TECHNOLOGY MANUAL

Maureen Petkewich University of South Carolina

Peter Flanagan-Hyde Phoenix Country Day School

Jack Morse University of Georgia

Debra Hydorn Mary Washington College

Jennifer Lewis Priestley Kennesaw State University

Michael Kowalski University of Alberta

SECOND EDITION

STATISTICS

THE ART AND SCIENCE OF LEARNING FROM DATA

Agresti • Franklin

Upper Saddle River, NJ 07458

Vice President and Editorial Director, Mathematics: Christine Hoag
Editor-in-Chief, Mathematics & Statistics: Deirdre Lynch
Editorial Assistant: Joanne Wendelken
Senior Managing Editor: Linda Mihatov Behrens
Project Manager, Production: Robert Merenoff
Supplement Cover Manager: Paul Gourhan
Supplement Cover Designer: Victoria Colotta
Operations Specialist: Ilene Kahn
Senior Operations Supervisor: Diane Peirano

Pearson Prentice Hall
Pearson Education, Inc.
Upper Saddle River, NJ 07458

Printed in the United States of America

10 9 8 7 6 5 4 3 2 1

ISBN-13: 978-0-13-603615-9

ISBN-10: 0-13-603615-5

Pearson Education Ltd., London
Pearson Education Singapore, Pte. Ltd.
Pearson Education, Canada, Ltd.
Pearson Education—Japan
Pearson Education Australia Pty, Ltd.
Pearson Education North Asia, Ltd.
Pearson Educación de Mexico, S.A. de C.V.
Pearson Education Malaysia, Pte. Ltd.
Pearson Education Upper Saddle River, New Jersey

EXCEL MANUAL

Jack Morse

University of Georgia

SECOND EDITION

STATISTICS

THE ART AND SCIENCE OF LEARNING FROM DATA

Agresti • Franklin

Upper Saddle River, NJ 07458

TABLE OF CONTENTS

INTRODUCTION TO EXCEL

INTRODUCTION: This lab session is designed to introduce you to the statistical aspects of Microsoft Excel. During this session you will learn how to enter and exit Excel, how to enter data and commands, how to print information, and how to save your work for use in subsequent sessions. As with any new skill, using this software will require practice and patience. Excel is a spreadsheet used for organizing data in columns and rows. It is an integrated part of Microsoft Office, and so data can be easily imported and exported into word processing documents, databases, graphics programs, etc. It offers a wide range of statistical functions and graphs and so is an alternative to specific statistical software.

BEGINNING AND ENDING AN EXCEL SESSION

To start Excel: Click on the Start button and choose Programs/Excel. If you have the Office shortcut bar installed, simply click on the Excel icon.

To exit Excel: To end a Excel session and exit the program, click the **Office Button** and then choose **Exit Excel**. A dialog box will appear, asking if you want to save the changes made to this worksheet. Click **Yes** or **No**.

You can also exit Excel by clicking the X in the upper right corner of the window.

THE EXCEL WINDOWS

The Document (sheet) Window: When you first start Excel you will be in a window titled "Book 1 - Microsoft Excel". Excel organizes itself in workbooks, each of which is made up of worksheets that are 65,536 rows by 256 columns. You can enter and edit data on several worksheets simultaneously and perform calculations based on data from multiple worksheets. When you create a chart, you can place the chart on the worksheet with its related data or on a separate chart sheet. Each of the cells within the sheet is identified by the intersection of its row and column, for example A2, or B7.

Note the three tabs at the bottom of the screen, called "Sheet1","Sheet2", and "Sheet3". The default is a workbook with three sheets, but the number of sheets in a workbook is limited only by available memory. To add a single worksheet, click Insert Worksheet after "Sheet3" at the bottom of the screen. To delete sheets from a workbook, select the sheets you want to delete, right click, and click Delete. To Rename a sheet, double-click the sheet tab, and type a new name over the current name.

The Application Window

When you first opened the workbook, there is one bar across the top of the screen. This bar consists of seven drop down menu items and is called the Menu Bar. It gives you access to all the Excel commands.

Analysis ToolPak : Microsoft Excel provides a set of data analysis tools — called the Analysis ToolPak — that you can use to save steps when you develop complex statistical or engineering analyses. You provide the data and parameters for each analysis; the tool uses the appropriate statistical or engineering macro functions and then displays the results in an output table. Some tools generate charts in addition to output tables. If the **Data Analysis** command is not on the **Data** menu, you need to install the Analysis ToolPak. To do this, click on the **Office Button** and click Excel Options. Click Add-Ins, and then in the Manage box, select Excel Add-ins and click Go. In the Add-Ins Available box, select the Analysis ToolPak and the Analysis ToolPak VBA, and then click OK. After you load the Analysis ToolPak, the Data Analysis command is available in the Analysis group on the **Data** tab.

In addition to the Data Analysis ToolPak, your textbook comes packaged with a program called PHStat2. Read the Readme file and follow the directions for loading the program on to your desktop. It contains additional macros for Excel to enable you to do advanced features.

You can choose additional toolbars by clicking the **Office Button**, selecting Excel Options, and selecting Customize to customize the toolbar you wish to display.

The Help Window in Excel

Information about Excel is stored in the program. If you forget how to use a command or need general information, you can ask Excel for help. Click on the ? in the top right hand corner and a drop down menu will appear.

ENTERING DATA

When a workbook is first opened, the cell A1 is outlined in black. This indicates the active cell. Move your cursor around the sheet, clicking into different cells to activate them. Note that the address changes in the box above A1. The address (row and column) of the active cell always appears here.

Let's enter data in the second column:

78 94 93 81 75 62 58 50 80 79

To do this press the down arrow key (↓) or enter key to move to the next entry position. Let’s fill the first column with the numbers 1 through 10. We can do it the same way, or we can let Excel do it for us. Enter a 1 in cell A1. Choose **Home> Editing > Fill > Series**. In the dialog box, select **columns**, **linear**, step **1**, stop value **10**. Then click **OK.**

Column 1 should now contain the integers 1 through 10.

While you are in the sheet window, fill columns 3 and 4 with a set of ten test scores each. You should now have four columns of data.

Changing a value entered

We can edit data directly in the cell or from the formula bar at the top of the sheet. If you have not hit the Enter key yet, you can simply back space and correct your mistake. If you have entered the data, click on the cell you wish to edit to make it active. You can either retype to overwrite the data, or click into the formula bar and edit the entry. Suppose we had inadvertently left out a value and we wish to enter it in a particular position. Place the cursor in the cell in which you wish to insert the new value. Choose **Home > Cells > Insert > Insert Cells** . A dialog box will appear, asking which way you wish to move the cells. A blank cell is created and the missing value can be entered. Entire rows and columns can be added the same way. You can take a short cut to this by right-clicking and selecting **Insert....**

A cell can be deleted by making the cell active, then Choose: **Home > Cells > Delete > Delete Cells** or by right-clicking and selecting **Delete....**

Copying Data

To copy the contents of one cell to another, simply activate the cell and choose **Home > Copy.** (**Control C** will also accomplish this.) Activate the cell that you want to paste the value into and choose **Home >Paste** (or **Control V**) This can also be done for a range of cells. Activate the upper left cell of the range. Press shift and click the lower right corner of the range. This should highlight the entire range. You can then copy and paste as above.

Cell References:

Previously, you entered four columns of data. Click on cell B11. Click Formulas > AutoSum > Sum. The ten values above it will be enclosed in a box. Press enter and the sum of the ten values will be in cell B11. Now activate cell B11, press **Control C**, highlight cells C11 and D11, and press **Control V.** This should give you the sums of columns C and D. Note what happened in the formula when you copied it. The references were changed to reflect the new column. This is called a relative reference.

If you need to preserve the value of a certain cell when copying a formula, you will have to use absolute referencing. This is accomplished by placing $ within the address. ($A$6 would keep the value in cell A6 to wherever it was copied within the worksheet.)

SAVING YOUR WORK

An Excel workbook contains all your work; the data, graphs, and all the sheets within the workbook. When you save a project, you save all of your work at once. When you open a project, you can pick up right where you left off.
The contents of each sheet can be saved and printed separately from the project, in a variety of formats. You can also delete a worksheet or graph, which removes the item from the project.

RETRIEVING A FILE

You can open a wide variety of files with Excel. Choose the **Office Button** and **Open** to select the appropriate one. There is an **Import Wizard** that will guide you through the process.

A CD ROM accompanies Agresti/Franklyn Statistics: The Art and Science of Learning from Data. This disk has data for many of the problems in the text. Follow the instructions that accompany the disk for use on your computer.
Let's use problem 1.19 to illustrate how to get the information from the CD into Excel. Choose the **Office Button** and click on **Open**. Use the drop down box to show the files on the CD in D drive. Open the Excel folder and click on the desired file. This should fill the worksheet with data.

PRINTING:

You have many options when it comes to printing from Excel. Choose the **Office Button** and click on **Print**. The Set Print Area choice allows you to select the range of cells you wish to print.

The **Print Preview** box has tabs that will help you customize your output. It allows you to play with your selections to get the best layout for your output before you commit it to paper.

Chapter 2 Example 3 How Much Electricity Comes From Renewable Energy Sources? Creating a Pie Chart and a Bar Graph

Copy the data from the text into the worksheet and click any cell in the data.

	A	B	C
1	Source	U. S. Percentage	Canada Percentage
2	Coal	51	16
3	Hydropower	6	65
4	Natural Gas	16	1
5	Nuclear	21	16
6	Petroleum	3	1
7	Other	3	1

Choose: **Insert > Chart > Pie > 2-D Pie**

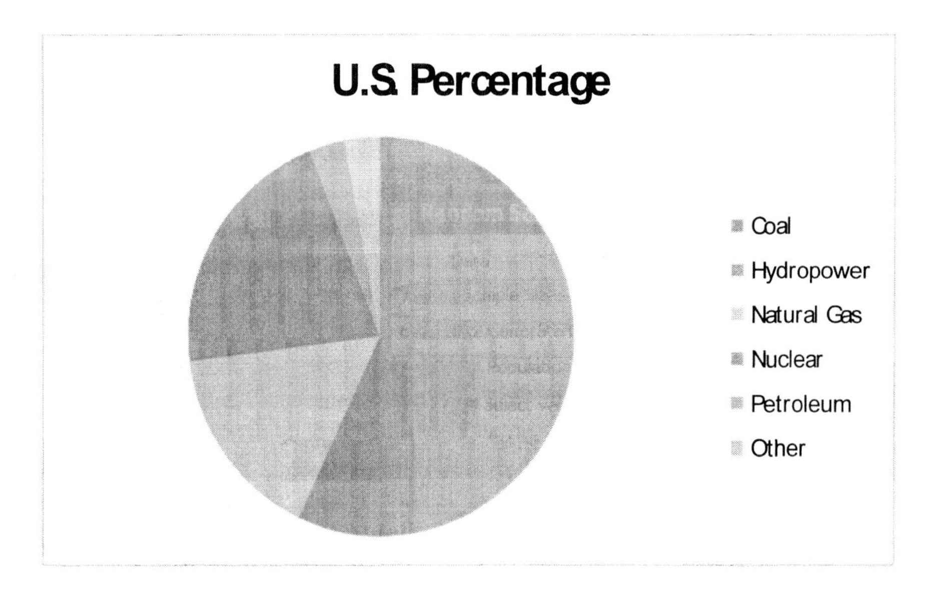

Click on the pie chart

Choose: **Chart Layouts > Layout 6**

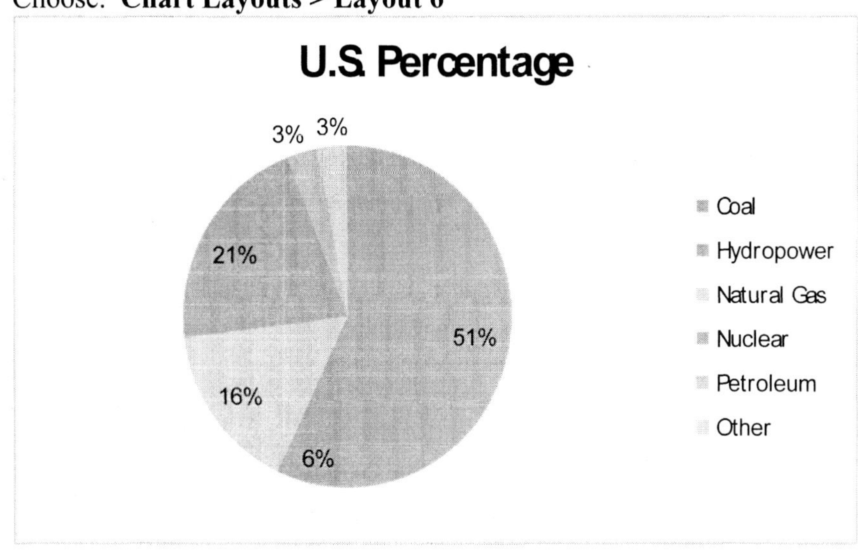

To get the bar chart:

Click any cell in the data.
Choose: **Insert > Chart > Column > 2-D Column**
Click on the bar chart
Choose: **Data > Select Data**
Click on "Canada Percentage" and click Remove.
Click OK.

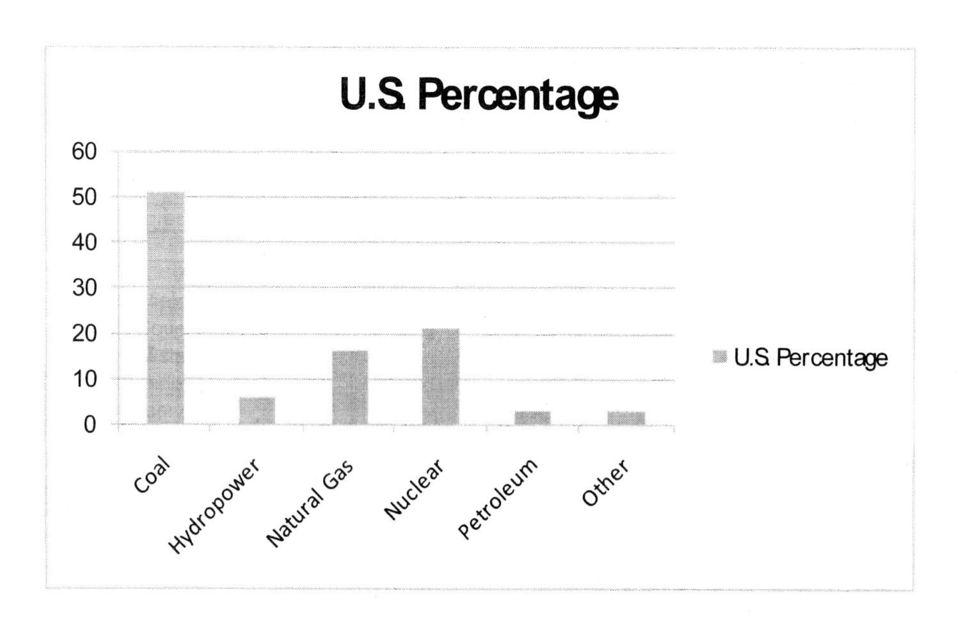

Click on smaller "U.S. Percentage" in the chart. Hit Delete.

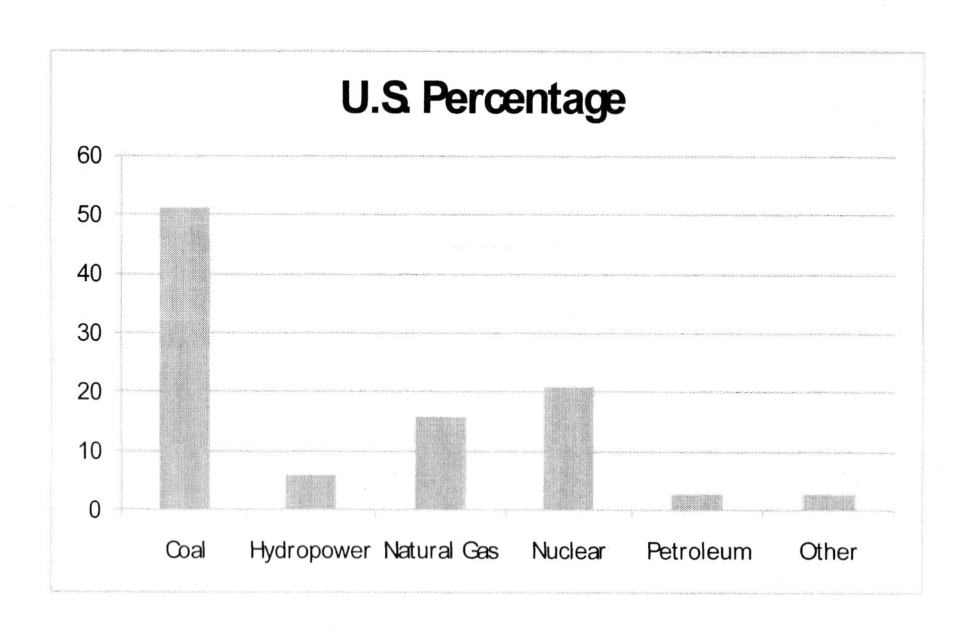

Chapter 2 Exercise 2.11 Weather Stations
Constructing a Pie Chart and Bar Graph

Use the data shown in the pie chart in the text
Click on any cell in the data:

	A	B
1	Region	Frequency
2	Southeast	67
3	Northeast	45
4	West	126
5	Midwest	121

Choose: **Insert > Chart > Pie > 2-D Pie**
Click on "Frequency" on the pie chart.
Delete "Frequency" text and enter "Regional Distribution of Weather Stations".
Click on the pie chart.
Choose: **Chart Layouts > Layout 6**

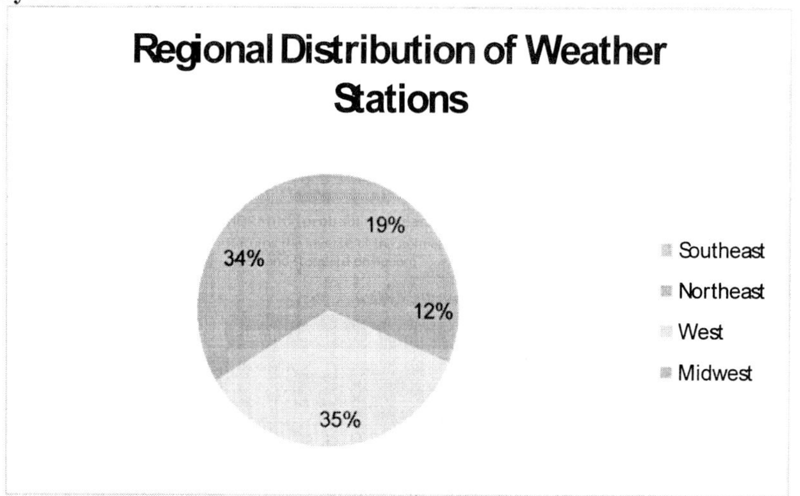

The corresponding bar chart:

Click on any cell in the data
Choose: **Insert > Chart > Column > 2-D Column**
Click on the larger "Frequency" on the bar chart.
Delete this "Frequency" text and enter "Regional Distribution of Weather Stations".
Click on smaller "Frequency" in the chart. Hit Delete.

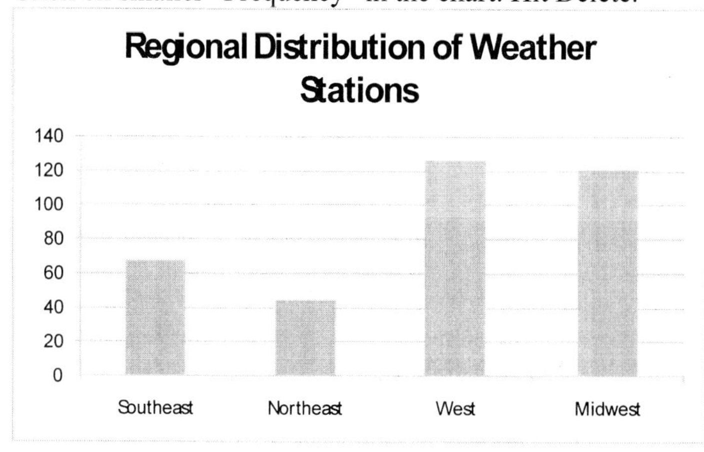

Chapter 2 Exercise 2.13 Shark Attacks Worldwide
Bar Chart and Pareto Chart

Enter the data from table 2.1 into the worksheet.

	A	B
1	Region	Frequency
2	Florida	365
3	Hawaii	60
4	California	40
5	Australia	94
6	Brazil	66
7	South Africa	76
8	Reunion Island	14
9	New Zealand	18
10	Japan	4
11	Hong Kong	6
12	Other	206

Sort the data alphabetically based on region and store the sorted columns in the original columns.
Click on a cell in the Region column.
Choose: **Editing > Sort & Filter > Sort A to Z**

To create the bar chart
Choose: **Insert > Chart > Column > 2-D Column**
Choose: **Chart Layouts > Layout 9**
Click on the larger "Frequency" on the bar chart.
Delete this "Frequency" text and enter "Shark Attacks".
Click on smaller "Frequency" in the chart. Hit Delete.
On the vertical "Axis Title", click and enter "Frequency".
On the horizontal "Axis Title", click and enter "Region".

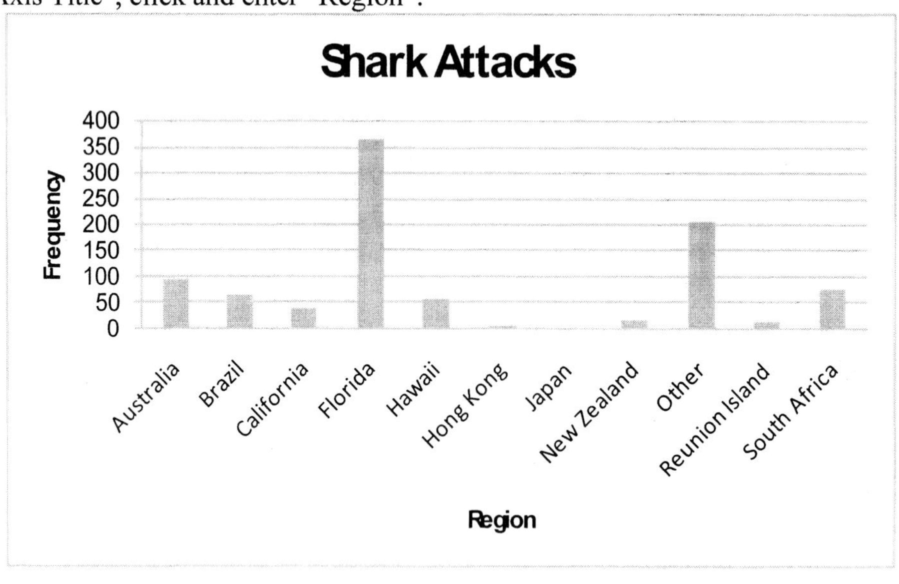

To obtain a Pareto chart for the same data, use the following menu choices

Click on a cell in the Frequency column of the data.
Choose: **Editing > Sort & Filter > Sort Largest to Smallest**

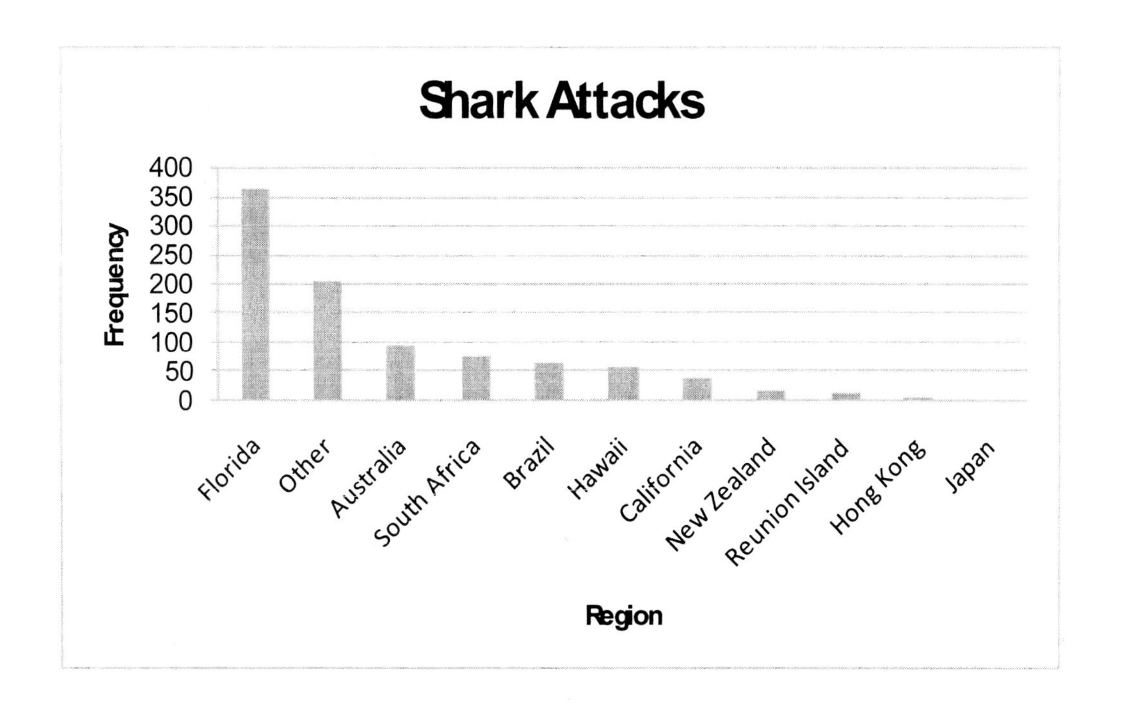

Chapter 2 Example 4 Exploring the Health Value of Cereals
Constructing a Dot Plot

Dot Plots are not available in Excel, but can be done using PHStat that comes with your text. Open the worksheet **cereal** from the Excel folder of the data disk.

Choose **PHStat > Descriptive Statistics > Dot Scale Diagram**

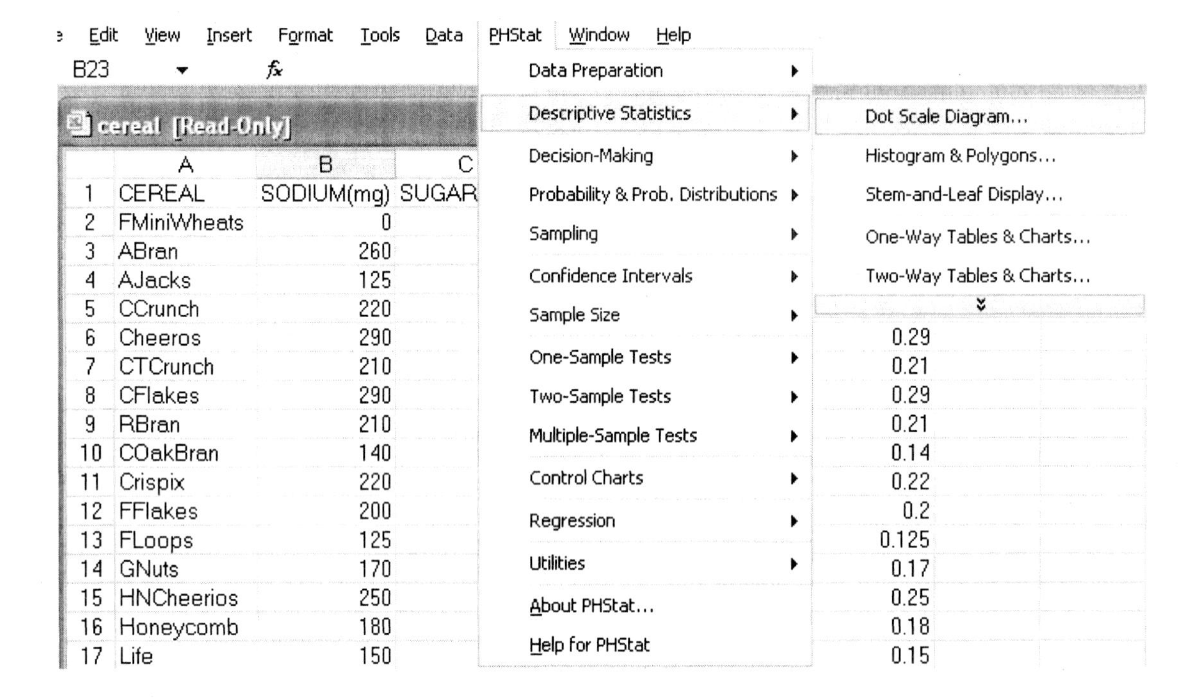

Enter the data by selecting column B.
Enter Title: **Sodium (mg)**

Click: **OK**

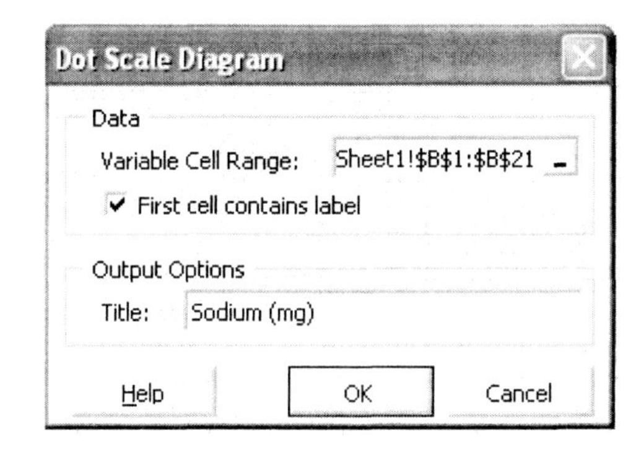

The output is shown below.

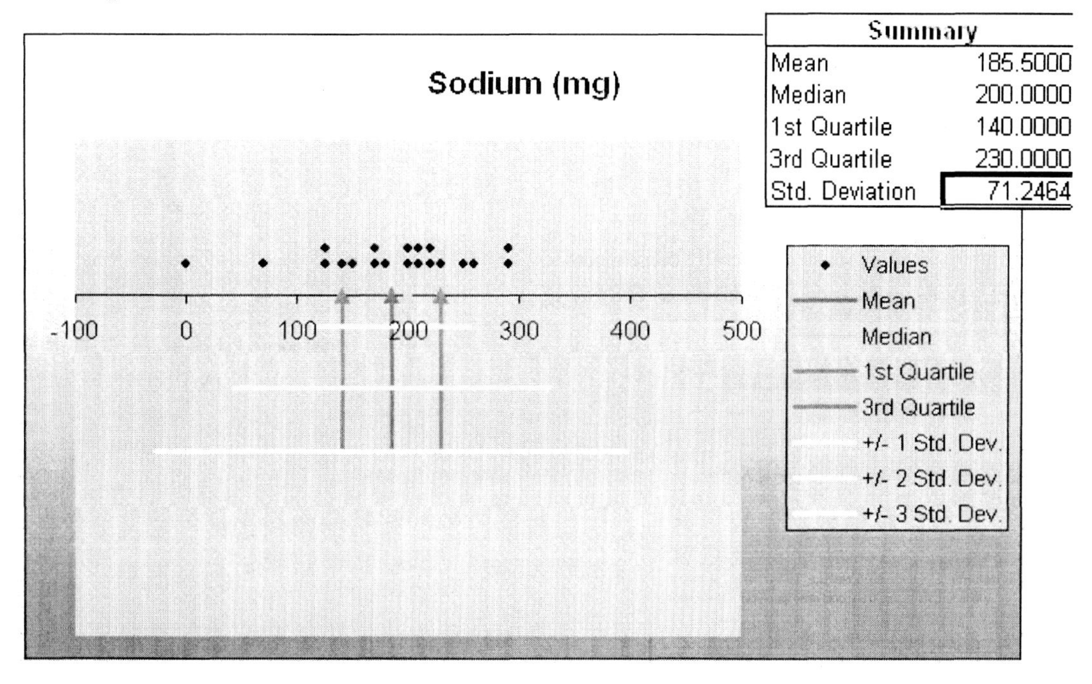

Summary	
Mean	185.5000
Median	200.0000
1st Quartile	140.0000
3rd Quartile	230.0000
Std. Deviation	71.2464

Chapter 2 Exercise 2.14 Sugar Dot Plot
Constructing a Dot Plot

Open the worksheet **cereal** from the Excel folder of the data disk.

Choose **PHStat > Descriptive Statistics > Dot Scale Diagram**

Enter the data by selecting column C.
Enter Title: **Sugar in Breakfast Cereals**
Click: **OK**

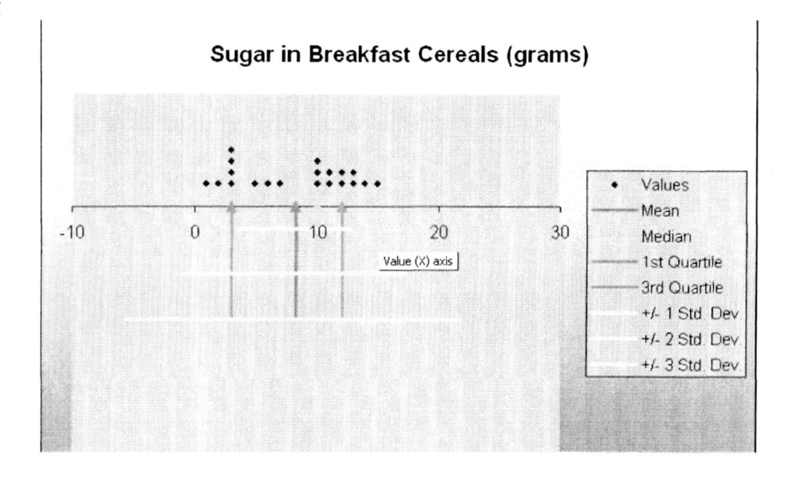

Chapter 2 Example 5 Exploring the Health Value of Cereals
Stem and Leaf Plots

Stem and Leaf Plots are not available in Excel, but can be done with PHStats.
Choose **PHStat > Descriptive Statistics > Stem-and-Leaf Display**

Enter: Variable Cell Range: **B1-B21**
Click: **First Cell contains label**
Click: **Autocalculate stem unit**
Click: **Summary Statistics**
Click: **OK**

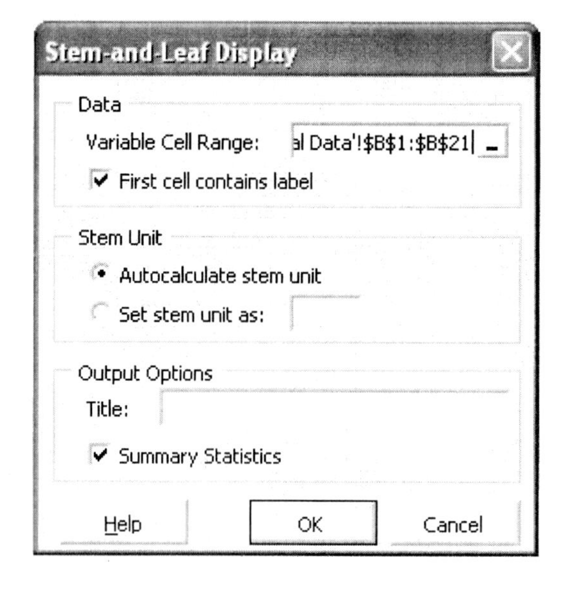

The results are shown below.

A	B	C	D	E
Stem-and-Leaf Display			Stem unit 10	
			0	0
Statistics			1	
Sample Size	20		2	
Mean	185.5		3	
Median	200		4	
Std. Deviation	71.24642		5	
Minimum	0		6	
Maximum	290		7	0
			8	
			9	
			10	
			11	
			12	5 5
			13	
			14	0
			15	0
			16	
			17	0 0
			18	0
			19	
			20	0 0
			21	0 0
			22	0 0
			23	0
			24	
			25	0
			26	0
			27	
			28	
			29	0 0

Chapter 2 Exercise 2.15 eBay Prices
Stem and Leaf Plot

Choose **PHStat > Descriptive Statistics > Stem-and-Leaf Display**
Enter: **A1 : A21**
Click **Autocalculate**
Click **Summary Statistics**
OK

	A	B	C	D	E
1				eBay Prices	
2					
3				Stem unit 10	
4					
5	Statistics			17	8
6	Sample Size	25		18	
7	Mean	232.16		19	9
8	Median	240		20	0
9	Std. Deviation	19.88022		21	0 0
10	Minimum	178		22	5 5 5 5 8
11	Maximum	255		23	2 5
12				24	0 0 0 5 6 6 6 9
13				25	0 0 0 5 5
14					
15					

Chapter 2 Example 7 Exploring the Health Value of Cereals
Constructing a Histogram

Open the **cereal** worksheet from the Excel folder of the data disk.

To do a histogram, you must use the Data Analysis Add-In. If Data Analysis does not show on the Tools menu:

Click the Office Button
Choose: **Excel Options > Add-Ins**
Click **GO** at the bottom of the page.
Select: **Analysis ToolPak**
Analysis ToolPak-VBA

To complete a Histogram, a Bin Range should be determined and entered into two columns of the worksheet. This is not required, though. If you omit the bin range, Excel creates a set of evenly distributed bins between the data's minimum and maximum values.

Choose: **Data > Data Analysis > Histogram**
Click: **OK**
Enter: Input Range: **select appropriate cells**
Bin Range:
Select: **Chart Output**

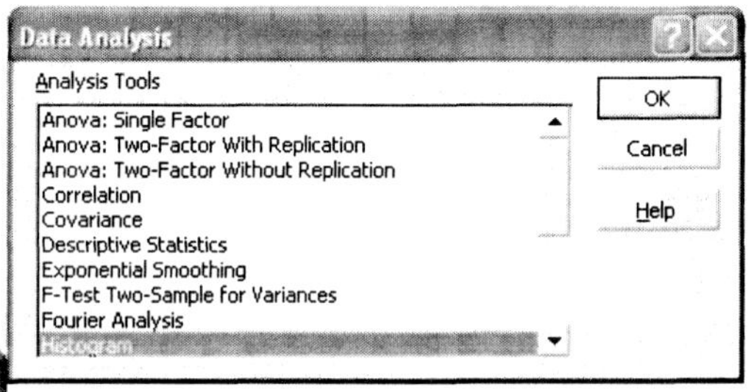

Data Analysis
Analysis Tools
Anova: Single Factor
Anova: Two-Factor With Replication
Anova: Two-Factor Without Replication
Correlation
Covariance
Descriptive Statistics
Exponential Smoothing
F-Test Two-Sample for Variances
Fourier Analysis
Histogram
OK
Cancel
Help

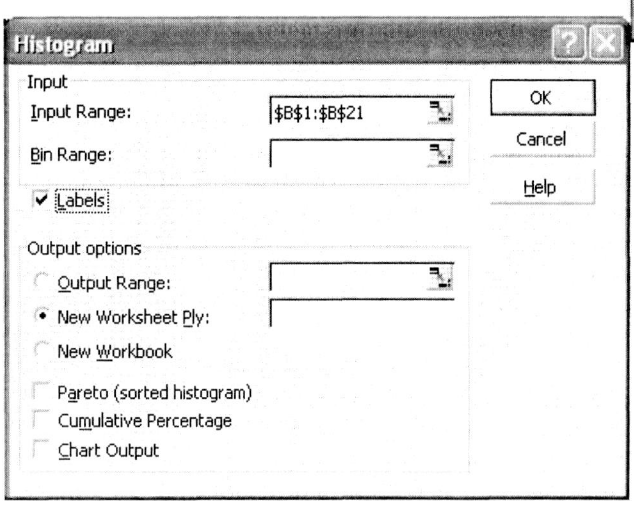

Histogram
Input
Input Range: $B$1:$B$21
Bin Range:
Labels
Output options
Output Range:
New Worksheet Ply:
New Workbook
Pareto (sorted histogram)
Cumulative Percentage
Chart Output
OK
Cancel
Help

This gives you the following chart:

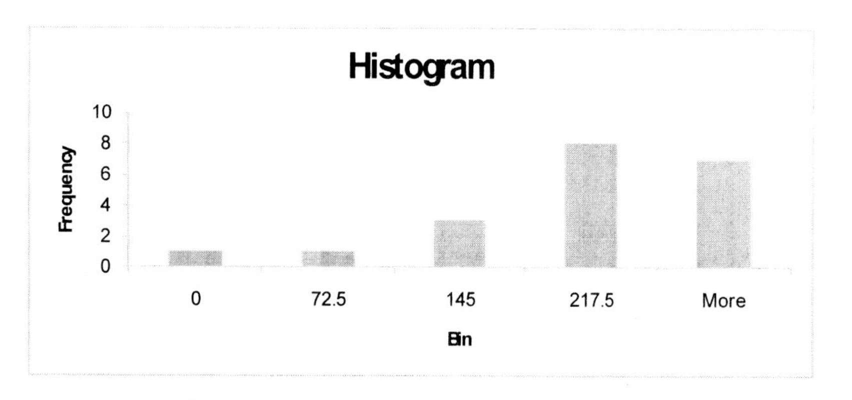

To edit the chart
Click: Anywhere within the chart on whatever it is you wish to edit

Enter an appropriate title, remove the legend, rename horizontal axis

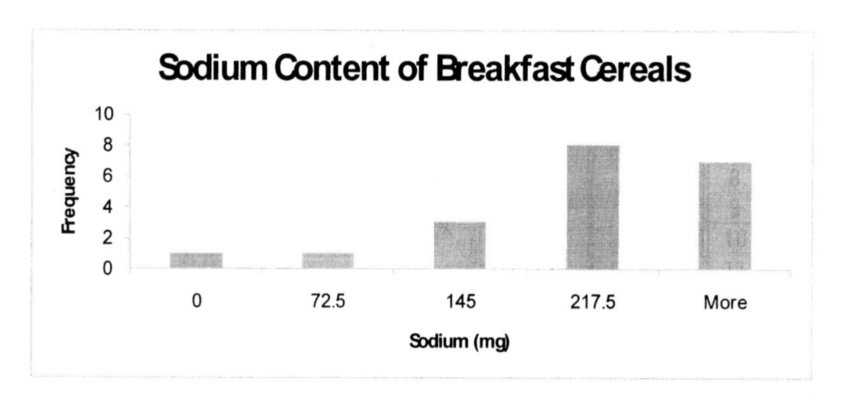

To remove gaps between bars

Right Click: **Any bar on graph**
Choose: **Format Data Series > Options**
Select: Gap width: **No Gap**
Click: **Close**

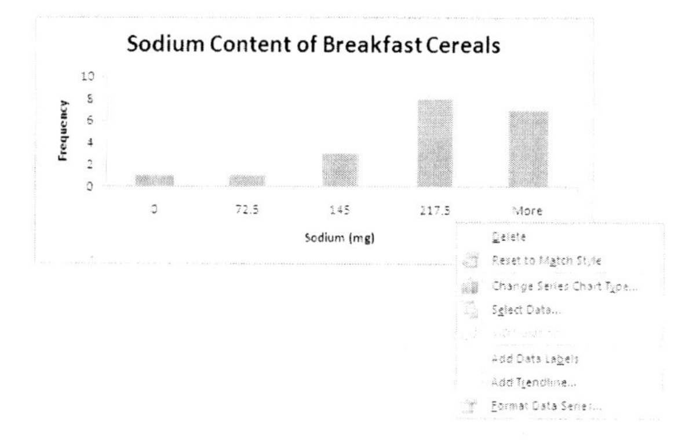

This gives you the finished chart.

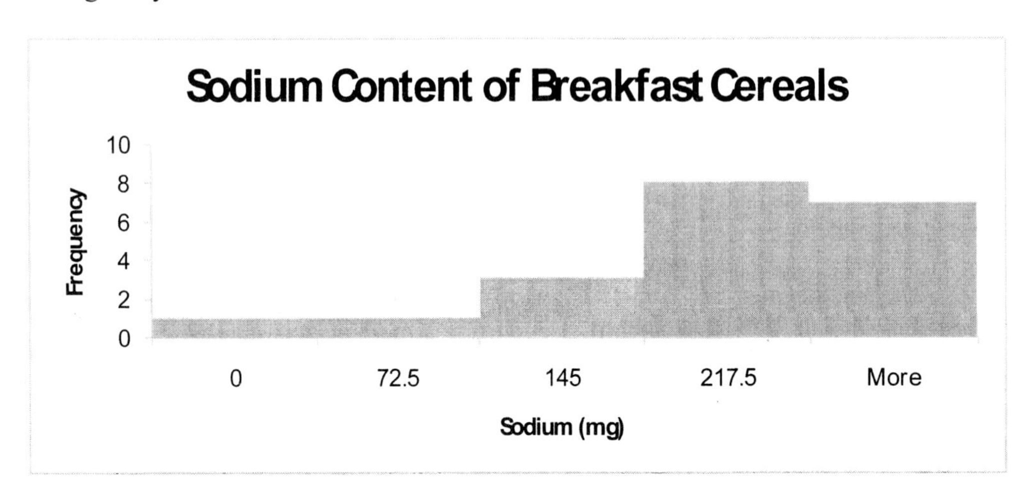

Chapter 2 Exercise 2.22 Sugar plots
Dot Plot, Stem-and-Leaf Plot and Histogram

Open the **cereal** worksheet from the Excel folder of the text data disk.

Choose **PHStat > Descriptive Statistics > Dot Scale Diagram**
Enter the data by selecting column **E**.
Enter Title: **Sugar (mg)**
OK
The chart is shown below.

Summary	
Mean	8200.0000
Median	10000.0000
1st Quartile	3000.0000
3rd Quartile	12000.0000
Std. Deviation	4560.7017

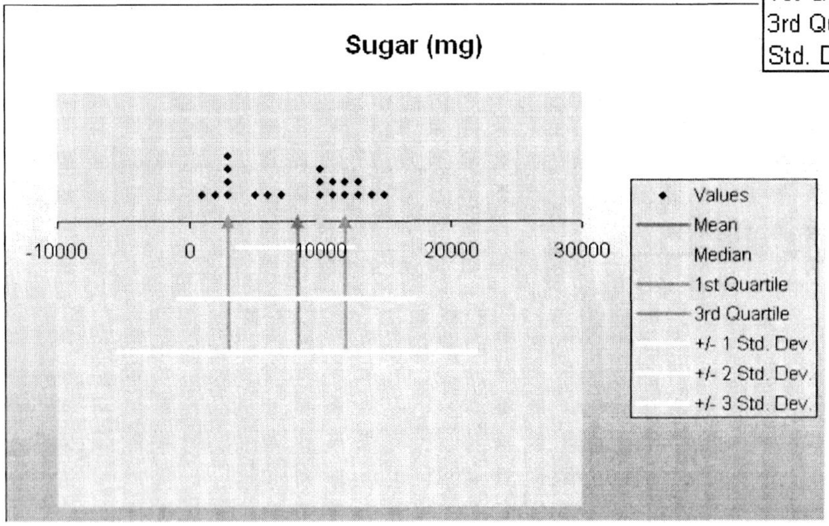

To construct the stem-and-leaf plot:
Choose **PHStat > Descriptive Statistics > Stem-and-Leaf Display**
Enter: Variable Cell Range: **E1-E21**
Click **First Cell contains label**
Click **Autocalculate**
Enter title**: Sugar (mg)**
Click **Summary Statistics**
OK

	A	B	C	D	E	F
1				Sugar(mg)		
2				Stem unit	1000	
3	Statistics			1	0	
4	Sample Size	20		2	0	
5	Mean	8200		3	0 0 0 0	
6	Median	10000		4		
7	Std. Deviation	4560.702		5	0	
8	Minimum	1000		6	0	
9	Maximum	15000		7	0	
10				8		
11				9		
12				10	0 0 0	
13				11	0 0	
14				12	0 0	
15				13	0 0	
16				14	0	
17				15	0	
18						

To construct the histogram:

Choose: **Data > Data Analysis > Histogram > OK**
Enter: Input Range: **E1 – E21**
Bin Range: **select cells - optional**
Select: **Chart Output**

Edit the chart

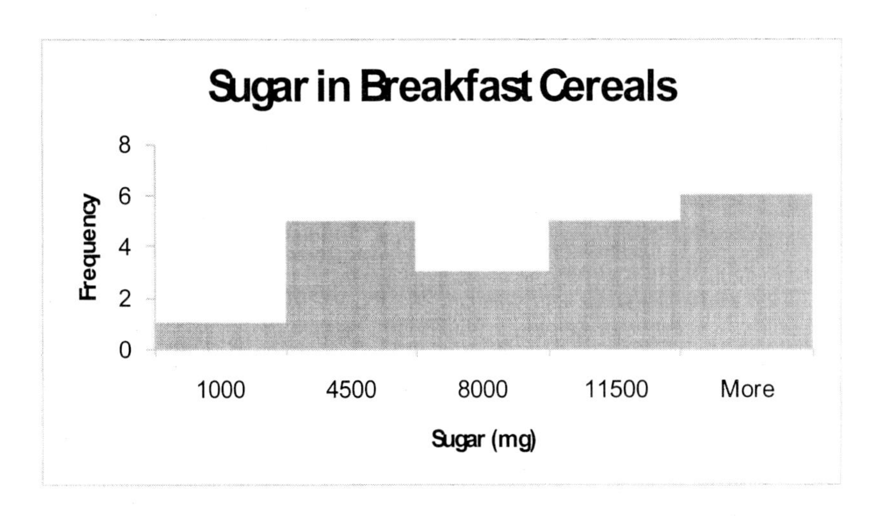

Chapter 2 Exercise 119a Temperatures in Central Park
Constructing a Histogram

Open the worksheet **central_park_yearly_temps** from the Excel folder of the data disk.

Choose: **Data > Data Analysis > Histogram > OK**
Enter: Input Range: **A2-A101**
Bin Range: **(select cells - optional)**
Select: **Chart Output**

To edit the chart
Click: Anywhere within the chart
To remove gaps between bars
Right Click: **Any bar on graph**
Choose: **Format Data Series > Options**
Select: Gap width: **No Gap**
Click: **Close**

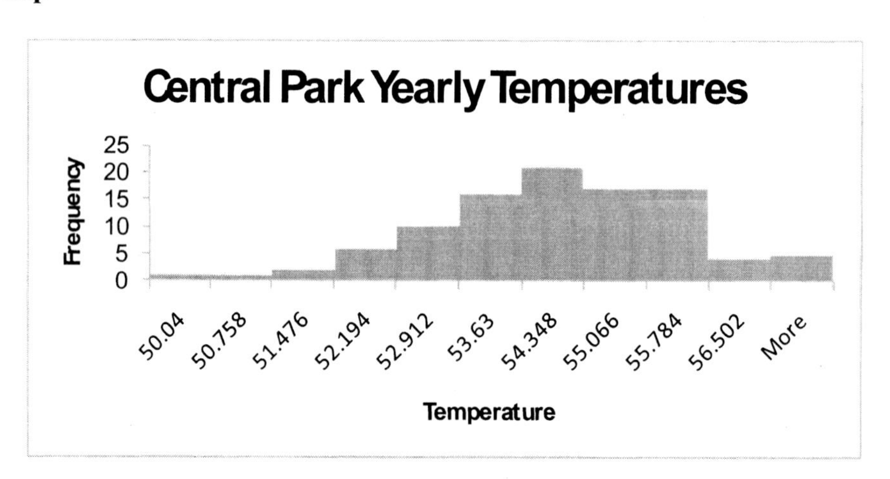

Chapter 2 Example 9 Is There a Trend Toward Warming in New York City?
Constructing a Time Plot

Open the worksheet **central_park_yearly_temps** from the Excel folder of the data disk. Excel does not do Time Plots, but we can simulate one using Insert > Chart > Scatter > Scatter with Straight Lines and Markers. Copy the data in the first column and paste it in the fourth column. Select the last two columns containing the data.

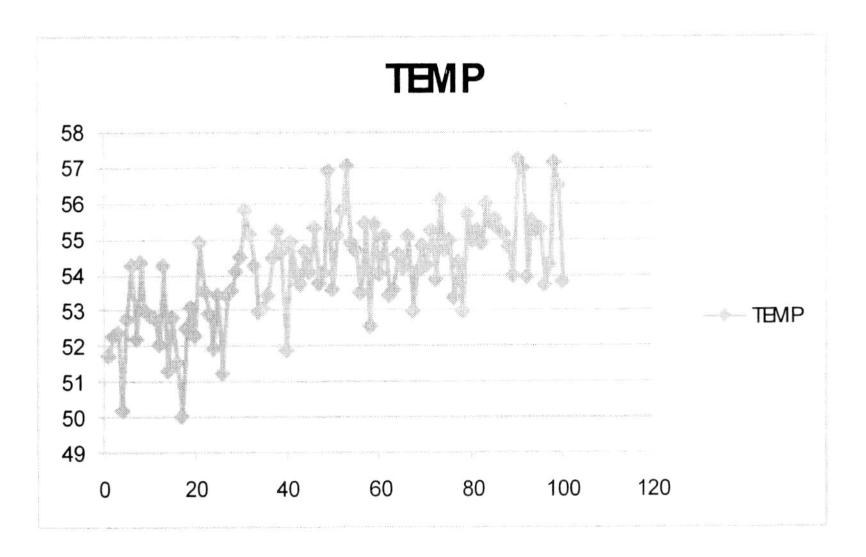

Add titles as appropriate. Place chart in new sheet.

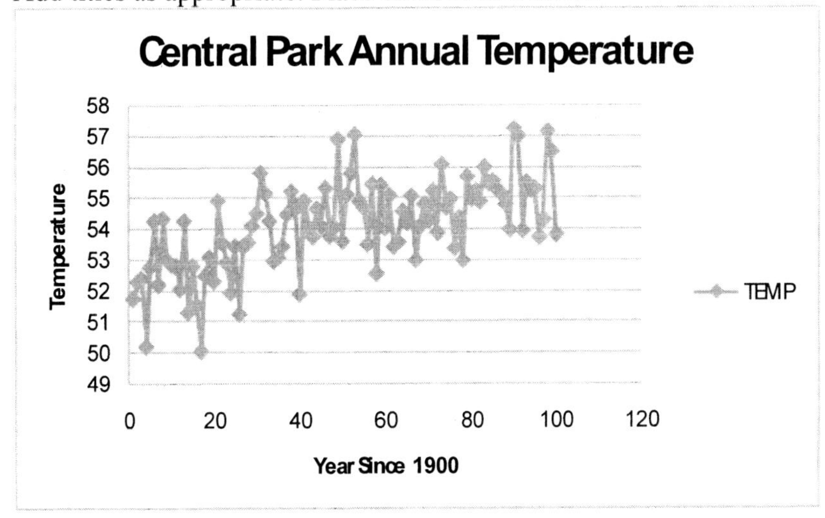

Format the graph, choosing proper axis scale and adding the trendline. You can do so by choosing Layout 9 under Chart Layouts.

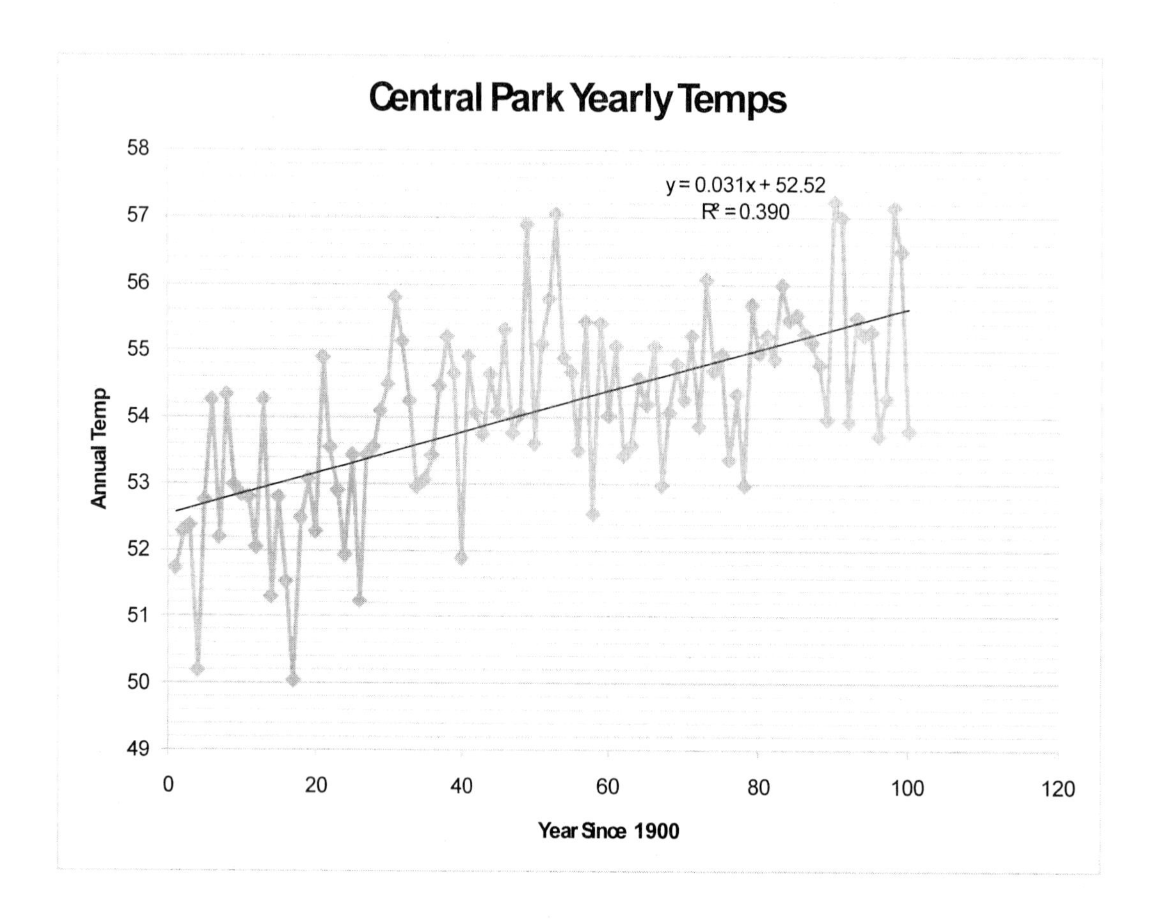

Chapter 2 Exercise 2.29 Warming in Newnan, Georgia?
Constructing a Time Plot

Open the worksheet **newnan_ga_temps** from the Excel folder of the data disk.
Using Insert > Chart > Scatter > Scatter with Straight Lines and Markers, graph a scatterplot with connected lines. Select the columns containing the data. Add titles as appropriate. Place chart in new sheet. Format the graph, choosing proper axis scale and adding the trendline.

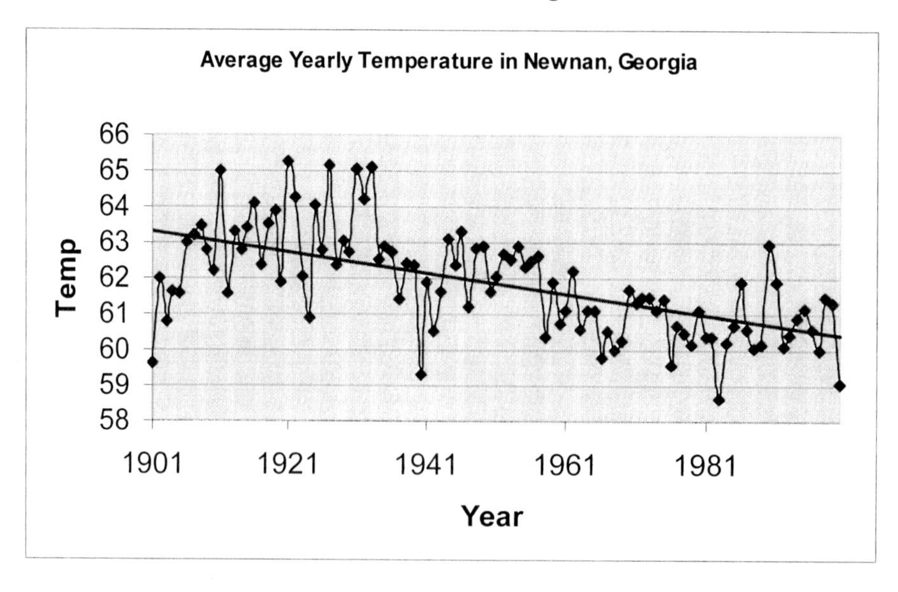

Chapter 2 Example 10 What is the Center of the Cereal Sodium Data? Determining Mean and Median

Open the Excel data file **cereal** from the Excel folder of the data disk.

Calculate the mean and median using the following commands.

Choose:
Data > Data Analysis > Descriptive Statistics > OK

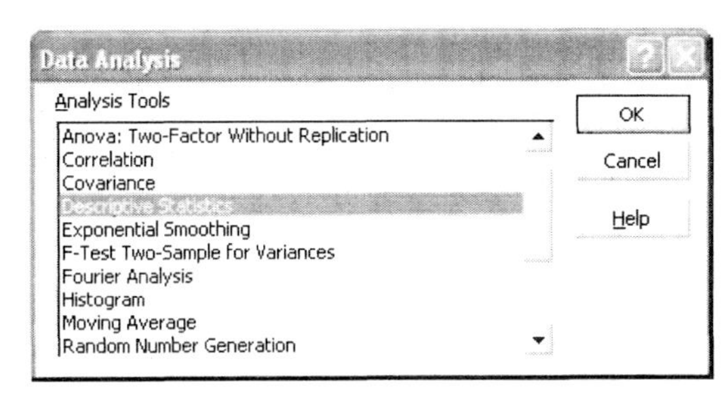

Enter: Input Range: **B1:B21**
Click: **Labels in First Row**
Choose: **Summary Statistics** and click **OK**

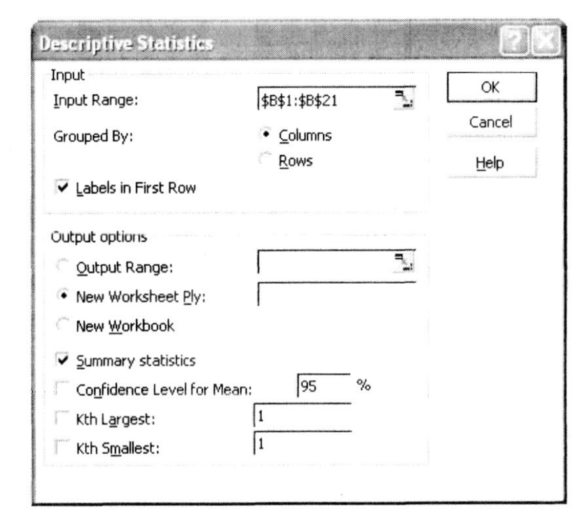

These commands generate the following output in a new worksheet ply.

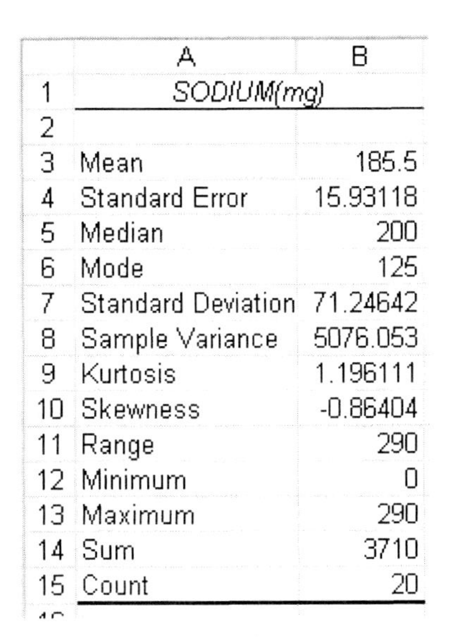

	A	B
1	*SODIUM(mg)*	
2		
3	Mean	185.5
4	Standard Error	15.93118
5	Median	200
6	Mode	125
7	Standard Deviation	71.24642
8	Sample Variance	5076.053
9	Kurtosis	1.196111
10	Skewness	-0.86404
11	Range	290
12	Minimum	0
13	Maximum	290
14	Sum	3710
15	Count	20

We can also find values by entering a formula and storing the result in a cell. For example, the midrange = (high + low)/2. To do this in Excel we would do the following:

> Select: **a particular cell**
> In the formula bar type in the expression: **=(MAX(B2:B21) + MIN(B2:B21)) /2**
> then press Enter.

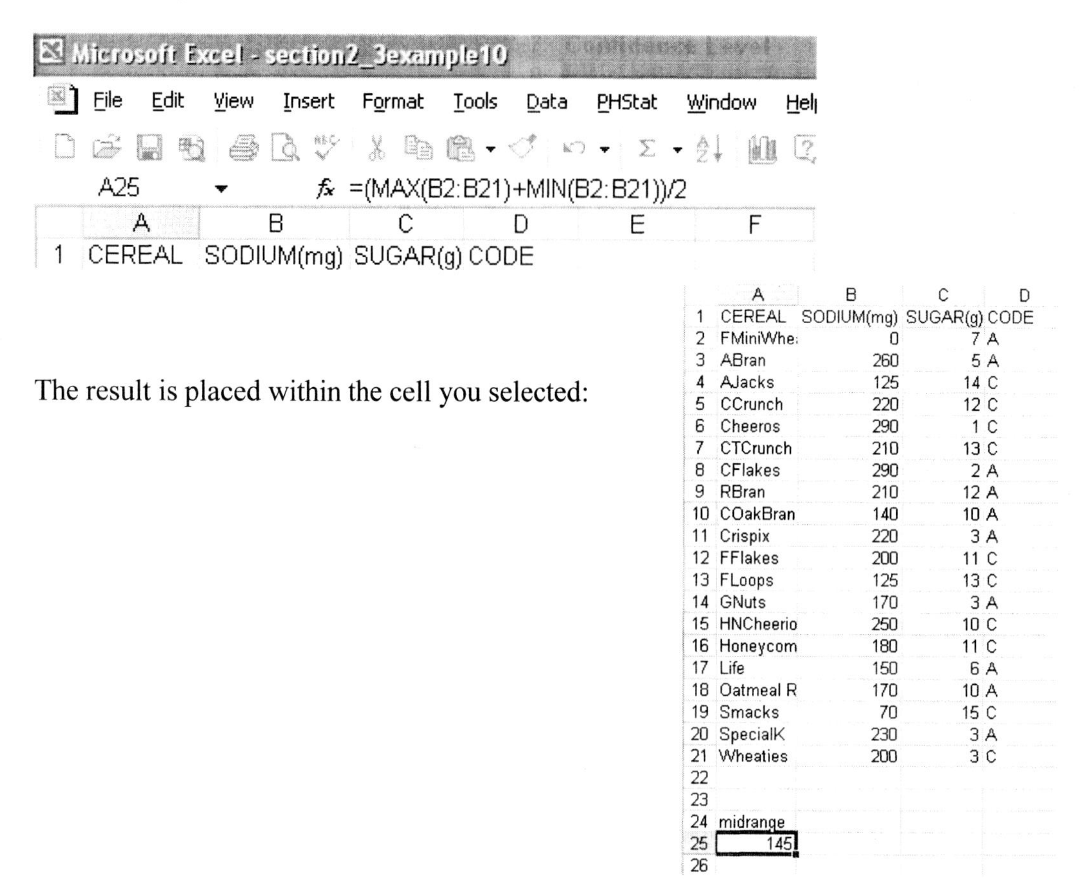

Microsoft Excel - section2_3example10

File Edit View Insert Format Tools Data PHStat Window Hel

A25 =(MAX(B2:B21)+MIN(B2:B21))/2

	A	B	C	D	E	F
1	CEREAL	SODIUM(mg)	SUGAR(g)	CODE		

The result is placed within the cell you selected:

	A	B	C	D
1	CEREAL	SODIUM(mg)	SUGAR(g)	CODE
2	FMiniWhe	0	7	A
3	ABran	260	5	A
4	AJacks	125	14	C
5	CCrunch	220	12	C
6	Cheeros	290	1	C
7	CTCrunch	210	13	C
8	CFlakes	290	2	A
9	RBran	210	12	A
10	COakBran	140	10	A
11	Crispix	220	3	A
12	FFlakes	200	11	C
13	FLoops	125	13	C
14	GNuts	170	3	A
15	HNCheerio	250	10	C
16	Honeycom	180	11	C
17	Life	150	6	A
18	Oatmeal R	170	10	A
19	Smacks	70	15	C
20	SpecialK	230	3	A
21	Wheaties	200	3	C
22				
23				
24	midrange			
25	145			
26				

Chapter 2 Exercise 2.29 More On CO_2 Emissions
Mean and Median

Enter the data (country and million metric tons of carbon equivalent) into the worksheet.

	A	B
1	Country	CO2 Emissions
2	U.S.	1892
3	China	1356
4	Russia	520
5	India	506
6	Japan	364
7	Germany	265
8	Brazil	230

Calculate the mean and median using the following commands.
Choose: **Data > Data Analysis > Descriptive Statistics > OK**
Enter: Input Range: **B1:B8**
Click: **Labels in First Row**
Choose: **Summary Statistics** and click **OK**

	A	B
1	*CO2 Emissions*	
2		
3	Mean	733.2857
4	Standard Error	240.8567
5	Median	506
6	Mode	#N/A
7	Standard Deviation	637.2469
8	Sample Variance	406083.6
9	Kurtosis	0.510528
10	Skewness	1.348203
11	Range	1662
12	Minimum	230
13	Maximum	1892
14	Sum	5133
15	Count	7

Chapter 2 Example 15 Describing Female College Student Heights Empirical Rule

	A	B	C
1	HEIGHT		GENDER
2	70		0
3	75		0
4	67		0
5	67		1
6	73		0
7	65		0
8	73		0
9	70		0
10	65		1
11	73		0
12	71		0
13	65.5		1

When we open the data file **heights** on the text CD, we find that the data includes male (indicated by 0) and females(indicated by 1). In this example we are only concerned with the heights of the females.

To generate the histogram of female student height data only, we must specify that the data to be used should be taken from only those rows where the gender is a 1. The histogram tool under **Data > Data Analysis > Histogram** does not provide for this. Instead, we will use the PHStat Add-in.

First, we must create the Bin Range, and bin midpoints.

D	E
Bin Cell Range	BIN midpoint
54.5	55
55.5	56
56.5	57
57.5	58
58.5	59
59.5	60
60.5	61
61.5	62
62.5	63
63.5	64
64.5	65
65.5	66
66.5	67
67.5	68
68.5	69
69.5	70
70.5	71
71.5	72
72.5	73
73.5	74
74.5	75
75.5	76
76.5	77
77.5	

To create the histogram of only the female heights:

Click: **Add-Ins > PHStat > Descriptive Statistics > Histogram & Polygons**

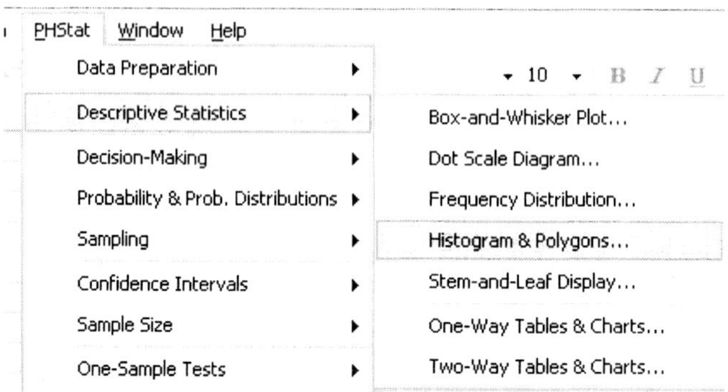

Complete the dialogue box as follows and click OK.

Histogram & Polygons
Data
Variable Cell Range: A1:A379
Bins Cell Range: Sheet1!$D$1:$D$25
Midpoints Cell Range: Sheet1!$E$1:$E$24
First cell in each range contains label
Input Options
Single Group Variable
Multiple Groups - Unstacked
Multiple Groups - Stacked
Grouping Variable Cell Range: $C$1:$C$379
Output Options
Title: Height
Histogram
Frequency Polygon
Percentage Polygon
Cumulative Percentage Polygon (Ogive)
Help OK Cancel

New worksheets are added, containing frequencies, percentages, and histograms. After some modification to the histogram, the sheet for the females appears as follows:

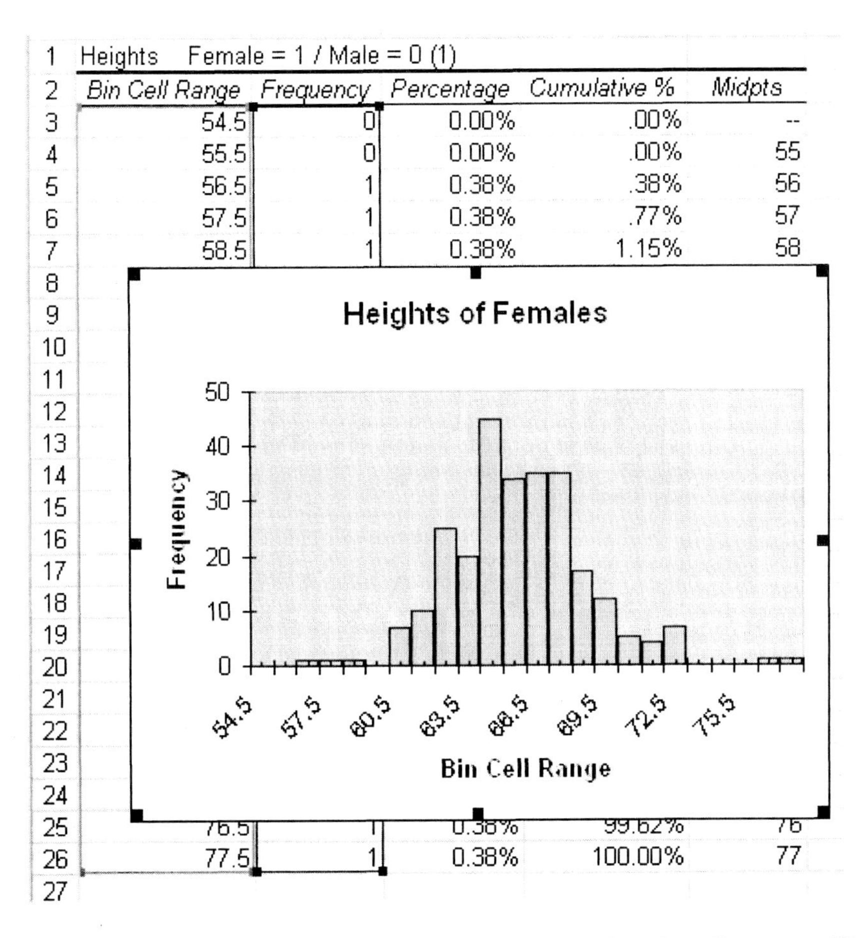

	Heights	Female = 1 / Male = 0 (1)			
2	*Bin Cell Range*	*Frequency*	*Percentage*	*Cumulative %*	*Midpts*
3	54.5	0	0.00%	.00%	--
4	55.5	0	0.00%	.00%	55
5	56.5	1	0.38%	.38%	56
6	57.5	1	0.38%	.77%	57
7	58.5	1	0.38%	1.15%	58
25	76.5	1	0.38%	99.62%	76
26	77.5	1	0.38%	100.00%	77

To generate the descriptive statistics for only the females, first, we will sort the data, by gender, and then by height:

Select columns A through C
Click: **Data > Sort**
Enter: Sort by **Column C** from the dropdown menu
Click **Add Level**
Then sort by **Column A** from the dropdown menu

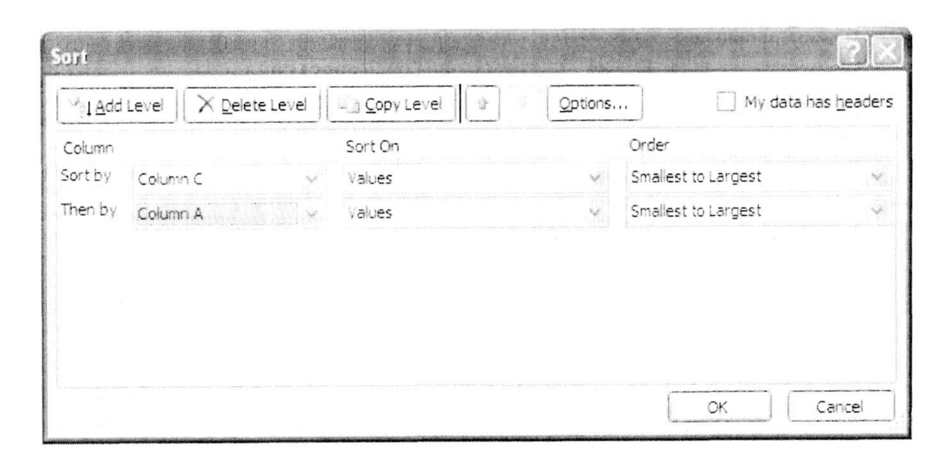

Now, we can perform our analysis on the female heights:

Choose: **Data > Data Analysis > Descriptive Statistics > OK**
Enter: Input Range: **A1:A262**
Click: **Labels in First Row**
Choose: **Summary Statistics** and click **OK**

	A	B
1	1	
2		
3	Mean	65.28352
4	Standard Error	0.182777
5	Median	65
6	Mode	64
7	Standard Deviation	2.952847
8	Sample Variance	8.719304
9	Kurtosis	1.330394
10	Skewness	0.40416
11	Range	21
12	Minimum	56
13	Maximum	77
14	Sum	17039
15	Count	261

	A	B	C
31	62		1
32	62		1
33	62		1
34	62		1
35	62		1
36	62		1
37	62		1
38	62		1
39	62		1
40	62		1
41	62		1
42	62		1
43	62		1
44	62		1
45	62		1
46	62.5		1
47	63		1
48	63		1
49	63		1
50	63		1
51	63		1
52	63		1
53	63		1
54	63		1
55	63		1

Now, with the lists sorted, counting the number of observations of female heights within each range is easier.

The heights between 62.3 and 68.3 lie from row 46 to row 232, which is 187 values, 72% of all female heights.

The heights between 59.3 and 71.3 lie from row 5 to row 252, which is 248 values, 95% of all female heights.

The heights between 56.3 and 74.3 lie from row 3 to row 260, which is 378 values, 99% of all female heights.

Chapter 2 Exercise 2.61 EU Data File
Empirical Rule

Open the **european_union_unemployment** data file on the text CD.

Generating the histogram, we see that the distribution is skewed, and not bell shaped.

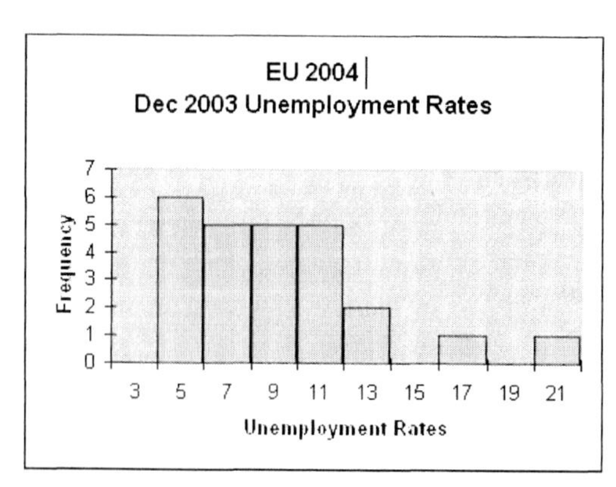

Generating basic statistics

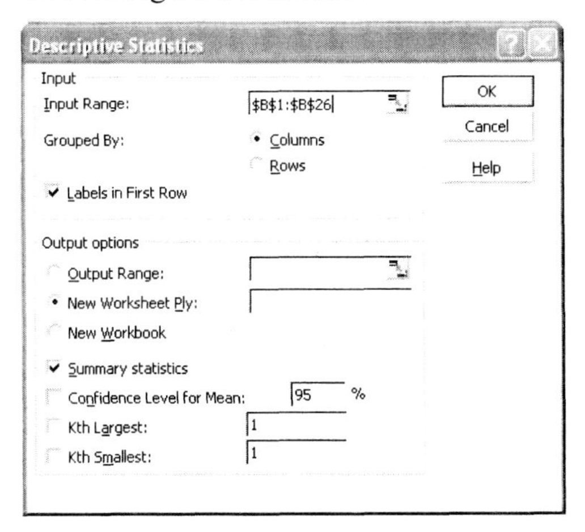

	A	B
1	*unemployment*	
2		
3	Mean	8.288
4	Standard Error	0.738948
5	Median	8.3
6	Mode	6
7	Standard Deviation	3.69474
8	Sample Variance	13.6511
9	Kurtosis	2.364485
10	Skewness	1.38482
11	Range	15.2
12	Minimum	3.9
13	Maximum	19.1
14	Sum	207.2
15	Count	25
16		

To determine the number of data values that fall in the ranges
$\bar{x} \pm 1 \cdot s = (4.593, 11.983), \bar{x} \pm 2 \cdot s = (.898, 15.678), \bar{x} \pm 3 \cdot s = (-2.797, 19.373)$
Sort the unemployment rates.

1	country	unemployment
2	Luxembourg	3.9
3	Netherlands	4.4
4	Austria	4.5
5	Ireland	4.6
6	Cyprus	4.7
7	United Kingdom	4.8
8	Hungary	5.9
9	Denmark	6
10	Sweden	6
11	Slovenia	6.4
12	Portugal	6.7
13	Czech Republic	8.1
14	Belgium	8.3
15	Italy	8.5
16	Malta	8.6
17	Finland	8.9
18	Germany	9.2
19	Greece	9.3
20	France	9.5
21	Estonia	9.6
22	Latvia	10.5
23	Spain	11.2
24	Lithuania	11.9
25	Slovakia	16.6
26	Poland	19.1

We determine that there are 20, 23, and 25 data values within the three ranges, respectively, which is 80%, 92 %, and 100%. Notice that theses percentages don't even begin to match the 68%, 95% and 99.7% of the Empirical Rule. This provides further evidence that the distribution of unemployment rates is not bell shaped.

Chapter 2 Example 16 What Are the Quartiles for the Cereal Sodium Data? Quartiles

Open the **cereal** data file on the CD.

In this example, we wish to determine the quartiles for the sodium values.
Choose: **Add-Ins > PHStats > Descriptive Statistics > Box-and-Whisker Plot**

Click: **Five-Number Summary**
Make selections as indicated and click **OK**

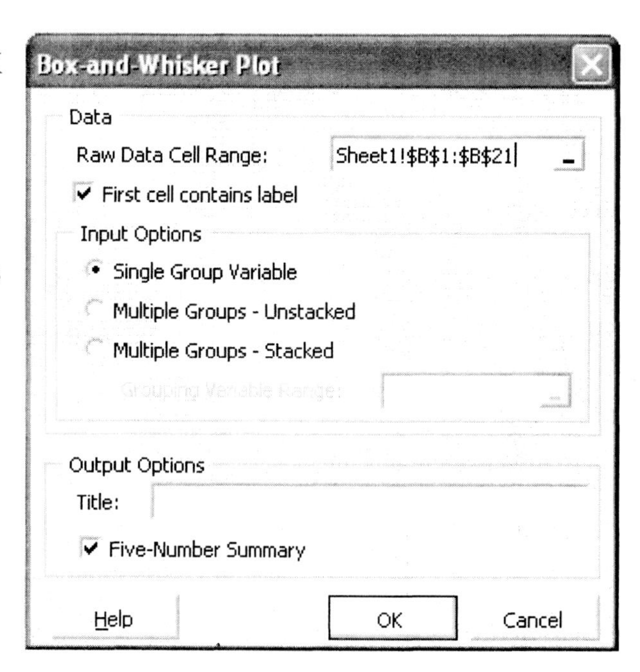

A new worksheet ply named FiveNumbers is created, containing the needed information.

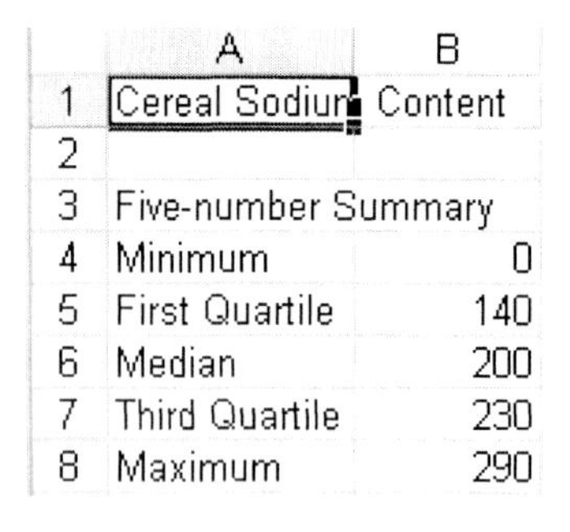

	A	B
1	Cereal Sodium	Content
2		
3	Five-number Summary	
4	Minimum	0
5	First Quartile	140
6	Median	200
7	Third Quartile	230
8	Maximum	290

Note that the results differ from those in the text. Not all statistical programs use the same method for determining the quartiles. You should explore how and why PHStat generates Q1=140 and Q3 = 230.

More general percentiles are very easy to generate with Excel. Let's say we're interested in the 30th, 65th and 98th percentiles.

Select a cell in the worksheet to store the response. When that cell is activated, you will type in the formula bar **PERCENTILE(array,k)** where **array** is the array or range of data that defines relative standing and **k** is the percentile value in the range 0..1, inclusive.

Arial 10

=PERCENTILE(B2:B21,0.3)

B	C	D	E
SODIUM(n	SUGAR(g)	CODE	
0	7	A	
260	5	A	
125	14	C	
220	12	C	
290	1	C	
210	13	C	
290	2	A	
210	12	A	
140	10	A	
220	3	A	
200	11	C	
125	13	C	
170	3	A	
250	10	C	
180	11	C	
150	6	A	
170	10	A	
70	15	C	
230	3	A	
200	3	C	

30th percentile
164

65th percentile
213.5

98th percentile
290

Chapter 2 Exercise 2.64 European Unemployment Quartiles

Enter the data into columns A and B of the worksheet. Be sure to label the columns.

	A	B
1	Country	Unemployment Rate
2	Belgium	7.8
3	Denmark	3.2
4	Germany	7.7
5	Greece	8.7
6	Spain	8.6
7	France	8.4
8	Portugal	7.2
9	Netherlands	3.6
10	Luxembourg	5
11	Ireland	4.4
12	Italy	6.7
13	Finland	7
14	Austria	4.5
15	Sweden	6
16	U.K.	5.4

Click **Add-Ins > PHStats > Descriptive Statistics > Box-and-Whisker Plot**
Enter: Raw Data Cell Range: **select appropriate cells**
Enter: Title: **an appropriate title**
Click: **Five-Number Summary > OK**

3	Five-number Summary	
4	Minimum	3.2
5	First Quartile	4.5
6	Median	6.7
7	Third Quartile	7.8
8	Maximum	8.7

Chapter 2 Example 17 Box Plot for the Breakfast Cereal Sodium Data
Box Plots

Open worksheet **cereal** from the Excel folder of the data disk. To construct a box plot (Box-and-Whisker Plot) for the breakfast cereal sodium data, use the following commands:

Click: **Add-Ins > PHStats > Descriptive Statistics > Box-and-Whisker Plot**

In the Box-and-Whisker Plot dialogue box, enter the **Raw Data Cell Range**, the **Title** and click **OK**.

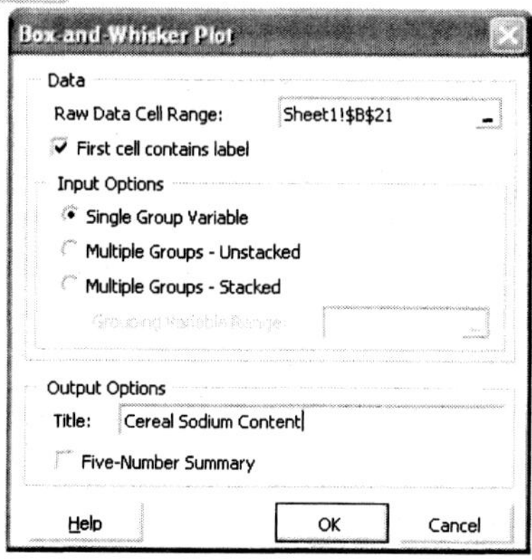

A new worksheet named BoxWhiskerPlot is created.

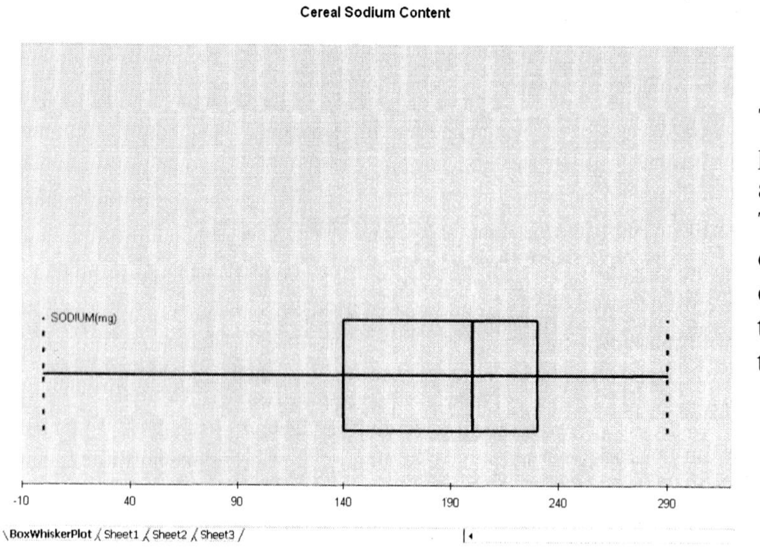

This macro does not highlight potential outliers with an asterisk, as MINITAB and the TI83/84 will do. Examination of whether values should be considered outliers would have to be carried out by you, using the 1.5 X IQR criterion.

Chapter 2 Exercise 2.77 European Union Unemployment Rates
Box Plots

Enter the data from exercise 2.64 into columns A and B.
Choose: **Add-Ins > PHStats > Descriptive Statistics > Box-and-Whisker Plot**

In the Box-and-Whisker Plot dialogue box
Enter: Raw Data Cell Range: **appropriate cells**
Title: **an appropriate title**
Click: **OK**.

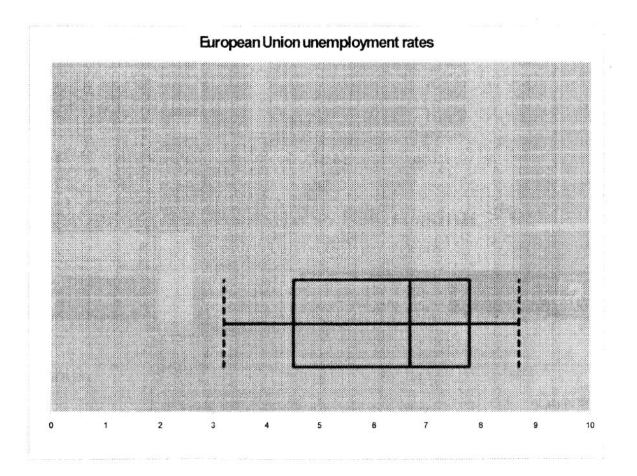

Chapter 2 Exercise 2.81 Florida Students Again
Box Plots

Open worksheet **fla_student_survey** from the Excel folder of the data disk.

Choose: **Add-Ins > PHStats > Descriptive Statistics > Box-and-Whisker Plot**

In the Box-and-Whisker Plot dialogue box
Enter: Raw Data Cell Range: **appropriate cells**
Title: **an appropriate title**
Click: **OK**

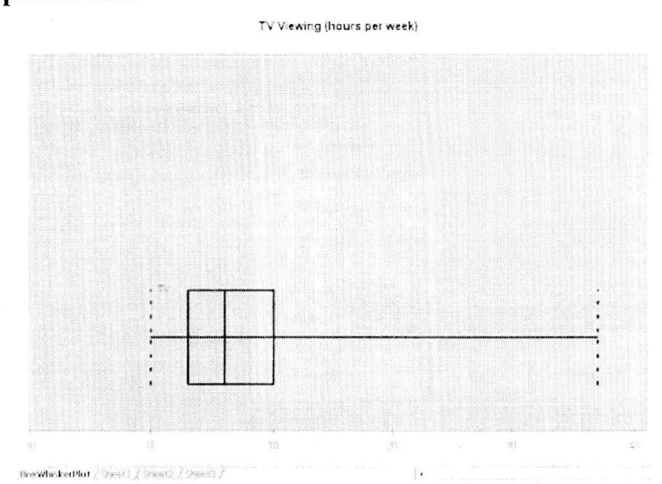

Chapter 2 Exercise 2.119b Temperatures in Central Park
Box Plots

Open worksheet **central_park_yearly_temps** from the Excel folder of the data disk.

Choose: **Add-Ins > PHStats > Descriptive Statistics > Box-and-Whisker Plot**

In the Box-and-Whisker Plot dialogue box
Enter: Raw Data Cell Range: **appropriate cells**
Title: **an appropriate title**
Click: **OK**

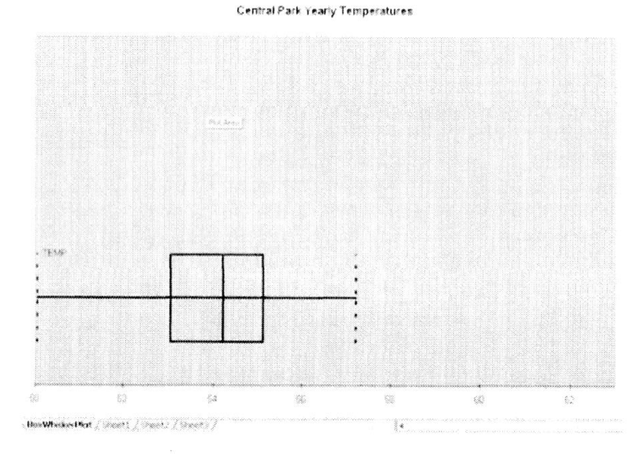

We can compare this to the histogram produced for part (a) of this exercise.

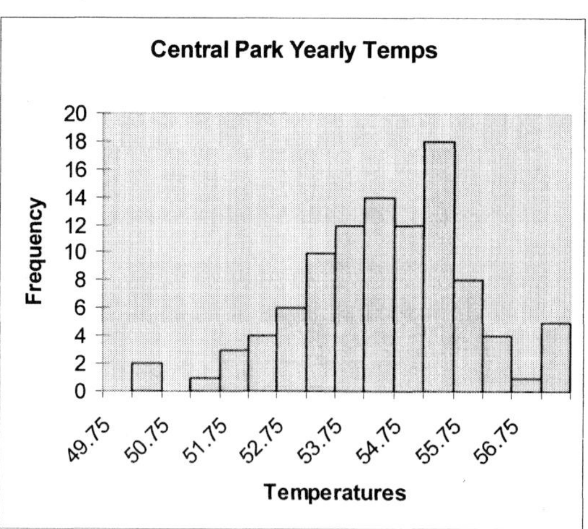

Chapter 2 Exercise 2.129 How Much is Spent on Haircuts?
Side-by-Side Plots

Open worksheet **georgia_student_survey** from the Excel folder of the data disk. For this exercise we wish to compare how much males and females spend on a haircut. First note that gender is indicated in column B as 0 for male and 1 for female. Many of the graphs illustrated in this chapter can be done as side-by-side plots, graphing with a separate plot for the males and the females, using the same scale, so that comparison of the distributions can be made.

	A	B	C	D	E
1	Height	Gender	Haircut	Job	Studytime
2	65	1	25	1	7
3	71	0	12	0	2
4	68	1	4	0	4
5	64	1	0	1	3.5
6	64	1	50	0	4.5
7	66	0	10	1	3

Generating histograms with the same scale (by using the same bin ranges)

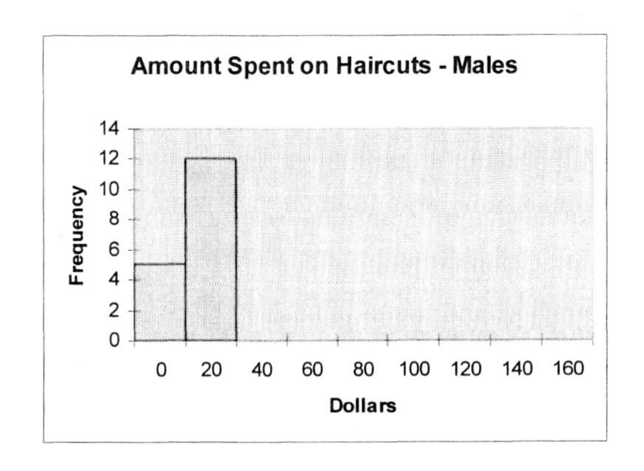

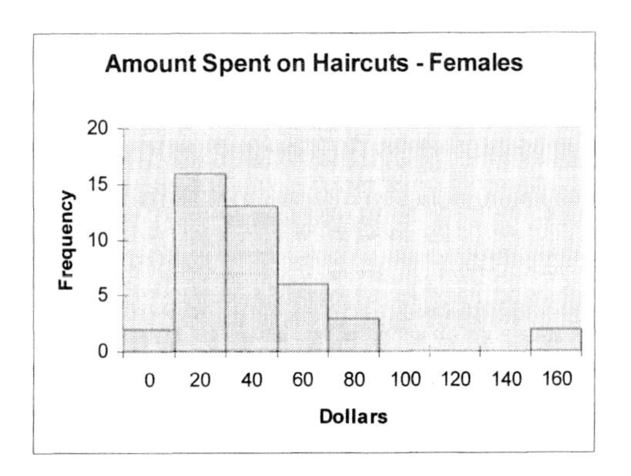

Side-by-side box plots:

Select: **Add-Ins > PHStat > Descriptive Statistics > Box-and-Whisker Plot**

Complete dialogue box as indicated.

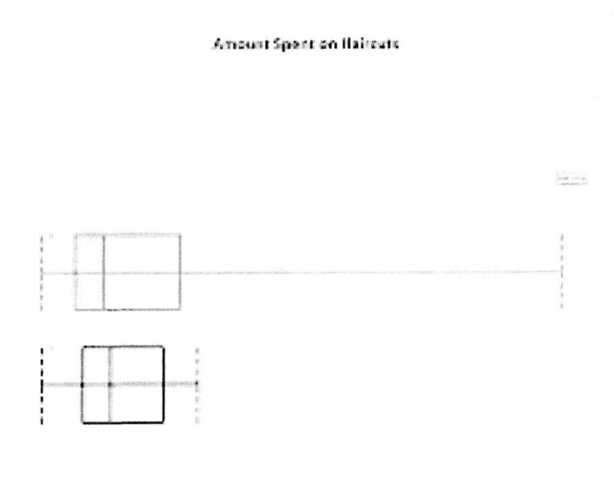

Chapter 3 Example 3 How Can We Compare Pesticide Residues For the Food Types Graphically?
Side-by-Side Bar Chart

Enter the conditional proportions (Table 3.2) into the worksheet.

	A	B	C
1	Food Type	Pesticide Present	Pesticide Not Present
2	Organic	0.23	0.77
3	Conventional	0.73	0.27
4			

To generate a single bar graph that shows side-by-side bars to compare the conditional proportion of pesticide residues in conventionally grown and organic foods:

Choose: **Insert > Chart > Column > 2D Column**
Choose: **Design > Chart Layouts > Layout 1**
Click on "Chart Title" and enter "Pesticide Status".

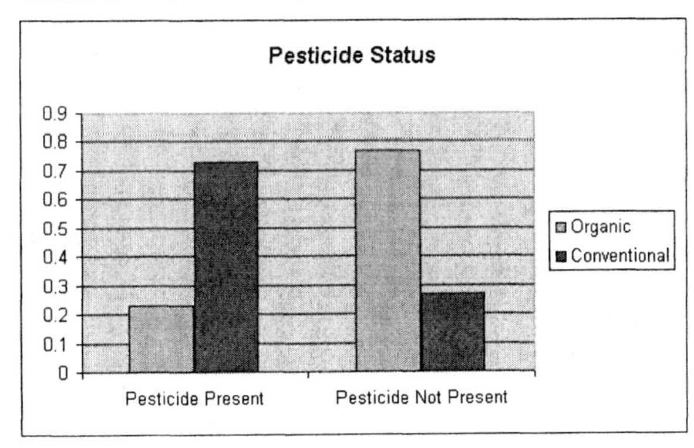

Chapter 3 Exercise 3.4 Religious Activities
Side-by-Side Bar Chart

Enter the data into the worksheet. Since we are comparing unequal numbers of males and females, we need to enter the data as proportions on hours of home religious activity within each category of gender.

	A	B	C	D	E	F
1	Gender	0	1 to 9	10 to 19	20 to 39	40 or more
2	Female	0.29896	0.38773	0.11488	0.13446	0.06397
3	Male	0.43533	0.38328	0.09306	0.06309	0.02524

Choose: **Insert > Chart > Column > 2D Column**
Choose: **Design > Chart Layouts > Layout 1**
Click on "Chart Title" and Enter "Hours of Religious Activity."

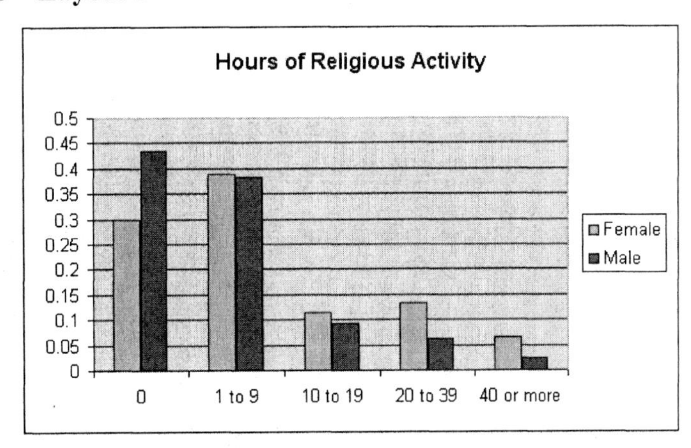

Chapter 3 Example 5 Constructing a Scatterplot for Internet Use and GDP
Constructing a Scatterplot

Open the worksheet **human_development**, which is found in the Excel folder of the data disk.

	A	B	C	D	E	F	G
1	C1-T	INTERNET	GDP	CO2	CELLULAR	FERTILITY	LITERACY
2	Algeria	0.65	6.09	3	0.3	2.8	58.3
3	Argentina	10.08	11.32	3.8	19.3	2.4	96.9
4	Australia	37.14	25.37	18.2	57.4	1.7	100
5	Austria	38.7	26.73	7.6	81.7	1.3	100
6	Belgium	31.04	25.52	10.2	74.7	1.7	100
7	Brazil	4.66	7.36	1.8	16.7	2.2	87.2
8	Canada	46.66	27.13	14.4	36.2	1.5	100

Since we are interested in how Internet use depends on GDP, column C GDP is the x-variable, and column B INTERNET is the y-variable.

To create a scatterplot of the data:
Highlight cells B1:C40
Choose: **Insert > Chart > Scatter > Scatter with only Markers**

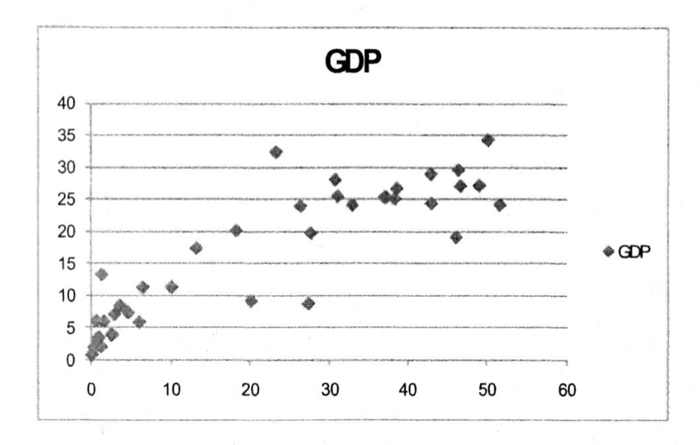

Note: by default, Column B Internet is selected as the x and column C GDP is selected as the y

We need to change this to
X Values: **Sheet1!$C$2:$C$40**
Y Values: **Sheet1!$B$2:$B$40**

Right click on the scatterplot and select “Select Data…”
Click **Edit.**
Enter: Series title: **GDP vs Internet Usage**
Series X Values: **=Sheet1!$C$2:$C$40**
Series Y Values: **=Sheet1!$B$2:$B$40**
Click OK. Click OK again.

Your graph should now look like this:

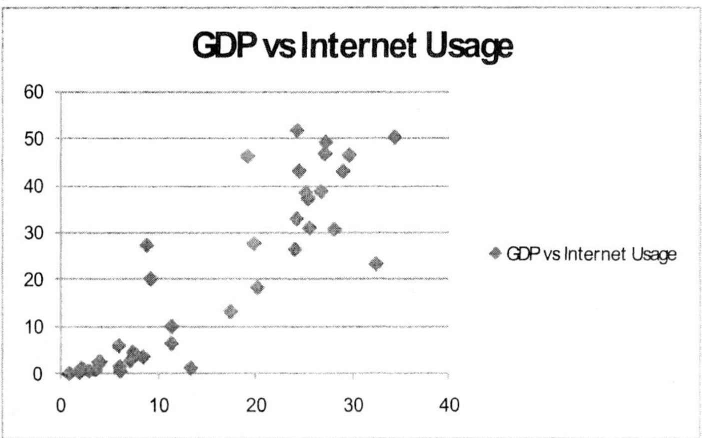

Lastly, we will add titles for the x and y axes.
Choose: **Design > Chart Layouts > Layout 1**
Click on the small “GDP vs Internet Usage” and click delete.
Click on the vertical “Axis Title” and enter: **% of adults who are internet subscribers**
Click on the horizontal “Axis Title” and enter: **Gross Domestic Product (thousands of $)**

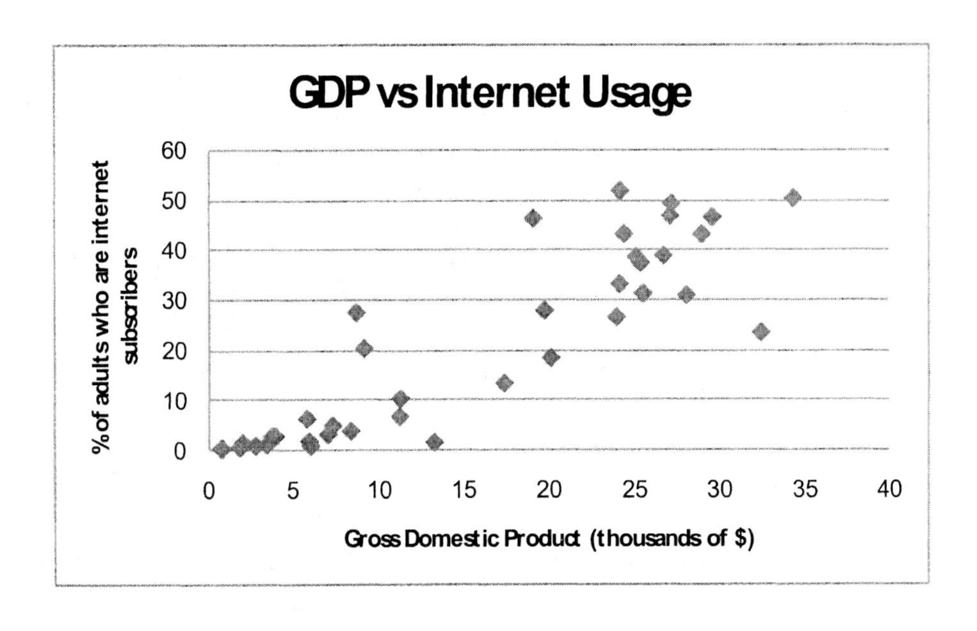

Chapter 3 Example 7 What's the Correlation Between Internet Use and GDP? Computing Correlation

Open the worksheet **human_development**, which is found in the Excel folder of the data disk.

Since we are interested in how Internet use depends on GDP, column C, GDP, is the x-variable, and Column B, Internet, is the y-variable.

To find the correlation:

Click: **within an empty cell**

	A	B	C	D	E	F	G	H	I
1		INTERNET	GDP	CO2	CELLULAF	FERTILITY	LITERACY		
2	Algeria	0.65	6.09	3	0.3	2.8	58.3		
3	Argentina	10.08	11.32	3.8	19.3	2.4	96.9		
4	Australia	37.14	25.37	18.2	57.4	1.7	100		
5	Austria	38.7	26.73	7.6	81.7	1.3	100		
6	Belgium	31.04	25.52	10.2	74.7	1.7	100		

Choose: **Formulas > Insert Function**

Select: Or select a category: **Statistical**

Select: **CORREL**

Click: **OK**

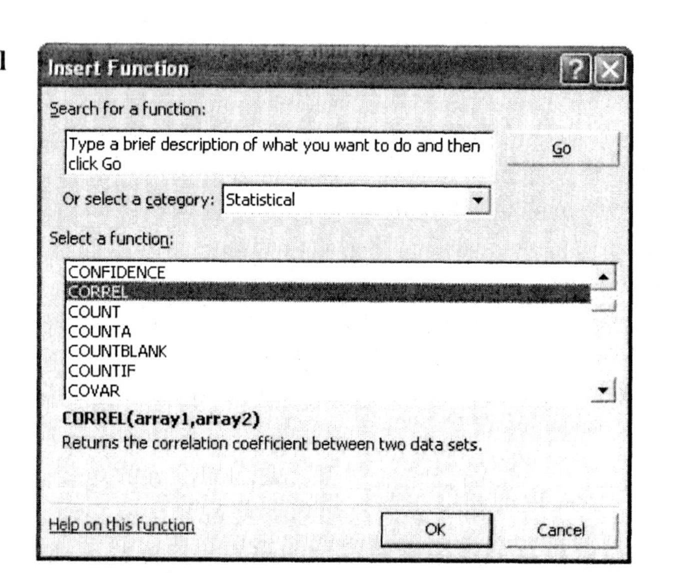

Complete the Function Arguments dialog box as shown. Click **OK**.

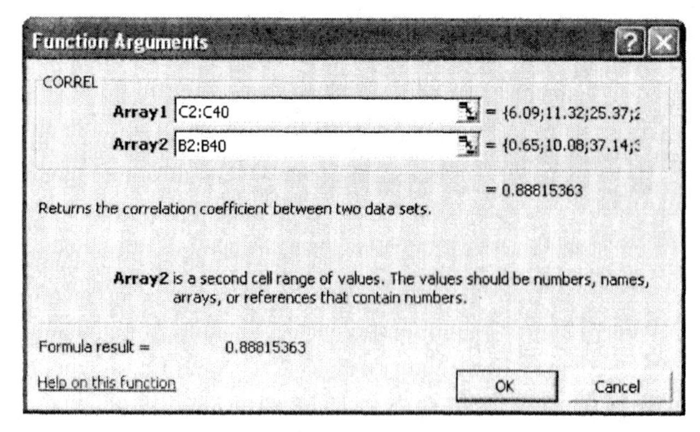

Your output should appear as shown.

	A	B	C	D	E	F	G	H	I
2	Algeria	0.65	6.09	3	0.3	2.8	58.3		0.888154
3	Argentina	10.08	11.32	3.8	19.3	2.4	96.9		
4	Australia	37.14	25.37	18.2	57.4	1.7	100		
5	Austria	38.7	26.73	7.6	81.7	1.3	100		
6	Belgium	31.04	25.52	10.2	74.7	1.7	100		

Chapter 3 Exercise 3.21 Which Mountain Bike to Buy?
Computing Correlation

Open the worksheet **mountain_bike**, which is found in the Excel folder of the data disk.

Since we are interested in whether and how weight affects the price, column B, weight, is the x-variable, and column A, price, is the y-variable.

To create a scatterplot of the data

Highlight cells A1:B13
Choose: **Insert > Chart > Scatter > Scatter with only Markers**

Notice, by default, that the values in column A (price) are being treated as the x-variable.
Right click on the scatterplot and select "Select Data…"
Click **Edit.**
Enter: Series title: **Mountain Bikes / Weight vs. Price**
Series X Values: **=Sheet1!$B$2:$B$13**
Series Y Values: **=Sheet1!$A$2:$A$13**
Click OK. Click OK again.

Now, we will add titles for the x and y axes.
Choose: **Design > Chart Layouts > Layout 1**
Click on the small "Mountain Bikes / Weight vs. Price" and click delete.
Click on the vertical "Axis Title" and enter: **Price ($)**
Click on the horizontal "Axis Title" and enter: **Weight (lb)**

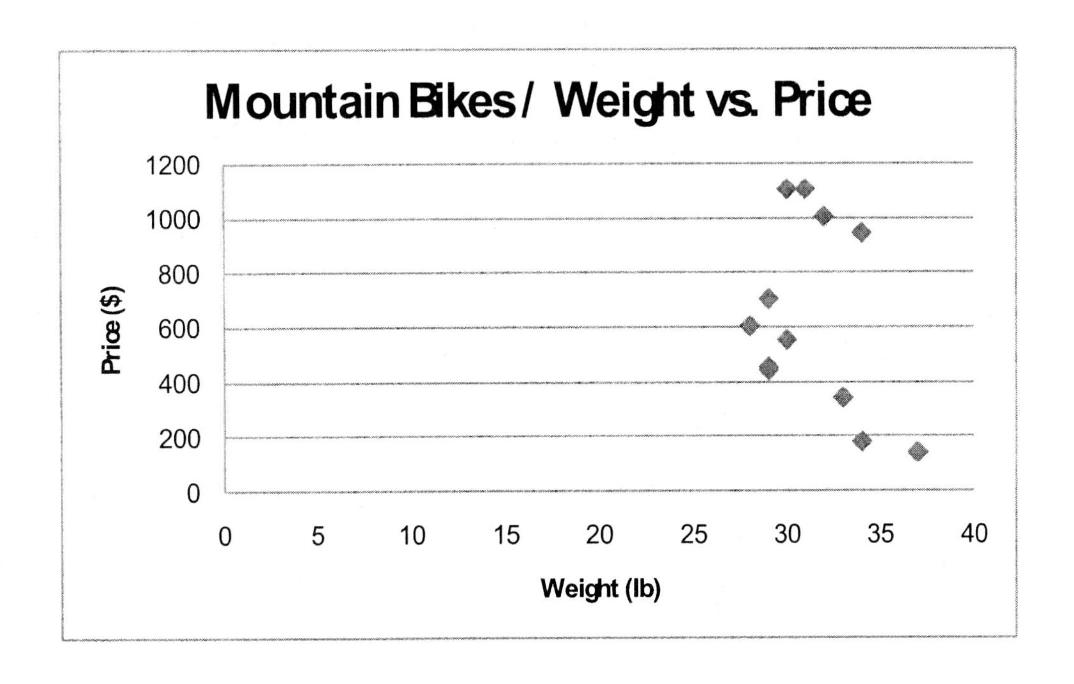

Notice that now we also need to adjust the scale on the x-axis. Place your cursor within the chart on the x-axis, so that the mouseover reads "Horizontal (Value) axis", and right-click. Choose "Format Axis…"

In the Axis Options window,

Enter: Fixed, Minimum: **27**

Click: **Close**

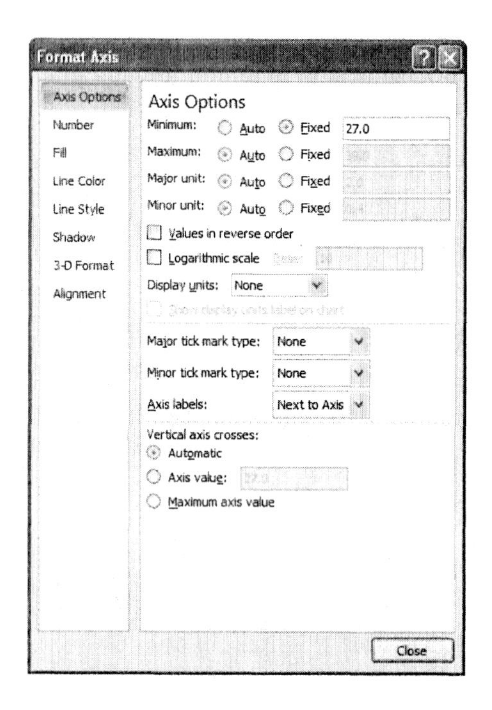

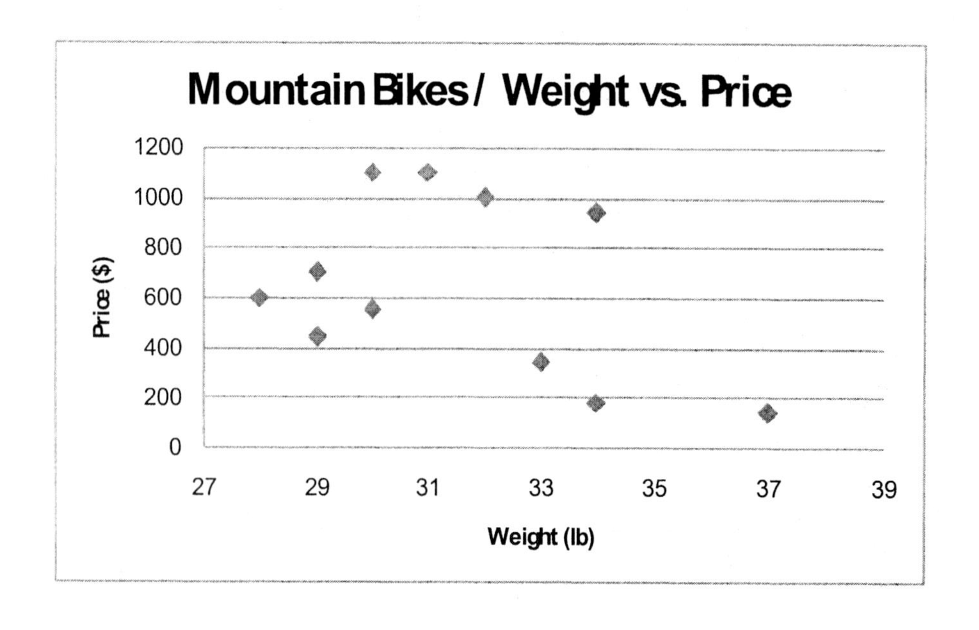

To find the correlation:

Click: **within an empty cell**
Choose: **Formulas > Insert Function**
Select: Or select a category: **Statistical**
Select: **CORREL**
Click: **OK**

Complete the Function Arguments dialog box as shown.
Click **OK**.

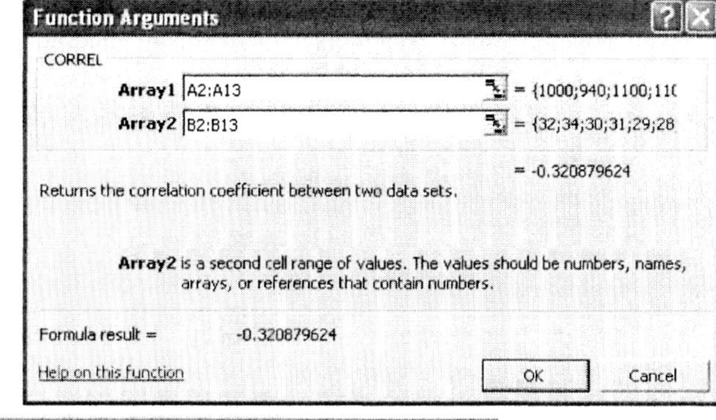

Your output should appear as shown.

	A	B	C	D
1	price	weight		
2	1000	32		-0.32088
3	940	34		
4	1100	30		
5	1100	31		

Chapter 3 Exercise 3.23 Buchanan Vote
Computing Correlation

Open the worksheet **Buchanan_and_the_butterfly_ballot** which is in the Excel folder of the data disk.

Creating a box plot of Gore and Buchanan:

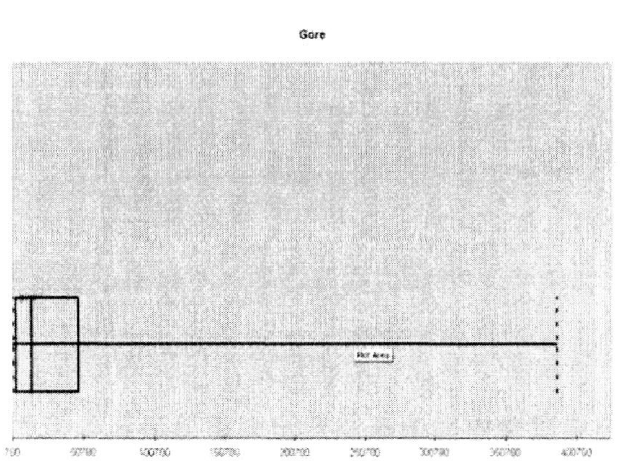

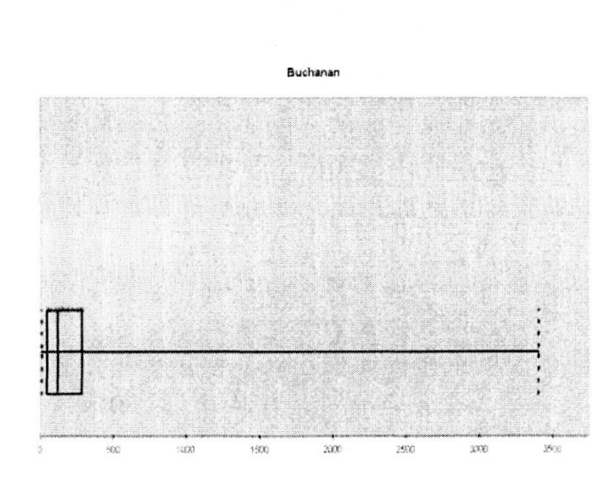

Note the different scales when comparing the two distributions

To create a scatterplot of the data
Highlight cells C1:D68
Choose: **Insert > Chart > Scatter > Scatter with only Markers**

Right click on the scatterplot and select "Select Data…"
Click **Edit.**
Enter: Series title: **The Gore Vote vs The Buchanan Vote**
Series X Values: =**Sheet1!$C$2:$C$68**
Series Y Values: =**Sheet1!$E$2:$E$68**
Click OK. Click OK again.

Now, we will add titles for the x and y axes.
Choose: **Design > Chart Layouts > Layout 1**
Click on the small "The Gore Vote vs The Buchanan Vote" and click delete.
Click on the vertical "Axis Title" and enter: **Buchanan**
Click on the horizontal "Axis Title" and enter: **Gore**

We will now adjust the scale on each axis. Place your cursor within the chart on the x-axis, so that the mouseover reads "Horizontal (Value) axis", and right-click.
Choose "Format Axis…"

In the Format Axis window,
Enter: Fixed, Maximum: **400000**
Click: **Close**

Place your cursor within the chart on the y-axis, so that the mouseover reads "Vertical (Value) axis", and right-click.
Choose "Format Axis…"

In the Format Axis window, on the Scale tab
Enter: Fixed, Maximum: **3500**
Click: **Close**

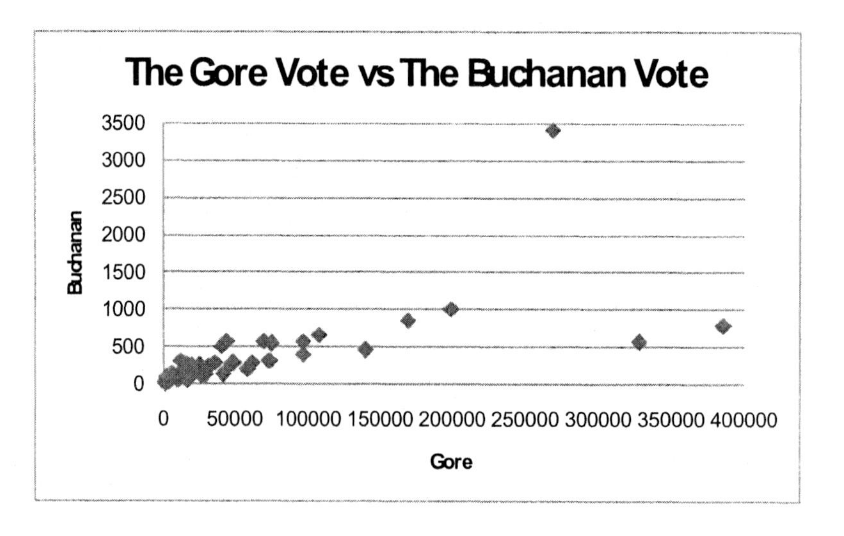

To find the correlation:

Click: **within an empty cell**
Choose: **Formulas > Insert Function**
Select: Or select a category: **Statistical**
Select: **CORREL**
Click: **OK**

Complete the Function Arguments dialog box as shown. Click **OK**.

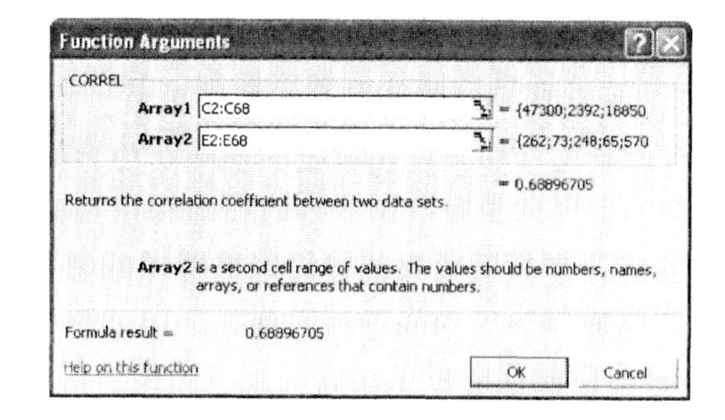

Your output should appear as shown.

	A	B	C	D	E	F	G
1	county	perot	gore	bush	buchanan		
2	Alachua	8072	47300	34062	262		0.688967
3	Baker	667	2392	5610	73		
4	Bay	5922	18850	38637	248		
5	Bradford	819	3075	5414	65		

Chapter 3 Example 9 How Can We Predict Baseball Scoring Using Batting Average? Generating the Regression Equation

Open the worksheet **al_team_statistics** which is in the Excel folder of the data disk.
Highlight cells D1:E15
Choose: **Insert > Chart > Scatter > Scatter with only Markers**
Choose: **Design > Chart Layouts > Layout 1**
Enter: Chart title: **Batting Average vs Team Scoring**
Value (x) axis: **Batting Average**
Value (y) axis: **Scoring Average**
Adjust the scale on each axis.

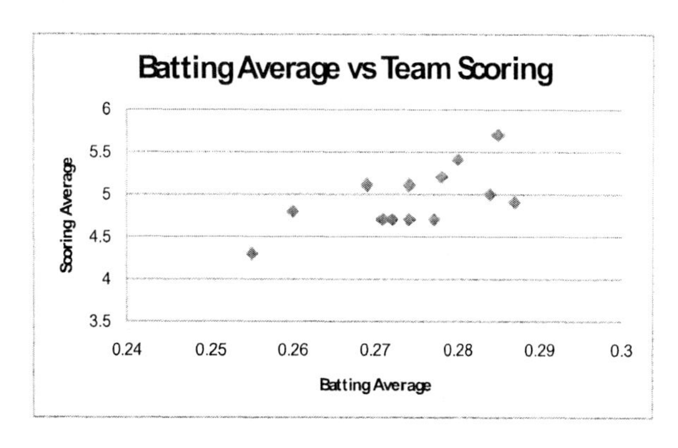

Now we will calculate the regression equation:

Choose: **Data >Data Analysis**
Select: **Regression**
Click: **OK**

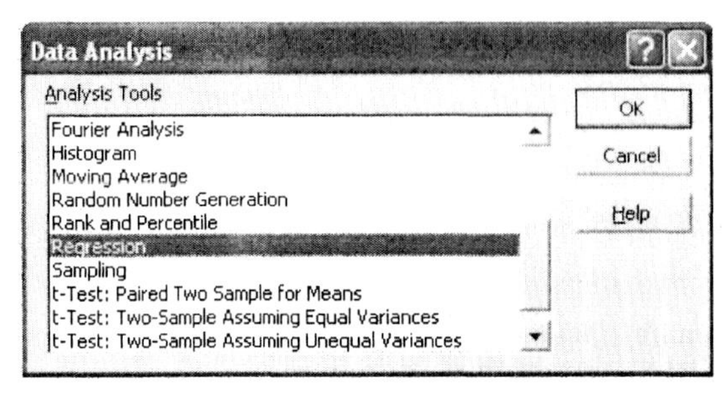

Indicate the location of the data, as appropriate:

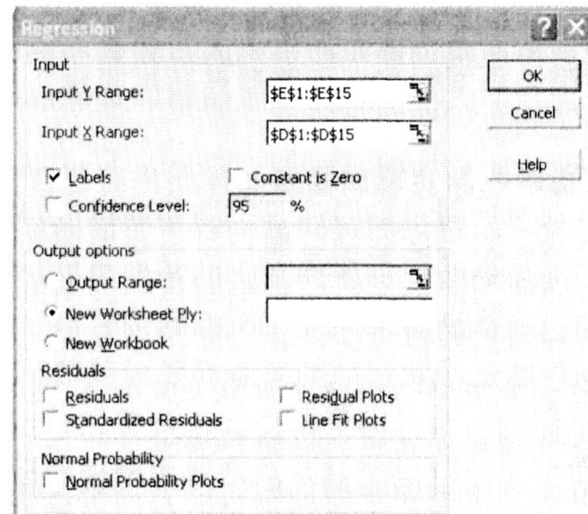

Here is the default output generated by the **Regression** command for Example 9.
Notice that a great deal of information is generated, but at this point we would need only the coefficients circled.

	A	B	C	D	E	F	G	H	I
1	SUMMARY OUTPUT								
2									
3	*Regression Statistics*								
4	Multiple R	0.652627							
5	R Square	0.425922							
6	Adjusted F	0.378082							
7	Standard E	0.290463							
8	Observatic	14							
9									
10	ANOVA								
11		*df*	*SS*	*MS*	*F*	*gnificance F*			
12	Regressic	1	0.751143	0.751143	8.903069	0.011405			
13	Residual	12	1.012428	0.084369					
14	Total	13	1.763571						
15									
16		*Coefficient*	*tandard Err*	*t Stat*	*P-value*	*Lower 95%*	*Upper 95%*	*ower 95.0%*	*pper 95.0%*
17	Intercept	-2.28361	2.435108	-0.93779	0.366846	-7.58926	3.022031	-7.58926	3.022031
18	BAT_AVG	26.43541	8.859641	2.983801	0.011405	7.131907	45.73891	7.131907	45.73891

The least squares line can be added to the plot, along with its equation and the value of r^2. Right click on one of the data points shown in the scatter plot. A drop-down menu will appear.

Select: **Add Trendline**
Select: Type: **Linear**
Check: **Display equation on chart** and **Display R-squared on chart > Close**

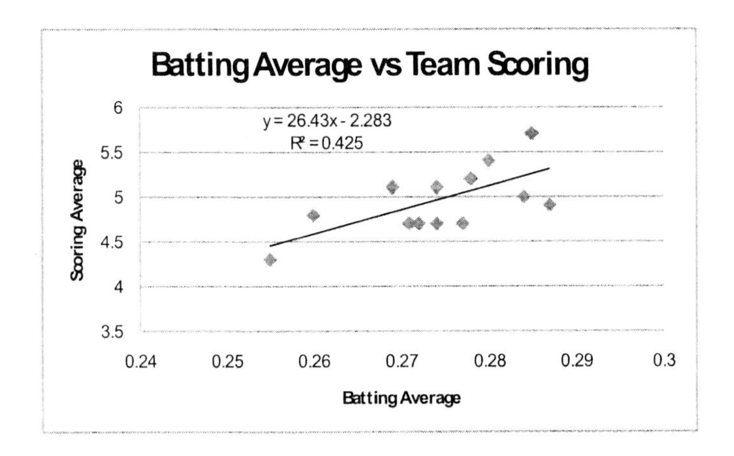

Use the regression equation to do the prediction.

Chapter 3 Exercise 3.41 Mountain Bikes Revisited
Generating a Regression Equation

Highlight cells A1:B13
Choose: **Insert > Chart > Scatter > Scatter with only Markers**

Right click on the scatterplot and select "Select Data…"
Click **Edit.**
Enter: Series title: **Weight vs. Price of Mountain Bikes**
Series X Values: **=Sheet1!$B$2:$B$13**
Series Y Values: **=Sheet1!$A$2:$A$13**
Click OK. Click OK again.

After formatting the chart and adding the trendline we get:

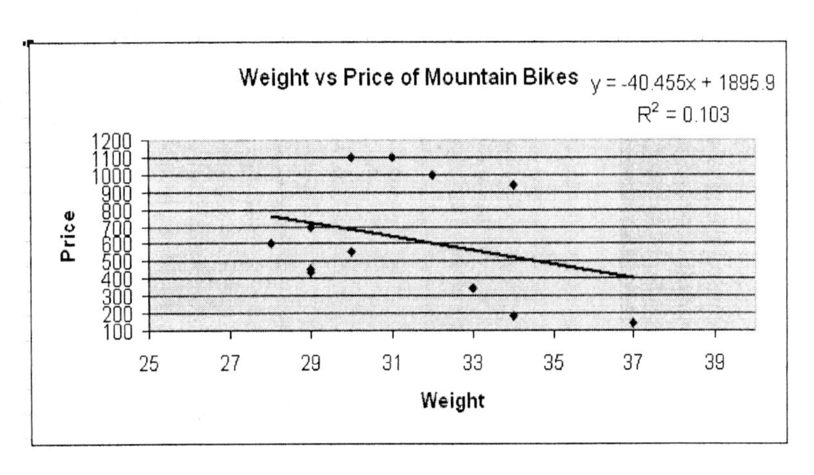

Chapter 3 Exercise 3.42 Mountain Bike and Suspension Type
Generating Regression Equations

Do a scatterplot designating front end and full suspensions as different series. Include both regression equations on the chart.
Highlight cells D1:E9
Choose: **Insert > Chart > Scatter > Scatter with only Markers**
Right Click on the chart and select "Select Data…"
Click Edit and enter the following information:

Series name: **FE**
Series X Values: **=Sheet1!$E$2:$E$9**
Series Y Values: **=Sheet1!$D$2:$D$9**

Click: **OK**
Click: **Add** and enter the following information:

Series name: **FU**
Series X Values: **=Sheet1!$G$2:$G$5**
Series Y Values: **=Sheet1!$F$2:$F$5**

Click: **OK**

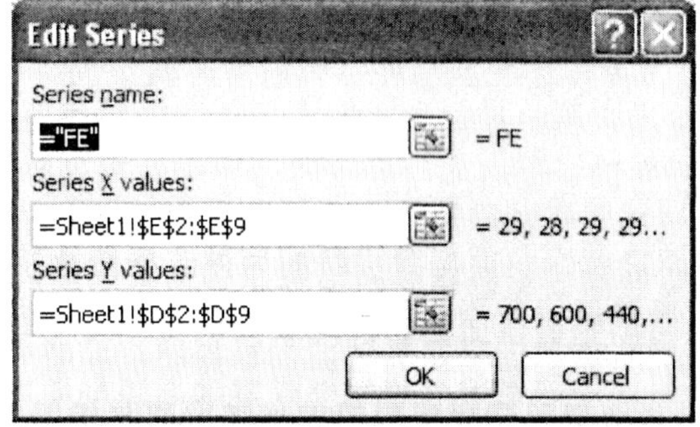

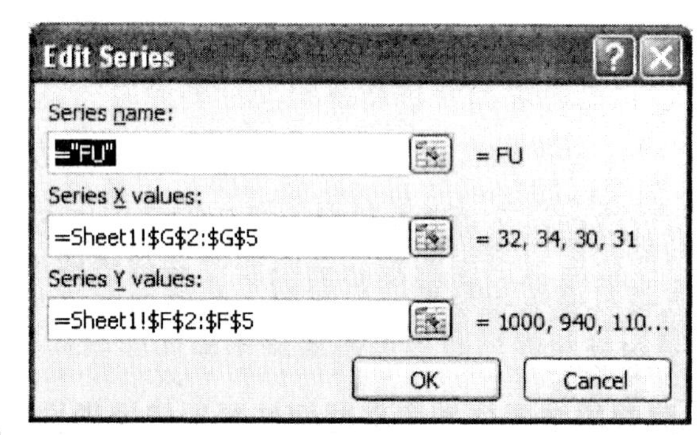

Choose: **Design > Chart Layouts > Layout 1**
Enter: Chart title: **Full Suspension & Front-end Suspension**
Value (x) axis: **Weight**
Value (y) axis: **Price**
After formatting the chart and adding the trendlines we get:

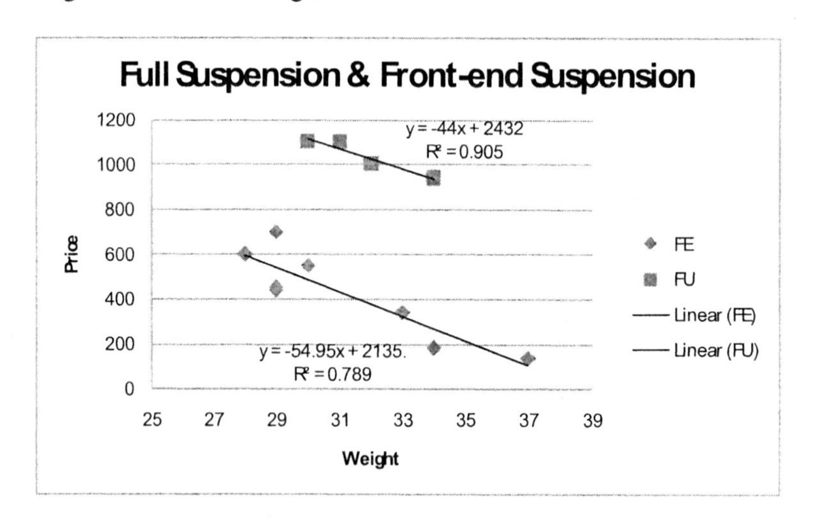

Find the correlations for the two types separately:

	A	B	C	D
1		weight_FU	price_FU	
2	weight_FU	1		
3	price_FU	-0.95178	1	
4				
5				

	A	B	C	D
1		weight_FE	price_FE	
2	weight_FE	1		
3	price_FE	-0.88842	1	
4				
5				

What conclusions can you draw?

Chapter 3 Exercise 3.89 High School Graduation Rates and Health Insurance Scatterplot, Correlation, and Regression

Open the worksheet **hs_graduation_rates** from the Excel folder of the data disk.

Choose: **Insert > Chart > Scatter > Scatter with only Markers**
Choose: **Design > Chart Layouts > Layout 1**
Enter: Chart title: **HS Grad rates vs Health Insurance**
Value (x) axis: **Grad rates**
Value (y) axis: **Health Ins**

After formatting the chart and adding the trendline we get:

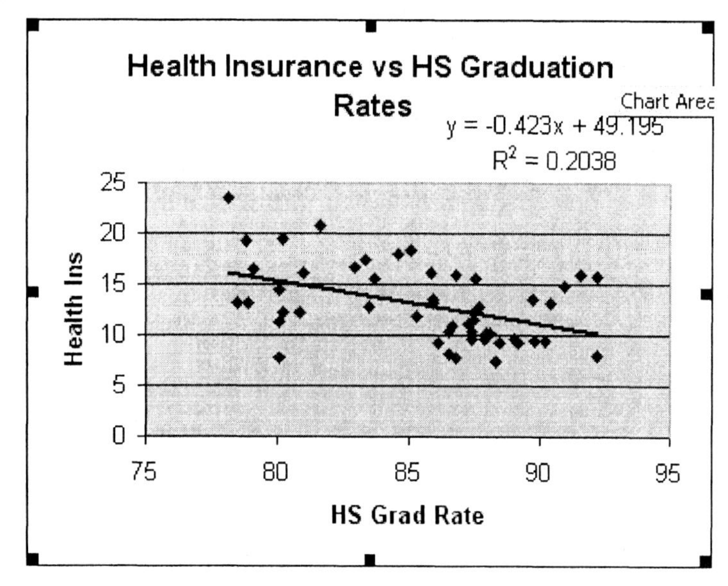

Computing correlation: Click: **within an empty cell**
Choose: **Formulas > Insert Function**
Select: Or select a category: **Statistical**
Select: **CORREL**
Click: **OK**

Complete the Function Arguments dialog box as appropriate. Click **OK**.

	A	B	C	D
1		HS Grad Rate	Health Ins	
2	HS Grad Rate	1		
3	Health Ins	-0.451437877	1	
4				

To do the regression: Choose: **Data > Data Analysis > Regression > OK**
Enter: Input Y Range: **D2:D52**
Input X Range: **C2:C52**
Click: **OK**

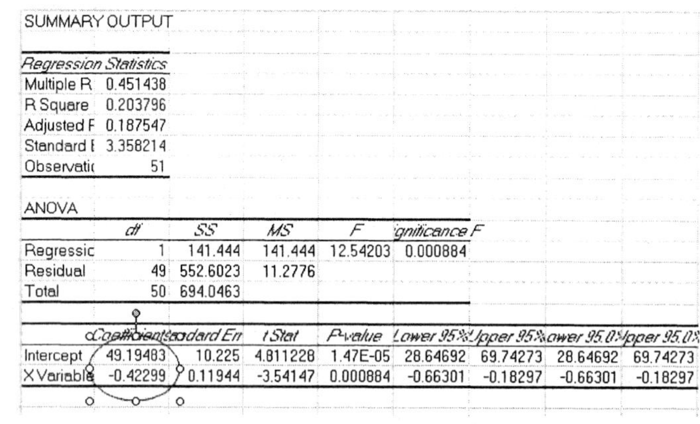

SUMMARY OUTPUT

Regression Statistics	
Multiple R	0.451438
R Square	0.203796
Adjusted F	0.187547
Standard I	3.358214
Observatio	51

ANOVA

	df	SS	MS	F	ignificance F
Regressic	1	141.444	141.444	12.54203	0.000884
Residual	49	552.6023	11.2776		
Total	50	694.0463			

	Coefficients	tandard Err	t Stat	P-value	Lower 95%	Upper 95%	ower 95.0%	pper 95.0%
Intercept	49.19483	10.225	4.811228	1.47E-05	28.64692	69.74273	28.64692	69.74273
X Variable	-0.42299	0.11944	-3.54147	0.000884	-0.66301	-0.18297	-0.66301	-0.18297

Chapter 3 Example 10 How Can We Detect an Unusual Vote Total? Looking at Residuals

Open the worksheet **Buchanan_and_the_butterfly_ballot** from the Excel folder of the data disk.

Select: **Data > Data Analysis > Regression**
Enter: Input Y Range: **select appropriate cells**
Input X Range: **select appropriate cells**
Select: **Residuals**

Click: **OK**

Regression
Input
Input Y Range: $E$2:$E$68
Input X Range: $B$2:$B$68
Labels
Constant is Zero
Confidence Level: 95 %
OK
Cancel
Help
Output options
Output Range:
New Worksheet Ply:
New Workbook
Residuals
Residuals
Standardized Residuals
Residual Plots
Line Fit Plots
Normal Probability
Normal Probability Plots

Next, you will copy the residuals to the worksheet containing the data so that you can use them in a histogram of the residuals. Click and drag over the residuals, so that the range is highlighted. Right click in the highlighted section and select copy. Paste into the data worksheet.

22	RESIDUAL OUTPUT		
23			
24	Observation	Predicted Y	Residuals
25	1	289.5312	-27.5312
26	2	24.91641	48.08359
27	3	212.7018	35.29824
28	4	30.34807	34.65193
29	5	903.3445	-333.344
30	6	1393.445	-604.445
31	7	23.59423	66.40577
32	8	279.2039	-97.2039
33	9	259.9429	10.05709
34	10	118.3267	67.67333
35	11	226.9241	-104.924
36	12	71.4786	17.5214
37	13	[illegible]	[illegible]

	A	B	C	D	E	F
1	county	perot	gore	bush	buchanan	residuals
2	Alachua	8072	47300	34062	262	-27.5312
3	Baker	667	2392	5610	73	48.08359
4	Bay	5922	18850	38637	248	35.29824
5	Bradford	819	3075	5414	65	34.65193
6	Brevard	25249	97318	115185	570	-333.344

Generating a histogram of the residuals:

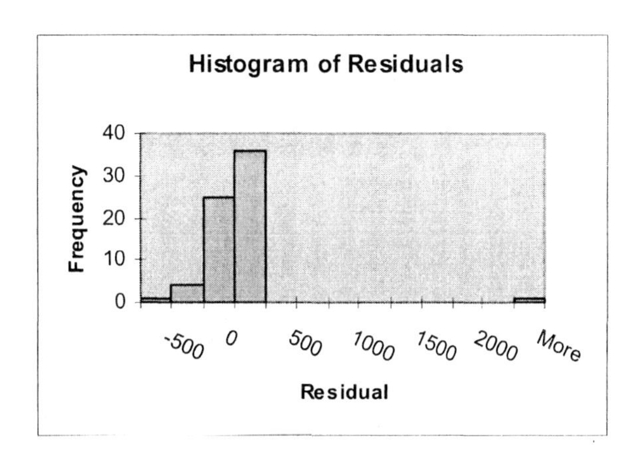

Chapter 3 Exercise 3.33 Regression Between Cereal Sodium and Sugar Residuals

Open the worksheet **CEREAL** from the Excel folder of the data disk. Perform the regression, checking Residuals in the dialog box.

Select: **Data > Data Analysis > Regression**
Enter: Input Y Range: **select appropriate cells**
Input X Range: **select appropriate cells**
Select: **Residuals**
Click: **OK**
Copy the residuals to the worksheet containing the data, and construct a histogram of the residuals.

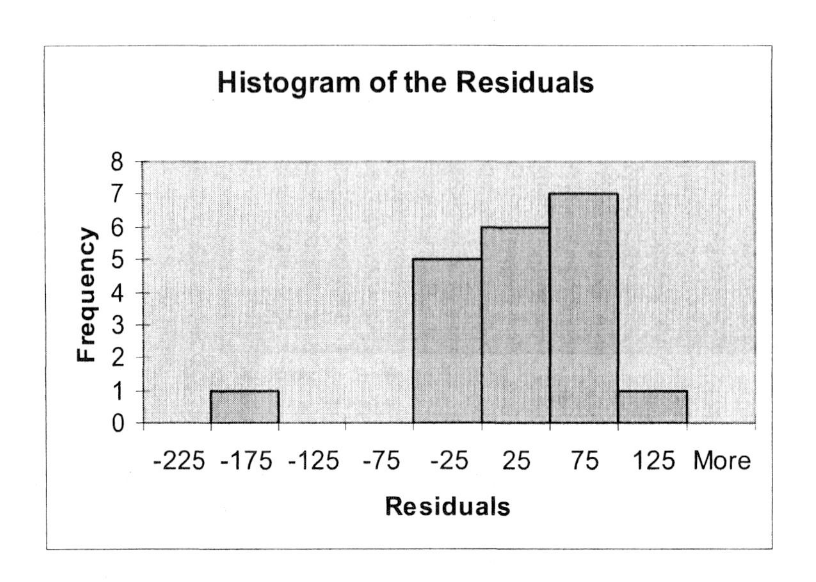

Chapter 3 Example 12 How Can We Forecast Future Global Warming? Regression on a Time Series Plot

Open the worksheet **central_park_yearly_temps** from the Excel folder of the data disk.

Because our x-variable is years, the scatterplot can be interpreted as a time plot.
Choose: **Insert > Chart > Scatter > Scatter with Straight Lines and Markers**

Select the columns containing the data.
Add titles as appropriate.

Format the graph, choosing proper axis scale and adding the trendline.

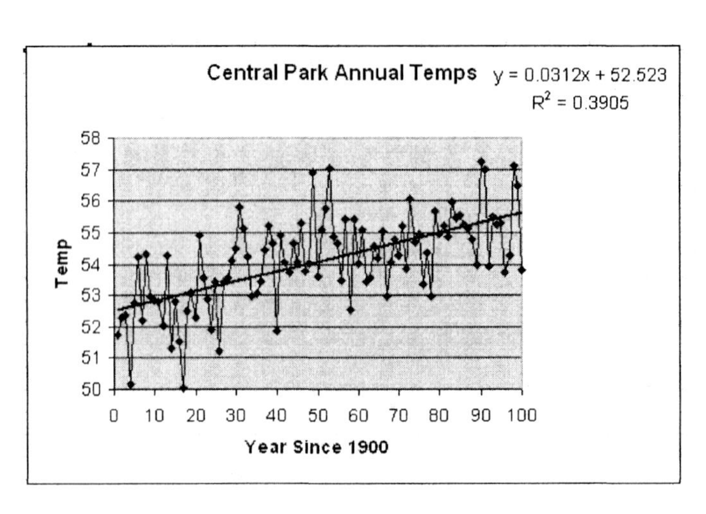

Chapter 3 Example 13 Is Higher Education Associated with Higher Murder Rates? Influential Outliers

Open the worksheet **us_statewide_crime** from the Excel folder of the data disk.

Choose: **Insert > Chart > Scatter > Scatter with only Markers**
Choose: **Design > Chart Layouts > Layout 1**
Enter: Chart title:
Value (x) axis: **% with College Education**
Value (y) axis: **Murder rate**

After formatting the chart and adding the trendline we get:

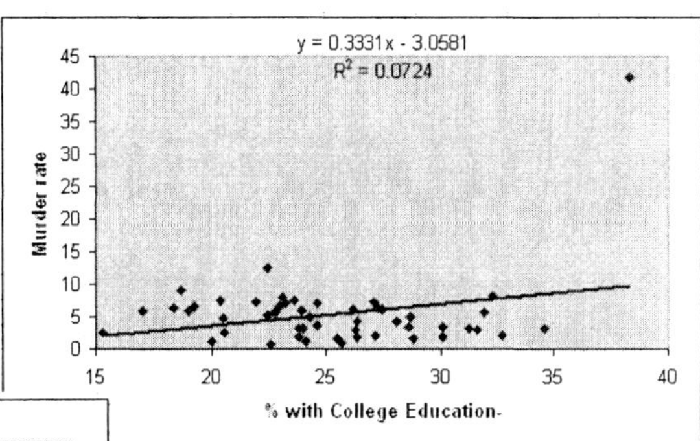

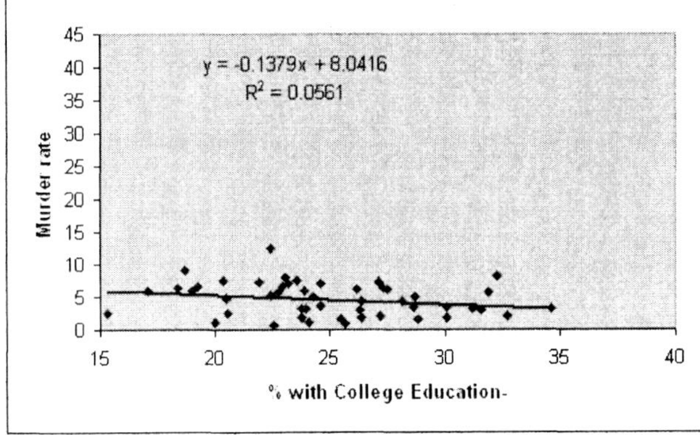

Redoing the entire process, excluding the data for D.C. we get:

Chapter 3 Exercise 3.47 Murder and Education Application of the Regression Model

Using the Excel results of the previous problem

a) In the equation y = 0.3331 x – 3.0581 we substitute 15 and 40 for x, we obtain 1.89 and 10.26 for y, respectively.

b) In the equation y = -.1379 x + 8.0416 we substitute 15 and 40 for x, we obtain 5.9 and 2.4 for y, respectively.

Chapter 3 Exercise 3.51 Regression Between Sodium and Sugar
Influential Outliers

Open the worksheet **CEREAL** from the Excel folder of the data disk. Let x = sodium (mg) and y = sugar (g). Using the chart wizard construct a scatterplot and include the regression line.

Choose: **Insert > Chart > Scatter > Scatter with only Markers**
Choose: **Design > Chart Layouts > Layout 1**
Enter: Chart title: **Sodium vs Sugar**
Value (x) axis: **Sodium (mg)**
Value (y) axis: **Sugar (mg)**

After formatting the chart and adding the trendline we get:

Note the outlier at (7000, 0).

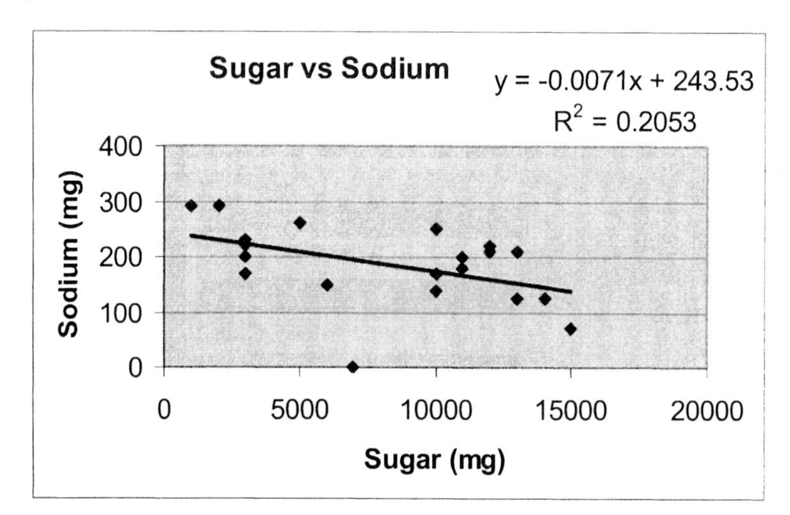

Computing correlation:
Click: **within an empty cell**
Choose: **Formulas > Insert Function**
Select: Or select a category: **Statistical**
Select: **CORREL**
Click: **OK**
Complete the Function Arguments dialog box as appropriate. Click **OK**

	A	B	C	D
1		SODIUM(mg	SUGAR(g)	
2	SODIUM(n	1		
3	SUGAR(g)	-0.45305	1	
4				

Then without the outlier:

	A	B	C	D
1		SODIUM(mg	SUGAR(g)	
2	SODIUM(n	1		
3	SUGAR(g)	-0.62255	1	
4				

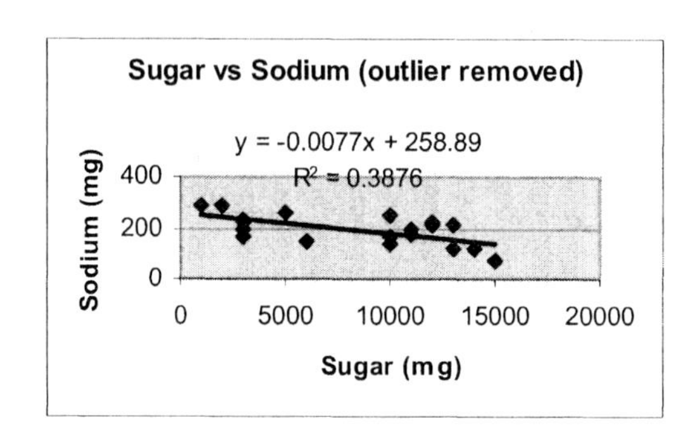

What conclusions can you draw?

Chapter 4 Example 5 Auditing the Accounts of a School District
Random Selection

To use random numbers within Excel to select 10 accounts to audit in a school district that has 60 accounts we will use the PHStat add-in.

Within a blank Excel worksheet,
Click: **Add-Ins > PHStat**
Select: **Sampling > Random Sample Generation**

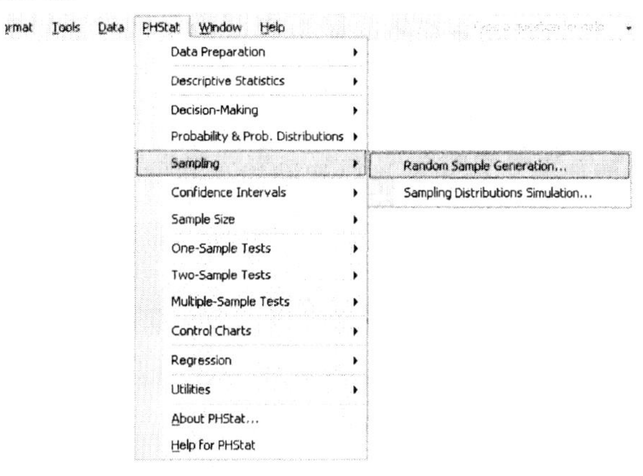

Complete the dialog box as shown:

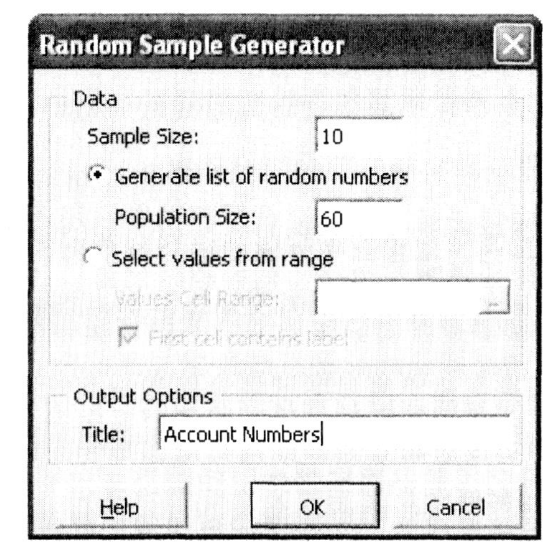

The output is displayed in a new worksheet.

	A
1	Account Numbers
2	51
3	50
4	30
5	54
6	52
7	7
8	18
9	43
10	42
11	26

Because these numbers were randomly generated, it is not likely that your output will be exactly the same.

Chapter 4 Exercise 4.18 Auditing Accounts
Random Selection

To use random numbers within Excel to select 10 accounts to audit in a school district that has 60 accounts we will use the PHStat add-in.

Within a blank Excel worksheet,
 Click: **Add-Ins > PHStat**
 Select: **Sampling >**
  Random Sample Generation

Complete the dialog box as shown:

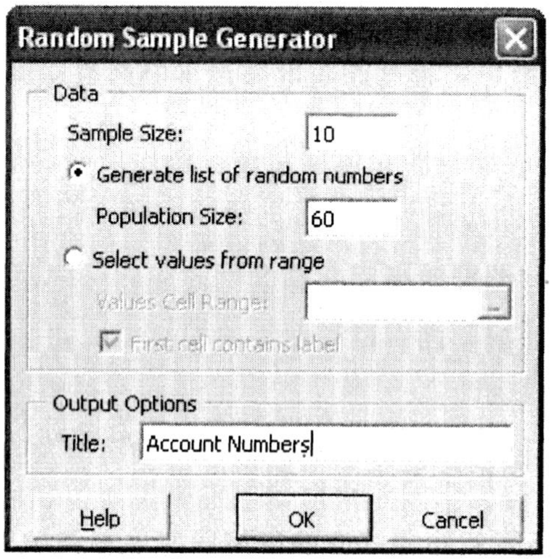

Random Sample Generator
Data
Sample Size: 10
Generate list of random numbers
Population Size: 60
Select values from range
Output Options
Title: Account Numbers
Help OK Cancel

The output is displayed in a new worksheet.

Because we are using a random number generator, each time this command is issued, different accounts will be selected.

	A
1	Account #'s
2	35
3	12
4	41
5	60
6	18
7	43
8	47
9	57
10	46
11	16

Chapter 6 Example 7 What IQ Do You Need to Get Into MENSA?
Determining Normal Probabilities

Stanford-Binet IQ scores are approximately normally distributed with $\mu = 100$ and $\sigma = 16$. To be eligible for MENSA, you must rank at the 98th percentile (i.e. you must score in the top 2%). Find the lowest IQ score that still qualifies for Mensa membership.

Choose: **Formulas > Insert Function**
Select: Or select a category: **Statistical**
Select: Select a function: **NORMINV**

Click: **OK**

Function Arguments
Enter: Probability: **.98**
(area to the left of the value we wish to find)
Mean: **100**
Standard deviation: **16**

Click: **OK**

The X-value should appear in the worksheet.

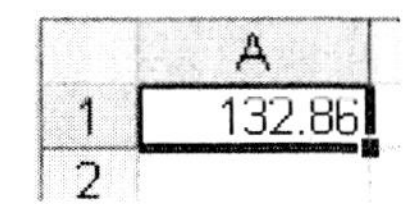

Notice that the test score needed to join Mensa is 133.

Chapter 6 Example 8 Finding Your Relative Standing on the SAT
Determining Normal Probabilities

SAT scores are approximately normally distributed with $\mu = 500$ and $\sigma = 100$. If one of your SAT scores was 650, what percentage of SAT scores was higher than yours?

Choose: **Formulas > Insert Function**
Select: Or select a category: **Statistical**
Select: Select a function: **NORMDIST**

Click: **OK**

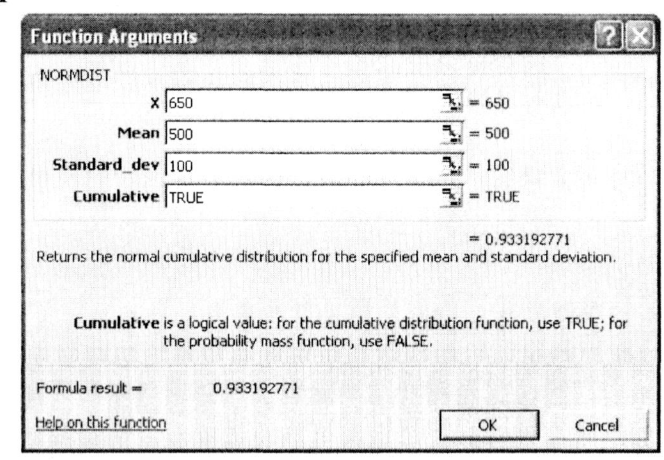

Enter: X **650**
Mean **500**
Standard_dev **100**
Click: **OK**

0.933193

Notice that the result displayed in the worksheet is the probability that X is less than 650.

Since we were asked to determine the percentage of SAT scores higher than 650, we must calculate 1 - 0.9332 = 0.0668. Only about 7% of SAT scores fall above 650.

Chapter 6 Example 9 What Proportion of Students Get a Grade of B?
Determining Normal Probabilities

On the midterm exam, an instructor always gives a grade of B to students who score between 80 and 90. One year, the scores on the exam have an approximately normal distribution with $\mu = 83$ and $\sigma = 5$. About what proportion of students get a B?

We will have to compute two cumulative probabilities, one for X = 80 and one for X = 90, and then subtract the two probabilities.

Choose: **Formulas > Insert Function > NORMDIST**
Enter: X **80** then repeat using **90**
Mean **83**
Standard_dev **5**
Click: **OK**

	A
1	0.274253
2	0.919243

It follows that about 0.9192 – 0.2743 = 0.6449, or about 64%, of the exam scores were in the B range.

Chapter 6 Exercise 6.23 Blood pressure
Determining Normal Probabilities

In Canada, systolic blood pressure readings has are normally distributed with $\mu = 121$ and $\sigma = 16$. A reading above 140 is considered to be high blood pressure. What proportion of Canadians suffers from high blood pressure?

Choose: **Formulas > Insert Function > NORMDIST**
Enter: X **140**
Mean **121**
Standard_dev **16**
Click: **OK**

0.882485 It follows that about 1 – 0.8825 = 0.1175, or about 12%, of Canadians suffer from high blood pressure.

We are also asked to determine the proportion of Canadians having systolic blood pressure in the range from 100 to 140.

We have already computed the $P(X < 140) = 0.8825$, so we just need to determine $P(X < 100)$ and subtract the two results.

0.094676

Approximately 0.8825 – 0.0947 = 0.7878, or 78% of Canadians have systolic blood pressures in the range from 100 to 140.

Chapter 6 Exercise 6.27 Mental Development Index
Determining Normal Probabilities

The Mental Development Index of the Bayley Scales of Infant Development is a standardized measure used in observing infants over time. MDI is approximately normally distributed with $\mu = 100$ and $\sigma = 16$.

What proportion of children has MDI of at least 120?

Choose: **Formulas > Insert Function > NORMDIST**
Enter: X **120**
Mean **100**
Standard_dev **16**
Click: **OK**

0.89435

It follows that about 1 – 0.8944 = 0.1056, or about 11%, of children have an MDI of at least 120.

We are also asked to determine the proportion of children having an MDI of at least 80.

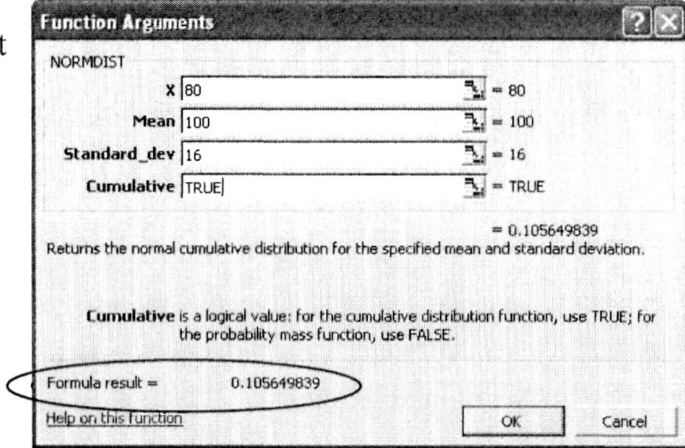

Note the Formula Result = 0.105649639

So, approximately 1.0 – 0.1057 = 0.8943, or 89% of children have an MDI of at least 80.

To determine the MDI score that is the 99th percentile:

Choose: **Formulas > Insert Function > NORMINV**
Enter: Probability **.99**
Mean **100**
Standard_dev **16**
Click: **OK**

Function Arguments
NORMINV
Probability .99 = 0.99
Mean 100 = 100
Standard_dev 16 = 16
= 137.221552
Returns the inverse of the normal cumulative distribution for the specified mean and standard deviation.
Standard_dev is the standard deviation of the distribution, a positive number.
Formula result = 137.221552
Help on this function
OK Cancel

An MDI of 138 is the 99th percentile.

To determine the MDI such that only 1% of the population has an MDI below it, we will use the probability **.01**

Function Arguments
NORMINV
Probability .01 = 0.01
Mean 100 = 100
Standard_dev 16 = 16
= 62.778448
Returns the inverse of the normal cumulative distribution for the specified mean and standard deviation.
Probability is a probability corresponding to the normal distribution, a number between 0 and 1 inclusive.
Formula result = 62.778448
Help on this function
OK Cancel

Approximately 1% of the population has an MDI less than 63.

Chapter 6 Example 12 Are Women Passed Over for Managerial Training?
Determining Binomial Probabilities

Let X denote the number of females selected in a random sample of ten employees (X can have any value 0, 1, 2, 3, …, 10). Enter column labels "x" and "P(X=x)" into cells A1 and B1.

Enter: **0** (in cell A2)
Choose: **Home > Editing > Fill > Series**
Select: **columns** **linear** step **1** stop value **10**
Click: **OK**

Select cell B2.
Select: **Formulas > Insert Function > BINOMDIST**
Click: **OK**

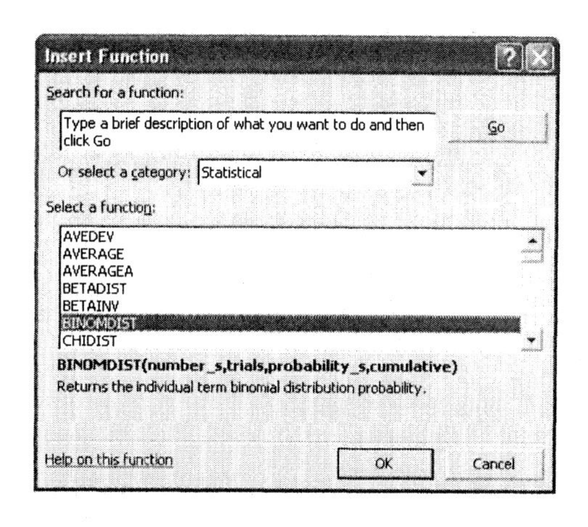

Enter: Number_s **A2**
Trials **10**
Probability_s **.5**
Cumulative: **FALSE**
Click: **OK**

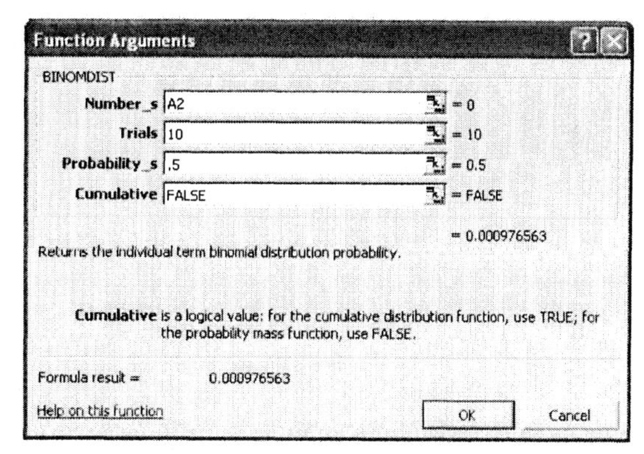

Drag to fill the remainder of B3 to B12

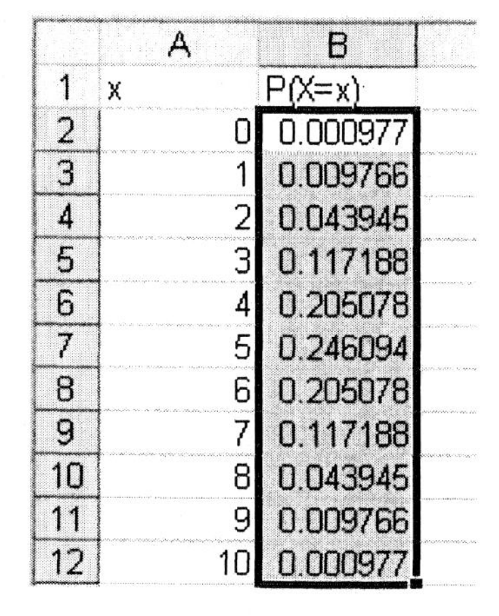

	A	B
1	x	P(X=x)
2	0	0.000977
3	1	0.009766
4	2	0.043945
5	3	0.117188
6	4	0.205078
7	5	0.246094
8	6	0.205078
9	7	0.117188
10	8	0.043945
11	9	0.009766
12	10	0.000977

This will generate the distribution in the column B of the worksheet.

Chapter 6 Example 14 How Can We Check for Racial Profiling?
Determining Binomial Probabilities

Let X denote the number of police car stops where the driver is African-American (X can have any value 0, 1, 2, 3, …, 262). Enter column labels "x" and "P(X=x)" into cells A1 and B1.

Enter: **0** (in cell A2)
Choose: **Home > Editing > Fill > Series**
Select: **columns** **linear** step **1** stop value **262**
Click: **OK**

Select cell B2.
Select: **Formulas > Insert Function > BINOMDIST**
Click: **OK**

Enter: Number_s **A2**
Trials **262**
Probability_s **.422**
Cumulative: **FALSE**
Click: **OK**

Drag to fill the remainder of B3 to B264.

Note the probabilities for X >=207.

x	P(X=x)
207	3.94491E-34
208	7.61589E-35
209	1.43666E-35
210	2.64724E-36
211	4.76321E-37
212	8.36601E-38
213	1.43382E-38
214	2.39696E-39
215	3.90704E-40
216	6.20693E-41
217	9.60637E-42
218	1.44777E-42
219	2.1237E-43
220	3.03056E-44
221	4.20499E-45
222	5.66997E-46
223	7.42541E-47
224	9.4389E-48

Chapter 7 Exercise 7.42 Exit Poll
Mean and Standard Deviation of the Binomial Random Variable X

The data is binary (vote for or against), the voters were randomly selected, and each voter is separate and independent from another voter

$n = 3000 \quad p = .5 \quad 1 - p = .5$

Recall that the mean and standard deviation for a general probability distribution is

$$\mu = \sum xP(x) \quad \text{and} \quad \sigma = \sqrt{\sum (x-u)^2 P(x)}$$

But, for the binomial distribution we can use the following expressions

$$\mu = n * p \text{ and } \quad \sigma = \sqrt{n \cdot p \cdot (1-p)}$$

Select an empty cell, and in the Function box, type in the required calculation.

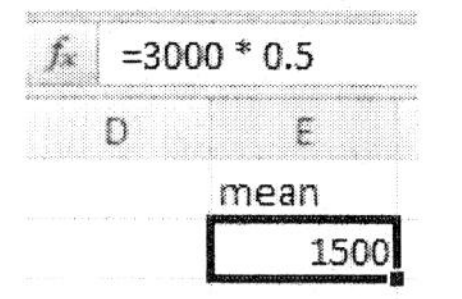

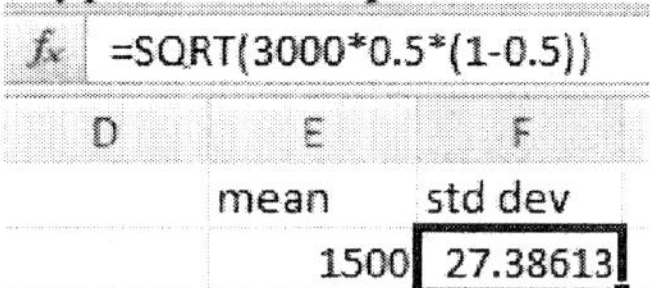

Chapter 7 Exercise 7.43 Jury Duty
Determining Binomial Probabilities

a) Check the three criteria for being binomial
b) Let n = 12, p = .4. Enter 0 through 12 into A2:A14, then
   Select cell **B2**
   Choose: **Formulas > Insert Function > BINOMDIST**

Enter: Number_s **A2**
Trials: **12**
Probability_s: **.4**
Cumulative: **FALSE**
Click: **OK**

Click and drag to fill B3:B14.

	A	B
1	x	P(X=x)
2	0	0.002177
3	1	0.017414
4	2	0.063852
5	3	0.141894
6	4	0.212841
7	5	0.22703
8	6	0.176579
9	7	0.100902
10	8	0.042043
11	9	0.012457
12	10	0.002491
13	11	0.000302
14	12	1.68E-05

Chapter 7 Exercise 7.5 Other Scenario for Exit Poll
The Sampling Distribution of Sample Proportion

Neither Excel nor PhStat can handle the calculations in determining the binomial probabilities for our sample size of 2705. See the Minitab section of this manual.

Chapter 7 Exercise 7.63 Class Exploration
Simulating a Sampling Distribution for a Sample Mean

Open the worksheet **heads_of_households** from the Excel folder of the data disk. Construct a histogram.

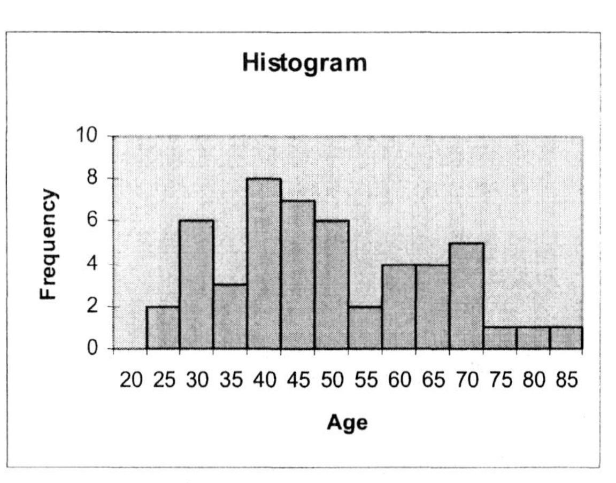

Lets assume a class size of 30 students. Each student can collect random samples of size 9. We will store the samples in nine columns of 30 cells each.

Click **Data > Data Analysis > Sampling > OK**

Enter: Input Range: **select appropriate cells**
Select: **Random**
Enter: Number of Samples: **30**
Enter: Output range**: enter a column**
Click: **OK**

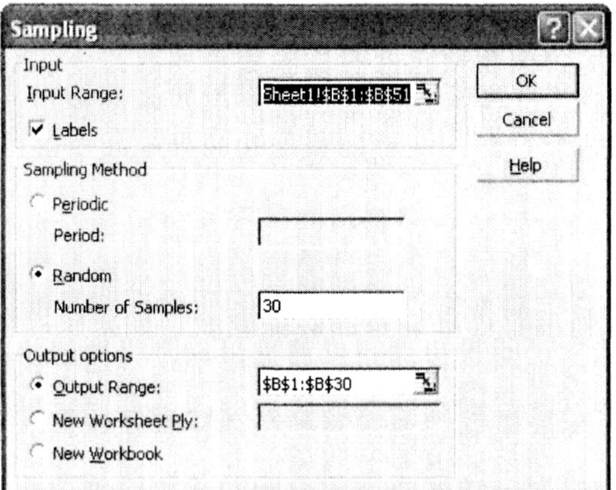

Note you will have to do this nine times, naming a new column each time.

To find the mean of each sample:

Select cell J1. Use formula bar to specify the mean of A1 through I1.

fx =AVERAGE(A1:I1)

	C	D	E	F	G	H	I	J
5	76	67	30	39	66	29	63	50.55556

Click and drag from J1 down to J30.

Now, find the mean and standard deviation of column J.
Select **Data > Data Analysis > Descriptive Statistics > OK**

Complete the dialogue box as indicated, specifying the column containing the sample means, and where you want the summary statistics placed.

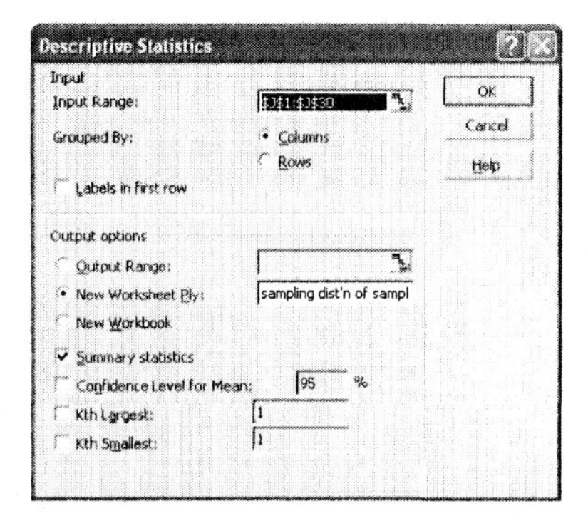

Note the mean of all thirty sample means is 48.5, and the standard deviation of all thirty sample means in 4.8547

	A	B
1	*Column1*	
2		
3	Mean	48.5
4	Standard Error	0.886343
5	Median	49.22222
6	Mode	47.22222
7	Standard Deviation	4.8547
8	Sample Variance	23.56811
9	Kurtosis	-0.6634
10	Skewness	-0.10158
11	Range	19.66667
12	Minimum	39.33333
13	Maximum	59
14	Sum	1455
		30

What conclusions can you draw?

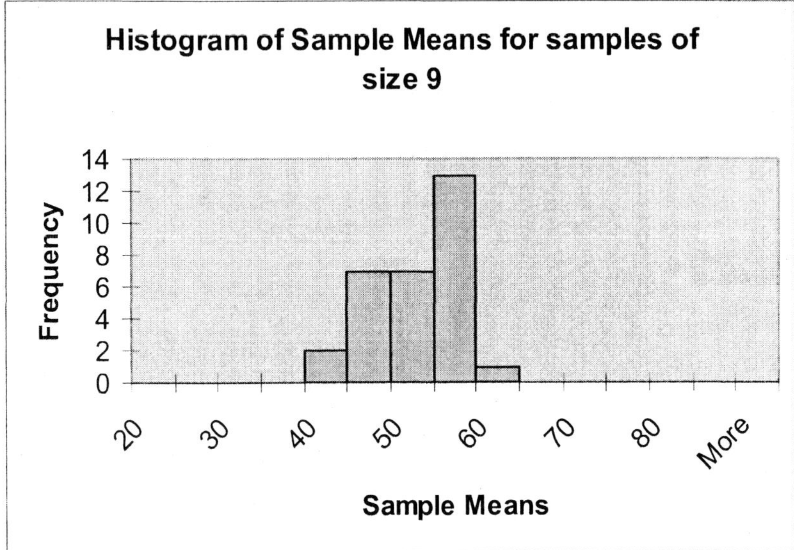

Compare this histogram to the histogram of the population that we did at the start of the exercise.

Chapter 8 Example 2 Should a Wife Sacrifice Her Career For Her Husband's? Constructing the Confidence Interval Estimate for a Population Proportion

From Chapter 6 we recall that 95% of a normal distribution falls within two standard deviations of the mean. So if we look at the mean $\pm$ 1.96 standard deviations, we would find 95% of a normal distribution. So in this problem, if we look at 1.96(.01) we get 0.02. We can then construct the interval as the sample proportion $\pm$ 0.02, or 0.19 $\pm$ 0.02 which gives us (0.17, 0.21).

To do this in Excel:
Choose **Add-Ins > PHStat > Confidence Interval > Estimate for the Proportion**

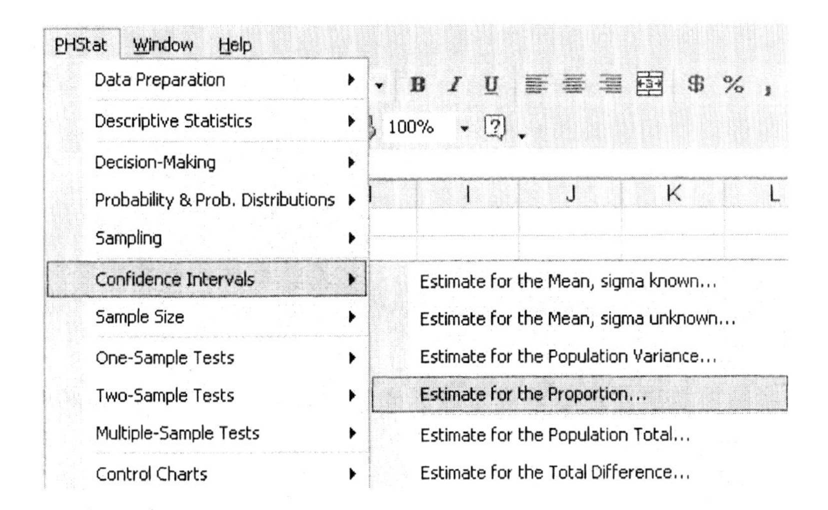

Enter Sample size **1823**
Number of Successes **346**
Confidence Level: **95**
Output Options:
Title: **C.I. for Proportion**
Click **OK**

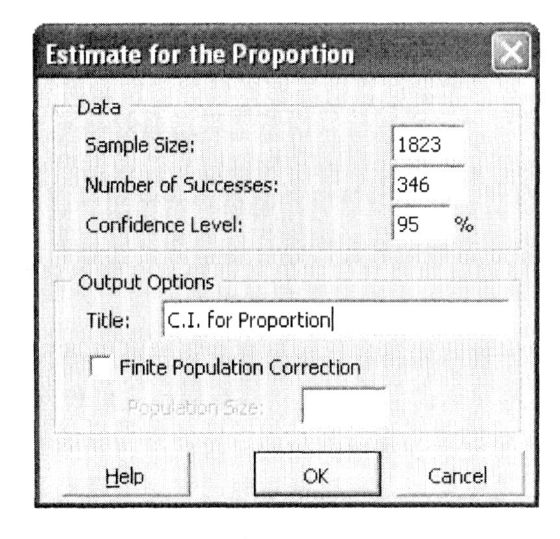

The output follows in a separate sheet:

	A	B
1	**C.I. for Proportion**	
2		
3	**Data**	
4	**Sample Size**	**1823**
5	**Number of Successes**	**346**
6	**Confidence Level**	**95%**
7		
8	Intermediate Calculations	
9	Sample Proportion	0.189797038
10	Z Value	-1.95996279
11	Standard Error of the Proportion	0.009184347
12	Interval Half Width	0.018000979
13		
14	**Confidence Interval**	
15	**Interval Lower Limit**	**0.171796059**
16	**Interval Upper Limit**	**0.207798017**

Note the point estimate (sample proportion) is 0.19 and the interval is (0.17, 0.21).
The margin of error is 0.02. Excel calls this the Interval Half Width.

Chapter 8 Exercise 8.7 Believe in Heaven?
Constructing the confidence interval estimate for a proportion

Choose **Add-Ins > PHStat > Confidence Interval > Estimate for the Proportion**
Enter Sample size **1158**
Number of Successes **996**
Confidence Level**: 95**

Note: This is 86% of the sample

Output Options:
Title: **C.I. for Proportion**
Click **OK**

The output appears in a separate sheet.

	A	B
1	**CI for Proportion**	
2		
3	**Data**	
4	**Sample Size**	**1158**
5	**Number of Successes**	**996**
6	**Confidence Level**	**95%**
7		
8	Intermediate Calculations	
9	Sample Proportion	0.860103627
10	Z Value	-1.95996279
11	Standard Error of the Proportion	0.010193524
12	Interval Half Width	0.019978927
13		
14	**Confidence Interval**	
15	**Interval Lower Limit**	**0.8401247**
16	**Interval Upper Limit**	**0.880082554**

Note the sample proportion, or point estimate = 0.86
The interval is (0.84, 0.88)
And the error is 0.02.

Chapter 8 Exercise 8.23 Exit Poll Predictions
Constructing the Confidence Interval Estimate for a Population Proportion

Choose **Add-Ins > PHStat > Confidence Interval > Estimate for the Proportion**
Enter: Sample size **1400**
Number of Successes **660**
Confidence Level**: 95**
Output Options:
Title: **95% C.I. for Proportions**
Click **OK**

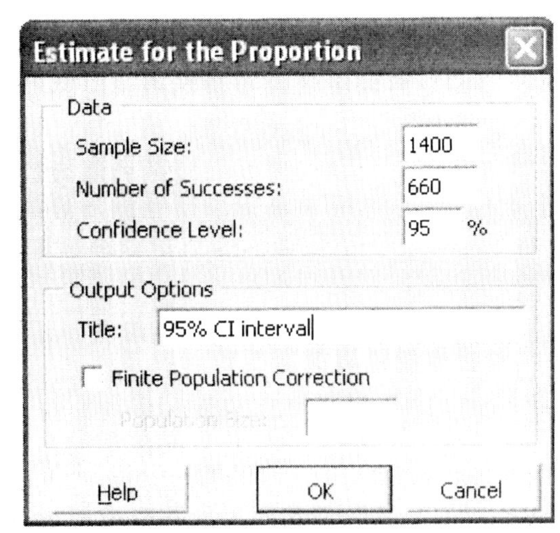

The output is in a separate sheet:

	A	B
1	**95% CI interval**	
2		
3	**Data**	
4	**Sample Size**	**1400**
5	**Number of Successes**	**660**
6	**Confidence Level**	**95%**
7		
8	Intermediate Calculations	
9	Sample Proportion	0.471428571
10	Z Value	-1.95996279
11	Standard Error of the Proportion	0.013341227
12	Interval Half Width	0.026148308
13		
14	**Confidence Interval**	
15	**Interval Lower Limit**	**0.445280263**
16	**Interval Upper Limit**	**0.49757688**

The interval is (0.45, 0.50) with a point estimate of 0.47.

Repeat the procedure for the 99% Confidence Interval to get a (0.44, 0.51) Interval with a point estimate of 0.47

What conclusions can you draw?

	A	B
1	**99% CI interval**	
2	**Data**	
3	**Sample Size**	**1400**
4	**Number of Successes**	**660**
5	**Confidence Level**	**99%**
6	Intermediate Calculations	
7	Sample Proportion	0.471428571
8	Z Value	-2.57583134
9	Standard Error of the Proportion	0.013341227
10	Interval Half Width	0.034364751
11	**Confidence Interval**	
12	**Interval Lower Limit**	**0.437063821**
13	**Interval Upper Limit**	**0.505793322**

Chapter 8 Example 7: eBay Auctions of Palm Handheld Computers Constructing a Confidence Interval Estimate for a Population Mean?

Open the data file **ebay_auctions** from the Excel folder of the data disk.
To get the descriptive statistics for the set of data
Choose **Data** > **Data Analysis** > **Descriptive Statistics** > **OK**

Enter Input Range **A1 – B19**
Click Grouped By **Columns**
Click **Summary Statistics**
Click **OK**

The output will appear in a new worksheet.

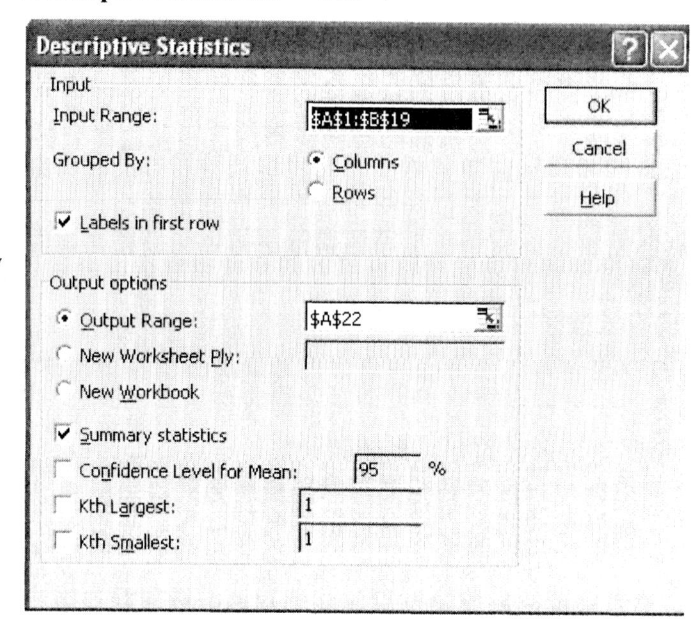

Note the mean and standard deviations.

22	*Buy-It-Now*		*Bidding*	
23				
24	Mean	233.5714286	Mean	231.6111111
25	Standard Error	5.532833352	Standard Error	5.170390112
26	Median	235	Median	240
27	Mode	225	Mode	246
28	Standard Deviation	14.63850109	Standard Deviation	21.93610746
29	Sample Variance	214.2857143	Sample Variance	481.1928105
30	Kurtosis	-0.65488889	Kurtosis	0.582289202
31	Skewness	-0.38824558	Skewness	-1.13783237
32	Range	40	Range	77
33	Minimum	210	Minimum	178
34	Maximum	250	Maximum	255
35	Sum	1635	Sum	4169
36	Count	7	Count	18

You can also compare the selling prices by doing a dotplot of the data values. Choose **Add-Ins > PHStat > Descriptive Statistics > Dot Scale Diagram** for each column of data. The output will also give you some descriptive statistics. Only the Buy It Now data is shown here.

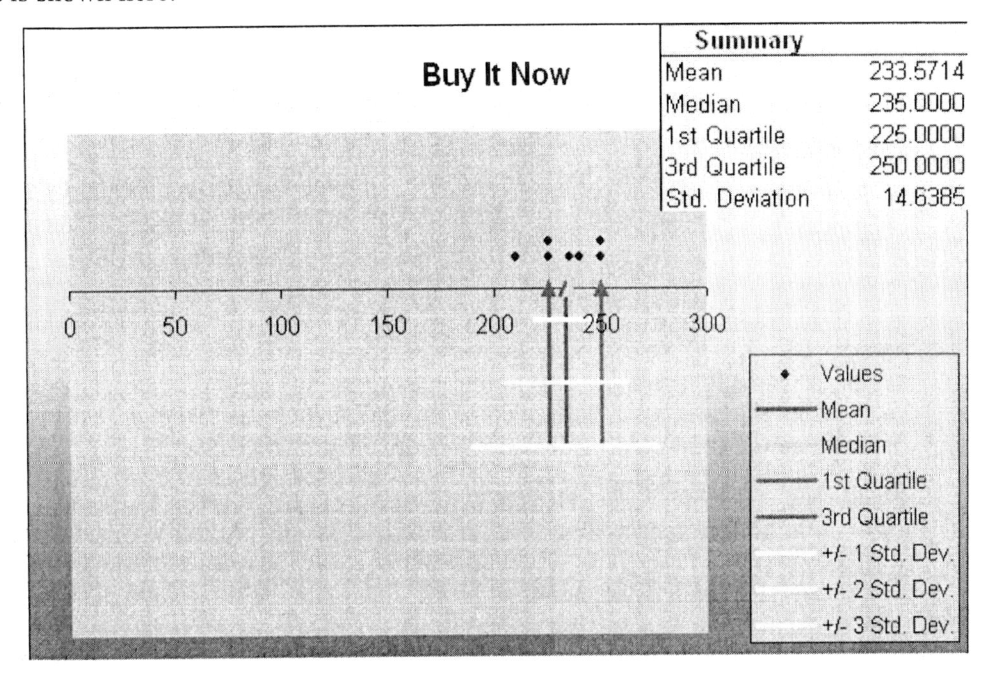

To get the Confidence interval around the mean
Choose **Add-Ins** > **PHStat** > **Confidence Intervals** > **Estimate for the Mean Sigma Unknown**
Click **Sample Statistics Unknown**
Sample Cell Range: **A1-A8**
Click: **First cell contains label**
Enter Title: **Buy It Now**
Click **OK**

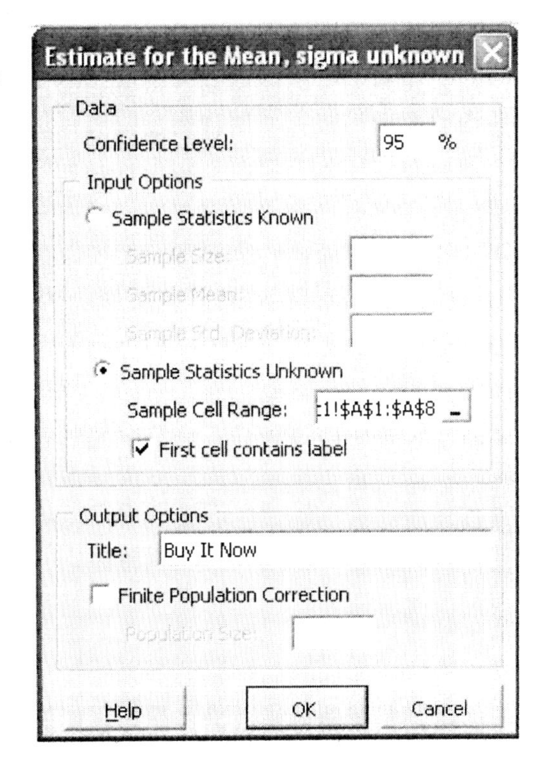

The output is placed in a separate sheet.

The interval is (220.03, 247.11)

	A	B
1	**Buy It Now**	
2		
3	**Data**	
4	**Sample Standard Deviation**	**14.63850109**
5	**Sample Mean**	**233.5714286**
6	**Sample Size**	**7**
7	**Confidence Level**	**95%**
8		
9	Intermediate Calculations	
10	Standard Error of the Mean	5.532833352
11	Degrees of Freedom	6
12	*t* Value	2.446913641
13	Interval Half Width	13.5383654
14		
15	**Confidence Interval**	
16	**Interval Lower Limit**	**220.03**
17	**Interval Upper Limit**	**247.11**

Using the same data, and now constructing the confidence interval for the Bidding column we get the following output. (This is exercise 7.29). Note that \$178 seems to be an outlier. Repeat the procedure after deleting \$178. Compare the two outputs.

	A	B
1	**Bidding**	
2		
3	**Data**	
4	**Sample Standard Deviation**	**21.93610746**
5	**Sample Mean**	**231.6111111**
6	**Sample Size**	**18**
7	**Confidence Level**	**95%**
8		
9	Intermediate Calculations	
10	Standard Error of the Mean	5.170390112
11	Degrees of Freedom	17
12	*t* Value	2.109818524
13	Interval Half Width	10.90858484
14		
15	**Confidence Interval**	
16	**Interval Lower Limit**	**220.70**
17	**Interval Upper Limit**	**242.52**

	A	B
1	**Bidding**	
2		
3	**Data**	
4	**Sample Standard Deviation**	**17.91831958**
5	**Sample Mean**	**234.7647059**
6	**Sample Size**	**17**
7	**Confidence Level**	**95%**
8		
9	Intermediate Calculations	
10	Standard Error of the Mean	4.345830838
11	Degrees of Freedom	16
12	*t* Value	2.119904821
13	Interval Half Width	9.212747743
14		
15	**Confidence Interval**	
16	**Interval Lower Limit**	**225.55**
17	**Interval Upper Limit**	**243.98**

Chapter 8 Exercise 8.55 Do You Like Tofu?
Confidence Interval Estimate of a Population Proportion with a Small Sample Size

In this problem, we have a very small sample of five students. All say they like tofu. We will use the "plus four" method in this situation, giving us seven successes out of nine trials. Using the formulas in the text:

We have N = 9, $\hat{p} = 7/9 = 0.7777$ and

$$se = \sqrt{\hat{p}(1-\hat{p})/n} = \sqrt{.7777(1-.7777)/9} = 0.1386.$$

The resulting interval = $\hat{p} \pm 1.96(se) = 0.7777 \pm (1.96)(0.1386)$.

This gives us the interval (0.506, 1.05). Note that the right end of the interval is 1.05. We cannot have a value greater than 1.0 for a proportion, so we would report the interval as (0.506, 1.00).

We can do this using Excel.
Choose **Add-Ins > PHStat > Confidence Intervals > Estimate for the Proportion** and enter the fields as shown.

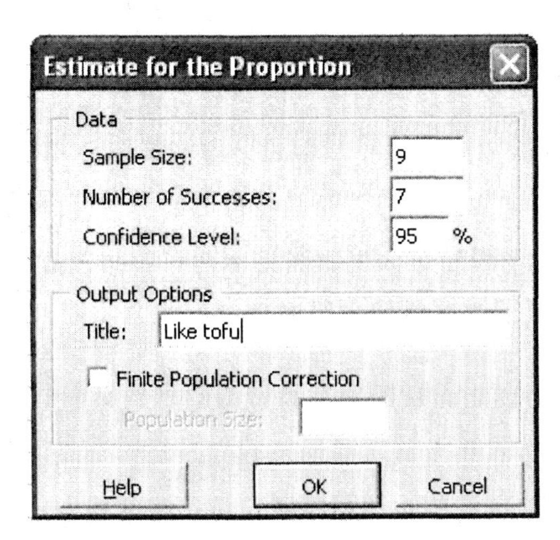

We are using the "plus 4" method, adding 4 to the sample size, 2 of which are considered successes.

The results are shown in a separate sheet.

Note that the upper limit is greater than 1, so we state the interval as (0.51, 1.0).

	A	B
1	**Like tofu**	
2		
3	**Data**	
4	**Sample Size**	9
5	**Number of Successes**	7
6	**Confidence Level**	95%
7		
8	Intermediate Calculations	
9	Sample Proportion	0.777777778
10	Z Value	-1.95996279
11	Standard Error of the Proportion	0.138579903
12	Interval Half Width	0.271611453
13		
14	**Confidence Interval**	
15	**Interval Lower Limit**	**0.506166324**
16	**Interval Upper Limit**	**1.049389231**

Chapter 9 Example 4 Dr. Dog: Can Dogs Detect Cancer By Smell? Hypothesis Testing For a Proportion

We are asked to test H_0: $p = 1/7$ vs. H_1: $p > 1/7$. The sample evidence presented is that in the total of 54 trials, the dogs made the correct decision 22 times.

Choose **Add-Ins > PHStat > One Sample Tests > Z Test for the Proportion**

Enter Null Hypothesis: **.143**
Level of Significance: **.05**
Number of Successes: **22**
Sample Size: **54**
Select: **Upper Tail Test**
Enter Title: **Dr. Dog Test for the Proportion**
Click **OK**

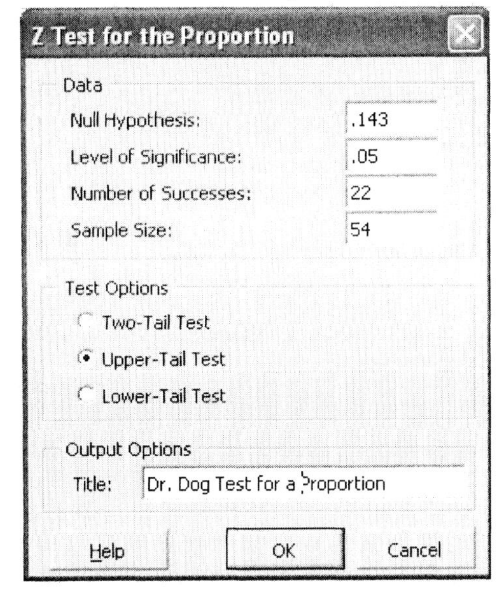

	A	B
1	**Dr. Dog Test for a Proportion**	
2		
3	**Data**	
4	**Null Hypothesis** *p*=	**0.143**
5	**Level of Significance**	**0.05**
6	**Number of Successes**	**22**
7	**Sample Size**	**54**
8		
9	Intermediate Calculations	
10	Sample Proportion	0.407407407
11	Standard Error	0.047638881
12	**Z Test Statistic**	**5.550243898**
13		
14	**Upper-Tail Test**	
15	**Upper Critical Value**	**1.644853476**
16	***p*-Value**	**1.43002E-08**
17	**Reject the null hypothesis**	

Note the test statistic ($z = 5.6$) and the p-value (0.000). With such a small p-value, we will reject the null hypothesis.

Chapter 9 Example 6 Can TT Practitioners Detect a Human Energy Field?
Hypothesis Testing For A Proportion

We are asked to test H_0: p = 0.5 vs. H_1: p > 0.5. The sample evidence presented is that in the total of 150 trials, the TT practitioners were correct with 70 of their predictions.

Choose **Add-Ins > PHStat > One Sample Tests > Z Test for the Proportion**

Enter Null Hypothesis: **.5**
Level of Significance: **.05**
Number of Successes: **70**
Sample Size: **150**
Select: **Upper Tail Test**
Enter Title: **TT Practitioners Test for the Proportion**
Click **OK**

The results appear in a separate worksheet.

With such a large p-value (p-value = 0.793), the decision is Fail To Reject.

	A	B
1	**TT Practitionors Test for the Proportio**	
2		
3	**Data**	
4	**Null Hypothesis** *p*=	**0.5**
5	**Level of Significance**	**0.05**
6	**Number of Successes**	**70**
7	**Sample Size**	**150**
8		
9	Intermediate Calculations	
10	Sample Proportion	0.466666667
11	Standard Error	0.040824829
12	Z Test Statistic	-0.816496581
13		
14	**Upper-Tail Test**	
15	**Upper Critical Value**	**1.644853476**
16	***p*-Value**	**0.792891972**
17	**Do not reject the null hypothesis**	

Chapter 9 Exercise 9.17 Another Test of Therapeutic Touch
Hypothesis Testing For A Proportion

We are asked to test H_0: p = 0.5 vs. H_1: p > 0.5. The sample evidence presented is that in the total of 130 trials, the TT practitioners were correct with 53 of their predictions.

Choose **Add-Ins > PHStat > One Sample Tests > Z Test for the Proportion**

Enter Null Hypothesis: **.5**
Level of Significance: **.05**
Number of Successes: **53**
Sample Size: **130**
Select: **Upper Tail Test**
Enter Title: **TT Practitioners Test for the Proportion**
Click **OK**

The results appear in a separate worksheet.

	A	B
1	**TT Practitionors Test for the Proportic**	
2		
3	**Data**	
4	**Null Hypothesis** *p*=	**0.5**
5	**Level of Significance**	**0.05**
6	**Number of Successes**	**53**
7	**Sample Size**	**130**
8		
9	Intermediate Calculations	
10	Sample Proportion	0.407692308
11	Standard Error	0.043852901
12	Z Test Statistic	-2.104939246
13		
14	**Upper-Tail Test**	
15	**Upper Critical Value**	**1.644853476**
16	***p*-Value**	**0.982351764**
17	**Do not reject the null hypothesis**	

Chapter 9 Exercise 9.19 Gender Bias In Selecting Managers
Hypothesis Testing For A Proportion

We are asked to test H_0: p = 0.6 vs. H_1: p $\neq$ 0.6. The sample evidence presented is that in the total of 40 employees chosen for management training, 28 were male.

Choose **Add-Ins > PHStat > One Sample Tests > Z Test for the Proportion**

Enter Null Hypothesis: **.6**
Level of Significance: **.05**
Number of Successes: **28**
Sample Size: **40**
Select: **Two Tail Test**
Enter Title: **Gender Bias Test for the Proportion**
Click **OK**

The results appear in a separate worksheet.

	A	B
1	Gender Bias Test for the Proportion	
2		
3	Data	
4	Null Hypothesis p=	0.6
5	Level of Significance	0.05
6	Number of Successes	28
7	Sample Size	40
8		
9	Intermediate Calculations	
10	Sample Proportion	0.7
11	Standard Error	0.077459667
12	Z Test Statistic	1.290994449
13		
14	Two-Tail Test	
15	Lower Critical Value	-1.959962787
16	Upper Critical value	1.959962787
17	p-Value	0.196705733
18	Do not reject the null hypothesis	

Chapter 9 Example 7 Mean Weight Change in Anorexic Girls
Significance Test About a Mean

Open the worksheet **anorexia,** which is found in the Excel folder of the data disk. We wish to perform a two-tailed test on H_0: $\mu = 0$ versus Ha: $\mu \neq 0$ using the weight gains found in Column G "cogchange".

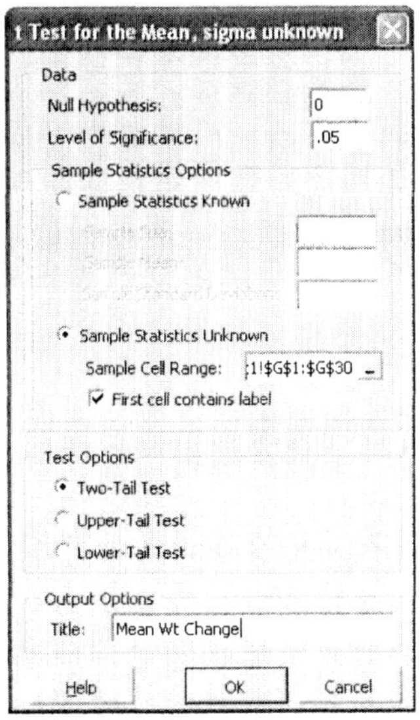

Choose **Add-Ins > PHStat >One Sample Tests > T test for the Mean, sigma unknown**
Enter:
Null Hypothesis: **0**
Level of Significance: **.05**
Choose **Sample Statistics Unknown**
Enter Sample Range: **G1 – G30**
Check: **First cell contains label**
Enter Title: **Mean Wt Change**
Click **OK**

t Test for the Mean, sigma unknown
Data
Null Hypothesis: 0
Level of Significance: .05
Sample Statistics Options
Sample Statistics Known
Sample Statistics Unknown
Sample Cell Range: ;1!$G$1:$G$30
First cell contains label
Test Options
Two-Tail Test
Upper-Tail Test
Lower-Tail Test
Output Options
Title: Mean Wt Change
Help OK Cancel

The output appears in a separate worksheet.

	A	B
1	Mean Wt Change	
2		
3	Data	
4	Null Hypothesis $\mu=$	0
5	Level of Significance	0.05
6	Sample Size	29
7	Sample Mean	3.006896552
8	Sample Standard Deviation	7.308504392
9		
10	Intermediate Calculations	
11	Standard Error of the Mean	1.357155195
12	Degrees of Freedom	28
13	*t* Test Statistic	2.215587844
14		
15	Two-Tail Test	
16	Lower Critical Value	-2.048409442
17	Upper Critical Value	2.048409442
18	*p*-Value	0.035022597
19	Reject the null hypothesis	

Chapter 9 Exercise 9.39 Anorexia in Teenage girls
Significance Test About a Mean

We wish to perform a two-tailed test on H_0: $\mu = 0$ versus Ha: $\mu \neq 0$ using the weight gains given. First we will plot the data with a box plot. There are no outliers indicated, and the data distribution is not highly skewed.

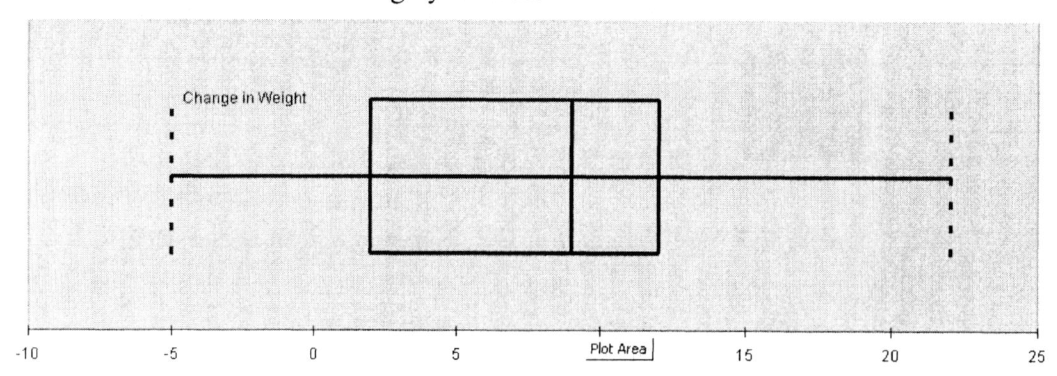

Performing the significance test about H_0: $\mu = 0$ versus Ha: $\mu \neq 0$
Choose **Add-Ins > PHStat >One Sample Tests > t test for the Mean, sigma unknown**
Enter:

Null Hypothesis: **0**
Level of Significance: **.05**

Choose **Sample Statistics Unknown**
Enter Sample Range: **A1-A18**
Check: **First cell contains label**
Enter Title: **Change in Weight**
Click **OK**

The output appears in a separate worksheet.

	A	B
1	Change in Weight	
2		
3	Data	
4	Null Hypothesis μ=	0
5	Level of Significance	0.05
6	Sample Size	17
7	Sample Mean	7.294117647
8	Sample Standard Deviation	7.183006908
9		
10	Intermediate Calculations	
11	Standard Error of the Mean	1.74213507
12	Degrees of Freedom	16
13	*t* Test Statistic	4.186884113
14		
15	Two-Tail Test	
16	Lower Critical Value	-2.119904821
17	Upper Critical Value	2.119904821
18	*p*-Value	0.000697365
19	Reject the null hypothesis	

Chapter 10 Example 4 Confidence Interval Comparing Heart Attack Rates for Aspirin and Placebo – Confidence Interval for Difference Between Two Proportions

To construct a confidence interval $p_1 - p_2$, note that for the placebo group X = 189 and n = 11034, and for the aspirin group X = 104 and n = 11037.

Select **Add-Ins > PHStat > Two Sample Tests > Z test for Differences in Two Proportions.**

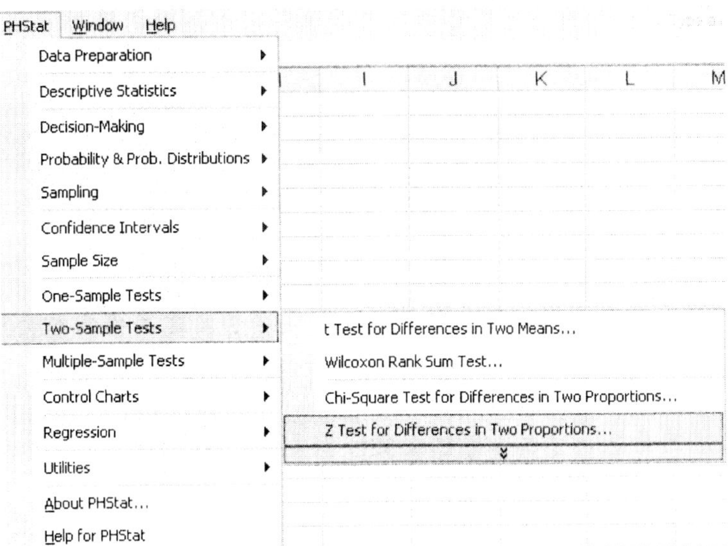

Enter Hypothesized Difference: **0**
Level of Significance: **.05**
Population 1 Sample
Number of Successes: **189**
Sample Size: **11034**
Population 2 Sample
Number of Successes: **104**
Sample Size: **11037**
Test Options
Click **Two Tail Test**
Output Options:
Title: **Placebo vs Aspirin**
Click **OK**

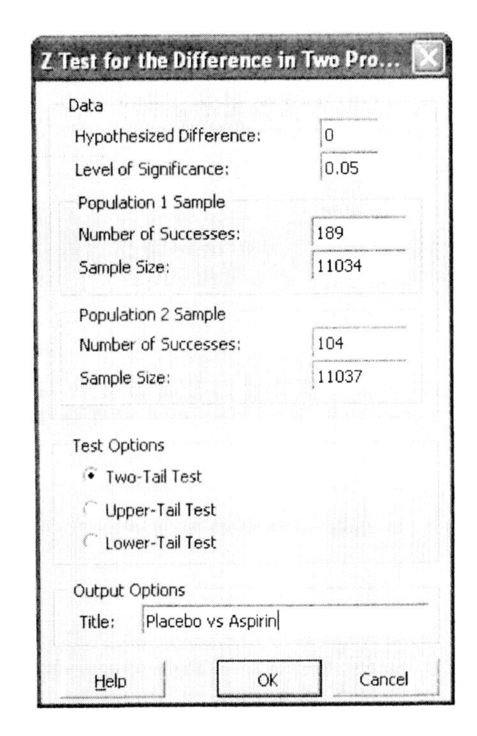

The results appear in a new worksheet.

This is not a confidence interval for a difference in two proportions. Excel and PHStat cannot produce a confidence interval. However by using this two-sided test with a level of significance = .05, we can find the z statistic and the point estimate. We would still need to calculate the margin of error to determine the interval.

	A	B
1	Placebo vs Aspirin	
2		
3	Data	
4	Hypothesized Difference	0
5	Level of Significance	0.05
6	Group 1	
7	Number of Successes	189
8	Sample Size	11034
9	Group 2	
10	Number of Successes	104
11	Sample Size	11037
12		
13	Intermediate Calculations	
14	Group 1 Proportion	0.017128874
15	Group 2 Proportion	0.00942285
16	Difference in Two Proportions	0.007706024
17	Average Proportion	0.013275339
18	Z Test Statistic	5.001388204
19		
20	Two-Tail Test	
21	Lower Critical Value	-1.959962787
22	Upper Critical Value	1.959962787
23	*p*-Value	5.70091E-07
24	Reject the null hypothesis	

Chapter 10 Example 5 Is TV Watching Associated With Aggressive Behavior? Significance Test for Difference Between Two Proportions

To perform a significance test for H_0: p1 – p2 = 0 versus H_a: p1 – p2 $\neq$ 0 note that for the "less than 1 hour of TV per day" group X = 5 and n = 88 and for the "at least 1 hour of TV per day" group X = 154 and n = 619.

Select **Add-Ins > PHStat > Two Sample Tests > Z test for Differences in Two Proportions.**

Enter Hypothesized Difference: **0**
Level of Significance: **.05**
Population 1 Sample
Number of Successes: **5**
Sample Size: **88**
Population 2 Sample
Number of Successes: **154**
Sample Size: **619**
Test Options
Click **Two Tail Test**
Output Options:
Title: **TV Watching with Aggressive Behavior**
Click **OK**

The results appear in a new worksheet.

	A	B
1	TV Watching with Agressive Behavior	
2		
3	Data	
4	Hypothesized Difference	0
5	Level of Significance	0.05
6	Group 1	
7	Number of Successes	5
8	Sample Size	88
9	Group 2	
10	Number of Successes	154
11	Sample Size	619
12		
13	Intermediate Calculations	
14	Group 1 Proportion	0.056818182
15	Group 2 Proportion	0.248788368
16	Difference in Two Proportions	-0.191970187
17	Average Proportion	0.224893918
18	Z Test Statistic	-4.035908513
19		
20	Two-Tail Test	
21	Lower Critical Value	-1.959962787
22	Upper Critical Value	1.959962787
23	*p*-Value	5.44182E-05
24	Reject the null hypothesis	

Chapter 10 Exercise 10.9 Drinking and Unplanned Sex
Significance Test for the Difference Between Two Populations

To perform a significance test for H_0: p1 – p2 = 0 versus H_a: p1 – p2 ≠ 0 note that for the 2005 group X = 146 (.3*485) and n = 485, and for the 1993 group X = 38 (.24*159) and n = 159.

Select **Add-Ins > PHStat > Two Sample Tests >**

Z test for Differences in Two Proportions.

Enter Hypothesized Difference: **0**
Level of Significance: **.05**
Population 1 Sample
Number of Successes: **146**
Sample Size: **485**
Population 2 Sample
Number of Successes: **38**
Sample Size: **159**
Test Options
Click **Two Tail Test**
Output Options:
Title: **Drinking and Unplanned Sex**
Click **OK**

Drinking and Unplanned Sex

Data	
Hypothesized Difference	0
Level of Significance	0.05
Group 1	
Number of Successes	146
Sample Size	485
Group 2	
Number of Successes	38
Sample Size	159

Intermediate Calculations	
Group 1 Proportion	0.301030928
Group 2 Proportion	0.238993711
Difference in Two Proportions	0.062037217
Average Proportion	0.285714286
Z Test Statistic	1.502715776

Two-Tail Test	
Lower Critical Value	-1.95996398
Upper Critical Value	1.959963985
p-Value	0.132912353
Do not reject the null hypothesis	

Chapter 10 Example 8 Nicotine – How Much More Addicted are Smokers than Ex-Smokers? -- Confidence Interval for Difference Between Population Means

To construct a confidence interval for $\mu_1 - \mu_2$, note that for the smoker group, $\bar{X}$ = 5.9 and s = 3.3 for n = 75 and for the ex-smoker group $\bar{X}$ = 1.0 and s = 2.3 for n = 257.

Choose **Add-Ins > PHStat> Two Sample Tests > t Test for Differences in Two Means**

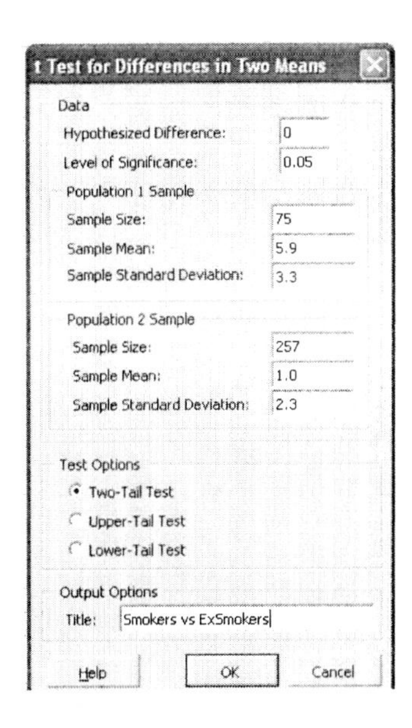

Enter
Hypothesized Difference **0**
Level of Significance **.05**
Population 1
Sample Size **75**
Sample Mean **5.9**

Sample Standard Deviation **3.3**
Population 2
Sample Size **257**
Sample Mean **1.0**
Sample Standard Deviation **2.3**
Choose **Two Tail Test**
Click **OK**

The results appear in a new worksheet. As noted before, Excel and PHStat cannot produce a confidence interval. However by using this two-sided test with a level of significance = .05, we can find the z statistic and the point estimate. We would still need to calculate the margin of error to determine the interval.

	A	B
1	Smokers vs ExSmokers	
2		
3	**Data**	
4	**Hypothesized Difference**	**0**
5	**Level of Significance**	**0.05**
6	**Population 1 Sample**	
7	**Sample Size**	**75**
8	**Sample Mean**	**5.9**
9	**Sample Standard Deviation**	**3.3**
10	**Population 2 Sample**	
11	**Sample Size**	**257**
12	**Sample Mean**	**1**
13	**Sample Standard Deviation**	**2.3**
14		
15	Intermediate Calculations	
16	Population 1 Sample Degrees of Freedom	74
17	Population 2 Sample Degrees of Freedom	256
18	Total Degrees of Freedom	330
19	Pooled Variance	6.545758
20	Difference in Sample Means	4.9
21	**_t_ Test Statistic**	**14.59299**
22		
23	**Two-Tail Test**	
24	**Lower Critical Value**	**-1.96718**
25	**Upper Critical Value**	**1.967178**
26	**_p_-Value**	**1.45E-37**
27	**Reject the null hypothesis**	

Chapter 10 Example 9 Does Cell Phone Use While Driving Impair Reaction Time? Significance Test for Comparing Two Population Means

To perform a significance test for H0: $\mu_1 - \mu_2 = 0$ versus: Ha: $\mu_1 - \mu_2 \neq 0$, note that for the "cell phone" group, $\bar{X}$ = 585.2 and s = 89.6 for n = 32 and for the "control" group $\bar{X}$ = 533.7 and s = 65.3 for n = 32.

Choose **Add-Ins > PHStat> Two Sample Tests > t Test for Differences in Two Means**

Enter
Hypothesized Difference **0**
Level of Significance **.05**
Population 1
Sample Size **32**
Sample Mean **585.2**
Sample Standard Deviation **89.6**
Population 2
Sample Size **32**
Sample Mean **533.7**
Sample Standard Deviation **65.3**
Choose **Two Tail Test**

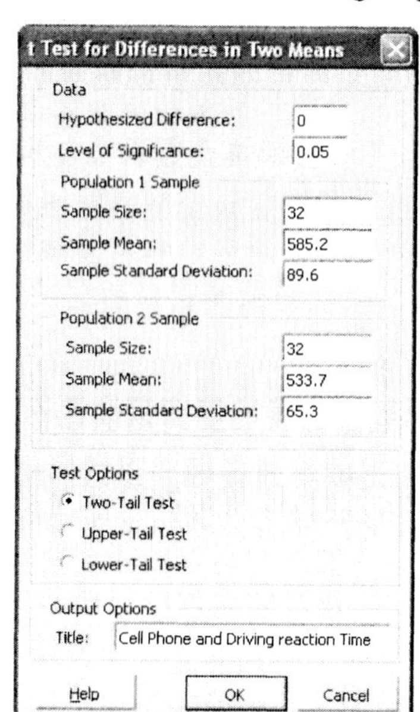

Click **OK**

The results appear in a new worksheet.

	A	B
1	Cell Phone and Driving reaction Time	
2		
3	Data	
4	Hypothesized Difference	0
5	Level of Significance	0.05
6	Population 1 Sample	
7	Sample Size	32
8	Sample Mean	585.2
9	Sample Standard Deviation	89.6
10	Population 2 Sample	
11	Sample Size	32
12	Sample Mean	533.7
13	Sample Standard Deviation	65.3
14		
15	Intermediate Calculations	
16	Population 1 Sample Degrees of Freedom	31
17	Population 2 Sample Degrees of Freedom	31
18	Total Degrees of Freedom	62
19	Pooled Variance	6146.125
20	Difference in Sample Means	51.5
21	*t* Test Statistic	2.627644
22		
23	Two-Tail Test	
24	Lower Critical Value	-1.99897
25	Upper Critical Value	1.998969
26	*p*-Value	0.010822
27	Reject the null hypothesis	

Chapter 10 Exercise 10.25 Females or Males More Nicotine Dependant Significance Test for Comparing Two Population Means

To perform a significance test for H$_0$: $\mu_1 - \mu_2 = 0$ versus: Ha: $\mu_1 - \mu_2 \neq 0$, note that for the females, $\overline{X} = 2.8$ and s = 3.6 for n = 150 and for the males $\overline{X} = 1.6$ and s = 2.9 for n = 182.

Choose **Add-Ins > PHStat> Two Sample Tests > t Test for Differences in Two Means**

Enter

Hypothesized Difference	**0**
Level of Significance	**.05**
Population 1	
Sample Size	**150**
Sample Mean	**2.8**
Sample Standard Deviation	**3.6**
Population 2	
Sample Size	**182**
Sample Mean	**1.6**
Sample Standard Deviation	**2.9**

Choose **Two Tail Test**
Click **OK**

	A	B
1	Nicotine dependance	
2		
3	Data	
4	Hypothesized Difference	0
5	Level of Significance	0.05
6	Population 1 Sample	
7	Sample Size	150
8	Sample Mean	2.8
9	Sample Standard Deviation	3.6
10	Population 2 Sample	
11	Sample Size	182
12	Sample Mean	1.6
13	Sample Standard Deviation	2.9
14		
15	Intermediate Calculations	
16	Population 1 Sample Degrees of Freedom	149
17	Population 2 Sample Degrees of Freedom	181
18	Total Degrees of Freedom	330
19	Pooled Variance	10.46439
20	Difference in Sample Means	1.2
21	*t* Test Statistic	3.363849
22		
23	Two-Tail Test	
24	Lower Critical Value	-1.96718
25	Upper Critical Value	1.967178
26	*p*-Value	0.000859
27	Reject the null hypothesis	

Chapter 10 Exercise 10.27 TV Watching and Gender
Significance Test for Comparing Two Population Means

To perform a significance test for H_0: $\mu_1 - \mu_2 = 0$ versus: Ha: $\mu_1 - \mu_2 \neq 0$, note that for the females, $\overline{X} = 2.99$ and s = 2.88 for n = 479 and for the males $\overline{X} = 2.72$ and s = 2.28 for n = 420.

Choose **Add-Ins > PHStat> Two Sample Tests > t Test for Differences in Two Means**

Enter

Hypothesized Difference	**0**
Level of Significance	**.05**
Population 1	
Sample Size	**479**
Sample Mean	**2.99**
Sample Standard Deviation	**2.88**
Population 2	
Sample Size	**420**
Sample Mean	**2.72**
Sample Standard Deviation	**2.28**

Choose **Two Tail Test**

Click **OK**

Note that we are using PHStat for this test. Data Analysis in Excel has a test for comparing two means, but not for summarized data.

TV Watching and Gender

Data	
Hypothesized Difference	0
Level of Significance	0.05
Population 1 Sample	
Sample Size	479
Sample Mean	2.99
Sample Standard Deviation	2.88
Population 2 Sample	
Sample Size	420
Sample Mean	2.72
Sample Standard Deviation	2.28

Intermediate Calculations	
Population 1 Sample Degrees of Freedom	478
Population 2 Sample Degrees of Freedom	419
Total Degrees of Freedom	897
Pooled Variance	6.848219
Difference in Sample Means	0.27
t Test Statistic	1.543432

Two-Tail Test	
Lower Critical Value	-1.96261
Upper Critical Value	1.962612
p-Value	0.123079
Do not reject the null hypothesis	

Chapter 10 Example 10 Is Arthroscopic Surgery Better Than Placebo?
Significance Test for Comparing Two Population Means Assuming Equal Population Standard Deviations

To perform a significance test for H_0: $\mu_1 - \mu_2 = 0$ versus: Ha: $\mu_1 - \mu_2 \neq 0$, note that for the "placebo" group, $\overline{X} = 51.6$ and s = 23.7 for n = 60 and for the "lavage arthroscopic" group $\overline{X} = 53.7$ and s = 23.7 for n = 61.

Choose **Add-Ins > PHStat> Two Sample Tests > t Test for Differences in Two Means**

Enter

Hypothesized Difference **0**
Level of Significance **.05**

Population 1

Sample Size **60**
Sample Mean **51.6**
Sample Standard Deviation **23.7**

Population 2

Sample Size **61**
Sample Mean **53.7**
Sample Standard Deviation **23.7**

Choose **Two Tail Test**
Click **OK**

Note that we are using PHStat for this test. Data Analysis in Excel has a test for comparing two means, but not for summarized data.

	A	B
1	Arthroscopic Surgery	
2		
3	Data	
4	Hypothesized Difference	0
5	Level of Significance	0.05
6	Population 1 Sample	
7	Sample Size	60
8	Sample Mean	51.6
9	Sample Standard Deviation	23.7
10	Population 2 Sample	
11	Sample Size	61
12	Sample Mean	53.7
13	Sample Standard Deviation	23.7
14		
15	Intermediate Calculations	
16	Population 1 Sample Degrees of Freedom	59
17	Population 2 Sample Degrees of Freedom	60
18	Total Degrees of Freedom	119
19	Pooled Variance	561.69
20	Difference in Sample Means	-2.1
21	*t* Test Statistic	-0.48733
22		
23	Two-Tail Test	
24	Lower Critical Value	-1.9801
25	Upper Critical Value	1.980097
26	*p*-Value	0.626924
27	Do not reject the null hypothesis	

Chapter 10 Example 13 Matched Pairs Analysis of Cell Phone Impact on Driver Reaction Time

Comparing Means With Matched Pairs

Enter the data from Table 10.9. Since neither Excel nor PHStat will do a matched pairs test, we will perform a t test on the difference column and get the same results.

Choose **Add-Ins > PHStat> One Sample Tests > t Test for the Mean Sigma Unknown**

Enter

Null Hypothesis: **0**
Level of Significance: **.05**
Select **Sample Statistics Unknown**
Sample Cell Range: **D1-D33**

Check **First cell contains label**
Check **Two Tail Test**
Enter Title: **Cell Phone Impact on Driver Reaction Time**
Click **OK**

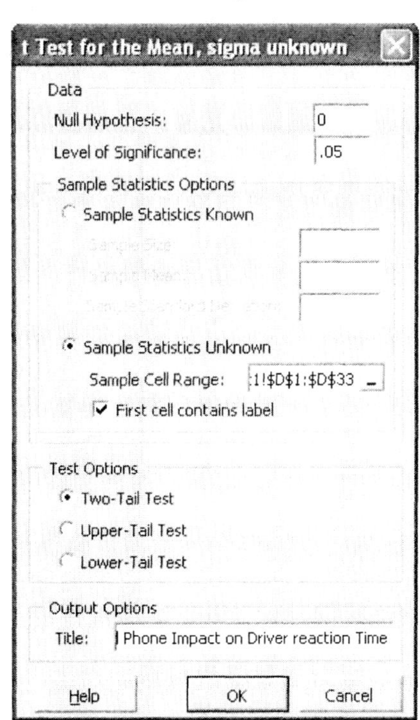

The results appear in a separate worksheet.

	A	B
1	Cell Phone Impact on Driver reaction Time	
2		
3	Data	
4	Null Hypothesis μ=	0
5	Level of Significance	0.05
6	Sample Size	32
7	Sample Mean	50.625
8	Sample Standard Deviation	52.48578917
9		
10	Intermediate Calculations	
11	Standard Error of the Mean	9.278264359
12	Degrees of Freedom	31
13	*t* Test Statistic	5.456300666
14		
15	Two-Tail Test	
16	Lower Critical Value	-2.039514584
17	Upper Critical Value	2.039514584
18	*p*-Value	5.80341E-06
19	Reject the null hypothesis	

Chapter 10 Exercise 10.49 Matched Pairs Analysis of Movies and Sports Comparing Means With Matched Pairs

Enter the data from table in the problem. Form the difference and place it in the next column. Since neither Excel nor PHStat will do a matched pairs test, we will perform a t test on the difference column and get the same results.

	A	B	C	D
1	Student	Movies	Sports	Difference
2	1	10	5	5
3	2	4	0	4
4	3	12	20	-8
5	4	2	6	-4
6	5	12	2	10
7	6	7	8	-1
8	7	45	12	33
9	8	1	25	-24
10	9	25	0	25
11	10	12	12	0

Choose **Add-Ins > PHStat> One Sample Tests > t Test for the Mean Sigma Unknown**
Enter

Null Hypothesis: **0**
Level of Significance: **.05**
Select **Sample Statistics Unknown**
Sample Cell Range: **D1-D33**

Check **First cell contains label**
Check **Two Tail Test**
Enter Title: **Movies VS Sports**
Click **OK**

	A	B
1	Movies vs Sports	
2		
3	Data	
4	Null Hypothesis μ=	0
5	Level of Significance	0.05
6	Sample Size	10
7	Sample Mean	4
8	Sample Standard Deviation	16.16580754
9		
10	Intermediate Calculations	
11	Standard Error of the Mean	5.112077203
12	Degrees of Freedom	9
13	*t* Test Statistic	0.782460796
14		
15	Two-Tail Test	
16	Lower Critical Value	-2.262158887
17	Upper Critical Value	2.262158887
18	*p*-Value	0.454037837
19	Do not reject the null hypothesis	

We can also get a confidence interval for the mean difference.

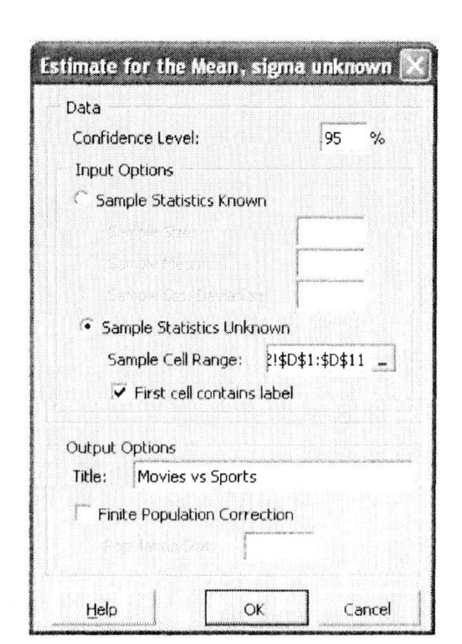

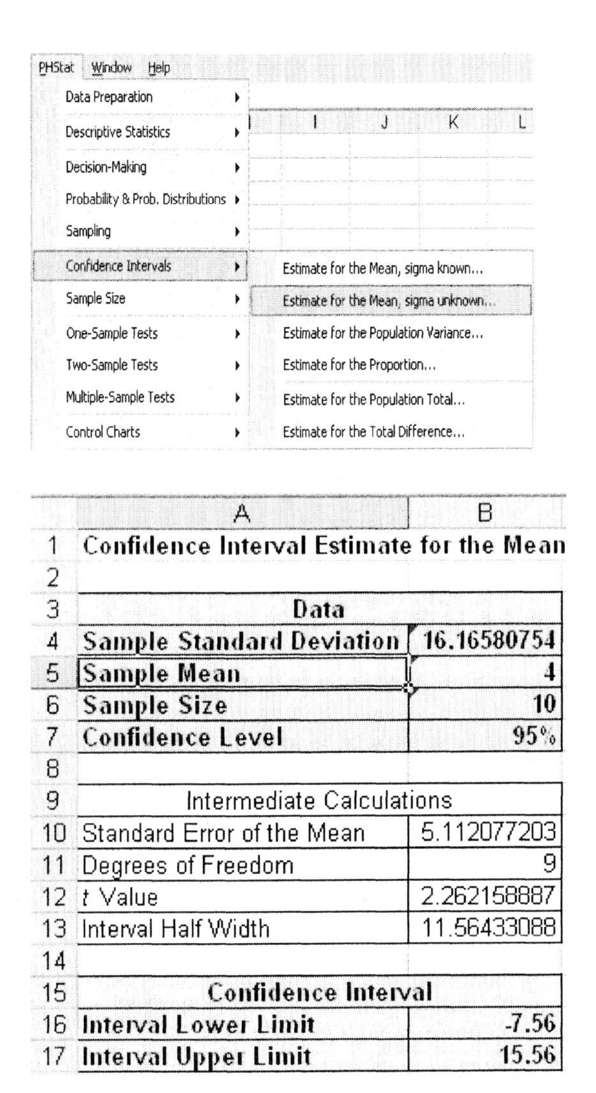

	A	B
1	**Confidence Interval Estimate for the Mean**	
2		
3	**Data**	
4	**Sample Standard Deviation**	**16.16580754**
5	**Sample Mean**	**4**
6	**Sample Size**	**10**
7	**Confidence Level**	**95%**
8		
9	Intermediate Calculations	
10	Standard Error of the Mean	5.112077203
11	Degrees of Freedom	9
12	*t* Value	2.262158887
13	Interval Half Width	11.56433088
14		
15	**Confidence Interval**	
16	**Interval Lower Limit**	**-7.56**
17	**Interval Upper Limit**	**15.56**

Chapter 10 Example 15 Inference Comparing Beliefs in Heaven and Hell Comparing Proportions with Dependent Samples

We will create the 95% confidence interval for population mean difference $p_1 - p_2$ using summary statistics.

First note that the sample mean of the 1120 difference scores equals

$[(0)(833) + (1)(125) + (-1)(2) + (0)(160)]/1120 = 123/1120 = 0.109821$

and the standard deviation of the 1120 difference scores equals

$$\sqrt{\{[(0-.109821)^2*(833)+(1-.109821)^2*(125)+(-1-.109821)^2*(2)+(0-.109821)^2*(160)]/(1120-1)\}}$$

$= 0.318469$

Choose **Add-Ins > PHStat > Confidence Intervals > Estimate for the Mean, Sigma Unknown**

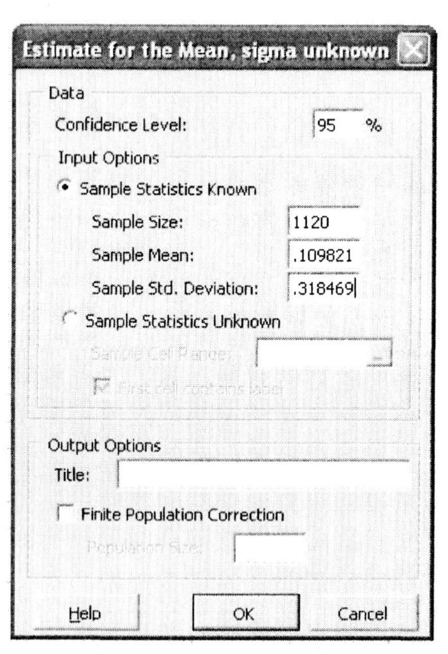

Enter: Confidence level **95**
Choose Sample Statistics Known

Enter: Sample Size **1120**
Sample Mean: **0.109821**
Sample Std. Deviation: **.318469**
Click **OK**
The results appear in a new worksheet.

	A	B
1	**Confidence Interval Estimate for the Mean**	
2		
3	**Data**	
4	**Sample Standard Deviation**	**0.318469**
5	**Sample Mean**	**0.109821**
6	**Sample Size**	**1120**
7	**Confidence Level**	**95%**
8		
9	Intermediate Calculations	
10	Standard Error of the Mean	0.009516081
11	Degrees of Freedom	1119
12	*t* Value	1.962084752
13	Interval Half Width	0.018671358
14		
15	**Confidence Interval**	
16	**Interval Lower Limit**	**0.09**
17	**Interval Upper Limit**	**0.13**

Chapter 10 Exercise 10.61 Heaven and Hell
Comparing Proportions for Dependent Samples

A point estimate for the difference between the population proportions believing in heaven and believing in hell is $\hat{p}_1 - \hat{p}_2 = .631 - .496 = .135$

Because we do not know the sample size used, but can assume dependent sampling, we'll use McNemar's test. Since McNemar's test is not included in Excel Data Analysis nor in PHStat, we will just do the computation in the worksheet.

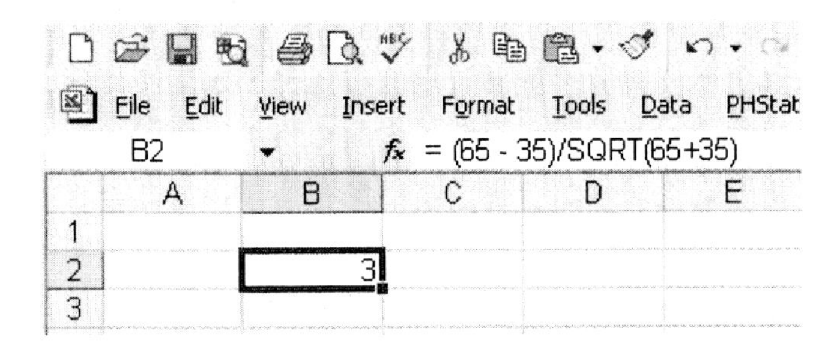

Chapter 11 Exercise 11.1 Gender Gap in Politics?
Contingency Tables

Excel does not normally do contingency tables, so all the work must be done manually. First enter the data in the spreadsheet. Use the Σ tool to enter the row and column totals.

	A	B	C	D	E
1	Gender	Democrat	Independent	Republican	Total
2	Female	885	992	593	2470
3	Male	551	859	539	1949
4	All	1436	1851	1132	4419

Now we will add the row and column percentages.

	A	B	C	D	E
1	Row Percent		Column Percent		
2	Gender	Democrat	Independent	Republican	Total
3	Female	885	992	593	2470
4		0.36	0.40	0.24	1.00
5		0.62	0.54	0.52	
6	Male	551	859	539	1949
7		0.28	0.44	0.28	1.00
8		0.38	0.46	0.48	
9	All	1436	1851	1132	4419
10		1.00	1.00	1.00	

Note the row percents are in rows 4 and 7 and the column percents are in rows 5 and 8.

To show bar graphs of this data, use the Chart menu as described in Chapter 2. Choose **Insert > Charts > Column > 2D Column** and choose the appropriate data.

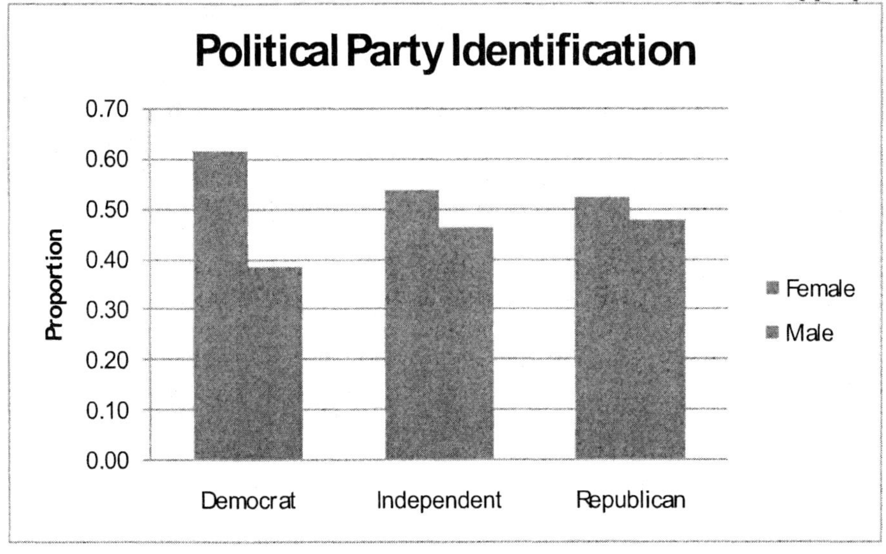

Chapter 11 Exercise 11.5 Marital Happiness and Income Contingency Table

Following the directions in this exercise, you can access the data in the GSS data base for this problem. Enter the fields as shown below.

SDA Frequencies/Crosstabulation Program
Help: General / Recoding Variables

REQUIRED Variable names to specify
Row: FINRELA(r:1-2; 3; 4-5)
OPTIONAL Variable names to specify
Column: HAPMAR
Control:
Selection Filter(s): YEAR(2006) *Example: age(18-50)*
Weight: No Weight

TABLE OPTIONS
Percentaging:
☐ Column ☑ Row ☐ Total
with 1 decimal(s)
☐ Confidence intervals Level: 95 percent
☐ Standard error of each percent
Sample Design: ⦿ Complex ○ SRS
☑ Statistics with 2 decimal(s)
☐ Question text ☐ Suppress table
☑ Color coding ☐ Show Z-statistic
☐ Include missing-data values

CHART OPTIONS
Type of chart: Stacked Bar Chart
Bar chart options:
Orientation: ⦿ Vertical ○ Horizontal
Visual Effects: ⦿ 2-D ○ 3-D
Show Percents: ☐ Yes
Palette: ⦿ Color ○ Grayscale
Size - width: 600 height: 400

Run the Table | Clear Fields

After you have accessed the data, enter it into an Excel worksheet as follows:

	A	B	C	D	E
1		Happily Married			
2	Income	very	pretty	not	Total
3	below	155	136	13	304
4	average	442	256	18	716
5	above	256	115	6	377

Using the Σ tool to enter the row and column totals and add the row and column percentages.

	A	B	C	D	E
1		Happily Married			
2	Income	very	pretty	not	Total
3	below	155	136	13	304
4		0.51	0.45	0.04	1.00
5		0.18	0.27	0.35	
6	average	442	256	18	716
7		0.62	0.36	0.03	1.00
8		0.52	0.50	0.49	
9	above	256	115	6	377
10		0.68	0.31	0.02	1.00
11		0.30	0.23	0.16	
12	Total	853	507	37	1397
13		1.00	1.00	1.00	

Chapter 11 Example 3 Chi-Squared for Happiness and Family Income

The table in the margin gives the data for this problem. To test the null hypothesis that Happiness and Family Income are independent, we will perform a Chi Square Test.
Choose **Add-Ins > PHStat > Multiple Sample tests> Chi-Square Test > OK**

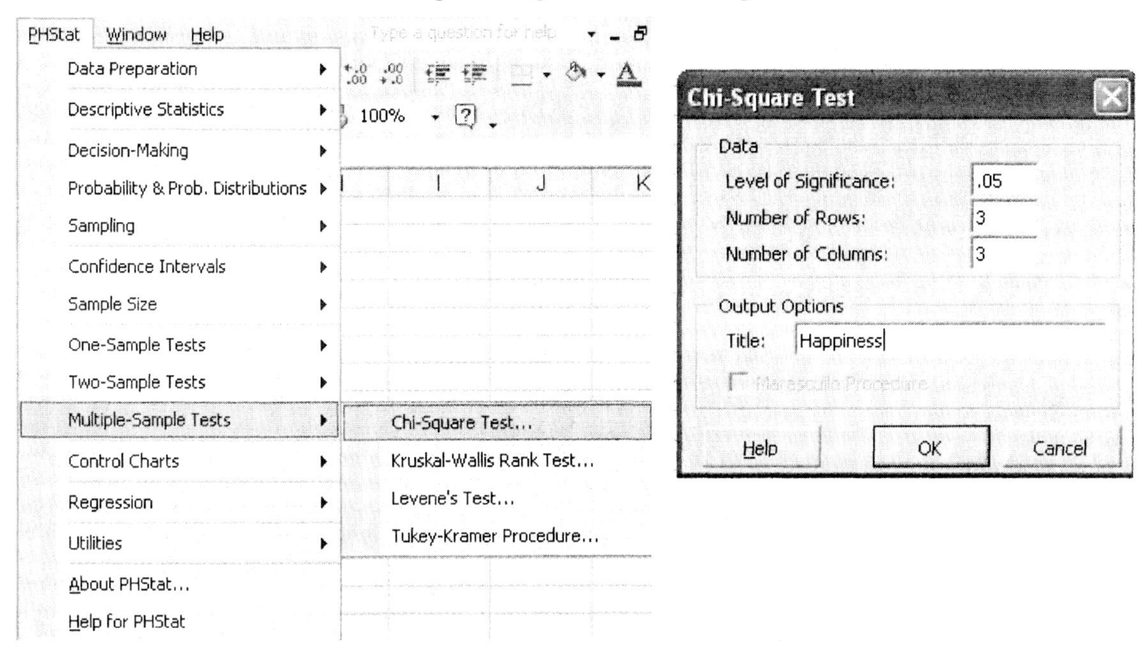

A chart template will pop up. Fill in the labels and counts in the appropriate columns. As you do, the rest of the chart will fill in according to pre-entered formulas.

Note the Chi-Square statistic and p-value in rows 27 and 28.

	A	B	C	D	E	F	G	H	I
1	**Happiness**								
2									
3		**Observed Frequencies**							
4			**Happiness**				Calculations		
5	**income**	**not**	**pretty**	**very**	**Total**			fo-fe	
6	**above**	21	159	110	290		-14.7709	-7.0793	21.85022
7	**average**	53	372	221	646		-26.6828	2.044053	24.63877
8	**below**	94	249	83	426		41.45374	5.035242	-46.489
9	**Total**	168	780	414	1362				
10									
11		Expected Frequencies							
12			Happimess						
13	income	not	pretty	very	Total			(fo-fe)^2/fe	
14	above	35.77093	166.0793	88.14978	290		6.099373	0.301762	5.416147
15	average	79.68282	369.9559	196.3612	646		8.935086	0.011294	3.091592
16	below	52.54626	243.9648	129.489	426		32.70286	0.103923	16.69042
17	Total	168	780	414	1362				
18									
19	**Data**								
20	**Level of Significance**	**0.05**							
21	Number of Rows	3							
22	Number of Columns	3							
23	Degrees of Freedom	4							
24									
25	**Results**								
26	**Critical Value**	9.487728							
27	**Chi-Square Test Statistic**	73.35246							
28	***p*-Value**	4.44E-15							
29	**Reject the null hypothesis**								
30									
31	*Expected frequency assumption*								
32	*is met.*								

Chapter 11 Example 4 Are Happiness and Income Independent?
Chi Squared Test of Independence

In the chart above we noted that the X^2 = 73.4 and the degrees of freedom = 4. The p-value reported in the table is 0 to very many decimal places. We therefore can reject H_0 and conclude that there is an association between happiness and family income.

Exercise 11.11 Life After Death and Gender
Contingency Tables

The data is that 891 of 1124 males and 1286 of 1505 females believe in life after death. Choose **Add-Ins > PHStat > Multiple Sample tests > Chi-Square Test**

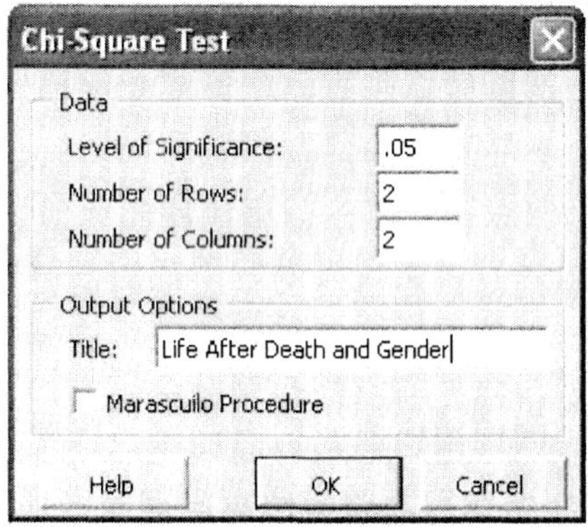

A chart template will pop up. Fill in the labels and counts in the appropriate columns. As you do, the rest of the chart will fill in according to pre-entered formulas. Note the Chi-Square statistic and p-value.

	A	B	C	D
1	**Life After Death and Gender**			
2				
3	**Observed Frequencies**			
4		**Life After Death**		
5	**Gender**	**Yes**	**No**	**Total**
6	**Male**	**891**	**233**	**1124**
7	**Female**	**1286**	**219**	**1505**
8	**Total**	**2177**	**452**	**2629**
9				
10	Expected Frequencies			
11		Life After Death		
12	Gender	Yes	No	Total
13	Male	930.7524	193.2476	1124
14	Female	1246.248	258.7524	1505
15	Total	2177	452	2629
16				
17	**Data**			
18	**Level of Significance**	**0.05**		
19	Number of Rows	2		
20	Number of Columns	2		
21	Degrees of Freedom	1		
22				
23	**Results**			
24	**Critical Value**	**3.841459**		
25	**Chi-Square Test Statistic**	**17.25037**		
26	***p*-Value**	**3.28E-05**		
27	**Reject the null hypothesis**			

Exercise 11.15 Help The Environment
Chi Squared Test for Different Levels

The data for this problem is given in the margin of the text. It is a 2 row by 5 column chart. Choose **Add-Ins > PHStats > Multiple Sample tests> Chi-Square Test > OK**
Enter the size of the chart and the title and the chart will pop up. Fill in the chart for a significance level of .05. Note the calculations on the right and the expected cell counts in rows 12 and 13.

	A	B	C	D	E	F	G	H	I	J	K	L	M
1	**Help the Environment**												
2			**Observed Frequencies**										
3				**Willing**					Calculations				
4	**Gender**	**vw**	**fw**	**nwu**	**nvw**	**nw**	**Total**				fo-fe		
5	**Female**	**34**	**149**	**160**	**142**	**168**	**653**		-1.71966	-7.2735	-14.1333	8.051282	15.07521
6	**Male**	**30**	**131**	**152**	**98**	**106**	**517**		1.719658	7.273504	14.13333	-8.05128	-15.0752
7	**Total**	**64**	**280**	**312**	**240**	**274**	**1170**						
8													
9			Expected Frequencies										
10				Willing									
11	Gender	vw	fw	nwu	nvw	nw	Total				(fo-fe)^2/fe		
12	Female	35.71966	156.2735	174.1333	133.9487	152.9248	653		0.08279	0.338534	1.147116	0.48394	1.486104
13	Male	28.28034	123.7265	137.8667	106.0513	121.0752	517		0.104568	0.427587	1.448872	0.611243	1.877032
14	Total	64	280	312	240	274	1170						
15	**Data**												
16	**Level of Significance**	**0.05**											
17	Number of Rows	2											
18	Number of Columns	5											
19	Degrees of Freedom	4											
20	**Results**												
21	**Critical Value**	**9.487728**											
22	**Chi-Square Test Statistic**	**8.007786**											
23	***p*-Value**	**0.091293**											
24	**Do not reject the null hypothesis**												
25	*Expected frequency assumption*												
26	*is met.*												

Repeat the chart for a significance level of .1 . Note the differences

	A	B	C	D	E	F	G	H	I	J	K	L	M
1	**Help the Environment**												
2			**Observed Frequencies**										
3				**Willing**					Calculations				
4	**Gender**	**vw**	**fw**	**nwu**	**nvw**	**nw**	**Total**				fo-fe		
5	**Female**	**34**	**149**	**160**	**142**	**168**	**653**		-1.71966	-7.2735	-14.1333	8.051282	15.07521
6	**Male**	**30**	**131**	**152**	**98**	**106**	**517**		1.719658	7.273504	14.13333	-8.05128	-15.0752
7	**Total**	**64**	**280**	**312**	**240**	**274**	**1170**						
8													
9			Expected Frequencies										
10				Willing									
11	Gender	vw	fw	nwu	nvw	nw	Total				(fo-fe)^2/fe		
12	Female	35.71966	156.2735	174.1333	133.9487	152.9248	653		0.08279	0.338534	1.147116	0.48394	1.486104
13	Male	28.28034	123.7265	137.8667	106.0513	121.0752	517		0.104568	0.427587	1.448872	0.611243	1.877032
14	Total	64	280	312	240	274	1170						
15	**Data**												
16	**Level of Significance**	**0.1**											
17	Number of Rows	2											
18	Number of Columns	5											
19	Degrees of Freedom	4											
20	**Results**												
21	**Critical Value**	**7.779434**											
22	**Chi-Square Test Statistic**	**8.007786**											
23	***p*-Value**	**0.091293**											
24	**Reject the null hypothesis**												
25	*Expected frequency assumption*												
26	*is met.*												

Chapter 12 Example 2 What Do We Learn From a Scatterplot in The Strength Study? Review of Producing a Scatterplot

Open the data file **high_school_female_athletes** from the Excel folder of the data disk.

To create a scatterplot of the data with column N, BP(60), as the x-variable, and column K, BP, as the y-variable:
Highlight Column K.
Click: **Insert > Chart > Scatter > Scatter with only Markers**
Right Click on the Scatter plot, choose "Select Data", and choose Edit.
On the Edit Series Tab, complete the dialogue box as illustrated.
Click: **OK**

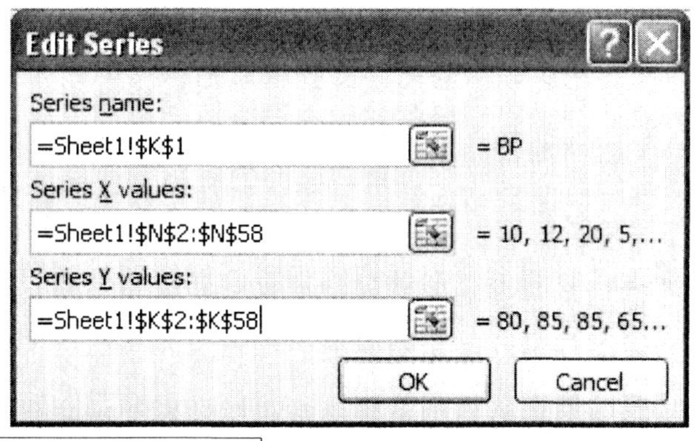

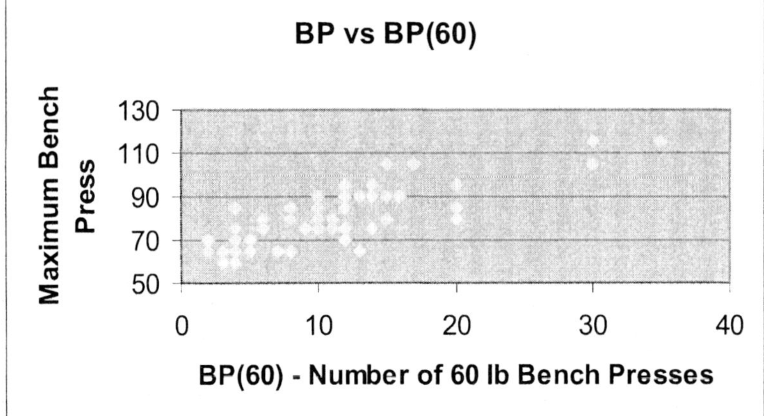

The scatterplot shows that female athletes with higher numbers of 60-pound bench presses also tended to have higher vales for the maximum bench press.

Chapter 12 Example 3 Which Regression Line Predicts Maximum Bench Press? Review of Generating a Regression Line

Open the data file **high_school_female_athletes** from the Excel folder of the text CD.

To generate the least squares regression line:

Click: **Data > Data Analysis > Regression**

Enter: Input Y Range: **L1:L58**
Input S Range: **O1:O58**
Check the **Labels** box
Click: **OK**

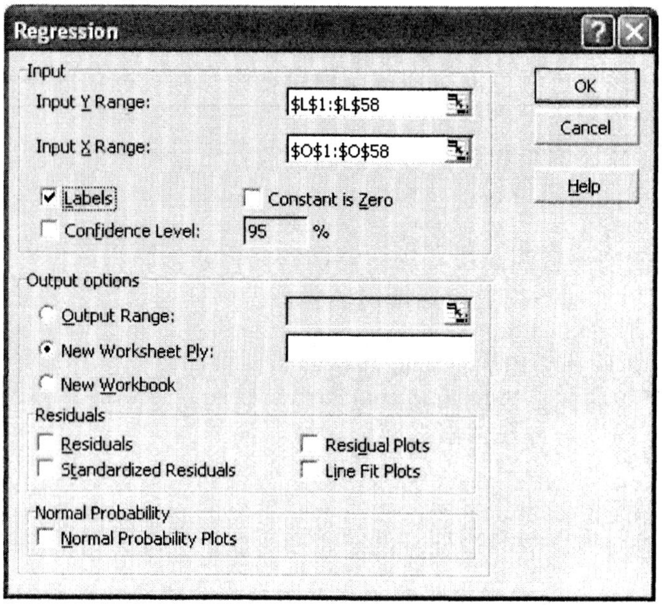

Regression

Input
Input Y Range: $L$1:$L$58
Input X Range: $O$1:$O$58
☑ Labels ☐ Constant is Zero
☐ Confidence Level: 95 %

Output options
○ Output Range:
◉ New Worksheet Ply:
○ New Workbook
Residuals
☐ Residuals ☐ Residual Plots
☐ Standardized Residuals ☐ Line Fit Plots
Normal Probability
☐ Normal Probability Plots

OK
Cancel
Help

The results are displayed in a new worksheet.

SUMMARY OUTPUT

Regression Statistics	
Multiple R	0.802025
R Square	0.643244
Adjusted F	0.636758
Standard E	8.003188
Observatic	57

ANOVA

	df	*SS*	*MS*	*F*	*gnificance F*
Regressic	1	6351.755	6351.755	99.16711	6.48E-14
Residual	55	3522.806	64.05103		
Total	56	9874.561			

	Coefficientstandard Err		*t Stat*	*P-value*	*Lower 95%*	*Upper 95%*	*Lower 95.0%*	*Upper 95.0%*
Intercept	63.53688	1.956469	32.47528	1.44E-37	59.61601	67.45771	59.61601	67.45771
BP (60)	1.491053	0.14973	9.958268	6.48E-14	1.190987	1.791119	1.190987	1.791119

To add the regression line to the scatterplot produced for chapter 11 example 2, right-click on one of the data points then select **Add Trendline, Linear, Display equation on chart** , and **Display R-squared on chart,** and Click **OK.**

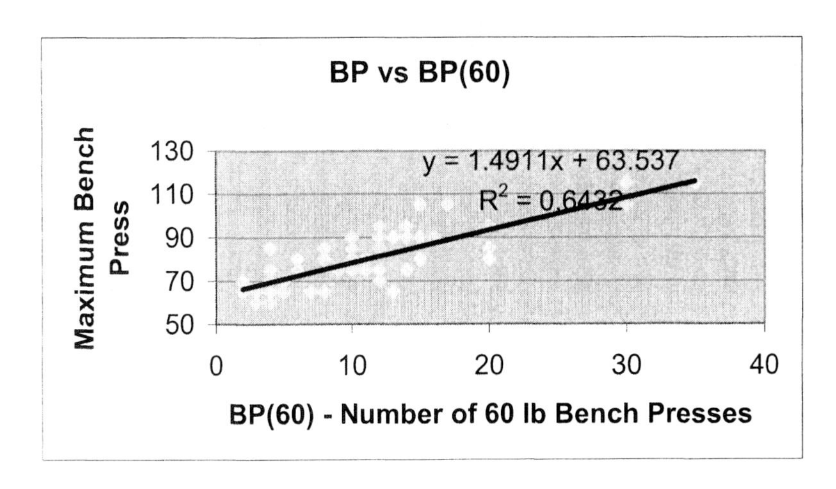

Chapter Exercise 12.1 Car Mileage and Weight
Generating a Regression Line

Open the data file **car_weight_and_mileage** from the Excel folder of the text CD.

To generate the least squares regression line:

Click: **Data > Data Analysis > Regression**

Enter: Input Y Range: **C1:C26**
Input S Range: **B1:B26**
Check the **Labels** box
Click: **OK**

SUMMARY OUTPUT

Regression Statistics	
Multiple R	0.866579
R Square	0.750959
Adjusted F	0.740131
Standard I	3.016149
Observati(	25

ANOVA

	df	*SS*	*MS*	*F*	*gnificance F*
Regressic	1	630.9254	630.9254	69.35415	2.14E-08
Residual	23	209.2346	9.097155		
Total	24	840.16			

	Coefficients	*Standard Err*	*t Stat*	*P-value*	*Lower 95%*	*Upper 95%*	*Lower 95.0%*	*Upper 95.0%*
Intercept	45.64536	2.602758	17.5373	8.3E-15	40.26115	51.02957	40.26115	51.02957
weight	-0.00522	0.000627	-8.32791	2.14E-08	-0.00652	-0.00392	-0.00652	-0.00392

Chapter Exercise 12.9 Predicting College GPA
Generating a Regression Line

Open the data file **georgia_student_survey** from the Excel folder of the text CD. We are interested in predicting college GPA based on high school GPA.

explanatory: high school GPA (column H)
response: college GPA (column I)

Highlight Column H.
To generate the scatterplot:
Click: **Insert > Chart > Scatter > Scatter with only Markers**
Right Click on the Scatter plot, choose "Select Data", and choose Edit.
On the Edit Series Tab, complete the dialogue box as illustrated.
Click: **OK**

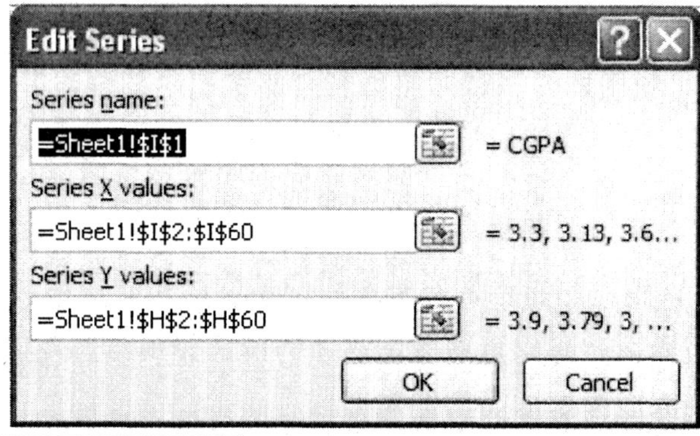

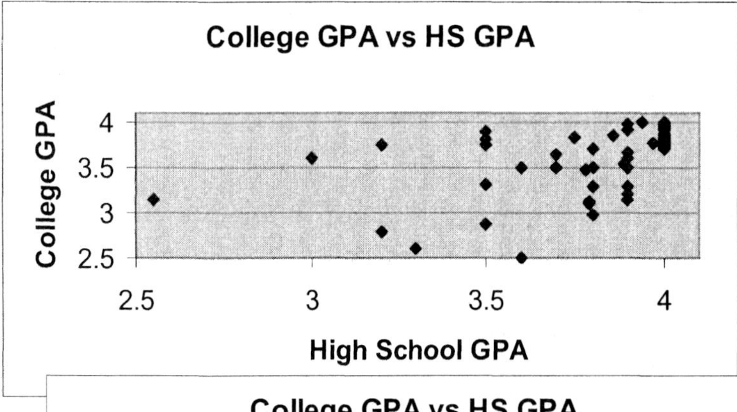

Add the least squares regression line to the scatterplot. Right-click on one of the data points then select **Add Trendline, Linear, Display equation on chart , and Display R-squared on chart,** and click **OK.**

College GPA vs HS GPA

y = 0.6369x + 1.1898
$R^2 = 0.2551$

College GPA

High School GPA

Chapter 12 Example 6 What's the Correlation for Predicting Strength?
Review of Determining Correlation

Open the data file **high_school_female_athletes** from the Excel folder of the text CD. We have already identified column K, BP, as the response variable and column N, BP(60), as the explanatory variable.

To determine the correlation using Excel:

Click: **within an empty cell**
Choose: **Formulas > Insert Function**
Select: Or select a category: **Statistical**
Select: **CORREL**
Click: **OK**

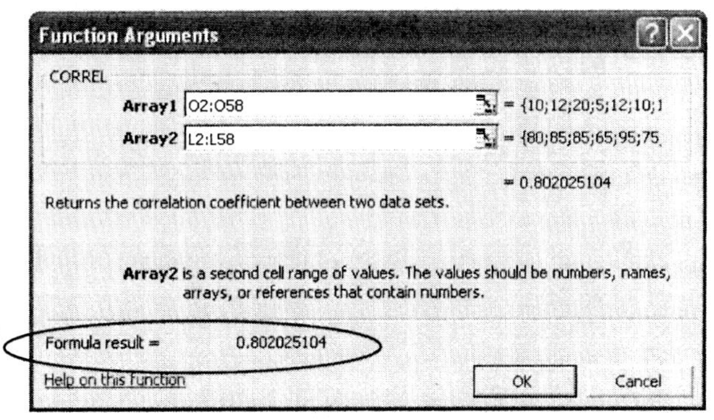

Complete the Functions Arguments dialog box as shown. Click **OK.**

The results are displayed within the dialog box. r = 0.802

Chapter 12 Example 9 What Does r^2 Tell Us In The Strength Study?
Determining r^2

Open the data file **high_school_female_athletes** from the Excel folder of the text CD. We have already identified column L, BP, as the response variable and column O, BP(60), as the explanatory variable.

To determine the value of r^2 using Excel:
Select: **Add-Ins > PHStat > Regression > Simple Linear Regression**
Enter: Y Variable Cell Range: **O1:O58**
X Variable Cell Range: **L1:L58**
Select: **Regression Statistics Table**
Click: **OK**

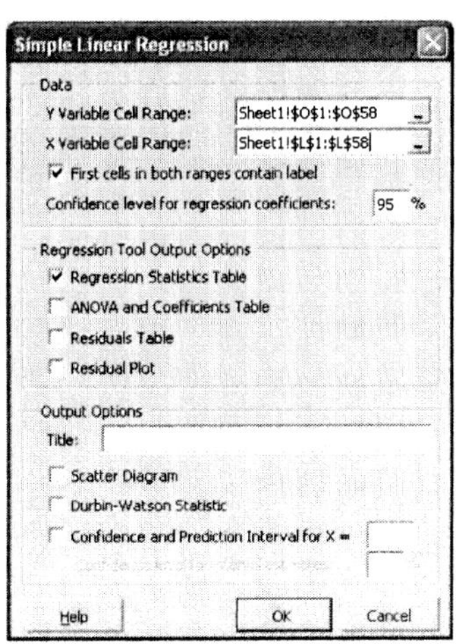

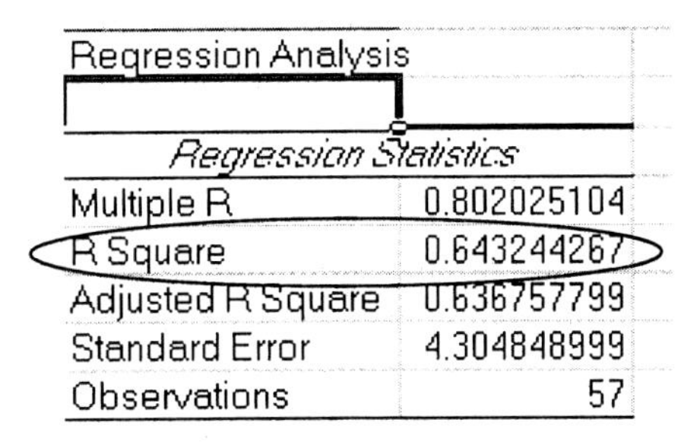

Regression Analysis	
Regression Statistics	
Multiple R	0.802025104
R Square	0.643244267
Adjusted R Square	0.636757799
Standard Error	4.304848999
Observations	57

Chapter 12 Exercise 12.15 Sit-ups and the 40-yard dash
Determining the Value of r^2

Open the data file **high_school_female_athletes** from the Excel folder of the text CD. For this exercise we have column H, SIT-UP, as the explanatory variable and column I, 40-YD (sec), as the response variable.

To determine the value of r^2 using Excel:
Select: **Add-Ins > PHStat > Regression > Simple Linear Regression**
Enter: Y Variable Cell Range: **O1:O58**
X Variable Cell Range: **L1:L58**
Select: **Regression Statistics Table**
Select: **ANOVA and Coefficients Table**
Click: **OK**

Regression Analysis

Regression Statistics	
Multiple R	0.459309491
R Square	0.210965208
Adjusted R Square	0.196619121
Standard Error	0.327208191
Observations	57

Note the value of R^2 = 21.1%

ANOVA

	df	*SS*	*MS*	*F*	*Significance F*
Regression	1	1.574438532	1.574438532	14.70541804	0.00032572
Residual	55	5.88858603	0.107065201		
Total	56	7.463024561			

	Coefficients	*Standard Error*	*t Stat*	*P-value*	*Lower 95%*	*Upper 95%*
Intercept	6.706527218	0.177891008	37.70020353	5.5637E-41	6.350025759	7.063028678
SIT-UP	-0.024346063	0.006348777	-3.834764405	0.00032572	-0.037069293	-0.011622833

Note the regression equation is $\hat{y} = 6.71 - 0.0243x$. Using this equation we can predict time in the 40-yard dash for any subject who can do a given number of sit-ups.

Chapter 12 Exercise 12.17 Student ideology
Determining the Value of r^2

Open the data file **fla_student_survey** from the Excel folder of the text CD. For this exercise we have column J, newspapers, as the explanatory variable and column N, political_ideology, as the response variable.

Doing a complete analysis of the association of these two variables, we start with the scatterplot:
To generate the scatterplot:
Highlight Column N.
Click: **Insert > Chart > Scatter > Scatter with only Markers**
Right Click on the Scatter plot, choose "Select Data", and choose Edit.
On the Edit Series Tab, complete the dialogue box as illustrated.
Click: **OK**
Complete the dialog box as indicated.

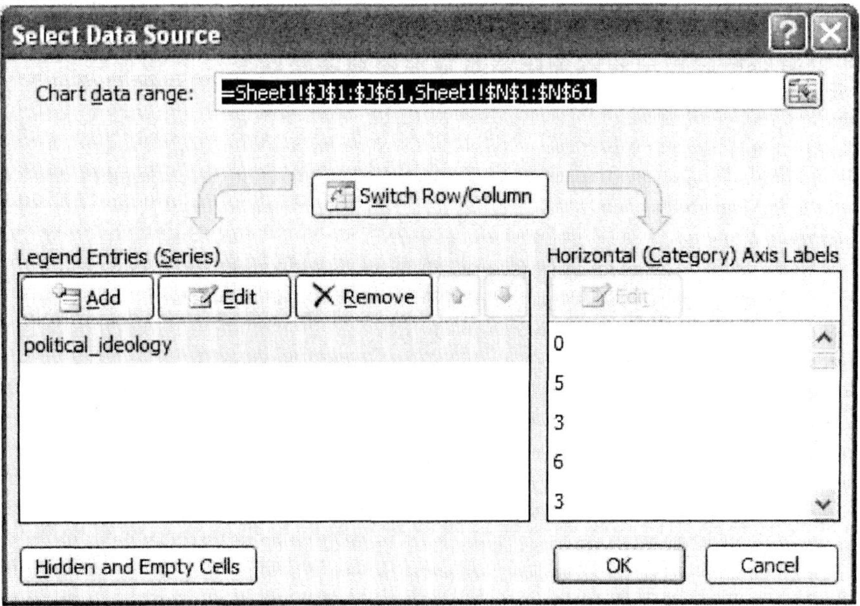

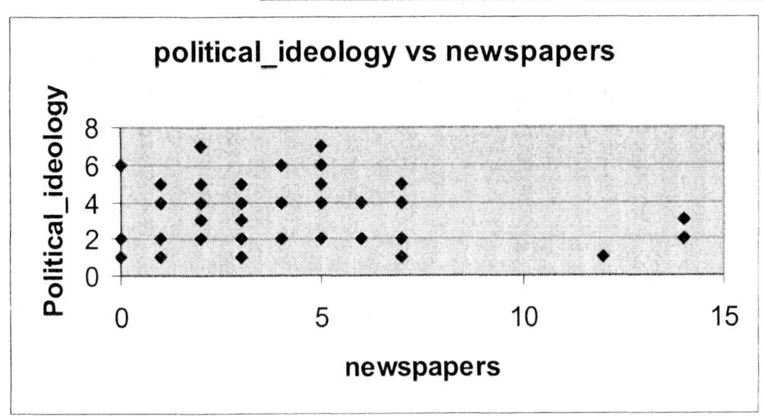

From this scatterplot, we can see that there is very little linear association.

Determining the correlation:

Select: **Formulas > Insert Function > CORR**

Complete the dialog box as indicated

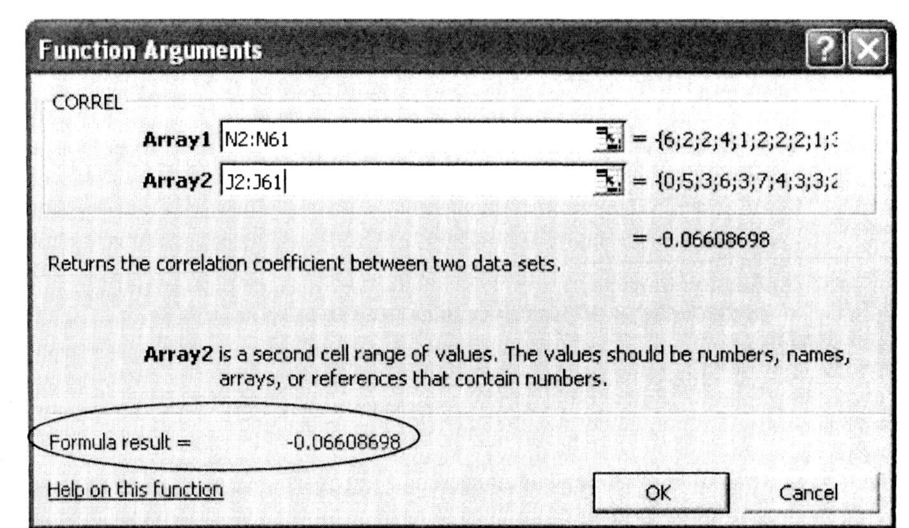

Again, based on the value of r = -0.066, we see very little evidence of a linear association between these two variables.

Determining the value of r^2:

Select: **Add-Ins > PHStat > Regression > Simple Linear Regression**

Enter: Y Variable Cell Range: **N1:N61**

X Variable Cell Range: **J1:J61**

Select: **Regression Statistics Table**

Select: **ANOVA and Coefficients Table**

A	B	C	D	E	F	G
Regression Analysis						
Regression Statistics						
Multiple R	0.06608698					
R Square	0.004367489					
Adjusted R Square	-0.012798589					
Standard Error	1.646540581					
Observations	60					
ANOVA						
	df	*SS*	*MS*	*F*	*Significance F*	
Regression	1	0.689772075	0.689772075	0.254425555	0.615886951	
Residual	58	157.2435613	2.711095884			
Total	59	157.9333333				
	Coefficients	*Standard Error*	*t Stat*	*P-value*	*Lower 95%*	*Upper 95%*
Intercept	3.180284775	0.360640175	8.818442854	2.64656E-12	2.458385572	3.902183979
newspapers	-0.035988108	0.071347482	-0.504406141	0.615886951	-0.178805504	0.106829288

Note the displayed value of R-Sq = .4%

Chapter 12 Example 11 Is Strength Associated with 60-pound Bench Presses? A Significance Test of Independence

Open the worksheet **high_school_female_student_athletes,** which is found in the Excel folder. The number of bench presses before fatigue, column N, BP(60), is the x-variable, and maximum bench press, column K, BP, is the y-variable. The regression output includes the t test statistic and the p-value.

Regression Analysis

Regression Statistics	
Multiple R	0.802025104
R Square	0.643244267
Adjusted R Square	0.636757799
Standard Error	8.003188444
Observations	57

ANOVA

	df	*SS*	*MS*	*F*	*Significance F*
Regression	1	6351.755014	6351.755014	99.16710914	6.48137E-14
Residual	55	3522.80639	64.05102527		
Total	56	9874.561404			

	Coefficients	*Standard Error*	*t Stat*	*P-value*	*Lower 95%*	*Upper 95%*
Intercept	63.53685646	1.956468584	32.4752756	1.43832E-37	59.61600676	67.45770615
BP (60)	1.491053006	0.149730149	9.95826838	6.48137E-14	1.190987157	1.791118856

Notice the t test statistic ($t = 9.96$) appears under the column heading "tStat" in the row for the predictor BP (60), and the p-value ($p = 0.000$) for the two-sided alternative H_a: $\beta \neq 0$ appears under the heading "P-value" in the row for the predictor BP (60).

Chapter 12 Exercise 12.33 Predicting House Prices Significance Test of Independence and a Confidence Interval for Slope

Open the worksheet **house_selling_prices**, which is found in the Excel folder. The size of the house is in column H, size, and selling price is in column G, price.

Select: **Add-Ins > PHStat > Regression > Simple Linear Regression**
Enter: Y Variable cell range: **G1:G101**
X Variable cell range: **H1:H101**
Select: **Regression Statistics Table**
Select: **ANOVA and Coefficients Table**
Click: **OK**

Regression Analysis

Regression Statistics	
Multiple R	0.761262112
R Square	0.579520003
Adjusted R Square	0.57522939
Standard Error	36730.2057
Observations	100

ANOVA

	df	*SS*	*MS*	*F*	*Significance F*
Regression	1	1.8222E+11	1.8222E+11	135.0669725	3.83804E-20
Residual	98	1.32213E+11	1349108011		
Total	99	3.14433E+11			

	Coefficients	*Standard Error*	*t Stat*	*P-value*	*Lower 95%*	*Upper 95%*
Intercept	9161.158864	10759.78616	0.851425737	0.396608629	-12191.28619	30513.60392
size	77.00769255	6.626123525	11.62183172	3.83804E-20	63.85836631	90.15701878

Notice the t test statistic ($t = 11.62$), and the p-value ($p = 0.000$) for the two-sided alternative H_a: $\beta \neq 0$. Since p-value = 0.000 is less than a 0.05 significance level, there is evidence that these two variables are not independent, and that the sample association between these two variables is not just random variation.

The 95% confidence interval for the population slope is (63.859, 90.157)

This interval does not support the builder's claim that selling price increases $100, on the average, for every extra square foot. (The interval does not contain 100.)

Chapter 12 Exercise 12.39 Advertising and Sales
A Significance Test of Independence

Enter the data into the Excel worksheet. Enter the x's into column A and name the column Advertising. Enter the y's into column B and name the column Sales.

Select: **Add-Ins > PHStat > Regression > Simple Linear Regression**
Enter: Y Variable cell range: **B1:B5**
X Variable cell range: **A1:A5**

Regression Analysis

Regression Statistics	
Multiple R	0.857142857
R Square	0.734693878
Adjusted R Square	0.602040816
Standard Error	1.362770288
Observations	4

ANOVA

	df	*SS*	*MS*	*F*	*Significance F*
Regression	1	10.28571429	10.28571429	5.538461538	0.142857143
Residual	2	3.714285714	1.857142857		
Total	3	14			

	Coefficients	*Standard Error*	*t Stat*	*P-value*	*Lower 95%*	*Upper 95%*
Intercept	5.285714286	0.997445717	5.299250068	0.033814111	0.994048759	9.577379812
Advertising	0.857142857	0.36421568	2.353393622	0.142857143	-0.709951822	2.424237536

Notice the t test statistic ($t = 2.35$), and the p-value ($p = 0.143$) for the two-sided alternative H_a: $\beta \neq 0$. Since p-value = 0.143 is not less than significance level 0.05, there is not evidence that these two variables are independent. A word of caution, though; were the assumptions validated? This is a very small sample, so results are suspect.

Chapter 12 Example 13 Detecting an Underachieving College Student
How Data Vary Around the Regression Line

Open the worksheet **georgia_student_survey**, found in the Excel folder. High school GPA is found in column H and college GPA is in column I.

As part of the regression analysis, PHStat can be instructed to include a table of the residuals. However, PHStat does not highlight any "unusual observations."

Select: **Add-Ins > PHStat > Regression > Simple Linear Regression**
Enter: Y Variable cell range: **I1:I60**
X Variable cell range: **H1:H60**
Select: **Regression Statistics Table**
Select: **ANOVA and Coefficients Table**
Select: **Residuals Table**
Click: **OK**

Here is a portion of the residual table produced.

Observation	*Predicted CGPA*	*Residuals*
1	3.673924508	-0.373924508
2	3.603860484	-0.473860484
3	3.100673406	0.499326594
4	3.673924508	-0.173924508
5	3.482840807	0.017159193
6	3.22806254	0.52193746
7	3.597491028	-0.127491028

Chapter 12 Example 16 Predicting Maximum Bench Press and Estimating Its Mean Confidence Interval for Population Mean of y and Prediction Interval for a Single y-value

Open the worksheet **high_school_female_athletes**, found in the Excel folder.

Proceed with the regular regression command:
Select: **Add-Ins > PHStat > Regression > Simple Linear Regression**

Enter: Y Variable Cell Range: **N1:N58**
X Variable Cell Range: **K1:K58**
Select: **Regression Statistics Table**
Enter: Output Options: Confidence and Prediction Interval for X = **11**
Enter: Confidence level for interval estimates: **95** %
Click: **OK**

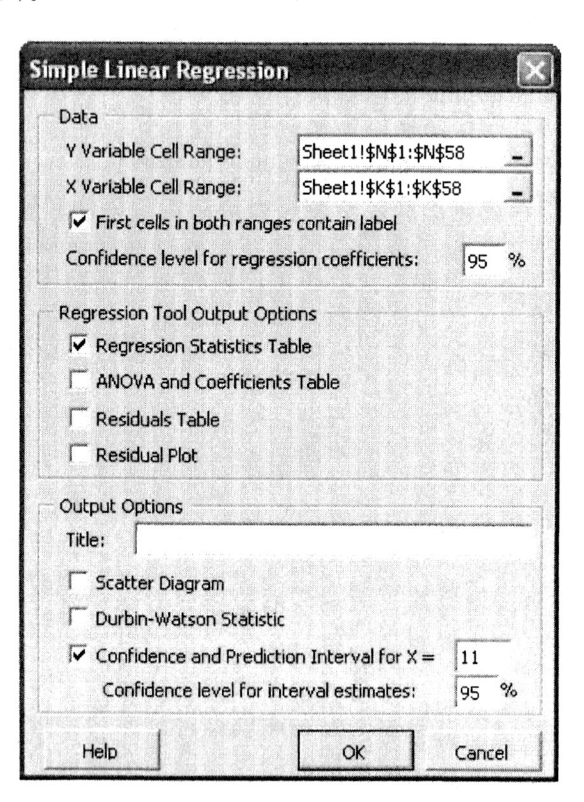

Data	
X Value	**11**
Confidence Level	**95%**
Intermediate Calculations	
Sample Size	57
Degrees of Freedom	55
t Value	2.004044
Sample Mean	10.98246
Sum of Squared Difference	2856.982
Standard Error of the Estimate	8.003188
h Statistic	0.017544
Predicted Y (YHat)	79.93844
For Average Y	
Interval Half Width	2.12439
Confidence Interval Lower Limit	**77.814**
Confidence Interval Upper Limi	**82.0628**
For Individual Response Y	
Interval Half Width	16.17882
Prediction Interval Lower Limit	**63.7596**
Prediction Interval Upper Limit	**96.1173**

Chapter 12 Exercise 12.43 Poor predicted strengths
How Data Vary Around the Regression Line

Open the worksheet **high_school_female_athletes**, found in the MINITAB folder.

Select: **Add-Ins > PHStat > Regression > Simple Linear Regression**
Enter: Y Variable Cell Range: **N1:N58**
X Variable Cell Range: **K1:K58**
Select: **Regression Statistics Table**
Select: **ANOVA and Coefficients Table**
Select: **Residuals Table**
Click: **OK**

Regression Analysis

Regression Statistics	
Multiple R	0.802025104
R Square	0.643244267
Adjusted R Square	0.636757799
Standard Error	8.003188444
Observations	57

ANOVA

	df	*SS*	*MS*	*F*	*Significance F*
Regression	1	6351.755014	6351.755014	99.16710914	6.48137E-14
Residual	55	3522.80639	64.05102527		
Total	56	9874.561404			

	Coefficients	*Standard Error*	*t Stat*	*P-value*	*Lower 95%*	*Upper 95%*
Intercept	63.53685646	1.956468584	32.4752756	1.43832E-37	59.61600676	67.45770615
BP (60)	1.491053006	0.149730149	9.95826838	6.48137E-14	1.190987157	1.791118856

A portion of the residuals table:

RESIDUAL OUTPUT

Observation	*Predicted BP*	*Residuals*
1	78.44738652	1.55261348
2	81.42949253	3.570507467
3	93.35791658	-8.357916585
4	70.99212149	-5.992121488

Chapter 12 Exercise 12.49 ANOVA Table for Leg Press

Open the worksheet **high_school_female_student_athletes,** which is found in the Excel folder. The number of leg presses before fatigue, column L, LP (200), is the x-variable, and maximum leg press, column M, LP, is the y-variable
Proceed with the regular regression command:

Select: **Add-Ins > PHStat > Regression > Simple Linear Regression**
Enter: Y Variable Cell Range: **L1:L58**
X Variable Cell Range: **M1:M58**
Select: **Regression Statistics Table**
Select: **ANOVA and Coefficients Table**
Select: **Residuals Table**
Click: **OK**

A	B	C	D	E	F	G
Regression Analysis						
Regression Statistics						
Multiple R	0.792504645					
R Square	0.628063612					
Adjusted R Square	0.621301132					
Standard Error	36.10702319					
Observations	57					
ANOVA						
	df	*SS*	*MS*	*F*	*Significance F*	
Regression	1	121082.4003	121082.4003	92.87474879	2.0619E-13	
Residual	55	71704.44179	1303.717123			
Total	56	192786.8421				
	Coefficients	*Standard Error*	*t Stat*	*P-value*	*Lower 95%*	*Upper 95%*
Intercept	233.8878981	13.06444129	17.90263303	1.32112E-24	207.7061792	260.0696171
LP (200)	5.271025893	0.546948359	9.6371546	2.0619E-13	4.174917157	6.367134629

From the ANOVA table, we can use SS Residual Error = 71704 to determine the residual standard deviation of y-values. Note that $df = 57 - 2 = 55$.

fx =SQRT(C13/B13)

B	C	D
		36.10702319

At any fixed value x of number of 200-pound leg presses, we estimate that the maximum leg press values have a standard deviation of 36.1 pounds.

For female athletes with $x = 22$, we would estimate the mean maximum leg press to be `LP = 234 + 5.27 (22)` = 349.94 pounds and the variability of their maximum leg press values to be 36.1 pounds.

If y-values are approximately normal, then 95% of the y-values would fall in the interval approximately $\hat{y} \pm 2s = 349.94 \pm 2(36.1) = (277.74, 422.14)$
Using Excel to generate the prediction interval, notice we get similar results.

For Individual Response Y	
Interval Half Width	72.99248
Prediction Interval Lower Limit	**276.858**
Prediction Interval Upper Limit	**422.843**

Chapter 12 Example 18 Explosion in Number of People Using the Internet Exponential Regression

Chapter 12 Exercise 12.61 Leaf Litter Decay Exponential Regression

Excel and PHStat do not include an exponential regression tool. See the MINITAB section.

Chapter 13 Example 2 Predicting Selling Price Using House and Lot Sizes Multiple Regression and Plotting the Relationships

Open the worksheet **house_selling_prices**, which is found in the Excel folder. The price is found in column G, house size is found in column H, and lot size is found in column I.

Select:
Add-Ins > PHStat > Regression > Multiple Regression

PHStat | Window | Help
Data Preparation
Descriptive Statistics
Decision-Making
Probability & Prob. Distributions
Sampling
Confidence Intervals
Sample Size
One-Sample Tests
Two-Sample Tests
Multiple-Sample Tests
Control Charts
Regression: Simple Linear Regression..., Multiple Regression..., Best Subsets..., Stepwise Regression...
Utilities
About PHStat...
Help for PHStat

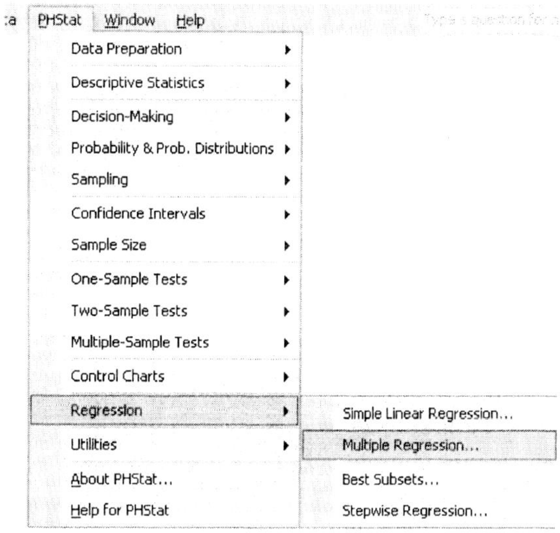

Enter:
Y Variable Cell Range: **select appropriate cells**
X Variables Cell Range: **select appropriate cells**
Select: **Regression Statistics Table**
Select: **ANOVA and Coefficients Table**
Click: **OK**

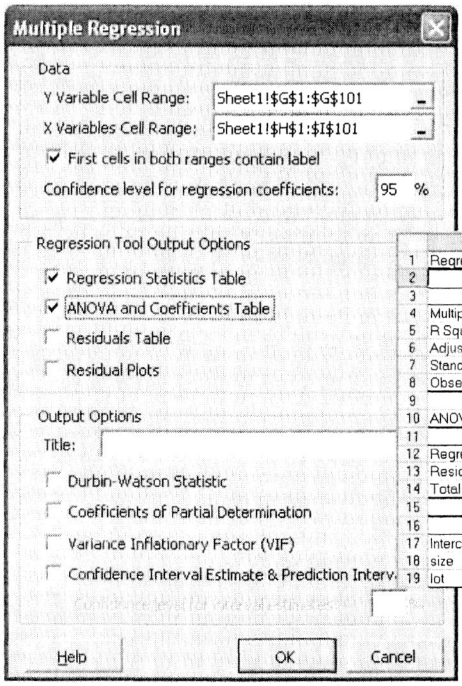

Multiple Regression
Data
Y Variable Cell Range: Sheet1!$G$1:$G$101
X Variables Cell Range: Sheet1!$H$1:$I$101
☑ First cells in both ranges contain label
Confidence level for regression coefficients: 95 %
Regression Tool Output Options
☑ Regression Statistics Table
☑ ANOVA and Coefficients Table
☐ Residuals Table
☐ Residual Plots
Output Options
Title:
☐ Durbin-Watson Statistic
☐ Coefficients of Partial Determination
☐ Variance Inflationary Factor (VIF)
☐ Confidence Interval Estimate & Prediction Interv...
Help | OK | Cancel

The results are displayed in a new worksheet.

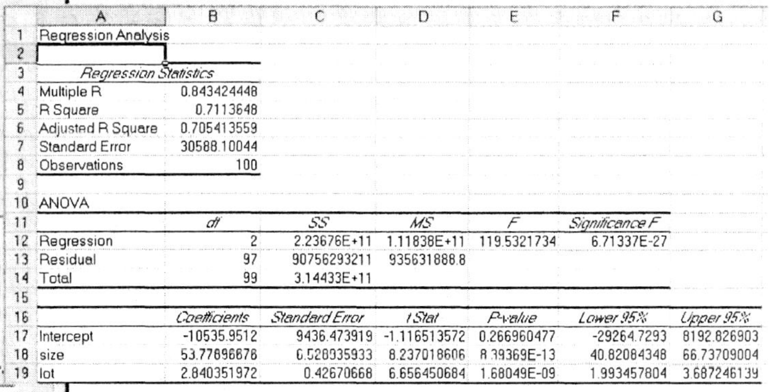

	A	B	C	D	E	F	G
1	Regression Analysis						
2							
3	*Regression Statistics*						
4	Multiple R	0.843424448					
5	R Square	0.7113648					
6	Adjusted R Square	0.705413559					
7	Standard Error	30588.10044					
8	Observations	100					
9							
10	ANOVA						
11		*df*	*SS*	*MS*	*F*	*Significance F*	
12	Regression	2	2.23676E+11	1.11838E+11	119.5321734	6.71337E-27	
13	Residual	97	90756293211	935631888.8			
14	Total	99	3.14433E+11				
15							
16		*Coefficients*	*Standard Error*	*t Stat*	*P-value*	*Lower 95%*	*Upper 95%*
17	Intercept	-10535.9512	9436.473919	-1.116513572	0.266960477	-29264.7293	8192.826903
18	size	53.77696676	6.528035933	8.237018606	8.39369E-13	40.82084348	66.73709004
19	lot	2.840351972	0.42670668	6.656450684	1.68049E-09	1.993457804	3.687246139

Excel and PHStat cannot be used to generate a scatterplot matrix. See MINITAB section.

Chapter 13 Exercise 13.5 Does more education cause more crime?
Multiple Regression

Open the worksheet **fla_crime**, which is found in the Excel folder. The crime rate is found in column B, education is found in column C, and urbanization is found in column D.

Select: **Add-Ins > PHStat > Regression > Multiple Regression**
Enter: Response: **B1:B68**
Predictors: **C1:C68, D1:D68** Select: **Results**
Select: **Regression Statistics Table**
Select: **ANOVA and Coefficients Table**
Click: **OK**

The results are displayed in the new worksheet.

	A	B	C	D	E	F	G
1	Regression Analysis						
2							
3	*Regression Statistics*						
4	Multiple R	0.686599834					
5	R Square	0.471419331					
6	Adjusted R Square	0.454901186					
7	Standard Error	20.81558241					
8	Observations	67					
9							
10	ANOVA						
11		*df*	*SS*	*MS*	*F*	*Significance F*	
12	Regression	2	24731.65726	12365.82863	28.53948225	1.37899E-09	
13	Residual	64	27730.46215	433.2884711			
14	Total	66	52462.1194				
15							
16		*Coefficients*	*Standard Error*	*t Stat*	*P-value*	*Lower 95%*	*Upper 95%*
17	Intercept	59.11806772	28.36531056	2.084167828	0.04114165	2.451896833	115.7842386
18	education	-0.583377348	0.472459127	-1.234767868	0.22143115	-1.527222104	0.360467407
19	urbanization	0.682501422	0.12321259	5.539218224	6.1108E-07	0.4363562	0.928646644

Using the regression equation generated by Excel,

```
crime = 59.1 - 0.583 education + 0.683 urbanization
```

we can predict crime rates for a county that has 0% in an urban environment by substituting 0 for "urbanization" and using the resulting equation crime = 59.1 – 0.583 education. With education set equal to 70% we determine the crime rate to be 59.1 – 0.583(70) = 18.29. For an 80% high school graduation rate, the crime rate is predicted to be 59.1 – 0.583(80) = 12.46.

Chapter 13 Example 3 How Well Can We Predict House Selling Prices? ANOVA and R

Open the worksheet **house_selling_prices**, which is found in the Excel folder. The price is found in column G, house size is found in column H, and lot size is found in column I. The regression command in PHStat includes output of the value of R-squared. We have done this many times.

Select: **Add-Ins > PHStat > Regression > Multiple Regression**
Enter: Y Variable Cell Range: **G1:G101**
X Variables Cell Range: **H1:H101, I1:I101**
Select: **Regression Statistics Table**
Select: **ANOVA and Coefficients Table**
Click: **OK**

	A	B
1	Regression Analysis	
2		
3	*Regression Statistics*	
4	Multiple R	0.843424448
5	R Square	0.7113648
6	Adjusted R Square	0.705413559
7	Standard Error	30588.10044
8	Observations	100

Note the value of R-squared is 71.1%. The multiple correlation between selling price and the two explanatory variables is $R = \sqrt{R^2} = \sqrt{.711} = 0.84$.

Chapter 13 Example 4 What Helps Predict a Female Athlete's Weight? Significance Test and Confidence Interval about a Multiple Regression Parameter β

Open the worksheet **college_athletes**, which is found in the Excel folder. The total body weight (TBW) is found in A, height (HGT) in inches is found in B, the percent of body fat (%BF) is found in C, and age (AGE) is found in K. PHStat will only allow selection of "X Variable Cell Ranges" that are contiguous. Copy columns HGT and %BF to columns L and M.

Select: **Add-Ins > PHStat > Regression > Multiple Regression**
Enter: Y Variable Cell Range: **A1:A65**
X Variable Cell Ranges: **K1:M65**
Select: **Regression Statistics Table**
Select: **ANOVA and Coefficients Table**
Click: **OK**

The results are displayed in the new worksheet.

	A	B
1	Regression Analysis	
2		
3	*Regression Statistics*	
4	Multiple R	0.818101194
5	R Square	0.669289564
6	Adjusted R Square	0.652754042
7	Standard Error	10.10860657
8	Observations	64

Note the value of R-squared is 66.9%. The predictive power is good.

ANOVA					
	df	*SS*	*MS*	*F*	*Significance F*
Regression	3	12407.94877	4135.982922	40.47586592	1.9772E-14
Residual	60	6131.03561	102.1839268		
Total	63	18538.98438			

	Coefficients	*Standard Error*	*t Stat*	*P-value*	*Lower 95%*	*Upper 9*
Intercept	-97.69378154	28.78521964	-3.393886959	0.001226151	-155.272775	-40.1147880
AGE	-0.960088375	0.648277817	-1.480982922	0.143843722	-2.256836658	0.33665990
HGT	3.428473996	0.367899254	9.319056661	2.87856E-13	2.692566158	4.16438183
%BF	1.364265481	0.312552565	4.364915328	5.09641E-05	0.739067469	1.98946349

To test whether age helps us to predict weight, if we already know height and percent body fat, we perform a significance test on $H_0 : \beta_3 = 0$ versus $H_a: \beta_3 \neq 0$. The *t* test statistic is reported in the table as -1.48 and the p-value is .144. This p-value does not give much evidence against the null hypothesis. Age does not significantly predict weight, if we already know height and percentage of body fat.

Chapter 13 Example 5 What's Plausible for the Effect of Age on Weight?
Confidence Interval about a Multiple Regression Parameter β

Open the worksheet **college_athletes**, which is found in the Excel folder. The total body weight (TBW) is found in A, height (HGT) in inches is found in B, the percent of body fat (%BF) is found in C, and age (AGE) is found in K.

Using the results from the previous example:

ANOVA					
	df	*SS*	*MS*	*F*	*Significance F*
Regression	3	12407.94877	4135.982922	40.47586592	1.9772E-14
Residual	60	6131.03561	102.1839268		
Total	63	18538.98438			

	Coefficients	*Standard Error*	*t Stat*	*P-value*	*Lower 95%*	*Upper 95%*
Intercept	-97.69378154	28.78521964	-3.393886959	0.001226151	-155.272775	-40.1147880
AGE	-0.960088375	0.648277817	-1.480982922	0.143843722	-2.256836658	0.33665990
HGT	3.428473996	0.367899254	9.319056661	2.87856E-13	2.692566158	4.16438183
%BF	1.364265481	0.312552565	4.364915328	5.09641E-05	0.739067469	1.98946349

The 95% confidence interval for the population slope β_3 is (-2.26, 0.34)
At fixed values of height and percent body fat, we infer that the population mean weight changes very little (and may not change at all, since this interval includes 0), making these results consistent with what we found in example 4.

Chapter 13 Example 7 The F Test For Predictors of Athletes' Weight
The F Test that All Multiple Regression Parameters $\beta = 0$

Open the worksheet **college_athletes**, which is found in the Excel folder. The total body weight (TBW) is found in A, height (HGT) in inches is found in B, the percent of body fat (%BF) is found in C, and age (AGE) is found in K.

Once again using the results from example 4:

Regression Analysis

Regression Statistics	
Multiple R	0.818101194
R Square	0.669289564
Adjusted R Square	0.652754042
Standard Error	10.10860657
Observations	64

ANOVA

	df	*SS*	*MS*	*F*	*Significance F*
Regression	3	12407.94877	4135.982922	40.47586592	1.9772E-14
Residual	60	6131.03561	102.1839268		
Total	63	18538.98438			

	Coefficients	*Standard Error*	*t Stat*	*P-value*	*Lower 95%*	*Upper 95%*
Intercept	-97.69378154	28.78521964	-3.393886959	0.001226151	-155.272775	-40.11478809
AGE	-0.960088375	0.648277817	-1.480982922	0.143843722	-2.256836658	0.336659908
HGT	3.428473996	0.367899254	9.319056661	2.87856E-13	2.692566158	4.164381834
%BF	1.364265481	0.312552565	4.364915328	5.09641E-05	0.739067469	1.989463493

The ANOVA table for the multiple regression output includes the value of the F test statistic ($F = 40.48$) and the p-value ($p = 0.000$) for testing H_0: $\beta_1 = \beta_2 = \beta_3 = 0$. We can reject H_0 and conclude that at least one predictor has an effect on weight.

Chapter 13 Exercise 13.20 Study time help GPA?
Hypothesis Test and Confidence Interval About β

Open the worksheet **georgia_student_survey**, which is found in the Excel folder.
Performing a multiple regression for college GPA based on high school GPA and study time:

Select: **Add-Ins > PHStat > Regression > Multiple Regression**
Enter: Y Variable Cell Range: **I1:I60**
X Variables Cell Range: Predictors: **H1:H60, E1:E60**
Select: **Regression Statistics Table**
Select: **ANOVA and Coefficients Table**
Click: **OK**

Regression Analysis

Regression Statistics	
Multiple R	0.508122099
R Square	0.258188067
Adjusted R Square	0.231694784
Standard Error	0.318824302
Observations	59

ANOVA

	df	*SS*	*MS*	*F*	*Significance F*
Regression	2	1.981222325	0.990611162	9.745416	0.000233468
Residual	56	5.692340387	0.101648935		
Total	58	7.673562712			

	Coefficients	*Standard Error*	*t Stat*	*P-value*	*Lower 95%*	*Upper 95%*
Intercept	1.126227293	0.568970902	1.979411054	0.05269364	-0.013557629	2.266012214
HSGPA	0.643428775	0.14576196	4.41424342	4.67363E-05	0.351432675	0.935424875
Studytime	0.007757836	0.016136202	0.480772131	0.632551508	-0.02456684	0.040082513

To test whether study time helps us to predict college GPA, if we already know high school GPA, we perform a significance test on H_0: $\beta_2 = 0$ versus H_a: $\beta_2 \neq 0$. The *t* test statistic is reported in the table as 0.48 and the p-value is 0.633. This p-value does not give evidence against the null hypothesis. Study time does not significantly predict college GPA if we already know high school GPA.

The 95% confidence interval for the population slope β_2: (-0.0247, 0.0399).
At a fixed value of high school GPA, we infer that the population mean college GPA changes very little (and may not change at all, since this interval includes 0), making these results consistent with the significance test we have already completed.

Chapter 13 Exercise 13.29 More predictors for house price
The F Test that All Multiple Regression Parameters β = 0

Open the worksheet **house_selling_prices**, which is found in the Excel folder. Performing multiple regression for predicting house selling price using size of home, lot size, and real estate tax (copy column B to column J):

Select: **Add-Ins > PHStat > Regression > Multiple Regression**
Enter: Y Variable Cell Range: **G1:G101**
X Variables Cell Range: **H1:J101**
Select: **Regression Statistics Table**
Select: **ANOVA and Coefficients Table**
Click: **OK**

Regression Analysis

Regression Statistics	
Multiple R	0.870016326
R Square	0.756928408
Adjusted R Square	0.749332421
Standard Error	28215.98487
Observations	100

ANOVA

	df	*SS*	*MS*	*F*	*Significance F*
Regression	3	2.38003E+11	79334302193	99.64845705	2.25232E-29
Residual	96	76429613021	796141802.3		
Total	99	3.14433E+11			

	Coefficients	*Standard Error*	*t Stat*	*P-value*	*Lower 95%*	*Upper 95%*
Intercept	6305.319269	9567.27338	0.65905081	0.511440892	-12685.58281	25296.22135
size	34.51179153	7.543276837	4.575172338	1.42238E-05	19.5384939	49.48508915
lot	1.594361818	0.491127242	3.246331463	0.001610374	0.619481202	2.569242433
Taxes	22.03549329	5.194517429	4.242067447	5.10402E-05	11.72444981	32.34653678

The ANOVA table for the multiple regression output includes the value of the F test statistic (F = 99.65) and the p-value (p = 0.000) for testing H_0: $\beta_1 = \beta_2 = \beta_3 = 0$. We can reject H_0 and conclude that at least one predictor has an effect on price.

The p-values of 0.000, 0.002, 0.000 for *t*-tests for each of the explanatory variables, respectively, indicate that each predictor (house size, lot size, and taxes) does significantly predict price, if we already know the values of the other two variables.

Chapter 13 Example 9 Another Residual Plot for House Selling Prices
Plots of Residuals Against Explanatory Variables

Open the worksheet **house_selling_prices** , which is found in the Excel folder. The price is found in column G, house size is found in column H, and lot size is found in column I.

To produce a histogram of standardized residuals for the multiple regression model predicting selling price by the house size and the lot size:

Within the **Add-Ins > PHStat > Regression > Multiple Regression** command, select **Residuals Table.** Although PHStat does not allow production of a histogram of the residuals, we can use the Histogram command from **Tools > Data Analysis > Histogram** to create a histogram of the residuals included in the Multiple Regression output, a portion of which is shown here.

RESIDUAL OUTPUT

Observation	*Predicted price*	*Residuals*
1	107766.7244	37233.27564
2	82070.99553	-14070.99553
3	107418.5374	7581.462637
4	90352.13611	-21352.13611
5	135017.8588	27982.1412
6	81020.78	-11120.78
7	63684.51979	-13684.51979
8	116402.7511	20597.24891
9	116410.7462	4889.25382
10	66145.24984	3854.750164

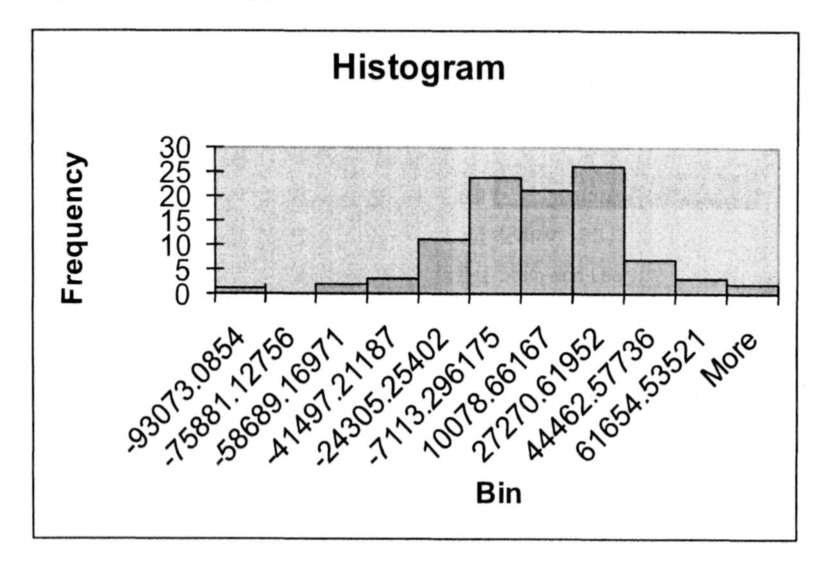

To plot the residuals against the explanatory variable house size:
Select: **Residual Plots** from within the PHStat Multiple Regression dialogue box.

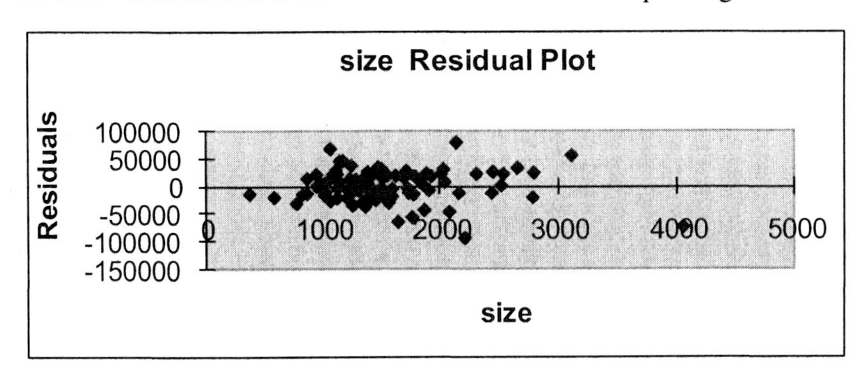

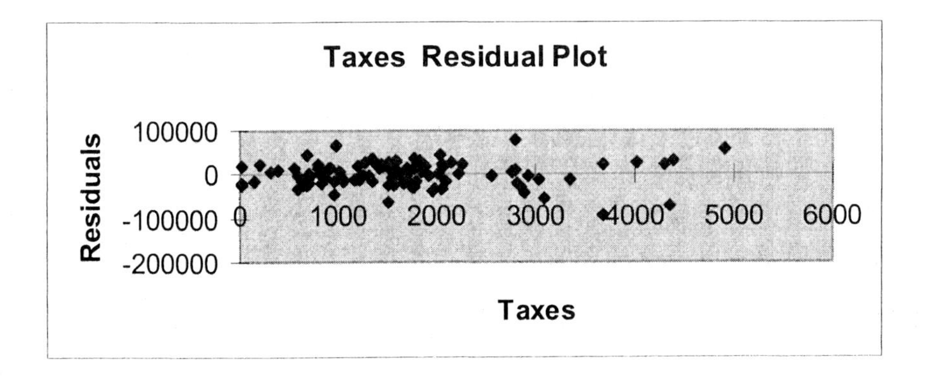

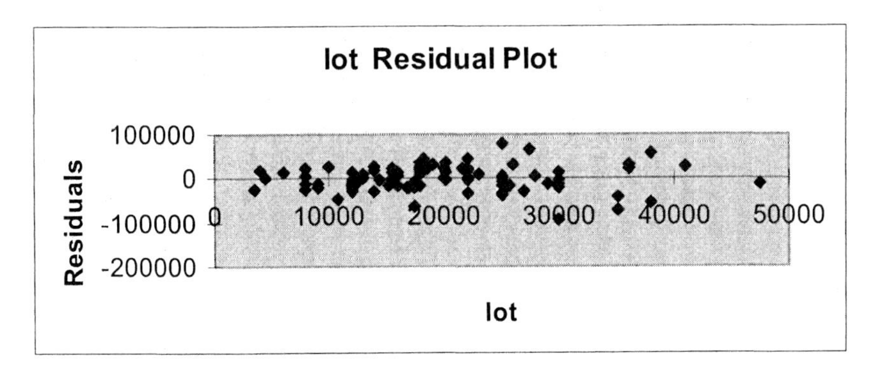

Chapter 13 Example 11 Comparing Winning High Jumps For Men and Women Including Categorical Predictors in Regression

Open the worksheet **high_jump**, which is found in the Excel folder. The following screen shot indicates the organization of the data file.

	A	B	C	D
1	Men_Mete	Women_M	Year_Men	Year_Wor
2	1.81	1.59	1896	1928
3	1.9	1.657	1900	1932
4	1.8	1.6	1904	1936
5	1.905	1.68	1908	1948
6	1.93	1.67	1912	1952
7	1.935	1.76	1920	1956
8	1.98	1.85	1924	1960

Note that 1928 is the first year that women participated in the high jump.

We will use the data in columns E, F and G in our multiple regression. Since we're only interested in years starting at 1928 we will not use the data for the men for prior years, so you can delete the data from E2 to G8. A screenshot of the data we are left with can be seen below.

E	F	G
r Winning H	Gender	Year
1.94	Men	1928
1.97	Men	1932
2.03	Men	1936
1.98	Men	1948
2.04	Men	1952

The last thing we need to do is in column F, replace any "Men" values with the number 1, and replace any "Women" values with the number 0, like so:

E	F	G
r Winning H	Gender	Year
1.94	1	1928
1.97	1	1932
2.03	1	1936
1.98	1	1948
2.04	1	1952
2.12	1	1956

Performing multiple regression for predicting winning height (in meters) as a function of number of years and gender (1 = male, 0 = female):

Select: **Add-Ins > PHStat > Regression > Multiple Regression**
Enter: Y Variable Cell Range:**E1:E37**
X Variables Cell Range: **F1:G37**
Select: **ANOVA and Coefficients Table**
Click: **OK**

A	B	C	D	E	F	G
Regression Analysis						
Regression Statistics						
Multiple R	0.982648584					
R Square	0.96559824					
Adjusted R Square	0.963513285					
Standard Error	0.044689426					
Observations	36					
ANOVA						
	df	*SS*	*MS*	*F*	*Significance F*	
Regression	2	1.849861861	0.92493093	463.1266263	7.13807E-25	
Residual	33	0.065905778	0.001997145			
Total	35	1.915767639				
	Coefficients	*Standard Error*	*t Stat*	*P-value*	*Lower 95%*	*Upper 95%*
Intercept	-11.13392071	0.643380553	-17.30534232	4.04757E-18	-12.44288828	-9.824953136
Gender	0.339055556	0.014896475	22.7607906	9.50162E-22	0.308748449	0.369362662
Year	0.006601975	0.000326767	20.20395168	3.75212E-20	0.005937163	0.007266786

To produce a scatterplot that includes the categorical distinction:
Highlight cells **E2:G37**
Select: **Insert > Charts > Scatter > Scatter with only Markers**
Right click on the plot and choose "Select Data"
Select "Series 1" and Edit and create a series for the men according to the output below.
Select "Series 2" and Edit and create a series for the women according to the output below.

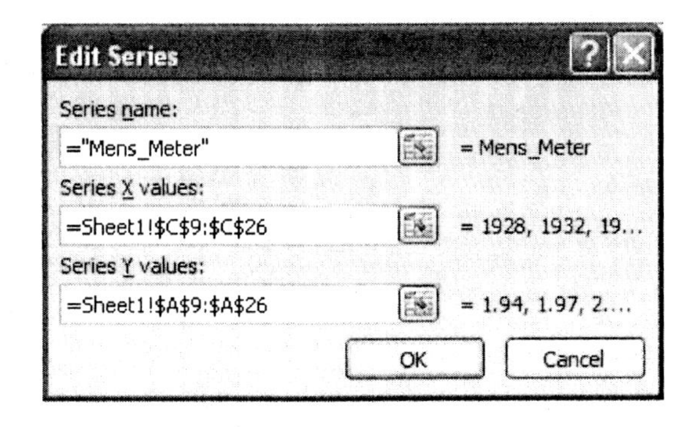

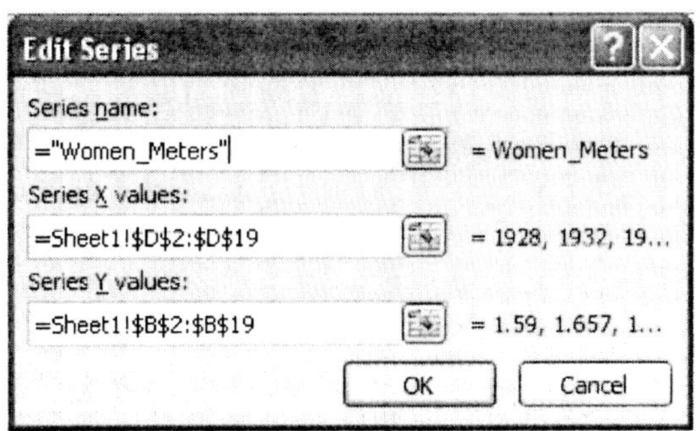

After modifying the axes:

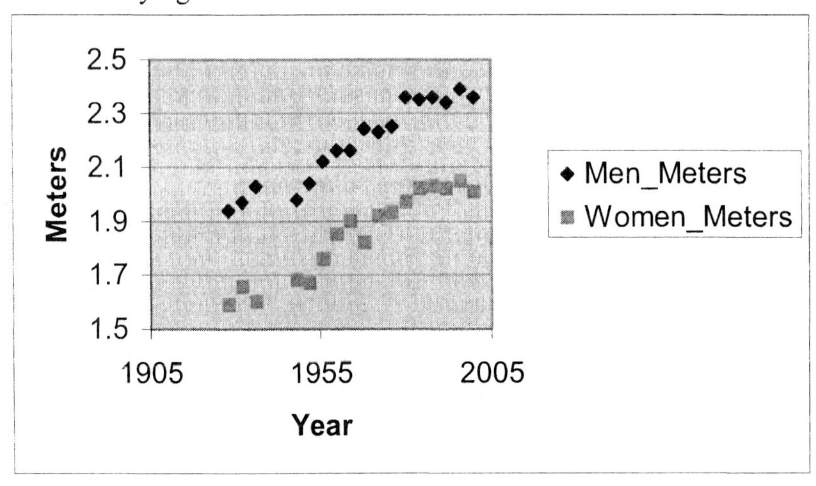

Chapter 13 Exercise 13.45 Houses, tax, and NW
Including Categorical Predictors in Regression

Open the worksheet **house_selling_prices** , which is found in the Excel folder. The following screen shot indicates the organization of the data file.

	A	B	C	D	E	F	G	H	I
1	House	Taxes	Bedrooms	Baths	Quadrant	NW	price	size	lot
2	1	1360	3	2	NW	1	145000	1240	18000
3	2	1050	1	1	NW	1	68000	370	25000
4	3	1010	3	1.5	NW	1	115000	1130	25000
5	4	830	3	2	SW	0	69000	1120	17000
6	5	2150	3	2	NW	1	163000	1710	14000
7	6	1230	3	2	NW	1	69900	1010	8000
8	7	150	2	2	NW	1	50000	860	15300

The data file includes column F, NW, where 1 = NW and 0 = other.

Performing multiple regression for predicting price as a function of real estate tax and whether the home is in the NW region:

Select: **Add-Ins > PHStat > Regression > Multiple Regression**
Enter: Y Variable Cell Range: **select appropriate cells**
X Variables Cell Range: **select appropriate cells**
Select: **ANOVA and Coefficients Table**
Click: **OK**

Regression Analysis

Regression Statistics	
Multiple R	0.827887418
R Square	0.685397577
Adjusted R Square	0.678910929
Standard Error	31934.40967
Observations	100

ANOVA

	df	SS	MS	F	Significance F
Regression	2	2.15511E+11	1.07756E+11	105.6628305	4.37986E-25
Residual	97	98921232511	1019806521		
Total	99	3.14433E+11			

	Coefficients	Standard Error	t Stat	P-value	Lower 95%	Upper 95%
Intercept	43014.39208	7830.66244	5.493071934	3.17562E-07	27472.70347	58556.08069
Taxes	45.3021067	3.215981488	14.08655705	3.41226E-25	38.91927725	51.68493616
NW	10814.17803	7458.343929	1.449943598	0.150299978	-3988.561816	25616.91787

From the coefficient for NW, we see that the selling price increases by $10,814 just because the house is located in the NW region.

Chapter 13 Example 12 Annual Income and Having a Travel Credit Card Logistic Regression

Chapter 13 Example 14 Estimating Proportion of Students Who Have Used Marijuana Logistic Regression

Logistic Regression is not available in Excel or PHStat. See the MINITAB section.

Chapter 14 Example 3 Customers' Telephone Holding Time ANOVA

Enter the data into three separate columns as found in Example 2:

	A	B	C
1	**Advertisement**	**Muzak**	**Classical**
2	5	0	13
3	1	1	9
4	11	4	8
5	2	6	15
6	8	3	7

Choose **Data > Data Analysis > ANOVA – Single Factor**
Enter Input Range: **A1 – C6**
Click Grouped by: **Columns**
Click **Labels in first row**
Enter Alpha: **.05**
Output range: **A10 – E15 or new worksheet**
Click **OK**

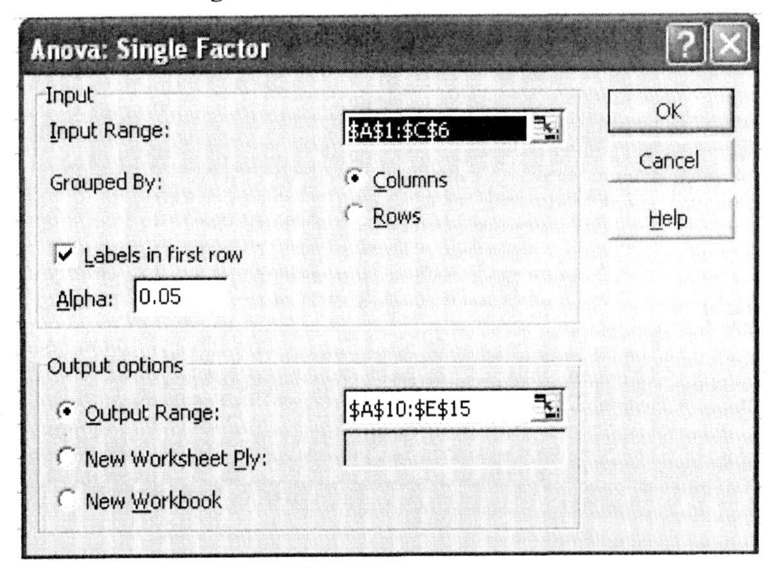

Note the p-value and the F statistic. The critical value is also given as 3.88.

10	Anova: Single Factor						
11							
12	SUMMARY						
13	*Groups*	*Count*	*Sum*	*Average*	*Variance*		
14	Advertisement	5	27	5.4	17.3		
15	Muzak	5	14	2.8	5.7		
16	Classical	5	52	10.4	11.8		
17							
18							
19	ANOVA						
20	*ource of Variatic*	*SS*	*df*	*MS*	*F*	*P-value*	*F crit*
21	Between Group	149.2	2	74.6	6.431034	0.012643	3.88529
22	Within Groups	139.2	12	11.6			
23							
24	Total	288.4	14				

Chapter 14 Exercise 14.3 What's The Best Way to Learn French? ANOVA

Enter the data given into three separate columns in a new worksheet.

	A	B	C
1	Group I	Group II	Group III
2	4	1	9
3	6	5	10
4	8		5

Choose **Data > Data Analysis > ANOVA – Single Factor**
Enter Input Range: **A1 – C4**
Click Grouped by**: Columns**
Click **Labels in first row**
Enter Alpha: **.05**
Output range: **A10 or you may choose new worksheet**
Click **OK**

The output is shown below.

10	Anova: Single Factor						
11	SUMMARY						
12	*Groups*	*Count*	*Sum*	*Average*	*Variance*		
13	Group1	3	18	6	4		
14	Group 2	2	6	3	8		
15	Group 3	3	24	8	7		
16							
17	ANOVA						
18	*Source of Variation*	*SS*	*df*	*MS*	*F*	*P-value*	*F crit*
19	Between Groups	30	2	15	2.5	0.1767767	5.786148449
20	Within Groups	30	5	6			
21	Total	60	7				

Chapter 14 Example 7 Regression Analysis of Telephone Holding Times

We will be using the data from Examples 1 – 4, but to do the regression analysis we will be setting it up a little differently. We will use two independent variables, x_1 and x_2. Let $x_1 = 1$ for advertisement and = 0 for Muzak and classical. Let $x_2 = 1$ for Muzak and = 0 for advertisement and classical. The dependent variable is the holding times. Observe the way the data is laid out.

	A	B	C	D	E	F	G	H
1	Advertisement	Muzak	Classical	Holding Time		x1	x2	
2	5	0	13	5		1	0	
3	1	1	9	1		1	0	
4	11	4	8	11		1	0	
5	2	6	15	2		1	0	
6	8	3	7	8		1	0	
7				0		0	1	
8				1		0	1	
9				4		0	1	
10				6		0	1	
11				3		0	1	
12				13		0	0	
13				9		0	0	
14				8		0	0	
15				15		0	0	
16				7		0	0	

To perform the regression analysis choose **Data > Data Analysis > Regression > OK**

Enter Input Y range: **D1 – D16**
Input X range: **F1 – G16**
Check **Labels**
Confidence Intervals **95%**
Choose Output Range: **A20**
Click **OK**

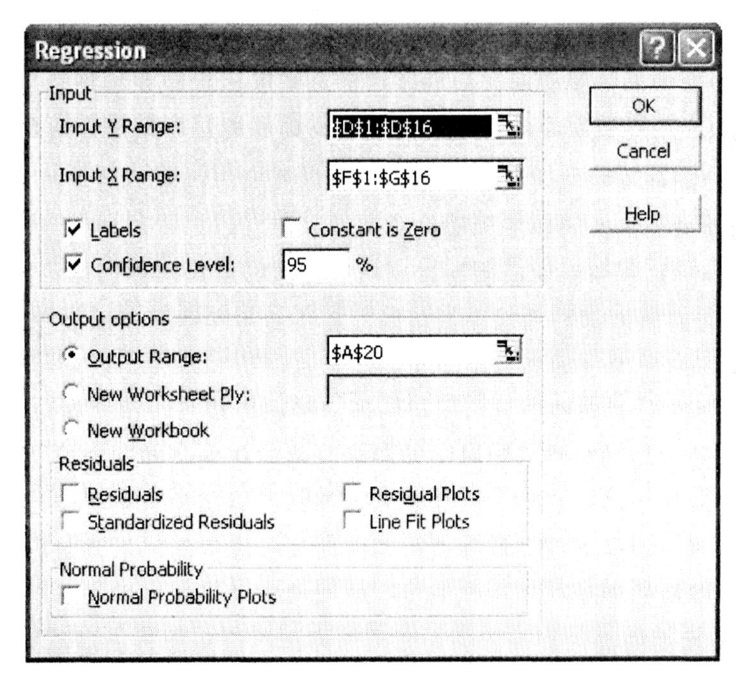

The output will be placed starting in cell A20.

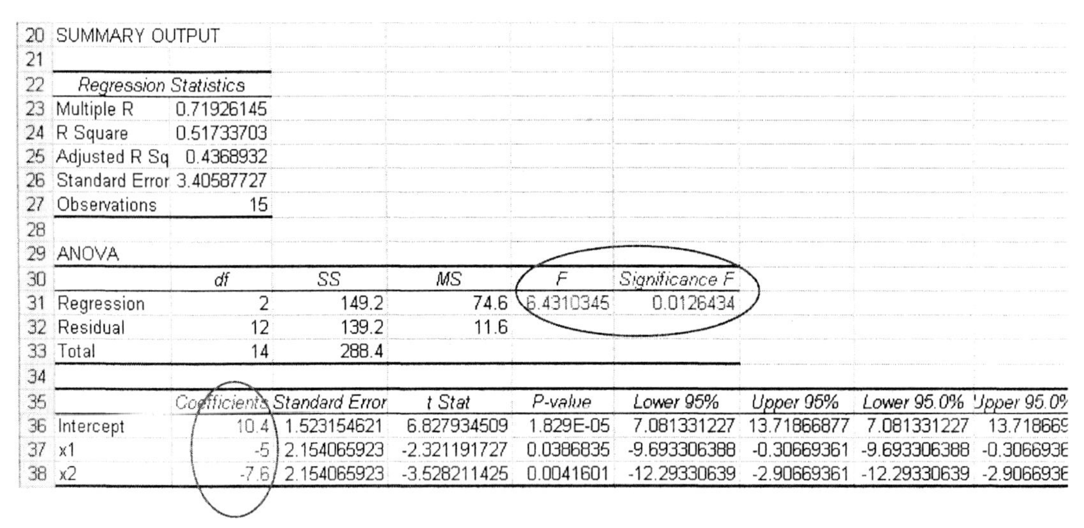

20	SUMMARY OUTPUT								
21									
22	*Regression Statistics*								
23	Multiple R	0.71926145							
24	R Square	0.51733703							
25	Adjusted R Sq	0.4368932							
26	Standard Error	3.40587727							
27	Observations	15							
28									
29	ANOVA								
30		*df*	*SS*	*MS*	*F*	*Significance F*			
31	Regression	2	149.2	74.6	6.4310345	0.0126434			
32	Residual	12	139.2	11.6					
33	Total	14	288.4						
34									
35		*Coefficients*	*Standard Error*	*t Stat*	*P-value*	*Lower 95%*	*Upper 95%*	*Lower 95.0%*	*Upper 95.0%*
36	Intercept	10.4	1.523154621	6.827934509	1.829E-05	7.081331227	13.71866877	7.081331227	13.718669
37	x1	-5	2.154065923	-2.321191727	0.0386835	-9.693306388	-0.30669361	-9.693306388	-0.3066936
38	x2	-7.6	2.154065923	-3.528211425	0.0041601	-12.29330639	-2.90669361	-12.29330639	-2.9066936

Note the coefficients, F statistic and P-value shown in the above chart.

These form the regression equation

$$\hat{y} = 10.4 - 5.1x_1 - 7.6x_2$$

Chapter 14 Example 11 Testing for Interaction with Corn Yield Data
Two Way ANOVA allowing for interaction

Enter the data in the worksheet as shown.

	A	B	C
1	Fertilizer		Manure
2		High	Low
3	High	13.7	16.4
4		15.8	12.5
5		13.9	14.1
6		16.6	14.4
7		15.5	12.2
8	Low	15	12.4
9		15.1	10.6
10		12	13.7
11		15.7	8.7
12		12.2	10.9

To perform a Two-Way ANOVA:
Choose **Data > Data Analysis > ANOVA: Two Factor with Replication**

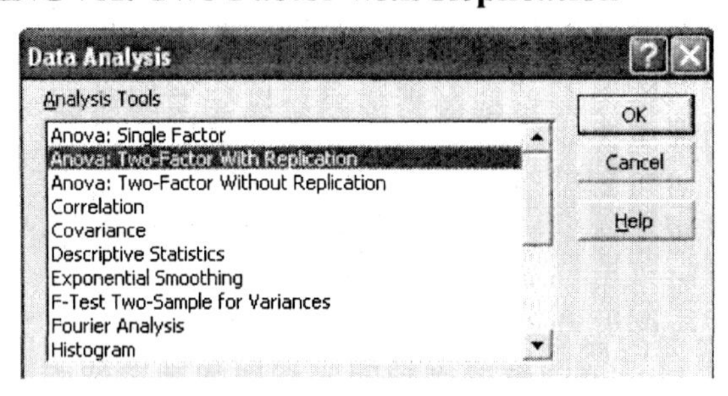

Enter Input Range: **A2-C12**
Rows per Sample: **5**
Alpha: **.05**
Output Range: **A17**
Click **OK**

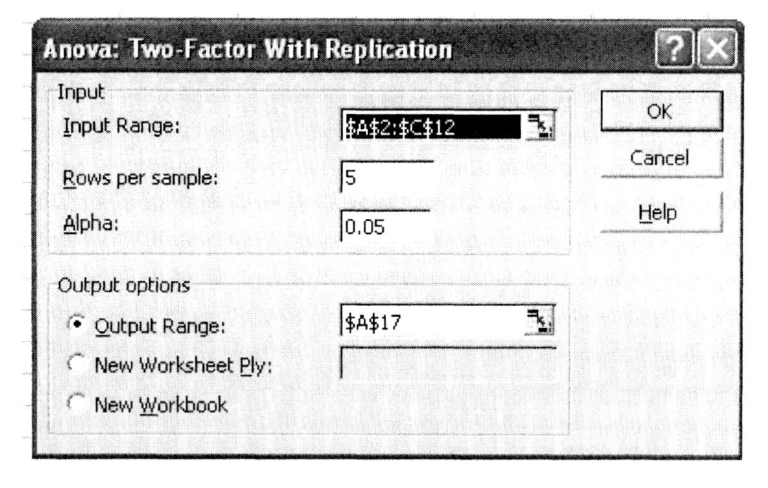

The output is extensive. Note the highlighted areas.

	A	B	C	D	E	F	G
17	Anova: Two-Factor With Replication						
18							
19	SUMMARY	High	Low	Total			
20	*High*						
21	Count	5	5	10			
22	Sum	75.5	69.6	145.1			
23	Average	15.1	13.92	14.51			
24	Variance	1.575	2.847	2.352111			
25							
26	*Low*						
27	Count	5	5	10			
28	Sum	70	56.3	126.3			
29	Average	14	11.26	12.63			
30	Variance	3.085	3.593	5.053444			
31							
32	*Total*						
33	Count	10	10				
34	Sum	145.5	125.9				
35	Average	14.55	12.59				
36	Variance	2.407222	4.827667				
37							
38							
39	ANOVA						
40	*Source of Variation*	*SS*	*df*	*MS*	*F*	*P-value*	*F crit*
41	Sample	17.672	1	17.672	6.368288	0.022578	4.493998
42	Columns	19.208	1	19.208	6.921802	0.018162	4.493998
43	Interaction	3.042	1	3.042	1.096216	0.310658	4.493998
44	Within	44.4	16	2.775			
45							
46	Total	84.322	19				

Chapter 14 Example 10: Regression Modeling to Estimate and Compare Mean Corn Yields

Enter the data in three columns, using 1 to represent high, and 0 to represent low for the levels of manure and fertilizer. The corn yield values will be the dependent variable and the manure and fertilizer levels will be the independent variables.

H	I	J
f	m	yield
1	1	13.7
1	1	15.8
1	1	13.9
1	1	16.6

To do the multiple regression we use PHStat.

Choose **Add-Ins > PHStat > Regression > Multiple Regression**
Enter Y Variable Cell Range: **J1 – J21**
X Variable Cell Range **H1 – I21**
Click **First cells contain labels**
Click **Regression Statistics Table**
Click **ANOVA and Coefficients Table**
Click **OK**

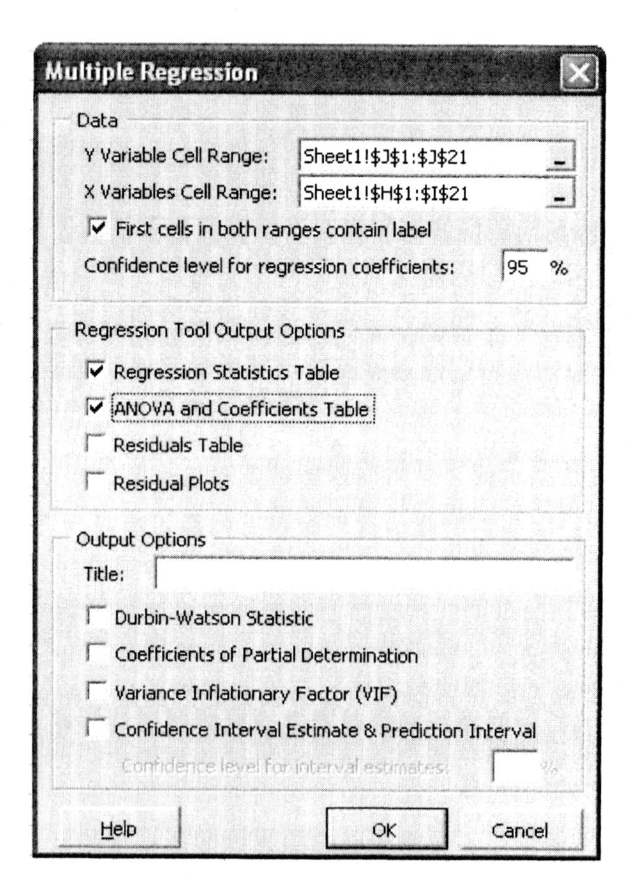

The output is extensive. Note the coefficients for the regression equation.

	A	B	C	D	E	F	G
1	Corn Yield						
2							
3	*Regression Statistics*						
4	Multiple R	0.661340328					
5	R Square	0.43737103					
6	Adjusted R Square	0.371179387					
7	Standard Error	1.670540596					
8	Observations	20					
9							
10	ANOVA						
11		*df*	*SS*	*MS*	*F*	*Significance F*	
12	Regression	2	36.88	18.44	6.607647232	0.007531609	
13	Residual	17	47.442	2.790705882			
14	Total	19	84.322				
15							
16		*Coefficients*	*Standard Error*	*t Stat*	*P-value*	*Lower 95%*	*Upper 95%*
17	Intercept	11.65	0.646997591	18.00624943	1.65275E-12	10.2849525	13.0150475
18	f	1.88	0.747088466	2.516435583	0.022188041	0.303778914	3.456221086
19	m	1.96	0.747088466	2.623517948	0.017792196	0.383778914	3.536221086
20							

Chapter 14 Exercise 14.33 Diet and Weight Gain
Two Way ANOVA, with and without interaction

Enter the data as presented in the problem. A small sample is shown to the right.

	A	B	C
1	Source	Level	
2		High	Low
3	Beef	73	90
4		102	76
5		118	90
6		104	64
7		81	86
8		107	51
9		100	72
10		87	90
11		117	95
12		111	78
13	Cereal	98	107
14		74	95
15		56	97
16		111	80

Choose **Data > Data Analysis > ANOVA: Two Factor without Replication**

Enter:
Input Range: **A2-C32**
Check Box for **Labels**
Output range: **M2**
Click **OK**

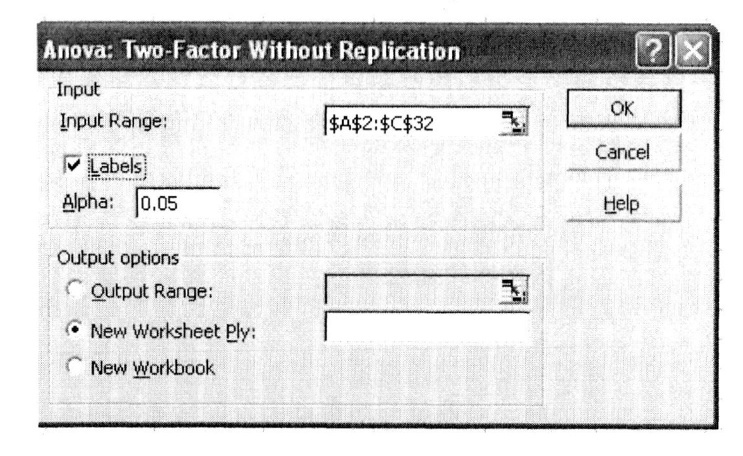

The output is extensive. Only part is shown here.

39	ANOVA						
40	*Source of Variation*	*SS*	*df*	*MS*	*F*	*P-value*	*F crit*
41	Rows	5597.933	29	193.0322	0.753146	0.775036	1.860812
42	Columns	3168.267	1	3168.267	12.3615	0.001462	4.182965
43	Error	7432.733	29	256.3011			
44							
45	Total	16198.93	59				

To repeat this and allow for interaction
Choose **Data > Data Analysis > ANOVA: Two Factor with Replication**

Enter
Input range**: A2 – C32**
Rows per sample **10**
Alpha: **.05**
Output range: **F2**
Click **OK**

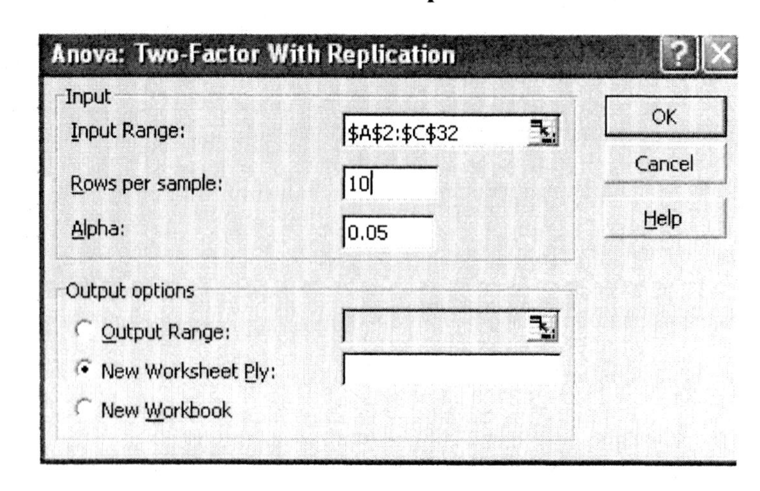

Again the output is extensive and only part is shown here.

	ANOVA						
30	*Source of Variation*	*SS*	*df*	*MS*	*F*	*P-value*	*F crit*
31	Sample	266.5333	2	133.2667	0.621129	0.541132	3.168246
32	Columns	3168.267	1	3168.267	14.76665	0.000322	4.01954
33	Interaction	1178.133	2	589.0667	2.74552	0.073188	3.168246
34	Within	11586	54	214.5556			
35							
36	Total	16198.93	59				

Note the interaction line in this output that does not appear in the one above. Take note of the other differences and similarities.

Chapter 15 Example 4 Estimating the Difference Between Median Reaction Times Wilcoxon Rank Sum Test

Excel does not provide Nonparametric tests, but we can perform some of them by using the add-in program PHStat. Open the worksheet **cell_phones** from the Excel folder of the data disk. A portion is shown below.

Microsoft Excel - Book1

File Edit View Insert Format Tools Data PHStat

E11 fx

	A	B	C	D	E
1	Cell phone	Control	rankcellph	rankcontrol	
2	636	557	27	20	
3	623	572	25	22	
4	615	457	24	5	
5	672	489	29	12	
6	601	532	22	17	

Choose **Add-Ins > PHStat > Two-Sample Tests > Wilcoxon Rank Sum Test**.

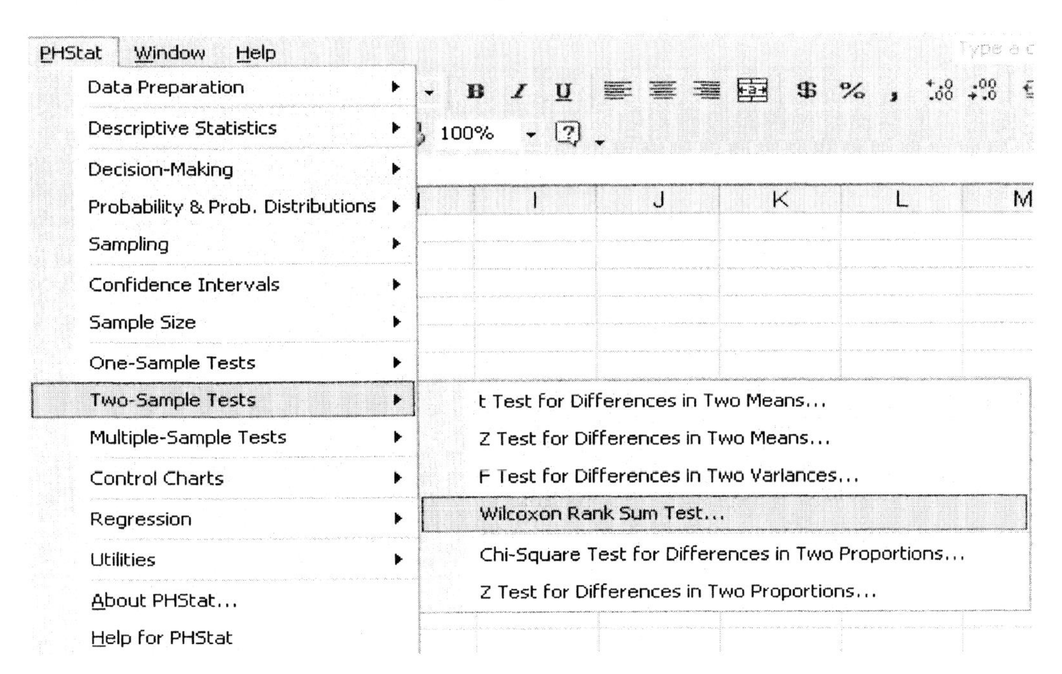

Enter: Level of Significance: **.05**
Population 1 Sample Cell Range: **B1:B33**
Population 2 Sample Cell Range: **A1:A33**
Check box for **First Cell contains label**
Check **Two Tail Test**
Enter Title: **Cell Phone Study of Reaction Time**
Click **OK**

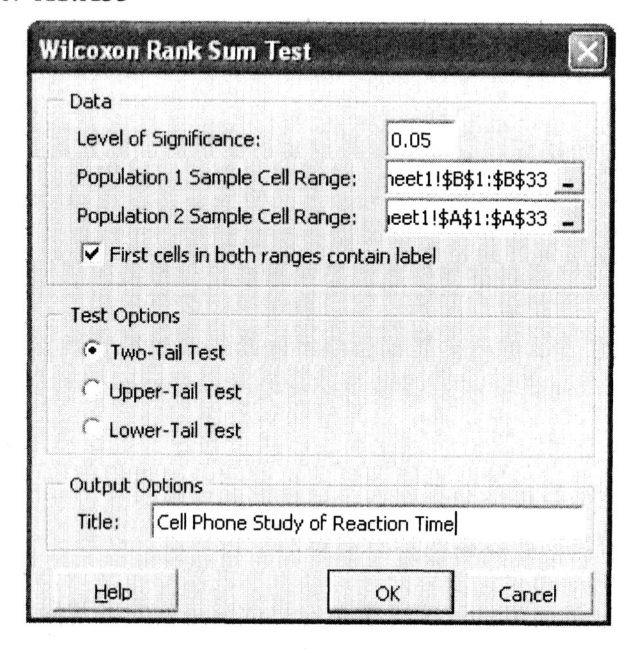

Notes:
PHStat performs the Wilcoxon rank sum test for differences between two medians in a new worksheet. This procedure also generates a second worksheet that contains the ranks of the sample data from the two populations.

	A	B	C
1	**Cell Phone Study of Reaction Time**		
2			
3	**Data**		
4	**Level of Significance**	**0.05**	
5			
6	Population 1 Sample		
7	Sample Size	32	
8	Sum of Ranks	864	
9	Population 2 Sample		
10	Sample Size	32	
11	Sum of Ranks	1216	
12			
13	Intermediate Calculations		
14	Total Sample Size n	64	
15	*T1* Test Statistic	864	
16	*T1* Mean	1040	
17	Standard Error of *T1*	74.4759469	
18	**Z Test Statistic**	**-2.363179084**	
19			
20	**Two-Tail Test**		
21	**Lower Critical Value**	**-1.959962787**	
22	**Upper Critical Value**	**1.959962787**	
23	***p*-Value**	**0.018118881**	
24	**Reject the null hypothesis**		

Chapter 15 Exercise 15.4 Anticipation of Hypnosis
One Sided Wilcoxon Rank Sum Test

Enter the data from the text into Columns A and B.
Choose **Add-Ins > PHStat > Two-Sample Tests > Wilcoxon Rank Sum Test**.
Enter: Level of Significance: **.05**
Population 1 Sample Cell Range**: A1:A9**
Population 2 Sample Cell Range: **B1:B9**
Check box for **First Cell contains label**
Check **Lower Tail Test**
Enter Title: **Anticipation of Hypnosis**
Click **OK**

	A	B	C
1	**Anticipation of Hypnosis**		
2			
3	**Data**		
4	**Level of Significance**	**0.05**	
5			
6	Population 1 Sample		
7	Sample Size	8	
8	Sum of Ranks	49	
9	Population 2 Sample		
10	Sample Size	8	
11	Sum of Ranks	87	
12			
13	Intermediate Calculations		
14	Total Sample Size n	16	
15	*T*1 Test Statistic	49	
16	*T*1 Mean	68	
17	Standard Error of *T*1	9.521905	
18	**Z Test Statistic**	**-1.9954**	
19			
20	**Lower-Tail Test**		
21	**Lower Critical Value**	**-1.64485**	
22	***p*-Value**	**0.023**	
23	**Reject the null hypothesis**		

Chapter 15 Exercise 15.7 Teenage Anorexia
Wilcoxon Rank Sum Test

Open the file **anorexia** from the Excel folder of the data disk.

Choose **Add-Ins > PHStat > Two-Sample Tests > Wilcoxon Rank Sum Test**.
Enter: Level of Significance: **.05**
Population 1 Sample Cell Range **G1:G30**
Population 2 Sample Cell Range: **H1:H27**
Check box for **First Cell contains label**
Check **Two Tail Test**
Enter Title: **Anorexia**
Click **OK**

Level of Significance	**0.05**
Population 1 Sample	
Sample Size	29
Sum of Ranks	907
Population 2 Sample	
Sample Size	26
Sum of Ranks	633
Intermediate Calculations	
Total Sample Size n	55
*T*1 Test Statistic	633
*T*1 Mean	728
Standard Error of *T*1	59.31835
Z Test Statistic	**-1.60153**
Two-Tail Test	
Lower Critical Value	**-1.95996**
Upper Critical Value	**1.959963**
***p*-Value**	**0.10926**
Do not reject the null hypothesis	

Chapter 15 Example 5 Does Heavy Dating Affect College GPA?
Kruskal-Wallis Test

PHStat performs the Kruskal-Wallis rank sum test of hypothesis for differences between medians from multiple independent sample groups and places the analysis in a new worksheet. It also generates a second worksheet that contains the ranks of the sample data from the multiple independent samples. Any non-numeric values in the sample data cell range (excluding the first cells of columns if First cells contain label is selected) are treated as zero values for the purposes of ranking.

Enter the data in three separate columns

	A	B	C	D
1	Rare	Occasional	Regular	
2	1.75	2	2.4	
3	3.15	3.2	2.95	
4	3.5	3.44	3.4	
5	3.68	3.5	3.67	
6		3.6	3.7	
7		3.71	4	
8		3.8		

Choose **Add-Ins > PHStat > Multiple-Sample Tests > Kruskal-Wallis Rank Test**

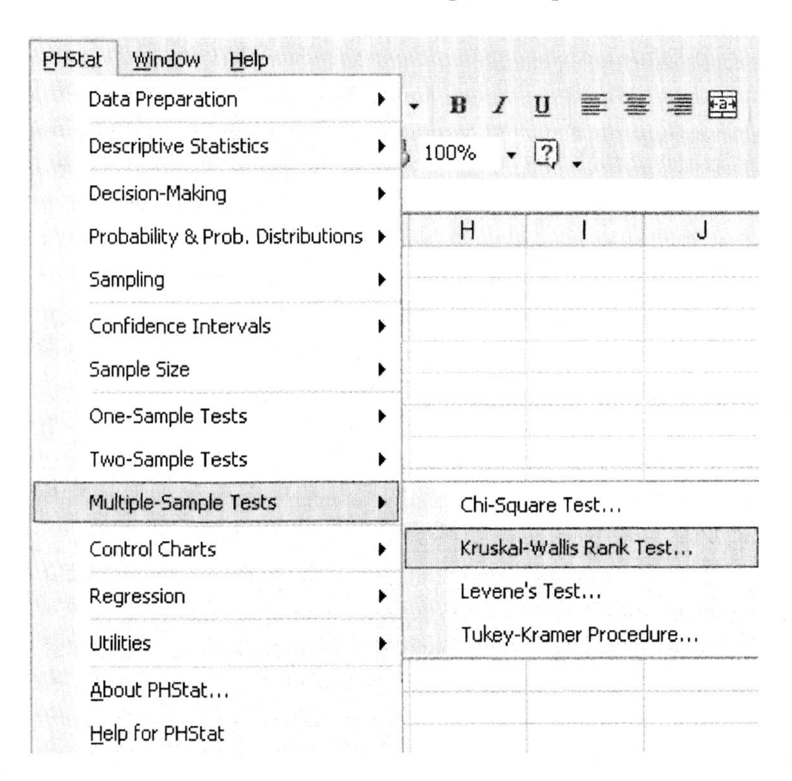

Enter Level of Significance: **.05**
Sample Data Cell Range: **A1:C8**
Check box for **First Cells contain labels**
Title: **GPA vs Dating**
Click **OK**

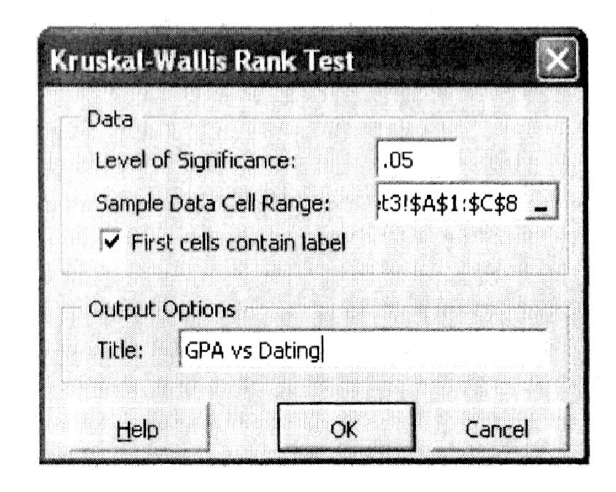

The results appear in a separate worksheet.

	A	B	C	D	E	F	G	H
1	GPA vs Dating							
2								
3	**Data**							
4	**Level of Significance**	**0.05**		Group	Sample Size	Sum of Ranks	Mean Ranks	
5				1	4	28.5	7.125	
6	Intermediate Calculations			2	7	67.5	9.64285714	
7	Sum of Squared Ranks/Sample Size	1395.455		3	6	57	9.5	
8	Sum of Sample Sizes	17						
9	Number of Groups	3						
10								
11	**Test Result**							
12	**H Test Statistic**	**0.723739**						
13	**Critical Value**	**5.991476**						
14	**p-Value**	**0.696373**						
15	**Do not reject the null hypothesis**							

Chapter 15 Example 6 Spend More Time Browsing the Internet or Watching TV? Sign Test for Matched Pairs

Excel does not have a nonparametric sign test, but we can do the calculations manually. Retrieve the data for the **georgia_student_survey** from the data disk. We are only interested in the columns for Internet and TV watching, so you can hide all the others. Create a column named DIFF for the differences between the two columns.
Use the COUNTIF function to count the number of entries greater than zero, which would be considered your successes.

Choose **Formulas > Insert Function > Statistical > COUNTIF**

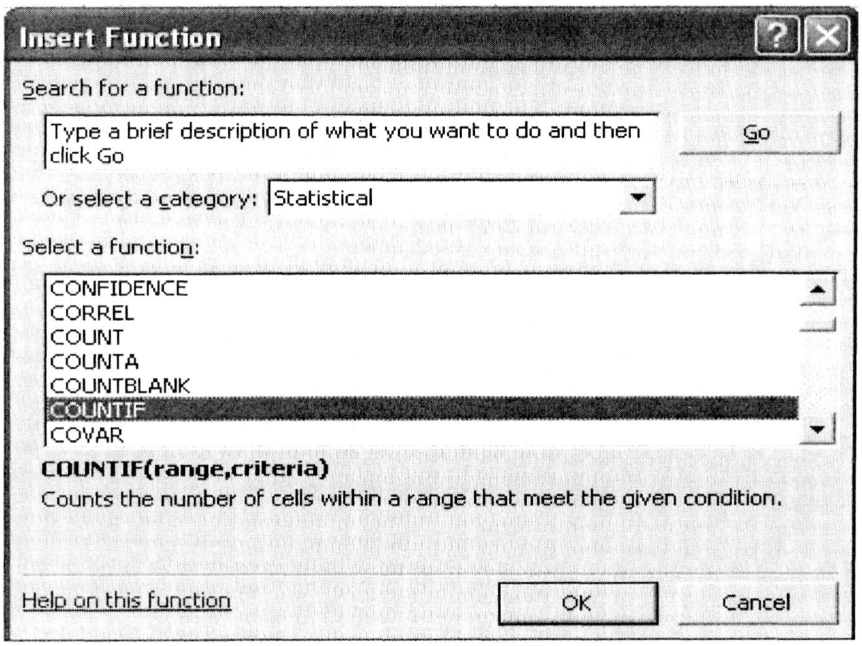

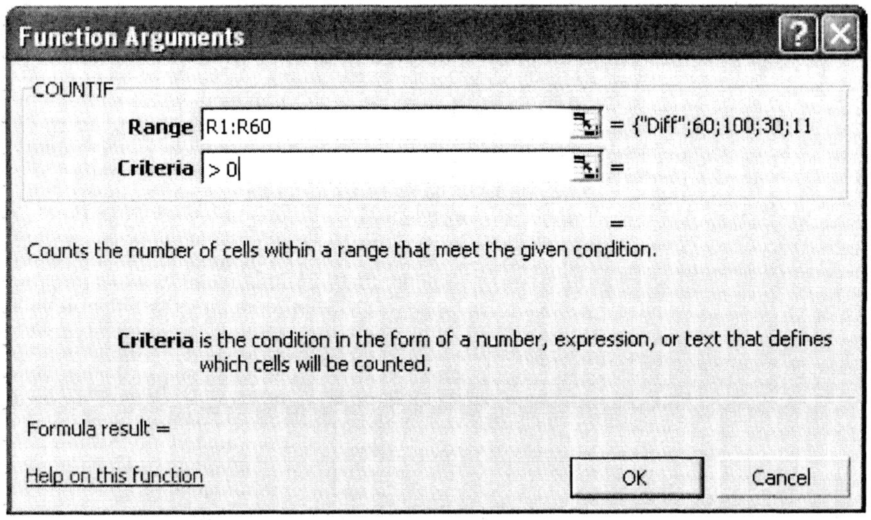

Insert Function

Search for a function:

Type a brief description of what you want to do and then click Go

Go

Or select a category: Statistical

Select a function:

CONFIDENCE
CORREL
COUNT
COUNTA
COUNTBLANK
COUNTIF
COVAR

COUNTIF(range,criteria)

Counts the number of cells within a range that meet the given condition.

Help on this function

OK Cancel

Function Arguments

COUNTIF

Range R1:R60 = {"Diff";60;100;30;11

Criteria > 0 =

=

Counts the number of cells within a range that meet the given condition.

Criteria is the condition in the form of a number, expression, or text that defines which cells will be counted.

Formula result =

Help on this function

OK Cancel

This will enter the number 35 into your worksheet. Then perform a z-test for the proportion by choosing **Add-Ins > PHStat > One-Sample test > Z test for the proportion.**

Enter Null Hypothesis: **.5**
Level of Significance **.05**
Number of Successes: **35**
Sample Size: **54**
Test Option: **Two Tail Test**
Title: **Internet vs TV**
Click **OK**

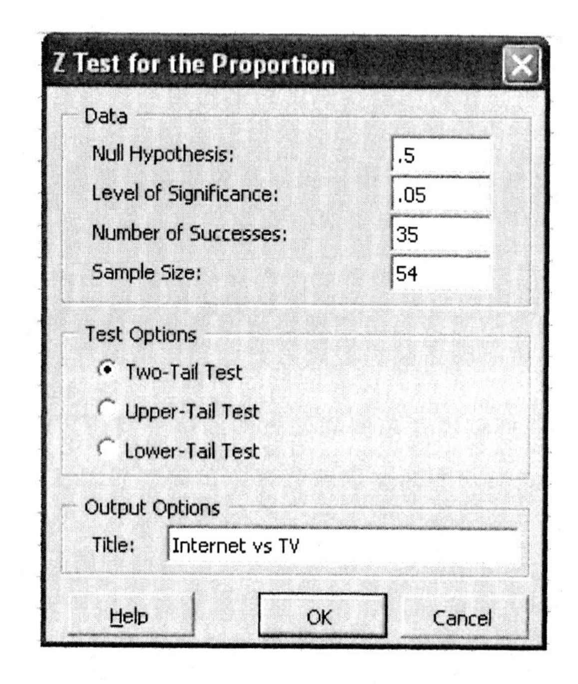

The result will appear in a new worksheet.

	A	B
1	**Internet vs TV**	
2		
3	**Data**	
4	**Null Hypothesis** *p=*	**0.5**
5	**Level of Significance**	**0.05**
6	**Number of Successes**	**35**
7	**Sample Size**	**54**
8		
9	Intermediate Calculations	
10	Sample Proportion	0.648148148
11	Standard Error	0.068041382
12	**Z Test Statistic**	**2.177324216**
13		
14	**Two-Tail Test**	
15	**Lower Critical Value**	**-1.959962787**
16	**Upper Critical value**	**1.959962787**
17	***p*-Value**	**0.029456281**
18	**Reject the null hypothesis**	

Chapter 15 Exercise 15.8 How Long Do You Tolerate Being Put On Hold? Kruskal-Wallis Test

Enter the data in three separate columns.

	A	B	C
1	Muzak	Advertisement	Classical
2	0	5	13
3	1	1	9
4	4	11	8
5	6	2	15
6	3	8	7

Choose **Add-Ins > PHStat > Multiple-Sample Tests > Kruskal Wallis Rank Test**
Enter Level of Significance: **.05**
Sample Data Cell Range: **A1:C6**
Check box for **First Cells contain labels**
Title: **Holding Time vs Group**
Click **OK**

The results appear in a separate worksheet.

	A	B	C	D	E	F	G
1	Holding Time vs Group						
2							
3	Data						
4	Level of Significance	0.05		Group	Sample Size	Sum of Ranks	Mean Ranks
5				1	5	22.5	4.5
6	Intermediate Calculations			2	5	37	7.4
7	Sum of Squared Ranks/Sample Size	1107.1		3	5	60.5	12.1
8	Sum of Sample Sizes	15					
9	Number of Groups	3					
10							
11	Test Result						
12	H Test Statistic	7.355					
13	Critical Value	5.991476					
14	p-Value	0.025286					
15	Reject the null hypothesis						

Chapter 15 Exercise 15.10 Sports vs TV
Sign Test

Open the **fla_student_survey** file. We are only interested in the Sports and TV columns, so hide the others. Create a Difference column. A small sample of the file is shown.

fla_student_survey [Read-Only]

	H	I	U
1	TV	sports	diff
2	3	5	-2
3	15	7	8
4	0		-4

Use the COUNTIF function to count the successes (ie > 0).
Choose **Formulas > Insert Function > Statistical > COUNTIF**
Enter: Range: **T1:T61**
Criteria**: >0**
Click **OK**
The result should be 30. This represents the number of successes.

Then perform a z-test for the proportion:
Choose **Add-Ins > PHStat > One-Sample test > Z test for the proportion**
Enter Null Hypothesis**: .5**
Level of Significance**: .05**
Number of Successes**: 30**
Sample Size**: 54**
Test Option: **Two Tail Test**
Title: **Sport vs TV**
Click **OK**

The result will appear in a new worksheet.

	A	B
1	**Sports vs TV**	
2		
3	**Data**	
4	**Null Hypothesis** *p*=	**0.5**
5	**Level of Significance**	**0.05**
6	**Number of Successes**	**30**
7	**Sample Size**	**54**
8		
9	Intermediate Calculations	
10	Sample Proportion	0.555555556
11	Standard Error	0.068041382
12	**Z Test Statistic**	**0.816496581**
13		
14	**Two-Tail Test**	
15	**Lower Critical Value**	**-1.959962787**
16	**Upper Critical value**	**1.959962787**
17	***p*-Value**	**0.414216057**
18	**Do not reject the null hypothesis**	

MINITAB MANUAL

Maureen Petkewich

University of South Carolina

SECOND EDITION

STATISTICS

THE ART AND SCIENCE OF LEARNING FROM DATA

Agresti • Franklin

Upper Saddle River, NJ 07458

TABLE OF CONTENTS

CHAPTER 1
INTRODUCTION TO MINITAB

INTRODUCTION: This lab is designed to introduce you to the statistical software MINITAB. During this session you will learn how to enter and exit MINITAB, how to enter data and commands, how to print information, and how to save your work for use in subsequent sessions.

BEGINNING AND ENDING A MINITAB SESSION

To start MINITAB

Double-click on the icon placed on your desktop when you installed the program.

If you didn't choose to have the icon installed, you may choose, from the taskbar, **Start > Programs > MINITAB 14 Student > MINITAB 14 Student.**

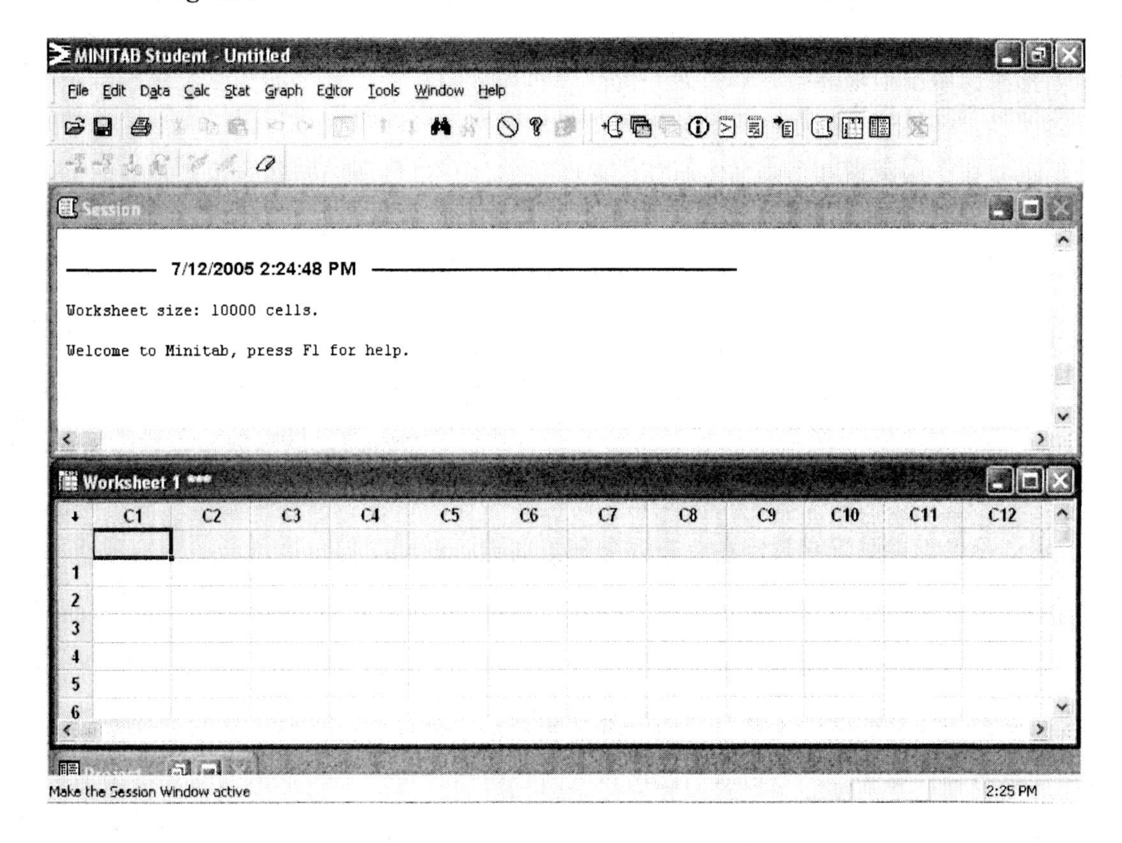

To exit MINITAB

To end a MINITAB session and exit the program, choose **File** from the menu bar and then choose **Exit**. A dialog box will appear, asking if you want to save the changes made to this worksheet. Click **Yes** or **No**.

It is also possible to exit MINITAB by clicking the X in the upper right corner of the window.

In MINITAB commands are executed with menus. When you click a menu selection, MINITAB performs an action or opens a dialog box.

MINITAB WINDOWS

The main MINITAB window opens when you first start MINITAB. You will be in a window titled "MINITAB - Untitled " within which a split window is shown; one titled "Session" and the other titled "Worksheet 1". The Session window displays text output such as tables of statistics. Data windows are where you enter, edit, and view the column data for each worksheet. Another window in the MINITAB environment that can be accessed through the Window menu is the Project Manager. The Project Manager summarizes each open worksheet. Within the Project manager, the History window records all the commands you have used. The graph window display graphs that were generated in the project.

Session Window

The data window is active when you first start MINITAB. To move to the Session Window just point the mouse to the Session Window and click. In older versions of MINITAB, whenever you issue a command from a menu, its corresponding Session command appears in the Session window. In version 14, the command will appear in the History folder within the Project Manager and will only appear in the session window if you have enabled the command language.

The Help Window in MINITAB

Information about MINITAB is stored in the computer. If you forget how to use a command or subcommand, or need general information, you can ask MINITAB for help. There are three methods for accessing Help: choose Help from the menu, select "?" from the toolbar, or press F1. It would be beneficial for you to read "How to use MINITAB Help" the first time you enter the program to help you understand the structures used in MINITAB.

Students: Practice using the HELP command by selecting the following and reading what is presented on the screen:

Choose: **Help > Help**
Select: **Index Help on**
Enter: **MEAN**

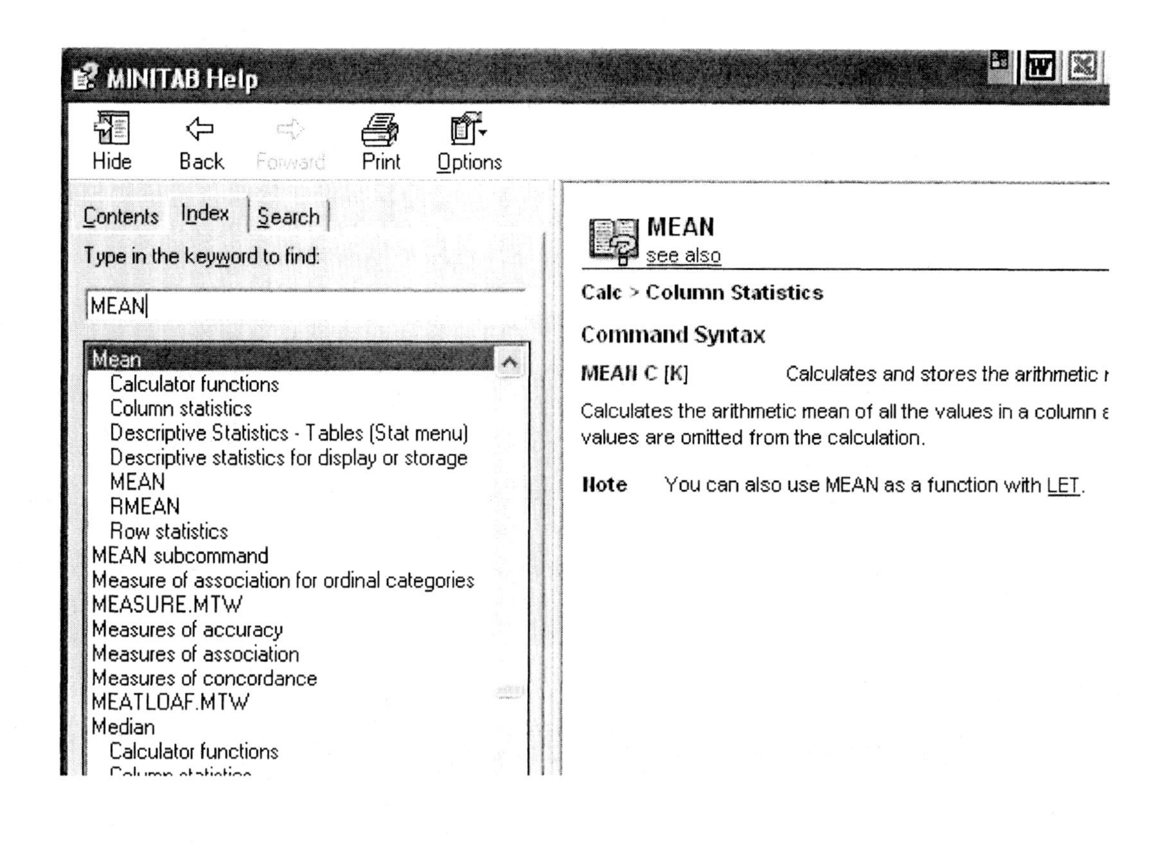
MINITAB Help
Hide
Back
Forward
Print
Options
Contents
Index
Search
Type in the keyword to find:
MEAN
Mean
Calculator functions
Column statistics
Descriptive Statistics - Tables (Stat menu)
Descriptive statistics for display or storage
MEAN
RMEAN
Row statistics
MEAN subcommand
Measure of association for ordinal categories
MEASURE.MTW
Measures of accuracy
Measures of association
Measures of concordance
MEATLOAF.MTW
Median
Calculator functions
MEAN
see also
Calc > Column Statistics
Command Syntax
MEAN C [K]
Calculates and stores the arithmetic
Calculates the arithmetic mean of all the values in a column
values are omitted from the calculation.
Note
You can also use MEAN as a function with LET.

The Data Window

Close Help and click in the worksheet.

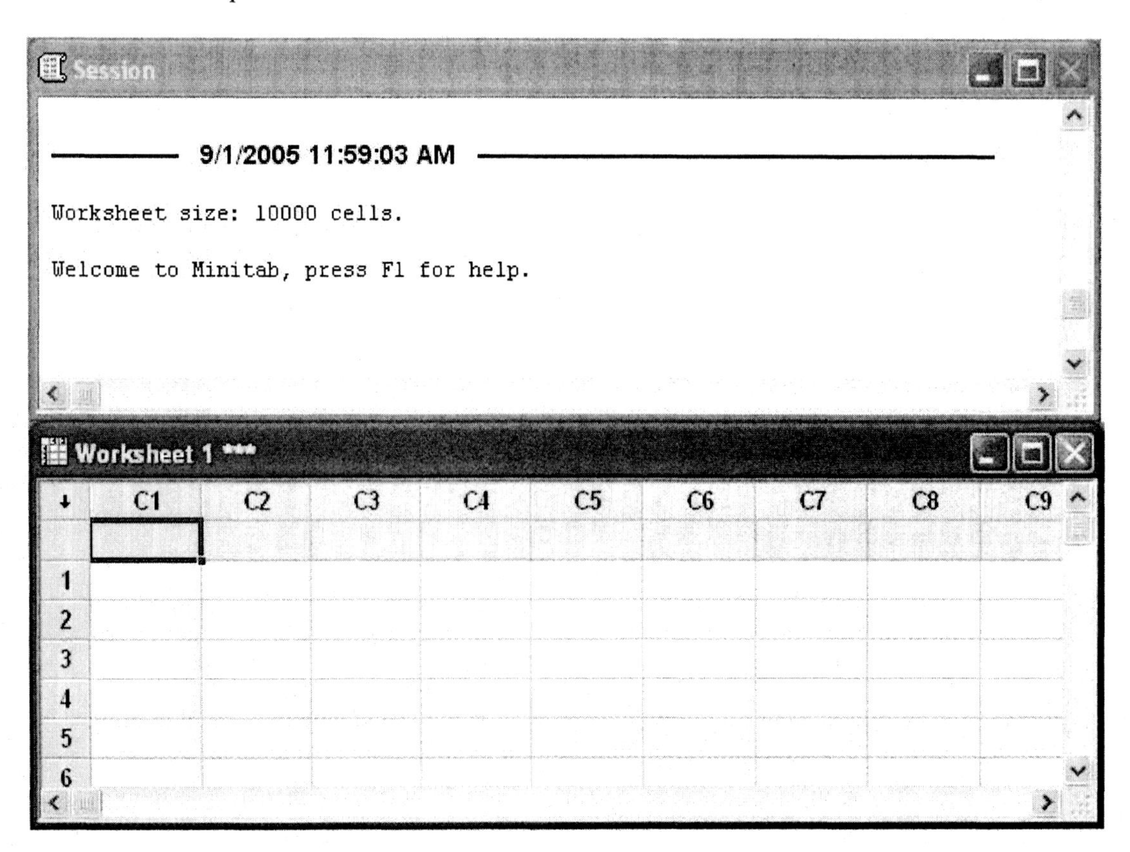

The worksheet is arranged by rows and columns. The columns C1, C2, C3, . . . , correspond to the variables in your data, the rows to observations. In general, a column contains all the data for one variable, and a row contains all the data for an individual subject or observation. You can refer to the columns as C1, C2, or by giving them descriptive names. Click into the column name cell (the blank space below the column number).

ENTERING DATA

As an example, consider a portion of the CEREAL data used in a number of future exercises:

Cereal	Sodium	Sugar	Type
Frosted Mini Wheats	0	7	A
Raisin Bran	210	12	A
All Bran	260	5	A

Now that we are in the data window, let's enter data in the first four columns. Name each column, appropriately.

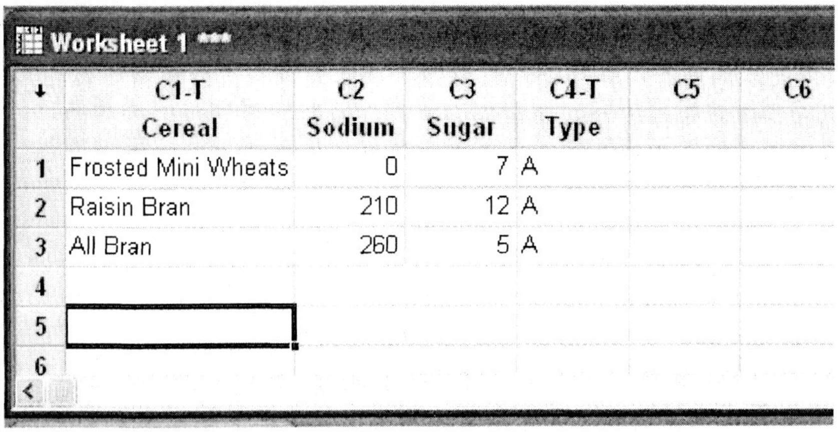

Worksheet 1 ***

↓	C1-T	C2	C3	C4-T	C5	C6
	Cereal	Sodium	Sugar	Type		
1	Frosted Mini Wheats	0	7	A		
2	Raisin Bran	210	12	A		
3	All Bran	260	5	A		
4						
5						
6						

Changing a value entered

We can edit data directly in the data window. Let's suppose we had incorrectly entered the third data item in the second column. It should have been a 620. Click cell C2 row 3 to make it active. Type in the correct value and press enter. Double-clicking allows insertion of new characters without retyping the entire entry.

Suppose we had inadvertently left out a single value and we wish to enter it in a particular position. Place the cursor in the cell in which you wish to insert the new value. Right click and choose Insert Cells. A blank cell is created and the missing value can be entered . There are buttons on the taskbar that are active when you are in the data window. They can be used to insert a cell, row, or column.

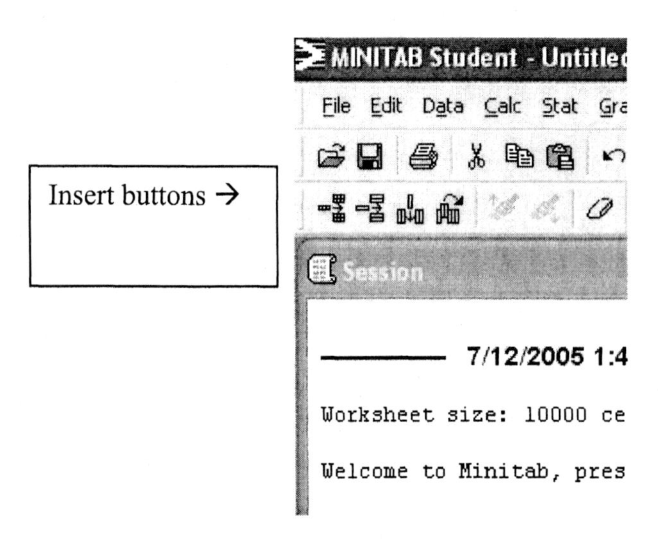

A cell can be deleted by making the cell active, then Choose: **Edit > Delete Cells** (or press the Del key).

Rows of values can also be inserted or deleted in a similar manner. The menu command to insert a row is only functional when the data window is active, and a row is active. To make a row active, click the row header (ie. the row number). An empty row will be added above the active row in the Data window and the remaining rows will be moved down.

Choose: **Editor > Insert Row**

To print your data choose **File > Print Worksheet,** make the appropriate selections and click **OK**

Suppose we wish to copy a column into another column. We can use the **COPY** command instead of reentering the data.

Choose: **Data> Copy> Copy columns to columns**
Enter: Copy from columns: **Cereal**
Select **Store Copied Data in columns** (choose from drop down arrow to select Column)
Click: **OK**

To erase an entire column we use the **ERASE** command.

Choose: **Data > Erase Variables**
Enter: Columns and constants: **select appropriate variable**
Click**: OK**

Entering Patterned Data

Suppose we wish to create a column that contains the integers 1 to 10. Although we could enter these numbers directly into the Data window by typing, there is a much easier way. Open a new worksheet by selecting **File > New > Worksheet**

Choose: **Calc > Make Patterned Data > Simple Set of Numbers**
Enter: Store patterned data in: **C1**
from first value: **1**
to last value: **10**
Click: **OK**

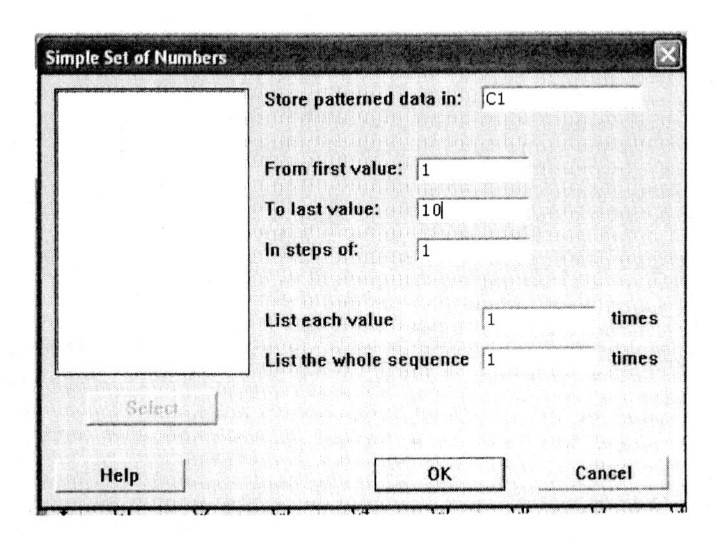

Column 1 should now contain the integers 1 through 10.

SAVING YOUR WORK

A MINITAB project contains all of your work; the data, text output from the commands, graphs, and more. When you save a project, you save all of your work at once. When you open a project, you can pick up right where you left off.
The project's many pieces can be handled individually. You can create data, graphs, and output from within MINITAB. You can also add data and graphs to the project by copying them from files. The contents of most windows can be saved and printed separately from the project, in a variety of formats. You can also *discard* a worksheet or graph, which removes the item from the project without saving it. Let's save the project and name it "Intro". Be sure to note where you are saving it.

To open, save, or close a project
To open a new project, choose **File** > **New**, click **Project**, and click **OK**.

To open a saved project, choose **File** > **Open Project**.

To save a project, choose **FILE** > **Save Project**.

To close a project, you must open a new project, open a saved project, or exit MINITAB.

RETRIEVING A FILE

To retrieve the project that we just saved:

Choose: **File** > **Open Project**

Click: Look in drop-down list arrow
Locate the file
Double-click: INTRO.MPJ
Click: **OPEN**

The data window now displays the test data you saved previously.

A CD ROM accompanies the Agresti/Franklin text, containing data sets for selected problems. They are all saved as MINITAB worksheets in the MINITAB folder.

Let's do exercise 1.21, which uses data from the CD ROM.

Click **File>Open Worksheet**

You must tell MINITAB where the file is located, usually the D drive.

Click the **appropriate drive > MINITAB > fla_student_survey.mtw**

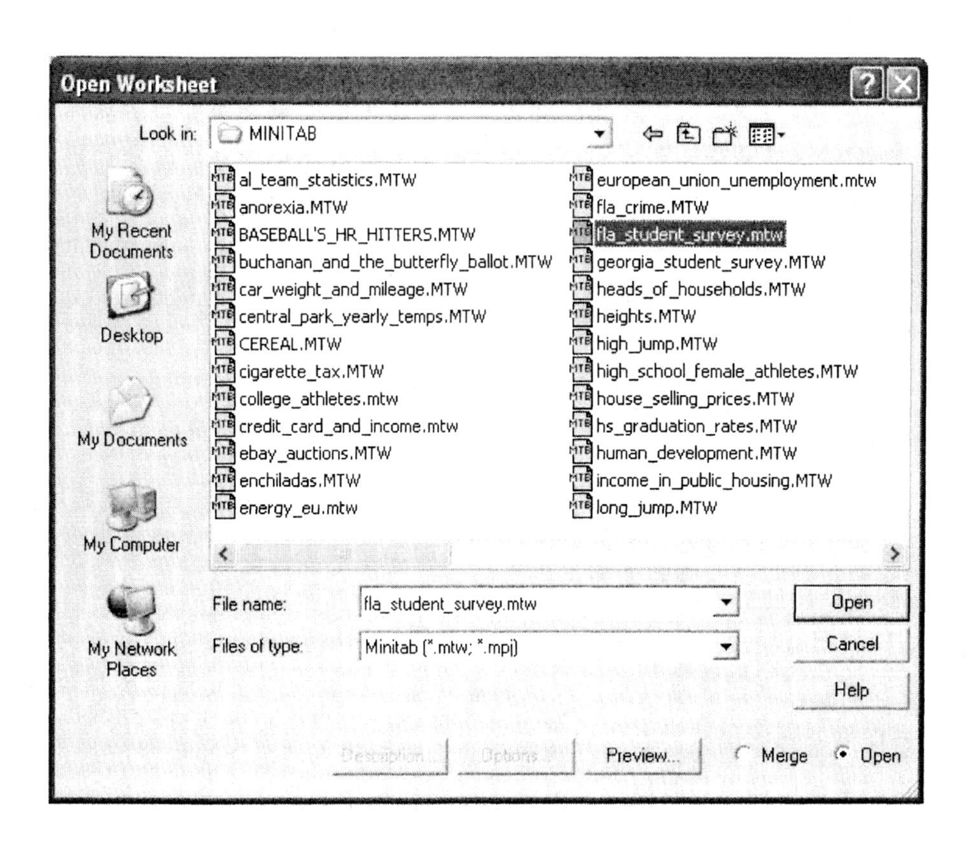

florida_student_survey.mtw ***

↓	C1-T	C2-T	C3	C4	C5	C6	C7	C8	
	subject	gender	age	high_sch_GPA	college_GPA	distance_home	distance_residence	TV	s
51	51	m	32	3.0	3.0	2000	5.00	5.0	
52	52	f	41	4.0	4.0	0	8.00	8.0	
53	53	f	29	3.0	3.9	300	3.70	2.0	
54	54	f	50	3.5	3.8	6	6.00	7.0	
55	55	f	22	3.4	3.7	80	7.00	10.0	
56	56	f	23	3.6	3.2	375	1.50	5.0	

Chapter 2 GRAPHIC PRESENTATIONS OF DATA

Graphically representing data is one of the most helpful ways to become acquainted with the sample data. In this lab you will use MINITAB to present data graphically. There are several ways to display a picture of the data. These graphical displays help us get acquainted with the data and to begin to get a feel for how the data is distributed and arranged. To see what is available to you, use the menu bar to select **Graph**. Note the different types of graphs that are listed there. We will use the menu bar to make our selections.

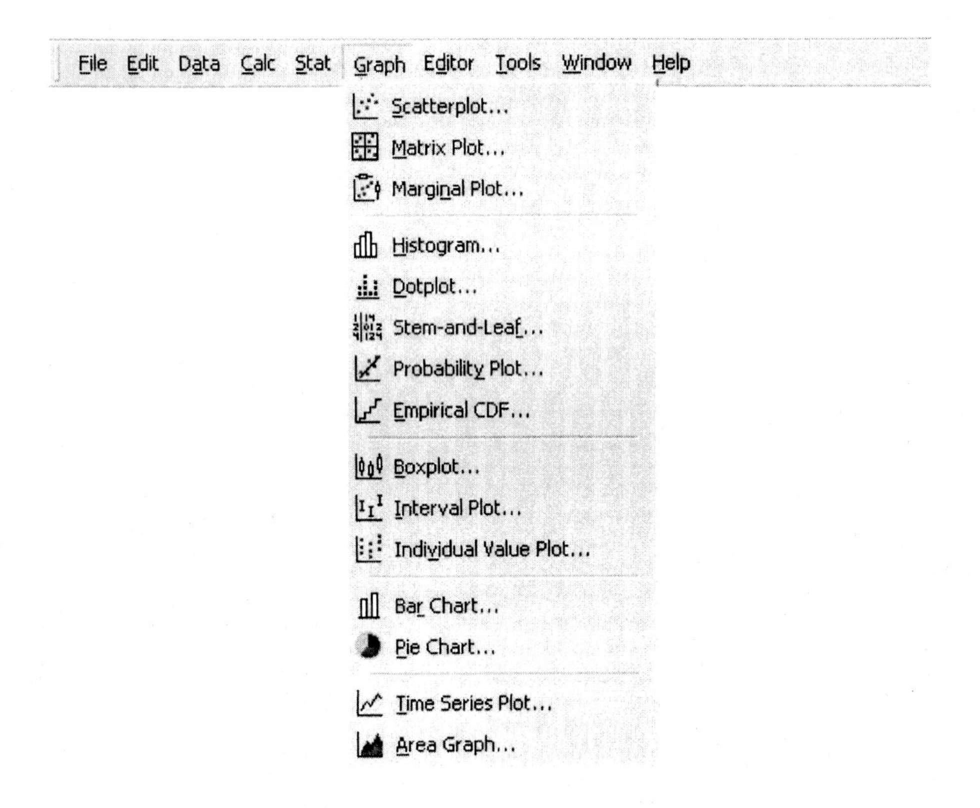

Chapter 2 Example 3 How Much Electricity Comes From Renewable Energy Sources? Constructing a Pie Chart

Copy the data from the text into the worksheet.

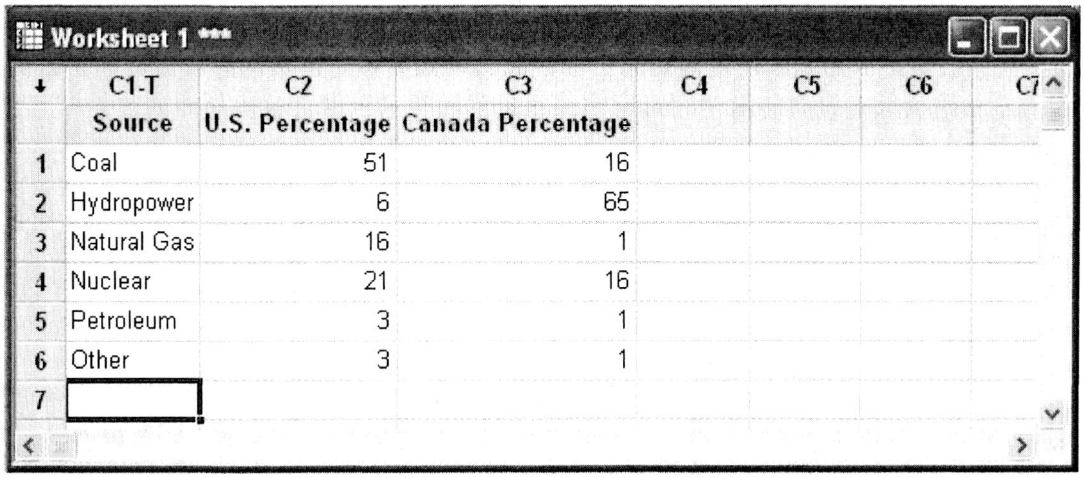

Worksheet 1 ***

↓	C1-T	C2	C3	C4	C5	C6	C7
	Source	U.S. Percentage	Canada Percentage				
1	Coal	51	16				
2	Hydropower	6	65				
3	Natural Gas	16	1				
4	Nuclear	21	16				
5	Petroleum	3	1				
6	Other	3	1				
7							

File Edit Data Calc Stat Graph Editor Tools Window Help

Scatterplot...
Matrix Plot...
Marginal Plot...
Histogram...
Dotplot...
Stem-and-Leaf...
Probability Plot...
Empirical CDF...
Boxplot...
Interval Plot...
Individual Value Plot...
Bar Chart...
Pie Chart...
Time Series Plot...
Area Graph...

Choose:
Graph > Pie Chart

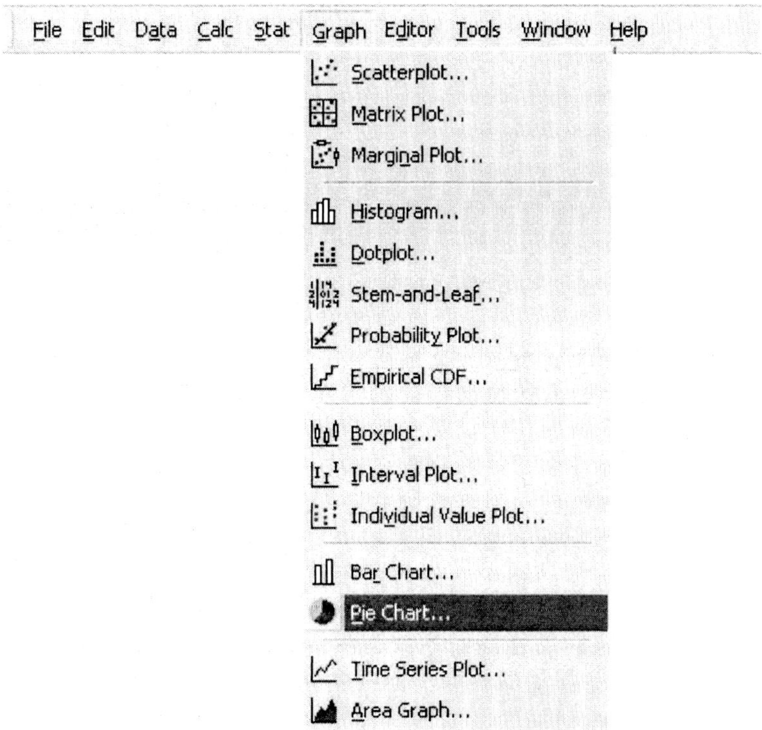

Select: **Chart values from a table**
Enter: Catagorical variable: **Source**
Summary variable: **US percentages**
Choose **labels**: Enter titles
Click: **OK**

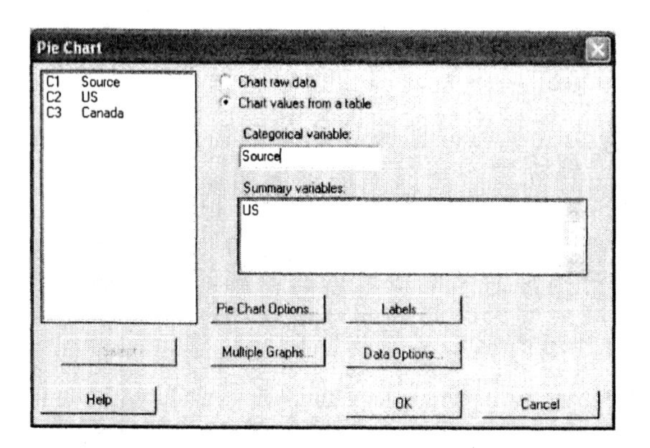

Use the following menu commands to get the Bar chart:

Choose: **Graph > Bar Chart**

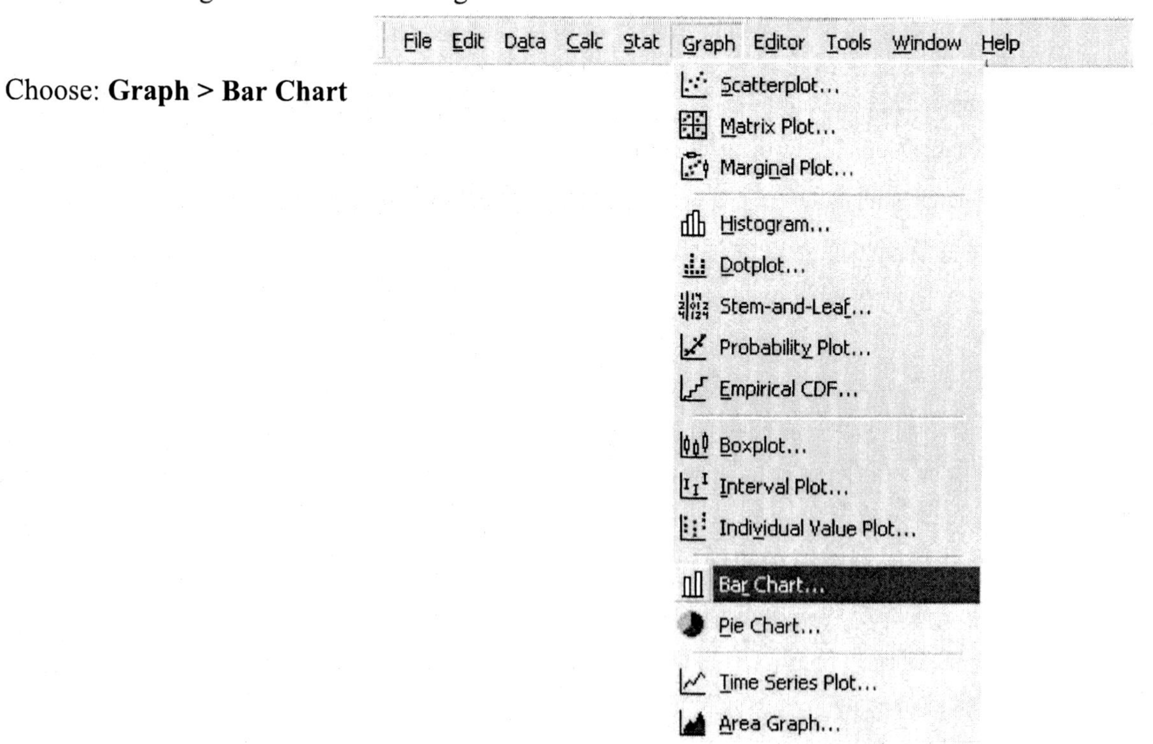

Choose: **Simple > OK**
Select: Bars represent:
Values from a table
Select: **Simple**
Click: **OK**

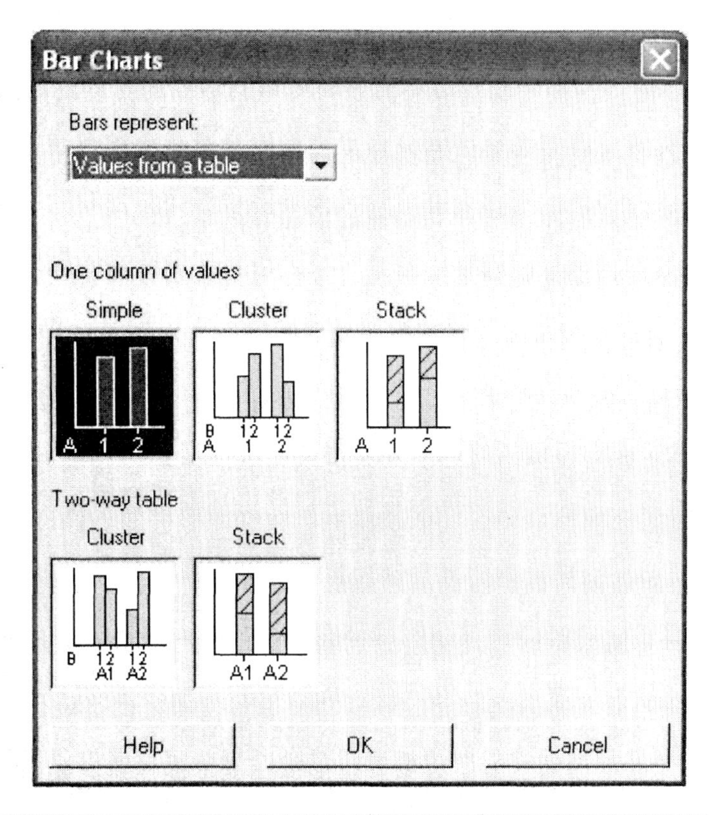

Enter: Graph variables: **C2**
Enter: Categorical variables: **C1**
Choose: **Labels**
Enter: **an appropriate title**
Click: **OK**

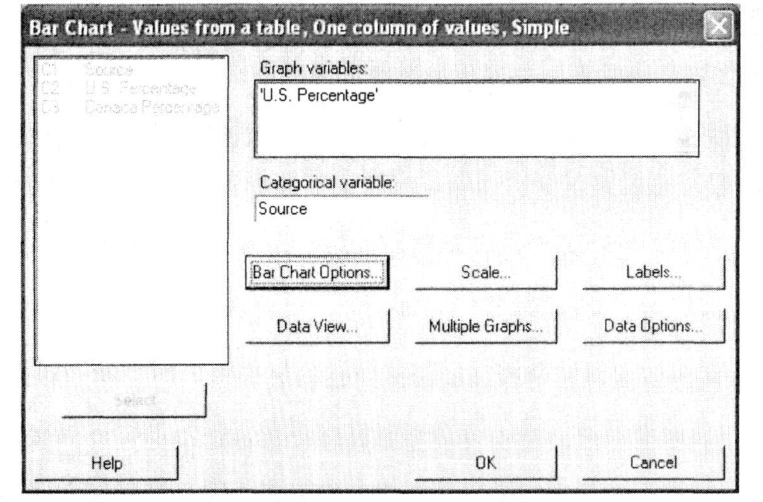

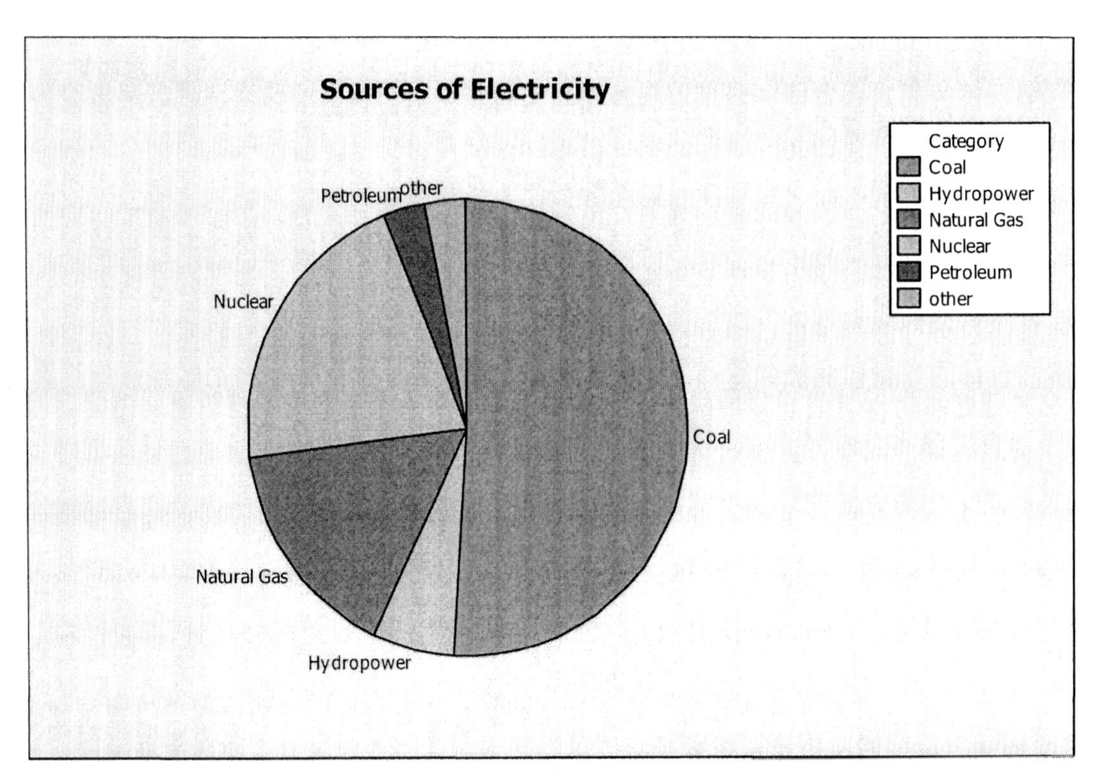
Sources of Electricity
Category
Coal
Hydropower
Natural Gas
Nuclear
Petroleum
other
Petroleum
other
Nuclear
Coal
Natural Gas
Hydropower

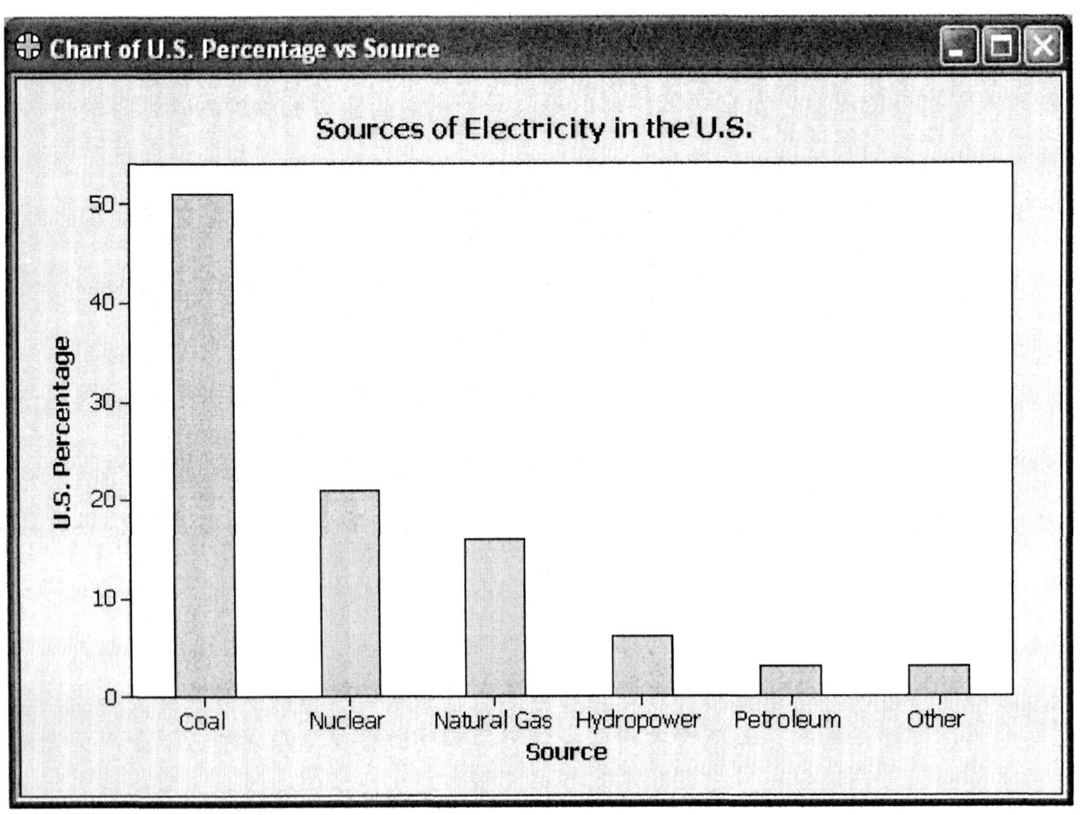
Chart of U.S. Percentage vs Source
Sources of Electricity in the U.S.
U.S. Percentage
50
40
30
20
10
0
Coal
Nuclear
Natural Gas
Hydropower
Petroleum
Other
Source

Chapter 2 Example 4 Exploring the Health Value of Cereals
Constructing a Dot Plot

Dot plots are a quick and efficient way to get a preliminary understanding of the distribution of your data. It results in a picture of the data as well as sorts the data into numerical order. Open the worksheet **CEREAL** from the MINITAB folder of the data disk

Choose: **Graph > Dotplot**

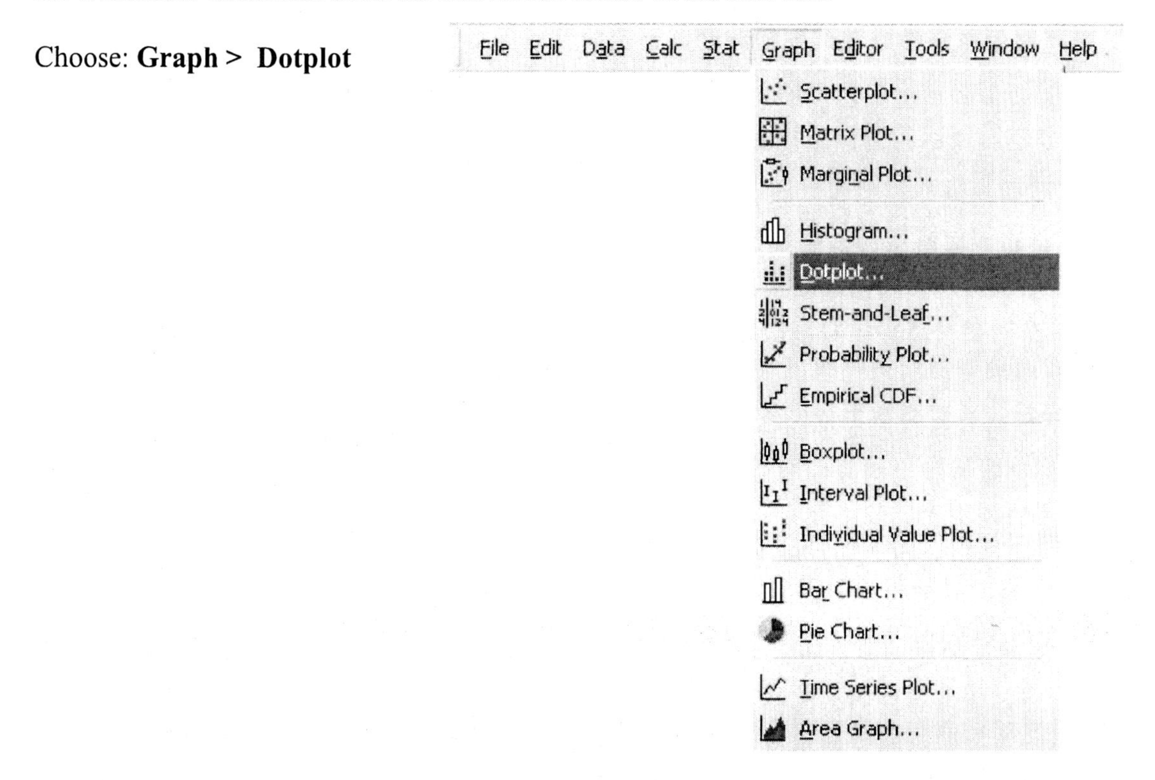

Choose: **One Y/Simple**
Click: **OK**

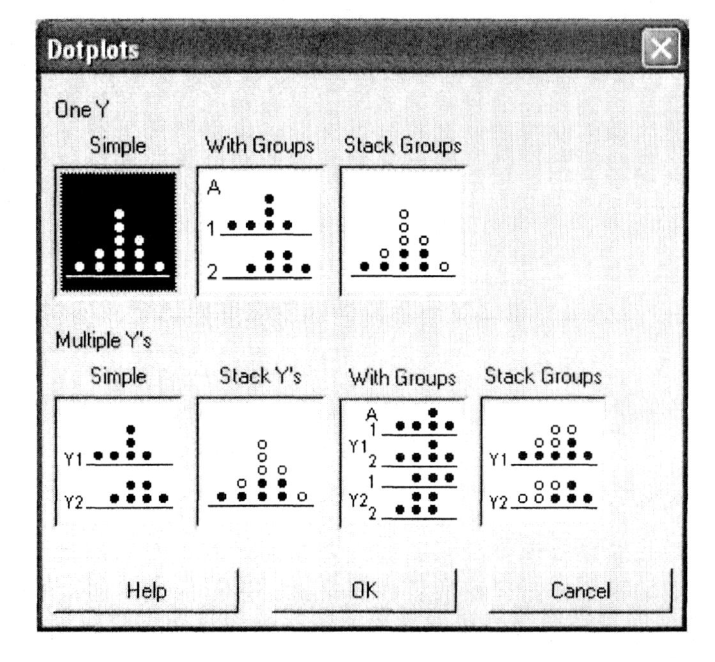

Enter: Graph variables: **C2**
Click: **OK**

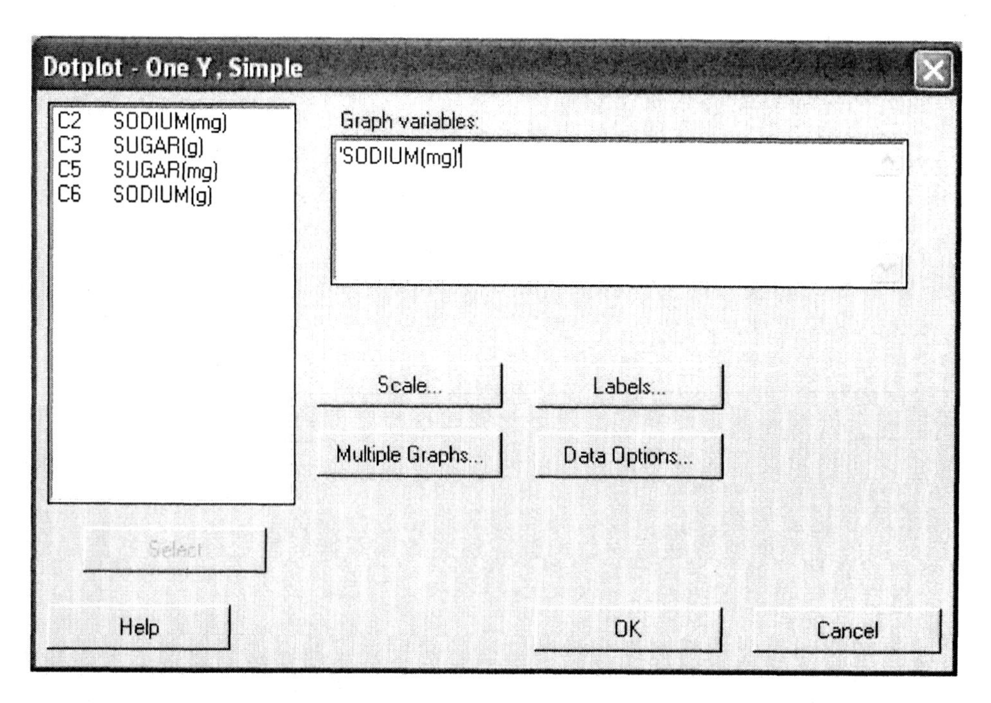

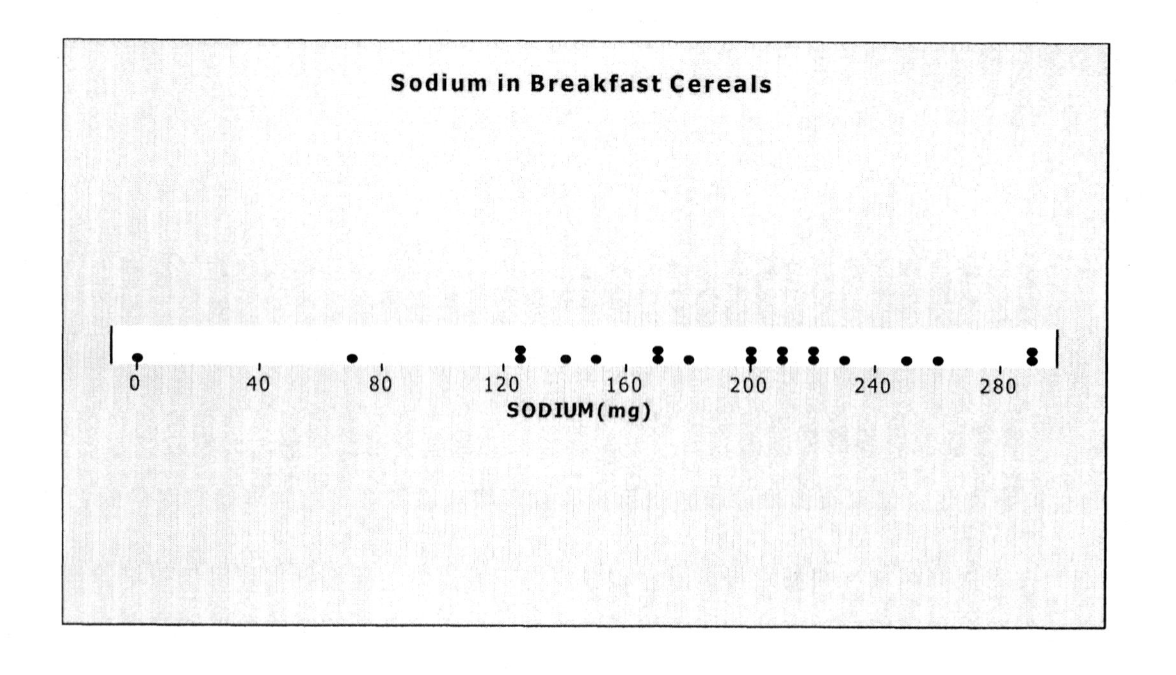

Chapter 2 Example 5 Exploring the Health Value of Cereals
Stem and Leaf Plots

To illustrate the commands necessary to construct a stem-and-leaf display, let's use the data from the **CEREAL** worksheet, sodium values in cereals.

Choose: **Graph >Stem-and-Leaf**

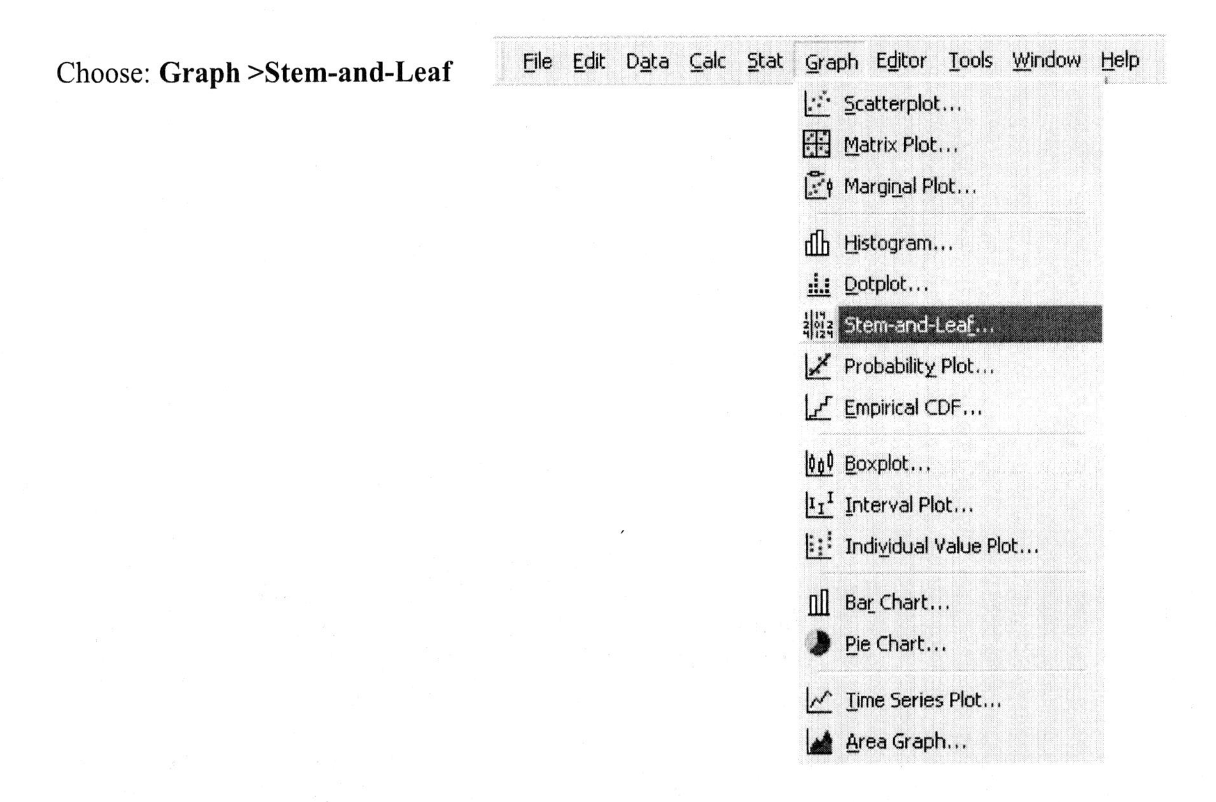

Enter: Variable: **C2**
Uncheck **"Trim Outliers"**
Click **OK**

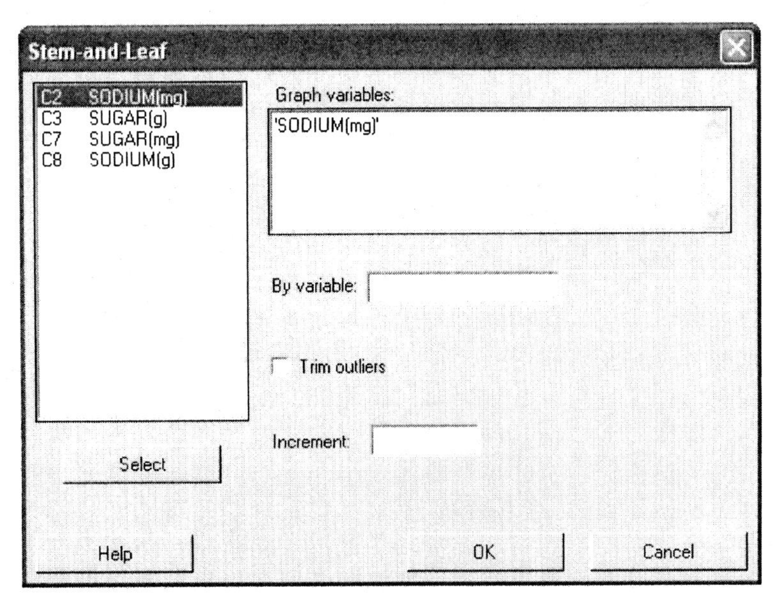

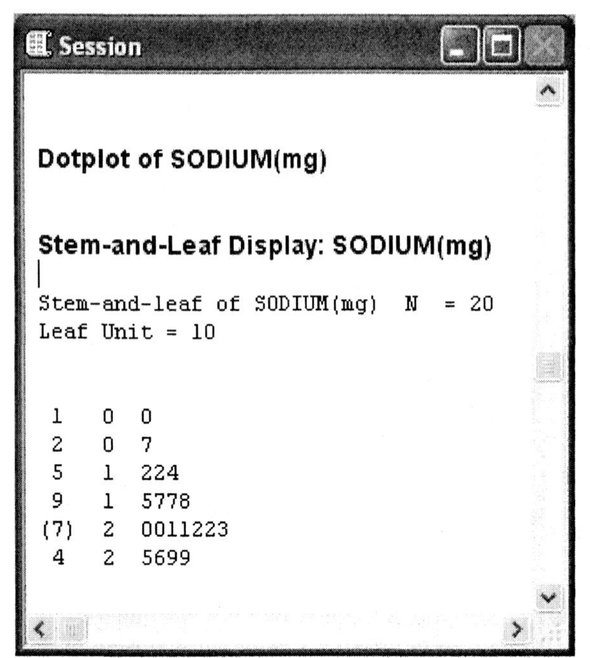

If we were to comment on the shape of the distribution, we can clearly see that it is skewed left.

We allowed MINITAB to choose the increment. We may choose to specify the increment ourselves. Try various values for "increment". Which Stem-and-Leaf display do you prefer?

Chapter 2 Example 7 Exploring the Health Value of Cereals
Constructing a Histogram

Open the **CEREAL** worksheet from the MINITAB folder of the data disk.

Choose: **Graph > Histogram**

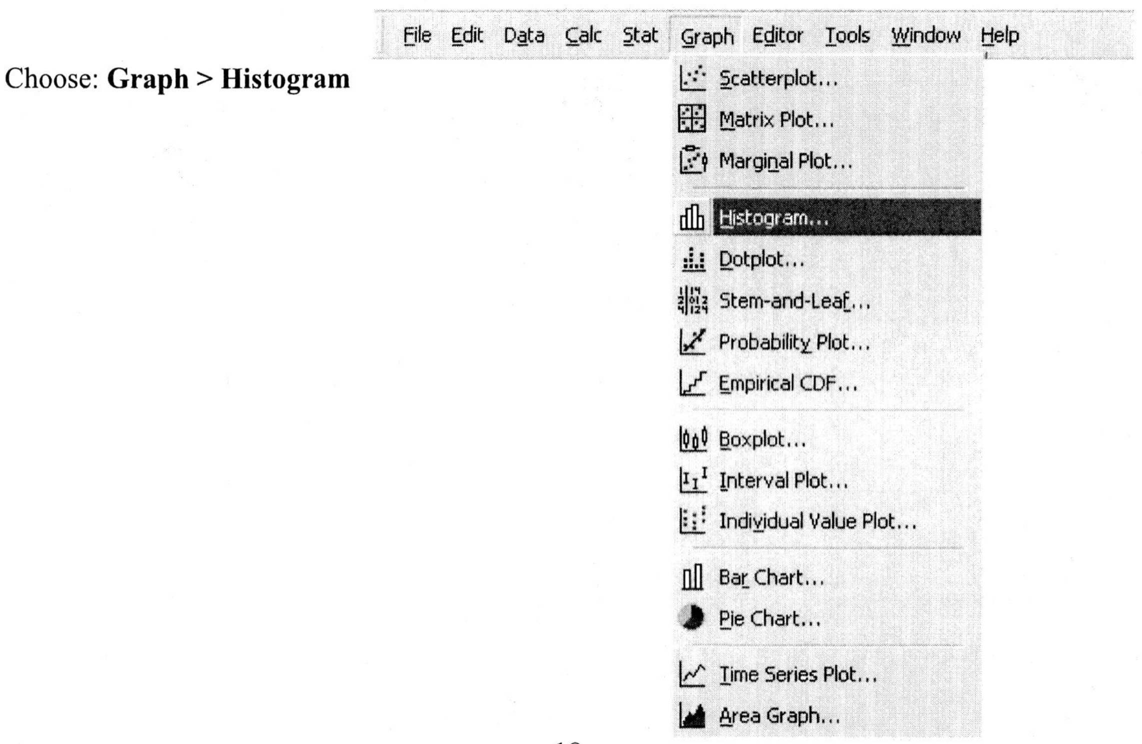

Choose: **Simple > OK**

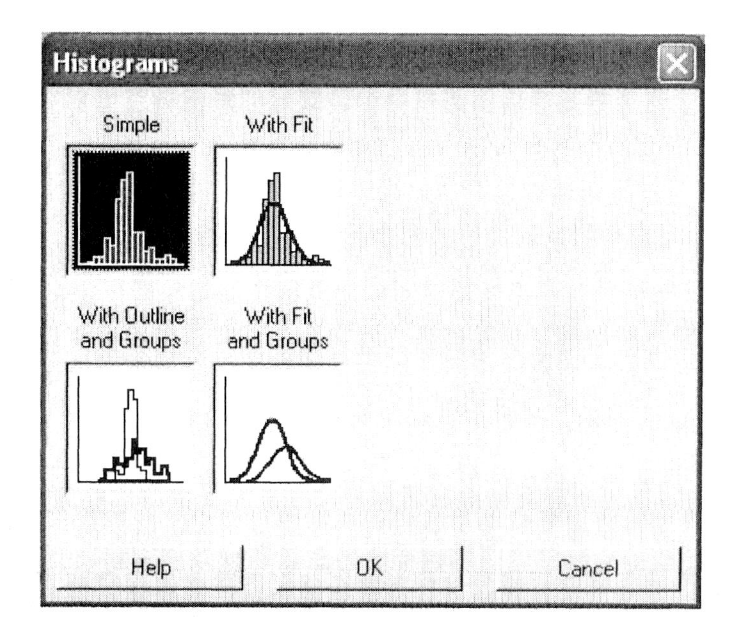

Enter: Graph variables: **C2**

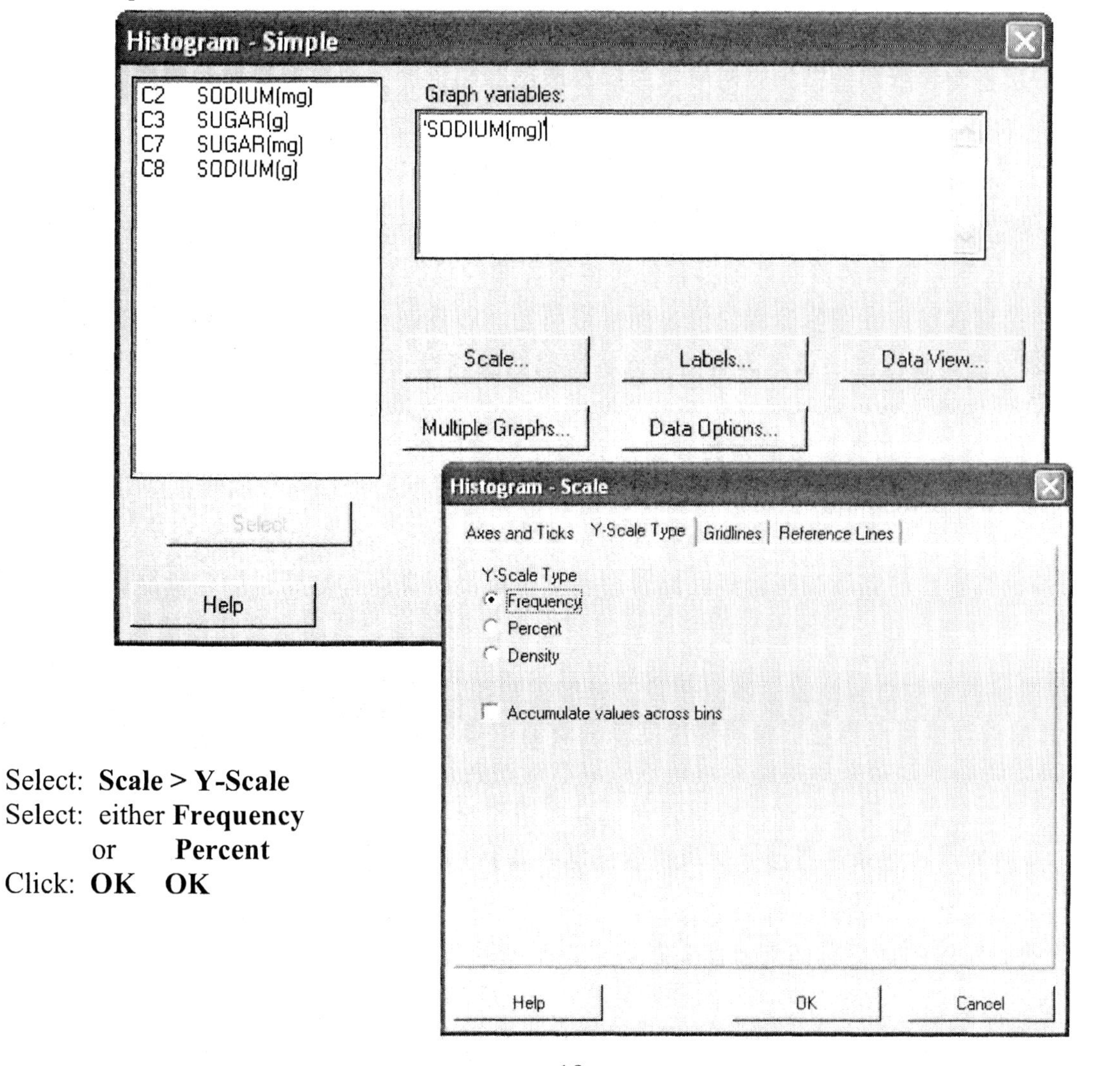

Select: **Scale > Y-Scale**
Select: either **Frequency**
or **Percent**
Click: **OK OK**

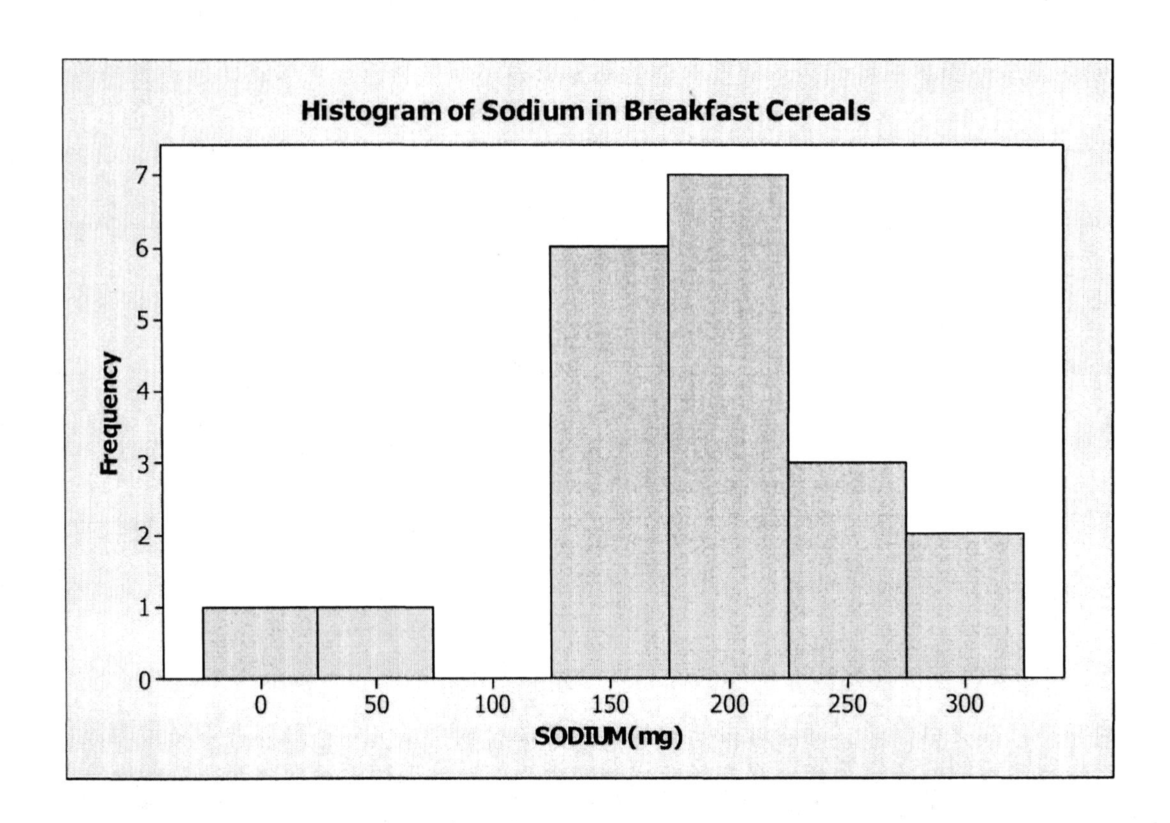

Chapter 2 Example 9 Is There a Trend Toward Warming in New York City? Constructing a Time Plot

For some variables, observations occur over a period of time. It is useful for showing trends in the data.

Open the worksheet **central_park_yearly_ temps** from the MINITAB folder of the data disk.

Select: **Graph > Time Series Plot**

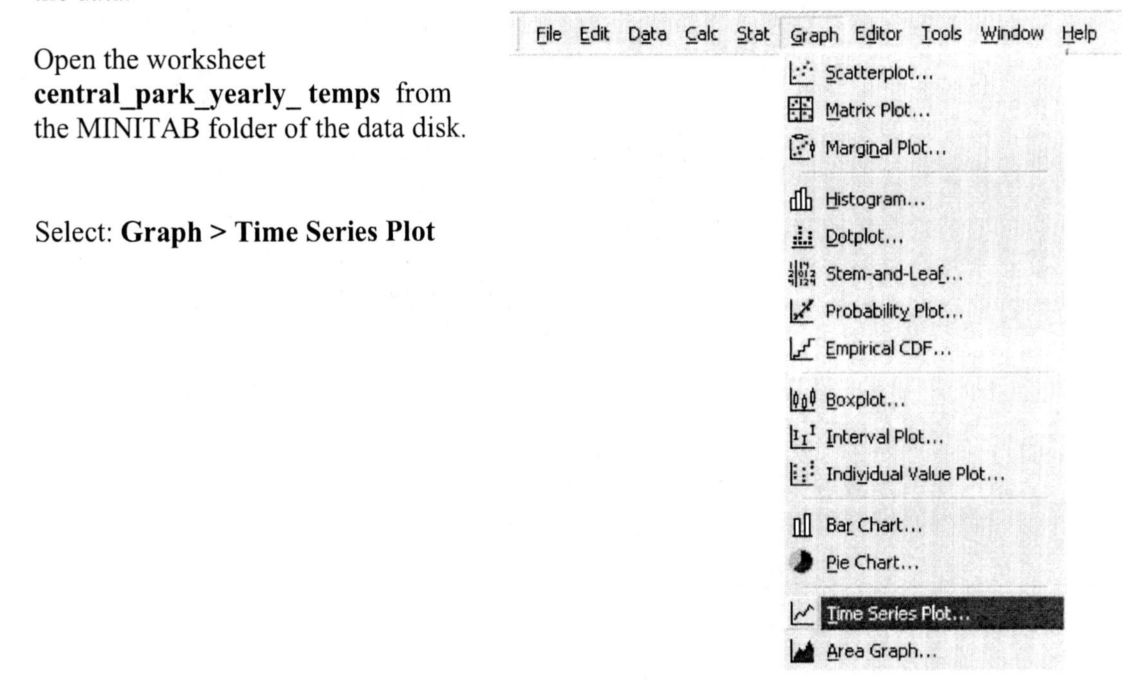

Select: **Simple**
Click: **OK**

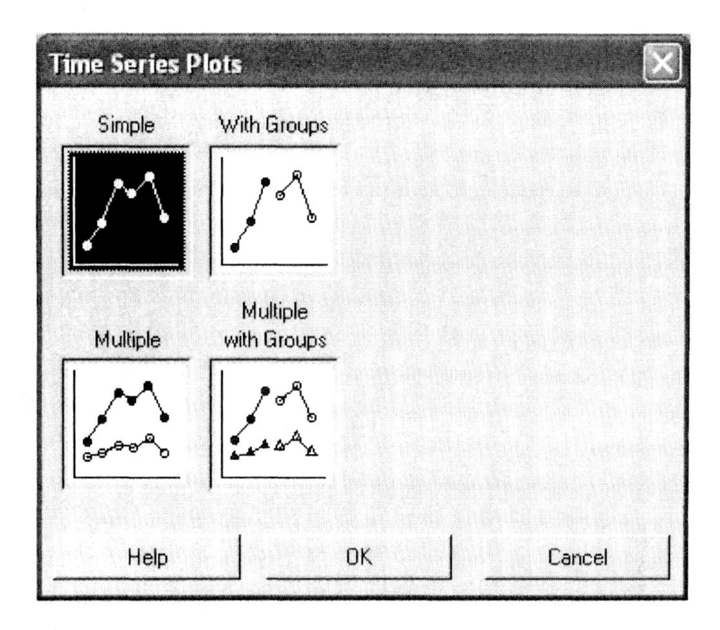

Enter:
  Series: **C1**

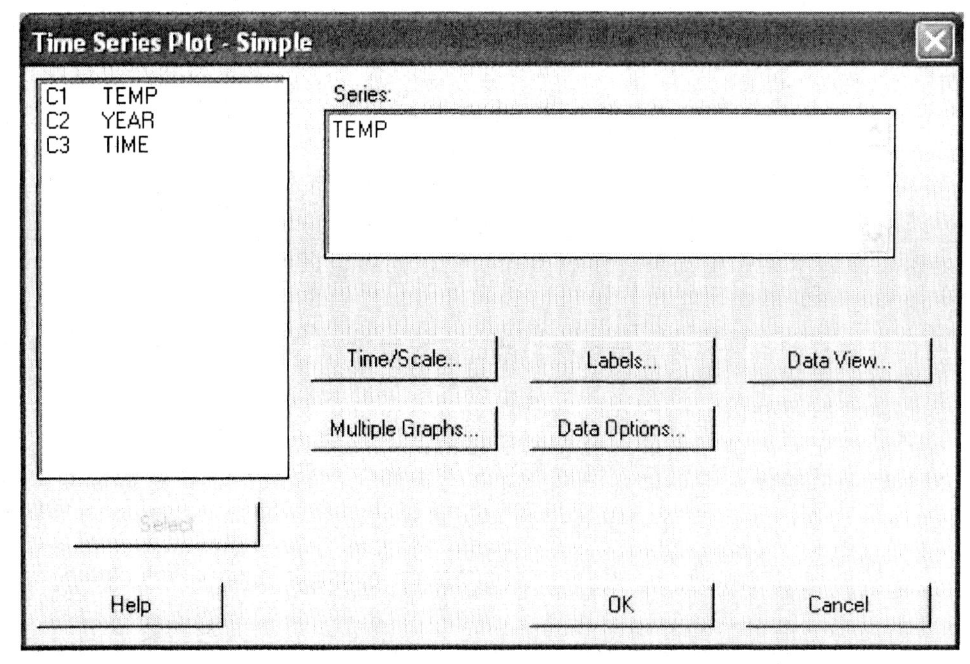

Click **Time/Scale**… **> Calendar > Year > OK**

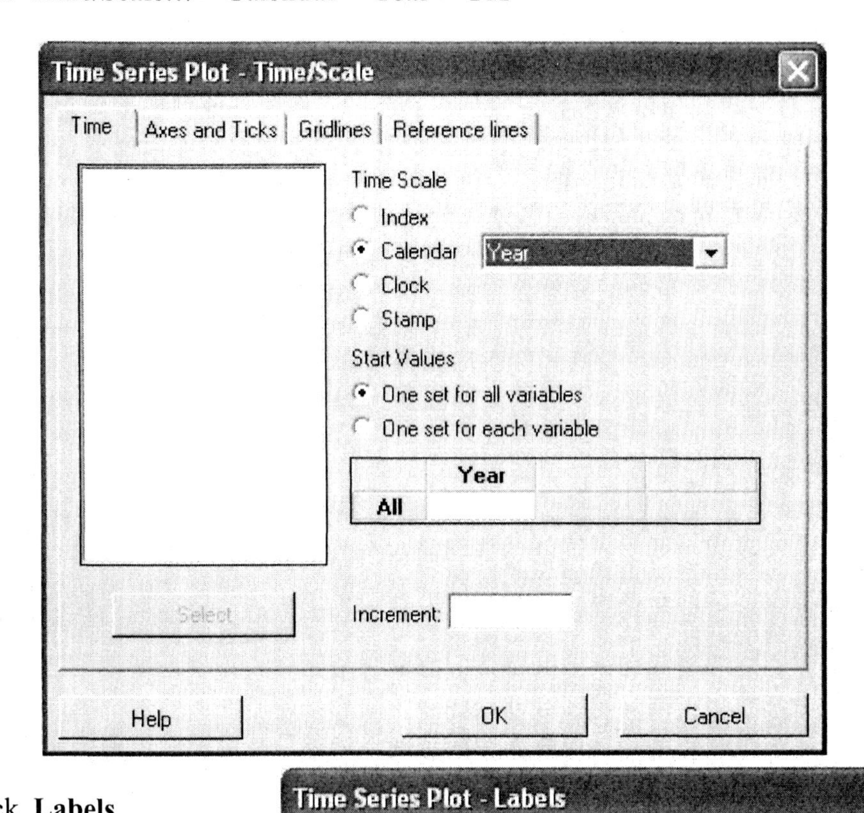

Click **Labels**
Enter: Title: **an appropriate title**
Click: **OK**

Time Series Plot - Labels
Titles/Footnotes | Data Labels
Title:
Average Yearly Temperatures in Central Park
Subtitle 1:
Subtitle 2:
Footnote 1:
Footnote 2:
Help OK Cancel

Click **Data View**
Select: **Symbols**
Connect Line
Click: **OK OK**

Select: **Smoother** tab

Select: Smoother: **Lowess**

(This adds a general trend line to the time series plot.)

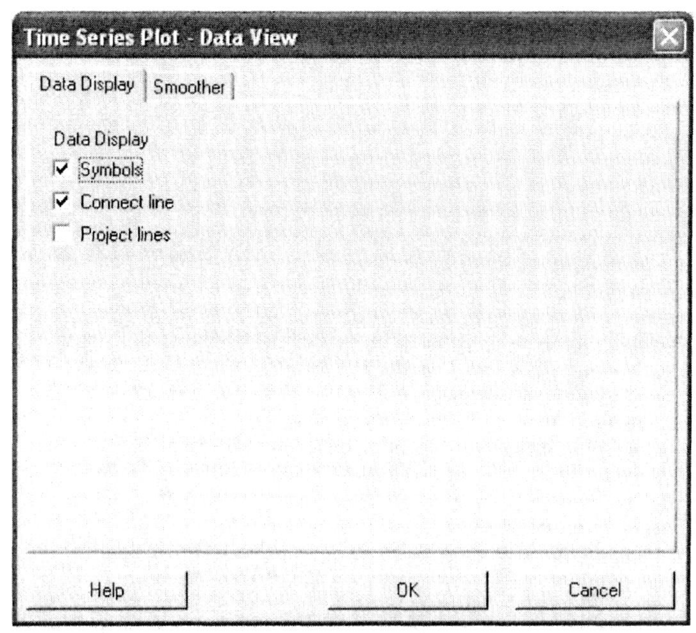

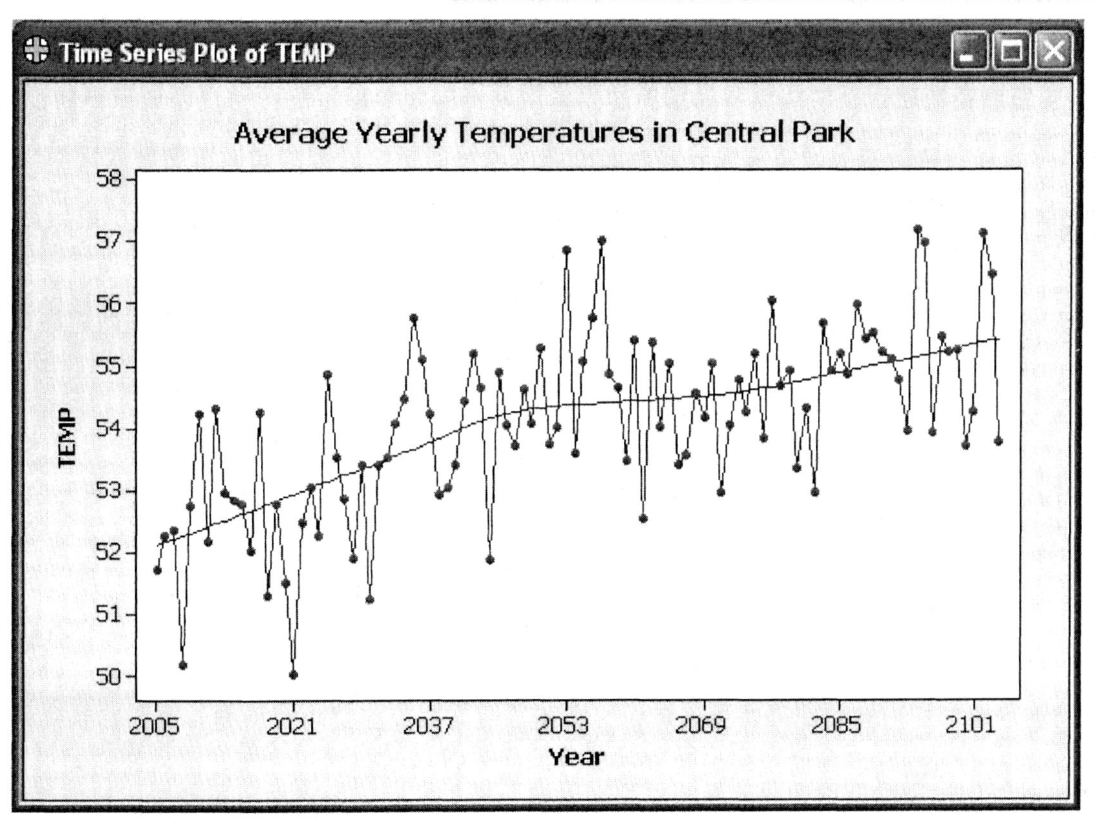

Chapter 2 Exercise 2.11 Weather Stations
Constructing a Pie Chart

Using the data shown in the pie chart in the text:

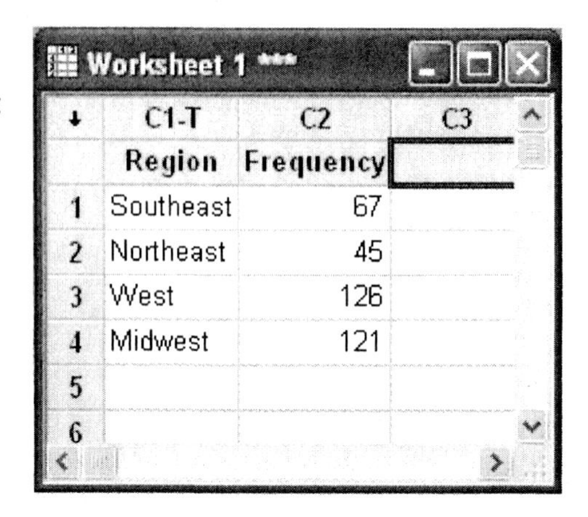

Worksheet 1 ***

↓	C1-T	C2	C3
	Region	Frequency	
1	Southeast	67	
2	Northeast	45	
3	West	126	
4	Midwest	121	
5			
6			

Choose: **Graph > Pie Chart**

Select**: Chart values from a table**
Enter: Categorical variable: **Region**
Summary variable: **Frequency**
Choose: **Labels**
Enter: **appropriate title**
Click: **OK**

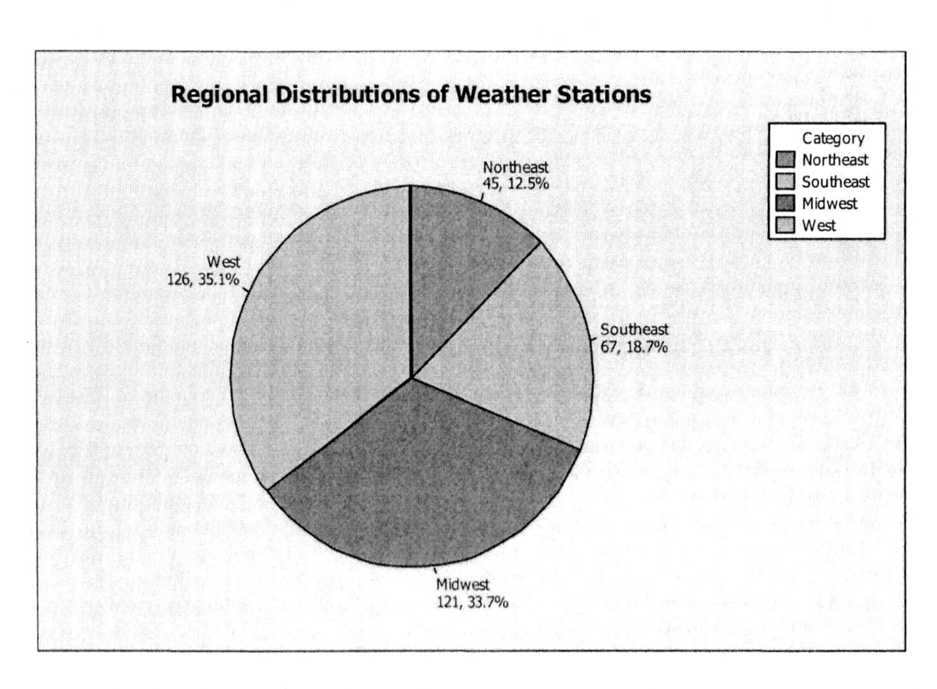

Note that by checking the boxes for category name, frequency and percent (on the Slice Labels tab), we get labels on each slice. If you compare the pie chart to the bar graph below, in which is it easier to identify the mode?

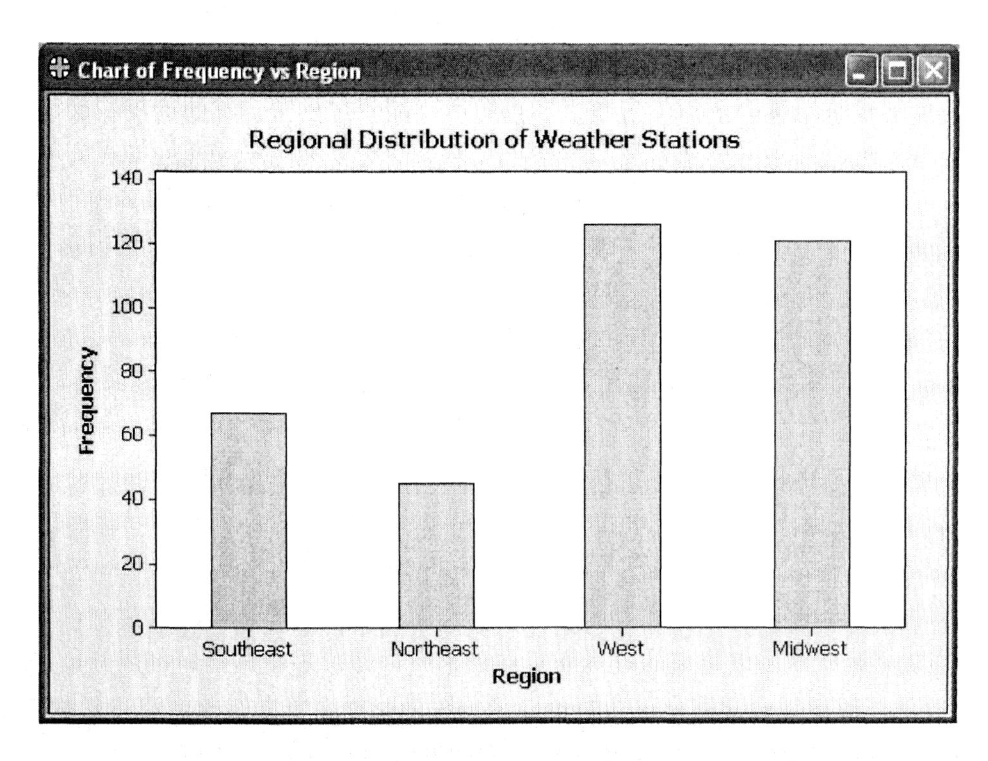

Chapter 2 Exercise 2.13 Shark attacks worldwide
Bar Chart and Pareto Chart

Enter the data from table 2.1 into the worksheet.

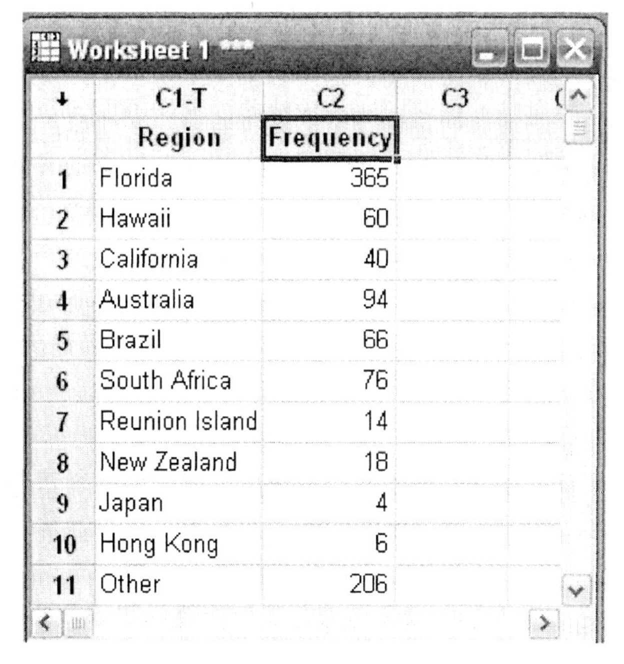

Worksheet 1 ***

↓	C1-T	C2	C3
	Region	Frequency	
1	Florida	365	
2	Hawaii	60	
3	California	40	
4	Australia	94	
5	Brazil	66	
6	South Africa	76	
7	Reunion Island	14	
8	New Zealand	18	
9	Japan	4	
10	Hong Kong	6	
11	Other	206	

Sort the data alphabetically based on region and store the sorted columns in original columns.

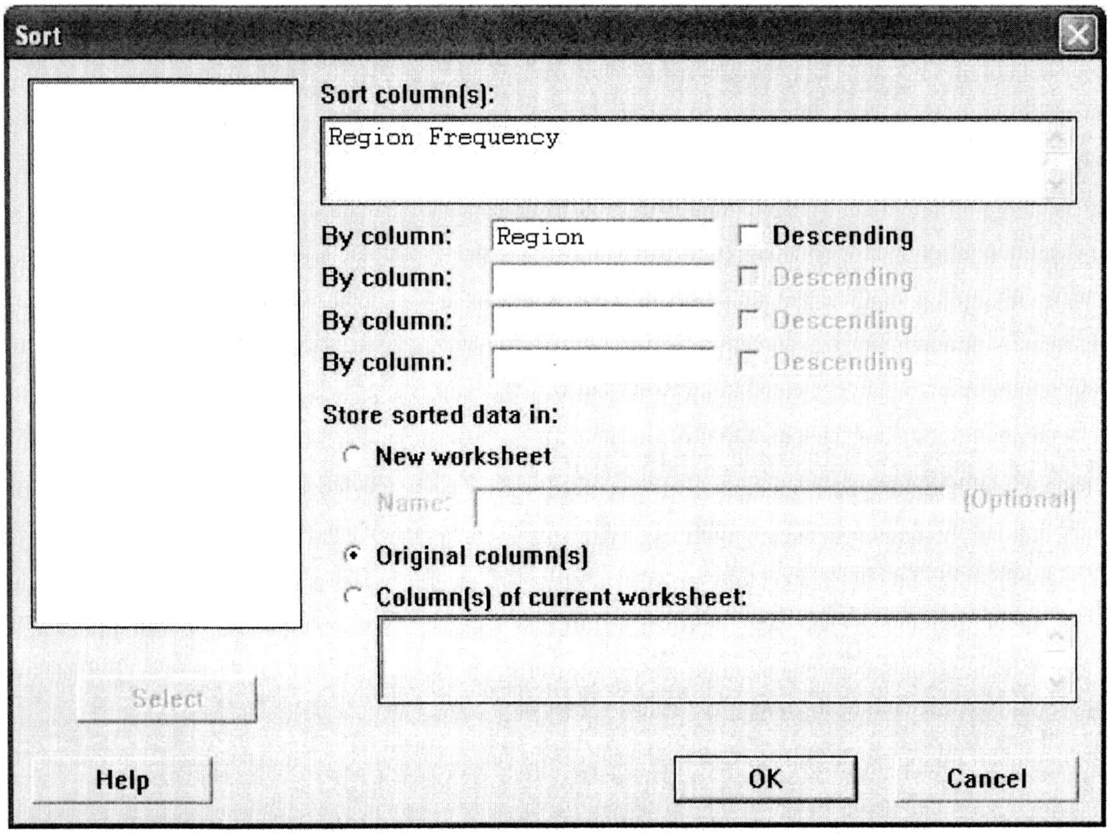

Sort

Sort column(s):
Region Frequency

By column: Region ☐ Descending
By column: ☐ Descending
By column: ☐ Descending
By column: ☐ Descending

Store sorted data in:
○ New worksheet
Name: (Optional)
◉ Original column(s)
○ Column(s) of current worksheet:

Select

Help   OK   Cancel

To create the bar chart

Choose: **Graph > Bar Chart**
Select: Bars represent: **Values from a table**
Select: **Simple > OK**
Enter: Graph variables: **C2**
Enter: Catagorical variables: **C1**
Choose: **Labels**
Enter: **an appropriate title**
Click: **OK**

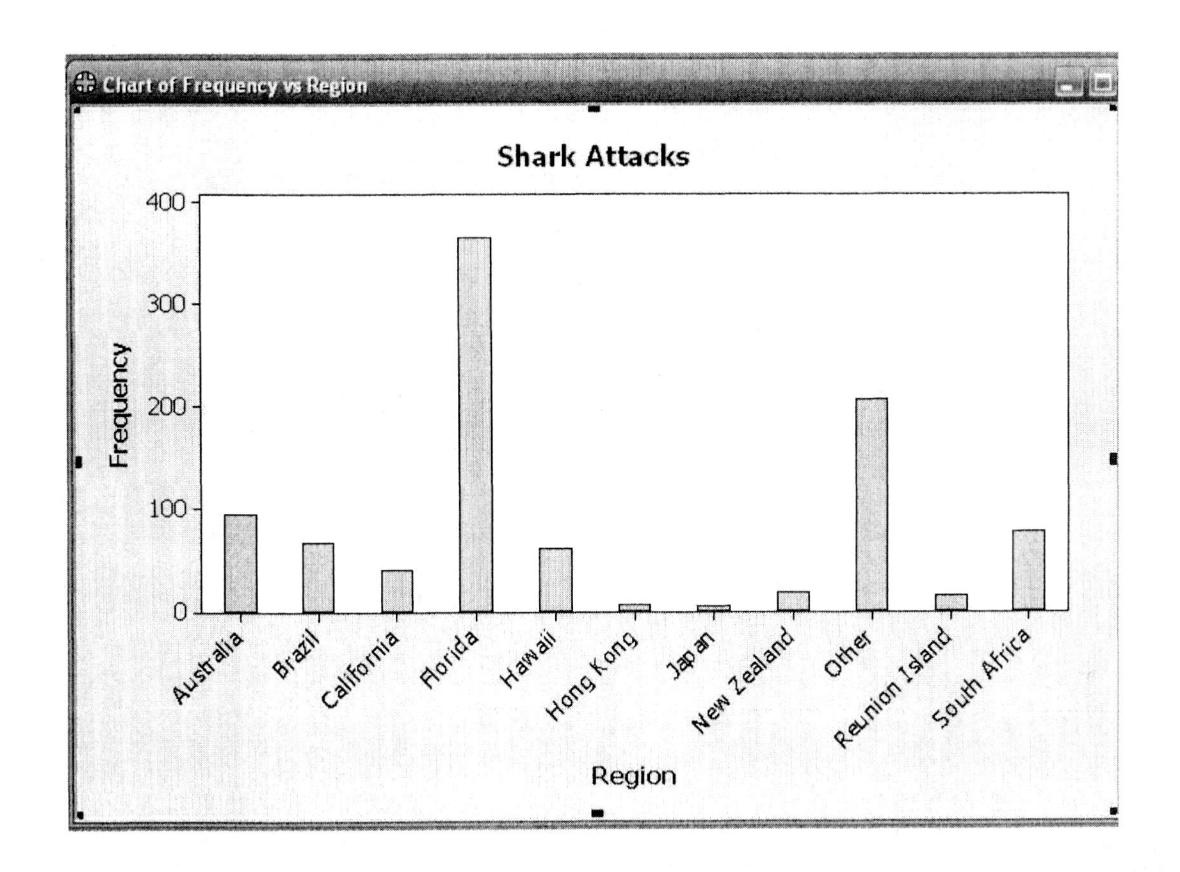

To obtain a Pareto Chart for the same data, use the following menu choices

Choose: **Stat > Quality Tools > Pareto Chart**
Click: Chart defects table
Enter: Labels in: **Region**
Frequencies in: **Frequency**
Combine all defects into one category after this percent: **99.9** %
Options: Title: **Shark Attacks**
Click: **OK** **OK**

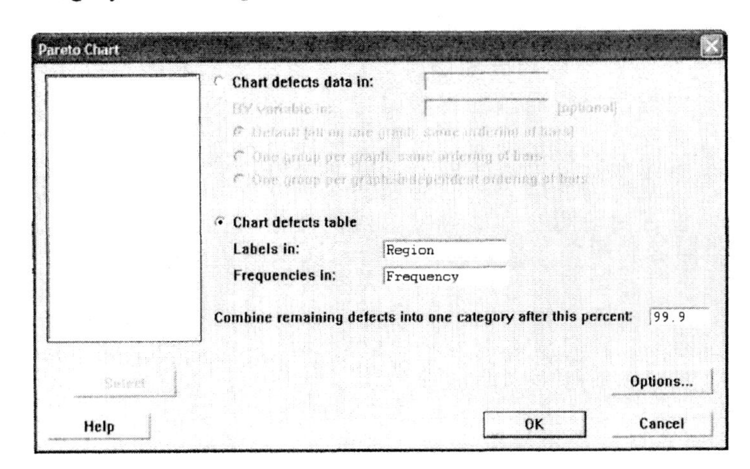

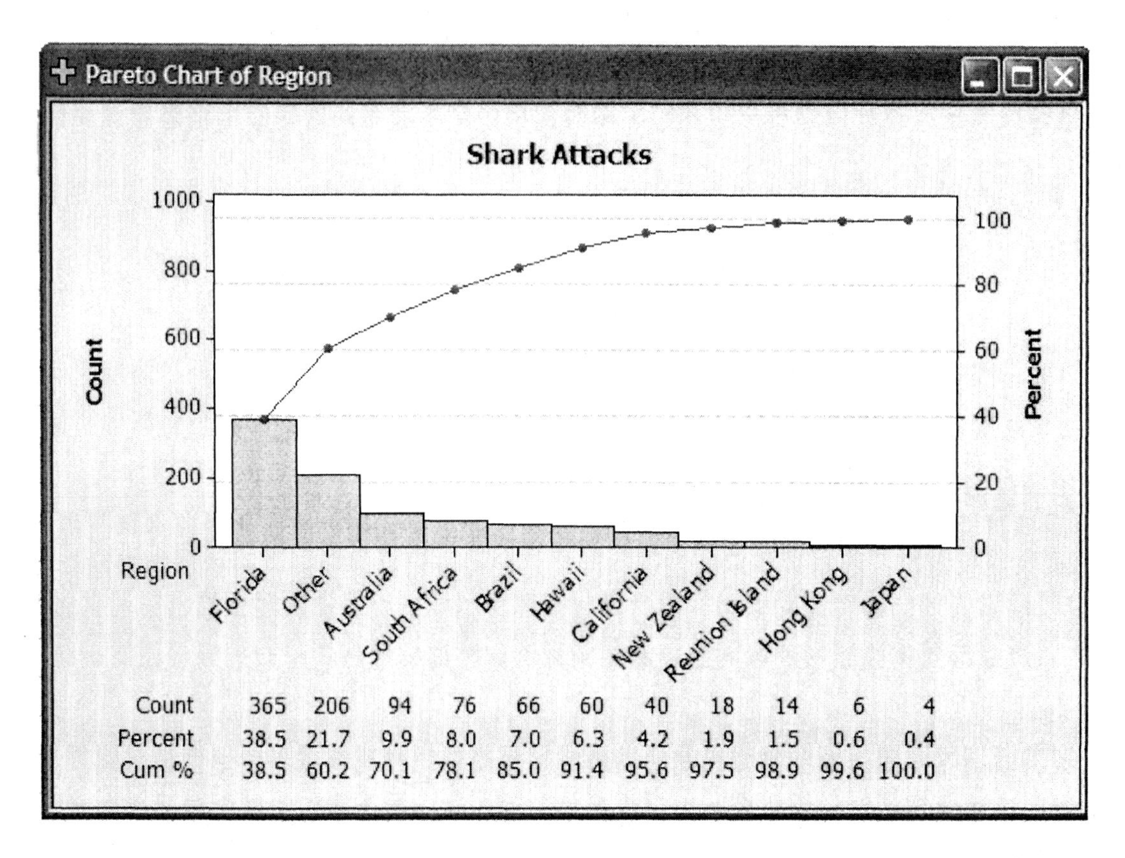

Chapter 2 Exercise 2.14 Sugar Dot Plot
Constructing a Dot Plot

Open the worksheet **CEREAL** from the MINITAB folder of the data disk

Choose: **Graph > Dotplot**
Choose: **One Y/Simple > OK**
Enter: Graph variables: **C3**
Click: **OK**

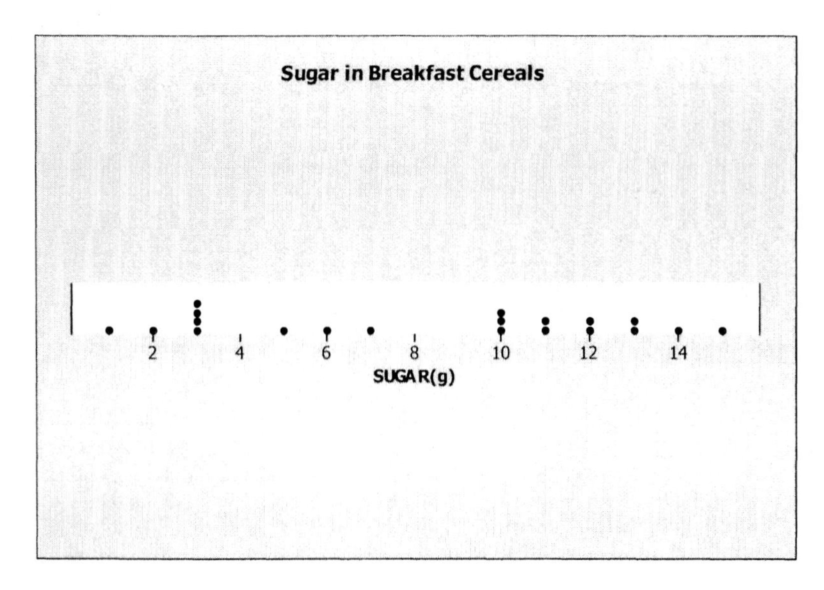

Chapter 2 Exercise 2.22 Sugar plots
Dot Plot, Stem-and-Leaf Plot and Histogram

Open the **CEREAL** worksheet from the MINITAB folder of the data disk.

Making the menu selections as previously demonstrated, we get the following graphical displays.

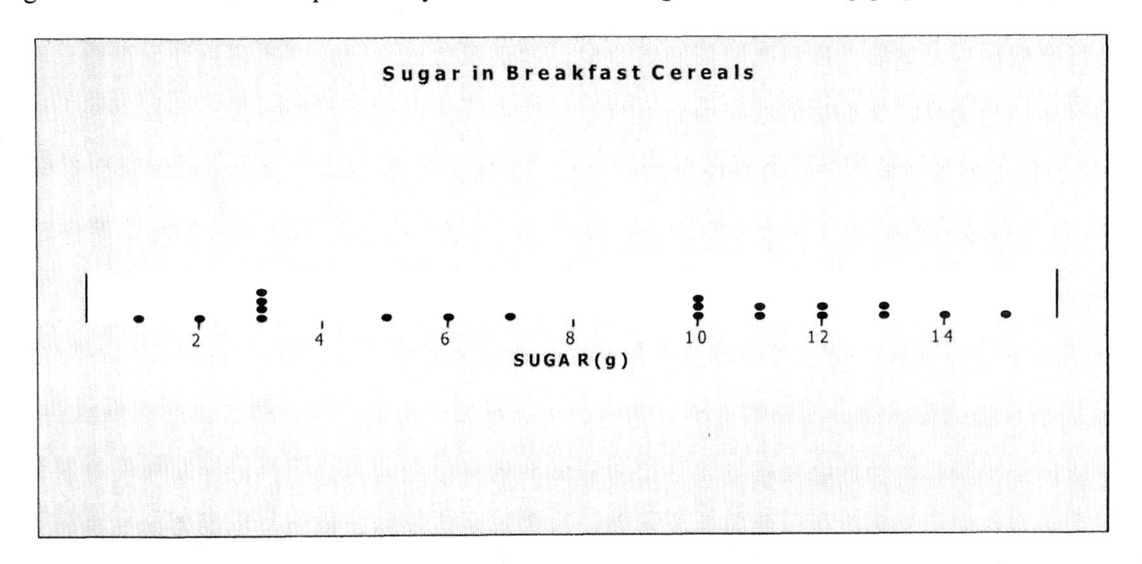

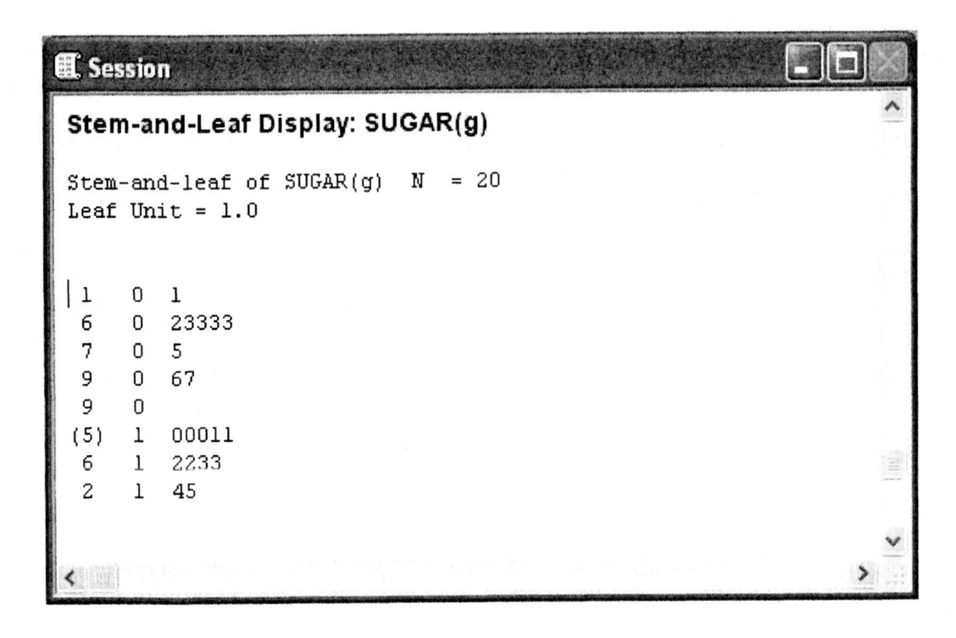
Session

Stem-and-Leaf Display: SUGAR(g)

```
Stem-and-leaf of SUGAR(g)  N  = 20
Leaf Unit = 1.0


  1    0  1
  6    0  23333
  7    0  5
  9    0  67
  9    0
 (5)   1  00011
  6    1  2233
  2    1  45
```

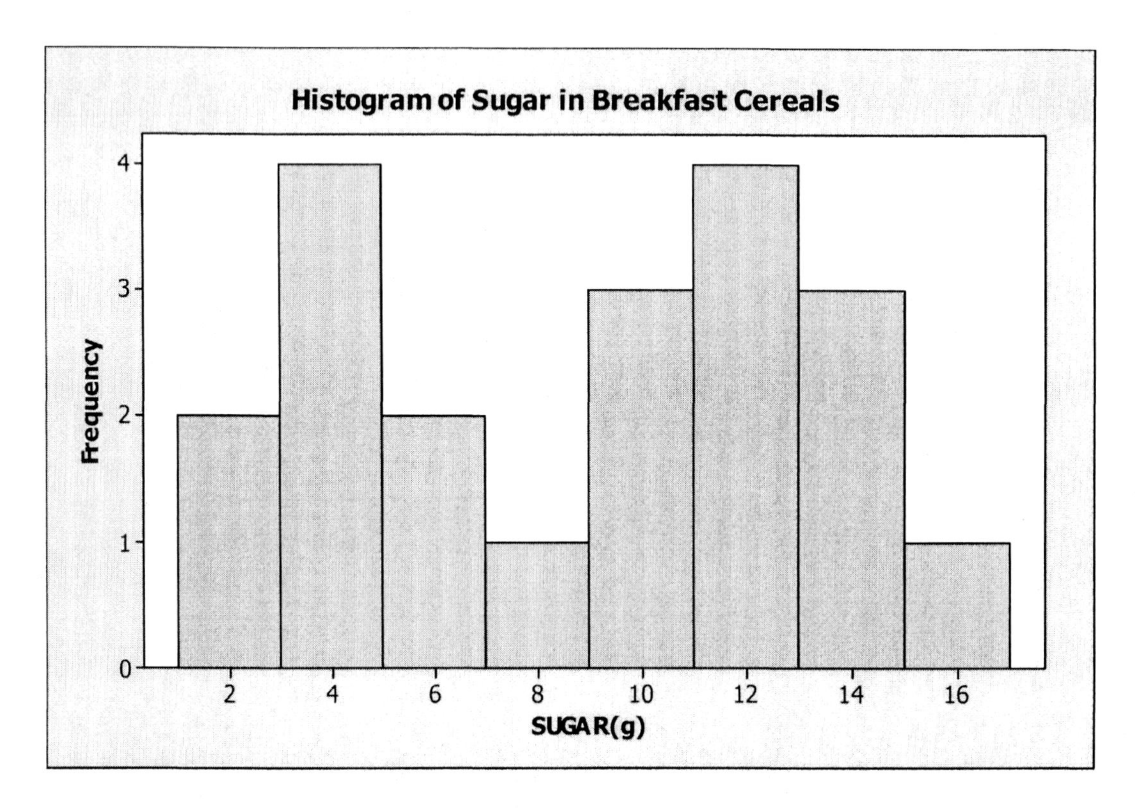

How would you explain each of these?

Chapter 2 Exercise 2.29 Warming in Newnan, Georgia?
Constructing a time plot

Open the worksheet **newnan_ga_temps** from the MINITAB folder of the data disk

Select: **Graph > Time Series Plot > Simple > OK**

Enter: Series: **C1**

Click **Time/Scale**... > **Calendar > Year >** Start Value **One set for all variables > 1901**

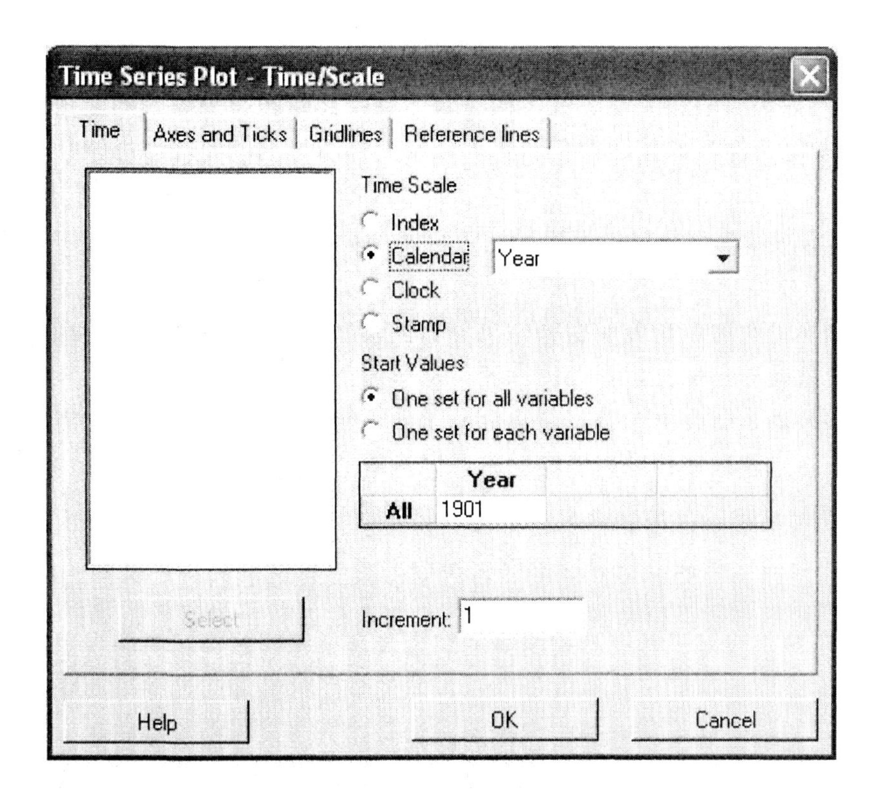

Click **Data View**
Select: **Symbols**
Connect Line
Click: **OK OK**

Select: **Smoother** tab

Select: Smoother: **Lowess**

(Note: These options remain as previously selected.)

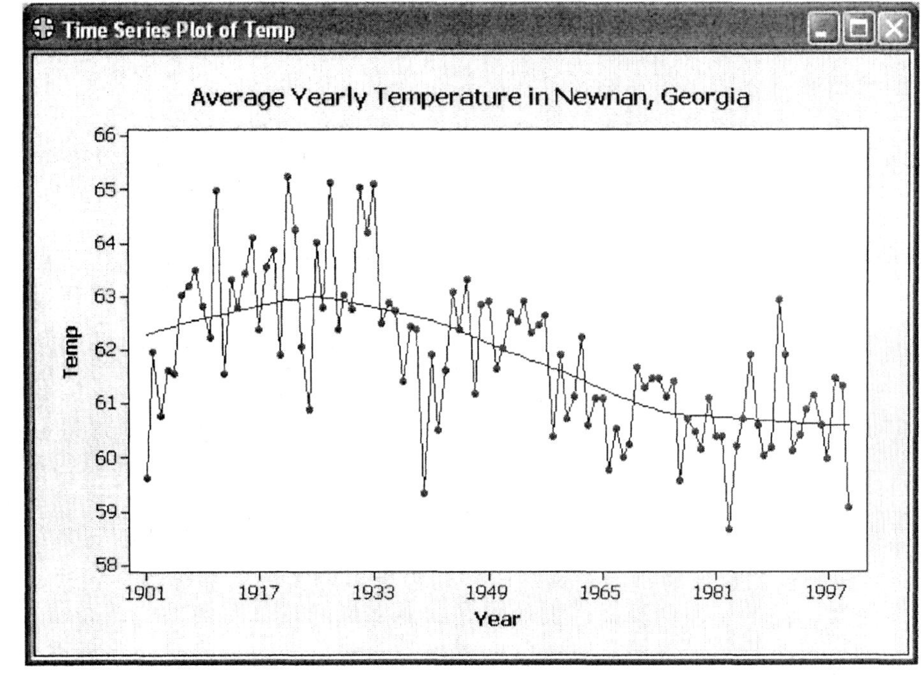

Chapter 2 Example 10 What is the Center of the Cereal Sodium Data?
Determining Mean and Median

Open the Minitab worksheet **cereal** from the MINITAB folder of the student data disk.

Calculate the mean and median using the following commands.

File Edit Data Calc Stat Graph Editor Tools Window Help

Choose: **Calc > Column Statistics**

Calculator...
Column Statistics...
Row Statistics...
Standardize...
Make Patterned Data
Make Indicator Variables...
Set Base...
Random Data
Probability Distributions
Matrices

Select: **Median**
Enter: Input variable: **C2**
Click: **OK**

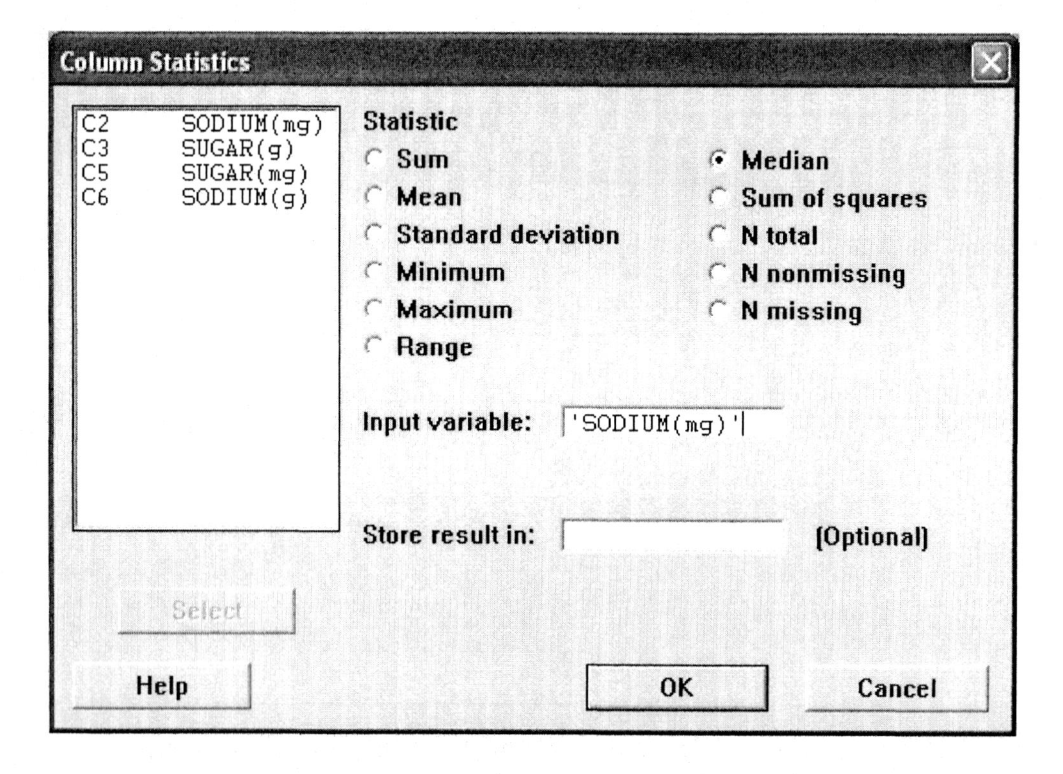

.The results are displayed in the Session Window.

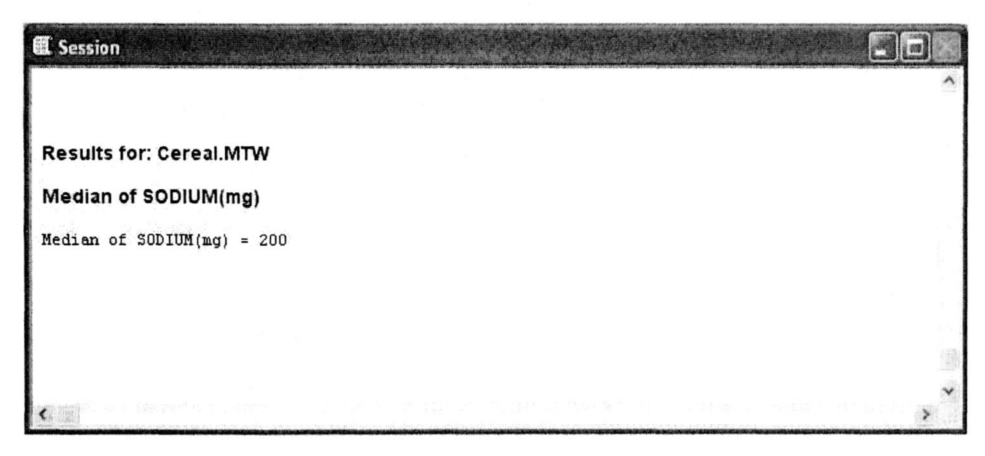

Use similar commands to find the mean:
Choose: **Calc > Column Statistics**
Select: **Mean**
Click: **OK**

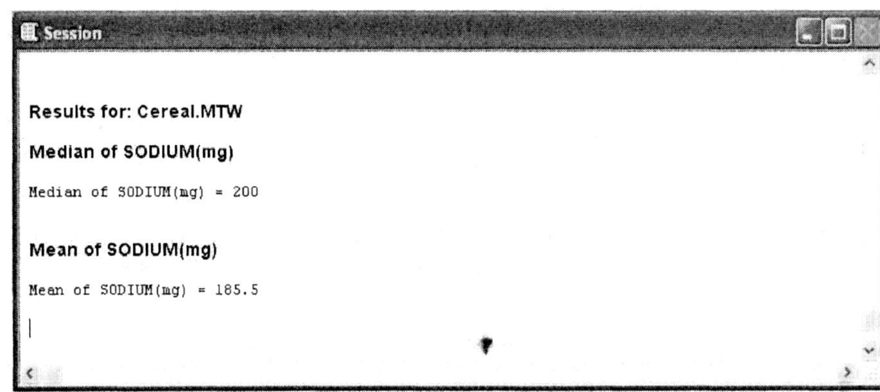

If you are interested in a variety of statistics, including median and mean, these values can be found more easily using the following:

Choose: **Stat > Basic Statistics > Display Descriptive Statistics**

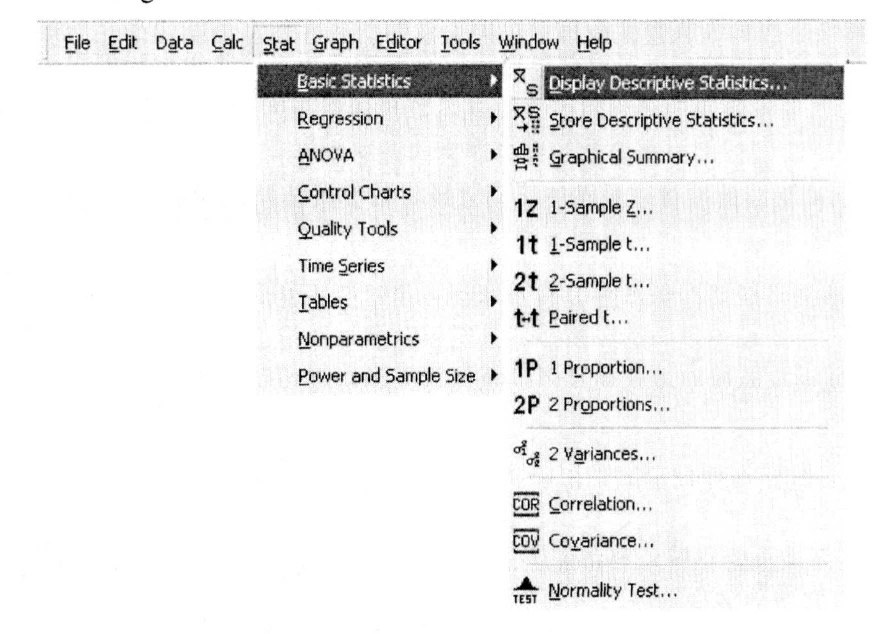

Enter: Variables: **C2**
Click: **OK**

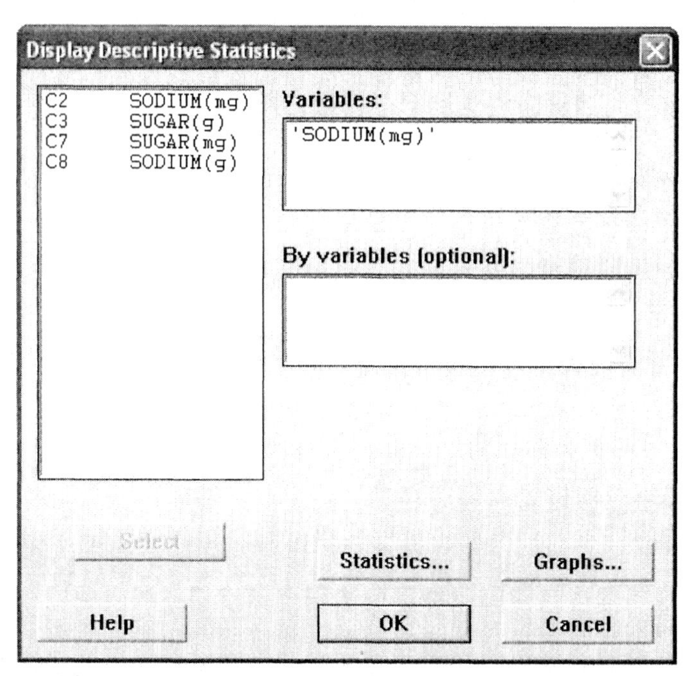

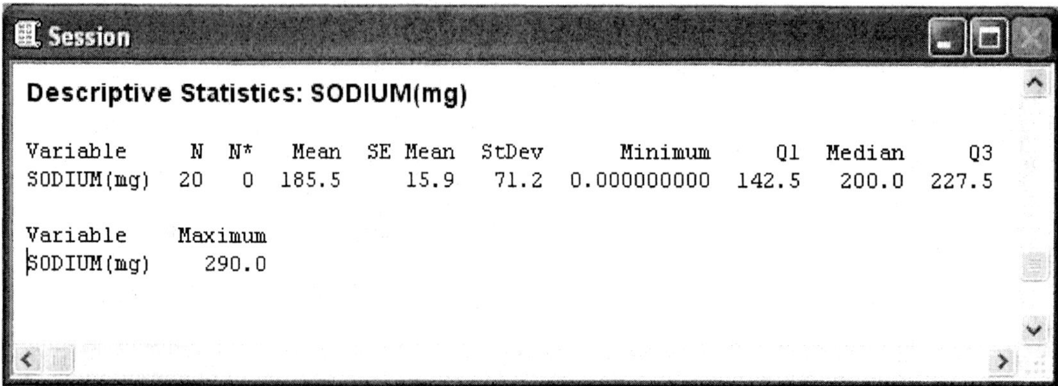

The only values we are interested in, at this point, are mean and median.

We can also find values by entering a formula and storing the result in a column. For example, the midrange = (high + low)/2. To do this in MINITAB we would do the following:

Select: **Calc>Calculator**

File Edit Data Calc Stat Graph Editor Tools Window Help

Calculator...
Column Statistics...
Row Statistics...
Standardize...
Make Patterned Data
Make Indicator Variables...
Set Base...
Random Data
Probability Distributions
Matrices

Store Result In: **midrange**
Type in the expression: **(MAX(C2) + MIN(C2)) /2**

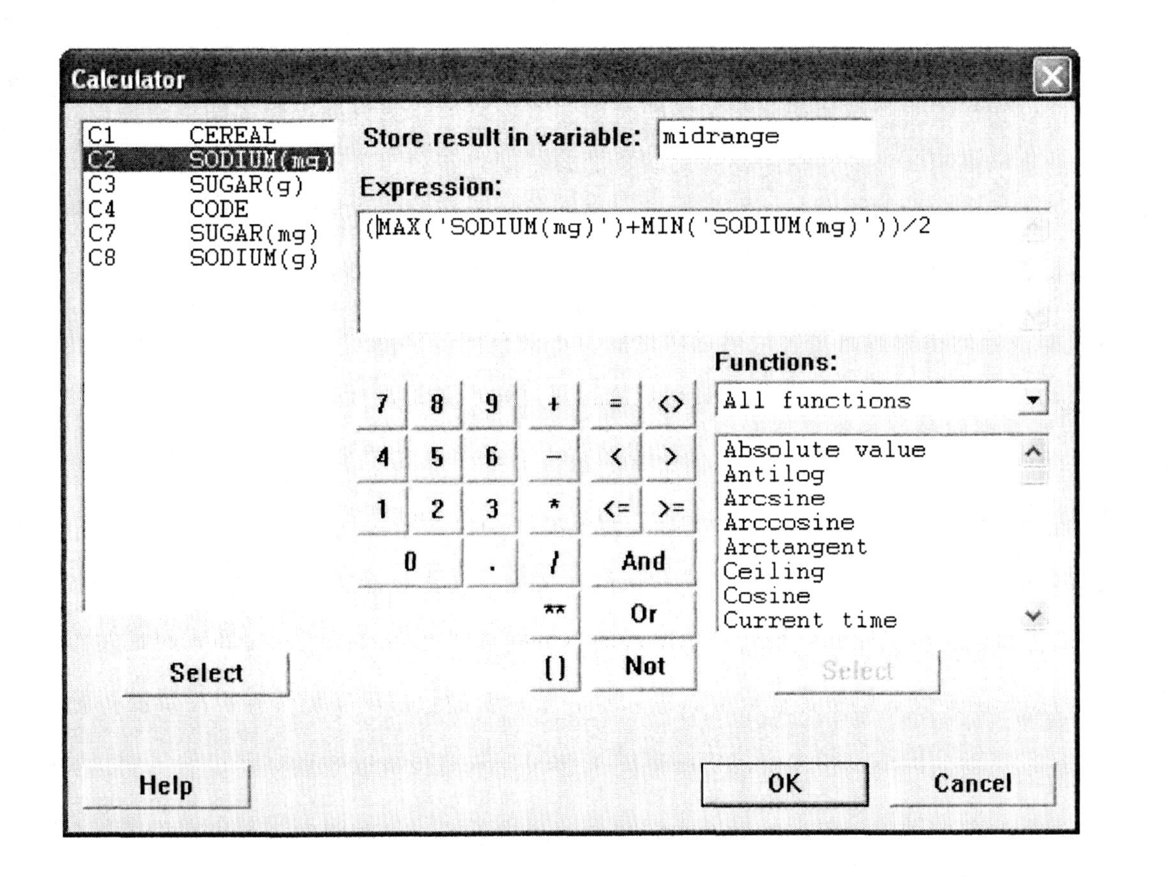

There is now a new column in the worksheet named midrange, which contains the result of the expression.

↓	C1-T	C2	C3	C4-T	C5	C6	C7	C8	C9
	CEREAL	SODIUM(mg)	SUGAR(g)	CODE			SUGAR(mg)	SODIUM(g)	midrange
1	FMiniWheats	0	7	A			7000	0.000	145
2	ABran	260	5	A			5000	0.260	
3	AJacks	125	14	C			14000	0.125	
4	CCrunch	220	12	C			12000	0.220	
5	Cheeros	290	1	C			1000	0.290	
6	CTCrunch	210	13	C			13000	0.210	
7	CFlakes	290	2	A			2000	0.290	
8	RBran	210	12	A			12000	0.210	
9	COakBran	140	10	A			10000	0.140	
10	Crispix	220	3	A			3000	0.220	

Chapter 2 Exercise 2.31 More On CO_2 Emissions
Mean and Median

Enter the data (country and million metric tons of carbon equivalent) into the worksheet.

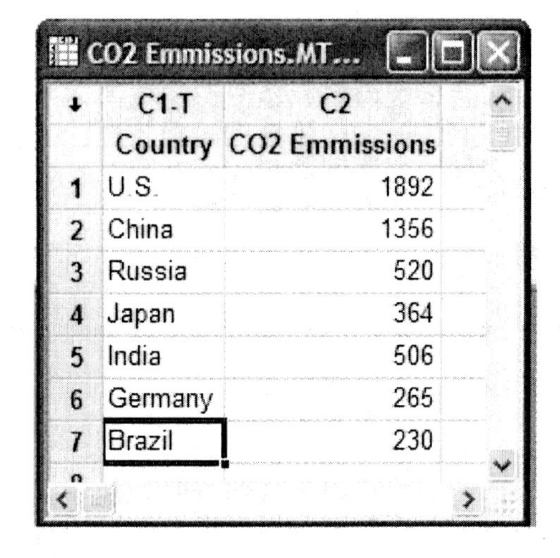
CO2 Emmissions.MT...

↓	C1-T	C2
	Country	CO2 Emmissions
1	U.S.	1892
2	China	1356
3	Russia	520
4	Japan	364
5	India	506
6	Germany	265
7	Brazil	230

Calculate the mean and median using the following commands.

Choose: **Calc > Column Statistics**
Select: **Mean** and then **Median**
Enter: Input variable: **C2**
Click: **OK**

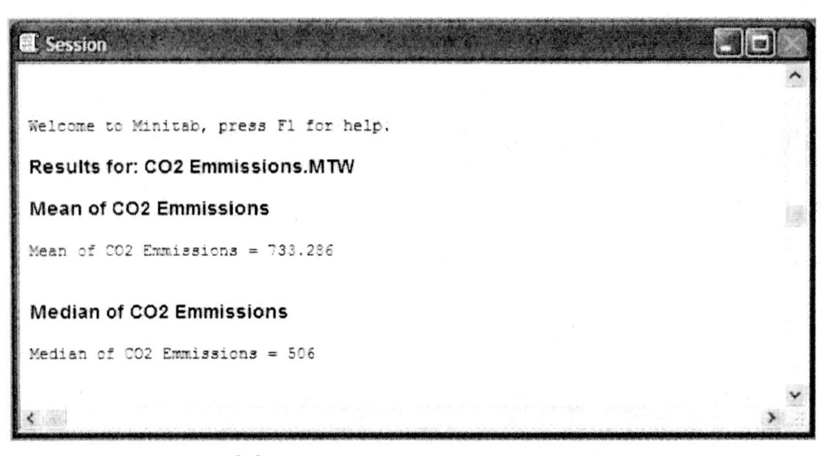
Session

Welcome to Minitab, press F1 for help.

Results for: CO2 Emmissions.MTW

Mean of CO2 Emmissions

Mean of CO2 Emmissions = 733.286

Median of CO2 Emmissions

Median of CO2 Emmissions = 506

Chapter 2 Example 15
Describing Female College Student Heights

Empirical Rule

When we open the data file **heights** on the text CD, we find that the data includes male (indicated by 0) and females(indicated by 1). In this example we are only concerned with the heights of the females.

heights.MTW ***

→	C1	C2	C3	C4	C5
	HEIGHT	GENDER			
1	70.0	0			
2	75.0	0			
3	67.0	0			
4	67.0	1			
5	73.0	0			
6	65.0	0			
7	73.0	0			
8	70.0	0			

To generate the histogram of female student height data only, we must specify that the data to be used should be taken from only those rows where the gender is a 1.

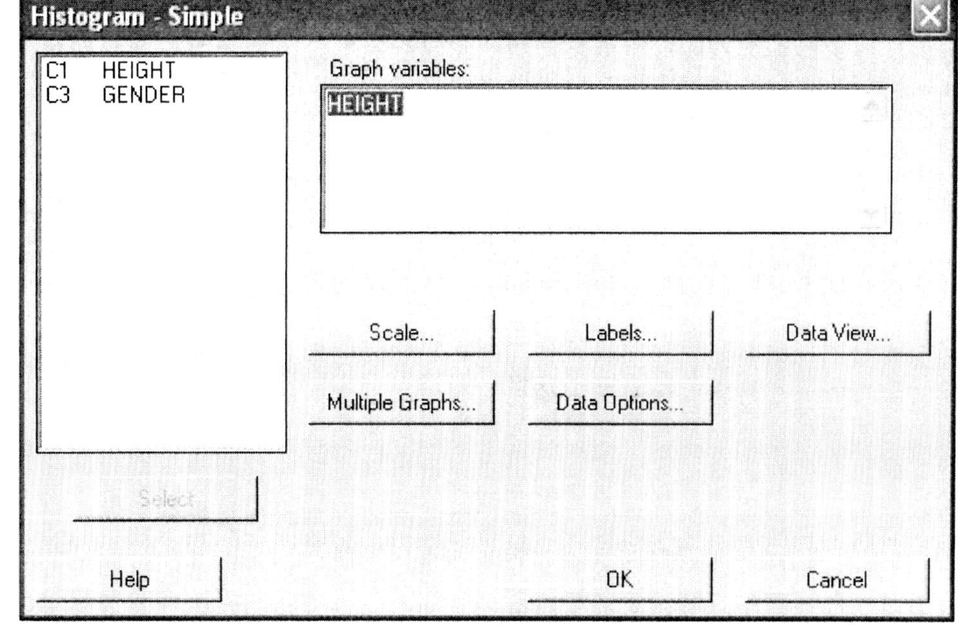

This can be accomplished by clicking the **Data Options** button within the **Graph>Histogram>Simple** task window.

In the **Data Options** window,
Click:**Specify Which Rows to Include > Rows that match > Condition** and within the Condition select **C3** and type "**=1**",

click **OK** for all three windows.

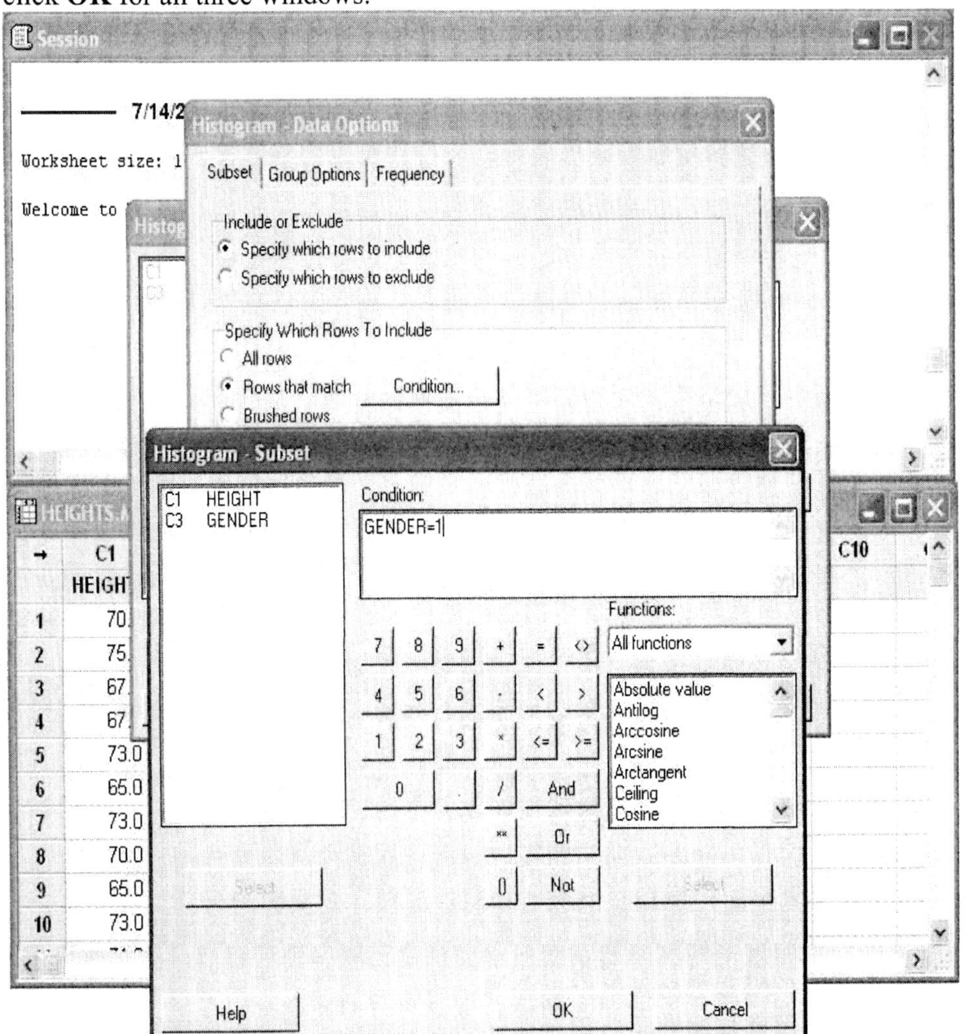

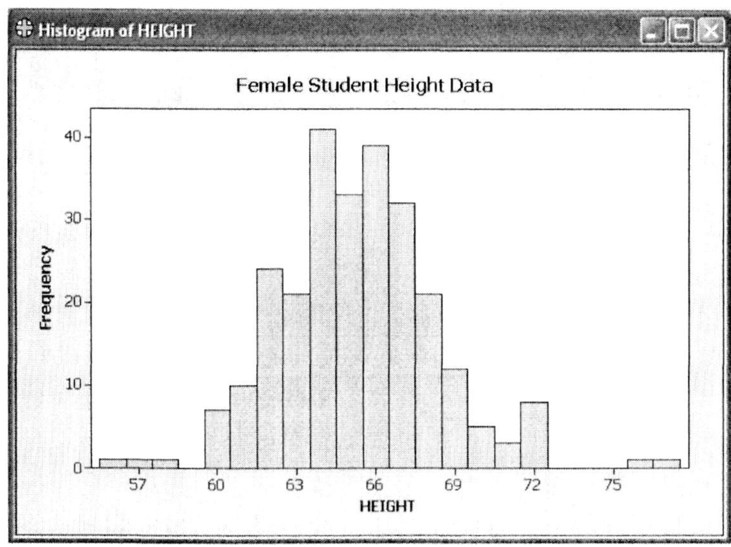

To generate the descriptive statistics for only the females, a similar selection is made from **Stat> Basic Statistics > Display Descriptive Statistics** within the By Variables box.

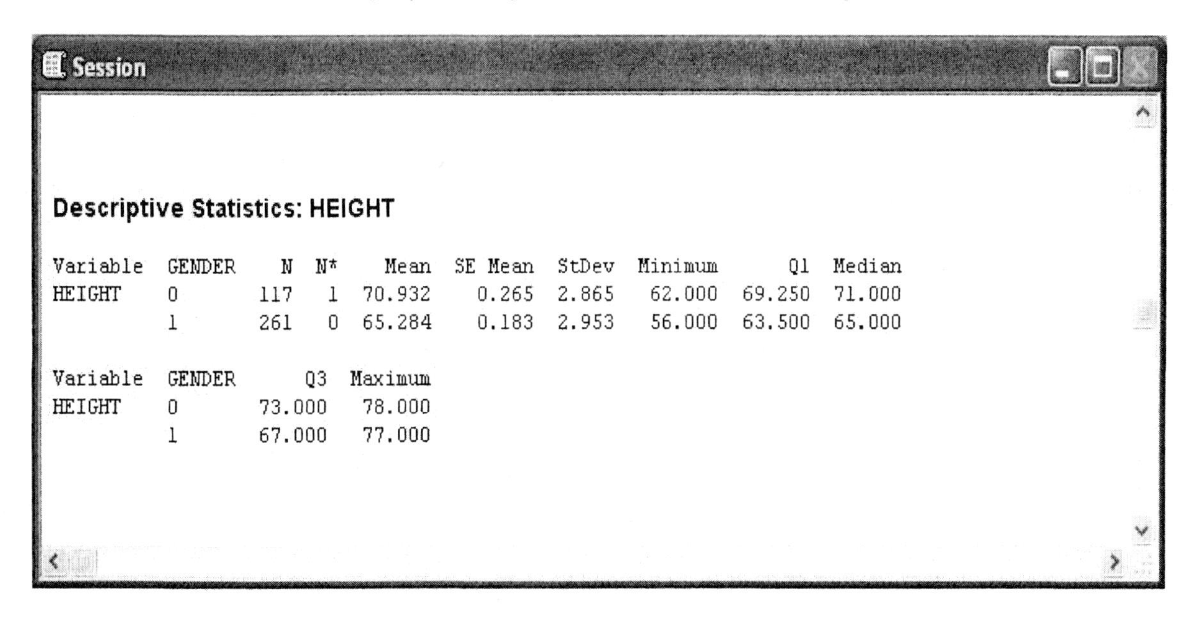
Session

Descriptive Statistics: HEIGHT

Variable	GENDER	N	N*	Mean	SE Mean	StDev	Minimum	Q1	Median
HEIGHT	0	117	1	70.932	0.265	2.865	62.000	69.250	71.000
	1	261	0	65.284	0.183	2.953	56.000	63.500	65.000

Variable	GENDER	Q3	Maximum
HEIGHT	0	73.000	78.000
	1	67.000	77.000

Notice this generates basic statistics for both the males and the females indicated by GENDER.

Now, to use MINITAB to count the number of observations within each of the ranges generated in the example, we will sort the data to facilitate the counting. Click **Data > Sort**

Enter both C1 and C3 into the **Sort Column(s)**, and C1 into the **By Column**. This way the correct gender identifier remains with the associated height. Also indicate into which **Column(s) of the current worksheet** you would like the sorted lists to be placed. Since we wish to handle the females separately from the males, we'll sort by both the Gender and the Height.

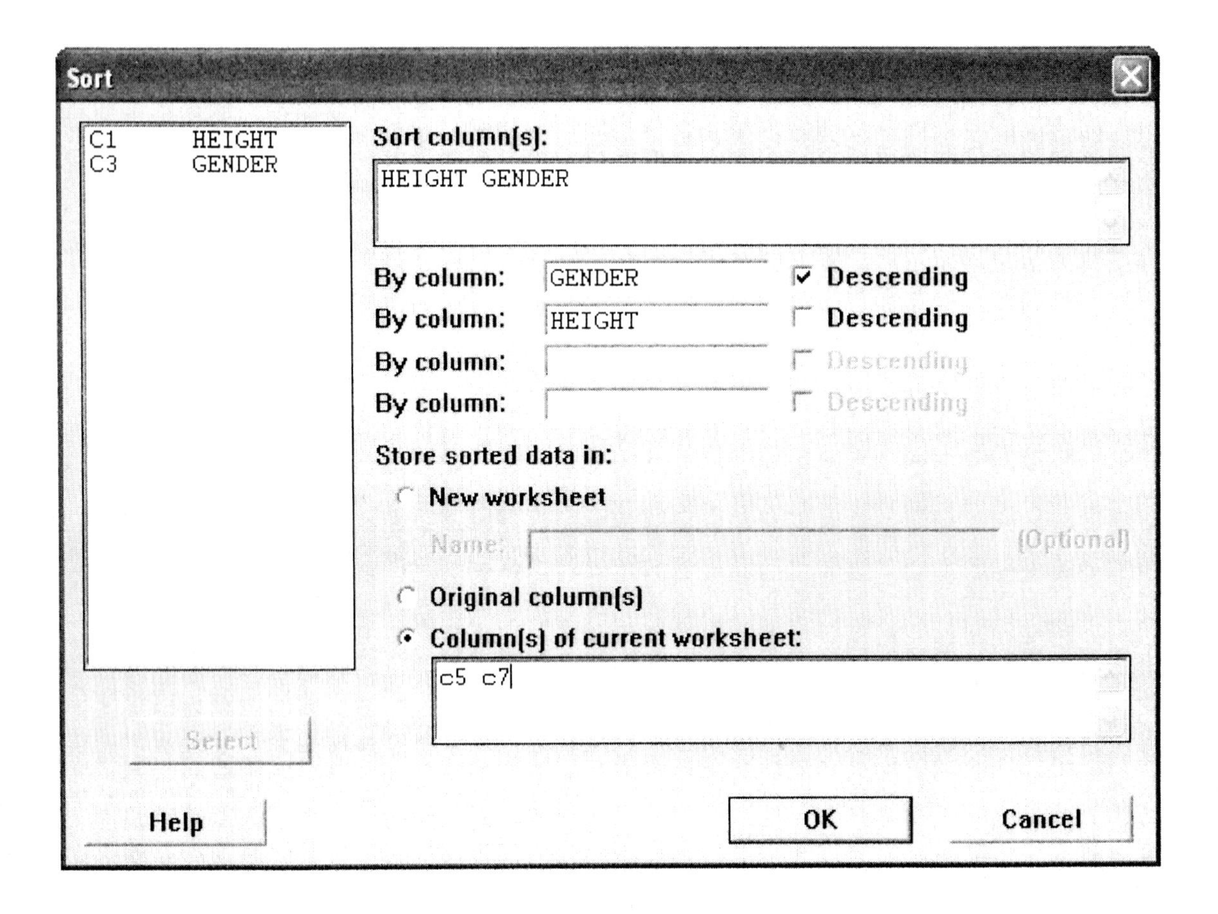

Now, with the lists sorted, counting the number of observations of female heights within each range is easier.

HEIGHTS.MTW ***

→	C1	C2	C3	C4	C5	C6	C7	C8	C9
	HEIGHT		GENDER						
43	64.0		1		62.0		1		
44	68.0		1		62.0		1		
45	63.0		1		62.5		1		
46	71.0		0		63.0		1		
47	68.0		1		63.0		1		
48	70.0		0		63.0		1		
49	60.0		1		63.0		1		
50	65.0		1		63.0		1		
51	62.0		1		63.0		1		
52	67.0		1		63.0		1		

The heights between 62.3 and 68.3 lie from row 45 to row 231, which is 187 values. 72% of all female heights.

The heights between 59.3 and 71.3 lie from row 4 to row 251, which is 248 values. 95% of all female heights.

The heights between 56.3 and 74.3 lie from row 2 to row 259,which is 378 values. 99% of all female heights.

Chapter 2 Exercise 2.61 EU data file
Empirical Rule

Open the **european_union_unemployment** data file on the text CD. Generating the histogram, we see that the distribution is skewed, and not bell shaped.

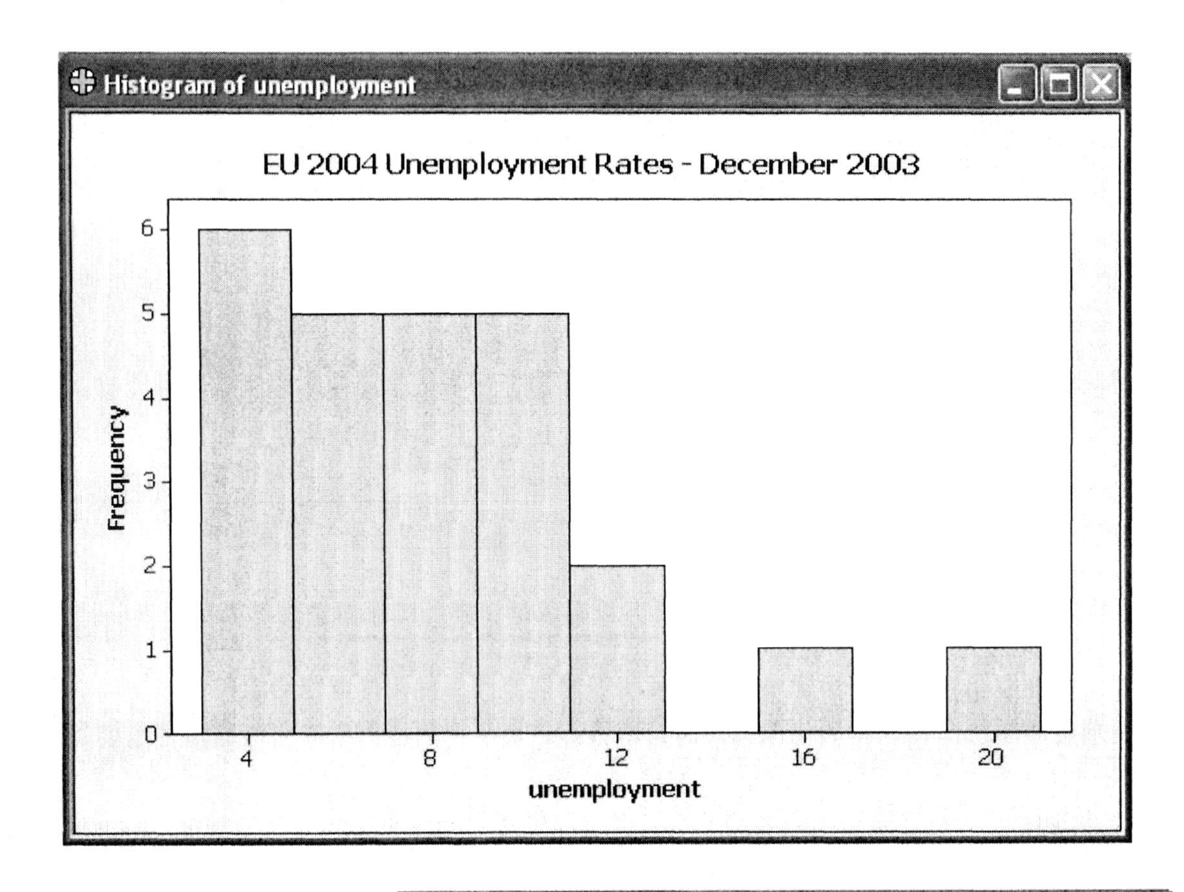

Generate the basic statistics.

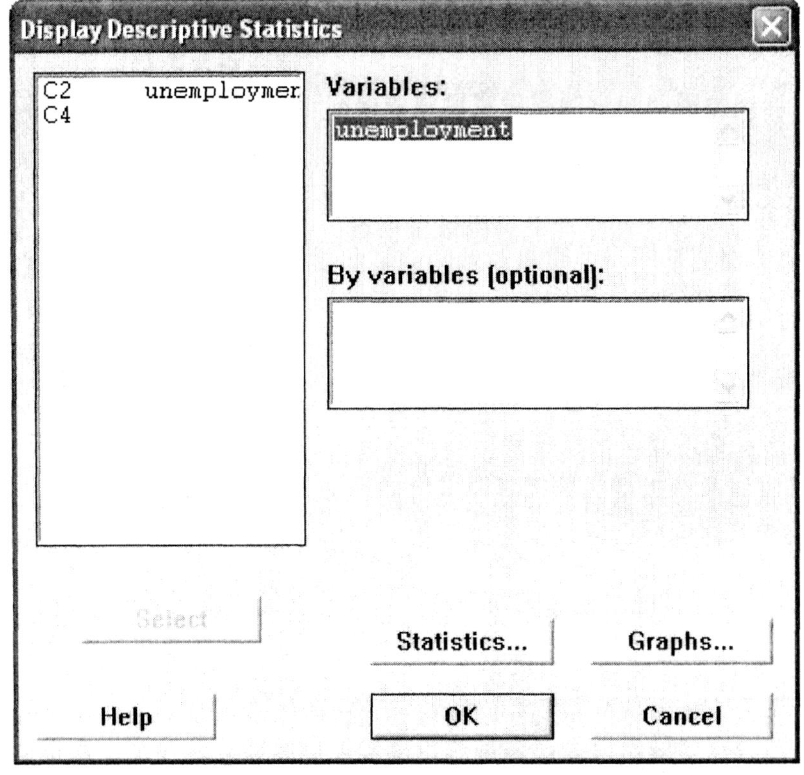

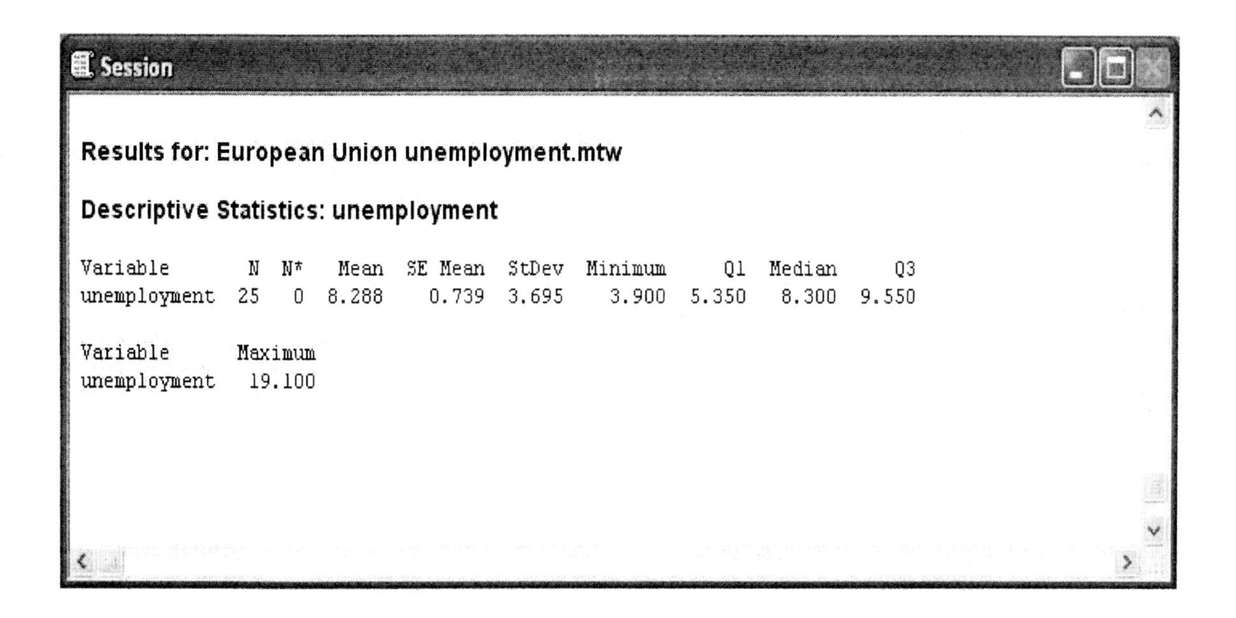

Session

Results for: European Union unemployment.mtw

Descriptive Statistics: unemployment

```
Variable        N  N*   Mean  SE Mean  StDev  Minimum     Q1  Median     Q3
unemployment   25   0  8.288    0.739  3.695    3.900  5.350   8.300  9.550

Variable      Maximum
unemployment   19.100
```

To determine the number of data values that fall in the ranges
$\overline{x} \pm 1 \cdot s = (4.593, 11.983), \overline{x} \pm 2 \cdot s = (.898, 15.678), \overline{x} \pm 3 \cdot s = (-2.797, 19.373)$,
sort the unemployment rates

European Union unemployment.mtw ***

↓	C1-T	C2	C3	C4	C5-T	C6	C7	C8	C9
	country	unemployment							
1	Belgium	8.3			Luxembourg		3.9		
2	Denmark	6.0			Netherlands		4.4		
3	Germany	9.2			Austria		4.5		
4	Greece	9.3			Ireland		4.6		
5	Spain	11.2			Cyprus		4.7		
6	France	9.5			United Kingdom		4.8		
7	Ireland	4.6			Hungary		5.9		
8	Italy	8.5			Denmark		6.0		
9	Luxembourg	3.9			Sweden		6.0		
10	Netherlands	4.4			Slovenia		6.4		

We determine that there are 20, 23, and 25 data values within the three ranges, respectively, which is 80%, 92 %, and 100%. Notice that theses percentages don't even begin to match the

68%, 95% and 99.7% of the Empirical Rule. This provides further evidence that the distribution of unemployment rates is not bell shaped.

Chapter 2 Example 16 What Are the Quartiles for the Cereal Sodium Data?
Quartiles

Open the **CEREAL** data file on the text CD.

In this example, we wish to determine the quartiles for the sodium values. Click **Stats > Basic Statistics > Display Descriptive Statistics > Statistics** and select First quartile, Median, and Third Quartile. Click **OK** for each window.

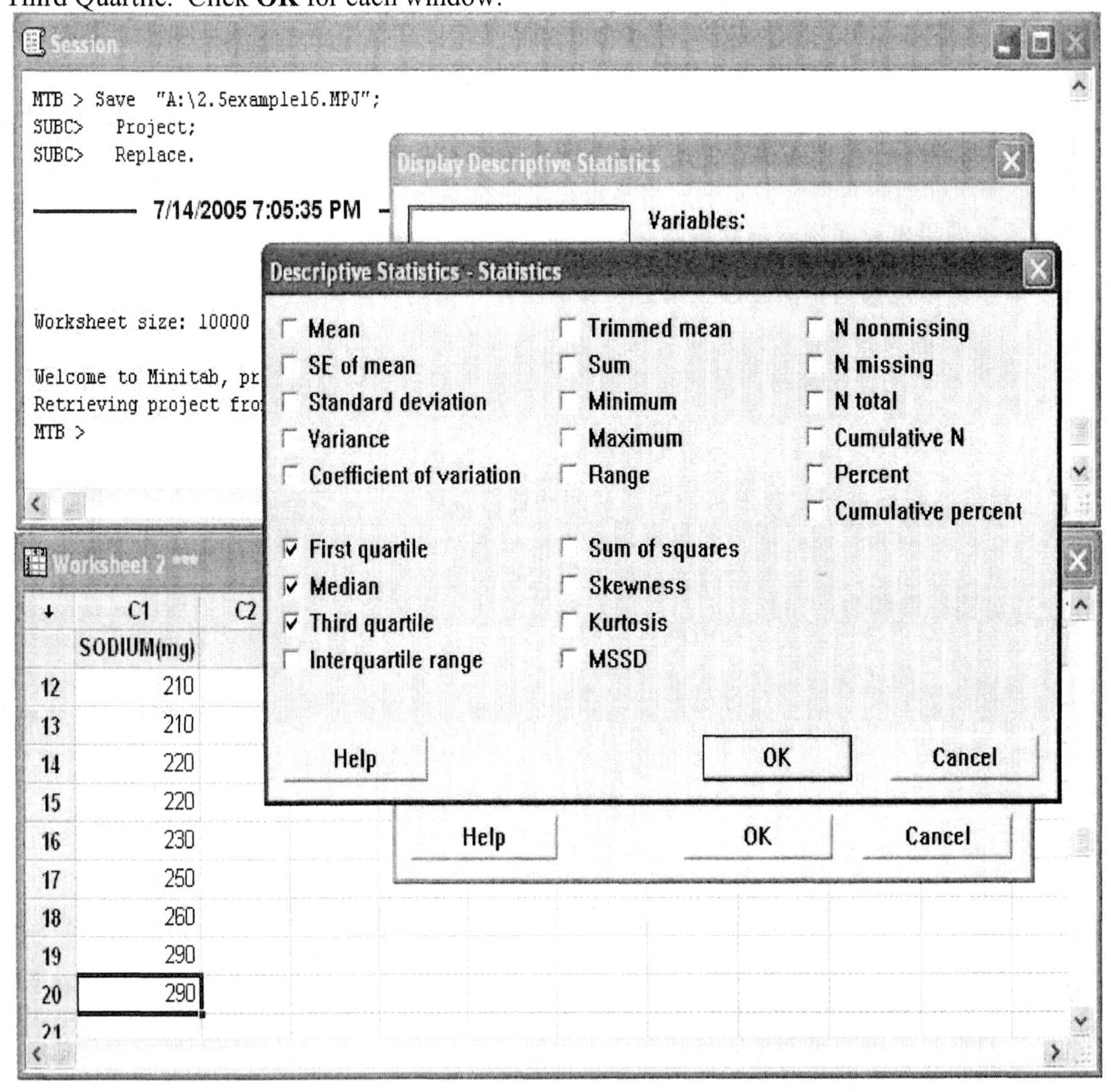

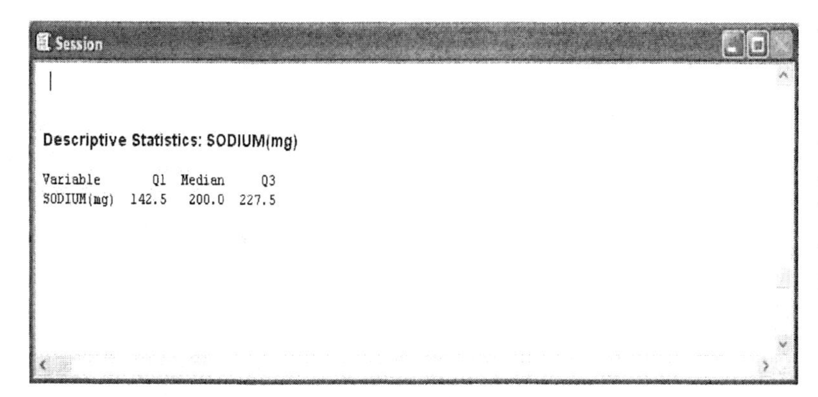

Note that the results differ from those in the text. Not all statistical programs use the same method for determining the quartiles. You should explore how and why MINITAB generates Q1=142.5 and Q3 = 227.5

Chapter 2 Example 17 Box Plot for the Breakfast Cereal Sodium Data
Box Plots

Open worksheet **CEREAL** from the MINITAB folder of the data disk. To construct a boxplot for the breakfast cereal sodium data, use the following commands:

Click: **Graph > Boxplot**

Select: **One Y/Simple > OK**

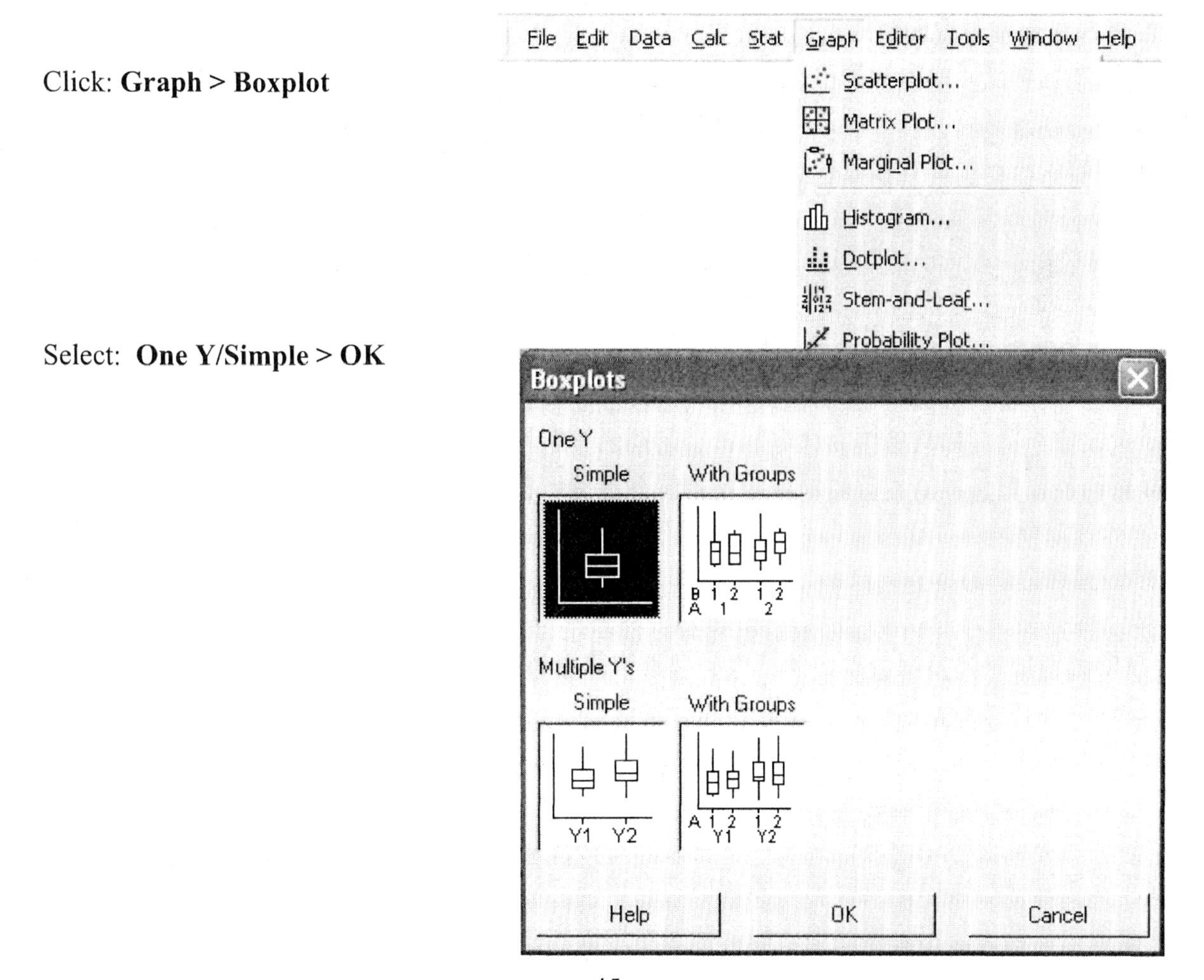

Enter:
Graph variable: **C2**
Click: **OK**

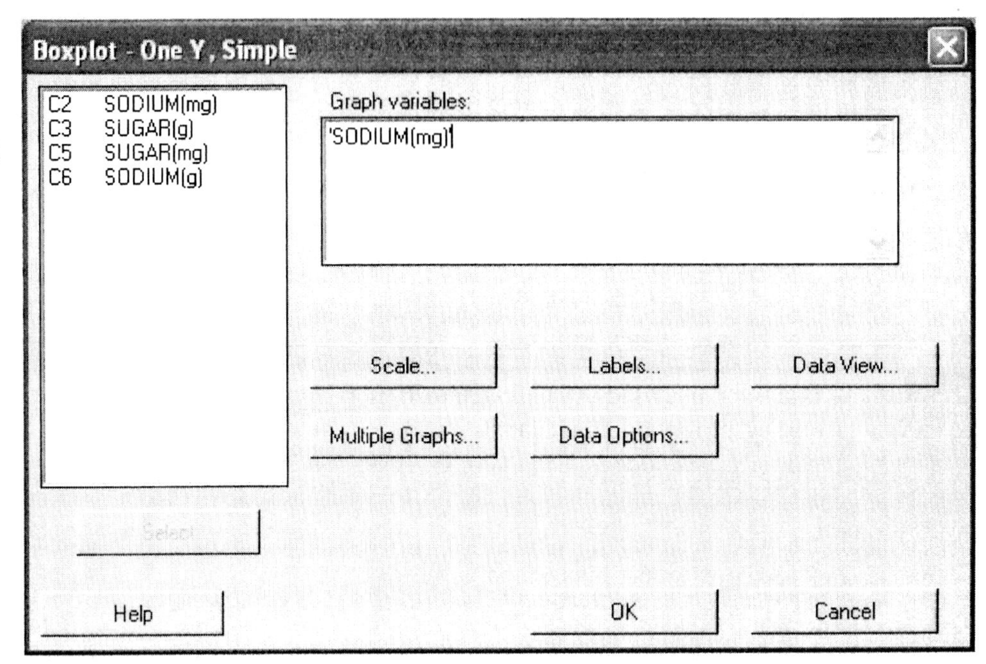

The default MINITAB boxplot plots vertically. This can be changed. Follow the commands above, and before clicking the second OK ,
Click: **Scale**
Select: **Transpose value and category scales**
Click: **OK** twice

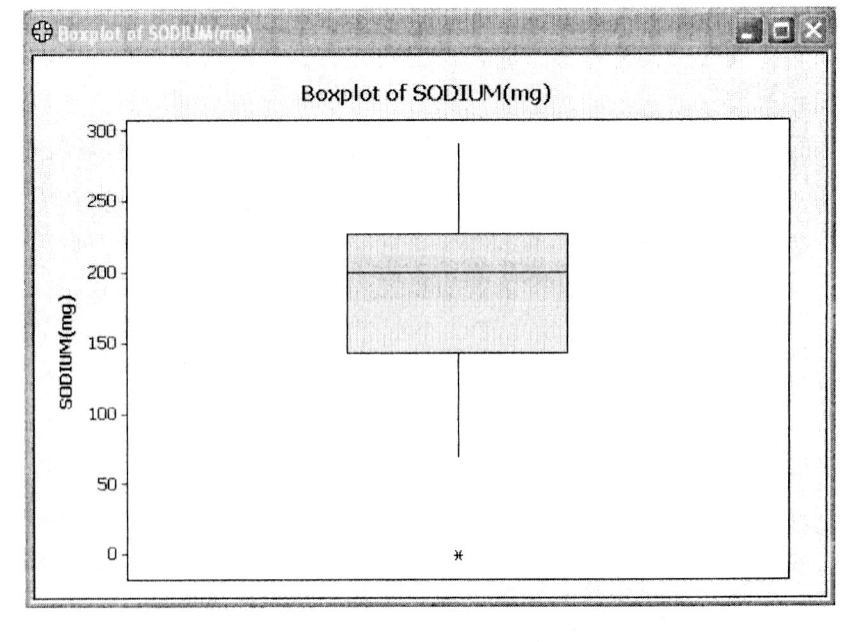

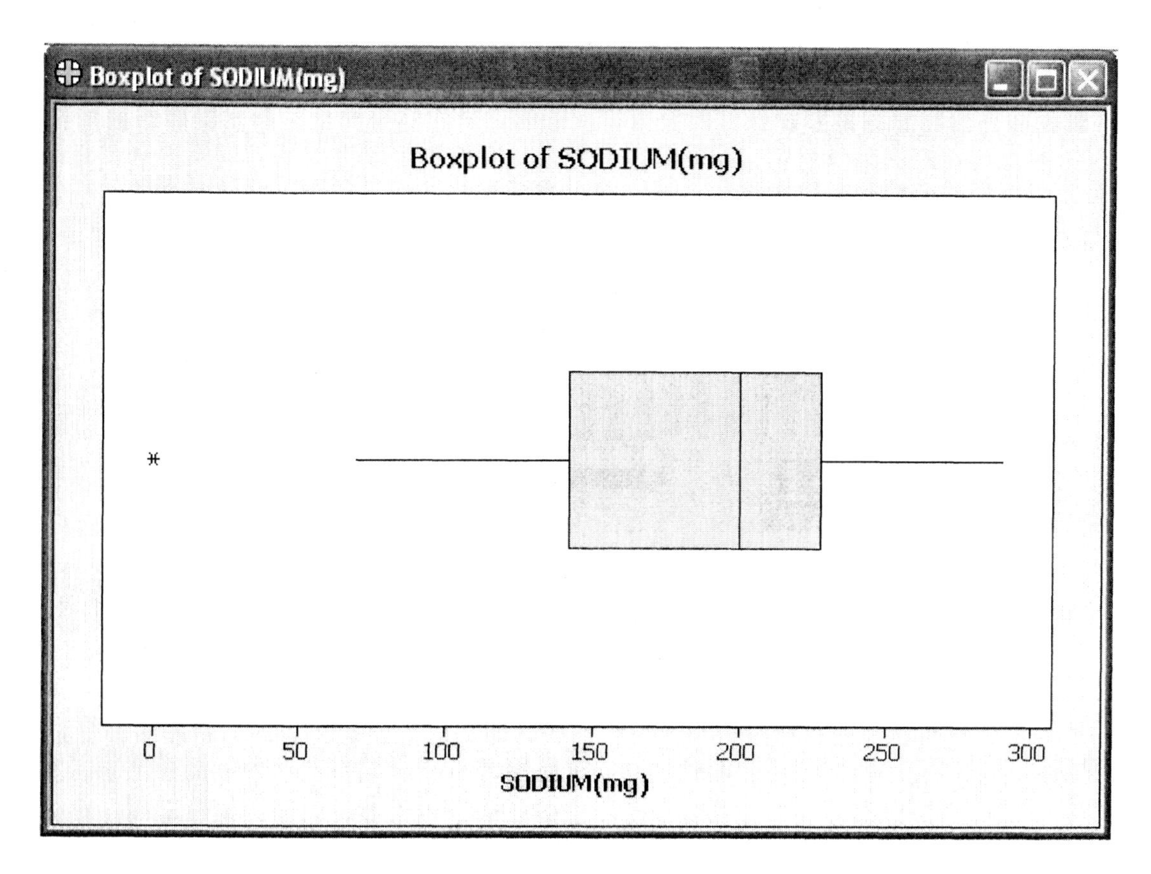

Chapter 2 Exercise 2.64 European Unemployment
Quartiles

Enter the data into C1 and C2 of the Data Window. Be sure to label the columns.

Click **Stats > Basic Statistics > Display Descriptive Statistics > Statistics** and select First quartile, Median, and Third Quartile. Click **OK** for each window. The selected descriptive statistics will be displayed in the Session Window.

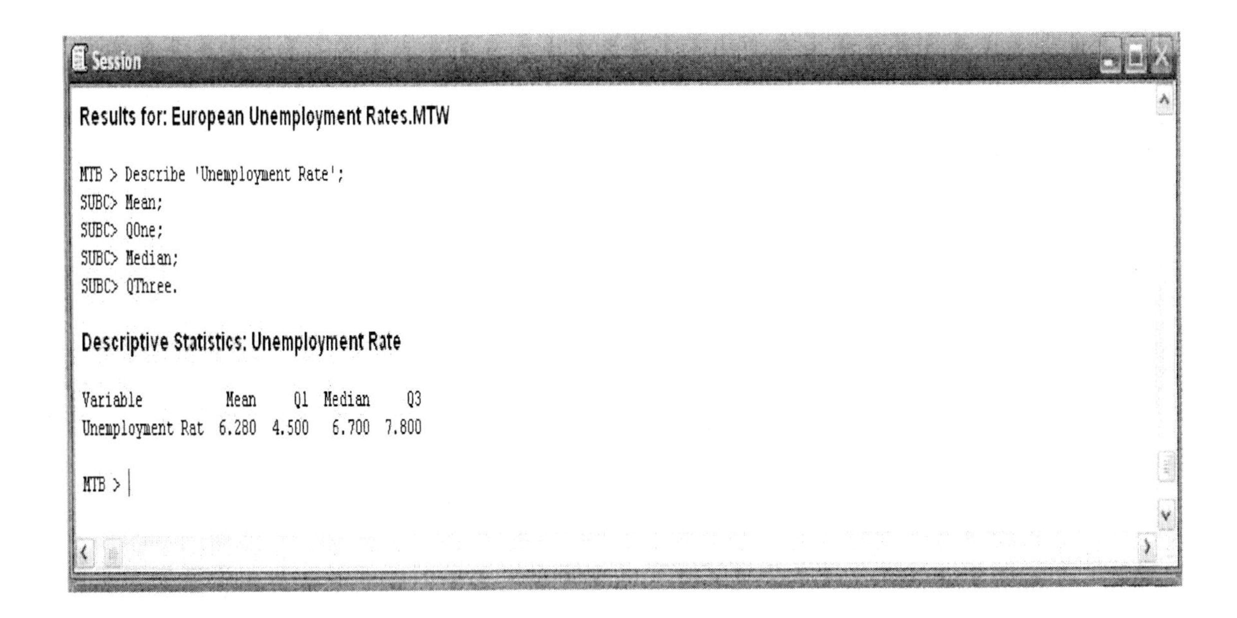

Chapter 2 Exercise 2.77 European Union Unemployment Rates
Box Plots

Enter the data from exercise 2.64 into C1 and C2.

Click: **Graph > Boxplot > Simple**
Click: **OK**
Enter: Graph variable: **C2**
Select: **Scale**
Click: **Transpose value and category scales**
Click: **OK OK**

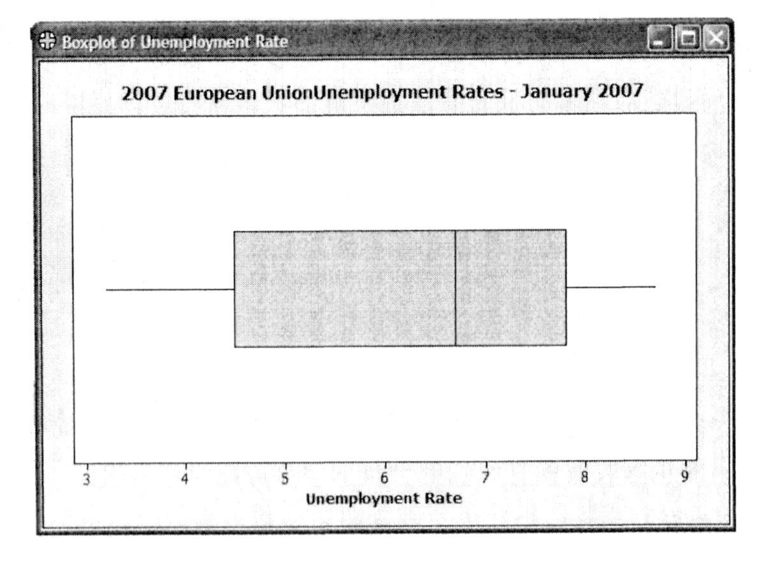

Chapter 2 Exercise 2.81 Florida Students Again
Box Plots

Open worksheet **fla_student_survey** from the MINITAB folder of the data disk.

Click: **Graph > Boxplot > Simple**
Click: **OK**
Enter: Graph variable: **C8**
Select: **Scale**
Click: **Transpose value and category scales**
Click: **OK**
Click: Mutiple Graphs
Select: type 'gender' under 'By variables with groups in separate panels
Click: OK

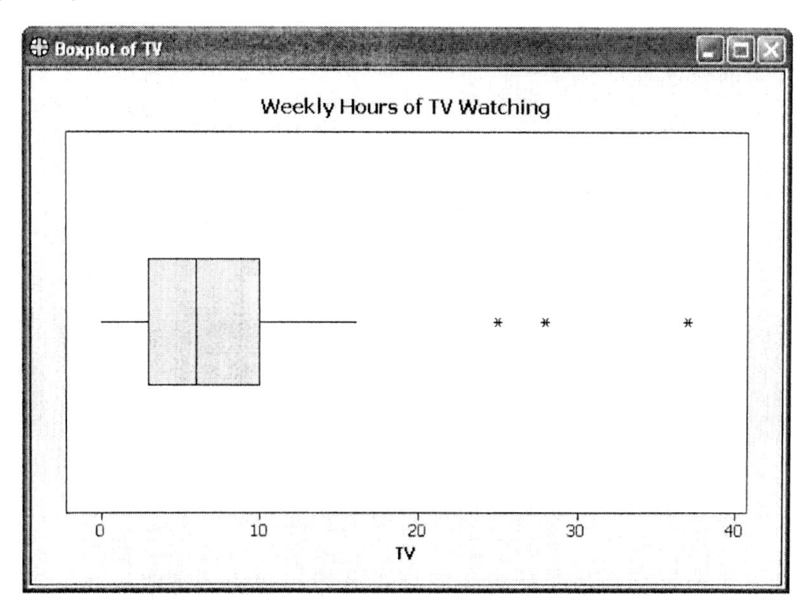

Chapter 2 Exercise 2.82 Females or males watch more T.V.?

Open worksheet **fla_student_survey** from the MINITAB folder of the data disk.

Click: **Graph > Boxplot > With Groups**
Click: **OK**
Enter: Graph variable: **C8**
Enter: Categorical variables for grouping: **C2**
Select: **Scale**
Click: **Transpose value and category scales**
Click: **OK OK**

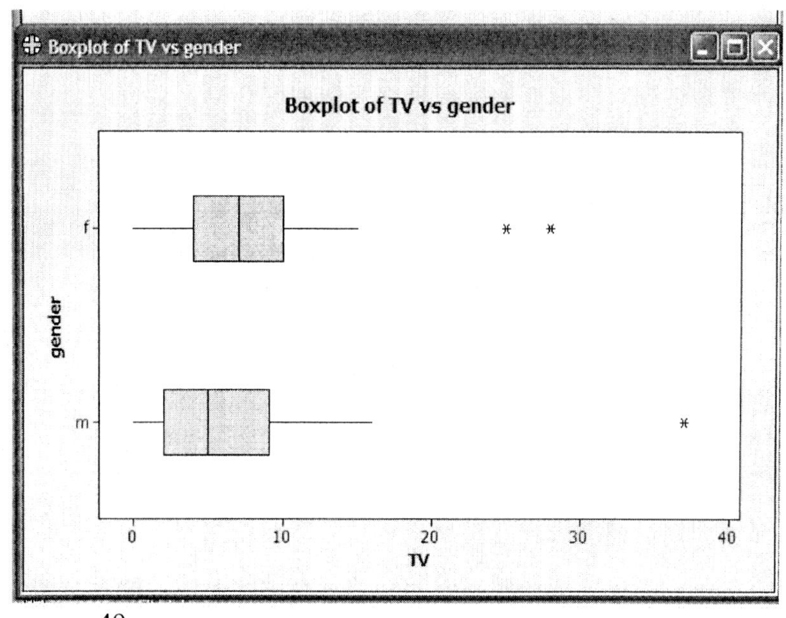

Chapter 2 Exercise 2.119a - Temperatures in Central Park
Constructing a Histogram

Open the worksheet **central_park_yearly_ temps** from the MINITAB folder of the data disk.

This time we will make the selection "With Fit"

Choose: **Graph > Histogram > With Fit > OK**
Select: **Label**
Enter: Title: **Central Park Annual Temperatures**
Select: **Frequency**
Click: **OK OK**

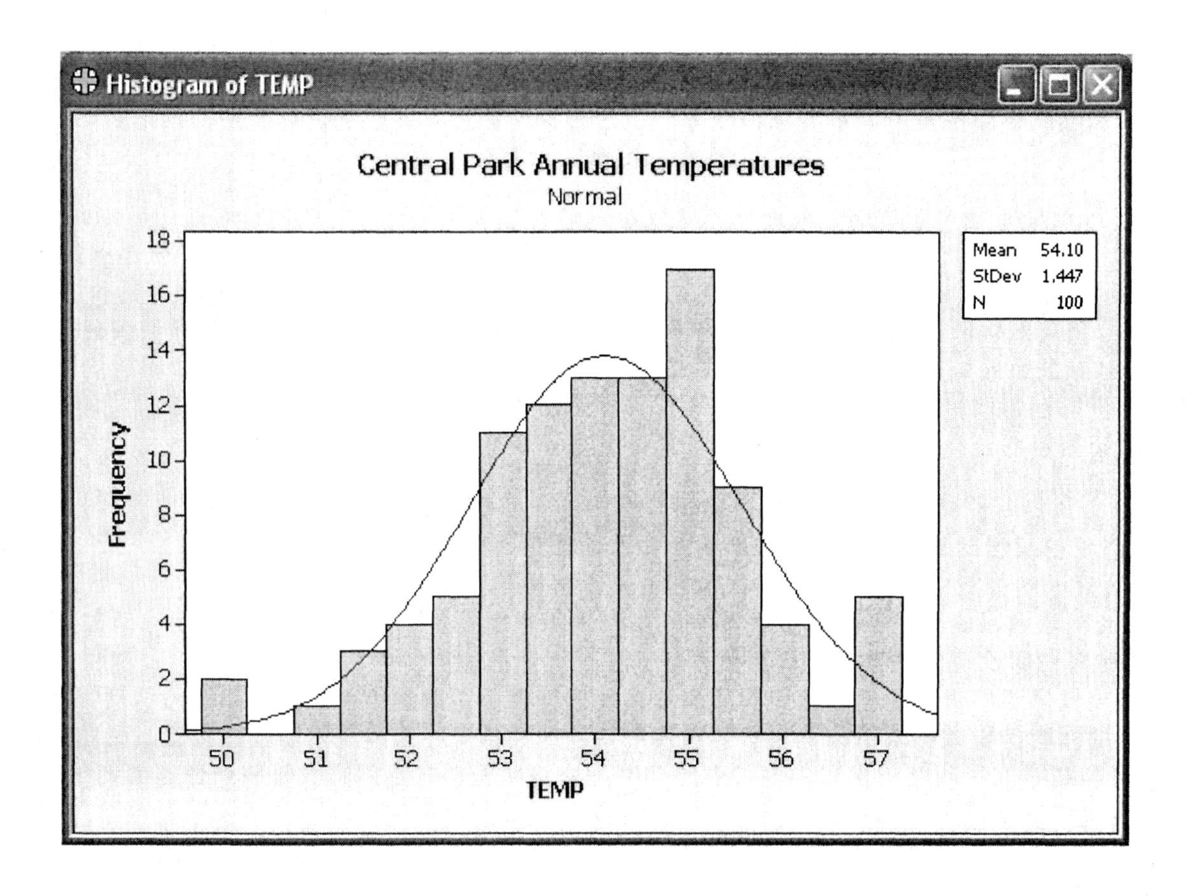

Chapter 2 Exercise 2.119b Temperatures in Central Park
Box Plot

Open worksheet **central_park_yearly_temps** from the MINITAB folder of the data disk.

Click: **Graph > Boxplot > Simple**
Click: **OK**
Enter: Graph variable: **C1**
Select: **Scale**
Click: **Transpose value and category scales**
Click: **OK OK**

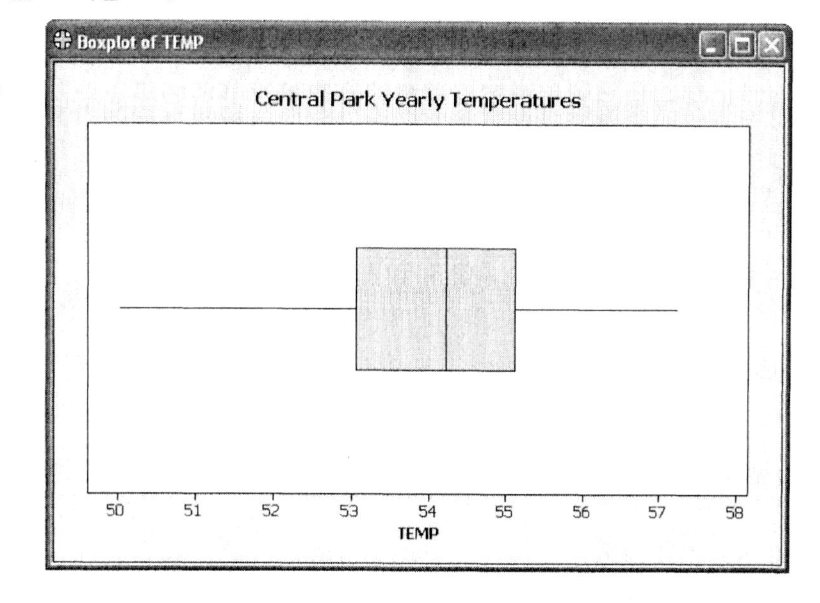

We can compare this to the histogram produced for part (a) of this exercise.

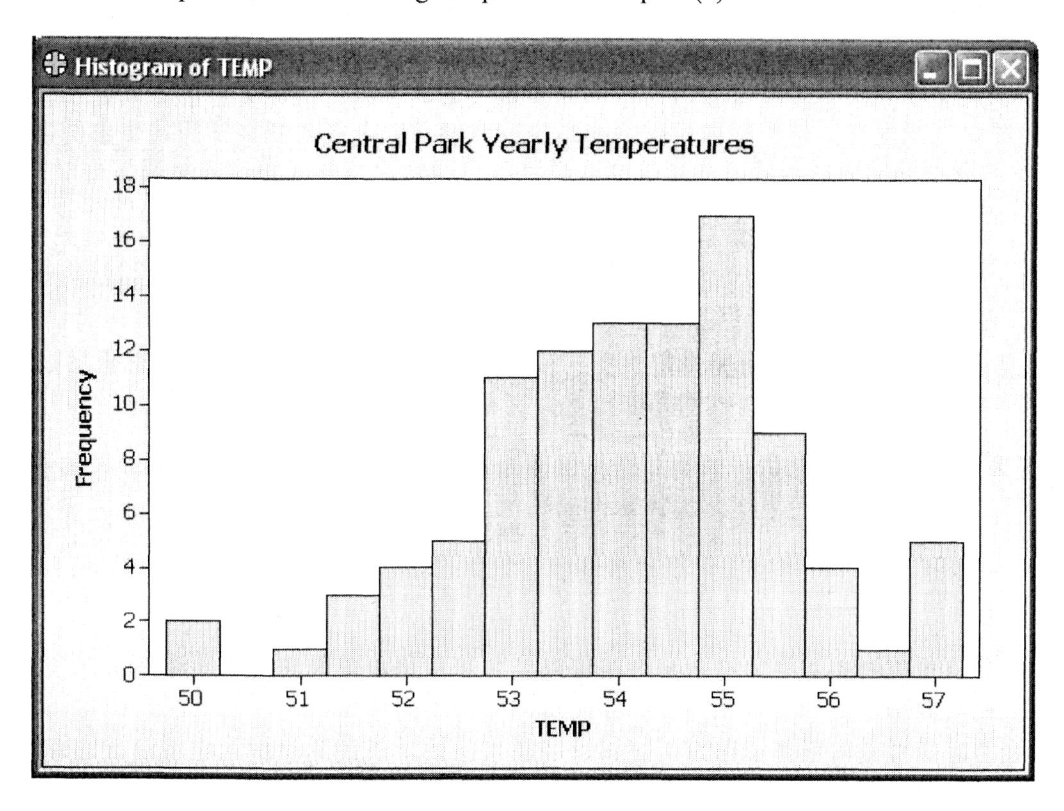

Chapter 2 Exercise 2.129 How Much is Spent on Haircuts? Side-by-Side Plots

Open worksheet **georgia_student_survey** from the MINITAB folder of the data disk. For this exercise we wish to compare how much males and females spend on a haircut. First note that gender is indicated in C2 as 0 for male and 1 for female. Many of the graphs illustrated in this chapter can be done as side-by-side plots, graphing with a separate plot for the males and the females, using the same scale, so that comparison of the distributions can be made.

georgia_student_survey.MTW ***

↓	C1	C2	C3	C4	C5	C6	C7	C8
	Height	Gender	Haircut	Job	Studytime	Smokecig	Dated	HSGPA
1	65	1	25	1	7.0	0	0	3.90
2	71	0	12	0	2.0	0	1	3.79
3	68	1	4	0	4.0	0	1	3.00
4	64	1	0	1	3.5	0	1	3.90
5	64	1	50	0	4.5	0	1	3.60
6	66	0	10	1	3.0	0	3	3.20

Click: **Graph > Boxplot (or Histogram or Dotplot) > One Y, With Groups**
Click: **OK**

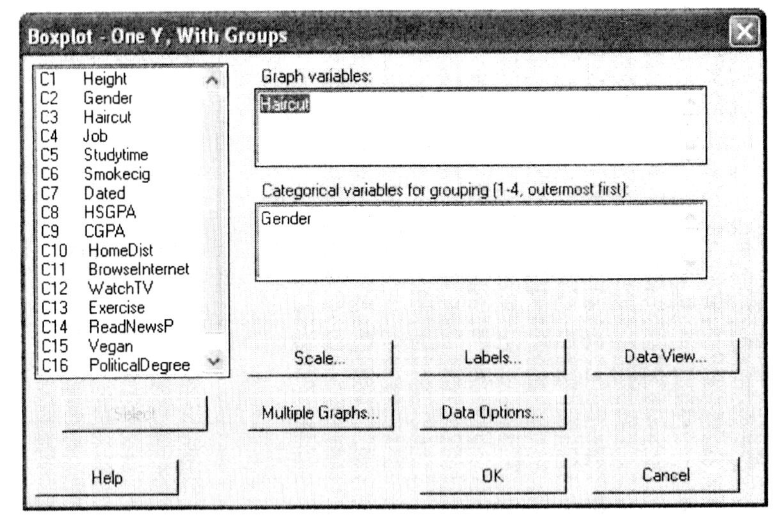

Select: Graph variable: **C3**
Select: Categorical variables for grouping: **G2**
Select: **Scale**
Click: **Transpose value and category scales**
Click: **OK OK**

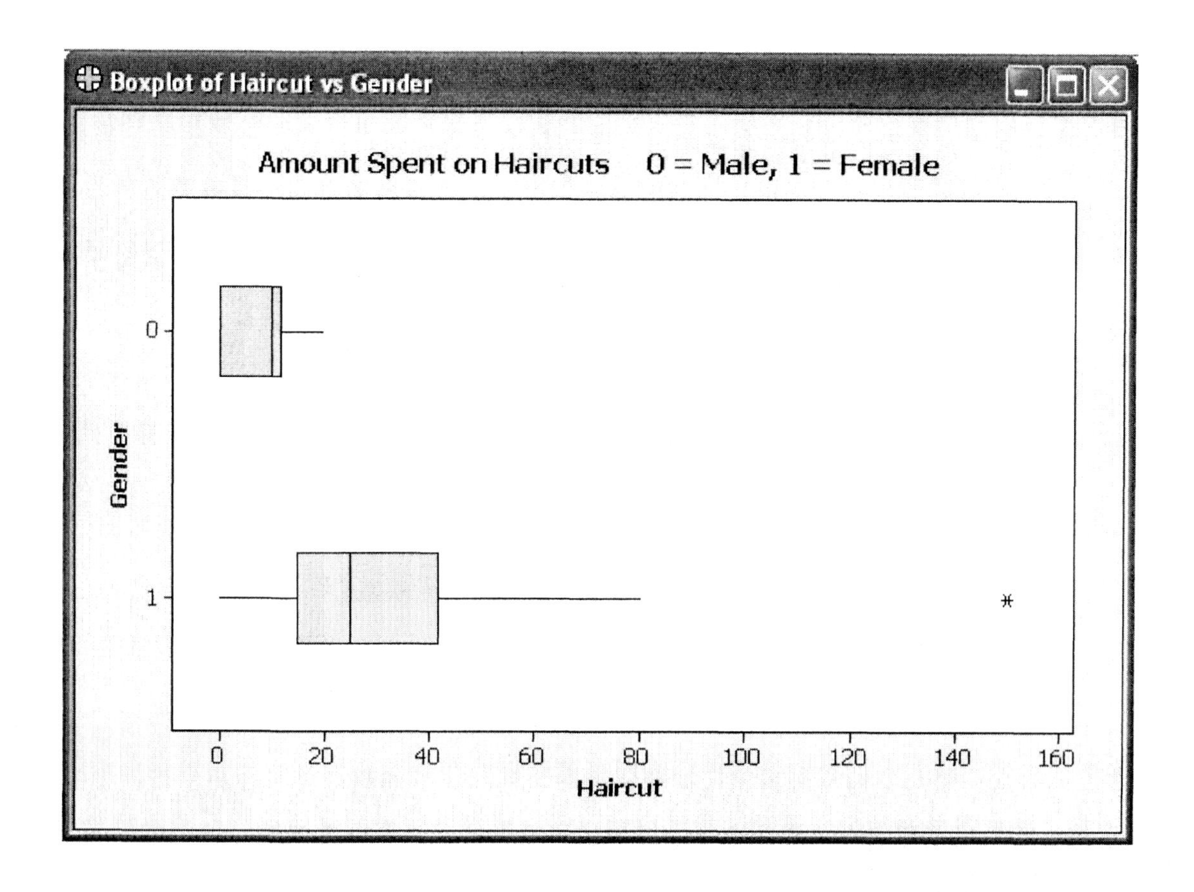

For comparing histograms, the results are more easily readable if we generate histograms based on gender in separate panels.

Click: **Graph > Histogram > Simple**
Click: **OK**
Click: **Labels**
Enter: Title: **an appropriate title**
Click: **OK**
Click: **Multiple Graphs**

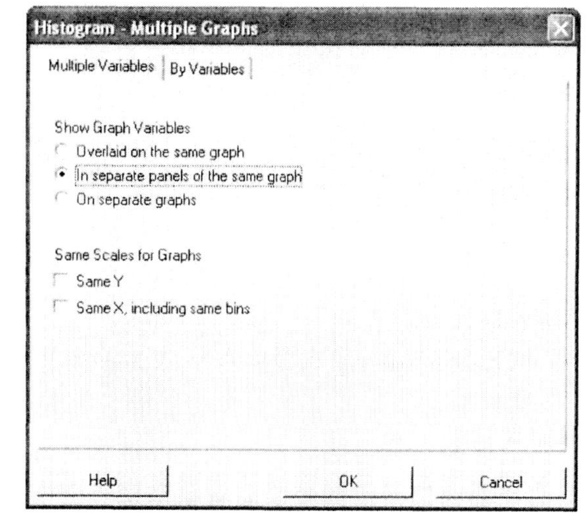

On the Multiple Variables tab
Select: **In separate panels of the same graph**

Click: the **By Variables** tab

Enter: By variables with groups in separate panels: **C2**

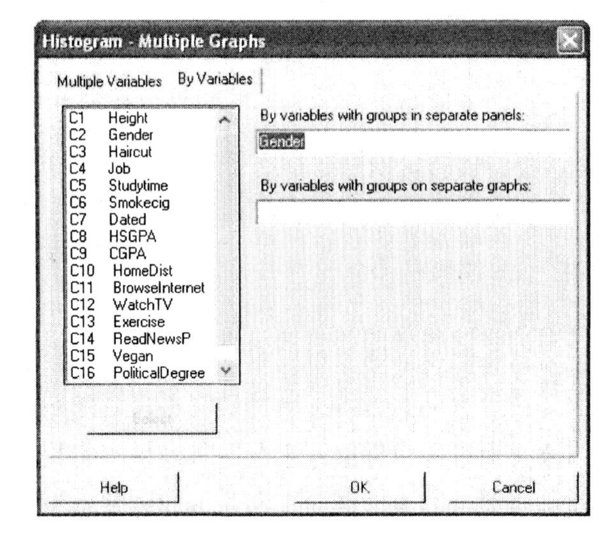

Then click **OK** for each window.

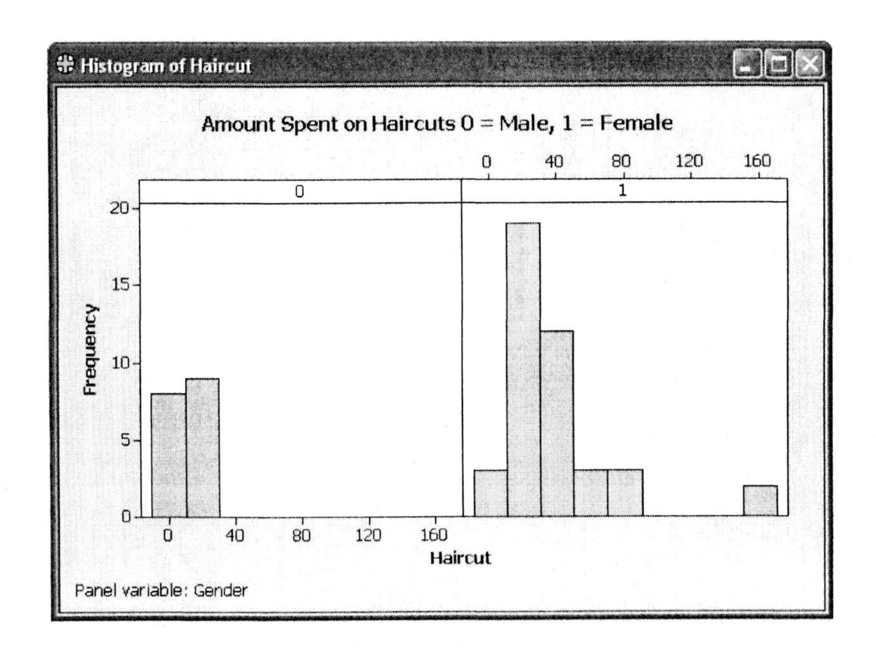

Chapter 3 Example 3 How Can We Compare Pesticide Residues For the Food Types Graphically?
Side-by-Side Bar Chart

Enter the conditional proportions (Table 3.2) into the worksheet.

Worksheet 1

↓	C1-T	C2	C3
	Food Type	Pesticide Present	Pesticide Not Present
1	Organic	0.23	0.77
2	Conventional	0.73	0.27

To generate a single bar graph that show side-by-side bars to compare the conditional proportion of pesticide residues in conventionally grown and organic foods:

Choose: **Graph > Bar Chart**
Select: Bars represent: **Values from a table**
Click: **Two-way table / Cluster > OK**
Enter: Graph Variables : **C2 C3**
Row labels: **C1**
Select: Table Arrangement: **Columns are outermost categories and rows are innermost**
Click: **OK**

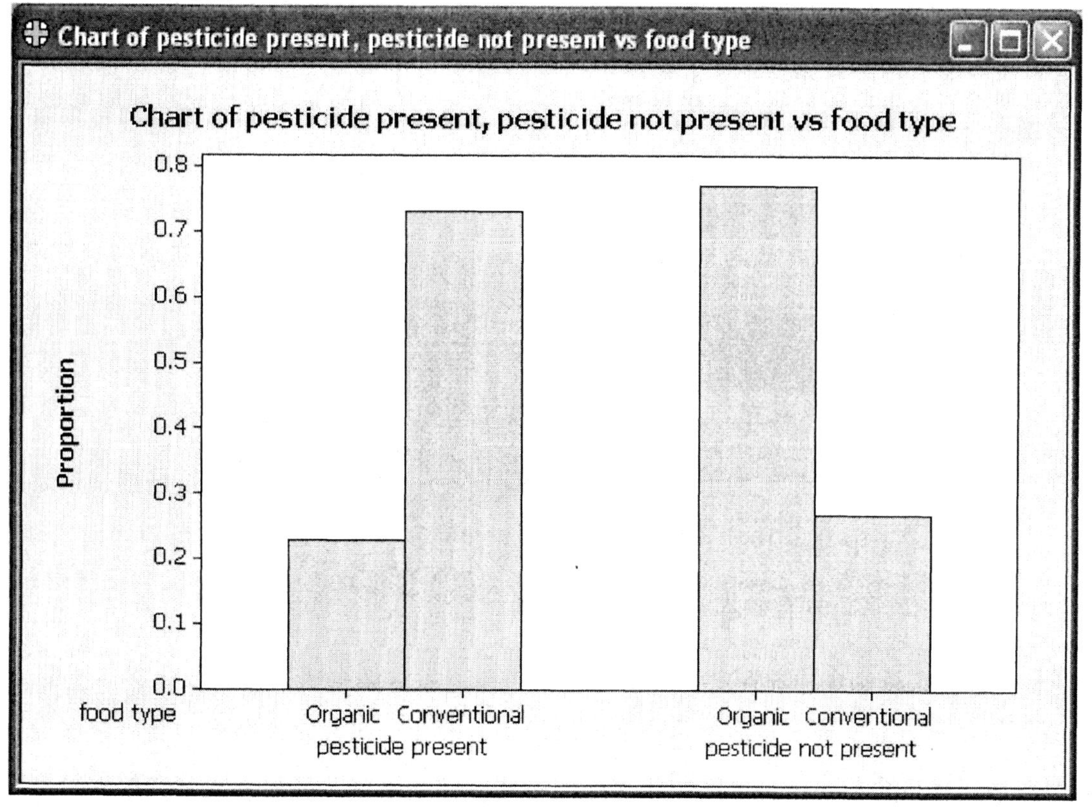

Making the selection of **Rows are outermost categories and columns are innermost** produces the following side-by-side bar chart.

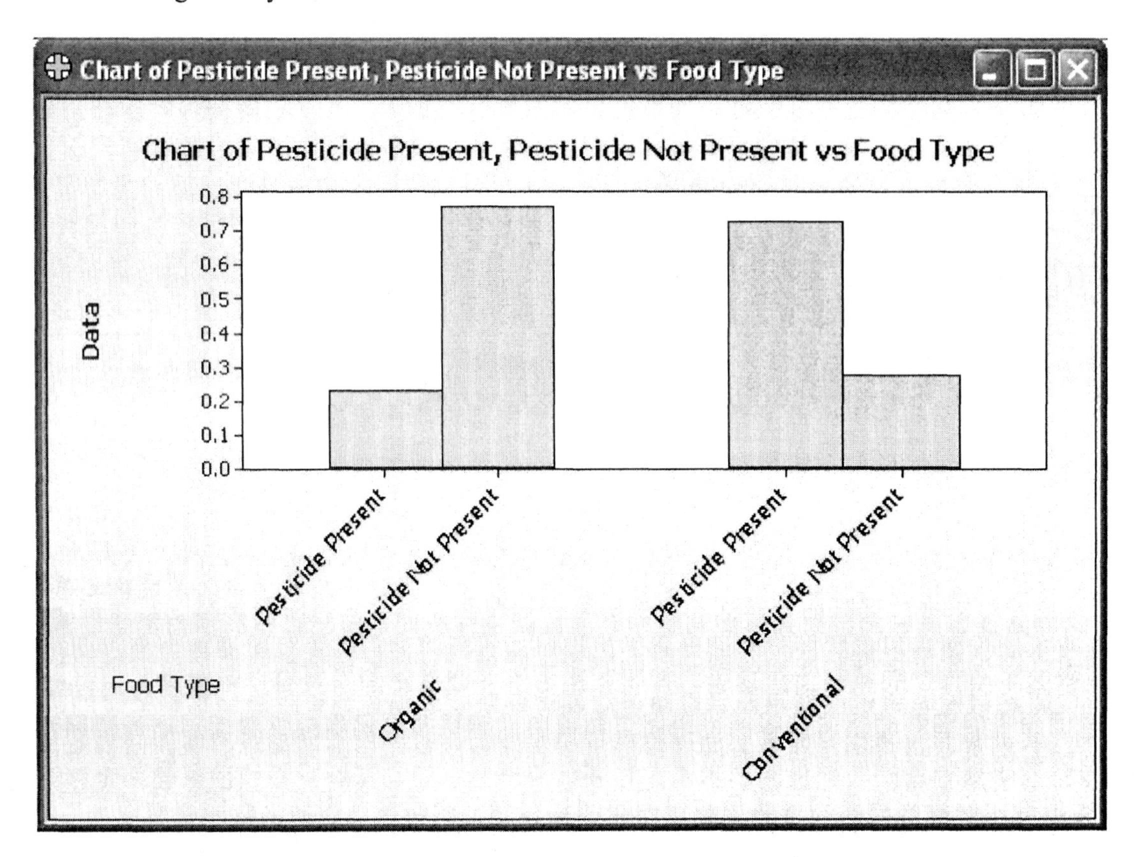

Chapter 3 Exercise 3.4 Religious Activities
Side-by-Side Bar Chart

Enter the data into the worksheet. Since we are comparing unequal numbers of males and females, we need to enter the data as proportions of hours of home religious activity within each category of gender.

↓	C1-T	C2	C3	C4	C5	C6
	Gender	0	1-9	10-19	20-39	40 or more
1	Female	0.29896	0.38773	0.114880	0.13446	0.06397
2	Male	0.43533	0.38328	0.093060	0.06309	0.02524

Choose: **Graph > Bar Chart**
Select: Bars represent: **Values from a table**
Click: **Two-way table / Cluster > OK**
Enter: Graph Variables : **C2 C3 C4 C5 C6**
Row labels: **C1**
Select: Table Arrangement: **Columns are outermost categories and rows are innermost**
Click: **OK**

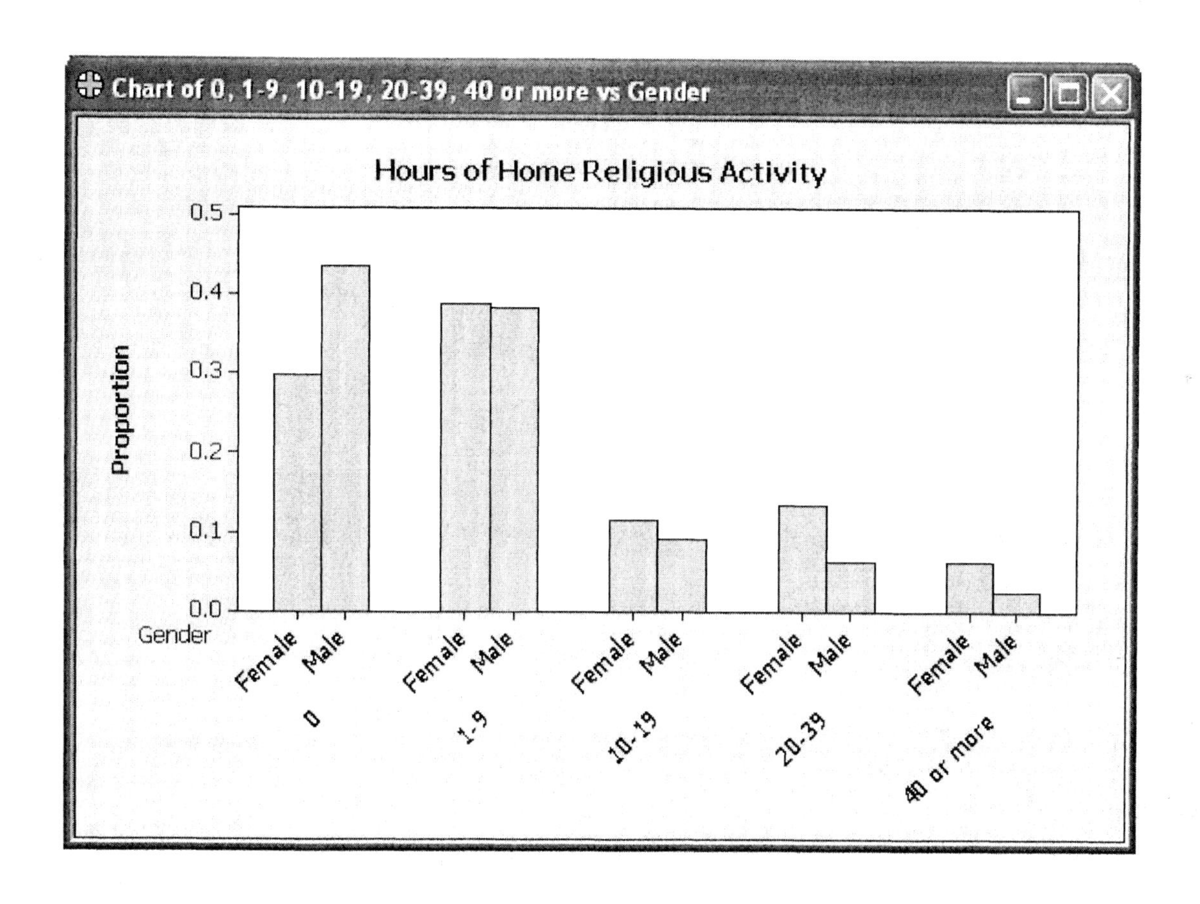

Chapter 3 Example 5 Constructing a Scatterplot for Internet Use and GDP
Constructing a Scatterplot

Open the worksheet **human_development**, which is found in the MINITAB folder on the data disk.

↓	C1-T	C2	C3	C4	C5	C6	C7
		INTERNET	GDP	CO2	CELLULAR	FERTILITY	LITERACY
1	Algeria	0.65	6.09	3.0	0.3	2.8	58.3
2	Argentina	10.08	11.32	3.8	19.3	2.4	96.9
3	Australia	37.14	25.37	18.2	57.4	1.7	100.0
4	Austria	38.70	26.73	7.6	81.7	1.3	100.0
5	Belgium	31.04	25.52	10.2	74.7	1.7	100.0
6	Brazil	4.66	7.36	1.8	16.7	2.2	87.2
7	Canada	46.66	27.13	14.4	36.2	1.5	100.0
8	Chile	20.14	9.19	4.2	34.2	2.4	95.7
9	China	2.57	4.02	2.3	11.0	1.8	78.7
10	Denmark	42.95	29.00	9.3	74.0	1.8	100.0

Since we are interested in how Internet use depends on GDP, C3 GDP is the x-variable, and C2 Internet is the y-variable.

To create a scatterplot of the data

Choose: **Graph > Scatterplot**

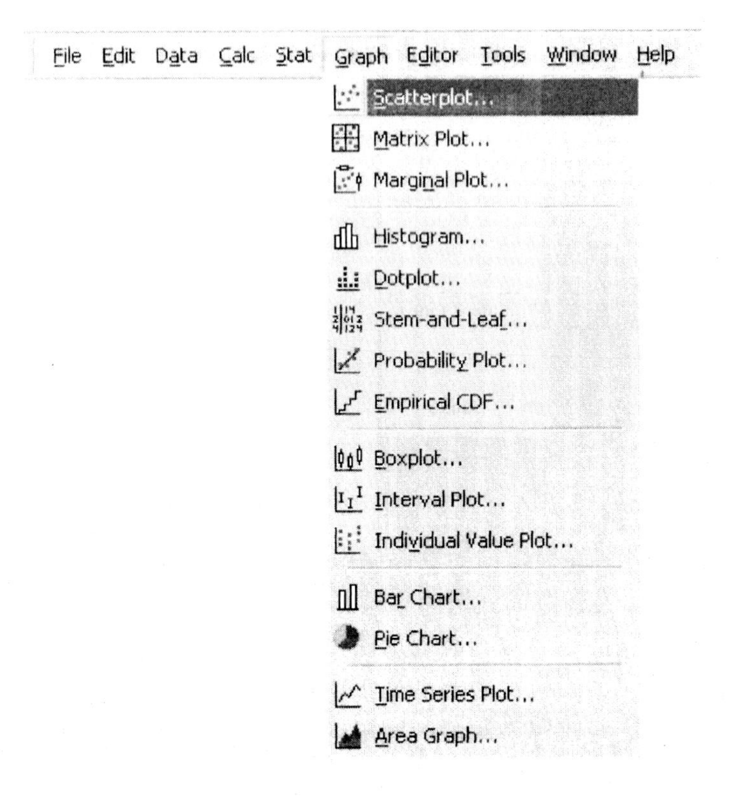

Select: **Simple > OK**

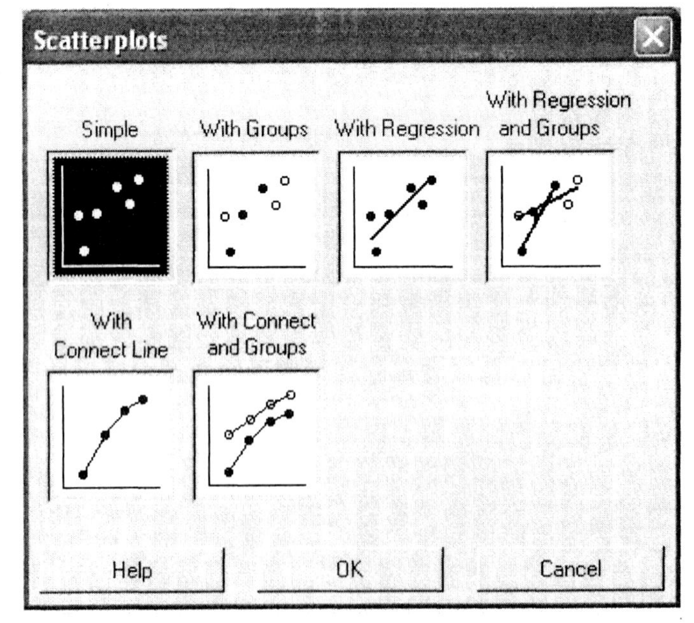

Enter: Y variables **C2**
X variables **C3**

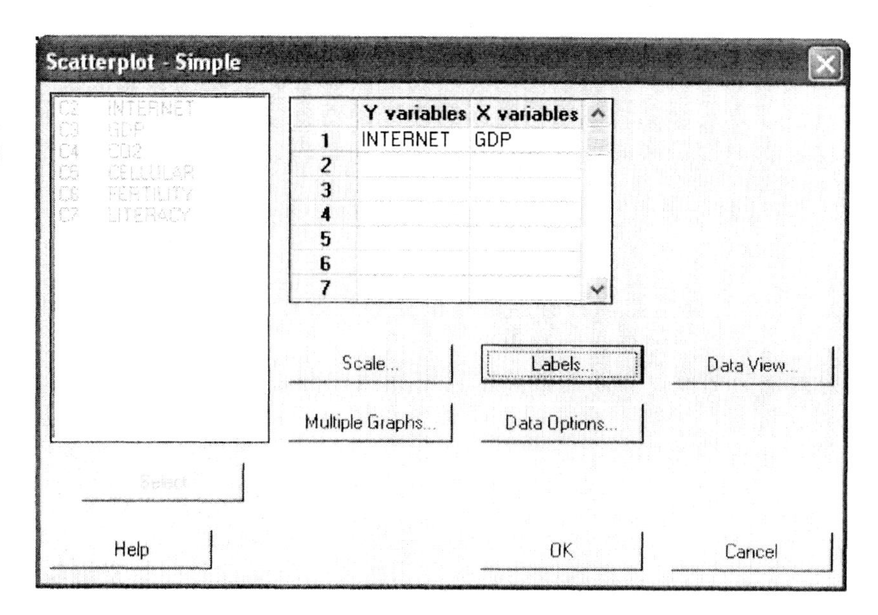

Select: Labels **enter an appropriate title**
Click: **OK OK**

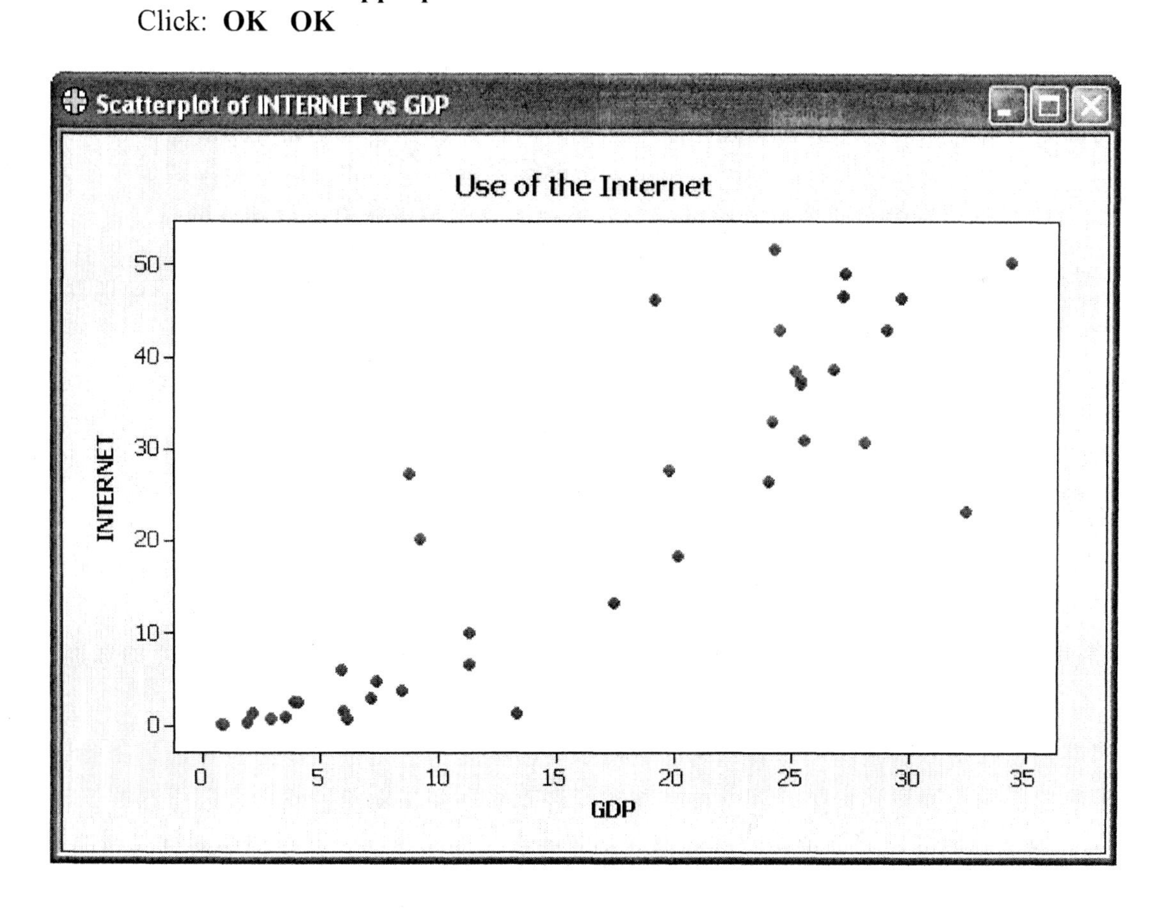

Chapter 3 Example 7 What's the Correlation Between Internet Use and GDP? Computing Correlation

Open the worksheet **human_development**, which is found in the MINITAB folder on the data disk.

↓	C1-T	C2	C3	C4	C5	C6	C7
		INTERNET	GDP	CO2	CELLULAR	FERTILITY	LITERACY
1	Algeria	0.65	6.09	3.0	0.3	2.8	58.3
2	Argentina	10.08	11.32	3.8	19.3	2.4	96.9
3	Australia	37.14	25.37	18.2	57.4	1.7	100.0
4	Austria	38.70	26.73	7.6	81.7	1.3	100.0
5	Belgium	31.04	25.52	10.2	74.7	1.7	100.0
6	Brazil	4.66	7.36	1.8	16.7	2.2	87.2
7	Canada	46.66	27.13	14.4	36.2	1.5	100.0
8	Chile	20.14	9.19	4.2	34.2	2.4	95.7
9	China	2.57	4.02	2.3	11.0	1.8	78.7
10	Denmark	42.95	29.00	9.3	74.0	1.8	100.0

Since we are interested in how Internet use depends on GDP, C3 GDP is the x-variable, and C2 Internet is the y-variable.

To find the correlation :

Click: **Stat > Basic Statistics > Correlation**

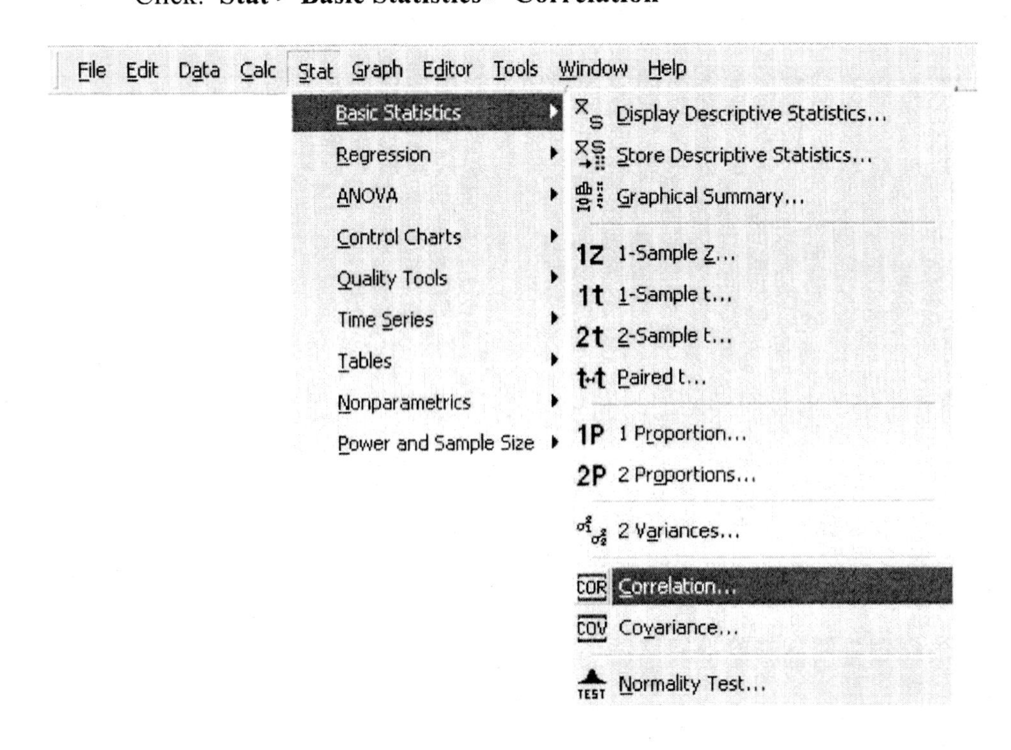

Enter: Variables: **C3 C2**
Click: **OK**

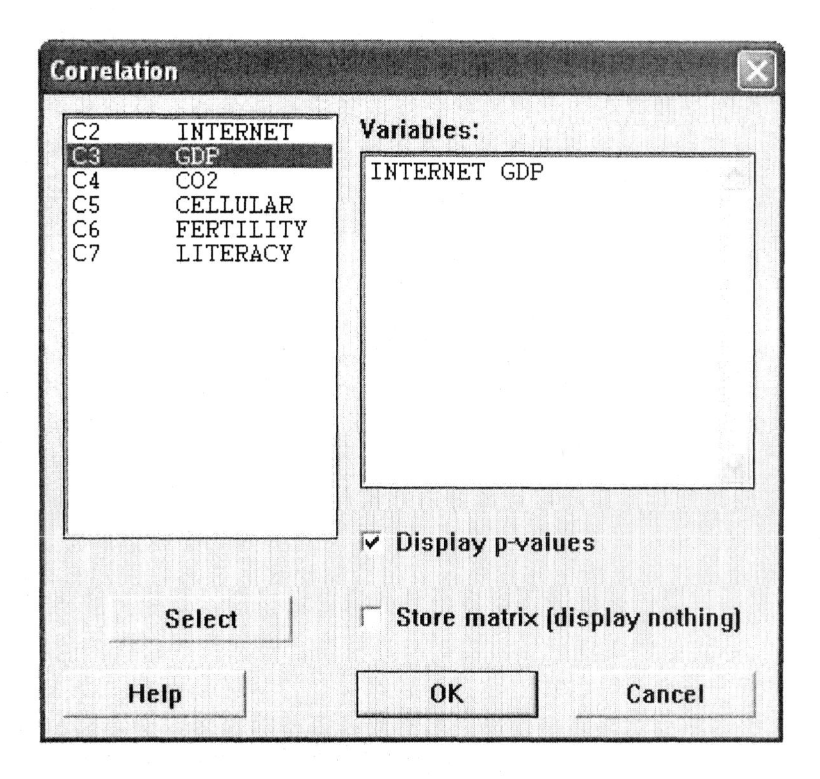

The correlation will be displayed in the Session Window.

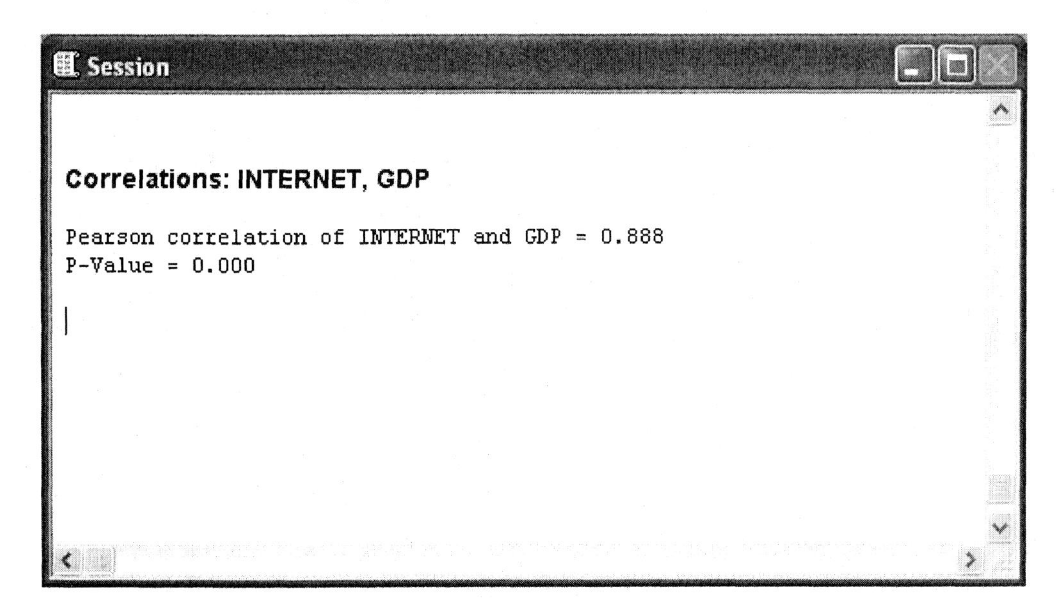

Chapter 3 Exercise 3.21 Which Mountain Bike to Buy?
Computing Correlation

Open the worksheet **mountain_bike**, which is found in the MINITAB folder of the data disk.

Since we are interested in whether and how weight affects the price, C2 weight is the x-variable, and C1 price is the y-variable.

To create a scatterplot of the data
Click: **Graph>Scatterplot > Simple > OK**
Select: Labels **enter an appropriate title**
Click: **OK OK**

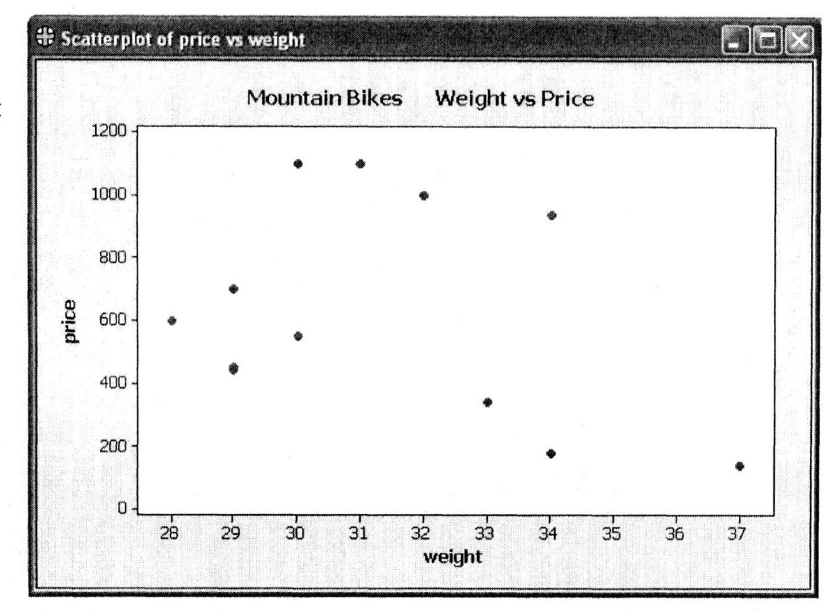

To find the correlation :

Click: **Stat > Basic Statistics > Correlation**
Enter: Variables: **C2 C1**
Click: **OK**

The correlation will be displayed in the Session Window.

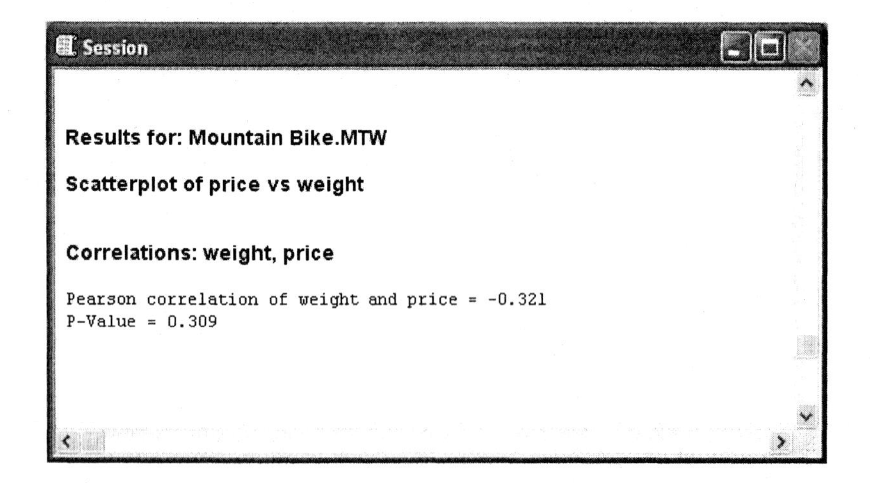

Chapter 3 Exercise 3.23 Buchanan vote
Computing Correlation

Open the worksheet **Buchanan_and_the_butterfly_ballot** which is in the MINITAB folder of the data disk.

Generating a box plot for Gore and Buchanan:

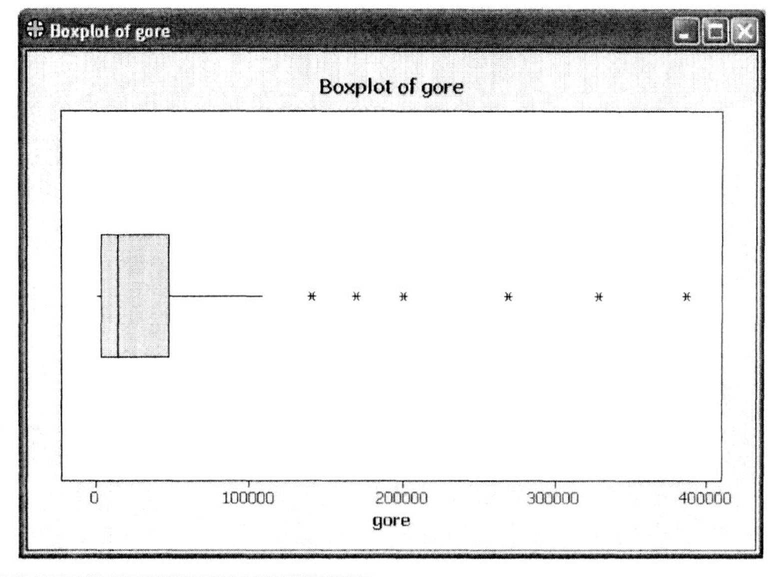

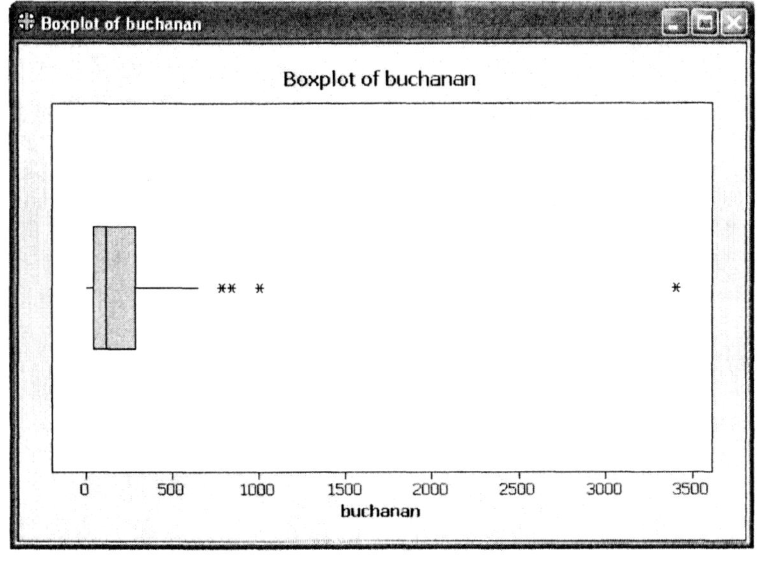

Note the different scales when comparing the two distributions.

To create a scatterplot of the data where C3 gore is the x-variable, and C5 buchanan is the y-variable

Click: **Graph > Scatterplot > Simple > OK**
Enter: Y Variables **C5**
X Variables **C3**
Select: Labels **enter an appropriate title**
Click: **OK OK**

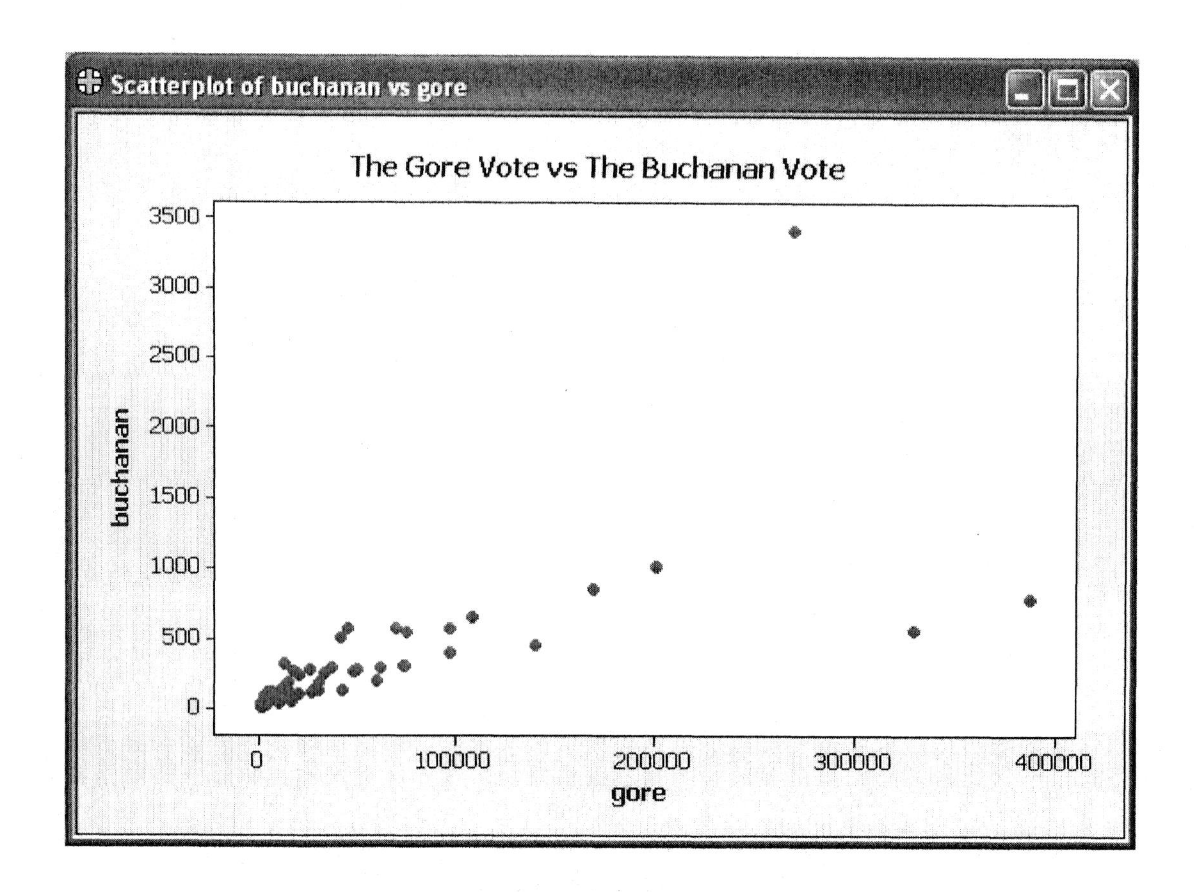

To find the correlation :

Click: **Stat > Basic Statistics > Correlation**
Enter: Variables: **C5 C3**
Click: **OK**

The correlation will be displayed in the Session Window.

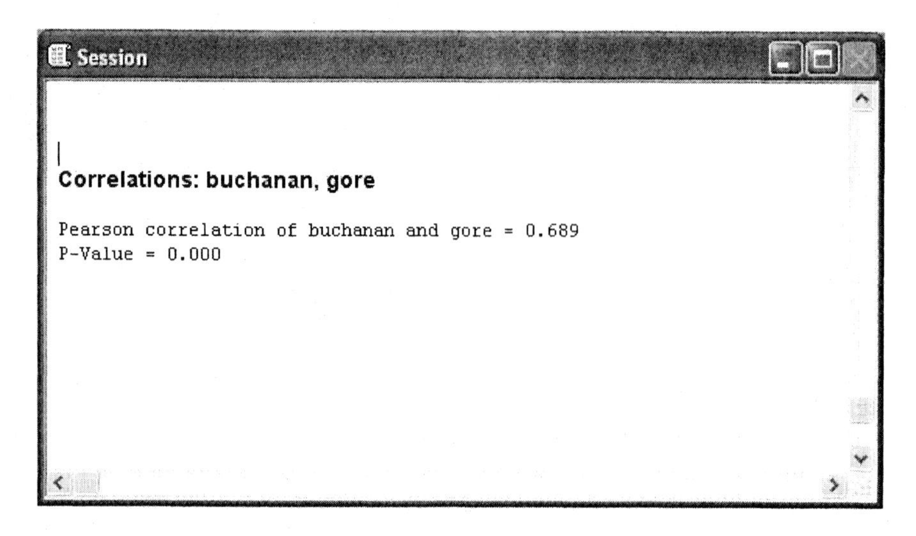

Chapter 3 Example 9 How Can We Predict Baseball Scoring Using Batting Average? Generating the Regression equation

Open the worksheet **al_team_statistics** from the MINITAB folder of the data disk.

We will first generate the scatterplot:

Choose: **Graph > Scatterplot**

Select: **Simple**

Enter: Y variables: **RUNS_AVG** X variables: **BAT_AVG**

Click: **Label:**

Title: **Predicting Scores Based on Batting Averages**

Click: **OK**

We get the following graph:

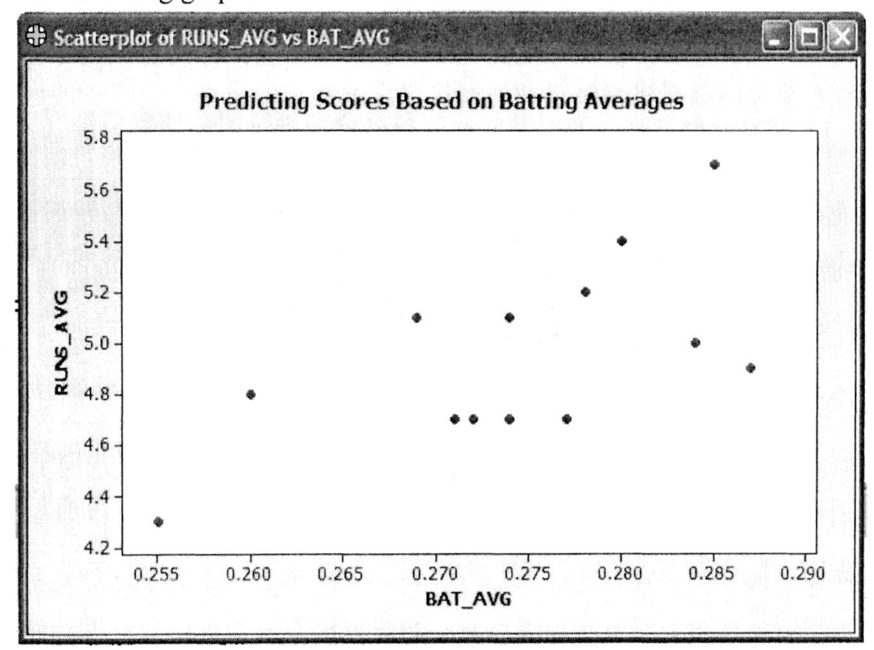

Now we will calculate the regression equation:

Choose:

Stat > Regression > Regression

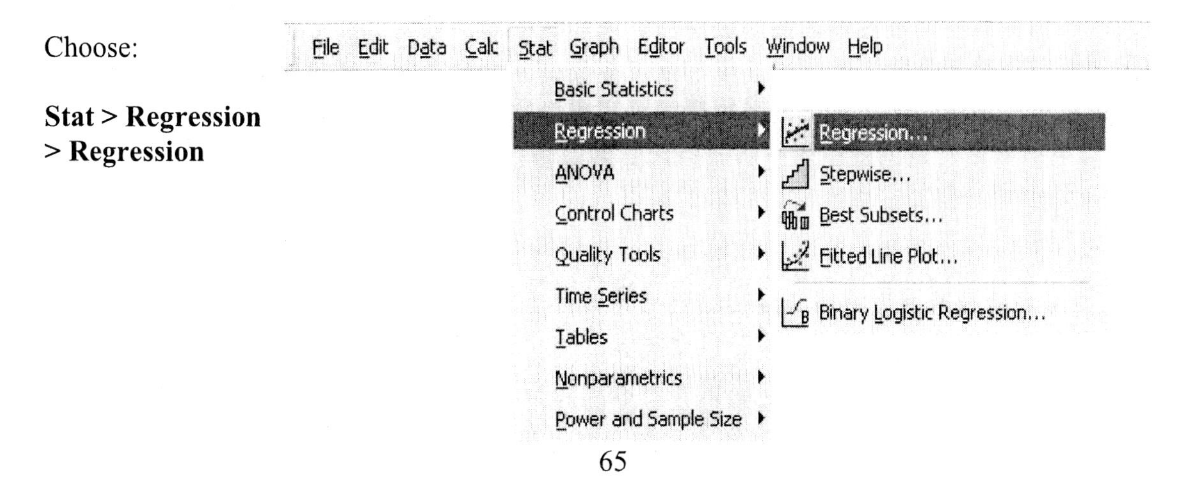

Enter:
Response: **C5**
Predictors: **C4**
Click: **OK**

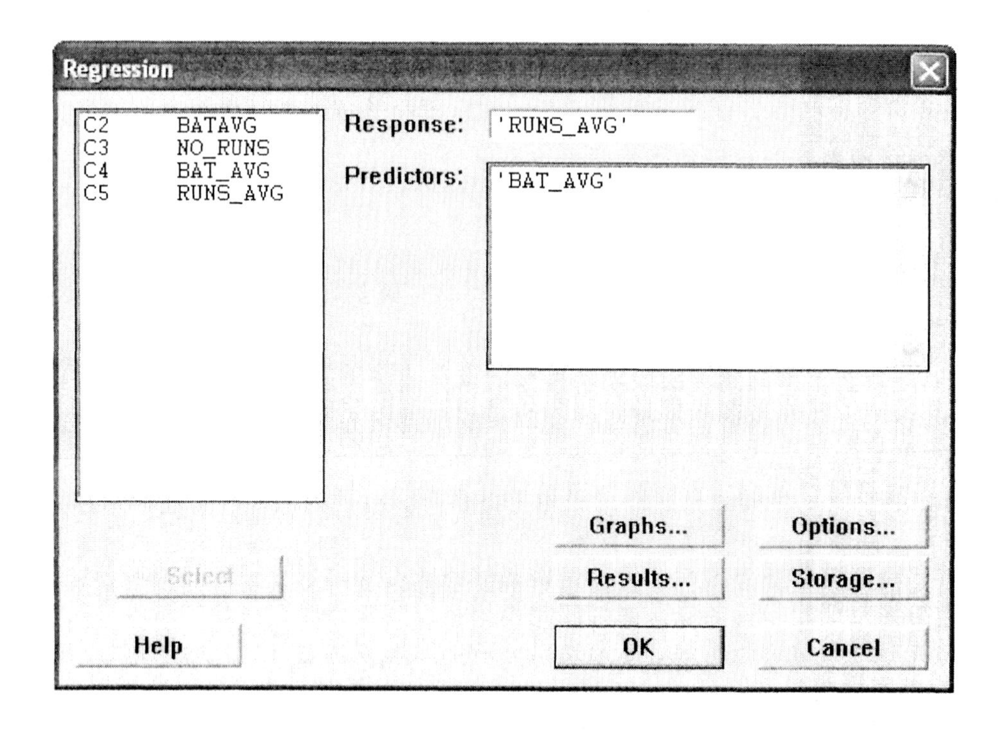

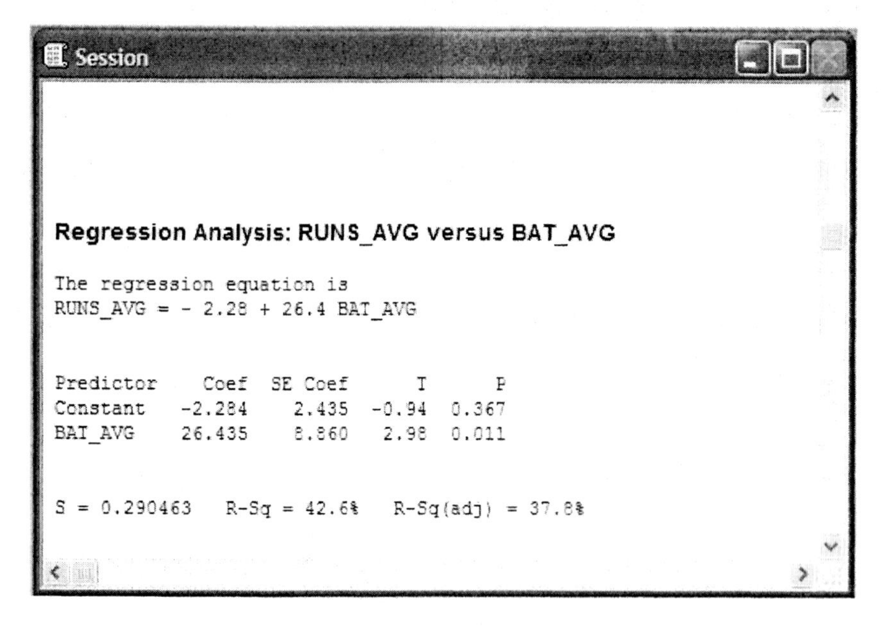

Session
Regression Analysis: RUNS_AVG versus BAT_AVG
The regression equation is
RUNS_AVG = - 2.28 + 26.4 BAT_AVG
Predictor Coef SE Coef T P
Constant -2.284 2.435 -0.94 0.367
BAT_AVG 26.435 8.860 2.98 0.011
S = 0.290463 R-Sq = 42.6% R-Sq(adj) = 37.8%

Chapter 3 Example 10 How Can We Detect An Unusual Vote Total? Looking at residuals

Open the worksheet **Buchanan_and_the_butterfly_ballot** from the MINITAB folder of the data disk.

Choose: **Stat > Regression > Regression**
Enter: Response: **C5**
Predictors: **C2**
Click: **OK**

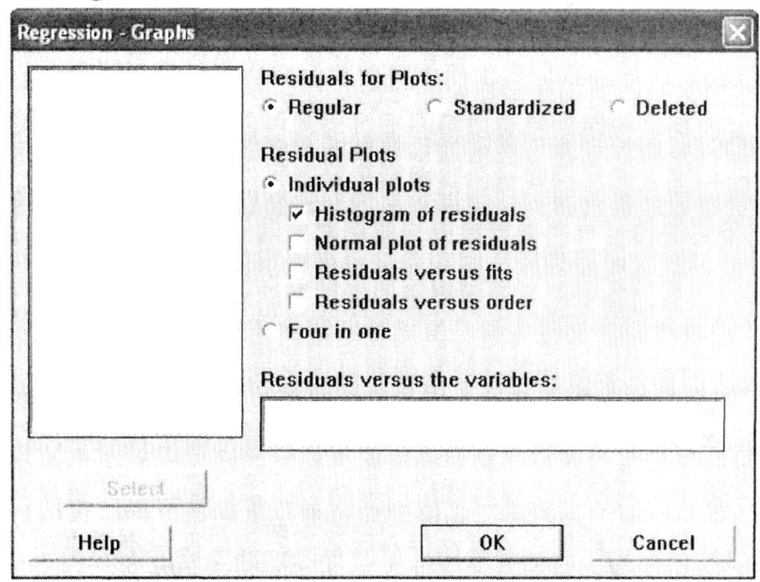

Click : Graphs
Select: Residual Plots / Individual Plots / **Histogram of residuals**

Click: **OK**

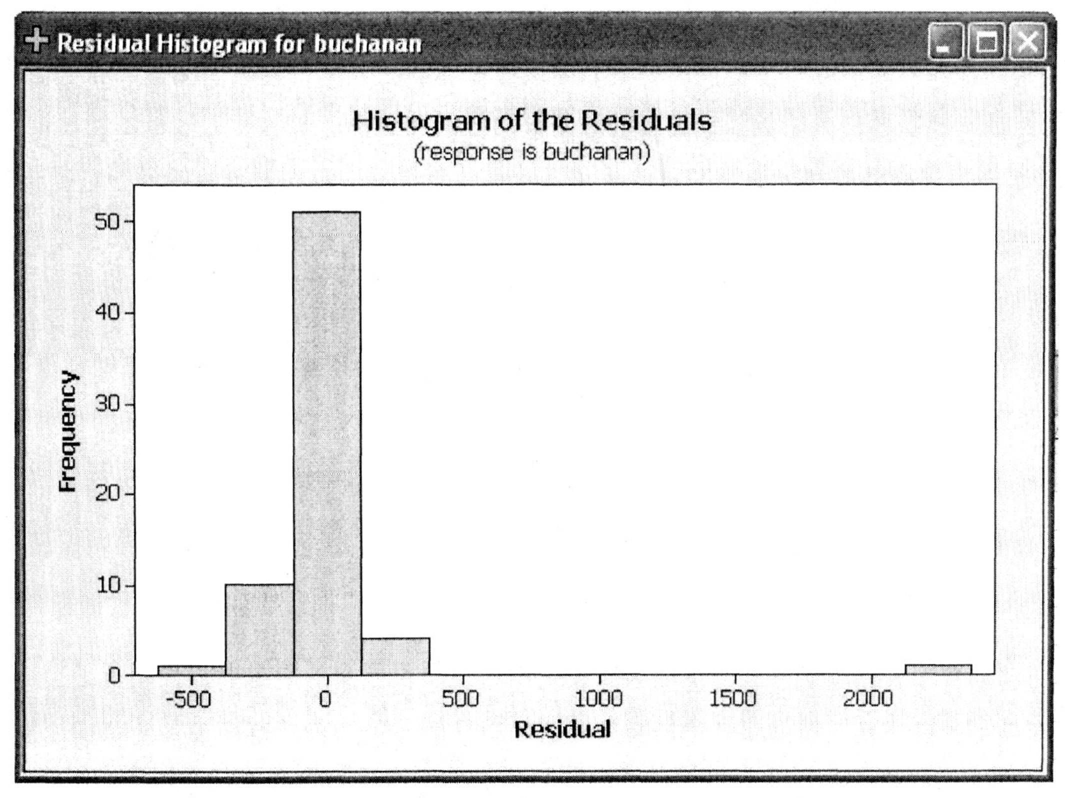

Chapter 3 Exercise 3.33 Regression between cereal sodium and sugar Residuals

Open the worksheet **CEREAL** from the MINITAB folder of the data disk.

Choose: **Graph > Scatterplot**
Select: **with regression**
Enter:
Y variables: **C2**
X variables: **C1**

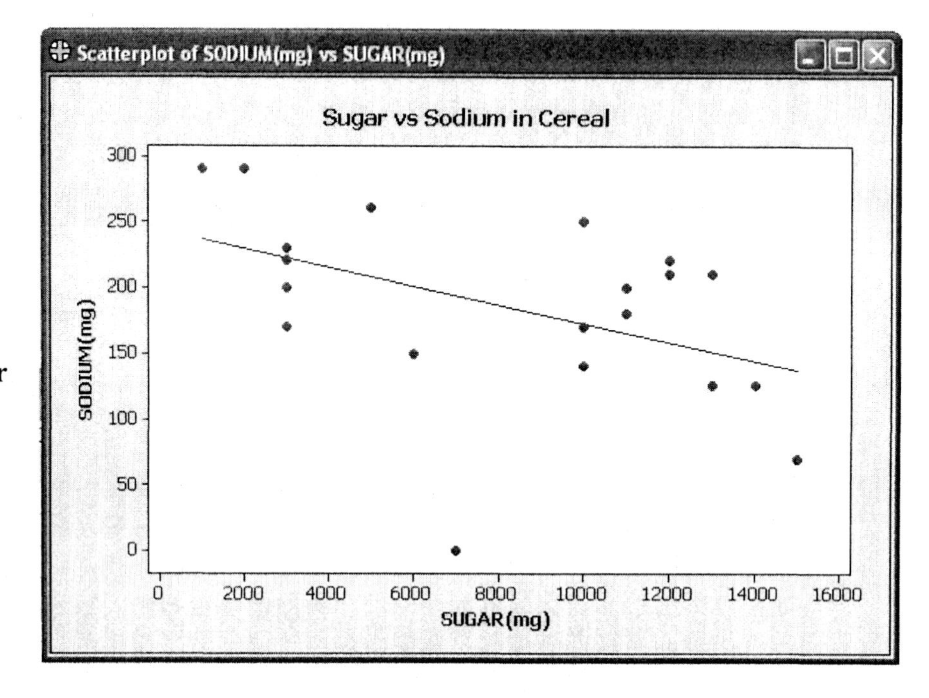

Click: **Label:**
Enter: Title: **Sugar vs Sodium in Cereal**
Click: **OK**

Histogram of the residuals

Choose: **Stat > Regression > Regression**
Enter: Response: **C2**
Predictors: **C7**
Click: **OK**
Click : Graphs
Select: Residual Plots / Individual Plots / **Histogram of residuals**

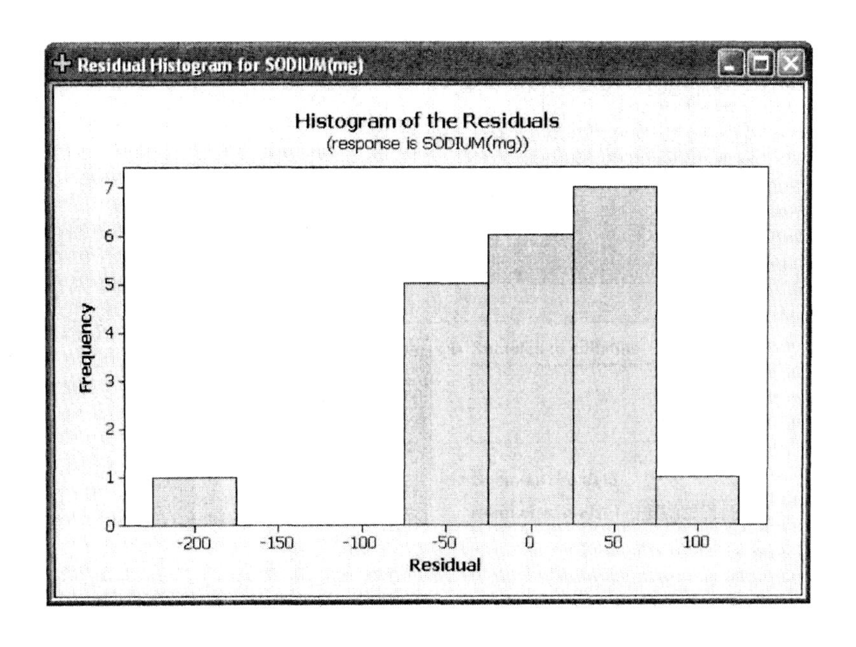

Chapter 3 Exercise 3.39 Study time and college GPA
Generating a regression Equation

Enter the study times and GPAs into a worksheet.

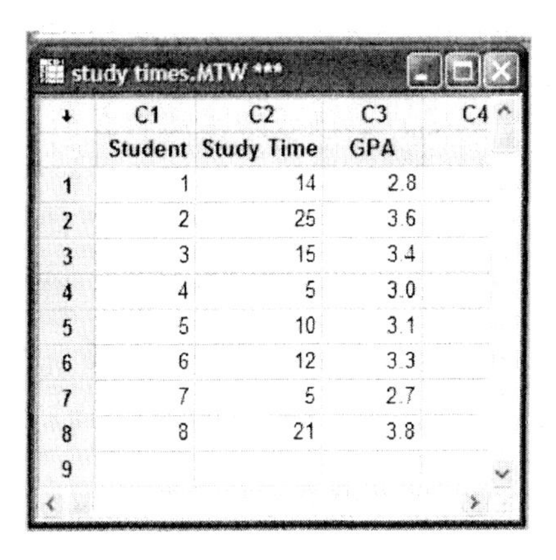
study times.MTW ***

↓	C1	C2	C3	C4
	Student	Study Time	GPA	
1	1	14	2.8	
2	2	25	3.6	
3	3	15	3.4	
4	4	5	3.0	
5	5	10	3.1	
6	6	12	3.3	
7	7	5	2.7	
8	8	21	3.8	
9				

Then get the scatterplot:

Choose: **Graph > Scatterplot**
Select: **With Regression**
Enter: Y variables: **GPA** X variables: **Study Time**
Click: **Label:**
Title: **GPA vs Study Time**
Click: **OK**

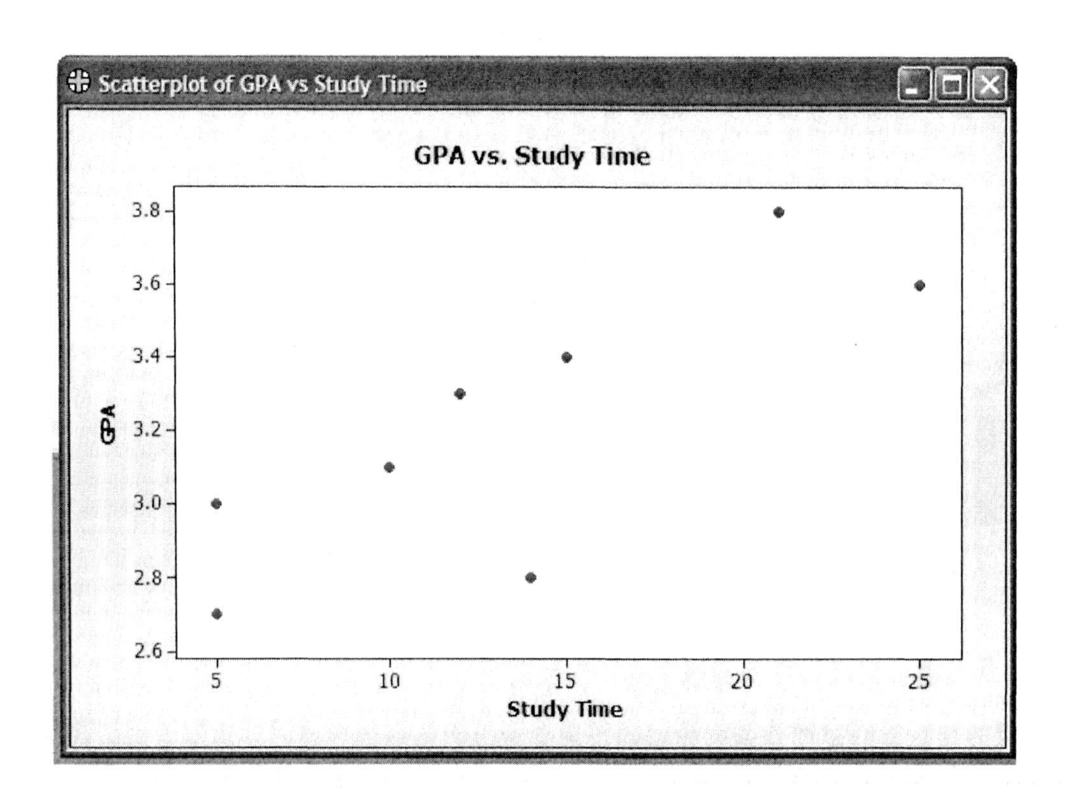

Find the correlation:

Choose: **Stat > Basic Statistics > Correlation**

Enter: **GPA 'Study Time'**

Click: **OK**

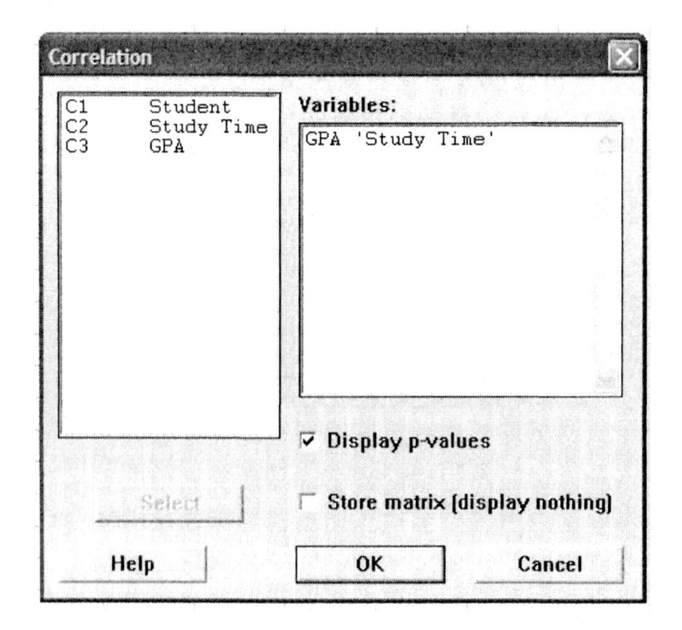

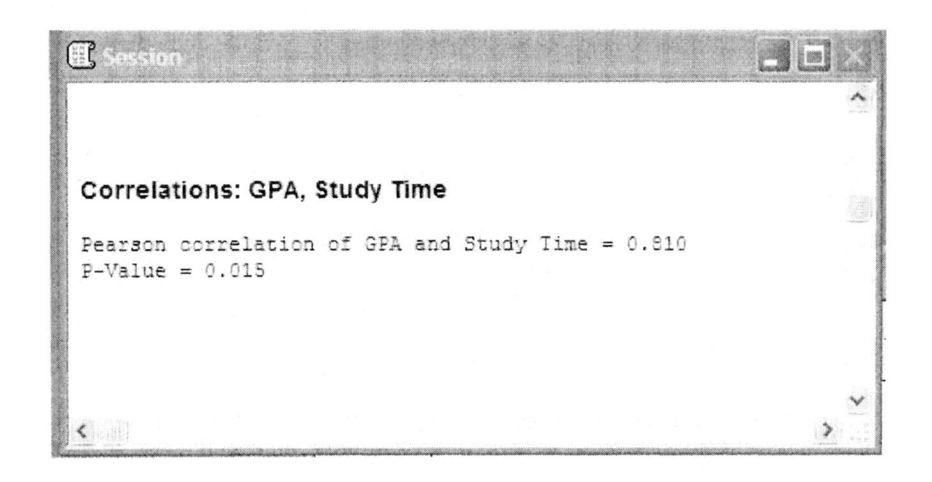

Session
Correlations: GPA, Study Time
Pearson correlation of GPA and Study Time = 0.810
P-Value = 0.015

Find the regression equation:

Choose: **Stat > Regression > Regression**

Enter: Response: **GPA**

Predictors: **Study Time**

Click: **OK**

Predicted GPA for a a student who studies 5 hours per week:
GPA=2.63 + .0439(5)= 2.85

Predicted GPA for a a student who studies 25 hours per week:
GPA=2.63 + .0439(25)= 3.73

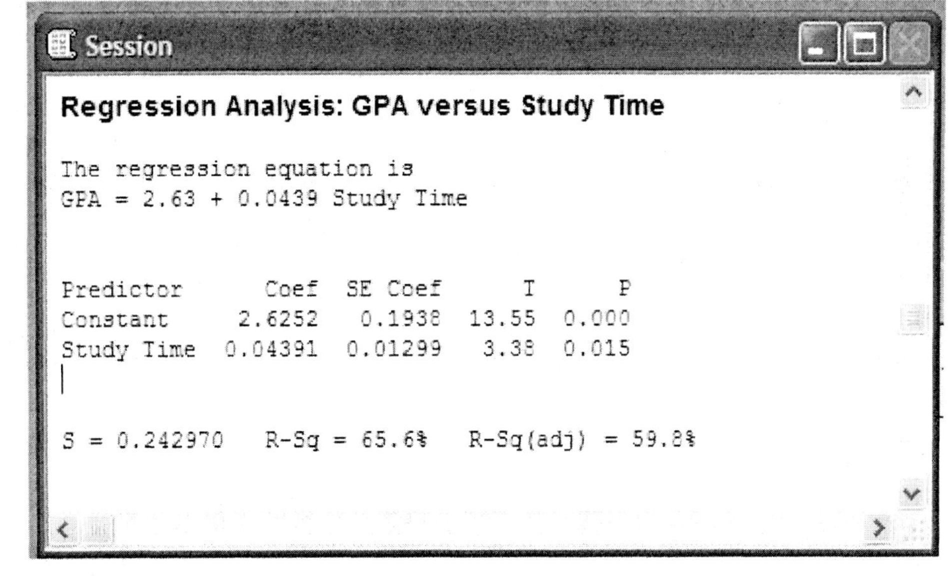

Session
Regression Analysis: GPA versus Study Time
The regression equation is
GPA = 2.63 + 0.0439 Study Time
Predictor Coef SE Coef T P
Constant 2.6252 0.1938 13.55 0.000
Study Time 0.04391 0.01299 3.38 0.015
S = 0.242970 R-Sq = 65.6% R-Sq(adj) = 59.8%

Chapter 3 Exercise 3.40 Oil and GDP
Generating a regression Equation

Enter the Oil and GDP data into a worksheet.

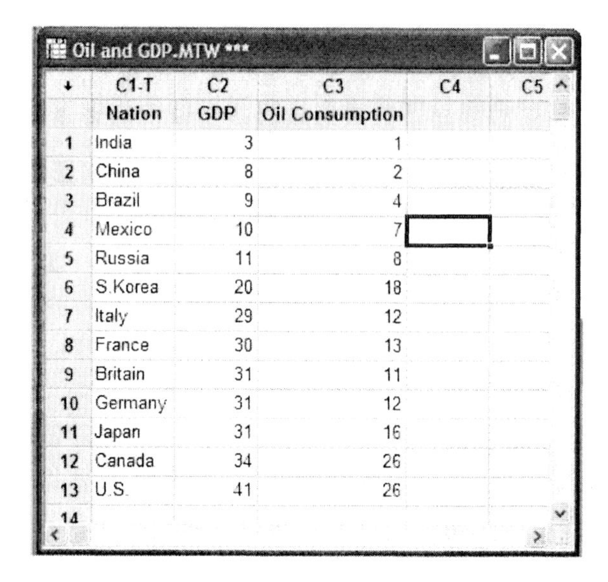

Oil and GDP.MTW ***

↓	C1-T	C2	C3	C4	C5
	Nation	GDP	Oil Consumption		
1	India	3	1		
2	China	8	2		
3	Brazil	9	4		
4	Mexico	10	7		
5	Russia	11	8		
6	S.Korea	20	18		
7	Italy	29	12		
8	France	30	13		
9	Britain	31	11		
10	Germany	31	12		
11	Japan	31	16		
12	Canada	34	26		
13	U.S.	41	26		
14					

Find the prediction equation.

Choose: **Stat > Regression > Regression**
Enter: Response: **Oil Consumption**
Predictors: **GDP**
Click: **OK**

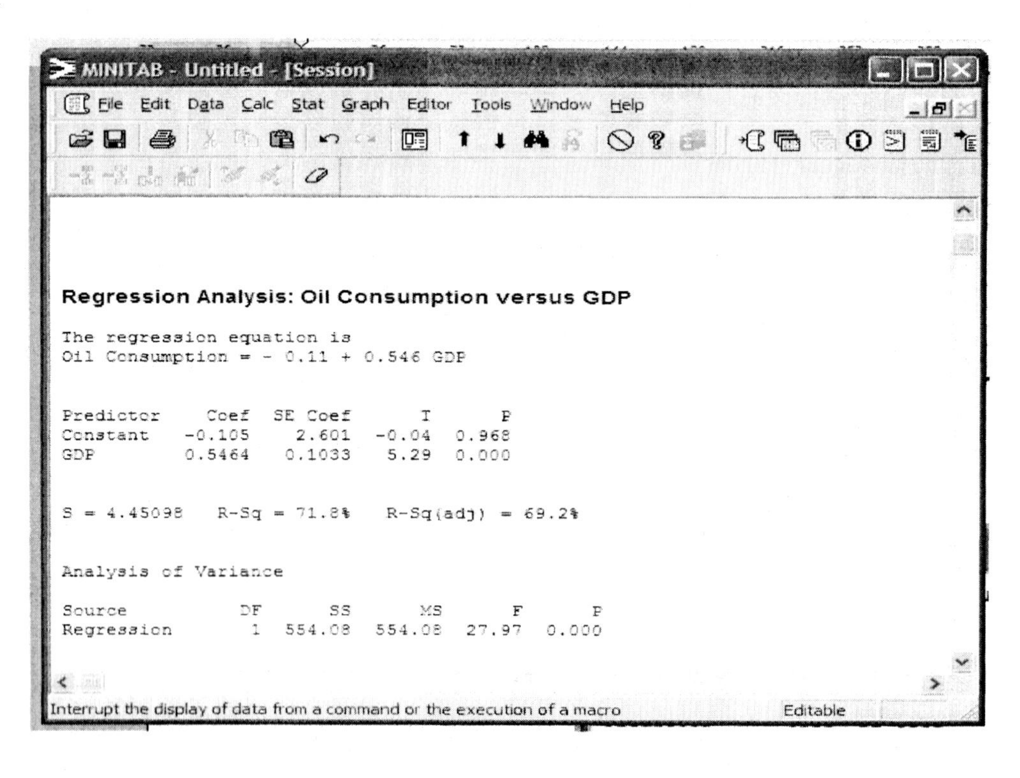

MINITAB - Untitled - [Session]

File Edit Data Calc Stat Graph Editor Tools Window Help

Regression Analysis: Oil Consumption versus GDP

```
The regression equation is
Oil Consumption = - 0.11 + 0.546 GDP


Predictor     Coef  SE Coef      T      P
Constant    -0.105    2.601  -0.04  0.968
GDP         0.5464   0.1033   5.29  0.000


S = 4.45098   R-Sq = 71.8%    R-Sq(adj) = 69.2%


Analysis of Variance

Source          DF      SS      MS      F      P
Regression       1  554.08  554.08  27.97  0.000
```

Interrupt the display of data from a command or the execution of a macro   Editable

Find the correlation:

Choose: **Stat > Basic Statistics > Correlation**

Enter: **GDP 'Oil Consumption'**

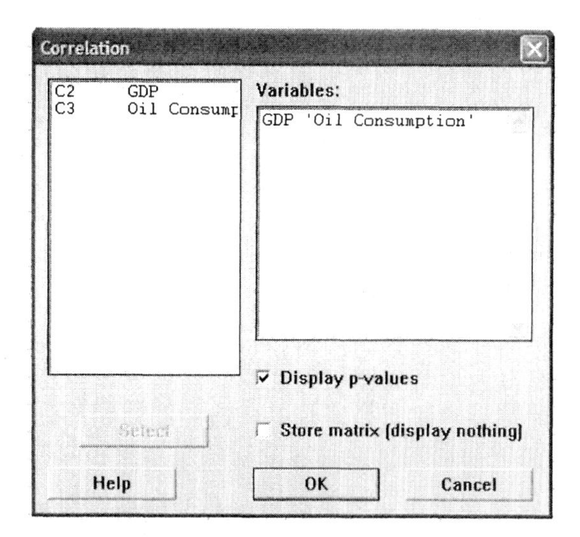

Click: **OK**

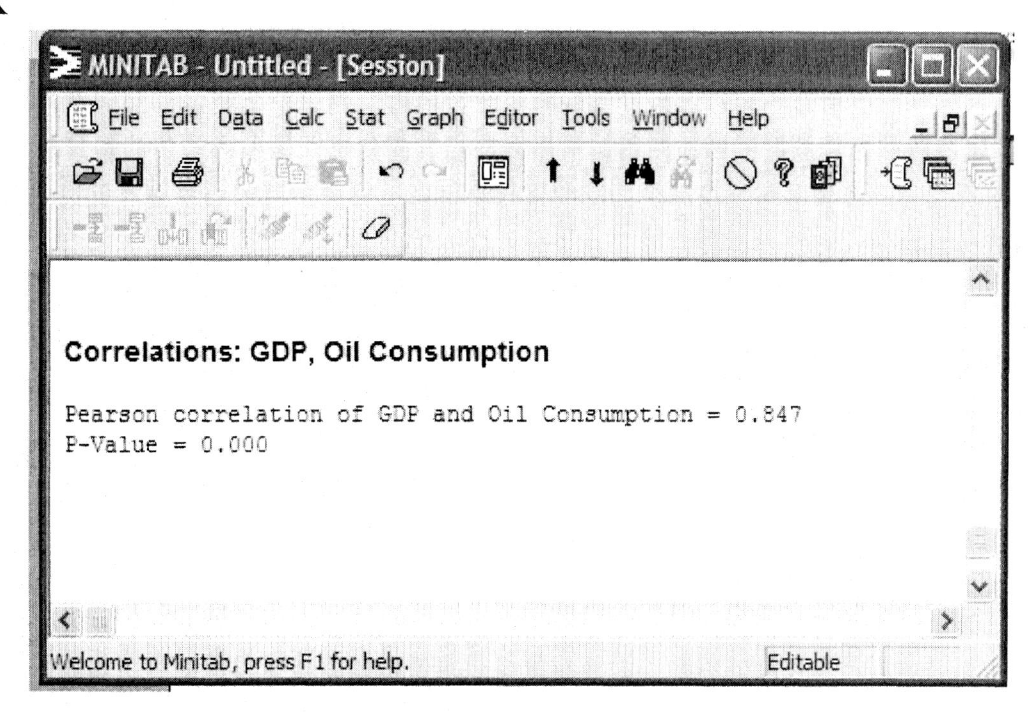

Chapter 3 Exercise 3.41 Mountain Bikes Revisited
Generating a regression equation

We first get the scatterplot:

Choose: **Graph > Scatterplot**
Select: **With Regression**
Enter: Y variables: **C1** X variables: **C2**
Click: **Label:**
Title: **Weight vs Price of Mountain Bikes**
Click: **OK**

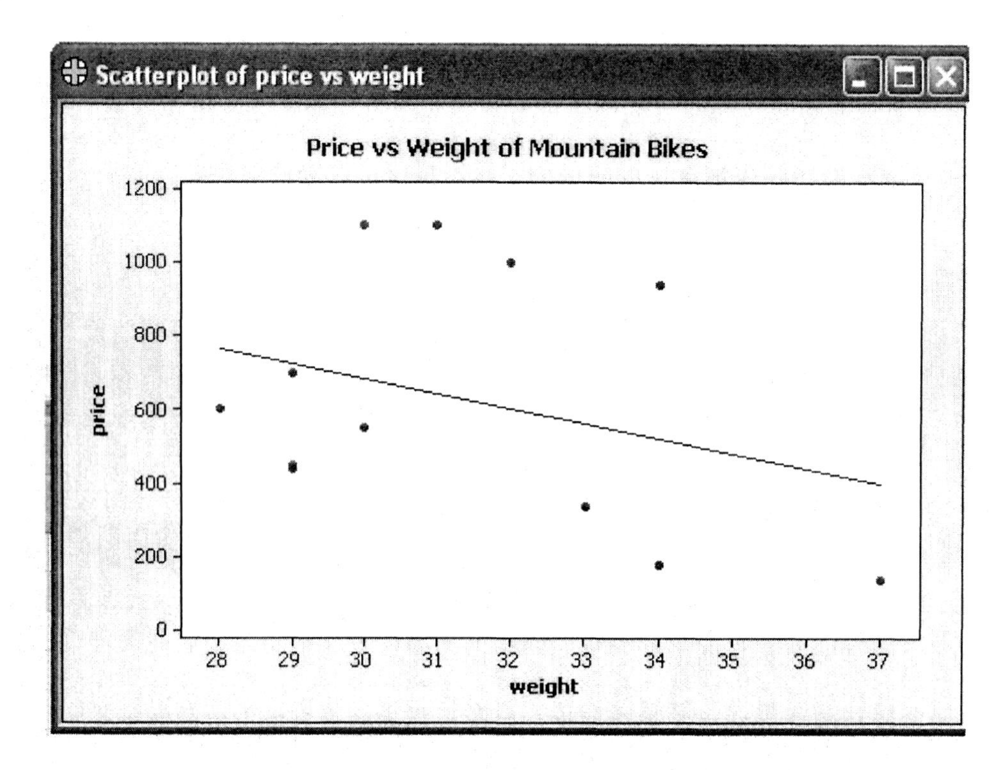

Find the regression equation:

Choose: **Stat > Regression > Regression**

Enter: Response: **price**

Predictors: **weight**

Click: **OK**

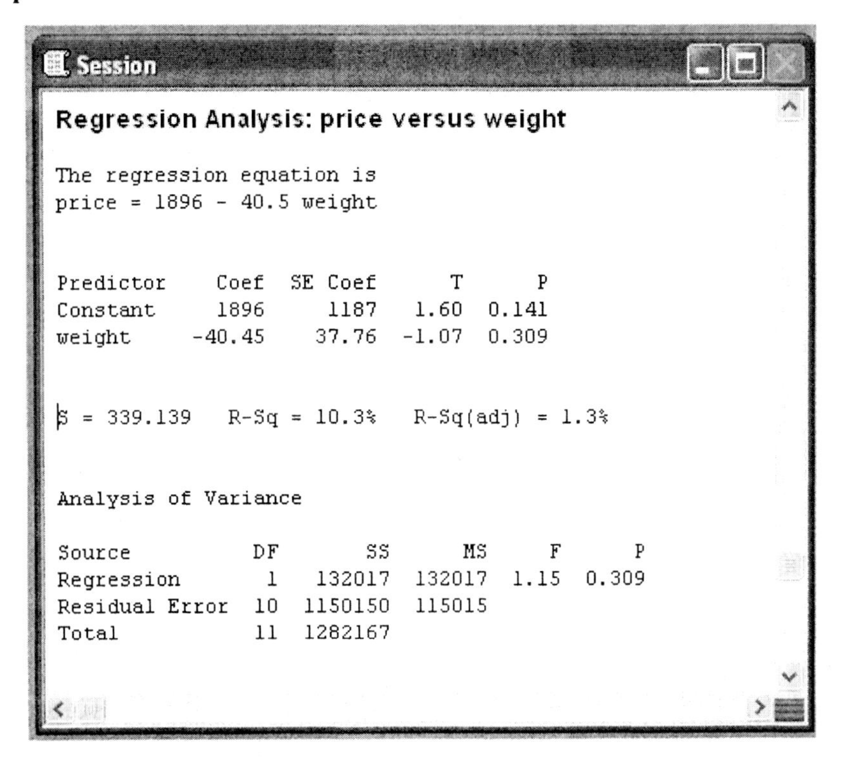

Session

Regression Analysis: price versus weight

The regression equation is
price = 1896 - 40.5 weight

Predictor	Coef	SE Coef	T	P
Constant	1896	1187	1.60	0.141
weight	-40.45	37.76	-1.07	0.309

S = 339.139   R-Sq = 10.3%   R-Sq(adj) = 1.3%

Analysis of Variance

Source	DF	SS	MS	F	P
Regression	1	132017	132017	1.15	0.309
Residual Error	10	1150150	115015		
Total	11	1282167			

Predicted price for a 30 lb bike:

price=1896-40.5(30)= 681

Chapter 3 Exercise 3.42 Mountain bike and suspension types
Generating Regression Equations

Choose: **Graph > Scatterplot**
Select: **with groups**

Enter: Y variables: **price**
X variables: **weight**
Enter: Categorical variables for grouping**: C3-T**

Click: **Label:**
Title: **Price vs Weight of Mountain Bikes**
Click: **OK**

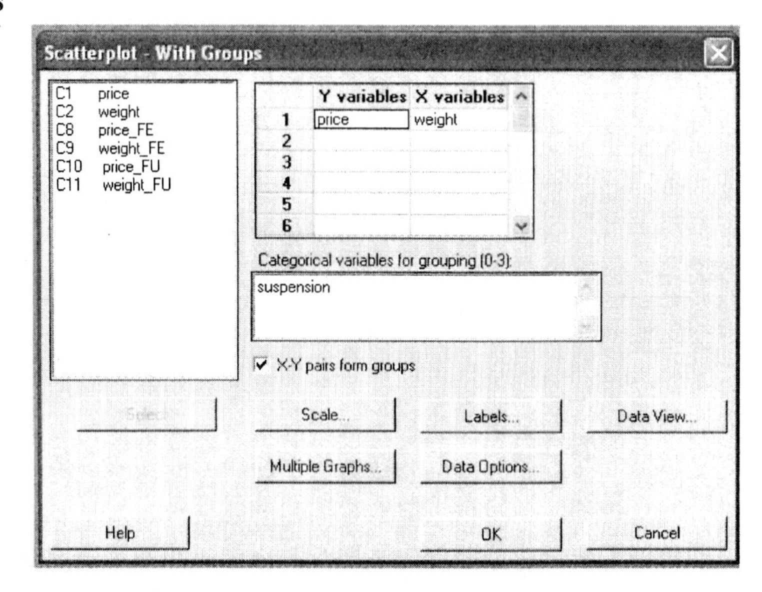

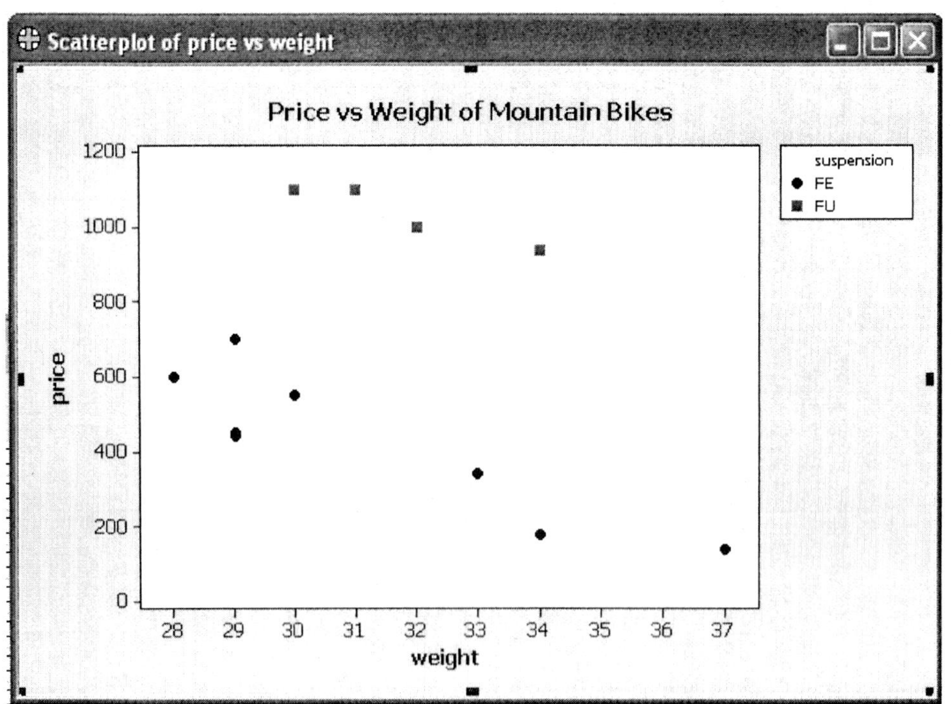

Now we do separate regression equations based on suspension type.

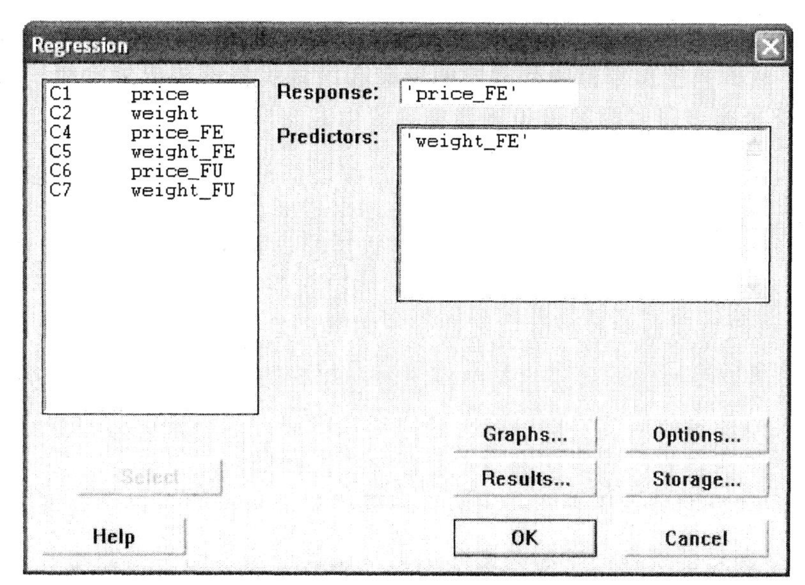

We get the following for front end suspension.

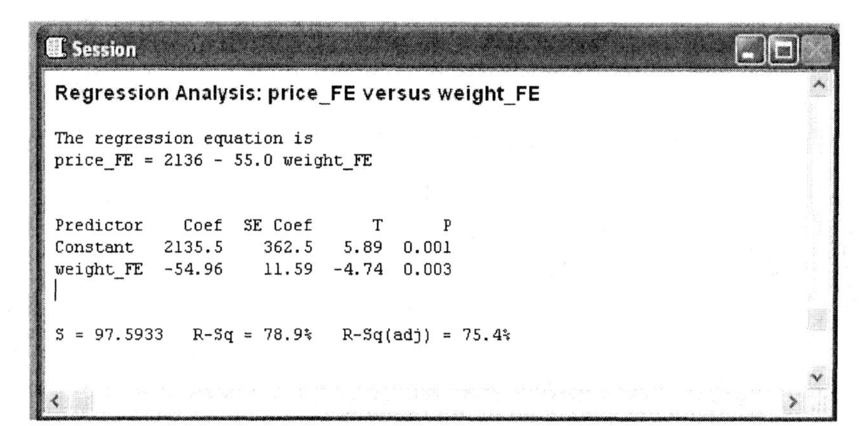
Session

Regression Analysis: price_FE versus weight_FE

The regression equation is
price_FE = 2136 - 55.0 weight_FE

Predictor	Coef	SE Coef	T	P
Constant	2135.5	362.5	5.89	0.001
weight_FE	-54.96	11.59	-4.74	0.003

S = 97.5933   R-Sq = 78.9%   R-Sq(adj) = 75.4%

Repeating the same steps for full suspension bikes we obtain:

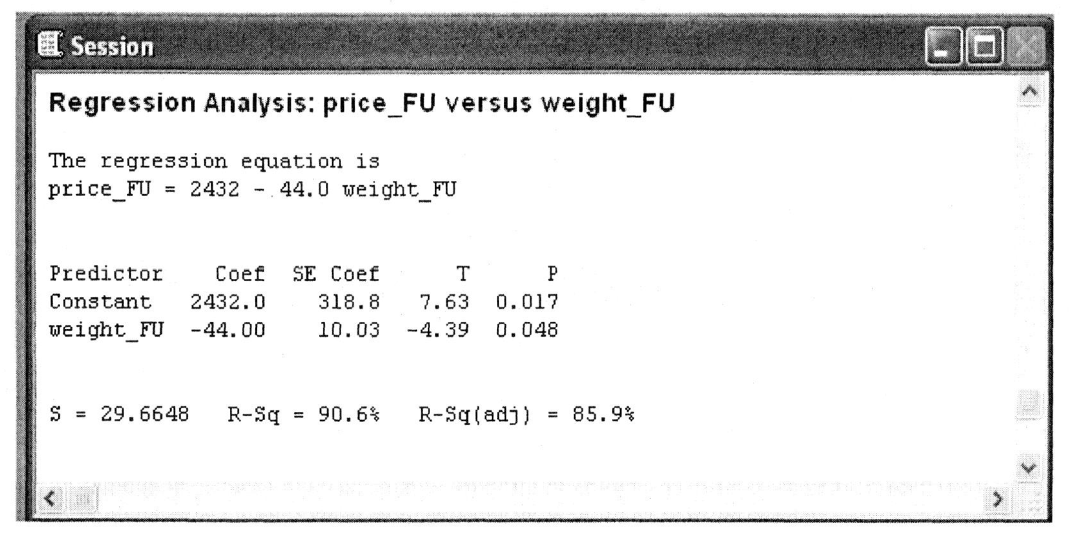
Session

Regression Analysis: price_FU versus weight_FU

The regression equation is
price_FU = 2432 - 44.0 weight_FU

Predictor	Coef	SE Coef	T	P
Constant	2432.0	318.8	7.63	0.017
weight_FU	-44.00	10.03	-4.39	0.048

S = 29.6648   R-Sq = 90.6%   R-Sq(adj) = 85.9%

Chapter 3 Example 12 How Can We Forecast Future Global Warming?
Regression on a time series plot

Open the worksheet **central_park_yearly_temps** from the MINITAB folder of the data disk.

In order to get a regression line printed on a time series plot we are going to use the scatterplot graph with regression, and choose to connect the data points.

Choose: **Graph > Scatterplot**
Select: **With Regression**
Enter: Y variables: **C1** X variables: **C3**
Click: **Label:**
Title: **Central Park Mean Annual Temperatures**
Click Data View>Data Display
    Check **symbols and connect line**
    Click: **OK**

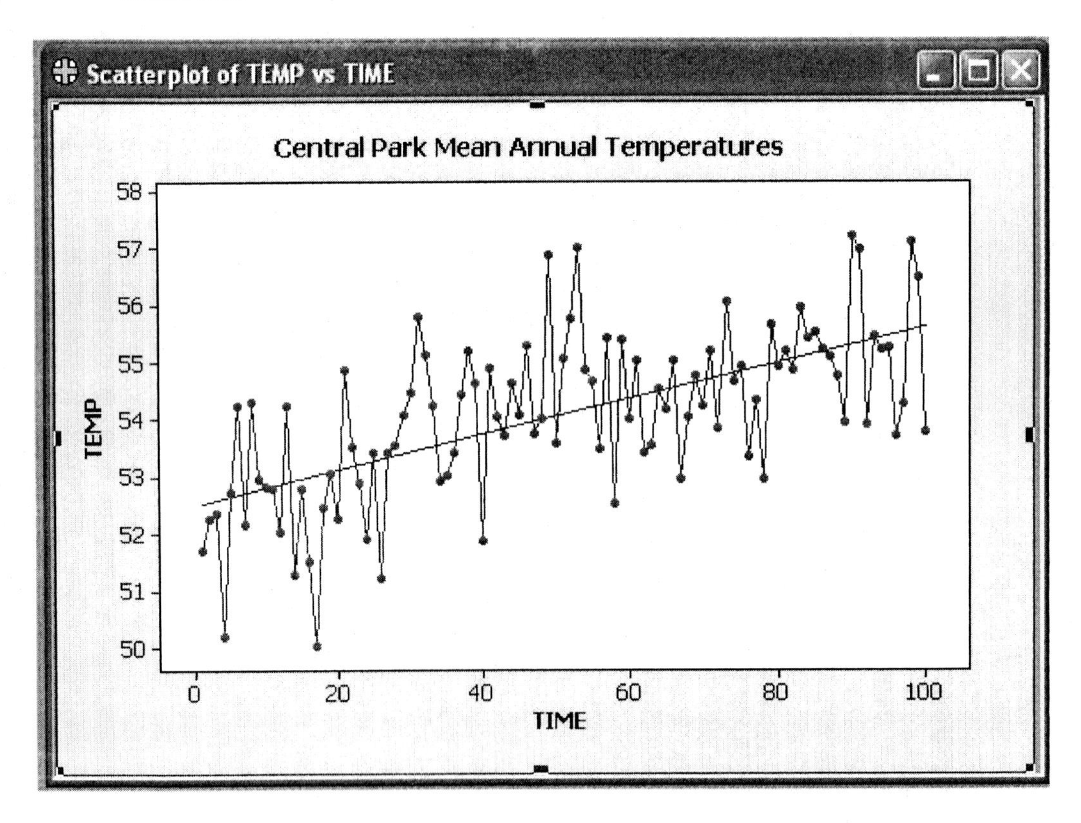

Chapter 3 Example 13 Is Higher Education Associated with Higher Murder Rates? Influential Outliers

Open the worksheet **us_statewide_crime** from the MINITAB folder of the data disk.

Scatterplot of murder rate vs college
Choose: **Graph > Scatterplot**
Select: **With Regression**
Enter: Y variables: **C3** X variables: **C6**
Click: **Label:**
Enter: Title: **Murder rate vs College education**
Click: **OK**

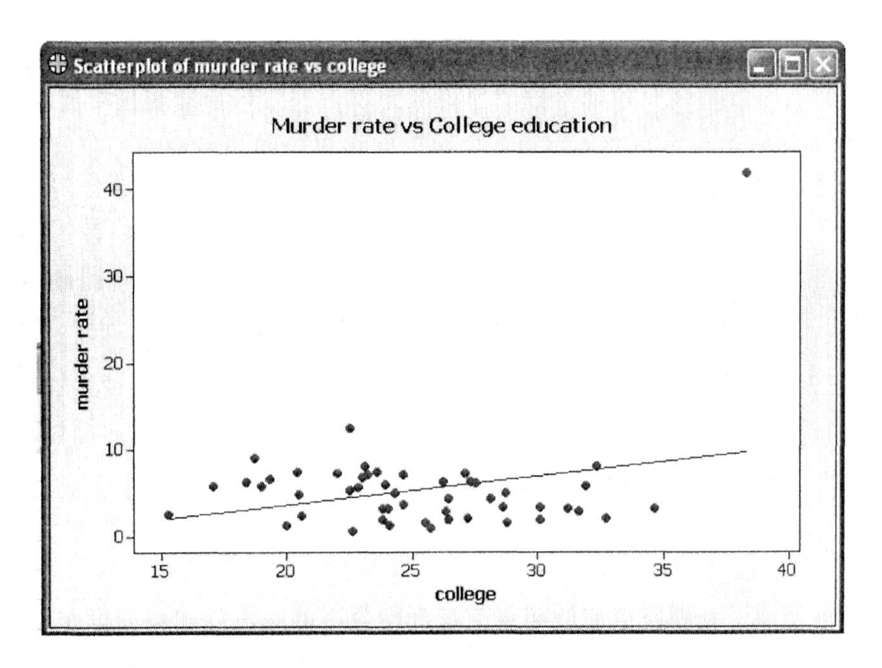

Now calculate the regression equation:

Choose: **Stat > Regression > Regression**
Enter: Response: **C3**
Predictors: **C6**
Click: **OK**

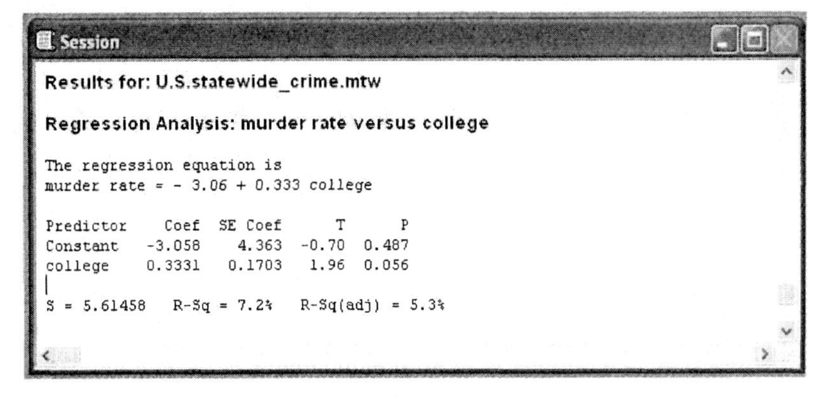

Session

Results for: U.S.statewide_crime.mtw

Regression Analysis: murder rate versus college

```
The regression equation is
murder rate = - 3.06 + 0.333 college

Predictor    Coef  SE Coef      T      P
Constant   -3.058    4.363  -0.70  0.487
college    0.3331   0.1703   1.96  0.056

S = 5.61458   R-Sq = 7.2%   R-Sq(adj) = 5.3%
```

Redoing the entire process, excluding the data for D.C. we get:

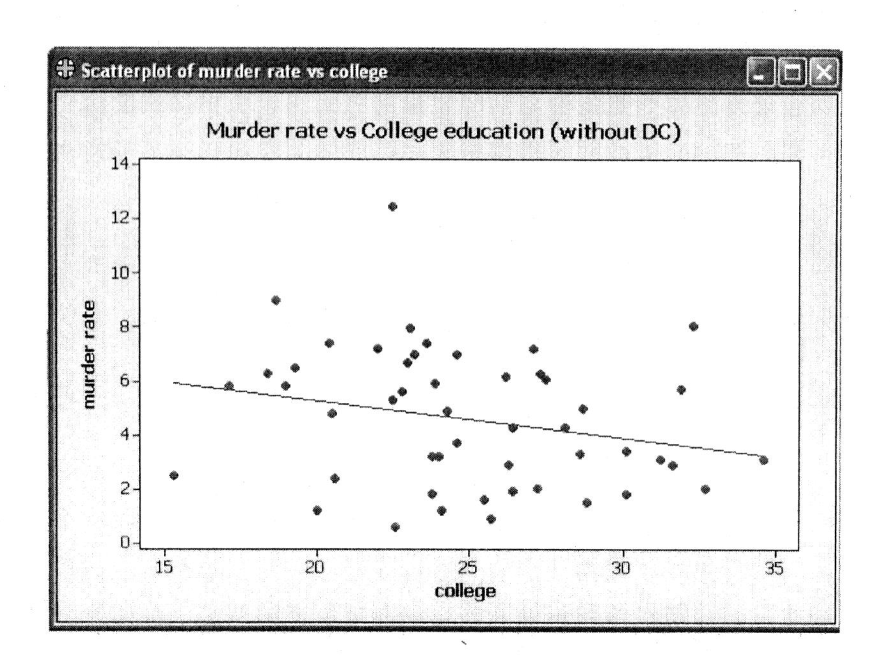

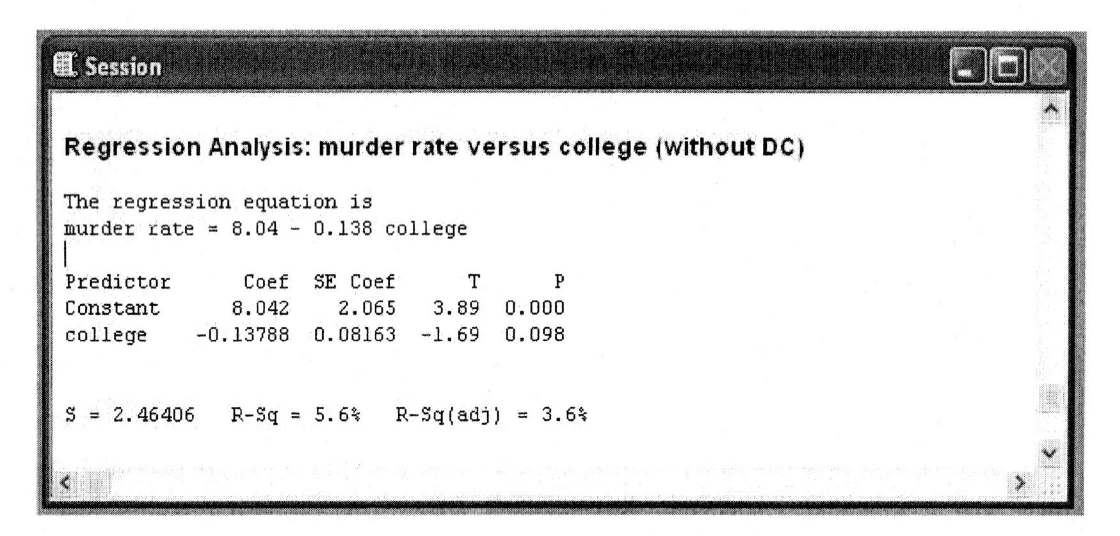
Session

Regression Analysis: murder rate versus college (without DC)

```
The regression equation is
murder rate = 8.04 - 0.138 college

Predictor       Coef   SE Coef      T      P
Constant       8.042     2.065   3.89  0.000
college     -0.13788   0.08163  -1.69  0.098


S = 2.46406   R-Sq = 5.6%   R-Sq(adj) = 3.6%
```

Chapter 3 Exercise 3.46 US Average Annual Temperatures
Trend Lines

Open the worksheet **us_temperatures** from the MINITAB folder of the data disk.

> Choose: **Stat > Regression>Fitted Line Plot**
> Enter: Response: **C2**
> Predictors: **C1**
> Click: **OK**

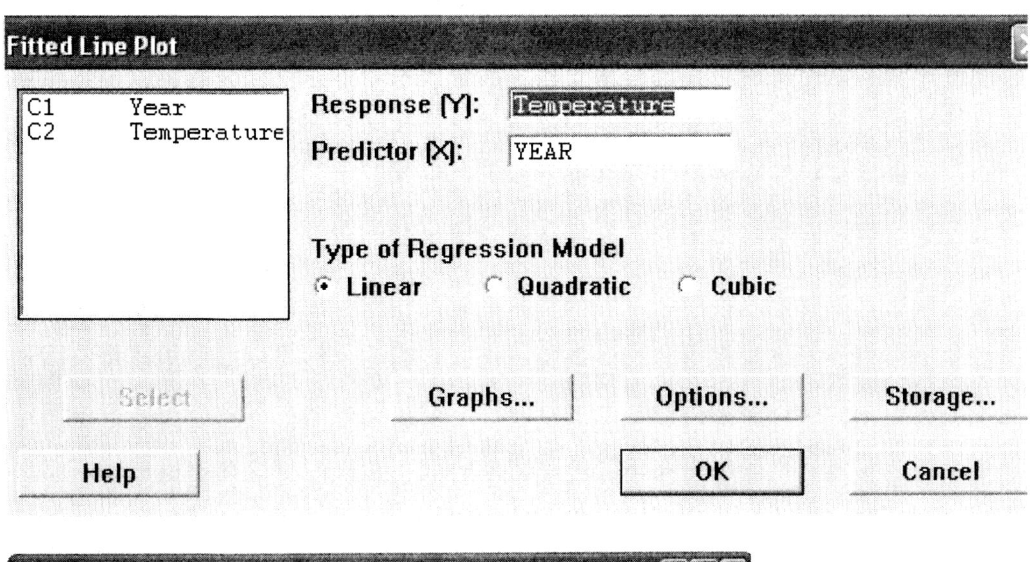

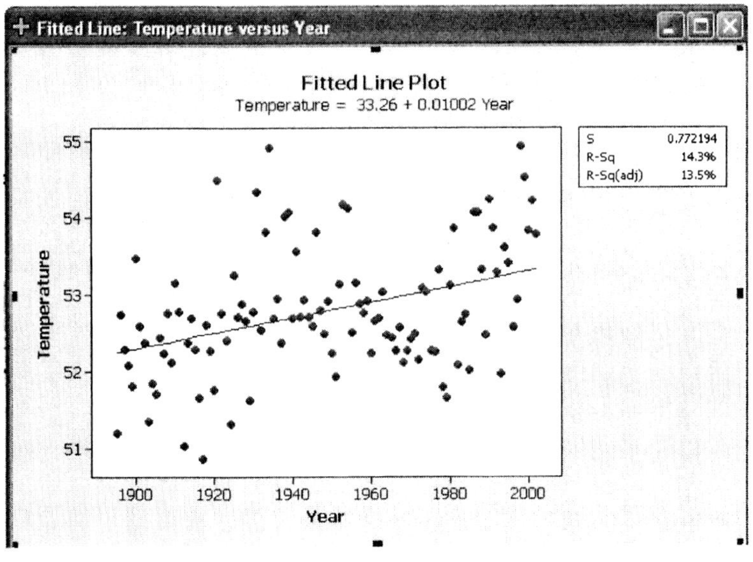

b) The equation, given at the top of the chart is: Temperature = 33.26 + 0.01002 Year.
Predict the temperature for a particular year by substituting that year into the equation.

Chapter 3 Exercise 3.47 Murder and Education
Application of the Regression Model

Using the MINITAB results of the previous problem

a) In the equation: murder rate = - 3.06 + 0.333 college we substitute 15% and 40% we obtain 1.89 and 10.26 respectively.

b) In the equation: murder rate = 8.04 - 0.138 college we substitute 15% and 40% we obtain 5.9 and 2.4 respectively.

Chapter 3 Exercise 3.51 Regression Between Sodium and Sugar
Scatterplot, Correlation, and Regression

Open the worksheet **CEREAL** from the MINITAB folder of the data disk.

a) Scatterplot:
Choose: **Stat > Regression>Fitted Line Plot**
Enter: Response: **C2** Predictor: **C5**

Note: an advantage to the fitted line plot versus the scatterplot is that the regression equation is given in the plot and is added to the scatterplot.

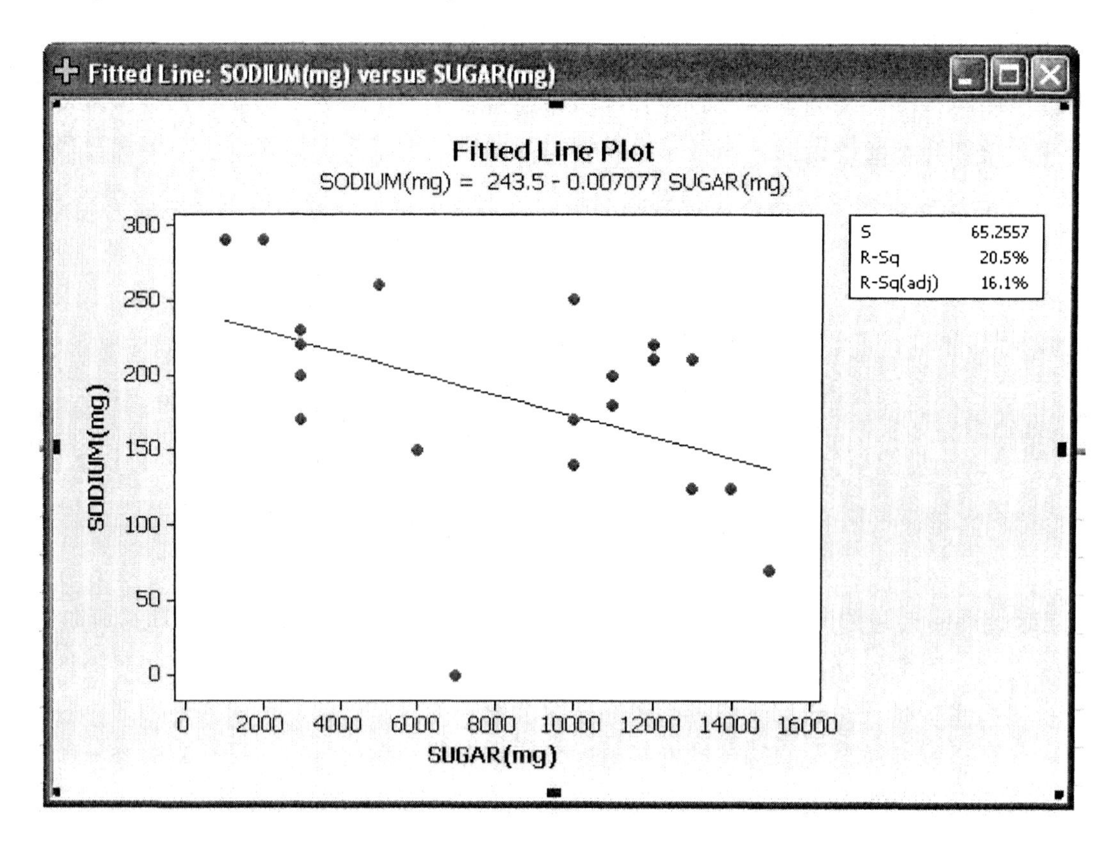

Note the outlier at (7000,0) for Frosted Mini Wheats. Repeating the regression analysis without that point we get:

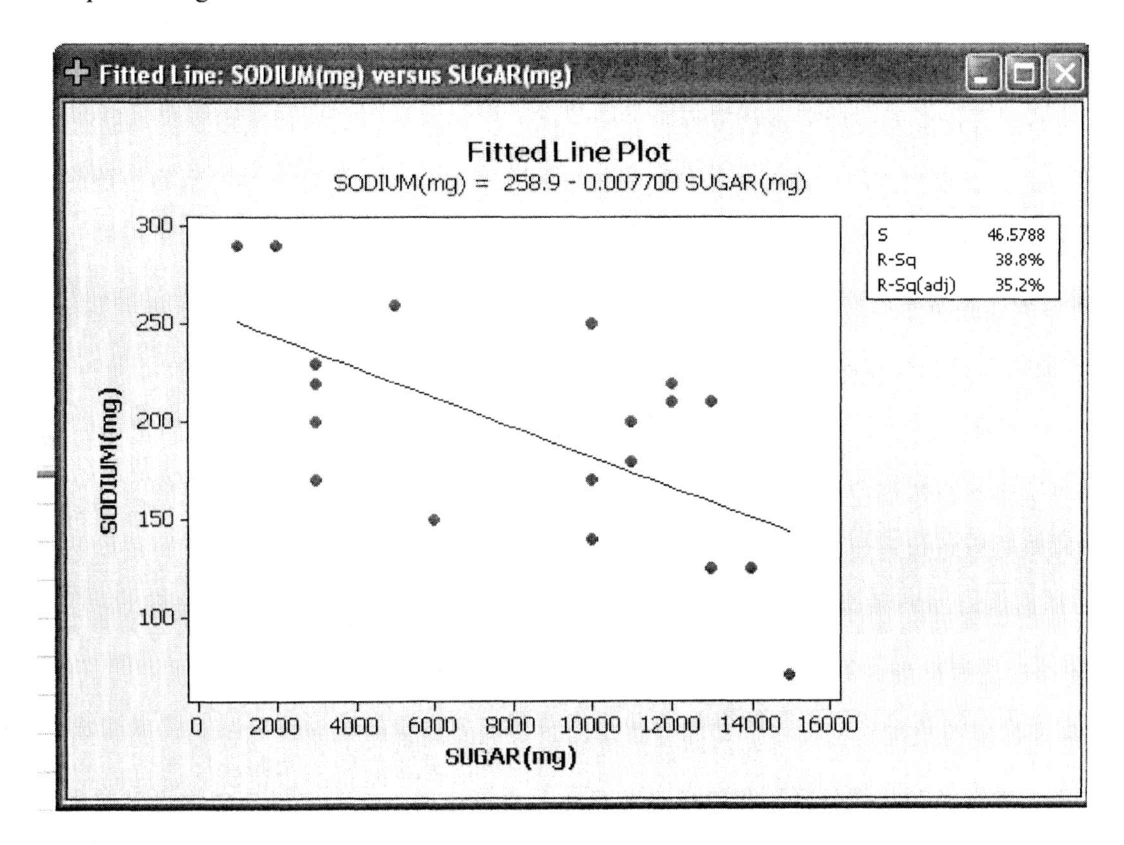

What are your conclusions?

Chapter 3 Exercise 3.57 Education Causes crime?
Scatterplot, Correlation, and Regression

Enter the data into a worksheet.

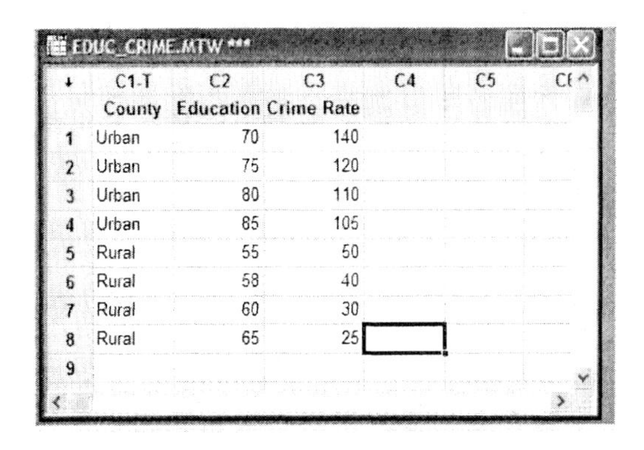

EDUC_CRIME.MTW ***

↓	C1-T	C2	C3	C4	C5
	County	Education	Crime Rate		
1	Urban	70	140		
2	Urban	75	120		
3	Urban	80	110		
4	Urban	85	105		
5	Rural	55	50		
6	Rural	58	40		
7	Rural	60	30		
8	Rural	65	25		
9					

Choose: **Graph > Scatterplot**
Select: **With Groups**
Enter: Y variables: **Crime Rate**
X variables: **Education**
Categorical variables for grouping: **County**
Click: **OK**

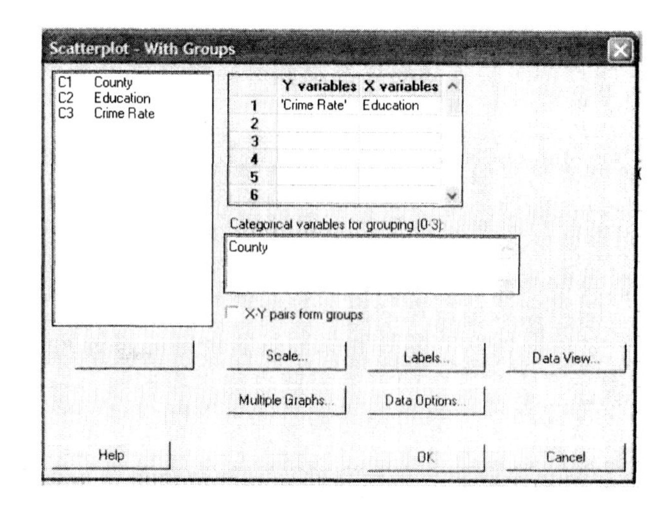

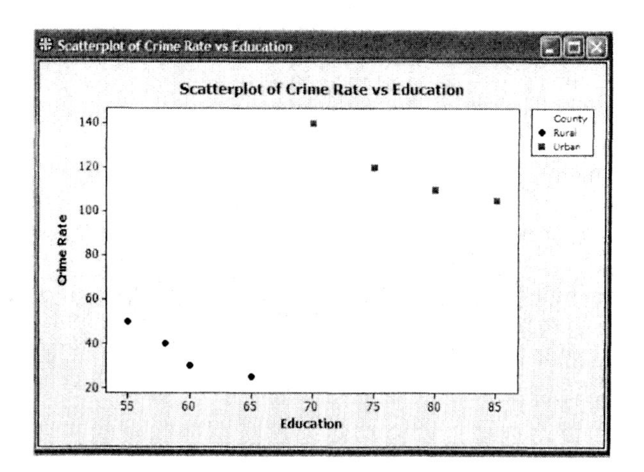

Choose: **Stat > Basic Statistics > Correlation**
Enter: **C2 C3**
Click: **OK**

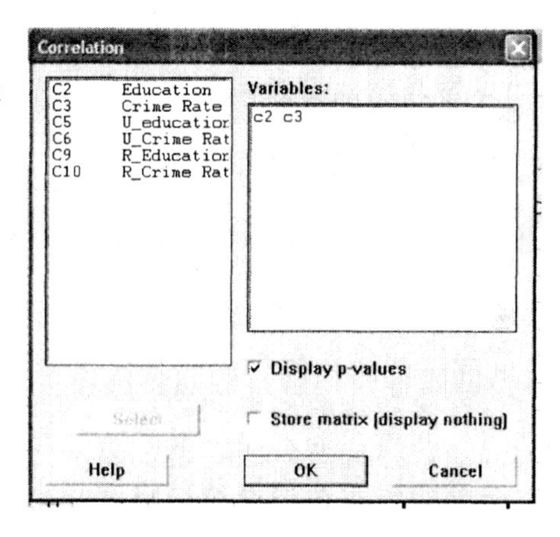

We can see that the correlation is .732 for all 8 data points.

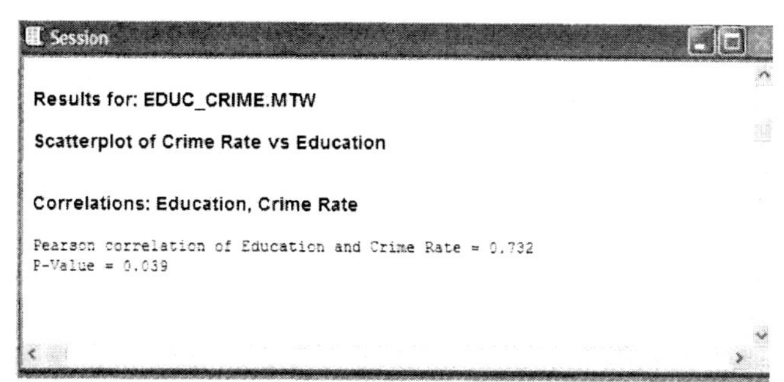

Repeating the same analysis above for rural counties and urban counties alone shows that the correlation of education and crime in urban counties is -9.59 and for rural counties, the correlation is -.948. As we saw in Example 14, urbanization affects the association of education and crime rate.

Chapter 3 Exercise 3.89 High School Graduation Rates and Health Insurance Scatterplot, Correlation, and Regression

Open the worksheet **hs_graduation_rates** from the MINITAB folder of the data disk.

Choose: **Graph > Scatterplot**
Select: **With Regression**
Enter: Y variables: **C4** X variables: **C3**
Click: **Label:**
Title: **Health Insurance vs HS Graduation Rate**
Click: **OK**

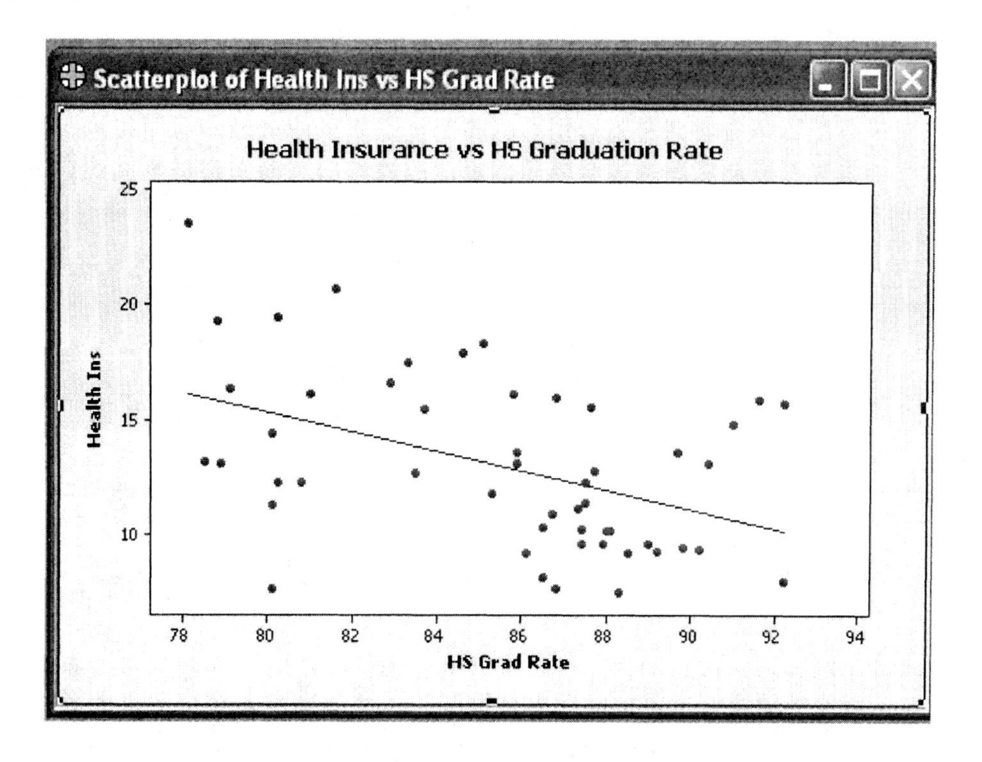

Computing correlation:

Choose: **Stat > Basic Statistics > Correlation**
Enter: Variables: **C3 C4**
Click: **OK**

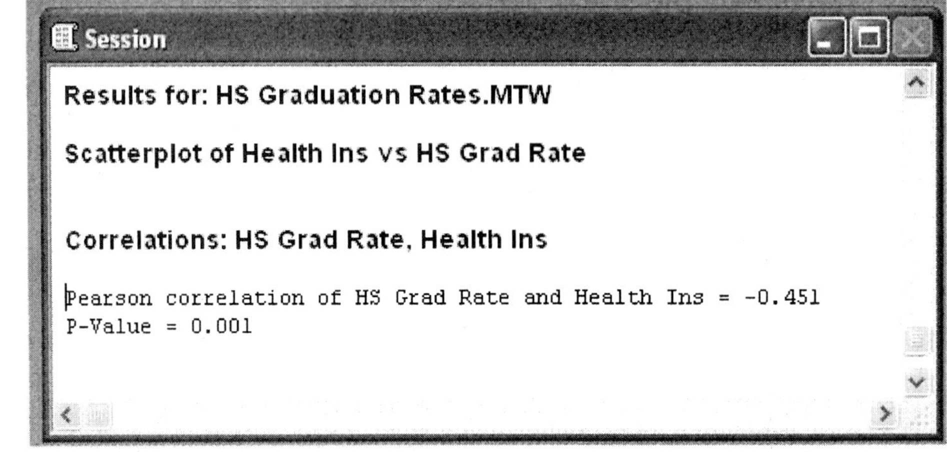

Session
Results for: HS Graduation Rates.MTW
Scatterplot of Health Ins vs HS Grad Rate
Correlations: HS Grad Rate, Health Ins
Pearson correlation of HS Grad Rate and Health Ins = -0.451
P-Value = 0.001

Generating the regression equation:

Choose: **Stat > Regression > Regression**
Enter: Response: **C4**
Predictors: **C3**
Click: **OK**

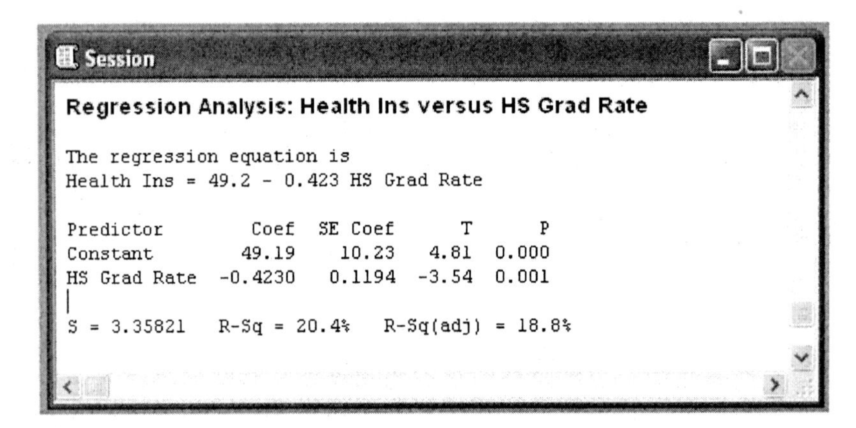

Session
Regression Analysis: Health Ins versus HS Grad Rate
The regression equation is
Health Ins = 49.2 - 0.423 HS Grad Rate
Predictor Coef SE Coef T P
Constant 49.19 10.23 4.81 0.000
HS Grad Rate -0.4230 0.1194 -3.54 0.001
S = 3.35821 R-Sq = 20.4% R-Sq(adj) = 18.8%

Chapter 4 Example 5 Auditing the Accounts of a School District
Random Selection

To use random numbers within MINITAB to select 10 accounts to audit in a school district that has 60 accounts:

Choose: **Calc > Random Data > Integer**

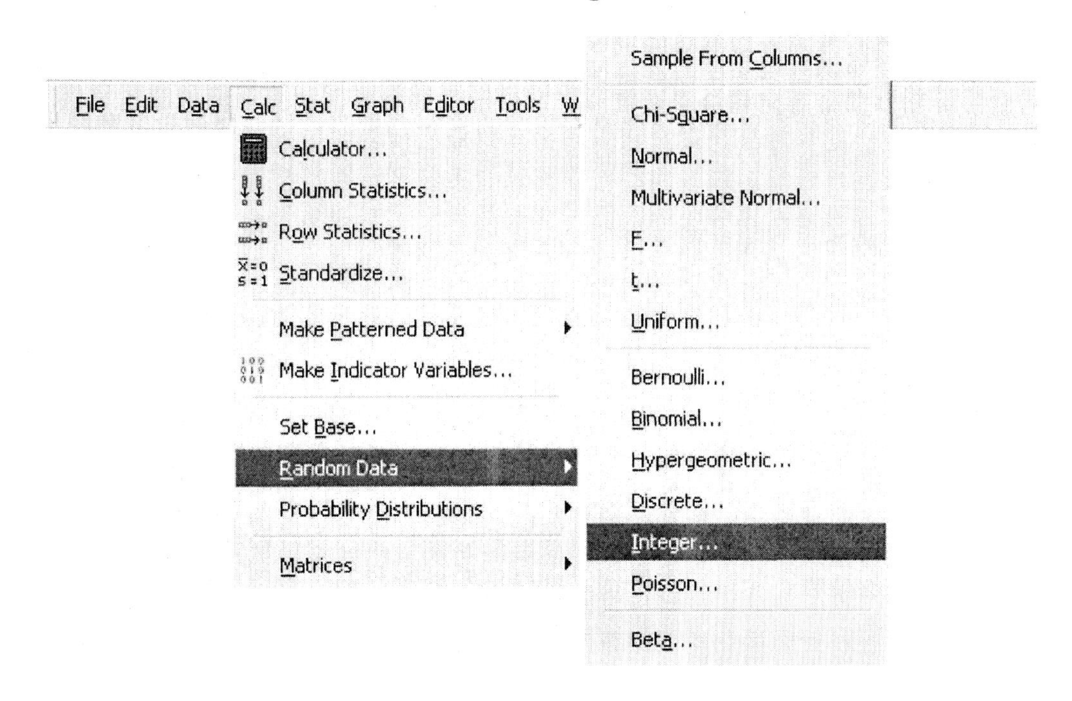

Enter: Generate **10** rows of data
Store in column(s): **C1**
Minimum value: **1**
Maximum value: **60**
Click: **OK**

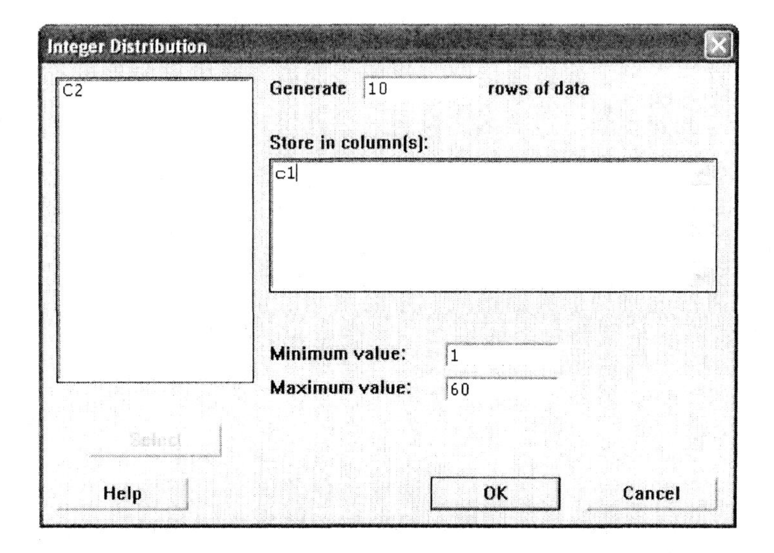

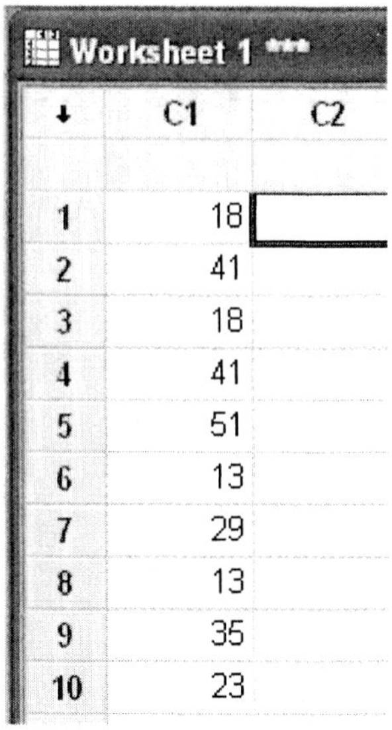

Worksheet 1 ***

↓	C1	C2
1	18	
2	41	
3	18	
4	41	
5	51	
6	13	
7	29	
8	13	
9	35	
10	23	

There is now a random sample of 10 accounts in C1.

Note the repetition of the number 18. We will need to generate additional random numbers to actually obtain a random sample of size 10.

Chapter 4 Exercise 4.18 Auditing Accounts
Random Selection

To use random numbers within MINITAB to select 10 accounts to audit in a school district that has 60 accounts:

Choose: **Calc > Random Data > Integer**
Enter: Generate **10** rows of data
Store in column(s): **C1**
Minimum value: **1**
Maximum value: **60**
Click: **OK**

There is now a random sample of 10 accounts in C1.

Issuing the same commands again, but storing in C2, we can see that a different random sample of size 10 has been selected. Since this is a random process, the results are different every time.

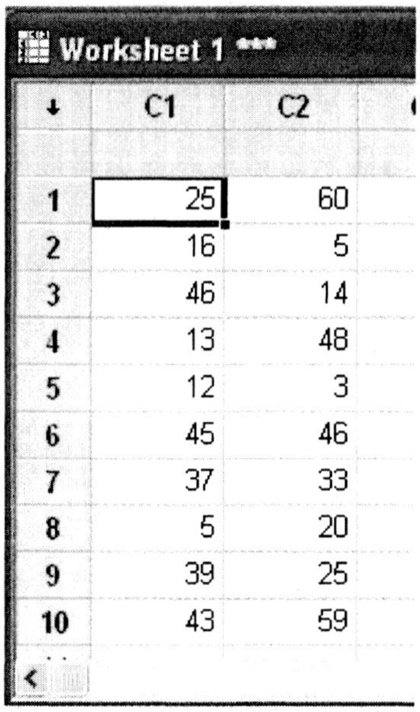

Worksheet 1 ***

↓	C1	C2
1	25	60
2	16	5
3	46	14
4	13	48
5	12	3
6	45	46
7	37	33
8	5	20
9	39	25
10	43	59

Chapter 6 Example 7 What IQ Do You Need to Get Into MENSA?
Determining Normal Probabilities

Stanford-Binet IQ scores are approximately normally distributed with $\mu = 100$ and $\sigma = 16$. To be eligible for MENSA, you must rank at the 98th percentile (i.e. you must score in the top 2%). Find the lowest IQ score that still qualifies for Mensa membership.

Choose:
Calc > Probability Distributions > Normal

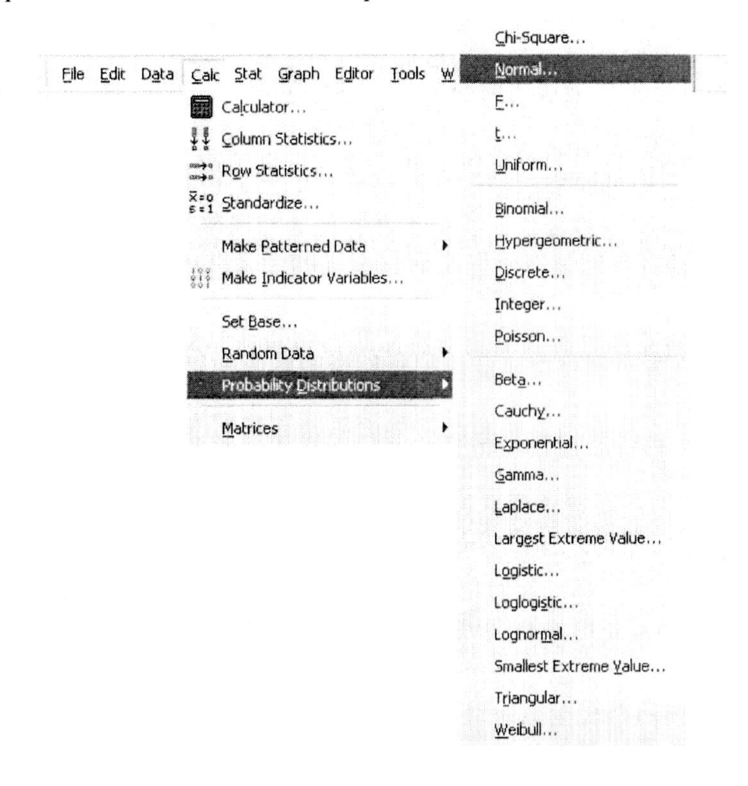

Select:
Inverse Cumulative Probability
Enter: Mean : **100**
Standard deviation: **16**
Enter: Input constant: **.98**
(area to the left of the value we wish to find)
Click: **OK**

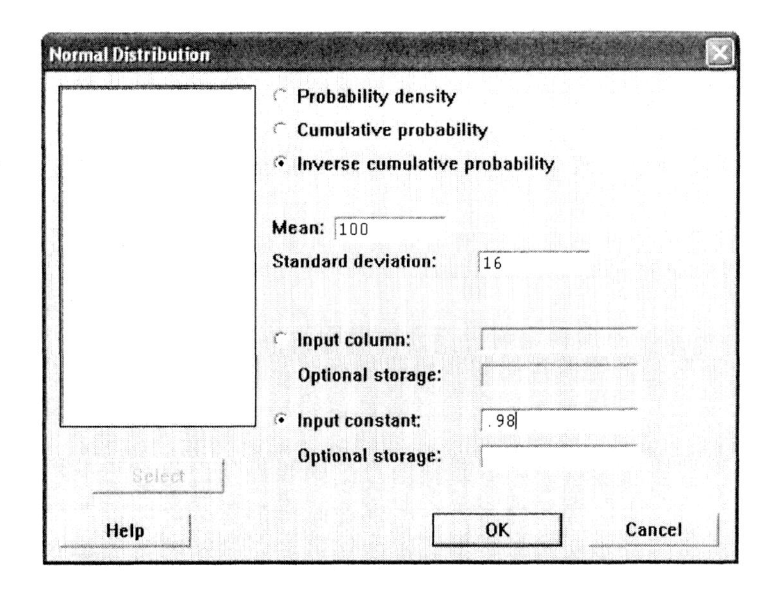

The X-value should appear in the Session window. Notice that the test score required to join Mensa is 133.

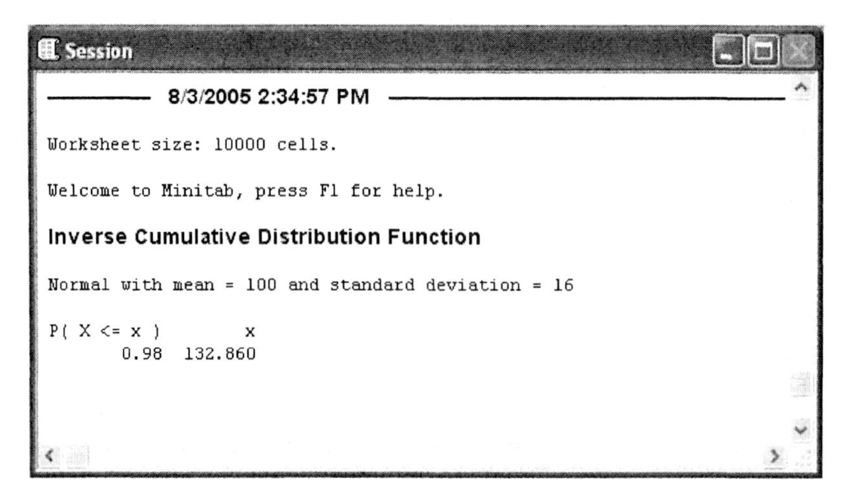

Chapter 6 Example 8 Finding Your Relative Standing on the SAT
Determining Normal Probabilities

SAT scores are approximately normally distributed with $\mu = 500$ and $\sigma = 100$. If one of your SAT scores was 650, what percentage of SAT scores were higher than yours?

Choose:
Calc > Probability Distributions > Normal
Select: **Cumulative probability**
Enter: Mean : **500**
Standard deviation: **100**
Enter: Input constant: **650**
Click: **OK**

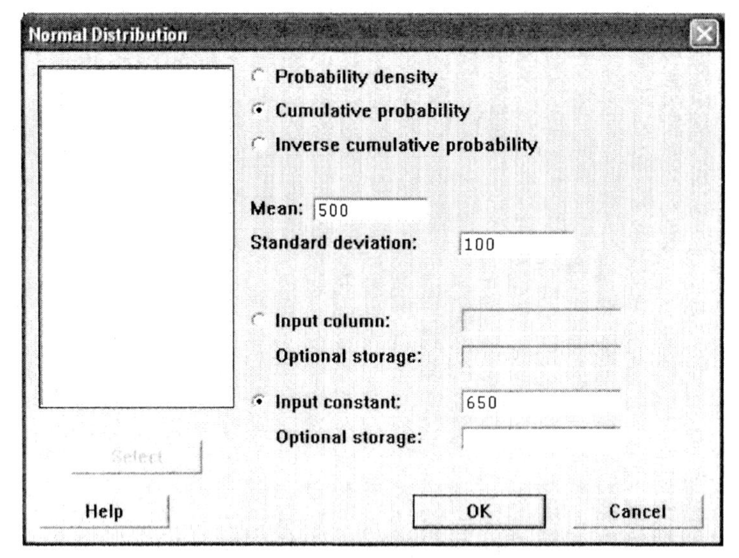

Notice that the result displayed in the Session window is the probability that X is less than 650.

Since we were asked to determine the percentage of SAT scores higher than 650 we must calculate 1 - 0.9332 = 0.0668. Only about 7% of SAT scores fall above 650.

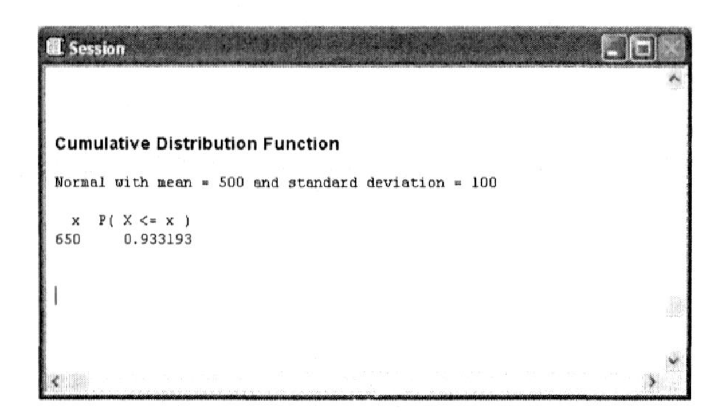

Chapter 6 Example 9 What Proportion of Students Get a Grade of B?
Determining Normal Probabilities

On the midterm exam, an instructor always gives a grade of B to students who score between 80 and 90. One year, the scores on the exam have an approximately normal distribution with $\mu = 83$ and $\sigma = 5$. About what proportion of students get a B?

We will have to compute two cumulative probabilities, one for X = 80 and one for X = 90, and then subtract the two probabilities.

Choose: **Calc > Probability Distributions > Normal**
Select: **Cumulative probability**
Enter: Mean : **83**
Standard deviation: **5**
Select: Input constant:
Enter: **80**
Click: **OK**

then repeat using Input constant: **90**

It follows that about 0.9192 – 0.2743 = 0.6449, or about 64%, of the exam scores were in the B range.

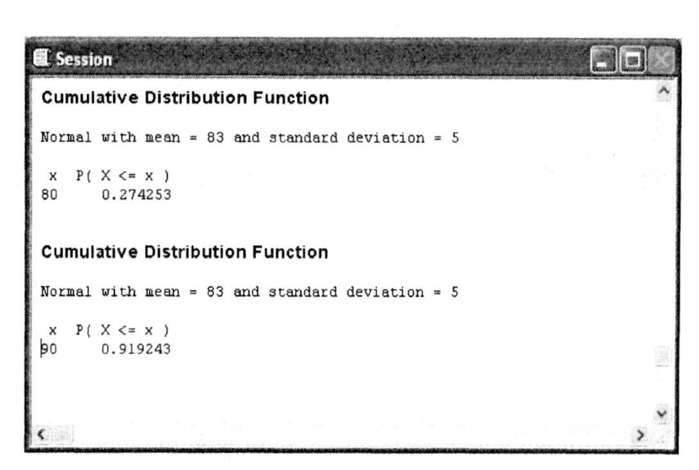

Chapter 6 Exercise 6.23 Blood pressure
Determining Normal Probabilities

In Canada, systolic blood pressure readings has are normally distributed with $\mu = 121$ and $\sigma = 16$. A reading above 140 is considered to be high blood pressure. What proportion of Canadians suffers from high blood pressure?

Choose: **Calc > Probability Distributions > Normal**

Select: **Cumulative probability**
Enter: Mean : **121**
Standard deviation: **16**

Select: Input constant:
Enter: **140**
Click: **OK**

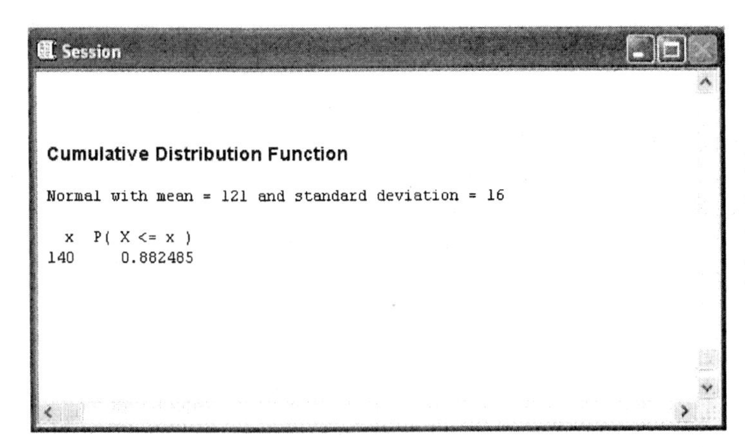

It follows that about 1 – 0.8825 = 0.1175, or about 12%, of Canadians suffer from high blood pressure.

We are also asked to determine the proportion of Canadians having systolic blood pressure in the range from 100 to 140.

We have already computed the P(X < 140) = 0.8825, so we just need to determine P(X < 100) and subtract the two results.

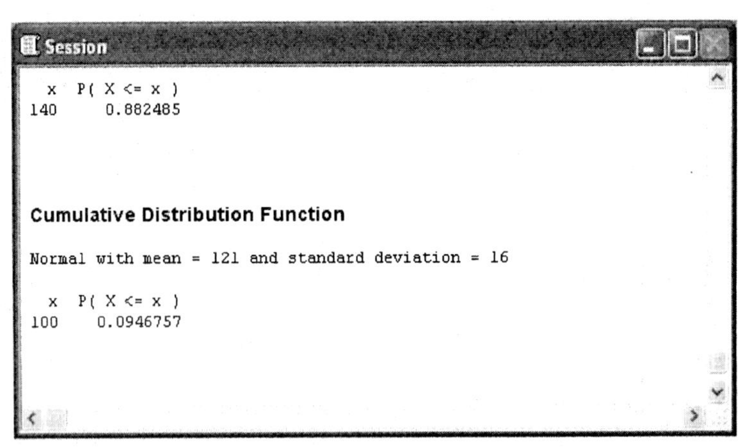

Approximately 0.8825 – 0.0947 = 0.7878, or 78% of Canadians have systolic blood pressures in the range from 100 to 140.

Chapter 6 Exercise 6.25 Energy Use
Determining Normal Probabilities

Suppose energy use is normally distributed with $\mu = 673$ kilowatt-hours and $\sigma = 556$ kilowatt-hours.

Find the probability that household electricity use was greater than 1000 kilowatt-hours.
Choose: **Calc > Probability Distributions > Normal**

Select: **Cumulative probability**
Enter: Mean : **673**
Standard deviation: **556**
Select: Input constant:
Enter: **1000**
Click: **OK**

```
Welcome to Minitab, press F1 for help.

Cumulative Distribution Function

Normal with mean = 673 and standard deviation = 556

   x  P( X <= x )
1000     0.721777
```

It follows that the probability that household electricity use is greater than 1000 kilowatt-hours is about 1 – 0.72177 = 0.27823, or 28%.

Chapter 6 Exercise 6.26 Apartment Rentals
Determining Normal Probabilities

Find the proportion of rentals that are at least than $1000.
Choose: **Calc > Probability Distributions > Normal**

Select: **Cumulative probability**
Enter: Mean : **700**
Standard deviation: **150**
Select: Input constant:
Enter: **1000**
Click: **OK**

```
1000     0.721777

Cumulative Distribution Function

Normal with mean = 700 and standard deviation = 150

   x  P( X <= x )
1000     0.977250
```

It follows that about 1 – 0.97725 = 0.02275, or about 2%, of rentals are at least $1000.

We are also asked to determine the proportion of rentals that are less than $500.

Choose: **Calc > Probability Distributions > Normal**

Select: **Cumulative probability**
Enter: Mean : **700**
Standard deviation: **150**
Select: Input constant:
Enter: **500**
Click: **OK**

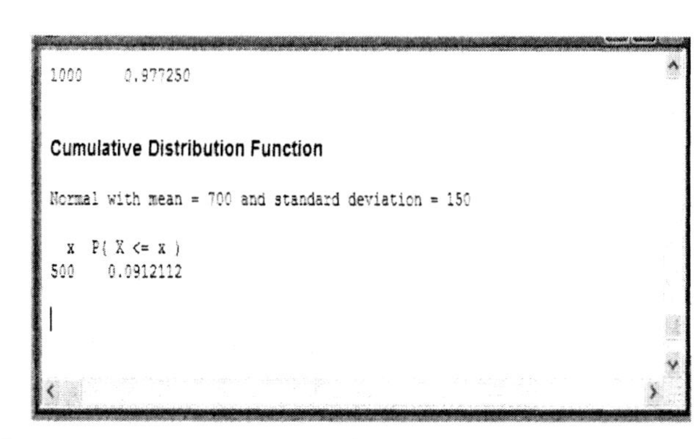

We see that .0912112, or about 9%, of rentals are less than $500.

We are also asked to determine the proportion of rentals that are between $500 and $1,000 a month. We have already computed the P(X <500) = 0.0912112 and that P(X<1000) = .977250, so we just need to subtract the two results: .977250-.0912112=.8860388. About 89% of the rentals are between $500 and $1000.

Chapter 6 Exercise 6.27 Mental Development Index
Determining Normal Probabilities

The Mental Development Index of the Bayley Scales of Infant Development is a standardized measure used in observing infants over time. MDI is approximately normally distributed with $\mu = 100$ and $\sigma = 16$.

What proportion of children has MDI of at least 120?

Choose: **Calc > Probability Distributions > Normal**

Select: **Cumulative probability**
Enter: Mean : **100**
Standard deviation: **16**
Select: Input constant:
Enter: **120**
Click: **OK**

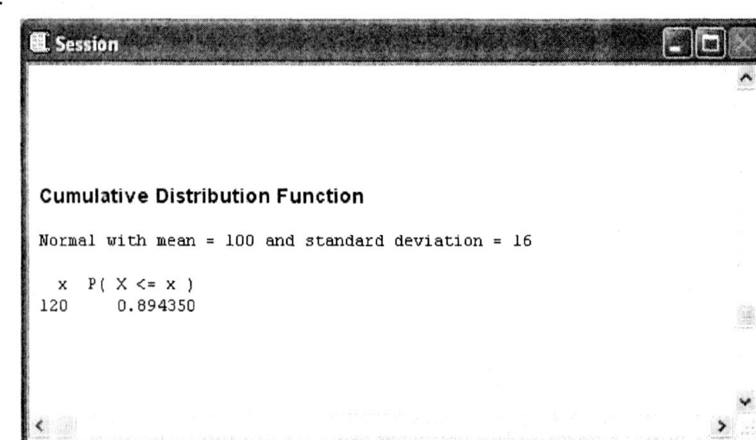

It follows that about 1 – 0.8944 = 0.1056, or about 11%, of children have an MDI of at least 120.

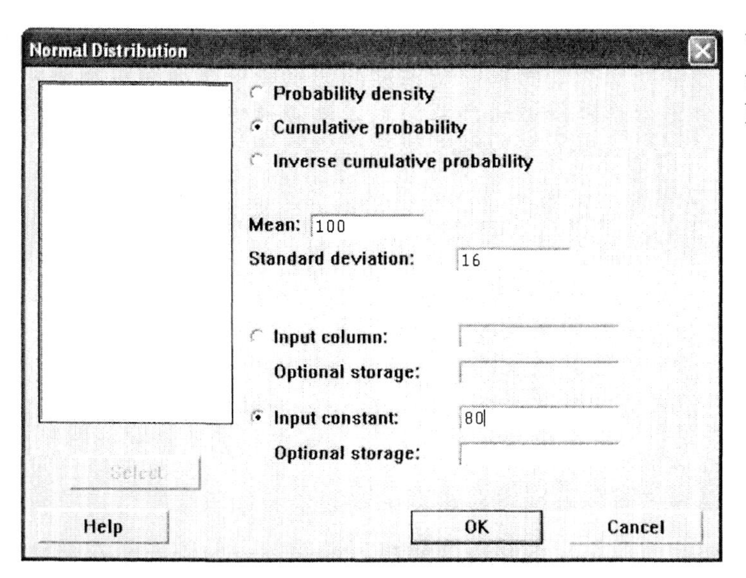

We are also asked to determine the proportion of children having an MDI of at least 80.

Approximately 1.0 – 0.1057 = 0.8943, or 89% of children have an MDI of at least 80.

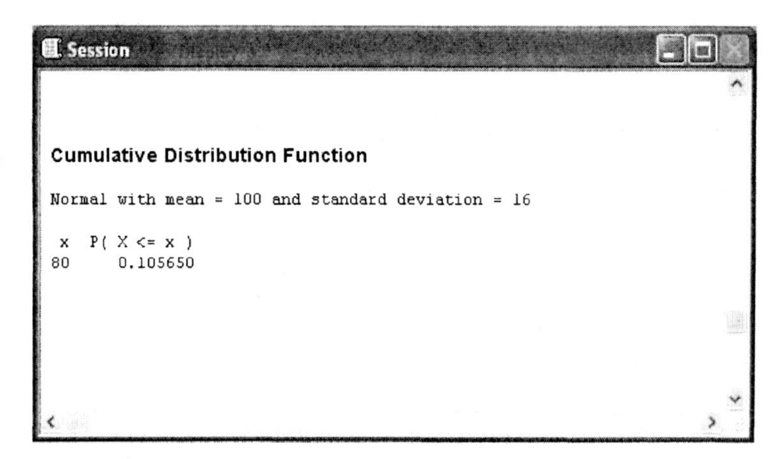

To determine the MDI score that is the 99th percentile:

Choose: **Calc >**
Probability Distributions > Normal
Select: **Inverse Cumulative Probability**
Enter: Mean : **100**
Standard deviation: **16**
Select: Input constant:
Enter: **.99**
Click: **OK**

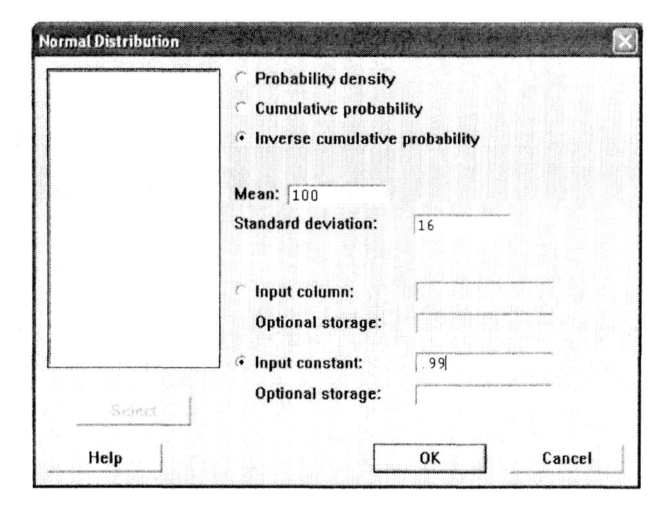

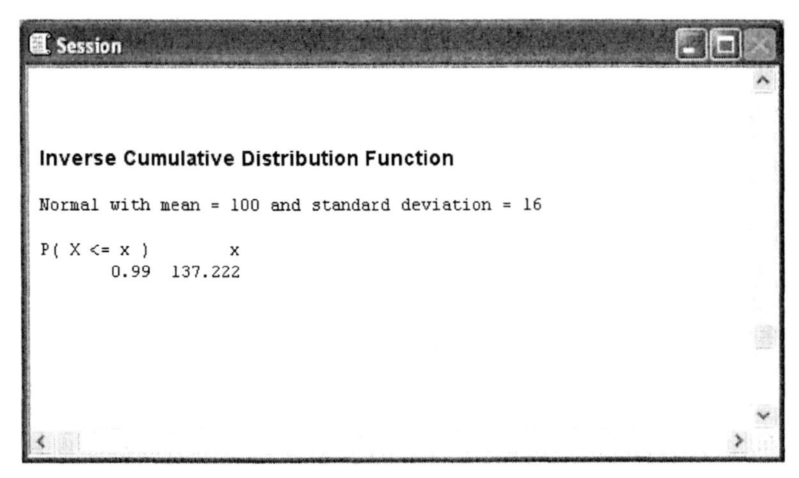

An MDI of 138 is the 99th percentile.

To determine the MDI such that only 1% of the population has an MDI below it, we will use the Input constant: **.01**

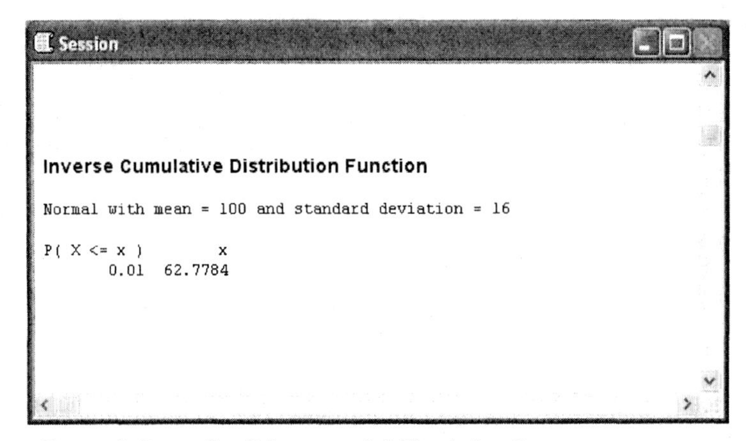

Approximately 1% of the population has an MDI less than 63.

Chapter 6 Example 12 Are Women Passed Over for Managerial Training? Determining Binomial Probabilities

Let X denote the number of females selected in a random sample of ten employees (X can have any value 0, 1, 2, 3, …, 10). We can do this using the following commands:

Choose: **Calc > Make Patterned Data > Simple Set of Numbers**
Enter: Store patterned data in: **C1**
From first value: **1**
To last value: **10**
List each value **1**
List whole sequence **1**
Click: **OK**

To find the binomial probability distribution:
Choose: **Calc > Probability Distributions > Binomial**

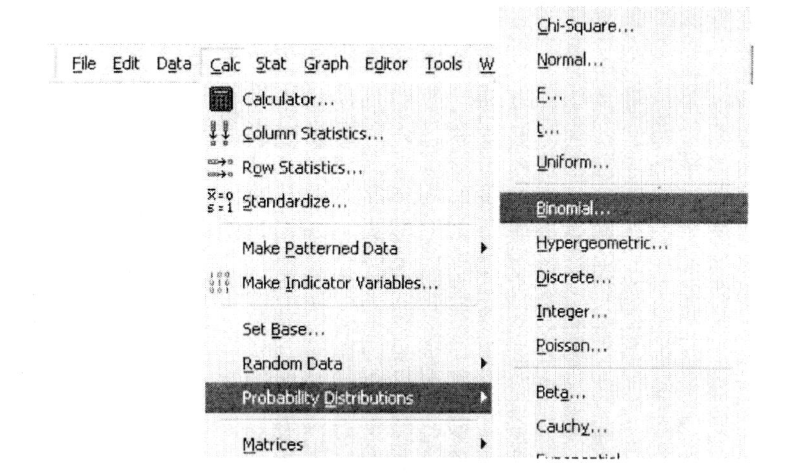

Select: **Probability**
Enter: Number of trials: **10**
Probability of success: **.5**
Input column: **C1**
Click: **OK**

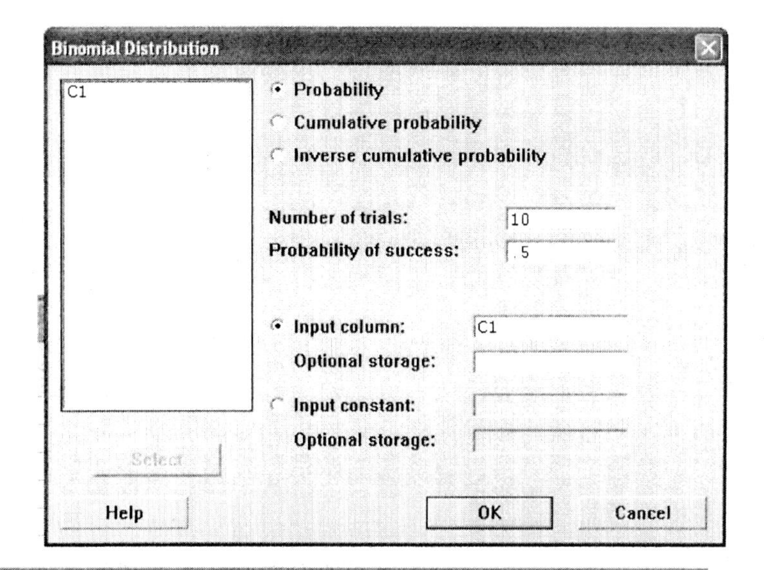

This will generate the distribution in the Session Window. We can see that if there were truly no gender bias, the probability of 0 females in 10 selections would be P(X = 0) = .00097.

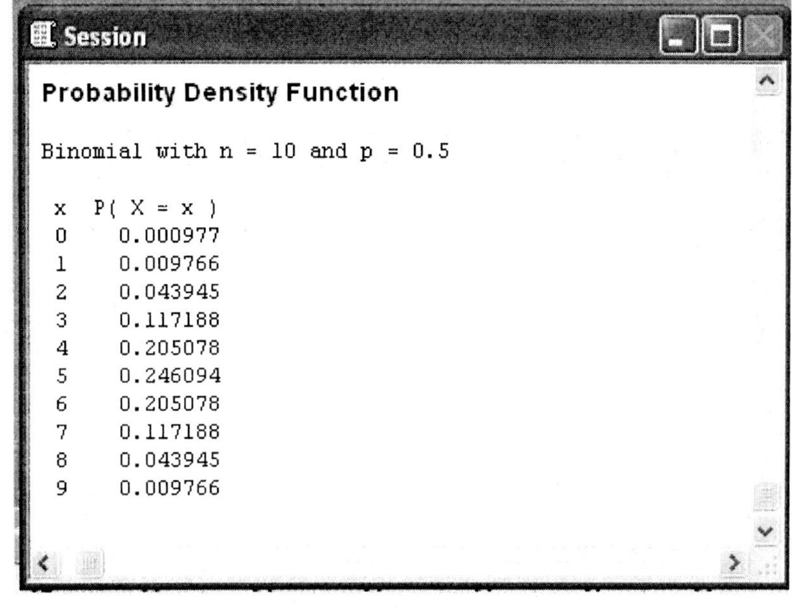

Session
Probability Density Function
Binomial with n = 10 and p = 0.5
x P(X = x)
0 0.000977
1 0.009766
2 0.043945
3 0.117188
4 0.205078
5 0.246094
6 0.205078
7 0.117188
8 0.043945
9 0.009766

Chapter 6 Example 14 How Can We Check for Racial Profiling? Determining Binomial Probabilities

In this problem, generate numbers 0 – 262 and store in C1.

Then generate the probability distribution:

Choose: **Calc > Probability Distributions > Binomial**
Select: **Probability**
Enter: Number of trials: **262**
Probability of success: **.422**
Input column: **C1**
Click: **OK**

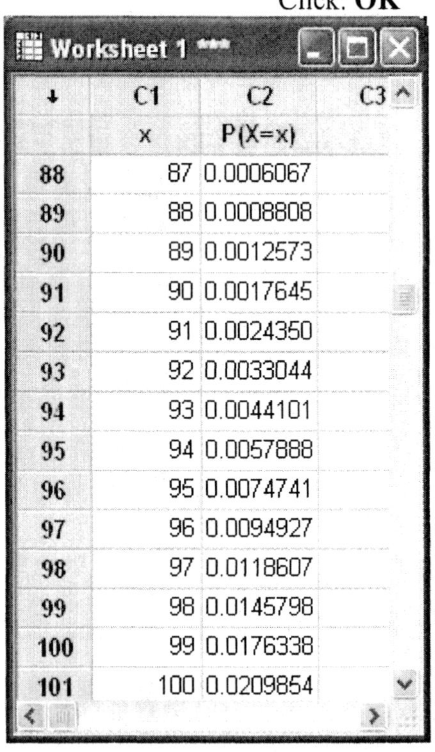

Worksheet 1 ***

↓	C1	C2	C3
	x	P(X=x)	
88	87	0.0006067	
89	88	0.0008808	
90	89	0.0012573	
91	90	0.0017645	
92	91	0.0024350	
93	92	0.0033044	
94	93	0.0044101	
95	94	0.0057888	
96	95	0.0074741	
97	96	0.0094927	
98	97	0.0118607	
99	98	0.0145798	
100	99	0.0176338	
101	100	0.0209854	

Here is part of the table generated. Notice where we begin to see results other than zero. What is the probability that $X \geq 207$? What conclusions can you draw from this?

Chapter 6 Exercise 6.42 Exit Poll
Mean and Standard Deviation of the Binomial Random Variable X

a) The data is binary (vote for or against), the voters were randomly selected, and each voter is separate and independent from another voter

N = 3000   p = .5   1-p = .5

Generate the probabilities.

First fill column 1 with numbers 0-3000.

Then generate the probability distribution:

Choose: **Calc > Probability Distributions > Binomial**
Select: **Probability**
Enter: Number of trials: **3000**
Probability of success: **.5**
Input column: **C1**
Optional storage: **C2**
Click: **OK**

b) Mean and Standard Deviation of the Binomial Distribution

Using the actual values for n, p and 1-p, the calculator within Calc>Calculator in MINITAB can be used to determine the mean and standard deviation for the binomial distribution

Recall that the mean and standard deviation for a general probability distribution is

$$\mu = \sum xP(x) \quad \text{and} \quad \sigma = \sqrt{\sum (x-u)^2 P(x)}$$

For the binomial distribution we can use the following expressions:

n * p   store result in "mean"
SQRT(n*p*(1-p))   store result in "std dev"

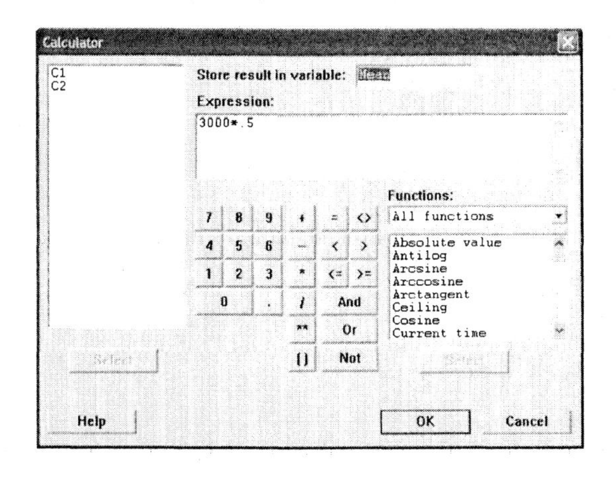

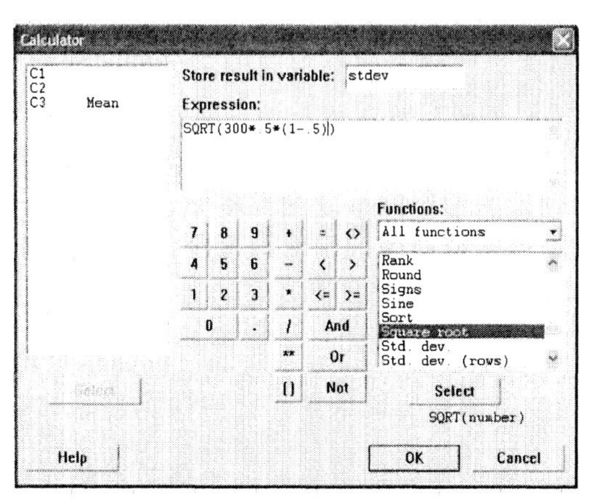

This will produce the following in the worksheet:

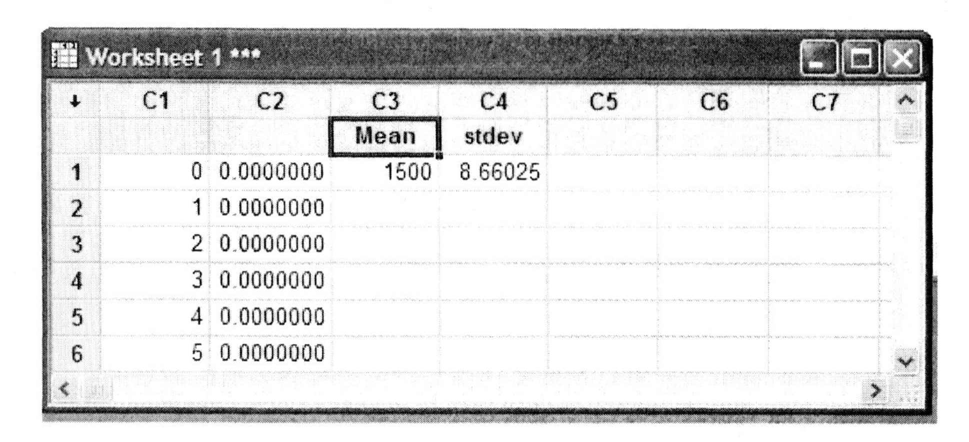

Worksheet 1 ***

↓	C1	C2	C3	C4	C5	C6	C7
			Mean	stdev			
1	0	0.0000000	1500	8.66025			
2	1	0.0000000					
3	2	0.0000000					
4	3	0.0000000					
5	4	0.0000000					
6	5	0.0000000					

Chapter 6 Exercise 6.43 Jury Duty
Determining Binomial Probabilities

a) Check the three criteria for being binomial
b) Let n = 12, p = .4 . Enter 0 through 12 into C1, then
   Choose: **Calc > Probability Distributions > Binomial**

   Select: **Probability**
   Enter: Number of trials: **12**
   Probability of success: **.4**
   Input column: **C1**
   Optional storage: **C2**
   Click: **OK**

Worksheet 1 ***

↓	C1	C2	C3	C4
1	0	0.002177		
2	1	0.017414		
3	2	0.063852		
4	3	0.141894		
5	4	0.212841		
6	5	0.227030		
7	6	0.176579		
8	7	0.100902		
9	8	0.042043		
10	9	0.012457		
11	10	0.002491		
12	11	0.000302		
13	12	0.000017		

Chapter 7 Example 2 Exit Poll of California Voters Revisited
The Mean and Standard Deviation of the Sampling Distribution of a Proportion

Considering the exit poll of 2705 voters, recall we had n = 2705 and p = .5
Enter patterned data into C1 and store the binomial distribution in C2.

Choose: **Calc > Make Patterned Data > Simple Set of Numbers**
Enter: Store patterned data in: **C1**
From first value: **0**
To last value: **2705**
List each value **1**
List whole sequence **1**
Click: **OK**

Choose: **Calc > Probability Distributions > Binomial**

Select: **Probability**
Enter: Number of trials: **2705**
Probability of success: **.5**
Input column: **C1**
Optional storage: **C2**
Click: **OK**

To calculate the mean:
Choose **Calc > Calculator**
Calculate the Mean using the formula
n * p

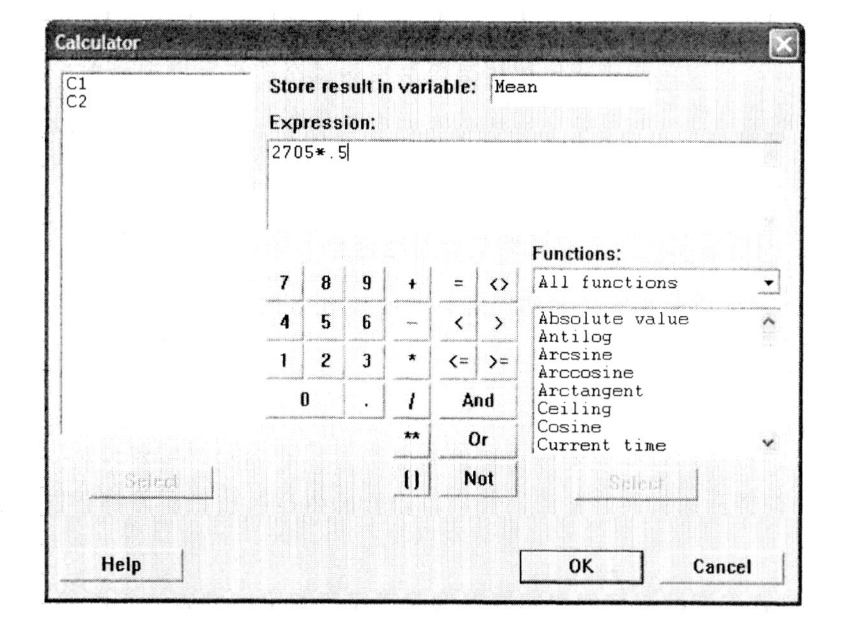

To calculate the standard deviation using Minitab's calculator

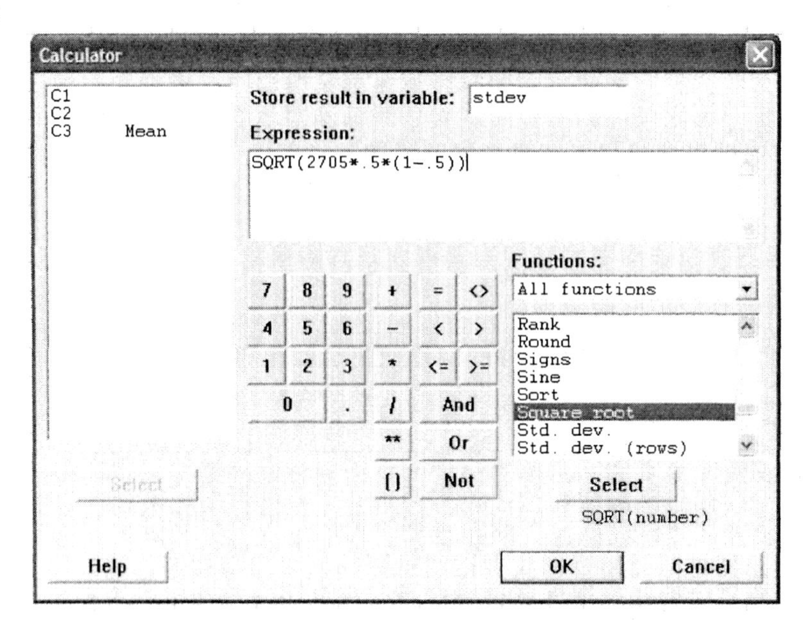

This gives the following results

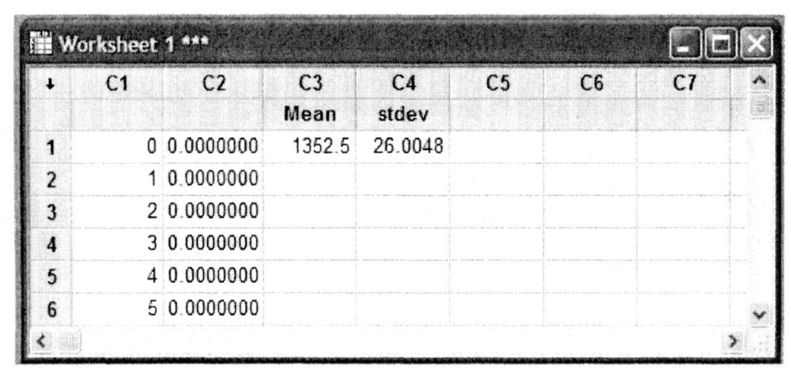

Worksheet 1 ***

↓	C1	C2	C3	C4	C5	C6	C7
			Mean	stdev			
1	0	0.0000000	1352.5	26.0048			
2	1	0.0000000					
3	2	0.0000000					
4	3	0.0000000					
5	4	0.0000000					
6	5	0.0000000					

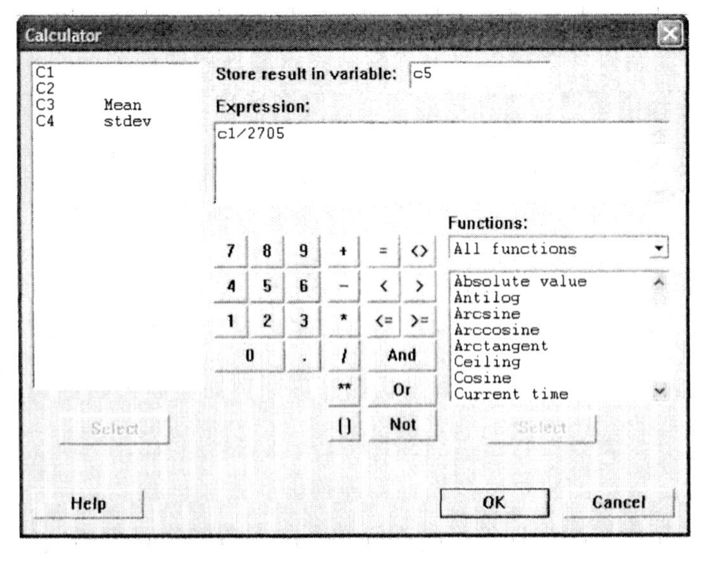

To get the proportions, divide the binomial random variable X(in C1) by 2705. Let Minitab do the work using the Calculator. Store results in C5.

Use the calculator to find the mean of the proportions we just stored in C5.

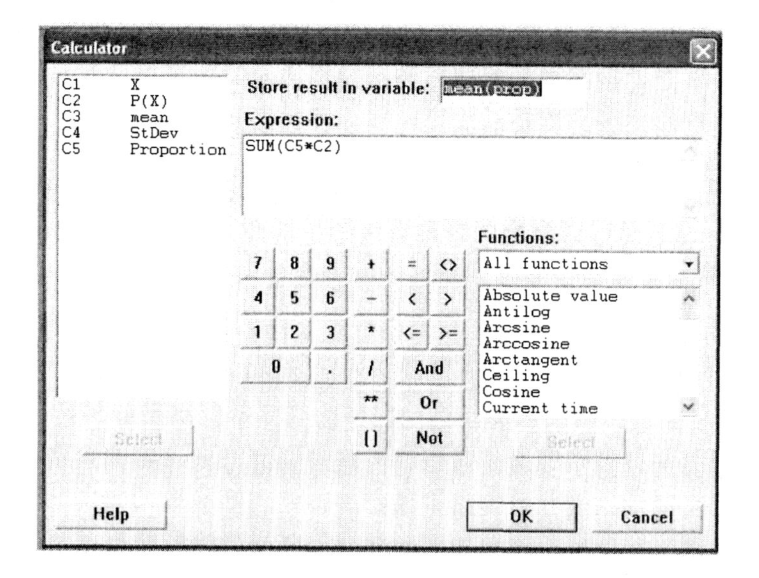

As you see in the following table, it is .5, just like the population proportion.

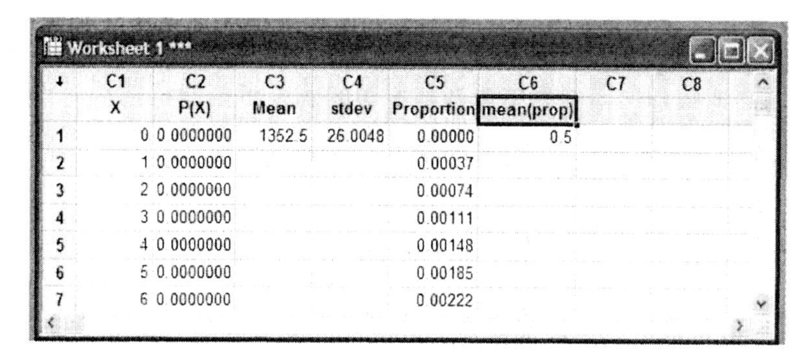

Worksheet 1 ***

↓	C1	C2	C3	C4	C5	C6	C7	C8
	X	P(X)	Mean	stdev	Proportion	mean(prop)		
1	0	0.0000000	1352.5	26.0048	0.00000	0.5		
2	1	0.0000000			0.00037			
3	2	0.0000000			0.00074			
4	3	0.0000000			0.00111			
5	4	0.0000000			0.00148			
6	5	0.0000000			0.00185			
7	6	0.0000000			0.00222			

Use the calculator to find the standard deviation of the proportions and store in C7.

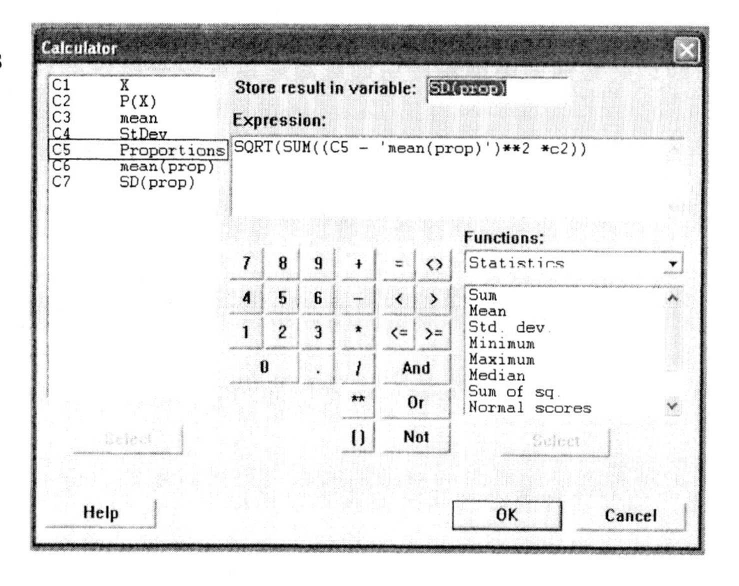

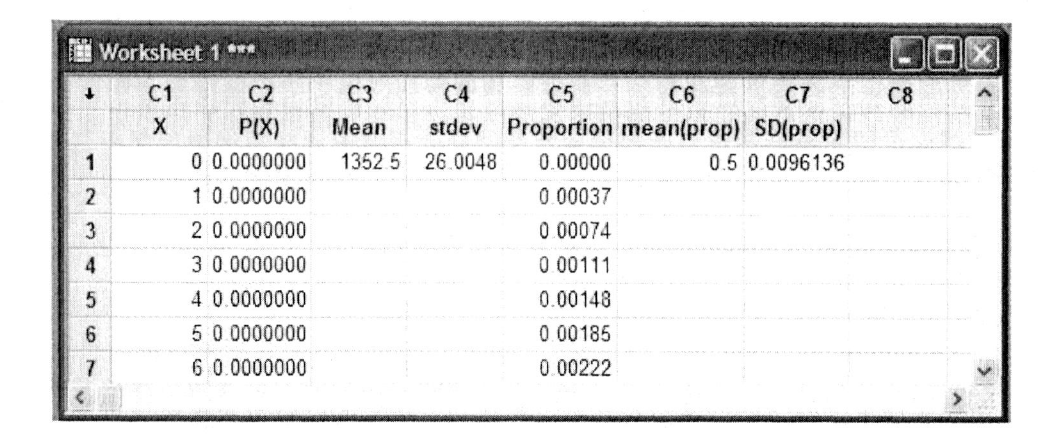

Worksheet 1 ***

↓	C1	C2	C3	C4	C5	C6	C7	C8
	X	P(X)	Mean	stdev	Proportion	mean(prop)	SD(prop)	
1	0	0.0000000	1352.5	26.0048	0.00000	0.5	0.0096136	
2	1	0.0000000			0.00037			
3	2	0.0000000			0.00074			
4	3	0.0000000			0.00111			
5	4	0.0000000			0.00148			
6	5	0.0000000			0.00185			
7	6	0.0000000			0.00222			

We got a standard deviation of .0096136 for the sampling distribution of the sample proportion.

This is the same as if we would use the formula $\sigma = \sqrt{\dfrac{p(1-p)}{n}}$

Chapter 7 Exercise 7.5 Other Scenario for Exit Poll
The Sampling Distribution of Sample Proportion

This problem repeats Example 2, but uses p = .559 instead of .5

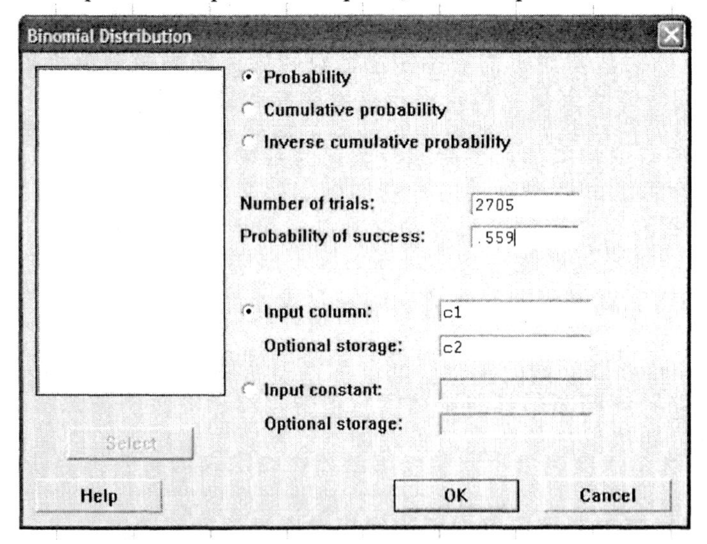

This will generate the sampling distribution for the number of voters who voted for Schwarzenegger. N = 2705 and p = .559 and store the binomial probabilities in C2.

To find the mean and standard deviation, use the **Calc> Calculator** as shown

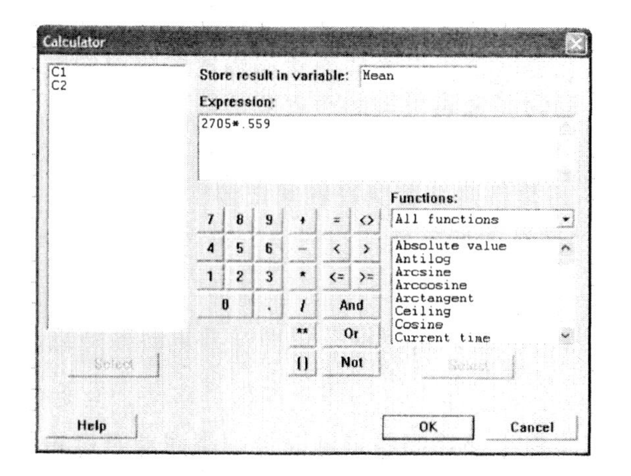

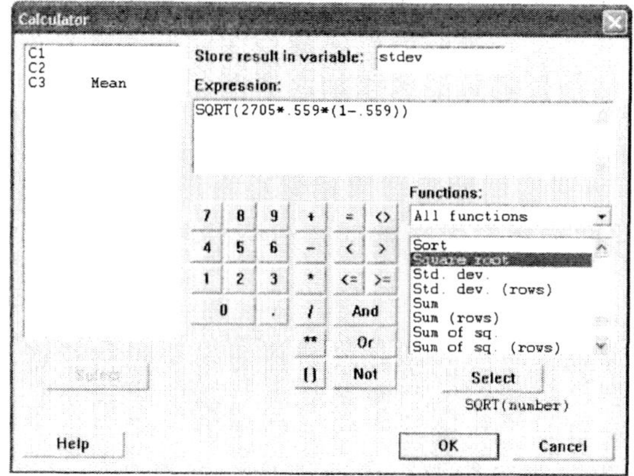

Note the answers stored in C3 and C4:

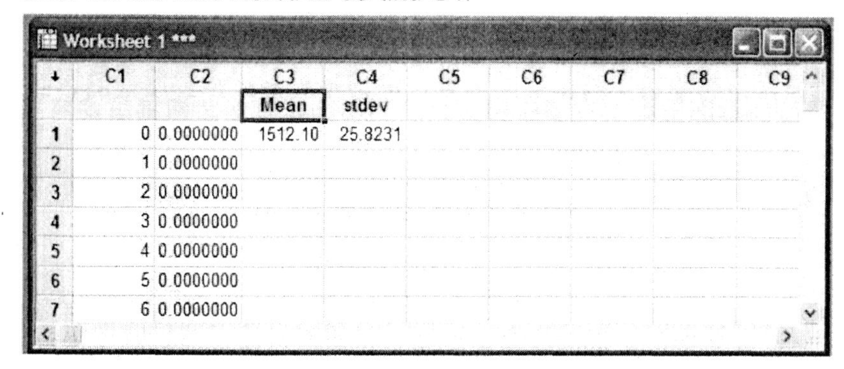
Worksheet 1 ***

↓	C1	C2	C3	C4	C5	C6	C7	C8	C9
			Mean	stdev					
1	0	0.0000000	1512.10	25.8231					
2	1	0.0000000							
3	2	0.0000000							
4	3	0.0000000							
5	4	0.0000000							
6	5	0.0000000							
7	6	0.0000000							

b) Now we will find the mean and standard deviation of sampling distribution of the proportion of the people who voted for the recall. Remember we have to divide by 2705.

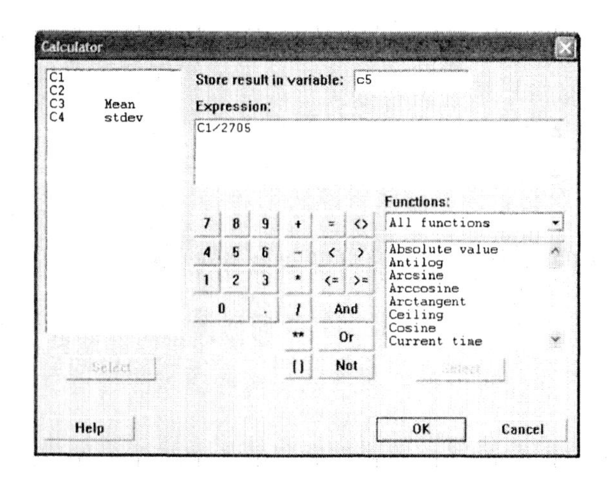

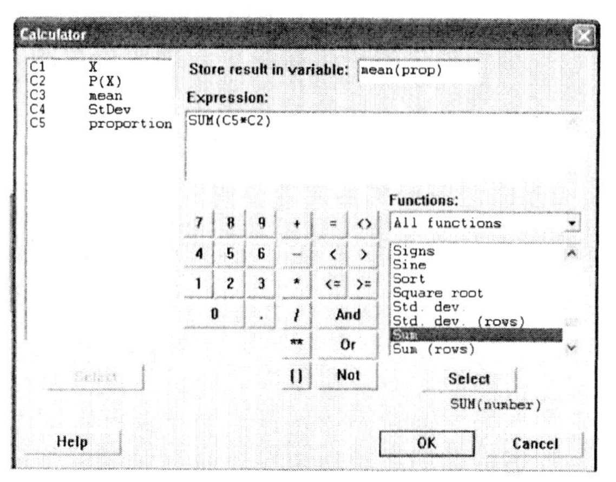

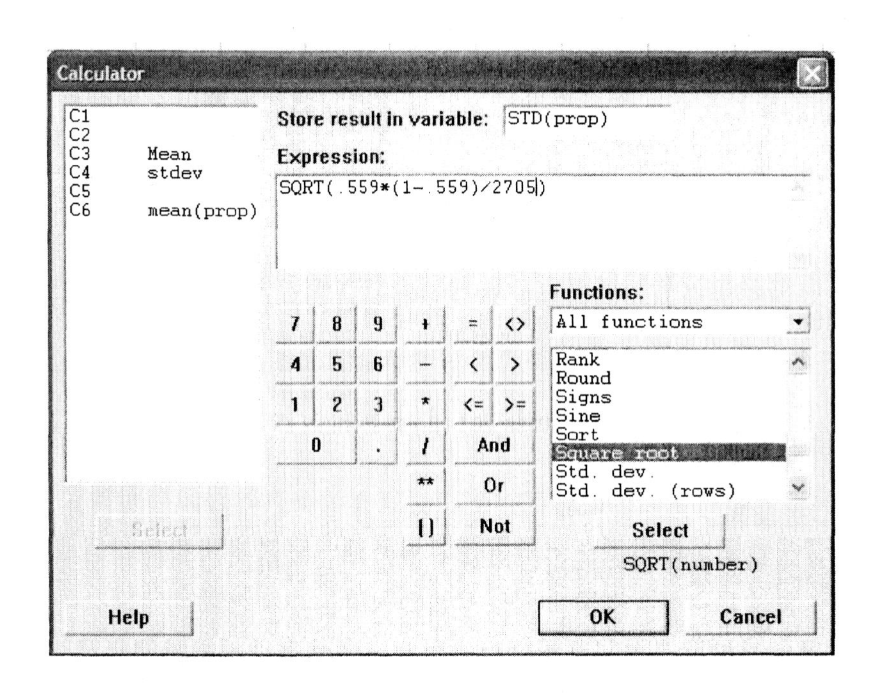

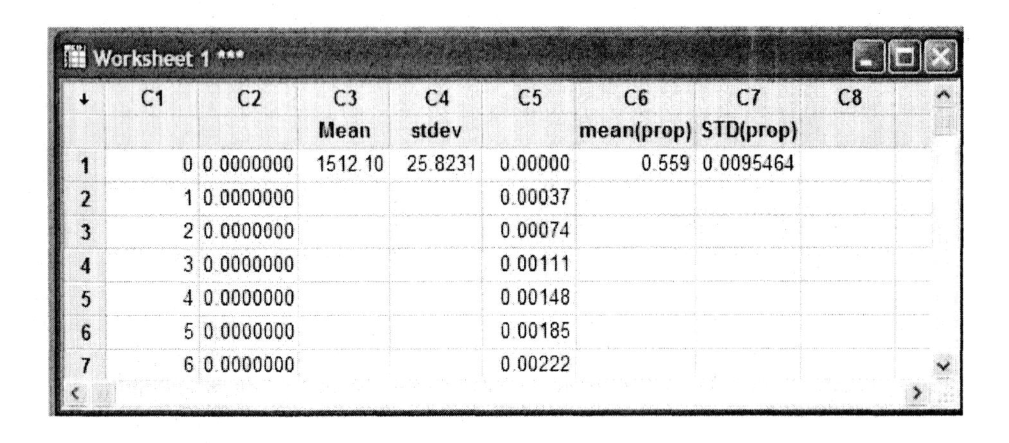

↓	C1	C2	C3	C4	C5	C6	C7	C8
			Mean	stdev		mean(prop)	STD(prop)	
1	0	0.0000000	1512.10	25.8231	0.00000	0.559	0.0095464	
2	1	0.0000000			0.00037			
3	2	0.0000000			0.00074			
4	3	0.0000000			0.00111			
5	4	0.0000000			0.00148			
6	5	0.0000000			0.00185			
7	6	0.0000000			0.00222			

Chapter 7 Class Exploration 7.63
Simulating a Sampling Distribution for a Sample Mean

Enter the data into the worksheet.
Construct a histogram.

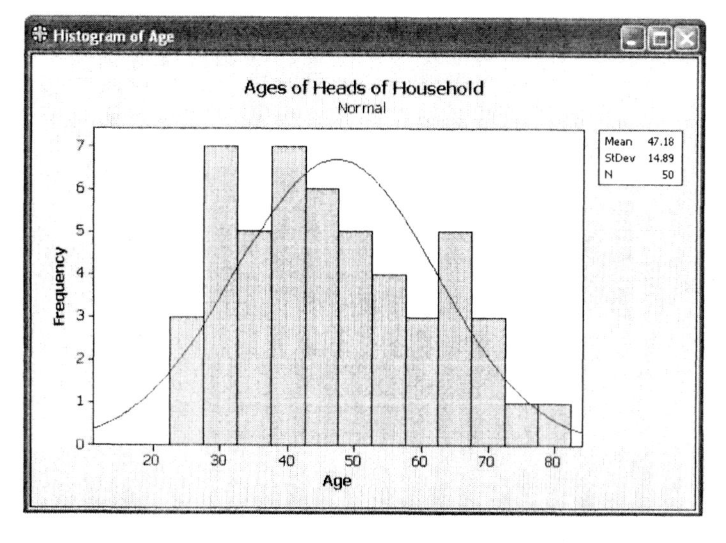

b) Lets assume a class size of 30 students. Each student can collect random samples of size 9. We will store the samples in nine columns of 30 cells each.

Calc > Random Data > Sample from Columns

Enter: Sample **30** Rows
Column **C3**
Store samples in **: enter a column**
Check box: Sample with replacement

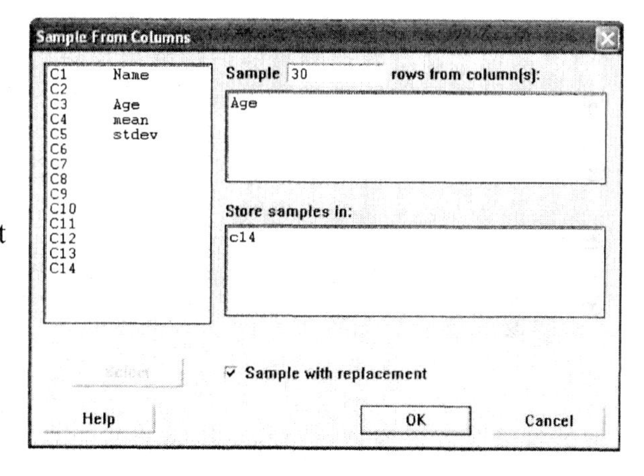

Note you will have to do this nine times, naming a new column each time.

To find the mean of each sample :

Choose: **Calc > Row Statistics**
Click **Mean**
Input variables: Enter **C6-C14**
Store in **C15**
OK

Find the mean and standard deviation of column 15. You can see these values by doing a histogram of column 15. What conclusions can you draw?

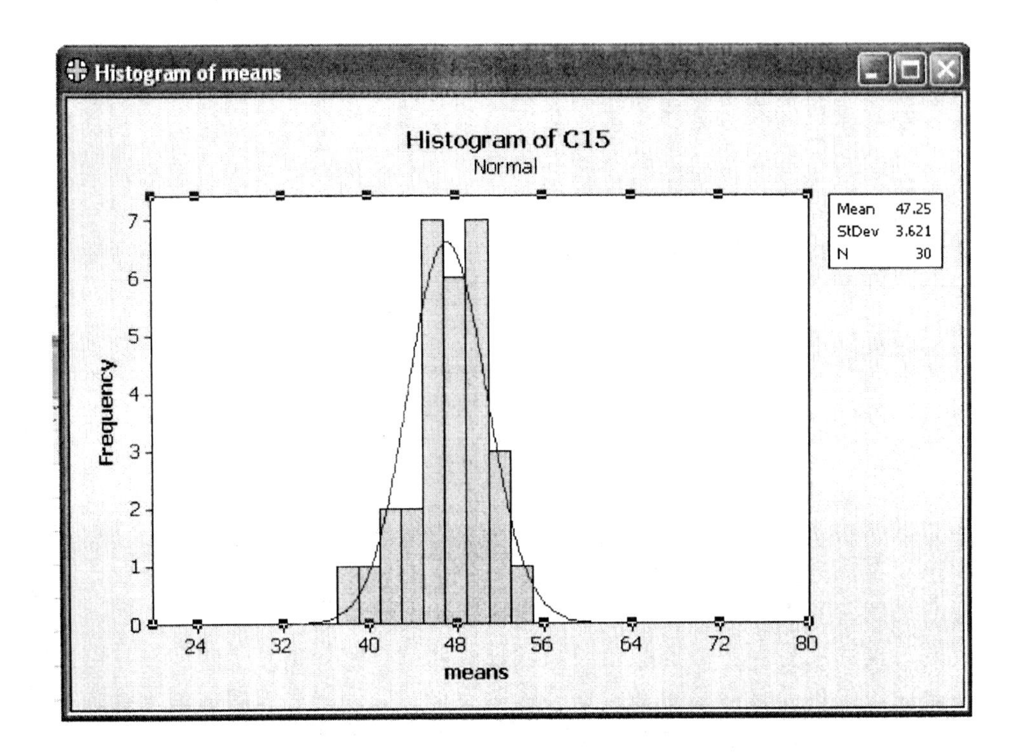

Chapter 8 Example 2 Should a Wife Sacrifice Her Career for Her Husband's? Constructing the Confidence Interval Estimate for a Population Proportion

From Chapter 6 we recall that 95% of a normal distribution falls within two standard deviations of the mean. So if we look at the (mean of the distribution) $\pm$ 1.96 (standard deviations for the distribution), we would find 95% of a normal distribution. So in this problem, if we look at 1.96(.01) we get 0.02. we can then construct the interval as the $\hat{p} \pm 0.02$, or $.19 \pm 0.02$ which gives us (0.17,0.21).

To do this in Minitab:

Choose: **Stat > Basic Statistics > 1-Proportion**

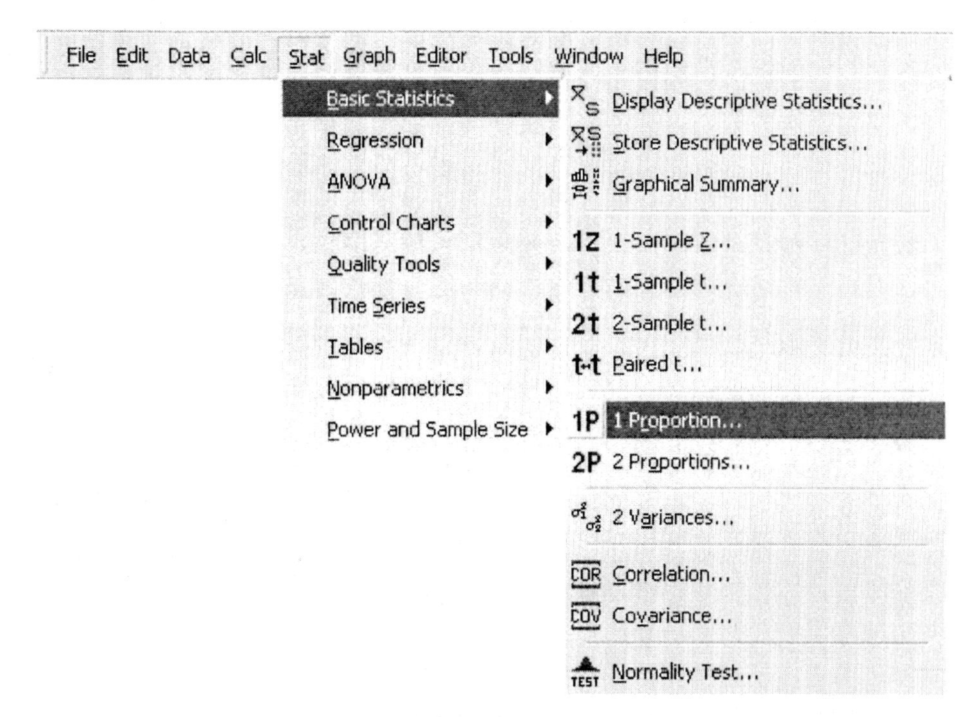

Select:
Summarized Data
Enter:
Number of trials **1823** Number of events **346** (Note: 19% of 1823)
Click: **OK**

Note: Minitab defaults to a 95% confidence interval.

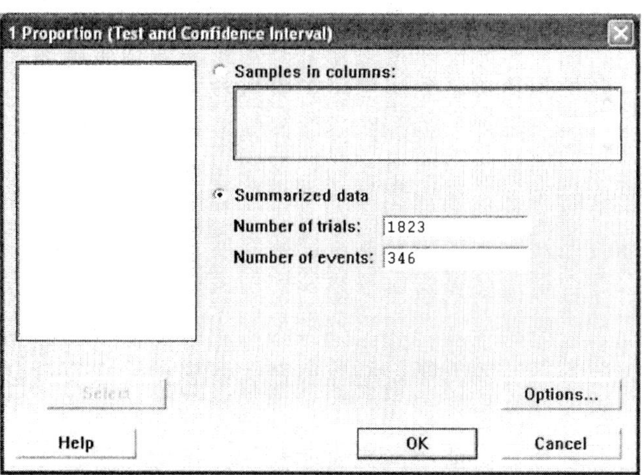

We get the following in the session window.

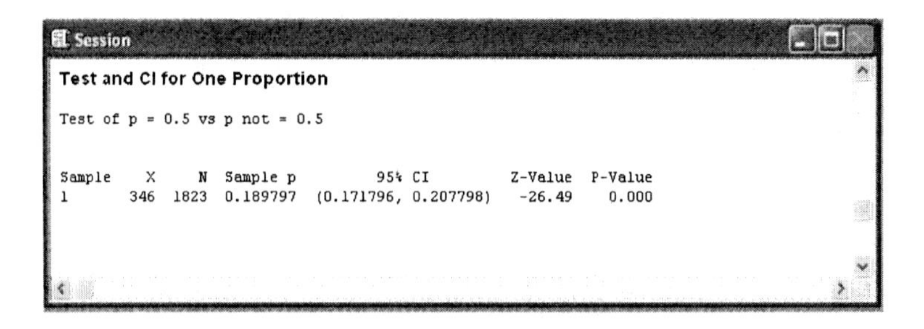

Session

Test and CI for One Proportion

Test of p = 0.5 vs p not = 0.5

Sample	X	N	Sample p	95% CI	Z-Value	P-Value
1	346	1823	0.189797	(0.171796, 0.207798)	-26.49	0.000

Given the interval in the above session window (0.17, 0.21) , and the point estimate of 0.19, do you see that the margin of error is 0.02?

Chapter 8 Exercise 8.7 Believe in heaven?
Constructing the Confidence Interval Estimate for a Proportion

In this exercise, n = 1158, and the sample proportion = 0.86

Choose: **Stat > Basic Statistics > 1-Proportion**
Select: Summarized Data
Enter: Number of trials **1156**
Number of events **996**
(Note: 86% of 1156)
Click: **OK**

1 Proportion (Test and Confidence Interval)
Samples in columns:
Summarized data
Number of trials: 1158
Number of events: 996
Select
Options...
Help
OK
Cancel

We get the following in the session window:

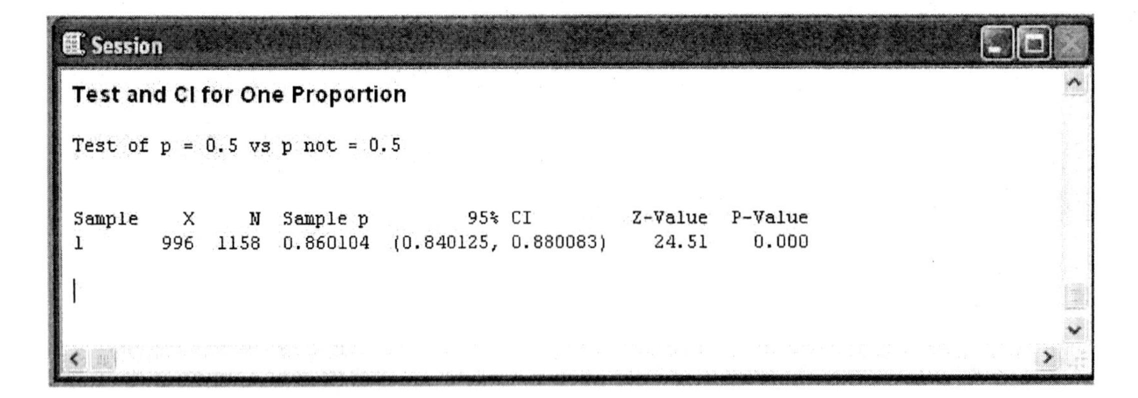

Session

Test and CI for One Proportion

Test of p = 0.5 vs p not = 0.5

Sample	X	N	Sample p	95% CI	Z-Value	P-Value
1	996	1158	0.860104	(0.840125, 0.880083)	24.51	0.000

Chapter 8 Exercise 8.23 Exit-poll Predictions
Constructing the Confidence Interval Estimate for a Proportion

In this problem we have n = 1400 voters, of whom 660 voted Democrat and 740 voted Republican. First using a 95% confidence interval:

Choose: **Stat > Basic Statistics > 1-Proportion**
Select: Summarized Data
Enter: Number of trials **1400**
Number of events **660**
Click: **OK**

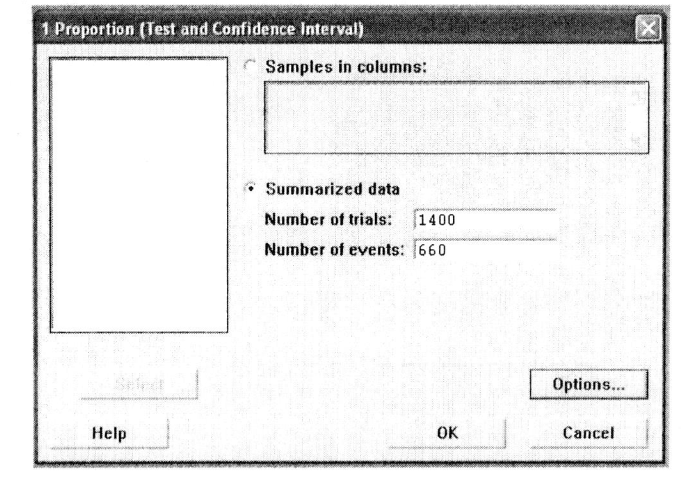

We get the following in the session window:

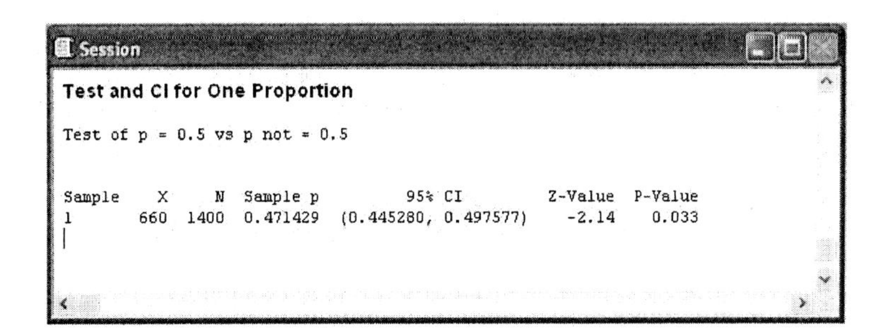
Session

Test and CI for One Proportion

Test of p = 0.5 vs p not = 0.5

Sample	X	N	Sample p	95% CI	Z-Value	P-Value
1	660	1400	0.471429	(0.445280, 0.497577)	-2.14	0.033

Now repeating this for a 99% confidence interval, we have to choose the Options tab to change the level.

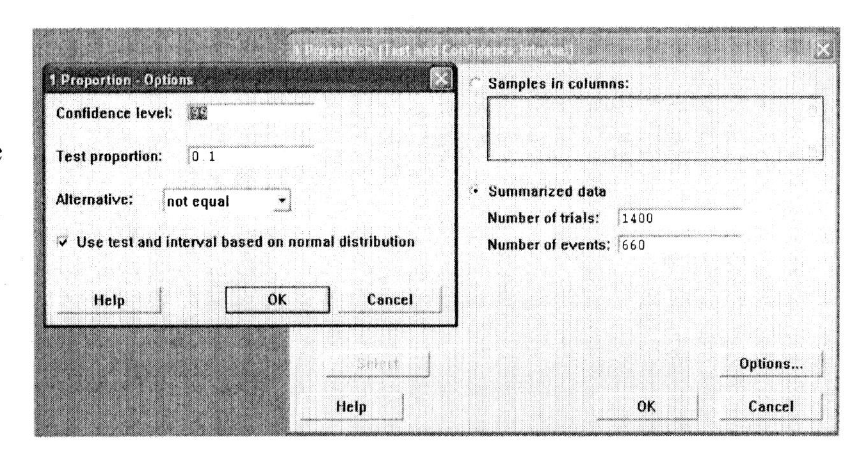

We now get these results:

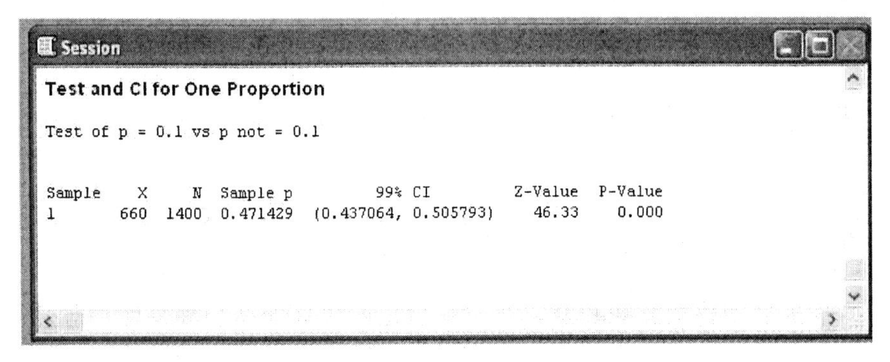

What conclusions can you draw?

Chapter 8 Example 7 eBay Auctions of Palm Handheld Computers Constructing a Confidence Interval Estimate for a Population Mean

Open worksheet **ebay_auctions** from the MINITAB folder of the data disk. To get the descriptive statistics

Choose **Stat> Basic Stat> Display Descriptive Statistics**
Click **Statistics** button (Choose the stats that you wish to see)
Click **OK OK**

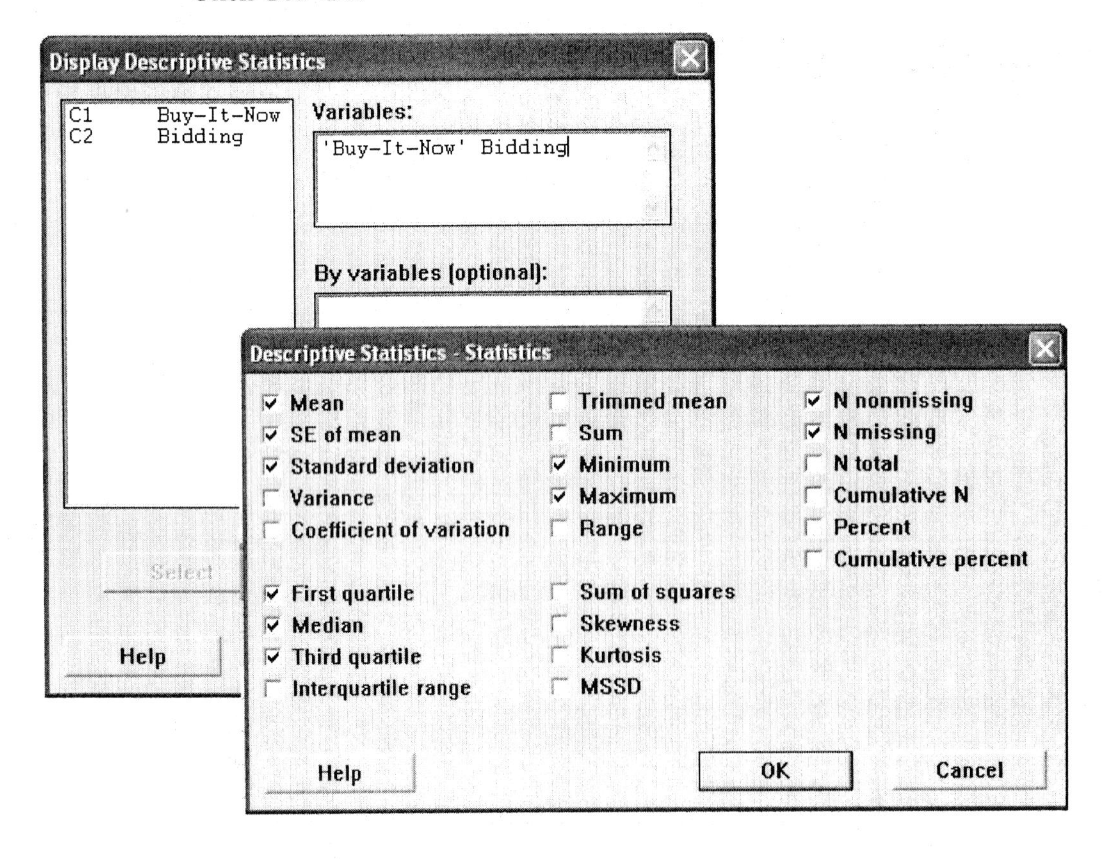

We then get the chart in the session window:

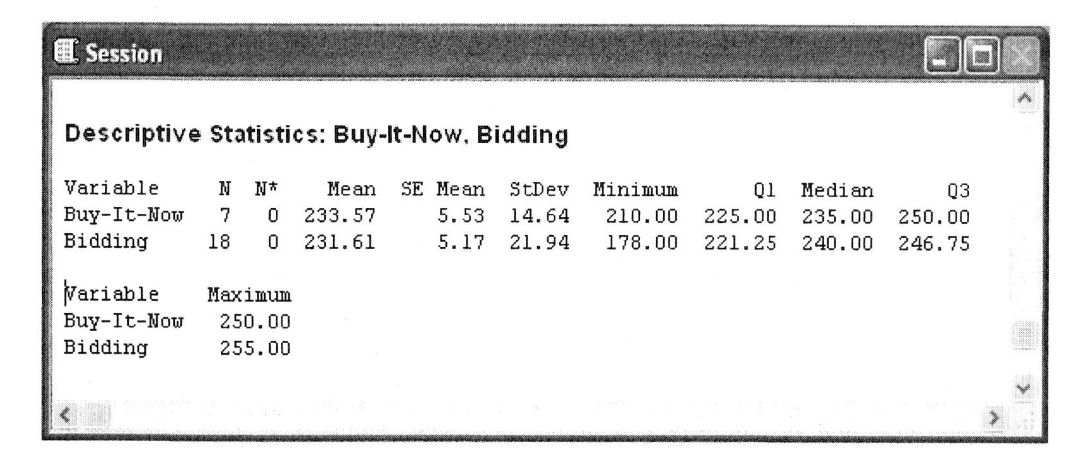

Session

Descriptive Statistics: Buy-It-Now, Bidding

Variable	N	N*	Mean	SE Mean	StDev	Minimum	Q1	Median	Q3
Buy-It-Now	7	0	233.57	5.53	14.64	210.00	225.00	235.00	250.00
Bidding	18	0	231.61	5.17	21.94	178.00	221.25	240.00	246.75

Variable	Maximum
Buy-It-Now	250.00
Bidding	255.00

We can do the dotplot for the data:
Choose: **Graph > DotPlot > Multiple y > Simple**

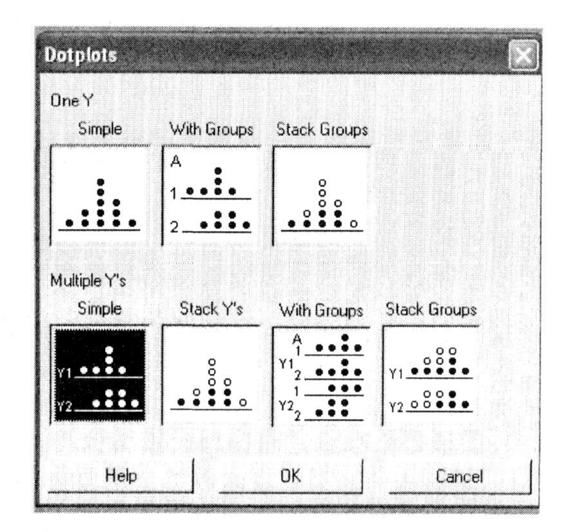

Enter Graph Variables: **C1 C2**
Enter : **appropriate title**
Click: **OK OK**

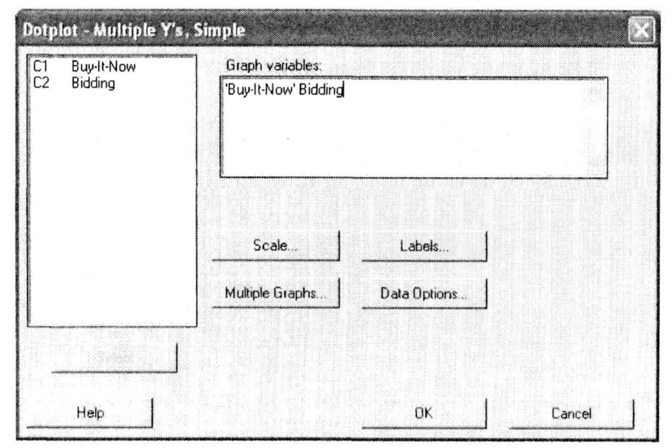

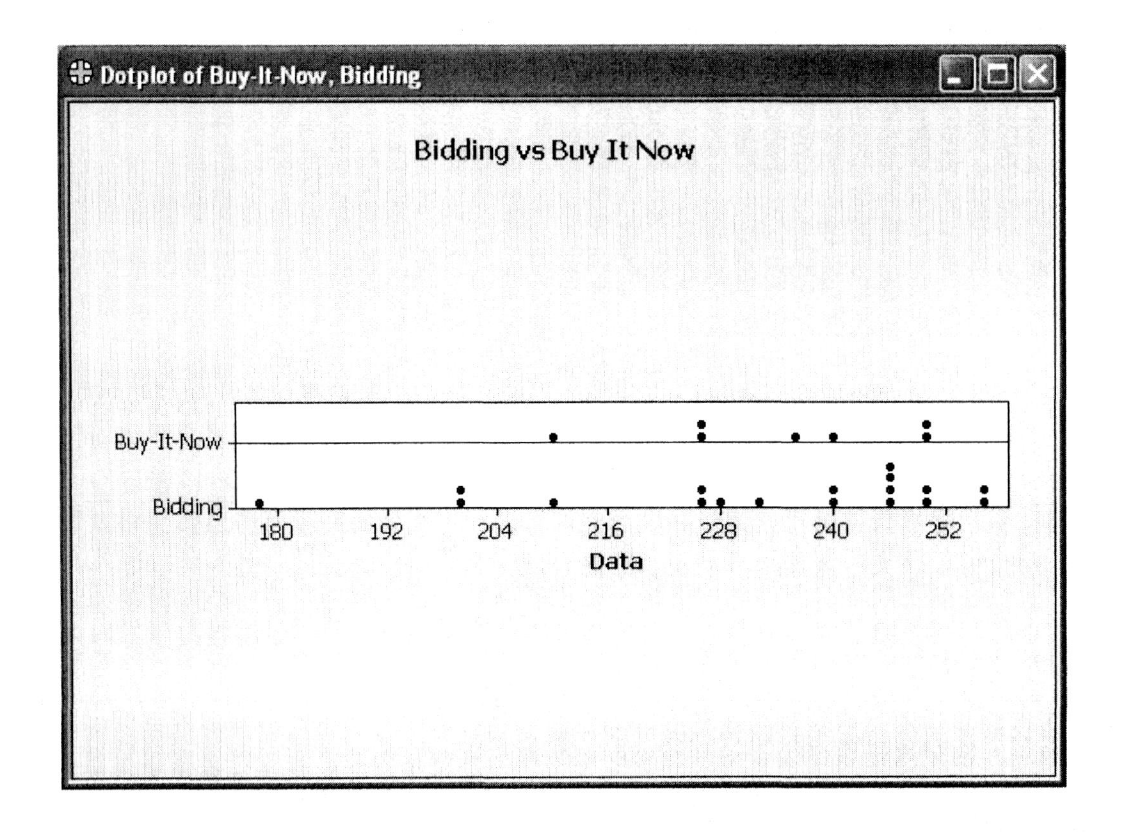

To obtain the confidence interval using the t-distribution:

Choose: **Stat>Basic Stat>1-sample t**

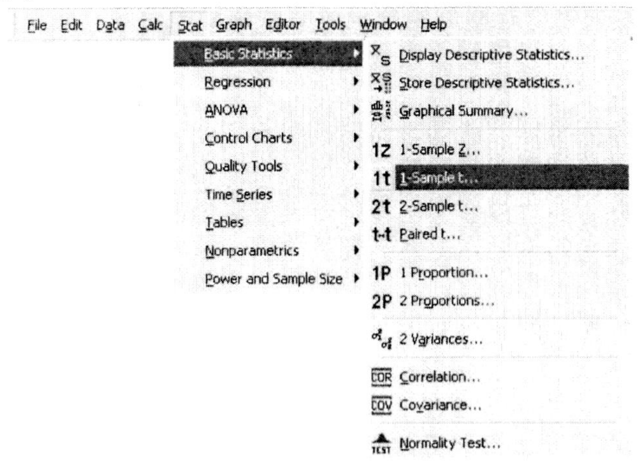

Enter: Samples in columns **C1**
Click : **OK**

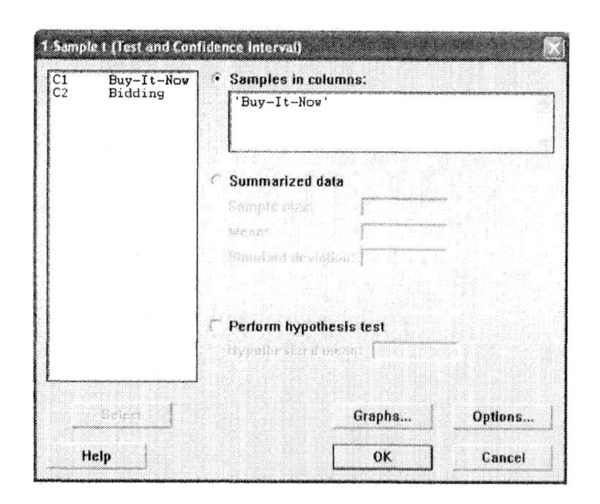

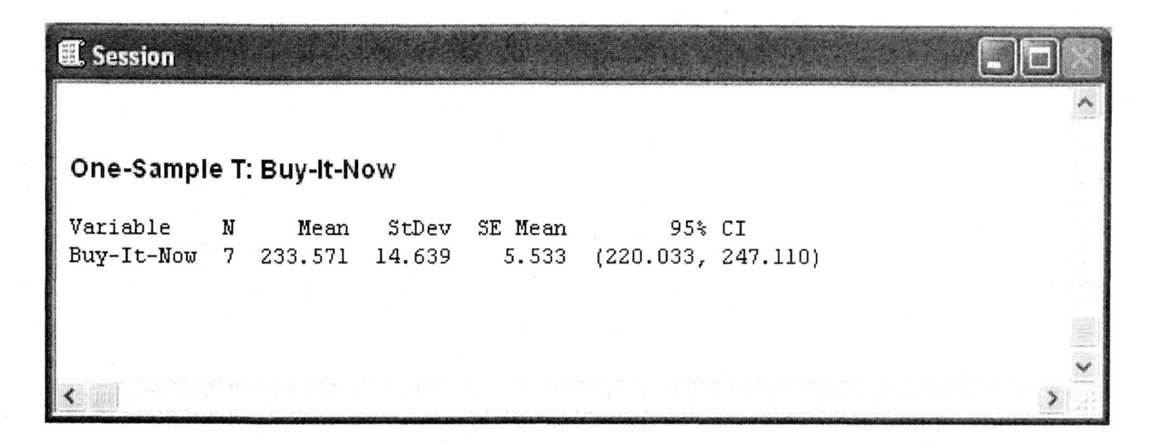

Session
One-Sample T: Buy-It-Now
Variable N Mean StDev SE Mean 95% CI
Buy-It-Now 7 233.571 14.639 5.533 (220.033, 247.110)

Using the same data, and now constructing the confidence interval for the Bidding column: (this is exercise 8.31)

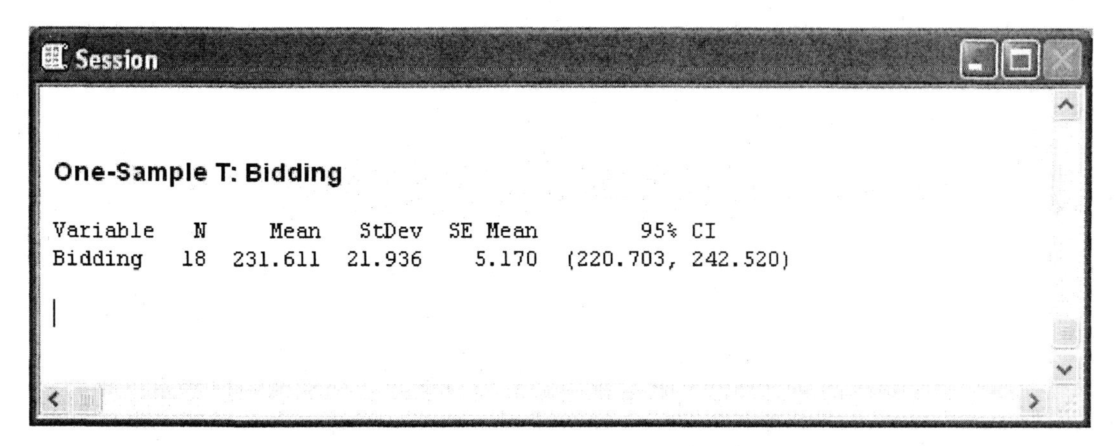

Session
One-Sample T: Bidding
Variable N Mean StDev SE Mean 95% CI
Bidding 18 231.611 21.936 5.170 (220.703, 242.520)

What conclusions can you draw?

Chapter 8 Exercise 8.55 Do You Like Tofu?

Confidence Interval Estimate of a Population Proportion with a Small Sample Size

In this problem we have a very small sample of five students. All say they like tofu. We will use the plus four method in this situation, giving us seven successes out of nine trials. Using the formulas in the text we get:

So we have N = 9, $\hat{p} = 7/9 = 0.7777$ and

$$se = \sqrt{\hat{p}(1-\hat{p})/n} = \sqrt{.7777(1-.7777)/9} = .1386.$$

The resulting interval = $\hat{p} \pm 1.96(se) = 0.7777 \pm (1.96)(0.1386)$.

This gives us the interval (0 .506, 1.05). Note that the right end of the interval is 1.05. We cannot have a value greater than 1.0 for a proportion, so we would report the interval as (0.506, 1.00).

Now we can do this using Minitab.

Choose **Stat > Basic Stat > 1 - Proportion**
Enter: Number of trials: **9**
Number of successes **7**

Click the Options tab and check the normal distribution box

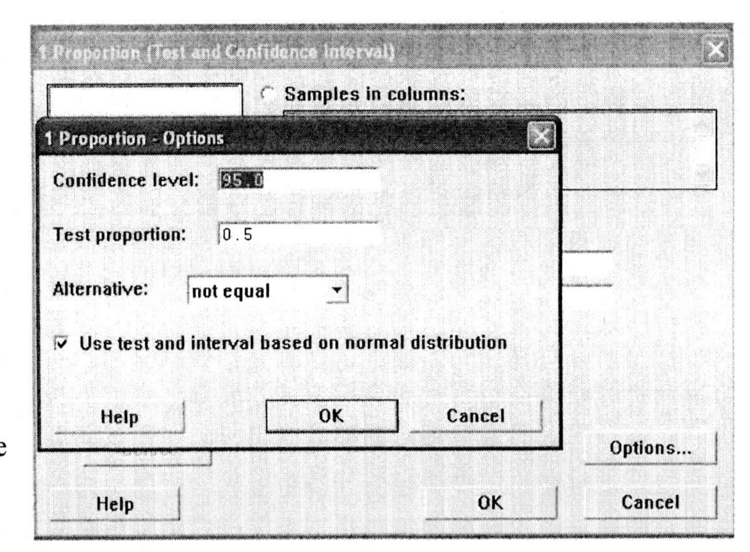

This will give you the same interval we calculated above.

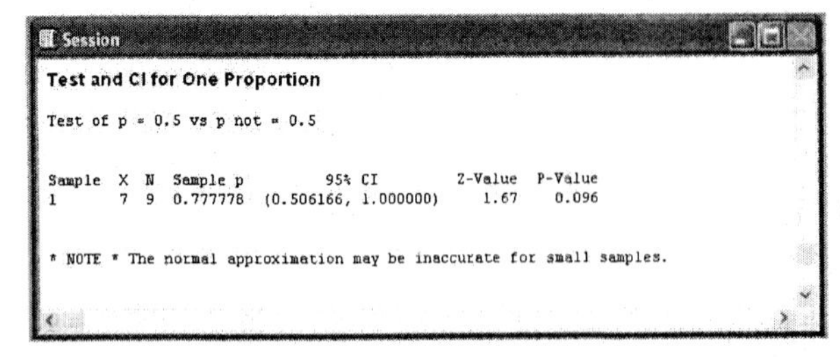

Chapter 9 Example 4 Dr Dog: Can Dogs Detect Cancer By Smell?
Hypothesis Testing For a Proportion

We are asked to test H_0: $p = 1/7$ vs H_1: $p > 1/7$. The sample evidence presented is that in the total of 54 trials, the dogs made the correct selection 22 times.

Choose: **Stat > Basic Statistics > 1-Proportion**
Select: **Summarized data**
Enter: Number of trials: **54**
Number of events: **22**

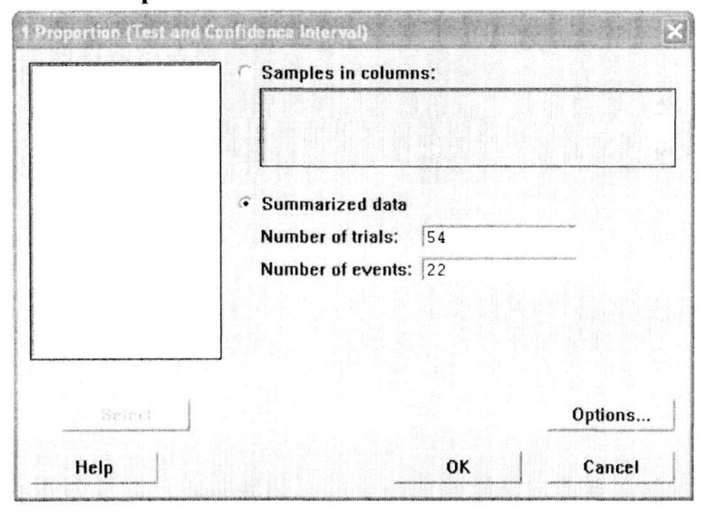

Click: **Options**
Enter: Test proportion: **.143**
Select: Alternative: **greater than**
Click: **Use test and interval based on normal distribution**
Click: **OK > OK**

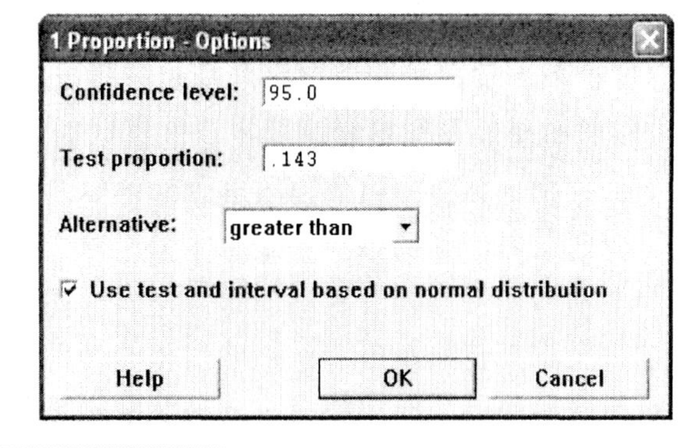

The results should be displayed in the Session window.

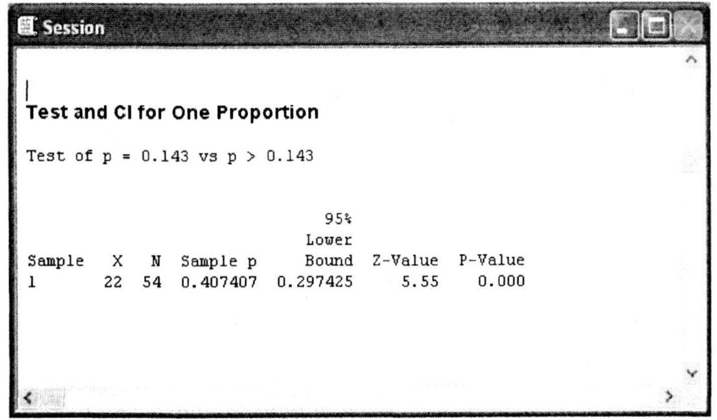

Test and CI for One Proportion

Test of p = 0.143 vs p > 0.143

Sample	X	N	Sample p	95% Lower Bound	Z-Value	P-Value
1	22	54	0.407407	0.297425	5.55	0.000

Note that the test statistic ($z = 5.55$) and the p-value ($p = 0.000$). With such a small p-value, we will reject the null hypothesis.

Chapter 9 Example 6 Can TT Practitioners Detect a Human Energy Field?
Hypothesis Testing For A Proportion

We are asked to test H_0: p = .50 vs H_1: p > .50. The sample evidence presented is that in the total of 150 trials, the TT practitioners were correct with 70 of their predictions.

Choose: **Stat > Basic Statistics > 1-Proportion**
Select: **Summarized data**
Enter: Number of trials: **150**
Number of events: **70**
Click: **Options**
Enter: Test proportion: **.50**
Select: Alternative: **greater than**
Click: **Use test and interval based on normal distribution**
Click: **OK > OK**

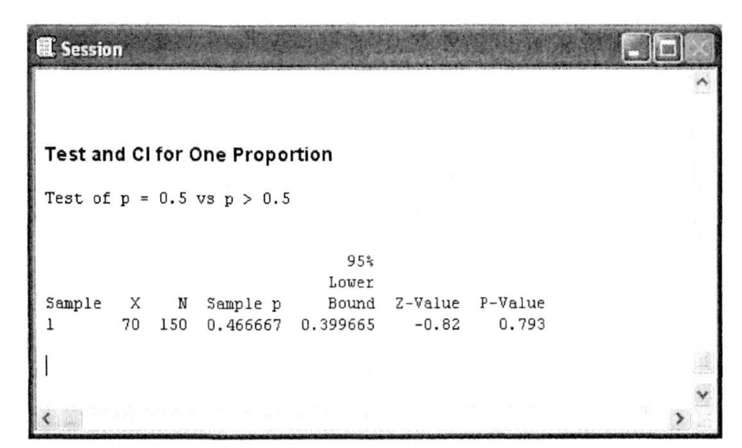
Session

Test and CI for One Proportion

Test of p = 0.5 vs p > 0.5

Sample	X	N	Sample p	95% Lower Bound	Z-Value	P-Value
1	70	150	0.466667	0.399665	-0.82	0.793

With such a large p-value (p-value = .793) the decision is Fail To Reject H_0.

Chapter 9 Exercise 9.17 Another Test of Therapeutic Touch
Hypothesis Testing For A Proportion

We are asked to test H_0: p = .50 vs H_1: p > .50. The sample evidence presented is that in the total of 130 trials, the TT practitioners were correct with 53 of their predictions.

Choose: **Stat > Basic Statistics > 1-Proportion**
Select: **Summarized data**
Enter: Number of trials: **130**
Number of events: **53**
Click: **Options**
Enter: Test proportion: **.50**
Select: Alternative: **greater than**
Click: **Use test and interval based on normal distribution**
Click: **OK > OK**

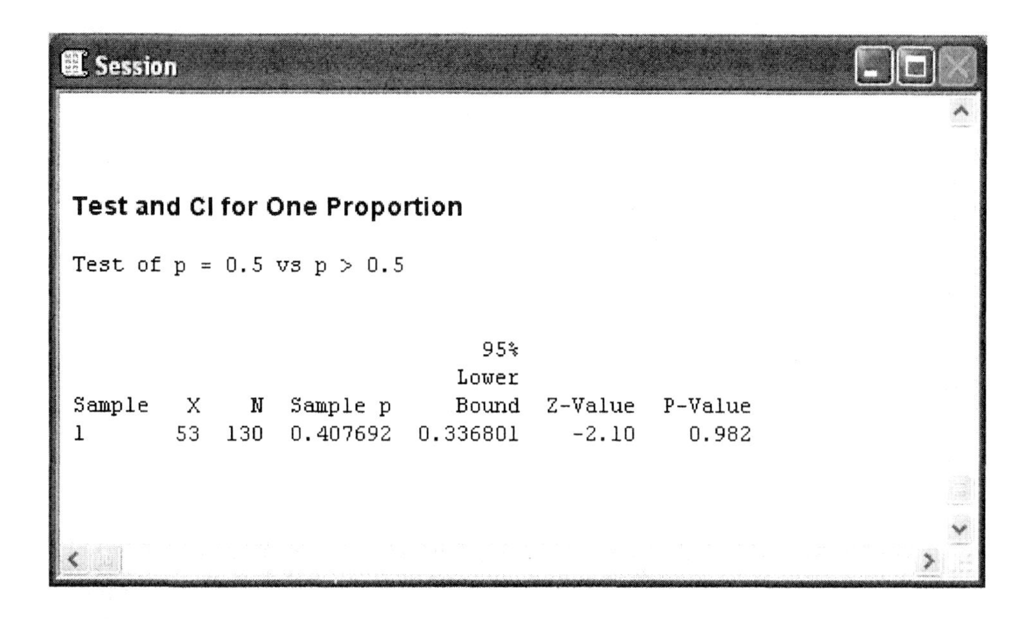

Test and CI for One Proportion

Test of p = 0.5 vs p > 0.5

Sample	X	N	Sample p	95% Lower Bound	Z-Value	P-Value
1	53	130	0.407692	0.336801	-2.10	0.982

Chapter 9 Exercise 9.19 Gender Bias in Selecting Managers
Hypothesis Testing For A Proportion

We are asked to test H_0: p = .60 vs H_1: p ≠ .60. The sample evidence presented is that in the total of 40 employees chosen for management training, 28 were male.

Choose: **Stat > Basic Statistics > 1-Proportion**
Select: **Summarized data**
Enter: Number of trials: **40**
Number of events: **28**
Click: **Options**
Enter: Test proportion: **.60**
Select: Alternative: **not equal**
Click: **Use test and interval based on normal distribution**
Click: **OK > OK**

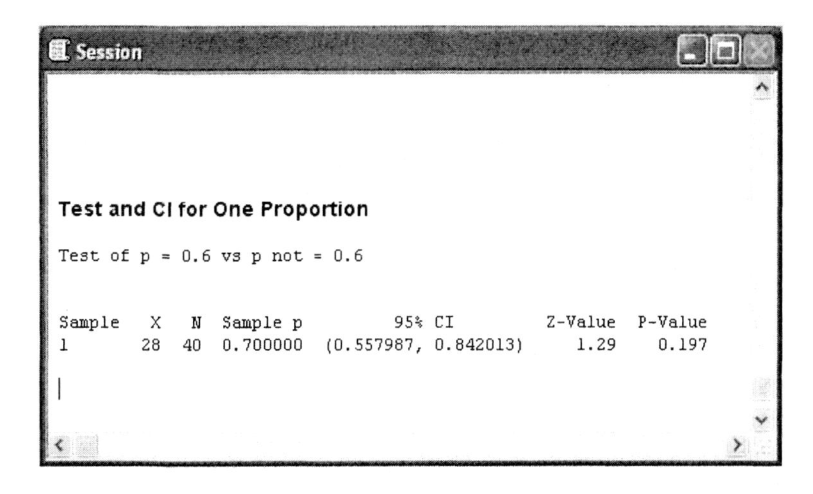

Test and CI for One Proportion

Test of p = 0.6 vs p not = 0.6

Sample	X	N	Sample p	95% CI	Z-Value	P-Value
1	28	40	0.700000	(0.557987, 0.842013)	1.29	0.197

Chapter 9 Example 7 Do Americans Work a 40 Hour Week?
Significance Test About a Mean

We wish to perform a two-tail test on H_0: $\mu = 40$ versus H_a: $\mu \neq 40$ using the summarized data:

Select: **Stat > Basic Statistics > 1-Sample t**

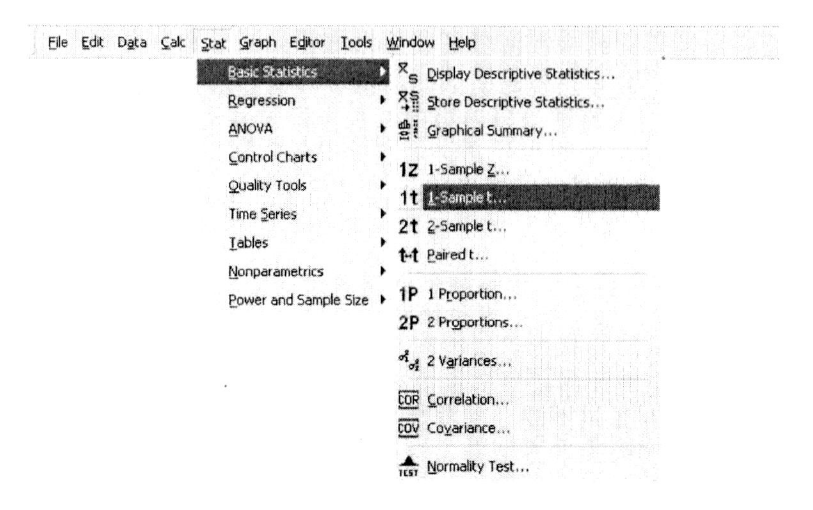

Select: **Summarized data:**
Enter:
Sample size: **868**
Mean: **39.1**
Standard Deviation: **14.6**
Test mean: **40**

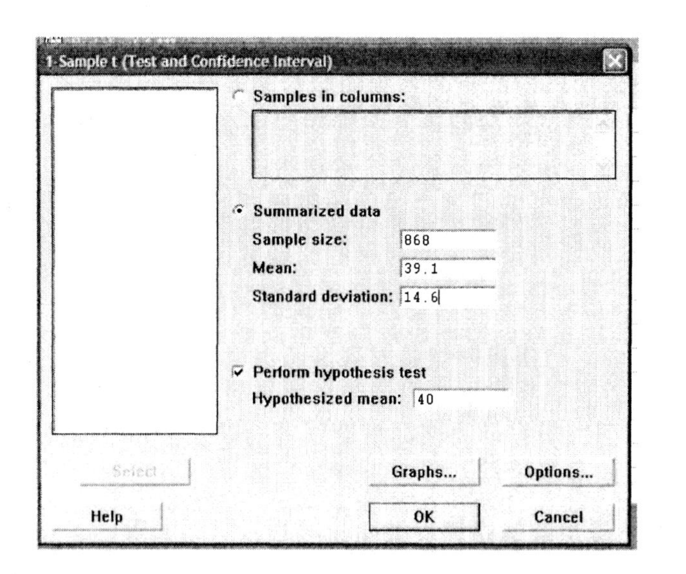

Select: **Options**
Select: Alternative: **not equal**
Click: **OK OK**

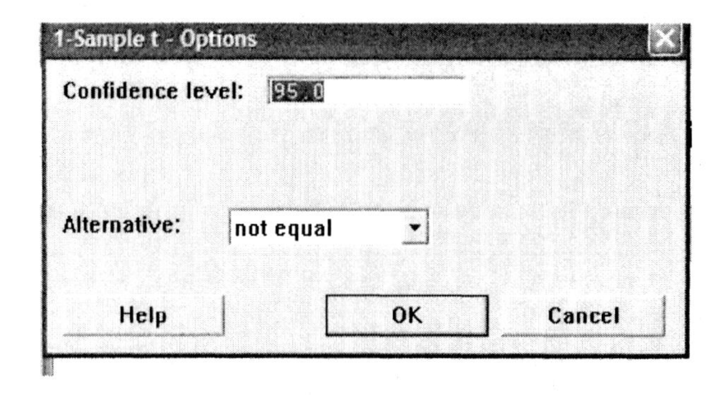

The results are displayed in the Session window. Note that the results are slightly different than those shown in the textbook due to rounding the summarized data.

One-Sample T

```
Test of mu = 40 vs not = 40

  N     Mean    StDev  SE Mean          95% CI              T      P
868  39.1000  14.6000   0.4956  (38.1274, 40.0726)  -1.82  0.070
```

Chapter 9 Exercise 9.39 Anorexia in teenage girls
Significance Test About a Mean

Enter the data into the MINITAB Data Window.

We wish to perform a two-tail test on H_0: $\mu = 0$ versus H_a: $\mu \neq 0$, using the weight gains given.

First, we will plot the data with a box plot. There are no outliers indicated, and the data distribution is not highly skewed.

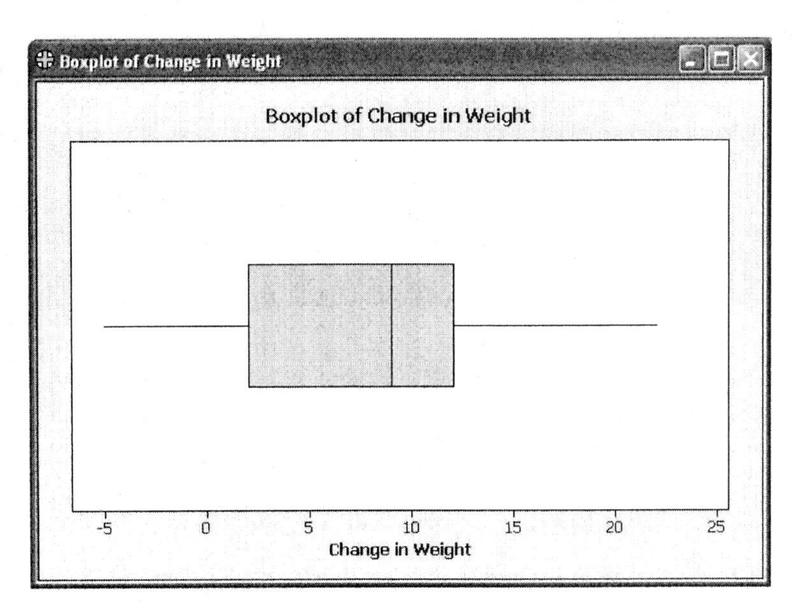

Performing the significance test about H_0: $\mu = 0$ versus H_a: $\mu \neq 0$

Select: **Stat > Basic Statistics > 1-Sample t**
Select: **Samples in columns:**

Enter: **C1**
Test mean: **0.0**

Select: **Options**
Select: Alternative: **not equal**
Click: **OK OK**

The results are displayed in the Session window.

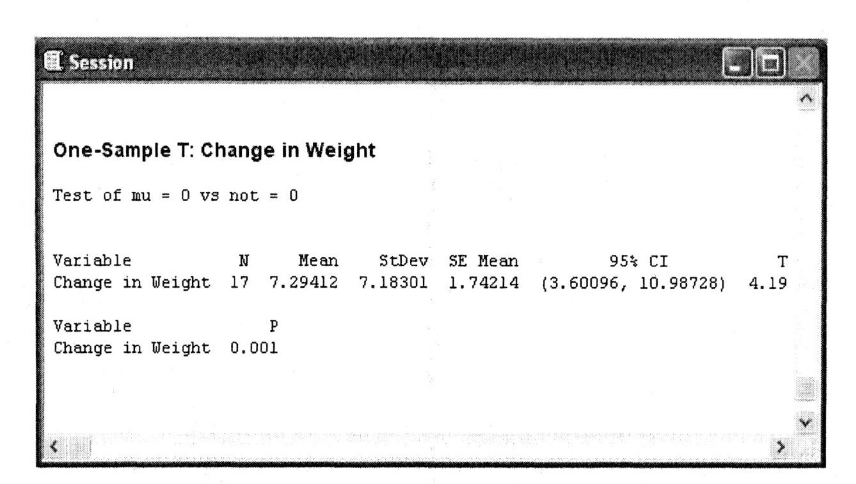

Session
One-Sample T: Change in Weight
Test of mu = 0 vs not = 0
Variable N Mean StDev SE Mean 95% CI T
Change in Weight 17 7.29412 7.18301 1.74214 (3.60096, 10.98728) 4.19
Variable P
Change in Weight 0.001

With a p-value of 0.001, the decision would be to reject H_0. There is significant evidence of a positive effect with the family therapy.

Chapter 9 Activity 2 Simulating the Performance of Significance Tests

To get a feel for the two possible errors in significance tests, we will simulate many samples from a population with a given proportion value, and perform a significance test for each sample. We will then check how often the tests make an incorrect inference.

For our first simulation, we will set the null hypothesis as H_0: $p = 1/3$ for a two-sided test using significance level $\alpha = 0.05$ with sample size 100.

To generate 100 samples of size 100:

Select: **Calc > Random Data > Bernoulli**

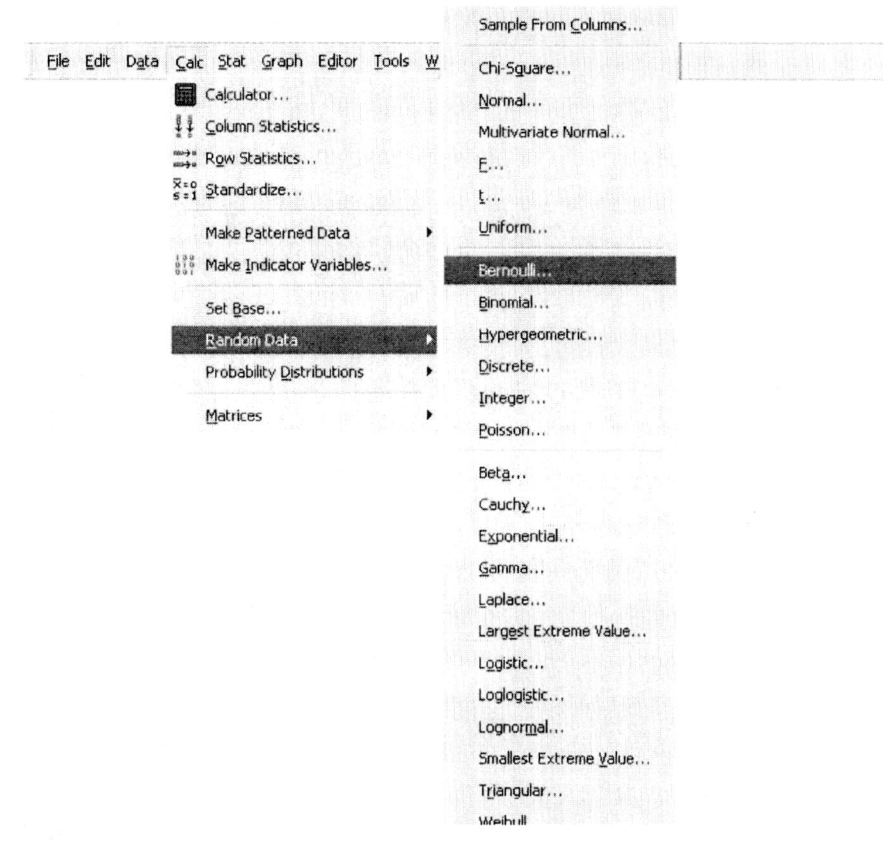

To simulate 100 samples of size 100, enter values as indicated. Click **OK**.

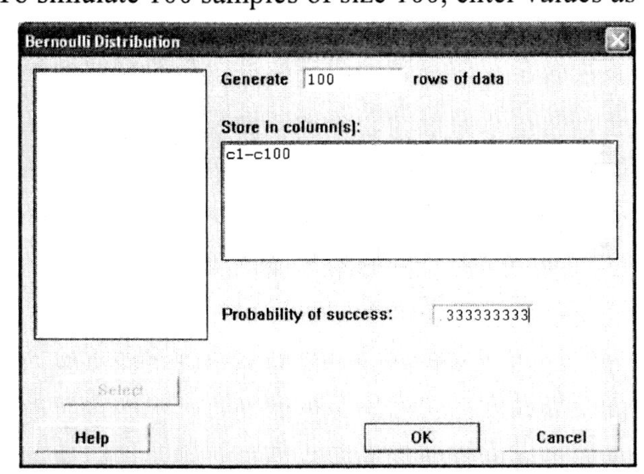

Your worksheet now contains 100 simulated samples from a binary trial where the true probability is p = 1/3.

To perform the significance test on each of the 100 samples:

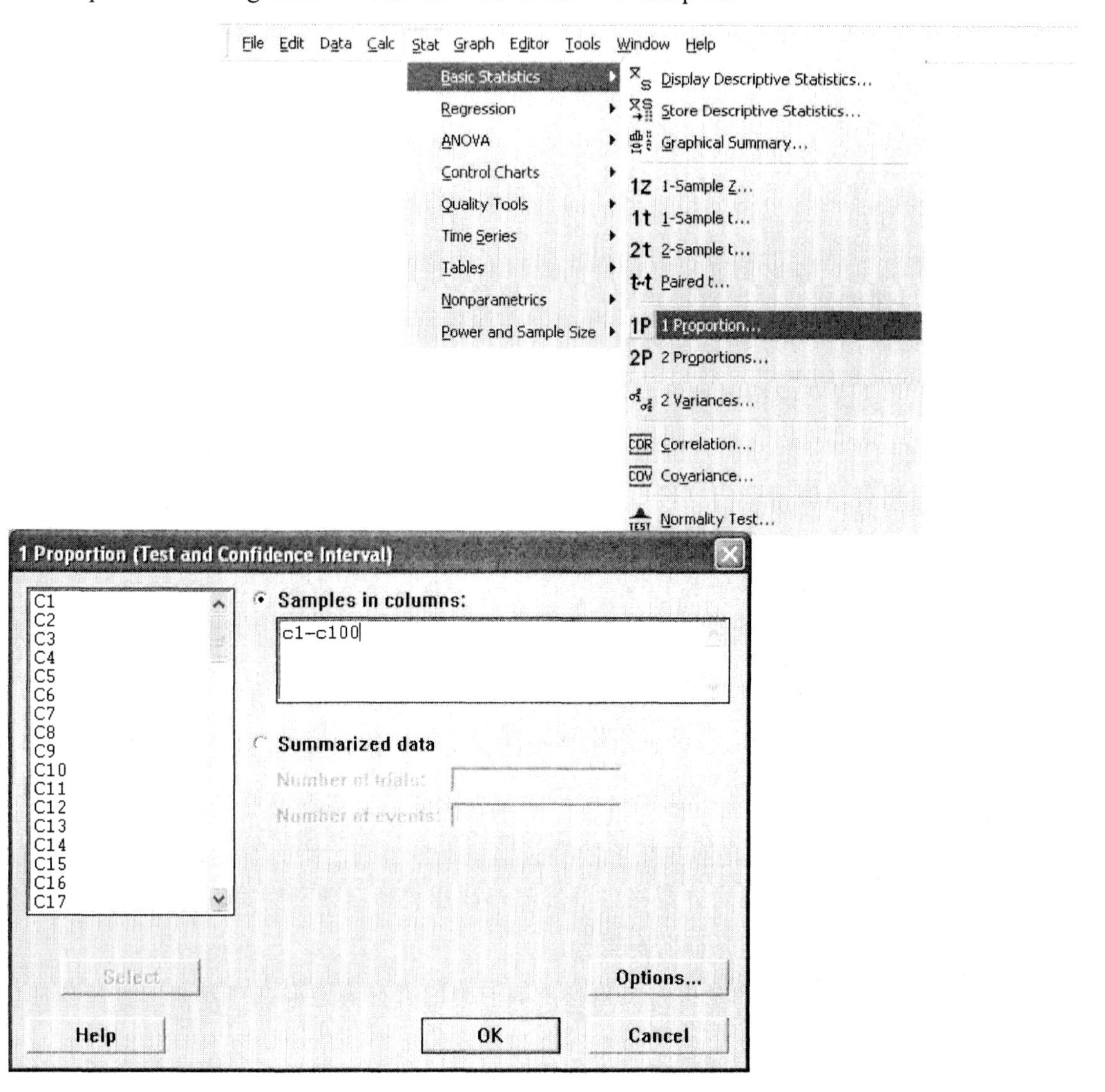

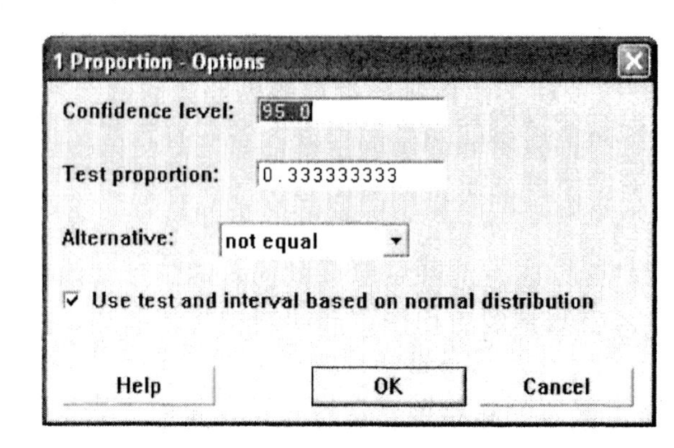

Analyzing the results

Counting the number of samples that would result in a p-value less than 0.05 (meaning we would reject H_0, thereby making an incorrect decision), we get approximately 5/100 or 5% of our samples lead to an incorrect decision. (Your results will differ slightly, the samples selected are random.)

Session

Test and CI for One Proportion: C1, C2, C3, C4, C5, C6, C7, C8, ...

Test of p = 0.333333 vs p not = 0.333333

Event = 1

Variable	X	N	Sample p	95% CI	Z-Value	P-Value
C1	34	100	0.340000	(0.247155, 0.432845)	0.14	0.888
C2	39	100	0.390000	(0.294403, 0.485597)	1.20	0.229
C3	35	100	0.350000	(0.256516, 0.443484)	0.35	0.724
C4	34	100	0.340000	(0.247155, 0.432845)	0.14	0.888
C5	32	100	0.320000	(0.228572, 0.411428)	-0.28	0.777
C6	31	100	0.310000	(0.219353, 0.400647)	-0.49	0.621
C7	34	100	0.340000	(0.247155, 0.432845)	0.14	0.888
C8	33	100	0.330000	(0.237840, 0.422160)	-0.07	0.944
C9	37	100	0.370000	(0.275372, 0.464628)	0.78	0.437

We will now redo the simulation in a new worksheet with 100 samples based on a Bernoulli distribution with p = 0.5, so that H_0 is actually false.

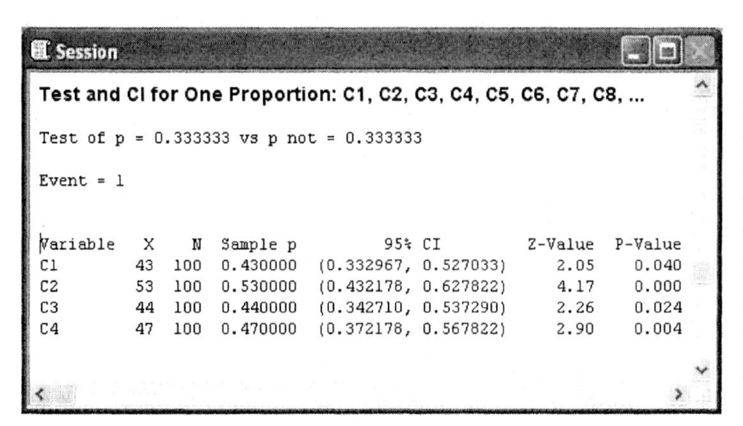

Session

Test and CI for One Proportion: C1, C2, C3, C4, C5, C6, C7, C8, ...

Test of p = 0.333333 vs p not = 0.333333

Event = 1

Variable	X	N	Sample p	95% CI	Z-Value	P-Value
C1	43	100	0.430000	(0.332967, 0.527033)	2.05	0.040
C2	53	100	0.530000	(0.432178, 0.627822)	4.17	0.000
C3	44	100	0.440000	(0.342710, 0.537290)	2.26	0.024
C4	47	100	0.470000	(0.372178, 0.567822)	2.90	0.004

Performing a significance test of H_0: p = 1/3, we will fail to reject our false Ho when the p-value is greater than alpha = 0.05. This occurs in 10 of the 100 samples. So, P(Type II) error is 0.10. Using the reasoning of example 13, we would expect P(type II) error to be 0.07.

Chapter 9 Exercise 9.61 Balancing Type I and Type II errors

Following the reasoning of Example 13, a Type II error occurs when the sample proportion falls less than z = 2.326 standard errors above the null hypothesis value of p = 1/3.

$$\hat{p} < 1/3 + 2.326\sqrt{[(1/3)(2/3)]/116} = .435$$

When p = 0.50, a Type II error has probability that $\hat{p} < 0.435$ when p = 0.50. This probability is

$$P(\hat{p} < 0.435) = P(z < (0.435 - 0.50)/\sqrt{[(.5)(.5)]/116}) = P(z < -1.40) = 0.0808$$

To see if we get approximately the same results using simulation, we will set the null hypothesis as H_0: p = 1/3 for a one-sided test using significance level $\alpha = 0.01$ with sample size 100 (Student version of Minitab 14 will not allow more than 10,000 pieces of data, so we cannot use sample size 116.)

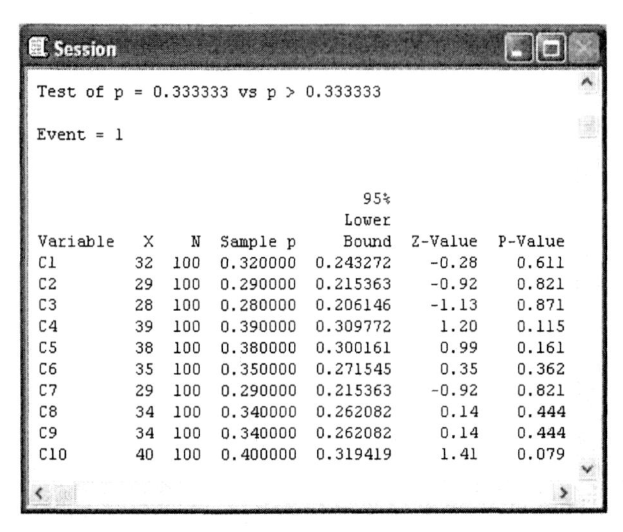
Session

Test of p = 0.333333 vs p > 0.333333

Event = 1

Variable	X	N	Sample p	95% Lower Bound	Z-Value	P-Value
C1	32	100	0.320000	0.243272	-0.28	0.611
C2	29	100	0.290000	0.215363	-0.92	0.821
C3	28	100	0.280000	0.206146	-1.13	0.871
C4	39	100	0.390000	0.309772	1.20	0.115
C5	38	100	0.380000	0.300161	0.99	0.161
C6	35	100	0.350000	0.271545	0.35	0.362
C7	29	100	0.290000	0.215363	-0.92	0.821
C8	34	100	0.340000	0.262082	0.14	0.444
C9	34	100	0.340000	0.262082	0.14	0.444
C10	40	100	0.400000	0.319419	1.41	0.079

In only one sample (of the 100 different samples generate) we would reject a true H_0. Thus, P(Type I error) = 1/100 = 0.01 (which is α).

Once again, redoing the simulation in a new worksheet with 100 samples based on a Bernoulli distribution with p = 0.5, so that H_0 is actually false, and performing the significance test of H_0: p = 1/3 versus H_a: p > 1/3, we will fail to reject our false H_0 when the p-value is greater than $\alpha = 0.01$.

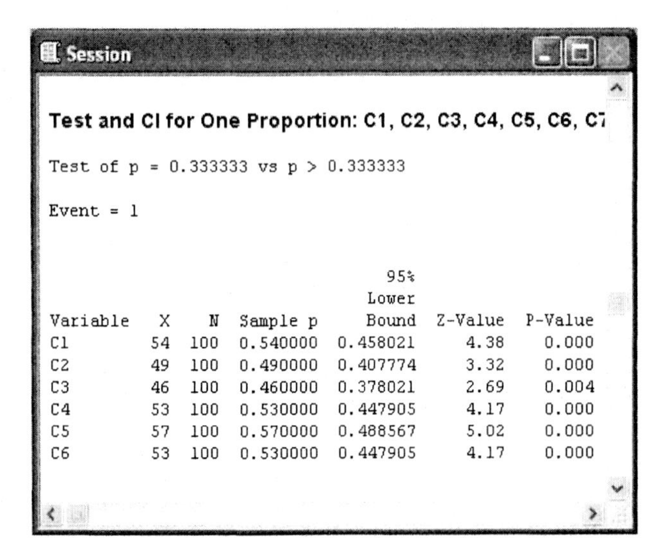
Session

Test and CI for One Proportion: C1, C2, C3, C4, C5, C6, C7

Test of p = 0.333333 vs p > 0.333333

Event = 1

Variable	X	N	Sample p	95% Lower Bound	Z-Value	P-Value
C1	54	100	0.540000	0.458021	4.38	0.000
C2	49	100	0.490000	0.407774	3.32	0.000
C3	46	100	0.460000	0.378021	2.69	0.004
C4	53	100	0.530000	0.447905	4.17	0.000
C5	57	100	0.570000	0.488567	5.02	0.000
C6	53	100	0.530000	0.447905	4.17	0.000

This occurs in 10 of the 100 samples. So, P(Type II) error is 0.10.

(This differs from the .0808 we calculated because it is based on these particular random samples.)

Chapter 10 Example 4 Confidence Interval Comparing Heart Attack Rates for Aspirin and Placebo - Confidence Interval for Difference Between Two Proportions

To construct a confidence interval $p_1 - p_2$, note that for the placebo group X = 189 and n = 11034, and for the aspirin group X = 104 and n = 11037.

Select: **Stat > Basic Statistics > 2 Proportions**

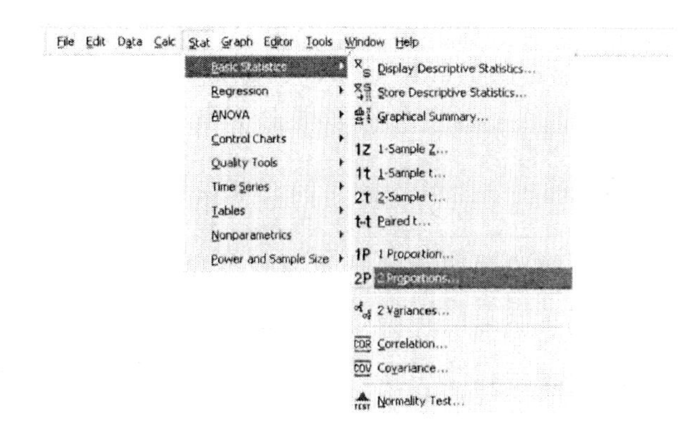

Select: **Summarized Data**
Enter: First: Trials: **11034** Events: **189**
Second: Trials: **11037** Events: **104**

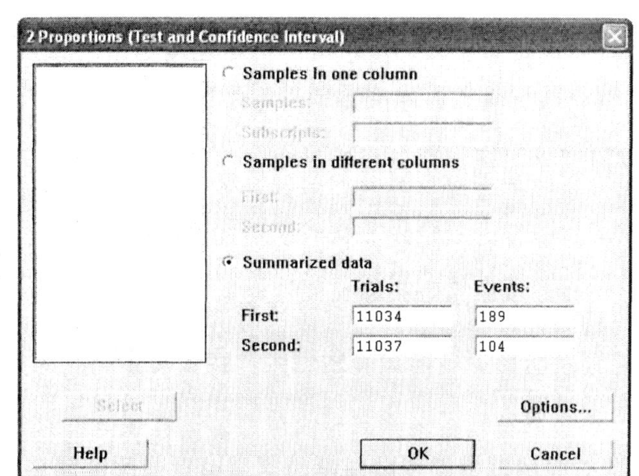

Select: **Options…**
Enter: Confidence level: **95**
Click: **OK OK**

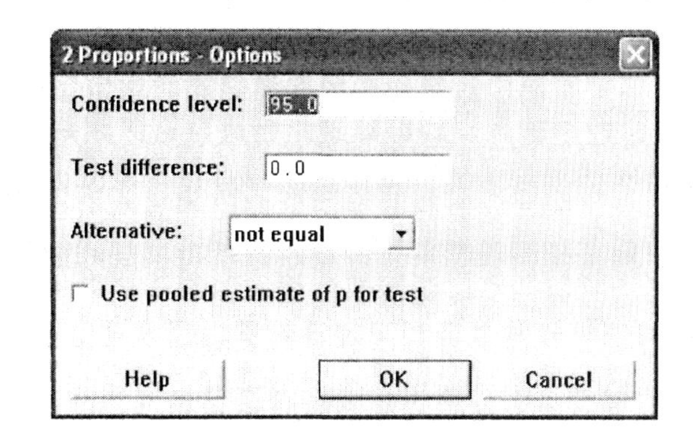

The results are displayed in the Session window.

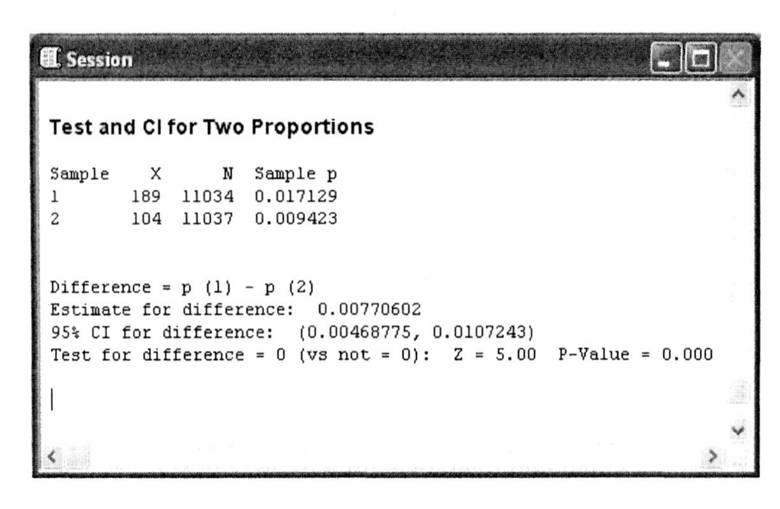

Session
Test and CI for Two Proportions
Sample X N Sample p
1 189 11034 0.017129
2 104 11037 0.009423
Difference = p (1) - p (2)
Estimate for difference: 0.00770602
95% CI for difference: (0.00468775, 0.0107243)
Test for difference = 0 (vs not = 0): Z = 5.00 P-Value = 0.000

Chapter 10 Example 5 Is TV Watching Associated with Aggressive Behavior? Significance Test for Difference Between Two Proportions

To perform a significance test for H_0: $p_1 - p_2 = 0$ versus H_a: $p_1 - p_2 \neq 0$ note that for the "less than 1 hour of TV per day" group X = 5 and n = 88, and for the "at least 1 hour of TV per day" group X = 154 and n = 619.

Select: **Stat > Basic Statistics > 2 Proportions**

Select: **Summarized Data**
Enter: First: Trials: **88** Events: **5**
Second: Trials: **619** Events: **154**

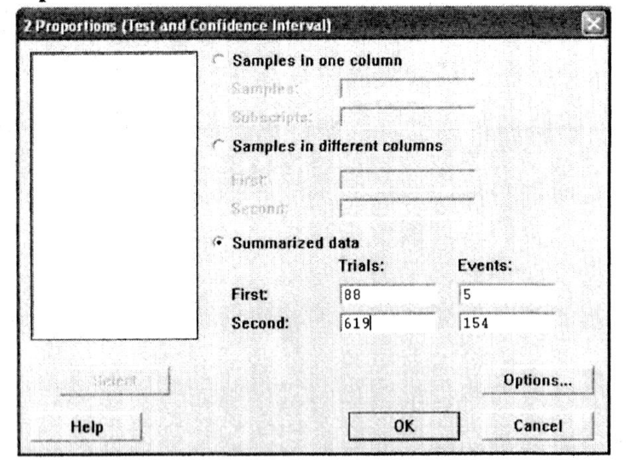

Select: **Options…**
Enter: Test difference: **0.0**
Select: Alternative: **not equal**
Select: **Use pooled estimate of p for test**
Click: **OK OK**

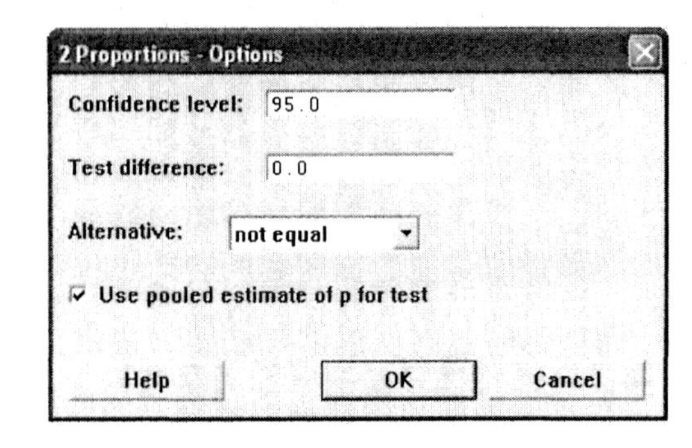

The results are displayed in the Session window.

```
Session

Test and CI for Two Proportions

Sample    X    N  Sample p
1         5   88  0.056818
2       154  619  0.248788


Difference = p (1) - p (2)
Estimate for difference:  -0.191970
95% CI for difference:  (-0.251124, -0.132816)
Test for difference = 0 (vs not = 0):  Z = -4.04  P-Value = 0.000
```

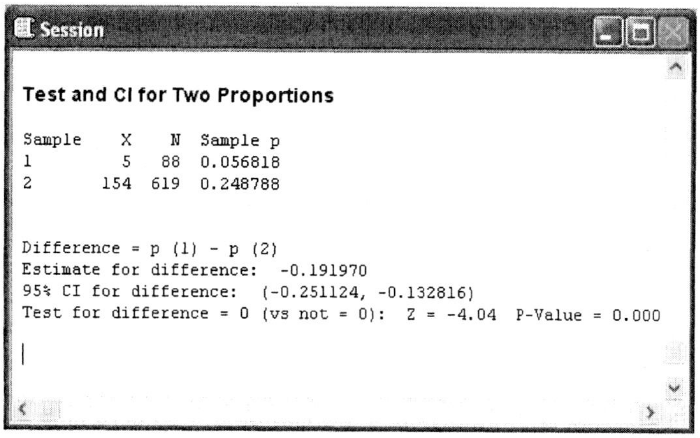

Chapter 10 Exercise 10.3 Binge Drinking
Confidence Interval for Difference Between Two Proportions

To form a 95% confidence interval for $p_1 - p_2$ note that for the 2005 group, X = 185 (.382 * 485) and n = 485, and for the 1993 group, = 50 (.312 * 159) and n = 159.

Select: **Stat > Basic Statistics > 2 Proportions**

Select: **Summarized Data**
Enter: First: Trials: **485** Events: **185**
Second: Trials: **159** Events: **50**
Select: **Options…**
Enter: Confidence level: **95.0**
Click: **OK OK**

The results are displayed in the Session window.

```
Test and CI for Two Proportions

Sample    X    N  Sample p
1       185  485  0.381443
2        50  159  0.314465


Difference = p (1) - p (2)
Estimate for difference:  0.0669779
95% CI for difference:  (-0.0171480, 0.151104)
Test for difference = 0 (vs not = 0):  Z = 1.56  P-Value = 0.119
```

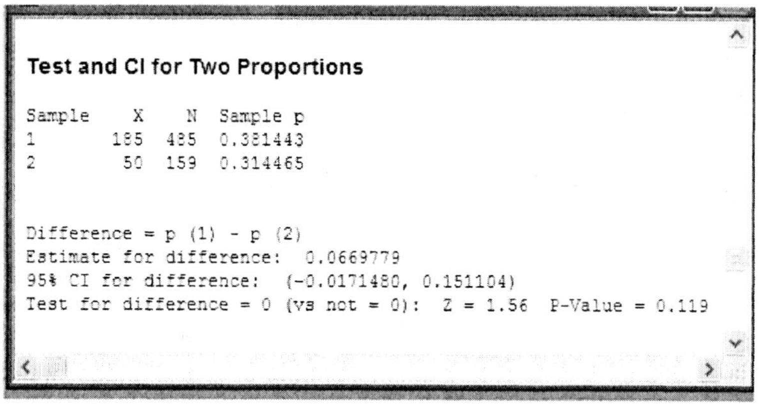

Chapter 10 Exercise 10.9 Drinking and Unplanned Sex
Significance Test for Difference Between Two Proportions

To perform a significance test for H0: $p_1 - p_2 = 0$ versus Ha: $p_1 - p_2 \neq 0$ note that for the 2005 group, X = 146 (.30 * 485) and n = 485, and for the 1993 group, X = 38 (.24 * 159) and n = 159.

Select: **Stat > Basic Statistics > 2 Proportions**

Select: **Summarized Data**
Enter: First: Trials: **485** Events: **146**
Second: Trials: **159** Events: **38**
Select: **Options…**

Enter: Test difference: **0.0**
Select: Alternative: **not equal**
Select: **Use pooled estimate of p for test**
Click: **OK OK**

The results are displayed in the Session window.

```
Sample    X    N  Sample p
1       146  485  0.301031
2        38  159  0.238994


Difference = p (1) - p (2)
Estimate for difference:  0.0620372
95% CI for difference:  (-0.0158133, 0.139888)
Test for difference = 0 (vs not = 0):  Z = 1.56  P-Value = 0.118
```

Chapter 10 Example 8 Nicotine – How Much More Addicted are Smokers than Ex-Smokers?
Confidence Interval for Difference Between Population Means

To construct a confidence interval for $\mu_1 - \mu_2$, note that for the smoker group, $\bar{x} = 5.9$ and s = 3.3 for n = 75, and for the ex-smoker group, $\bar{x} = 1.0$ and s = 2.3 for n = 257.

Select: **Stat > Basic Statistics > 2-Sample t…**

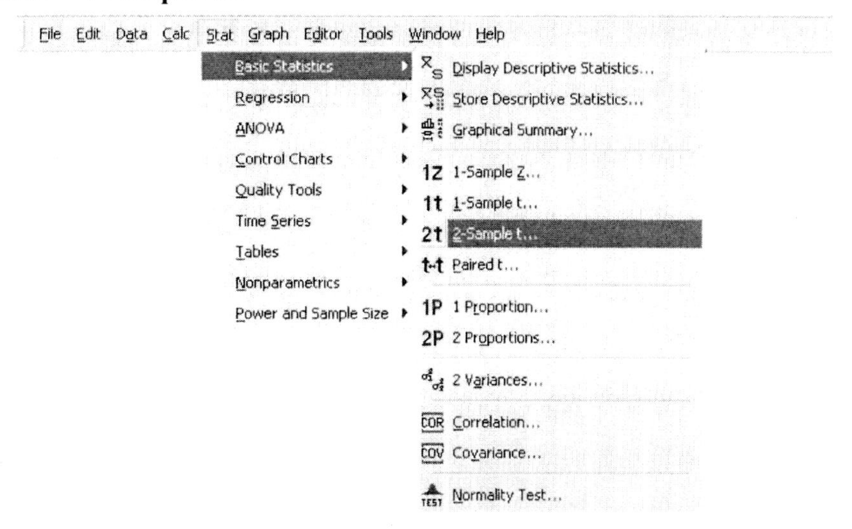

Select: **Summarized Data**

Enter:
First: Sample size:**75** Mean: **5.9** Standard deviation: **3.3**
Second: Sample size: **257** Mean: **1.0** Standard deviation: **2.3**

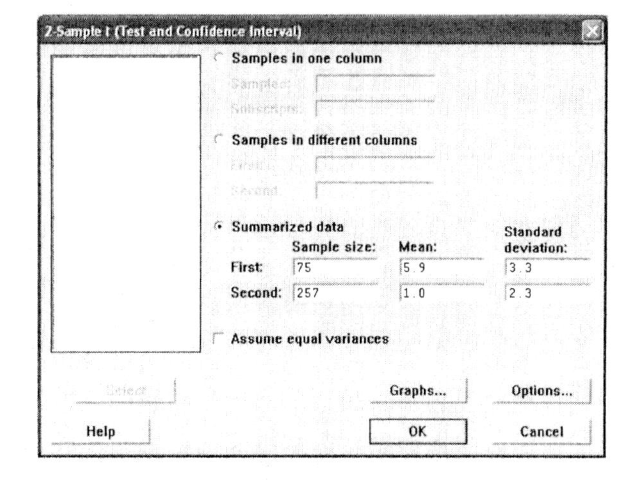

Select: **Options…**
Enter: Confidence level: **95**
Click: **OK OK**

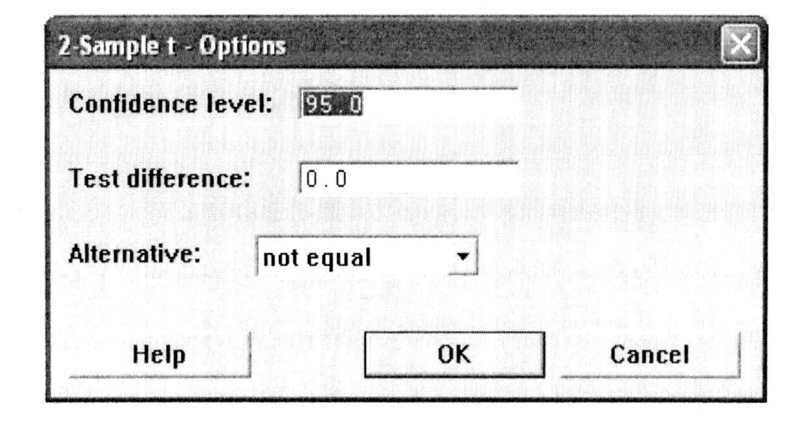

The results are displayed in the Session window.

```
Session

Two-Sample T-Test and CI

Sample    N  Mean  StDev  SE Mean
1        75  5.90   3.30     0.38
2       257  1.00   2.30     0.14


Difference = mu (1) - mu (2)
Estimate for difference:  4.90000
95% CI for difference:  (4.09167, 5.70833)
T-Test of difference = 0 (vs not =): T-Value = 12.03  P-Value = 0
```

Chapter 10 Example 9 Does Cell Phone Use While Driving Impair Reaction Times? Significance Test for Comparing Two Population Means

To perform a significance test for H_0: $\mu_1 - \mu_2 = 0$ versus H_a: $\mu_1 - \mu_2 \neq 0$ note that for the "cell phone" group, $\bar{x} = 585.2$ and s = 89.6 for n = 32, and for the "control" group, $\bar{x} = 533.7$ and s = 65.3 for n = 32.

Select: **Stat > Basic Statistics > 2-Sample t…**

Select: **Summarized Data**
Enter: First: Sample size: **32** Mean: **585.2** Standard deviation: **89.6**
Second: Sample size: **32** Mean: **533.7** Standard deviation: **65.3**

Select: **Options…**
Enter: Test difference: **0.0**
Select: Alternative: **not equal**
Click: **OK OK**

The results are displayed in the Session window.

```
Session
Two-Sample T-Test and CI

                             SE
Sample   N   Mean  StDev   Mean
1       32  585.2   89.6     16
2       32  533.7   65.3     12


Difference = mu (1) - mu (2)
Estimate for difference:  51.5000
95% CI for difference:  (12.2379, 90.7621)
T-Test of difference = 0 (vs not =): T-Value = 2.63  P-Value = 0.011  DF = 56
```

Chapter 9 Exercise 10.23 Some Smoked But Didn't Inhale
Confidence Interval for Difference Between Population Means

To construct a 95% confidence interval for $\mu_1 - \mu_2$, note that for the "Inhalers" group, $\bar{x} = 2.9$ and s = 3.6 for n = 237, and for the "Non-inhalers" group, $\bar{x} = 0.1$ and s = 0.5 for n = 95.

Select: **Stat > Basic Statistics > 2-Sample t...**

Select: **Summarized Data**
Enter: First: Sample size: **237** Mean: **2.9** Standard deviation: **3.6**
Second: Sample size: **95** Mean: **0.1** Standard deviation: **0.5**

Select: **Options...**
Enter: Confidence level: **95**
Click: **OK OK**

The results are displayed in the Session window.

```
Session

Two-Sample T-Test and CI

Sample    N   Mean  StDev  SE Mean
1       237   2.90   3.60     0.23
2        95  0.100  0.500    0.051


Difference = mu (1) - mu (2)
Estimate for difference:  2.80000
95% CI for difference:  (2.32855, 3.27145)
T-Test of difference = 0 (vs not =): T-Value = 11.70  P-Value = 0.000  DF = 257
```

Chapter 10 Exercise 10.25 Females or Males More Nicotine Dependent?
Significance Test for Comparing Two Population Means

To perform a significance test for H_0: $\mu_1 - \mu_2 = 0$ versus H_a: $\mu_1 - \mu_2 \neq 0$ note that for the "females", $\bar{x} = 2.8$ and s = 3.6 for n = 150, and for the "males", $\bar{x} = 1.6$ and s = 2.9 for n = 182.

Select: **Stat > Basic Statistics > 2-Sample t...**

Select: **Summarized Data**
Enter: First: Sample size: **150** Mean: **2.8** Standard deviation: **3.6**
Second: Sample size: **182** Mean: **1.6** Standard deviation: **2.9**

Select: **Options...**
Enter: Confidence level: **95**
Click: **OK OK**

The results are displayed in the Session window.

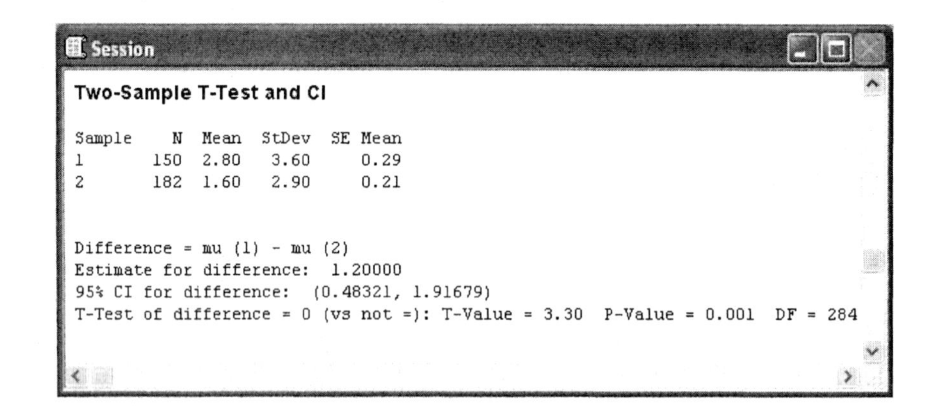

```
Session

Two-Sample T-Test and CI

Sample    N  Mean  StDev  SE Mean
1       150  2.80   3.60     0.29
2       182  1.60   2.90     0.21


Difference = mu (1) - mu (2)
Estimate for difference:  1.20000
95% CI for difference:  (0.48321, 1.91679)
T-Test of difference = 0 (vs not =): T-Value = 3.30  P-Value = 0.001  DF = 284
```

Chapter 10 Exercise 10.27 TV Watching and Gender
Significance Test for Comparing Two Population Means

To perform a significance test for H_0: $\mu_1 - \mu_2 = 0$ versus H_a: $\mu_1 - \mu_2 \neq 0$ note that for the "females", $\bar{x}$ = 2.99 and s = 2.34 for n = 1179, and for the "males", $\bar{x}$ = 2.86 and s = 2.22 for n = 870.

Select: **Stat > Basic Statistics > 2-Sample t…**

Select: **Summarized Data**
Enter: First: Sample size: 1117 Mean: **2.99** Standard deviation: **2.34**
Second: Sample size: 870 Mean: **2.86** Standard deviation: **2.22**

Select: **Options…**
Enter: Test difference: **0.0**
Alternative: **not equal**
Click: **OK OK**

The results are displayed in the Session window.

```
———— 11/30/2007 3:14:33 PM ————————————————

Welcome to Minitab, press F1 for help.

Two-Sample T-Test and CI

Sample     N  Mean  StDev  SE Mean
1       1117  2.99   2.34    0.070
2        870  2.86   2.22    0.075


Difference = mu (1) - mu (2)
Estimate for difference:  0.130000
95% CI for difference:  (-0.071603, 0.331603)
T-Test of difference = 0 (vs not =): T-Value = 1.26  P-Value = 0.206  DF = 1910
```

Chapter 10 Example 10 Is Arthroscopic Surgery Better Than Placebo?
Significance Test for Comparing Two Population Means Assuming Equal Population Standard Deviations

To perform a significance test for H_0: $\mu_1 - \mu_2 = 0$ versus H_a: $\mu_1 - \mu_2 \neq 0$ note that for the "placebo" group, $\bar{x} = 51.6$ and s = 23.7 for n = 60, and for the "lavage arthroscopic" group, $\bar{x} = 53.7$ and s = 23.7 for n = 61.

Select: **Stat > Basic Statistics > 2-Sample t…**

Select: **Summarized Data**
Enter: First: Sample size: **60** Mean: **51.6** Standard deviation: **23.7**
Second: Sample size: **61** Mean: **53.7** Standard deviation: **23.7**
Select: **Assume equal variances**

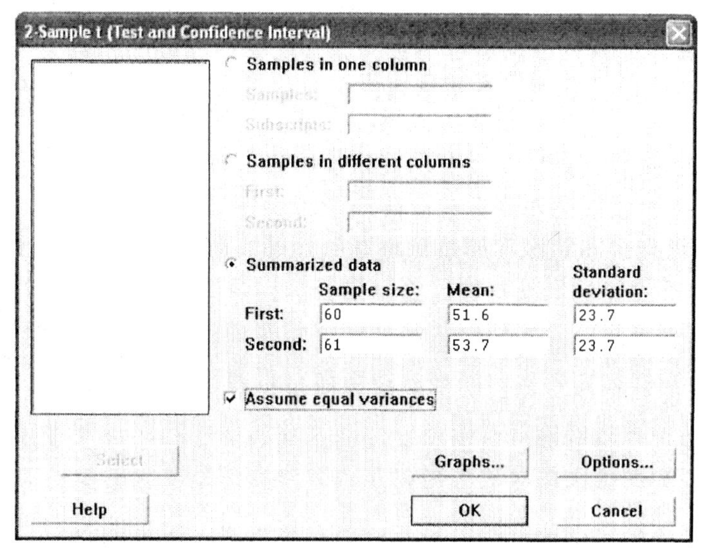

Select: **Options…**
Enter: Test difference: **0.0**
Select: Alternative: **not equal**
Click: **OK OK**

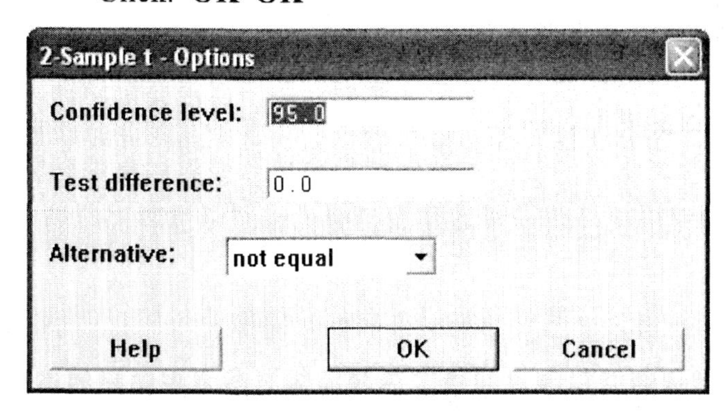

The results are displayed in the Session window.

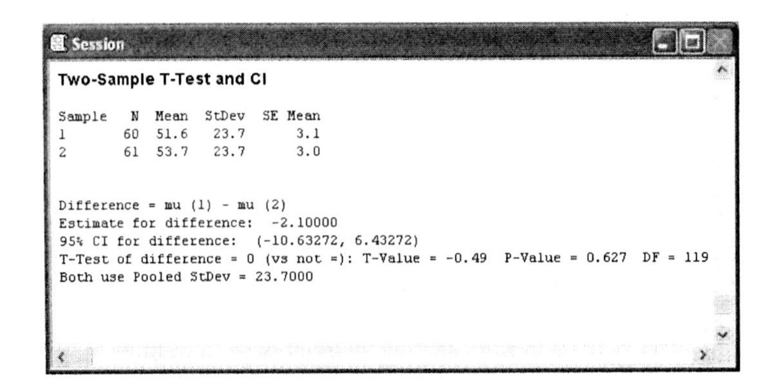

```
Two-Sample T-Test and CI

Sample   N  Mean  StDev  SE Mean
1       60  51.6   23.7      3.1
2       61  53.7   23.7      3.0


Difference = mu (1) - mu (2)
Estimate for difference:  -2.10000
95% CI for difference:  (-10.63272, 6.43272)
T-Test of difference = 0 (vs not =): T-Value = -0.49  P-Value = 0.627  DF = 119
Both use Pooled StDev = 23.7000
```

Chapter 10 Exercise 10.39 Vegetarians More Liberal?
Significance Test for Comparing Two Population Means Assuming Equal Population Standard Deviations

To perform a significance test for H_0: $\mu_1 - \mu_2 = 0$ versus H_a: $\mu_1 - \mu_2 \neq 0$ assuming equal variances note that for the "non-vegetarian" group, $\bar{x} = 3.18$ and s = 1.72 for n = 51, and for the "vegetarian" group, $\bar{x} = 2.22$ and s = .67 for n = 9.

Select: **Stat > Basic Statistics > 2-Sample t...**
Select: **Summarized Data**
Enter: First: Sample size: **51** Mean: **3.18** Standard deviation: **1.72**
Second: Sample size: **9** Mean: **2.22** Standard deviation: **.67**
Select: **Assume equal variances**
Select: **Options...**
Enter: Test difference: **0.0**
Select: Alternative: **not equal**
Click: **OK OK**

The results are displayed in the Session window.

```
Session

Two-Sample T-Test and CI

Sample   N   Mean  StDev  SE Mean
1       51   3.18   1.72     0.24
2        9  2.220  0.670     0.22


Difference = mu (1) - mu (2)
Estimate for difference:  0.960000
95% CI for difference:  (-0.209716, 2.129716)
T-Test of difference = 0 (vs not =): T-Value = 1.64  P-Value = 0.106  DF = 58
Both use Pooled StDev = 1.6162
```

Chapter 10 Example 13 Matched Pairs Analysis of Cell Phone Impact on Driver Reaction Time Comparing Means with Matched Pairs

Enter the data from Table 10.9.
To perform a significance test for H_0: $\mu_d = 0$ versus H_a: $\mu_d \neq 0$

Select: **Stat > Basic Statistics > Paired t…**

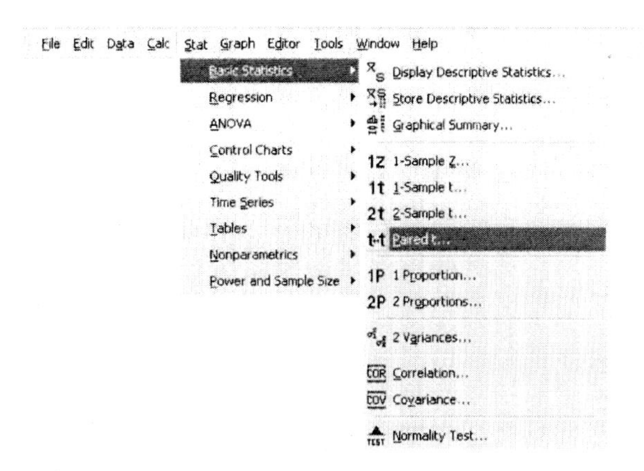

Select: **Samples in columns**
Enter: First sample: **Yes**
Second sample: **No**

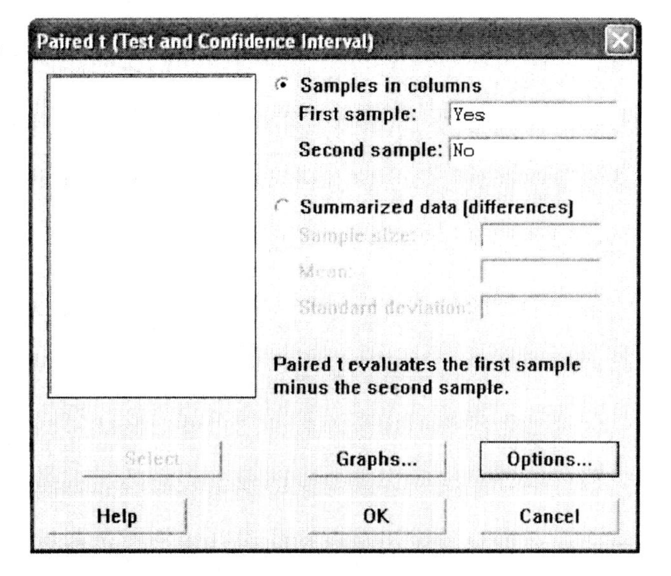

Select: **Options...**
Enter: Confidence level: **95**
Enter: Test difference: **0.0**
Select: Alternative: **not equal**
Click: **OK OK**

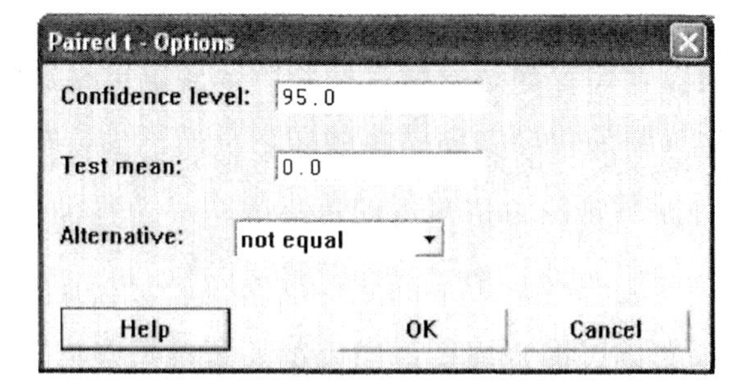

The results are displayed in the Session window.

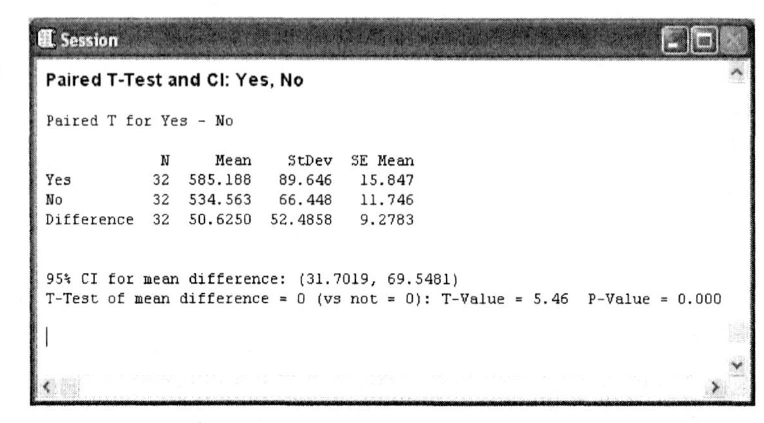

Session
Paired T-Test and CI: Yes, No
Paired T for Yes - No
N Mean StDev SE Mean
Yes 32 585.188 89.646 15.847
No 32 534.563 66.448 11.746
Difference 32 50.6250 52.4858 9.2783
95% CI for mean difference: (31.7019, 69.5481)
T-Test of mean difference = 0 (vs not = 0): T-Value = 5.46 P-Value = 0.000

Chapter 10 Example 15 Inference Comparing Beliefs in Heaven and Hell Comparing Proportions with Dependent Samples

We will create the 95% confidence interval for population mean difference $p_1 - p_2$ using summary statistics.

First note that the sample mean of the 1120 difference scores equals

[(0)(833) + (1)(125) + (-1)(2) + (0)(160)] / 1120 = 123/1120 = 0.109821

and the sample standard deviation of the 1120 difference scores equals

$\sqrt{\{ [(0 - .109821)^2 * (833) + (1- .109821)^2 *(125)+(-1- .109821)^2 *(2) + (0- .109821)^2 *(160)] / (1120-1)\}}$
= 0.318469

Select: **Stat > Basic Statistics > Paired t…**
Select: **Summarized data (differences)**
Enter: Sample size: **1120**
Mean: **.109821**
Standard deviation: **.318469**

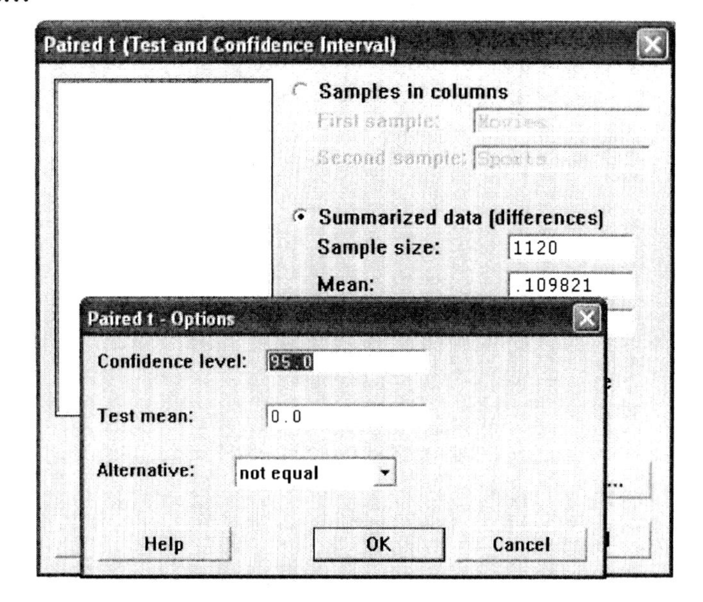

Select: **Options**
Enter: Confidence level: **95**
Click: **OK OK**

The results are displayed in the Session window.

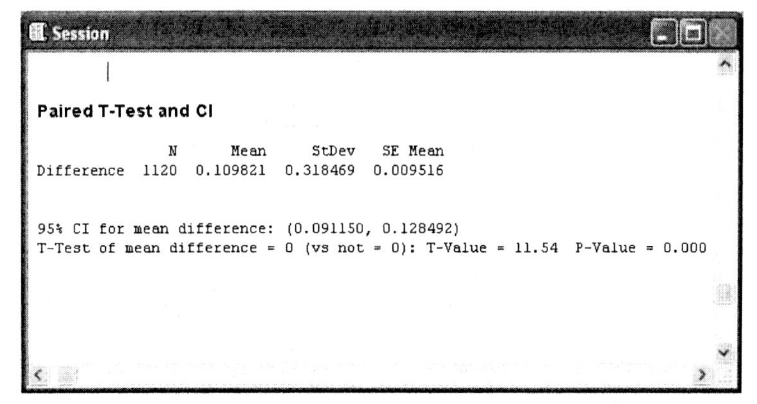

Session
Paired T-Test and CI
N Mean StDev SE Mean
Difference 1120 0.109821 0.318469 0.009516
95% CI for mean difference: (0.091150, 0.128492)
T-Test of mean difference = 0 (vs not = 0): T-Value = 11.54 P-Value = 0.000

Chapter 10 Exercise 10.49 MINITAB Output for Inferential Analysis: Paired t for movies - sports
Comparing Means with Matched Pairs

Enter the data into a worksheet. We wish to perform inference on “movies – sports”.

To perform a significance test for H_0: $\mu_d = 0$ versus H_a: $\mu_d \neq 0$ and to generate a 95% confidence interval

Select: **Stat > Basic Statistics > Paired t...**

Select: **Samples in columns**
Enter: First sample: **Movies**
Second sample: **Sports**

Select: **Options...**
Enter: Confidence level: **95**
Enter: Test difference: **0.0**
Select: Alternative: **not equal**
Click: **OK OK**

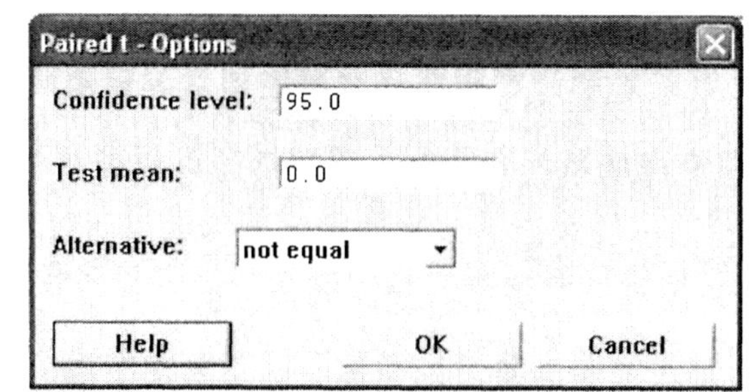

The results are displayed in the Session window.

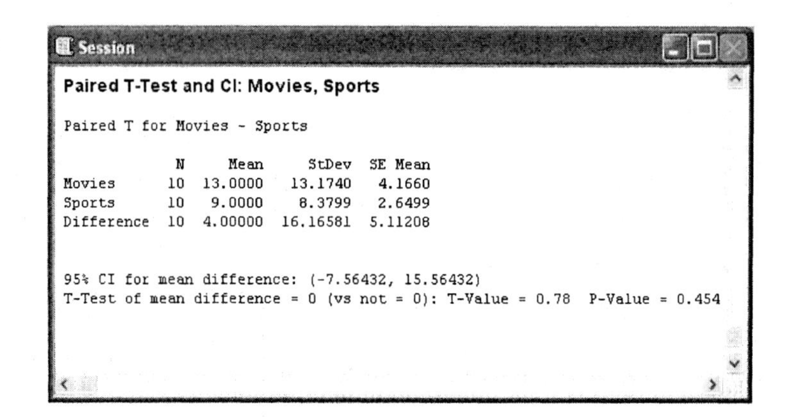

Session

Paired T-Test and CI: Movies, Sports

Paired T for Movies - Sports

	N	Mean	StDev	SE Mean
Movies	10	13.0000	13.1740	4.1660
Sports	10	9.0000	8.3799	2.6499
Difference	10	4.00000	16.16581	5.11208

95% CI for mean difference: (-7.56432, 15.56432)
T-Test of mean difference = 0 (vs not = 0): T-Value = 0.78 P-Value = 0.454

Chapter 10 Exercise 10.61 Heaven and Hell
Comparing Proportions for Dependent Samples

A point estimate for the difference between the population proportions believing in heaven and believing in hell is $\hat{p}_1 - \hat{p}_2 = .631 - .496 = .135$

Because we do not know the sample size used, but can assume dependent sampling, we'll use McNemar's test. Since McNemar's test is not a procedure included in MINITAB, we'll just use MINITAB calculator to do the computation.

Determining the z-score:

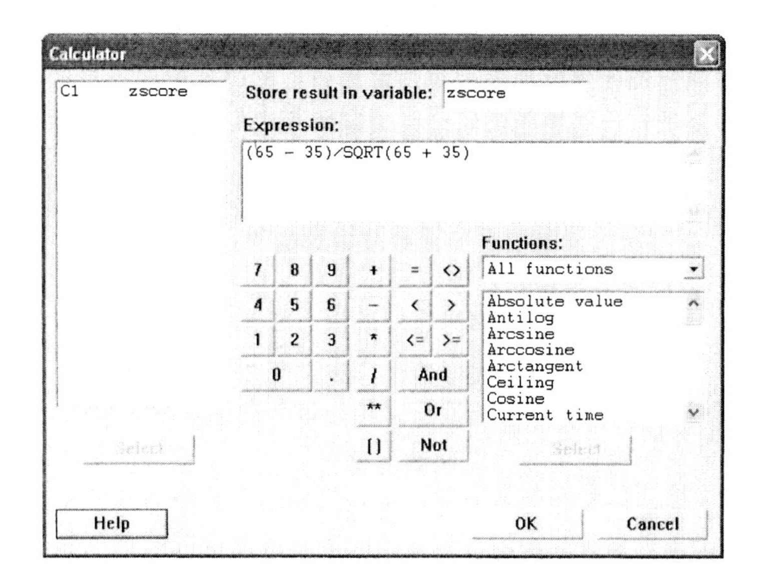

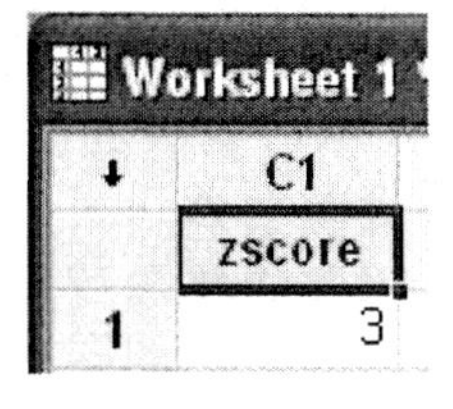

For z = 3, the two-sided p-value equals 0.0026. The sample gives extremely strong evidence that the proportion p_1 definitely believing in heaven is higher than the proportion p_2 definitely believing in hell.

Chapter 11 Exercise 11.1 Gender Gap in Politics?
Contingency Tables

Enter the data into the worksheet.

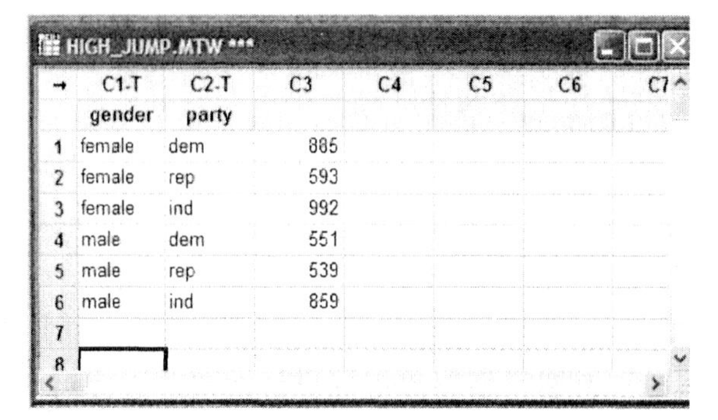

To construct the basic table choose **Stat > Tables > Cross Tabulation and Chi Square**

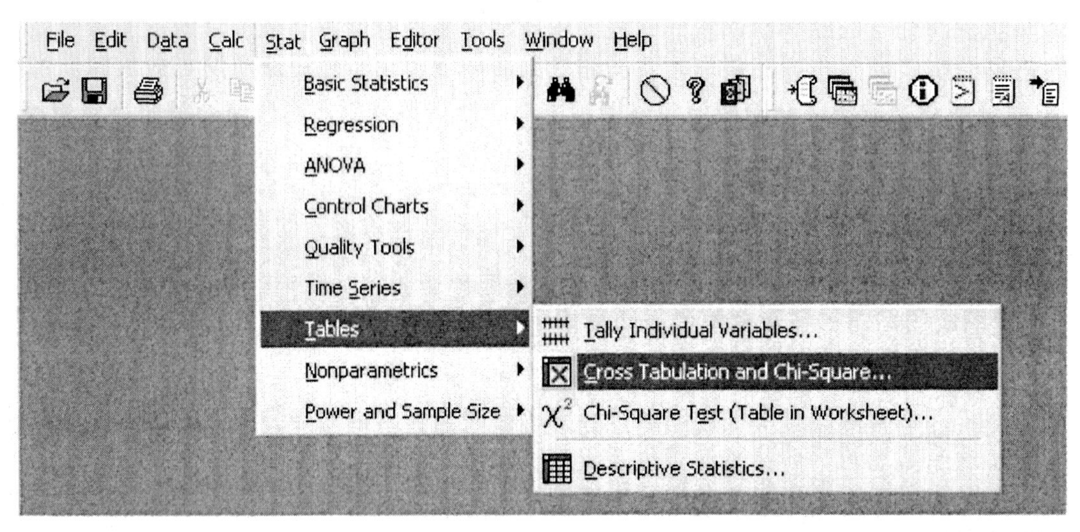

Enter the variables as shown

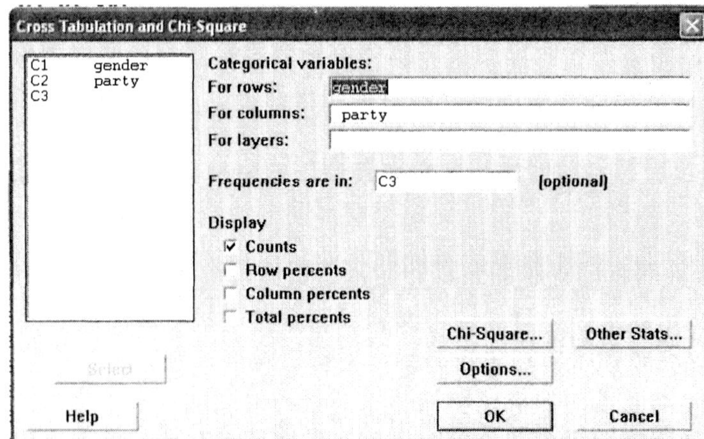

This will generate the basic table.

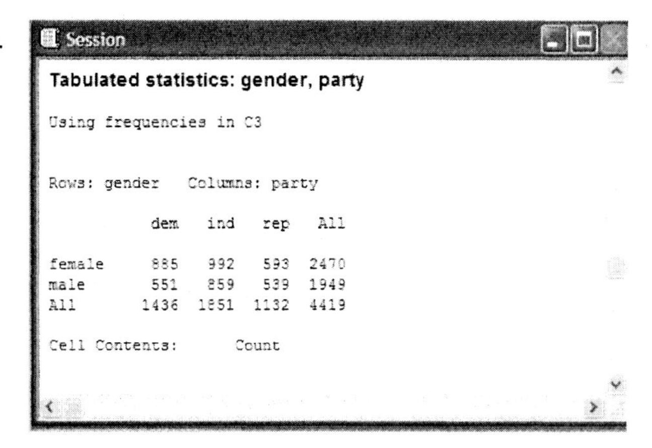

Now to show the percentages, click on Counts, Row Percents and Column Percents.

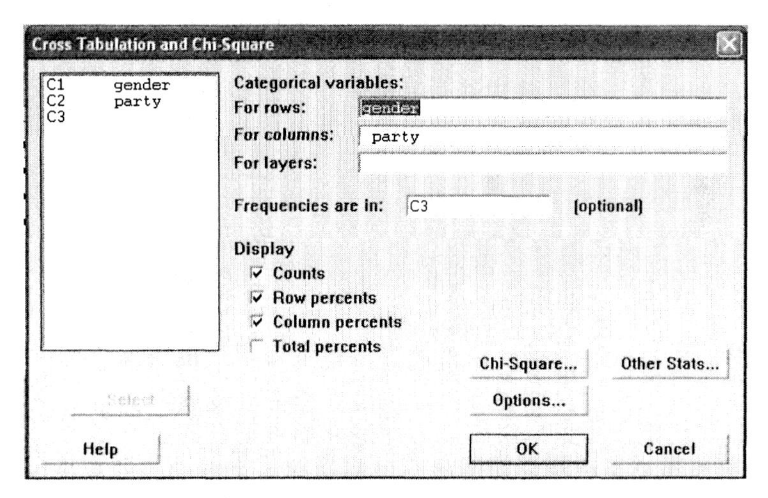

This gives the following in the session window.

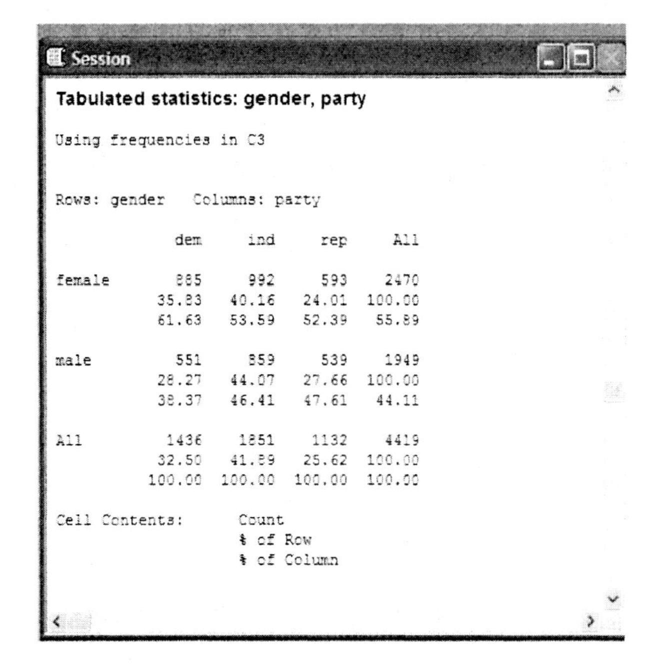

We now enter the chart into the worksheet, so that we can graph it.

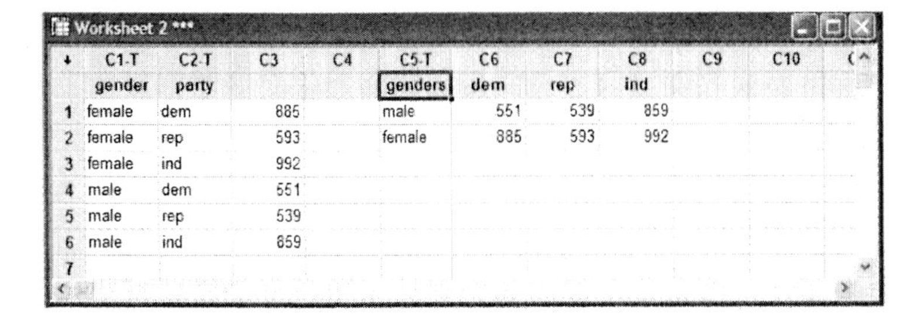

Worksheet 2 ***

↓	C1-T	C2-T	C3	C4	C5-T	C6	C7	C8	C9	C10
	gender	party			genders	dem	rep	ind		
1	female	dem	885		male	551	539	859		
2	female	rep	593		female	885	593	992		
3	female	ind	992							
4	male	dem	551							
5	male	rep	539							
6	male	ind	859							
7										

To create the bar chart
Choose: **Graph > Bar Chart > Values from a table > Two way table cluster**

Enter the values as shown.

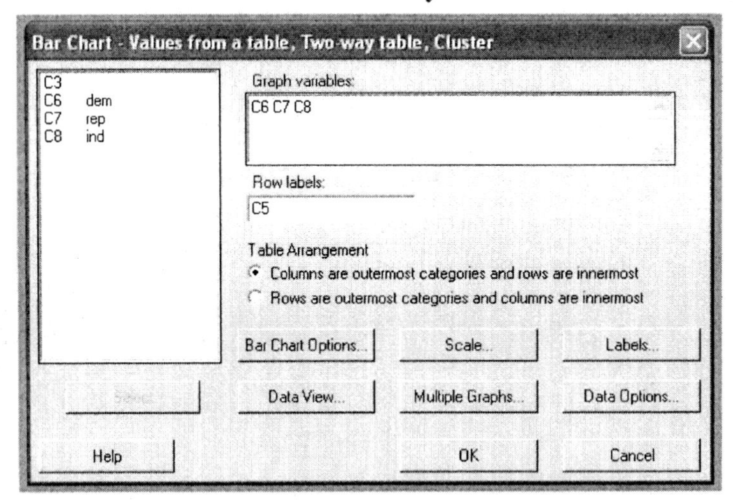

This gives you the following bar chart

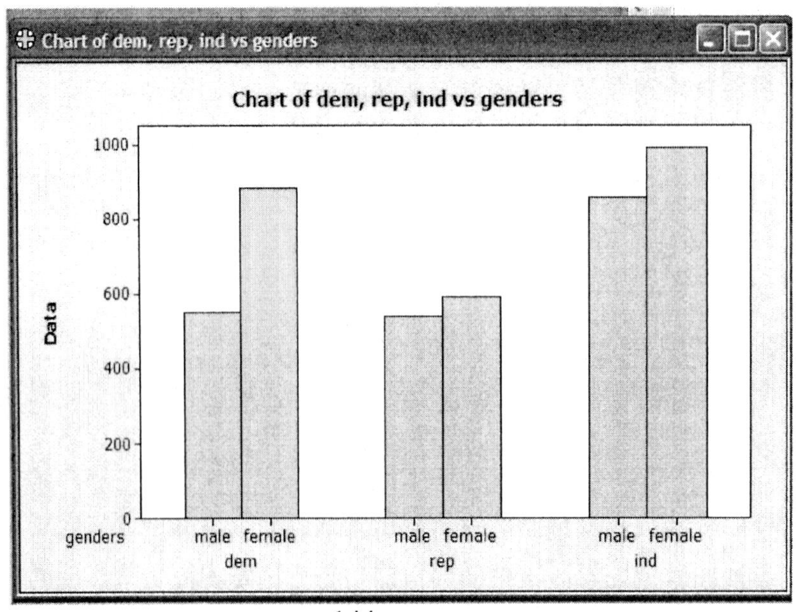

If you had chosen rows to be the outermost category, you would get the next chart.

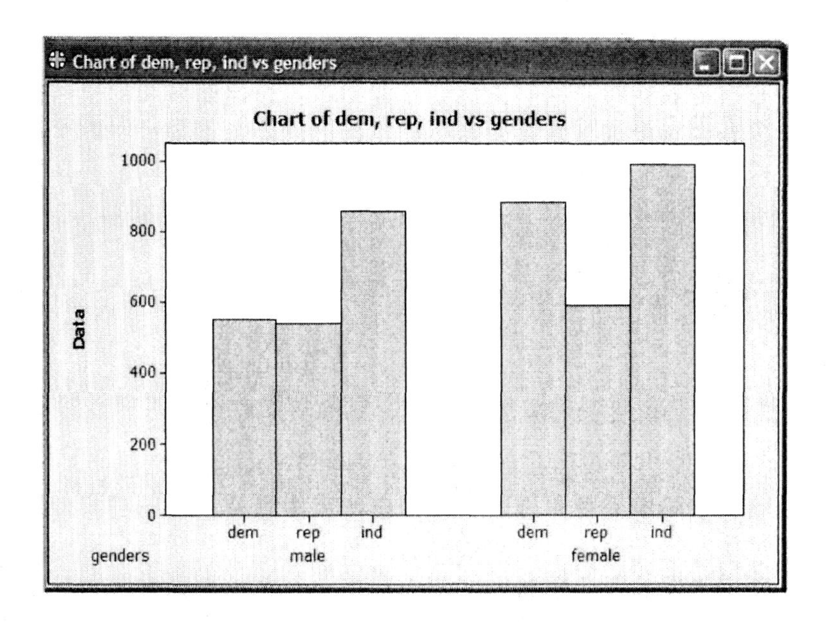

Exercise 11.5 Marital Happiness and Income
Contingency Table and the Chi Squared Statistic

Following the directions in Activity2 at the end of section 11.2, you can access the data in the GSS data base for this problem. Enter the fields as shown below.

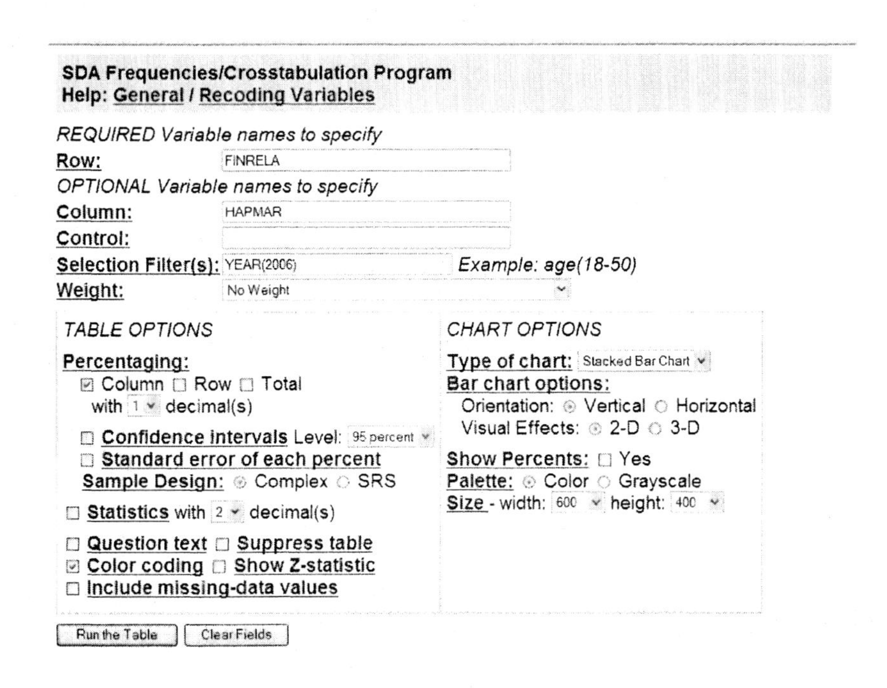

After you have accessed the data, enter it into a Minitab worksheet as follows:

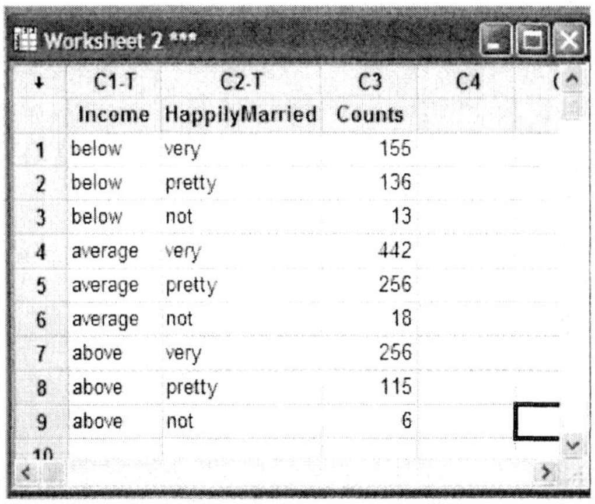

	C1-T Income	C2-T HappilyMarried	C3 Counts	C4
1	below	very	155	
2	below	pretty	136	
3	below	not	13	
4	average	very	442	
5	average	pretty	256	
6	average	not	18	
7	above	very	256	
8	above	pretty	115	
9	above	not	6	
10				

Choose **Stat > Tables > Cross Tabulation and Chi Square**

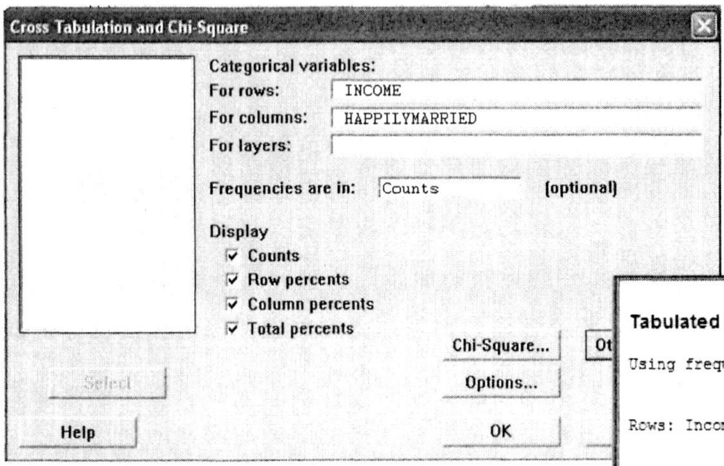

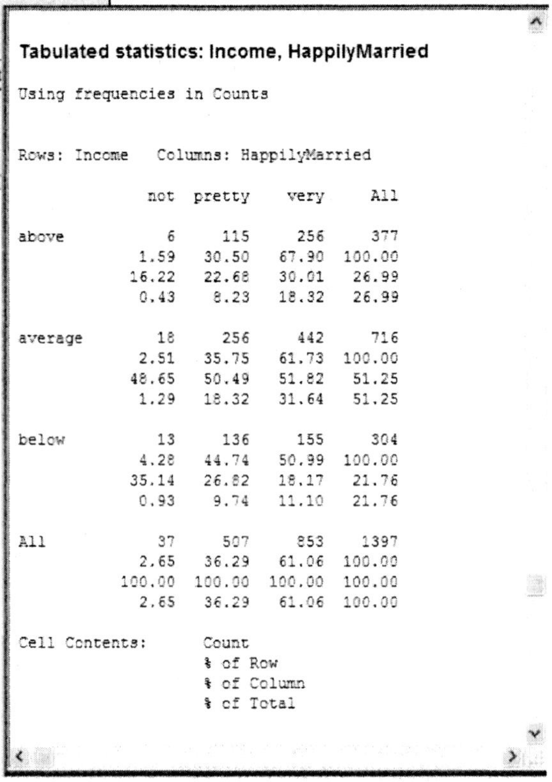

Tabulated statistics: Income, HappilyMarried

Using frequencies in Counts

Rows: Income Columns: HappilyMarried

	not	pretty	very	All
above	6	115	256	377
	1.59	30.50	67.90	100.00
	16.22	22.68	30.01	26.99
	0.43	8.23	18.32	26.99
average	18	256	442	716
	2.51	35.75	61.73	100.00
	48.65	50.49	51.82	51.25
	1.29	18.32	31.64	51.25
below	13	136	155	304
	4.28	44.74	50.99	100.00
	35.14	26.82	18.17	21.76
	0.93	9.74	11.10	21.76
All	37	507	853	1397
	2.65	36.29	61.06	100.00
	100.00	100.00	100.00	100.00
	2.65	36.29	61.06	100.00

Cell Contents: Count
% of Row
% of Column
% of Total

This gives us the basic contingency table.

We now enter the chart into the worksheet, so that we can graph it.

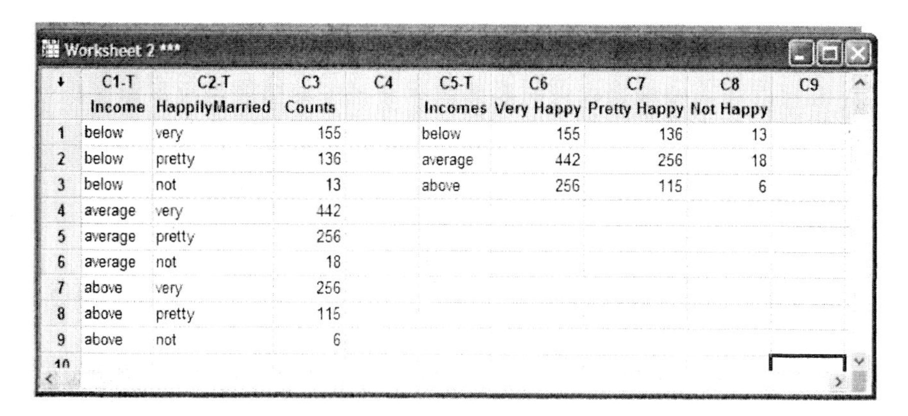

Worksheet 2 ***

↓	C1-T	C2-T	C3	C4	C5-T	C6	C7	C8	C9
	Income	HappilyMarried	Counts		Incomes	Very Happy	Pretty Happy	Not Happy	
1	below	very	155		below	155	136	13	
2	below	pretty	136		average	442	256	18	
3	below	not	13		above	256	115	6	
4	average	very	442						
5	average	pretty	256						
6	average	not	18						
7	above	very	256						
8	above	pretty	115						
9	above	not	6						
10									

To create the bar chart choose **Graph > Bar Chart > Values from a table > Two way table cluster.** Enter the values as shown.

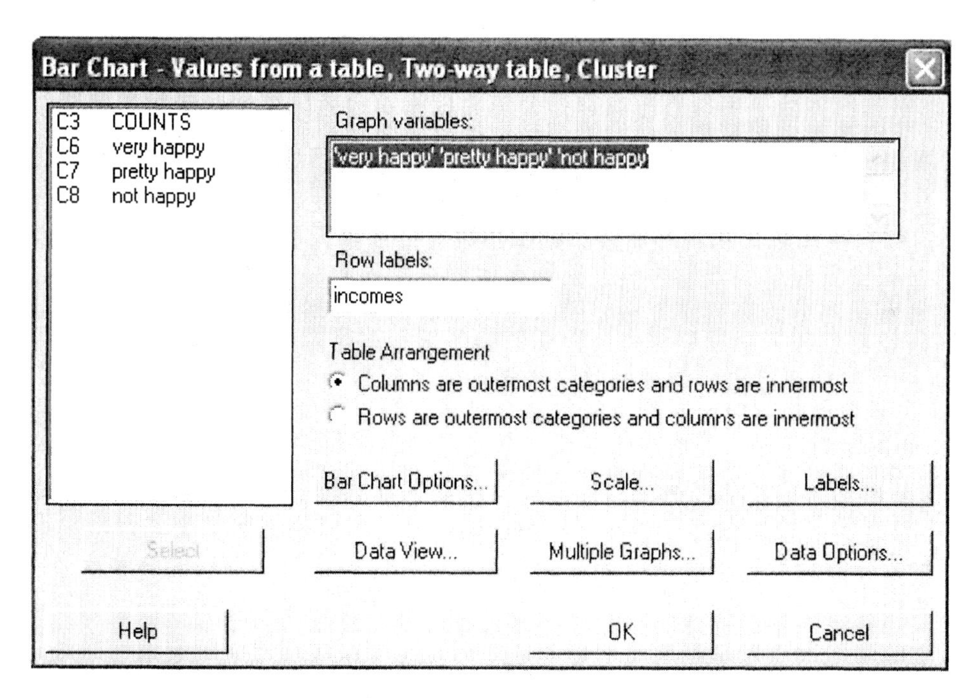

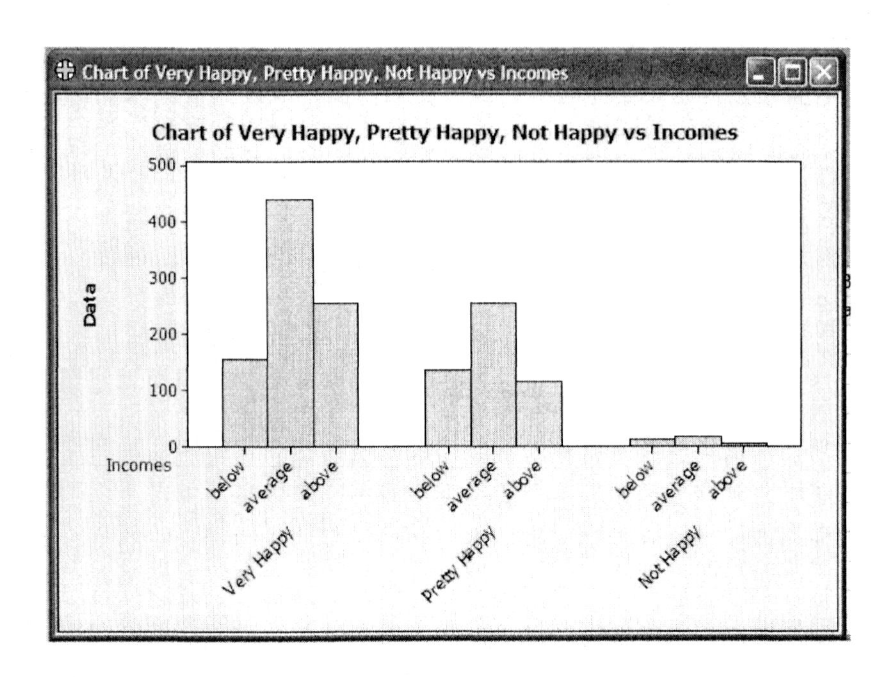

Chapter 11 Example 3 Chi Squared Test for Happiness and Family Income
Chi Squared Test of Independence

In this problem we are using the data summarized in the margin of the text, originally retrieved from the GSS. We can create a data file using it by manually entering the data.

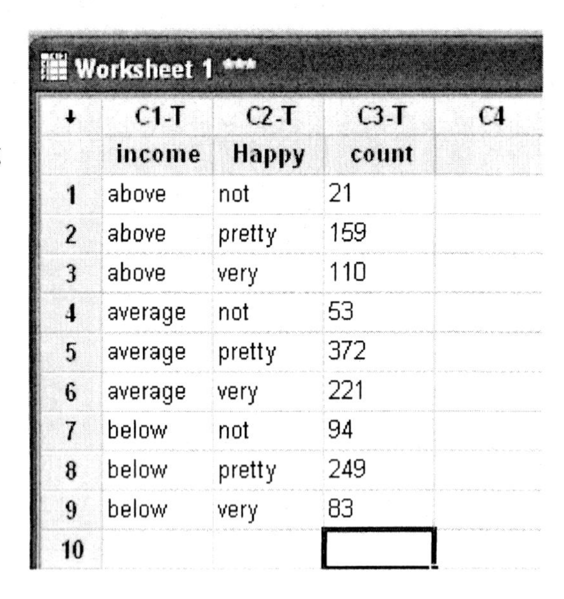

Worksheet 1 ***

↓	C1-T	C2-T	C3-T	C4
	income	Happy	count	
1	above	not	21	
2	above	pretty	159	
3	above	very	110	
4	average	not	53	
5	average	pretty	372	
6	average	very	221	
7	below	not	94	
8	below	pretty	249	
9	below	very	83	
10				

To create the table as shown in the text :
Choose **Stat > Tables > Cross Tabulation and Chi Squared**

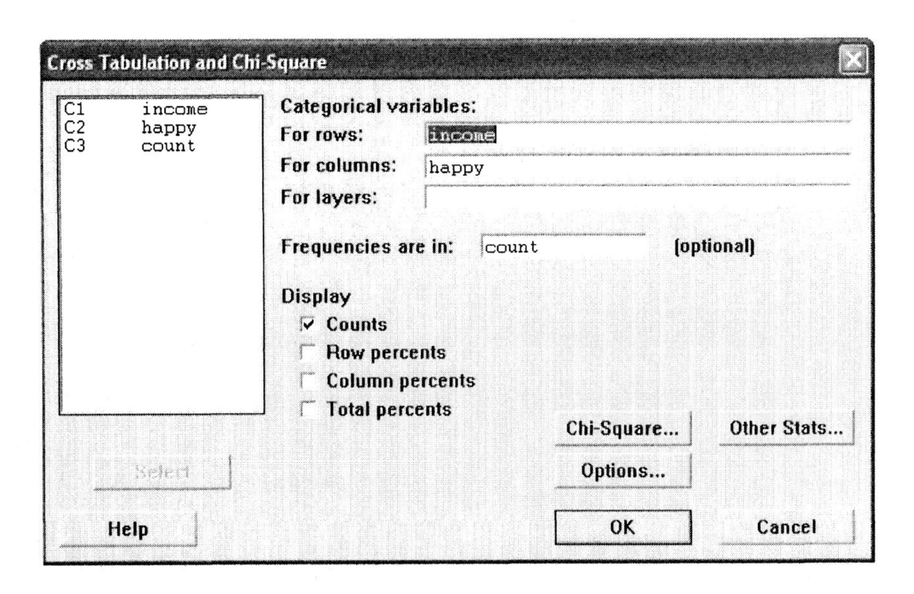

Click on the **Chi-Square** box and check the desired boxes as shown:

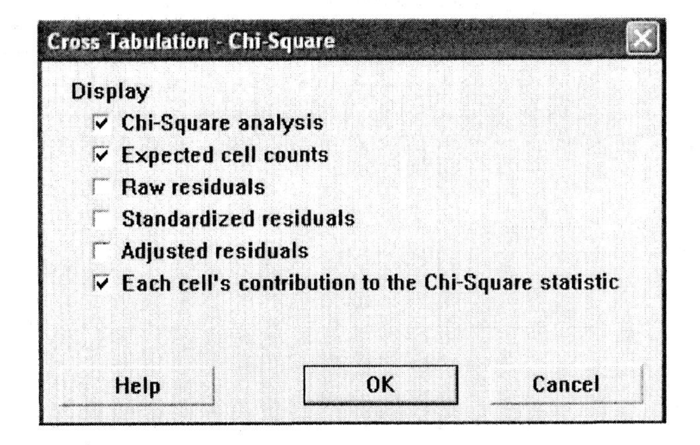

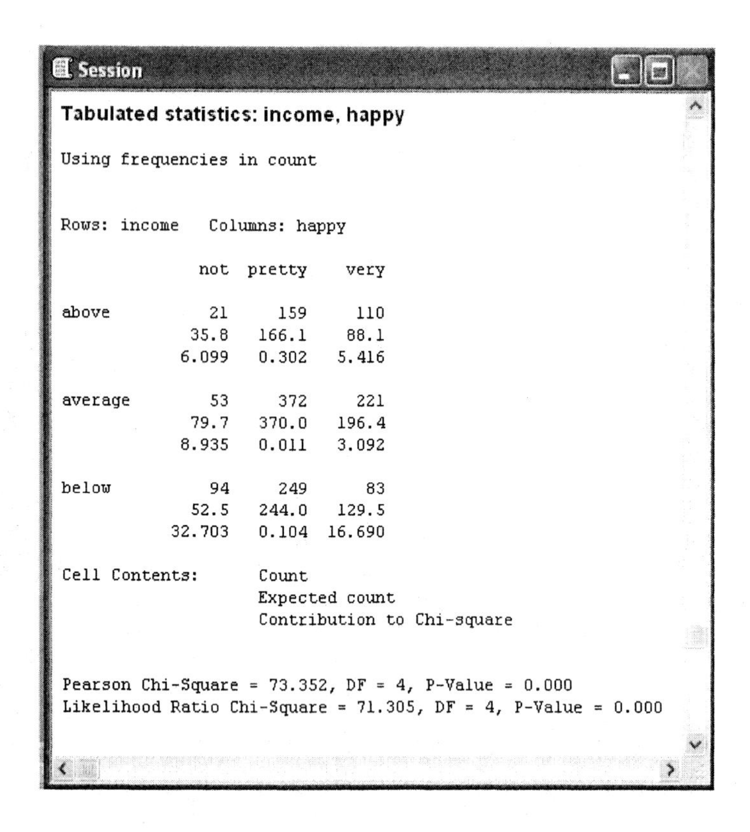

Session

Tabulated statistics: income, happy

```
Using frequencies in count


Rows: income    Columns: happy

              not  pretty    very

above          21     159     110
             35.8   166.1    88.1
            6.099   0.302   5.416

average        53     372     221
             79.7   370.0   196.4
            8.935   0.011   3.092

below          94     249      83
             52.5   244.0   129.5
           32.703   0.104  16.690

Cell Contents:      Count
                    Expected count
                    Contribution to Chi-square


Pearson Chi-Square = 73.352, DF = 4, P-Value = 0.000
Likelihood Ratio Chi-Square = 71.305, DF = 4, P-Value = 0.000
```

Chapter 11 Example 4 Are Happiness and Income Independent?
Chi Squared Test of Independence

To create the table as shown in the text :

Choose **Stat > Tables > Cross Tabulation and Chi Squared**
Click on the **Chi-Square** box and check the desired boxes

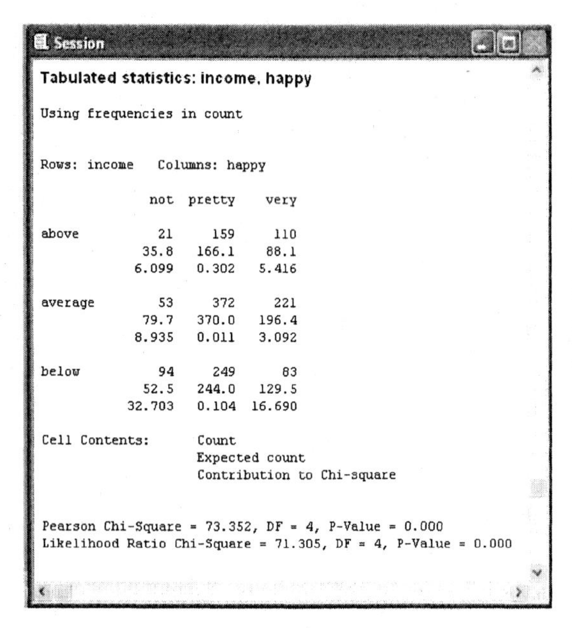

Session

Tabulated statistics: income, happy

```
Using frequencies in count


Rows: income    Columns: happy

              not  pretty    very

above          21     159     110
             35.8   166.1    88.1
            6.099   0.302   5.416

average        53     372     221
             79.7   370.0   196.4
            8.935   0.011   3.092

below          94     249      83
             52.5   244.0   129.5
           32.703   0.104  16.690

Cell Contents:      Count
                    Expected count
                    Contribution to Chi-square


Pearson Chi-Square = 73.352, DF = 4, P-Value = 0.000
Likelihood Ratio Chi-Square = 71.305, DF = 4, P-Value = 0.000
```


Notice the p-value given as P = 0.000, degrees of freedom = 4 and Chi-Square = 73.3

Since the p-value is below 0.05, we can reject the null hypothesis that happiness and family income are independent.

Chapter 11 Exercise 11.11 Life After Death and Gender
Contingency Table and the Chi Squared Statistic

a) Construct a contingency table by entering the data into the worksheet.

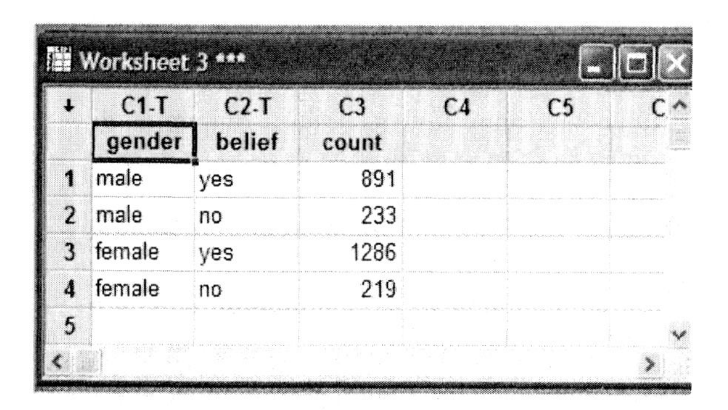

Worksheet 3 ***

↓	C1-T	C2-T	C3	C4	C5	C
	gender	belief	count			
1	male	yes	891			
2	male	no	233			
3	female	yes	1286			
4	female	no	219			
5						

To create the table Choose **Stat > Tables > Cross Tabulation and Chi Squared**

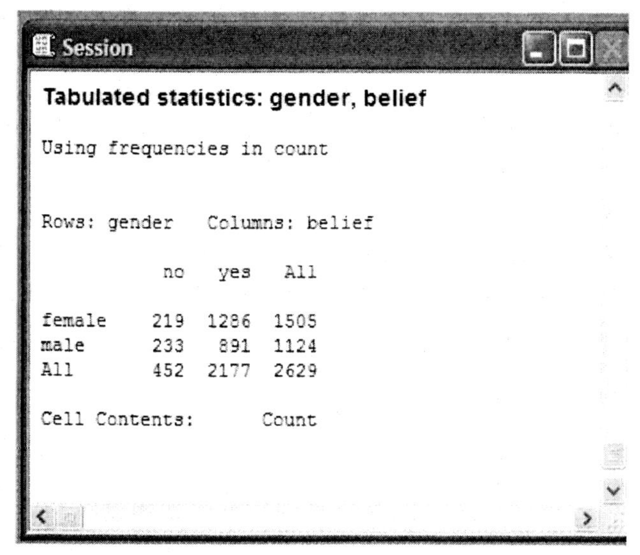

Session

Tabulated statistics: gender, belief

```
Using frequencies in count


Rows: gender   Columns: belief

           no   yes   All

female    219  1286  1505
male      233   891  1124
All       452  2177  2629

Cell Contents:      Count
```

To obtain the expected counts for each cell, its contribution to the Chi-Square and the analysis, check the appropriate boxes.

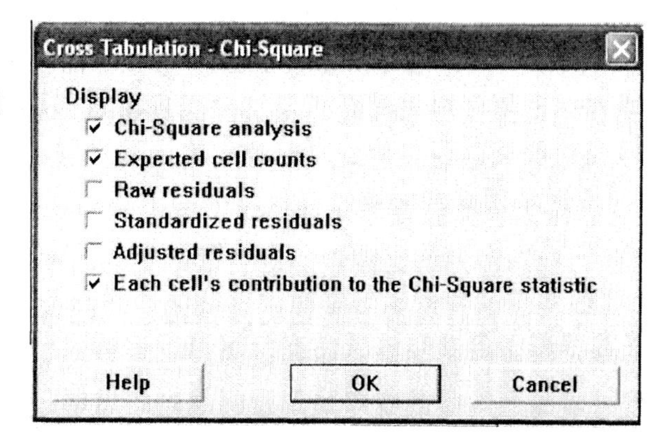

This gives the following in the session window:

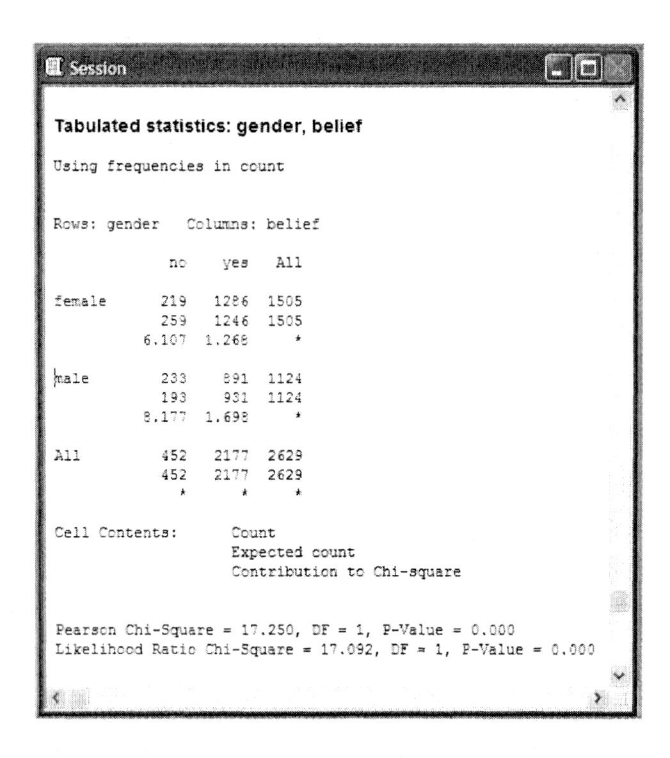

```
Session

Tabulated statistics: gender, belief

Using frequencies in count


Rows: gender   Columns: belief

            no    yes   All

female     219   1286  1505
           259   1246  1505
         6.107  1.268     *

male       233    891  1124
           193    931  1124
         8.177  1.698     *

All        452   2177  2629
           452   2177  2629
             *      *     *

Cell Contents:      Count
                    Expected count
                    Contribution to Chi-square


Pearson Chi-Square = 17.250, DF = 1, P-Value = 0.000
Likelihood Ratio Chi-Square = 17.092, DF = 1, P-Value = 0.000
```

Chapter 11 Exercise 11.15 Help the Environment
A Chi-Squared test

First enter the data into the worksheet.

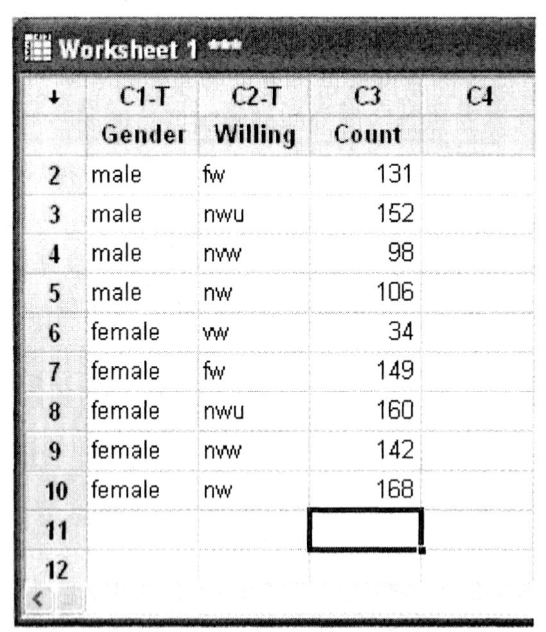

Worksheet 1 ***

↓	C1-T	C2-T	C3	C4
	Gender	Willing	Count	
2	male	fw	131	
3	male	nwu	152	
4	male	nvw	98	
5	male	nw	106	
6	female	vw	34	
7	female	fw	149	
8	female	nwu	160	
9	female	nvw	142	
10	female	nw	168	
11				
12				

Create the two-way table: Choose **Stat > Tables > Cross Tabulation and Chi Squared**
Click on the **Chi-Square** box and check the desired boxes

Note the degrees of freedom, the Chi Square statistic and the p-value are all given at the bottom of the chart.

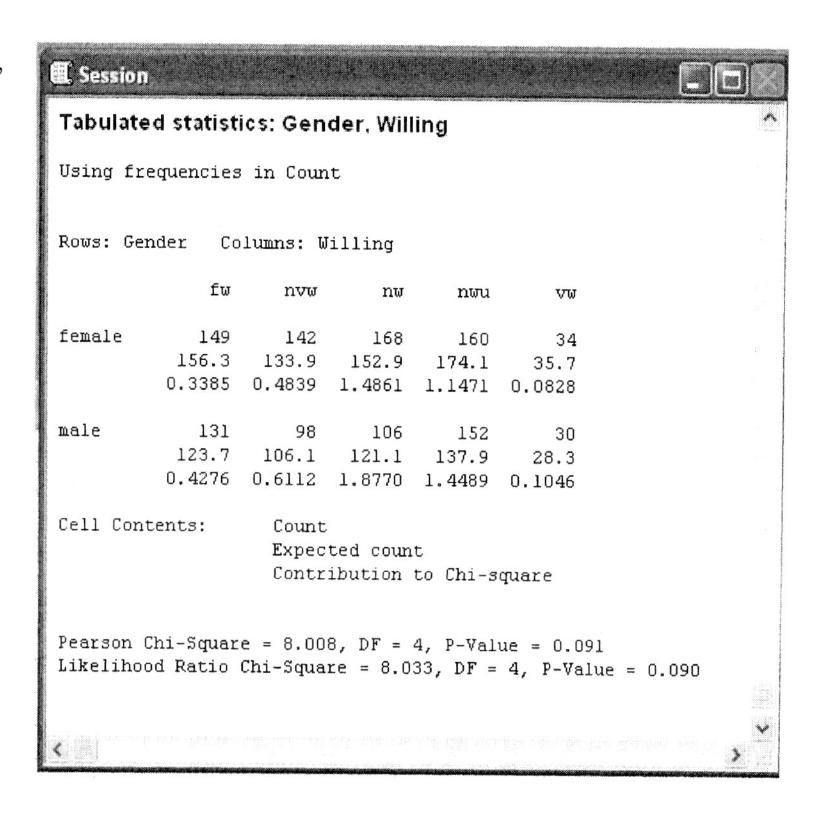
Session

Tabulated statistics: Gender, Willing

Using frequencies in Count

Rows: Gender   Columns: Willing

	fw	nvw	nw	nwu	vw
female	149	142	168	160	34
	156.3	133.9	152.9	174.1	35.7
	0.3385	0.4839	1.4861	1.1471	0.0828
male	131	98	106	152	30
	123.7	106.1	121.1	137.9	28.3
	0.4276	0.6112	1.8770	1.4489	0.1046

Cell Contents: Count
Expected count
Contribution to Chi-square

Pearson Chi-Square = 8.008, DF = 4, P-Value = 0.091
Likelihood Ratio Chi-Square = 8.033, DF = 4, P-Value = 0.090

Chapter 11 Example 9 Standardized Residuals for Religiosity and Gender

We can duplicate the chart in Example 9 by entering the data into the worksheet.

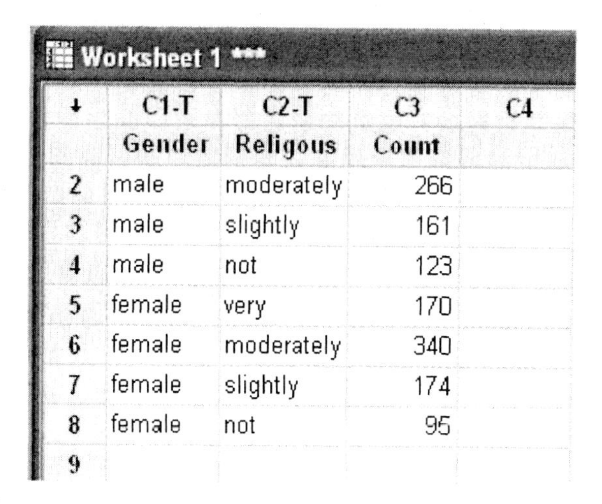
Worksheet 1 ***

↓	C1-T	C2-T	C3	C4
	Gender	Religous	Count	
2	male	moderately	266	
3	male	slightly	161	
4	male	not	123	
5	female	very	170	
6	female	moderately	340	
7	female	slightly	174	
8	female	not	95	
9				

Choose **Stat > Tables > Cross Tabulation and Chi Squared**

To show the residuals, check the adjusted residuals box.

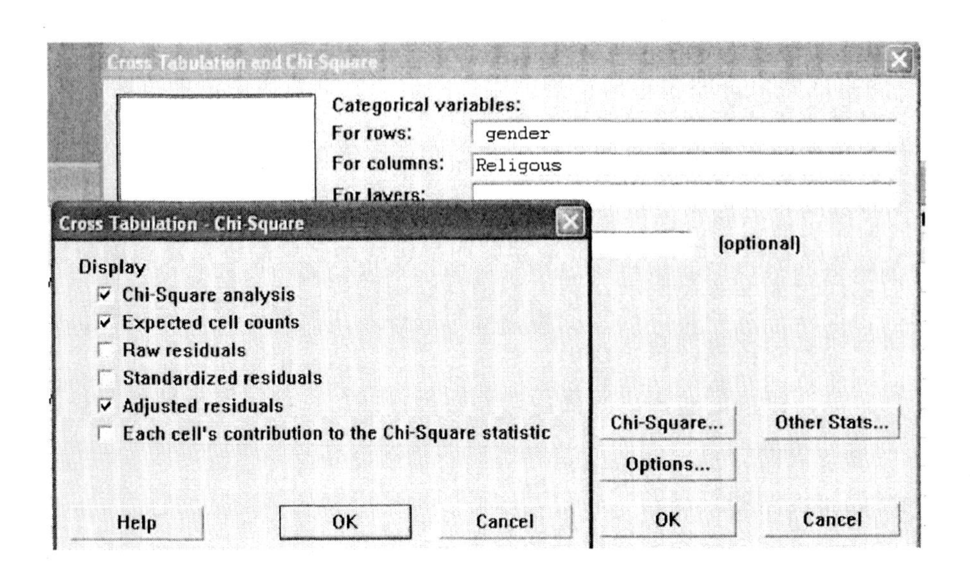

This will give you Table 11.17 in the text.

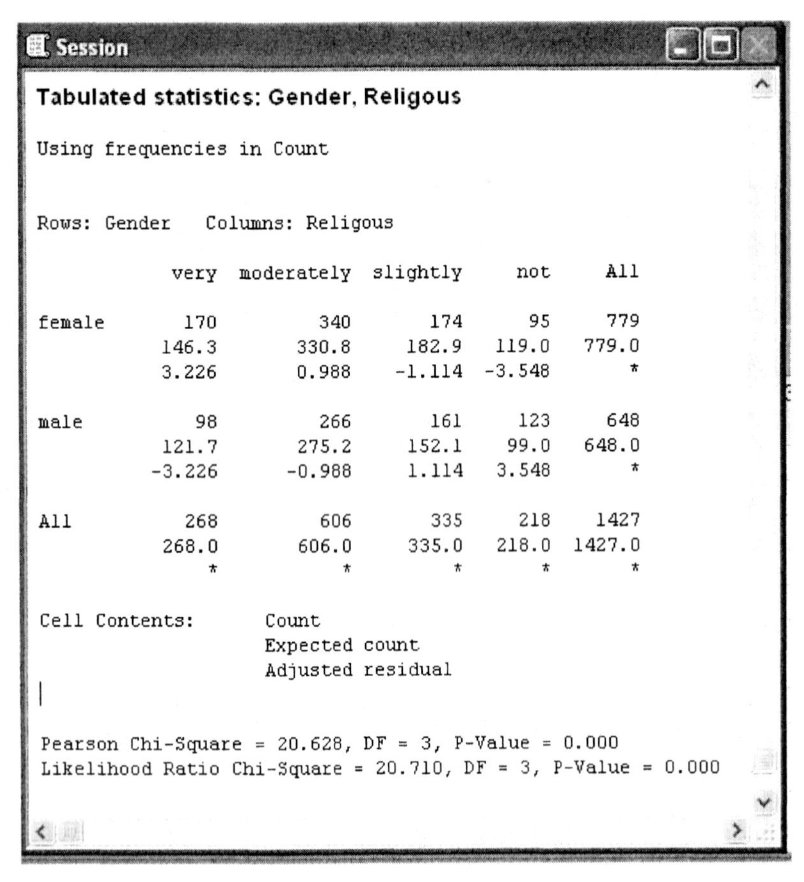

Session

Tabulated statistics: Gender, Religous

Using frequencies in Count

Rows: Gender   Columns: Religous

	very	moderately	slightly	not	All
female	170	340	174	95	779
	146.3	330.8	182.9	119.0	779.0
	3.226	0.988	-1.114	-3.548	*
male	98	266	161	123	648
	121.7	275.2	152.1	99.0	648.0
	-3.226	-0.988	1.114	3.548	*
All	268	606	335	218	1427
	268.0	606.0	335.0	218.0	1427.0
	*	*	*	*	*

Cell Contents: Count
Expected count
Adjusted residual

Pearson Chi-Square = 20.628, DF = 3, P-Value = 0.000
Likelihood Ratio Chi-Square = 20.710, DF = 3, P-Value = 0.000

When you look at this chart, note the entry for females that are very religious. The difference between the observed and expected counts is more than three standard deviations. What conclusion can you draw? What do the negative residuals mean?

Chapter 11 Exercise 11.33 Standardized residuals for Happiness and Income

We again have this worksheet of raw data:

Worksheet 1 ***

↓	C1-T Income	C2-T Happy	C3 Count	C4
1	above	not	21	
2	above	pretty	159	
3	above	very	110	
4	average	not	53	
5	average	pretty	372	
6	average	very	221	
7	below	not	94	
8	below	pretty	249	
9	below	very	83	
10				
11				

We can generate the chart by choosing

Stat > Tables > Cross Tabulation and Chi Squared

Be sure to check the appropriate boxes in the Chi Square tab.

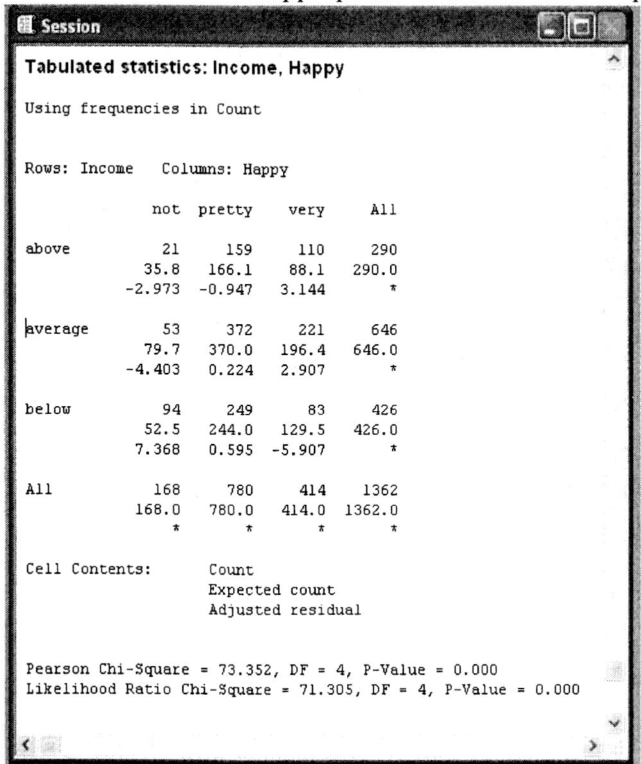

Session

Tabulated statistics: Income, Happy

Using frequencies in Count

Rows: Income   Columns: Happy

	not	pretty	very	All
above	21	159	110	290
	35.8	166.1	88.1	290.0
	-2.973	-0.947	3.144	*
average	53	372	221	646
	79.7	370.0	196.4	646.0
	-4.403	0.224	2.907	*
below	94	249	83	426
	52.5	244.0	129.5	426.0
	7.368	0.595	-5.907	*
All	168	780	414	1362
	168.0	780.0	414.0	1362.0
	*	*	*	*

Cell Contents: Count
Expected count
Adjusted residual

Pearson Chi-Square = 73.352, DF = 4, P-Value = 0.000
Likelihood Ratio Chi-Square = 71.305, DF = 4, P-Value = 0.000

Notice the high residuals in certain cells. How would you interpret these?
Notice the negative residuals in certain cells. How would you interpret these?

Chapter 11 Example 10 Tea Tastes Better with Milk Poured First? Small Sample Size – Fisher's Exact Test

We start by creating the worksheet.

Worksheet 1 ***

↓	C1-T	C2-T	C3	C4
	Actual	Predicted	Counts	
1	Milk	Milk	3	
2	Milk	Tea	1	
3	Tea	Milk	1	
4	Tea	Tea	3	
5				

To create the table as shown in the text :
Choose **Stat > Tables > Cross Tabulation and Chi Squared**

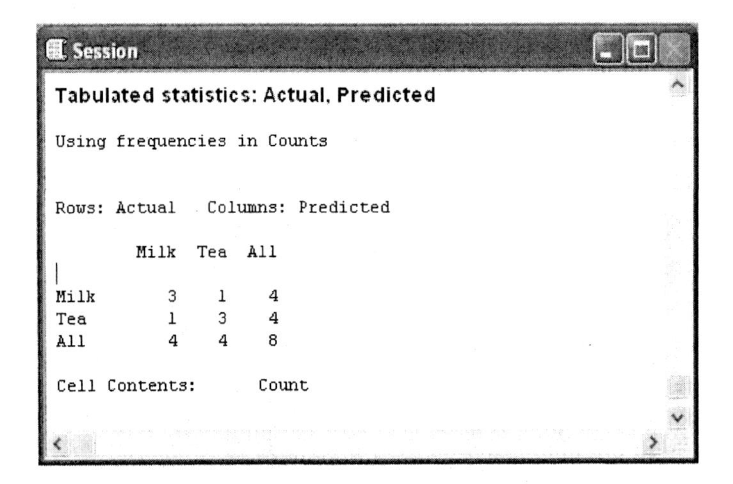

Session

Tabulated statistics: Actual, Predicted

Using frequencies in Counts

Rows: Actual   Columns: Predicted

	Milk	Tea	All
Milk	3	1	4
Tea	1	3	4
All	4	4	8

Cell Contents: Count

In the Chi Square tab, check the Chi Square Analysis and Expected Cell Counts boxes.

In the Other Stats tab, check the box for Fishers Exact Test for 2x2 Tables.

This will give you the following chart:

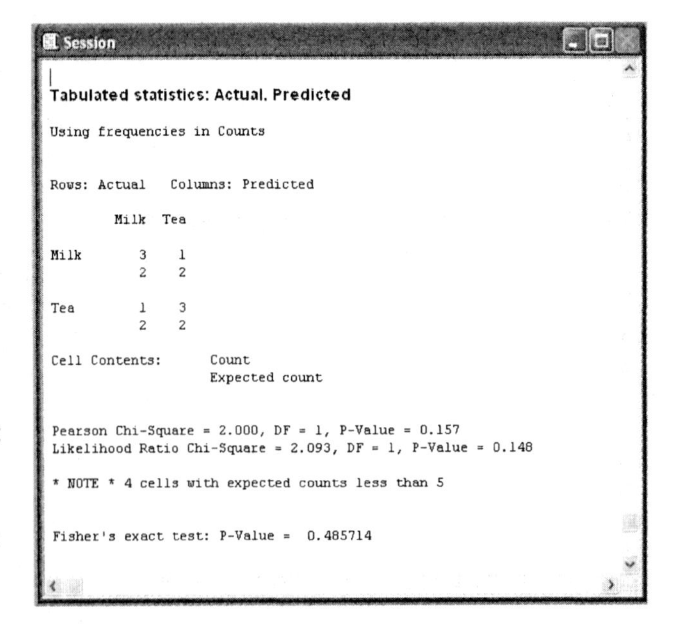

Session

Tabulated statistics: Actual, Predicted

Using frequencies in Counts

Rows: Actual   Columns: Predicted

	Milk	Tea
Milk	3 2	1 2
Tea	1 2	3 2

Cell Contents: Count
Expected count

Pearson Chi-Square = 2.000, DF = 1, P-Value = 0.157
Likelihood Ratio Chi-Square = 2.093, DF = 1, P-Value = 0.148

* NOTE * 4 cells with expected counts less than 5

Fisher's exact test: P-Value = 0.485714

Fisher's exact test is a test based on an exact distribution rather than on the approximate chi-square distribution used for Pearson's and likelihood ratio tests. Fisher's exact test is useful when the expected cell counts are low and chi-square approximation is not very good.

Chapter 11 Exercise 11.41 Claritin and Nervousness
Small Sample Size – Fisher's Exact Test

We enter the data into the worksheet as follows:

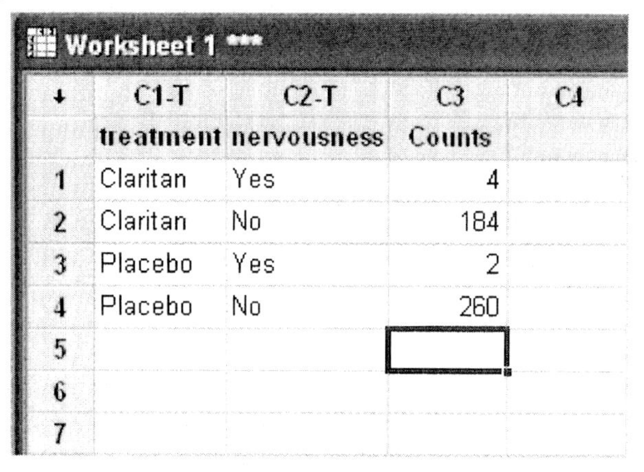

↓	C1-T	C2-T	C3	C4
	treatment	nervousness	Counts	
1	Claritan	Yes	4	
2	Claritan	No	184	
3	Placebo	Yes	2	
4	Placebo	No	260	
5				
6				
7				

To create the table Choose **Stat > Tables > Cross Tabulation and Chi Squared**
In the Chi Square tab, check the Chi Square Analysis and Expected Cell Counts boxes.
In the Other Stats tab, check the box for Fishers Exact Test for 2x2 Tables.

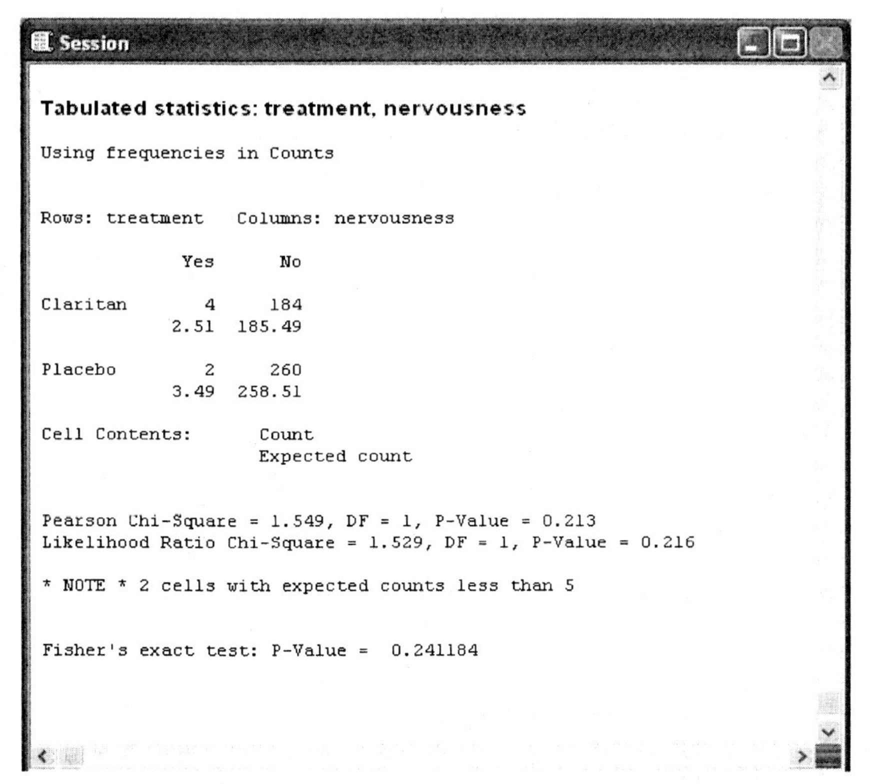

```
Tabulated statistics: treatment, nervousness

Using frequencies in Counts


Rows: treatment   Columns: nervousness

             Yes      No

Claritan       4     184
            2.51  185.49

Placebo        2     260
            3.49  258.51

Cell Contents:       Count
                     Expected count


Pearson Chi-Square = 1.549, DF = 1, P-Value = 0.213
Likelihood Ratio Chi-Square = 1.529, DF = 1, P-Value = 0.216

* NOTE * 2 cells with expected counts less than 5


Fisher's exact test: P-Value =  0.241184
```

Note the p-value, the Chi Square statistic.

Chapter 12 Example 2 What Do We Learn From a Scatterplot in The Strength Study? Review of Producing a Scatterplot

Open the data file **high_school_female_athletes** from the MINITAB folder of the data disk.

To create a scatterplot of the data with C15 BP(60) as the x-variable, and C12 BP as the y-variable

Click: **Graph > Scatterplot > Simple > OK**
Enter: Y Variables **C12**
X Variables **C15**
Select: Labels **enter an appropriate title**
Click: **OK OK**

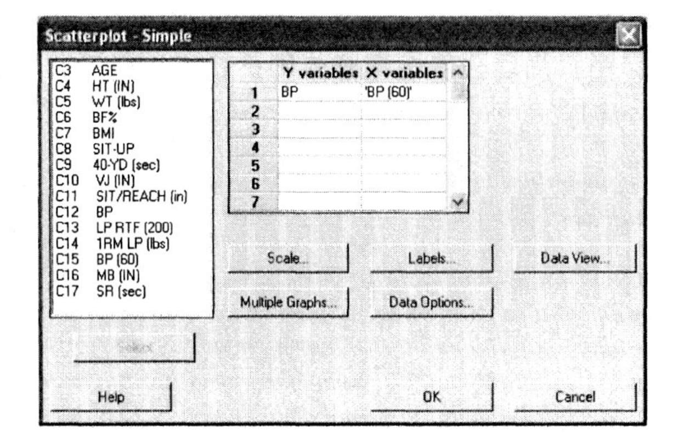

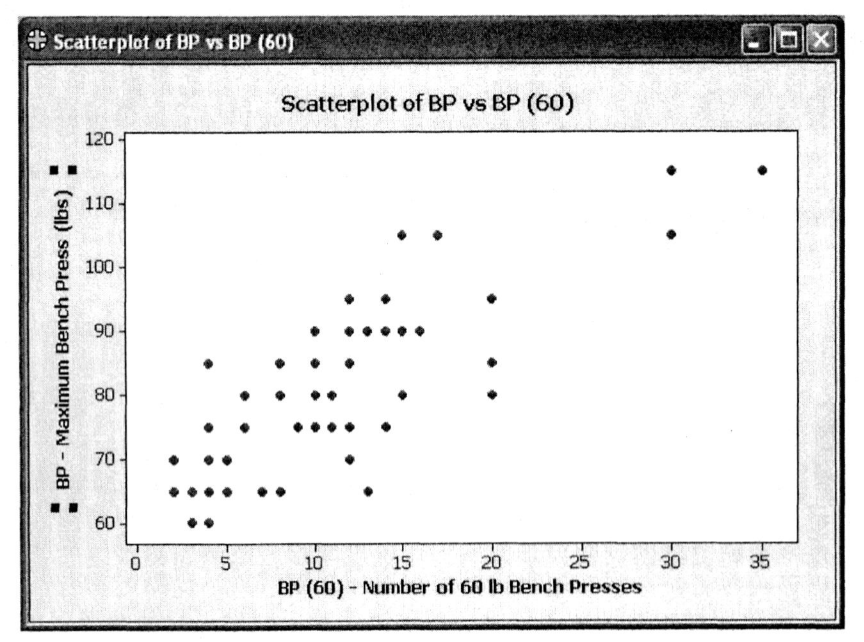

The scatterplot shows that female athletes with higher numbers of 60-pound bench presses also tended to have higher vales for the maximum bench press.

Chapter 12 Example 3 Which Regression Line Predicts Maximum Bench Press? Review of Generating a Regression Line

Open the data file **high_school_female_athletes** from the MINITAB folder of the text CD.

To generate the least squares regression line

Click: **Stat > Regression > Regression**

Enter: Response: **C12**
Predictors: **C15**
Click: **OK**

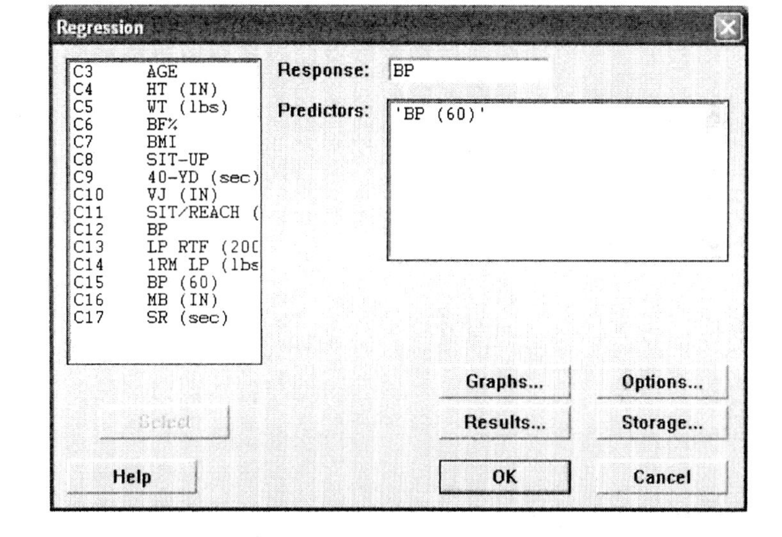

The results are displayed in the Session window.

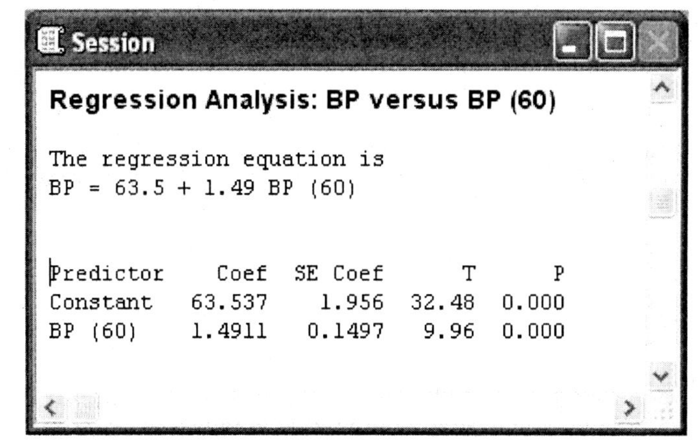

To create a scatterplot of the data with the regression line added:

Click: **Graph > Scatterplot > With Regression > OK**
Enter: Y Variables **C12**
X Variables **C15**
Select: Labels **enter an appropriate title**
Click: **OK OK**

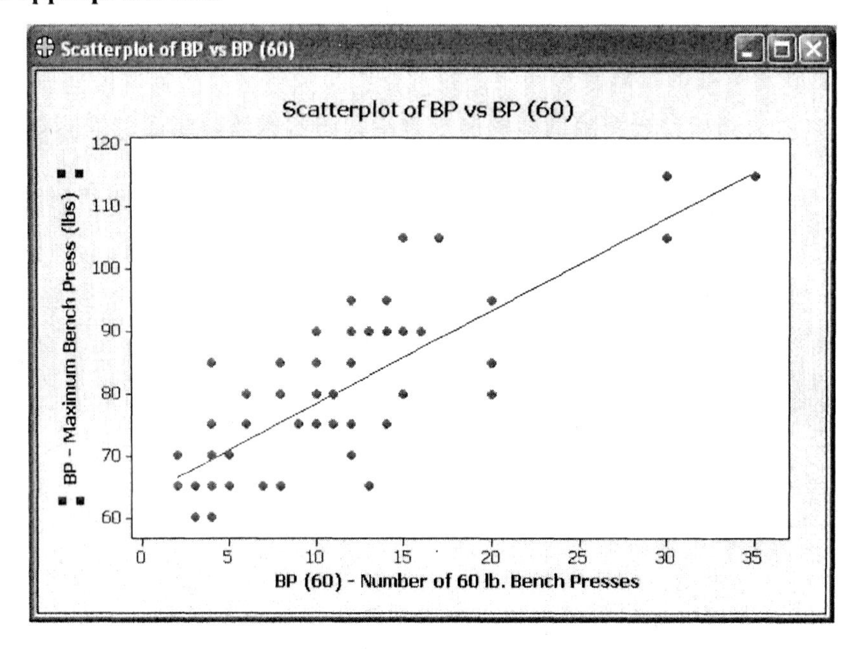

Chapt Exercise 12.1 Car Mileage and Weight
Generi ing a Regression Line

Open tl data file **car_weight_and_mileage** from the MINITAB folder of the text CD.

To gen ate the least squares regression line
lick: **Stat > Regression > Regression**

Enter: Response: **mileage**
Predictors: **weight**
Click: **OK**

The res lts are displayed in the Sessio indow.

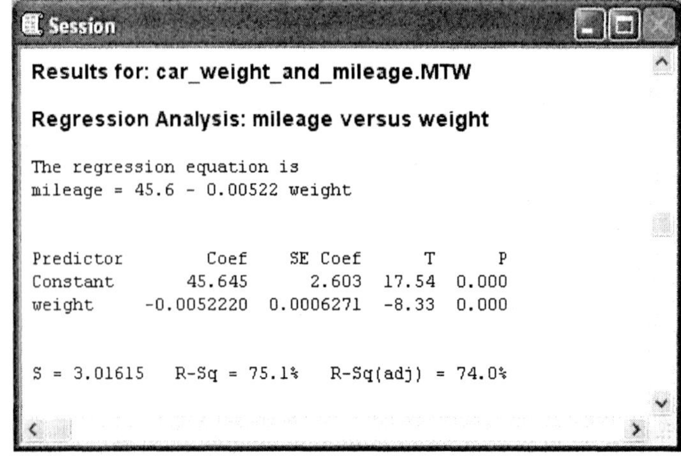

Session

Results for: car_weight_and_mileage.MTW

Regression Analysis: mileage versus weight

```
The regression equation is
mileage = 45.6 - 0.00522 weight


Predictor        Coef    SE Coef      T      P
Constant       45.645      2.603  17.54  0.000
weight     -0.0052220  0.0006271  -8.33  0.000


S = 3.01615   R-Sq = 75.1%   R-Sq(adj) = 74.0%
```

Chapter 12 Exercise 12.7 Study time and college GPA.

Generating a Scatterplot and Regression Line

Enter the data in to a worksheet.

12.7 ***

↓	C1	C2	C3	C4	C5	C6	C7
	Student	Study Time	GPA	Skipped			
1	1	14	2.8	9			
2	2	25	3.6	0			
3	3	15	3.4	2			
4	4	5	3.0	5			
5	5	10	3.1	3			
6	6	12	3.3	2			
7	7	5	2.7	12			
8	8	21	3.8	1			
9							

Current Worksheet: 12.7

Choose: **Graph > Scatterplot> Simple**
Click: **OK**
Enter: Y variable: **C3**
X Variable: **C2**

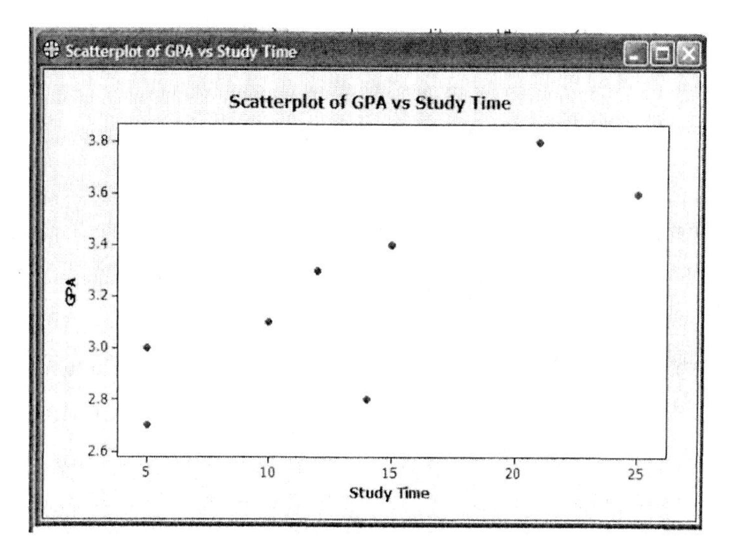

The scatterplot shows a fairly linear, positive association between study time and GPA.

Choose: Stat > **Regression > Regression**
Enter: Response: **C3**
Predictors: **C2**
Click: **OK**

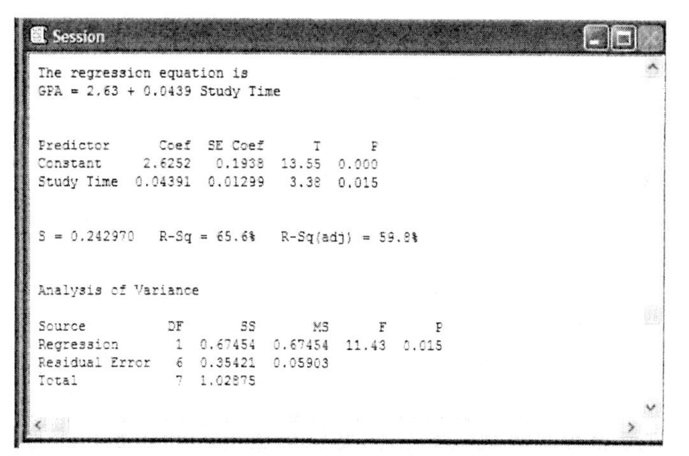

Session

```
The regression equation is
GPA = 2.63 + 0.0439 Study Time


Predictor      Coef  SE Coef      T      P
Constant     2.6252   0.1938  13.55  0.000
Study Time  0.04391  0.01299   3.38  0.015


S = 0.242970   R-Sq = 65.6%   R-Sq(adj) = 59.8%


Analysis of Variance

Source          DF       SS       MS      F      P
Regression       1  0.67454  0.67454  11.43  0.015
Residual Error   6  0.35421  0.05903
Total            7  1.02875
```

The regression equations is
GPA = 2.63 + 0.0439 Study Time

The slope of .0439 tells us that the predicted GPA increases by about .04 points for each additional hour of study time per week.

The predicted GPA for a student who studies 25 hours is GPA = 2.63 + .0439*25 = 3.7275.
The residual for Student 2, who reported x = 25, is 3.6 - 3.7275 = -.1275.

Chapter 12 Exercise 12.8 GPA and skipping class.
Generating a Scatterplot and Regression Line

Use the data in the worksheet from exercise 12.7.

12.7 ***

↓	C1	C2	C3	C4	C5	C6	C7
	Student	Study Time	GPA	Skipped			
1	1	14	2.8	9			
2	2	25	3.6	0			
3	3	15	3.4	2			
4	4	5	3.0	5			
5	5	10	3.1	3			
6	6	12	3.3	2			
7	7	5	2.7	12			
8	8	21	3.8	1			
9							

Current Worksheet: 12.7

Choose: **Graph > Scatterplot> Simple**
Click: **OK**
Enter: Y variable: **C3**
X Variable: **C4**

The scatterplot shows a negative association between the number of classes skipped and GPA.

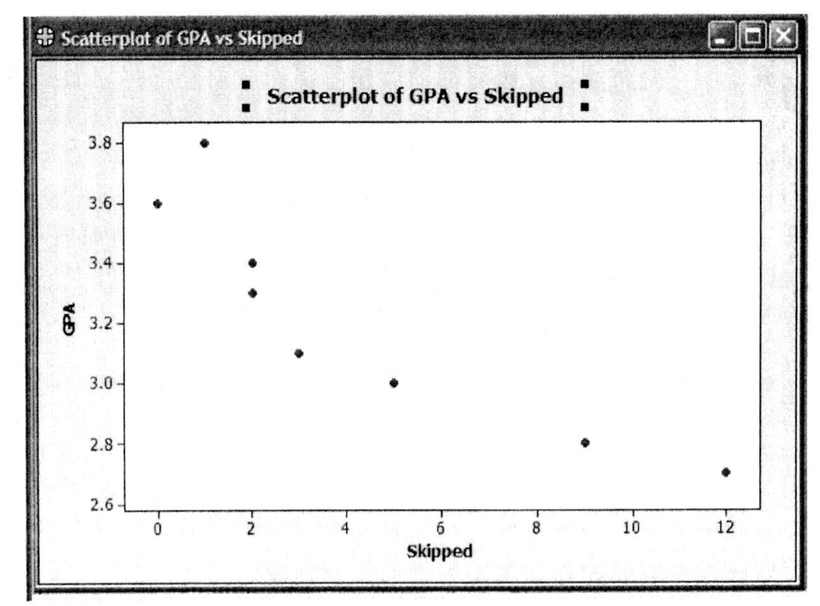

Choose: **Stat > Regression > Regression**
Enter: Response: **C3**
Predictors: **C4**
Click: **OK**

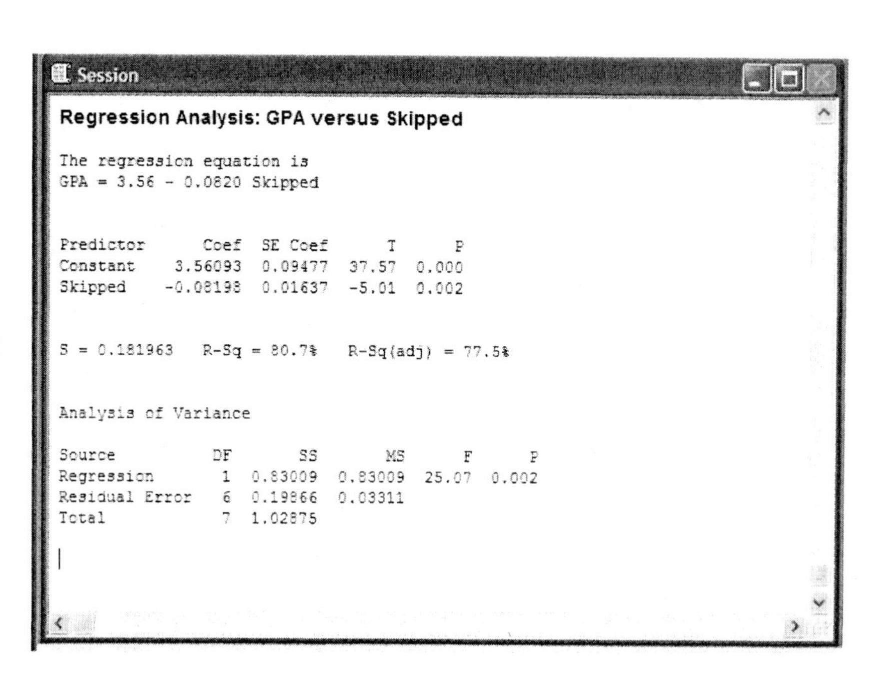

Session
Regression Analysis: GPA versus Skipped
The regression equation is
GPA = 3.56 - 0.0820 Skipped
Predictor Coef SE Coef T P
Constant 3.56093 0.09477 37.57 0.000
Skipped -0.08198 0.01637 -5.01 0.002
S = 0.181963 R-Sq = 80.7% R-Sq(adj) = 77.5%
Analysis of Variance
Source DF SS MS F P
Regression 1 0.83009 0.83009 25.07 0.002
Residual Error 6 0.19866 0.03311
Total 7 1.02875

The regression equations is
GPA= 3.56 + 0.0820 Skipped

The predicted GPA for Student 1, who skipped 9 classes, is GPA = 3.56 - 0.0820*9 = 2.822.

Chapter Exercise 12.9 Predicting College GPA
Generating a Regression Line

Open the data file **georgia_student_survey** from the MINITAB folder of the text CD. We are interested in predicting college GPA based on high school GPA.

explanatory: high school GPA
response: college GPA

To generate the scatterplot:

Select: **Graph > Scatterplot > Simple > OK**

Enter: Y variables: **CGPA** X variables: **HSGPA**

Click: **OK**

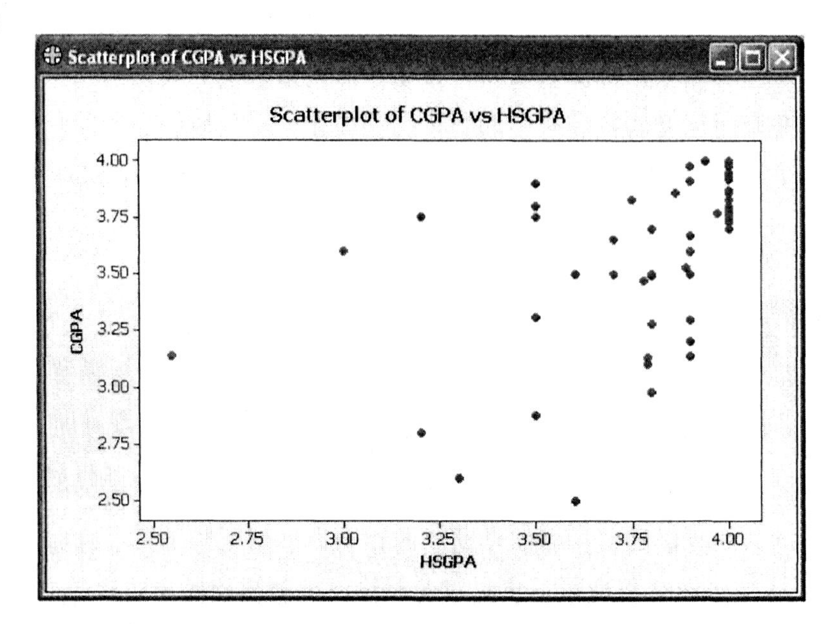

To generate the least squares regression line

Click: **Stat > Regression > Regression**

Enter: Response: **CGPA**

Predictors: **HSGPA**

Click: **OK**

The results are displayed in the Session window.

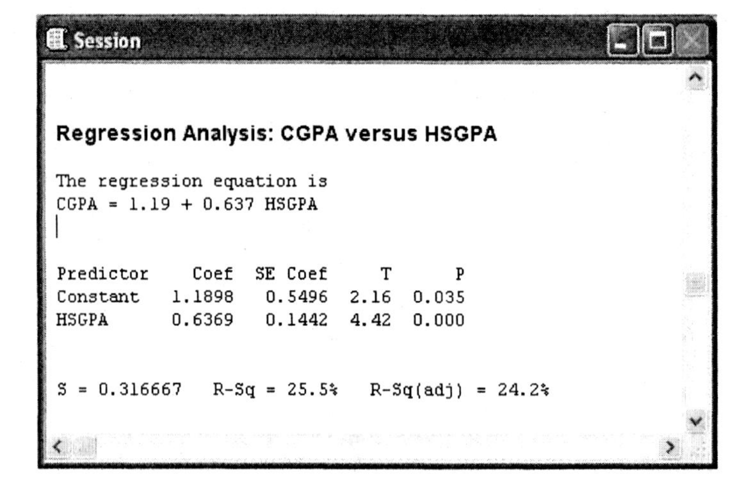

Chapter 12 Example 6 What's the Correlation for Predicting Strength?
Review of Determining Correlation

Open the data file **high_school_female_athletes** from the MINITAB folder of the text CD. We have already identified C12 BP as the response variable and C15 BP(60) as the explanatory variable.

To determine the correlation using MINITAB:

Select: **Stat > Basic Statistics > Correlation**
Enter: Variables: **C15 C12**
Click: **OK**

The results are displayed in the Session window.

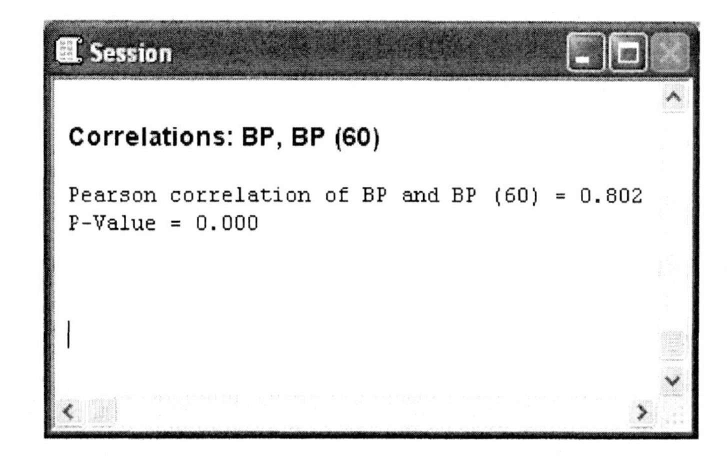

Session
Correlations: BP, BP (60)
Pearson correlation of BP and BP (60) = 0.802
P-Value = 0.000

Chapter 12 Example 9 What Does r^2 Tell Us In The Strength Study?
Determining r^2

Open the data file **high_school_female_athletes** from the MINITAB folder of the text CD. We have already identified C12 BP as the response variable and C15 BP(60) as the explanatory variable.

To determine the value of r^2 using MINITAB:

Select: **Stat > Regression > Regression**
Enter: Variables: **C15 C12**
Click: **OK**

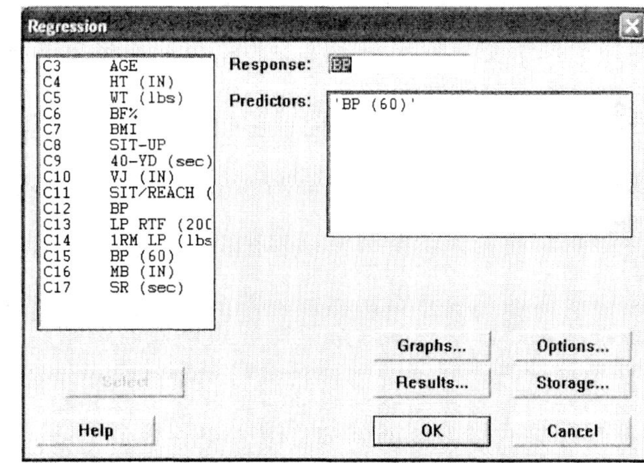

Select: **Results**

Select: **Regression equation, table of coefficients, s, R-squared, and basic analysis of variance**

Click: **OK OK**

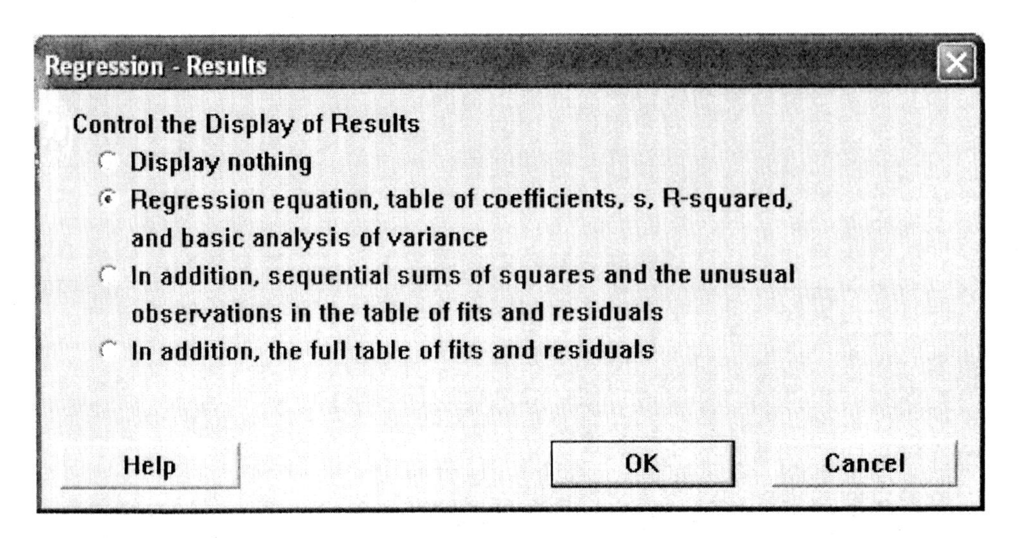

The results are displayed in the Session window.

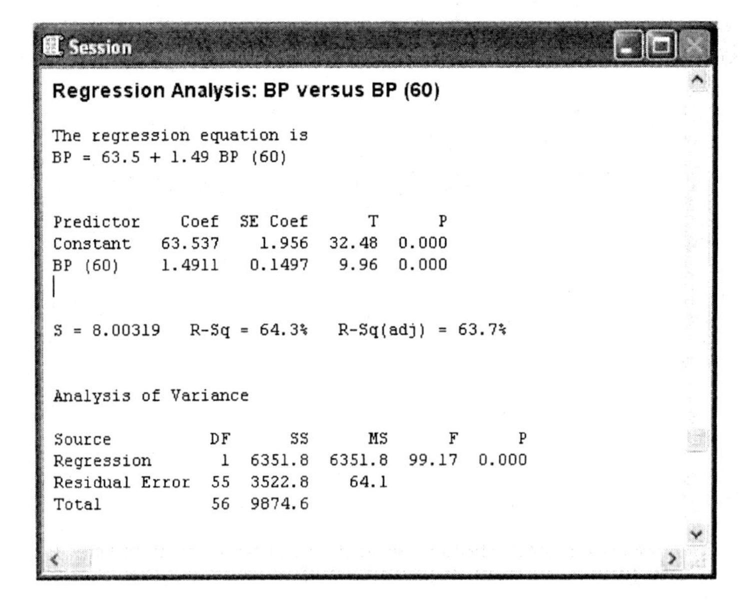

Chapter 12 Exercise 12.15 Sit-ups and the 40-yard dash
Determining the Value of r^2

Open the data file **high_school_female_athletes** from the MINITAB folder of the text CD. For this exercise we have C8 SIT-UP as the explanatory variable and C9 40-YD (sec) as the response variable.

To determine the value of r^2 using MINITAB:

Select: **Stat > Regression > Regression**
Enter: Variables: **C8 C9**
Select: **Results**
Select: **Regression equation, table of coefficients, s, R-squared, and basic analysis of variance**
Click: **OK OK**

The results are displayed in the Session window.

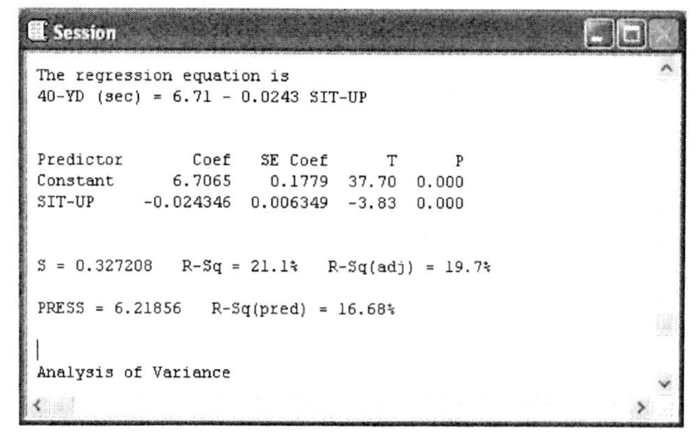

Note the regression equation is $\hat{y} = 6.71 - 0.0243x$. Using this equation we can predict time in the 40-yard dash for any subject who can do a given number of sit-ups.

Note the displayed value of R-Sq = 21.1%

Chapter 12 Exercise 12.17 Student ideology
Determining the Value of r^2

Open the data file **fla_student_survey** from the MINITAB folder of the text CD. For this exercise we have C10 newspapers as the explanatory variable and C14 political_ideology as the response variable.

Doing a complete analysis of the association of these two variables, we start with the scatterplot:

Select: **Graph > Scatterplot > Simple > OK**

Enter: Y variables: **C14** X variables: **C10**
Click: **OK**

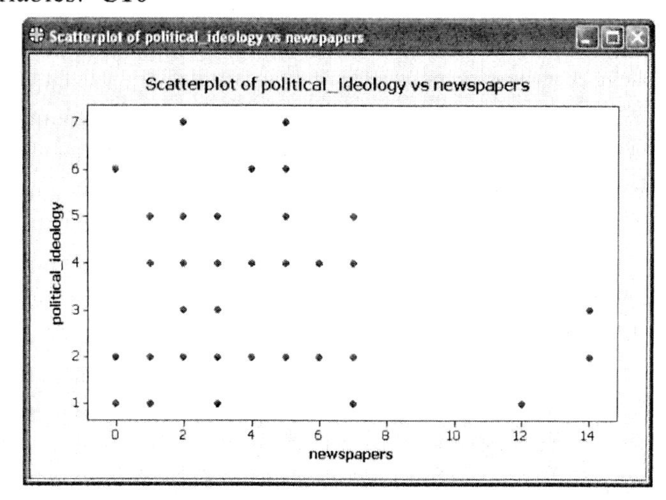

From this scatterplot, we can see that there is very little linear association. Determining the correlation:
Select: **Stat > Basic Statistics > Correlation**
Enter: **C10 C14**
Click: **OK**

Again, based on the value of r = -0.066, we see very little evidence of a linear association between these two variables.

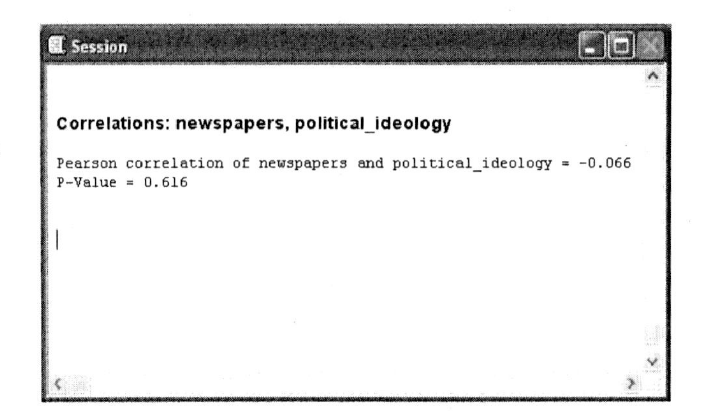
Correlations: newspapers, political_ideology

Pearson correlation of newspapers and political_ideology = -0.066
P-Value = 0.616

Determining the value of r^2:

Select: **Stat > Regression > Regression**
Enter: Response: **C14**
Predictors: **C10**

Select: **Results**
Select: **Regression equation, table of coefficients, s, R-squared, and basic analysis of variance**
Click: **OK OK**

The results are displayed in the Session window.

Note the displayed value of R-Sq = .4%

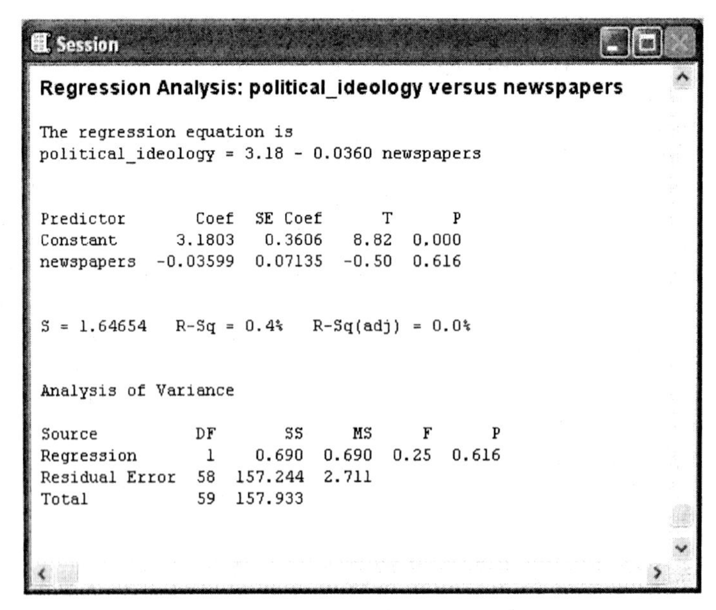
Regression Analysis: political_ideology versus newspapers

The regression equation is
political_ideology = 3.18 - 0.0360 newspapers

Predictor	Coef	SE Coef	T	P
Constant	3.1803	0.3606	8.82	0.000
newspapers	-0.03599	0.07135	-0.50	0.616

S = 1.64654 R-Sq = 0.4% R-Sq(adj) = 0.0%

Analysis of Variance

Source	DF	SS	MS	F	P
Regression	1	0.690	0.690	0.25	0.616
Residual Error	58	157.244	2.711		
Total	59	157.933			

Chapter 12 Exercise 12.21 GPA and study time.
Determining the Value of r^2

Use the worksheet from exercise 12.7.

Choose: **Stat > Basic statistics > correlation.**
Enter: Variables: **C2 C3**
Click: **OK**

The output shows that the correlation is **.810**.

$r^2 = (.810)^2 = .6561 = 66\%$ (Note that this value is also given in the regression analysis from exercise 12.7.)

Both r and r^2 indicate a strong, positive association between GPA and study time.

Chapter 12 Exercise 12.22 GPA and skipping class.
Determining the Value of r^2

Use the worksheet from exercise 12.7.

Choose: **Stat > Basic statistics > Display descriptive statistics.**
Enter: Variables: **C3 C4**
Click: **OK**

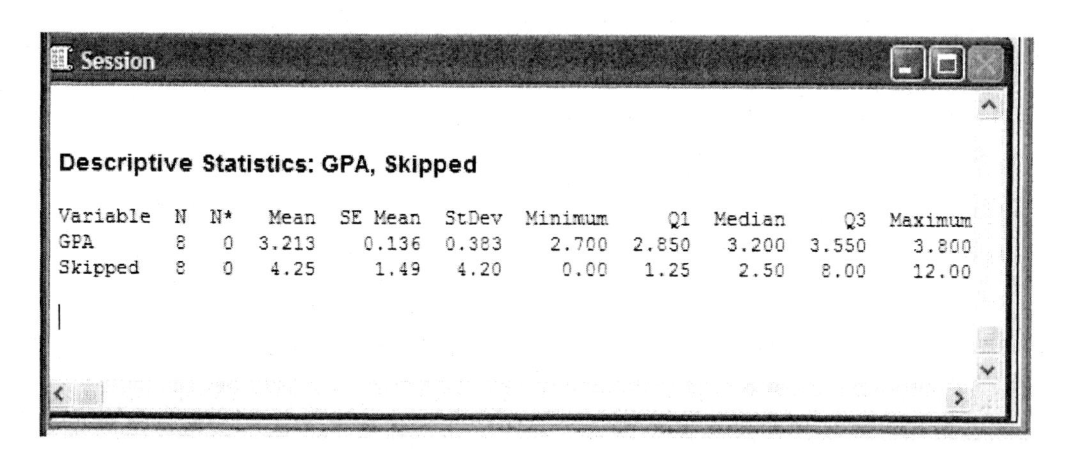
Session

Descriptive Statistics: GPA, Skipped

Variable	N	N*	Mean	SE Mean	StDev	Minimum	Q1	Median	Q3	Maximum
GPA	8	0	3.213	0.136	0.383	2.700	2.850	3.200	3.550	3.800
Skipped	8	0	4.25	1.49	4.20	0.00	1.25	2.50	8.00	12.00

We know from exercise 12.8 that the slope of the prediction equation is -0.082.
To find the correlation:
Choose: **Stat > Basic statistics > correlation.**
Enter: Variables: **C3 C4**
Click: **OK**

The output shows that the correlation is **-.898.**

Alternatively, we can use the formula **$r = b(s_x/s_y)$** to find the correlation.

$r = b(s_x/s_y)$= -0.082(4.2/.383) = .899

Chapter 12 Example 11 Is Strength Associated with 60-pound Bench Presses?
A Significance Test of Independence

Open the worksheet **high_school_female_student_athletes,** which is found in the MINITAB folder.
The number of bench presses before fatigue, C15 BP(60), is the x-variable, and maximum bench press, C12 BP, is the y-variable. The regression output includes the t test statistic and the p-value.

Select: **Stat > Regression > Regression**
Enter: Response: **C12**
Predictors: **C15**
Click: **OK**

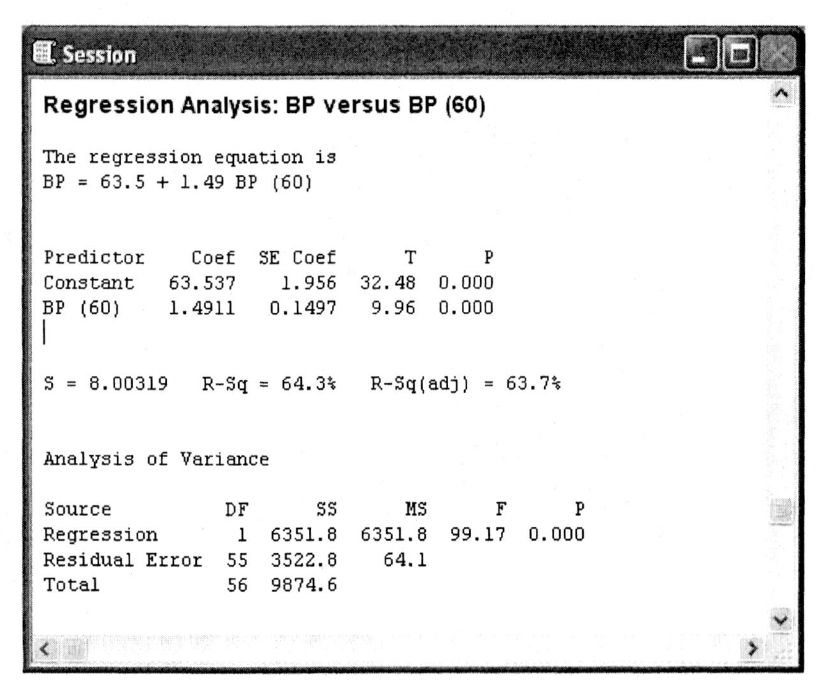

```
Session

Regression Analysis: BP versus BP (60)

The regression equation is
BP = 63.5 + 1.49 BP (60)


Predictor     Coef  SE Coef      T      P
Constant    63.537    1.956  32.48  0.000
BP (60)     1.4911   0.1497   9.96  0.000


S = 8.00319   R-Sq = 64.3%   R-Sq(adj) = 63.7%


Analysis of Variance

Source          DF      SS      MS      F      P
Regression       1  6351.8  6351.8  99.17  0.000
Residual Error  55  3522.8    64.1
Total           56  9874.6
```

Notice the *t* test statistic (*t* = 9.96) appears under the column heading "T" in the row for the predictor BP (60), and the p-value (p = 0.000) for the two-sided alternative H_a: $\beta \neq 0$ appears under the heading "P" in the row for the predictor BP (60).

Chapter 12 Exercise 12.33 Predicting house prices
Significance Test of Independence and a Confidence Interval for Slope

Open the worksheet **house_selling_prices**, which is found in the MINITAB folder.

The size of the house is in C8 size, and selling price is in C7 price.

> Select: **Stat > Regression > Regression**
> Enter: Response: **price**
> Predictors: **size**
> Click: **OK**

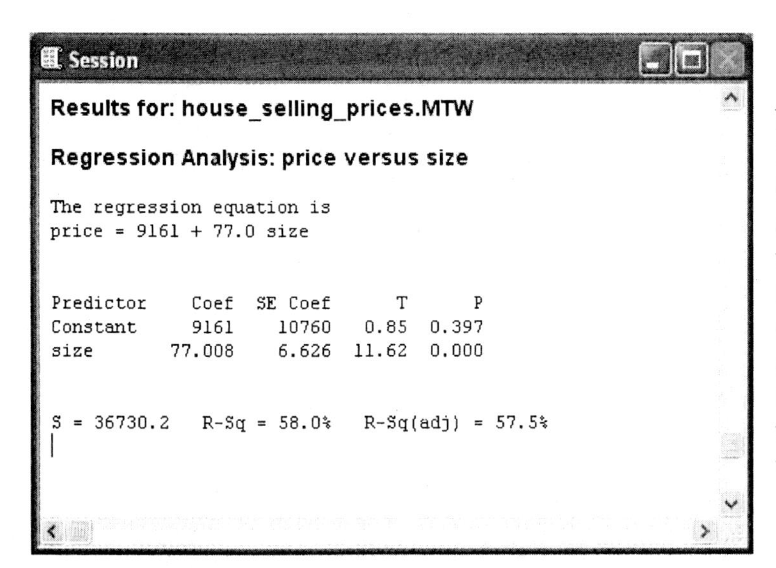

Session

Results for: house_selling_prices.MTW

Regression Analysis: price versus size

```
The regression equation is
price = 9161 + 77.0 size


Predictor     Coef  SE Coef      T      P
Constant      9161    10760   0.85  0.397
size        77.008    6.626  11.62  0.000


S = 36730.2   R-Sq = 58.0%   R-Sq(adj) = 57.5%
```

Notice the t test statistic (t = 11.62), and the p-value (p = 0.000) for the two-sided alternative H_a: $\beta \neq 0$. Since p-value = 0.000 is less than a 0.05 significance level, there is evidence that these two variables are not independent, and that the sample association between these two variables is not just random variation.

To compute a 95% confidence interval for the population slope, note that the $t_{.025}$ value for $df = n - 2 = 98$ is t = 1.9845, and that b = 77.008, and se = 6.626. (MINITAB does not include a t-interval estimate for a population slope.)

$$b \pm t_{.025}(se) = 77.008 \text{ +/- } 1.9845\ (6.626)$$
$$= 77.008 \text{ +/- } 13.149$$
$$(\ 63.859,\ 90.157)$$

This interval does not support the builder's claim that selling price increases $100, on the average, for every extra square foot. (The interval does not contain 100.)

Chapter 12 Exercise 12.39 Advertising and Sales
A Significance Test of Independence

Enter the data into the MINITAB Data Window. Enter the x's into C1 and name the column Advertising. Enter the y's into C2 and name the column Sales.

> Select: **Stat > Regression > Regression**
> Enter: Response: **Sales**
> Predictors: **Advertising**

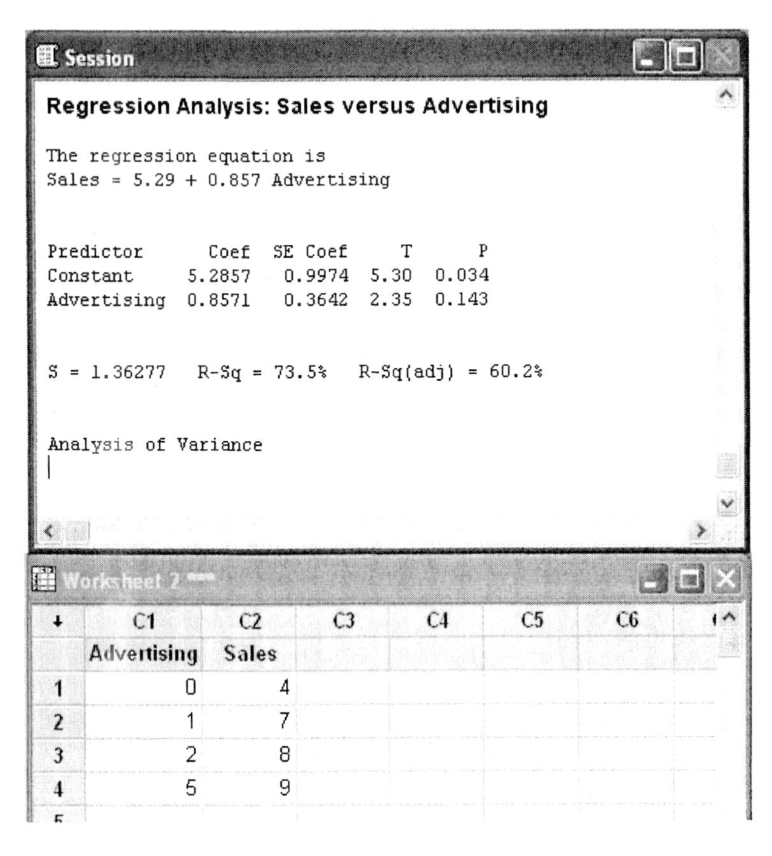

Notice the t test statistic (t = 2.35), and the p-value (p = 0.143) for the two-sided alternative H_a: $\beta \neq 0$. Since p-value = 0.143 is not less than significance level 0.05, there is not evidence that these two variables are independent. A word of caution though, were the assumptions validated? This is a very small sample, so results are suspect.

Chapter 12 Example 13 Detecting an Underachieving College Student
How Data Vary Around the Regression Line

Open the worksheet **georgia_student_survey**, found in the MINITAB folder. High school GPA is found in column C8 and college GPA is in column C9.

As part of the regression analysis, MINITAB highlights observations that have standardized residuals with absolute value larger than 2. To generate this table of "unusual observations":

Select: **Stat > Regression > Regression**
Enter: Response: **CGPA**
Predictors: **HSGPA**

Select: **Results**
Select: **In addition, sequential sums of squares and the unusual observations in the table of fits and residuals**
Click: **OK OK**

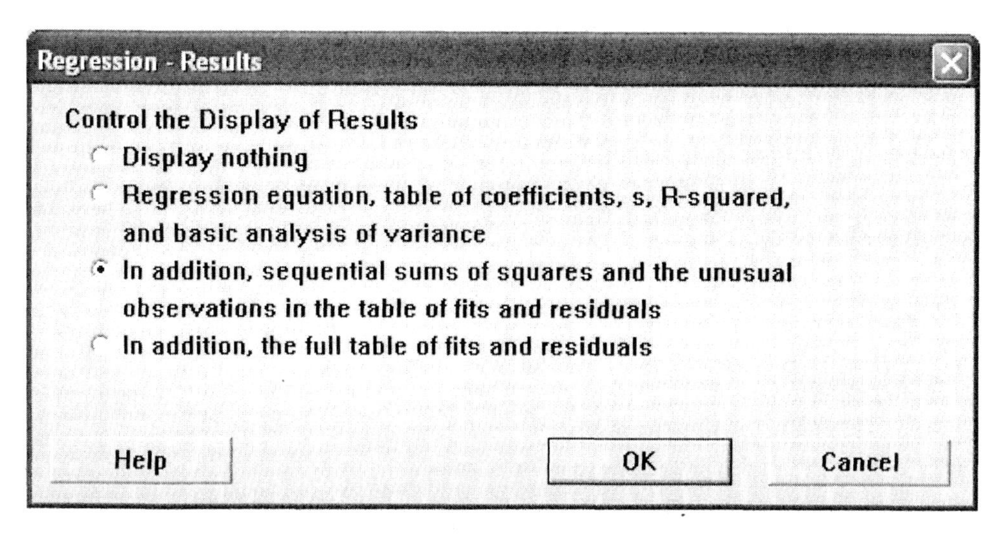

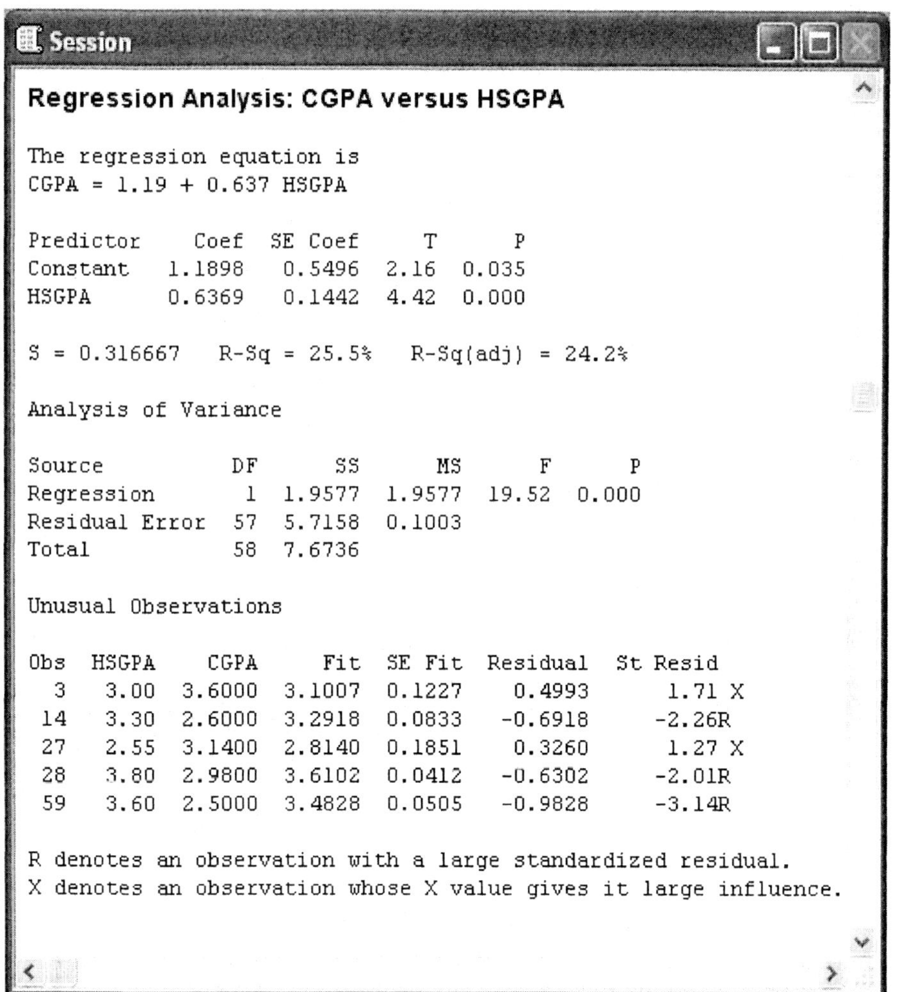

Session

Regression Analysis: CGPA versus HSGPA

The regression equation is
CGPA = 1.19 + 0.637 HSGPA

Predictor	Coef	SE Coef	T	P
Constant	1.1898	0.5496	2.16	0.035
HSGPA	0.6369	0.1442	4.42	0.000

S = 0.316667   R-Sq = 25.5%   R-Sq(adj) = 24.2%

Analysis of Variance

Source	DF	SS	MS	F	P
Regression	1	1.9577	1.9577	19.52	0.000
Residual Error	57	5.7158	0.1003		
Total	58	7.6736			

Unusual Observations

Obs	HSGPA	CGPA	Fit	SE Fit	Residual	St Resid
3	3.00	3.6000	3.1007	0.1227	0.4993	1.71 X
14	3.30	2.6000	3.2918	0.0833	-0.6918	-2.26R
27	2.55	3.1400	2.8140	0.1851	0.3260	1.27 X
28	3.80	2.9800	3.6102	0.0412	-0.6302	-2.01R
59	3.60	2.5000	3.4828	0.0505	-0.9828	-3.14R

R denotes an observation with a large standardized residual.
X denotes an observation whose X value gives it large influence.

Chapter 12 Exercise 12.40 GPA and study time
One-Sided Significance Test About a Population Slope β

Recall the regression output from exercise 12.7.

```
Session

The regression equation is
GPA = 2.63 + 0.0439 Study Time

Predictor      Coef   SE Coef      T      P
Constant     2.6252    0.1938  13.55  0.000
Study Time  0.04391   0.01299   3.38  0.015

S = 0.242970    R-Sq = 65.6%    R-Sq(adj) = 59.8%

Analysis of Variance

Source          DF        SS        MS      F      P
Regression       1   0.67454   0.67454  11.43  0.015
Residual Error   6   0.35421   0.05903
Total            7   1.02875
```

Ha: $\beta > 0$ predicts a positive association. Notice the slope is .04391, the standard error is .01299, t test statistic ($t = 3.38$), and the p-value = 0.015/2 = .0075 for the one sided alternative. Since 0.0075 is less than significance level 0.05, there is evidence of a positive association between these variables. This is a small sample, so results are suspect.

Chapter 12 Exercise 12.41 GPA and skipping class
Confidence Interval for a Population Slope β

Recall the regression output from exercise 12.8.

```
Session

Regression Analysis: GPA versus Skipped

The regression equation is
GPA = 3.56 - 0.0820 Skipped

Predictor      Coef   SE Coef      T      P
Constant    3.56093   0.09477  37.57  0.000
Skipped    -0.08198   0.01637  -5.01  0.002

S = 0.181963    R-Sq = 80.7%    R-Sq(adj) = 77.5%

Analysis of Variance

Source          DF        SS        MS      F      P
Regression       1   0.83009   0.83009  25.07  0.002
Residual Error   6   0.19866   0.03311
Total            7   1.02875
```

To compute a 90% confidence interval for the population slope, note that the $t_{.05}$ value for $df = n - 2 = 6$ is $t =$ 1.943, and that b = -.08198, and se = .01637. (MINITAB does not include a t-interval estimate for a population slope.)

$$b \pm t_{.05}(se) = -.08198 +/- 1.943\ (.01637)$$
$$= -.08198 +/- .03181$$
$$(-.114, -.0502)$$

We are 90% confident that the population slope β falls between -.0502 and -.114.
On average, college GPA decreases by between .05 and .11 points for each additional class that is skipped.

Chapter 12 Example 16 Predicting Maximum Bench Press and Estimating Its Mean Confidence Interval for Population Mean of y and Prediction Interval for a Single y-value

Open the worksheet **high_school_female_athletes**, found in the MINITAB folder.

Proceed with the regular regression command:
Select: **Stat > Regression > Regression**
Enter: Response: **BP**
Predictors: **BP (60)**

Now, to find both the confidence interval and the prediction interval:
Select: **Options**
(Fit intercept is selected by default)
Enter: Prediction intervals for new observations: **11**
Enter: Confidence level: **95**
Select: **Confidence limits**
Select: **Prediction limits**
Click: **OK OK**

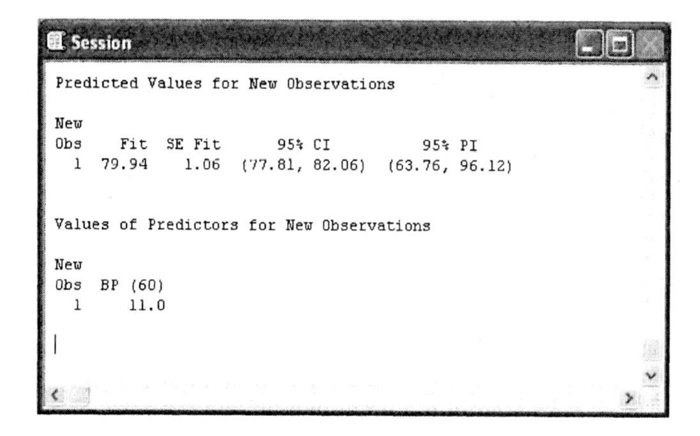
Session

Predicted Values for New Observations

New Obs	Fit	SE Fit	95% CI	95% PI
1	79.94	1.06	(77.81, 82.06)	(63.76, 96.12)

Values of Predictors for New Observations

New Obs	BP (60)
1	11.0

Chapter 12 Exercise 12.43 Poor predicted strengths
How Data Vary Around the Regression Line

Open the worksheet **high_school_female_athletes**, found in the MINITAB folder.

Select: **Stat > Regression > Regression**
Enter: Response: **BP**
Predictors: **BP (60)**

Select: **Results**
Select: **In addition, sequential sums of squares and the unusual observations in the table of fits and residuals**
Click: **OK OK**

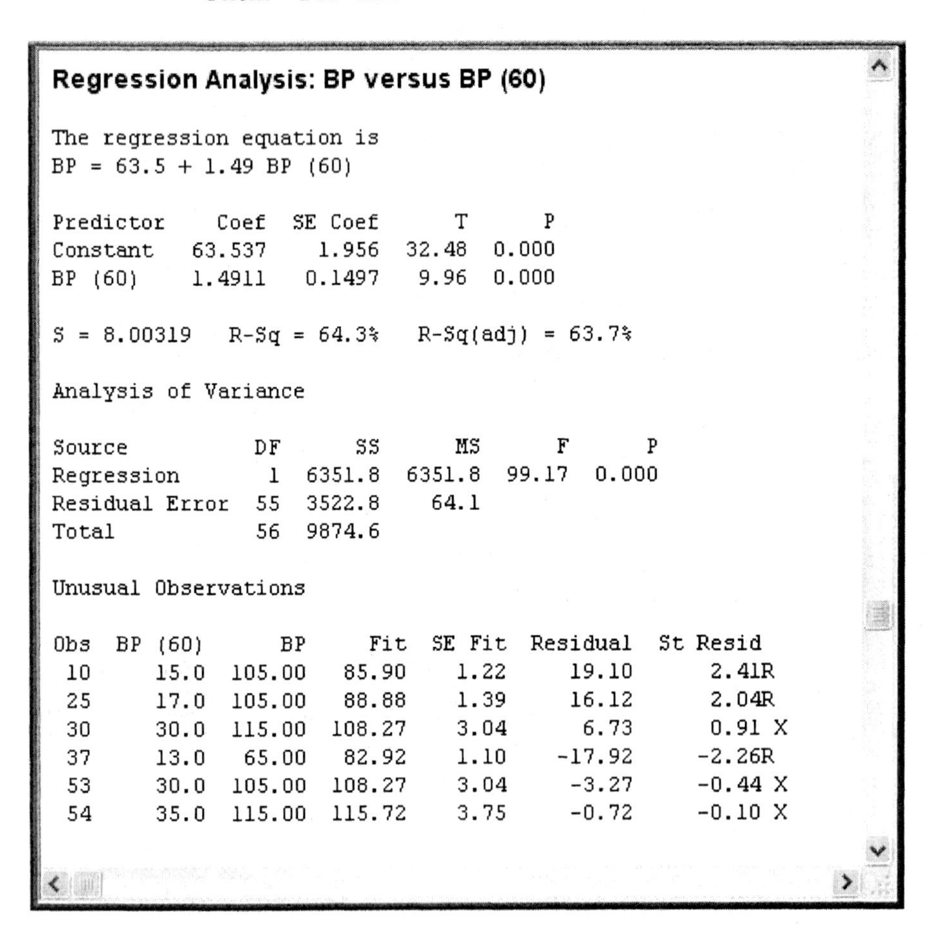

Regression Analysis: BP versus BP (60)

```
The regression equation is
BP = 63.5 + 1.49 BP (60)

Predictor     Coef  SE Coef      T      P
Constant    63.537    1.956  32.48  0.000
BP (60)     1.4911   0.1497   9.96  0.000

S = 8.00319   R-Sq = 64.3%   R-Sq(adj) = 63.7%

Analysis of Variance

Source          DF      SS      MS      F      P
Regression       1  6351.8  6351.8  99.17  0.000
Residual Error  55  3522.8    64.1
Total           56  9874.6

Unusual Observations

Obs  BP (60)      BP     Fit  SE Fit  Residual  St Resid
 10     15.0  105.00   85.90    1.22     19.10     2.41R
 25     17.0  105.00   88.88    1.39     16.12     2.04R
 30     30.0  115.00  108.27    3.04      6.73     0.91 X
 37     13.0   65.00   82.92    1.10    -17.92    -2.26R
 53     30.0  105.00  108.27    3.04     -3.27    -0.44 X
 54     35.0  115.00  115.72    3.75     -0.72    -0.10 X
```

Chapter 12 Exercise 12.49 ANOVA Table for Leg Press

Open the worksheet **high_school_female_student_athletes,** which is found in the MINITAB folder.
The number of leg presses before fatigue, C13 LP (200), is the x-variable, and maximum leg press, C14 LP , is the y-variable

Proceed with the regular regression command:
Select: **Stat > Regression > Regression**
Enter: Response: **LP**
Predictors: **LP (200)**

Now, to find to generate the ANOVA table
Select: **Results**
Select: **Regression equation, table of coefficients, s, R-squared, and basic analysis of variance**
Click: **OK OK**

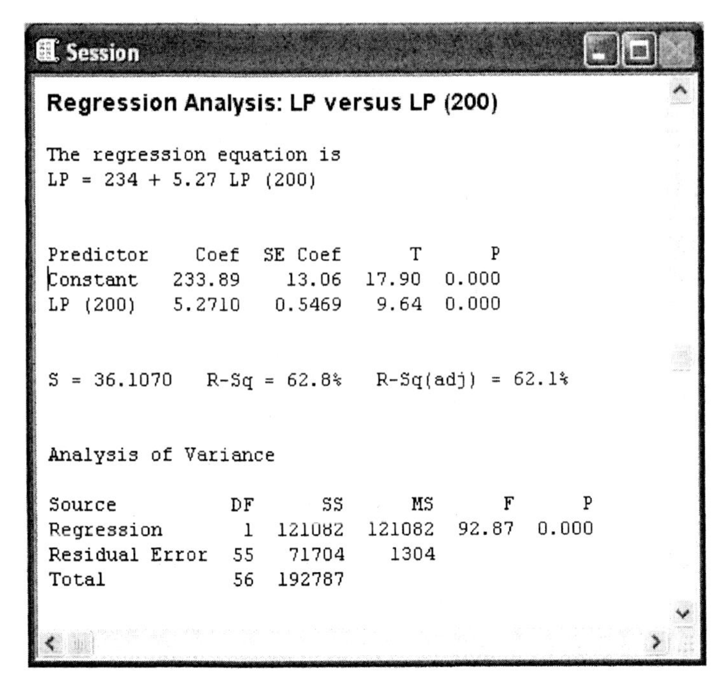

Session

Regression Analysis: LP versus LP (200)

```
The regression equation is
LP = 234 + 5.27 LP (200)


Predictor      Coef  SE Coef      T      P
Constant     233.89    13.06  17.90  0.000
LP (200)     5.2710   0.5469   9.64  0.000


S = 36.1070   R-Sq = 62.8%   R-Sq(adj) = 62.1%


Analysis of Variance

Source          DF      SS      MS      F      P
Regression       1  121082  121082  92.87  0.000
Residual Error  55   71704    1304
Total           56  192787
```

From the ANOVA table, we can use SS Residual Error = 71704 to determine the residual standard deviation of y-values. Note that $df = 57 - 2 = 55$.

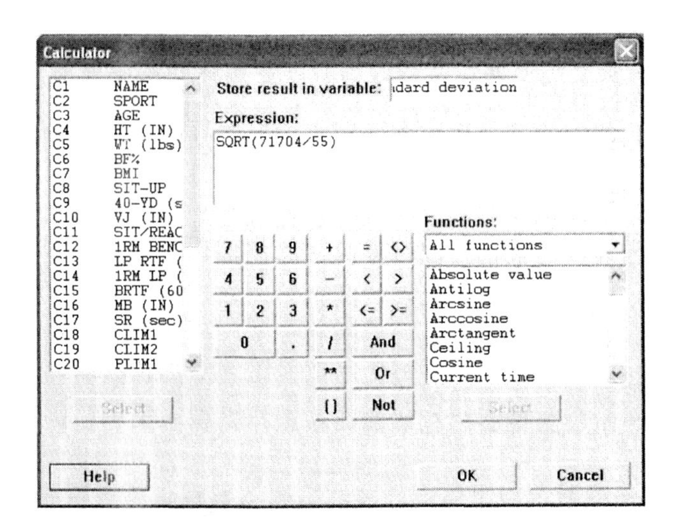

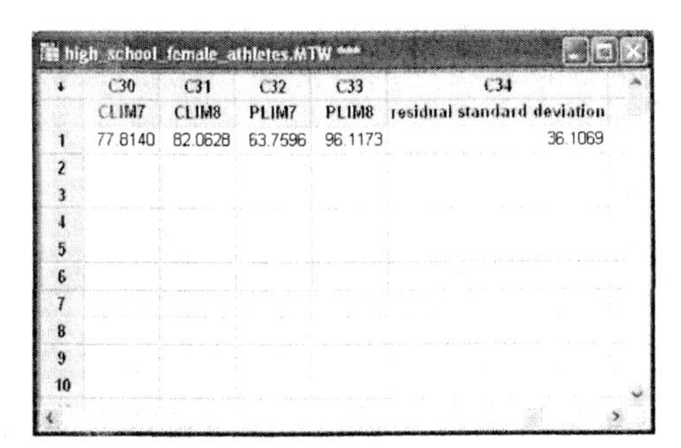

	C30	C31	C32	C33	C34
	CLIM7	CLIM8	PLIM7	PLIM8	residual standard deviation
1	77.8140	82.0628	63.7596	96.1173	36.1069

At any fixed value x of number of 200-pound leg presses, we estimate that the maximum leg press values have a standard deviation of 36.1 pounds.

For female athletes with $x = 22$, we would estimate the mean maximum leg press to be `LP = 234 + 5.27 (22)` = 349.94 pounds and the variability of their maximum leg press values to be 36.1 pounds.

If y-values are approximately normal, then 95% of the y-values would fall in the interval approximately $\hat{y} \pm 2s = 349.94 \pm 2(36.1) = (277.74, 422.14)$.

Using MINITAB to generate the prediction interval, notice we get similar results.

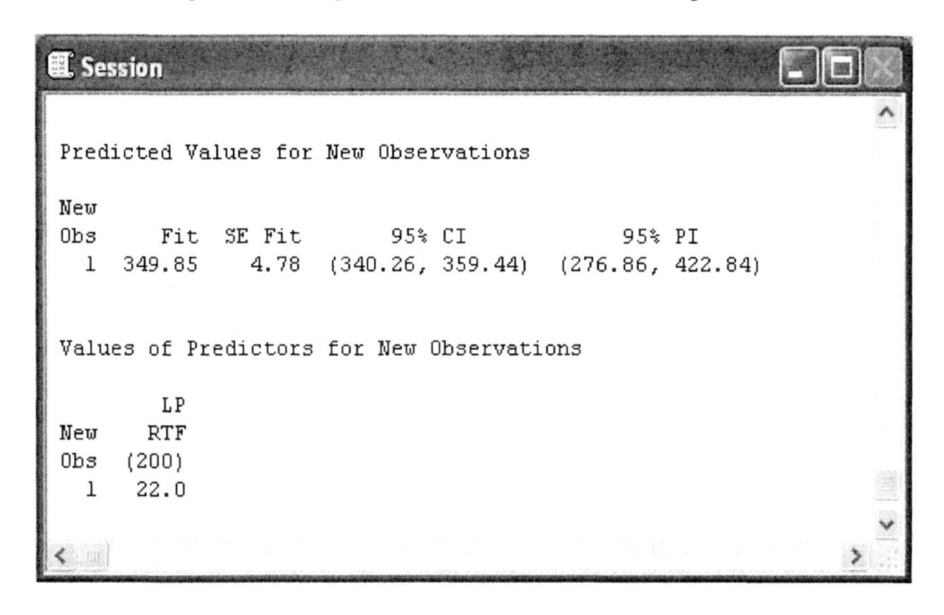

Session

```
Predicted Values for New Observations

New
Obs     Fit  SE Fit       95% CI              95% PI
  1  349.85    4.78  (340.26, 359.44)  (276.86, 422.84)


Values of Predictors for New Observations

         LP
New     RTF
Obs   (200)
  1    22.0
```

Chapter 12 Example 18 Explosion in Number of People Using the Internet Exponential Regression

Enter data into the Data Window, using C1 for number of years since 1995, and C2 for number of people (in millions).

internet_use.MTW ***

↓	C1	C2	C3	C4	C5
	No. Years Since 1995	Number People			
1	0	16			
2	1	36			
3	2	76			
4	3	147			
5	4	195			
6	5	369			
7	6	513			
8					
9					
10					

Producing a scatterplot, we see that the number of people increases over time, and the amount of increase from one year to the next seems itself to increase over time.

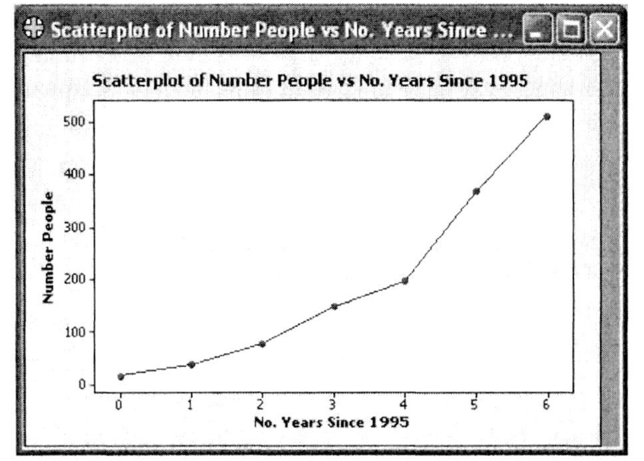

Generating the logarithm of the Number People:

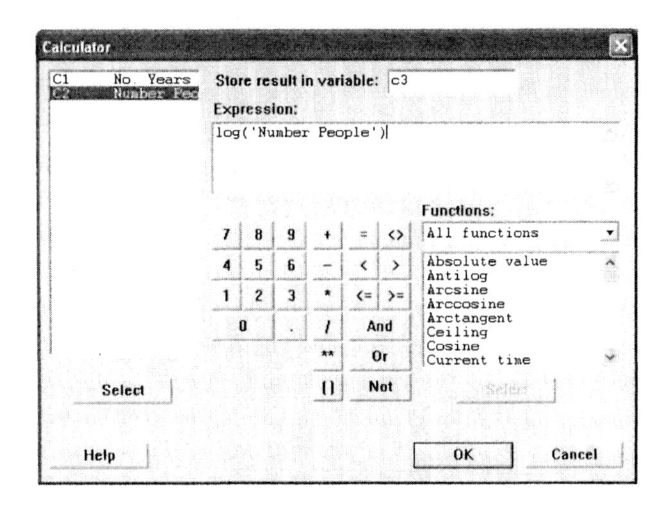

And then producing a scatterplot of C1 versus the logarithms:

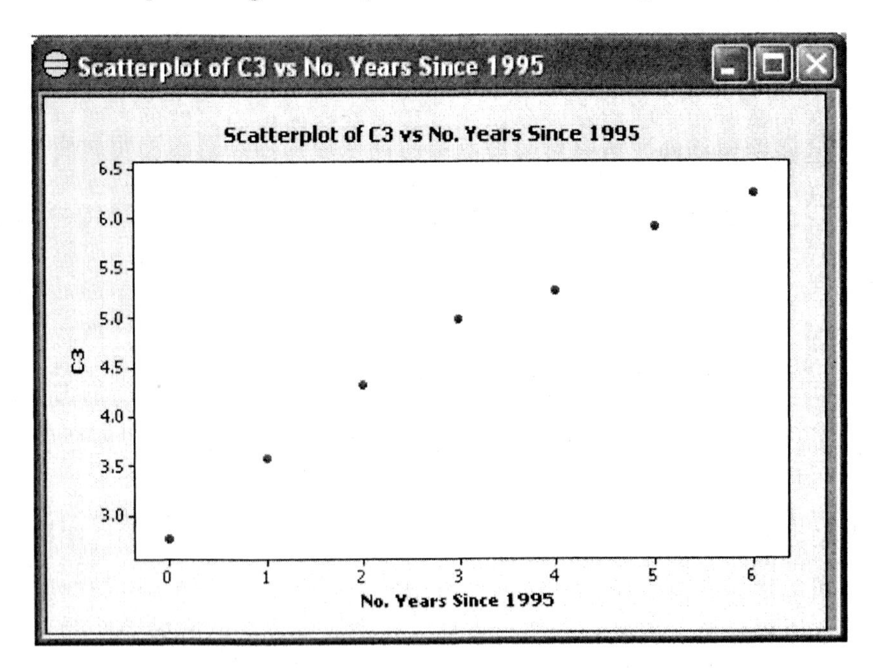

Determining the correlation between the logarithms and number of years since 1995

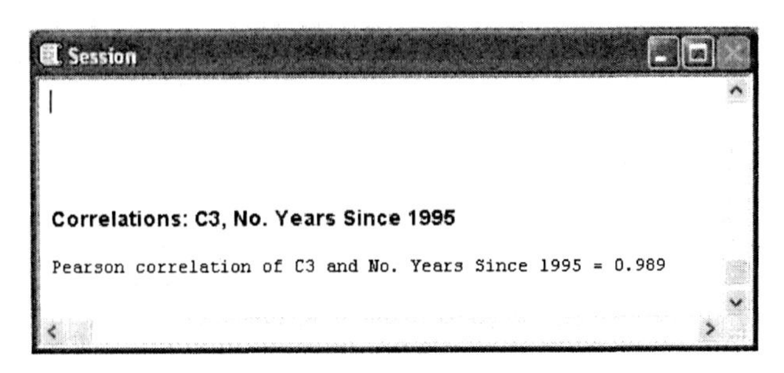

This value of 0.99 suggests that growth in Internet users over this time period was approximately exponential.

To generate the exponential regression model:

Select: **Stat > Time Series > Trend Analysis**

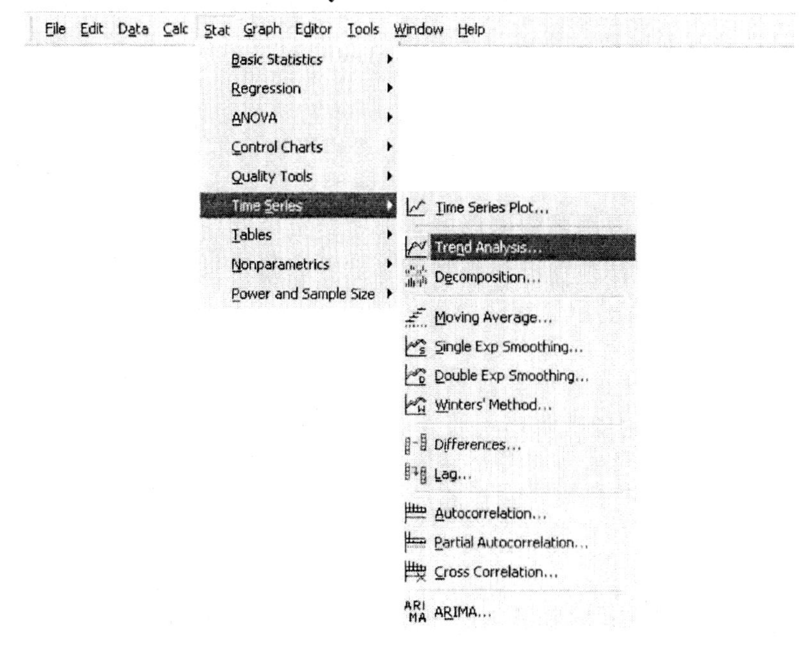

Enter: Variable: **Number People**
Select: **Exponential growth**
Click: **OK**

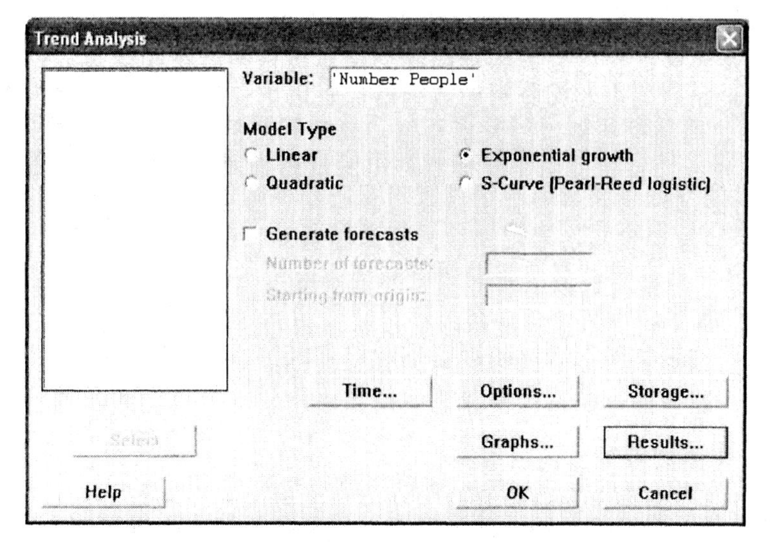

MINITAB produces a trend analysis plot, which includes the exponential regression model,

Yt = 11.5083 * (1.77079**t)

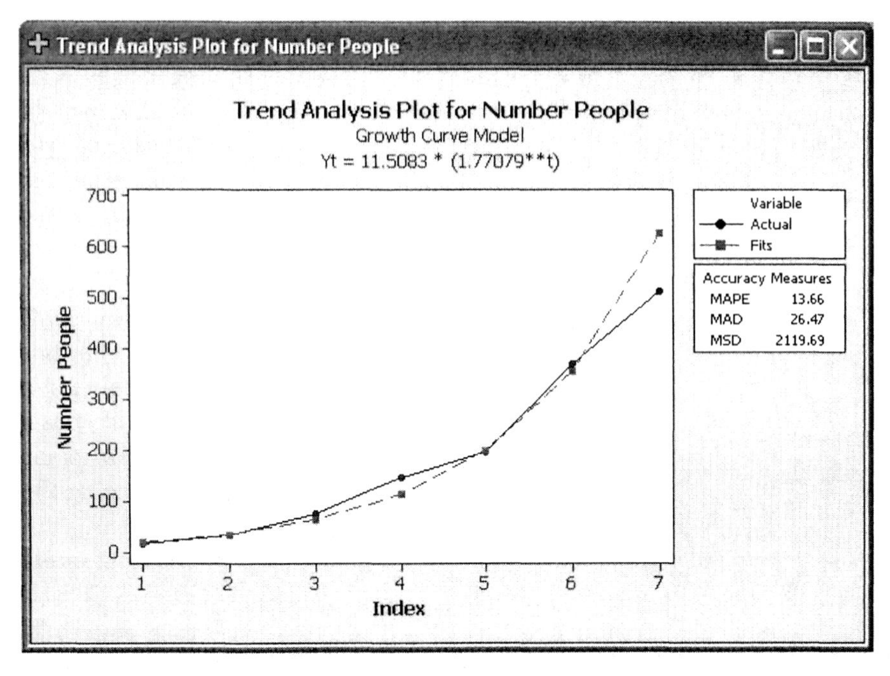

Note that Index = 1 means 1995, Index = 2 means 1996, and so on.

For the year 2000, Index = 6, and the predicted $\hat{y} = 11.5083 * 1.77079^6 = 354.8$ million.

Chapter 12 Exercise 12.61 Leaf Litter Decay Exponential Regression

Worksheet 1 ***

↓	C1	C2	C3
	weeks	weight	
1	0	75.0	1.87506
2	1	60.9	1.78462
3	2	51.8	1.71433
4	3	45.2	1.65514
5	4	34.7	1.54033
6	5	34.6	1.53908
7	6	26.2	1.41830
8	7	20.4	1.30963
9	8	14.0	1.14613
10	9	12.3	1.08991
11	10	*	*
12	11	8.2	0.91381
13	12	*	*
14	13	*	*
15	14	*	*
16	15	3.1	0.49136
17	16	*	*
18	17	*	*
19	18	*	*
20	19	*	*
21	20	1.4	0.14613

Enter data into the Data Window, using C1 for number of weeks, and C2 for the weight.

Note the inclusion of weeks 10, 12, 13, 14, 16, 17, 18, and 19, even though we do not have weights for these particular weeks. The time series trend analysis assumes that the variable values (weight) are given at time intervals having an increment of one.

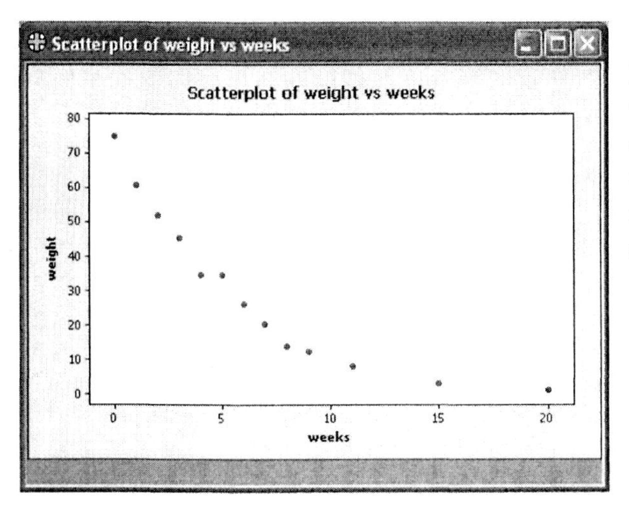

Producing a scatterplot, we see that the weight of the organic mass in the bag decreases over time, and the amount of decrease from one week to the next seems itself to decrease over time. Thus, a straight-line model is inappropriate.

Generating the logarithm of the weight and then producing a scatterplot of C1 versus the logarithms:

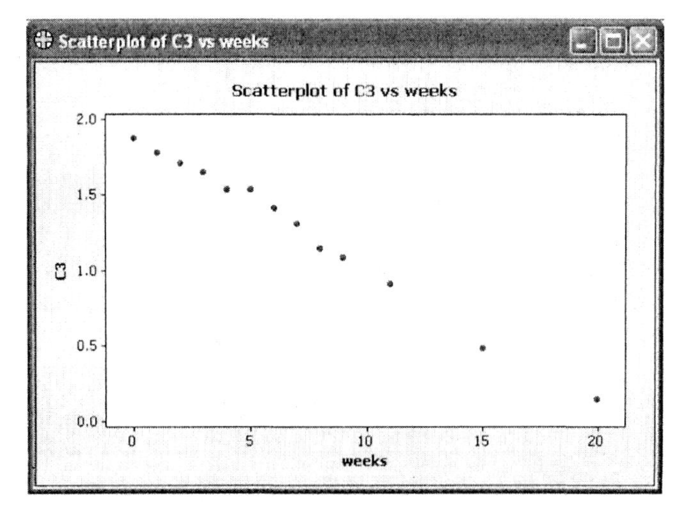

Determining the correlation between the logarithms and number of weeks

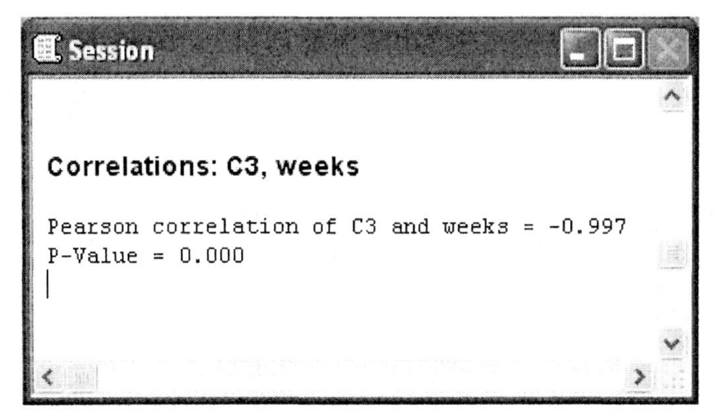

Session
Correlations: C3, weeks
Pearson correlation of C3 and weeks = -0.997
P-Value = 0.000

This value of -0.997 suggests that the decay of the organic mass over this time period was approximately exponential.

To generate the exponential regression model:

Select: **Stat > Time Series > Trend Analysis**

Enter: Variable: **weight**
Select: **Exponential growth**
Click: **OK**

MINITAB produces a trend analysis plot, which includes the exponential regression model,
Yt = 99.1144 * (0.813443**t)

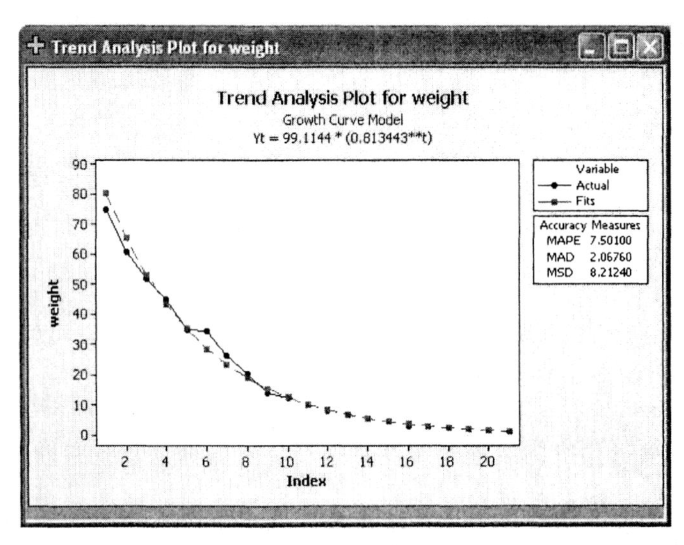

Note that Index = 1 means "initially", Index = 2 means "after one week", and so on.

For after the 20th week Index = 21, and the predicted $\hat{y} = 99.1144 * 0.813443^{21} = 1.30$ kg.

Chapter 13 Example 2 Predicting Selling Price Using House and Lot Sizes
Multiple Regression and Plotting the Relationships

Open the worksheet **house_selling_prices** , which is found in the MINITAB folder. The price is found in C7, house size is found in C8, and lot size is found in C9.

Select: **Stat > Regression > Regression**
Enter: Response: **price**
Predictors: **size lot**
Select: **Results**
Select: **Regression equation, table of coefficients, s, R-squared, and basic analysis of variance**
Click: **OK OK**

The results are displayed in the Session Window.

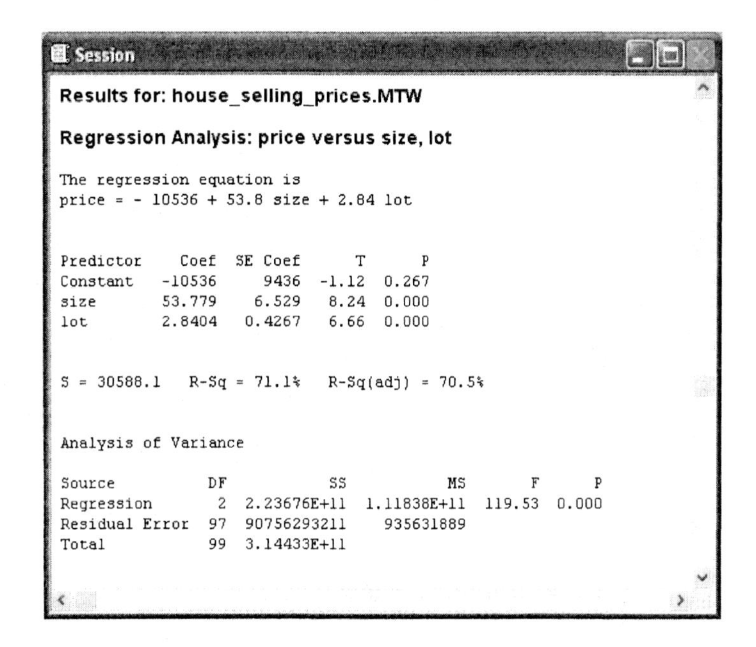
Session

Results for: house_selling_prices.MTW

Regression Analysis: price versus size, lot

```
The regression equation is
price = - 10536 + 53.8 size + 2.84 lot


Predictor      Coef  SE Coef      T      P
Constant     -10536     9436  -1.12  0.267
size         53.779    6.529   8.24  0.000
lot          2.8404   0.4267   6.66  0.000


S = 30588.1   R-Sq = 71.1%   R-Sq(adj) = 70.5%


Analysis of Variance

Source          DF           SS           MS       F      P
Regression       2  2.23676E+11  1.11838E+11  119.53  0.000
Residual Error  97  90756293211    935631889
Total           99  3.14433E+11
```

To use MINITAB to generate a scatterplot matrix

Select: **Graph > Matrix plot**

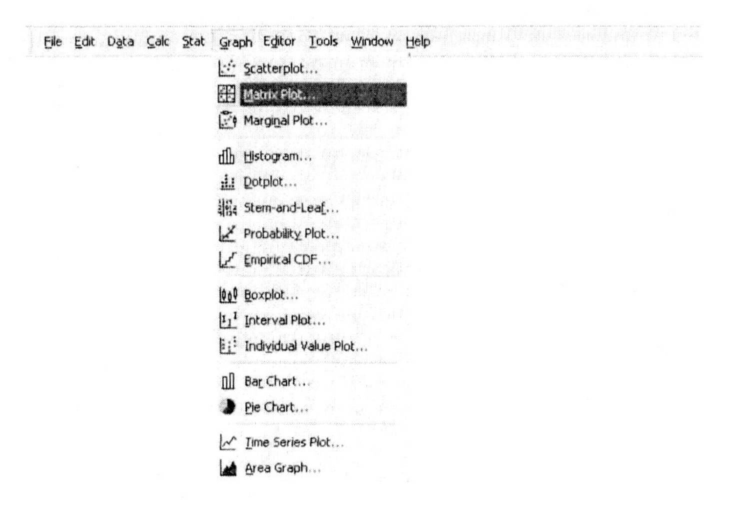

Select: **Matrix of plots / Simple > OK > OK**

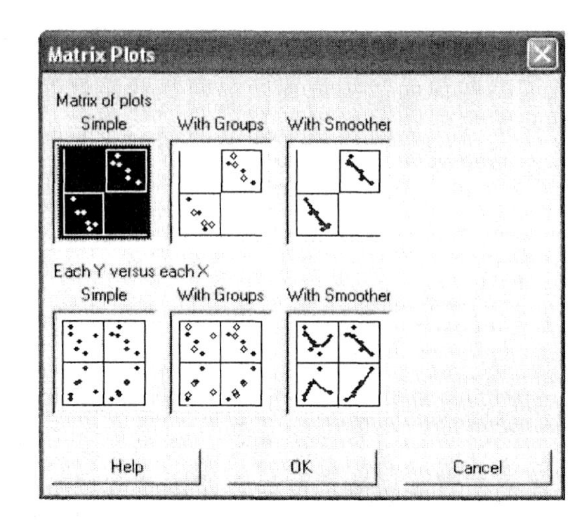

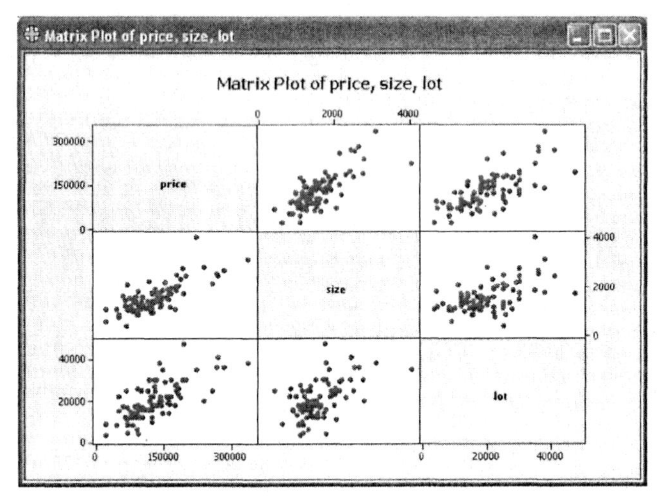

Since we are truly interested in the association between price and house size, and price and lot size, another option is

Select: **Graph > Matrix of plots**
Select: **Each Y versus each X, Simple**
Click: **OK**

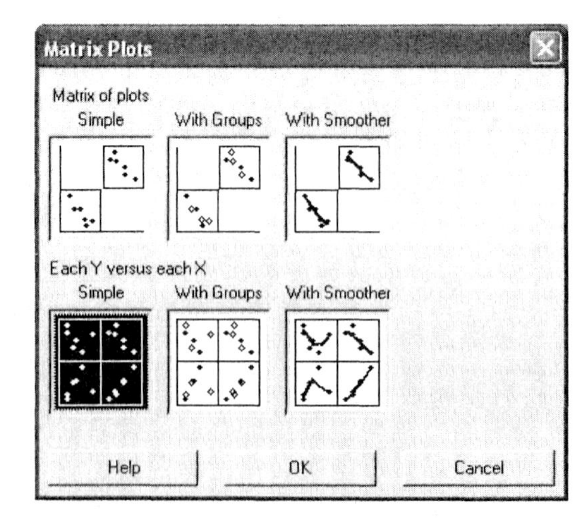

Enter: Y variables: **price**
X variables: **size lot**
Click: **OK**

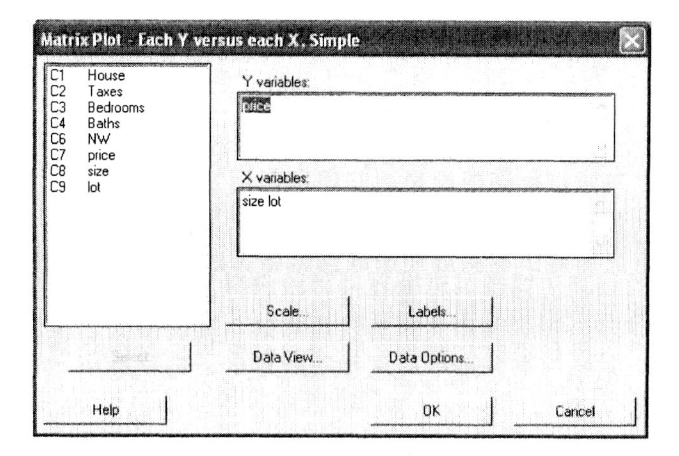

These commands produce the following graph window.

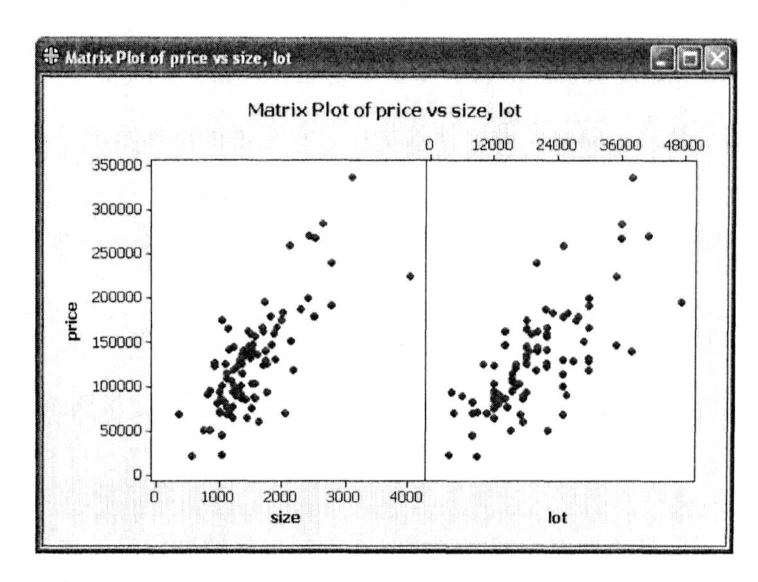

Chapter 13 Exercise 13.5 Does more education cause more crime?
Multiple Regression

Open the worksheet **fla_crime**, which is found in the MINITAB folder. The crime rate is found in C2, education is found in C3, and urbanization is found in C4.

Select: **Stat > Regression > Regression**
Enter: Response: **crime**
Predictors: **education urbanization**
Select: **Results**
Select: **Regression equation, table of coefficients, s, R-squared, and basic analysis of variance**
Click: **OK OK**

The results are displayed in the Session Window.

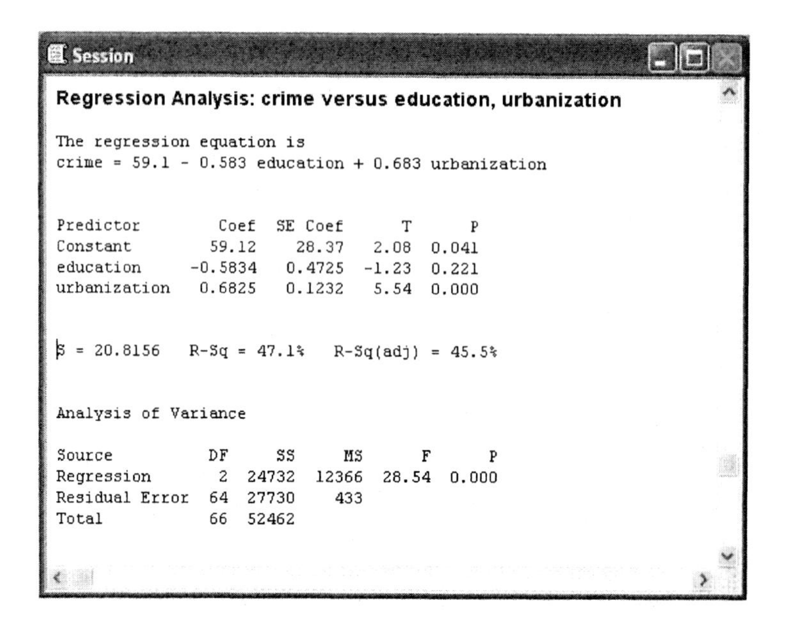

Regression Analysis: crime versus education, urbanization

```
The regression equation is
crime = 59.1 - 0.583 education + 0.683 urbanization


Predictor         Coef  SE Coef      T      P
Constant         59.12    28.37   2.08  0.041
education      -0.5834   0.4725  -1.23  0.221
urbanization    0.6825   0.1232   5.54  0.000


S = 20.8156   R-Sq = 47.1%   R-Sq(adj) = 45.5%


Analysis of Variance

Source          DF     SS     MS      F      P
Regression       2  24732  12366  28.54  0.000
Residual Error  64  27730    433
Total           66  52462
```

To generate a scatterplot matrix

Select: **Graph > Matrix plot**

Select: **Matrix of plots / Simple > OK > OK**

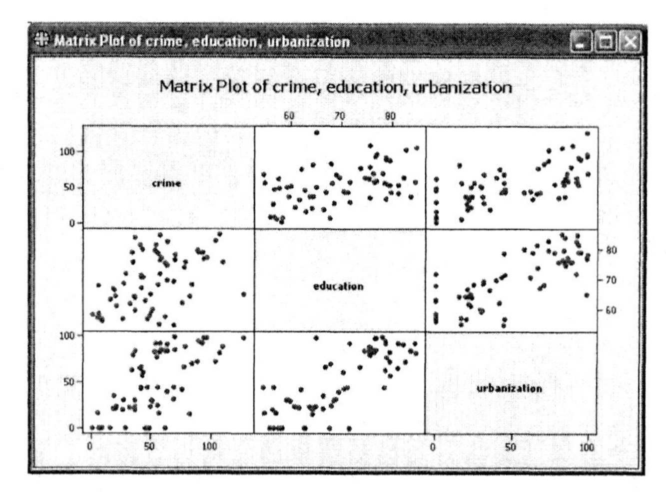

Since we are truly interested in the association between crime rate and education, and crime rate and urbanization, another option is

Select: **Graph > Matrix of plots**
Select: **Each Y versus each X, Simple**
Click: **OK**

Enter: Y variables: **crime**
X variables: **education urbanization**
Click: **OK**

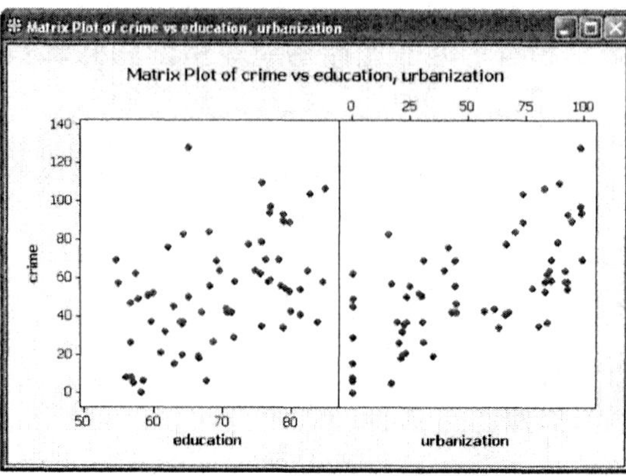

Using the regression equation generated by MINITAB,

```
crime = 59.1 - 0.583 education + 0.683 urbanization
```

we can predict crime rates for a county that has 0% in an urban environment by substituting 0 for "urbanization" and using the resulting equation crime = 59.1 – 0.583 education . With education set equal to 70% we determine the crime rate to be 59.1 – 0.583(70) = 18.29. For an 80% high school graduation rate, the crime rate is predicted to be 59.1 – 0.583(80) = 12.46.

Chapter 13 Example 3 How Well Can We Predict House Selling Prices? ANOVA and R

Open the worksheet **house_selling_prices** , which is found in the MINITAB folder. The price is found in C7, house size is found in C8, and lot size is found in C9. The regression command can be tailored to include the value of R-squared. We have done this numerous times.

Select: **Stat > Regression > Regression**
Enter: Response: **price**
Predictors: **size lot**
Select: **Results**
Select: **Regression equation, table of coefficients, s, R-squared, and basic analysis of variance**
Click: **OK OK**

The results are displayed in the Session Window.

Note the value of R-squared is 71.1%. The multiple correlation between selling price and the two explanatory variables is $R = \sqrt{R^2} = \sqrt{.711} = 0.84$.

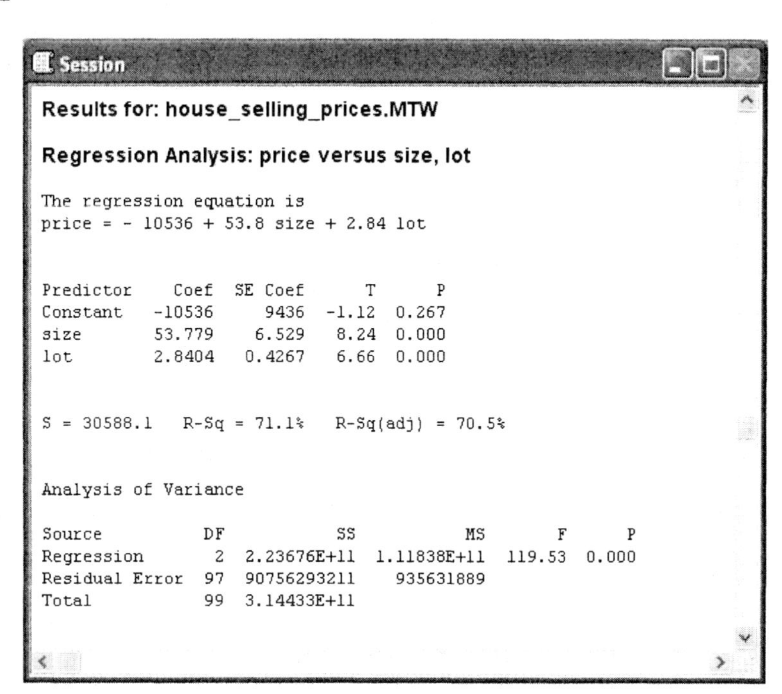
Session

Results for: house_selling_prices.MTW

Regression Analysis: price versus size, lot

```
The regression equation is
price = - 10536 + 53.8 size + 2.84 lot


Predictor    Coef  SE Coef      T      P
Constant   -10536     9436  -1.12  0.267
size       53.779    6.529   8.24  0.000
lot        2.8404   0.4267   6.66  0.000


S = 30588.1   R-Sq = 71.1%   R-Sq(adj) = 70.5%


Analysis of Variance

Source          DF           SS           MS       F      P
Regression       2  2.23676E+11  1.11838E+11  119.53  0.000
Residual Error  97  90756293211    935631889
Total           99  3.14433E+11
```

Chapter 13 Example 4 What Helps Predict a Female Athlete's Weight?
Significance Test and Confidence Interval about a Multiple Regression Parameter β

Open the worksheet **college_athletes**, which is found in the MINITAB folder. The total body weight (TBW) is found in C1, height (HGT) in inches is found in C2, the percent of body fat (%BF) is found in C3, and age (AGE) is found in C11.

Select: **Stat > Regression > Regression**
Enter: Response: **TBW**
Predictors: **HGT %BF AGE**
Select: **Results**
Select: **Regression equation, table of coefficients, s, R-squared, and basic analysis of variance**
Click: **OK OK**

The results are displayed in the Session Window.

Note the value of R-squared is 66.9%. The predictive power is good.

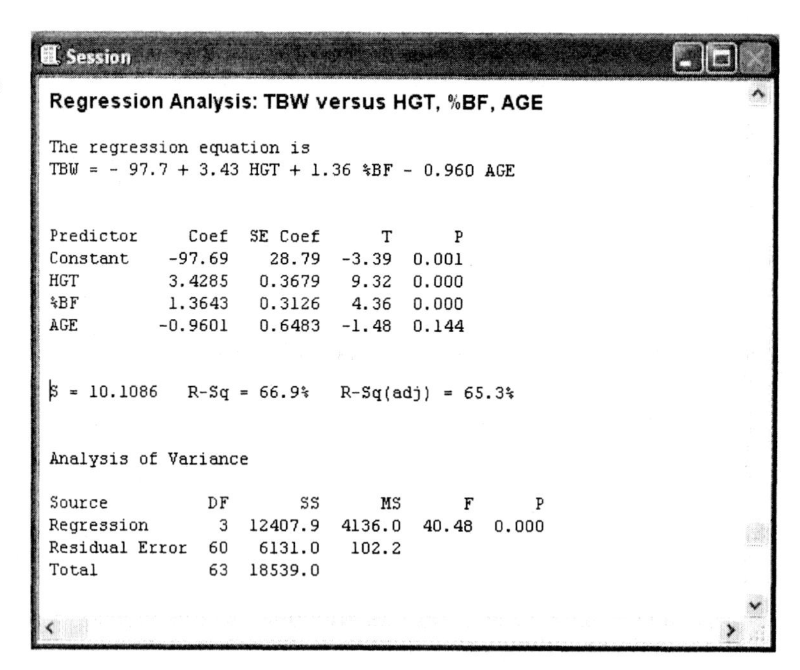
Session

Regression Analysis: TBW versus HGT, %BF, AGE

The regression equation is
TBW = - 97.7 + 3.43 HGT + 1.36 %BF - 0.960 AGE

Predictor	Coef	SE Coef	T	P
Constant	-97.69	28.79	-3.39	0.001
HGT	3.4285	0.3679	9.32	0.000
%BF	1.3643	0.3126	4.36	0.000
AGE	-0.9601	0.6483	-1.48	0.144

S = 10.1086 R-Sq = 66.9% R-Sq(adj) = 65.3%

Analysis of Variance

Source	DF	SS	MS	F	P
Regression	3	12407.9	4136.0	40.48	0.000
Residual Error	60	6131.0	102.2		
Total	63	18539.0			

To test whether age helps us to predict weight, if we already know height and percent body fat, we perform a significance test on $H_0 : \beta_3 = 0$ versus H_a: $\beta_3 \neq 0$. The *t* test statistic is reported in the table as -1.48 and the p-value is .144. This p-value does not give much evidence against the null hypothesis. Age does not significantly predict weight, if we already know height and percentage of body fat.

Chapter 13 Example 5 What's Plausible for the Effect of Age on Weight?
Confidence Interval about a Multiple Regression Parameter β

Open the worksheet **college_athletes**, which is found in the MINITAB folder. The total body weight (TBW) is found in C1, height (HGT) in inches is found in C2, the percent of body fat (%BF) is found in C3, and age (AGE) is found in C11.

Select: **Stat > Regression > Regression**
Enter: Response: **TBW**
Predictors: **HGT %BF AGE**
Select: **Results**
Select: **Regression equation, table of coefficients, s, R-squared, and basic analysis of variance**
Click: **OK OK**

The results are displayed in the Session Window.

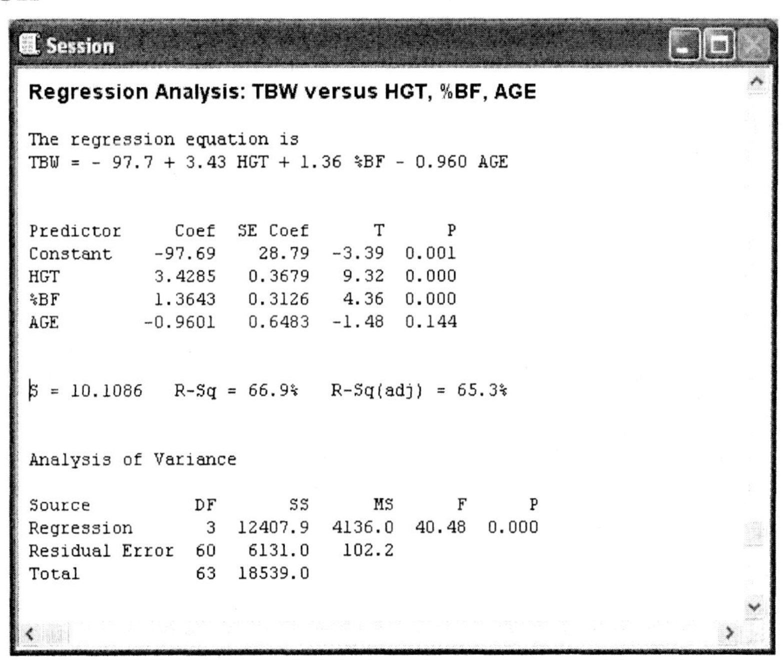

Session

Regression Analysis: TBW versus HGT, %BF, AGE

```
The regression equation is
TBW = - 97.7 + 3.43 HGT + 1.36 %BF - 0.960 AGE


Predictor      Coef  SE Coef      T      P
Constant     -97.69    28.79  -3.39  0.001
HGT          3.4285   0.3679   9.32  0.000
%BF          1.3643   0.3126   4.36  0.000
AGE         -0.9601   0.6483  -1.48  0.144


S = 10.1086   R-Sq = 66.9%   R-Sq(adj) = 65.3%


Analysis of Variance

Source          DF       SS      MS      F      P
Regression       3  12407.9  4136.0  40.48  0.000
Residual Error  60   6131.0   102.2
Total           63  18539.0
```

To compute a 95% confidence interval for the population slope β_3, note that the $t_{.025}$ value for $df = n - 2 = 60$ is $t = 2.00$, and that the estimate of β_3 is -0.9601, and $se = 0.6483$. (MINITAB does not include a t-interval estimate for a population slope.)

$$b \pm t_{.025}(se) = -0.9601 \pm 2.00\,(0.6483) = -0.9601 \pm 1.2966$$

$$(-2.26,\ 0.34)$$

At fixed values of height and percent body fat, we infer that the population mean weight changes very little (and may not change at all, since this interval includes 0), making these results consistent with what we found in example 4.

Chapter 13 Example 7 The F Test For Predictors of Athletes' Weight
The F Test that All Multiple Regression Parameters β = 0

Open the worksheet **college_athletes**, which is found in the MINITAB folder. The total body weight (TBW) is found in C1, height (HGT) in inches is found in C2, the percent of body fat (%BF) is found in C3, and age (AGE) is found in C11.

Select: **Stat > Regression > Regression**
Enter: Response: **TBW**
Predictors: **HGT %BF AGE**
Select: **Results**
Select: **Regression equation, table of coefficients, s, R-squared, and basic analysis of variance**
Click: **OK OK**

The results are displayed in the Session Window.

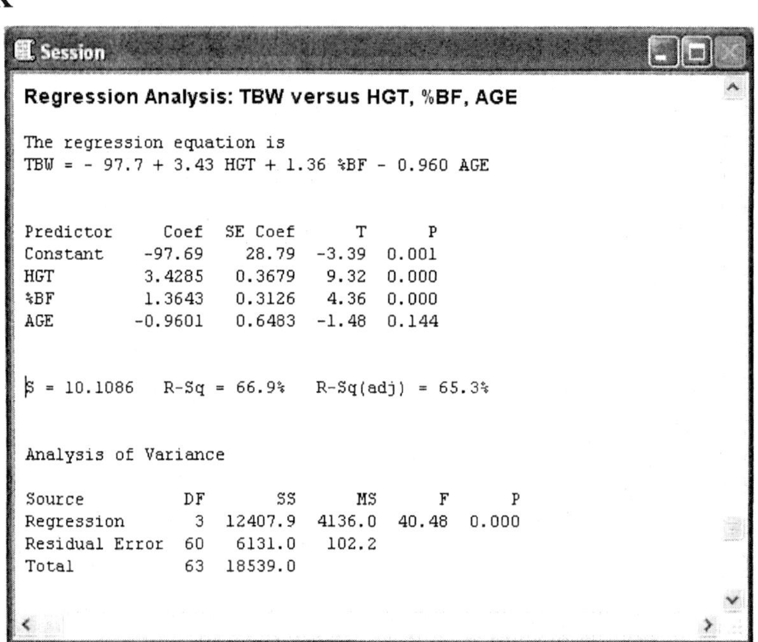
Session

Regression Analysis: TBW versus HGT, %BF, AGE

The regression equation is
TBW = - 97.7 + 3.43 HGT + 1.36 %BF - 0.960 AGE

Predictor	Coef	SE Coef	T	P
Constant	-97.69	28.79	-3.39	0.001
HGT	3.4285	0.3679	9.32	0.000
%BF	1.3643	0.3126	4.36	0.000
AGE	-0.9601	0.6483	-1.48	0.144

S = 10.1086 R-Sq = 66.9% R-Sq(adj) = 65.3%

Analysis of Variance

Source	DF	SS	MS	F	P
Regression	3	12407.9	4136.0	40.48	0.000
Residual Error	60	6131.0	102.2		
Total	63	18539.0			

The ANOVA table for the multiple regression output includes the value of the F test statistic (F = 40.48) and the p-value (p = 0.000) for testing H_0: $\beta_1 = \beta_2 = \beta_3 = 0$. We can reject H_0 and conclude that at least one predictor has an effect on weight.

Chapter 13 Exercise 13.20 Study time help GPA?
Hypothesis Test and Confidence Interval About β

Open the worksheet **georgia student survey**, which is found in the MINITAB folder. Performing a multiple regression for college GPA based on high school GPA and study time:

Select: **Stat > Regression > Regression**
Enter: Response: **CGPA**
Predictors: **HSGPA Studytime**
Select: **Results**
Select: **Regression equation, table of coefficients, s, R-squared, and basic analysis of variance**
Click: **OK OK**

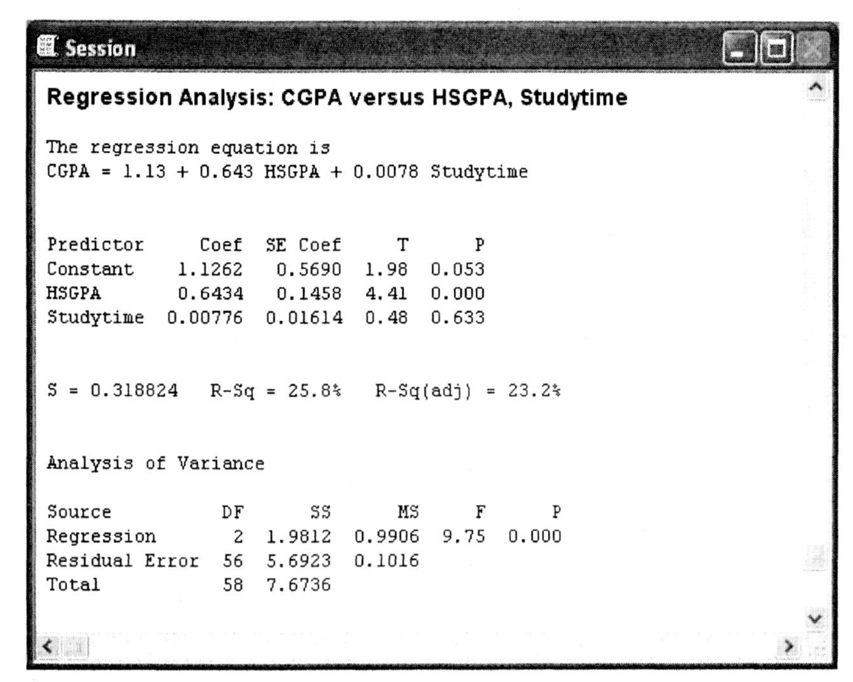

Session

Regression Analysis: CGPA versus HSGPA, Studytime

```
The regression equation is
CGPA = 1.13 + 0.643 HSGPA + 0.0078 Studytime


Predictor      Coef   SE Coef     T      P
Constant     1.1262    0.5690  1.98  0.053
HSGPA        0.6434    0.1458  4.41  0.000
Studytime   0.00776   0.01614  0.48  0.633


S = 0.318824   R-Sq = 25.8%   R-Sq(adj) = 23.2%


Analysis of Variance

Source          DF       SS      MS     F      P
Regression       2   1.9812  0.9906  9.75  0.000
Residual Error  56   5.6923  0.1016
Total           58   7.6736
```

To test whether study time helps us to college GPA, if we already know high school GPA, we perform a significance test on $H_0 : \beta_2 = 0$ versus H_a: $\beta_2 \neq 0$. The t test statistic is reported in the table as 0.48 and the p-value is 0.633. This p-value does not give evidence against the null hypothesis. Study time does not significantly predict college GPA if we already know high school GPA.

To compute a 95% confidence interval for the population slope β_2, note that the $t_{.025}$ value for $df = n - 2 = 57$ is $t =$ 2.0025, and that the estimate of β_2 is 0.00776, and $se = 0.01614$. (MINITAB does not include a t-interval estimate for a population slope.)

$$b \pm t_{.025}\,(se) = 0.00776 \pm 2.0025\,(0.01614)$$
$$= 0.00776 \text{ +/- } 0.0323$$
$$(-0.0247,\ 0.0399)$$

At a fixed value of high school GPA, we infer that the population mean college GPA changes very little (and may not change at all, since this interval includes 0), making these results consistent with the significance test we have already completed.

Chapter 13 Exercise 13.29 More predictors for house price
The F Test that All Multiple Regression Parameters β = 0

Open the worksheet **house_selling_prices**, which is found in the MINITAB folder. Performing multiple regression for predicting house selling price using size of home, lot size, and real estate tax:

Select: **Stat > Regression > Regression**
Enter: Response: **price**
Predictors: **size lot Taxes**
Select: **Results**
Select: **Regression equation, table of coefficients, s, R-squared, and basic analysis of variance**
Click: **OK OK**

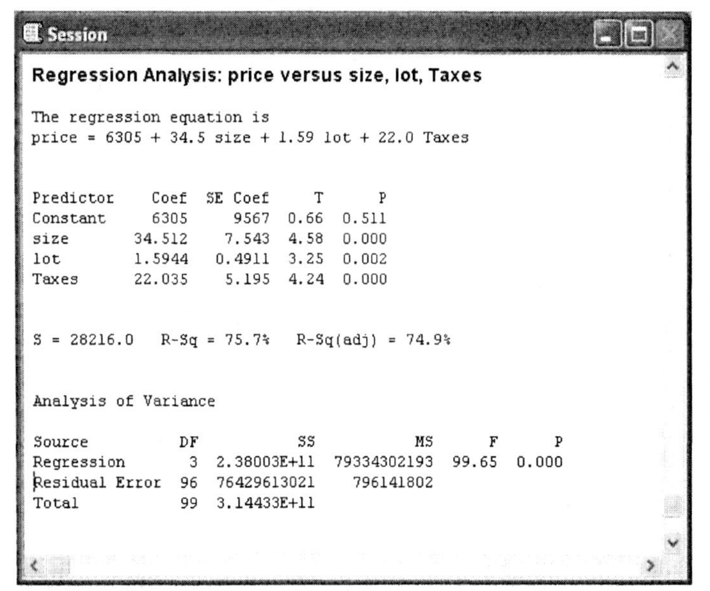

Session

Regression Analysis: price versus size, lot, Taxes

The regression equation is
price = 6305 + 34.5 size + 1.59 lot + 22.0 Taxes

Predictor	Coef	SE Coef	T	P
Constant	6305	9567	0.66	0.511
size	34.512	7.543	4.58	0.000
lot	1.5944	0.4911	3.25	0.002
Taxes	22.035	5.195	4.24	0.000

S = 28216.0   R-Sq = 75.7%   R-Sq(adj) = 74.9%

Analysis of Variance

Source	DF	SS	MS	F	P
Regression	3	2.38003E+11	79334302193	99.65	0.000
Residual Error	96	76429613021	796141802		
Total	99	3.14433E+11			

The ANOVA table for the multiple regression output includes the value of the F test statistic (F = 99.65) and the p-value (p = 0.000) for testing H_0: $\beta_1 = \beta_2 = \beta_3 = 0$. We can reject H_0 and conclude that at least one predictor has an effect on price.

The p-values of 0.000, 0.002, 0.000 for *t*-tests for each of the explanatory variables, respectively, indicate that each predictor (house size, lot size, and taxes) does significantly predict price, if we already know the values of the other two variables.

Chapter 13 Example 9 Another Residual Plot for House Selling Prices Plots of Residuals Against Explanatory Variables

Open the worksheet **house_selling_prices** , which is found in the MINITAB folder. The price is found in C7, house size is found in C8, and lot size is found in C9.

To produce a histogram of standardized residuals for the multiple regression model predicting selling price by the house size and the lot size:

Select: **Stat > Regression > Regression**
Enter: Response: **price**
Predictors: **size lot**
Select: **Graphs**

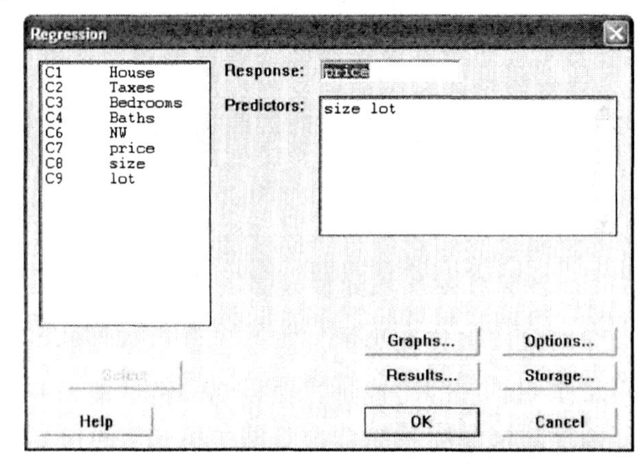

Select: Residuals for Plots: **Standardized**
Residual Plot: **Histogram of residuals**
Click: **OK OK**

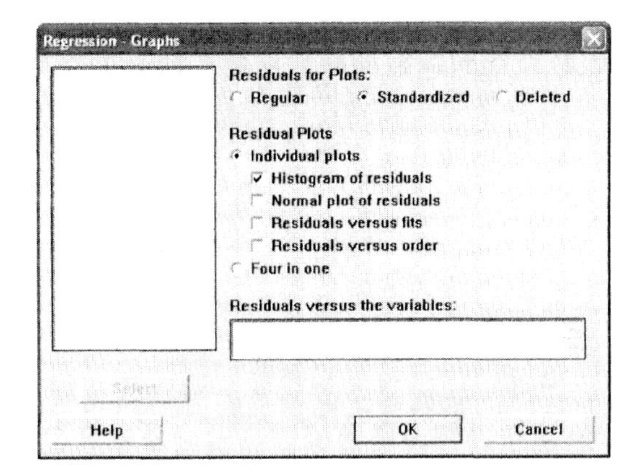

The histogram is displayed in a Graph Window.

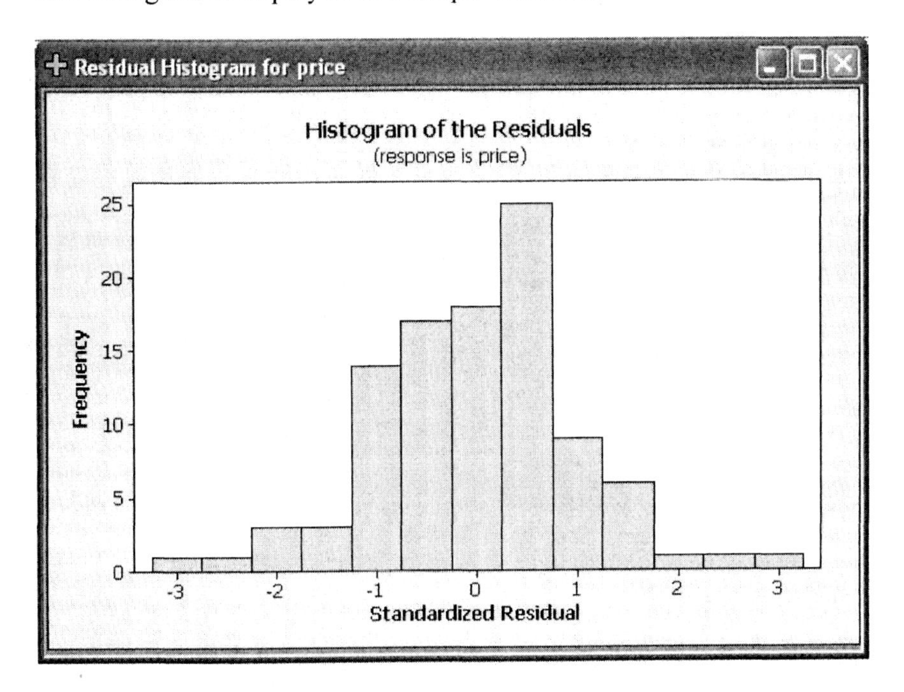

To plot the residuals against the explanatory variable house size:

Select: **Stat > Regression > Regression**
Enter: Response: **price**
Predictors: **size lot**
Select: **Graphs**
Select: Residuals for Plots: **Standardized**
Enter: Residuals versus the variables: **size**
Click: **OK OK**

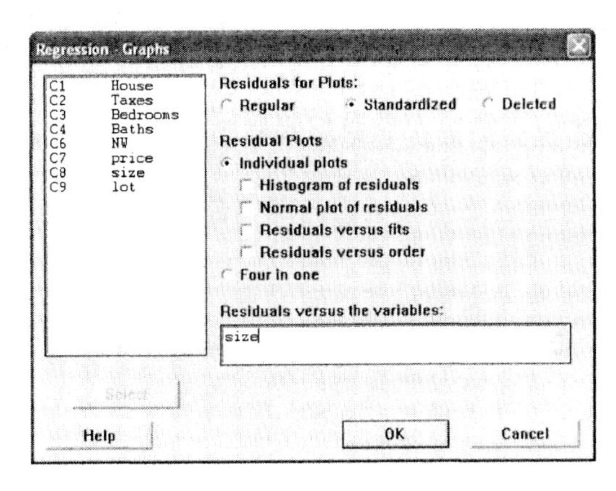

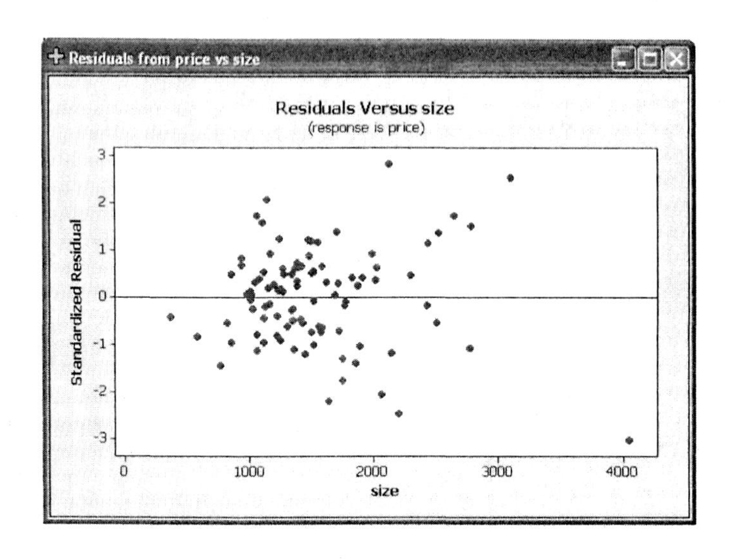

Chapter 13 Example 11 Comparing Winning High Jumps For Men and Women Including Categorical Predictors in Regression

Open the worksheet **high_jump** , which is found in the MINITAB folder. The following screen shot indicates the organization of the columns used for the regression in this example.

↓	C5	C6	C7-T	C8
		Winning Height (m)	Gender	Year
16		2.240	Men	1968
17		2.230	Men	1972
18		2.250	Men	1976
19		2.360	Men	1980
20		2.350	Men	1984
21		2.360	Men	1988
22		2.340	Men	1992
23		2.390	Men	1996
24		2.360	Men	2000
25		2.360	Men	2004
26		1.590	Women	1928
27		1.657	Women	1932
28		1.600	Women	1936
29		1.680	Women	1948
30		1.670	Women	1952
31		1.760	Women	1956

Note that 1928 is the first year that women participated in the high jump.

We need to add columns to indicate x_1 = number of years since 1928 and x_2 = gender (1 = male, 0 = female). Since we're only interested in years starting at 1928 we will not use the data for the men for prior years for the regression.

HIGH_JUMP.MTW ***

↓	C1 Men_Meters	C2 Women_Meters	C3 Year_Men	C4 Year_Women	C5	C6 Winning Height (m)	C7-T Gender	C8 Year	C9 x1	C10 x2
1	1.810	1.590	1896	1928		1.940	Men	1928	0	1
2	1.900	1.657	1900	1932		1.970	Men	1932	4	1
3	1.800	1.600	1904	1936		2.030	Men	1936	8	1
4	1.905	1.680	1908	1948		1.980	Men	1948	20	1
5	1.930	1.670	1912	1952		2.040	Men	1952	24	1
6	1.935	1.760	1920	1956		2.120	Men	1956	28	1
7	1.980	1.850	1924	1960		2.160	Men	1960	32	1
8	1.940	1.900	1928	1964		2.160	Men	1964	36	1
9	1.970	1.820	1932	1968		2.240	Men	1968	40	1
10	2.030	1.920	1936	1972		2.230	Men	1972	44	1
11	1.980	1.930	1948	1976		2.250	Men	1976	48	1
12	2.040	1.970	1952	1980		2.360	Men	1980	52	1
13	2.120	2.020	1956	1984		2.350	Men	1984	56	1
14	2.160	2.030	1960	1988		2.360	Men	1988	60	1
15	2.160	2.020	1964	1992		2.340	Men	1992	64	1
16	2.240	2.060	1968	1996		2.390	Men	1996	68	1

Select: **Stat > Regression > Regression**
Enter: Response: **Winning Height (m)**
Predictors: **x1 x2**
Click: **OK**

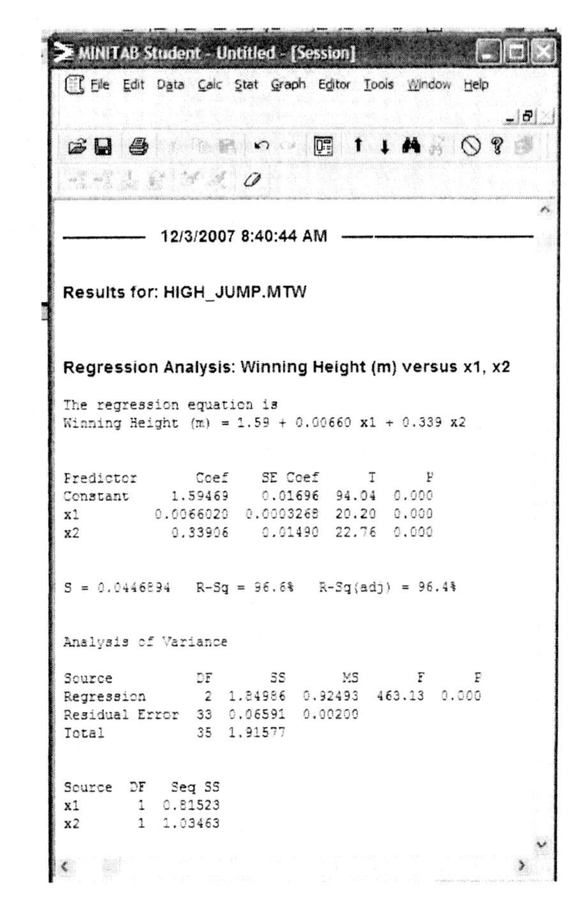
MINITAB Student - Untitled - [Session]

File Edit Data Calc Stat Graph Editor Tools Window Help

12/3/2007 8:40:44 AM

Results for: HIGH_JUMP.MTW

Regression Analysis: Winning Height (m) versus x1, x2

```
The regression equation is
Winning Height (m) = 1.59 + 0.00660 x1 + 0.339 x2

Predictor        Coef     SE Coef      T      P
Constant      1.59469     0.01696  94.04  0.000
x1          0.0066020   0.0003268  20.20  0.000
x2            0.33906     0.01490  22.76  0.000

S = 0.0446894   R-Sq = 96.6%   R-Sq(adj) = 96.4%

Analysis of Variance

Source          DF       SS       MS       F      P
Regression       2  1.84986  0.92493  463.13  0.000
Residual Error  33  0.06591  0.00200
Total           35  1.91577

Source  DF   Seq SS
x1       1  0.81523
x2       1  1.03463
```

To produce a scatterplot that includes the categorical distinction:
Select: **Graph > Scatterplot > With Groups > OK**
Select: **With Groups**

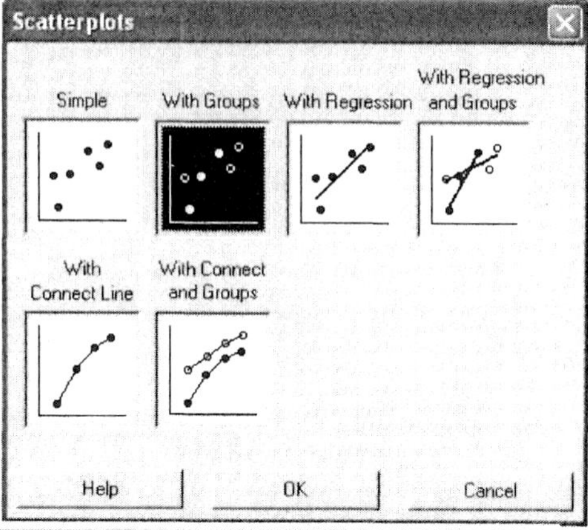

Enter the appropriate variables and select **X-Y pairs form groups > OK**

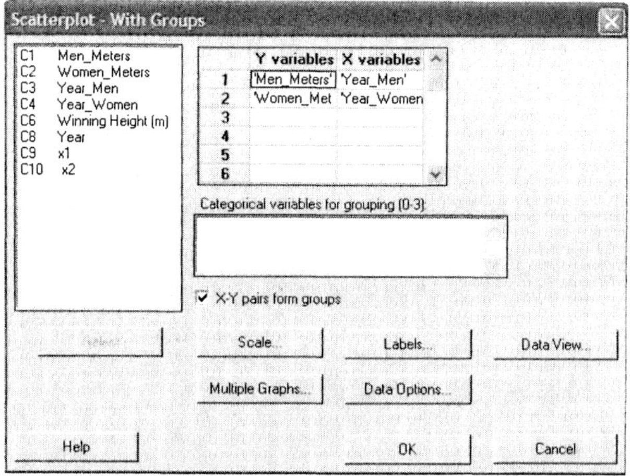

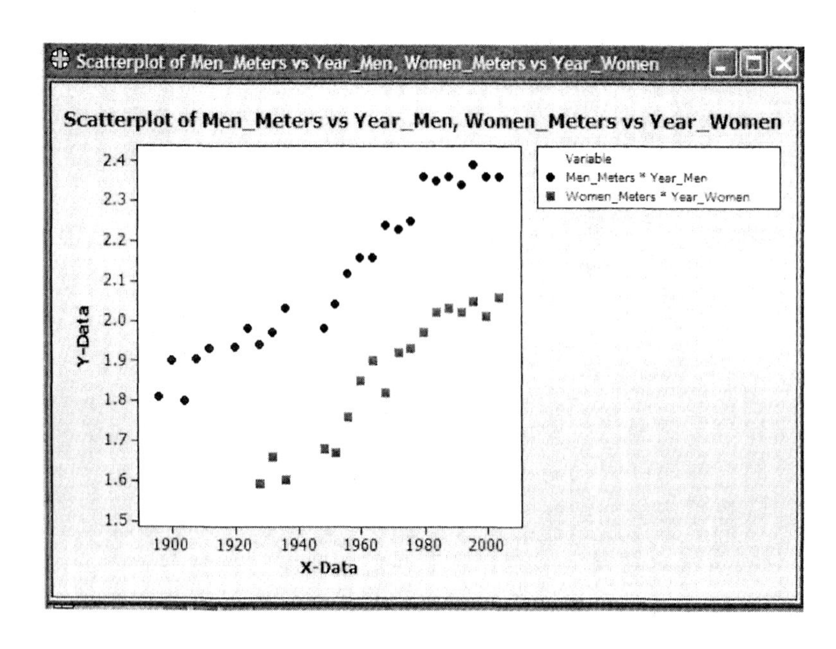

Chapter 13 Exercise 13.45 Houses, tax, and NW
Including Categorical Predictors in Regression

Open the worksheet **house_selling_prices** , which is found in the MINITAB folder. The following screen shot indicates the organization of the data file.

house_selling_prices.MTW ***

↓	C1 House	C2 Taxes	C3 Bedrooms	C4 Baths	C5-T Quadrant	C6 NW	C7 price	C8 size	C9 lot
1	1	1360	3	2.0	NW	1	145000	1240	18000
2	2	1050	1	1.0	NW	1	68000	370	25000
3	3	1010	3	1.5	NW	1	115000	1130	25000
4	4	830	3	2.0	SW	0	69000	1120	17000
5	5	2150	3	2.0	NW	1	163000	1710	14000
6	6	1230	3	2.0	NW	1	69900	1010	8000
7	7	150	2	2.0	NW	1	50000	860	15300
8	8	1470	3	2.0	NW	1	137000	1420	18000
9	9	1850	3	2.0	NW	1	121300	1270	16000
10	10	320	3	2.0	NW	1	70000	1160	8000

The data file includes C6 NW where 1 = NW and 0 = other.

Performing multiple regression for predicting price as a function of real estate tax and whether the home is in the NW region:

Select: **Stat > Regression > Regression**
Enter: Response: **price**
Predictors: **Taxes NW**
Click: **OK**

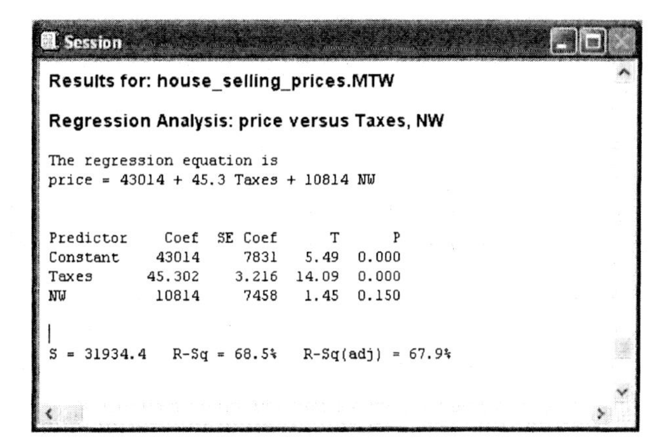

Session

Results for: house_selling_prices.MTW

Regression Analysis: price versus Taxes, NW

```
The regression equation is
price = 43014 + 45.3 Taxes + 10814 NW


Predictor     Coef  SE Coef      T      P
Constant     43014     7831   5.49  0.000
Taxes       45.302    3.216  14.09  0.000
NW           10814     7458   1.45  0.150


S = 31934.4   R-Sq = 68.5%   R-Sq(adj) = 67.9%
```

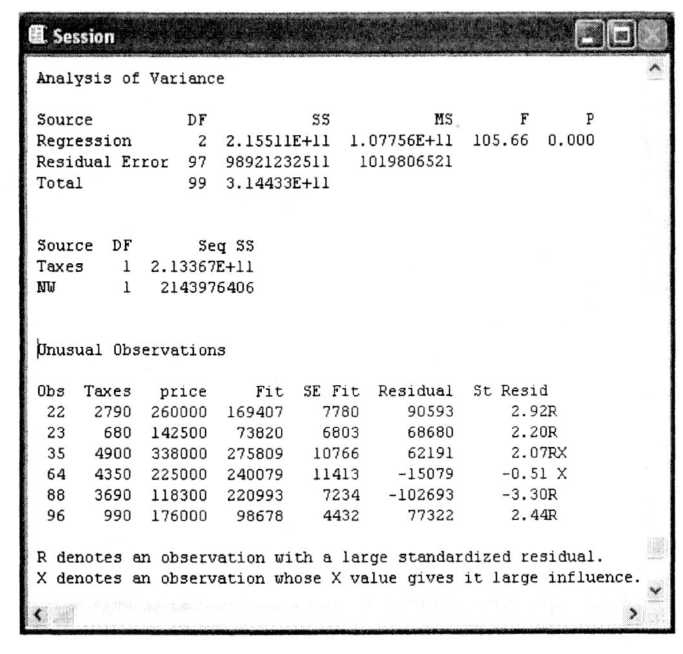

Session

```
Analysis of Variance

Source          DF           SS          MS       F      P
Regression       2  2.15511E+11  1.07756E+11  105.66  0.000
Residual Error  97  98921232511   1019806521
Total           99  3.14433E+11


Source  DF       Seq SS
Taxes    1  2.13367E+11
NW       1   2143976406


Unusual Observations

Obs  Taxes   price     Fit  SE Fit  Residual  St Resid
 22   2790  260000  169407    7780     90593     2.92R
 23    680  142500   73820    6803     68680     2.20R
 35   4900  338000  275809   10766     62191     2.07RX
 64   4350  225000  240079   11413    -15079    -0.51 X
 88   3690  118300  220993    7234   -102693    -3.30R
 96    990  176000   98678    4432     77322     2.44R

R denotes an observation with a large standardized residual.
X denotes an observation whose X value gives it large influence.
```

From the coefficient for NW we see that the sclling price increases by $10,814 just because the house is located in the NW region.

Chapter 13 Example 12 Annual Income and Having a Travel Credit Card Logistic Regression

Open the data file **credit_card_and_income** from the MINITAB folder of the text CD.

Choose: **Stat > Regression > Binary Logistic Regression**

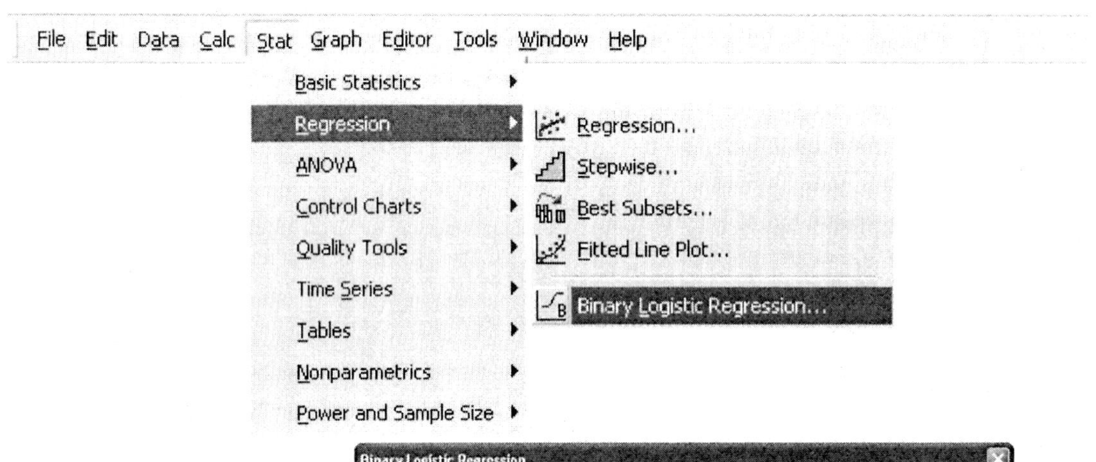

Enter: Response: **y**
Model: **income**
Click: **OK**

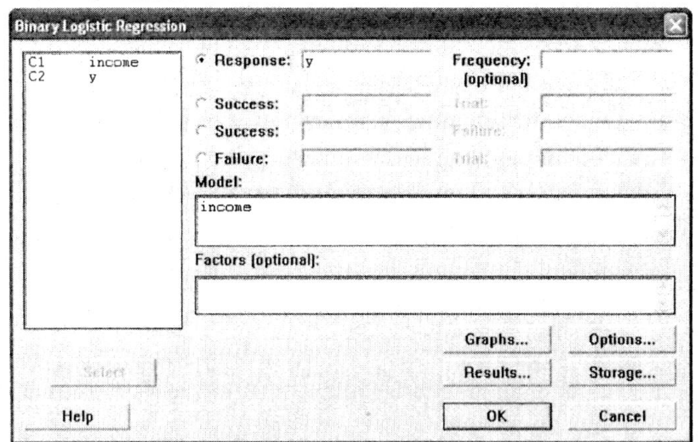

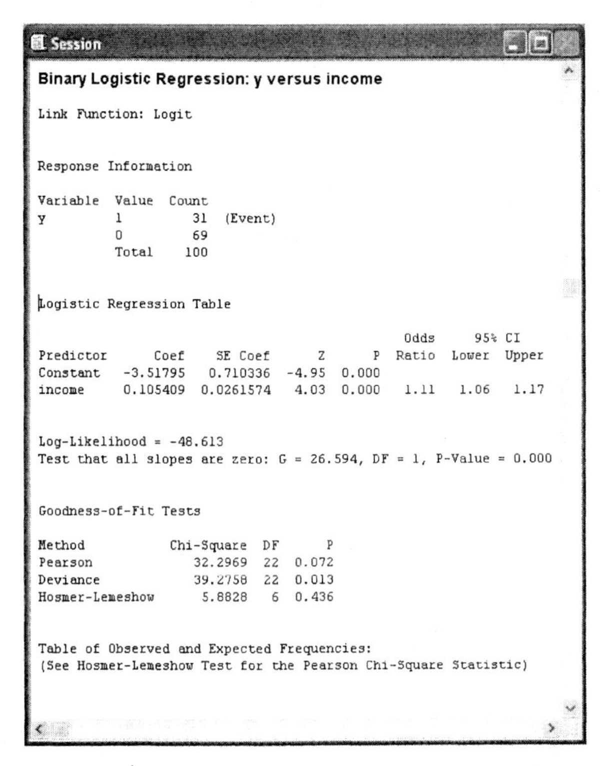

Session

Binary Logistic Regression: y versus income

```
Link Function: Logit


Response Information

Variable  Value  Count
y         1         31  (Event)
          0         69
          Total    100


Logistic Regression Table

                                                   Odds     95% CI
Predictor      Coef    SE Coef      Z      P  Ratio  Lower  Upper
Constant  -3.51795   0.710336  -4.95  0.000
income    0.105409  0.0261574   4.03  0.000   1.11   1.06   1.17


Log-Likelihood = -48.613
Test that all slopes are zero: G = 26.594, DF = 1, P-Value = 0.000


Goodness-of-Fit Tests

Method           Chi-Square  DF      P
Pearson             32.2969  22  0.072
Deviance            39.2758  22  0.013
Hosmer-Lemeshow      5.8828   6  0.436


Table of Observed and Expected Frequencies:
(See Hosmer-Lemeshow Test for the Pearson Chi-Square Statistic)
```

Quite a bit of information is displayed in the Session Window.

The values we need are α and β. They are found under the Coef heading.
$\alpha = -3.51795$ and $\beta = 0.105409$

So, the equation for estimated probability $\hat{p}$ of possessing a travel credit card is

$$\hat{p} = \frac{e^{-3.52 + .105x}}{1+e^{-3.52 + .105x}}$$

Chapter 13 Example 14 Estimating Proportion of Students Who Have Used Marijuana Logistic Regression

For this problem we will need to enter the data. The response variable "marijuana use" will be coded 1 = yes and 0 = no. Likewise we will code the indicator variables, alcohol use, and cigarette use, 1 = yes and 0 = no. We also need to include a column with the corresponding frequencies.

Worksheet 3 ***

↓	C1 Alcohol Use	C2 Cigarette Use	C3 Marijuana Use	C4 Count
1	1	1	1	911
2	1	1	0	538
3	1	0	1	44
4	1	0	0	456
5	0	1	1	3
6	0	1	0	43
7	0	0	1	2
8	0	0	0	279
9				
10				

Choose: **Stat > Regression > Binary Logistic Regression**
Enter: Response: **Marijuana Use**
Model: **Alcohol Use   Cigarette Use**
Click: **OK**

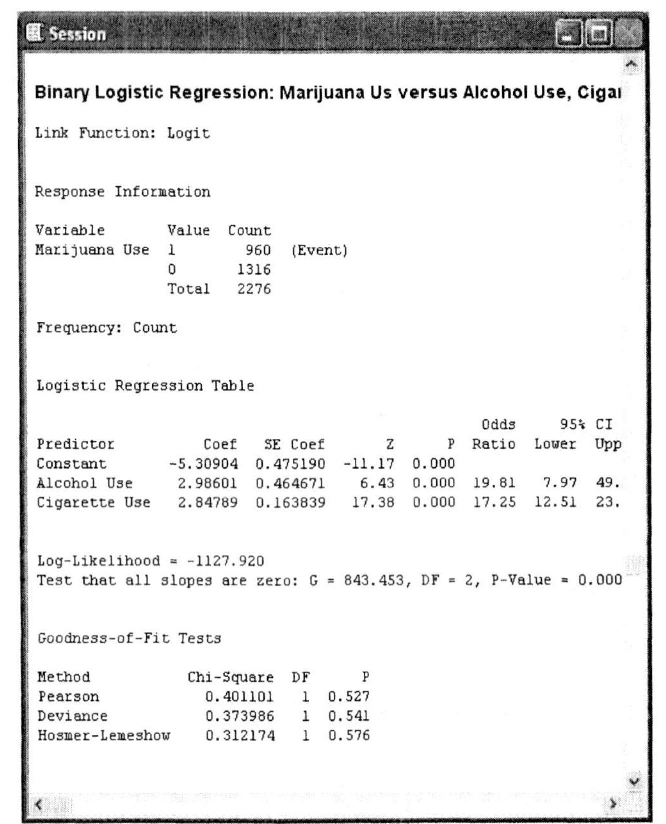
Session

Binary Logistic Regression: Marijuana Us versus Alcohol Use, Cigai

Link Function: Logit

Response Information

Variable	Value	Count	
Marijuana Use	1	960	(Event)
	0	1316	
	Total	2276	

Frequency: Count

Logistic Regression Table

Predictor	Coef	SE Coef	Z	P	Odds Ratio	95% CI Lower	Upp
Constant	-5.30904	0.475190	-11.17	0.000			
Alcohol Use	2.98601	0.464671	6.43	0.000	19.81	7.97	49.
Cigarette Use	2.84789	0.163839	17.38	0.000	17.25	12.51	23.

Log-Likelihood = -1127.920
Test that all slopes are zero: G = 843.453, DF = 2, P-Value = 0.000

Goodness-of-Fit Tests

Method	Chi-Square	DF	P
Pearson	0.401101	1	0.527
Deviance	0.373986	1	0.541
Hosmer-Lemeshow	0.312174	1	0.576

Quite a bit of information is displayed in the Session Window.

The values we need are α and β_1 and β_2. They are found under the Coef heading. $\alpha = -5.30904$ and $\beta 1 = 2.98601$ and $\beta_2 = 2.84789$

So, the equation for estimated probability $\hat{p}$ of using marijuana is

$$\hat{p} = \frac{e^{-5.31 + 2.99x_1 + 2.85x_2}}{1 + e^{-5.31 + 2.99x_1 + 2.85x_2}}$$

Chapter 13 Exercise 13.57 Death Penalty and Race Logistic Regression

Enter the data into the worksheet as indicated.

Worksheet 4 ***

↓	C1 d	C2 v	C3 y	C4 count
1	1	1	1	53
2	1	1	0	414
3	1	0	1	0
4	1	0	0	16
5	0	1	1	11
6	0	1	0	37
7	0	0	1	4
8	0	0	0	139
9				
10				

Choose: **Stat > Regression > Binary Logistic Regression**

Enter: Response: **y**
Model: **d v**
Click: **OK**

The results are displayed in the Session Window.

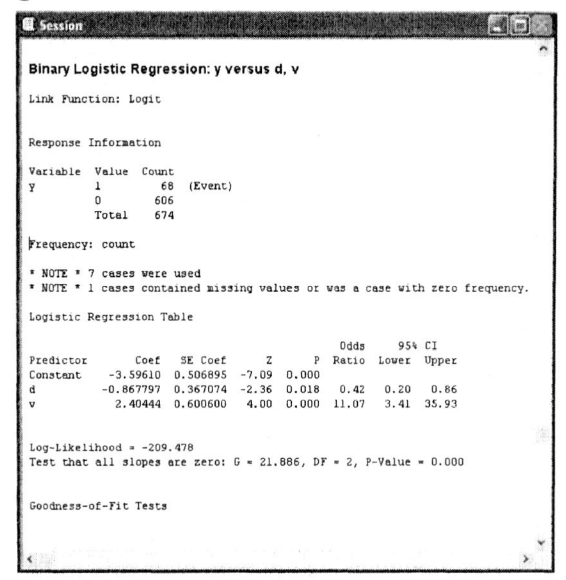

Session

Binary Logistic Regression: y versus d, v

Link Function: Logit

Response Information

Variable	Value	Count	
y	1	68	(Event)
	0	606	
	Total	674	

Frequency: count

* NOTE * 7 cases were used
* NOTE * 1 cases contained missing values or was a case with zero frequency.

Logistic Regression Table

Predictor	Coef	SE Coef	Z	P	Odds Ratio	95% CI Lower	95% CI Upper
Constant	-3.59610	0.506895	-7.09	0.000			
d	-0.867797	0.367074	-2.36	0.018	0.42	0.20	0.86
v	2.40444	0.600600	4.00	0.000	11.07	3.41	35.93

Log-Likelihood = -209.478
Test that all slopes are zero: G = 21.886, DF = 2, P-Value = 0.000

Goodness-of-Fit Tests

The values we need are α and β_1 and β_2. They are found under the Coef heading.
$\alpha = -3.59610$ and $\beta_1 = -0.867797$ and $\beta_2 = 2.40444$

So, the equation for estimated probability $\hat{p}$ of death penalty verdict is

$$\hat{p} = \frac{e^{-3.596 - 0.868d + 2.404v}}{1+e^{-3.596 - 0.868d + 2.404v}}$$

Chapter 14 Example 3 Customers' Telephone Holding Time ANOVA

We first must enter the data into the spreadsheet. The data is found in Example 2 of the previous section.

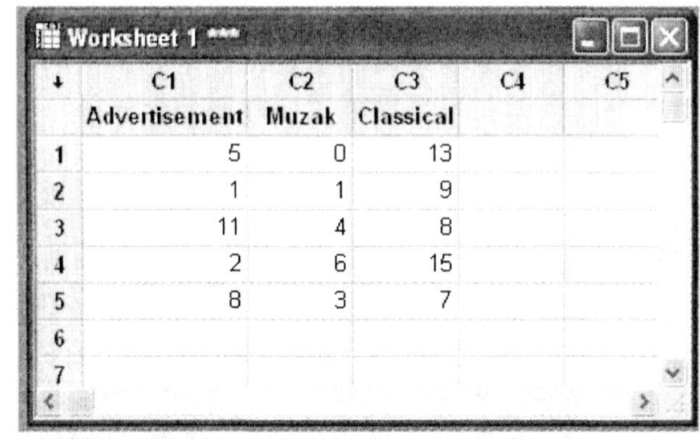

↓	C1	C2	C3	C4	C5
	Advertisement	Muzak	Classical		
1	5	0	13		
2	1	1	9		
3	11	4	8		
4	2	6	15		
5	8	3	7		
6					
7					

If we do a dotplot on this data some interesting things can be shown.

Choose: Graph > Dotplot> Multiple y's Simple

Enter the variables as shown:

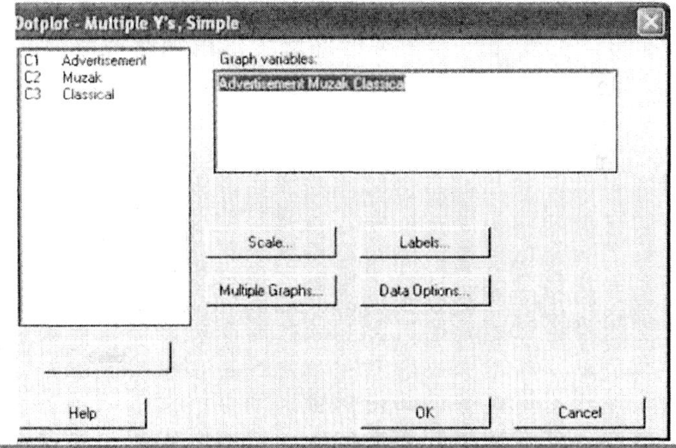

Notice where the plots overlap and where they do not.

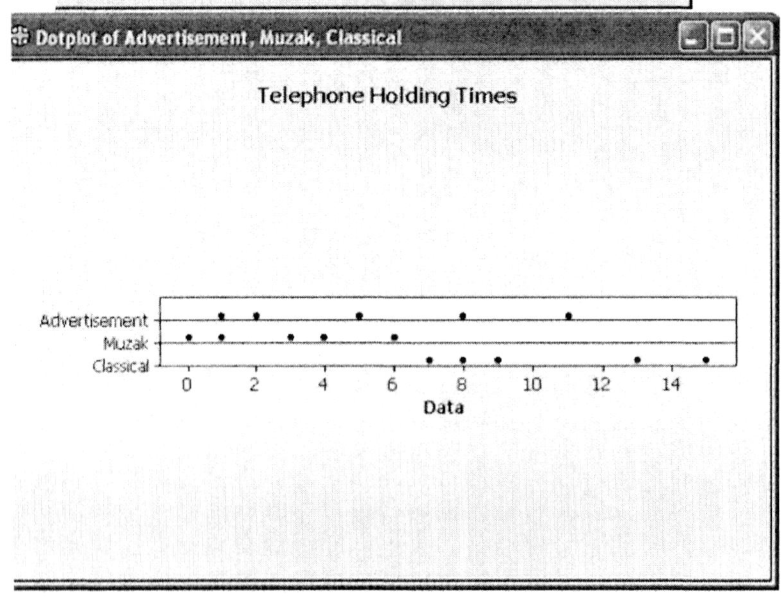

To do the ANOVA analysis

Choose: **Stat > ANOVA >Oneway [Unstacked]**

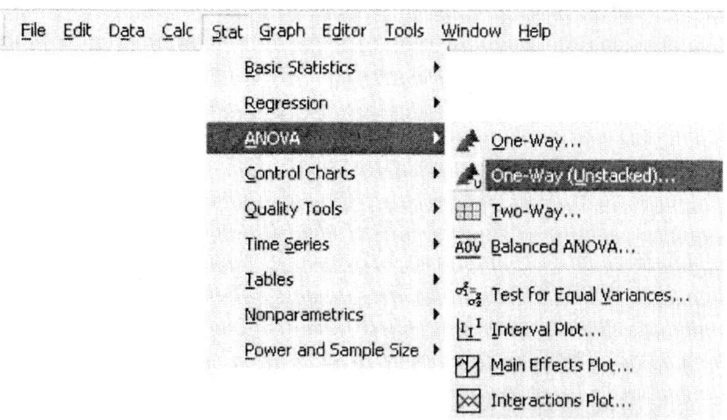

Enter the variables:

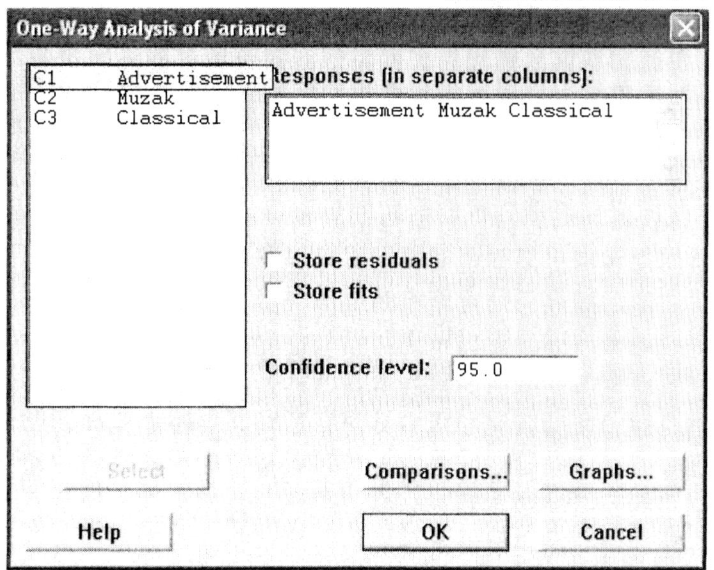

This gives you the following output in the Session window.

Notice the chart lists the degrees of freedom, the F statistic and the p-value.

Session

One-way ANOVA: Advertisement, Muzak, Classical

```
Source  DF     SS    MS     F      P
Factor   2  149.2  74.6  6.43  0.013
Error   12  139.2  11.6
Total   14  288.4

S = 3.406   R-Sq = 51.73%   R-Sq(adj) = 43.69%


                                Individual 95% CIs For Mean Based on
                                Pooled StDev
Level          N    Mean  StDev  --+---------+---------+---------+-------
Advertisement  5   5.400  4.159        (--------*-------)
Muzak          5   2.800  2.387  (-------*-------)
Classical      5  10.400  3.435                     (-------*-------)
                                 --+---------+---------+---------+-------
                                 0.0       4.0       8.0      12.0

Pooled StDev = 3.406
```

Chapter 14 Exercise 14.3 What's the best way to learn French?
ANOVA analysis

We are given data of quiz scores for three groups that are students studying French:
Group I: never studied a foreign language but have good English skills
Group II: never studied a foreign language but have poor English skills
Group III: Studied at least one other foreign language.

We enter the data as follows:

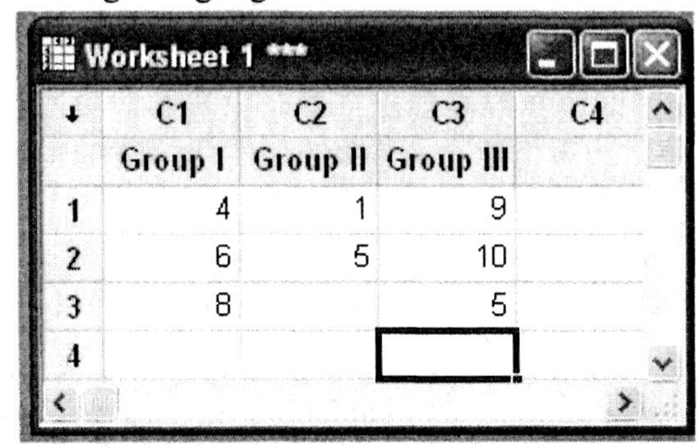

↓	C1	C2	C3	C4
	Group I	Group II	Group III	
1	4	1	9	
2	6	5	10	
3	8		5	
4				

Step 1: Assumptions: independent random samples, normal population distributions with equal standard deviations.

Step 2: Hypothesis: H_0: $\mu_1 = \mu_2 = \mu_3$

H_a: at least 2 of the population means are unequal

We then give the following commands to do the ANOVA analysis to find F and p:
Choose: **Stat > ANOVA >Oneway [Unstacked]**

Enter the variables as shown:

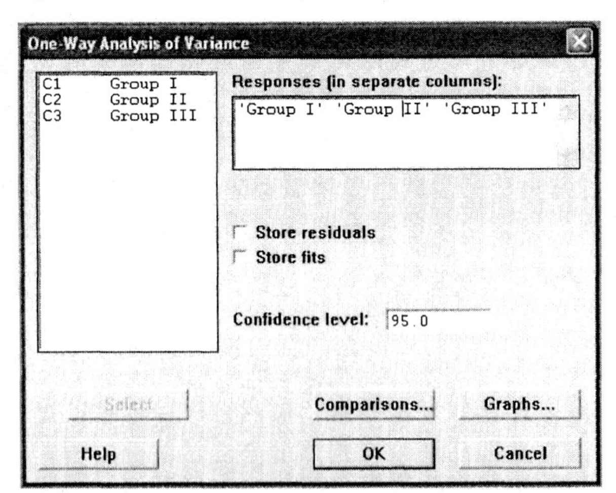

Step 3: F = 2.5

Step 4: p = 0.177

Step 5: Interpret.

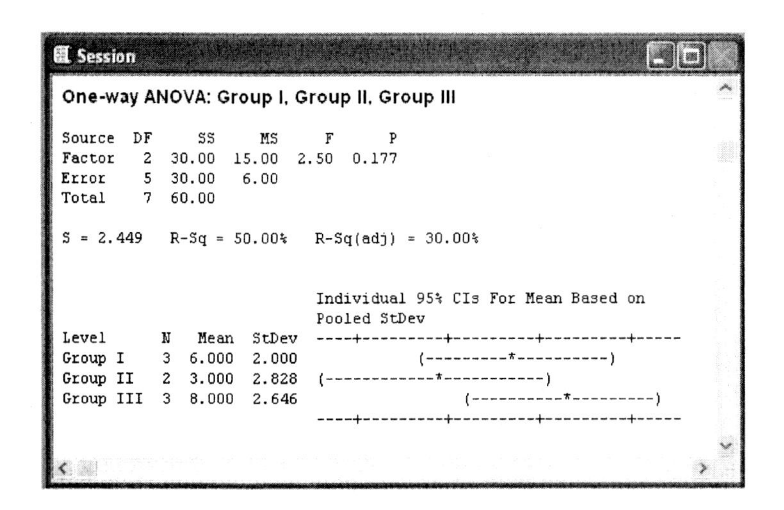

```
One-way ANOVA: Group I, Group II, Group III

Source  DF     SS     MS     F      P
Factor   2  30.00  15.00  2.50  0.177
Error    5  30.00   6.00
Total    7  60.00

S = 2.449   R-Sq = 50.00%   R-Sq(adj) = 30.00%

                                 Individual 95% CIs For Mean Based on
                                 Pooled StDev
Level      N   Mean  StDev  ----+---------+---------+---------+-----
Group I    3  6.000  2.000             (---------*----------)
Group II   2  3.000  2.828  (------------*-----------)
Group III  3  8.000  2.646                  (----------*---------)
                            ----+---------+---------+---------+-----
```

Chapter 14 Exercise 14.15 Telephone Holding Time Comparisons
The ANOVA F Test   Tukey method

We first must enter the data into the spreadsheet. The data is found in Example 2.

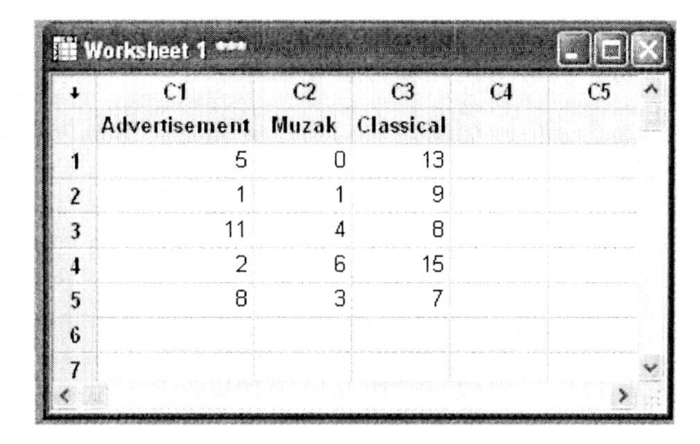

↓	C1 Advertisement	C2 Muzak	C3 Classical	C4	C5
1	5	0	13		
2	1	1	9		
3	11	4	8		
4	2	6	15		
5	8	3	7		
6					
7					

To do the ANOVA analysis   Choose:
Stat > ANOVA >Oneway [Unstacked]

Enter the variables:

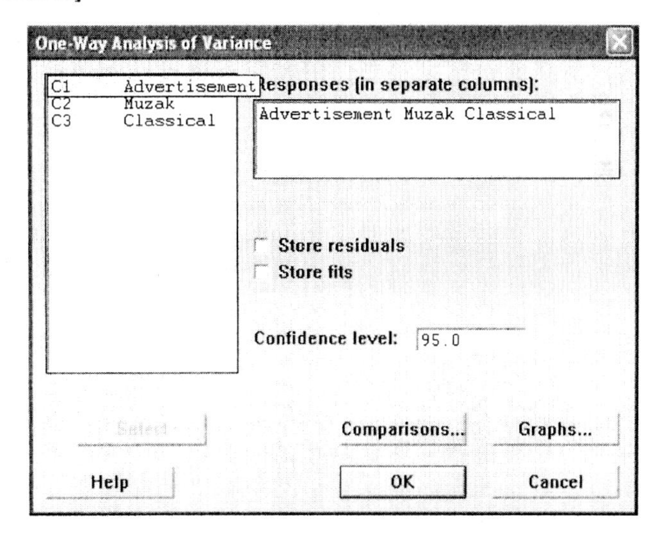

Select: **Comparisons**

Select: **Tukey's, family error rate**

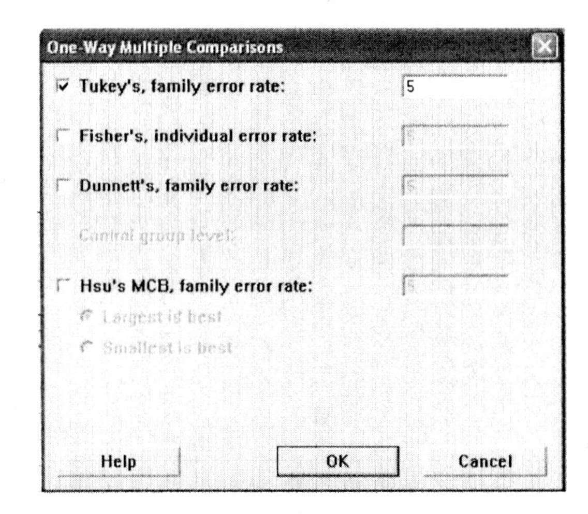

In the session window you will see the results.

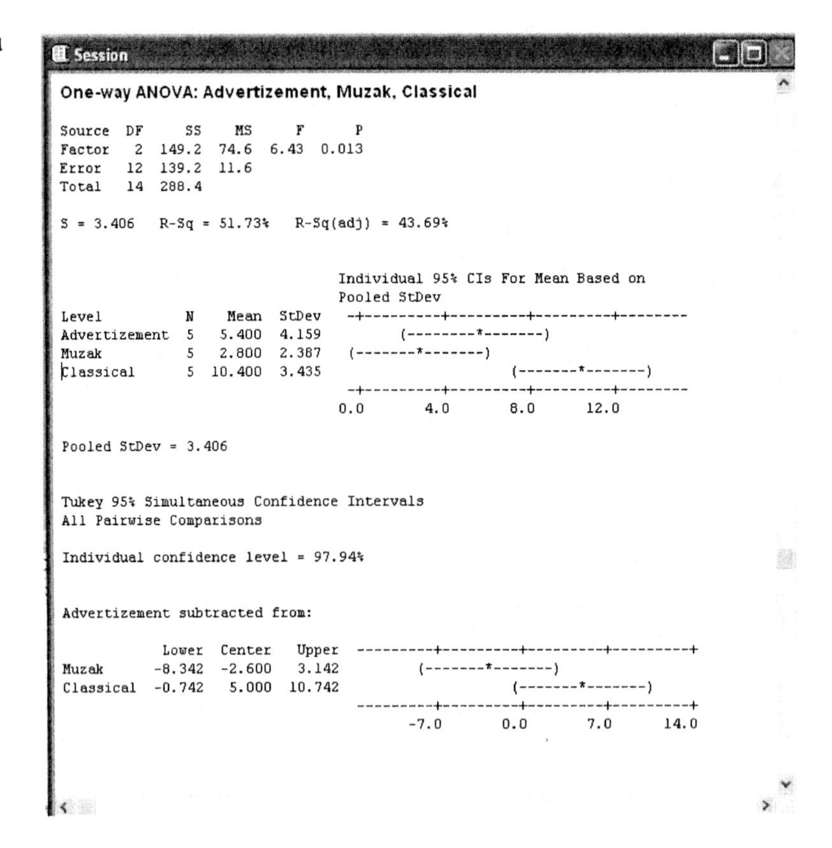

```
Session

One-way ANOVA: Advertizement, Muzak, Classical

Source  DF     SS    MS     F      P
Factor   2  149.2  74.6  6.43  0.013
Error   12  139.2  11.6
Total   14  288.4

S = 3.406   R-Sq = 51.73%   R-Sq(adj) = 43.69%


                                  Individual 95% CIs For Mean Based on
                                  Pooled StDev
Level          N    Mean  StDev   -+---------+---------+---------+--------
Advertizement  5   5.400  4.159          (--------*-------)
Muzak          5   2.800  2.387    (-------*-------)
Classical      5  10.400  3.435                      (-------*-------)
                                  -+---------+---------+---------+--------
                                  0.0      4.0       8.0      12.0

Pooled StDev = 3.406


Tukey 95% Simultaneous Confidence Intervals
All Pairwise Comparisons

Individual confidence level = 97.94%


Advertizement subtracted from:

             Lower  Center   Upper  ---------+---------+---------+---------+
Muzak       -8.342  -2.600   3.142         (-------*-------)
Classical   -0.742   5.000  10.742                    (-------*-------)
                                    ---------+---------+---------+---------+
                                          -7.0       0.0       7.0      14.0
```

Chapter 14 Exercise 14.21 French ANOVA
Comparing Means using Fisher and Tukey methods

We enter the data as shown:

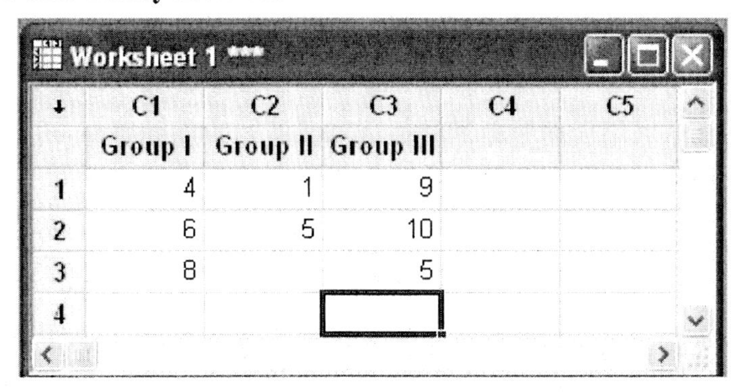

Worksheet 1 ***

↓	C1	C2	C3	C4	C5
	Group I	Group II	Group III		
1	4	1	9		
2	6	5	10		
3	8		5		
4					

We can do the ANOVA as before, checking the Fisher box under the Comparisons screen
Choose: **Stat > ANOVA >Oneway [Unstacked**

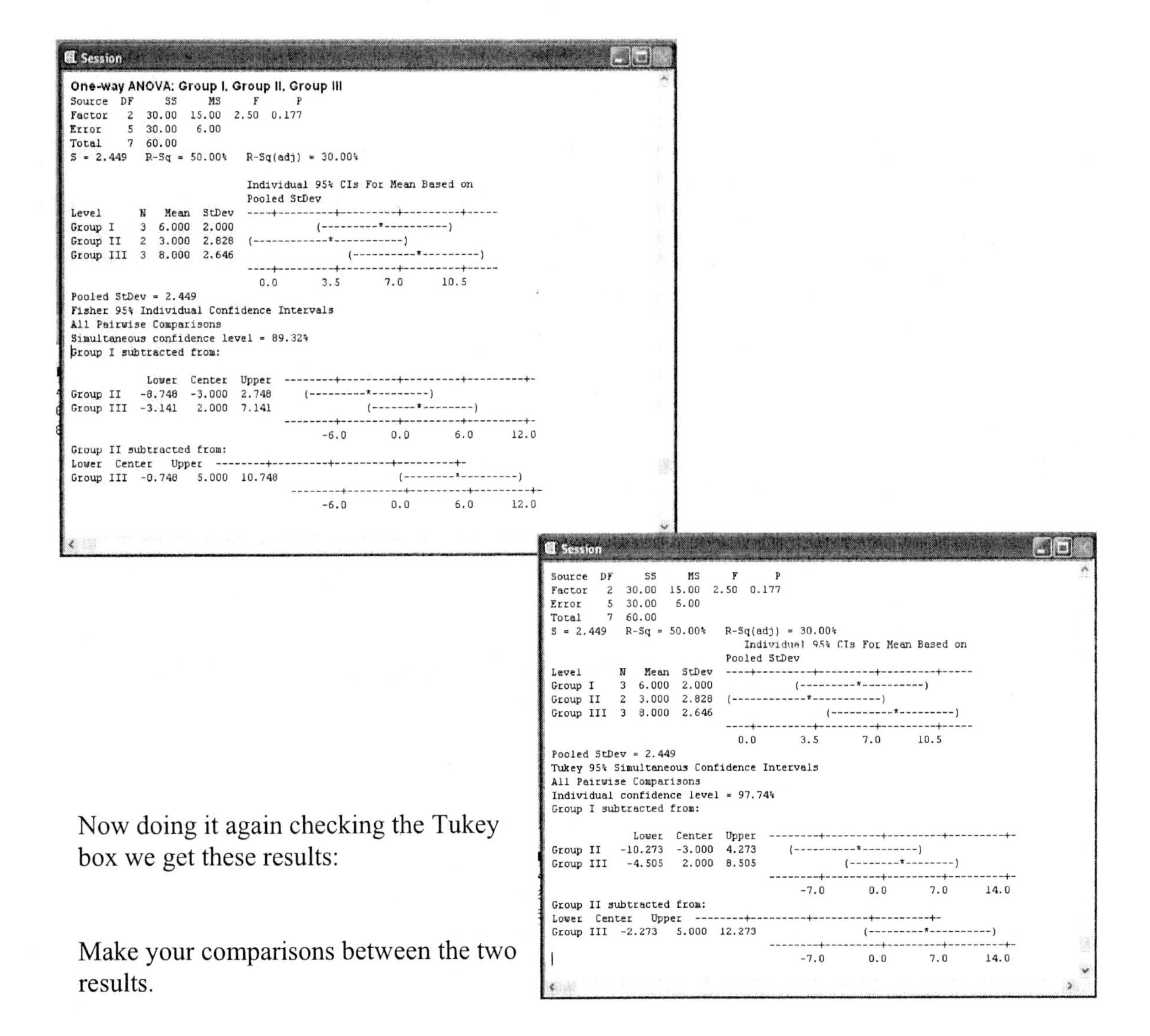

```
Session
One-way ANOVA: Group I, Group II, Group III
Source  DF     SS     MS     F      P
Factor   2  30.00  15.00  2.50  0.177
Error    5  30.00   6.00
Total    7  60.00
S = 2.449   R-Sq = 50.00%   R-Sq(adj) = 30.00%

                            Individual 95% CIs For Mean Based on
                            Pooled StDev
Level      N   Mean  StDev  ----+---------+---------+---------+-----
Group I    3  6.000  2.000              (---------*----------)
Group II   2  3.000  2.828  (------------*-----------)
Group III  3  8.000  2.646                   (----------*---------)
                            ----+---------+---------+---------+-----
                              0.0       3.5       7.0      10.5
Pooled StDev = 2.449
Fisher 95% Individual Confidence Intervals
All Pairwise Comparisons
Simultaneous confidence level = 89.32%
Group I subtracted from:

             Lower  Center  Upper  --------+---------+---------+---------+-
Group II    -8.748  -3.000  2.748    (---------*---------)
Group III   -3.141   2.000  7.141          (-------*--------)
                                   --------+---------+---------+---------+-
                                        -6.0       0.0       6.0      12.0
Group II subtracted from:
Lower  Center   Upper   --------+---------+---------+---------+-
Group III  -0.748   5.000  10.748             (--------*---------)
                                   --------+---------+---------+---------+-
                                        -6.0       0.0       6.0      12.0
```

```
Session
Source  DF     SS     MS     F      P
Factor   2  30.00  15.00  2.50  0.177
Error    5  30.00   6.00
Total    7  60.00
S = 2.449   R-Sq = 50.00%   R-Sq(adj) = 30.00%
                            Individual 95% CIs For Mean Based on
                            Pooled StDev
Level      N   Mean  StDev  ----+---------+---------+---------+-----
Group I    3  6.000  2.000              (---------*----------)
Group II   2  3.000  2.828  (------------*-----------)
Group III  3  8.000  2.646                   (----------*---------)
                            ----+---------+---------+---------+-----
                              0.0       3.5       7.0      10.5
Pooled StDev = 2.449
Tukey 95% Simultaneous Confidence Intervals
All Pairwise Comparisons
Individual confidence level = 97.74%
Group I subtracted from:

              Lower  Center  Upper  --------+---------+---------+---------+-
Group II    -10.273  -3.000  4.273    (----------*---------)
Group III    -4.505   2.000  8.505          (--------*--------)
                                    --------+---------+---------+---------+-
                                         -7.0       0.0       7.0      14.0
Group II subtracted from:
Lower  Center   Upper   --------+---------+---------+---------+-
Group III  -2.273   5.000  12.273             (---------*----------)
                                    --------+---------+---------+---------+-
                                         -7.0       0.0       7.0      14.0
```

Now doing it again checking the Tukey box we get these results:

Make your comparisons between the two results.

Chapter 14 Example 9 Testing the Main Effects for Corn Yield
Two Way ANOVA

We first enter the data as shown:

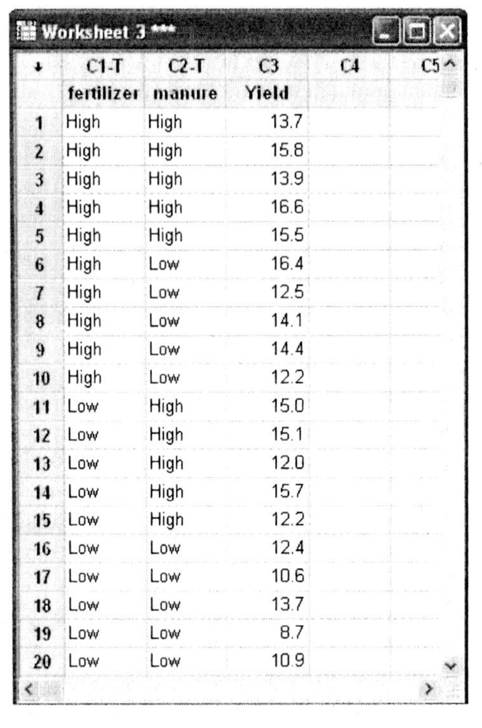

Worksheet 3 ***

↓	C1-T fertilizer	C2-T manure	C3 Yield	C4	C5
1	High	High	13.7		
2	High	High	15.8		
3	High	High	13.9		
4	High	High	16.6		
5	High	High	15.5		
6	High	Low	16.4		
7	High	Low	12.5		
8	High	Low	14.1		
9	High	Low	14.4		
10	High	Low	12.2		
11	Low	High	15.0		
12	Low	High	15.1		
13	Low	High	12.0		
14	Low	High	15.7		
15	Low	High	12.2		
16	Low	Low	12.4		
17	Low	Low	10.6		
18	Low	Low	13.7		
19	Low	Low	8.7		
20	Low	Low	10.9		

The two factors we are considering are fertilizer levels (High, Low) and Manure levels (High, Low). We are given five sample yields for each level, for a total of twenty data points.

To perform a two-way ANOVA ,
Choose **Stat > ANOVA > Two Way**

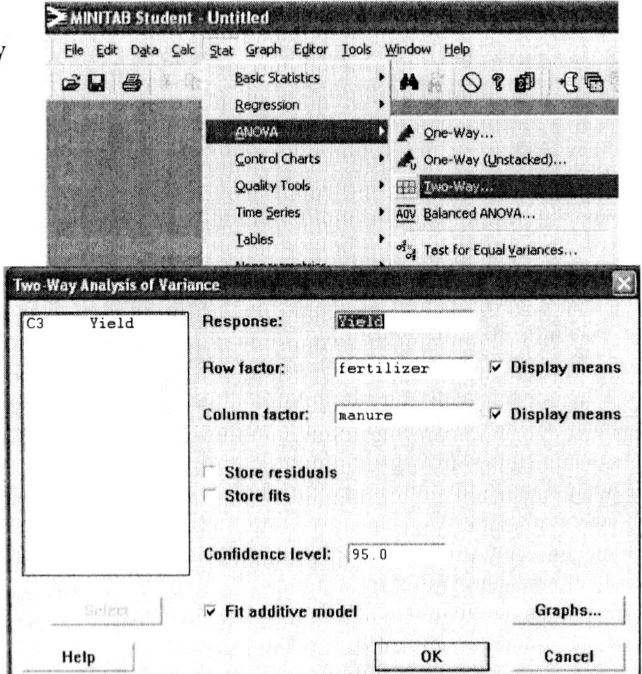

Enter the response, Row Factor, Column Factor as indicated.
Check the Fit Additive model box.

The results of the ANOVA table will be in the session window.

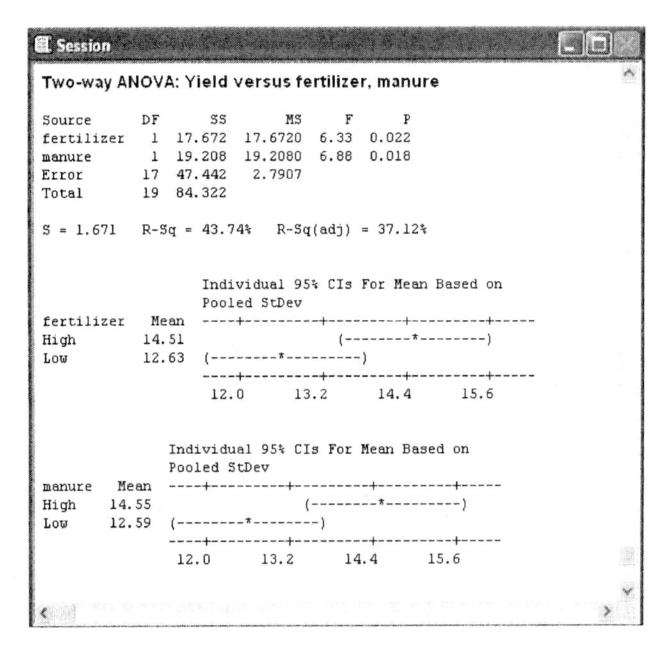

Session

Two-way ANOVA: Yield versus fertilizer, manure

```
Source      DF      SS       MS     F      P
fertilizer   1  17.672  17.6720  6.33  0.022
manure       1  19.208  19.2080  6.88  0.018
Error       17  47.442   2.7907
Total       19  84.322

S = 1.671   R-Sq = 43.74%   R-Sq(adj) = 37.12%

                   Individual 95% CIs For Mean Based on
                   Pooled StDev
fertilizer  Mean   ----+---------+---------+---------+-----
High       14.51                   (--------*--------)
Low        12.63   (--------*---------)
                   ----+---------+---------+---------+-----
                     12.0      13.2      14.4      15.6

               Individual 95% CIs For Mean Based on
               Pooled StDev
manure  Mean   ----+---------+---------+---------+-----
High   14.55                   (--------*---------)
Low    12.59   (--------*--------)
               ----+---------+---------+---------+-----
                 12.0      13.2      14.4      15.6
```

Note that this example is assuming no interaction. To see if this is a valid assumption, redo the ANOVA table to allow for interaction in assessing the effects of fertilizer level and manure level in the mean corn yield. We can do this simply by unchecking **the Fit Additive model box**. This tests the hypothesisH_0: There is no interaction. This results in the following chart:

Based on the test statistic F = 1.10, and the reported p-value = 0.31, there is not much evidence of interaction so we would not reject H_0 .

In doing a Two Way ANOVA in Minitab Student, you must have the same number of observations in each cell of the contingency table. In this example we had five observations in each of the four cells of the table.

Session

Two-way ANOVA: Yield versus fertilizer, manure

```
Source       DF      SS      MS     F      P
fertilizer    1  17.672  17.672  6.37  0.023
manure        1  19.208  19.208  6.92  0.018
Interaction   1   3.042   3.042  1.10  0.311
Error        16  44.400   2.775
Total        19  84.322

S = 1.666   R-Sq = 47.34%   R-Sq(adj) = 37.47%

                   Individual 95% CIs For Mean Based on
                   Pooled StDev
fertilizer  Mean   ----+---------+---------+---------|-----
High       14.51                   (--------*--------)
Low        12.63   (--------*---------)
                   ----+---------+---------+---------+-----
                     12.0      13.2      14.4      15.6

               Individual 95% CIs For Mean Based on
               Pooled StDev
manure  Mean   ----+---------+---------+---------+-----
High   14.55                   (--------*---------)
Low    12.59   (--------*--------)
|              ----+---------+---------+---------+-----
```

Chapter 14 Example 10 Regression Modeling to Estimate and Compare Mean Corn Yields

To do the regression analysis, we must use numeric data to represent High (1) and Low (0). The worksheet now looks like this:

Worksheet 3 ***

↓	C1-T fertilizer	C2-T manure	C3 Yield	C4 F	C5 M
1	High	High	13.7	1	1
2	High	High	15.8	1	1
3	High	High	13.9	1	1
4	High	High	16.6	1	1
5	High	High	15.5	1	1
6	High	Low	16.4	1	0
7	High	Low	12.5	1	0
8	High	Low	14.1	1	0
9	High	Low	14.4	1	0
10	High	Low	12.2	1	0
11	Low	High	15.0	0	1
12	Low	High	15.1	0	1
13	Low	High	12.0	0	1
14	Low	High	15.7	0	1
15	Low	High	12.2	0	1
16	Low	Low	12.4	0	0
17	Low	Low	10.6	0	0
18	Low	Low	13.7	0	0
19	Low	Low	8.7	0	0
20	Low	Low	10.9	0	0

To perform the regression Choose: **Stat > Regression > Regression**

Enter Response: **Yield**
Predictors: **F M**
OK

The chart will appear in the session window.

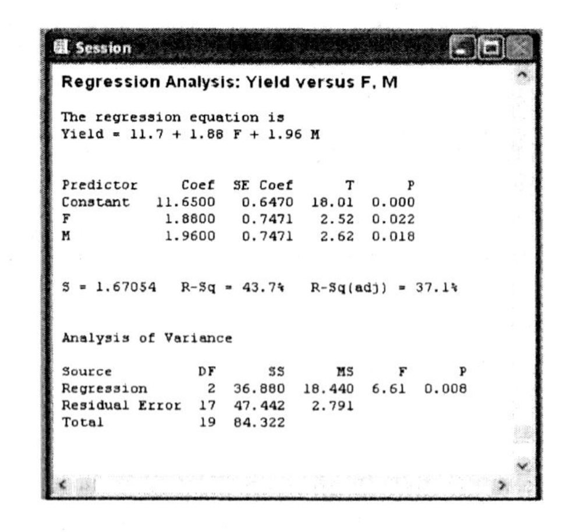
Session

Regression Analysis: Yield versus F, M

The regression equation is
Yield = 11.7 + 1.88 F + 1.96 M

Predictor	Coef	SE Coef	T	P
Constant	11.6500	0.6470	18.01	0.000
F	1.8800	0.7471	2.52	0.022
M	1.9600	0.7471	2.62	0.018

S = 1.67054 R-Sq = 43.7% R-Sq(adj) = 37.1%

Analysis of Variance

Source	DF	SS	MS	F	P
Regression	2	36.880	18.440	6.61	0.008
Residual Error	17	47.442	2.791		
Total	19	84.322			

Chapter 14 Exercise 14.33 Diet and Weight Gain
Two Way ANOVA, with and without interaction.

Enter the data into the worksheet either by hand or opening the file **protein_and_weight_gain** worksheet from the MINITAB folder. A portion is shown here. The file is 60 rows by 3 columns. Note that there are ten observations in each cell, so we have a balanced design.

protein_and_weight_gain.M...

↓	C1-T	C2-T	C3
	source	level	weight_gain
1	beef	high	73
2	beef	high	102
3	beef	high	118
4	beef	high	104
5	beef	high	81
6	beef	high	107
7	beef	high	100
8	beef	high	87
9	beef	high	117
10	beef	high	111

a) Conduct an ANOVA without interaction. Choose **Stat > ANOVA > Two Way ...** and enter the variables as shown

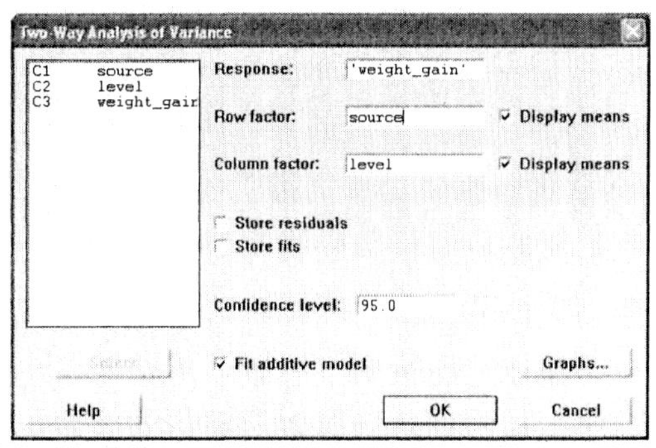

The results will be in the session window.

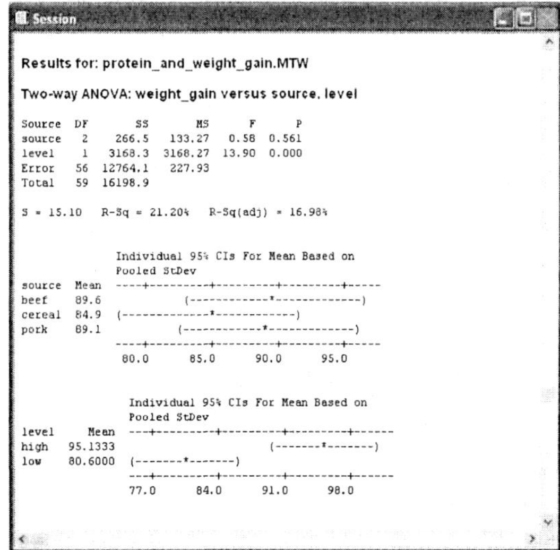

Session
Results for: protein_and_weight_gain.MTW
Two-way ANOVA: weight_gain versus source, level
Source DF SS MS F P
source 2 266.5 133.27 0.58 0.561
level 1 3168.3 3168.27 13.90 0.000
Error 56 12764.1 227.93
Total 59 16198.9
S = 15.10 R-Sq = 21.20% R-Sq(adj) = 16.98%
Individual 95% CIs For Mean Based on Pooled StDev
source Mean
beef 89.6
cereal 84.9
pork 89.1
80.0 85.0 90.0 95.0
Individual 95% CIs For Mean Based on Pooled StDev
level Mean
high 95.1333
low 80.6000
77.0 84.0 91.0 98.0

b) Conduct an ANOVA with interaction.
Choose **Stat > ANOVA > Two Way …** and enter the variables as appropriate.
This time uncheck the box for Fit Additive Model. Again the results will be in the session window.

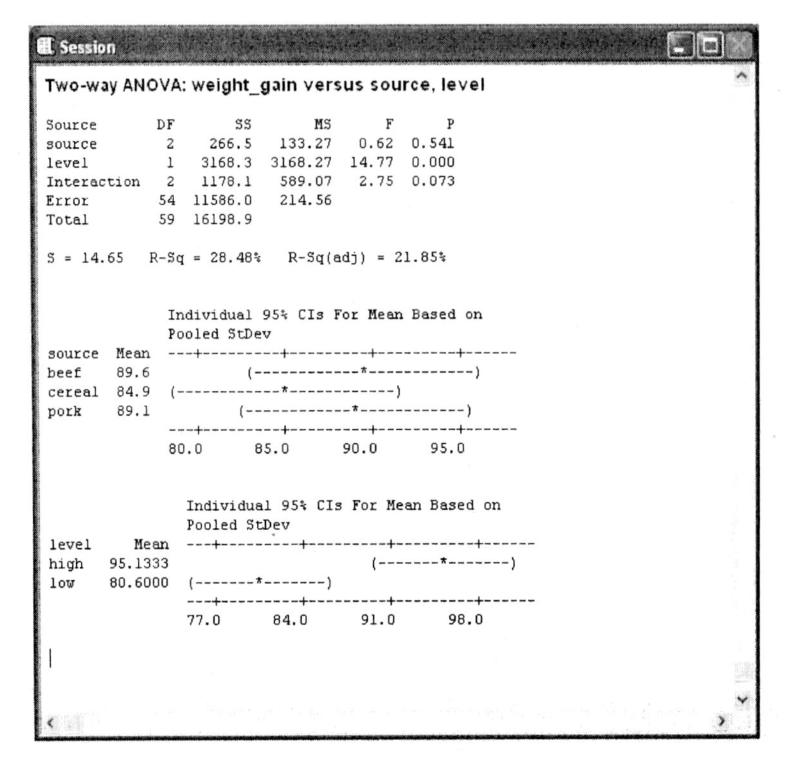

Chapter 15 Example 4 Estimating the Difference between Median Reaction Times Two sample Wilcoxon Test (Mann-Whitney)

You can perform a two-sample Wilcoxon rank sum test of the equality of two population medians, and calculate the corresponding point estimate and confidence interval. MINITAB calls this the Mann-Whitney test.

The hypotheses are: H_0: $\eta_1 = \eta_2$ versus H_1: $\eta_1 \neq \eta_2$, where η is the population median.

Open the worksheet entitled **cell_phones** from the MINITAB folder of the data disk.

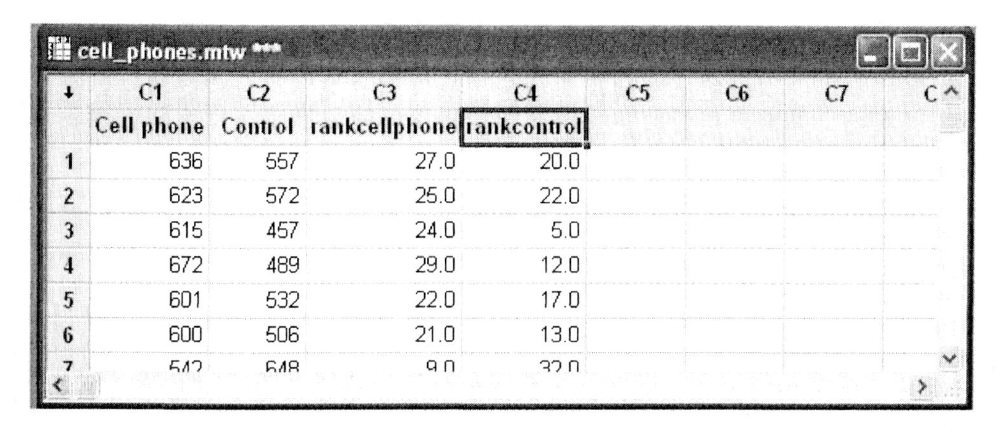

cell_phones.mtw ***

↓	C1	C2	C3	C4	C5	C6	C7
	Cell phone	Control	rankcellphone	rankcontrol			
1	636	557	27.0	20.0			
2	623	572	25.0	22.0			
3	615	457	24.0	5.0			
4	672	489	29.0	12.0			
5	601	532	22.0	17.0			
6	600	506	21.0	13.0			

To perform the test Choose **Stat > Nonparametrics > Mann- Whitney**

Enter the variables as shown:

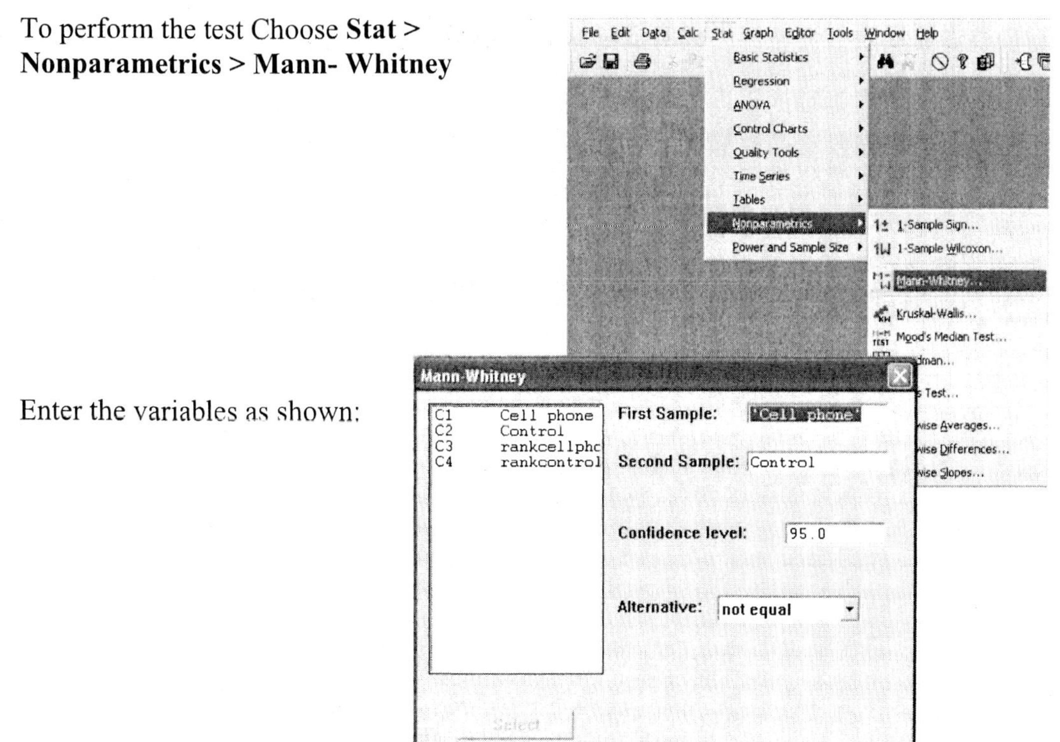

The results are in the session window.

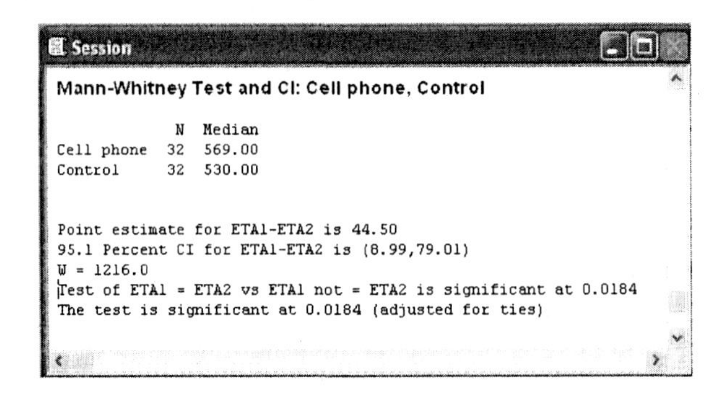

Chapter 15 Exercise 15.5 Estimating Hypnosis Effect
Two sample Wilcoxon Test (Mann-Whitney)

Enter the data into the worksheet:

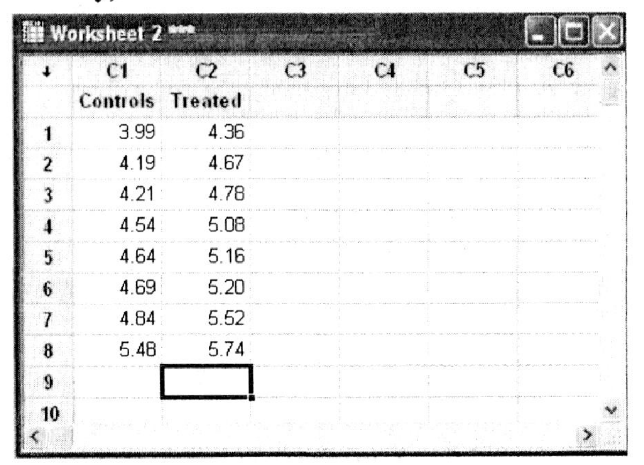

	C1 Controls	C2 Treated
1	3.99	4.36
2	4.19	4.67
3	4.21	4.78
4	4.54	5.08
5	4.64	5.16
6	4.69	5.20
7	4.84	5.52
8	5.48	5.74

To perform the test Choose **Stat > Nonparametrics > Mann- Whitney**

Enter the columns, leaving the default confidence level and alternatives.

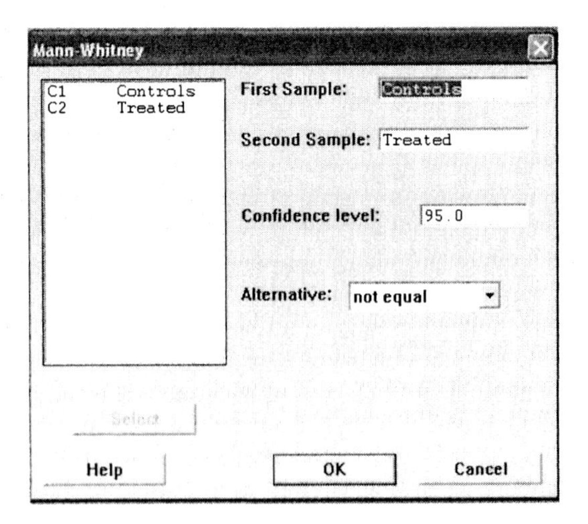

The results appear in the Session window.

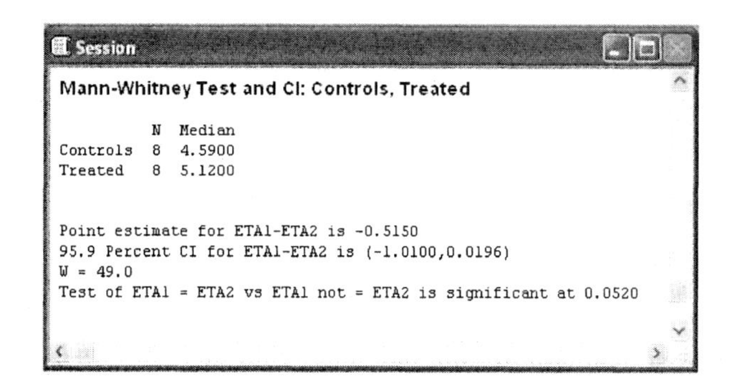

Chapter 15 Exercise 15.7 Teenage Anorexia
Two – Way Wilcoxon (Mann-Whitney)

Open the worksheet entitled **anorexia** from the MINITAB folder of the data disk.

To perform the test: Choose **Stat > Nonparametrics > Mann-Whitney**
And choose the columns appropriately:

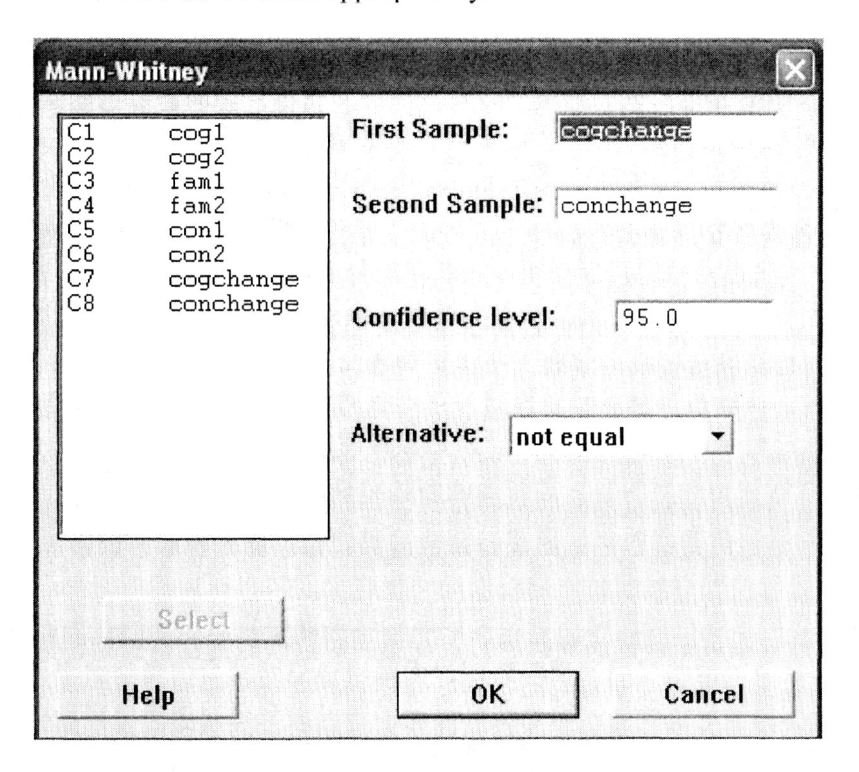

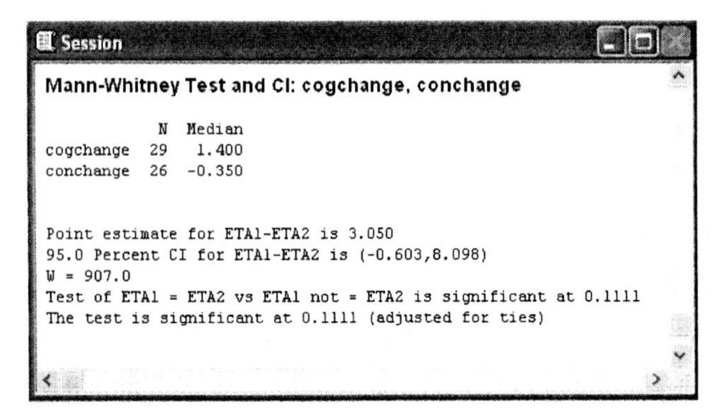

Note the point estimate = 3.050 and the confidence interval is (-0.603, 8.098).

The hypotheses are: H_0: $\eta_1 = \eta_2$ versus H_1: $\eta_1 \neq \eta_2$, where η is the population median. The test statistic W = 907 has a p-value of 0.111 when adjusted for ties. What conclusions can you draw?

Chapter 15 Example 5 Does Heavy Dating Affect College GPA? Kruskal-Wallis Test

The Kruskal-Wallis hypotheses are:
H_0: the population medians are all equal versus H_1: the medians are not all equal

First enter the GPA data in column 1. Then you can rank the data by choosing **Data > Rank** and entering the ranks into Column 2. Enter the dating factors into column 3.

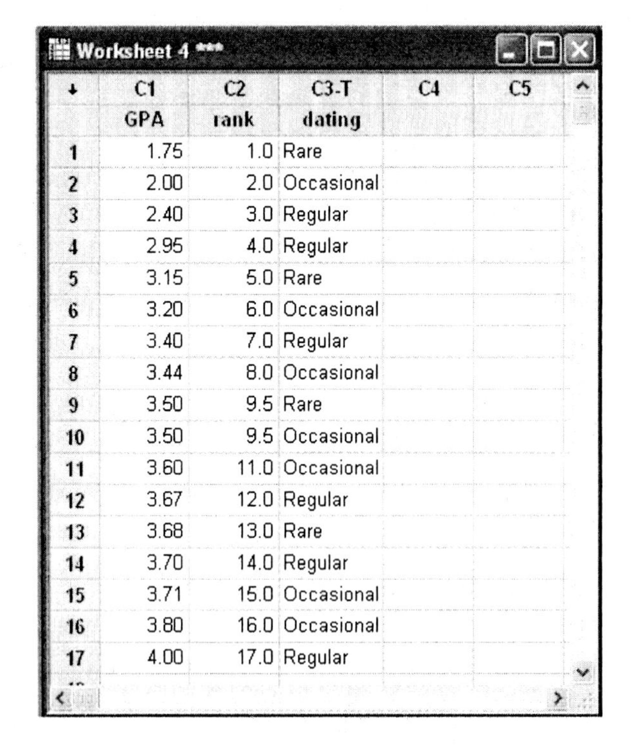
Worksheet 4 ***

	C1 GPA	C2 rank	C3-T dating	C4	C5
1	1.75	1.0	Rare		
2	2.00	2.0	Occasional		
3	2.40	3.0	Regular		
4	2.95	4.0	Regular		
5	3.15	5.0	Rare		
6	3.20	6.0	Occasional		
7	3.40	7.0	Regular		
8	3.44	8.0	Occasional		
9	3.50	9.5	Rare		
10	3.50	9.5	Occasional		
11	3.60	11.0	Occasional		
12	3.67	12.0	Regular		
13	3.68	13.0	Rare		
14	3.70	14.0	Regular		
15	3.71	15.0	Occasional		
16	3.80	16.0	Occasional		
17	4.00	17.0	Regular		

To perform the test choose **Stats> Nonparametric> Kruskal-Wallis** and enter the response and Factor as shown:

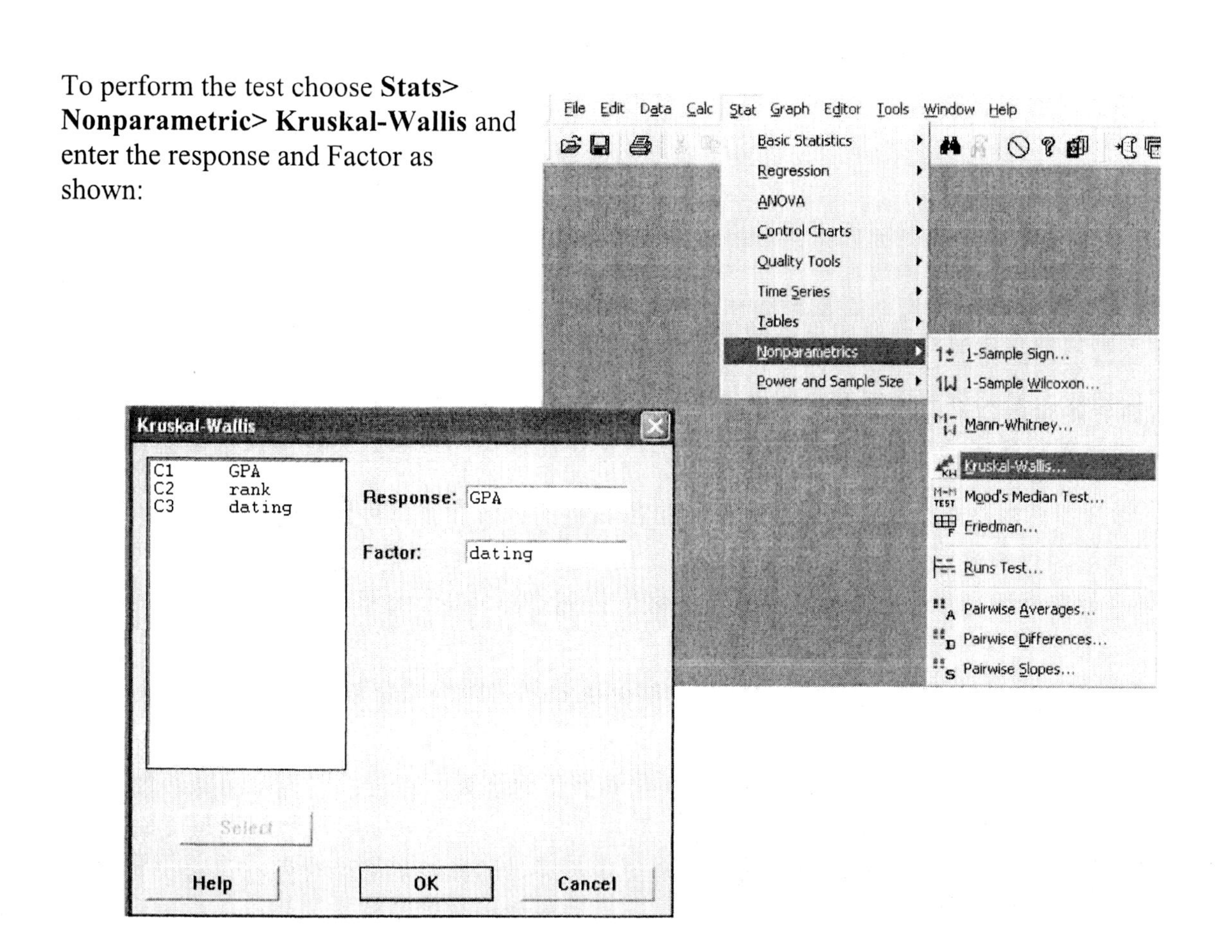

The results are shown in the session window.

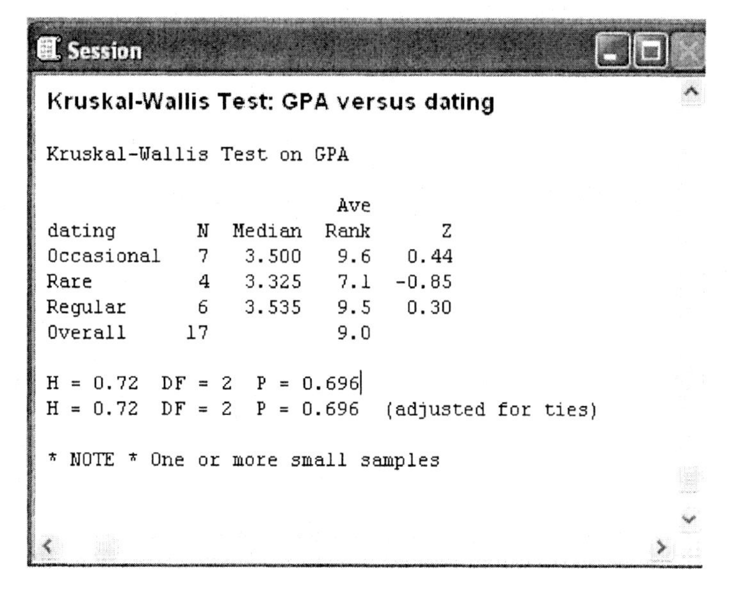

Chapter 15 Example 6 Spend More Time Browsing the Internet or Watching TV? Redone with The Wilcoxin Signed Ranks Test

Open the worksheet for the **georgia_student_survey** from the MINITAB folder of the data disk. This is a large data set, but we are only interested in two columns: "Watching TV" and "Browse Internet". We want to look at the paired differences between these two columns. To create the column of paired differences choose **Calc> Calculator** and enter the dialog boxes as shown. A section of the worksheet is also shown.

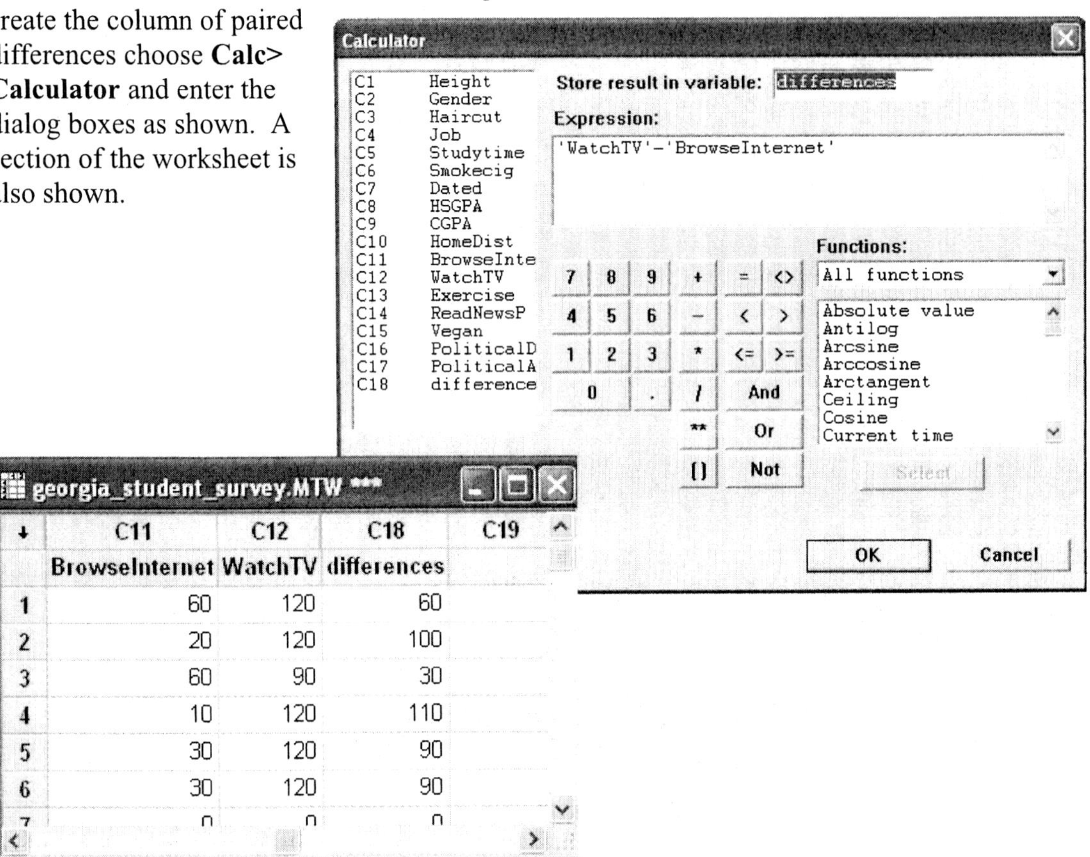

To perform the test choose

Stat > Nonparametrics > One-Sample Wilcoxon

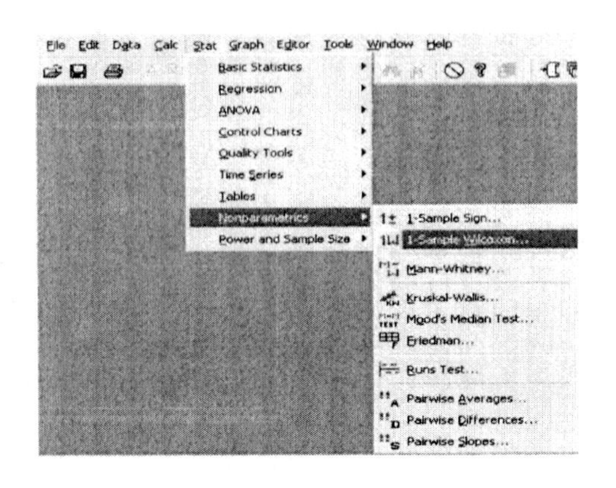

Enter the dialog boxes as shown.

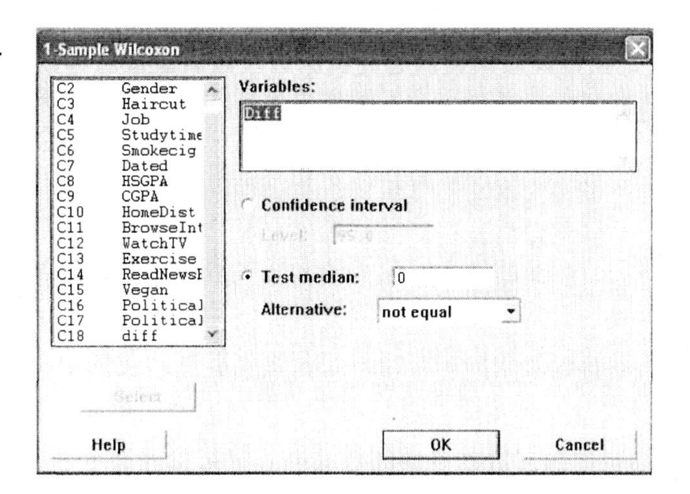

The results as in the session window.

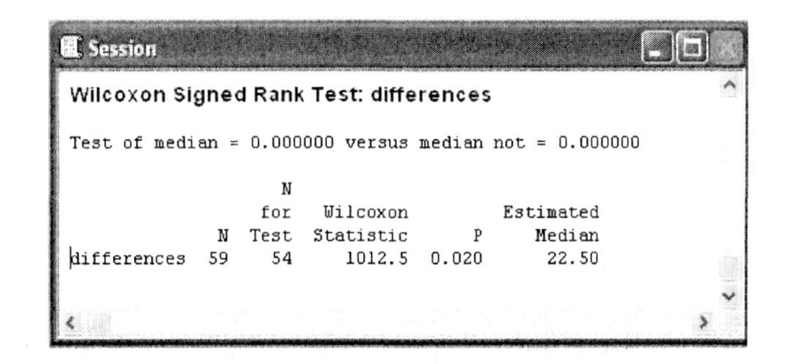

```
Wilcoxon Signed Rank Test: differences

Test of median = 0.000000 versus median not = 0.000000

                      N
                     for   Wilcoxon              Estimated
                N   Test  Statistic       P         Median
differences    59     54     1012.5   0.020          22.50
```

Chapter 15 Exercise 15.8 How Long do You Tolerate Being Put on Hold? Kruskal Wallis Test

The Kruskal-Wallis hypotheses are
H_0: the population medians are all equal versus H_1: the medians are not all equal

First enter the GPA data in column 1.

You can rank the data by choosing **Data > Rank** and entering the ranks into Column 2. Enter the message types into column 3.

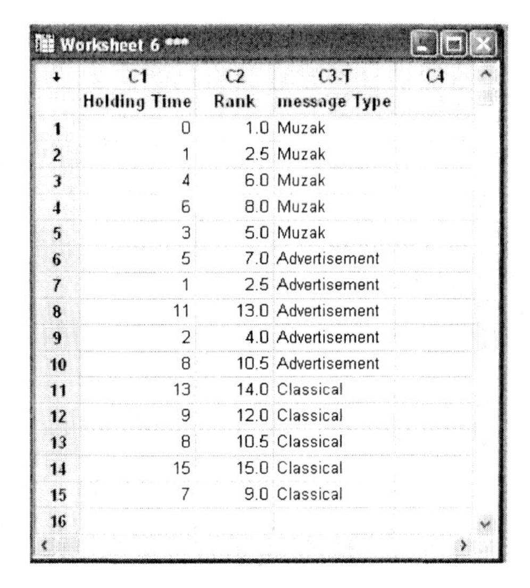

↓	C1	C2	C3-T	C4
	Holding Time	Rank	message Type	
1	0	1.0	Muzak	
2	1	2.5	Muzak	
3	4	6.0	Muzak	
4	6	8.0	Muzak	
5	3	5.0	Muzak	
6	5	7.0	Advertisement	
7	1	2.5	Advertisement	
8	11	13.0	Advertisement	
9	2	4.0	Advertisement	
10	8	10.5	Advertisement	
11	13	14.0	Classical	
12	9	12.0	Classical	
13	8	10.5	Classical	
14	15	15.0	Classical	
15	7	9.0	Classical	
16				

To perform the test choose

Stats> Nonparametric> Kruskal-Wallis

and enter the response and Factor as shown

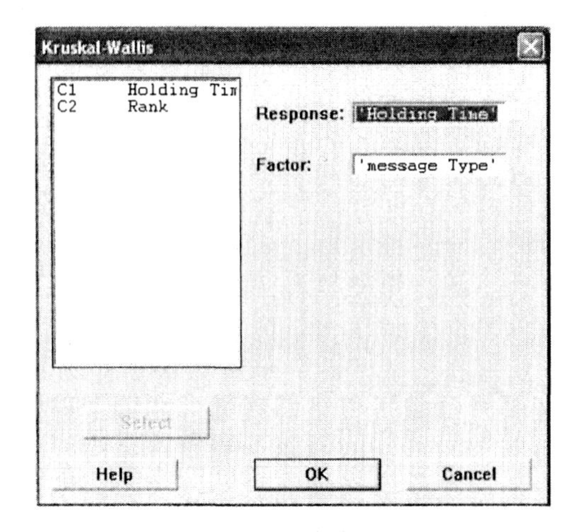

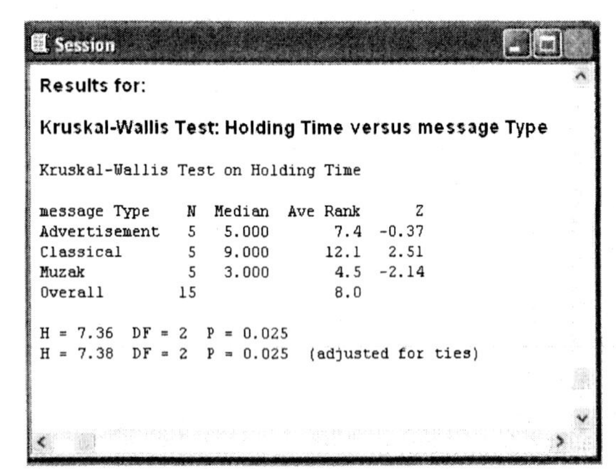

Session

Results for:

Kruskal-Wallis Test: Holding Time versus message Type

```
Kruskal-Wallis Test on Holding Time

message Type    N  Median  Ave Rank      Z
Advertisement   5   5.000       7.4  -0.37
Classical       5   9.000      12.1   2.51
Muzak           5   3.000       4.5  -2.14
Overall        15               8.0

H = 7.36  DF = 2  P = 0.025
H = 7.38  DF = 2  P = 0.025  (adjusted for ties)
```

The sample medians for the three treatments were calculated 5.0,9.0,3.0. The test statistic (H) had a p-value of 0.0125 both unadjusted and adjusted for ties, indicating that the null hypothesis can be rejected at levels higher than 0.025 in favor of the alternative hypothesis of at least one difference among the treatment groups.

Chapter 15 Exercise 15.16 More on cell phones
Wilcoxon Signed Rank Test

Using the data in Example 12 from Chapter 10, create the worksheet. A small segment of the worksheet is shown here:

REACTION_TIME_CELL_PHONE_USE.MTW ***

↓	C1	C2	C3	C4	C5
	Student	No	Yes	Difference	
1	1	604	636	32	
2	2	556	623	67	
3	3	540	615	75	
4	4	522	672	150	
5	5	459	601	142	
6	6	544	600	56	
7	7	513	542	29	

To perform the test on this data choose

Stat> Nonparametrics > 1 - Sample Wilcoxon… and enter the dialog boxes as shown:

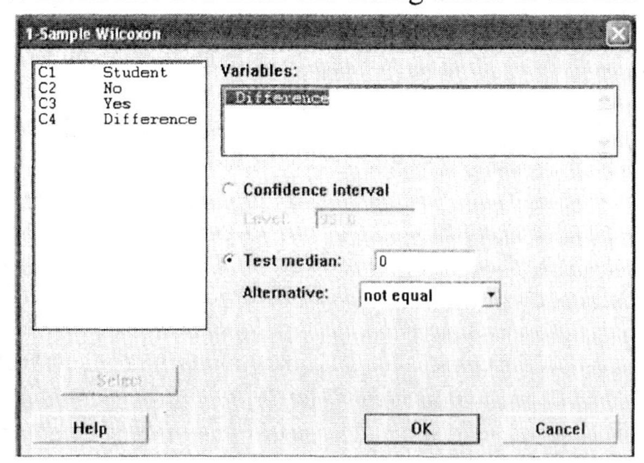

The results are shown in the session window.

Note the hypothesis in the box. It is testing whether the population median of difference scores is 0.

The Wilcoxon test statistic is the sum of ranks for the positive differences. Note the P-Value and the Estimated Median.

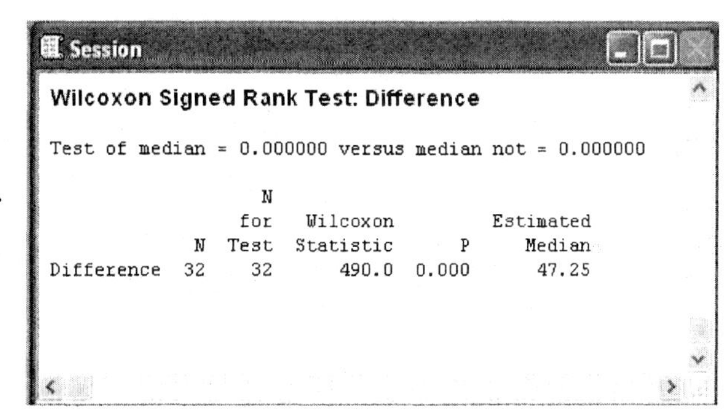

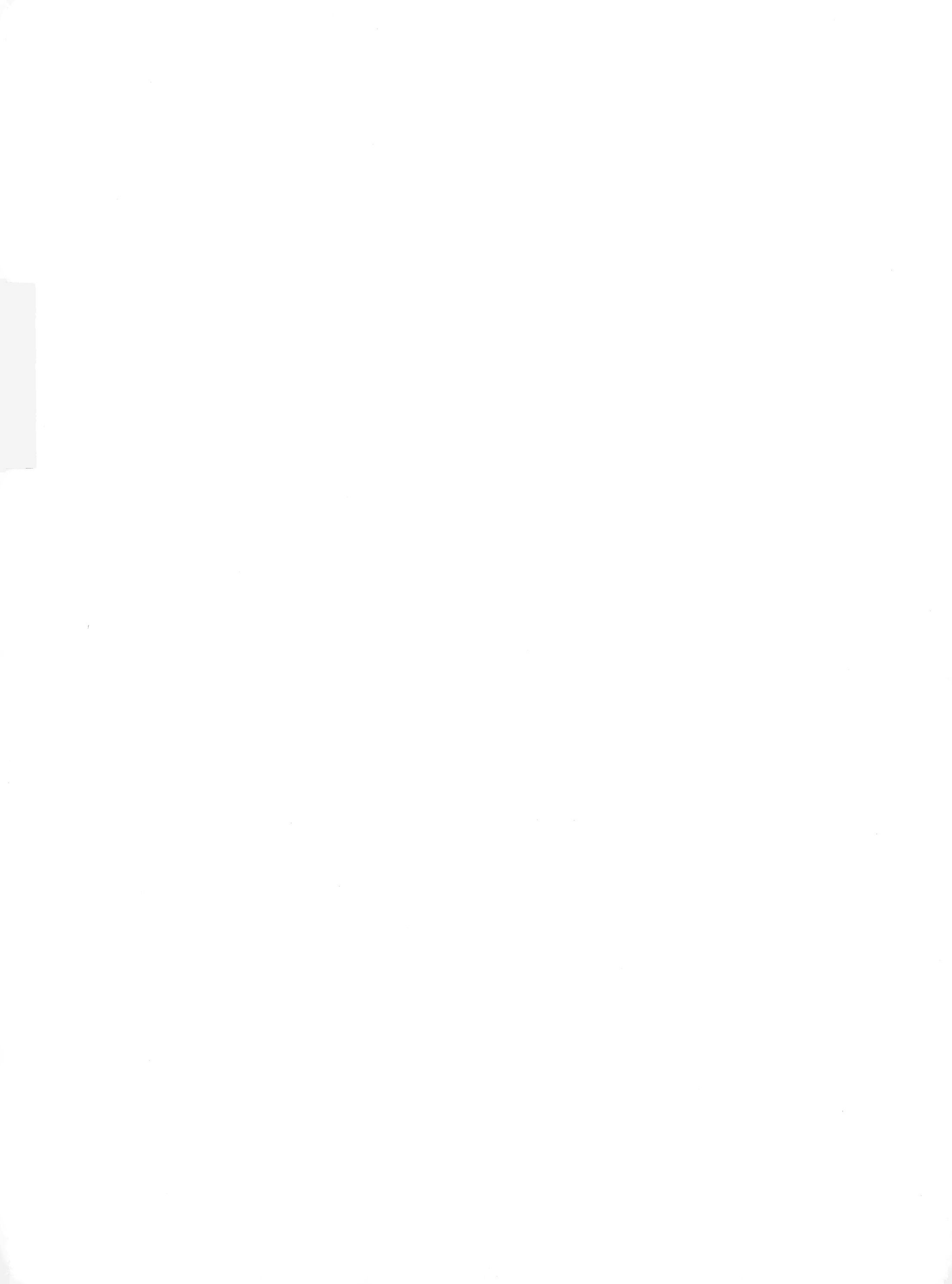

TI-83/84 MANUAL

Peter Flanagan-Hyde

Phoenix Country Day School

SECOND EDITION

STATISTICS

THE ART AND SCIENCE OF LEARNING FROM DATA

Agresti • Franklin

Upper Saddle River, NJ 07458

Table of Contents

Chapter One: Introduction to the TI-83 and TI-84

This chapter covers some basic information to get you started using your TI-83 or TI-84 calculator. Subsequent chapters will use examples and exercises from *Statistics: The Art and Science of Learning from Data*, Second Edition, by Alan Agresti and Chris Franklin, to show how to implement each of the statistical methods with your calculator.

Introduction to the TI-83 and TI-84 families of calculators

The calculators in these two families include the TI-83, TI-83 Plus, TI-83 Plus Silver Edition, TI-84 Plus, the TI-84 Plus Silver Edition, and the TI Inspire with TI-84 Plus keypad. These are all keystroke-compatible and can readily exchange data. The "Plus" calculators include additional memory space (archive) that allows data and programs to be set aside, built-in APPS (application programs), and improved statistical software. The "Silver Edition" calculators have larger memory capacity and faster processors. The 84 series has a few additional statistical options and APPS, and it adds a USB connection port for better connectivity. In this manual the calculator will be assumed to be a TI-83 Plus or better. All screenshots in this manual were done with a TI-84 Plus Silver Edition. Users of a TI-83 (no longer sold) will be alerted when a feature is used that is not present on that calculator.

This manual supplements the Agresti/Franklin text, covering the topics from the textbook. The guidebook that came with your calculator includes more information about other aspects of your calculator and provides additional examples. Free downloads of calculator guidebooks are available from education.ti.com/us/global/guides.html.

Other models

Statistics applications for the TI-89 family of calculators are available with the addition of the Stats/List Editor APP, available at education.ti.com/us/product/apps/statsle.html. Once this is loaded, the TI-89 will be able to perform all the statistical functions of the TI-83 family. Most of the instructions in this manual can be followed with the TI-89, but no specific screenshots or instructions will be provided.

Other TI calculators, such as the TI-85 and TI-86, were not designed with statistics in mind. There are third-party solutions that enable features similar to those of the TI-83 family, but they are not described in this manual.

Keyboard Basics

Once the calculator is turned on, the cursor will appear as a blinking solid square on the screen. The primary key functions are printed on the buttons themselves. Touching most of the keys on the keyboard types the character or function on the screen. Some of the keys, such as **MATH**, **PGRM**, or **VARS**, produce a menu with more options instead of a character or function.

The secondary key functions are above each of the button to the left, printed in blue (TI-84) or yellow (TI-83). Press the **2nd** key to access these functions or menus—you'll see the cursor change to an up-arrow. The **ALPHA** key functions are above most of the buttons to the right, printed in green. Press the **ALPHA** key to access these functions—you'll see the cursor change to include an inset **A**.

Once you've typed on the screen, you can move the cursor back and forth with the left and right cursor keys. Typing with the blinking square cursor overwrites any characters to the right of the cursor. Pressing **DEL** deletes the character at the cursor. You may insert characters without overwriting by pressing **INS** (**2nd DEL**). In this mode, the cursor will appear as a blinking underline. Pressing **CLEAR** will remove the active line of text or clear the entire screen if there is nothing on the active line.

Home Screen, Graph Screen, and other screens

The home screen is the default screen where calculations are performed. From any other screen, pressing **QUIT** (**2nd MODE**) or **CLEAR** from other menus and screens will get you back to the home screen.

The graph screen is where the calculator displays graphs of functions, statistical graphical displays, and other drawings. The function set up screen (**Y=**) allows you to enter a function to be graphed, and the **STAT PLOT** screen (**2nd Y=**) allows you to set up statistical graphs. The graph window (**WINDOW**) lets you specify the parameters for the graph screen. Other screens, such as **MODE**, **TABLE**, **ZOOM**, and **FORMAT** allow you to control other aspects of the calculator. As these are encountered their use will be described in this manual.

Miscellaneous Tips

- The calculator distinguishes between a negative sign and the subtraction arithmetic operator. For subtracting a value or expression from another, use the – key, found on the right side of the keypad, between **+** and **×**. To designate a value or expression as negative, you must use the negative sign key **(-)**, found to the left of **ENTER**. The two keys are not interchangeable.
- **ENTRY** (**2nd ENTER**) can be used to call up previous expressions when using the home screen. This can save you many keystrokes. Repeatedly pressing **2nd ENTER** takes you back through your most recent entries.
- **ANS** (**2nd (-)**) can be used to call up the most recently calculated value. This not only saves keystrokes – it also minimizes rounding error.
- The lightness and darkness of the screen is controlled by **2nd** (**up arrow**) and **2nd** (**down arrow**). Up is darker, down is lighter.
- **CATALOG (2nd 0)** lists all calculator functions in alphabetical order. Pressing any of the alphabetic keys jumps to that portion of the list.
- The **APP** key accesses applications, such as **ProbSim** that allows you to roll dice, flip coins, and produce other random numbers for simulations, and **CtlgHelp** that displays the sequence of function inputs needed for various commands.

Managing Data and Memory

The TI-83 Plus stores statistical data in lists. The data can be put into a list either by directly typing it, by copying a list from another calculator (using a link cable), or by copying a list from a computer (using a GraphLink or USB cable and TI-Connect software). There are six default lists, labeled **L1** through **L6**, that are directly accessible from the keyboard (**2nd 1**, etc.). Other lists can be given names that indicate what they represent, using the **ALPHA** keys.

The number of lists that can be held by the calculator is limited by its memory (longer lists, with up to 999 numbers, take up more memory). All calculators in the TI-83 Plus family have 24k of active memory (RAM) and additional, larger, archive memory. A list must be in the active memory to be used or processed. Lists can be moved back and forth between the active memory and the archived memory through the memory management menu (**2nd MEM 2:Mem Mgmt/Del**). The screens below indicate that this calculator has a total of 15141 bytes of active memory free and much more archive memory free, 412495 bytes.

```
MEMORY
1:About
2:Mem Mgmt/Del...
3:Clear Entries
4:ClrAllLists
5:Archive
6:UnArchive
7↓Reset...
```

```
RAM FREE    15141
ARC FREE   412495
1:All...
2:Real...
3:Complex...
4:List...
5:Matrix...
6↓Y-Vars...
```

Here are some of the lists that are stored on this calculator: the default lists **L4**, **L5**, and **L6**, and the named lists **CERAL**, **SHARK**, and **TVFRQ**. The asterisk beside **SHARK** indicates that it is in archive memory. The arrow pointing to **CERAL** indicates that this will be the object of any action. In the last screen, **CERAL** has been deleted by pressing **DEL**. The active memory has increased by 196, the size of the list **CERAL**.

```
RAM FREE    15158
ARC FREE   412495
  L4           84
  L5           84
  L6           12
▸ CERAL       196
 *SHARK       115
  TVFRQ       169
```

```
RAM FREE    15354
ARC FREE   412495
  L4           84
  L5           84
  L6           12
▸*SHARK       115
  TVFRQ       169
  TVHR        168
```

It is a good habit to always remove or archive lists that you are not using. This ensures that you have plenty of active memory available and do not exhaust the active memory. Doing so makes the calculator unusable until you clear some space.

If the situation gets desperate you can **RESET** the calculator's memory, removing all lists, programs, and other data stored in active memory. This is done by **2nd MEM**

If the situation gets desperate you can **RESET** the calculator's memory, removing all lists, programs, and other data stored in active memory. This is done by **2nd MEM 7:Reset**. After reset, the active memory is much larger, but the list **SHARK**, which was archived, is still there.

Using Groups

It is hoped that you will make extensive use of your calculator as you work through *Statistics: The Art and Science of Learning from Data.* Therefore you will be constantly moving data in and out of your calculator, and given the limited active memory this can frequently result in the message "**ERR MEMORY**," indicating that you have tried to exceed the limits of the active memory. One feature of the TI-83 Plus family of calculators that can help manage the memory limitation is the use of **groups**. A group collects together any combination of lists, programs, or other data and holds it in archive memory. When you need the contents of the group, you can "ungroup" the data, copying it into active memory. The group itself remains in archive, ready if you need to refresh a list. Most of the data that is used in this manual is available on the textbook CD in groups by topic or chapter. The lists are also available separately on the CD, since the TI-83 calculators do not include the grouping capability.

Transferring Data from a Computer

To transfer data from a computer to your calculator, you need two things: software on the computer to manage the transfer and a cable to connect the computer to the calculator. This setup also allows you to upload data from your calculator to your computer, and you can make backups of the calculator's memory.

The software is called TI-Connect™, for Windows and Mac, and is available for free from the TI website at

education.ti.com/educationportal/sites/US/productDetail/us_ti_connect.html

The part of the program that transfers data is called the Device Explorer, and lists, groups, or other data can be drag-and-dropped from the computer onto the calculator that appears in this window. A window from the TI Device Explorer software is shown below.

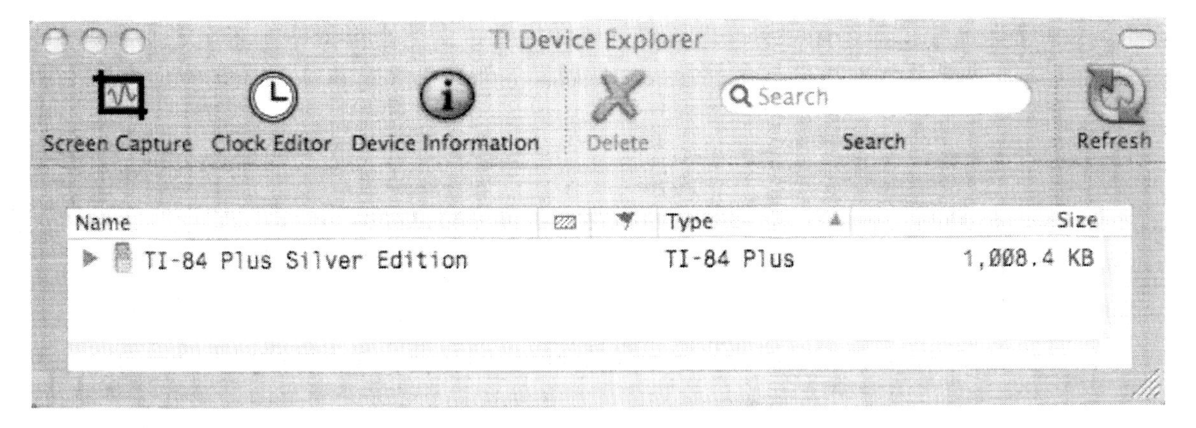

The cable needed to transfer data comes in the package with a TI-84 Plus or Silver Edition. It can be purchased separately, but it is a USB Standard A-to-miniB cable that is also used by other devices such as digital cameras and cell phones.

Entering Data

Data can be stored into a list directly from the Home Screen. For example, suppose we wish to store the numbers 12, 16, and 35 in the first default list, **L1**. We could enter the numbers, separated by commas and enclosed in braces, and use the key labeled **STO>** to put them in **L1**. Note that the calculator displays the list enclosed in braces separated by a space.

```
{12,16,35}→L1
        {12 16 35}
█
```

An easier way to enter a list of data is to use the list editor. Pressing **STAT EDIT 1:Edit…** shows a table with the default lists. You should see, by default, lists **L1-L3**. If you scroll to the right, you'll see **L4-L6**.

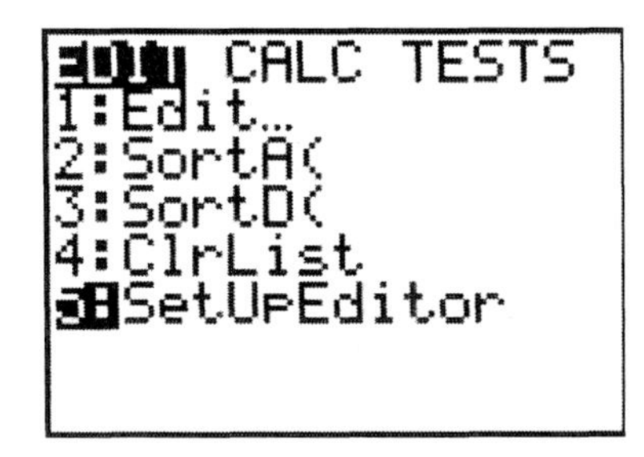

If this is not what you see – if any of the six default lists are missing, for example – you can reset them in the editor by pressing **STAT EDIT 5:SetUpEditor**. As with any numbered menu, you can either scroll down to 5, or just press 5 on the keyboard. Press **ENTER** in the home screen. This has no effect on any data that you have, it just sets up what appears in the editor.

Other lists can be specified to appear in the list editor, but in this manual the convention will be to always copy other named lists to the default lists. This allows you to choose the lists from the keyboard, and keeps your original data safe from inadvertent changes.

Removing data from a list

Since you will be using the default lists for many different problems, you will need to remove old data from a list. There are two easy ways to do this. One way is the following: in the list editor, move the cursor up to highlight the list name, press **CLEAR**, then **ENTER**.

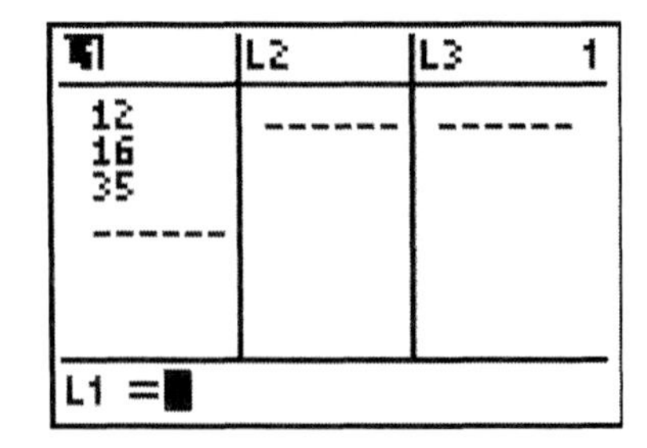

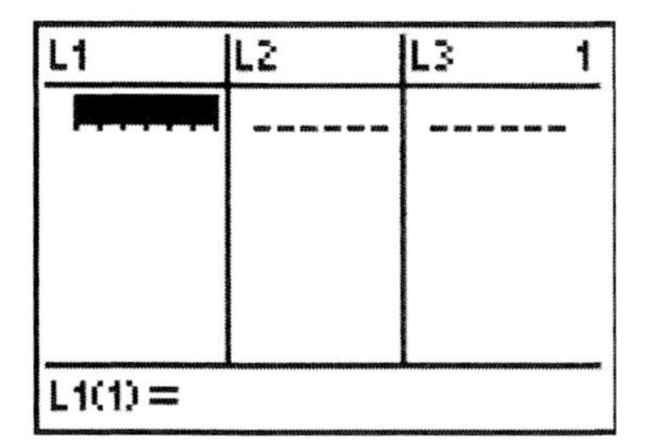

A second way it to press **STAT EDIT 4:ClrList**, then specify which list is to be cleared.

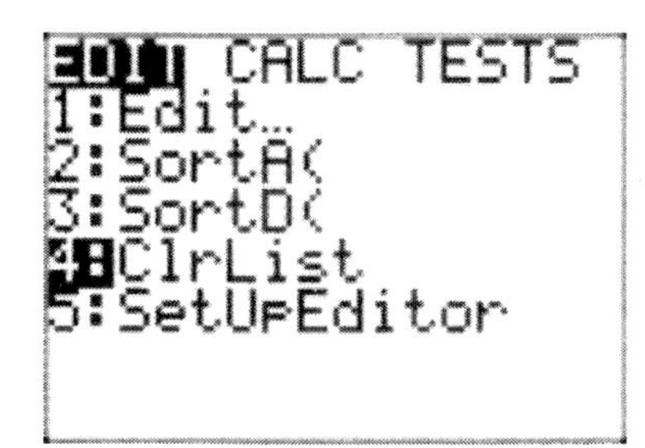

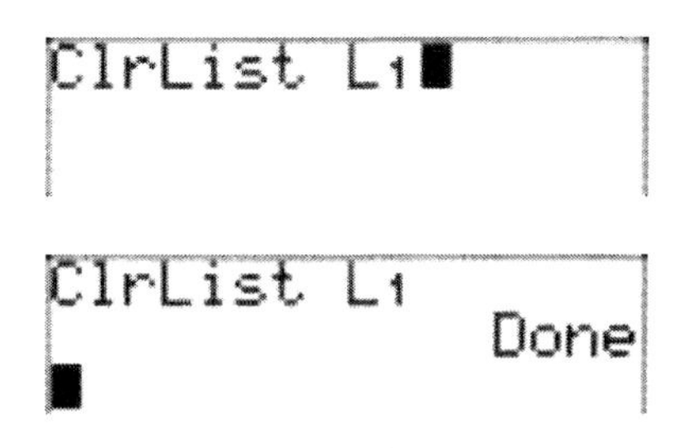

Getting Data From the Textbook CD

Most of the data used in this manual is on the CD that comes with your textbook. With the setup described above, you can copy either an individual list to active memory or copy a **group** of lists to archive memory. With a group you get all the lists for each chapter bundled together. This manual will assume you have the means to copy the lists to your calculator, but unless noted otherwise you can also type the data from your textbook directly into a list using the list editor. All the lists for each chapter are stored in a group named AF followed by the chapter number. For example, lists for Chapter 2 are in the group **AF02**. Copy this group to your calculator, and we will show you how to access the lists that are included.

Press **2nd MEM** to access the memory management menu. Scroll down to find **8:Group...** , then choose **UNGROUP** and choose the group **AF02**. This will take the lists from the group and put them in active memory. The lists that are copied are on page 1 of this chapter.

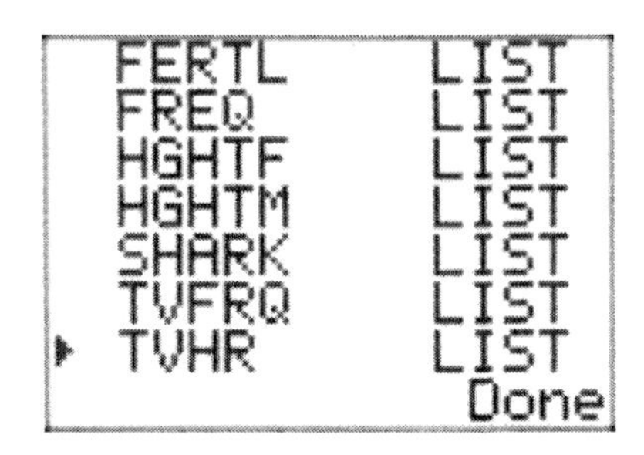

Once the lists are copied from the group you can access them through the **LIST** menu (**2nd LIST**). The **NAMES** submenu shows lists that are available, the six default lists followed by the named lists. To get the data about the shark attacks on p. 28, scroll down to the list **SHARK**.

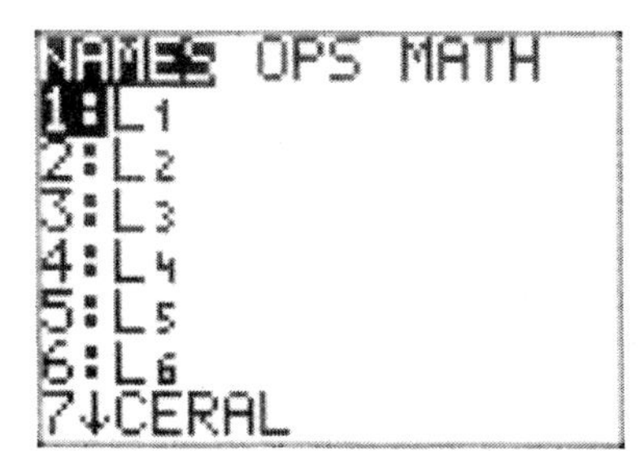

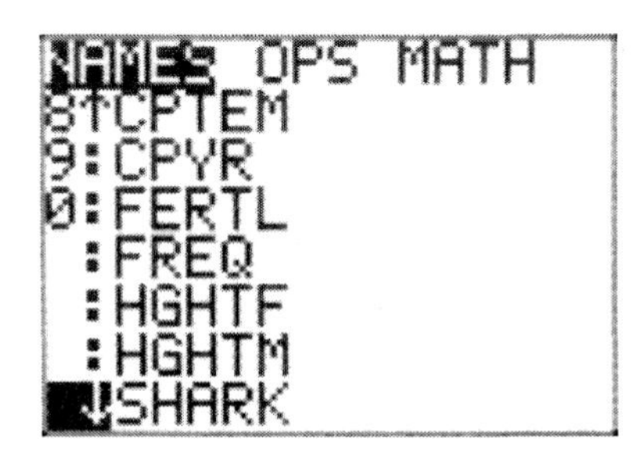

Copy the data in **SHARK** to **L1** by pressing the key **STO>**, then **ENTER**.

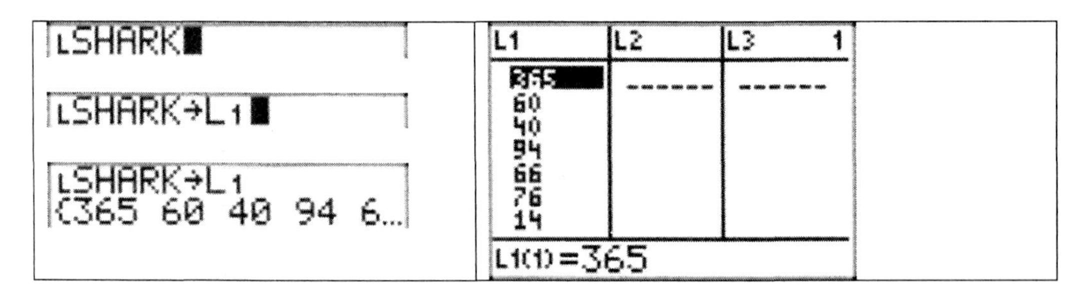

Troubleshooting: It's possible that you might not be able to copy the lists into your active memory, if there isn't enough free space. If this is the case you get the message **ERR MEMORY**. The solution is to delete some of the lists or programs stored in your calculator. Do this by accessing the memory management menu (**2nd MEM 2:Mem Mgmt/Del...**). You can then select either **4: List**, **7: Prgm**, or other data type, deleting unneeded items using the **DEL** key.

A Note on Group Format: There are two formats that are used for groups, ending in **.8xo** or **.8xg**. In general the **.8xo** is preferred, since it keeps the group together; the **.8xg** immediately ungroups the data on some calculators. However, the **.8xo** may not work on all calculators.

Agresti/Franklin Data **Computer File to TI List Index**

Computer File Name	TI-83/84 Lists	TI-83/84 Group
AL_team_statistics_07	ALBAT, ALRUN	ALTEAMS
anorexia	ANCG1, ANCG2, ANFM1, ANFM2, ANCN1, ANCN2, ANCHG	ANOREXIA
baseball's_hr_hitters_07	HRBR, HRRM, HRHA, HRMM, HRSS, HRBB	HOMERUNS
buchanan_and_the_butterfly_ballot	BPERO, BGORE, BBUSH, BBUCH	BUCHANAN
car_weight_and_mileage	CARWT, CARMI	CARWTMI
central_park_yearly_temps	CPYR, CPTEM	CENTRALP
cereal_07	CERSO, CERSG	CEREAL
cigarette_tax	CIGTX, CIGRG	CIGARETT
college_athletes	CLHGT,CLWGT, CLBFT, CLLBM, CLRP5, CLRP7, CL1RM, CLAGE	COLLATHL
credit_card_and_income	CCINC, CCYES	CREDITCR
ebay_auctions	EBNOW, EBBID	EBAY
ebay_regression_07	ERNOW, ERPRC, ERBID, ERSEL	EBAYREG
enchiladas	ENCST, ENSOD	ENCHILAD
energy_eu	EUENR	EUROPE
European_union_unemployment	EUUNP	EUROPE
fl_crime	FLCRI, FLEDU, FLURB, FLINC, FLCRT, FLIRT, FLHS4, FLRDY, FLDNS	FLCRIME
fl_student_survey	FLAGN, FLAGE, FLAHS, FLACG, FLADH, FLADR, FLATV, FLASP, FLANP, FLAAD, FLAVG, FLAPI, FLAPL, FLARL, FLAAL, FLAAA, FLALD, FLNFE, FLNMA	FLASTUDT
georgia_student_survey	GAHGT, GAGEN, GAHAR, GAJOB, GASTT, GACIG, GADAT, GAHSG, GACGG, GADIH, GABRO, GATV, GAEXR, GANWS, GAVEG, GAPDE, GAPAF	GASTUDNT
heads_of_households	HHAGE	(no group)
heights	HGHT, HGEN	HEIGHTS
high_jump_07	HJMEN, HJMYR, HJWOM, HJWYR	HIGHJUMP
high_school_female_athletes	HSAGE, HSHGT, HSWGT, HSBFT, HSBMI, HSSIT, HS40Y, HSVJ, HSRCH, HSBNH, HSREP, HSLP, HSRTF, HSMB, HSSR	HSFEMATH
house_selling_prices	HPTAX, HPBED, HPBTH, HPQD, HPNW, HPPRI, HPSIZ, HPLOT	HOUSPRIC
hs_graduation_rates	GRSAT, GRRAT, GRHLT	GRADRATE

human_development_07	HDINT, HDGDP, HDCO2, HDCEL, HDFRT, HDLIT, HDFEC	HUMANDEV
income_in_public_housing	PHINC	(no group)
long_jump_07	LJUMP, LJYR	LONGJUMP
mental_health	MHMEN, MHLIF, MHSES	MENTHLTH
mountain_bike	MBPRI, MBWGT, MBSUS	MTNBIKE
newnan_ga_temps	NGAYR, NGATP	NEWNANGA
NL_team_statistics	NLWIN, NLBAT, NLRUN, NLSCR, NLERA	NLTEAMS
protein_and_weight_gain	PRSRC, PRLEV, PRGN	PROTEIN
quality_and_productivity	QCPLT, QCDEF, QCTIM	QUALPROD
sharks_07	SHARK, SHPRP, SHYR, SHWLA, SHWLF, SHWLN, SHFLA, SHFLF, SHFLN	SHARKS
softball	SBGAM, SBDIF, SBRUN, SBHIT, SBERR, SBSTA	SOFTBALL
tv_europe	TVIMP, TVPRV	TVEURO
us_statewide_crime	CRIME, CRMUR, CRPOV, CRHS, CRCOL, CRSPA, CRUNM, CRMET	USCRIME
us_temperatures	USTYR, USTMP	USTEMPS
whooping_cough	WCYR, WCRAT	WHOOPCGH

Agresti/Franklin Data — Technology Manual Lists and Groups

For convenience, the lists used in some chapters of this manual have been grouped together. The table below lists these chapter groups.

Tech Manual Chapter	TI-83/84 Lists	TI-83/84 Group
2	SHARK, ELEUS, VISIT, TVHRS, TVFRQ, CERSO, CERSG, WCRAT, WCYR, CPTEM, CPYR, HGHTF, HGHTM	AF02
3	HDINT, HDGDP, MBPRI, MBWGT, ALBAT, ALRUN, CERSO, CERSG, CPTEM, CPYR	AF03
8	EBNOW, ANCHG	AF08
10	EBNOW, EBBID, FLNFE, FLNMA, CELLY, CELLN, CELLD, STAGR, STANG	AF10
12	HSBNH, HSRTF, LEAFW, LEAFR	AF12

Chapter Two: Exploring Data with Graphs and Numerical Summaries

All of these lists referred to in this Chapter are on the textbook CD with the names below. They are also collected together in the group AF02 on the CD.

The process for copying lists and groups from a computer to your calculator is described in Chapter 1 of this manual, beginning on page 4.

SHARK	Number of shark attacks
ELEUS	Source of electricity in the U.S.
VISIT	Most visited countries
TVFRQ	Frequency of TV hours
TVHR	Daily hours watching TV
CERSO	Sodium amounts in breakfast cereals
CERSG	Sugar amounts in breakfast cereals
WCRAT	Incidence rate for whooping cough
WCYR	Year of each rate for whooping cough
CPTEM	Average annual temperature in Central Park, NY
CPYR	Year in which Central Park temperature occurred
HGHTF	Heights of female college students
HGHTM	Heights of male college students

2.1: What Are the Types of Data?

A **variable** is any characteristic that is observed for subjects in a study.

A variable is called **categorical** if each observation belongs to one of a set of categories.
A variable is called **quantitative** if observations on it take numerical values that represent different magnitudes of the variable.

A quantitative variable is **discrete** if its possible values form a set of separate numbers, such as 0, 1, 2, 3, ...
A quantitative variable is **continuous** if its possible values form an interval.

The values of a variable are stored as a **list** in the calculator. All lists on the TI-83 Plus family of calculators are stored in the same way, as numerical values, with no distinction between discrete and continuous data. To work with categorical data, it must be coded as numerical values.

EXAMPLE 1: How many shark attacks in different locations around the world have occurred recently? (pp. 27-29 in Agresti/Franklin)

Table 2.1 in your textbook (p. 28) displays a frequency table of shark attacks in various regions for 1990 - 2002. This is categorical data; for each of the shark attacks the **region** is recorded, and the table lists how many times shark attacks occurred in each of the regions.

Either type the data directly into L1 or get the data from the textbook CD as described in Chapter 1 of this manual, copying the list **SHARK** to **L1**. The data in **L1** is the **frequency table** for the **SHARK** data. Proportions and percentages can be added to the data by operations on the original list. Create the proportions by dividing each of the values by the total number of attacks, 949.

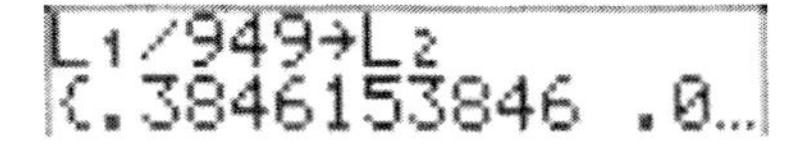

L1	L2	L3 2
365	[illegible]	------
60	.06322	
40	.04215	
94	.09905	
66	.06955	
76	.08008	
14	.01475	

L2(1)=.3846153846...

The total number of shark attacks is listed in the textbook, but in many cases you might not know this sum. You can also add up the values in a list using the **sum(** command. This is accessed by pressing **2nd LIST MATH 5:sum(**. Once you have the value calculated, you can divide by this value, as shown. Notice the use of the "**ANS**" key (**2nd (-)**) that always uses the most recently calculated value, from the line above. This is handy to reduce the number of values you need to enter manually. The values stored in **L2** are the same as in the previous screens.

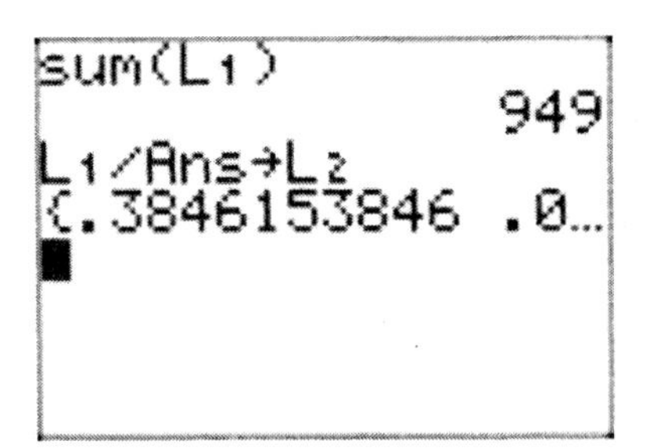

The other way to operate on a list is from the list editor. Push the arrow keys to move the cursor to the top of the column **L3**. The entry panel at the bottom of the screen changes from element editing (single value) to list editing (entire list). Type **L2*100** to create the percentages, which are automatically placed in **L3**. To see the other values in the list, scroll down using the cursor keys. Note that the editing pane shows the position of the cursor, indicating both the list and the row number, and the value. In the last screen below, the cursor is in **L3** at row **11**. The 206 shark attacks that took place in "other" locations are about 21.7% of all attacks.

L1	L2	L3 3
365	.38462	------
60	.06322	
40	.04215	
94	.09905	
66	.06955	
76	.08008	
14	.01475	

L3 =

L1	L2	L3 3
365	.38462	------
60	.06322	
40	.04215	
94	.09905	
66	.06955	
76	.08008	
14	.01475	

L3 =L2*100

L1	L2	L3 3
365	.38462	[illegible]
60	.06322	6.3224
40	.04215	4.215
94	.09905	9.9052
66	.06955	6.9547
76	.08008	8.0084
14	.01475	1.4752

L3(1)=38.46153846...

L1	L2	L3 3
66	.06955	6.9547
76	.08008	8.0084
14	.01475	1.4752
18	.01897	1.8967
4	.00421	.4215
6	.00632	.6322[illegible]
206	.21707	[illegible]

L3(11) =21.7070600...

Try it! Do Exercise 2.8, p. 30

Use the methods we've just described to do problem 2.8. You'll need to clear out the existing list, type in the data, then add the proportions and percentages in each category. The final screen that you should get is shown below.

L1	L2	L3 3
1216	.2704	[illegible]
710	.15788	15.788
1147	.25506	25.506
738	.16411	16.411
386	.08584	8.5835
139	.03091	3.0909
83	.01846	1.8457

L3(1)=27.04024905...

2.2 How Can We Describe Data Using Graphical Summaries?

Graphs for categorical variables include **pie charts** and **bar graphs**. The TI-83 Plus family does not produce pie charts, except by using an application such as CellSheet. This isn't described in this manual. Bar graphs can be created using one of the graphical options in the **STAT PLOT** menu. The next example demonstrates this.

EXAMPLE 3: How much electricity comes from renewable energy resources? (pp. 30-32 of Agresti/Franklin)

The best we can do for the categories in this example (coal, hydropower, etc.) is to enter them as numbers 1 through 6 in **L1**, then use **L2** to enter the U.S. percentages for sources of electricity in the order in which they appear in table 2.2. (The U.S. values are in the list **ELEUS**).

L1	L2	L3
1	51	------
2	6	
3	16	
4	21	
5	3	
6	3	

L2(7) =

All graphical displays are set up in the **STAT PLOT** menu (**2nd Y=**). There are three plots that can be set up and each can be turned on and off. In most cases you will only use one at a time, so the other two should be set to **OFF**. You can quickly turn all three of the plots off by using the **PlotsOff** function in the **STAT PLOT** menu, then turn on the plot you want (typically using **Plot1**).

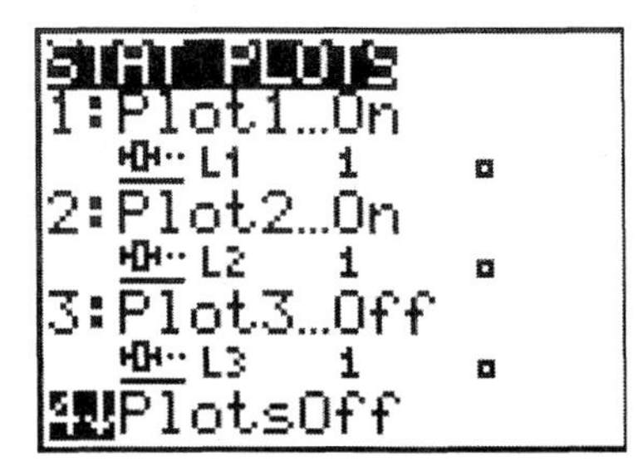

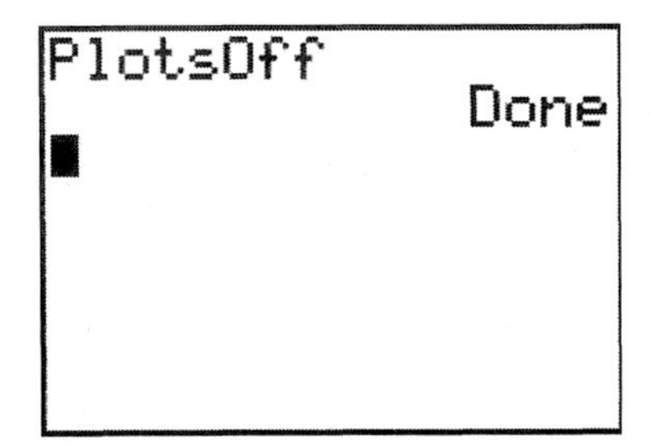

Go back to the **STAT PLOT** menu. Choose **1:Plot1**, then turn it on by choosing **On** and then **ENTER**. There are six possible types of plots; the third icon on the top row is for a histogram ( ), a type of quantitative graphical display, but this can also be used to create a bar graph. Change the type to the histogram icon (so it looks like) by moving the cursor to the icon and pressing **ENTER**. The **Xlist** is for the categories, so use **L1**, and the frequencies are in **L2**, for **Freq**.

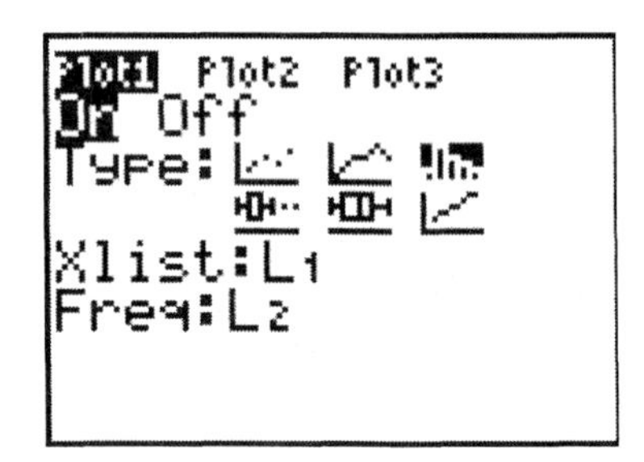

The next step in setting up the graph is to specify the values that determine the "window" for the graph screen. Press the **WINDOW** key (the second key from the left immediately below the screen). The values for the left and right endpoints of the window are specified by **Xmin** and **Xmax**, respectively, and for the top and bottom by **Ymin** and **Ymax**.

We have coded the numerical values 1 through 6 to represent the categories, so **Xmin = 0** and **Xmin = 7** will include these values. The value **Xscl** determines the width of the bars in the graphs, so this is set equal to **1** since each number is a different category. **Xscl** also is used in other settings to determine the tick marks along the x-axis. **Xres** is not used for statistical graphs, only in function graphing. Leaving this at 1 is fine.

Ymax must be large enough to include the height of the tallest bar, which from the data above is 51. A little extra room is usually a good idea, so set this at 60. For statistical graphs the calculator can add some helpful text to the graph at the bottom of the screen. To leave room for this, it's a good rule of thumb to have **Ymin** be about 1/4 of the size of **Ymax**, but of course negative; so set **Ymin** to -15. Make sure you use the negative sign, **(-)**, not the subtraction key, –. This positions the x-axis nicely on the graph. **Yscl** determines the tick marks on the y-axis, set this value at 5 to make each mark on the y-axis represent 5 percentage points.

You can have functions graphed at the same time as statistical graphs, but for now make sure that there are no functions in the **Y=** menu or that they are turned off. Press **GRAPH** to show the plot.

WINDOW
Xmin=0
Xmax=7
Xscl=1
Ymin=-15
Ymax=60
Yscl=5
Xres=1

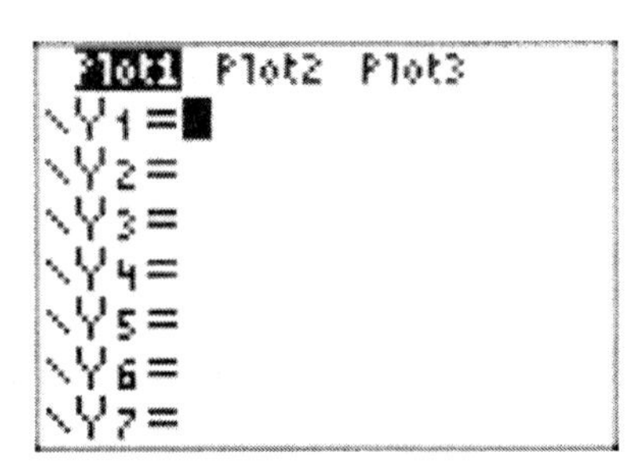

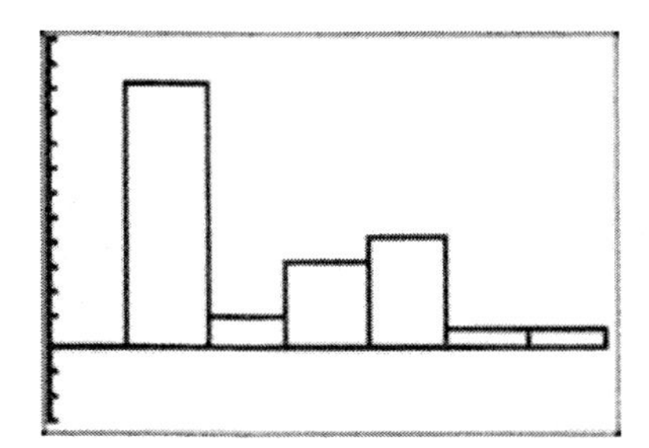

When viewing the graph, pressing **TRACE** will allow you to get more information. The image below shows the first bar for the first category, coal, with the interval including 1, but not including 2. The height of the bar is given by **n = 51**, but you need to keep in mind that this is actually a percentage. Scroll over to the right with the cursor keys to verify the other percentages in table 2.2. The second image shows natural gas, category 3, is 16% of the total electricity.

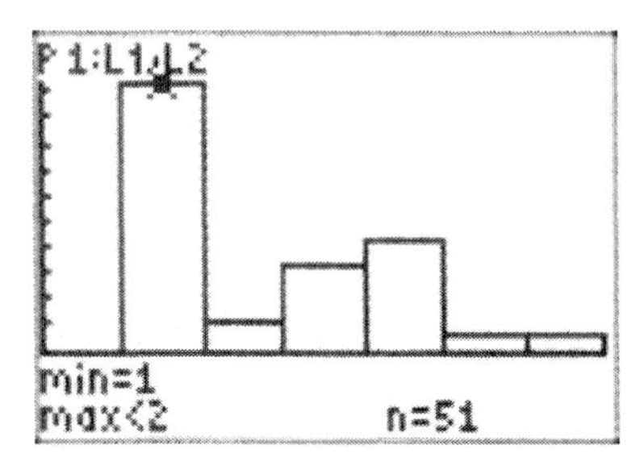

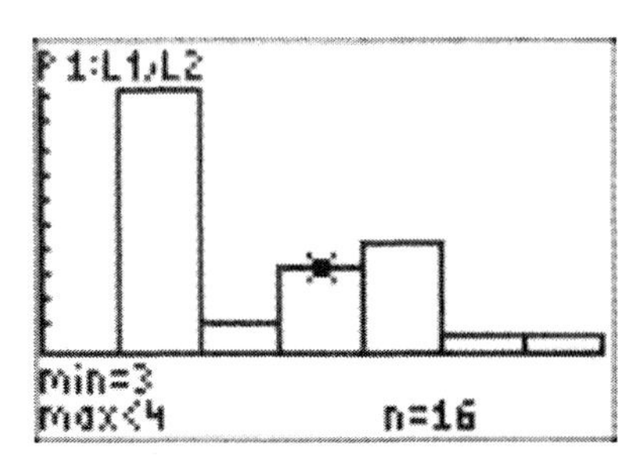

You can turn this graph into a Pareto chart by ordering the data from the largest to smallest percentages (tallest to shortest bar). The calculator can sort the list into the correct order for you. Press **STAT EDIT**, then choose **3:SortD**. Provide **L2** as the list to be sorted. Since **L1** is just a dummy list of numbers, it remains as is. You can verify that the list was sorted in the list editor.

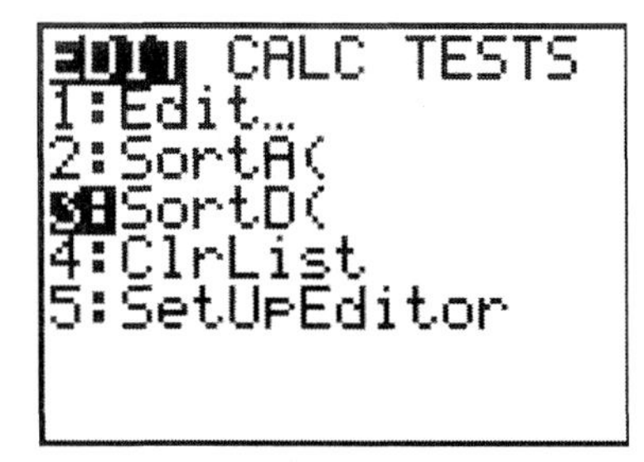

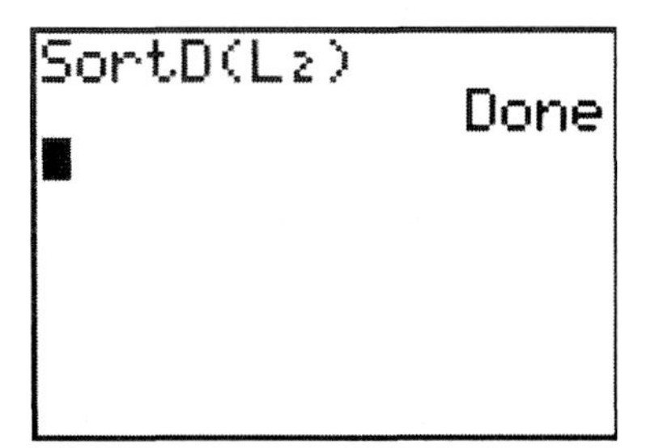

L1	L2	L3
1	51	------
2	21	
3	16	
4	6	
5	3	
6	3	

L2(1)=51

Press **GRAPH** to see the Pareto chart. Use TRACE to see that the smallest category, other renewable resources, represents only 3% of U.S. electricity production.

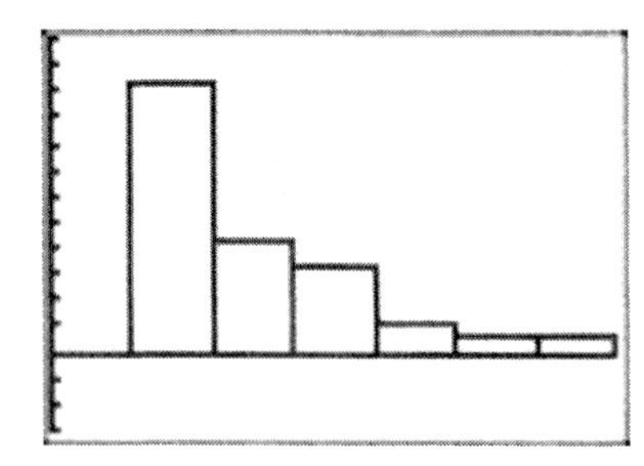

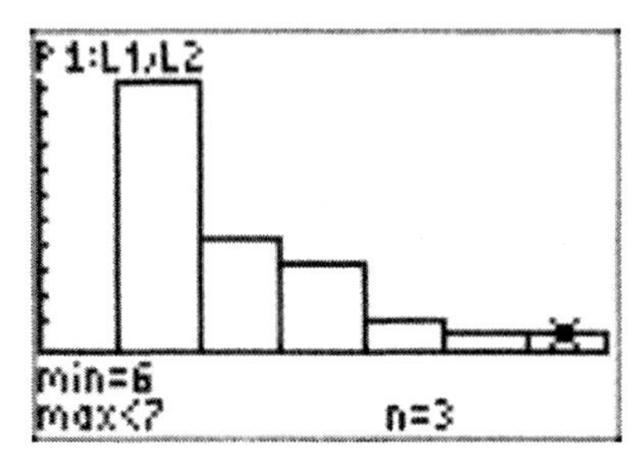

Try it! Do Exercise 2.12, p. 45

Use the methods we've just described to do problem 2.12. Put the categories 1 – 8 in **L1** and either type the values into **L2** or copy the list **VISIT**. The results are below.

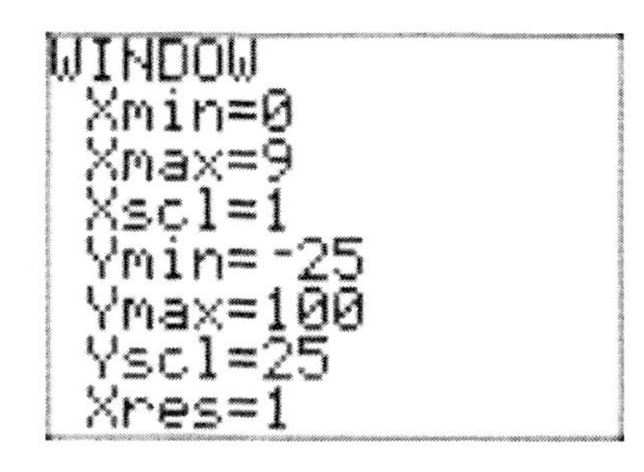

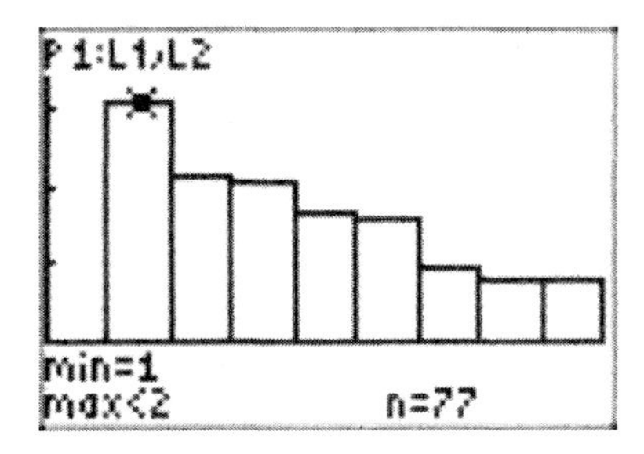

Dot Plots and Stem-and-Leaf Plots

Neither dot plots nor stem-and-leaf plots are available with the TI-83 family, but histograms can be set up for either discrete or continuous quantitative data.

Histograms for Discrete Quantitative Data

EXAMPLE 6: Histogram of TV Watching (pp. 37-38 of Agresti/Franklin)

A histogram uses bars to portray the frequencies or relative frequencies of the outcomes of a quantitative variable. For discrete quantitative data this is similar to making a bar chart. The big difference is that the categories now represent meaningful quantities, rather than arbitrary labels.

The data for this example is from the General Social Survey, summarized in the frequency table at the right. This is available on the CD as lists **TVHRS** and **TVFREQ**. Either type in the values from the table or copy these two lists to **L1** and **L2** respectively.

TV Hours	Frequency
0	29
1	215
2	226
3	166
4	103
5	71
6	37
7	7
8	24
9	2
10	6
11	2
12	8
13	5

For discrete data, it's usually most appropriate to have the bar that represents the frequency or percentage be centered over the value. This is how the graph on page 37 of the textbook is constructed. To have your TI-83 Plus calculator reproduce this, you will need to make some modifications to the window settings, so that the intervals for the bars start and end at "half-integers" such as 1.5, 2.5, etc.

Set up the histogram as before, using the **STAT PLOT** menu, with **Xlist = L1** and **Freq = L2**. Here is a window that will let you center the bars over each value for TV hours, then the graph is shown. Tracing reveals the tallest bar, at 2 hours, was reported by 226 people.

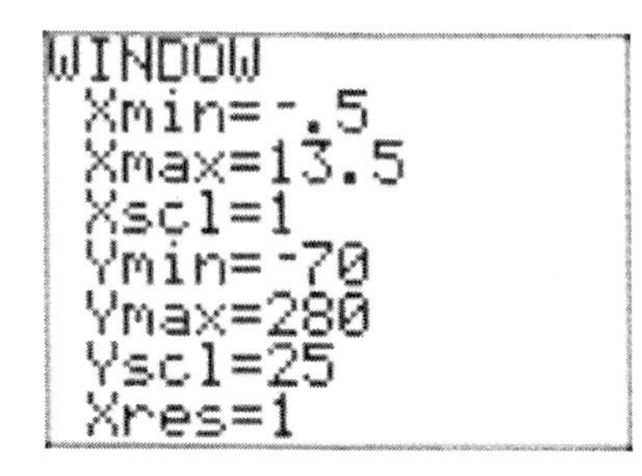

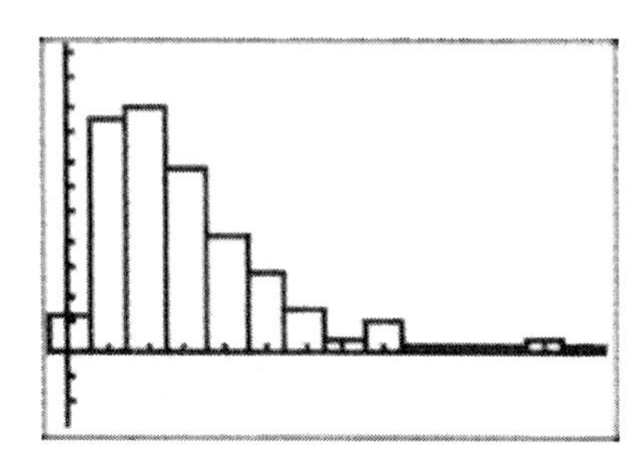

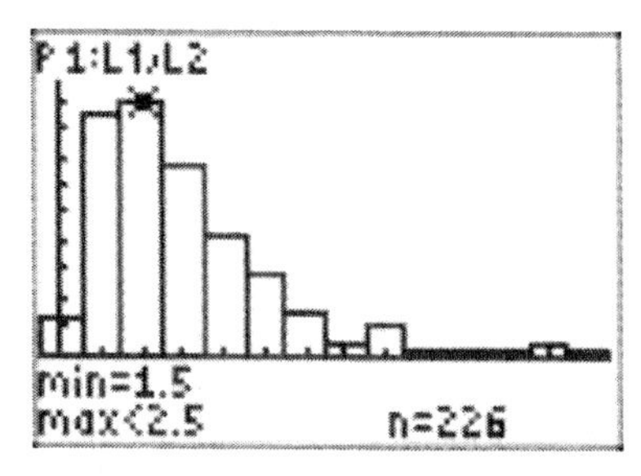

You can change the histogram to show percentages, as is the case in the textbook. The first screen shows how you can use the list functions to create **L3** to have the percentages, then an appropriate window is shown.

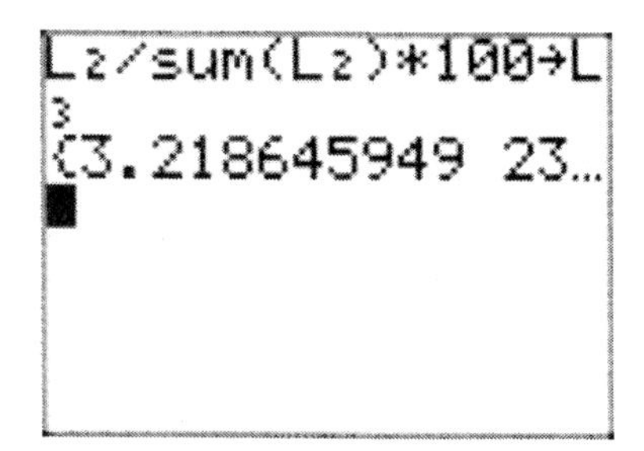

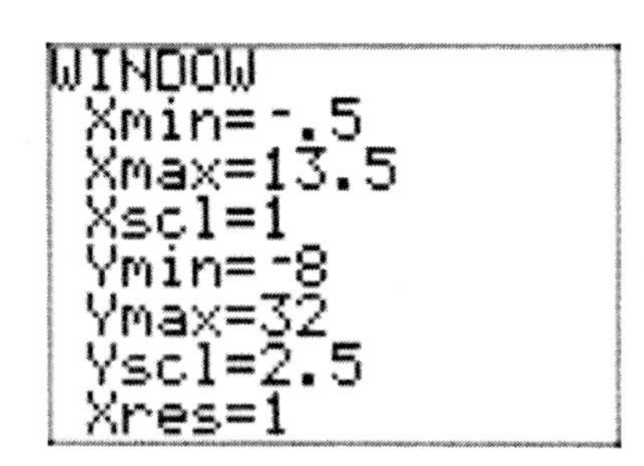

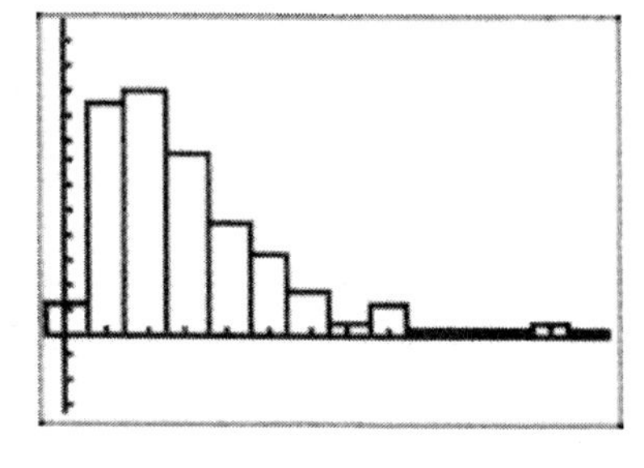

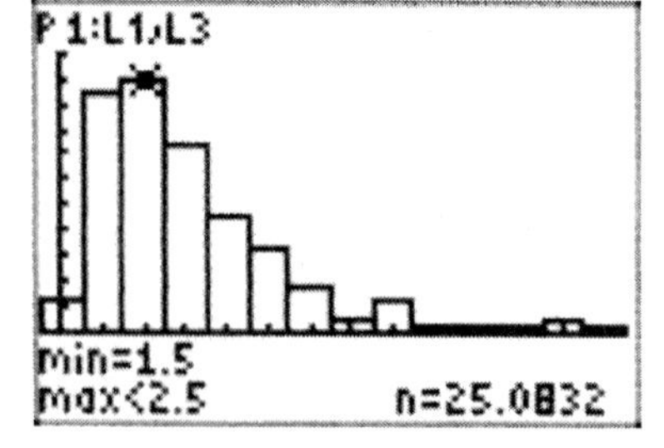

The graph at the left above, with no scale, is essentially identical to the frequency histogram. Use **TRACE** to see that 2 hours includes just over 25% of the recorded responses.

EXAMPLE 7: Exploring the health value of cereals (pp. 38-39 of Agresti/Franklin)

With continuous quantitative variables, it is much less likely that you will have multiple instances of the same value. Thus you will typically not have it set up with a list of values and the corresponding frequency, usually you just have a list with each of the values appearing as one item in the list. To create a histogram of the cereal sodium values, type in or copy the list **CERSO** into **L1**. The values are in Table 2.3, p. 34 of the textbook.

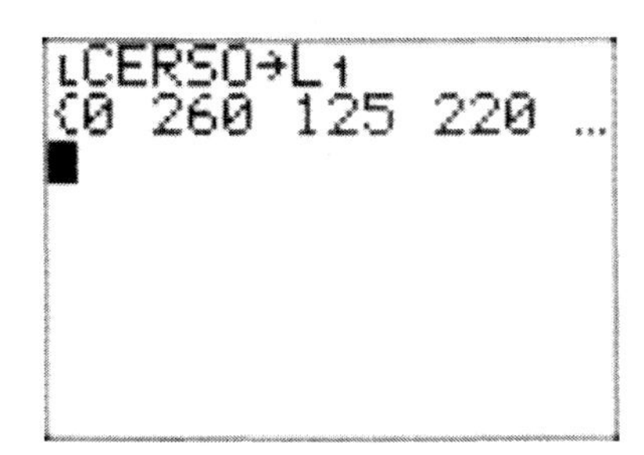

From the **STAT PLOT** menu, make sure that **Plot1** is turned on and the other two plots are off. To have each of the 20 values in **L1** count as 20 individual values, change **Freq** to **1**. This is the most common situation, so this is the default setting on the calculator.

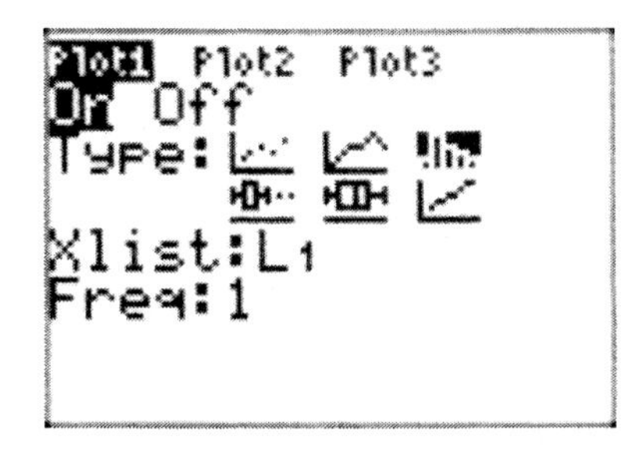

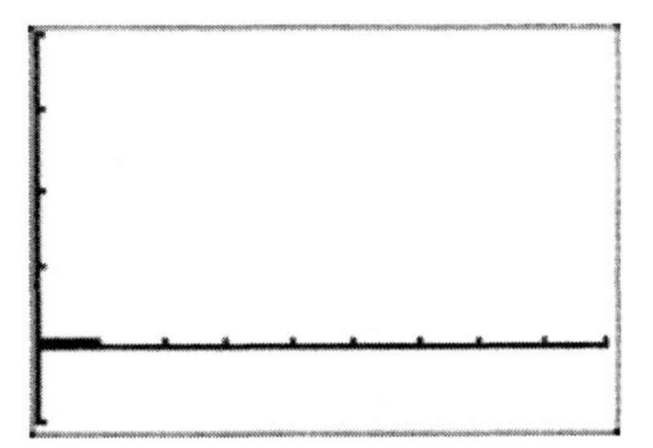

The **STAT PLOT** is set up correctly, above on the left, but if you just press **GRAPH** you will probably get a graph like that on the right—not very satisfactory. Specifying the window dimensions for the bar graph or discrete histogram was relatively easy, since you had a list of values (for x) and a list of frequencies (for y). But setting up the **WINDOW** values is more complicated with continuous data, since the range of the values in the list might not be known, the best value for the width of the bars is not given, and the height of the tallest bar is hard to predict. But here's some good news: you can use the **ZoomStat** function to automatically set the window dimensions to fit the data. Press **ZOOM**, then choose **9:ZoomStat**.

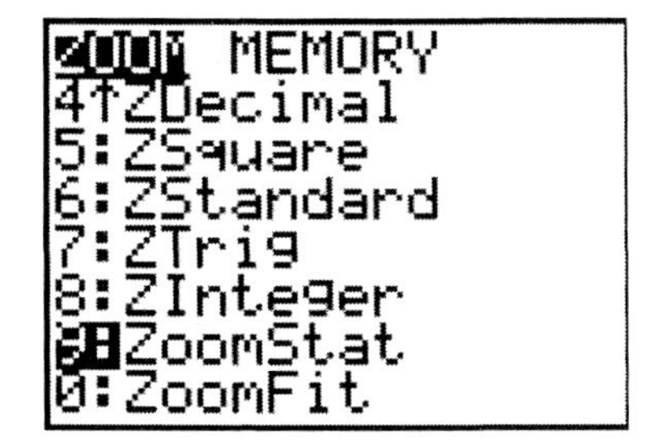

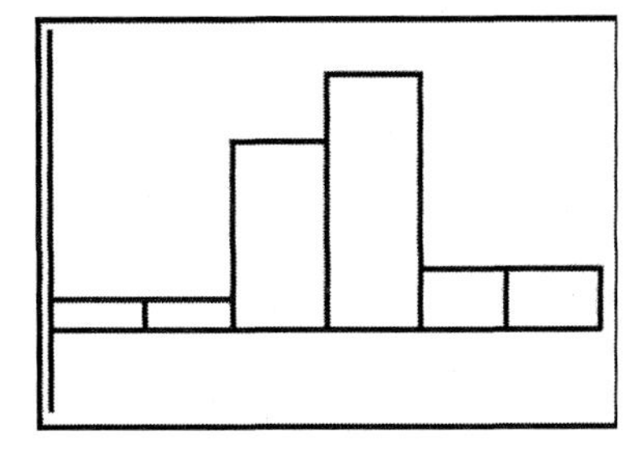

Viola! Now we can view our histogram. Notice that it doesn't look exactly like the histogram on p. 39 of Agresti/Franklin. It has fewer bars, six instead of eight, and the pattern isn't exactly the same. This is because the interval width isn't the same as was chosen in the book, 40. You can see the window parameters by pressing the **WINDOW** key, at the left below. The interval width is 58, a seemingly arbitrary number. You can change the window to match the interval width, 40, and the range of the data to be 0 to 320. Since the calculator can only make reasonable guess at an appropriate interval width, the graph that is produced by **ZoomStat** is often best thought of as a rough draft of the histogram, and you should consider adjusting the interval widths. You may also need to adjust the values for **Ymin**, **Ymax**, and **Yscl**, as is done here.

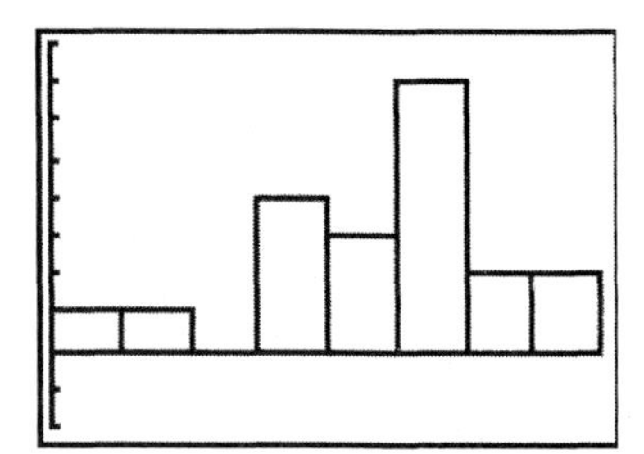

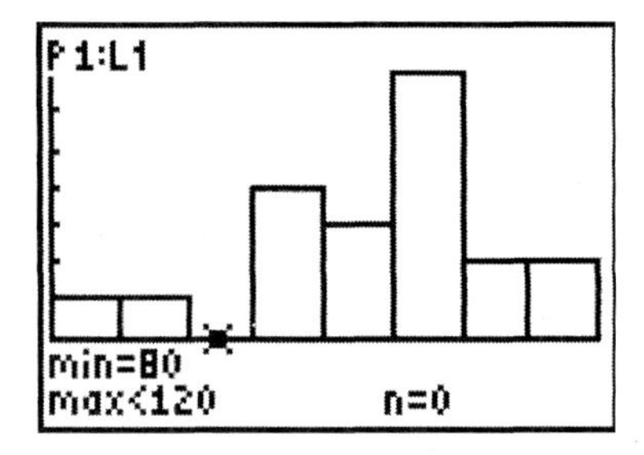

Press **GRAPH** to show the new histogram, matching the one in the textbook. The **TRACE** function is used below to show that no sodium values fell between 80 and 120 mg.

P1:L1
min=80
max<120 n=0

Try it! Do Exercise 2.21, p. 46

Use the methods we've just described to do problem 2.21. Enter the data from Table 2.3, p. 34, or copy the list **CERSG** to **L1**. A window and a histogram that matches the figure in the textbook are shown below.

```
WINDOW
 Xmin=-1
 Xmax=17
 Xscl=2
 Ymin=-1.5
 Ymax=5
 Yscl=1
 Xres=1
```

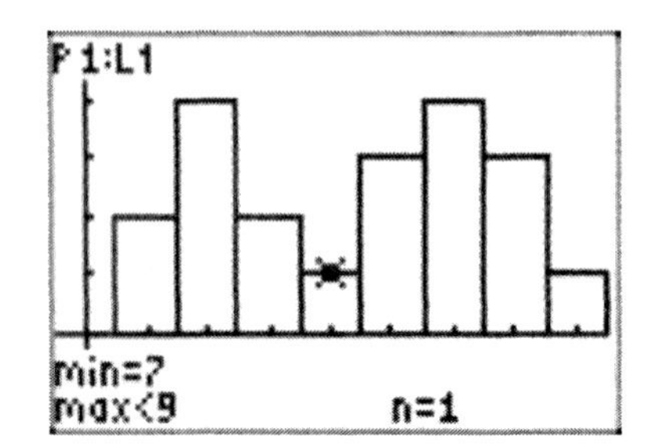

The Shape of a Distribution

A **distribution** of data is a frequency table or a graph that shows the values a variable takes and how often they occur.

One aspect that is important in describing a distribution is its **shape**. Making a histogram for quantitative data allows you to see the shape of the distribution at a glance.

A distribution is **unimodal** if it has a single peak or mound.
A distribution is bimodal if it has two distinct peaks or mounds.
A distribution is **symmetric** if the side of the distribution below a central value is a mirror image of the side above the central value.
A distribution is **skewed to the left** if the left tail is longer than the right tail.
A distribution is **skewed to the right** if the right tail is longer than the left tail.

- The distribution of TV watching times is **skewed to the right**. Most people watch a couple hours of TV each day, but a few people watch much more.

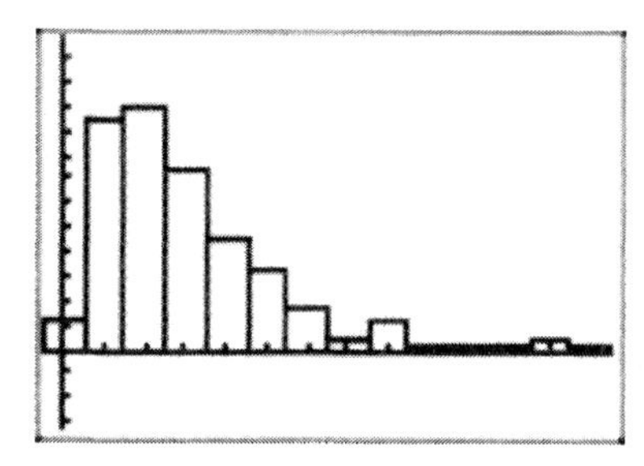

- The distribution of sodium content in cereals is **skewed to the left**. Most cereals have sodium levels around 200 mg, but a few have much less than this.

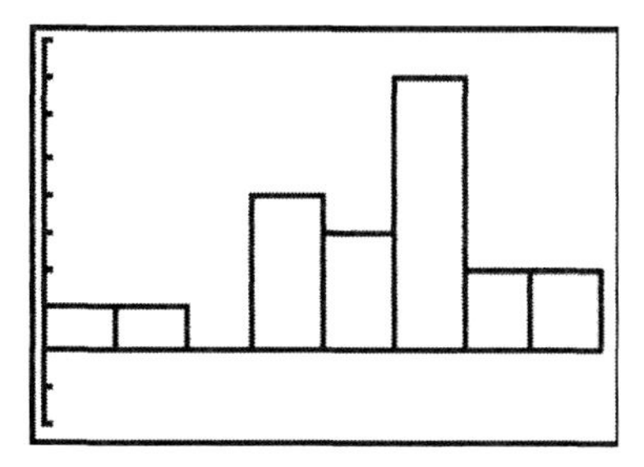

- The distribution of sugar content of cereals is **bimodal**. There is a peak at around 4 and another at about 12. This may indicate cereals that are directed at adults and children.

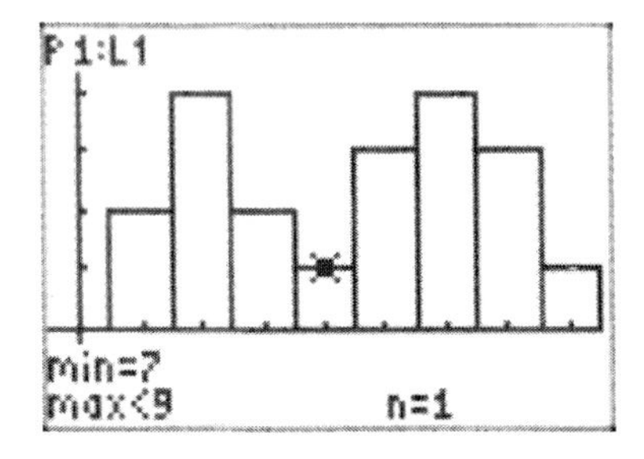

Time Plots

Exercise 2.28: Is whooping cough close to being eradicated? (p. 48 of Agresti/Franklin)

To make a time plot of a time series, the TI-83 Plus family has another graph called an **xyLine** plot. Using two lists of equal length, this plots a series of points on a graph as you might do in Algebra class, and connects them in the order that they appear in the lists. Use the **STAT PLOT** menu to set this up, selecting the second icon in the Type: menu. The **Xlist** is a list of times or dates, and the **Ylist** is the corresponding data value, in this case the incidence rate of whooping cough per 100,000 people. You can copy the data from the textbook to **L1** and **L2** or copy the lists **WCYR** and **WCRAT** to lists **L1** and **L2**. In the menu the Mark determines the symbol that will be drawn at each point, and this is up to you. It only really makes a difference if you are drawing two plots simultaneously and need to distinguish the points in one plot from the points in another.

L1	L2	L3 2
1940	139.6	
1945	101	
1950	80.1	
1955	38.2	
1960	8.3	
1965	3.5	
1970		

L2(10) =2.1

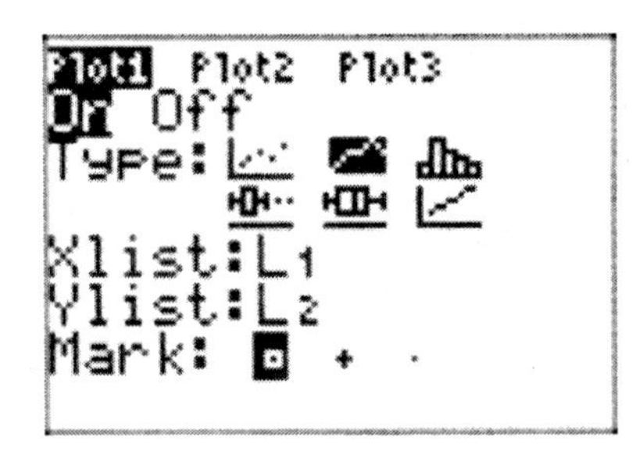

You can use **ZoomStat** to get a decent graph, and then polish it up a bit with the **WINDOW** values. In particular, **ZoomStat** does not adjust the tick marks on the axes, so these have been reset to 5 years on the x-axis and 25 on the y-axis. Below is the graph, and **TRACE** can be used to scroll through the years and verify, for example, that the rate was 38.2 per 100,000 in 1955.

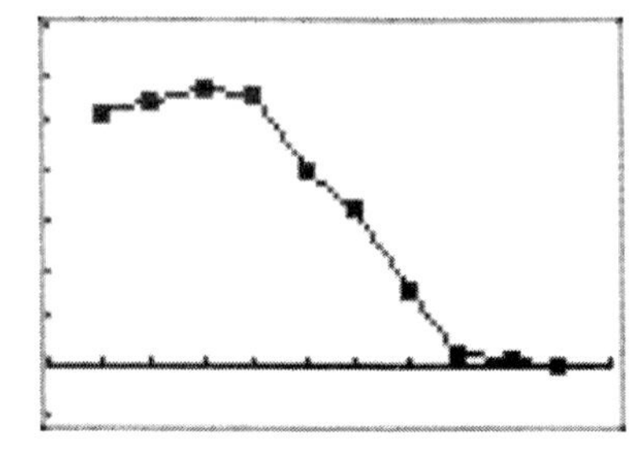

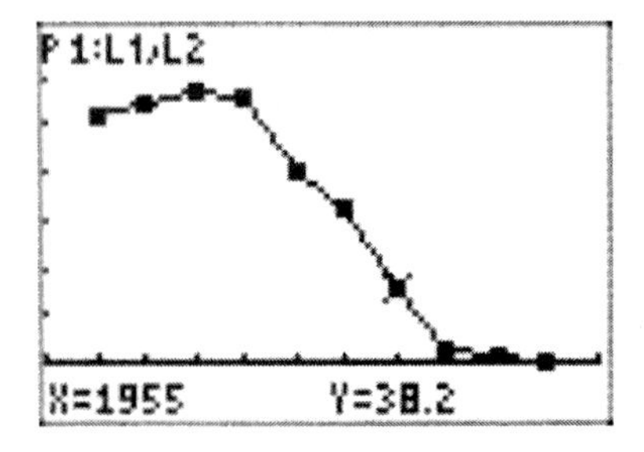

Try it!

Use the lists **CPYR** and **CPTEM** to see if you can reproduce the time plot shown on page 43, a display of the average yearly temperature in Central Park, New York, for each year in the twentieth century.

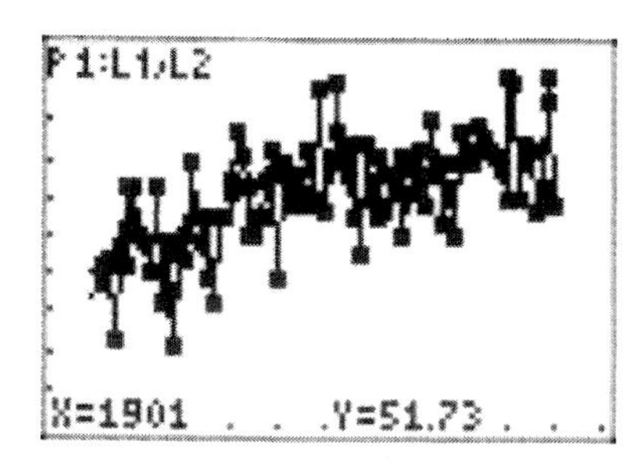

2.3 How Can We Describe the Center of Quantitative Data?

> The **mean** is the sum of the observations divided by the number of observations. The **median** is the midpoint of the observations when they are ordered from smallest to largest (or largest to smallest).

EXAMPLE 10: What's the center of the cereal sodium data? (p. 49 of Agresti/Franklin)

The sodium levels of 20 different cereals first appeared in Table 2.3 on p. 34 (and these are in the list **CERSO**). On page 48 the same numbers appear in order from smallest to largest. There are times when it is useful to have the data in order, and the **SortA** function accomplishes this – **A** is for *ascending*. This is found in two places, in the **STAT EDIT** menu and also in the **LIST OPS** menu. Type in the original data or copy **CERSO** to **L1**, then apply the **SortA** function to do an *ascending* sort.

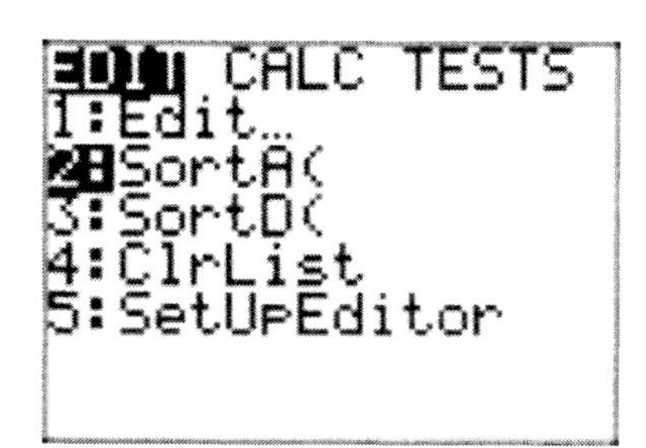

```
SortA(L1)
              Done
```

```
L1      L2      L3      1
0       ------  ------
70
125
125
140
150
170
L1(1)=0
```

You can scroll through the sorted values in **L1** to find that the 10th and 11th values are both 200, so this is the median. If these values weren't the same, then the median would be the average of the two numbers.

```
L1      L2      L3      1
150
170
170
180
200
200
210
L1(10) =200
```

You can find the median and other numerical measures from a list using functions that are available in the **LIST** menu. Press **LIST** (**2nd STAT**). **NAMES** allows us to access lists by name. The screen below shows the default lists **L1**-**L6**, plus the list **CERSO** that is in active memory. The **OPS** menu provides various list operations, including the sort functions we used from the **STAT EDIT** menu before. **MATH** contains various functions, including median.

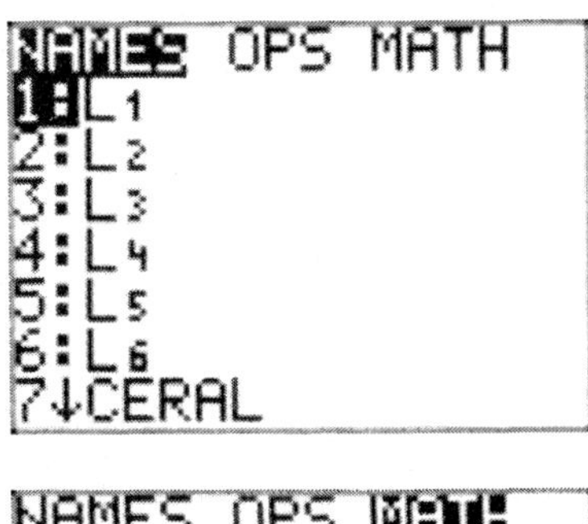

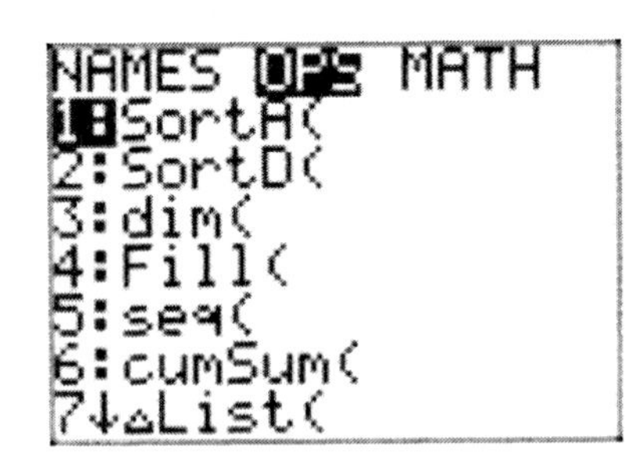

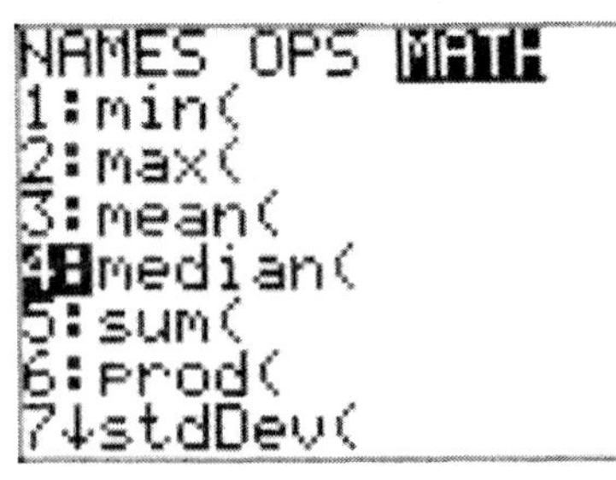

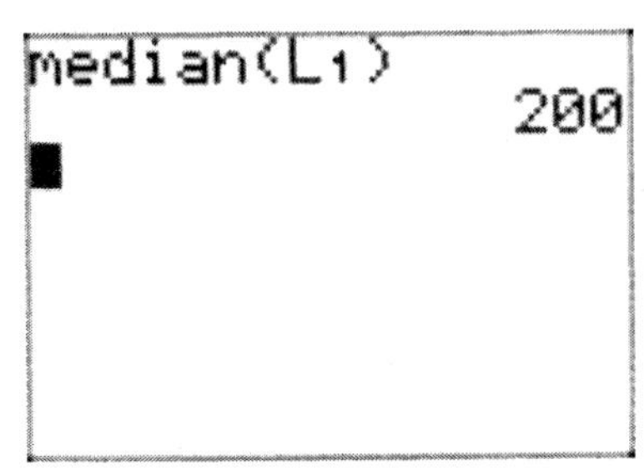

You can also use this menu to find the mean sodium level in the cereal, choosing **LIST MATH 3:mean(** in the menu above. You can also use the **LIST** functions to apply the formula for the mean, $\bar{x} = \frac{\sum x}{n}$. The numerator is accessed by **LIST MATH 5:sum**. It's less obvious that the denominator is also in these menus. The length of a list is sometimes called its "dimension," and you can have the calculator figure out the length of a list with **LIST OPS 3:dim(**. The screen below demonstrates that these give the same result.

```
mean(L1)
                185.5
sum(L1)/dim(L1)
                185.5
■
```

In the section above on the shapes of distributions, we described the sodium data as skewed to the left. Thus it is no surprise that the mean ends up less than the median in this case, 185.5 vs. 200.

2.4 How Can We Describe the Spread of Quantitative Data?

The **range** is the difference between the largest and the smallest observations.

You can find the range of a list using the **LIST** menus. The range of the cereal sodium data in **L1** is 290.

```
NAMES OPS MATH
1:min(
2:max(
3:mean(
4:median(
5:sum(
6:prod(
7↓stdDev(
```

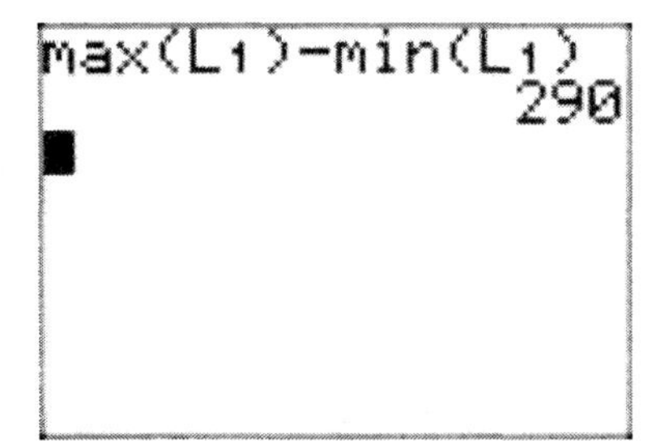

The **variance s^2** of n observations is $s^2 = \frac{\sum (x - \bar{x})^2}{n-1}$.

The **standard deviation s** of n observations is $s = \sqrt{\frac{\sum (x - \bar{x})^2}{n-1}}$.

The standard deviation is the square root of the variance.

The standard deviation describes the *typical* deviation of individual observations from the mean. You can find the standard deviation of a list with **LIST MATH 7:stdDev(**.

As with the mean, you can use the **LIST** functions to work through the steps of the formula. In practice you don't often do this, but it's good experience in working with lists in the list editor. As you work through the steps, move the cursor up to the heading of each list. **L2** ends up with the **deviations from the mean** and **L3** the **sum of squared deviations**. The last screen completes the calculation.

L1	L2	L3
0	------	------
210		
260		
125		
220		
290		
210		

L2 =(L1-mean(L1))

L1	L2	L3
0	-185.5	------
210	24.5	
260	74.5	
125	-60.5	
220	34.5	
290	104.5	
210	24.5	

L2(1)= -185.5

L1	L2	L3
0	-185.5	------
210	24.5	
260	74.5	
125	-60.5	
220	34.5	
290	104.5	
210	24.5	

L3 =L2²

L1	L2	L3
0	-185.5	34410
210	24.5	600.25
260	74.5	5550.3
125	-60.5	3660.3
220	34.5	1190.3
290	104.5	10920
210	24.5	600.25

L3(1)=34410.25

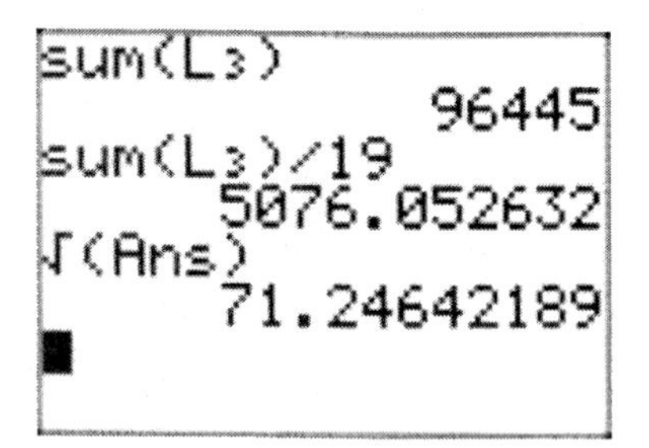

While you can work through the menus to make the individual calculations of the mean, median, or standard deviation, there is a very useful shortcut that the TI-83 Plus family of calculators provides to quickly generate a report that lists all the usual numerical summaries. This is accessed from the **STAT** menu, but push over to the **CALC** option and choose **1:1-Var Stats**. This produces even more than we have discussed at the present time, but includes the mean ($\bar{x}$), the standard deviation (**Sx**), the minimum, the maximum, and the median. There are actually two “pages” of numerical output, so you have to scroll down to see all the values.

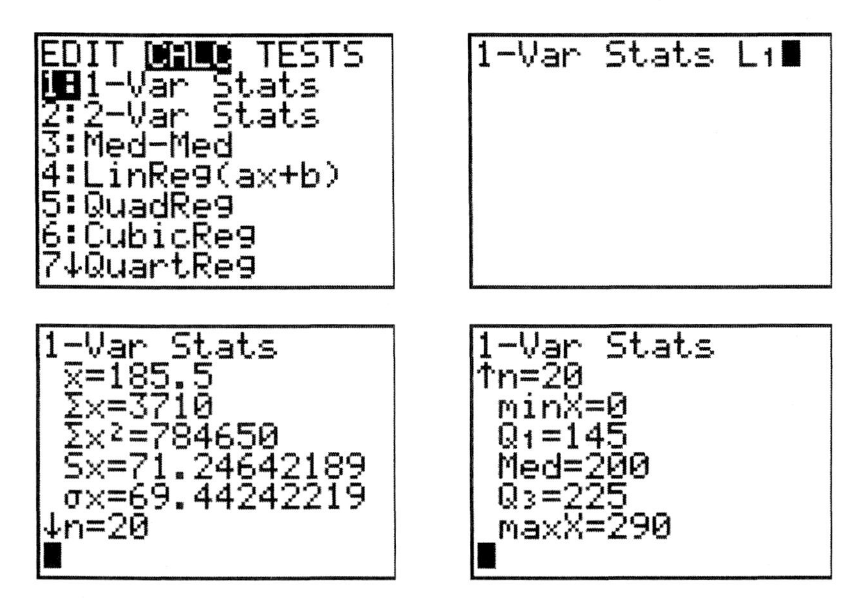

Try it! Heights of Female Students

Copy the list **HGHTF** to **L1**, and reproduce the histogram in figure 2.13, p. 62. Use **1-VarStats** to display the summary statistics.

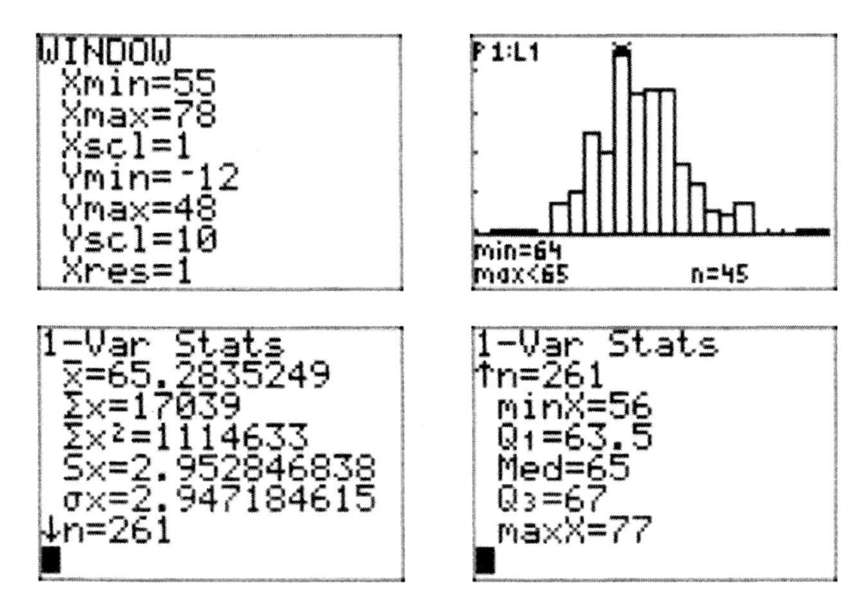

2.5 How Can Measures of Position Describe Spread?

The **quartiles** of a distribution split the distribution into four parts. The **first quartile** coincides with the 25th percentile and the **third quartile** is at the 75th percentile. The **1-Var Stats** report lists the quartiles as **Q1** and **Q3**. Above, **Q1 = 63.5** and **Q3 = 67** for the heights of female college students. This represents the middle portion of the distribution of heights; half of the college students are between 63.5 inches and 67 inches tall.

The difference between the values Q3 and Q1 is a measure of spread called the **interquartile range**, often abbreviated as **IQR**, with **IQR = Q3 – Q1**. In the example above, the IQR is 67 – 63.5 = 3.5 inches.

Finding Outliers

An observation is a potential outlier if it falls more than 1.5 x IQR below Q1 or 1.5 x IQR above Q3.

```
1.5*3.5
                5.25
63.5-5.25
               58.25
67+5.25
               72.25
```

Any value that is less than 58.25 or more than 72.25 is a potential outlier by the 1.5 x IQR criterion. If you sort the values in **L1**, these are easy to find, at the beginning or end of the list.

```
L1      L2      L3     1
56      ------  ------
57
58
60
60
60
60
L1(1)=56
```

```
L1      L2      L3     1
72
72
72
72
76
77

L1(262) =
```

There are three potential outliers on the low side, 56, 57, and 58. There are two on the high side, 76 and 77.

EXAMPLE 16: What are the quartiles and the IQR for the cereal sodium data? (p. 67 of Agresti/Franklin)

Copy the list **CERSO** to **L1** and use the **1-Var Stats** function. You should find that **Q1 = 145**, and **Q3 = 225**. This means that the IQR = 225 – 145 = 80 mg. A potential outlier would be a value that is less than 145 – 1.5 x 80 = 25. There is one value less than this, the lowest value of 0. A potential outlier on the high side would be more than 225 + 1.5 x 80 = 345. This is greater than the maximum sodium value, so there are no outliers on the high side.

```
1-Var Stats
↑n=20
 minX=0
 Q1=145
 Med=200
 Q3=225
 maxX=290
```

The Five-Number Summary and Boxplots

The five numbers at the end of the 1-Var Stats report, the minimum, Q1, median, Q3, and the maximum, are often called the **five-number summary**. A **boxplot** is a graphical display of the five-number summary, and can also be used to identify potential outliers graphically. The boxplot has a "box" that stretches from Q1 to Q3, with the median marked as a line across the box. In lists where there are no outliers, a "whisker" from Q1 is extended to the minimum value, and another whisker is drawn from Q3 to the maximum. If potential outliers are present, then the whiskers stop at the most extreme value that is not an outlier, and the potential outliers are each marked individually.

EXAMPLE 17: Boxplot for the breakfast cereal sodium data (p. 70 of Agresti/Franklin)

With the cereal sodium data, **CERSO**, in **L1**, you begin setting up a boxplot in the **STAT PLOT** menu. Choose the first icon in the second row of types of graphs. Choose **L1** for **Xlist** and 1 for **Freq**. The **Mark** is of your choosing, used to indicate potential outliers.

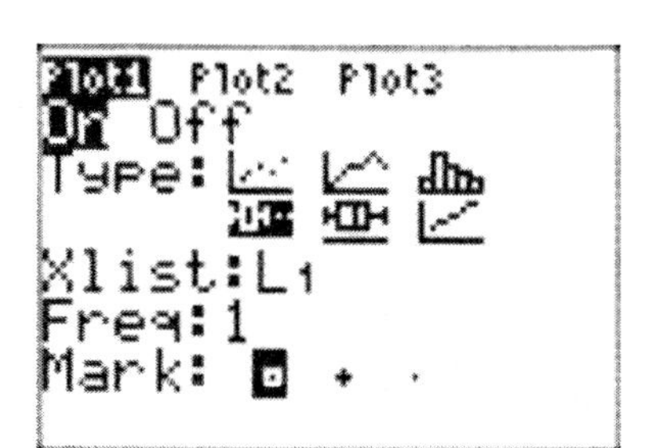

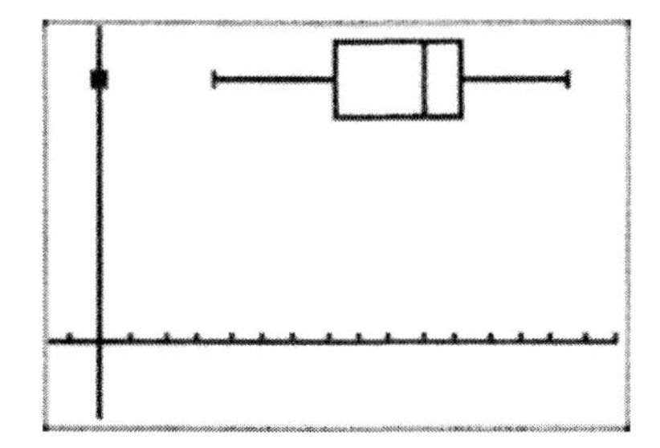

The one outlier is shown, but it is easy to miss, partially obscured by the vertical axis.

It is also worth noting that the calculator provides a second boxplot option, the icon in the middle of the second row. This simplified boxplot just shows the five-number summary and makes no note of potential outliers. This is usually not the preferred choice, since it makes evaluating the overall shape of the distribution harder in most cases, with the outliers included in the whisker.
This is a good opportunity, however, to use two different **STAT PLOT**s at the same time.

Leave **Plot1** as is. Turn **Plot2** on, choose the simple boxplot icon, and specify the same list of data, the sodium values in **L1**. Now press **GRAPH**. No need to adjust the window or use **ZoomStat**; the window dimensions are set for the sodium data.

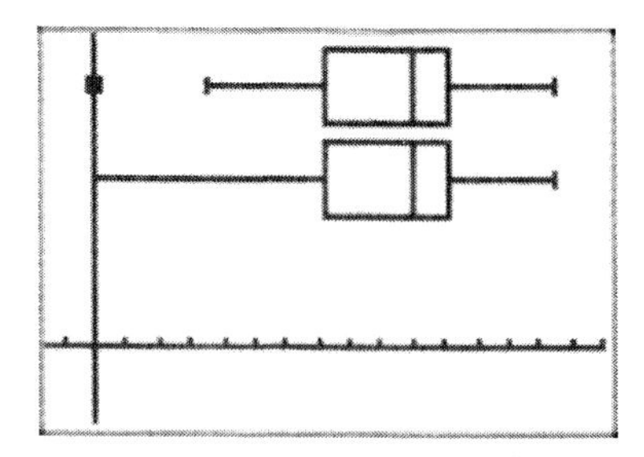

The lower boxplot, **Plot2**, is the simple boxplot. Notice that the left tail extends all the way to the minimum value of zero. This is the only difference in the plots – all other values are exactly the same.

Comparing Distributions with Boxplots
Boxplots are especially effective in making side-by-side comparisons of two distributions. Here is an example, comparing the heights of college female and male students, that uses the data is in lists **HGHTF** and **HGHTM**.

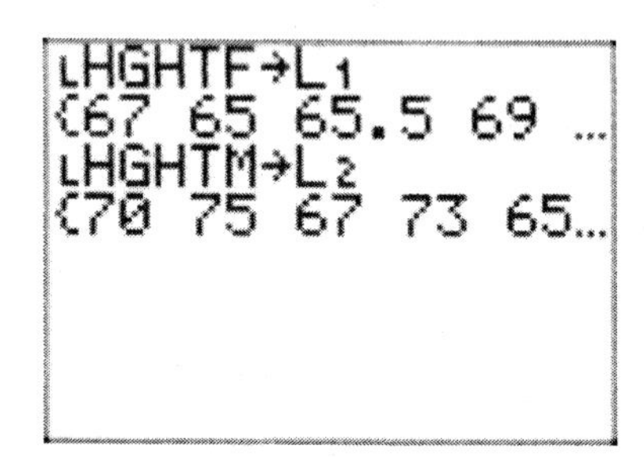

The setup for **Plot2** is changed so that it shows the data that is in **L2**. After **ZoomStat**, this produces the graph below.

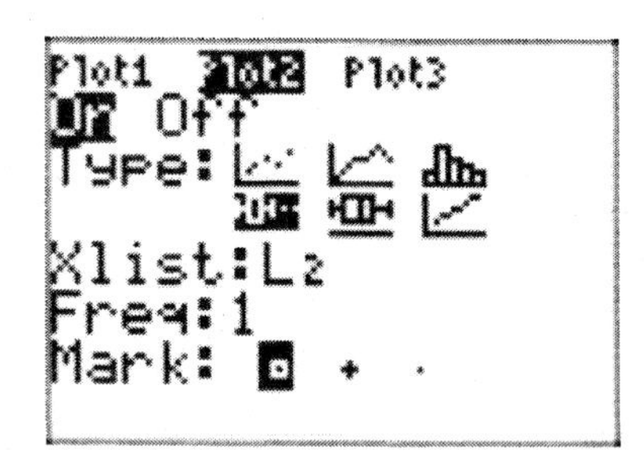

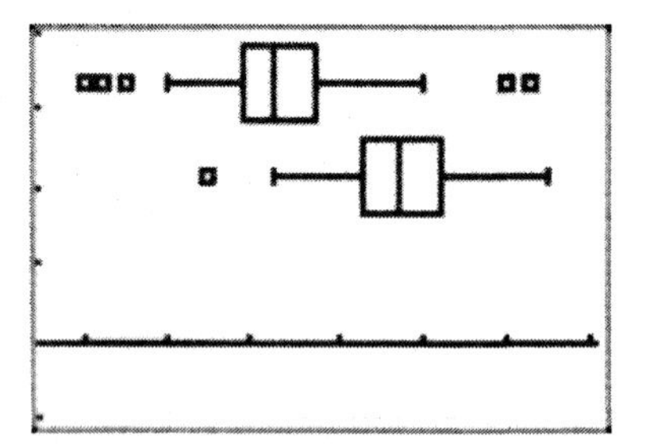

Comparisons of two distributions should address shape, center, and spread, and each of these characteristics is easy to see in a boxplot. Both the female and male distributions are roughly symmetric, the center of the male distribution is higher than the female distribution by about 6 inches, and the spread of the two distributions in similar.

Using z-scores

The **z-score** for and observation is the number of standard deviations that it falls from the mean. For sample data, the z-score is calculated as

$$z = \frac{\text{observation - mean}}{\text{standard deviation}}.$$

The z-score allows you to make easy comparisons of the relative positions of observations in different distributions. For example the heights of the male and female students revealed that the shortest male and the shortest female were both outliers. You can use **TRACE** with the boxplots above to find that the shortest female was 56 inches tall and the shortest male is 62 inches tall. The z-scores of each of these students can be calculated using the **LIST MATH** menu to find **mean(** and **stDev(**. Relative to their groups, the shortest students are similar; the shortest female student is about 3.14 standard deviations below the mean of the female group, and the shortest male is about 3.12 standard deviations below the mean of the male group

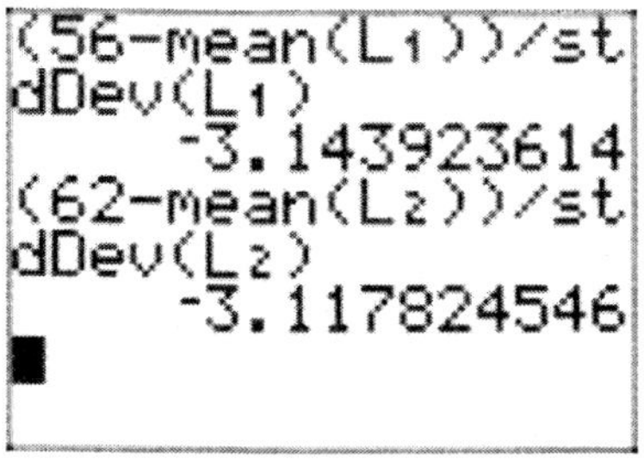

```
(56-mean(L1))/st
dDev(L1)
     -3.143923614
(62-mean(L2))/st
dDev(L2)
     -3.117824546
```

The z-score formula can be applied all at once to the data in lists **L1** and **L2**. This allows you to compare the two distributions on a common scale, shown below. Change the **STAT PLOT**s to show the distributions in **L3** and **L4**. The **WINDOW** parameters have been changed to z-score units, measuring with the "ruler" of the standard deviation in each distribution.

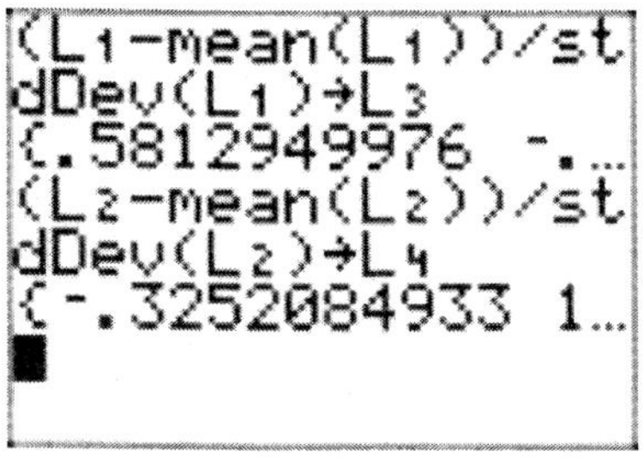

```
(L1-mean(L1))/st
dDev(L1)→L3
{.5812949976 -.…
(L2-mean(L2))/st
dDev(L2)→L4
{-.3252084933 1…
```

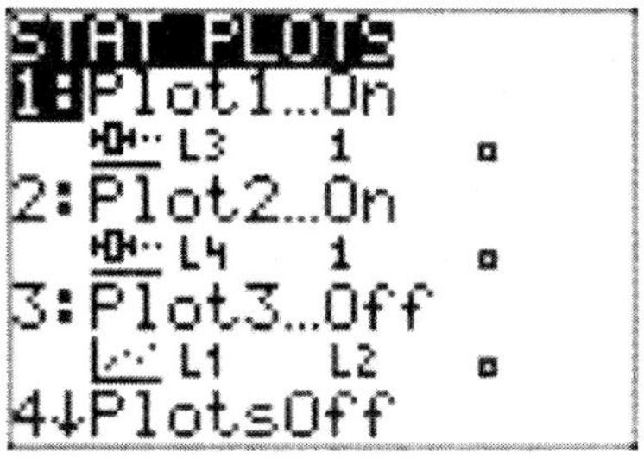

```
STAT PLOTS
1:Plot1…On
    L3   1
2:Plot2…On
    L4   1
3:Plot3…Off
    L1   L2
4↓PlotsOff
```

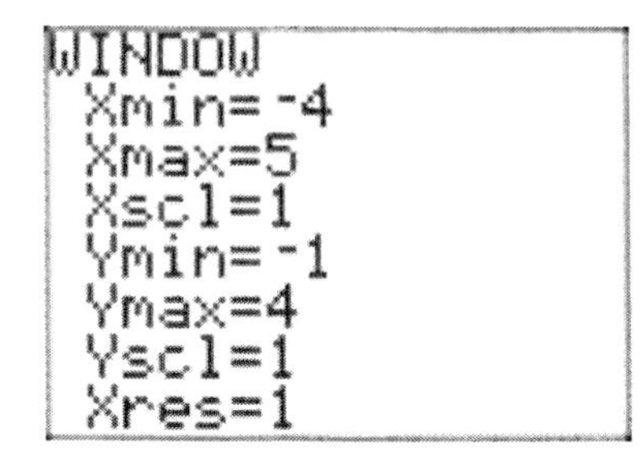

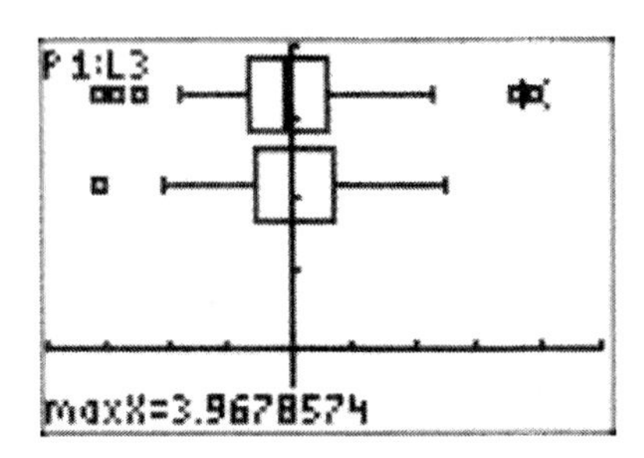

You can easily see with the comparative boxplots that the two shortest students are similarly placed in their respective distributions. It is also clear that the tallest female students are much taller, relative to their female peers, than are the tallest male students. The graph above uses **TRACE** to show that the tallest female student is nearly 4 standard deviations above the mean of her group.

2.6 How Can Graphical Summaries Be Misused?

The good advice in the textbook includes the requirement that both axes of a graph should be labeled and a heading included on all graphs. Since the TI-83 Plus family has a small screen and does not include text, this requirement shifts to the user to be sure that he or she understands the scales used. The **WINDOW** provides this information and should be a common button-press to make sure that the graph is correctly interpreted.

Chapter Three: Association: Contingency, Correlation, and Regression

All of these lists referred to in this Chapter are on the textbook CD with the names below. They are also collected together in the group AF03 on the CD.

The process for copying lists and groups from a computer to your calculator is described in Chapter 1 of this manual, beginning on page 4.

HDINT	Internet use, by country
HDGDP	Gross domestic product, by country
MBPRI	Mountain bike prices
MBWGT	Mountain bike weights
ALBAT	American league team batting averages, 2006
ALRUN	American league team runs scored per game, 2006
CERSO	Sodium amounts in breakfast cereals
CERSG	Sugar amounts in breakfast cereals
CPTEM	Average annual temperature in Central Park, NY
CPYR	Year in which Central Park temperature occurred

3.1 How Can We Explore the Association between Two Categorical Variables?

EXAMPLE 3: How can we compare pesticide residues for the food types graphically? (pp. 97-98 of Agresti/Franklin)

This will be quite similar to the bar graph we created for example 3 of chapter 2. Again, since the TI-83 Plus family does not deal easily with categorical data, we are "tricking" the calculator to give us a bar graph by supplying the percentages. This uses the histogram option under **STAT PLOT**. Our goal is to create a side-by-side bar graph like figure 3.2 on p. 98 of Agresti/Franklin. First enter the categories as numbers in **L1**. We'll need the numbers 1-5 to cover the two food types (twice), plus a "spacer" between the two levels of pesticide status. Put the percentages from table 3.2 (p. 97) in **L2**.

L1	L2	L3 2
1	27	------
2	77	
3	0	
4	73	
5	23	

L2(6) =

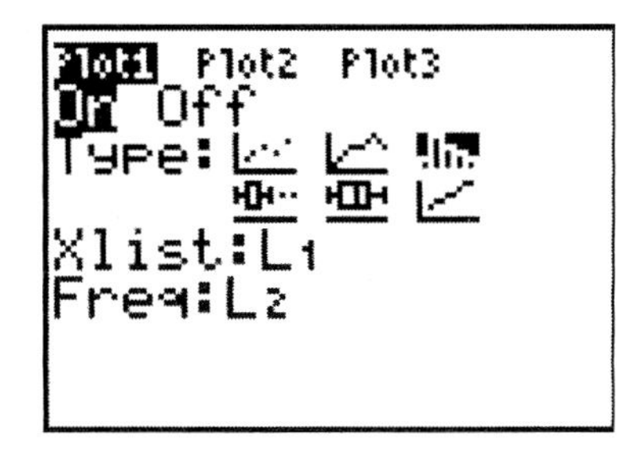

In the table above, "1" represents (no pesticide, conventional), "2" represents (no pesticide, organic), "3" inserts a space, "4" represents (pesticide, conventional), and "5" represents (pesticide, organic). Note that a zero is entered on the third line so that no bar appears at that location. Next press **STAT PLOT**. Select the histogram icon for **Plot1**, specify **L1** for **Xlist**, and **L2** for **Freq**. Because we need to be careful to make the x-values correspond to the categories, **ZoomStat** isn't useful here – we need to specify the **WINDOW** dimensions manually in order to produce the intended graph.

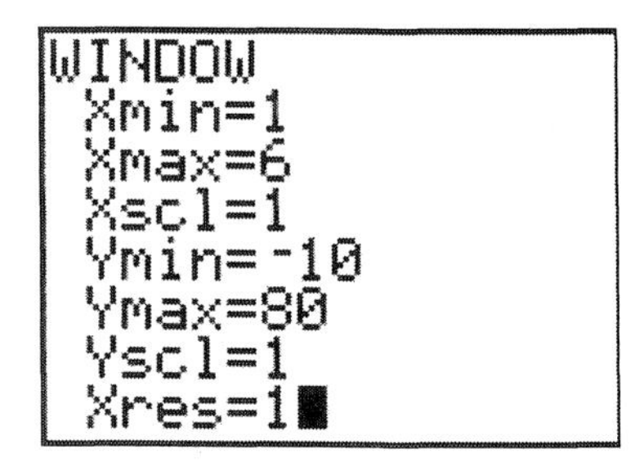

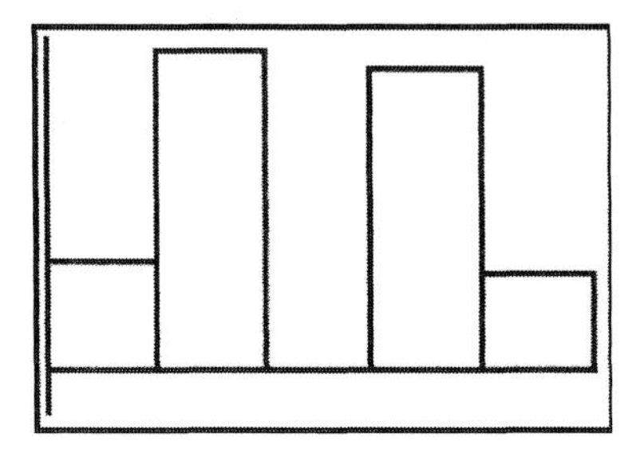

The choices for **Xmin** and **Xmax** accommodate the five categories; **Xscl** should be one. **TRACE** tells us that, for example, 73% of the conventional foods had pesticide residue.

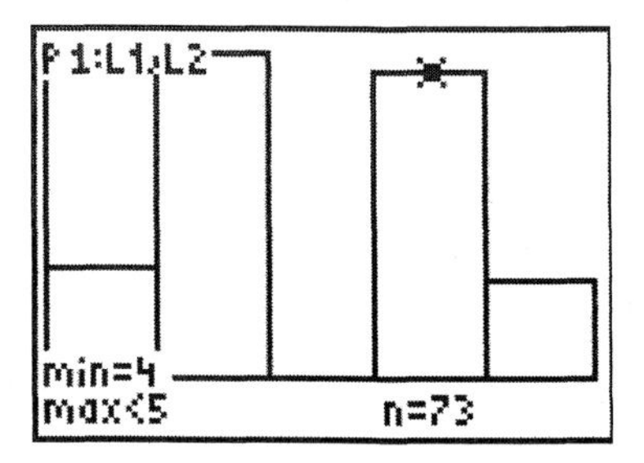

3.2 How Can We Explore the Association between Two Quantitative Variables?

EXAMPLE 5: Constructing a scatterplot for Internet use and GDP (pp. 104-105 of Agresti/Franklin)

The goal here is to create a graph to illustrate what relationship, if any, exists between Internet use (in % for each country) and GDP (per capita, in thousands of US dollars). The data appear in table 3.4 on p. 103 of Agresti/Franklin, from the computer file "human development" on the textbook CD. Enter or download the lists **HDGDP** and **HDINT** into lists **L1** and **L2**.

```
LHDGDP→L1
{6.09 11.32 25....
LHDINT→L2
{.65 10.08 37.1...
```

L1	L2	L3 1
6.09	.65	------
11.32	10.08	
25.37	37.14	
26.73	38.7	
25.52	31.04	
7.36	4.66	
27.13	46.66	

L1(1)=6.09

As in the textbook, you can make boxplots to illustrate the distribution of values in each of these lists.

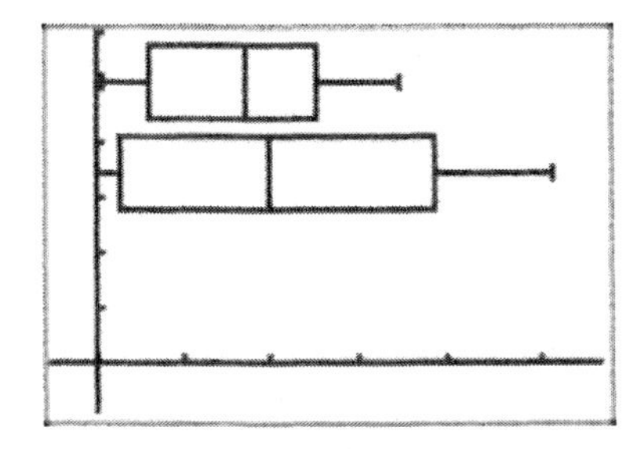

Our goal, though, is to show the relationship between the values for Internet use and GDP in each of the countries. This is done in a scatterplot. Begin the set up of the scatterplot by pressing **STATPLOT**. Choose the first icon in **Type** to make a scatterplot, with **L1** as the **Xlist** (explanatory variable, GDP) and **L2** as **Ylist** (response variable, Internet use). Using the information in the boxplots, you can make choices about a reasonable **WINDOW**, but **ZoomStat** is very helpful, too, with scatterplots. You might want to first try **ZoomStat**, then go back to the window to make the tick marks on the graph meaningful. These are determined by **Xscl** and **Yscl**.

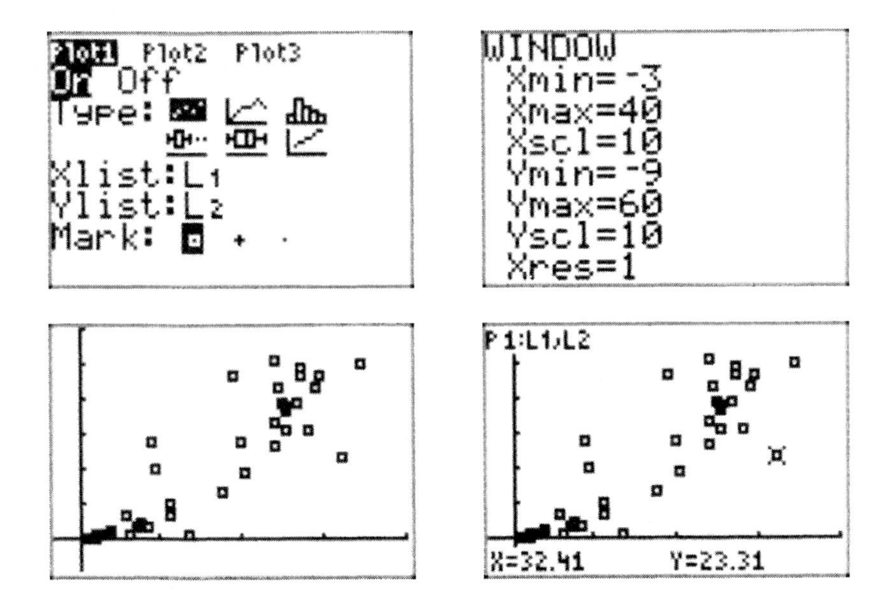

Use **TRACE** to see the coordinates of any of the individual points. The last screen shows the information for Ireland, (32.41, 23.31), which is the (GDP, Internet use %).

EXAMPLE 7: The correlation between Internet use and GDP (pp. 109 - 111 of Agresti/Franklin)

The correlation coefficient measures the strength of a linear association between two quantitative variables. The idea of correlation is illustrated graphically on p. 111 of Agresti/Franklin, with a scatterplot divided into four quadrants by a horizontal line at the mean y-value and a vertical line at the mean x-value. This can be reproduced on your calculator, using the **DRAW** menu (**2nd PRGM**). From the Home Screen, press **DRAW**, then select either the **Horizontal** or **Vertical** choices, then specify the mean of the appropriate list to make the line appear.

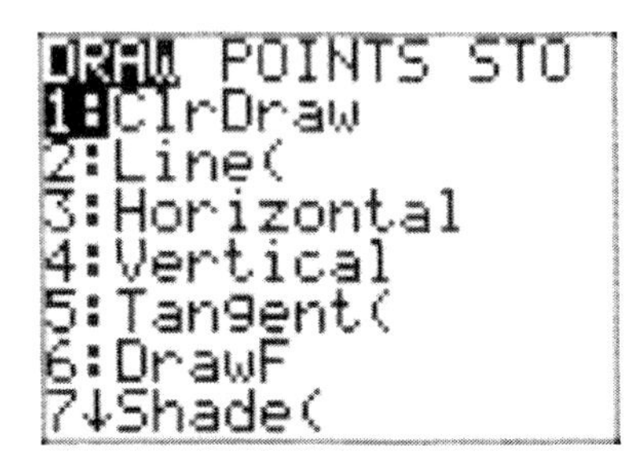

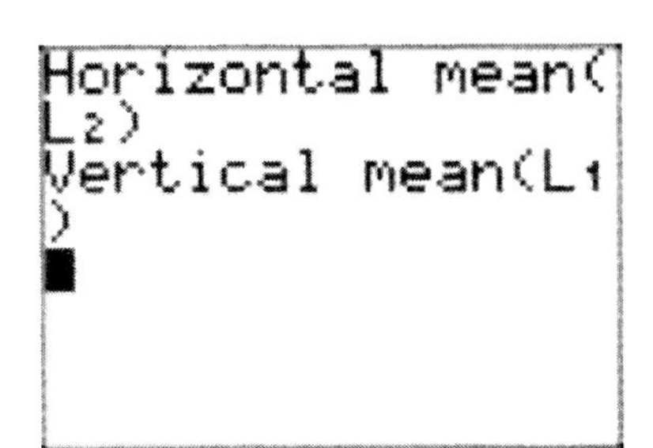

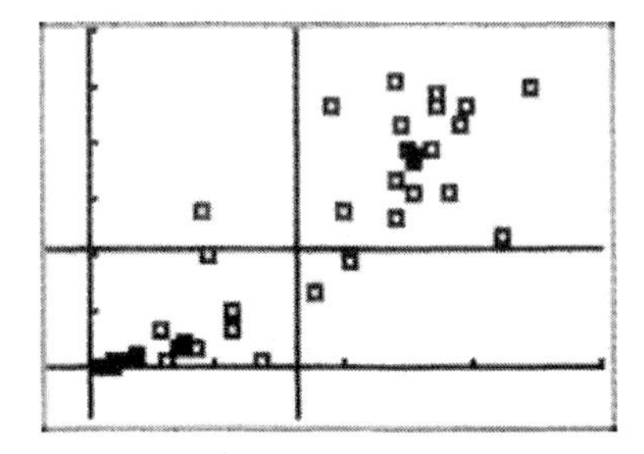

The points in the "first quadrant" have larger than average GDP and higher than average Internet use, while those in the "third quadrant" have smaller than average GDP and lower than average Internet use. Both of these contribute to a positive correlation coefficient, while the three points in the second and fourth quadrants take away from the correlation coefficient.

You can implement the formula shown on page 106 to calculate the value of the correlation coefficient. Calculate the z-scores for L1 and L2 and store them in L3 and L4, then multiply them together in L5. Finally add them up and divide by $n - 1 = 38$ to calculate the correlation coefficient, approximately 0.888.

```
(L1-mean(L1))/st
dDev(L1)→L3
{-.9345296894 -...
(L2-mean(L2))/st
dDev(L2)→L4
{-1.109526277 -...
L3*L4→L5
```

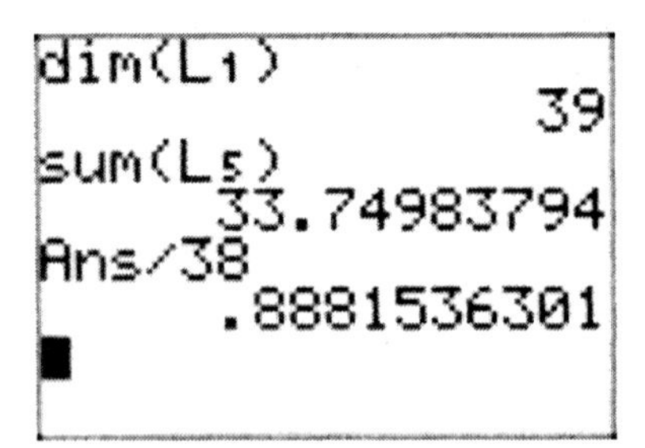

This procedure is awkward! As stated in the textbook, it's more important to understand how the correlation describes association, in terms of how it reflects, the relative numbers of points in the four quadrants, than to be able to calculate it. In the next section you'll see that the correlation coefficient is produced when you make a linear regression model, so there is no need to go through this calculation every time you need a correlation.

3.3 How Can We Predict the Outcome of a Variable?

EXAMPLE 9: How can we predict baseball scoring using batting average? (pp. 117-119 of Agresti/Franklin)

The team batting average and mean runs scored per game for each of the teams in the American League is presented in Table 3.5 on p. 117. Copy this data or use the lists **ALBAT** and **ALRUN** to enter the data into **L1** and **L2**. Since we want to predict team scoring from batting average, begin with a scatterplot with **Xlist** as **L1**, the batting averages, and **Ylist** as **L2**, the scoring averages. Set up the scatterplot as before to see the relationship between the teams' batting average and runs scored. From the graph it is evident that there is a positive association: teams with higher average score more runs, on average.

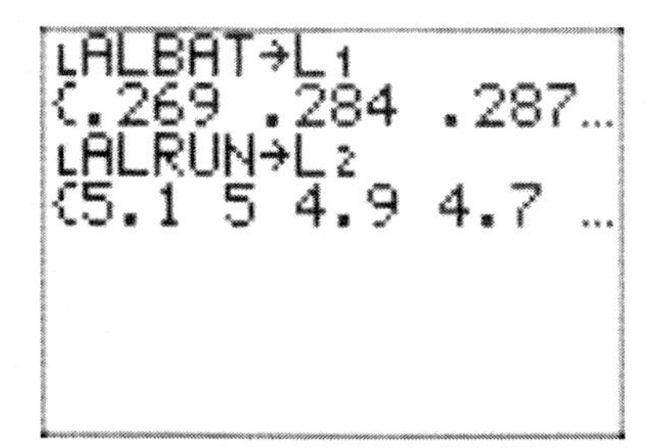

L1	L2	L3
.269	5.1	------
.284	5	
.287	4.9	
.271	4.7	
.272	4.7	
.285	5.7	
.274	4.7	

L1(1)=.269

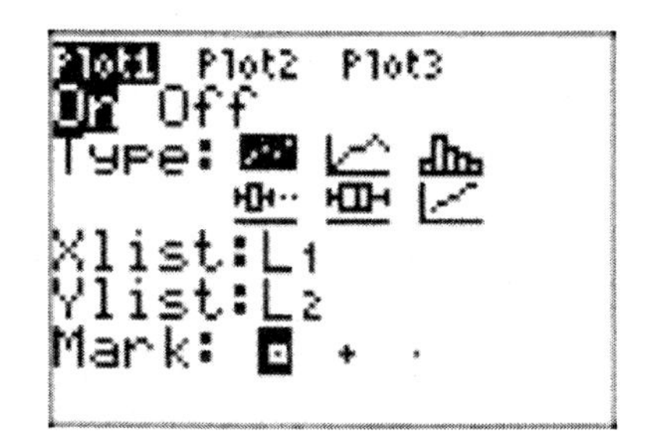

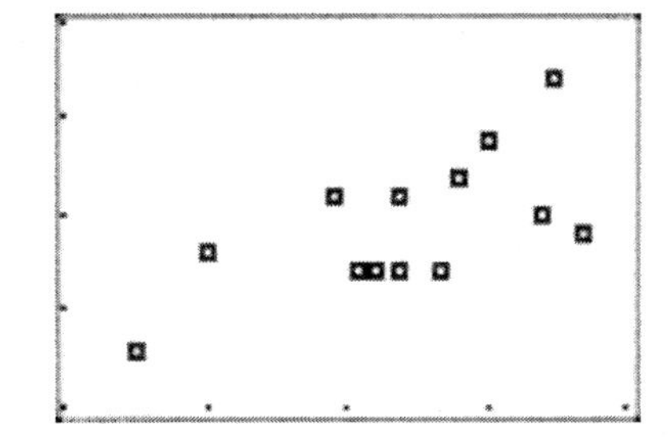

By making a linear model, called a **regression line**, you can answer the question of *how many* more runs are predicted to be scored for a given improvement in batting average.

> The **regression line** predicts the value for the response variable *y* as a straight-line function of the value *x* of the explanatory variable. Let $\hat{y}$ denote the **predicted value** of *y*. The equation for the regression line has the form $\hat{y} = a + bx$. In this formula, *a* denotes the ***y*-intercept** and *b* denotes the **slope**.

To calculate the regression line for this data, use press the **STAT** key then **CALC**. The choice **8:LinReg(a+bx)** is what you need, then specify the lists for the explanatory variable and response variable, **L1** and **L2**, respectively. The regression line for this data is shown: $\hat{y} = -2.28 + 26.4x$. A team batting .275 is predicted to score about 5 runs per game.

```
EDIT CALC TESTS
2↑2-Var Stats
3:Med-Med
4:LinReg(ax+b)
5:QuadReg
6:CubicReg
7:QuartReg
8:LinReg(a+bx)
```

```
LinReg(a+bx) L1,
L2
```

```
LinReg
 y=a+bx
 a=-2.28361244
 b=26.4354067
```

```
-2.28+26.4(.275)
                4.98
```

The regression report above, however, didn't produce the correlation coefficient as was promised in the previous section. This is because, by default, the calculator produces a simplified report. To get the full report you need to turn the "diagnostics" on. This term indicates that the full report will be useful in diagnosing whether the linear model is appropriate for this setting. To turn this on, choose **CATALOG** (**2nd 0**). This lists, alphabetically, every command and function on the calculator, including those functions like the diagnostics that don't appear in any of the menus. Since the Alpha indicator appears in the top right corner, you can jump to the **D** section of this long list by pressing the **D** key (**ALPHA** $\mathbf{x^{-1}}$). Scroll to find diagnostics. If you repeat the **LinReg(a+bx)** command, you get the full report, with *r* indicating the correlation coefficient.

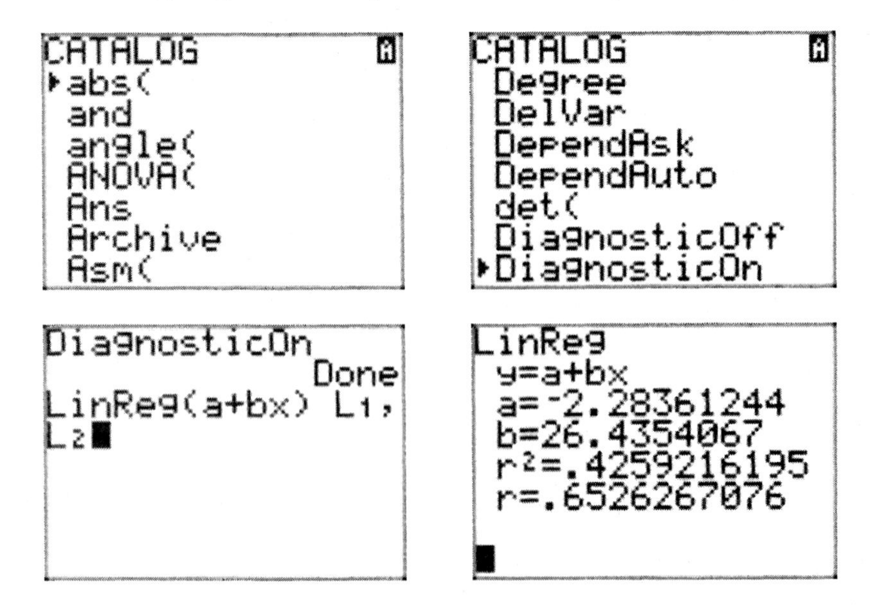

The regression line can be drawn on the scatterplot, and this is very useful in many situations. To do so, you could just type the equation into the **Y=** menu. However, the calculator is happy to do this automatically as part of the regression calculation. You need to add a third argument to the **LinReg** command to indicate which of the built in functions **Y1**, **Y2**, etc. will have the regression equation. After choosing the **LinReg** function and supplying the **L1** and **L2** lists, press the **VARS** button, right arrow to **Y-VARS**, and choose **1:Function**. Pressing **ENTER** chooses **Y1** and pastes this choice back to the home screen.

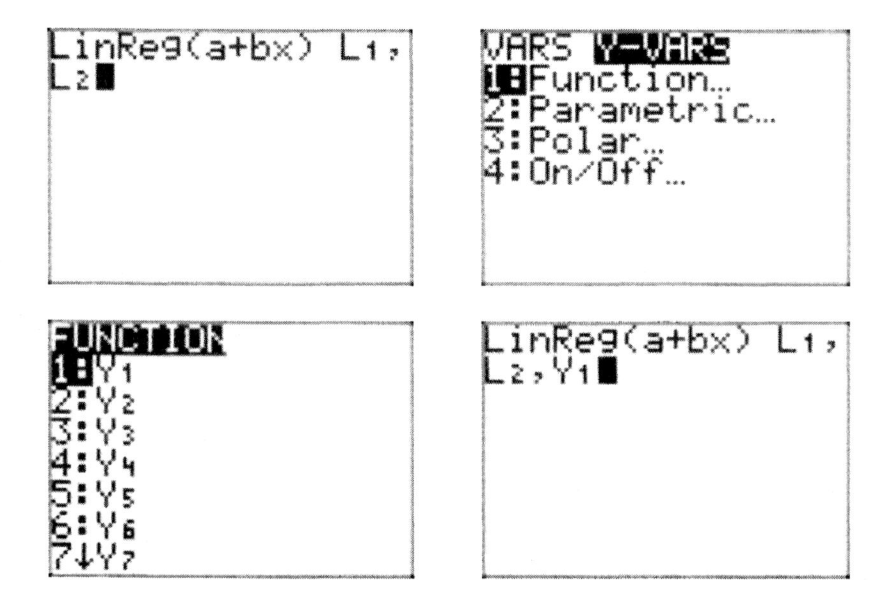

Press **Y=** to see the equation, and **GRAPH** to see it on the scatterplot.

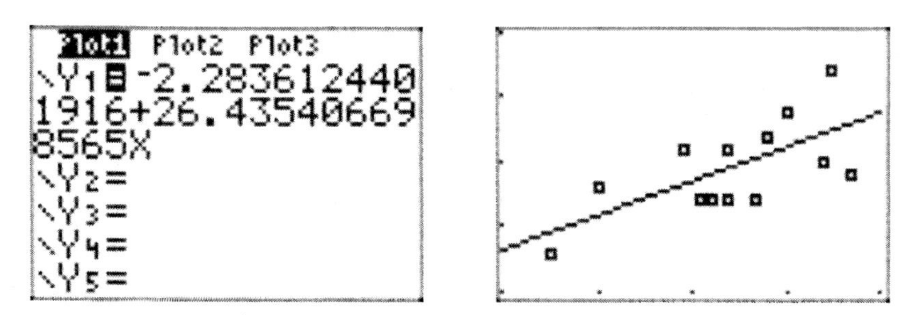

You can graphically use the regression line to make predictions of runs scored for a particular batting average with the **CALC** command (**2nd TRACE**). Select value from the **CALC** menu and enter .275. This shows the point on the regression line that corresponds to a prediction of about 5 runs scored for an average of .275. You can also scroll along the regression line with **TRACE**.

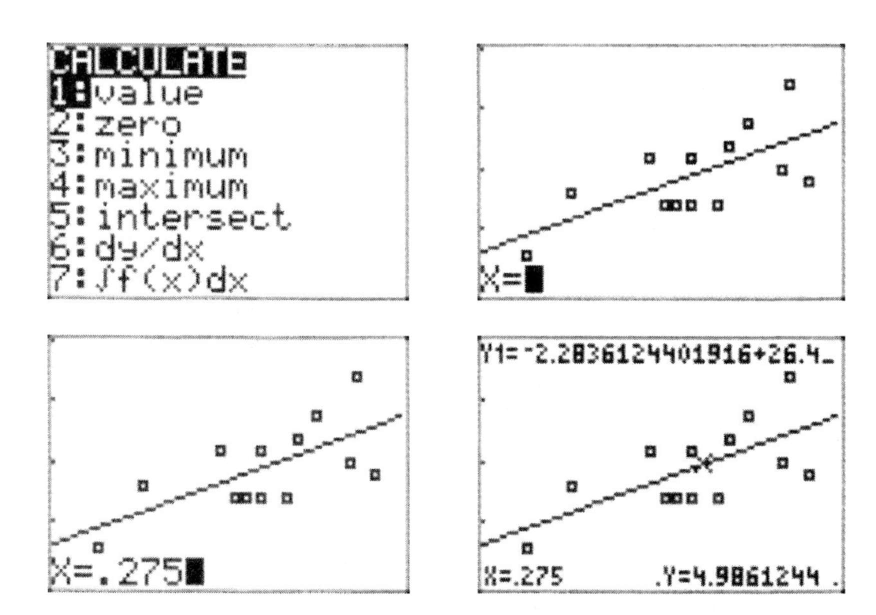

Exercise 3.41 (p. 128 of Agresti/Franklin)

Mountain bikes vary in price and weight, and this exercise examines the relationship between these two quantities. Enter the price data in **L1**, and the weight data in **L2** (from exercise 3.21), or copy the lists **MBPRI** and **MBWGT**. Can you predict the price if you know the weight of a bike is 30 pounds? Note that in the following the **Xlist** is **L2** and **Ylist** is **L1**.

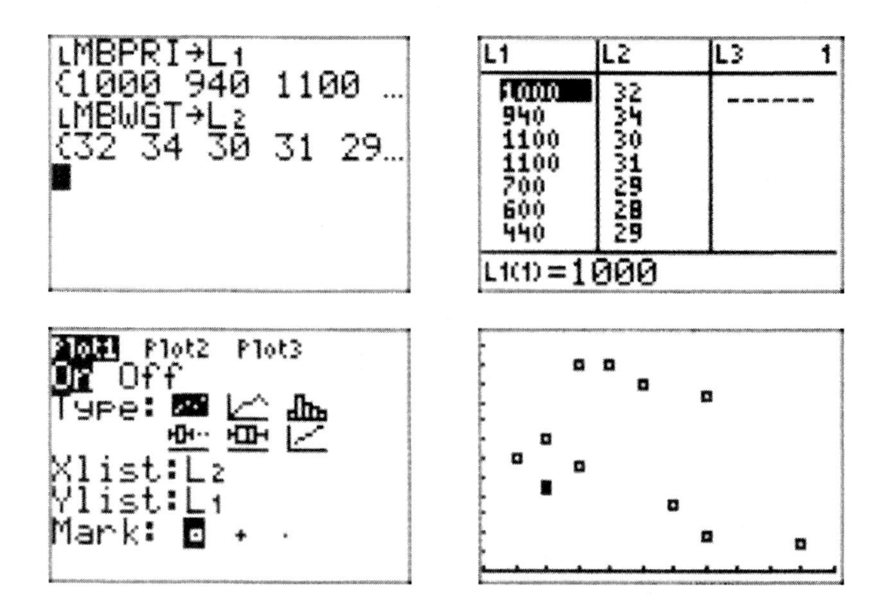

The **TRACE** function allows us to explore the "solid square", actually two points that are nearly on top of each other.

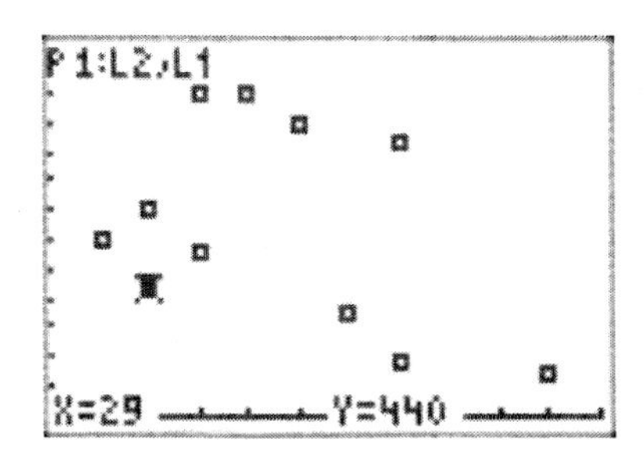

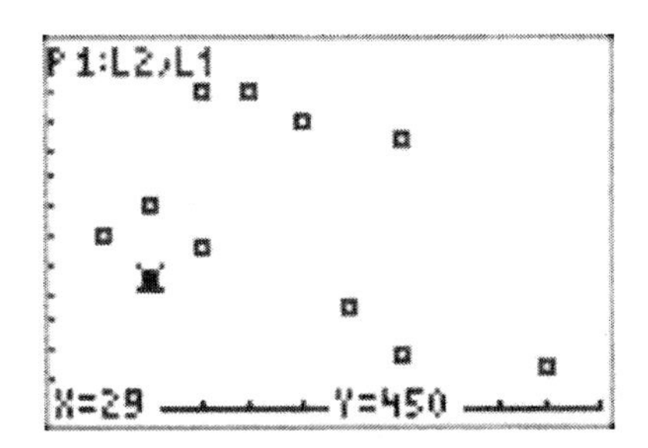

The Haro Escape A7.1 and the Giant Yukon SE weigh the same (29 lbs), and cost nearly the same ($440 and $450, respectively). Next let's get the linear model and correlation coefficient between price and weight. We should expect a negative slope, based on the scatterplot. The correlation is negative, too, indicating a negative association. Heavier bikes cost less, not more.

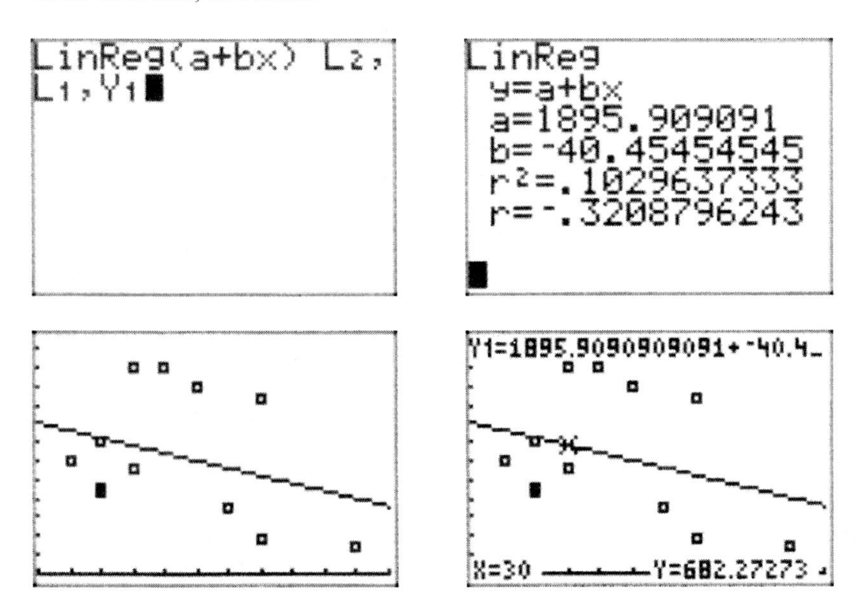

Using **CALC**, the predicted price of a mountain bike that weighs 30 pounds is $682.27. It doesn't seem, though, that this is a very helpful prediction. So let's consider the other variable that is given on p. 113, the type of suspension. This may help make more useful predictions.

We can create a scatterplot of price vs. weight like the one above, with different symbols for FU=full suspension and FE=front end suspension bikes. To accomplish this, we need to move the data points for the four full suspension bikes to say, **L3** and **L4**, leaving the front suspension pairs of data (the last eight bikes) in **L1** and **L2**. Then we can use **Plot1** and **Plot2**, both turned on at the same time, and using different symbols for the points.

L1	L2	L3
700	29	1000
600	28	1100
440	29	940
450	29	1100
550	30	
340	33	
180	34	

L3(5) =

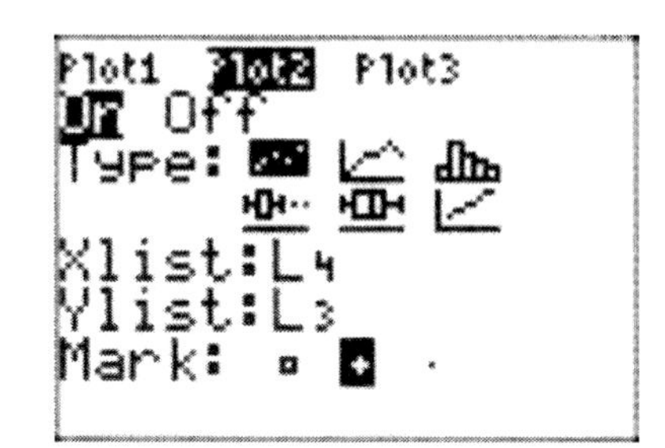

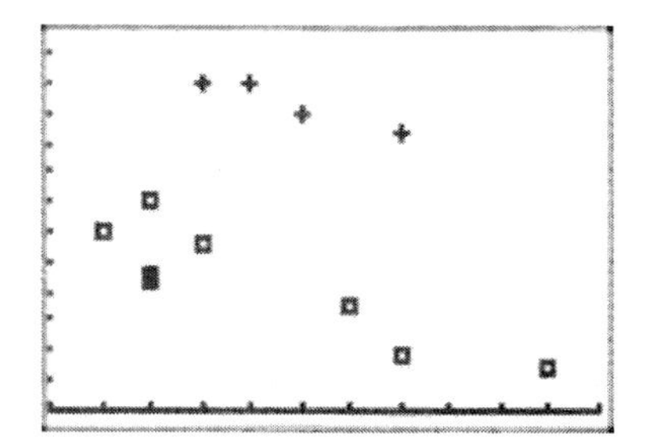

A regression line can be drawn for each of the two groups.

```
LinReg(a+bx) L2,
L1,Y1
```

```
LinReg
 y=a+bx
 a=2135.502646
 b=-54.95590829
 r²=.7892819424
 r=-.888415411
```

```
LinReg(a+bx) L4,
L3,Y2
```

```
LinReg
 y=a+bx
 a=2432
 b=-44
 r²=.9058823529
 r=-.9517785209
```

Notice that the correlation coefficients for each of these two separate regression lines indicate a much stronger correlation between weight and price among the bikes in each type, -.888 and -.952 for the FE and FU bikes, respectively. The graph below shows the two regression lines.

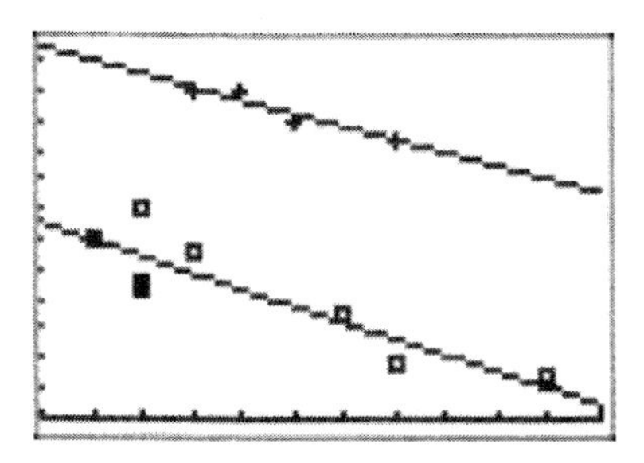

Using **CALC** you can make predictions of the price of a bike that weighs 30 pounds: \$486.82 if it is a front suspension bike, and \$1112 for the pricier full suspension model. These predicted prices are likely to be much more accurate than the single predicted price. Knowing the type of suspension provides very useful information.

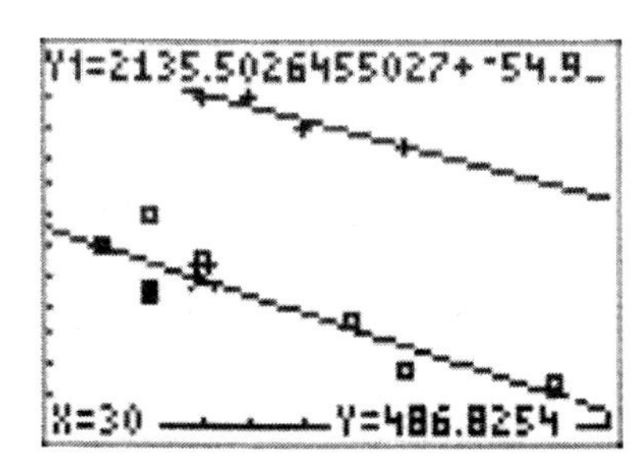

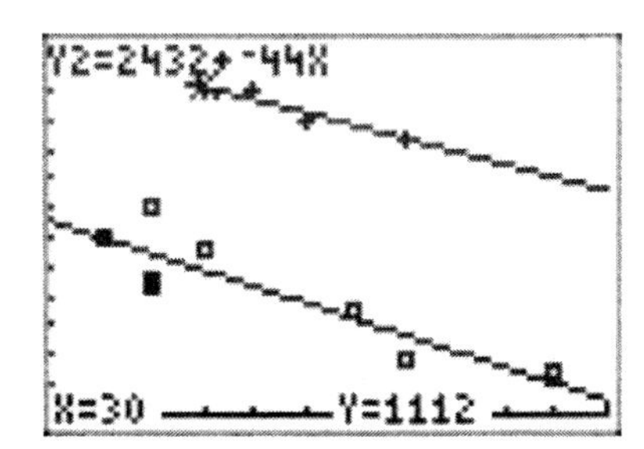

Exercise 3.33 (p. 126 of Agresti/Franklin)

The sugar and sodium content data for 20 breakfast cereals appears in table 2.3 on p. 34 of Agresti/Franklin and is available in the lists **CERG** and **CERSO**. Copy the sugar data into **L1** and the sodium data into **L2**. Here are the scatterplot and regression line.

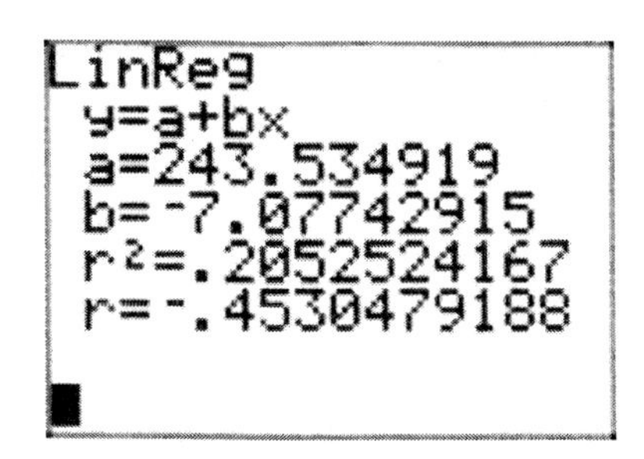

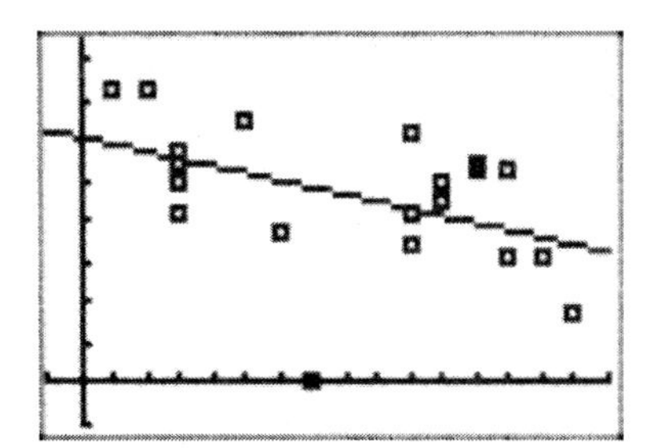

> The **prediction error** for an observation, which is the difference $y - \hat{y}$ between the actual value and the predicted value of the response variable, is called the **residual**.

Use **TRACE** to find the coordinates of the point for the sugar and sodium content of All Bran, (5, 260). Use **CALC** to find the predicted value of sodium for a cereal with 5 grams of sugar: 208.15.

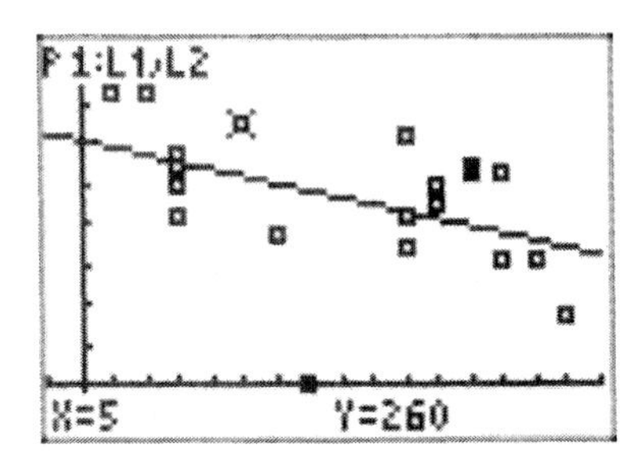

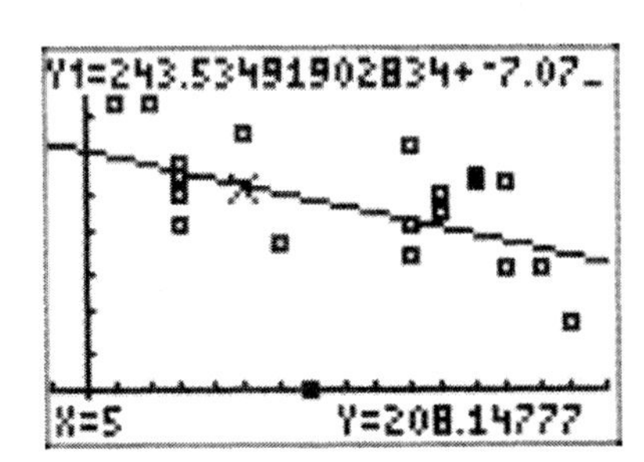

The difference between these values is the **residual** for All Bran, **260 – 208.15 = 51.85**. You recreate the graph that appears on p. 126, a histogram of all the residuals. Your calculator will make a list of all the residuals (**RESID**) as part of the regression calculation. (This only happens when Diagnostics are turned on.) You can find this in the **LIST** menu, and create a histogram.

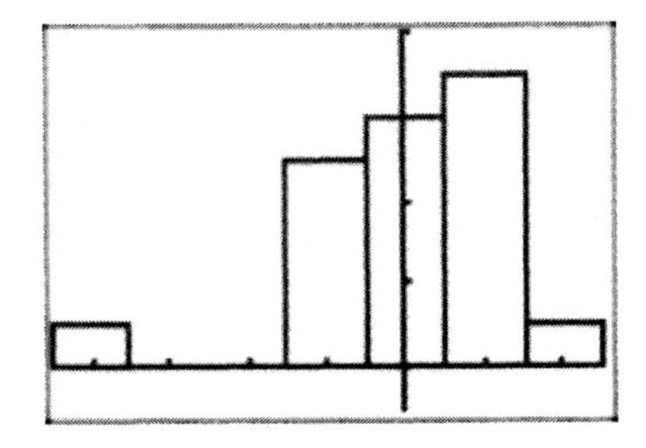

There is one residual that is quite different from the others. You can add a boxplot in **Plot2**, choosing the **RESID** list, to show that this very low value is an outlier among the residuals.

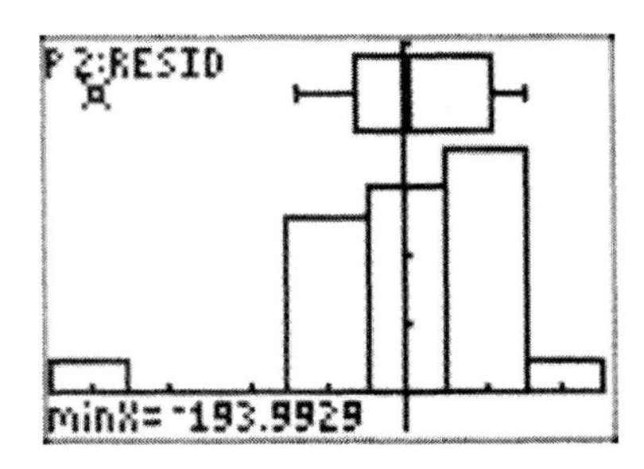

This residual value corresponds to Frosted Mini Wheats, which have an average amount of sugar (middle of the x values) but no sodium at all ($y = 0$).

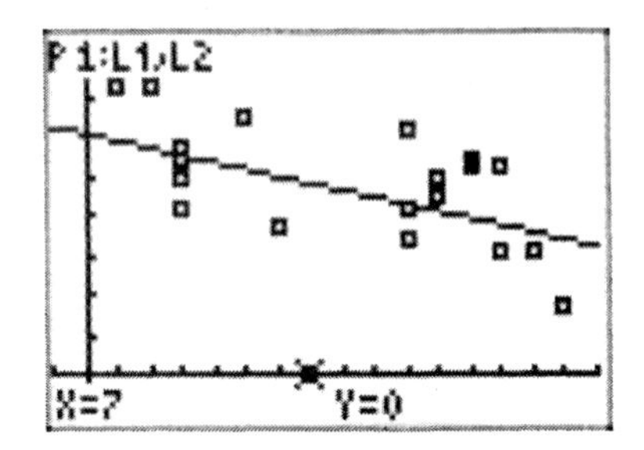

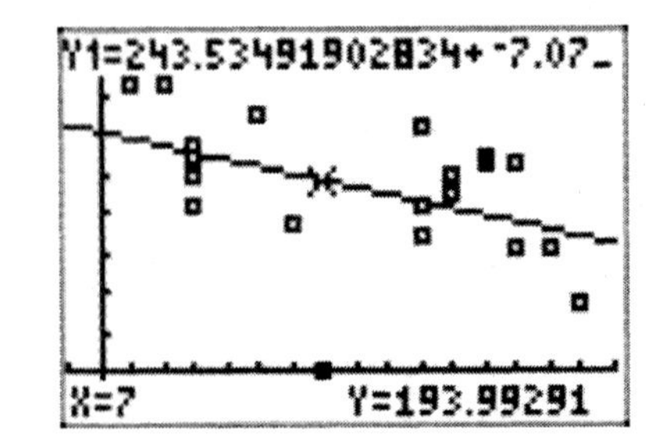

An observation is said to be **influential** when its presence or absence makes a big difference in the results of a regression analysis. Does the Frosted Mini Wheats, with its large residual, qualify? You can delete the values for this point from **L1** and **L2** (they are the first entries; delete them with the **DEL** key). The results of the subsequent linear regression is shown, along with the new regression line on the graph – the new line is slightly higher and slightly steeper, but not really much different from the original.

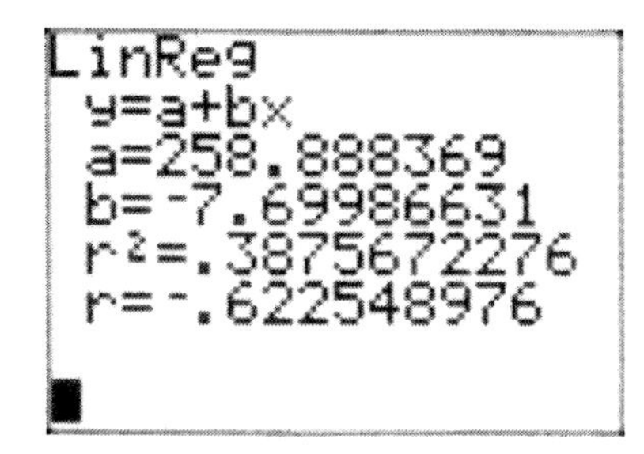

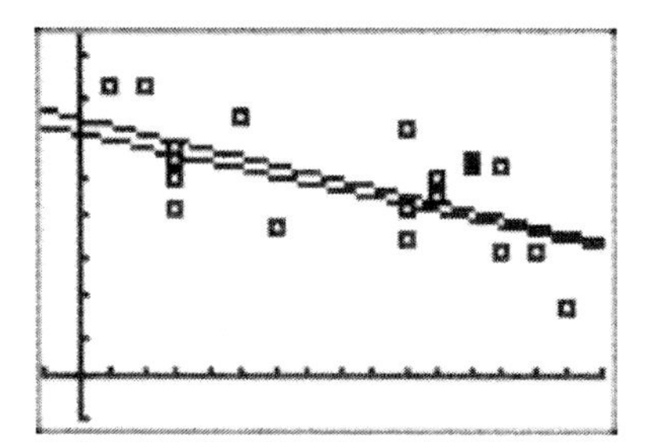

Since the "before" and "after" lines are not very different, the Frosted Mini Wheats are not influential in this regression. Generally, influential points have large residuals *and* are located either to the far left or far right of the data. Frosted Mini Wheats have a large residual, but are located in the center of the x-values.

Chapter Four: Gathering Data

There is no data from the CD that is used in this chapter.

4.2 What Are Good Ways and Poor Ways to Sample?

EXAMPLE 5: Auditing the accounts of a school district (pp. 166-167 of Agresti/Franklin)

In this example, the authors use a table of random numbers to generate a simple random sample of size 10 from a set of accounts that are numbered from 1 to 60. Your calculator can easily make random numbers that can be used for random sampling. We will use the **randInt** function, which can be found in the **MATH PRB** menu. Choose **5:randInt.** This function randomly generates an integer (or a list of several integers) in a certain range. So we need to supply a lower and upper bound for these integers. In this example, the lower bound is 1, and the upper bound is 60. We can also give a third argument for how many random integers we would like in this range.

MATH NUM CPX PRB
1:rand
2:nPr
3:nCr
4:!
5:randInt(
6:randNorm(
7:randBin(

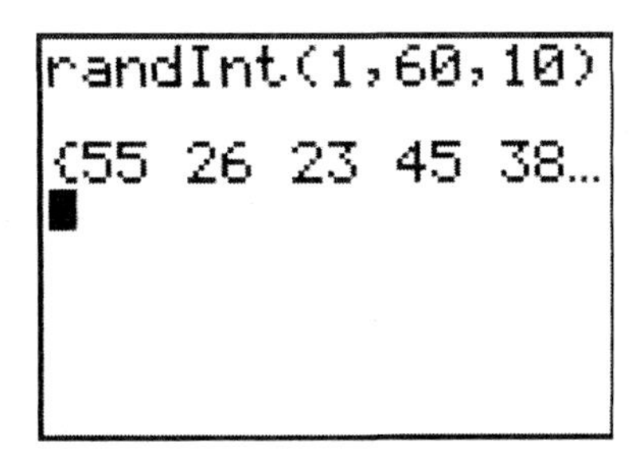

NOTE: You will get different results on your calculator! That's part of being random!

Above, you can see that 10 integers were requested. The first five are visible: 55, 26, 23, 45, and 38. Scroll right to see the rest. We can't select account 33 and 38 twice, so at this point we have actually only selected eight accounts. Let's use the **randInt** function to generate integers one-at-a-time until we have a total of ten unique account numbers from 1 to 60.

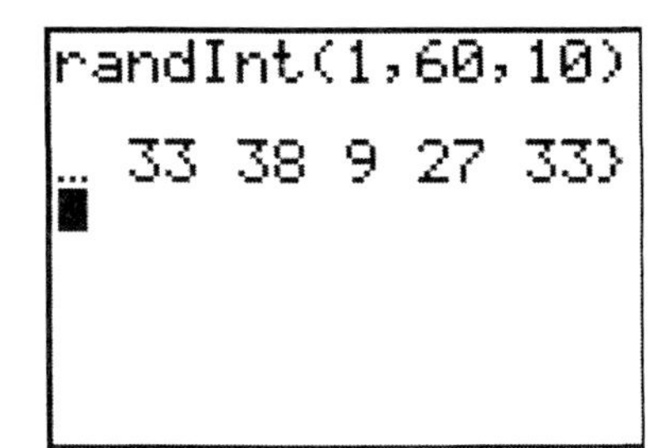

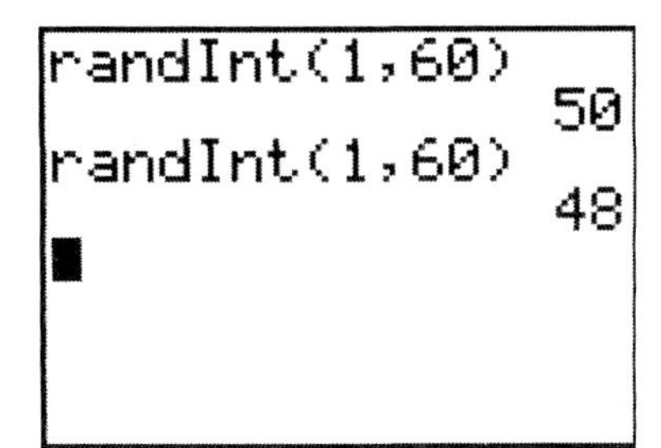

Twice more did the trick; now we have our simple random sample. We should select accounts 55, 26, 23, 45, 38, 33, 9, 27, 50 and 48.

Exercise 4.15: Sample students (p. 173 of Agresti/Franklin)

To randomly sample 3 students from 50, we will use a lower integer bound of 1 (same as last time), and an upper integer bound of 50. Remember that you will get different integers from your calculator!

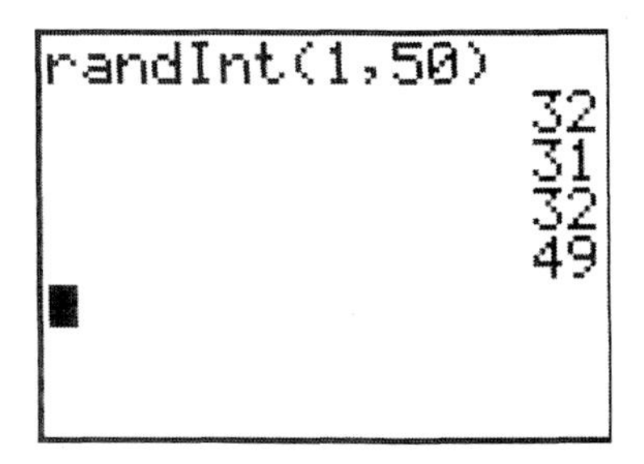

The method above is maybe the best way to select a simple random sample using the TI-83. The **randInt** function was used, along with the appropriate lower and upper integer bounds. After generating the first integer, you can just press **ENTER** again, and another integer will be generated. You can see above that since 32 appeared a second time, **ENTER** was pressed four times in order to get three unique numbers. The random sample of three students from the class is: students #32, 31, and 49.

Chapter Five: Probability in Our Daily Lives

There is no data from the CD that is used in this chapter.

5.1 How Can Probability Quantify Randomness?

EXAMPLE 2: Is a die fair? (pp. 214-216 of Agresti/Franklin)

We can simulate the repeated rolling of a fair die with the TI-83, adapting the **randInt** function that we used in Chapter 4. To create a graph like that shown in Figure 5.1, we need two lists. The trial number will be stored in **L1**, and the cumulative proportion will end up in **L5**. Let's start with the list of trial number, 1, 2, …, 100. This would be tedious to type in manually, so instead let's use the sequence function (**seq**), which creates a list of values according to a formula or pattern. Here the pattern is very simple: consecutive integers. Find the **seq** function in the **LIST OPS** menu. The format for this it **seq(formula, variable, start, end)**. Here the formula is just the same as the variable, so that explains the two X's in the expression. The consecutive numbers are put in **L1**.

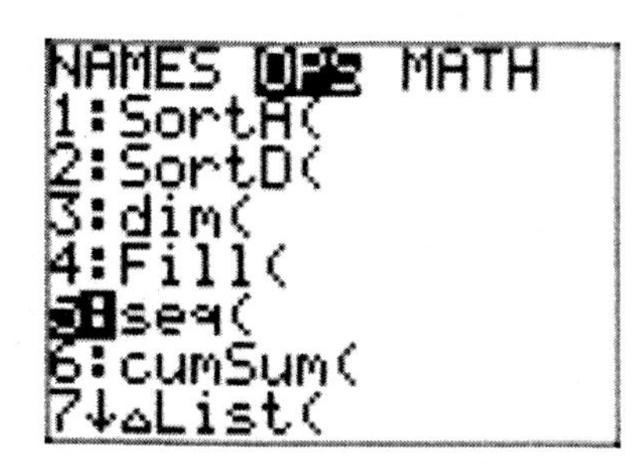

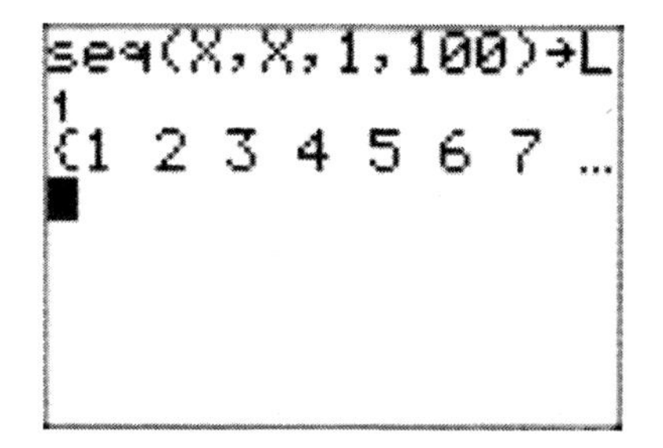

The next task is to create the rolls of the die, 100 times, using the **randInt** function. Did a six appear on each of the rolls? This question is answered by using the **TEST** (**2nd MATH**) menu to check for whether each of the values in **L2** is a six or not. Enter the list **L2**, then choose the equal sign from the **TEST** menu.

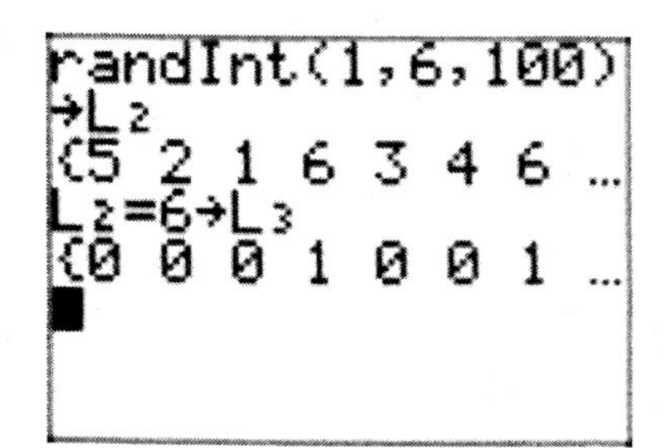

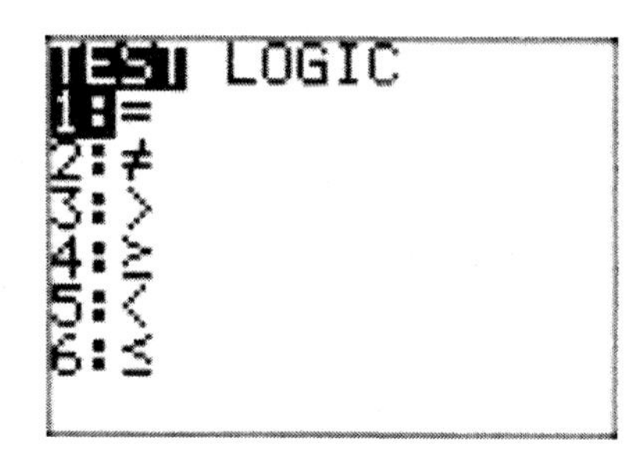

This test either produces a **0** (when the roll is not a six) or a **1** (when the roll is a six).

The cumulative proportion for the instances of six is done in two steps: first a cumulative sum (in the **LIST OPS** menu), then dividing by the number of trials to get a cumulative proportion.

NAMES OPS MATH
1:SortA(
2:SortD(
3:dim(
4:Fill(
5:seq(
6:cumSum(
7↓ΔList(

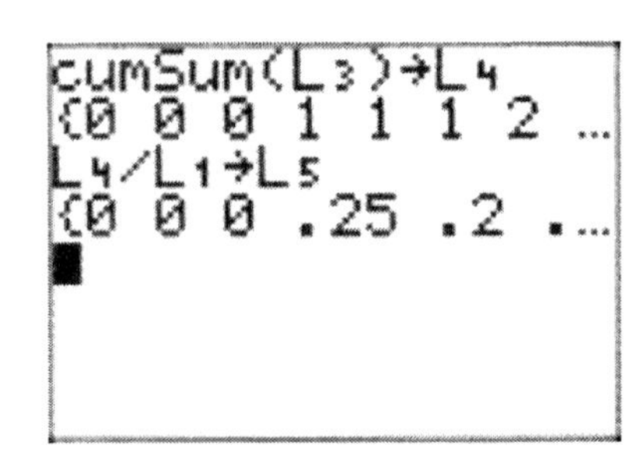

Now we are ready to illustrate the long run behavior of rolling a die, producing a graph similar to figure 5.1 on p. 216 of Agresti/Franklin. Note that since these are independently generated random trials, the shape of the graph won't match the textbook, and if you follow along your graph will be different still. Set up the **STAT PLOT** with **Xlist** = **L1**, **Ylist** = **L5**. Use the second **Type** icon, the **xyLine** plot. Since there are many points, it will look better with the smallest plotting symbol for the **Mark**. Press **ZoomStat** to easily scale your graph.

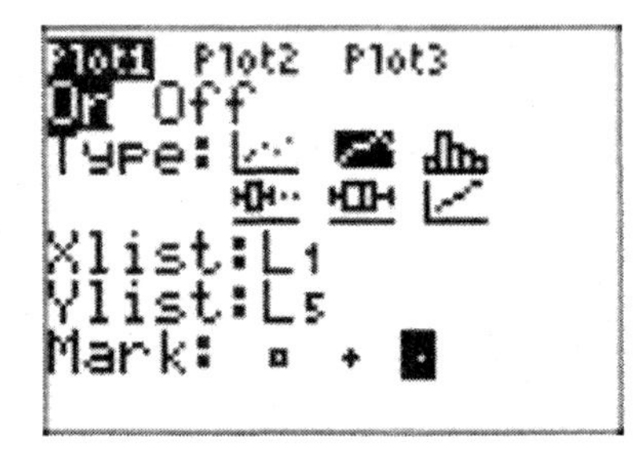

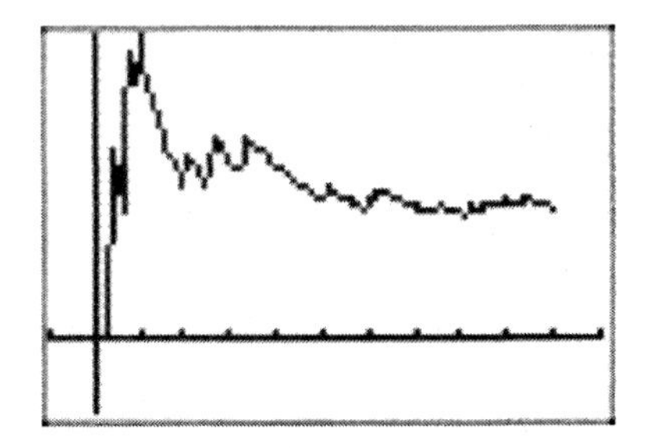

Note that while this isn't exactly the same as figure 5.1, the overall pattern is the same: large variability at the beginning, then settling down to more consistent values. You can use **TRACE** to find that at the end of our 100 trials, the cumulative proportion was .17. You can also scroll in the **STAT** editor to see this value.

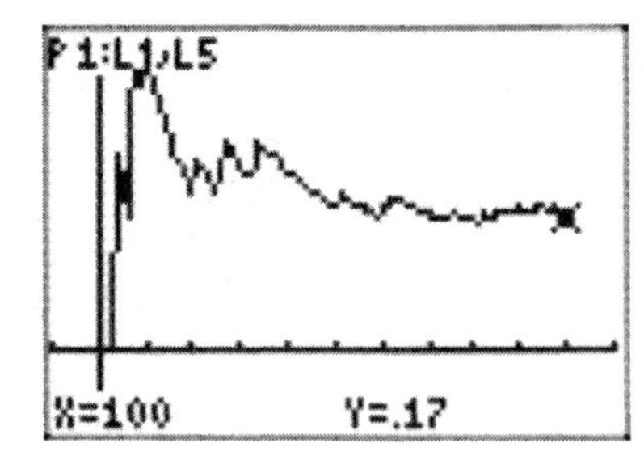

L3	L4	L5
0	17	.17895
0	17	.17708
0	17	.17526
0	17	.17347
0	17	.17172
0	17	

L5(100) =.17

Your results may vary from this. To illustrate the kind of variability that is present from sample to sample, here are two more examples of 100 trials, ending with cumulative proportions of .13 and .18. What was your value?

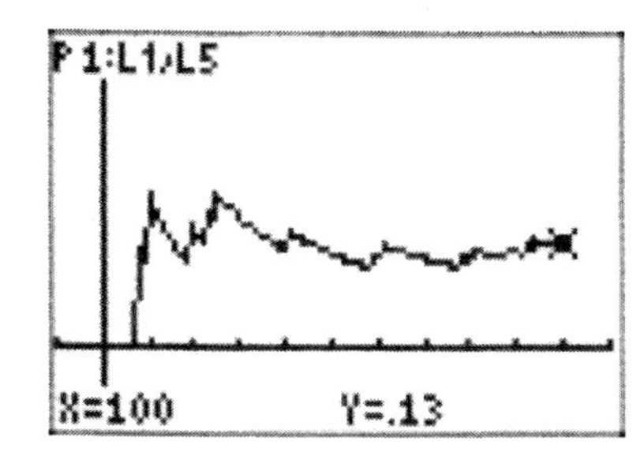
P1:L1/L5
X=100
Y=.13

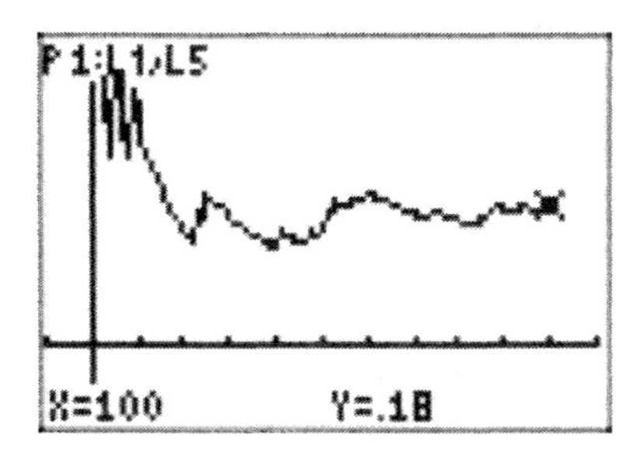
P1:L1/L5
X=100
Y=.18

Chapter Six: Probability Distributions

The only list used in this Chapter is HGHTF, the heights of a random sample of female students. This is on the textbook CD.

The process for copying lists and groups from a computer to your calculator is described in Chapter 1 of this manual, beginning on page 4.

6.1 How Can We Summarize Possible Outcomes and Their Probabilities?

> A **random variable** is a numerical measure of the outcome of a random phenomenon.

The outcomes of a random variable are summarized in a **probability distribution**. The Agresti/Franklin textbook shows you how to use statistical tables, graphs, and formulas to find probabilities and percentiles for certain probability distributions. This chapter presents methods to use your TI-83 Plus family calculator to do the same, often with greater efficiency.

EXAMPLE 2: How Many Home Runs Will the Red Sox Hit in a Game? (pp. 269-270 of Agresti/Franklin)

Table 6.1 presents a probability distribution for a discrete random variable, the number of home runs hit in a game by the Boston Red Sox in 2004. Enter the values into lists **L1** and **L2**. You can use the **sum** function, **2nd LIST MATH 5:sum(**, to verify that this is a legitimate probability distribution, with the probabilities adding up to 1.

L1	L2	L3 2
0	.23	------
1	.38	
2	.22	
3	.13	
4	.03	
5	.01	

L2(7) =

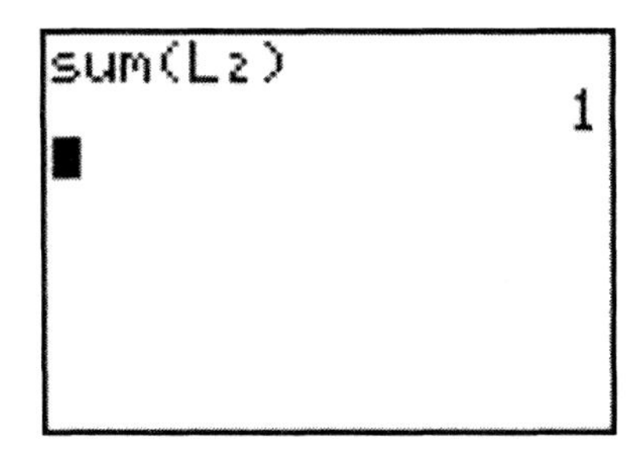

You can make a plot of the probability distribution similar to Figure 6.1. Set up a histogram with **Xlist** = **L1** and **Freq** = **L2**. This is the method to specify the values and the corresponding probabilities. For the most part, **ZoomStat** will produce an unacceptable histogram, but it isn't difficult to predict window parameters to make the graph fit nicely. Below the **Xmin** and **Xmax** are chosen with half-integer values to center the bars of the histogram over the integer values. The value **Xscl** = 1 makes a bar for each integer, and the **Ymax** must be large enough to fit the largest probability, 0.38.

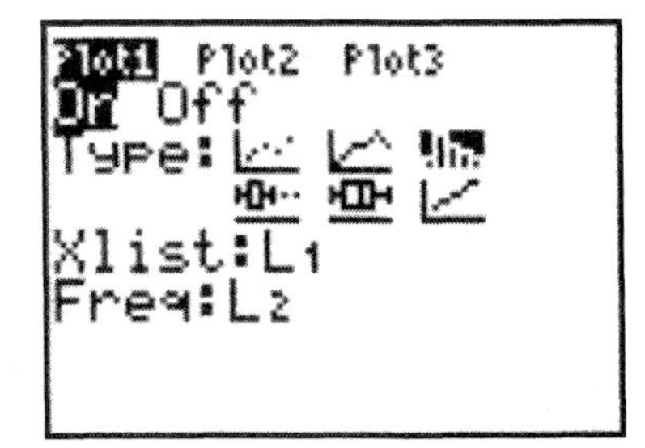

WINDOW
Xmin=-.5
Xmax=5.5
Xscl=1
Ymin=-.1
Ymax=.4
Yscl=1
Xres=1

The graph can be traced to show the value of any of the probabilities.

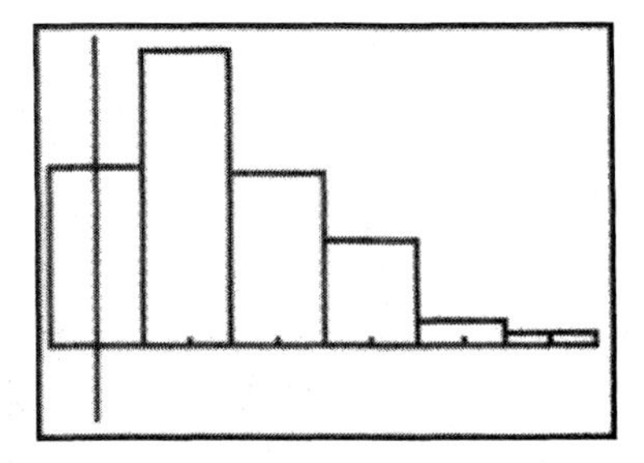

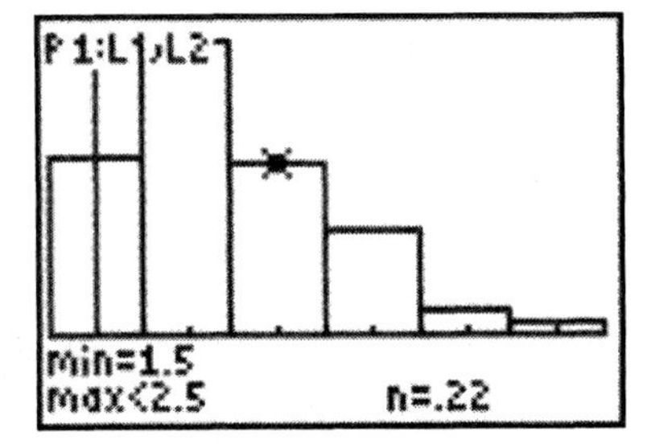

EXAMPLE 3: What's the Expected Number of Home Runs in a Baseball Game? (pp. 271-272 of Agresti/Franklin)

You can implement the formula for the mean of a discrete random variable, $\mu = \sum x \cdot P(x)$, using lists. **L1** has the values x an **L2** has the corresponding probability. Multiply these together, store the products in **L3**, and then sum **L3** to find the mean number of home runs, 1.38 per game.

L1	L2	L3
0	.23	------
1	.38	
2	.22	
3	.13	
4	.03	
5	.01	

L3 =L1*L2

L1	L2	L3
0	.23	0
1	.38	.38
2	.22	.44
3	.13	.39
4	.03	.12
5	.01	.05

L3(1)=0

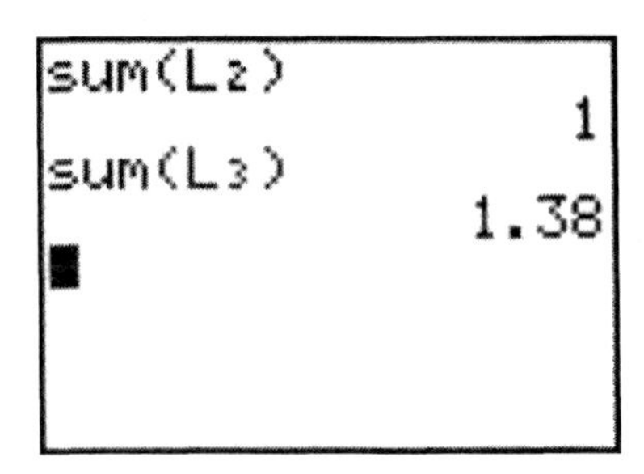

You can get a more complete numerical summary of the probability distribution using **1-Var Stats**. This is done by specifying the values and their corresponding probabilities, **L1** and **L2**.

```
1-Var Stats L1,L
2
```

```
1-Var Stats
 x̄=1.38
 Σx=1.38
 Σx²=3.16
 Sx=
 σx=1.120535586
↓n=1
```

Note in the report that $n = 1$, which is the sum of the probabilities. There is no value for the sample standard deviation **Sx**, since $n - 1$ would work out to be zero in this case. It does calculate a value for σ_x, the standard deviation of the probability distribution, which is discussed in more detail in sections 6.2 and 6.3. See Exercise 6.85 for the formula for this calculation.

You can scroll down in the report to find the quartiles of the probability distribution, too, which correspond to where the cumulative probabilities reach .25, .50, and .75. You can calculate the cumulative probabilities of the distribution of home runs using the **cumSum** function, in the **LIST OPS** menu. Do you see how the cumulative probabilities help you find the values in the report?

```
1-Var Stats
↑n=1
 minX=0
 Q1=1
 Med=1
 Q3=2
 maxX=5
```

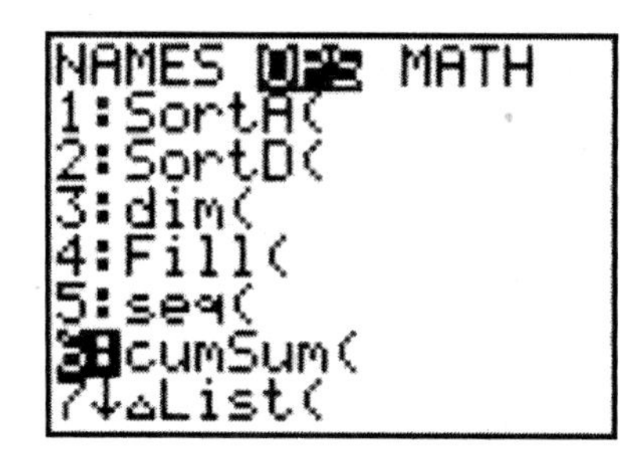

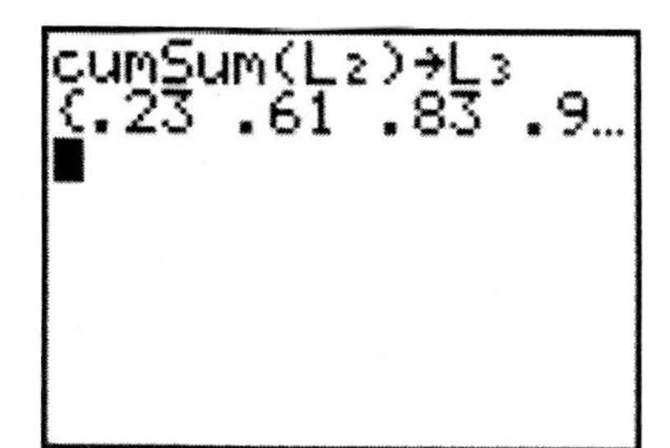

L1	L2	L3
0	.23	.23
1	.38	.61
2	.22	.83
3	.13	.96
4	.03	.99
5	.01	1

L3(1)=.23

Nearly 40% of the baseball games have 1 home run, so it is not surprising that this is **Q1** and also the median value.

These techniques can be used with a categorical variable with only two categories by designating them with the numerical codes 0 and 1.

Try it! Do Exercise 6.1, p. 277

Use the methods we've just described to do problem 6.1. The mean roll of two dice is 7, and the standard deviation of the rolls is 2.415. Type the probabilities in **L2** as fractions, 1/36, etc. The calculator will convert these to decimal equivalents.

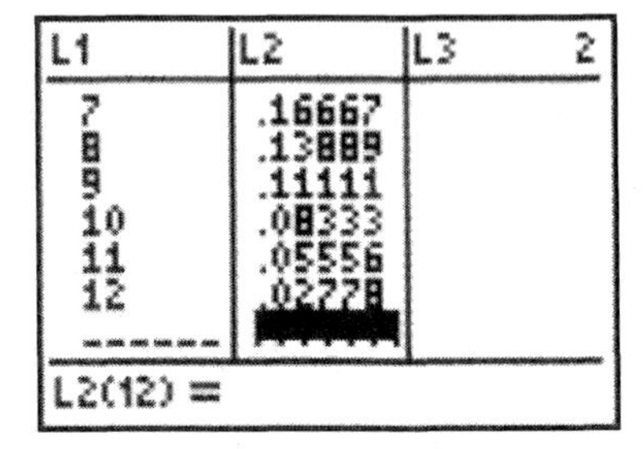

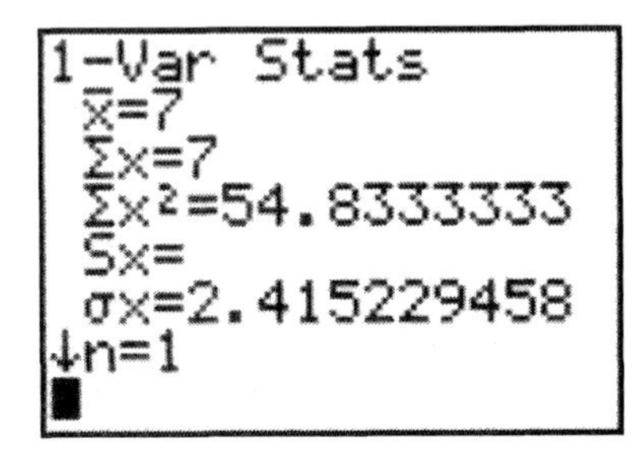

A **continuous random variable** has possible values that form an interval. Its **probability distribution** is specified by a curve that determines the probability that the random variable falls in any particular interval of values.

- Each interval has probability between 0 and 1. This is the area under the curve, above that interval.
- The interval containing all possible values has probability equal to 1, so the total area under the curve equals 1.

In practice, many continuous variables such as heights are measured with a rounded value. Here is a histogram of the heights of female students at the University of Georgia, available in the list **HGHTF** on the CD. To make this into a probability distribution, **L2** must be set up with the same number of values, 261. Each recorded height is then "weighted" by $1/n$, stored in **L2** using the **Fill** command (**2nd LIST OPS**).

```
LHGHTF→L1
{67 65 65.5 69 ...
dim(L1)→dim(L2)
                261
Fill(1/261,L2)
               Done
```

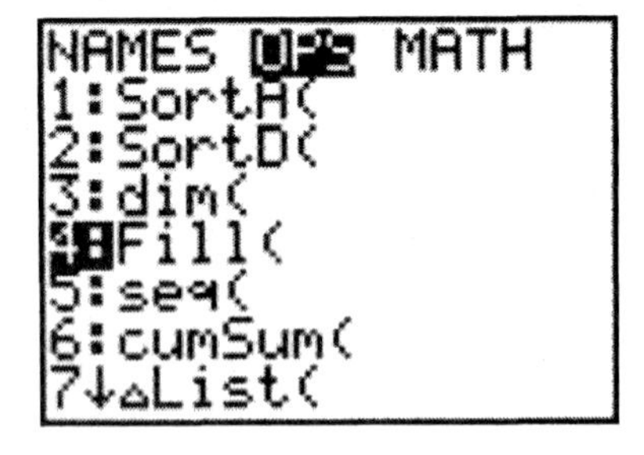

Make a probability distribution histogram as before, with **Xlist** = **L1** and **Freq** = **L2**, and scroll to find the probability of a height occurring in any of the intervals. These intervals are each 1 inch wide, the median height is 65, and the graph shows that the probability of a height being between 64.5 and 65.5 inches is about 0.126.

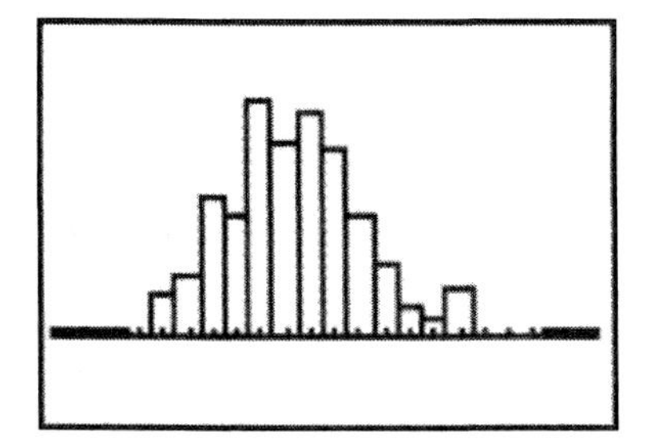

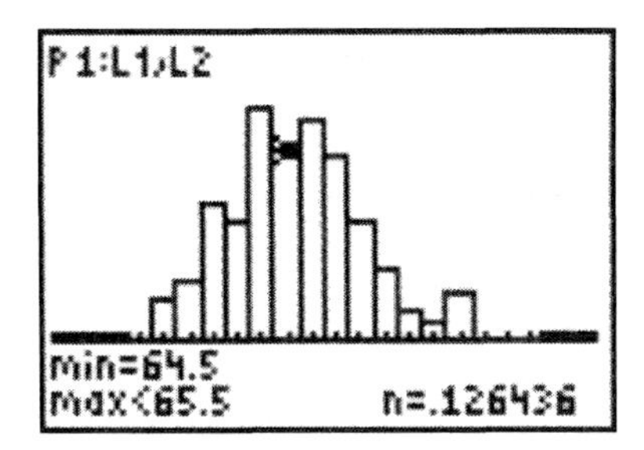

The useful functions for some specific probability distributions can be found under the distributions (**DISTR**) menu by pressing **2nd VARS**. Scroll down to view them all. Seven probability distributions are represented, and this chapter will cover two of them: Normal and Binomial, in sections 6.2 and 6.3. Some of the others, t, χ^2 (pronounced "kai" squared), and F will be covered in later chapters. If you take another statistics course (lucky you) you may see the remaining distributions, Poisson and Geometric.

```
DISTR DRAW
1:normalpdf(
2:normalcdf(
3:invNorm(
4:tpdf(
5:tcdf(
6:X²pdf(
7↓X²cdf(
```

```
DISTR DRAW
9↑Fcdf(
0:binompdf(
A:binomcdf(
B:poissonpdf(
C:poissoncdf(
D:geometpdf(
E:geometcdf(
```

6.2 How Can we Find Probabilities for Bell-Shaped Distributions?

The distributions that are labeled with "normal" or "norm" are an idealized form of a bell shaped curve called the **normal distribution**, a continuous probability distribution. The mathematical function that describes this is complex and not described in your book, but it is in the function **normalpdf** used to draw this curve on a graph. Here is an example that reproduces the graph in figure 6.3 on p. 276. The normal probability distribution function is drawn at each X value, with the specified mean and standard deviation taken from the heights of the Georgia students, 65.3 and 2.95, respectively. You can see that the pattern in the histogram roughly follows the curved graph of the normal distribution.

```
1-Var Stats
 x̄=65.2835249
 Σx=65.2835249
 Σx²=4270.62452
 Sx=
 σx=2.947184615
↓n=1
```

```
Plot1 Plot2 Plot3
\Y1=normalpdf(X,
65.3,2.95)
\Y2=
\Y3=
\Y4=
\Y5=
\Y6=
```

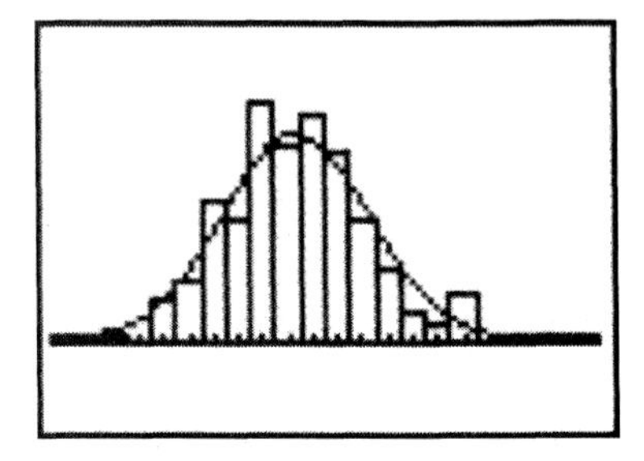

The second function in the distribution menu, **normalcdf**, is much more commonly used, since it calculates **probabilities** over a specified interval. This is the accumulated area under the normal curve between two z-scores. This will be one of the most commonly used functions. To find the probability in the normal distribution above, for example, specify the endpoints of the interval, 64.5 and 65.5 inches, and the mean and standard deviation are 65.3 and 2.95.

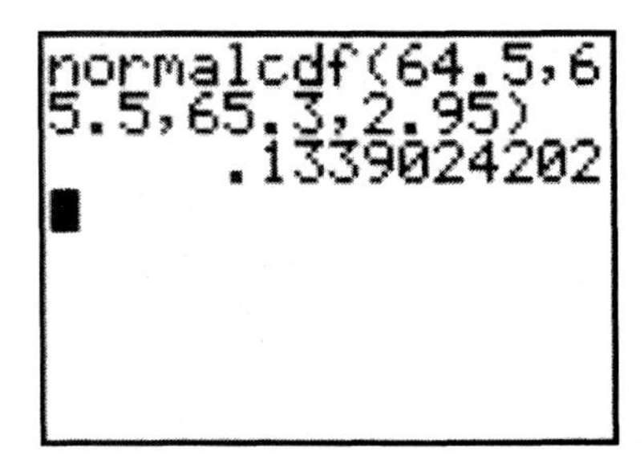

This value, 0.134, is a little bigger than the area included in the rectangle, 0.126. This makes sense because on the graph above you can see that the smooth curve is just a bit higher than the bar at that point.

This can be illustrated graphically with the **ShadeNorm** function (**2nd DISTR DRAW**). The interval and mean and standard deviation are specified exactly the same as with **normalcdf**.

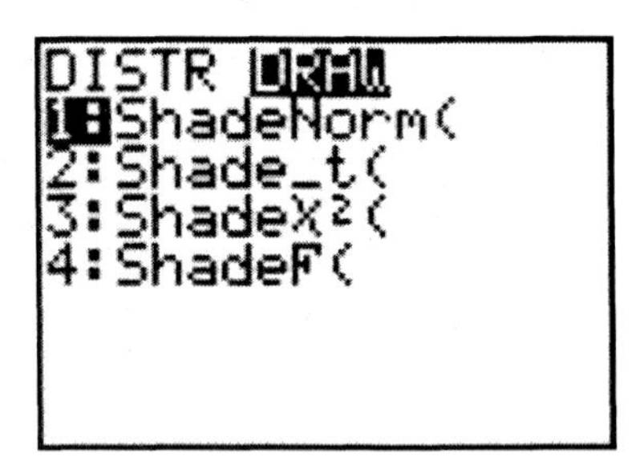

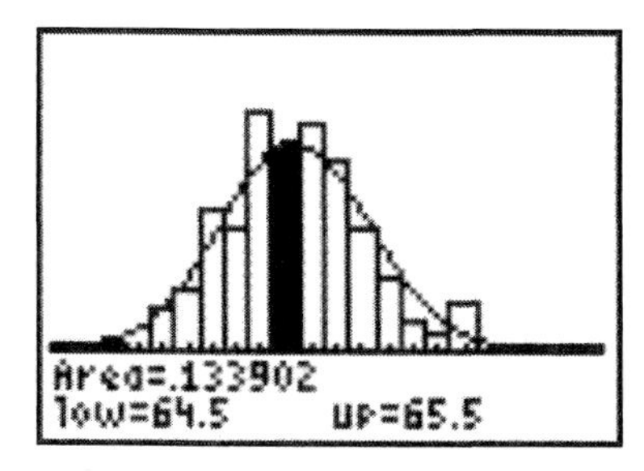

The **standard normal distribution** is the normal curve with mean 0 and standard deviation 1. Any normal distribution can be converted to the standard normal distribution by replacing the data values with z-scores. This is why the textbook can have one table for all normal distributions: they are all fundamentally the same shape and all can be standardized.

The values in the empirical rule can be verified using your calculator, with either **normalcdf** or **ShadeNorm**. Here are some examples. Notice that these are a little different from above in that they only specify two numbers. Since it is the most commonly used, for the standard normal distribution, you don't have to specify the mean and standard deviation. If you supply only two numbers, the calculator assumes you are using 0 and 1 for the mean and standard deviation – the standard normal distribution.

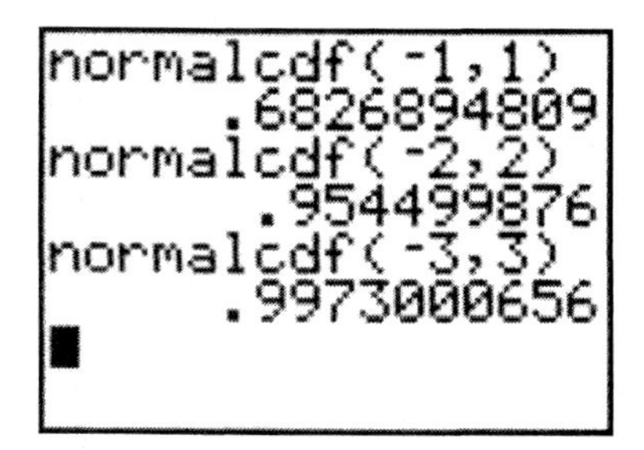

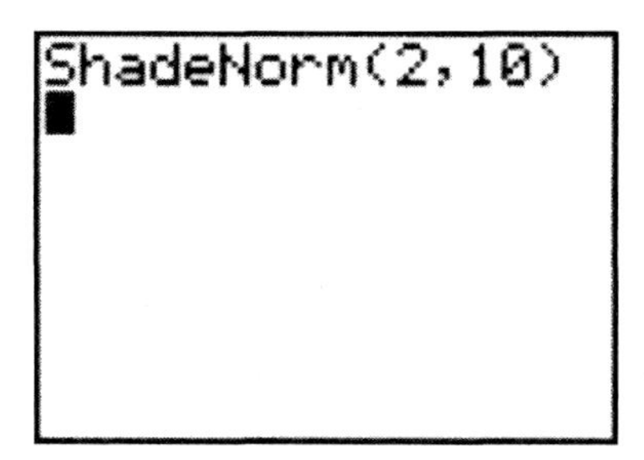

In the **ShadeNorm** example, the upper limit of 10 is enough to include all reasonable values, so the interpretation of this is that the probability of a value randomly selected from a normal distribution with value 2 or more. This is a calculation of a "tail probability." An appropriate window is shown for the standard normal distribution; particularly note that .4 is just a bit higher than the maximum of the standard normal distribution.

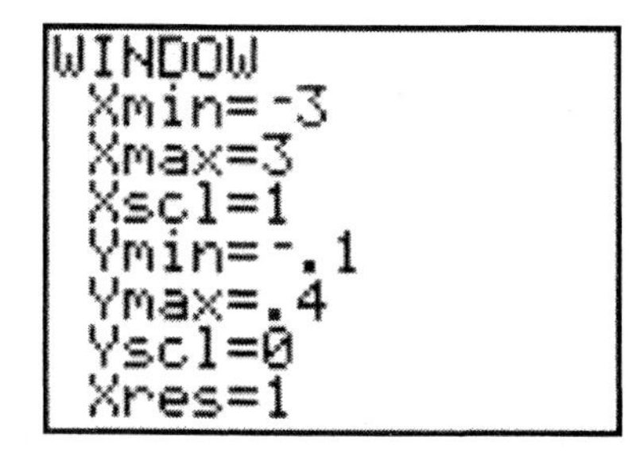

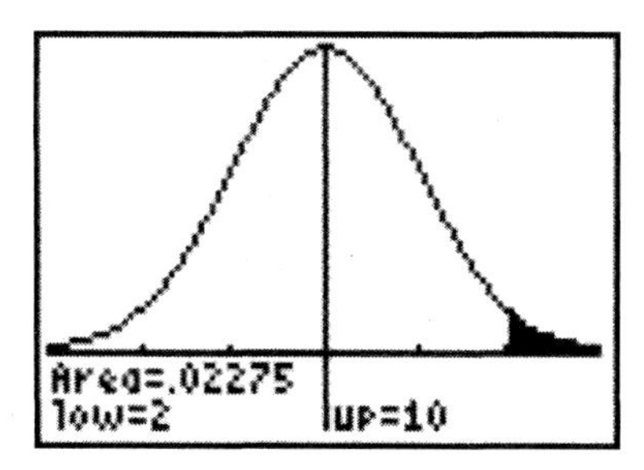

The third function in the distribution menu that uses the normal distribution is **invNorm**, short for "inverse normal". This function takes a **percentile** (or cumulative probability) value and returns the corresponding z-score. You can think of the percentile as the cumulative area that starts from the extreme low side of the distribution and continues until the area equals the specified value.

EXAMPLE 7: What IQ do you need to get into MENSA? (pp. 284-285 of Agresti/Franklin)

We are told that the IQ scores are normally distributed with mean 100 and standard deviation 16. What is the z-score corresponding to the 98th percentile of the standard normal distribution? Choose the **invNorm** function from the **DISTR** menu. Supply just one argument, the desired percentile. This measures standard deviations from the mean, so you can calculate the IQ score that corresponds to this as well.

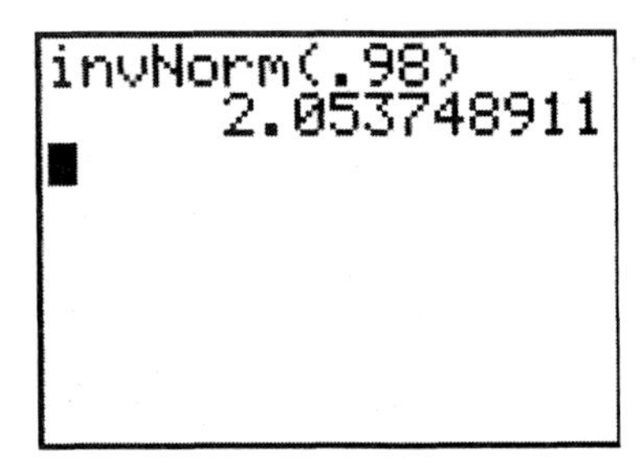

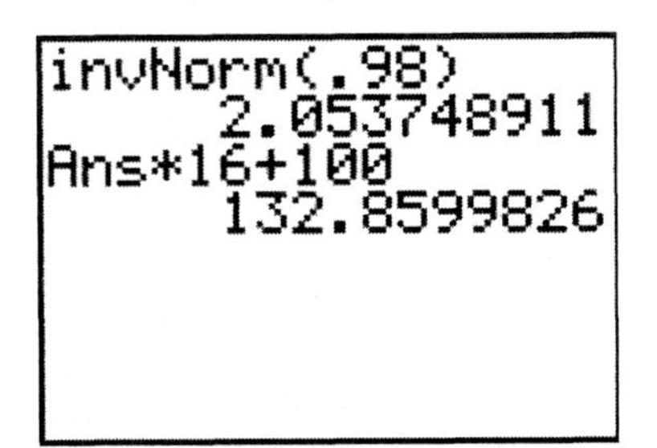

If you like, you can skip a step by specifying the mean and standard deviation of the IQ scores in the **invNorm** function. The arguments follow the form **invNorm(percentile, mean, stdev)** where the standard normal distribution, mean=0 and stdev=1, is assumed if the last two arguments are not supplied.

```
invNorm(.98,0,1)
          2.053748911
invNorm(.98,100,
16)
          132.8599826
```

EXAMPLE 9: What proportion of students gets a grade of B? (pp. 286-287 of Agresti/Franklin)

Now we are working with a normal distribution that has $\mu = 83$ and $\sigma = 5$. What is the proportion of students that score between 80 and 90? Let's first find the z-scores.

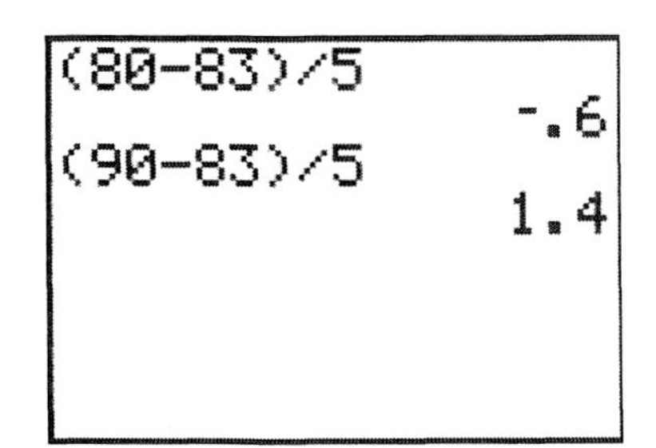

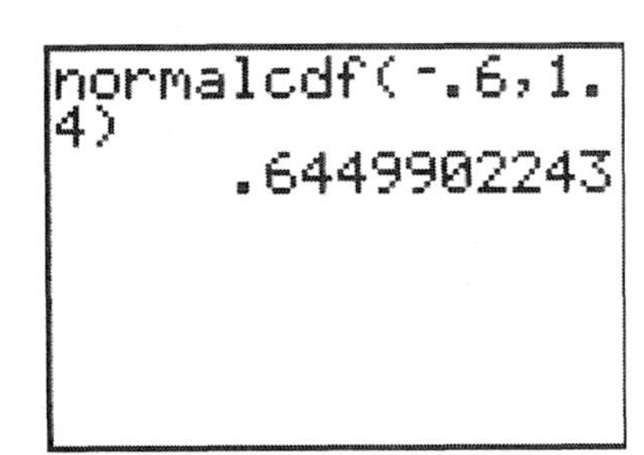

The area under the standard normal distribution curve between –0.6 and 1.4 is about .65, so 65% of the students get a B. Note that since you can specify both ends of the interval, it is not necessary to find two probabilities and subtract one from the other as you do when using the table in the textbook, Table A. You can do the same thing with **ShadeNorm**, which has the added benefit of a picture of the probability as an area.

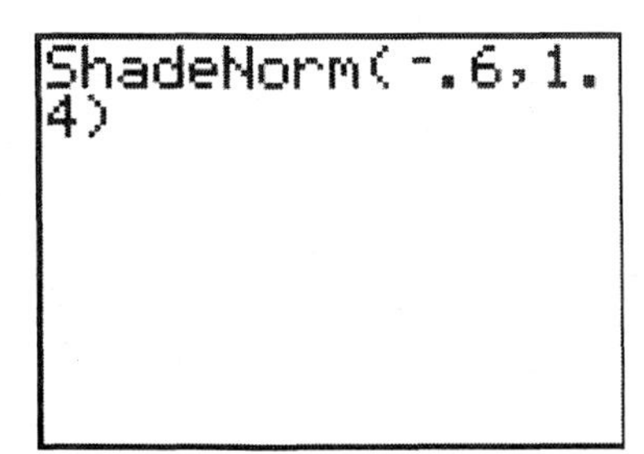

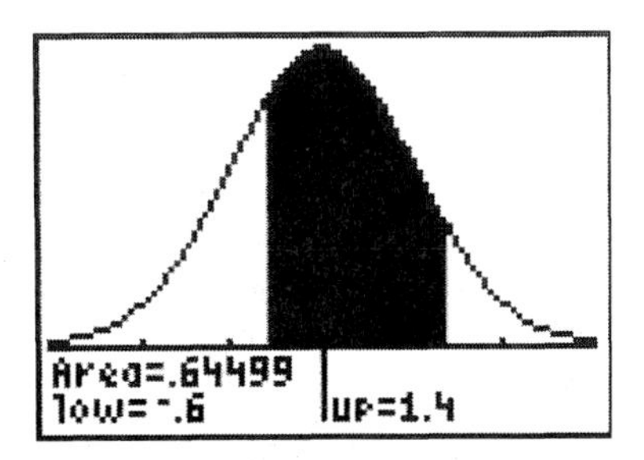

Using z-scores and the standard normal distribution has another advantage – it helps you begin to think in units of standard deviations, the most natural units when comparing different distributions, especially those that are reasonably bell-shaped.

Try it! Do Exercise 6.23 (p. 290 of Agresti/Franklin)

The systolic blood pressure readings are normal with $\mu = 121$ and $\sigma = 16$. Convert to a z-score, then use either **normalcdf** or **ShadeNorm**.

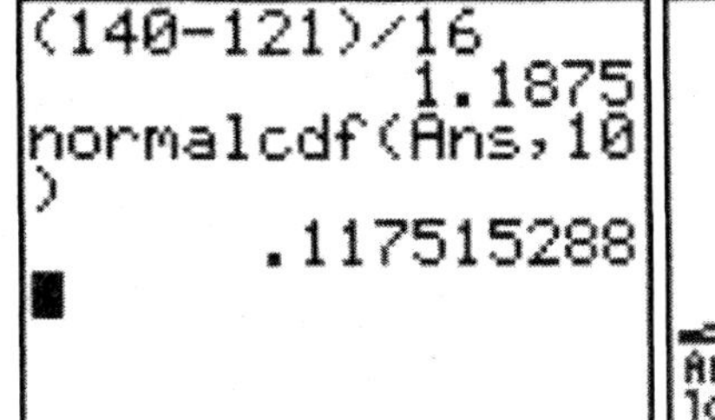

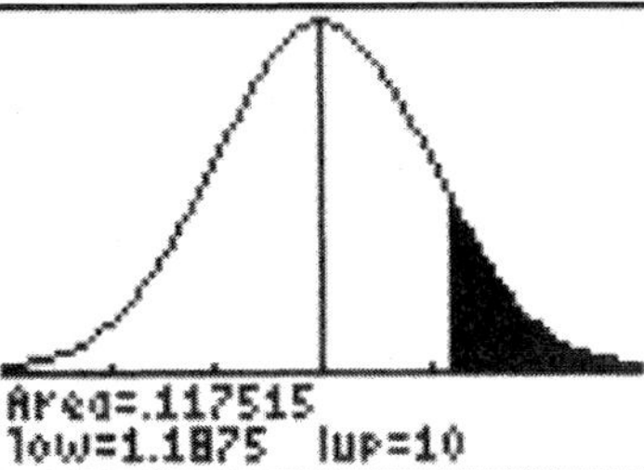

One additional note with **ShadeNorm**: if you use it in succession, it doesn't erase old shadings. This is great if that's what you want, but it can be confusing if you've moved on to a new problem. Use **ClrDraw**, in the **DRAW** menu, to remove shadings you don't need anymore.

For example, part *c* of this problem asks for the proportion of Canadians with blood pressure readings between 100 and 140. If you don't clear the drawing screen, then the shading won't correspond to the calculated percentages, since the shading from the previous part remains.

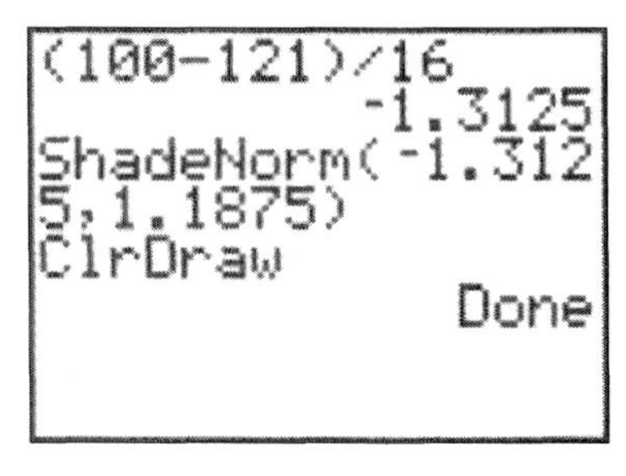

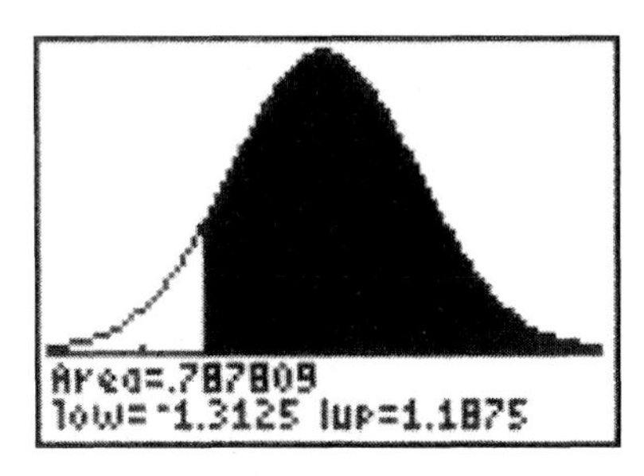

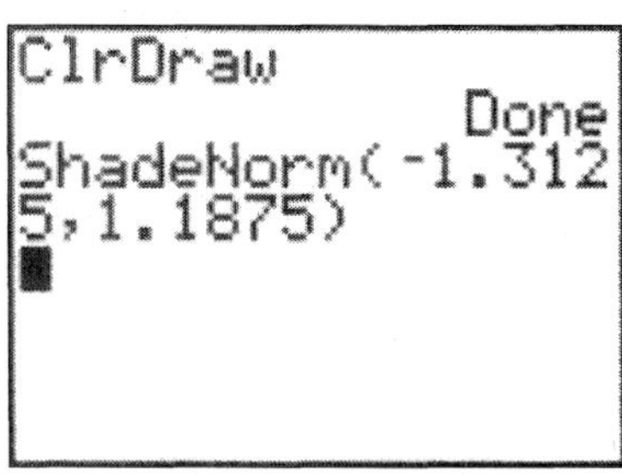

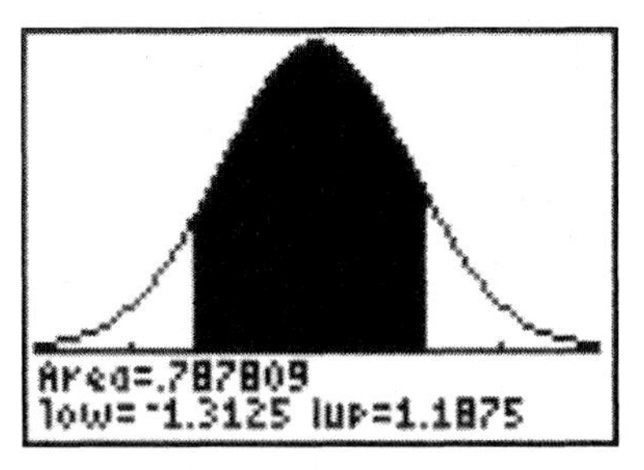

6.3 How Can We Find Probabilities When Each Observation has Two Possible Outcomes?

The **binomial distribution** is an important discrete probability distribution. The conditions needed to have a binomial distribution are

- Each of *n* trials has two possible outcomes. The outcome of interest is called a "success" and the other outcome is called a "failure."
- Each trial has the same probability of success, denoted *p*, so the probability of a failure is $1 - p$.
- The *n* trials are independent. That is, the result of one trial does not depend on the results of other trials.

The **binomial random variable** *X* is the number of successes in *n* trials.

There are two functions that relate to this in the **DISTR** menu. **binompdf** calculates probabilities for exactly *X* successes, while **binomcdf** calculates the probability that there are *X* or fewer successes.

EXAMPLE 12: Are women passed over for managerial training? (pp. 294-295 of Agresti/Franklin)

Note that the conditions for a binomial experiment are satisfied. If *X* is the number of females selected, then *X* is binomial with n = 10 and p = .50. *X* can take any value from 0, 1, 2, …, 10. The probability that zero females are chosen (see calculation on p. 295) is shown below. Use the **binompdf** function, since this gives the probability that X is equal to a specific value. The arguments are (**n, p, x**), where **x** = 0 represents 0 women chosen.

```
binompdf(10,.5,0
)
        9.765625E-4
```

This is a very small probability, expressed by the calculator in scientific notation. The probability that no females are chosen is .0009765625, or about .001. It is very unlikely that, if all employees are equally likely to be chosen to be a manager, that all would end up male. The probability of this occurring is about 1 in a thousand.

If you use the **binompdf** function with only two numbers, n and p, then you will get the distribution of all probabilities for the entire range of X values, as in the table on p. 295. It's easiest to save this to a list rather than scroll through on the home screen. You can type the values for X into list **L1** (or use the **seq** command, shown below) and then put the corresponding probabilities in **L2**.

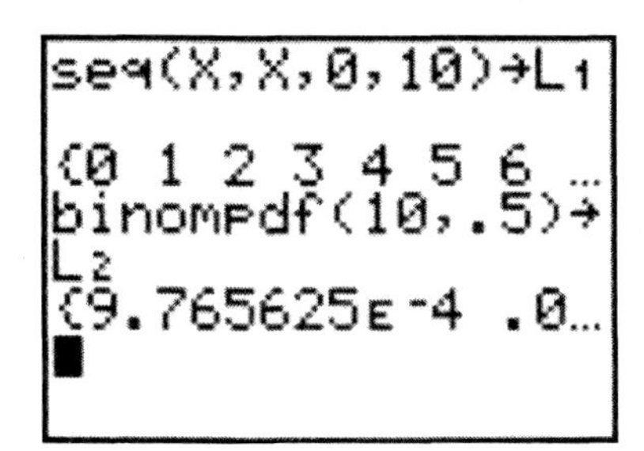

L1	L2	L3
0	9.8E-4	
1	.00977	
2	.04395	
3	.11719	
4	.20508	
5	.24609	
6	.20508	

L3(1)=

You can also make a graphical display of this probability distribution, using the histogram in **STAT PLOT**. Specify **L1** as the **Xlist**, and **L2** for **Freq**.

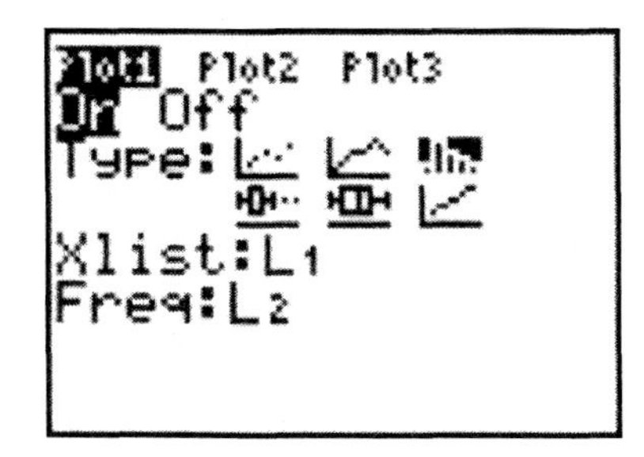


```
WINDOW
 Xmin=-.5
 Xmax=10.5
 Xscl=1
 Ymin=-.075
 Ymax=.3
 Yscl=0
 Xres=1
```

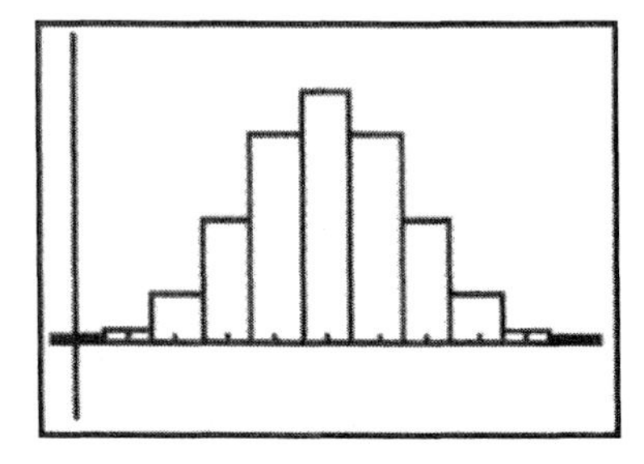

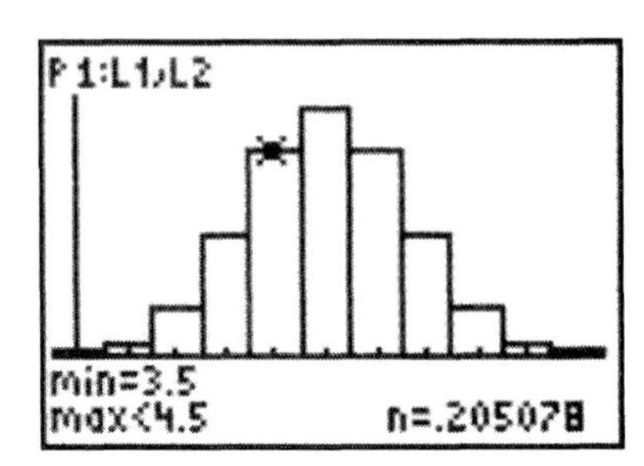

EXAMPLE 14: How can we check for racial profiling? (pp. 298-299 of Agresti/Franklin)

In this example, a binomial distribution is used to assess the probability of 207 of the 262 traffic stops involving an African-American motorist. We can use the same methods as above:

```
seq(X,X,0,262)→L
1
{0 1 2 3 4 5 6 ...
binompdf(262,.42
2)→L2
{4.217874624E-6...
```

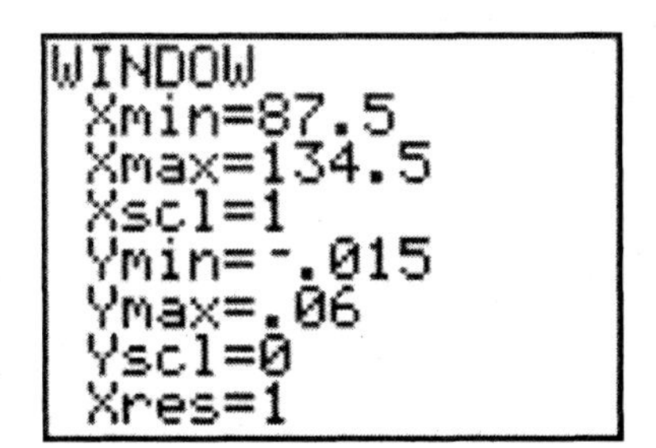

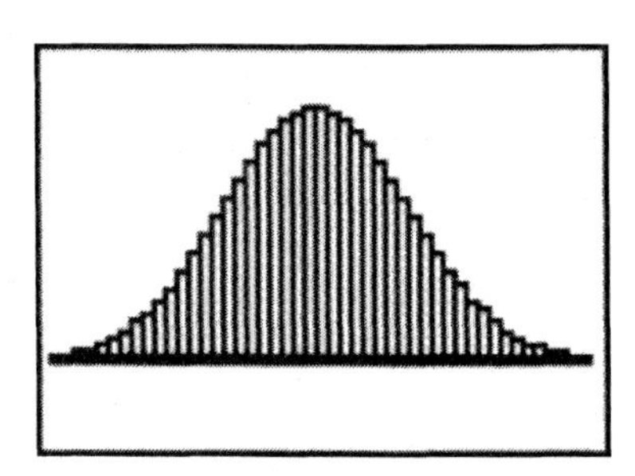

This has many bars (263), so only the middle section of the graph is shown. In fact, due to the technical limitations of the calculator screen it can draw no more than 47 bars (that's what is shown here). But this is certainly where the bulk of the probability distribution falls. You should notice that the shape of this distribution looks like a classic normal distribution. For large number of trials – when np is at least 15 and $n(1-p)$ is at least 15 – this is true.

You can see a graphical reinforcement of this by plotting a normal distribution that has the same mean and standard deviation as this binomial distribution. From p. 297 of the textbook, the mean of a binomial distribution is $\mu = np$ and the standard deviation is $\sigma = \sqrt{np(1-p)}$.

```
262(.422)
                110.564
√(262*.422*.578)

            7.994122341
```

Here is the normal distribution that has the same mean and standard deviation as the binomial, plotted using the **normalpdf** function.

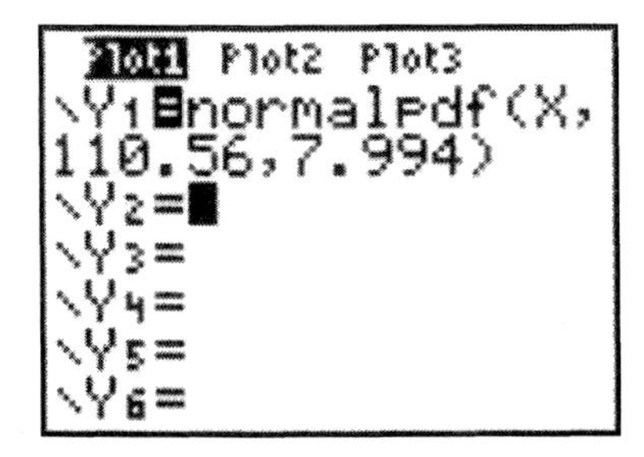

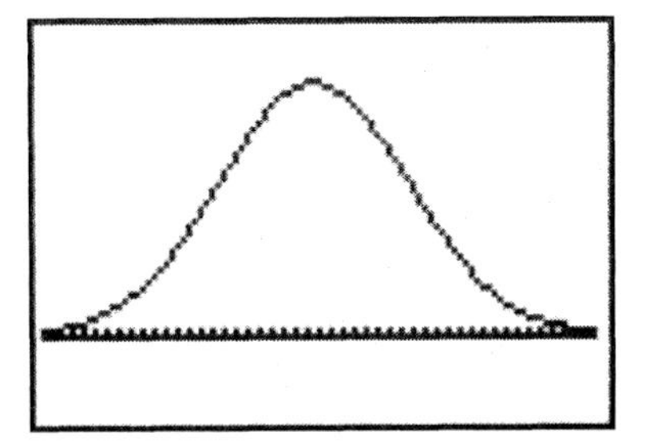

If this is plotted over the histogram in the same window, you can see that the two match up very well. You'll need to look very closely to see the difference between this graph and the previous graph that had only the histogram.

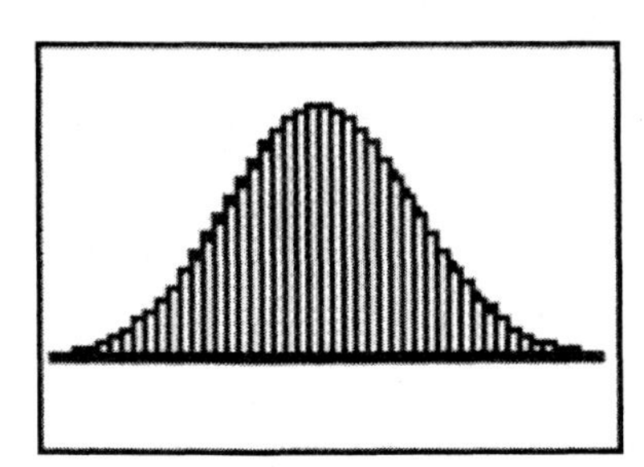

The normal distribution can be used effectively to make very good estimates of cumulative binomial probabilities. This will be very useful in the next several chapters.

Chapter Seven: Sampling Distributions

The lists referred to in this Chapter are the output of simulations. They are on the textbook CD with the names below. They are also collected together in the group AF07 on the CD.

The process for copying lists and groups from a computer to your calculator is described in Chapter 1 of this manual, beginning on page 4.

BATAV	Results of the batting average simulation
DISIM	Results of the dice simulation

7.1 How Likely Are the Possible Values of a Statistic? The Sampling Distribution

A **sampling distribution** of a statistic is the probability distribution that specifies probabilities for the possible values that the statistic can take.

A sampling distribution shows haw a statistic varies from sample to sample. Through simulations, the TI-83 Plus calculators can estimate a sampling distribution and reinforce the fact that for sample proportions and sample means, the sampling distribution is approximately normal for large samples.

EXERCISE 7.7: Random Variability in Baseball (pp. 320 of Agresti/Franklin)

In baseball, the batting average is a sample proportion of the number of successes (hits) in a given number of at bats. Lets simulate what happens when a player whose ability indicates that he has a probability of .300 of getting a hit plays a season and has 500 at bats. (Batting averages are usually expressed as proportions with three decimal places.) This is modeled as a binomial situation, with $n = 500$, $p = .3$. The calculator has a function **randBin** (in the **MATH PRB** menu) that runs one trial of a simulation of this scenario.

The screen below shows two examples of this, the first producing 163 hits out of 500 at bats for a batting average of .326. The second simulation produced a batting average of .298, and the two steps in the operation were combined in one line – simulating the number of hits, then dividing by 500.

```
MATH NUM CPX PRB
1:rand
2:nPr
3:nCr
4:!
5:randInt(
6:randNorm(
7:randBin(
```

```
randBin(500,.3)
                163
Ans/500
               .326
randBin(500,.3)/
500
               .298
```

The sampling distribution expresses the variability in these numbers. Like any distribution, we'll want to be able to describe its shape, center, and spread.

You could do this over and over again and write down the observed sample proportions, but the calculator can automate this process. If you add a third argument to the **randBin** function, it will repeat the binomial process the specified number of times (100 in the example below). These 100 batting averages are stored **L1**. (Be patient if you are following along, as this takes several minutes to complete.)

L1	L2	L3 1
.288	------	------
.26		
.304		
.282		
.276		
.3		
.326		

L1(1)=.288

Set up an appropriate window and you can graph the approximate sampling distribution.

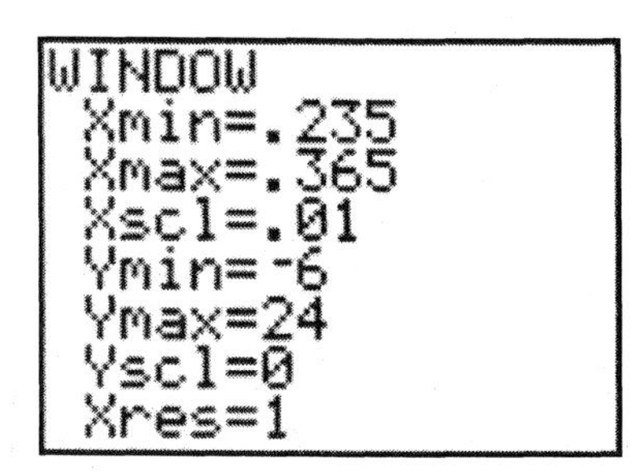

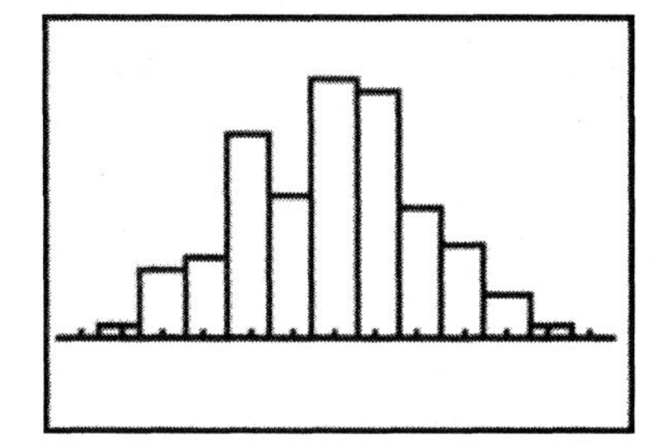

Use **TRACE** to answer the question posed in the exercise: is a season batting average (sample proportion) of .280 or .320 unusual for a .300 hitter? Looks like the answer is no, these are not uncommon occurrences among the 100 trials that we have.

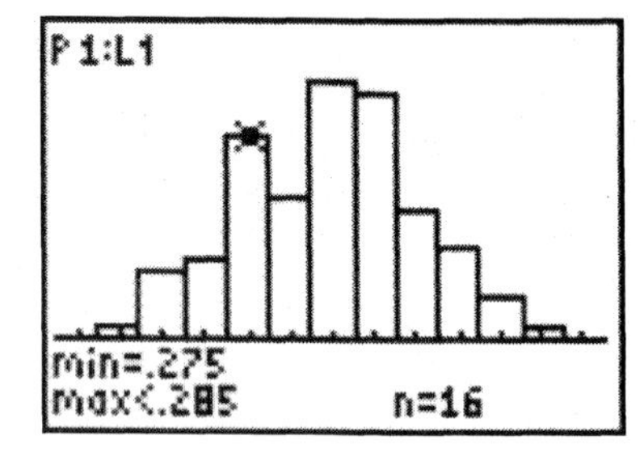

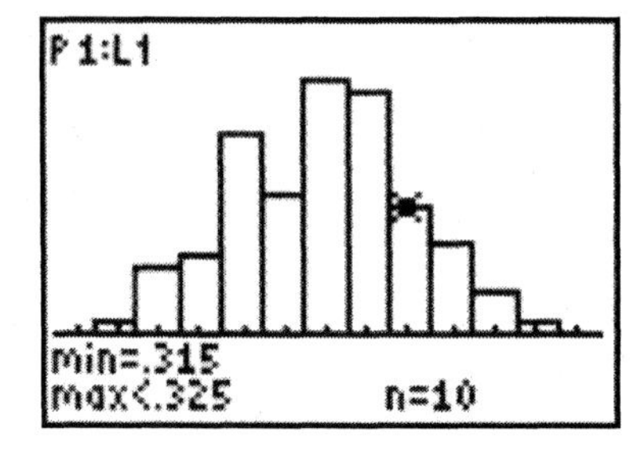

The binomial distribution that is behind this simulation is well approximated by a normal distribution, as we saw in chapter 6. You can demonstrate this by superimposing the correct normal curve on the graph above. The mean of this sampling distribution should be exactly the population mean, .300. The standard error is the standard deviation of this sampling distribution, determined by the formula $SE = \sqrt{\frac{p(1-p)}{n}}$. This works out to approximately 0.020, so the values of .280 and .320 are each about one standard error from the mean. Thus they mark the boundaries of typical variability, and you would expect about 68% of the sample proportions to fall between these values using the empirical rule. The calculator can get you a more exact value, using **normalcdf**, specifying the mean and standard error. The maximum value of the sample batting averages was .352, a value that is more than 2.5 standard errors above the mean, measured by its z-score.

```
√(.3*.7/500)
       .0204939015
normalcdf(.280,.
320,.300,Ans)
       .6708860218
```

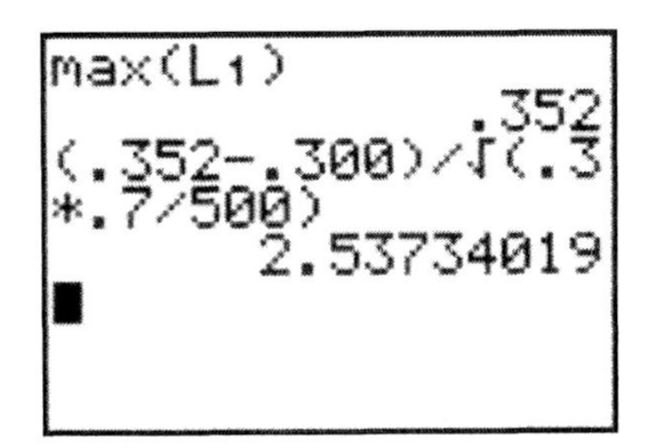

The same mean and standard error can be specified to sketch a normal curve over the histogram. The curve fits the histogram fairly well, and with more of the binomial trials it would fit even better.

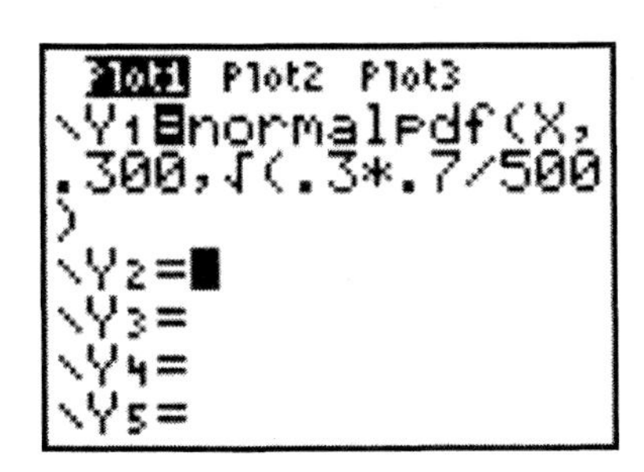

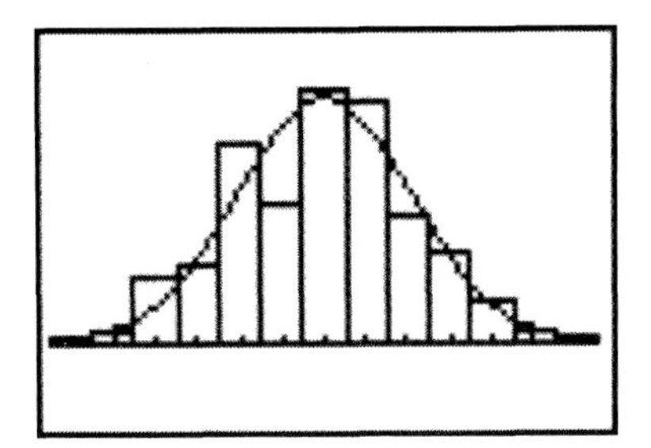

Drawing the normal curve over a histogram is sometimes difficult due to scaling issues. The area enclosed by the normal curve is exactly one square unit, but the histogram might not have this value as the total of areas in the bars. In this case it does, since the total height of the bars is 100 (the number of trials) and the width of the bars is .01. These multiply to exactly 1. If this isn't the case, then you need to rescale the normal curve so that it has the same area as the area of the bars.

7.2 How Close Are Sample Means to Population Means?

EXAMPLE: Maximum value of the roll of a die. (pp. 326-327 of Agresti/Franklin)

Even for a skewed distribution, the sampling distribution of a sample mean is approximately normal. This example uses the distribution of highest values in a roll of two dice that was introduced in Exercise 6.2, p. 278.

In order to simulate a sampling distribution, we need first to have a population distribution from which to draw values. This is expressed in the probability distribution histogram on p. 326. There are several ways to enter this probability distribution in you calculator; for the purposes of the simulation we're about to do it is easiest to enter values in **L1** so that the number present corresponds to the probability. The sample space has 36 possible outcomes, so we'll enter 36 numbers in **L1**, each corresponding to the maximum of the particular pair of dice. There is one pair in which 1 is the maximum, so enter this value once, three pairs where 2 is the maximum, so enter 2 three times, and so forth: 3 is entered 5 times, 4 entered 7 times, 5 entered 9 times, and finally 6 entered 11 times.

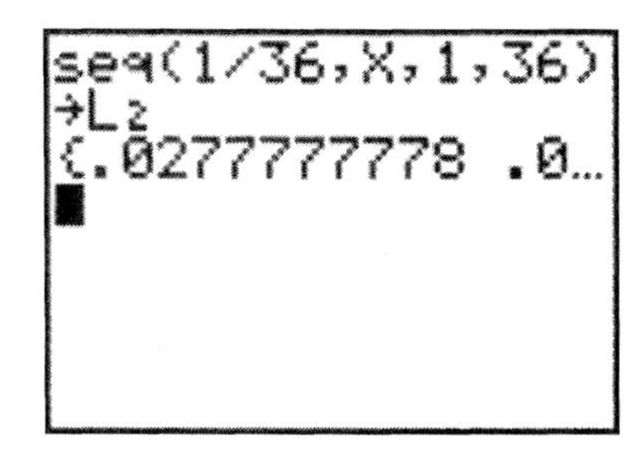

Entering the weight 1/36 in each entry from **L2** (or, use **seq**, shown above). You can verify, using **TRACE**, that the probability of having the maximum roll equal to 5 is 0.25.

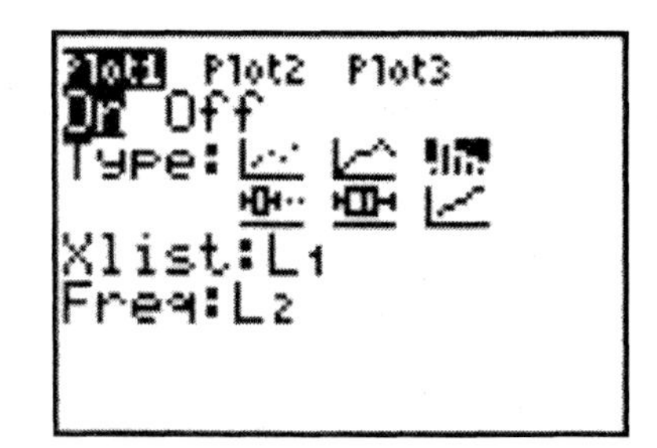

You can create the probability histogram to check your work, making sure it matches the graph on p. 326.

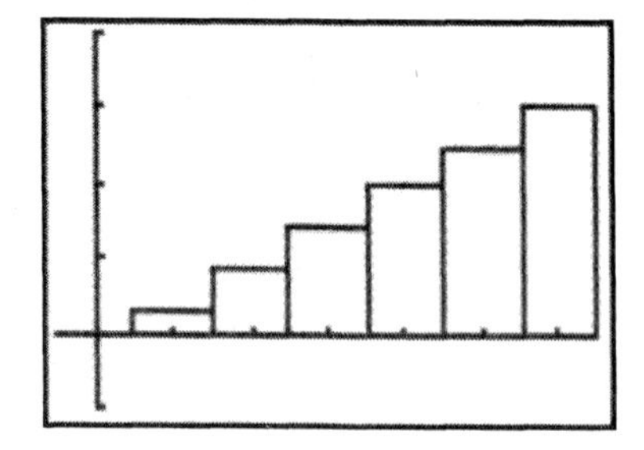

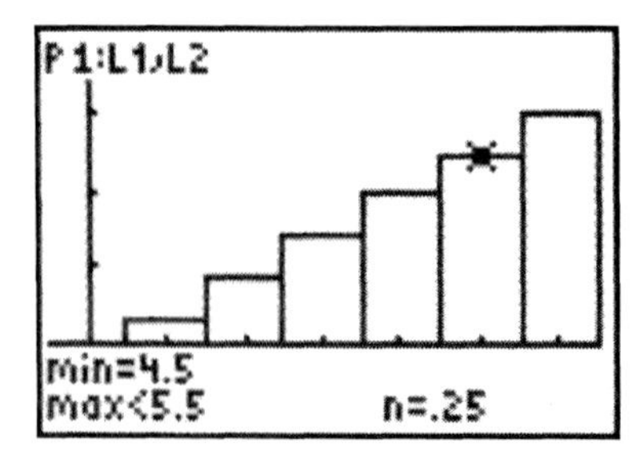

To sample from this population distribution is a little trickier. There is no built-in function as was the case with the proportions using **randBin**, so to do this you need to do a little programming. This isn't really anything more than gathering together a couple of calculator commands and having them repeated.

In the example below, the calculator will calculate a sample mean (3rd line) from a random sample of 30 taken from the values in **L1**. These results are stored as 200 values in **L3**. You can enter this program by pressing **PRGM**, then **NEW**. When you are typing in the program, pressing the **PRGM** button brings up a menu of programming controls. This is where you find **For** and **End**. The others are entered by using the keyboard, just as you would from the home screen. You can also find them in the **Catalog**.

Once you've entered the program, go back to the home screen. Then press **PRGM** again, select the program and press **ENTER**. This pastes it to the home screen. Press enter again to run the program, which will fill values into **L3**. A sample run of these values is included in the list **DISIM** on the CD. Be especially careful with the parentheses – if you run the program and it says **ERR:SYNTAX**, misplaced parentheses are a likely cause. Press **PRGM**, then **EDIT** to fix your program if need be.

```
PROGRAM:SAMPMEAN
:ClrList L3
:For(I,1,200)
:mean(seq(L1(ran
dInt(1,36)),X,1,
30))→L3(I)
:End
:
```

L1	L2	L3 1
1	.02778	4.2
2	.02778	4.4667
2	.02778	4.6
2	.02778	4.6333
3	.02778	4.7667
3	.02778	4.9
3	.02778	4.6

L1(1)=1

To display a histogram of the sample means, set up **Plot2** with **Xlist** as **L3**. The window below will be a good match for the picture shown in Figure 7.7. This is a frequency histogram of the 200 sample means.

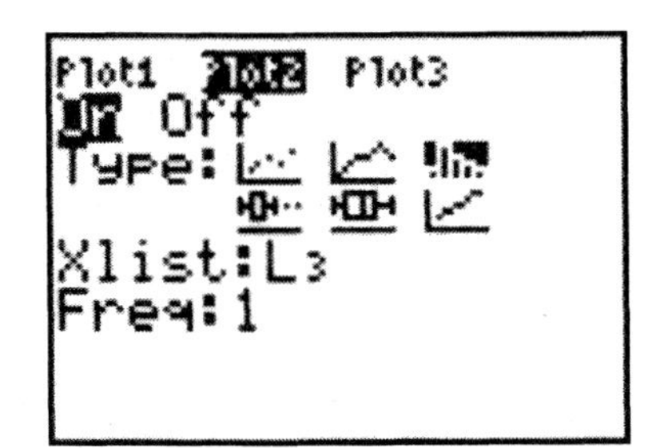

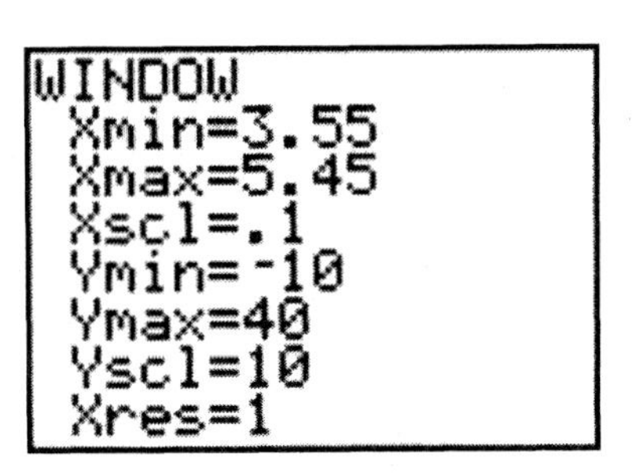

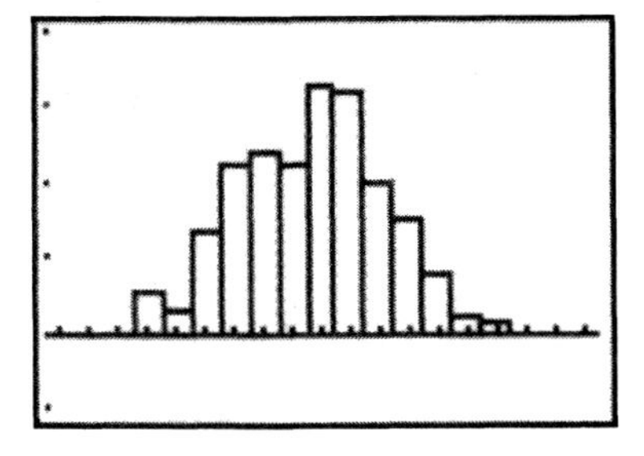

Does this look approximately normal? According to the textbook, the sample size of 30 is large enough so that the sampling distribution is approximately normal, with the same mean as the population distribution, and standard error $SE = \frac{\sigma}{\sqrt{n}}$. You can find the population distribution mean μ and standard deviation σ using **1-Var Stats**.

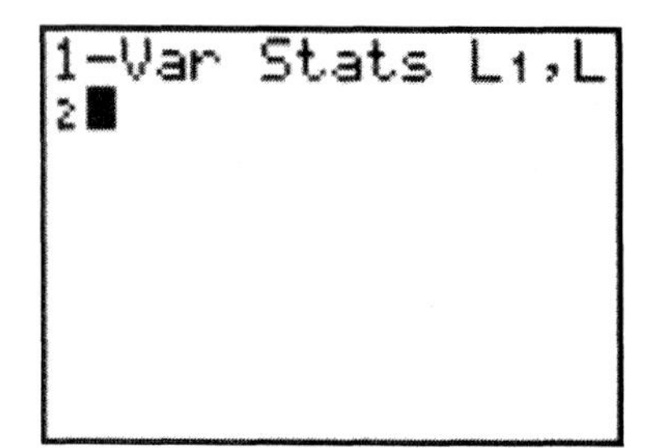

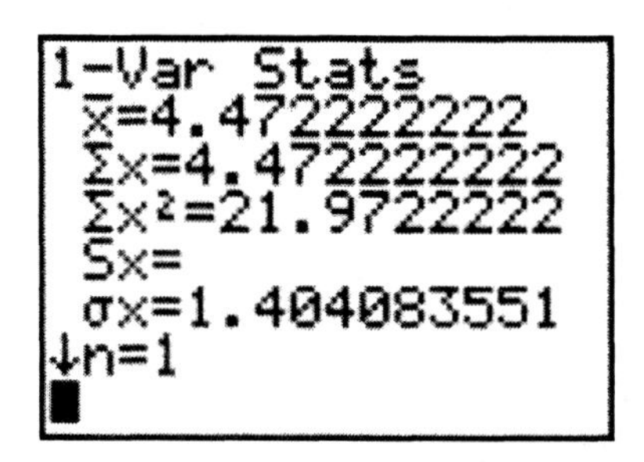

Now, you could write these values down, $\mu = 4.472$ and $\sigma = 1.404$, but once you do the **1-Var Stats**, the calculator stores these values. You can find them by pressing the **VARS** key, then **5:Statistics**. Finally choose $\bar{x}$ and σ_x from the list. Note that the calculator always uses $\bar{x}$ for the mean, as it doesn't know whether this is a sample mean or a population mean.

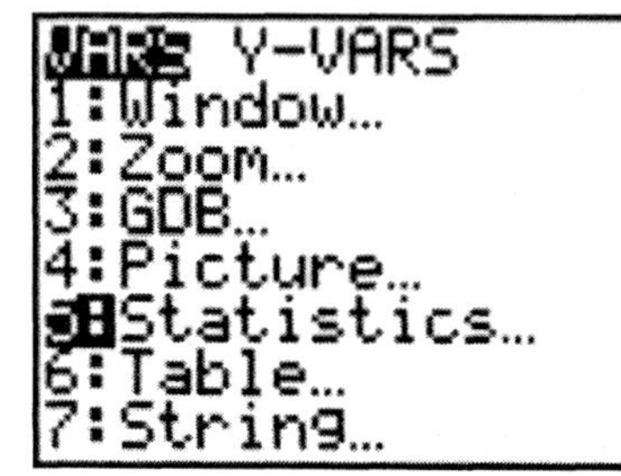

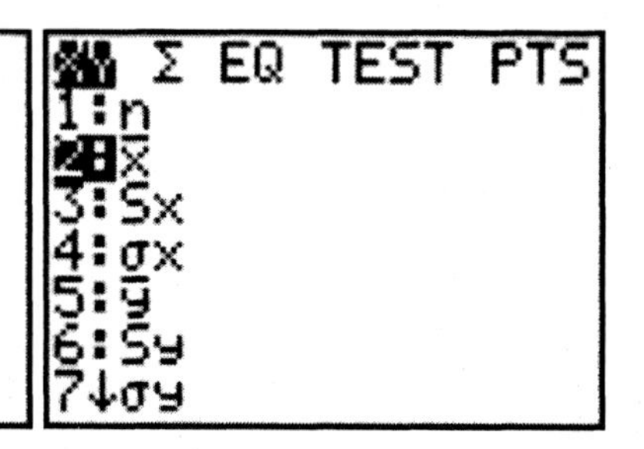

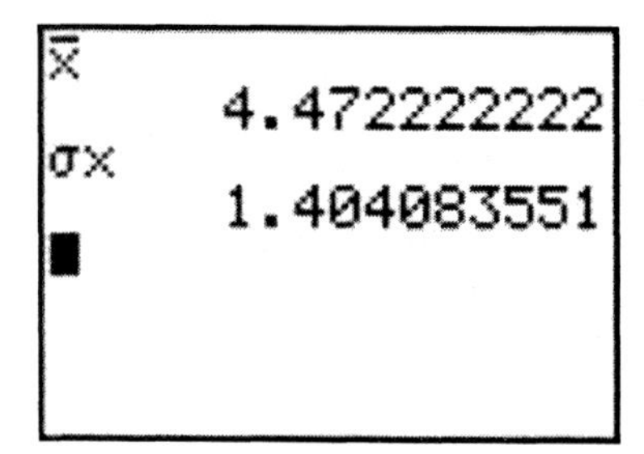

Now you can use these to plot the theoretical sampling distribution, even entering the variables in the **y=** menu. Note that the histogram above is not a probability distribution, since the bars represent counts, not relative frequencies. This means that the normal distribution, which encloses area = 1, must be scaled to match. The histogram has bars with total height 200 (200 samples). The width of each bar is 0.1, so the total area of the bars is (200)(0.1) = 20. This is the number to use to scale the normal curve to fit.

```
Plot1 Plot2 Plot3
\Y1=20normalpdf(
X,x̄,σx/√(30))
\Y2=
\Y3=
\Y4=
\Y5=
\Y6=
```

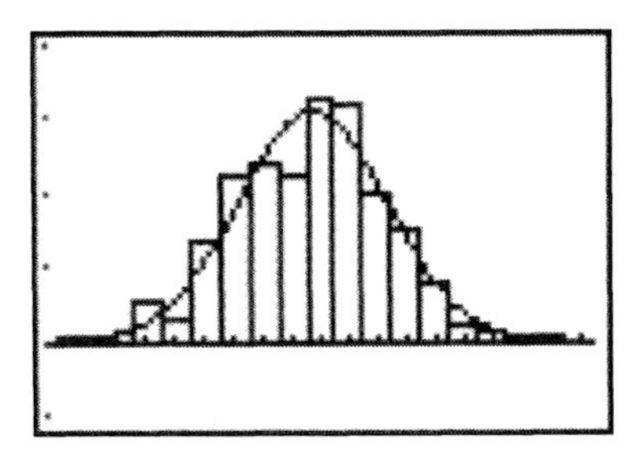

This fits well. You can now uses **normalcdf** to answer questions such as "What is the probability that a sample mean is between 4.0 and 4.5?" or "What is the probability it is greater than 5?" These values are about .51 and .20, respectively, calculated below. The Central Limit Theorem says that the sampling distribution is approximately normal, so these values are only approximately correct.

```
normalcdf(4,4.5,
x̄,σx/√(30))
        .5104146864
normalcdf(5,10,x̄
,σx/√(30))
        .0197555852
```

Chapter Eight: Statistical Inference – Confidence Intervals

All of the lists referred to in this Chapter are on the textbook CD with the names below. They are also collected together in the group AF08 on the CD.

The process for copying lists and groups from a computer to your calculator is described in Chapter 1 of this manual, beginning on page 4.

EBNOW	Prices of Palm devices on eBay with the "buy now" option
ANCHG	Change in weights for anorexia patients

8.1 What Are Point and Interval Estimates of Population Parameters?

The previous chapter concerned the sampling distribution of a sample proportion or mean, where the pattern of many, many values of a statistic was explored. The more common and more practical problem, however, is what to do with the *one value*, a single sample proportion, $\hat{p}$, or sample mean, $\bar{x}$, that is drawn from a population with unknown parameters.

> A **confidence interval** is an interval containing the most believable values for a parameter. The probability that this method produces an interval that contains the parameter is called the **confidence level**. This is a number chosen close to 1, most commonly 0.95.

EXAMPLE 2: Should a wife sacrifice her career for her husband's? (p. 360 of Agresti/Franklin)

In this example, we have a sample proportion of 0.19, a point estimate that has a standard error of about 0.01. To construct a confidence interval, we first must determine the **margin of error**.

> The **margin of error** measures how accurate the point estimate is likely to be in estimating a parameter. It is a multiple of the standard error of the sampling distribution of the estimate, such as 1.96 x (standard error) when the sampling distribution is close to normal.

The number 1.96 comes from an approximate percentile of the standard normal distribution such that the middle 95% of the distribution falls between. You can use the **invNorm** command, found in the **DISTR** menu, to confirm these percentile values. Using the approximate value 1.96 and the standard error estimate 0.01, we can find the margin of error and the confidence interval.

```
DISTR DRAW
1:normalpdf(
2:normalcdf(
3:invNorm(
4:invT(
5:tpdf(
6:tcdf(
7↓X²pdf(
```

```
invNorm(0.025)
      -1.959963986
invNorm(0.975)
       1.959963986
■
```

```
1.96*.01
                .0196
.19-.0196
                .1704
.19+.0196
                .2096
```

You might wonder how the standard error estimate of 0.01 was made. In Chapter 7 this was shown to be $\sqrt{\frac{p(1-p)}{n}}$, but it isn't possible to figure this out without knowing *p*, the very quantity we are trying to estimate. Replacing *p* with the sample proportion $\hat{p}$ produces a reasonable approximation to the standard error. In this case $SE \approx \sqrt{\frac{.19(.81)}{1823}} = 0.0092$, very close to 0.01.

8.2 How Can We Construct a Confidence Interval to Estimate a Population Proportion?

EXAMPLE 3: Would you pay higher prices to protect the environment? (p. 364 of Agresti/Franklin)

The sample proportion and standard error of $\hat{p}$ are calculated below from the numbers in the textbook. Remember that you can call up the most recently calculated value via **ANS** (**2nd (-)**). You can also store results into a variable, using the **ALPHA** key. Both of these techniques can save time typing, and produces a more accurate result with less rounding error. Recall from example 2 above that we must multiply the standard error by 1.96 to achieve 95% confidence. The interval produced is from 0.4202 to 0.4776.

```
518/1154
         .4488734835
Ans→P
         .4488734835
√(P(1-P)/1154)
         .0146414714
Ans*1.96→M
```

```
         .0146414714
Ans*1.96→M
          .028697284
P-M
         .4201761995
P+M
         .4775707675
```

This procedure is very common, and so your calculator has automated the process. This is true of many of the inference procedures that we will see in the next several chapters, and these automated procedures are available in the **STAT TESTS** menu. Press **STAT**, then move right to **TESTS**. Scroll down to **A:1-PropZInt**. This is an abbreviation of the procedure for a one sample confidence interval for a population proportion using a normal approximation (z-scores).

```
EDIT CALC TESTS
6↑2-PropZTest...
7:ZInterval...
8:TInterval...
9:2-SampZInt...
0:2-SampTInt...
A:1-PropZInt...
B↓2-PropZInt...
```

```
1-PropZInt
 x:518
 n:1154
 C-Level:.95
 Calculate
```

The **1-PropZInt** procedure needs three inputs. First we enter x, the number of "successes", which is the number of people willing to pay higher prices to protect the environment in this example. Then we enter *n*, the sample size, and *C*, the confidence level. After supplying this information, move the cursor down to **Calculate** and press **ENTER**.

```
1-PropZInt
 (.42018,.47757)
 p̂=.4488734835
 n=1154
```

The display shows $\hat{p}$, the point estimate, along with the interval – the same interval as calculated above. As a general rule, we'll work out the point estimate and margin of error, then use **1-PropZInt** to check our work. In practice, the **1-PropZInt** is a terrific shortcut.

EXAMPLE 5: Should a Husband Be Forced to Have Children? (p. 367-368 of Agresti/Franklin)

We used **invNorm** function to find 1.96, the normal percentile needed for 95% confidence intervals. For other confidence levels, the percentile is different. For a 99% confidence interval, only 0.005 is left in each tail (.99 in the middle). The percentiles for a 99% confidence interval are shown below, along with the percentiles for a 90% confidence interval. These values are shown in Table 8.2.

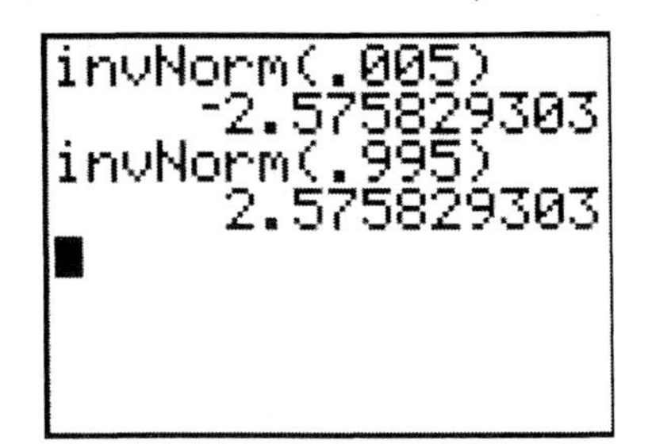

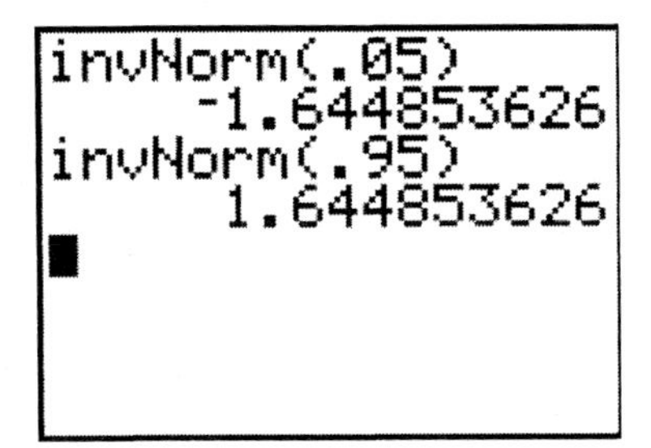

Of course, you only need to find one of the above percentiles. By the symmetry of the normal distribution, the two values are always opposites of one another. Note also that the greater the confidence level, the higher the *z* percentiles. To have higher confidence in your interval, it must be wider – larger margin of error. You can have any confidence interval that you want – the procedure is to divide the tail area (1 – confidence level) in half. Here are the values for 98% and 99.9% confidence intervals.

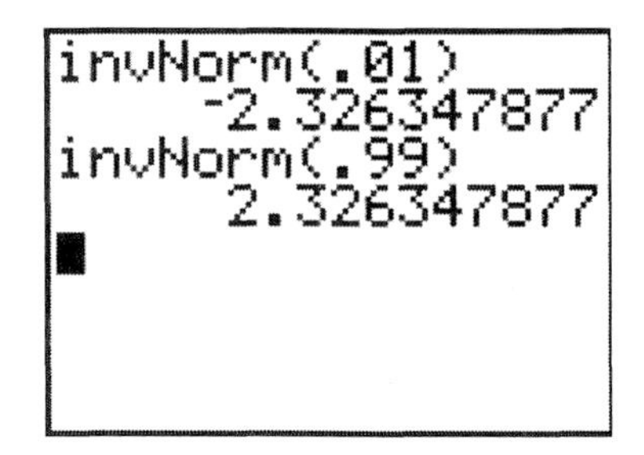


```
invNorm(.0005)
        -3.290526729
invNorm(.9995)
         3.290526729
```

Exercise 8.19: Instant Messaging (p. 372 of Agresti/Franklin)

In a random sample of 1100 teens, 75% said they regularly use instant messaging. A 95% confidence interval is shown below.

```
√(.75*.25/1100)
        .0130558242
Ans*1.96
        .0255894154
Ans→M
        .0255894154
```

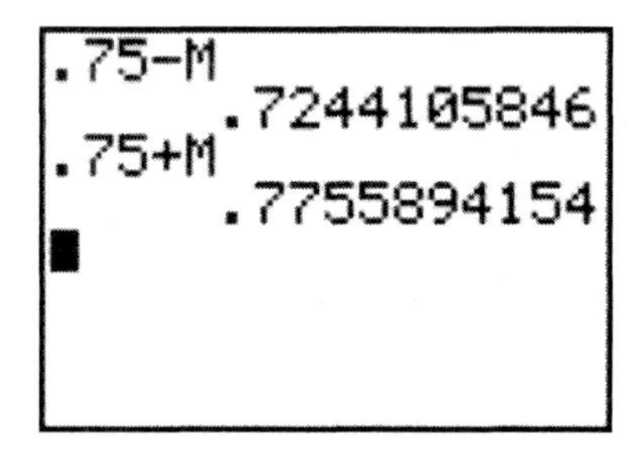

The value for $\hat{p}$ is 0.75, but the study doesn't report the *number* of teens who said that they regularly use IM. This value is needed to confirm the interval with **1-PropZInt**, but you can enter a calculation for x – it's .75 of the total, so .75*1100. Then go to **Calculate**.

```
1-PropZInt
 x:.75*1100
 n:1100
 C-Level:.95
 Calculate
```

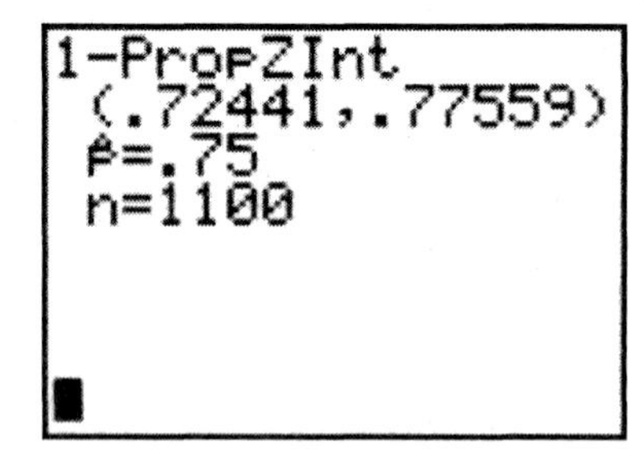

Of course, you could find .75 x 1100 = 825 manually and enter the value, too. There also would have been a problem had this not worked out to be a whole number – after all, it represents some number of teenagers, and so no fractions are allowed. If you try to proceed with a fraction, you'll get the error message **ERR:DOMAIN**.

8.3 How Can We Construct a Confidence Interval to Estimate a Population Mean?

EXAMPLE 7: eBay auctions of palm handheld computers (pp. 378-380 of Agresti/Franklin)

We have a sample of seven final prices using the "buy it now" option with which to estimate the true mean price. Enter these data into **L1** (or copy the list **EBNOW**).

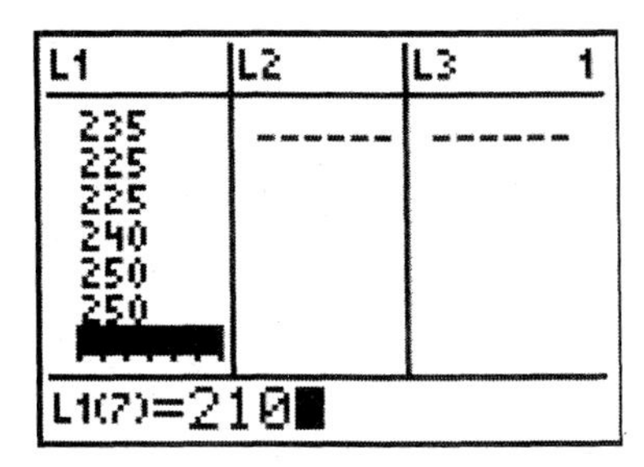

The procedure for a confidence interval for a population mean has the same strategy as with a proportion: add or subtract a multiple of the estimated standard error from the sample statistic. However, with means there is an additional complication: not only is the population meaning an unknown, but also so is the population standard deviation. This is needed to estimate the standard error.

The only option we have is to use the sample standard deviation instead, but this introduces additional uncertainty in the interval. The t distribution accounts for this by including more area in the tails. This makes the calculations more difficult, and so the built-in procedures are especially helpful. We'll demonstrate the formula by working through all the details, then see that our calculations match the built-in procedures.

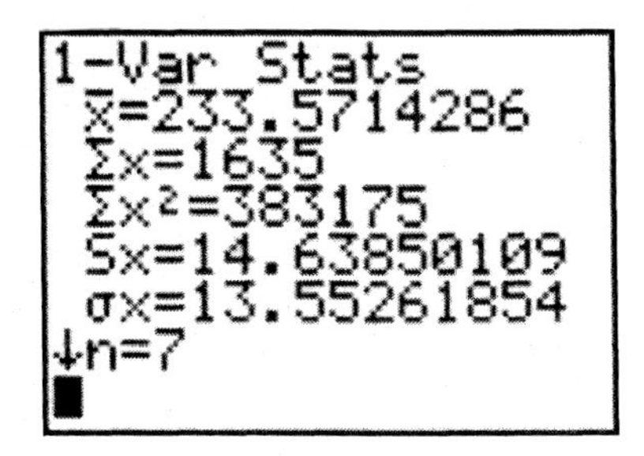

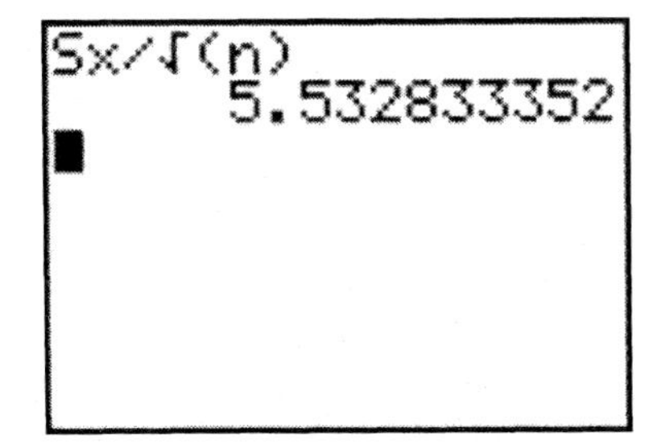

Here we see that $\bar{x} = 233.57$, $s = 14.64$, and the estimated standard error is 5.53. Now all we need is the multiplier, the t critical value for 95% confidence and $n - 1 = 6$ degrees of freedom. The newer calculators have this included in the **DISTR** menu under the command **invT**. (If you have an older calculator, you can find this value from Table B on Appendix page A3.)

As with the normal distribution, specify the percentiles that leave half the tail area on either side of the area that corresponds to the confidence level.

```
DISTR DRAW
1:normalpdf(
2:normalcdf(
3:invNorm(
4:invT(
5:tpdf(
6:tcdf(
7↓X²pdf(
```

```
invT(.025,6)
     -2.446911839
invT(.975,6)
      2.446911839
Ans*Sx/√(n)→M
      13.53835543
```

The margin of error is about 13.54, so the 95% confidence interval can be calculated. Given the values in this sample, I am 95 confident that the population mean of prices is between $220.03 and $247.11.

```
233.57-M
      220.0316446
233.57+M
      247.1083554
```

The easiest way to construct a confidence interval is to use the built-in procedures that automate this process, found in the **STAT TEST** menu. Choose **8:TInterval** to begin.

```
EDIT CALC TESTS
2↑T-Test…
3:2-SampZTest…
4:2-SampTTest…
5:1-PropZTest…
6:2-PropZTest…
7:ZInterval…
8:TInterval…
```

There are two different ways that you can enter the information for this calculation, either as raw data or as summary statistics. This is specified on the **Inpt** line, as **Data** or **Stats**. We'll try it both ways.

First, using **Data**: Since we have the eBay data entered in **L1**, enter this as the **List**. The **Freq**: value will be **1**, since we merely have 7 data observations in **L1**. Specify a 95% confidence level, and press **ENTER** with the cursor on **Calculate**.

```
TInterval
 Inpt:Data Stats
 List:L1
 Freq:1
 C-Level:.95
 Calculate
```

```
TInterval
 (220.03,247.11)
 x̄=233.57
 Sx=14.64
 n=7
```

To use the Stats option, recall that the sample mean was 233.57 and the standard deviation about 14.64. Enter these values, move the cursor to **Calculate**, and press **ENTER**.

```
TInterval
 Inpt:Data Stats
 x̄:233.57
 Sx:14.64
 n:7
 C-Level:.95
 Calculate
```

```
TInterval
 (220.03,247.11)
 x̄=233.57
 Sx=14.64
 n=7
```

Exercise 8.30: Anorexia in teenage girls (p. 383 of Agresti/Franklin)

Enter the 17 weight changes into **L1**, or use the list **ANCHG**.

```
L1    L2     L3    1
11    ------ ------
11
6
9
14
-3
0
L1(7)=0
```

The *t* interval procedures are quite robust, but you should check that the data isn't severely skewed. This is best done graphically, with a histogram, a boxplot, or both. Request a modified boxplot for **Plot1**, in order to check for outliers. A histogram is set up in **Plot2**, which allows a closer examination of the shape of the data distribution.

```
Plot1 Plot2 Plot3
On Off
Type:
Xlist:L1
Freq:1
Mark:
```

```
Plot1 Plot2 Plot3
On Off
Type:
Xlist:L1
Freq:1
```

Since a rough sketch should suffice, use **ZoomStat** to quickly format the graph.

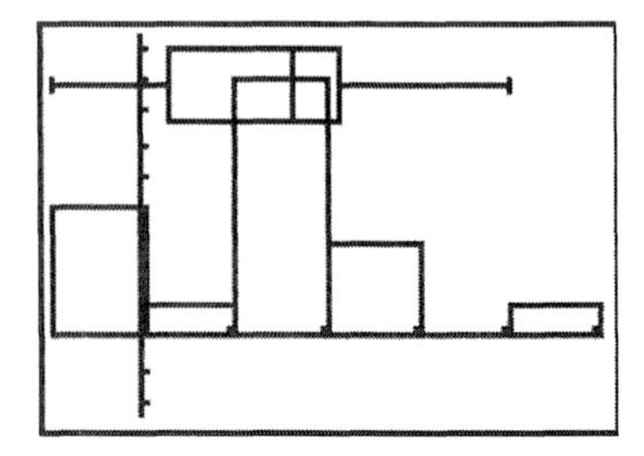

The data are not strongly skewed, and there are no outliers, so the t methods should work well.

Use the **1-Var Stats** function to verify the sample mean and standard deviation.

```
1-Var Stats
 x̄=7.294117647
 Σx=124
 Σx²=1730
 Sx=7.183006908
 σx=6.96854028
↓n=17
```

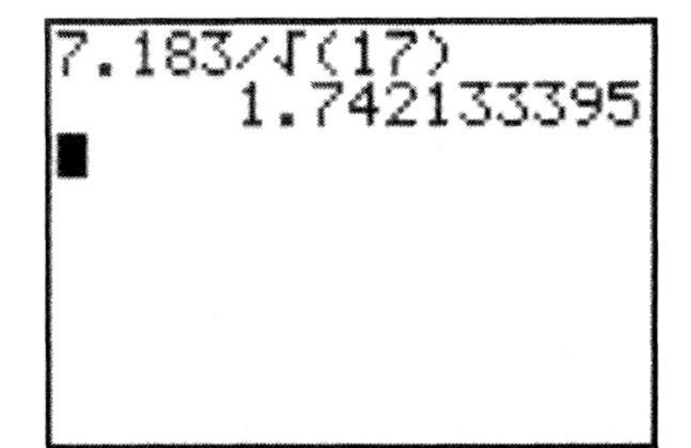

You can calculate the estimated standard error using the formula $SE = \frac{\sigma}{\sqrt{n}}$, which works out to be 1.742 (above).

Use **InvT** to show that $t = 2.120$ is the approximate t percentile for a 95% interval with $df = 16$.

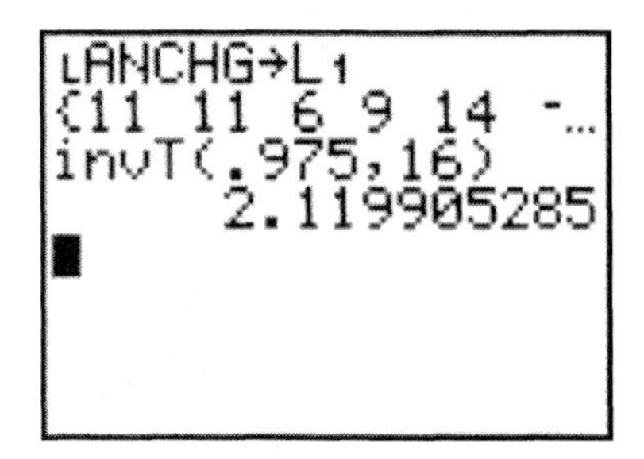

We can get the confidence interval estimate the "long way"...

```
Ans*1.7421→M
            3.693086997
7.294-M
            3.600913003
7.294+M
              10.987087
```

… or using the more efficient **STAT TESTS 7:TInterval**.

```
TInterval
 Inpt:Data Stats
 List:L1
 Freq:1
 C-Level:.95
 Calculate
```

```
TInterval
 (3.601,10.987)
 x̄=7.294117647
 Sx=7.183006908
 n=17
```

Chapter Nine: Statistical Inference – Significance Tests about Hypotheses

There is no data from the CD that is used in this chapter.

This chapter explores another type of statistical inference: significance tests. Just like with confidence intervals, we'll demonstrate the methods with a series of calculations and distributions, but also show you that the procedures are built into the **STAT TESTS** menus that efficiently carry out the necessary calculations.

9.2 Significance Tests about Proportions

EXAMPLE 4: Dr. Dog: Can dogs detect cancer by smell? (pp. 417-418 of Agresti/Franklin)

The sample proportion and standard error calculations are verified below (left).

```
22/54
          .4074074074
√((1/7)*(6/7)/54
)
          .0476190476
```

```
√((1/7*(6/7)/54)
          .0476190476
(22/54-1/7)/Ans
          5.555555556
```

The *z* test statistic in this case is 5.56. This measures standard deviations, so you should be thinking that this is very unusual – if the null hypothesis is true – and in fact 5.56 is not even included in Table A of standard normal probabilities at the back of the textbook. The p-value measures this unlikelihood, and the **normalcdf** function to calculate this as the probability of observing a result that is more than 5.56 standard deviations above the mean. Press **2nd VARS** to access the **DISTR** menu, then **2:normalcdf**.

```
DISTR DRAW
1:normalpdf(
2:normalcdf(
3:invNorm(
4:invT(
5:tpdf(
6:tcdf(
7↓χ²pdf(
```

```
normalcdf(5.56,1
0)
   1.352364232E-8
Ans*2
   2.704728464E-8
```

Each tail probability is a little more than .00000001 (reported in scientific notation), and the p-value, twice this for the two tails, is about .00000003. P-values this small are often reported by software as "p-value < .0001". We have very convincing evidence to reject H_0.

There is a built-in function available for tests such as this. Press **STAT**, then move right to **TESTS**. Select **1-PropZTest**.

```
1-PropZTest
 p0:1/7
 x:22
 n:54
 prop≠p0 <p0 >p0
 Calculate Draw
```

We can enter the null value of p_0 as 1/7. The calculator will convert it to a decimal. Remember that $\hat{p}=\frac{x}{n}$, where x is the number of successes (correct selections by the dog), and n is the sample size. Choose the **≠p0** option for a two-tailed test, then **Calculate**. The results list the alternative hypothesis, with the value **prop≠** as the decimal .14286, test statistic (**z**), the p-value (**p**), sample proportion($\hat{p}$), and sample size(**n**).

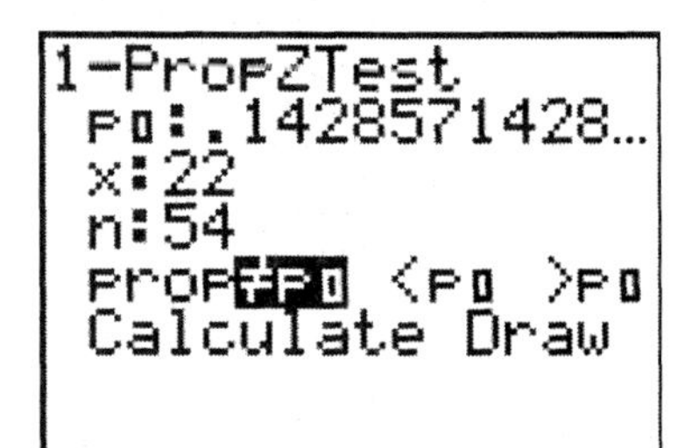

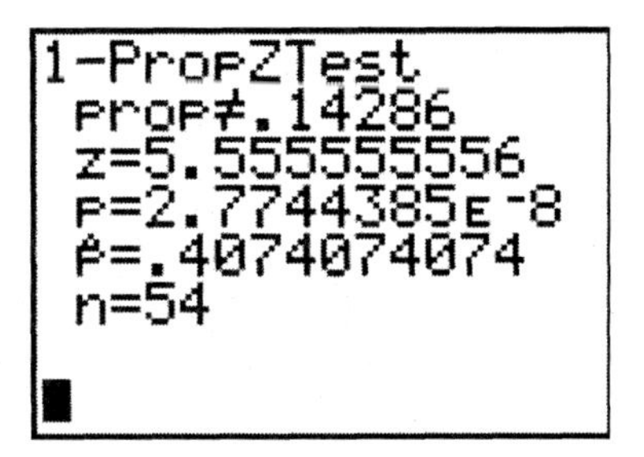

We can also choose **Draw**, but at first glance this isn't so informative. Our test statistic, 5.56, is too large to show up on the screen, which shows z-values from -3.5 to +3.5. Notice that the p-value is reported as 0.

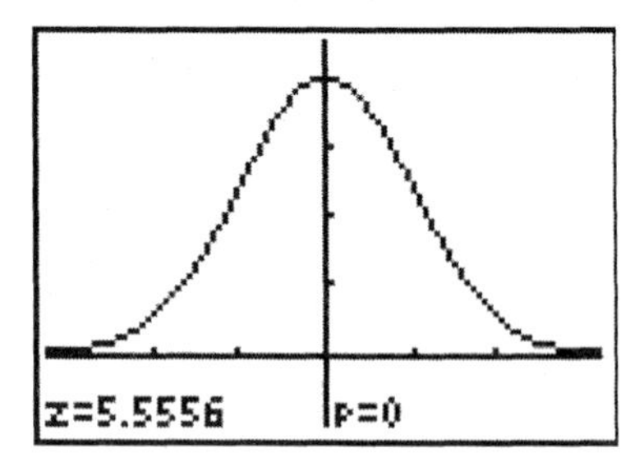

Exercise 9.21: Garlic to repel ticks (p. 427 of Agresti/Franklin)

Here we wish to test H_0: $p = 0.5$ vs. H_a: $p \neq 0.5$. The sample size is $n = 37 + 29 = 66$. Let's call garlic being more effective as the "success". Thus we have $\hat{p}=\frac{37}{66}=.56$. The test statistic calculation is shown below (left). We can use **normalcdf** (from the **DISTR** menu) to get the p-value. Since this is a two-sided test, the tail area is doubled to get the p-value.

```
37/66
          .5606060606
(Ans-.5)/√(.5*.5
/66)
          .9847319278
```

```
normalcdf(Ans,10
)
          .1623778823
Ans*2
          .3247557646
```

This is a large p-value, consistent with random variation being a plausible explanation for the observed difference. We do not have sufficient evidence of a real difference between garlic and placebo.

Here's the same procedure, automated with the **1-PropZTest**. Press **STAT**, then **TESTS**, to access this function. We can double check our work above.

```
EDIT CALC TESTS
1:Z-Test...
2:T-Test...
3:2-SampZTest...
4:2-SampTTest...
5:1-PropZTest...
6:2-PropZTest...
7↓ZInterval...
```

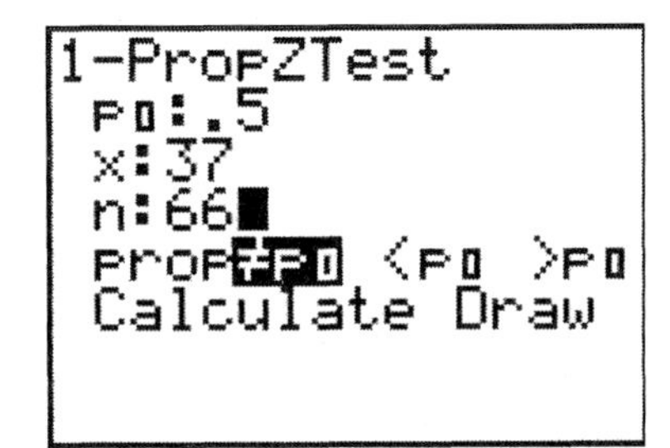

Pressing **ENTER** with the cursor on **Calculate** to verify our previous results. Pressing **ENTER** with the cursor on **Draw** will give a nice visual picture of the large, two-sided p-value.

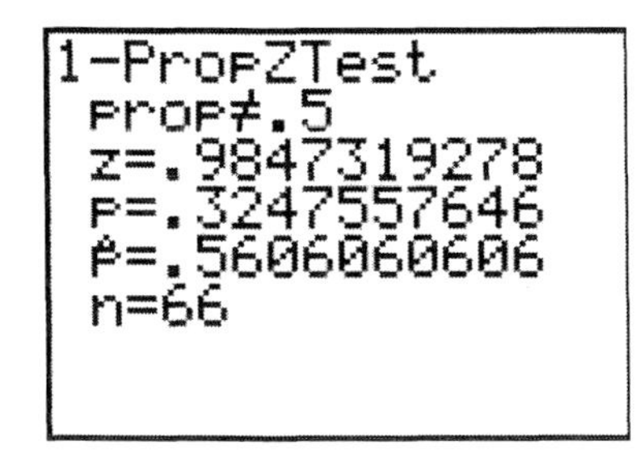

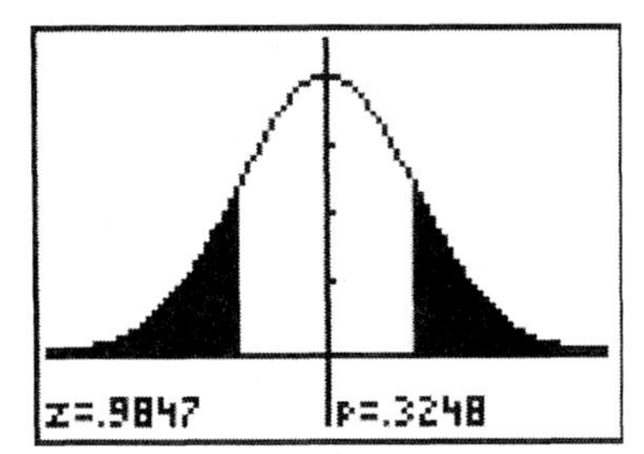

Exercise 9.16: Another test of astrology (p. 426 of Agresti/Franklin)

This problem illustrates a one-sided test, H_0: p = 1/3 vs. H_a: p > 1/3. The sample proportion and test statistic calculation are shown below (left). Note that we do not need to double the tail probability, since this is a one-tailed test.

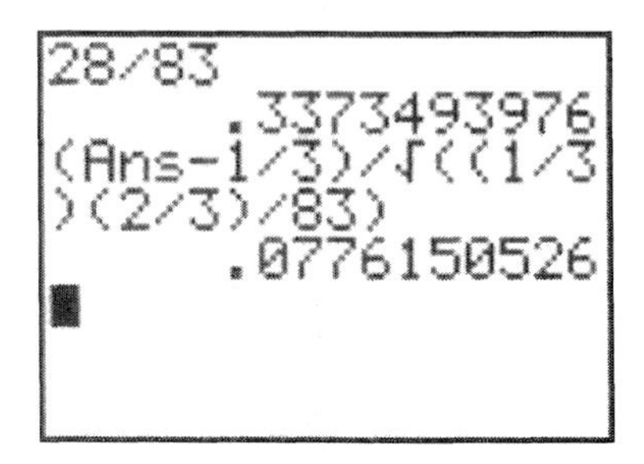

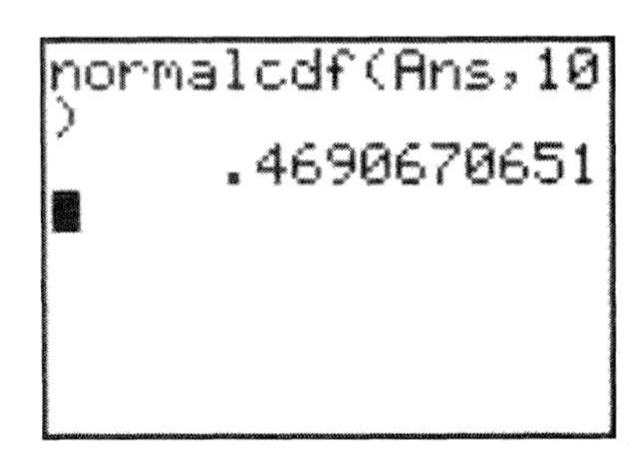

This is a large p-value, consistent with random guessing. We do not have sufficient evidence that people are more likely to select their "correct" horoscope.

Let's double-check our work using **1-PropZTest**. Note that we are choosing the "right-tail" alternative.

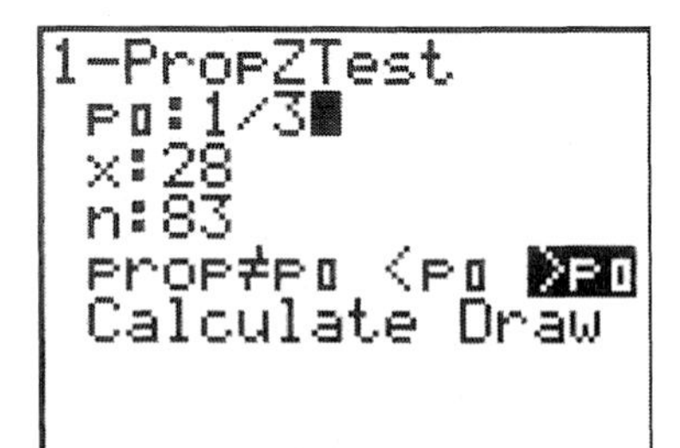

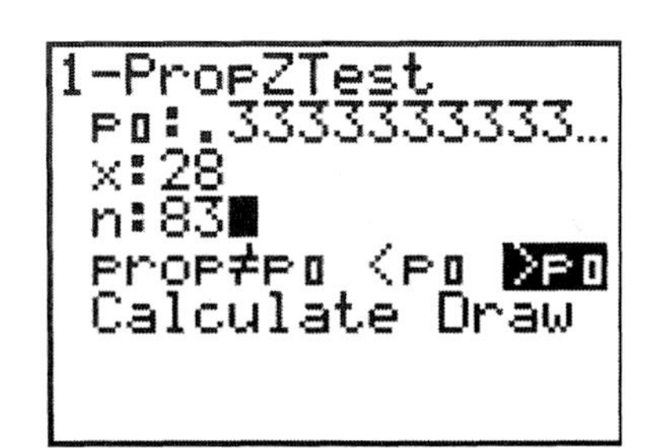

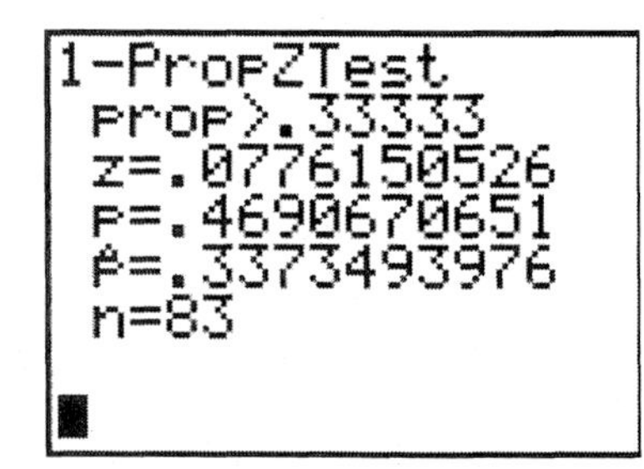

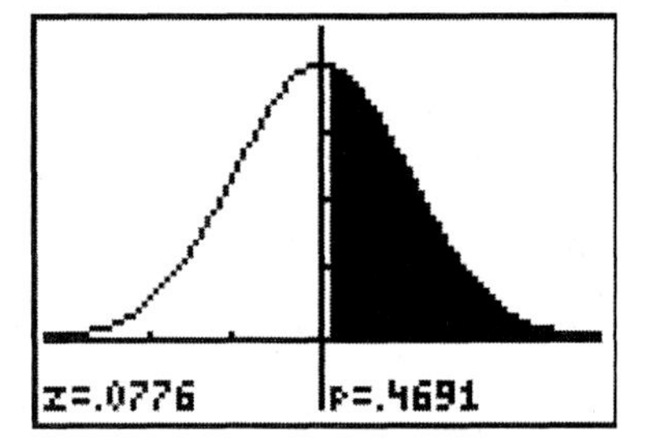

Calculate verifies our work, and **Draw** shows how close the p-value is to .5.

9.3 Significance Tests about Means

In this section, we consider hypothesis tests about one mean. This method makes use of the *t*-distribution, so we'll use the **tcdf** function from the **DISTR** menu. There is also a very useful function under **STAT TESTS** called **T-Test**.

Exercise 9.30: Low Carbohydrate Diet (p. 439 of Agresti/Franklin)

This study found that, for the 41 participants in the study, a mean weight change of -9.7 kg and a standard deviation of 3.4 kg. Is this enough evidence to say that the diet is effective in the entire population? This is a significance test of H_0: $\mu = 0$ vs. H_a: $\mu < 0$, where μ is the population mean weight change.

As previously done, we estimate the standard error and calculate a test statistic, but then we need to use the **tcdf** function to find the p-value.

```
3.4/√(41)
          .5309907904
-9.7/Ans
          ⁻18.2677368
tcdf(-100,Ans,40
)
                    0
```

The test statistic, t = -18.27, is extremely small. Remember, like z it is counting number of standard deviations from the mean. The reported p-value is 0, indicating that there is no way these weight losses happened by chance. The **STAT TESTS 2:T-Test** confirms these values.

```
T-Test
 Inpt:Data Stats
 μ0:0
 x̄:⁻9.7
 Sx:3.4
 n:41
 μ:≠μ0 <μ0 >μ0
 Calculate Draw
```

```
T-Test
 μ<0
 t=⁻18.2677368
 p=0
 x̄=⁻9.7
 Sx=3.4
 n=41
```

Exercise 9.33: Lake pollution (p. 440 of Agresti/Franklin)

The t procedures can be used with even small samples, as long as there is no skewness or outliers. This problem has a very small sample, n = 4! Enter the data into L1, then calculate the summary statistics with **1VarStats**.

```
L1      L2      L3     1
2000    ------  ------
1000
3000
2000

L1(5)=
```

```
1-Var Stats
 x̄=2000
 Σx=8000
 Σx²=18000000
 Sx=816.4965809
 σx=707.1067812
↓n=4
```

```
816.5/√(4)
               408.25
(2000-1000)/Ans
          2.449479486
tcdf(Ans,100,3)
         .0458598729
```

The standard error is 408.25 and the test statistic is $z = 2.449$. The p-value is 0.0459, and we can reject H_0 at the 0.05 level (but it's a close call).

Let's double-check our work using the **T-Test** function. Press **STAT**, then cursor right to **TESTS**. Choose **T-Test**, specifying **Data** this time, since we have it stored in **L1**. Enter the null value of the population mean ($\mu = 1000$), supply the list where the data are stored (**L1**), leave **Freq:1** as is (since each data value simply represents one observation), and specify a right-sided alternative.

```
EDIT CALC TESTS
1:Z-Test...
2:T-Test...
3:2-SampZTest...
4:2-SampTTest...
5:1-PropZTest...
6:2-PropZTest...
7↓ZInterval...
```

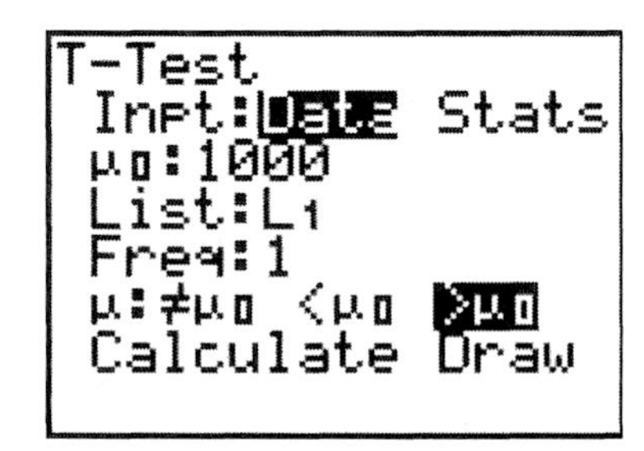

Calculate produces the results shown below left. **Draw** gives the picture shown below right.

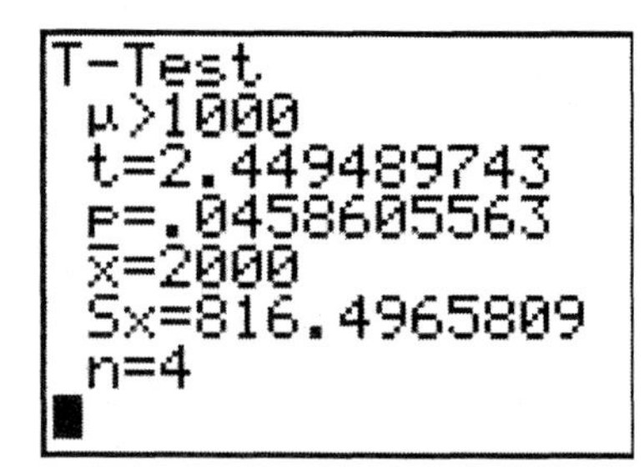

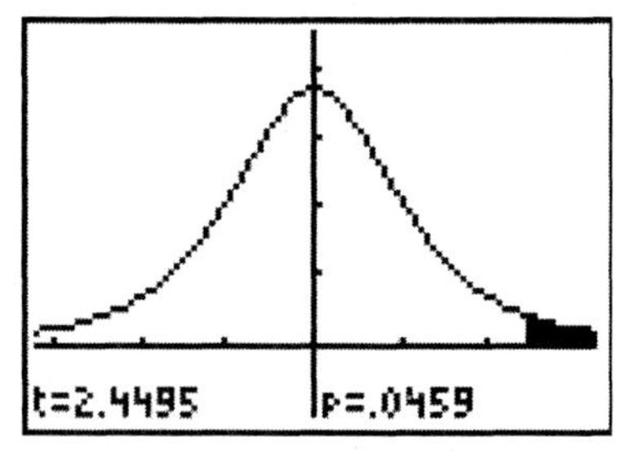

These confirm our previous results.

Chapter Ten: Comparing Two Groups

All of these lists referred to in this Chapter are on the textbook CD with the names below. They are also collected together in the group AF10 on the CD.

The process for copying lists and groups from a computer to your calculator is described in Chapter 1 of this manual, beginning on page 4.

EBNOW	Prices of Palm devices on eBay with the "buy now" option
EBBID	Prices of Palm devices on eBay through auction bidding
FLNFE	Number of times newspaper read, female Florida students
FLNMA	Number of times newspaper read, male Florida students
CELLY	Cell phone use group
CELLN	Cell phone control group
CELLD	Differences in cell phone reaction times

10.1 Categorical Response: How Can We Compare Two Proportions?

EXAMPLE 3: Standard error for heart attack rates for aspirin and placebo (pp. 471 of Agresti/Franklin)

The proportion of heart attacks with placebo and the heart attack rates for aspirin differed in the study examined in this example. To help understand this difference, the **standard error of the difference** is calculated. Each of the sample proportions is an independent random variable, so this is done by adding the *variances* of each of the proportions, then taking the square root. It is convenient in this two-sample situation to utilize the calculator's ability to store a numerical result into a variable for later use in the problem. Pressing **ALPHA** and then the key accesses the calculator variables. In the example below, pressing **ALPHA 8** accesses the variable **P**. Here is the difference of the two proportions, about 0.0077. The estimated standard error, 0.00154, is stored as **S**.

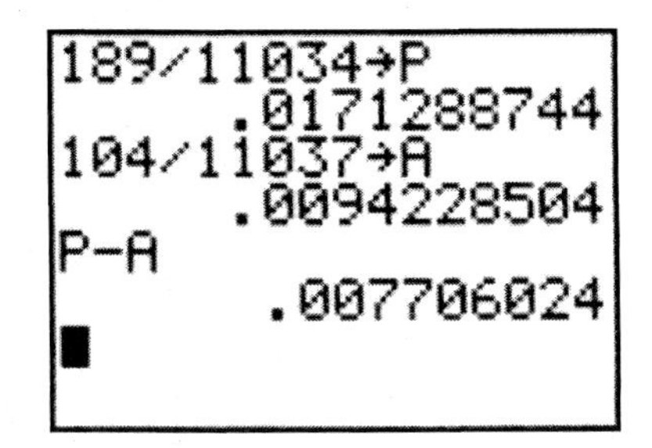

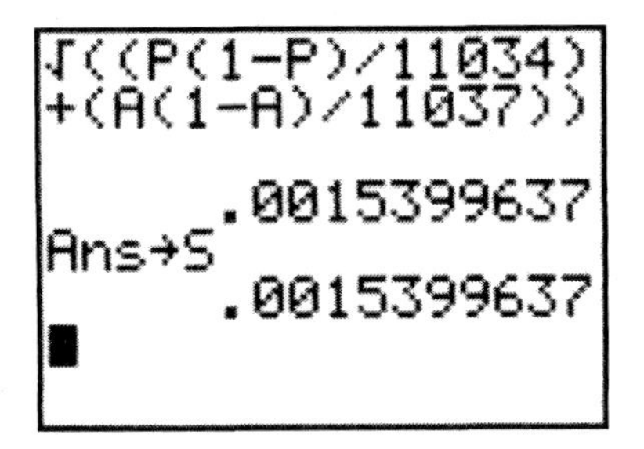

EXAMPLE 4: Confidence interval comparing heart attack rates for aspirin and placebo (pp. 473-474 of Agresti/Franklin)

Once you know the standard error, you can make a confidence interval estimate of the true difference in heart attack rates between the placebo and aspirin.

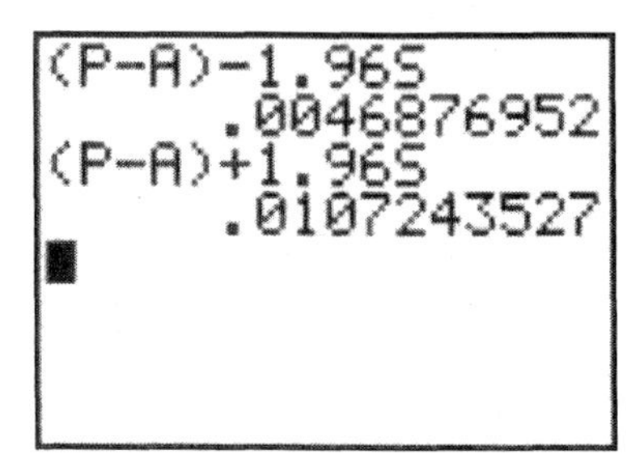

I am 95% confident that the true difference in heart attack rates between the placebo and aspirin is between 0.00469 and 0.01072. This is a small but meaningful difference – especially note that 0 is not in the interval, so there is evidence of a genuine difference between the treatments. The heart attack rate with aspirin is lower.

As with the previous chapter, figuring out the appropriate standard error is a somewhat cumbersome procedure, and you need to be very careful with all the details. Using the **STAT TESTS** menu can automate this calculation. Press **STAT**, then cursor right to **TESTS**. Choose **B:2-PropZInt**. Recall that group 1, the placebo group, had 189 heart attacks out of 11034 doctors. Recall that group 2, the aspirin group, had 104 heart attacks out of 11037 doctors. Specify 95% confidence, and press **ENTER** with the cursor on Calculate.

```
EDIT CALC TESTS
0↑2-SampTInt...
A:1-PropZInt...
B:2-PropZInt...
C:χ²-Test...
D:2-SampFTest...
E:LinRegTTest...
F:ANOVA(
```

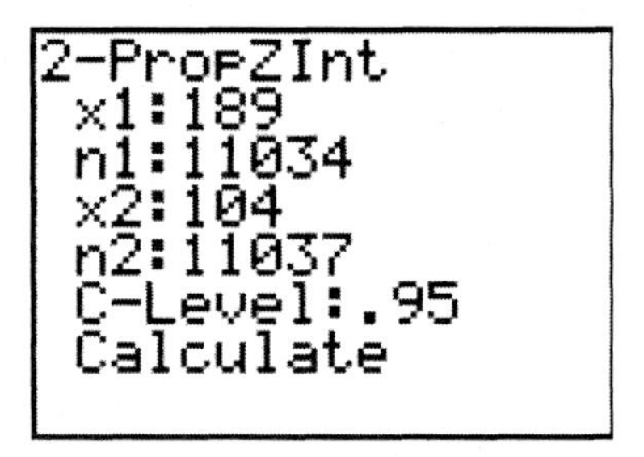


```
2-PropZInt
 (.00469,.01072)
 p̂1=.0171288744
 p̂2=.0094228504
 n1=11034
 n2=11037
```

The book reported this interval as 0.005 to 0.011, rounded to three decimal places. This is often preferred, as it lets you focus on the most significant aspect of the information present.

Exercise 10.4: Decline in Smoking (p. 480 of Agresti/Franklin)

In order to use the automated procedure for calculating a confidence interval for a difference of proportions, you need to know the number of successes and the sample size.

Sometimes this isn't reported, as in this exercise, where it says that 21.4% of a sample of 33,326 adults indicated that they are smokers in 2003 vs. 25.6% of 42,000 adults in 1993. The actual number who indicated themselves to be smokers isn't stated. You can enter a calculation in the entry for **X1** or **X2** in the **2-PropZInt** procedure. However, since **X1** and **X2** are both counts, they must be whole numbers. With rounding of percentages, often they do not work out this way, so you must round the calculated value to the nearest integer. In the example below, the value for **X1** is rounded to 7132.

```
2-PropZInt
 x1:.214*33326█
 n1:0
 x2:0
 n2:0
 C-Level:.95
 Calculate
```

```
2-PropZInt
 x1:7131.764
 n1:█
 x2:0
 n2:0
 C-Level:.95
 Calculate
```

```
2-PropZInt
 x1:7132
 n1:█
 x2:0
 n2:0
 C-Level:.95
 Calculate
```

```
2-PropZInt
 x1:7132
 n1:33326
 x2:10752
 n2:42000
 C-Level:.99
 Calculate
```

```
2-PropZInt
 (-.05,-.034)
 p̂1=.2140070816
 p̂2=.256
 n1=33326
 n2=42000
█
```

Note that the requested interval was a 99% confidence interval, so the **C-Level** is changed to .99. Also note that the reported interval in the textbook was (0.034, 0.050), whereas the interval calculated above is (-0.05, -0.034), the exact opposite. In most cases it is not important which is "group 1" and which is "group 2," but your interpretation of the interval should make the direction of the difference clear. In this case you can be 99% confident that the smoking rate is between 0.034 and 0.050 lower in 2003 than it was in 1991.

EXAMPLE 5: Is TV watching associated with aggressive behavior? (pp. 476-478 of Agresti/Franklin)

The calculator also has a built-in procedure for completing a two-proportion significance test, as in Example 5. The standard error used in a significance test is different than that used in a confidence interval calculation, since the null hypothesis claims that the two proportions are equal. Let's go through the calculation of the standard error, then show the built-in method.

In this example, 5 out of 88 teens who reported less than 1 hour a day of TV later committed aggressive acts, whereas 154 of 619 teens who reported 1 or more hours of TV a day later committed aggressive acts. Overall, then, 159 of the total of 707 teens in the study later committed an aggressive act. If the null hypothesis is true, then this is the best estimate of the true proportion of all teens who will later commit an aggressive act, since TV viewing is not associated with this behavior. The test statistic, z is calculated and then the normal probability.

```
(5+154)/(88+619)
→P
          .224893918
√(P(1-P)(1/88+1/
619))
        .0475655446
```

```
(5/88-154/619)/A
ns
       -4.035908513
2*normalcdf(-10,
Ans)
     5.441824398E-5
```

Here a two-sided test was used. It is robust against violations of the condition that you should have at least 10 successes and 10 failures, and with only 5 "successes" this is the most appropriate choice. Since the calculated P-value is about 0.0000544, this is evidence against the null hypothesis, and since the test statistic is negative, we can say that more TV watching is associated with a higher incidence of later aggressive acts. Be cautious in interpreting this, however, as it is not correct to say that TV viewing *causes* increased likelihood to commit aggressive acts, given that this is an observational study.

These calculations can be automated with **STAT TESTS 6:2-PropZTest**:

```
2-PropZTest
 x1:5
 n1:88
 x2:154
 n2:619
 p1:≠p2 <p2 >p2
 Calculate Draw
```

```
2-PropZTest
 p1≠p2
 z=-4.035908513
 p=5.4418244E-5
 p̂1=.0568181818
 p̂2=.2487883683
↓p̂=.224893918
```

```
2-PropZTest
 p1≠p2
↑p̂1=.0568181818
 p̂2=.2487883683
 p̂=.224893918
 n1=88
 n2=619
```

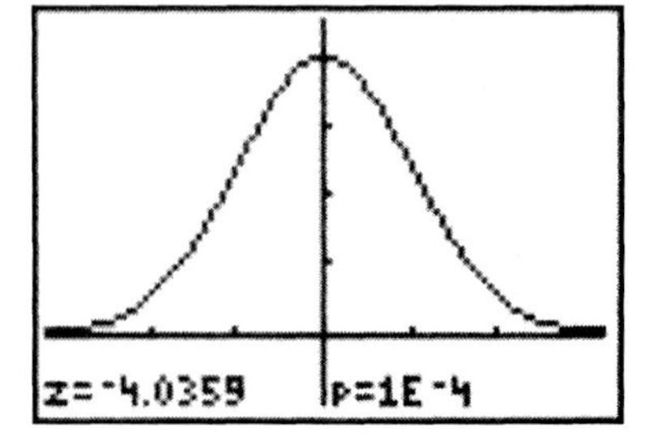

The report with the **Calculate** option reports test statistic and P-value, along with the individual proportions, the pooled proportion, and the sample sizes. The draw option shades the area probability, but in this case the shaded areas are miniscule and off the left and right sides of the window. Be especially careful with the P-values that are reported in scientific notation.

10.2 Quantitative Response: How Can We Compare Two Means?

EXAMPLE 7: How can we compare nicotine dependence for smokers and ex-smokers? (pp. 483-484 of Agresti/Franklin)

Using the values for the two groups from the textbook, the standard error can be calculated. The method involves squaring each of the two separate standard errors, adding the squares, then taking the square root: $se = \sqrt{\left[se(\bar{x}_1)\right]^2 + \left[se(\bar{x}_2)\right]^2}$. This is appropriate for two independent random samples, as is the case here. This calculation may remind you of the Pythagorean Theorem that you might have learned in Geometry, and if you take another course in Statistics you may learn that this connection is not an accident.

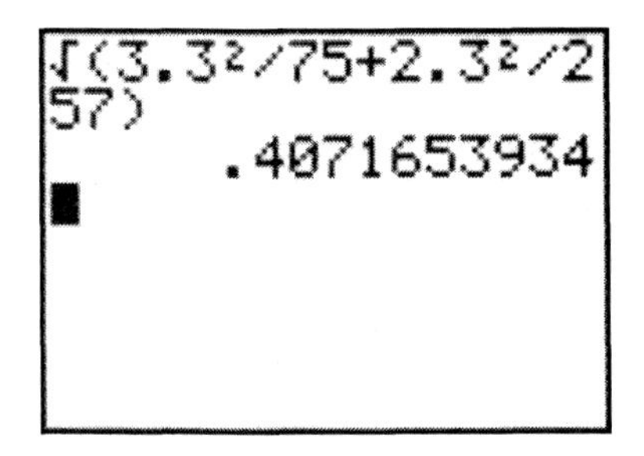

The interpretation of this value is that in repeated samples of the same sizes, we expect the observed difference in the means of the two groups to vary, with the standard error 0.407 quantifying this variability.

This standard error can be used to create confidence intervals or complete significance tests. With the differences of means, however, the calculations are more complicated than with proportions. This is due to the difficulty of estimating the correct degrees of freedom to use. In practice, then, technology is used is to complete the calculations.

EXAMPLE 8: Nicotine – How much more addicted are smokers than ex-smokers? (pp. 485-486 of Agresti/Franklin)

We can use **STAT TESTS 0:2-SampTTest** to calculate a confidence interval for the difference of the two means. Choose the **Stats** option, then specify the sample mean, standard deviation, and sample size for each of the two groups.

```
EDIT CALC TESTS
4↑2-SampTTest...
5:1-PropZTest...
6:2-PropZTest...
7:ZInterval...
8:TInterval...
9:2-SampZInt...
0:2-SampTInt...
```

```
2-SampTInt
 Inpt:Data Stats
 x̄1:5.9
 Sx1:3.3
 n1:75
 x̄2:1
 Sx2:2.3
↓n2:257
```

Specify 95% confidence, and choose **No** in the **Pooled** option. Pooling is another method of combining the two individual standard errors that may be appropriate in some circumstances. This is discussed in section 10.3, but most commonly pooling is not recommended when comparing two independent samples.

```
2-SampTInt
↑n1:75
 x̄2:1
 Sx2:2.3
 n2:257
 C-Level:.95
 Pooled:No Yes
 Calculate
```

```
2-SampTInt
 (4.0918,5.7082)
 df=95.91055203
 x̄1=5.9
 x̄2=1
 Sx1=3.3
↓Sx2=2.3
```

The confidence interval matches the one shown on p. 486. Note that the reported degrees of freedom is 95.91, from the complicated formula on page 484.

Exercise 10.20: eBay auctions (pp. 492 of Agresti/Franklin)

A confidence interval calculation can also be completed from the raw data if it is stored in two lists. Either type the data into **L1** and **L2** or copy the data from the lists **EBNOW** and **EBBID**. These are small samples, and one of the conditions for using the two-sample *t* methods is that the two population distributions are approximately normal. You should check the reasonableness of this assumption by looking at the distribution of each of the samples. We are especially concerned about the presence of outliers, so a boxplot is a good choice for a display.

Let's use **Plot1** to display the distribution of group 1 and **Plot2** for group 2. Once you have set up the two boxplots, using **ZoomStat** is effective in creating the display.

```
STAT PLOTS
1:Plot1...Off
   L1   1
2:Plot2...Off
   L1   1
3:Plot3...Off
   L4   1
4↓PlotsOff
```

```
Plot1 Plot2 Plot3
On Off
Type:
Xlist:L1
Freq:1
Mark: □ + ·
```

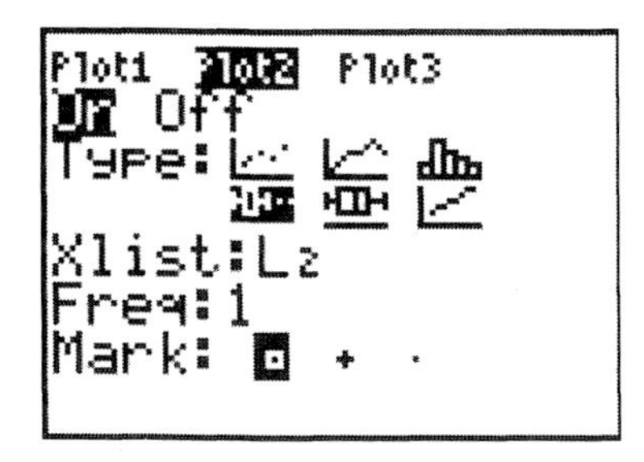

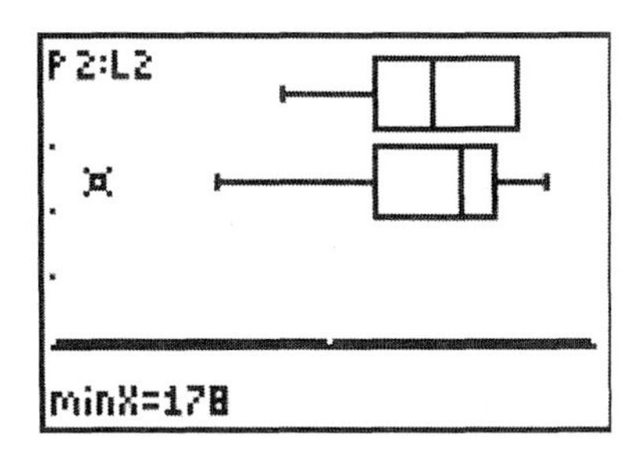

Uh-oh, an outlier in the bidding-only group. The **TRACE** function was used to find that the outlier is the minimum of group 2, a price of merely $178. This may affect the interpretation of the interval we create.

From **STAT TESTS**, choose **0:2-SampTInt**. Choose **Data**, and specify lists **L1** and **L2**. Leave **Freq1** and **Freq2** equal to **1**, since each price in the data set represents one auction. We want 95% confidence, and do not want to pool the standard deviations together. With the cursor blinking on **Calculate**, press **ENTER**.

```
EDIT CALC TESTS
4↑2-SampTTest...
5:1-PropZTest...
6:2-PropZTest...
7:ZInterval...
8:TInterval...
9:2-SampZInt...
0:2-SampTInt...
```

```
2-SampTInt
 Inpt:Data Stats
 List1:L1
 List2:L2
 Freq1:1
 Freq2:1
 C-Level:.95
↓Pooled:No Yes
```

```
2-SampTInt
 (-14.05,17.967)
 df=16.58973553
 x̄1=233.5714286
 x̄2=231.6111111
 Sx1=14.6385011
↓Sx2=21.9361075
```

```
2-SampTInt
 (-14.05,17.967)
↑x̄2=231.6111111
 Sx1=14.6385011
 Sx2=21.9361075
 n1=7
 n2=18
```

The results would indicate that we are 95% confident that the true mean difference between prices using these two options is between -14.05 and 19.97 – no significant difference one way or the other. Given the presence of the outlier we may be cautious about the exact values of the endpoints of the interval. But this doesn't change the main conclusion, that there is no evidence that the type of sale affects the sales price.

EXAMPLE 9: Does Cell-Phone Use while Driving Impair Reaction Times? (pp. 488-490 of Agresti/Franklin)

We are not given the data for this example, so we are not able to replicate the boxplots shown in figure 9.6 with the calculator. However, the presence of the outlier in the cell phone group is a concern. The summary statistics are presented both with the outlier included and with the outlier omitted, and with the **STAT TESTS 4:2-SampTTest** it's easy enough to try both and see if the outlier affects the conclusions. We are given the following sample statistics. For group 1 (cell phone), we have $\bar{x}_1 = 585.2$, $s_1 = 89.6$, and $n_1 = 32$. For group 2 (control), we have $\bar{x}_2 = 533.7$, $s_2 = 65.3$, and $n_2 = 32$. Enter these values and we're off.

```
EDIT CALC TESTS
1:Z-Test...
2:T-Test...
3:2-SampZTest...
4:2-SampTTest...
5:1-PropZTest...
6:2-PropZTest...
7↓ZInterval...
```

```
2-SampTTest
 Inpt:Data Stats
 x̄1:585.2
 Sx1:89.6
 n1:32
 x̄2:533.7
 Sx2:65.3
↓n2:32
```

Indicate a two-tailed test, no pooling, and then **Calculate**. The report shows the test statistic, p-value, and calculated degrees of freedom, **df**.

```
2-SampTTest
↑n1:32
 x̄2:533.7
 Sx2:65.3
 n2:32
 μ1:≠μ2 <μ2 >μ2
 Pooled:No Yes
 Calculate Draw
```

```
2-SampTTest
 μ1≠μ2
 t=2.627644003
 p=.0110413561
 df=56.68480423
 x̄1=585.2
↓x̄2=533.7
```

Scrolling down further just confirms our sample statistics. If you go back to **STAT TESTS** and choose **Draw** from **2-SampTTest**, you'll see the picture below right. The test statistic and two-tailed p-value just barely show up on-screen.

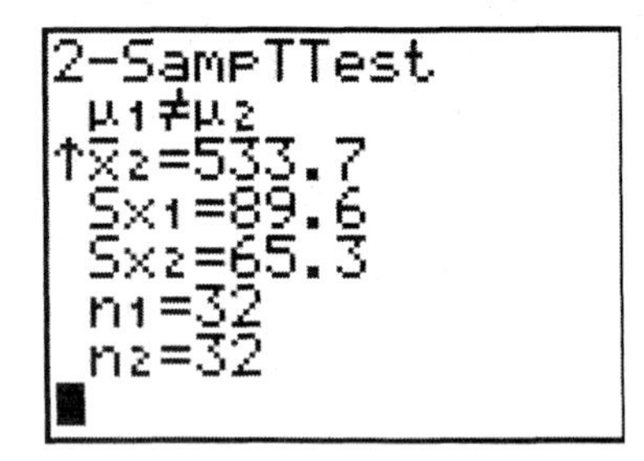

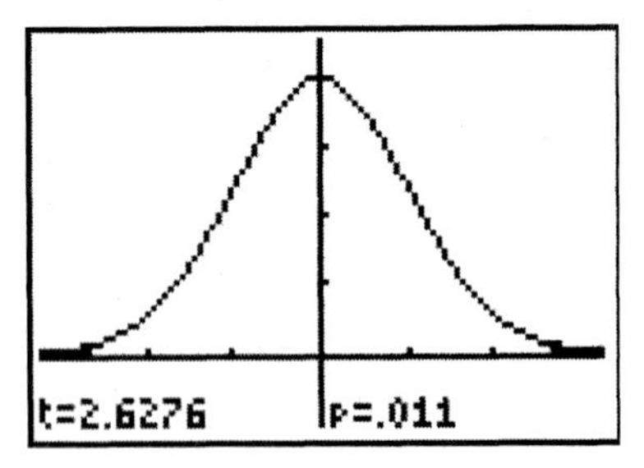

Remember that we were concerned about the presence of an outlier in the bidding-only group. Did our conclusion swiing one way or the other due to this outlier? Repeating this procedure with the summary statistics for the lists with the outlier deleted produces the following:

```
2-SampTTest
 Inpt:Data Stats
 x̄1:573.1
 Sx1:58.9
 n1:31
 x̄2:533.7
 Sx2:65.3
↓n2:32
```

```
2-SampTTest
 μ1≠μ2
 t=2.5163377
 p=.014518401
 df=60.69685004
 x̄1=573.1
↓x̄2=533.7
```

```
2-SampTTest
 μ1≠μ2
↑x̄2=533.7
 Sx1=58.9
 Sx2=65.3
 n1=31
 n2=32
```

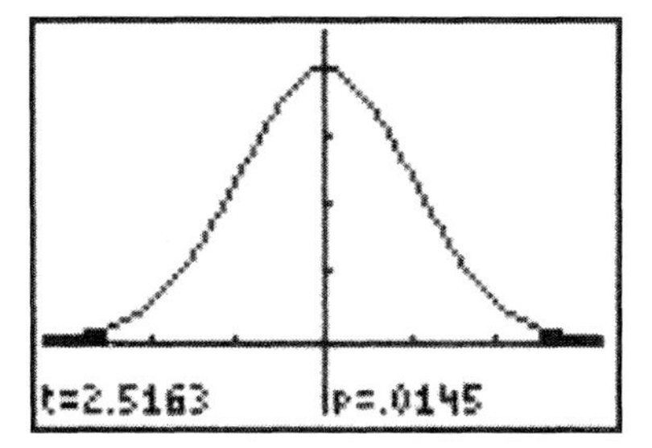

These results aren't much different than the previous, so the outlier doesn't change the conclusion we reached: there is good evidence that the mean reaction times for cell phone users is higher than for non-cell phone users.

Try It! Do Exercise 10.28, p. 493.

Complete a significance test for population means. Use the lists **FLNFE** and **FLNMA** from the textbook CD or enter the data from the textbook.

10.3 Other Ways of Comparing Means and Comparing Proportions

EXAMPLE 10: Is arthroscopic surgery better than placebo? (pp. 495-497 of Agresti/Franklin)

We are comparing group 1 (placebo) with group 2 (lavage arthroscopic surgery). We have $n_1 = 60$, $\bar{x}_1 = 51.6$, $s_1 = 23.7$ and $n_2 = 61$, $\bar{x}_2 = 53.7$, $s_2 = 23.7$. Use **STAT TESTS 0:2-SampTInt**, indicate **Stats**, since we only have the summary statistics, and enter the statistics for each group.

```
EDIT CALC TESTS
4↑2-SampTTest...
5:1-PropZTest...
6:2-PropZTest...
7:ZInterval...
8:TInterval...
9:2-SampZInt...
0:2-SampTInt...
```

```
2-SampTInt
 Inpt:Data Stats
 x̄1:51.6
 Sx1:23.7
 n1:60
 x̄2:53.7
 Sx2:23.7
↓n2:61
```

Specify 95% a confidence level, and this time request the **Pooled** procedure. **Calculate**. The interval is (–10.63, 6.43).

```
2-SampTInt
↑n1:60
 x̄2:53.7
 Sx2:23.7
 n2:61
 C-Level:.95
 Pooled:No Yes
 Calculate
```

```
2-SampTInt
 (-10.63,6.4327)
 df=119
 x̄1=51.6
 x̄2=53.7
 Sx1=23.7
↓Sx2=23.7
```

Note that **df** = $n_1 + n_2 - 2 = 60 + 61 - 2 = 119$. The pooled standard deviation **Sxp** is 23.7.

```
2-SampTInt
 (-10.63,6.4327)
↑Sx1=23.7
 Sx2=23.7
 Sxp=23.7
 n1=60
 n2=61
```

To confirm the significance test results, choose **4:2-SampTTest** from the **STAT TESTS** menu. Since we have just done the confidence interval with the same statistics, the means, standard deviations and sample sizes are already ready to go!

```
EDIT CALC TESTS
1:Z-Test...
2:T-Test...
3:2-SampZTest...
4:2-SampTTest...
5:1-PropZTest...
6:2-PropZTest...
7↓ZInterval...
```

```
2-SampTTest
 Inpt:Data Stats
 x̄1:51.6
 Sx1:23.7
 n1:60
 x̄2:53.7
 Sx2:23.7
↓n2:61
```

Indicate a two-tailed test and pooled procedure. The test statistic is $t = -0.487$, and P-value = .6269.

```
2-SampTTest
↑n1:60
 x̄2:53.7
 Sx2:23.7
 n2:61
 μ1:≠μ2 <μ2 >μ2
 Pooled:No Yes
 Calculate Draw
```

```
2-SampTTest
 μ1≠μ2
 t=-.4873251288
 p=.6269244633
 df=119
 x̄1=51.6
↓x̄2=53.7
```

The **Draw** option produces the image below. The heavier shading indicates the large P-value. There is no evidence in this study to conclude that the arthroscopic surgery with lavage is any different in effectiveness from the placebo.

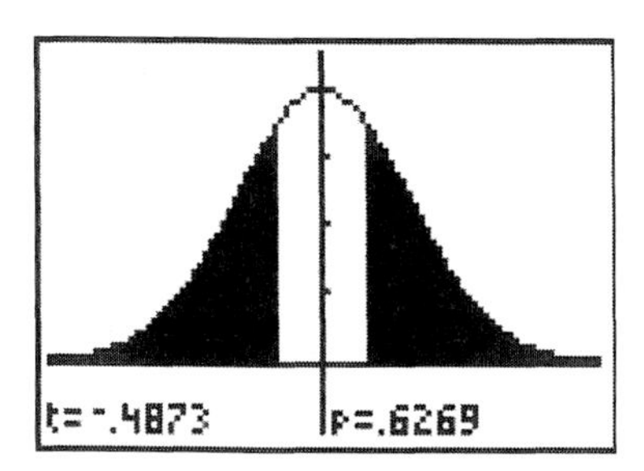

10.4 How Can We Analyze Dependent Samples?

EXAMPLE 12: Matched-pairs designs for cell phones and driving study (pp. 502-504 of Agresti/Franklin)

Table 10.9 shows the data from the study of reaction times with cell phones in which each participant was tested under both conditions. These results can be entered into **L1** and **L2** or copied from the lists **CELLN** and **CELLY**. The differences can either be calculated from the lists, shown below, or copied from the list **CELLD** to **L3**.

L1	L2	L3
604	636	------
556	623	
540	615	
522	672	
459	601	
544	600	
513	542	

L3 =L2−L1

L1	L2	L3
604	636	32
556	623	67
540	615	75
522	672	150
459	601	142
544	600	56
513	542	29

L3(1)=32

Let's replicate the comparison boxplots given in the textbook. Press **STAT PLOT**. Define **Plot1** to be a boxplot for **L1**, and **Plot2** to be a boxplot for **L2**.

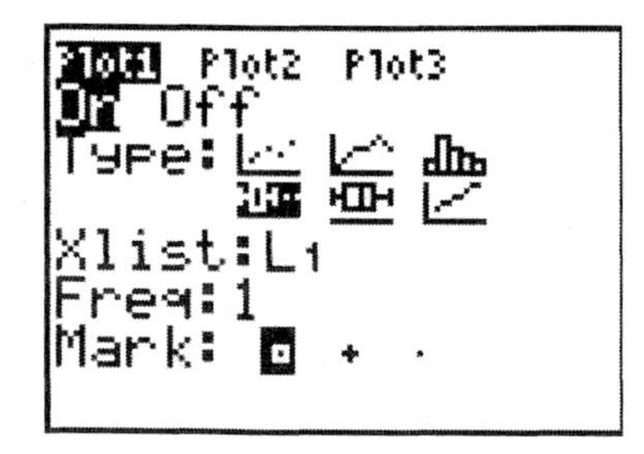

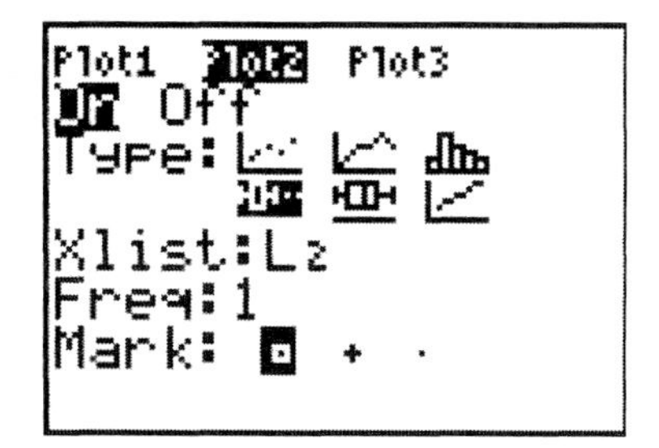

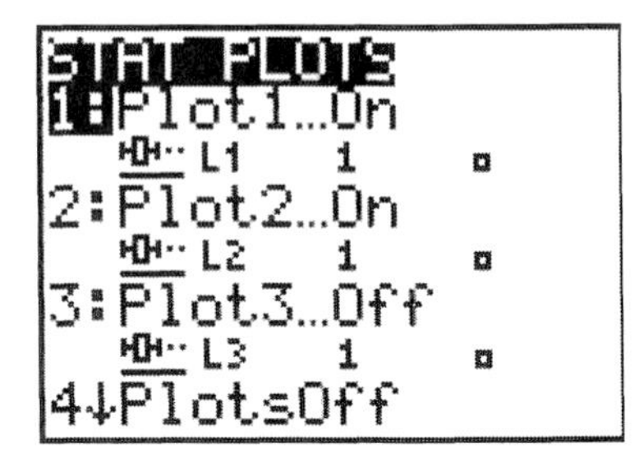

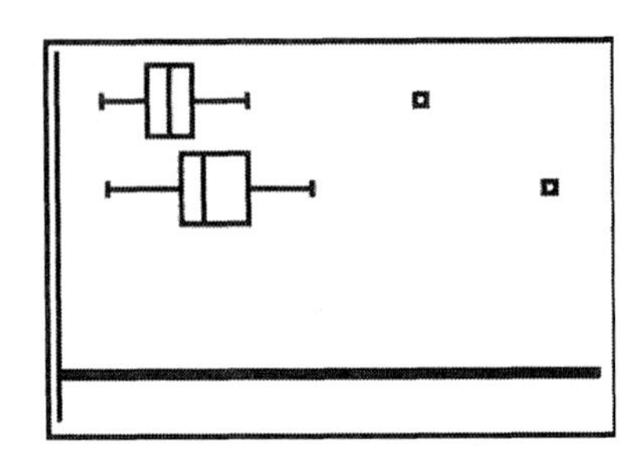

Two outliers are visible in the above graph. Using the TRACE function, can you determine which students they belong to?

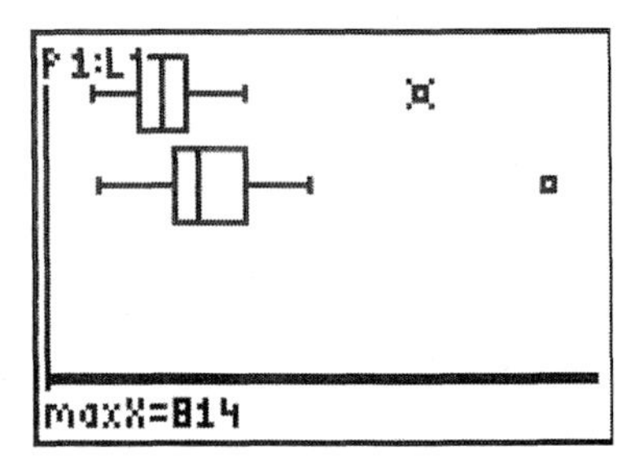

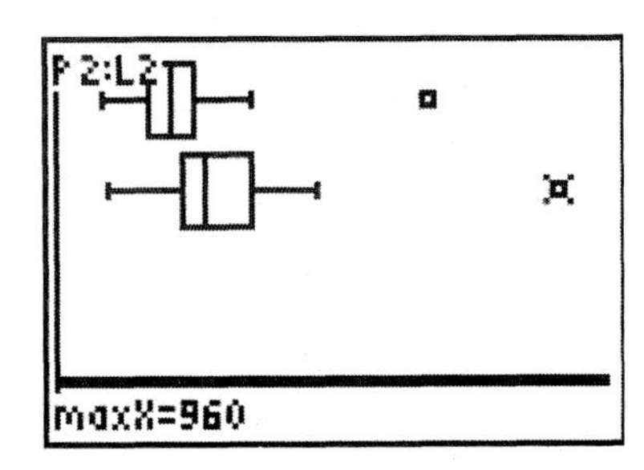

This is the same student, #28, who seems to be slow to react regardless of whether or not he or she was using a cell phone. The differences between cell phone "No" and "Yes" for each student are really what we're interested in here. These are *not* two independent samples, they are the same subjects in each group. Thus the appropriate approach is a paired-difference t-test, not a two-sample t-test. This is continued in Example 13, which examines the list of differences that are in **L3**.

EXAMPLE 13: Matched-pairs analysis of cell phone impact on driver reaction time (pp. 505-507 of Agresti/Franklin)

Assuming you just worked Example 12, turn off **Plot1** and **Plot2** and define **Plot3** to be a boxplot for the differences, **L3**. You'll need to change the window with **ZoomStat** in order to view the new graph.

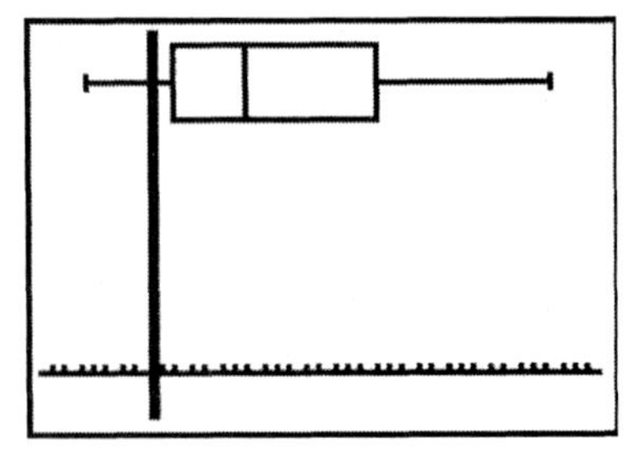

Let's find the mean and standard deviation of the differences using **1-Var Stats**. Press **STAT CALC** and specify **L3**.

```
1-Var Stats L3
```

```
1-Var Stats
 x̄=50.625
 Σx=1620
 Σx²=167410
 Sx=52.48578917
 σx=51.65918965
↓n=32
```

We can see that $\bar{x}_d = 50.6$ and $s_d = 52.5$.

For a paired difference significance test, the hypotheses are H_0: $\mu_d = 0$ vs. H_0: $\mu_d \neq 0$. Press **STAT TESTS 2:T-Test**. The differences are in **L3**, so choose **Data**. Supply a zero for μ_0 (to match H_0) and specify a two-tailed alternative. Choose **Calculate** or **Draw**.

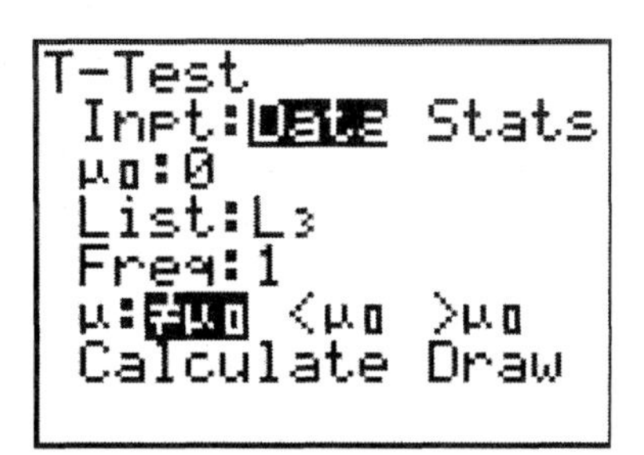

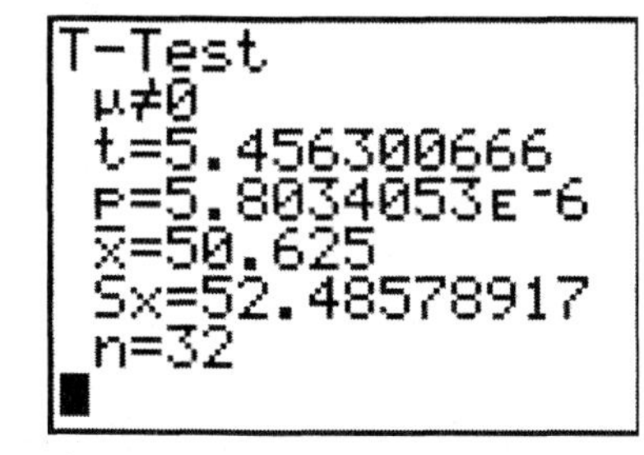

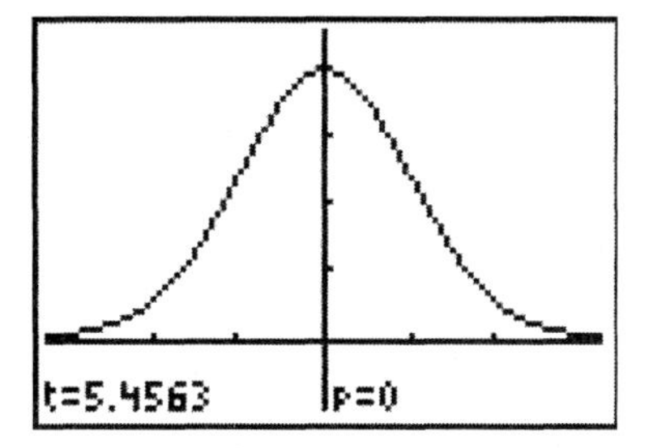

The test statistic is $t = 5.46$ and the P-value is .000006, essentially 0. We have strong evidence that the reaction times are different. The reactions times are slower on average with a cell phone.

With strong evidence of a difference, the natural question is how large is the difference? A confidence interval answers this question, and is usually an appropriate follow-up whenever a significant difference is observed. The differences are in **L3** and note again that the procedure is really a one-sample situation, using the one set of differences.

```
EDIT CALC TESTS
2↑T-Test...
3:2-SampZTest...
4:2-SampTTest...
5:1-PropZTest...
6:2-PropZTest...
7:ZInterval...
8:TInterval...
```

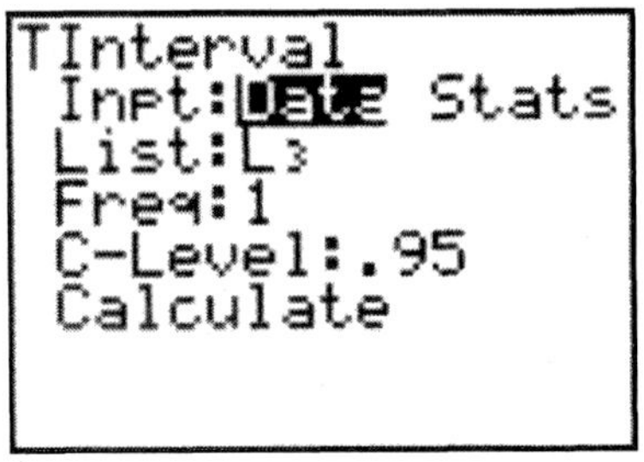

```
TInterval
 Inpt:Data Stats
 List:L3
 Freq:1
 C-Level:.95
 Calculate
```

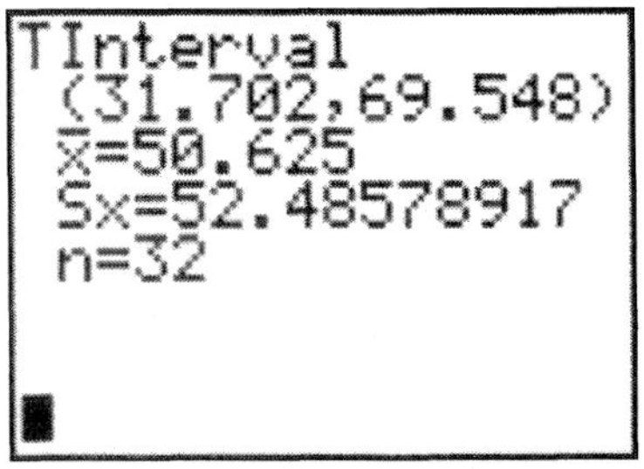

```
TInterval
 (31.702,69.548)
 x̄=50.625
 Sx=52.48578917
 n=32
```

I am 95% confident that the true mean difference in reaction times is between 31.7 and 69.5 milliseconds slower when using a cell phone.

Chapter Eleven: Analyzing the Association between Categorical Variables

There is no data from the CD that is used in this chapter.

11.2 How Can We Test Whether Categorical Variables are Independent?

In this section, we will use the **STAT TESTS** function **χ^2-Test**, and the **χ^2cdf** command from the **DISTR** menu.

EXAMPLE 4: Are happiness and income independent? (pp. 555-556 of Agresti/Franklin)

In the contingency table shown in Example 4 (Table 11.6) the conditional distributions across the three rows are not identical, so there is an association. The χ^2 test statistic is a measure of the degree to which these conditional distributions differ and can be used to test whether the observed association is statistically significant. The form of the χ^2 test statistic is $\chi^2 = \sum \frac{(O-E)^2}{E}$, where O is the observed count in a cell and E is the expected count in a cell. The expected counts are the unique values in each cell that would have the same row and column totals, but for which the conditional distributions are identical. This involves a bunch of arithmetic, so for practical reasons we will use the built-in procedures to demonstrate this.

First, the counts in the contingency table must be entered into the calculator. This is done using the matrix feature, accessed with **2nd MATRIX** (or, on the original TI-83, the key labeled **MATRX**) and then **EDIT**. This matrix needs to have three rows and three columns to correspond to the contingency table; these are specified and then the values are typed in, pressing **ENTER** after each entry.

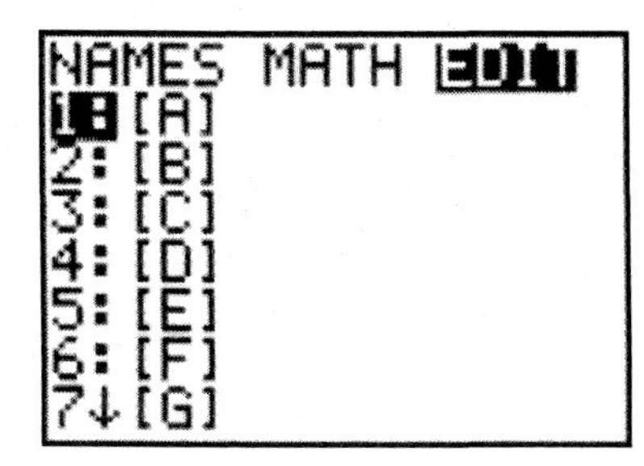

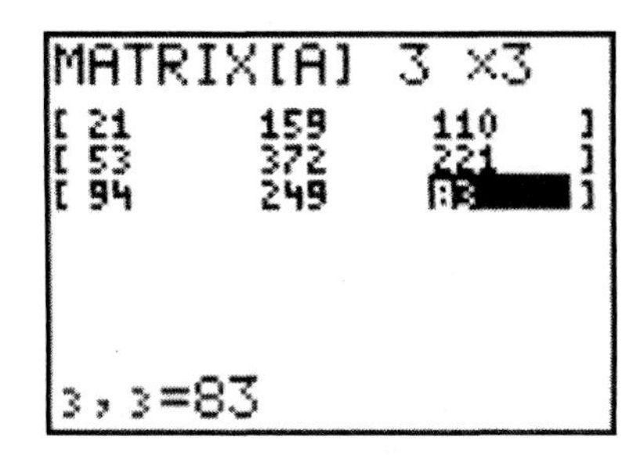

To get the χ^2 test, choose **STAT TESTS C:χ^2-Test**. This asks you to specify the matrix in which the **Observed** counts have been typed, and the matrix in which the calculator will leave the **Expected** counts. You can use any of the ten defined matrices, but the default choices are matrices **[A]** and **[B]** for these roles.

```
EDIT CALC TESTS
7↑ZInterval...
8:TInterval...
9:2-SampZInt...
0:2-SampTInt...
A:1-PropZInt...
B:2-PropZInt...
C:χ²-Test...
```

```
χ²-Test
 Observed:[A]
 Expected:[B]
 Calculate Draw
```

Choosing the Calculate option produces the report on the left. Like the *t* distribution, there is a value for the degrees of freedom associated with a particular χ^2 distribution. In this case the degrees of freedom is 4, calculated with the formula
(#rows – 1)(#columns – 1) = (3 – 1)(3 – 1) = 4.

```
χ²-Test
 χ²=73.35246138
 p=4.44409E-15
 df=4
```

The P-value for this test is incredibly small: 4×10^{-15}, essentially 0. We reject H_0 and conclude that happiness and family income are *not* independent.

You can see the expected counts that are stored in matrix B. Press **MATRIX NAMES 2:[B]**. Scroll to the right to see all the values in the matrix, which agree with the values quoted in the textbook.

```
NAMES MATH EDIT
1:[A] 3×3
2:[B] 3×3
3:[C]
4:[D]
5:[E]
6:[F]
7↓[G]
```

```
[B]
[[35.77092511 1...
 [79.68281938 3...
 [52.54625551 2...
```

```
[B]
... 166.0792952 8...
... 369.9559471 1...
... 243.9647577 1...
```

```
[B]
...2 88.14977974]
...1 196.3612335]
...7 129.4889868]]
```

Choosing the **Draw** option produces a graph of the χ^2 distribution for **df** = 4. The χ^2 value, 73.4 is far to the right (the tick marks on the axis are each 1 unit); the area beyond 73.4, measured by the P-value is essentially 0.

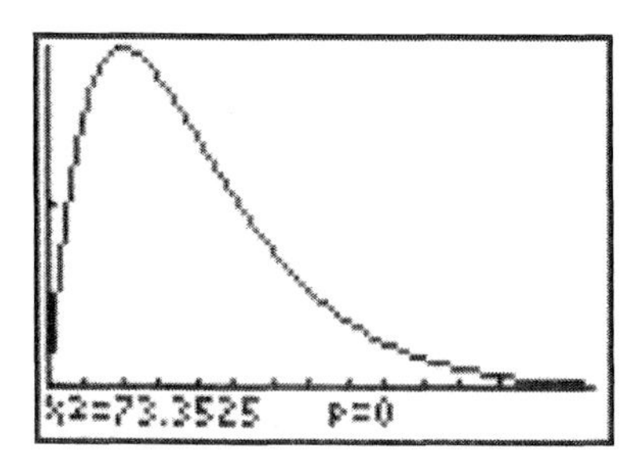

Exercise 11.9: Are happiness and gender independent? (p. 562 of Agresti/Franklin)

Enter the values from the problem into a 2x3 matrix. Press **STAT TESTS C:χ^2-Test**. Choose the **Calculate** option to produce the screen at the right below. The χ^2 test statistic is much smaller than the previous at about 3.5, and the P-value is 0.17. This is not sufficient evidence to reject the null hypothesis that happiness and gender are independent.

MATRIX[A] 2 ×3
[86 394 235]
[94 344 184]
2,3=184

χ²-Test
χ²=3.498694438
p=.1738874168
df=2

The **Draw** option produces the χ^2 distribution with **df** = 2. Notice that this shape is different than the previous example, showing that each different value for the degrees of freedom has a different χ^2 distribution. The χ^2 value and P-value are shown, indicated by the shading.

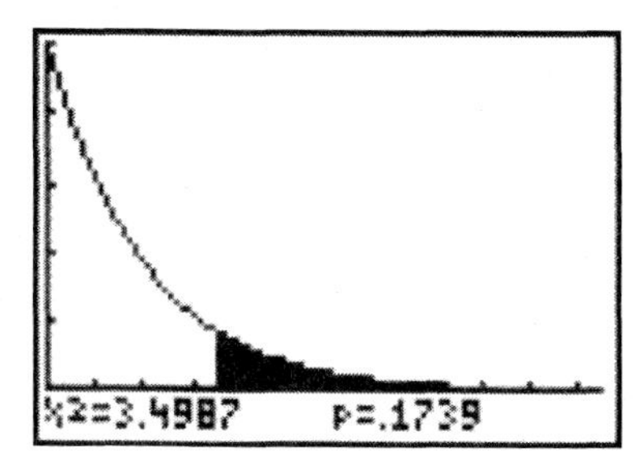

Exercise 11.21: Is genetic theory correct? (p. 564 of Agresti/Franklin)

The χ^2 test statistic can also be used in other circumstances, anytime there is an observed count that is matched up with an expected count by a hypothesis. With a single categorical variable the expected counts might come from a scientific theory, as is the case in this example. This setting is called a "goodness of fit" test, since we are testing whether the observed results fit the predictions of the theory. Put the observed counts in **L1** and the expected counts, determined by the theory that predicts that 75% of the seedlings should be green, in **L2**.

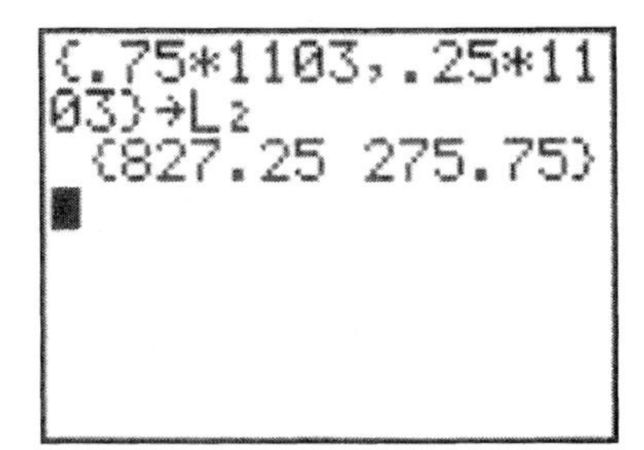

The χ^2 formula can be implemented with the lists **L1** and **L2** replacing the observed counts and expected counts, respectively. The components are in **L3** and the sum is the χ^2 test statistic, 3.712.

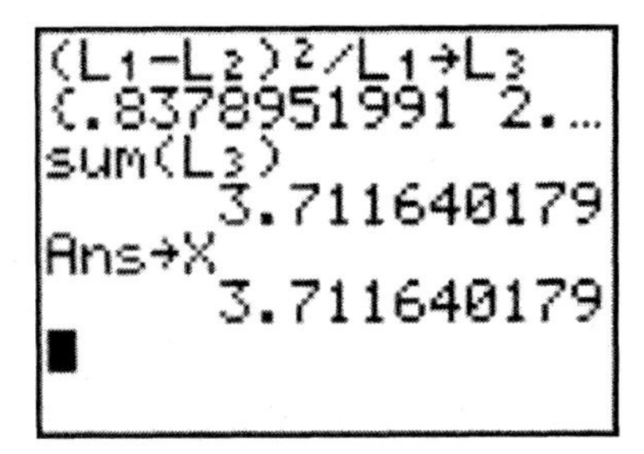

The P-value is calculated using the χ^2cdf from the **DISTR** menu. This calculates the probability of getting a χ^2 test statistic of 3.712 or larger if the genetic theory is correct. The lower bound has been stored as **X**; the upper bound is a little harder to specify since, unlike *z* or *t* distributions, the χ^2 units don't have a natural scale. You can always use 10^99, essentially positive infinity, as the upper bound.

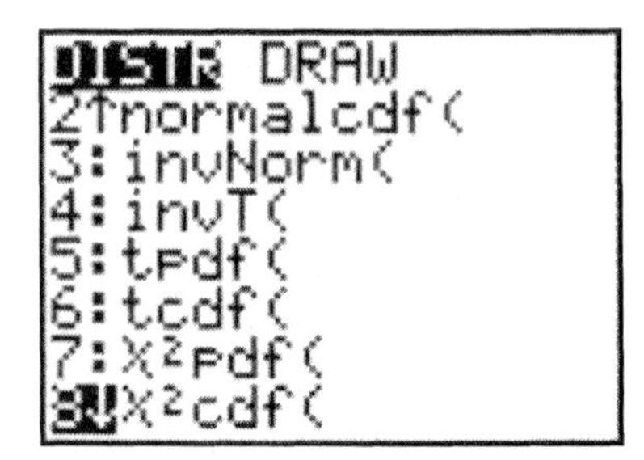

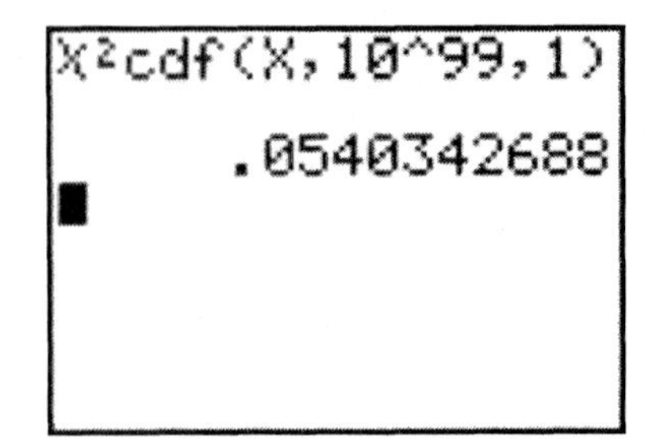

The probability is 0.054, so there is not sufficient evidence to reject the null hypothesis. The degrees of freedom is 1, one less than the number of categories.

Exercise 11.22: Is the roulette wheel fair? (p. 564 of Agresti/Franklin)

In this example the individual data is not reported, just the χ^2 test statistic, 34.4. This is a much larger value than the previous example, but it is important to understand that what constitutes a large value of χ^2 depends greatly on the degrees of freedom. Since the European roulette wheel has 37 pockets, the degrees of freedom is 36. The χ^2-Test can determine if 34.4 is an unusually high value or a typical value for this distribution.

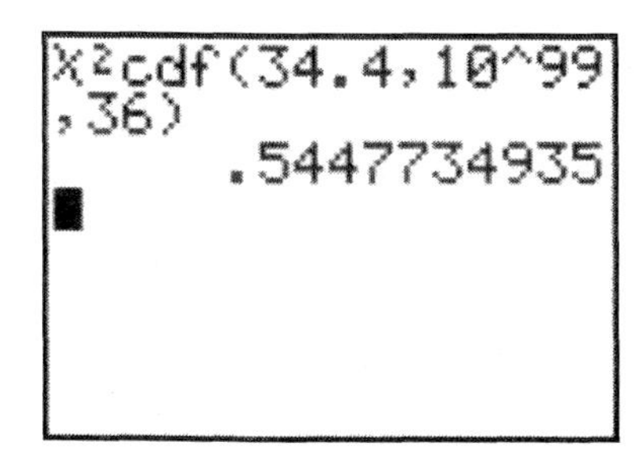

The P-value of 0.5448 reveals that 34.4 is not an unusually high value for a fair roulette wheel. Since there are 37 individual components to the χ^2 test statistic, it is not surprising that this will typically produce a higher total than the previous example with only two categories.

In the **DISTR** menu, the calculator has an option for **Shade**χ^2, with the same format as χ^2**cdf**.

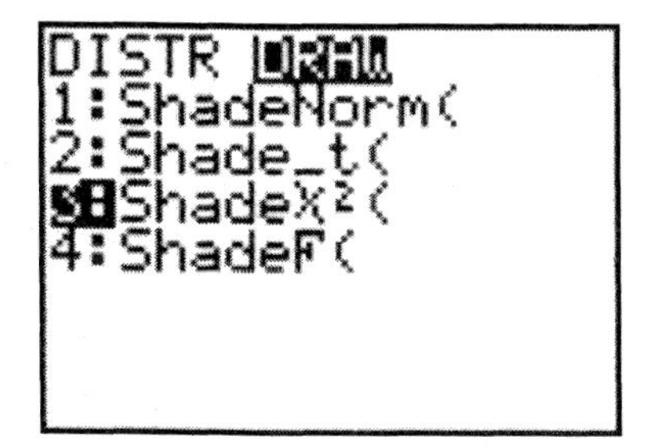

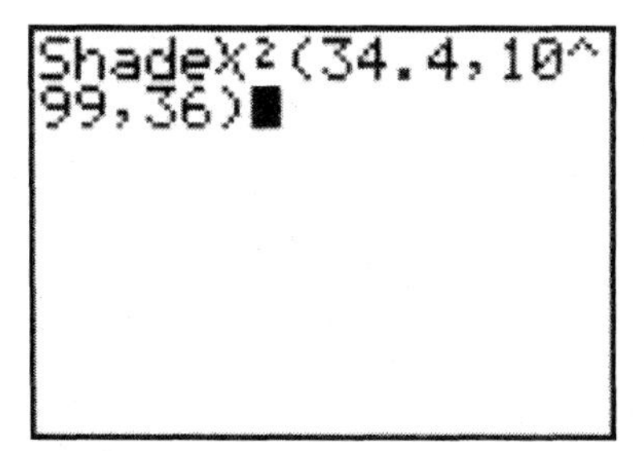

Unfortunately, this doesn't necessarily scale the window appropriately, so if you use this you will have to set the window manually, which may require some trial and error. Here is a window that works in this case, and you can see that the shading of the distribution illustrates the P-value of just over 0.5.

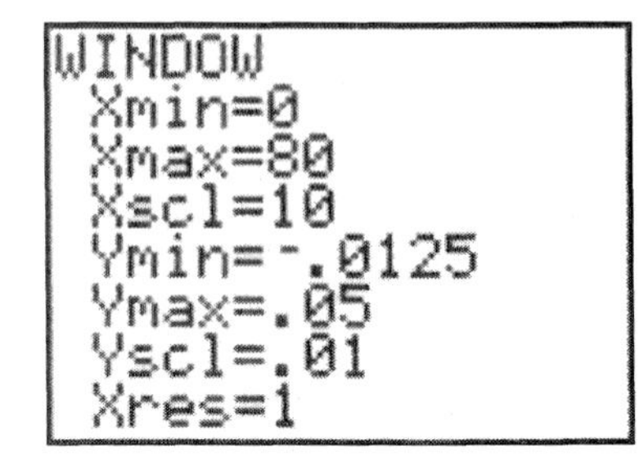

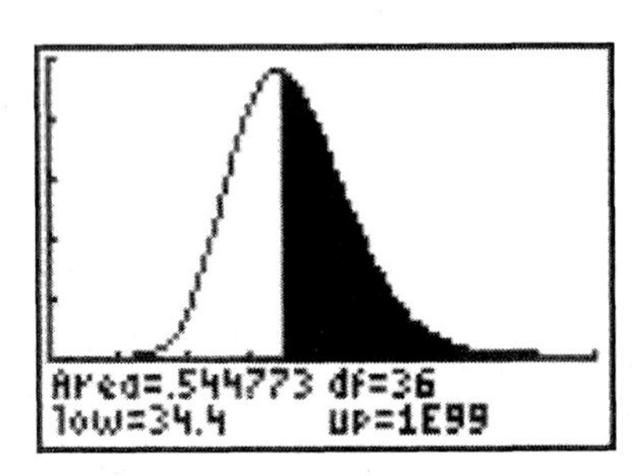

You can see another feature of the family of χ^2 distributions in this graph as well. All the χ^2 distributions are skewed, but as the degrees of freedom increases, the skewness decreases. In fact, this graph is hardly skewed at all and doesn't look too different from a normal distribution

Another Example: Birth Days

A study was done of the days of the week on which students were born, with a sample of 175 students taken from a large population, and the day of the week on which they were born was recorded. The null hypothesis is that a student would be equally likely to be born on any day of the week, each day having probability 1/7. The alternate hypothesis is that these probabilities are not all 1/7.

The table at the right reports the results of the survey. The observed counts are in **L1**, the expected counts are all 25 (175 divided by 7) and stored in **L2**. The χ^2 formula puts the components in **L3**.

Day	Count
Mon	29
Tue	30
Wed	33
Thu	27
Fri	26
Sat	19
Sun	11
Total	175

L1	L2	L3 3
29	25	------
30	25	
33	25	
27	25	
26	25	
19	25	
11	25	

L3 =(L1−L2)²/L2

L1	L2	L3 3
29	25	.64
30	25	1
33	25	2.56
27	25	.16
26	25	.04
19	25	1.44
11	25	7.84

L3(1)=.64

The χ^2 test statistic, the sum of **L3**, is 13.68, and the P-value is calculated to be 0.0334. This is small, sufficient evidence to reject the null hypothesis. Births are not equally likely to occur on the seven days of the week. Examining the χ^2 components makes it clear that births on the weekend, particularly on Sunday, are less likely, probably due to scheduled Caesarian deliveries or induced labors.

```
sum(L3)
                13.68
χ²cdf(Ans,10^99,
6)
        .0334223255
```

The goodness of fit setting makes the term “degrees of freedom” perhaps more easy to understand. In this example, once the numbers of births on Monday through Saturday are recorded, the number of Sunday births is predetermined, given that there are 175 in the sample.

Chapter Twelve: Analyzing Association between Quantitative Variables: Regression Analysis

All of these lists referred to in this Chapter are on the textbook CD with the names below. They are also collected together in the group AF12 on the CD.

The process for copying lists and groups from a computer to your calculator is described in Chapter 1 of this manual, beginning on page 4.

HSBNH	Maximum bench press of a sample of high school female athlete
HSRTF	Number of bench press repetitions at 60 pounds for female athletes
LEAFR	Kilograms of leaf litter remaining in a forest
LEAFW	Number of weeks since the start of the leaf litter study

Recall that in chapter 3 we explored the association between pairs of quantitative variables. We created scatterplots, fit regression lines, and calculated correlations. Since then, we have learned the art of significance testing. We can apply that knowledge to regression analysis to determine if a linear relationship that exists between two quantitative variables is statistically significant or is the result of random variation.

12.1 How Can We "Model" How Two Variables Are Related

In this section, we will revisit scatterplots and regression lines. All of the examples here use the strength study data.

EXAMPLE 2: What do we learn from a scatterplot in the strength study? (pp. 588-589 of Agresti/Franklin)

The strength study is introduced in Example 1. The two variables are the number of repetitions that an athlete can complete using a 60-lb weight on the bench press, and the maximum possible bench press (pounds) that the athlete can perform. Copy the data from lists **HSRTF** and **HSBNH** to **L1** and **L2** or type the data in to your calculator. (The data isn't presented in the textbook, but it's on the CD in a number of formats that you can read, including Excel, if you don't have the cable to connect you calculator directly to a computer).

Let's first confirm the summary statistics on p. 588, $\bar{x} = 11.0$ reps, $s_x = 7.1$ reps, and $\bar{y} = 79.9$ lbs., $s_y = 13.3$ lbs. This can be done all at once by using the command **2-VarStats** in the **STAT CALC** menu. These do confirm the values in the book, with additional decimal places. You need to scroll down to see the complete display.

```
LHSRTF→L1
{10 12 20 5 12 ...
LHSBNH→L2
{80 85 85 65 95...
```

```
EDIT CALC TESTS
1:1-Var Stats
2:2-Var Stats
3:Med-Med
4:LinReg(ax+b)
5:QuadReg
6:CubicReg
7↓QuartReg
```

```
2-Var Stats
 x̄=10.98245614
 Σx=626
 Σx²=9732
 Sx=7.142656639
 σx=7.079724517
↓n=57
```

```
2-Var Stats
↑ȳ=79.9122807
 Σy=4555
 Σy²=373875
 Sy=13.27898541
 σy=13.16198767
↓Σxy=54285
```

```
2-Var Stats
↑σy=13.16198767
 Σxy=54285
 minX=2
 maxX=35
 minY=60
 maxY=115
```

Create a scatterplot to examine the relationship between the number of repetitions and the maximum bench press weight. There clearly appears to be a positive association.

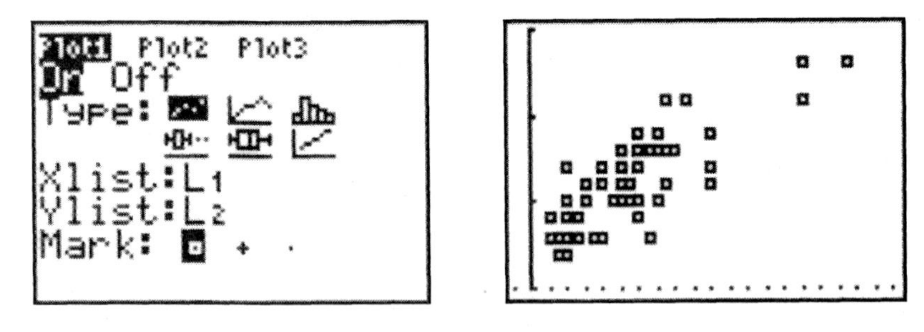

TRACE can be used to examine the points of the graph, including the point in the extreme top right, the athlete with the largest number of repetitions, 35, and the maximum bench press strength, 115.

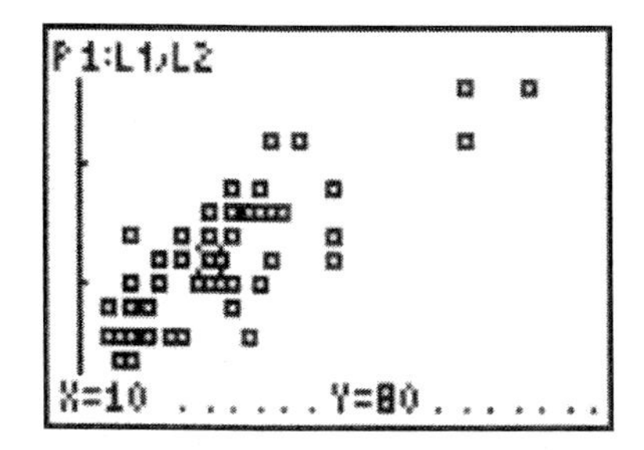

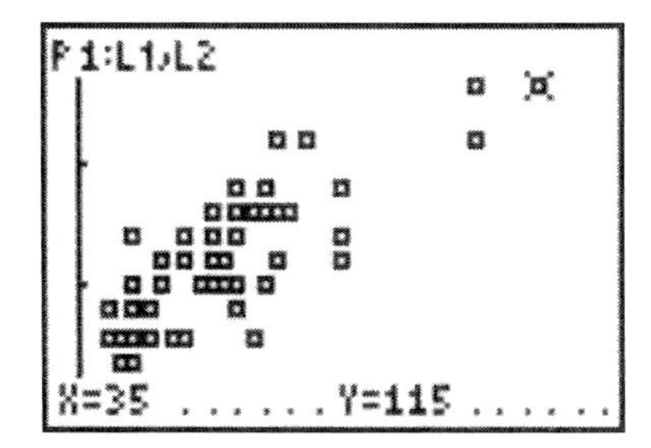

EXAMPLE 3: Which regression line predicts maximum bench press strength? (pp. 590 of Agresti/Franklin)

The regression equation models how you might predict the maximum bench press strength if you knew the number of repetitions that an athlete can do at 60 pounds.

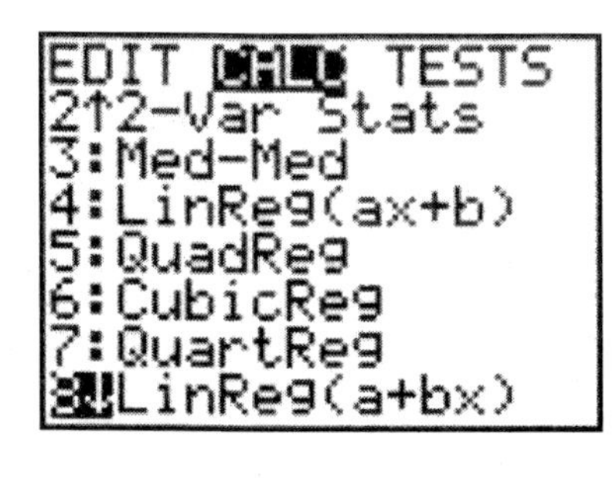

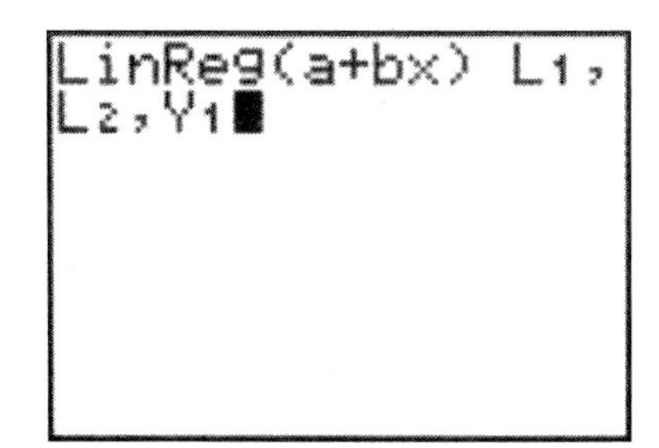

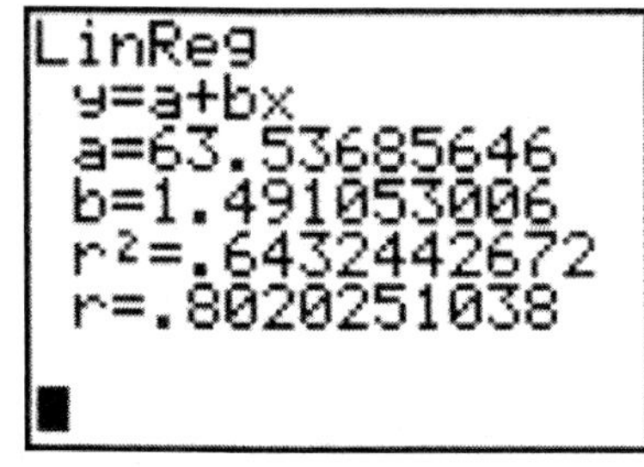

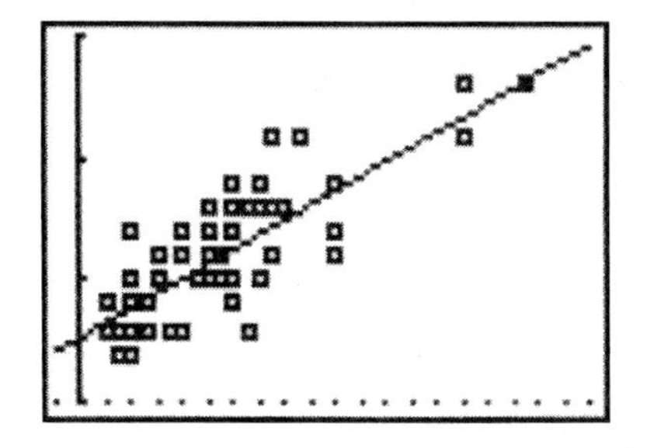

The prediction equation has a slope of about 1.49, meaning that for each additional 60-lb bench press completed, the estimate for the maximum bench press strength increases by about a pound and a half.

When you make a prediction with this line, you are really finding the *mean* maximum weight for all athletes that can do a certain number of repetitions. For example, if the number of repetitions is 15, the predicted value is 63.54 + 1.49(15) = 85.89. Thus, among the group of all athletes who can complete 15 repetitions, the predicted mean maximum bench press strength is 85.89 lbs. This is why the textbook writes the **population regression equation** as $\mu_y = \alpha + \beta x$. The sample regression equation $\hat{y} = a + bx$ provides estimates, *a* and *b*, of the population intercept and slope, α and β.

12.2: How Can We Describe Strength of Association?

EXAMPLE 6: What's the correlation for predicting strength? (pp. 599-600 of Agresti/Franklin)

One feature of the TI-83 Plus family of calculators is that once you have requested **2-VarStats** or done a linear regression with **LinReg(a+bx)**, the values that are produced in the process are stored as variables. Pressing VARS, then 5:Statistics, can access these variables. For example, you can find the sample correlation coefficient, *r*, under the **EQ** submenu. Using the other available variables you can confirm the relationship between the correlation, the slope, and the standard deviations of *x* and *y* that is shown on p. 599.

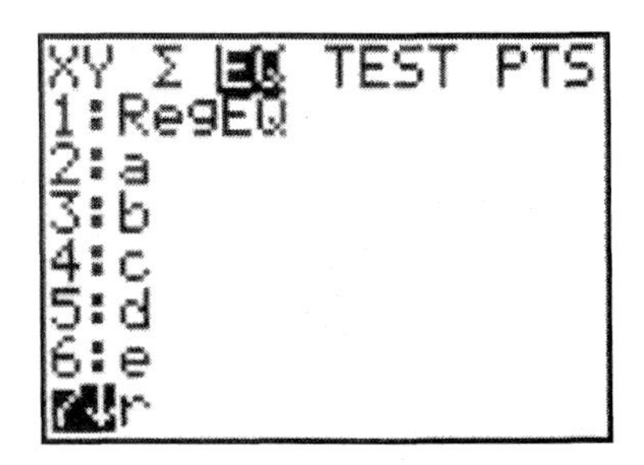

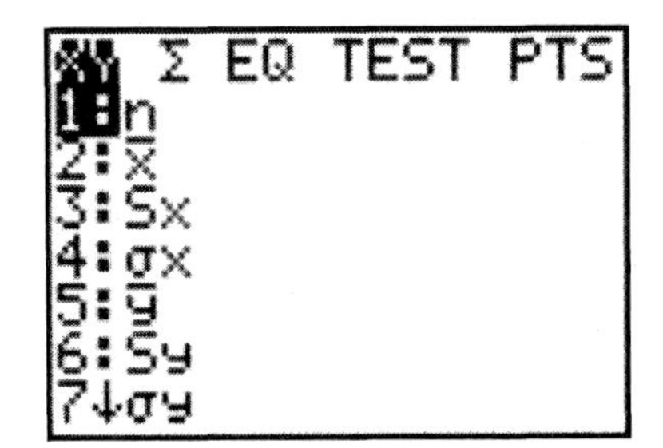


```
r
          .8020251038
b(Sx/Sy)
          .8020251038
```

You can also demonstrate that the correlation coefficient does not depend on the units of measure that are used. For example, convert the bench press strengths to kilograms by dividing by 2.2, the number of pounds in a kilogram. If you repeat the linear regression, the intercept and slope of the line are completely different but the correlation remains exactly the same.

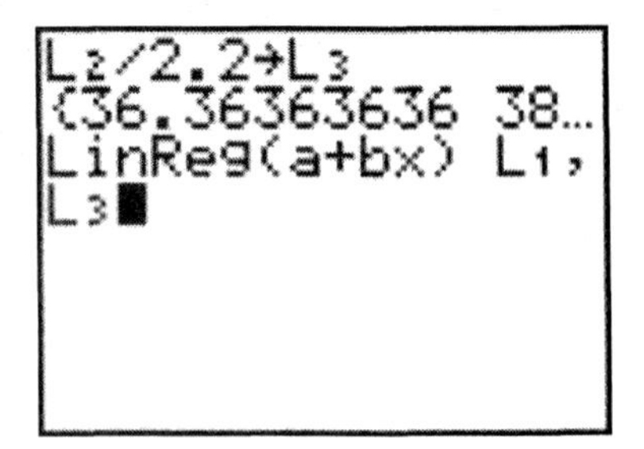


```
LinReg
 y=a+bx
 a=28.8803893
 b=.677513666
 r²=.6432442672
 r=.8020251038
```

The term "regression" means that values of the response variable, *y*, tend to be less extremely removed from the mean than the explanatory variable, *x*. If an *x* value is a certain number of standard deviations from its mean, then the predicted *y* is *r* times that many standard deviations from its mean. This can be confirmed using the stored values in the variables. First run the regression again with the original lists—this is a good lesson that the values stored in the variables are always from the most recently completed regression.

You can see that the y-value calculated from the regression equation can be found using the idea of regression to the mean. One standard deviation above the mean of x is 18.13. One standard deviation above the mean of y is 93.19, but the predicted value for a point whose x-value is one standard deviation above the mean is only 90.56.

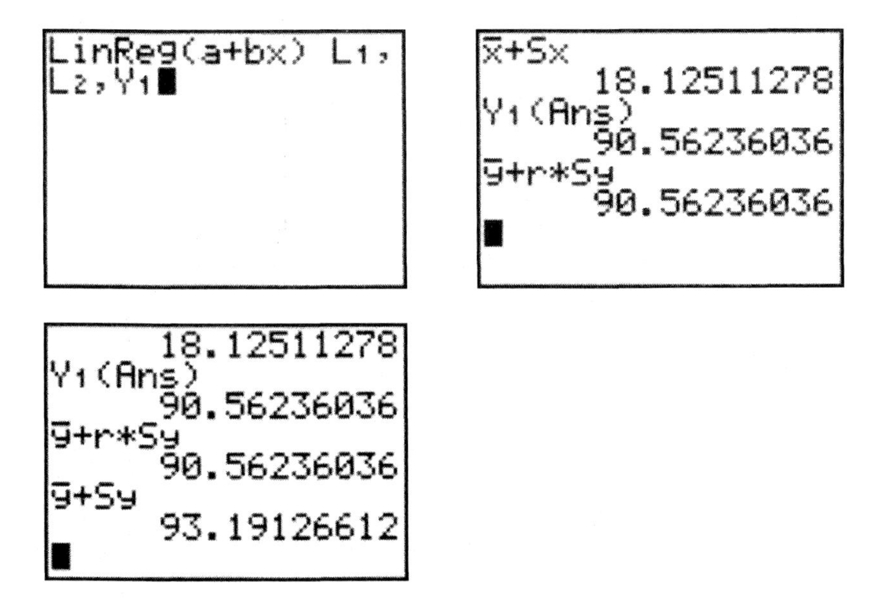

EXAMPLE 9: What does r^2 tell us in the strength study? (pp. 603-604 of Agresti/Franklin)

The predictive power of a linear model is measured by the proportional reduction in prediction error, r^2. The formulas on p. 603 demonstrate the meaning of r^2, and can be implemented on the calculator. This isn't the easy way to get the numerical value of r^2; just doing a linear regression produces the value, but it illustrates the meaning of what r^2 measures.

First, you can calculate the value of the **total sum of squares,** $\sum(y-\bar{y})^2$. Having just done the regression with number of repetitions and maximum bench press strength, above, the value for $\bar{y}$ is available in the **VARS** menu. The y-values are stored in **L2**, so that replaces y in the formula above. The sum command is in the list menu, and the value for the total sum of squares is stored to the variable **S** for later use.

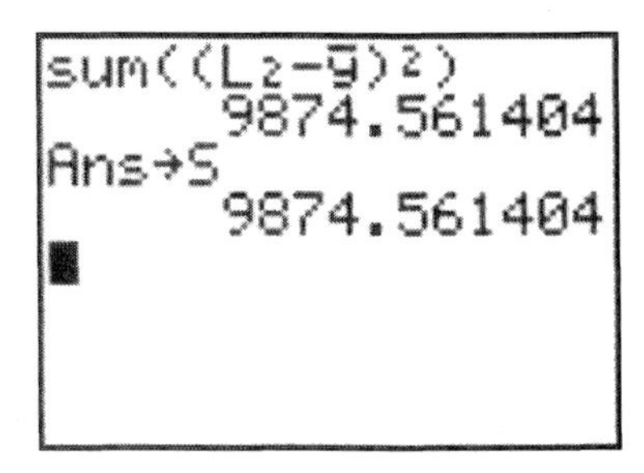

In the process of completing the linear regression the calculator also creates a list of the residuals about the line. These residuals are stored as the list **RESID**, which is available in the **LIST** menu. The **residual sum of squares,** $\sum(y-\hat{y})^2$, is the sum of the values in the **RESID** list.

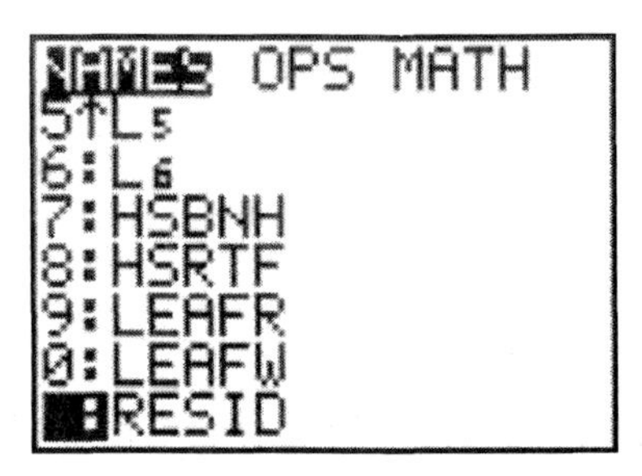


```
        9874.561404
Ans→S
        9874.561404
sum(ʟRESID²)
         3522.80639
Ans→R
         3522.80639
```

The value of the residual sum of squares, about 3523, is stored as **R**. This is smaller than the total sum of squares, 9875, with the difference of 6352 due to the benefit of the regression model. The proportional reduction is $r^2 = 0.643$, matching the calculation shown on p. 603.

```
S-R
        6351.755014
(S-R)/S
        .6432442672
r²
        .6432442672
```

12.3: How Can We Make Inferences about the Association?

EXAMPLE 11: Is strength associated with 60-pound bench presses? (pp. 611-613 of Agresti/Franklin)

The TI-83 Plus family of calculators has a built-in procedure for conducting a significance test for the slope of a regression line. The assumptions in this test are 1) that in the population the variables x and y are related by the regression equation $\mu_y = \alpha + \beta x$, 2) the data comes from a random process, and 3) that the conditional distribution of y at any given x is normally distributed with the same standard deviation at each x value. To access this procedure, press **STAT TESTS**, then scroll down to **LinRegTTest** (it's at **E** on the older calculators, **F** on the newer TI-84).

Specify the lists where the data is stored, for the **Xlist** and **Ylist**. The **Freq:1** indicates that each data value is a distinct value, rather than using a summary table, and this is a two-sided test. Note that this is testing whether the population slope, β, is zero, but that this is equivalent to testing whether the population correlation coefficient, ρ, is zero. Optionally, you can specify which function will be used to store the resulting regression equation.

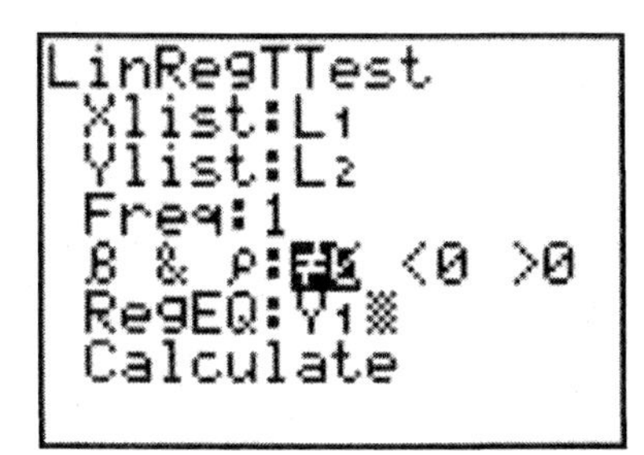

The results reproduce the work on p. 612, indicating very strong evidence that the population slope is not zero. The value for the t test statistic is 9.958, and this is counting in units of estimated standard error. This is huge, and the resulting P-value is essentially zero. The calculator displays this in scientific notation, **6.481372E⁻14**, the calculator syntax for 6.481372×10^{-14}.

The degrees of freedom for the t test statistic in linear regression is $n - 2$ or 55 in this case. In general the degrees of freedom is determined by subtracting the number of parameters that are being estimated from the sample size n. The estimated parameters in linear regression are α and β, the intercept and slope of the line.

There is more information available if you scroll down. The values of the estimates of the intercept and slope are provided, as is s, the **residual standard deviation**. The formula for this is presented on p. 619, and the value can be verified. The numerator of the formula is the residual sum of squares, which we had previously stored as **R**. This indicates that the typical distance from the points to the regression line is about 8 lbs.

```
LinRegTTest
 y=a+bx
 β≠0 and ρ≠0
↑b=1.491053006
 s=8.003188444
 r²=.6432442672
 r=.8020251038
```

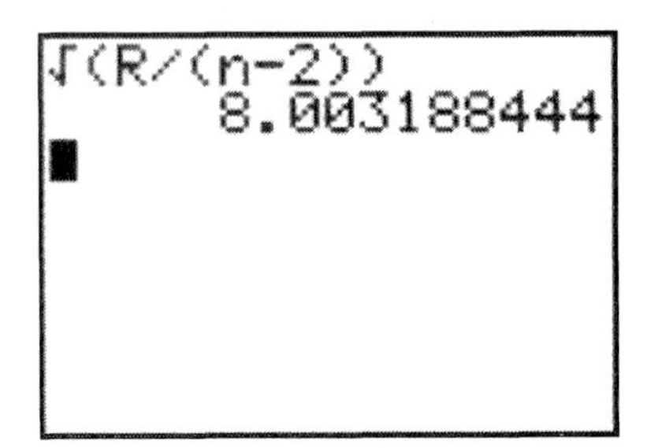

EXAMPLE 12: Estimating the Slope for Predicting Maximum Bench Press Strength. (pp. 613-614 of Agresti/Franklin)

The significance test above only demonstrates that a female athlete who can complete a higher number of repetitions at 60 pounds than another athlete does in fact have a greater expected maximum strength. The more interesting question of how much stronger the athlete is likely to be is answered by the slope. Our data revealed a point estimate of 1.49, meaning that for each additional repetition completed, the expected strength increases by nearly one and a half pounds.

However, this is based on a sample and may not be exactly the correct value. The plausible values for this slope are included in a confidence interval for the slope.

To calculate this confidence interval you need the standard error of the slope. This is not included directly in the regression report above, but it is easily calculated. The test statistic t is the ratio of the slope estimate to its standard error, $t = \frac{b}{SE}$. This also means that you can calculate the standard error if you know t and b, which are included in the regression report. The standard error of the slope is about 0.150. A confidence interval will extend about 2 standard errors on either side of the point estimate b, from about 1.19 to 1.79.

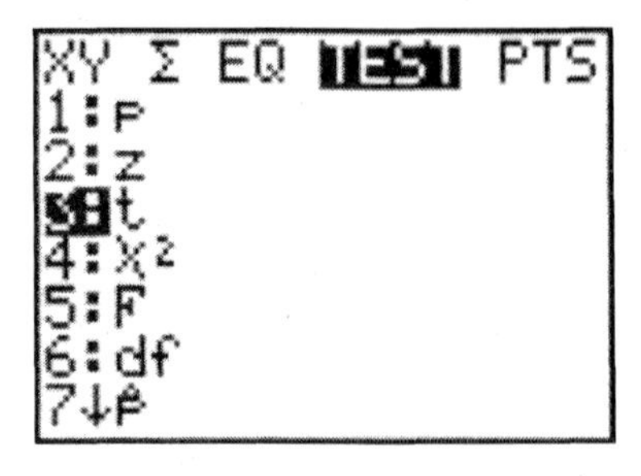

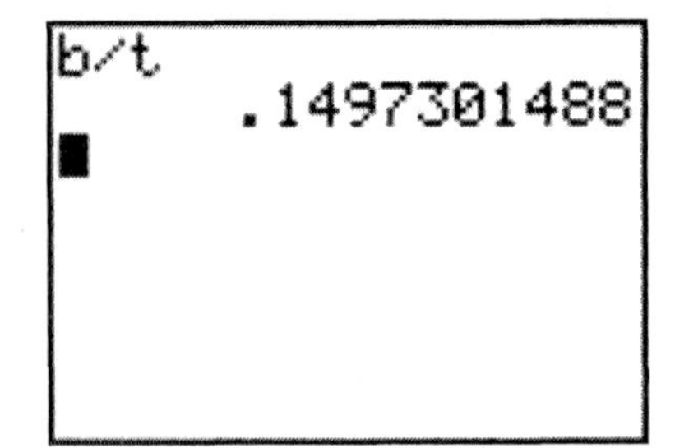


```
      .1497301488
Ans→E
      .1497301488
b+2.00E
      1.790513304
b-2.00E
      1.191592709
```

The newer TI-84 calculators also have the confidence interval as a built in function, via **STAT TESTS G:LinRegTInt**. The setup is the same as the significance test, and the results agree with our calculation above and that in the textbook.

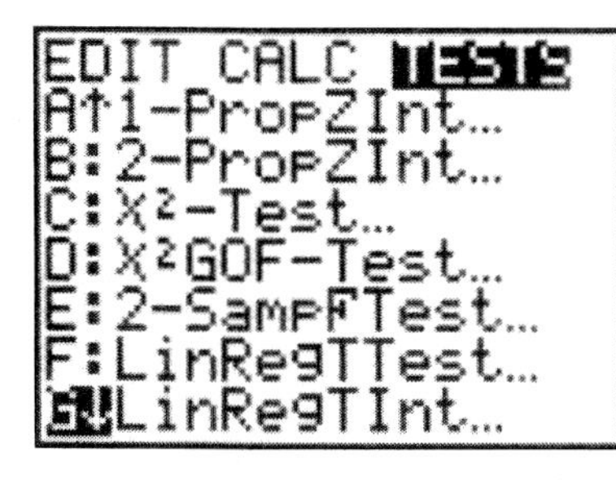


```
LinRegTInt
 Xlist:L1
 Ylist:L2
 Freq:1
 C-Level:.95
 RegEQ:Y1
 Calculate
```

```
LinRegTInt
 y=a+bx
 (1.191,1.7911)
 b=1.491053006
 df=55
 s=8.003188444
↓a=63.53685646
```

```
LinRegTInt
 y=a+bx
↑df=55
 s=8.003188444
 a=63.53685646
 r²=.6432442672
 r=.8020251038
```

12.4: What Do We Learn from How the Data Vary around the Regression Line?

EXAMPLE 15: How variable are athletes' strengths? (pp. 619-620 of Agresti/Franklin)

One of the assumptions for inference for regression is the residuals are normally distributed with the same standard deviation at each x value. The estimate of the residual standard deviation of y is reported as **s** with the **LinRegTTest** or **LinRegTInt**. In the report above, the value is about 8.0, matching the calculation on p. 620.

You can check the shape of the overall distribution of residuals with a histogram:

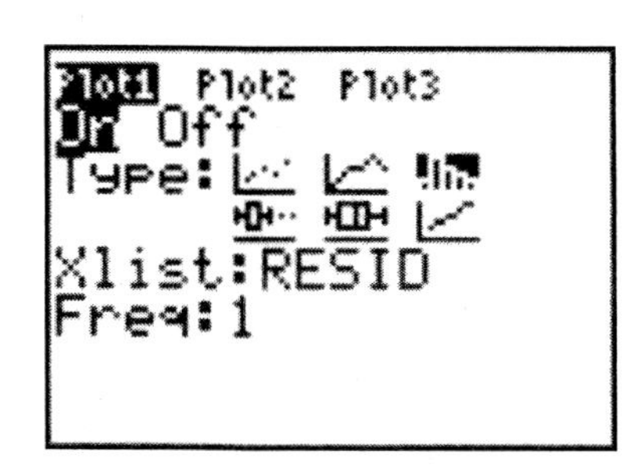

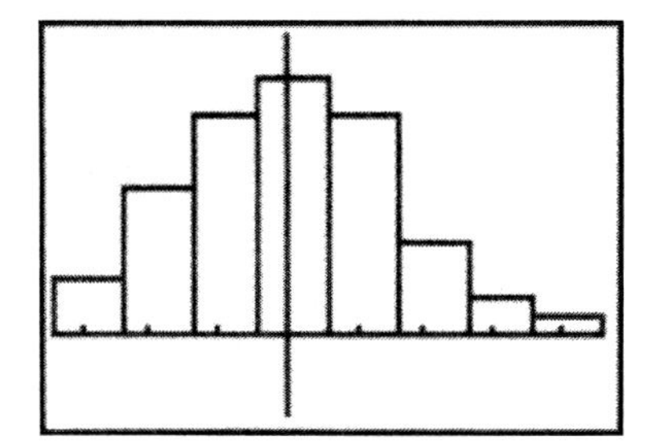

This looks good, mound shaped and symmetric.

You can also examine the residuals individually by making a scatterplot of the residuals vs. the explanatory variable. This is often called a residual plot, and residual plots are done in Chapter 13 in the context of multiple regression. However, a residual plot provides useful information about a linear regression model, too. This shows that the residuals are evenly distributed above and below the line. The assumption that the standard deviation of the residuals is the same at each x is reasonable for the left side of the graph, but the data is too sparse on the right to make much of a judgment. Overall, though, the conditions appear to be reasonably satisfied.

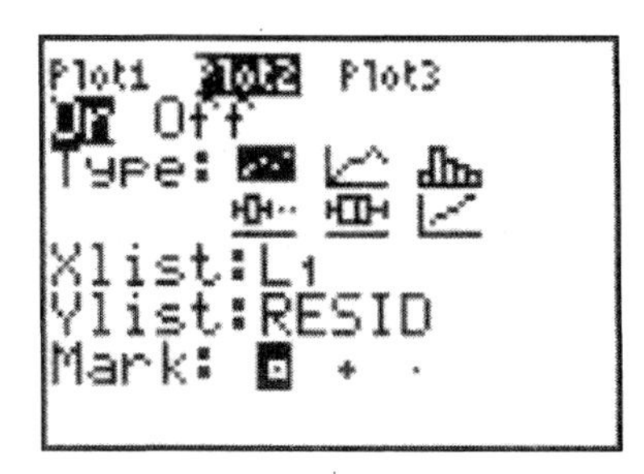

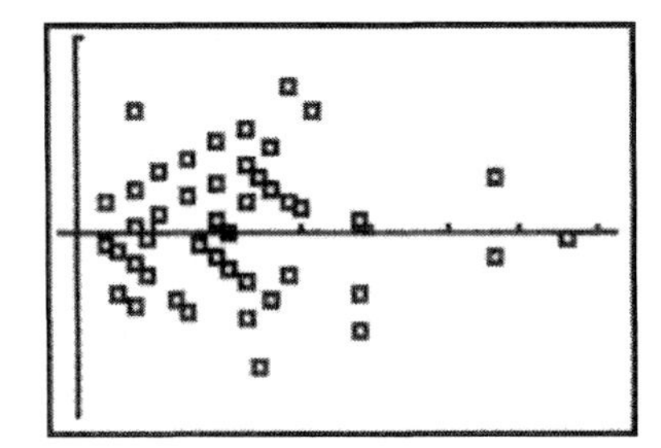

EXAMPLE 16: Predicting maximum bench press and estimating its mean. (pp. 621-623 of Agresti/Franklin)

Among all athletes who can complete 11 repetitions at 60 pounds, the mean bench press maximum is about 80 pounds, according to our linear regression model. For a given individual who can complete 11 repetitions, however, you would expect that they would have a maximum strength that might be a little more than 80 or a little less. A **prediction interval** is an estimate of the plausible values of a response variable y for a given value of x, and in this case it extends about 2 standard deviations on either side of the point estimate of 80:

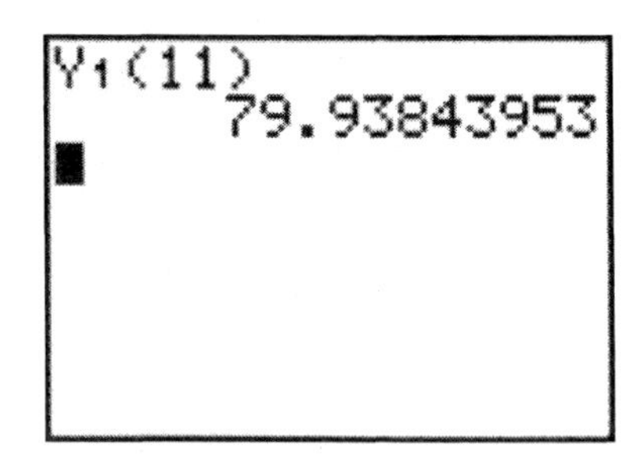

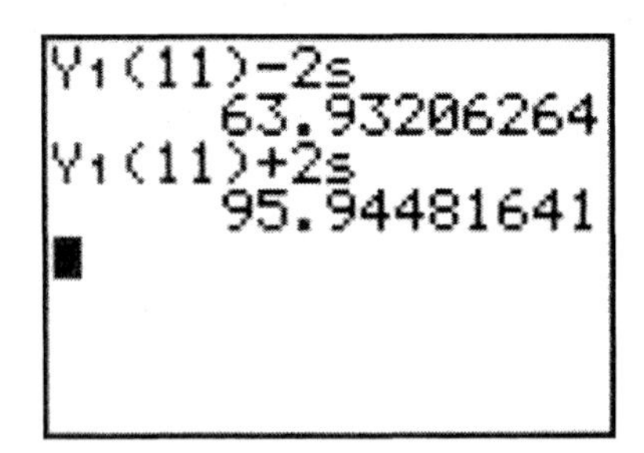

I am 95% confident that *an individual* female athlete who can do 11 repetitions at 60 pounds will have a maximum strength between 64 and 96 pounds. It would be surprising if someone who completed 11 repetitions had a maximum strength less than about 64 or more than about 96 pounds.

The *mean* bench press strength of all athletes who can perform 11 repetitions is also subject to some uncertainty, but not as much as the uncertainty for one specific randomly chosen individual. A **confidence interval for μ_y** expresses this uncertainty, dividing the residual standard deviation by $\sqrt{n}$.

```
Y1(11)-2(s/√(n))
          77.81834312
Y1(11)+2(s/√(n))
          82.05853594
```

Given the data in this study, I am 95% confident that the mean bench press maximum among *all* female athletes who can complete 11 repetitions at 60 pounds is between about 78 and 82 pounds.

The value of the **regression sum of squares** for this example is calculated on p. 623. This is the sum of the squares of the differences of the predicted values $\hat{y}$ from the mean $\bar{y}$. Using the regression equation, the predicted values are the value of the prediction equation, stored in **Y1**, at each of the x values in **L1**. This agrees with our previous calculation of the difference between the total sum of squares and the residual sum of squares, which we had previously stored as **S** and **R**.

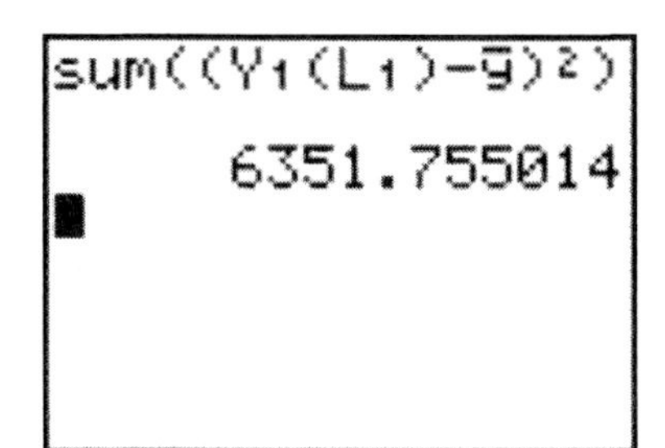

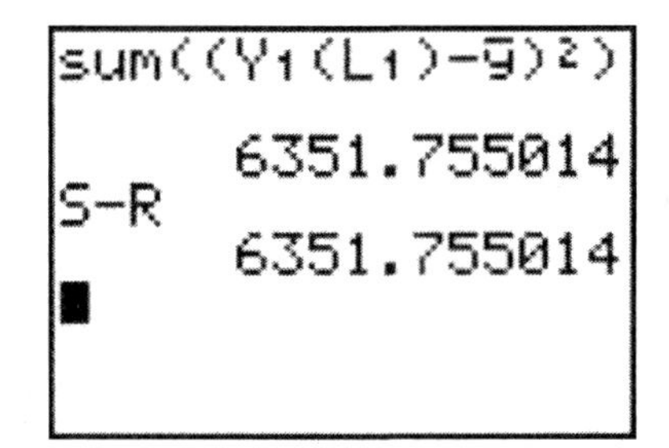

The **mean square error** (or **MSE**) is an estimate of the variance of the conditional distributions of points about the regression line. It is the residual sum of squares divided by the degrees of freedom, $n - 2$ or 55 in this case. The **VARS** menu has the value for **df** as well, and this can be used to calculate this value, 64.05. You can also confirm that the value for s is the square root of this. Some software will report the estimate s as "**RMSE**" or **root mean square error**.

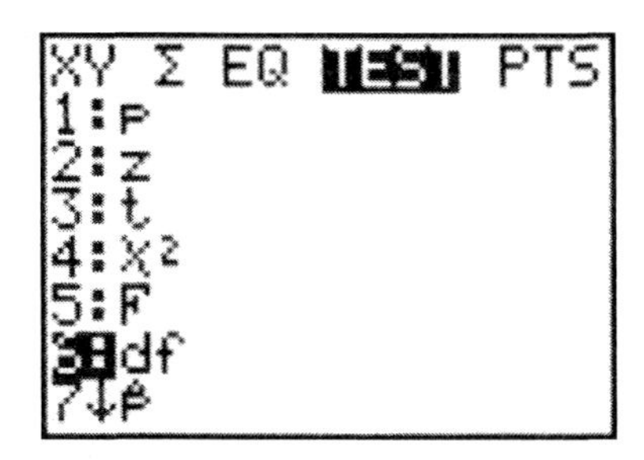

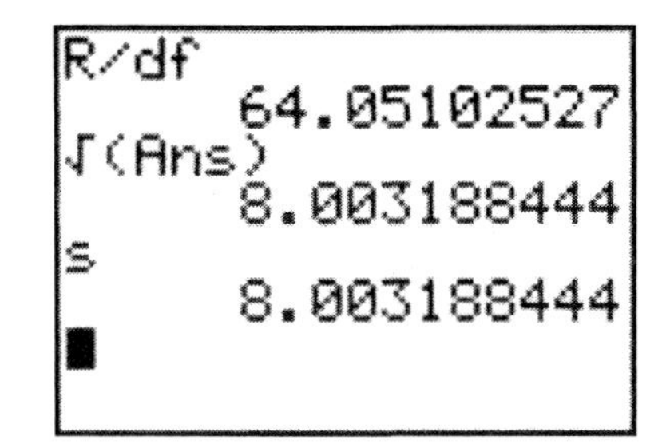

12.5: Exponential Regression – A Model for Nonlinearity

Exercise 12.61: Leaf litter decay (p. 631)

Sometimes there is a clear association between two variables, but the relationship is not linear. That is, for each increase in x the corresponding increase or decrease in y is ***not*** consistent. this is demonstrated in exercise 12.61. Copy the list **LEAFW** and **LEAFR** to **L1** and **L2**, or type the data in from the textbook.

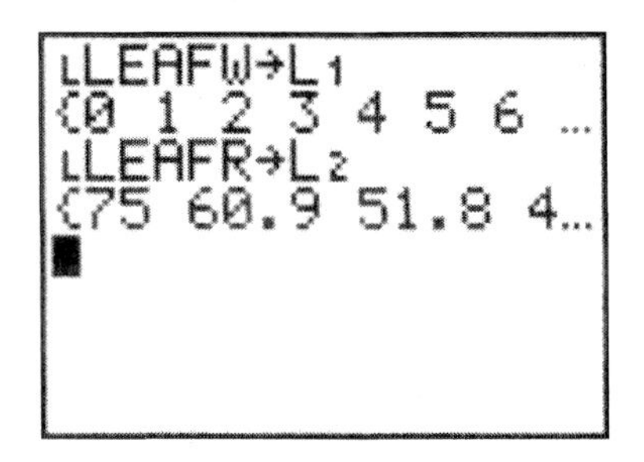

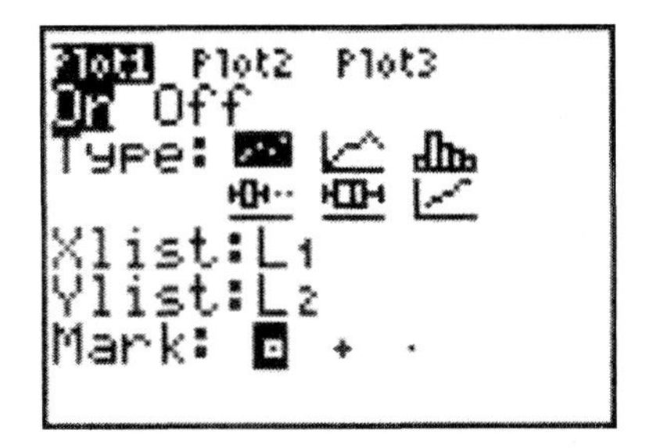

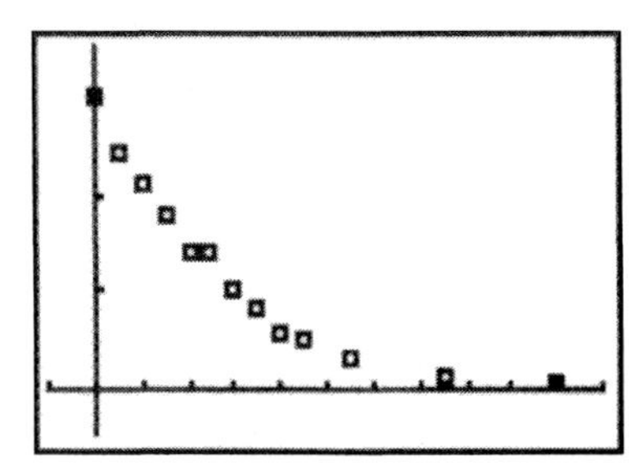

A scatterplot clearly shows an association, but a regression line doesn't fit well since the points seem to fall along a curved path rather than a straight line.

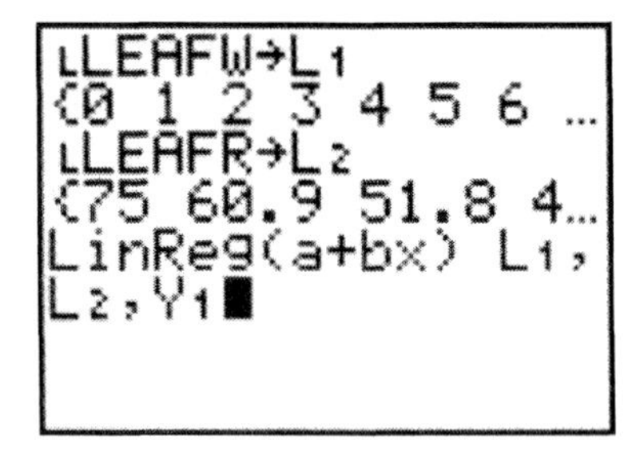

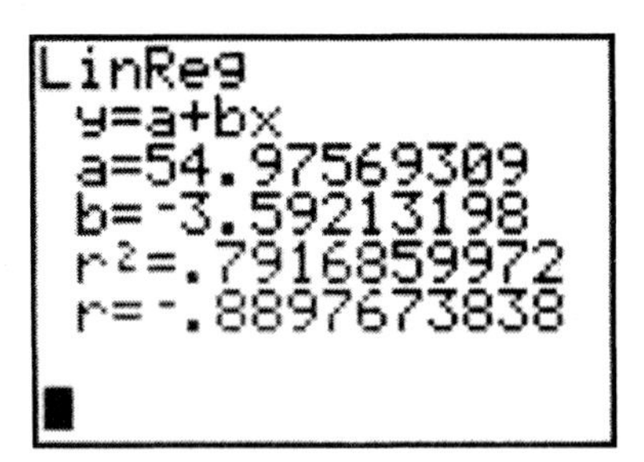

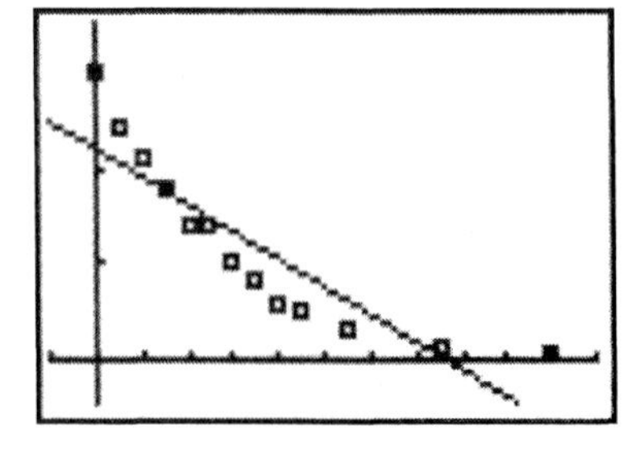

One solution to this is to **transform** one or both of the lists of data by applying a function to each of the values. The hope is that a scatterplot with the transformed values will show a linear pattern.

There are many different types of nonlinear relationships. One that occurs in a variety of circumstances is one of **exponential growth** or **exponential decay**. This occurs when the values of y don't differ but a constant value for increments of x, but rather have a common **ratio** over equal increments of x. For example, if a population of rabbits is growing exponentially with year 1 showing 100 rabbits and year 2 having 150 rabbits, the expected number of rabbits in year 3 would be 225. The ratios 150/100 and 225/150 are both equal to 3/2. Exponential decay is how radioactive isotopes lose their radioactivity. An isotope with a half-life of 10 years will decrease to half its potency after 10 years, and be at 1/4 of its original potency after 20 years, 1/8 after 30 years, etc.

If two variables do have an exponential relationship, then transforming the y values with a logarithm will produce a linear relationship between x and the logarithm of y. This is because the logarithm is the **inverse function** for exponential growth or decay. Use the **LOG** button to complete the transformation of the values in **L2**, which are then stored in **L3**.

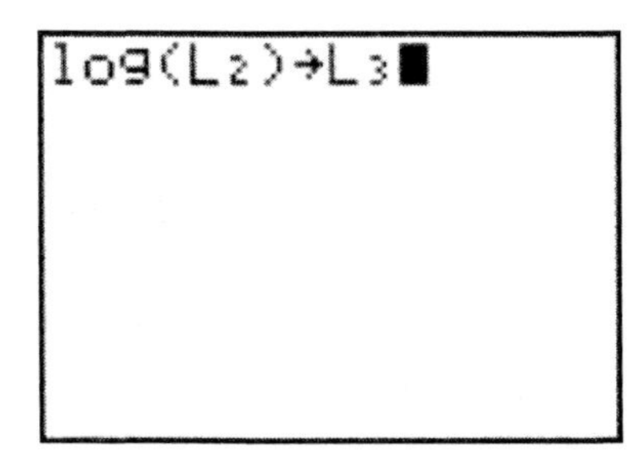

A scatterplot of the original x values and the transformed y values is apparently linear.

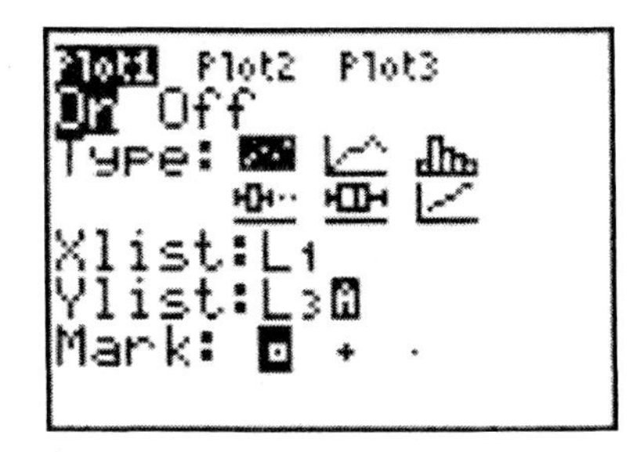

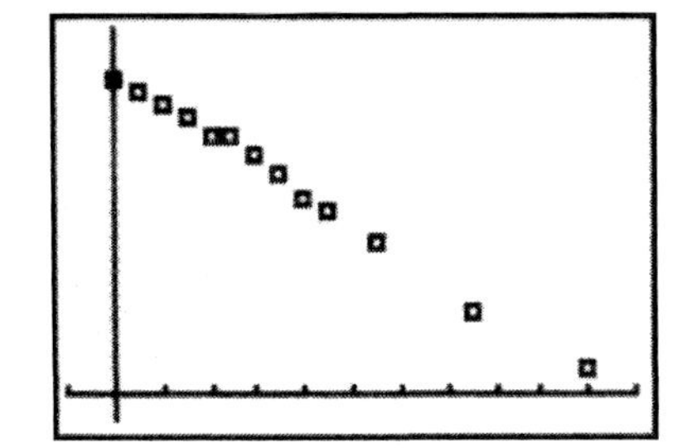

Do a linear regression to find the line that relates x and log(y).

```
log(L2)→L3
{1.875061263 1....
LinReg(a+bx) L1,
L3,Y2
```

This fits the scatterplot very well, much more closely than the previous scatterplot with the original data. The values of r and r^2 also indicate that the predicted values fit the data values extremely closely.

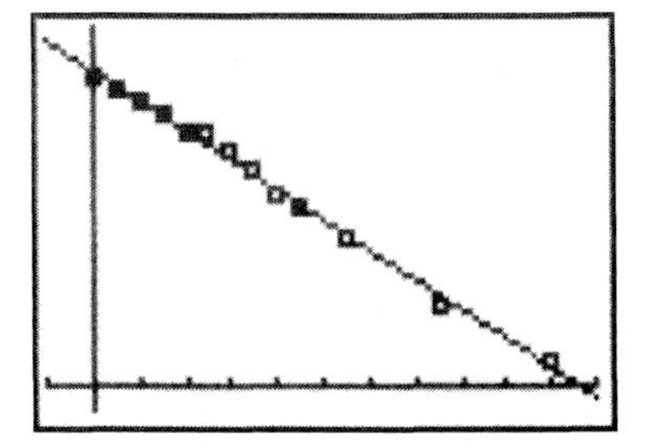

```
LinReg
 y=a+bx
 a=1.906463997
 b=-.0896726532
 r²=.9935636516
 r=-.9967766308
```

Plotting the residuals against the explanatory variable shows that the variation from the line has no pattern and is more or less random. This is the best evidence that the exponential model we have created is as good as can be: if the variation of the residuals shows no pattern, then it is unlikely that any other model for this data will be much better.

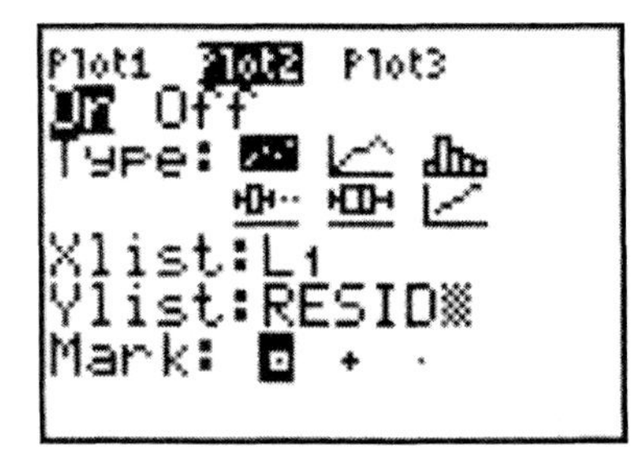

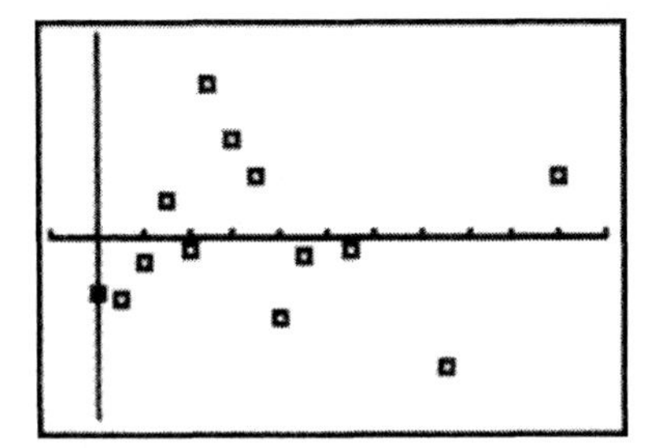

Making a prediction using the transformed model is a little tricky. Since the model relates x to $\log(y)$, the predicted values are actually the logs of the values we actually want. To go from the logarithm to the actual value, use the exponential function $y = 10^x$, the inverse of the log function. The function **10^X** is the second function on the **LOG** key since they are inverses.

```
Y2(13)
        .7407195047
10^(Ans)
        5.504520642
```

```
Y2(16)
        .471701545
10^(Ans)
        2.962794604
Y1(16)
        -2.498418586
```

The predicted value after 13 months is that there will be about 5.5 kg of leaf litter remaining.

Note that there are two *different* log buttons on your calculator, base 10 (common logs) or base *e* (natural logs). This example was done with the common logs. To use the natural logs, use the button **LN** and its inverse, e^x, **2nd LN**. If you work through this example with these values you will get a different linear model, but the predicted values for a given *x* will be exactly the same. So use the log function that you are most comfortable with.

If you like an algebra challenge, you can show that the linear regression we calculated above can be turned into an exponential function (where the variable *x* appears in the exponent). The slope and intercept we found relate the log(*y*) to *x*, or $\log(y) = 1.906 - .08967x$.

To get back to just *y*, apply the inverse function 10^x to both sides: $10^{\log(y)} = 10^{1.906-.08967x}$.

Apply the rules for exponents to the left and the right sides: $y = 10^{1.906} \times 10^{-.08967x}$.

Simplify the right side a little more: $y = 10^{1.906} \times \left(10^{-.08967}\right)^x$.

Calculate the powers of 10: $y = 80.54 \times (.8134)^x$.

Your calculator can go right to this last form of the exponential model using the **ExpReg** option in the **STAT CALC** menu.

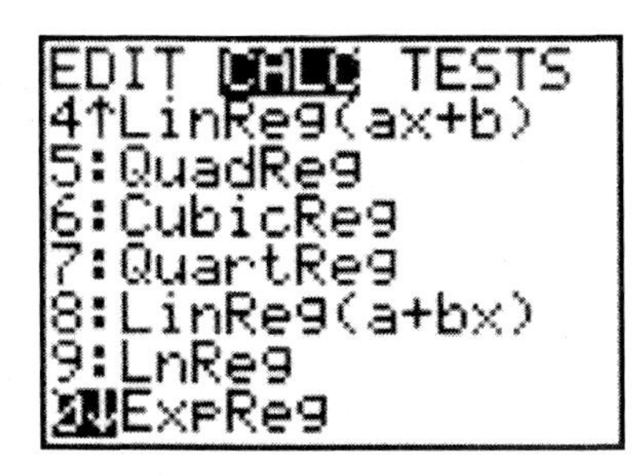

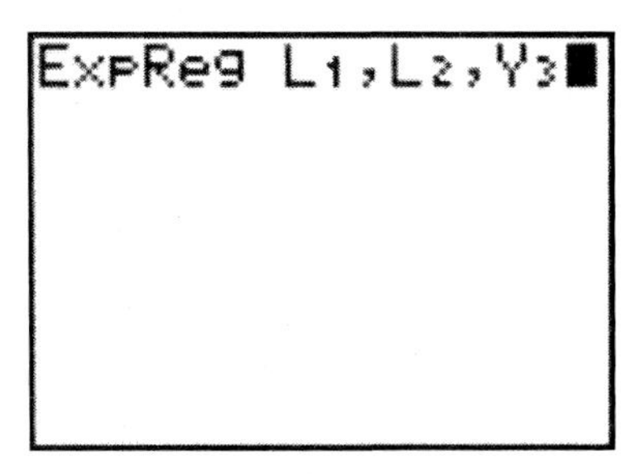

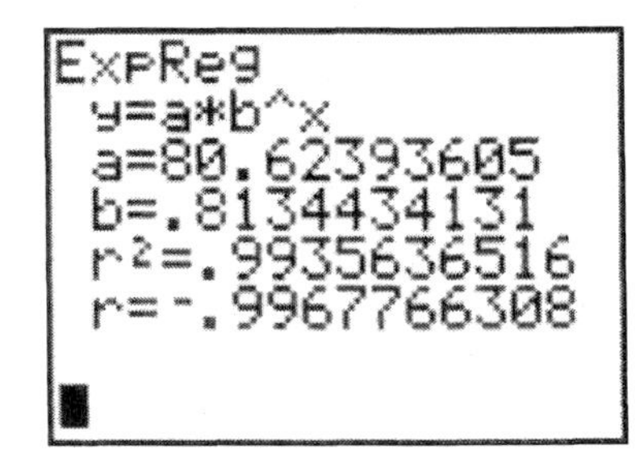

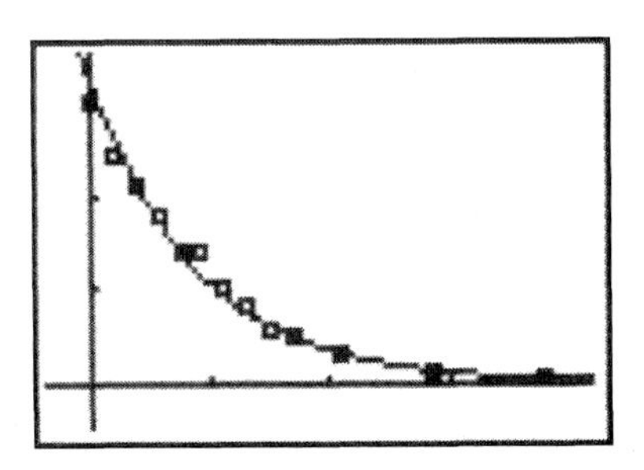

This matches the equation $y = 80.54 \times (.8134)^x$ that we found above. The scatterplot shows the graph of this exponential decay function, a decreasing curved line that fits the original data very well.

However, using the **ExpReg** shortcut doesn't let you evaluate the appropriateness of this model in the same way as making the transformation. Using transformations is also a very important concept in more advanced work in statistics and allows you to work with a much wider variety of curved patterns.

It's also important to understand that the values for r and r^2 that are reported with **ExpReg** are actually from the transformed variables. You can see that in the report above – these values are identical to those we got earlier with the transformed values. These are from evaluating the squares of the residuals about the transformed linear equation, not the squares of the residuals that you would get with the curved exponential. This is not evident in just looking at the report or the graph, so even if you use **ExpReg**, make sure you are thinking about the transformations when you interpret r or r^2.

Chapter Thirteen: Multiple Regression

All of these lists referred to in this Chapter are on the textbook CD with the names below.

The process for copying lists and groups from a computer to your calculator is described in Chapter 1 of this manual, beginning on page 4.

HPSIZ	Size of houses in real estate study, in square feet
HPLOT	Lot sizes in real estate study
HPPRI	Prices of houses in real estate study
HPNW	Indicator variable for northwest region in real estate study
CCINC	Income of participants in credit card study
CCYES	Indicator variable for travel credit card in study

Unfortunately, the TI-83 Plus family of calculators does not include functions to create a multiple regression model. However, a number of the computations that are shown in the textbook that use the multiple regression models can be confirmed with the calculator.

13.2: Extending the Correlation and *R*-Squared for Multiple Regression

EXAMPLE 3: How well can we predict house selling prices? (pp. 649-650 in Agresti/Franklin)

The calculation of R^2 in multiple regression has the same form as in linear regression, so the same method from Chapter 12 can be applied here to confirm the value calculated on p. 649. The data for this example is in lists **HPSIZ**, **HPLOT**, and **HPPRI** (separately or in the group **HOUSPRIC**). First, enter the two explanatory variables, house size and lot size, into **L1** and **L2**. The response variable, selling price, is in **L3**. Using the multiple regression model that is on p. 643, the predicted prices are put in **L4**. Each of the two price variables is changed to units of $1000, as is suggested in the textbook.

```
ʟHPSIZ→L1
{1240 370 1130 ...
ʟHPLOT→L2
{18000 25000 25...
```

```
ʟHPPRI/1000→L3
{145 68 115 69 ...
(-10536+53.779L1
+2.8404L2)/1000→
L4
{107.27716 80.3...
```

The residuals are the difference between the actual selling prices and the predicted selling prices. These are stored in **L5**, shown in the first screen below. The second screen shows the calculated values of the sums of squares: the total sum of squares, $\sum(y-\bar{y})^2 = 314433$, the residual sum of squares, $\sum(y-\hat{y})^2 = 90756$, and the difference between these is the numerator of the formula for R^2, 223676.

```
{145 68 115 69 ...
(-10536+53.779L1
+2.8404L2)/1000→
L4
{107.27716 80.3...
L3-L4→L5
{37.72284 -12.3...
```

```
sum((L3-mean(L3)
)²)→S
         314432.5196
sum(L5²)→R
         90756.29332
S-R
         223676.2263
```

Here is the calculation that confirms the value $R^2 = 0.711$, as shown in the textbook. The value of the difference can also be calculated directly as the regression sum of squares, shown on p. 650, as $\sum(\hat{y}-\bar{y})^2$.

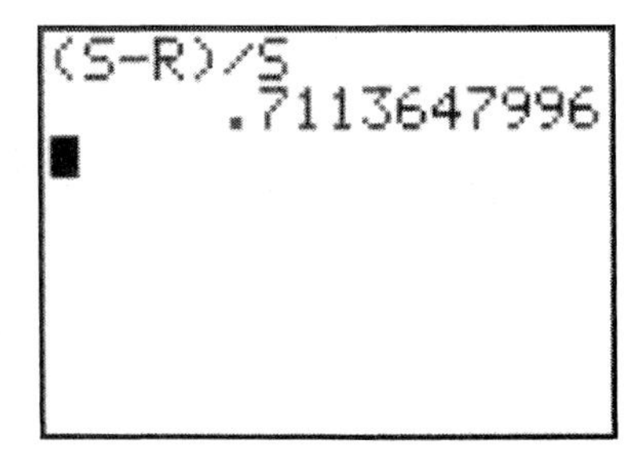

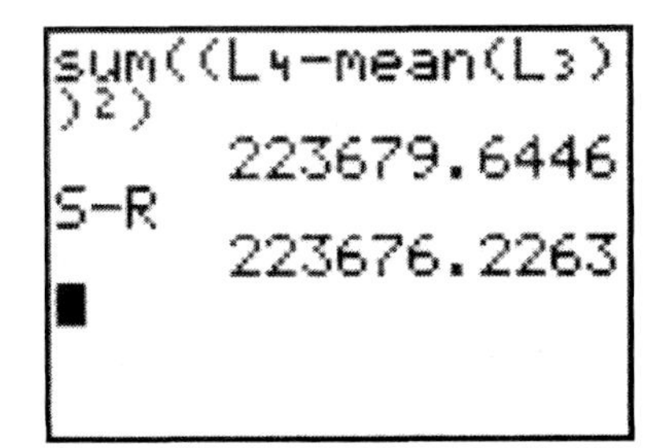

On the calculator output above you'll note that the value doesn't agree exactly with the previously calculated value. This is because the coefficients of the multiple regression model are rounded values that were typed in manually, the best we could do without access to the procedures for having the calculator compute the model for us. It's convincingly close, however.

13.4: Checking a regression model using residual plots

EXAMPLE 8: Residuals for house selling prices. (pp. 664 in Agresti/Franklin)

The residuals are in **L5**, so making a histogram of the residuals to examine the shape of the distribution is not difficult. This helps to reassure us that the assumptions of multiple regression are satisfied in the problem we are working, to the extent that the data can confirm this.

The calculator does not, however, calculate the standardized residuals, so the *x*-axis is scaled in the units of the response ($1000), not the standardized units in the graph shown in the textbook. However, the main point that the shape of the distributions of the residuals does not depart from normality very much is still clear. The distribution is roughly bell shaped. Here is the histogram and the window that produced it.

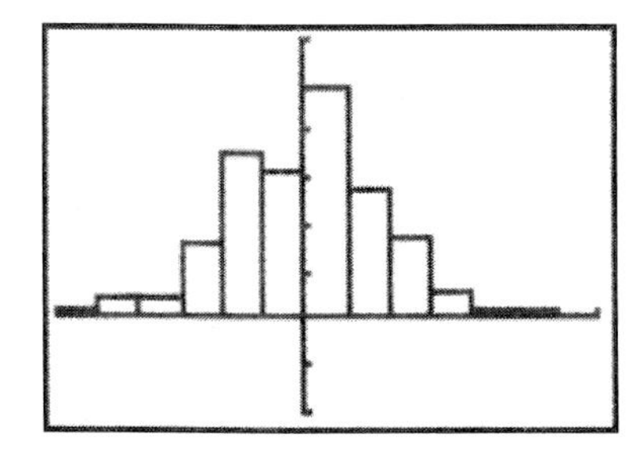

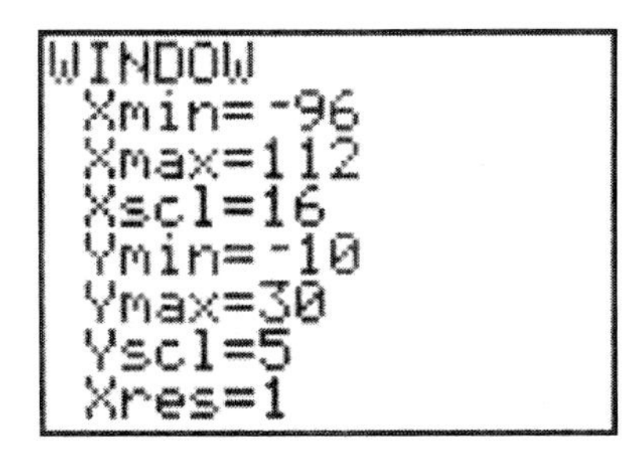

EXAMPLE 9: Another residual plot for house selling prices. (pp. 665-666 in Agresti/Franklin)

The residuals plotted in a scatterplot against each of the explanatory variables also checks for potential problems with the regression model. Either a nonconstant standard deviation of the residuals or a nonlinear pattern would be a problem. The first graph below reproduces the scatterplot on p. 666 for the residuals vs. house size, and the second examines the residuals vs. the other explanatory variable, lot size. Both show a slight increase in variability for increasing values of the response variable. This doesn't invalidate the model, but it does prompt greater caution when making predictions using this model.

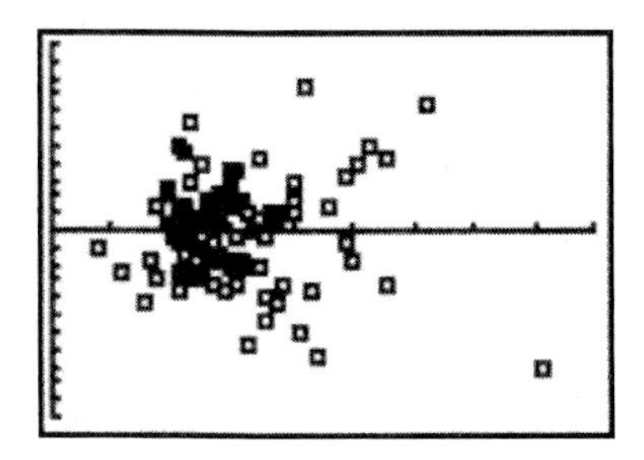

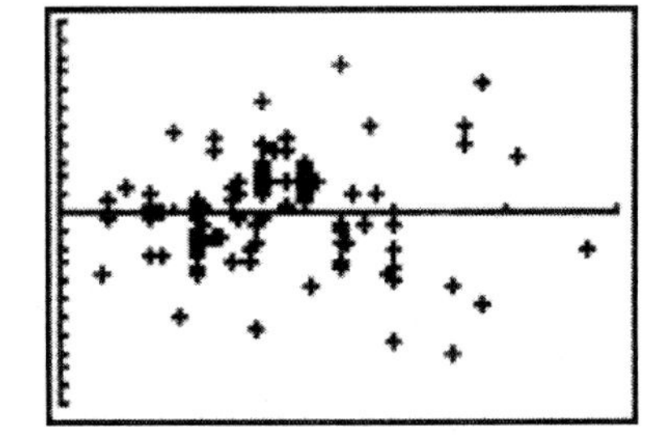

13.5: How Can Regression Include Categorical Predictors?

EXAMPLE 10: Including region in regression for house selling price. (pp. 669-672 in Agresti/Franklin)

An **indicator variable** is a categorical indicator variable that is set to either 1 (condition met) or 0 (condition not met). In this example there is an indicator variable, **HPNW**, that indicates whether a house is located in the more desirable northwest region of the market. With this variable included with the house size, a multiple regression model is given on p. 669 as $\hat{y} = -15258 + 78.0x_1 + 30569x_2$, where x_1 represents house size and x_2 represents the indicator variable, NW.

The graph on p. 670, showing the two lines that make up the multiple regression model, can be reproduced on your calculator with one function. Enter the model parameters into your calculator using **X** for x_1 and the short list {0, 1} for x_2. This produces a graph with two lines, one for each of the values of the indicator variable.

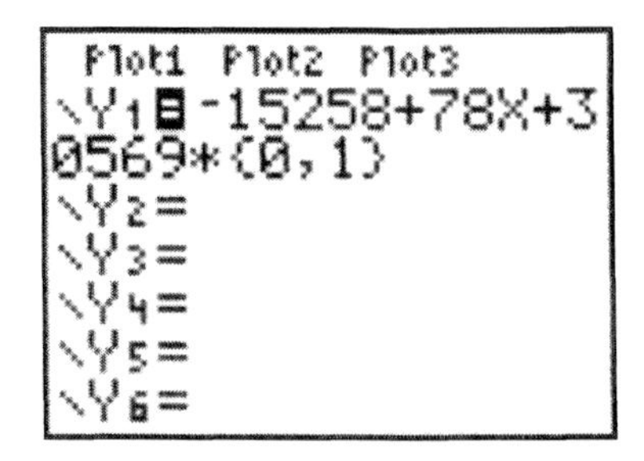

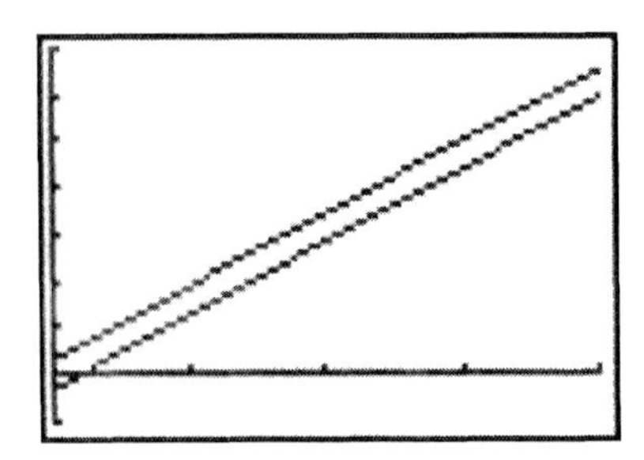

Of course, these lines are parallel, both with slope 78, since the multiple regression model assumes that the relationship between house size and price is exactly the same for all values of the other response variables, in this case the indicator variable NW. Is this too simplistic? Examining the data for an **interaction** between the explanatory variables can help answer this question.

This is done by performing separate regressions for the houses from the northwest and the houses from the other regions. The TI-83 Plus calculators can do this, since it's a simple linear regression

To accomplish this, however, the data needs to be broken up according to the value of the indicator variable. This is a little tricky. First, clear the lists from the previous work to make sure we have a clean slate for the new explanatory variables. Then load the data with the house size, region, and price.

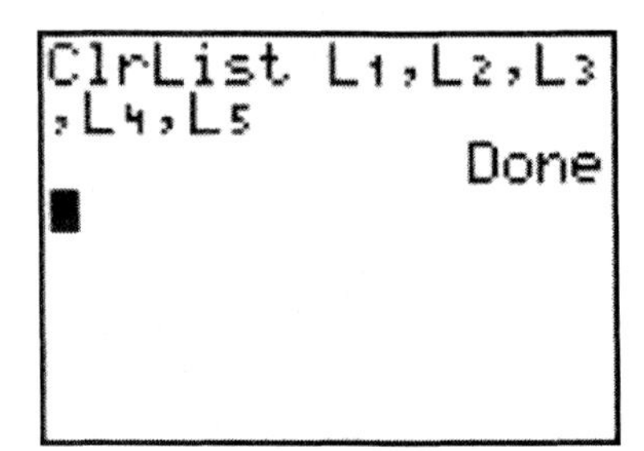

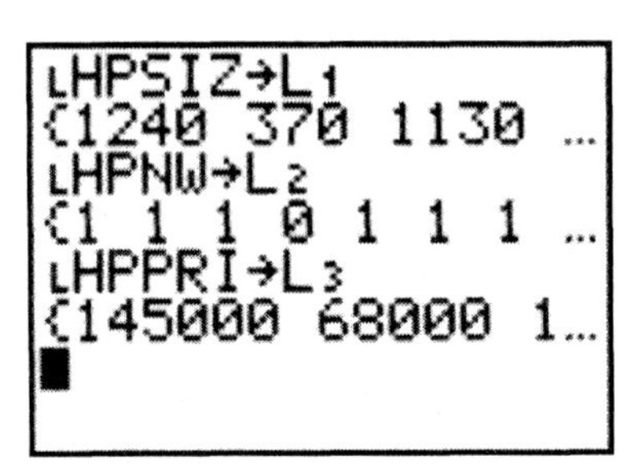

Notice that the indicator variable is either 1 (NW) or 0. To get just the northwest houses, sort the lists according to the value of this indicator. In the **SortD** command, the first list determines the order, and the other lists follow along, making sure the house sizes and prices remain together. SortD is used to put the houses with NW = 1 at the top of the lists and those with NW = 0 at the bottom.

```
SortD(L2,L1,L3)
                Done
```

L1	L2	L3
370	1	68000
1130	1	115000
1340	1	127200
1710	1	163000
1010	1	69900
860	1	50000
1420	1	137000

L2(1)=1

Scroll down to find where the last house that is in the NW = 1 group falls. It's at position 75 in the list. Next, the dim command 9short for "dimension" can be used to shorten the lists so they only include the houses from the northwest.

L1	L2	L3
2030	1	183000
1390	1	140000
1880	1	160000
1240	1	145000
1520	0	103000
2070	0	70000
1020	0	70700

L2(75)=1

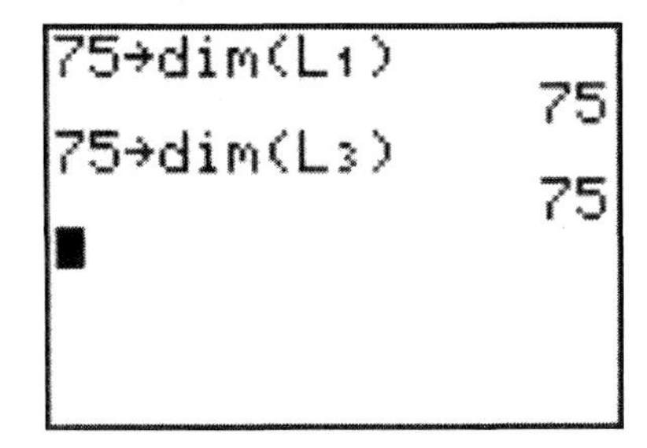

A linear regression done with the remaining data produces a linear regression model for the northwest houses, which has a slope of 83.7.

```
LinReg(a+bx) L1,
L3,Y2
```

```
LinReg
 y=a+bx
 a=6619.708462
 b=83.72915557
 r²=.6746989181
 r=.8214005832
```

This process has to be repeated for the houses with NW = 0, from other regions. This time, however, use the SortA command (ascending) to put the houses with NW = 0 at the top. Scroll down to find the last house in the non-NW group (it shouldn't be a surprise that it's at 25, since there are 100 houses in all, and you could skip the step of scrolling down).

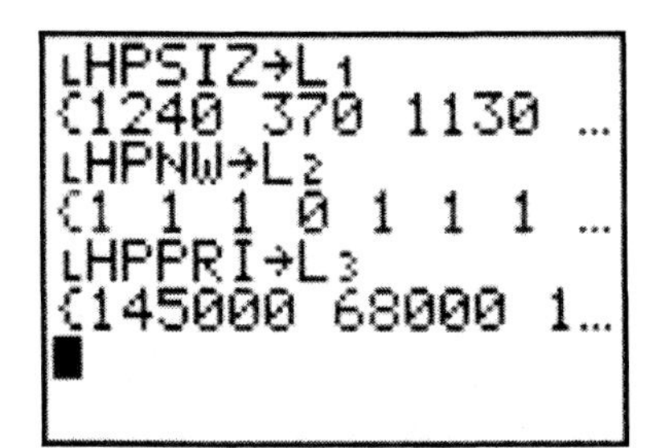


```
SortA(L2,L1,L3)
                Done
```

L1	L2	L3
1170	0	142500
1350	0	86500
1450	0	65000
1560	0	88400
1250	1	118600
1760	1	140000
1550	1	148000

L2(25)=0

```
25→dim(L1)
                  25
25→dim(L3)
                  25
LinReg(a+bx) L1,
L3,Y3
```

The linear regression produces a model with slope 64.6 for the non-NW houses. Showing the two linear models together demonstrates that the lines are **not** parallel, indicating some interaction between the region and house size. In the NW region, house prices increase *more* for a given increases in size than they do in the other regions.

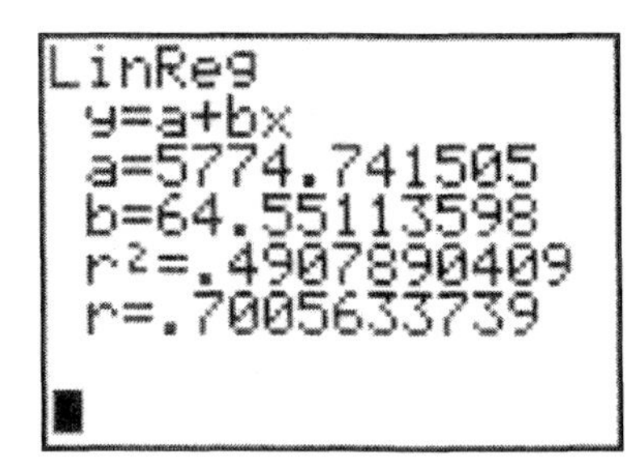

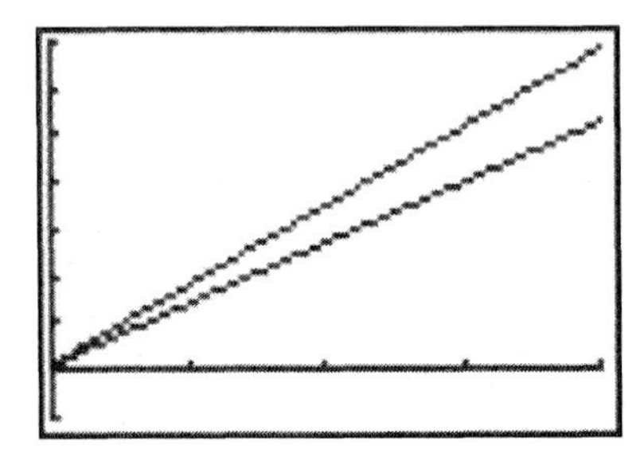

13.6: How Can We Model a Categorical Response?

EXAMPLE 13: Interpreting effect of income on credit card use. (pp. 676-679 in Agresti/Franklin)

Logistic regression creates a model that predicts one of two values (success/failure) of a binary categorical variable. In this example, the explanatory variable, income, is used to predict whether or not an individual carries at least one travel credit card, such as American Express or Diners Club. The data for a sample of 100 individuals is available on the CD in lists **CCINC** and **CCYES** (together in group **CREDITCR**). Move **CCINC** to **L1** and **CCYES** to **L2**. You can visualize this by making a scatterplot of **L1** vs. **L2**. The axes have been turned off in the graphs below to allow you to see the points that fall on the x-axis. (Do this with **2nd FORMAT AxesOff**). Also, the small markers have been used for the points.

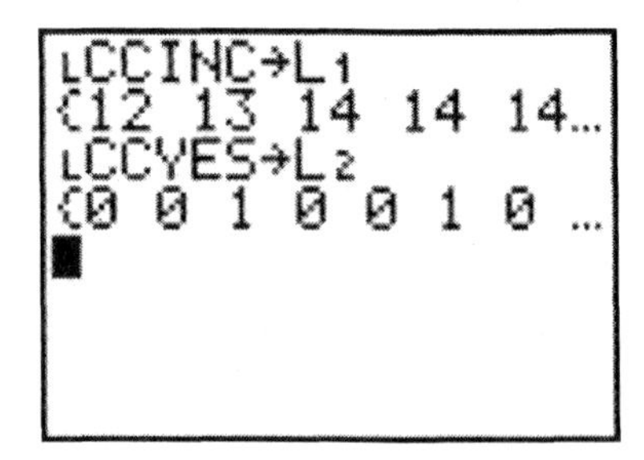

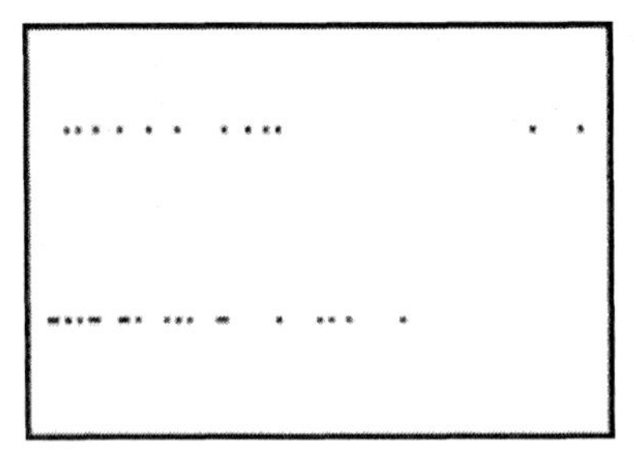

Now, this is potentially deceptive, since there are 100 data points, but the graph doesn't appear to show 100 points. That's because there are a number of repeated values, which you can see if you TRACE the points on the graph. There are many duplicates, but the graph still gives a good idea of the relative density of points along each of the two vertical lines, and y =0 and y = 1.

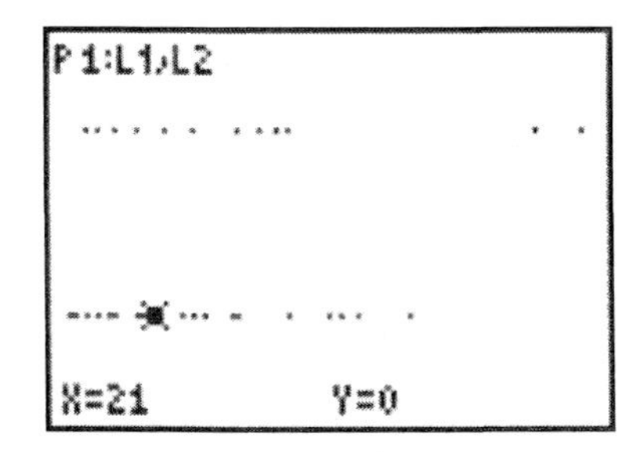

The logistic regression model is given in the textbook on p. 677: $\hat{p} = \dfrac{e^{-3.52+0.105x}}{1+e^{-3.52+0.105x}}$.

This equation is entered in the function menu, and then it is graphed.

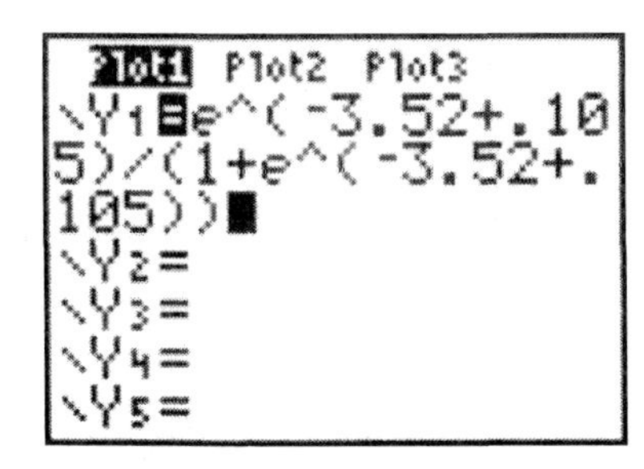

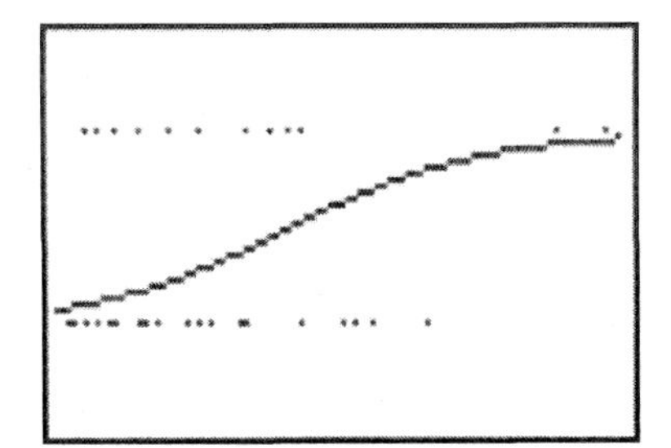

Here's one way to imagine this logistic model: At each x-value, imagine a little tug-of-war between the dots at $y = 0$ and $y = 1$. On the left side of the graph the points are much denser on the $y = 0$ side, so the line is close to those points. In the middle the points are sparser on the $y = 0$ side, so the line shifts towards a higher value. Finally, on the far right, the points on the $y = 1$ line dominate, and the line ends up very close to $y = 1$.

You can find the estimated proportion of individuals with a travel credit card at any particular income by finding a value of the logistic model at that value. You can do this visually on the graph by choosing **2nd CALC** and then **VALUE**. Type in 30 to produce the graph below.

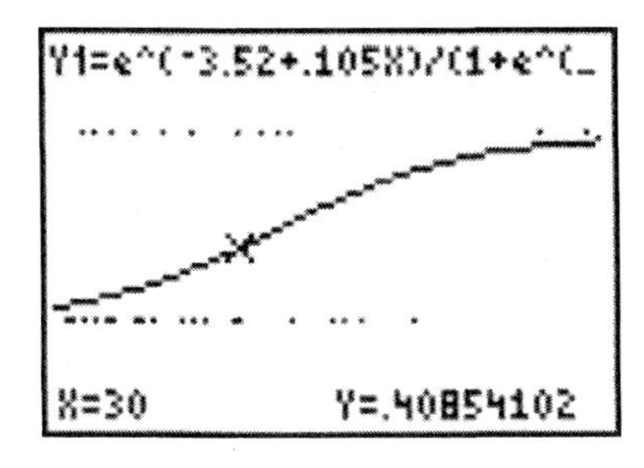

The form of the logistic equation guarantees that the curve will always be between 0 and 1. It's always greater than 0 since both the numerator and denominator are both positive for all x-values (powers of e are always positive, no matter what the exponent). It's always less than 1 because the denominator is always larger than the numerator (exactly 1 larger, in fact).

A linear regression model made with the values in this study doesn't have this property, which you can see below. The linear model will predict proportions larger than 1 at the right side with the larger incomes in this study.

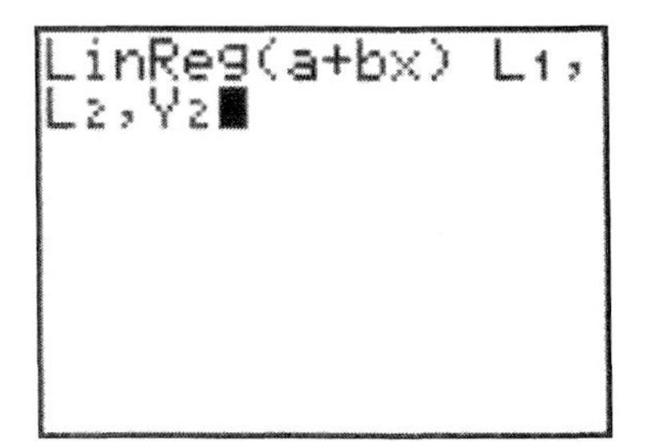


```
LinReg
 y=a+bx
 a=-.15944954
 b=.0188458266
 r²=.2597270321
 r=.509634214
```

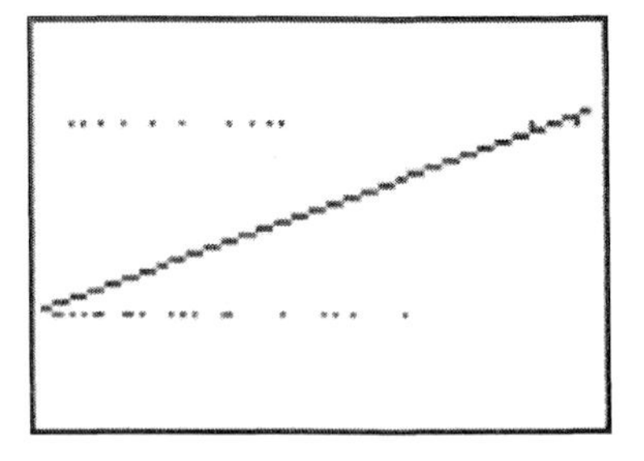

Special Note: Don't be confused by the calculator's option for "Logistic"

You may have noticed that there is an option "**Logistic**" in the **STAT CALC** menu that presents options for regression. This, however, does ***not*** do a logistic regression, the little tug-of-war that is described above. Trying it with the credit card data produces an error message.

```
EDIT CALC TESTS
7↑QuartReg
8:LinReg(a+bx)
9:LnReg
0:ExpReg
A:PwrReg
B:Logistic
C↓SinReg
```

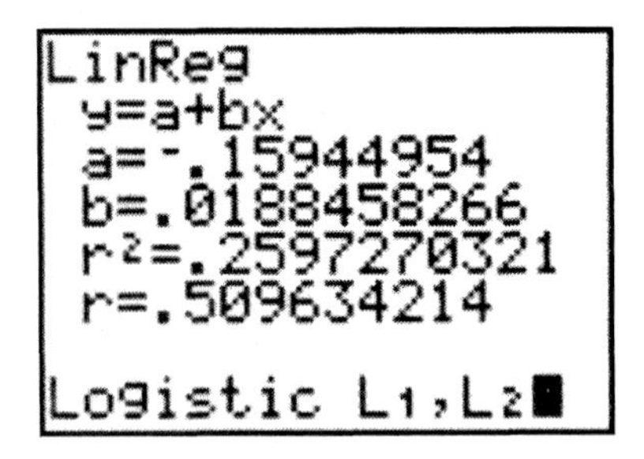


```
ERR:DOMAIN
1:Quit
2:Goto
```

What this command does instead is to fit a curve similar to the logistic function above if you provide a few sample points. So, for instance, if you had made estimates of points on the curve above such as (15, 0.12), (30, 0.4), (45, 0.77), and (60, 0.94), the logistic option would try to find a curve that fits these values. The points above a very close to the logistic function from the textbook, and the graph below shows that they are similar enough to appear as one line.

```
{15,30,45,60}→L5
   {15 30 45 60}
{.12,.4,.77,.94}
→L6
{.12 .4 .77 .94}
Logistic L5,L6,Y
3
```

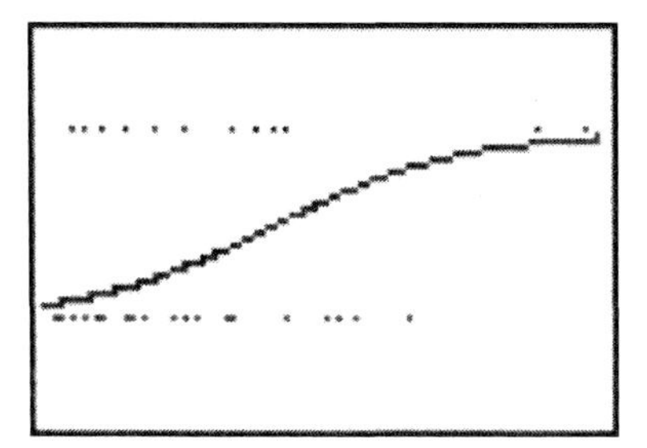

Note also that the form of this function is quite different from the logistic function in the textbook, and doesn't have the guarantees that it produces only values that are proportions, larger than 0 and less than 1.

```
Logistic
 y=c/(1+ae^(-bx))
 a=37.08775534
 b=.1073436818
 c=.9962792282
```

This logistic function does have uses in a variety of curve-fitting applications, but it doesn't necessarily model logistic regression the way a statistician thinks of it.

Chapter Fourteen: Comparing Groups: Analysis of Variance Methods

There is no data from the CD that is used in this chapter.

14.1: How Can We Compare Several Means? One-Way ANOVA

Analysis of Variance, or ANOVA, is used to compare the means of three or more groups. This is an extension of the methods of section 10.2 that compared two means. The **ANOVA** function from the **STAT TESTS** menu provides this on the TI-83 Plus family. Graphically, boxplots are used to illustrate the distribution of values in each of the groups.

EXAMPLE 2: How long will you tolerate being put on hold? (pp. 693-694 of Agresti/Franklin)

The data for this study appears on p. 693. Enter the five holding times for the advertisement group into **L1**, the times for the muzak group into **L2**, and the times for the classical music group into **L3**.

L1	L2	L3
5	0	13
1	1	9
11	4	8
2	6	15
8	3	7

```
L3(6) =
```

The first stage in an ANOVA analysis is to examine the data, numerically and/or graphically. The assumptions for what will follow includes that the population distributions from each group are normal and have the same standard deviations. Here are the means and standard deviation of the two groups, entered as a small list with the braces { and }. Then there is a set boxplots that completes the comparison of the three groups.

```
{mean(L1),stdDev
(L1)}
{5.4 4.15932686...
{mean(L2),stdDev
(L2)}
{2.8 2.38746727...
```

```
{5.4 4.15932686...
{mean(L2),stdDev
(L2)}
{2.8 2.38746727...
{mean(L3),stdDev
(L3)}
{10.4 3.4351128...
```

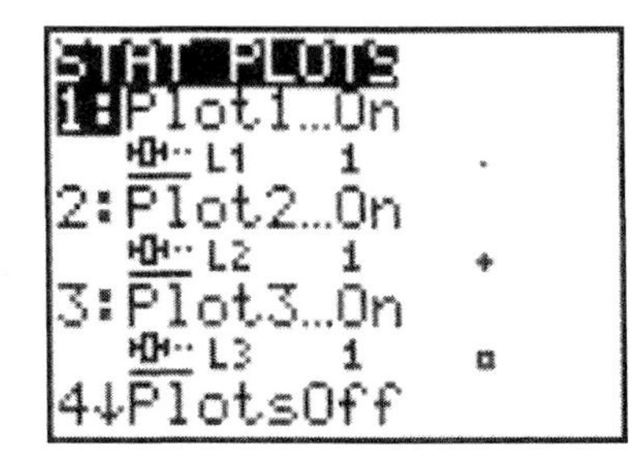

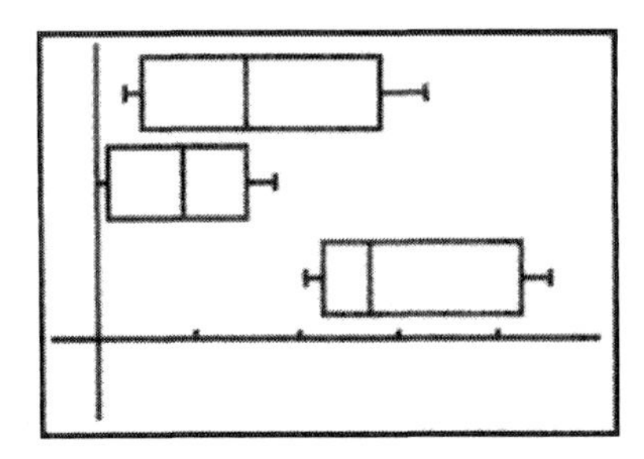

You can see by examining either the values in the table or the boxplots that there are differences in the centers of each of the two groups, some difference in the spread, and no evidence of striking difference from a normal distribution in each of the three cases. These are parts of the assumptions for continuing with an ANOVA analysis that are listed on p. 695.

From the boxplots it is visually clear that there is variability *within* each of the three groups **and** there is variability *between* the groups. Comparing these two types of variability is the key to the inference procedure for ANOVA. In this procedure a test statistic, F, is constructed that is the ratio of the between-group variability to the within-group variability. From the distribution of F statistics a P-value is then calculated.

Each of the estimates of variability has an associated degrees of freedom. For the between-group variability it is one ness than the number of groups, $g-1$ (works out to be 2 in this case). For the within-group variability it is the total sample size minus the number of groups, $N-g$ (12 in this case).

Formulas for estimating the two types of variability are on p. 697. The overall mean is 6.2, and the between-group variability adds up the differences from this for each of the group means, squared. Five members in each group and $df = 2$ completes the calculation of the estimate, 74.6.

The within-group estimate is just the average of the three variances, about 11.6. The calculated F-ratio, the quotient of the values, is about 6.43, the same as in the textbook.

```
(5.4-6.2)²+(2.8-
6.2)²+(10.4-6.2)
²
                29.84
Ans*5
                149.2
```

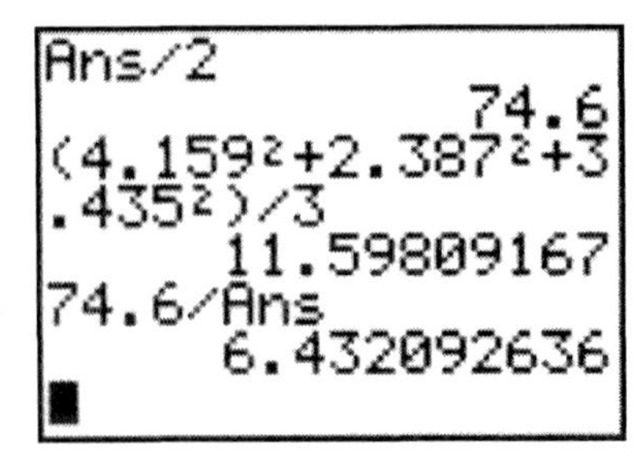

Is 6.43 an unusually high value for this F-ratio? This is where the inference comes in. To calculate the P-value, you need a distribution of F values that are typical if you have three groups of five measurements each <u>and</u> the null hypothesis of no difference between the means is true. The calculator has just this distribution, in the **DISTR** menu (**2nd VARS**). The format of the command includes an interval (from 6.43 to infinity, or a very

large value) and the two degrees of freedom from the numerator and denominator of the F-ratio, 2 and 12 in this problem.

```
DISTR DRAW
5↑tpdf(
6:tcdf(
7:X²pdf(
8:X²cdf(
9:Fpdf(
0:Fcdf(
A↓binompdf(
```

```
Fcdf(6.432,1000,
2,12)
           .0126375095
```

This P-value agrees with the computer output table on p. 697. The conclusion is to reject the null hypothesis with this low P-value: there is a difference in how long customers are willing to wait on hold depending on the music or message they listen to.

In practice, the calculator automates these calculations. There is a built-in **ANOVA** procedure, in the **STAT TESTS** menu. The data is entered just as we did above, into as many lists as there are groups. The sample size in each list need not be the same. Then the test is called up, with the lists indicated.

```
EDIT CALC TESTS
B↑2-PropZInt...
C:X²-Test...
D:X²GOF-Test...
E:2-SampFTest...
F:LinRegTTest...
G:LinRegTInt...
H:ANOVA(
```

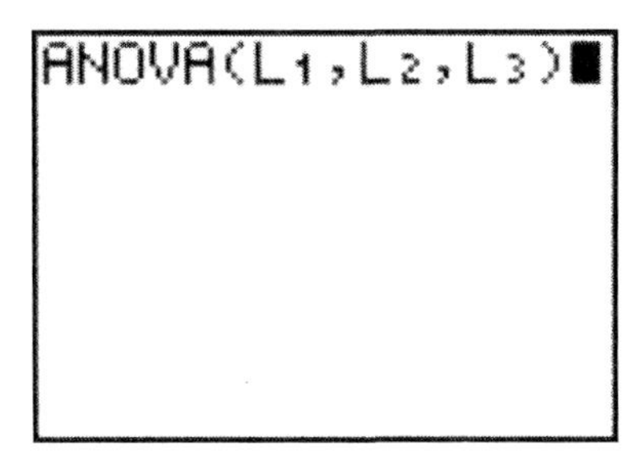


```
One-way ANOVA
 F=6.431034483
 p=.0126433999
 Factor
  df=2
  SS=149.2
↓ MS=74.6
```

```
One-way ANOVA
↑ MS=74.6
 Error
  df=12
  SS=139.2
  MS=11.6
 Sxp=3.40587727
```

The output above shows almost exactly the same information that the computer output table in the textbook displays, with estimates for the two forms of variability identified with MS or **mean squares**. The group estimate of variability is under the heading "Factor" rather than "Group". The F test statistic and P-value are calculated, and an estimate is provided for the pooled standard deviation of the responses, identified as **Sxp**, about 3.4.

This of course is just the same as the estimate of the within-group variability expressed as a standard deviation: $\sqrt{11.6} = 3.40587727$. While not included in the computer printout in the textbook, it is present in most software. Remember that one of the assumptions of the ANOVA procedure is the within-group variability is the same in all groups, and this is the best estimate of this value from the data we have.

With the P-value of 0.126, there is good evidence that the mean holding times are not all the same for the different presentations. But this is the extent of the conclusion we can make at this time. The overall ANOVA provides no information about the nature of the differences. For any practical situation, the obvious next task is to quantify the differences that we have demonstrated do exist.

14.2: How Should We Follow Up an ANOVA F Test?

Exercise 14.14: Comparing telephone holding times (p. 710 of Agresti/Franklin)

This exercise takes this ext step, using the formula for a 95% confidence interval for the difference of two group means, $\mu_i - \mu_j$, that appears on p. 702:

$$\left(y_i - y_j\right) \pm t_{0.25}\, s\sqrt{\frac{1}{n_1} + \frac{1}{n_2}}.$$

The value, *s*, that appears here is residual standard deviation, from the estimated within-group variability. This is **Sxp** in the ANONA report above. The *t* critical value is for the *t*-distribution with $df = 12$, $N - g$. This value can be gotten from the **invT** command with a newer TI-84 or the table. In this case it's about 2.18, and the value is stored in the variable **T** since we'll need it for each of the three intervals, one for each comparison of means.

```
DISTR DRAW
1:normalpdf(
2:normalcdf(
3:invNorm(
4:invT(
5:tpdf(
6:tcdf(
7↓X²pdf(
```

```
invT(.975,12)
        2.178812801
Ans→T
        2.178812801
■
```

The next screens show these intervals, confirming the values that are presented in Exercise 14.14.

```
(10.4-2.8)-T*Sxp
*√(1/5+1/5)
        2.906693594
(10.4-2.8)+T*Sxp
*√(1/5+1/5)
        12.29330641
■
```

```
(10.4-5.4)-T*Sxp
*√(1/5+1/5)
        .3066935939
(10.4-5.4)+T*Sxp
*√(1/5+1/5)
        9.693306406
■
```

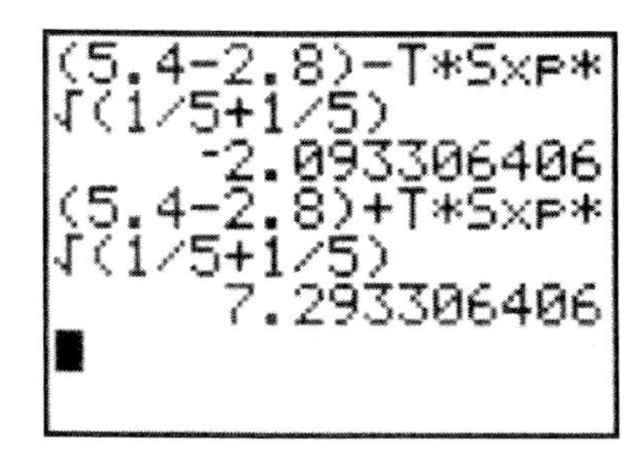

Two of these intervals indicate a statistically significant difference between the means. These are the difference between classical and muzak, with confidence interval (2.9, 12.3), and between classical and advertisement, confidence interval (0.3, 9.7). In neither case is 0 in the interval of plausible differences.

The difference between advertisement and muzak, however, is not statistically significant since 0 is in the interval (-2.1, 7.3). It is plausible to believe that there is no difference between the mean wait time of all customers with either muzak or advertisements.

Part (b) of this question is about the margin of error of these intervals, which are all equal to $t_{0.25}\, s\sqrt{\frac{1}{n_1}+\frac{1}{n_2}}$. In each case this is about 4.7.

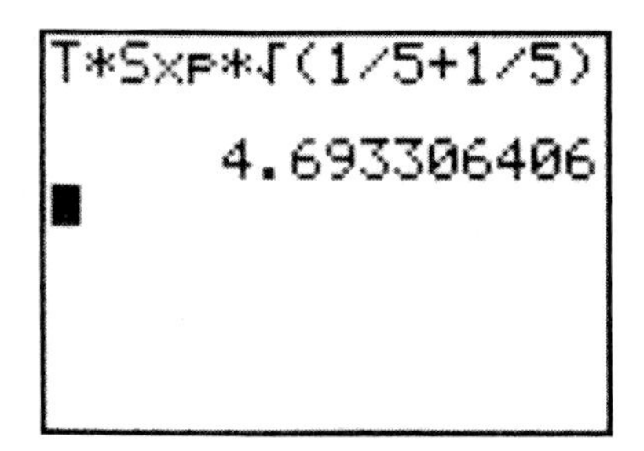

The other methods presented in the chapter, the equivalence between ANOVA and multiple regression and two-way ANOVA, are not easily implemented on the calculator.

Chapter Fifteen: Nonparametric Statistics

The capabilities of the TI-83 Plus family of calculators do not include working with nonparametric statistics.

There are programs available on the Internet that can be downloaded to perform these tests, but these are beyond the scope of this manual.

SPSS MANUAL

Debra Hydorn

Mary Washington College

SECOND EDITION

STATISTICS

THE ART AND SCIENCE OF LEARNING FROM DATA

Agresti • Franklin

Upper Saddle River, NJ 07458

Table of Contents

Chapter 1 Getting Started with SPSS

SPSS is a Windows-based program with drop-down menus for you to access commands for manipulating and analyzing your data. In SPSS, your data is entered and manipulated in one window, the Data Editor, and your output appears in another window, the Output Viewer. To produce a graph or data analysis of your data, you select the appropriate command from one of the menus and typically fill in the details about the analysis you want in a dialog box. If you have used drop-down menus and dialog boxes before then you will find SPSS very easy to use, and, if you haven't used them before, don't worry – this manual will provide step-by-step instructions! The purpose of this manual is to get you started with using SPSS for your data analyses. SPSS includes commands for conducting many more advanced methods than will be covered here. And, for most of the commands demonstrated in this manual you will find some options that are not discussed as they are beyond the scope of an introductory statistics course. The examples for this manual were produced using version 15.0 of SPSS; if you are using an earlier version you might notice some differences in the screens or output that you encounter.

Depending on the computer you are using, you may access SPSS through "All Programs" found on that computer's **Start** button, or by double clicking on an SPSS icon on the desk top. If you access it through the Start button, make sure to select SPSS for Windows (and not an SPSS Production Facility, for example). When you first start SPSS you will see the following dialog box that appears for you to indicate what you want to do:

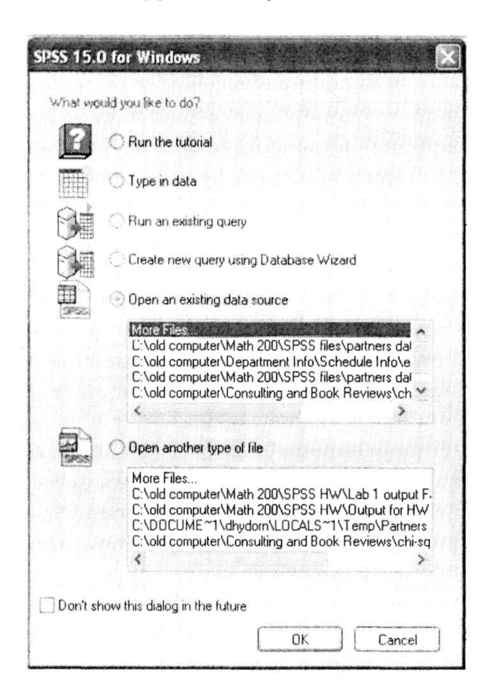

The default setting in this dialog box is to "Open an existing data source." The CD that comes with your textbook includes a number of SPSS data files that you will access later on using this option. This dialog box also has an option for accessing output files that have been saved. For a simple introduction to SPSS, though, click on "Type in data" and then on the **OK** button. You should now see the following display:

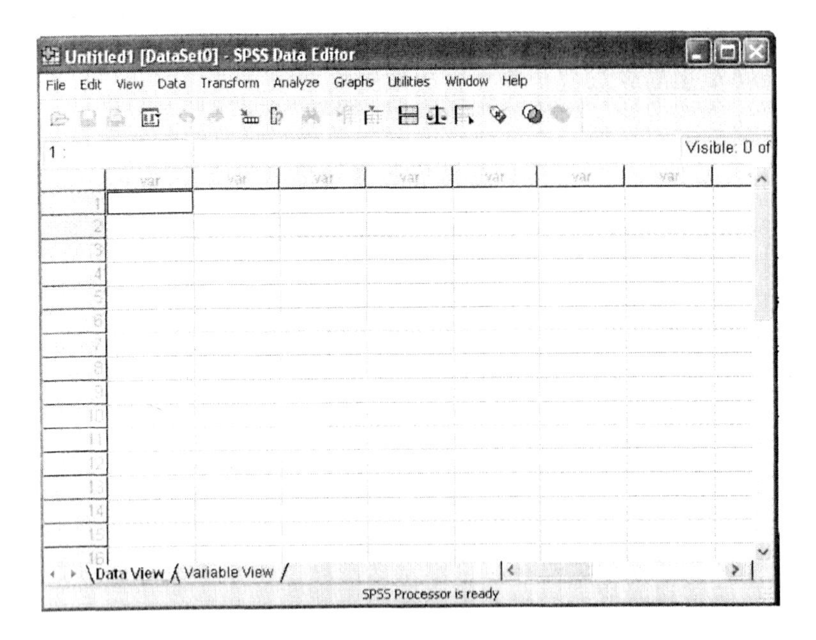

This is the first of the two windows that you will be using in SPSS. This is the **SPSS Data Editor** window, and it is here that you will enter new data to analyze. It resembles spreadsheets you have probably seen before and it has some of the same properties. Unlike some other spreadsheets, however, the Data Editor in SPSS is for data only. All output you produce will appear in the **Output Viewer** window, which is the second SPSS window you will be using. To see the Output Viewer you will have to first enter some data, but let's explore the Data Editor a little more first.

The menu bar is immediately below the title bar. Once you enter some data and save it you will see the name on that bar change. The menus you will be using most are:

File – commands for opening and saving data files and for exiting SPSS
Edit – commands for modifying SPSS data files or output (e.g., copy and paste)
Data – commands for adding or deleting variables and for manipulating the data set (e.g., select cases and split file)
Transform – commands for creating or re-coding new variables using variables that are already in the data set
Analyze – commands for producing univariate and bivariate descriptive statistics as well as for conducting inferential statistics
Graphs – commands for producing various graphical displays

Below the menu bar is a series of icons representing a few common commands, for those who prefer that approach. The directions in this manual use the drop-down menus only. Notice the two tabs at the bottom of the window, **Data View** and **Variable View**. The Data View is for data entry and has columns for the variables and rows for the observations. The columns are currently each labeled "var" but these labels will change when you enter data. The Variable View is for defining and labeling your variables. You can enter the data in the Data View and then move to the Variable View to name the variables, or you can define your variables first in the Variable View and then enter the data in the Data View. If you click on the Variable View tab you will see that the labels on the columns of the spreadsheet change, as shown below.

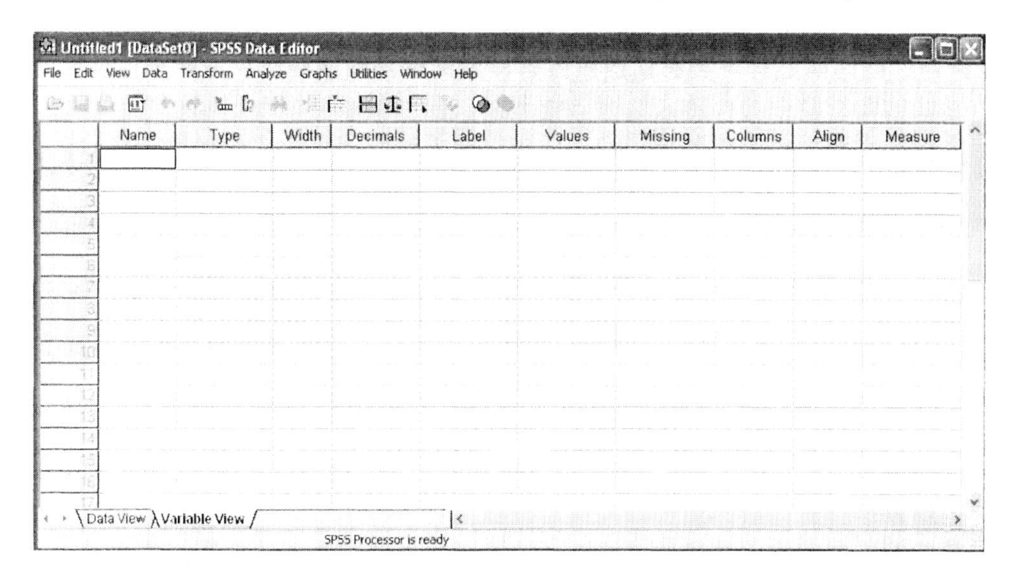

Information about the variables in your data set is entered using the columns in this new spreadsheet, some of which have obvious functions. In the "Name" column you can enter a one word name for your variable; SPSS does not allow the use of spaces or some special characters in this column. For more descriptive names you can use the underscore _ symbol or enter longer descriptions in the "Label" column. Try to avoid using variable names such as var1 or X as these names do not provide any information about the data that could be useful when interpreting the output you produce. If you click on the far right side of the cells under some of the columns a dialog box will appear with options for your variables. For example, if you click on the right side of the first cell under "Type" the following dialog box will appear.

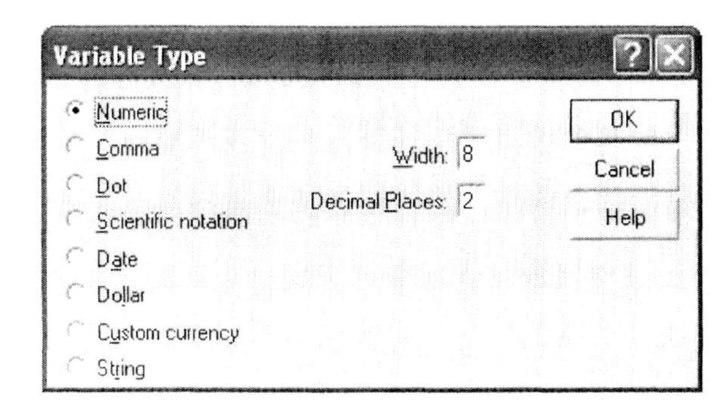

Notice that "Numeric" is the default variable type; if you want to enter character data you will need to change the type to "String." You can increase or decrease the total number of digits or characters by changing the entry under "Width," while the number of decimals to the right of the decimal point can be altered under "Decimals." For both of these variable characteristics up and down arrows will appear when you click on the right side of a cell in those columns. If you have categorical data that you wish to code numerically, this can be done using the "Values" column. This option will be demonstrated in the example below.

Entering Data

To introduce you to data entry in SPSS, let's start with the University of Florida Student Survey example in the first chapter of your text. The complete data set is available on the text CD but you will enter only a portion of it here. In addition to a student identification number, the variables you will enter are:

Gender (f=female, m=male)
Racial-ethnic group (b=black, h=Hispanic, w=white)
Age (in years)
College GPA (scale 0 to 4)
Average number of hours per week watching TV
Whether a vegetarian (yes, no)
Political party (dem=Democrat, rep=Republican, ind=Independent)
Marital status (1=married, 0=unmarried)

Some of these variables are categorical and some are numeric. For entering categorical variables you have two choices: you can enter it as a string variable or you can enter it as a numeric variable with codes to identify the categories. Of the categorical variables, all but "Marital status" will be entered as string variables. Marital status is entered using a numerical coding scheme (but the variable is still categorical). How you enter categorical data will determine which analyses will be available in SPSS for you to use. You will learn more about this later on.

Let's start by defining the variables in the Variable View and then entering the data in the Data View. The following display shows the variable names entered and which variables are numeric and which are string. It also shows the number of decimals for the numerical variables decreased to match the data to be entered. Notice the up and down arrows on the right side of the cell in the "Decimals" column of the "married" variable. These were used to decrease the number of decimals from the default setting of 2 to 0. To match the variable names as they are given in the text, variable labels have been included.

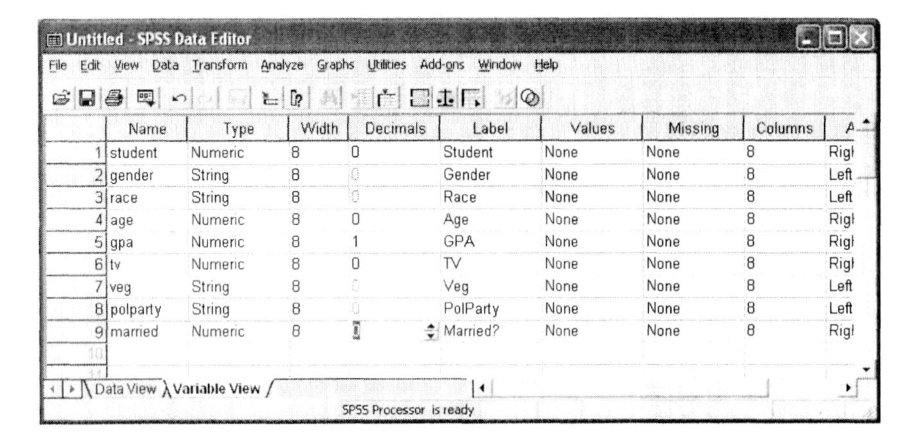

	Name	Type	Width	Decimals	Label	Values	Missing	Columns	A
1	student	Numeric	8	0	Student	None	None	8	Rigl
2	gender	String	8	0	Gender	None	None	8	Left
3	race	String	8	0	Race	None	None	8	Left
4	age	Numeric	8	0	Age	None	None	8	Rigl
5	gpa	Numeric	8	1	GPA	None	None	8	Rigl
6	tv	Numeric	8	0	TV	None	None	8	Rigl
7	veg	String	8	0	Veg	None	None	8	Left
8	polparty	String	8	0	PolParty	None	None	8	Left
9	married	Numeric	8	0	Married?	None	None	8	Rigl

The last thing you need to do is enter the coding for the "married" variable. Click on the right hand side of the cell under the "Values" column in the "married" variable row. The following dialog box will appear:

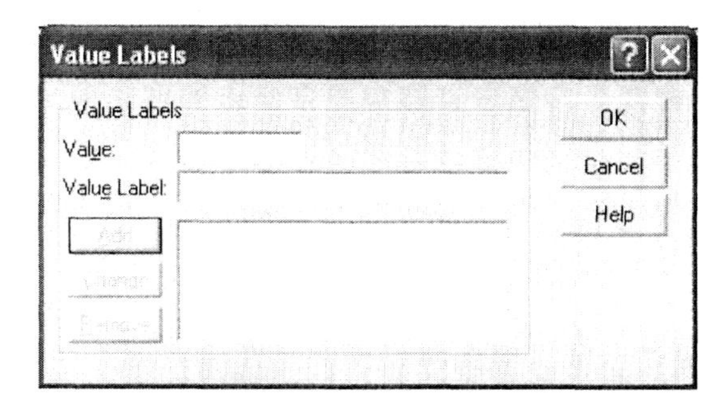

Enter the value "1" in the "Value" box and then click on the "Value Label" box and enter "Married" (but without the quotes). Notice that the "Add" button becomes highlighted as soon as you start entering a label. Click on **Add** so that the dialog box looks like:

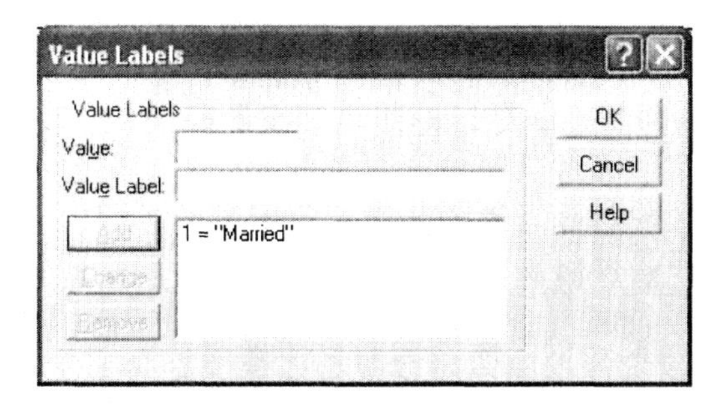

Now enter "0" in the "Value" box and "Unmarried" in the "Value Label" box and click on **Add** again. Click on **OK** to complete the process and close this dialog box. The Variable View should now look like:

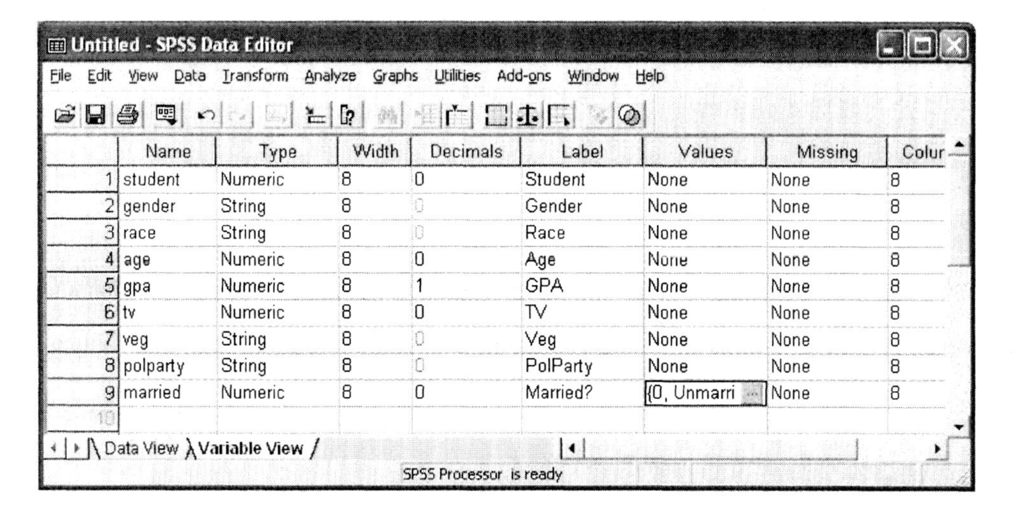

	Name	Type	Width	Decimals	Label	Values	Missing	Colur
1	student	Numeric	8	0	Student	None	None	8
2	gender	String	8	0	Gender	None	None	8
3	race	String	8	0	Race	None	None	8
4	age	Numeric	8	0	Age	None	None	8
5	gpa	Numeric	8	1	GPA	None	None	8
6	tv	Numeric	8	0	TV	None	None	8
7	veg	String	8	0	Veg	None	None	8
8	polparty	String	8	0	PolParty	None	None	8
9	married	Numeric	8	0	Married?	{0, Unmarri	None	8
10								

If you wish, you can add "value" labels for the string variables, Gender and Racial-ethnic group, following the same process. That is, you can the Gender variable coded so that the "value" m will appear as "male" in your output and f will appear as "female." Now click on the Data View tab. Notice that the variable names now replace "var" at the top of each column. The data to be entered is shown in the following table and represents just a portion of the data set available on your text CD in the "FL student survey" data file.

Student	Gender	Race	Age	GPA	TV	Veg	PolParty	Married?
1	m	w	32	3.5	3	no	rep	1
2	f	w	23	3.5	15	yes	dem	0
3	f	w	27	3.0	0	yes	dem	0
4	f	h	35	3.2	5	no	ind	1
5	m	w	23	3.5	6	no	ind	0
6	m	w	39	3.5	4	yes	dem	1
7	m	b	24	3.7	4	no	ind	0
8	f	h	31	3.0	4	no	ind	1

Enter the data so that your Data Editor appears as shown below.

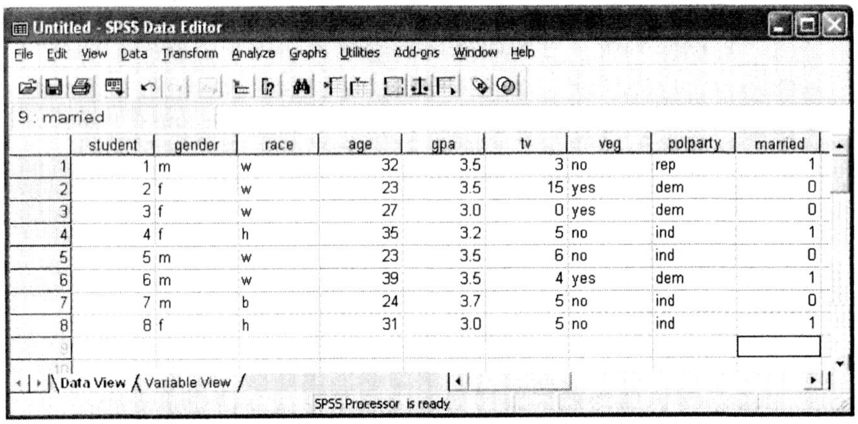

Untitled - SPSS Data Editor

File Edit View Data Transform Analyze Graphs Utilities Add-ons Window Help

9 : married

	student	gender	race	age	gpa	tv	veg	polparty	married
1	1	m	w	32	3.5	3	no	rep	1
2	2	f	w	23	3.5	15	yes	dem	0
3	3	f	w	27	3.0	0	yes	dem	0
4	4	f	h	35	3.2	5	no	ind	1
5	5	m	w	23	3.5	6	no	ind	0
6	6	m	w	39	3.5	4	yes	dem	1
7	7	m	b	24	3.7	5	no	ind	0
8	8	f	h	31	3.0	5	no	ind	1
9									

Data View / Variable View

SPSS Processor is ready

To view the value labels for the "married" variable, instead of the values 0 and 1, click on the "View" menu and select "Value labels." A check mark will appear next to "Value labels" and the labels you entered for each value should now appear in the Data Editor in place of the values 0 and 1. Once you have the data entered it is important that you now check for any data entry errors. Possible errors to look for are misplaced decimals or omitted cases. You can easily correct any mistakes by using the arrow keys on your keyboard to move to the cell you need to correct or by clicking on that cell with the mouse. Make a correction and press **Enter**.

Note: If you first enter data in the Data View before defining your variables in the Variable View, the variable names at the top of each column you use will automatically change to "var00001" and "var00002," for example, which you can then change to appropriate names in the Variable View.

Note: If a data set has "missing" values, these observations are entered in SPSS as a "." (without the quotes). Many of the SPSS commands you will be using include a table indicating if there were any missing values omitted from an analysis.

Saving Data

Now that you have entered this data set in SPSS (and it is free from errors), let's see how to save it. Select the "Save as..." command under the "File" menu. A dialog box will appear in which you can enter a name for the data file and where you want the data file to be saved. In the display below, the data file is being saved into a folder called "Intro Statistics" with the name "fla student survey" (again, without the quotes). Note that it is OK to have spaces in the names of SPSS data files.

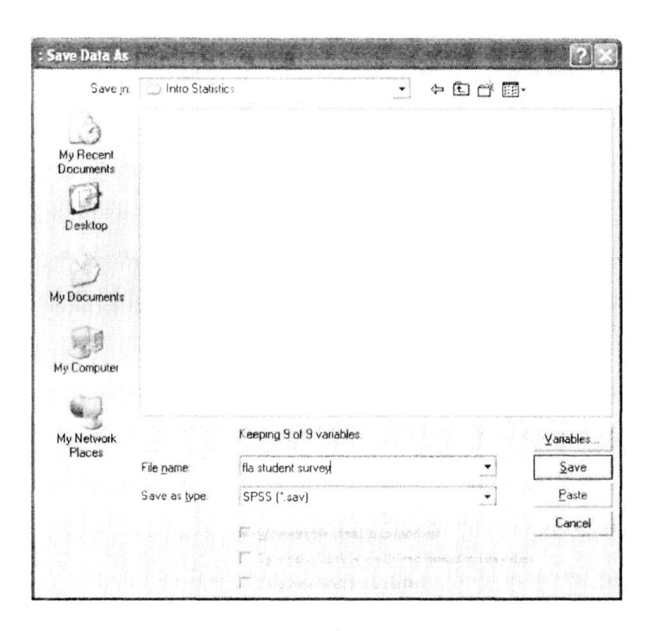

Click on **Save** to complete the process. The data file will be saved with the ".sav" file-type extension. Notice that once the data has been saved with a name, the title in the bar at the top of the Data Editor now shows that data file name. SPSS will produce a log file in the Output Viewer to indicate that the file has been saved.

```
SAVE OUTFILE='C:\old computer\Consulting and Book Reviews\Agresti-Franklin
SPSS
   manual\Intro Statistics\FL student survey.sav'
 /COMPRESSED.
```

A log file will be produced for each command you use, including when you open and close data files. These files have been omitted from most of the SPSS output displayed in this manual. With the appearance of this log file, SPSS now has the Output Viewer as the "active" window, since it is on top of the Data Editor. To switch between the Output Viewer and the Data Editor you can use the tabs at the bottom of the computer screen. You can also use a command from one of the menus, as described below.

Producing Output

To demonstrate how SPSS works, let's print a display of this data and then find some descriptive statistics and produce a graph for the "age" variable. The command to print a table of your data is the "Case Summaries" option under the "Reports" command from the Analyze menu. Click on the "Analyze" menu and then scroll down to "Reports." Then click on "Case Summaries" from the list of options that appears. Let's abbreviate this action with select **Analyze → Reports → Case Summaries...** (menu name → command name → subcommand or option). Using this convention, the command for changing between the data and output windows would be **Window → Data Editor** if the Output Viewer is the active window. A triangle to the right of a command name indicates that there are several options from which you can choose. The ellipsis (…) following a command or subcommand indicates that a dialog box will appear for you to enter information about the analysis you want SPSS to do. Once you have selected the Case Summaries command, the following dialog box will appear. All of the variables in the data set are listed in the box on the left. Notice the different symbols that SPSS uses to identify the variable type.

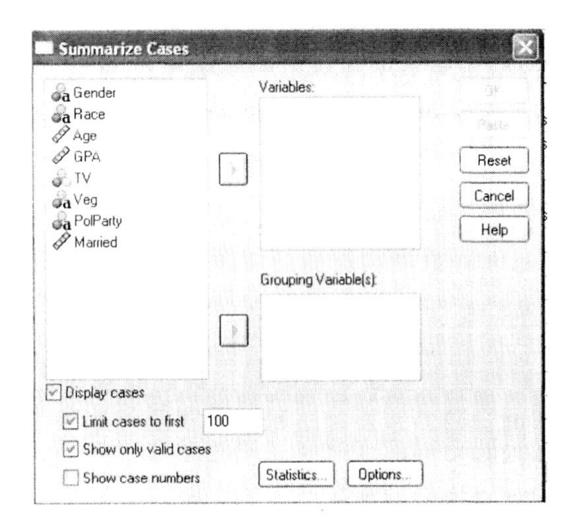

To select a variable to be included in the output from this command, click once on the variable name and then once on the triangle button between the box containing the list of variables and the box labeled "Variables:". Or, just double click on the variable name. This double-clicking option doesn't work all of the time, but it does work when there is only one place to enter a variable name. You can select more than one variable at the same time by pressing the Shift or Ctrl keys while clicking on variable names. Using Shift will select all of the variables between the ones you click on, while using the Ctrl key will select just the ones you click on. You can also select all of the variables at once by holding down the mouse button and scrolling down the list of variables. The dialog box below shows all of the variables selected.

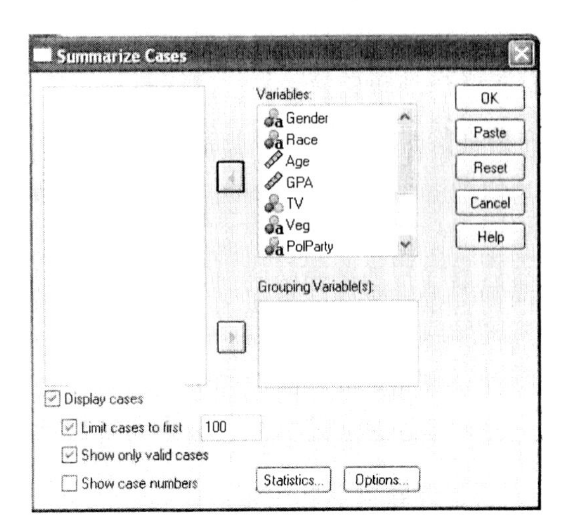

To un-select a variable, click on it once in the "Variables" box and then on the triangle button. In the dialog box below, the variable gender has been selected. Notice that the triangle button is pointing the other way. If you click on the triangle button, the gender variable will be returned to the (now empty) variable list on the left.

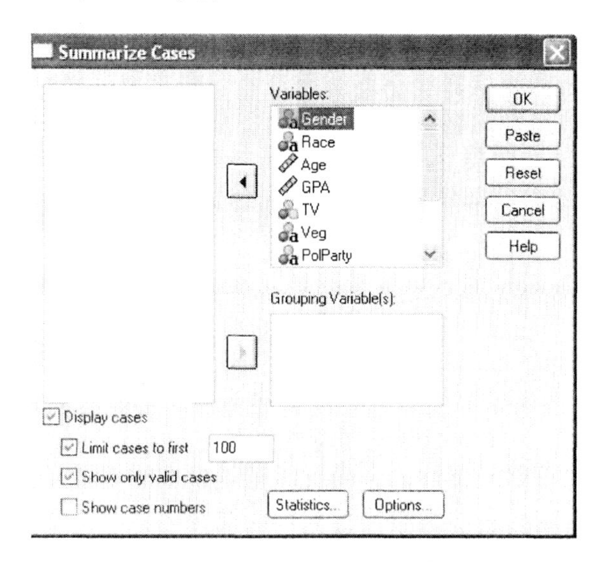

Notice also that the **OK** button has become highlighted. This is how you will know when you have entered enough information in a dialog box for SPSS to be able to execute the command. Click on **OK** to produce the output. Notice that the Output Viewer also has a list of menus at the top and is divided into two parts. On the right you will see the actual output produced, in this case, a table showing the data that has been entered. A flow chart, or **Output Navigator**, showing the commands you have used and the parts of the output produced, is shown on the left. The list of menus in the menu bar of the Output Viewer is slightly different from those available in the Data Editor window, but the Analyze and Graphs menus are in both menu bars so you can produce additional output while either the Data Editor or the Output Viewer is the active window.

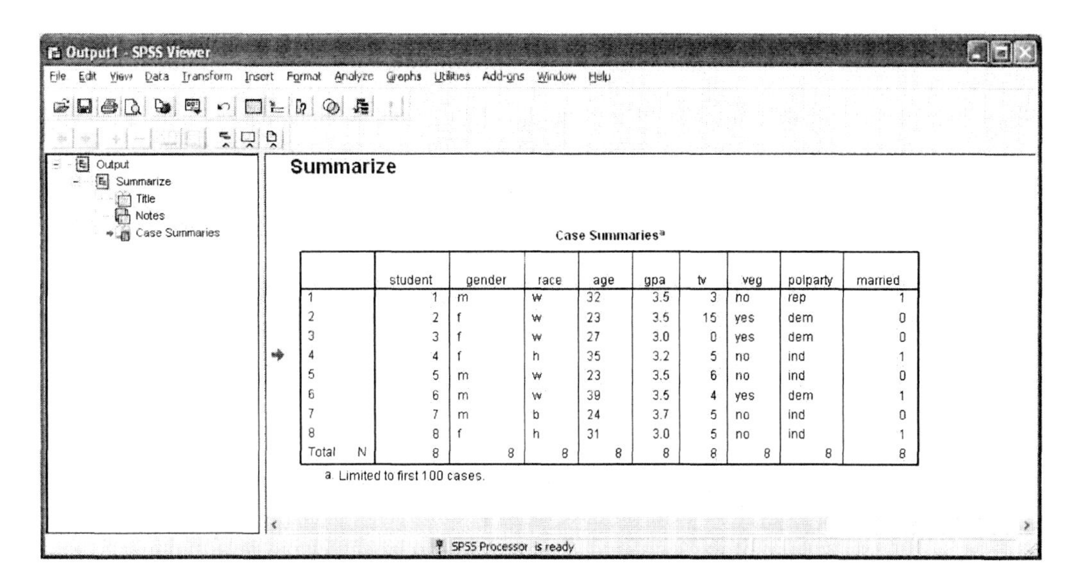

Case Summaries[a]

		student	gender	race	age	gpa	tv	veg	polparty	married
1		1	m	w	32	3.5	3	no	rep	1
2		2	f	w	23	3.5	15	yes	dem	0
3		3	f	w	27	3.0	0	yes	dem	0
4		4	f	h	35	3.2	5	no	ind	1
5		5	m	w	23	3.5	6	no	ind	0
6		6	m	w	39	3.5	4	yes	dem	1
7		7	m	b	24	3.7	5	no	ind	0
8		8	f	h	31	3.0	5	no	ind	1
Total	N	8	8	8	8	8	8	8	8	8

a. Limited to first 100 cases.

This output is typical of what SPSS produces. It includes a log, a title, some notes that are currently "hidden," the name of the active data set, a "Case Processing Summary," and the actual output. The log, the data set name and the "Case Processing Summary" were omitted in the output above. To delete a piece of output just click on it once and press the "Delete" key on your keyboard. You can also "hide" a piece of output by double clicking on its icon in the Output Navigator. To "un-hide" it just double click on the icon again. Notice that the title of the output is the name of the command you used. You will see how to modify the title so that it is more descriptive after making a histogram for "age." Within the output, SPSS uses the variable names and labels that you entered in the Variable View of the Data Editor. If you forget to name a variable, you will see "var00001" in heading of the column for that variable. You will need to go into the Variable View to name the variable and then redo the output. Unlike some other data analysis programs, in SPSS any changes you make to the data after producing output are not incorporated into output that has already been produced. This is also true for correcting any data entry errors – SPSS does not automatically update output.

One SPSS command that finds a small collection of descriptive statistics is the "Descriptive Statistics" command under the "Analyze" menu. Select **Analyze → Descriptive Statistics → Descriptives...** and the following dialog box will appear.

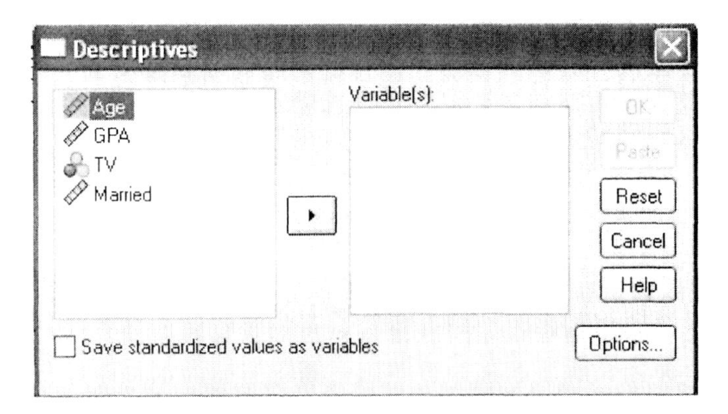

Notice that only the numeric variables are listed on the left. Some SPSS dialog boxes, like this one, allow you to select more than one variable to analyze at the same time. Others only allow you to select and analyze one variable at a time. Select the "age" variable so that it appears in the "Variable(s)" list as shown below.

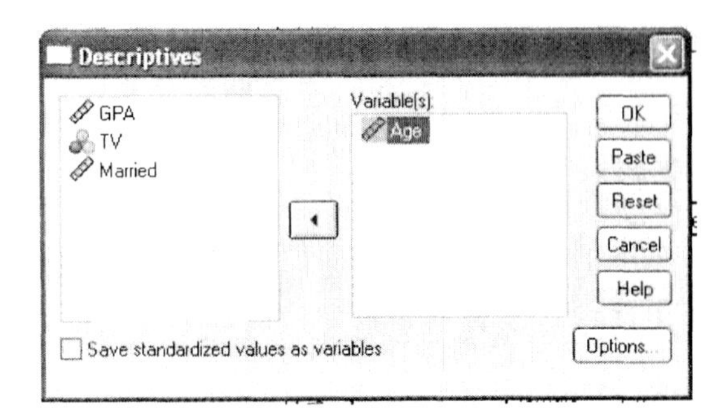

Click on **OK** to produce the output

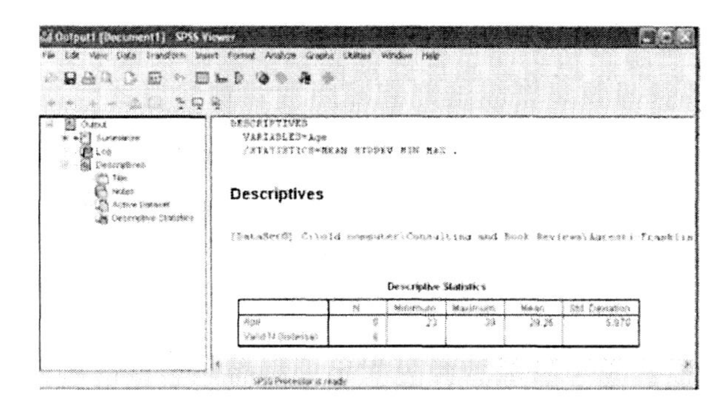

In this output, the Log and Active Dataset information were left intact.

To make a histogram, you have two options. Click on the **Graphs** menu to reveal the list of graphing commands. You can use the "Legacy Dialogs" option under the Graphs menu, or you can use the "Interactive option." The Legacy Dialogs are the dialog boxes from earlier versions of SPSS. If you have learned SPSS before you might prefer to use those. If you have an earlier version of SPSS then the graphing commands you have are similar to the Legacy Dialog commands described in this manual.

FL student survey.sav [DataSet0] - SPSS Data Editor

File Edit View Data Transform Analyze Graphs Utilities Window Help

Chart Builder...
Interactive
Legacy Dialogs

13 : Gender Visible: 8 of 8

	Gender	Race	Ag		TV	Veg	PolParty	Married
1	m	w	32	3.50	3	no	rep	1
2	f	w	23	3.50	15	yes	dem	0
3	f	w	27	3.00	0	yes	dem	0
4	f	h	35	3.20	5	no	ind	1
5	m	w	23	3.50	6	no	ind	0
6	m	w	39	3.50	4	yes	dem	1
7	m	b	24	3.70	5	no	ind	0
8	f	h	31	3.00	5	no	ind	1

Data View / Variable View

SPSS Processor is ready

We will demonstrate the Interactive commands first. Select **Graphs → Interactive→ Histogram….** The following dialog box will appear:

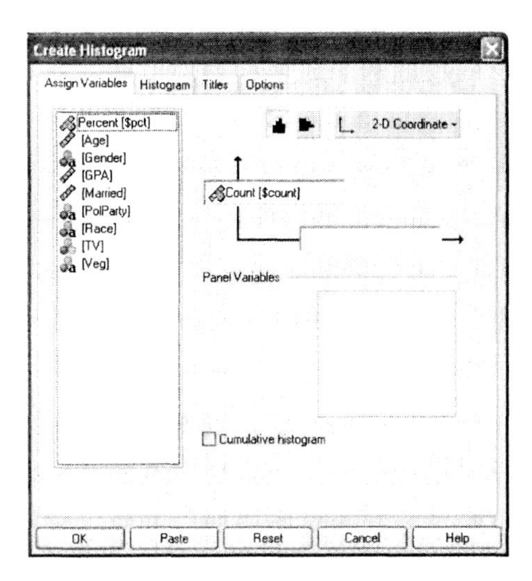

This dialog box has four tabs. The Assign Variables tab includes a space for you to indicate which variable to graph. The "Panel Variables" section of this dialog box provides the option for creating a histogram for one variable for each value of another variable. For example, we could produce separate histograms of the age variable for each gender using this option. The Histogram tab allows you to specify the placement and the number of intervals for SPSS to use to make the histogram. It also includes an option for overlaying a normal curve on the histogram. The Titles tab has options for entering a title, subtitle and caption to the graph, and the Options tab includes additional options for customizing the output.

To make a histogram of the age variable, "grab" that variable from the list on the left by clicking on it and holding down the mouse button. Then drag it to the box on the horizontal axis displayed in the dialog box.

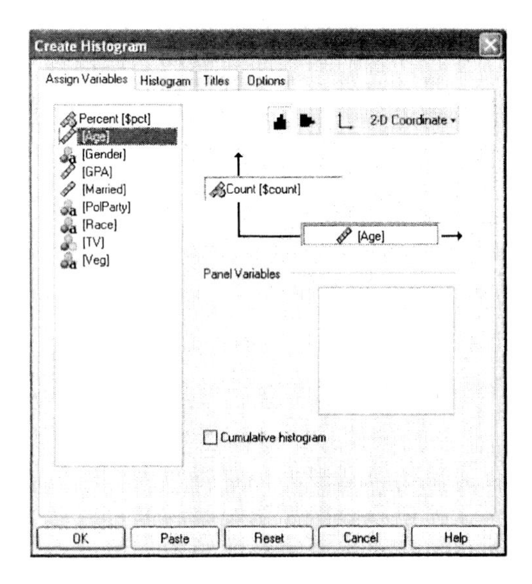

Click on OK to make the graph.

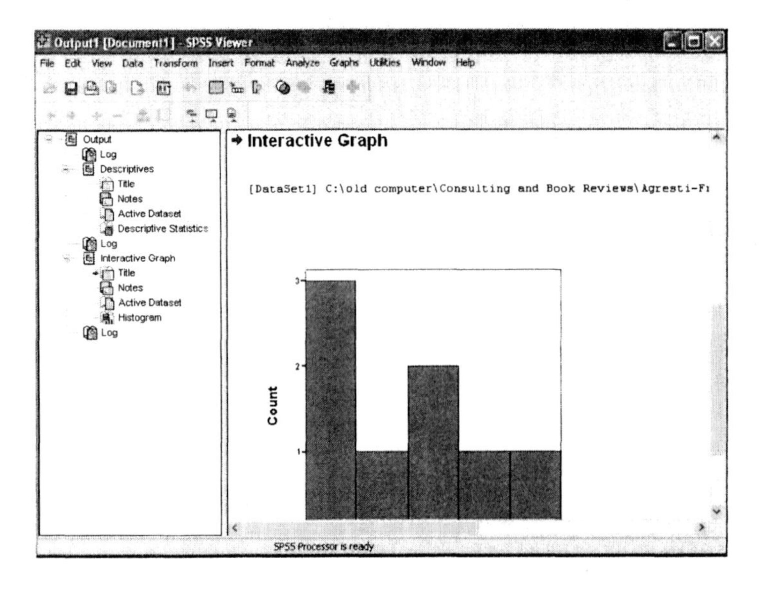

To view the entire histogram you may have to use the scroll bar on the right and increase the size of the window. Notice that the Output Navigator shows the additional output and its parts, and that the title for the histogram is just "Interactive Graph," which like the "Descriptives" title is not very informative. But, it is very easy to modify SPSS output so that it is more useful!

Now let's see how the Legacy Dialog command for a histogram works. Select **Graphs → Legacy Dialog → Histogram...** In the dialog box that appears, click on the age variable and then on the arrow button to the left of the Variable box. The dialog box should appear as the one below.

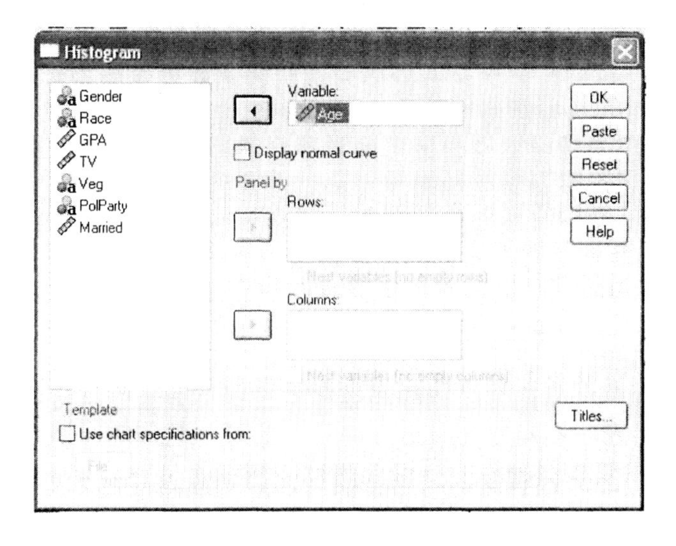

Click on OK to produce the output.

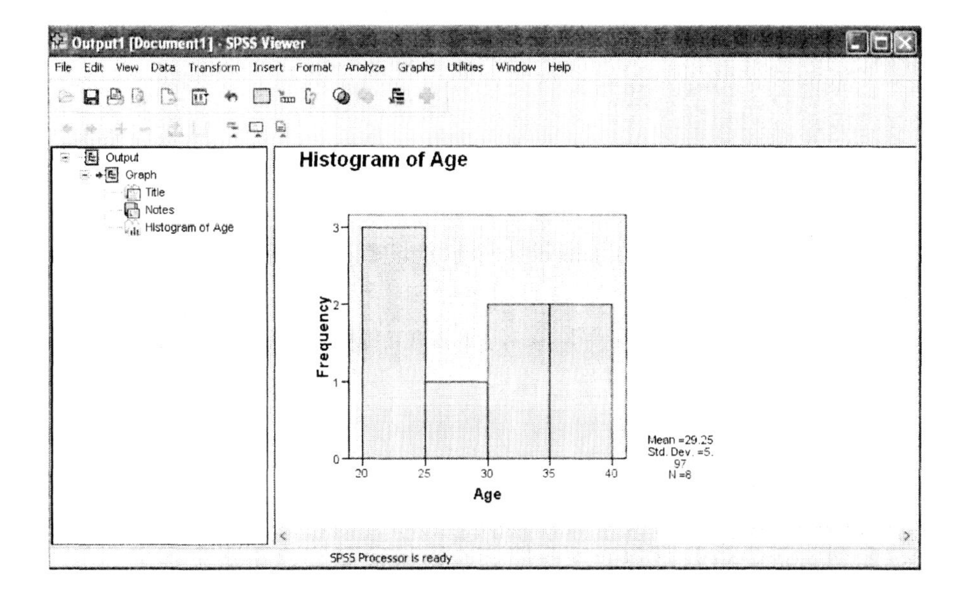

The output appears slightly different from the graph produced using the Interactive histogram command. You also need to follow a different procedure for modifying this type of output.

Modifying output

To modify any part of any output produced by SPSS all you have to do is double click on it to open an Editor. For example, double click on the "Descriptives" title and the following text editor window will appear.

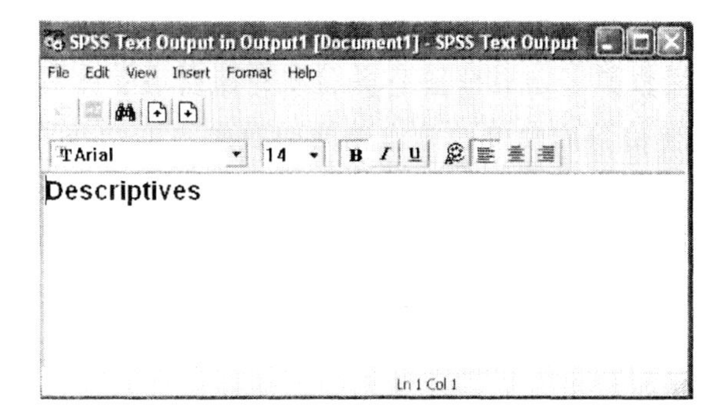

Enter a new title, such as "Descriptive Statistics for Age" and then close this window. For longer titles, press **Enter** when you want to start a new line. SPSS should automatically increase the size of the title box. If it doesn't, follow the process described below for changing the size of output. Using the same process, change the title of the histogram to something like "Histogram for Age."

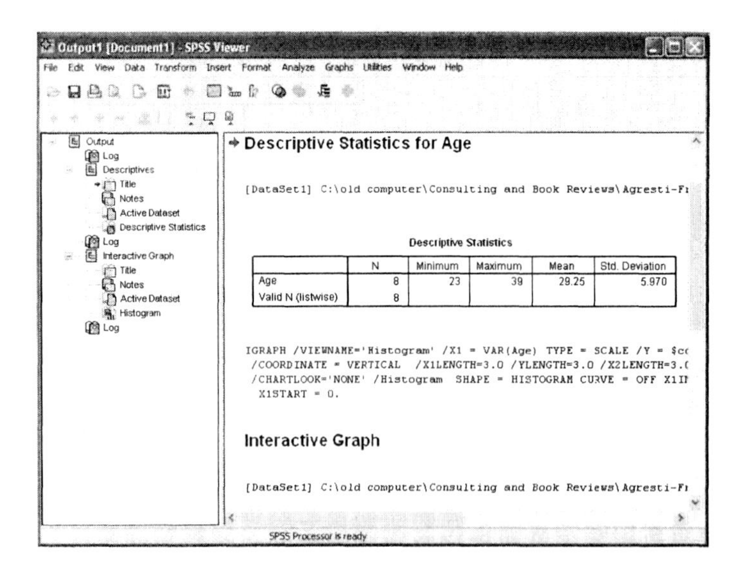

The "Descriptives" output is actually a text table, but some of the horizontal and vertical bars are hidden. You can change the height and width of the cells in the table so that the table takes up less space. Just double click on the output to open a text Editor. Two boxes appear, one called "Pivoting Trays" and the other "Formatting Toolbar," but you don't need them. Move the cursor around within the table until it changes to a double arrow. Then, drag the column or row to the desired width or height. Double click anywhere outside that output to close the Editor.

You will see how to modify a Histogram and other graphs made using the Interactive commands using the Chart Editor in the next chapter, but one thing you might want to do is resize it so that the entire graph appears in the Output Viewer window. To do this, double click once on the histogram to open a Chart Editor. Notice that menu bars form above and to the right of the graph. Now, double click within the box containing the histogram. A dialog box should appear that allows you to resize the graph. The example below shows the height and width reduced from 3.00 to 2.50. Click on **OK** and then anywhere in the right side of the window off of any output to close the chart editor. The histogram will then appear in the output window in the smaller size.

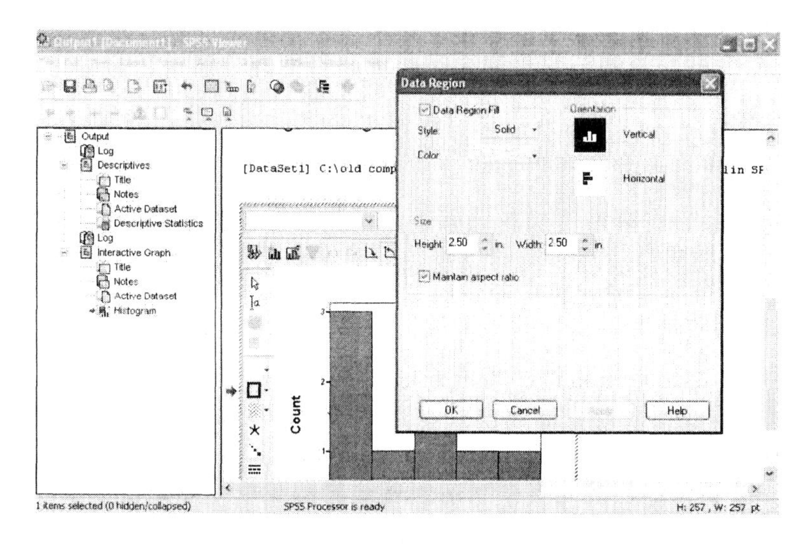

To resize a graph made using one of the Legacy Dialog commands, click on the graph once. Size boxes should appear at the corners and midway between the corners.

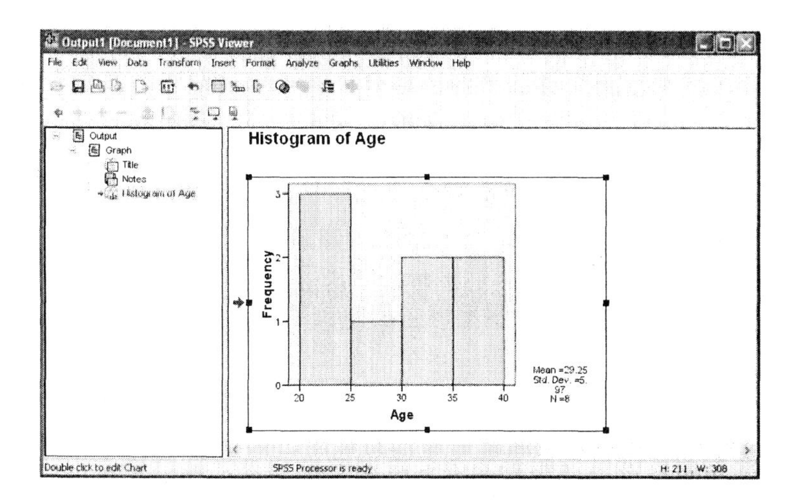

Using the mouse button grab and drag one of these boxes to resize the graph.

Printing Output

Before selecting the "Print" command from the "File" menu make sure that all of the output you want to print has been selected. For example, after you have changed a title or another piece of output, only that piece of output is selected unless you click once somewhere else in the output (but not on another piece of output). To determine which pieces of output, if any, are currently selected look at the Output Navigator. In the display below, notice that there is a shaded box around Histogram indicated that it is currently selected and only it would be printed if the print command was executed.

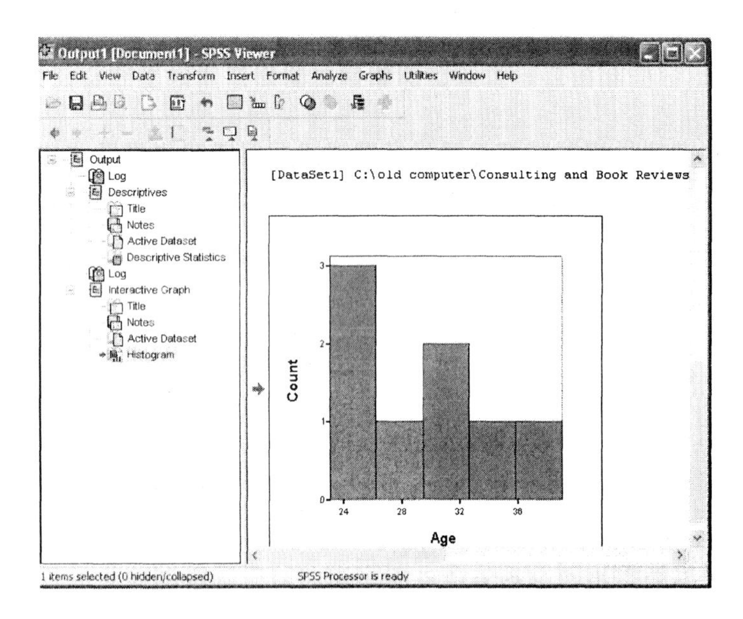

To print all of the output you can either have nothing selected or everything selected. Either way, SPSS will automatically print all of your output. To select everything click on the word "Output" at the top of the Output Navigator. All of the pieces of output in the Output Viewer will then be selected. To select nothing click anywhere in the output window off of the actual output. If you only want to print a portion of your output, just select those items in the Output Navigator. If the pieces of output you want to print are not next to each other you can hide the output in between them, or press the "Ctrl" button on your keyboard while selecting output you want to print. To print your output, select **File → Print**. Another option for printing output is described below under Saving Output.

Saving Output

You have two options for saving SPSS output. The first is to select **File → Save As...** which will save your output as an SPSS output file with the ".spo" file-type extension. If you resize graphs and insert page breaks (available under the "Insert" menu) so that titles are not separated from their output, this approach works well for producing output that might appear as an appendix to a report, for example. Another option is to save the output into a word processing document. This is a good option for when you want to include output as part of your report, rather than in an appendix, and is also the second option for printing output. Open a word processing document along with your SPSS output, then click on a piece of output to copy. You can use either the "Copy" command or "Copy Objects" commands. For graphical output you won't notice much difference, but for text output the "Copy" command will recreate the output as a text table in the document, while with "Copy Object" the output will appear as a single object that you won't be able to manipulate. After selecting "Copy" or "Copy Objects" move to the word processing document and click where you want the output to appear and select "Paste" from the "Edit" menu. For example, the output from the "Descriptives" command produced, above, along with its title, has been copied to this document below using this process. With this option you can save your SPSS output and print it later, even on a computer that doesn't have SPSS installed on it. Output saved in a .spo file can only be printed from a computer that has SPSS installed on it.

Descriptive Statistics for Age

Descriptive Statistics

	N	Minimum	Maximum	Mean	Std. Deviation
Age	8	23	39	29.25	5.970
Valid N (listwise)	8				

Exiting SPSS

Before you leave SPSS make sure that you have saved any data or output that you might want to have access to later on. This is especially important for larger data sets! To quit this or any SPSS session select **File → Exit**. If you haven't saved your data or output at this point a dialog box will appear asking if you want to save it now.

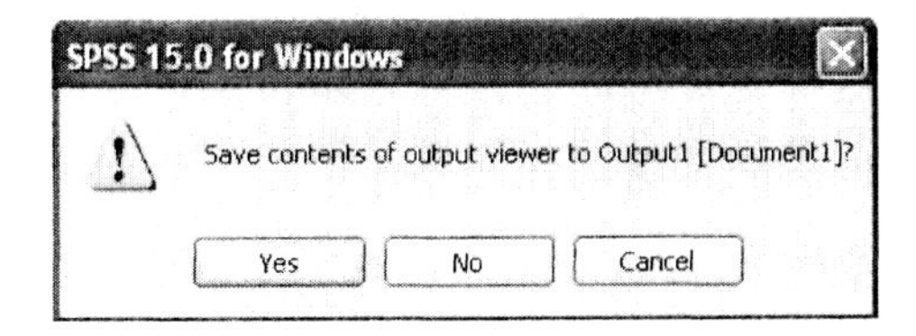

Additional Comments

In the example provided above you saw how typical SPSS commands operate and the output those commands produce. Every command you might need for an introductory statistics course is found in one of the drop-down menus and most commands have dialog boxes similar to those shown above. The rest of this manual will continue to use the convention used above for identifying the menu, command and option, that is, select **menu → command → subcommand**.

Because there are different options for entering data, some SPSS commands include options for producing the same or similar output. For example, the "married" variable in the above example could have been entered as a string variable as were the other categorical variables in that data set. As another example, if we wanted to compare the ages of the male and female students in a class, we could enter the data as two separate variables, one for the ages of the male students and the other for the ages of the female students. Or, we could enter the data as in the example data set, using one variable for the ages and another for the students' gender. Where appropriate, the different options for data entry will be described in this manual and the corresponding command options will be demonstrated.

Hints for Effective Data Analysis

The quality of the output you produce will go a long way toward helping you to communicate the information available in your data. Producing descriptive statistics and graphs is only half of the process involved in data analysis. The second half involves presenting that output to others and describing what it shows. Any output you produce should be created by keeping in mind that others will need to be able to examine it and understand what is being displayed. Here are some hints as you begin learning how to analyze data using SPSS:

1. Always check for data entry errors; any observations that appear as outliers in a histogram or box plot should be double checked with the recorded data.
2. Use descriptive variable names and variable labels, where appropriate.
3. Use value labels for numerically coded categories.
4. Include descriptive titles on all output that is part of a report or project.
5. Provide meaningful labels on all axes of graphs and include units and a starting point.
6. Provide a legend if needed.
7. Avoid any features or unnecessary output that distract from the data, such as unusual colors or shading patterns in graphs.
8. Use graphs that are appropriate for the type of data and how it was collected.
9. Use the same scales on graphs that will be compared, or place the graphs on the same axes.
10. Use proportions instead of frequencies for comparing groups that are of different sizes.
11. Use multiple statistics and graphs to showcase your data – each statistic and graph provides different information about the data.

Chapter 2 Exploring Data with Graphs and Numerical Summaries

Working with Categorical Data

Recall from the previous chapter that you have two options for entering categorical data. You can enter the data as a "String" variable, or you can enter it as a coded "Numeric" variable. In this section you will explore both options.

Pie Charts for Tabulated Data

When data has already been tabulated, as in Exercises 2.11 and 2.13, it is very easy to produce a pie or bar chart using SPSS. The key is to use a "weight" variable to let SPSS know how many observations are in each category. The data for Example 2.11 is in the form of a pie chart that shows the regional distribution of weather stations and can be summarized in the following table:

Region	Count	Percent
Southeast	67	18.7
Northeast	45	12.5
West	126	35.1
Midwest	121	33.7

We will enter the data in SPSS in two columns, one for the region and the other for the count in each region.

Data Entry Option 1: Category as a string variable

In a new SPSS Data Editor create two variables, "region" as a string variable and "count" as a numeric variable. You will have to increase the Width for "region" from 8 to 9 (under the Variable View) if you want to type "Southeast" and "Northeast" entirely. Enter the data from the table above so that your Data Editor appears as the one in the following display.

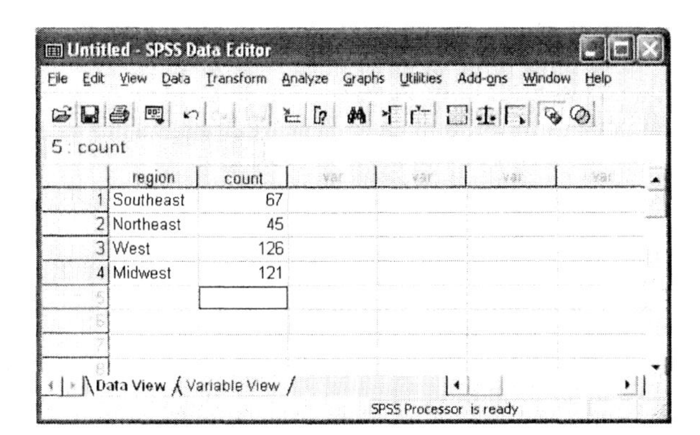

To let SPSS know that the variable "count" represents the number of observations in each category defined by the variable "region," select **Data → Weight Cases....** In the dialog box that appears, click on "Weight cases by" and enter the "count" variable in the "Frequency Variable" box.

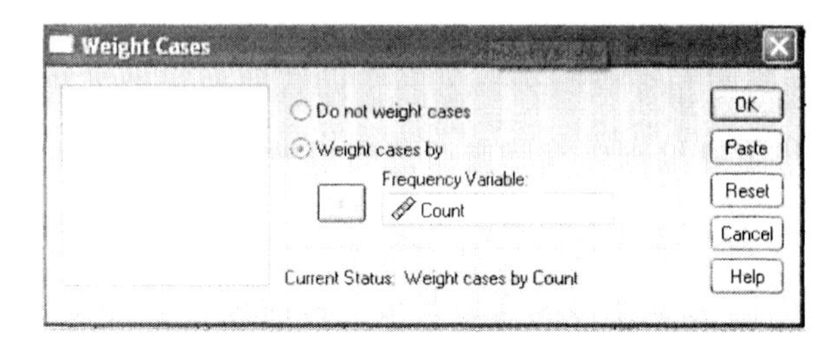

Click on **OK** to complete the process. To produce a pie chart for this data, select **Graphs → Interactive → Pie → Simple…** and in the dialog box that appears drag Region to the "Slice by" box. SPSS has a default or built-in variable called "Count($count)" in the Slice Summary box. This is not the same variable as the Count variable we entered.

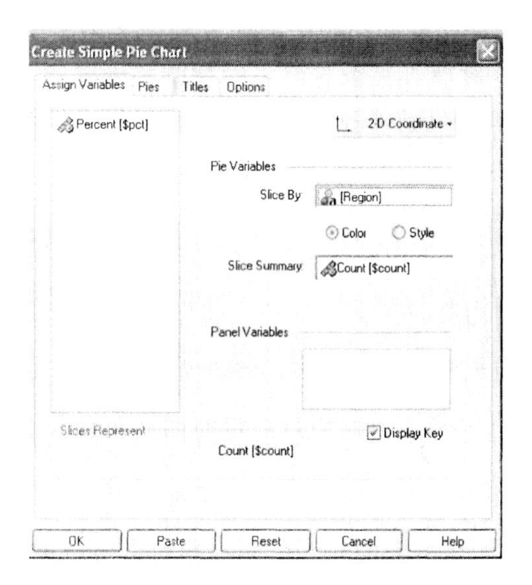

Click on **OK** to produce the graph.

The Output View will show the following pie chart.

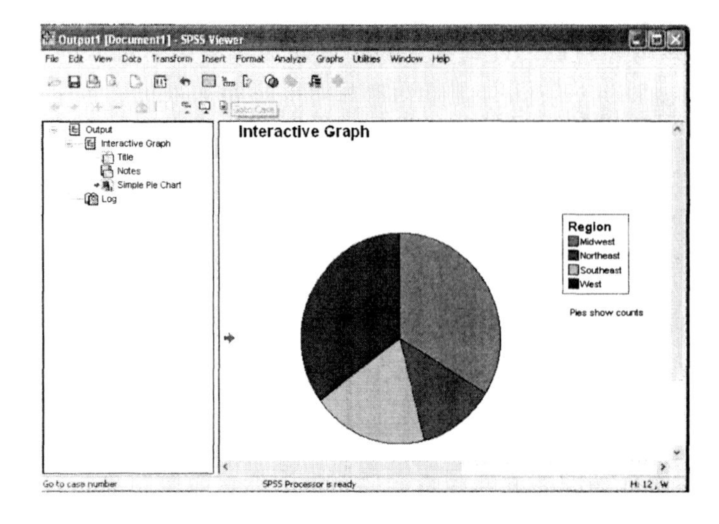

Notice that the slices are arranged alphabetically. If you don't like the color scheme selected by SPSS it is very easy to change. Double click anywhere on the pie chart to open the **Chart Editor**. Then, double click on one of the colored boxes in the legend. A dialog box will appear allowing you to change the color of each slice. Click on each of the other categories in this dialog box to change the color of that category.

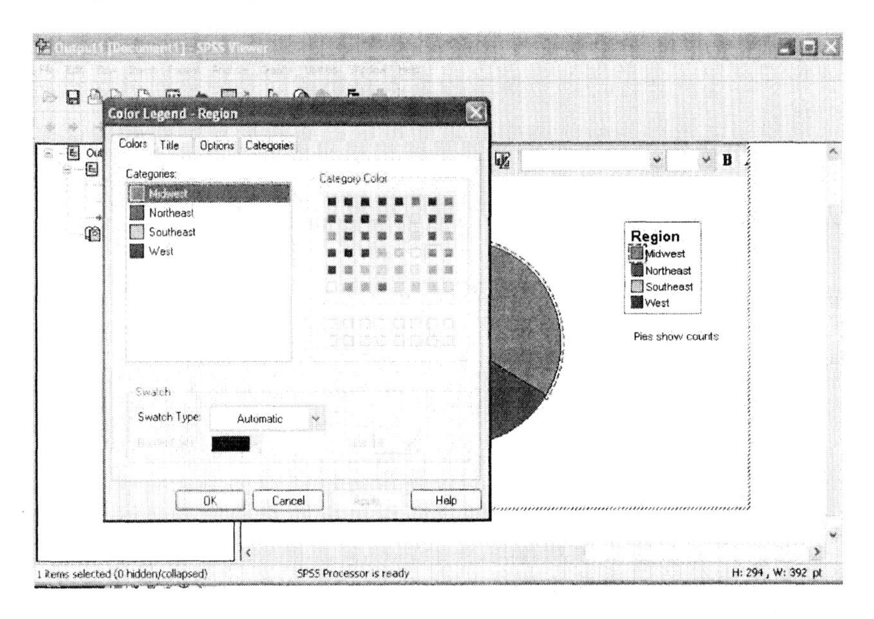

You can also change the location of the legend using the Title tab of this dialog box. If you double click on one of the slices in the pie chart a dialog box will appear that provides options for changing the order in which the slices are presented in the chart and where the angle of the first slice starts.

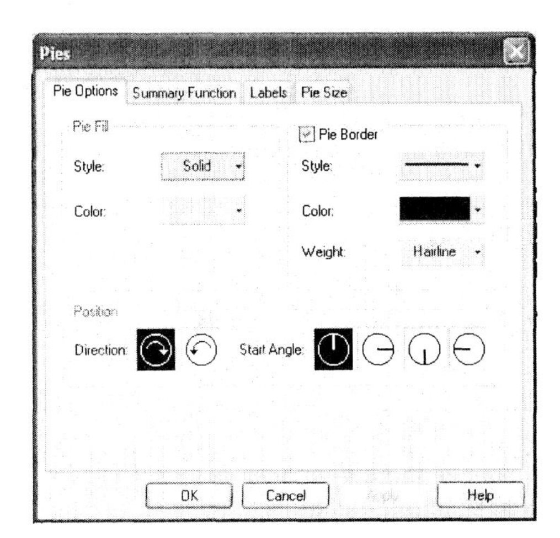

The Labels tab of this dialog box provides options for labeling each slice with its category, count and percent values. To resize the pie chart, follow the directions given in Chapter 1 for resizing a histogram. When you are done making changes, click anywhere off of the graph to close the Chart Editor.

If you don't want to use Count as a weighting variable, there is another option for producing this same output. Select **Data → Weight Cases...** and click on "Do not weight cases" and then on **OK**. Select **Graphs → Interactive → Pie → Simple...** and place Region in the Slice By box and Count (the variable we entered in the Data Editor) in the Slice Summary box. Then click on **OK** to produce the output.

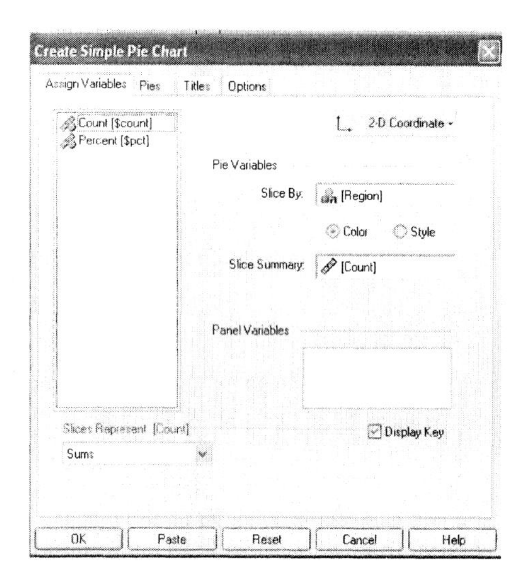

Data Entry Option 2: Category as a coded numeric variable

To code "region" numerically, use the "Numeric" type in the Variable View and enter the following value labels following the process described in the first chapter of this manual.

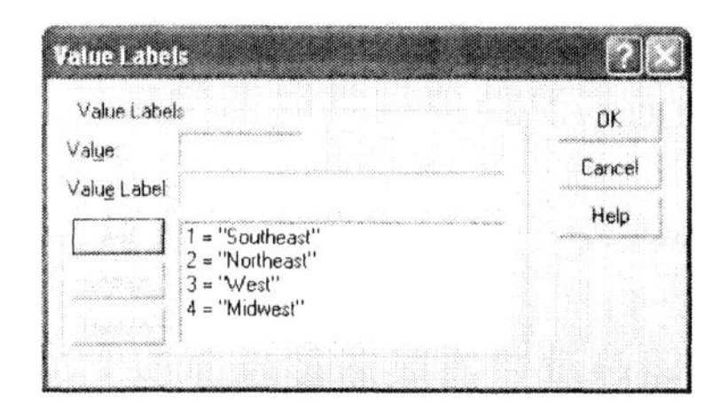

Click on the Data View tab and enter the data the same as above, except enter the values 1, 2, 3 and 4 for each region. Select **View → Value Labels** so that the Data View appears as follows:

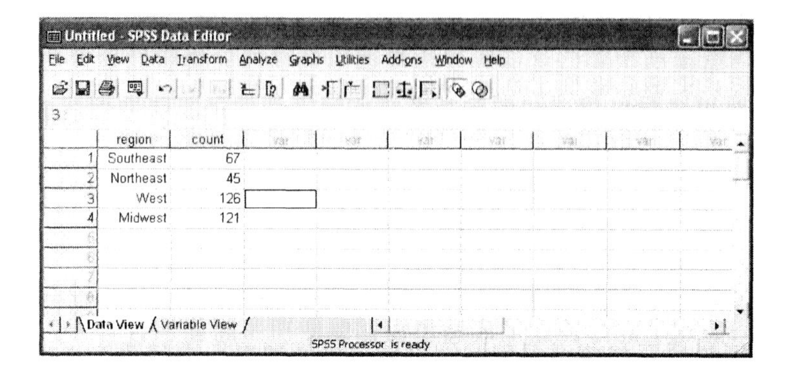

The data set appears exactly the same as it did for Option 1 but the process for producing a pie chart is slightly different. Select **Graph → Legacy Dialog → Pie…** and then select the "Values of individual cases" option.

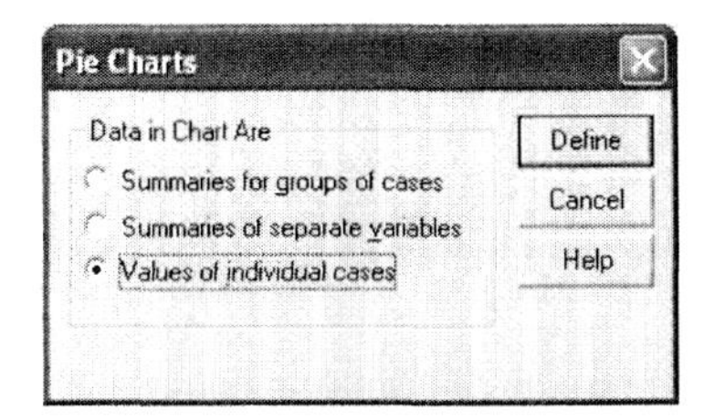

Click on **Define**. Enter "Count" in the "Slices Represent:" box and then click on "Variable" under "Slice Labels." Enter "region" in the box below "Variable" and click on **OK**.

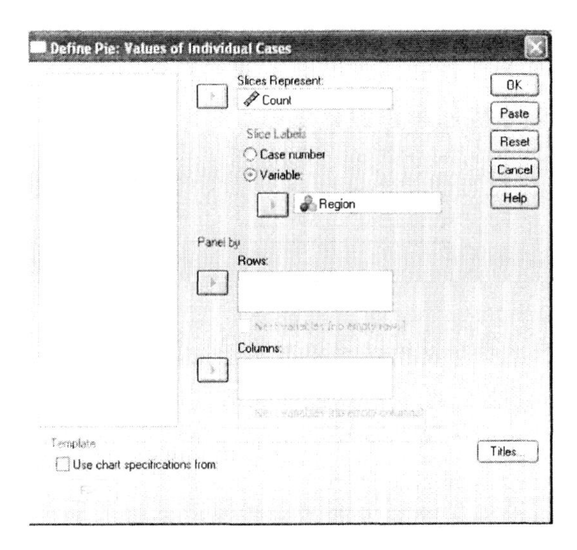

The resulting pie chart differs from the first pie chart in the order that the slices are presented. When "Region" was a string variable, SPSS ordered the slices alphabetically, but when it is coded numerically the slice coded 1 (Southeast) appears first.

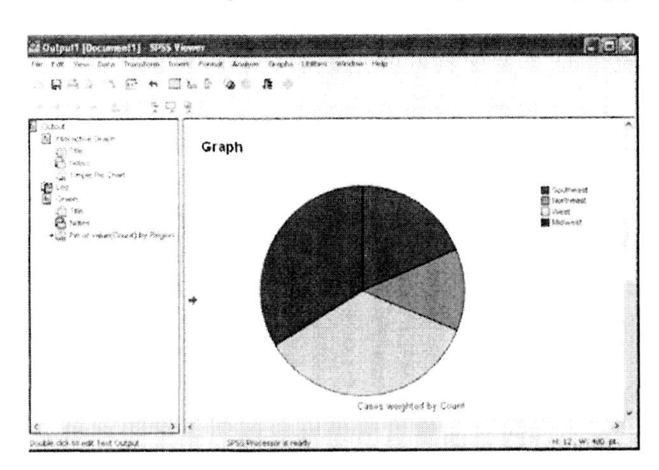

To modify this pie chart double click on it to open a Chart Editor.

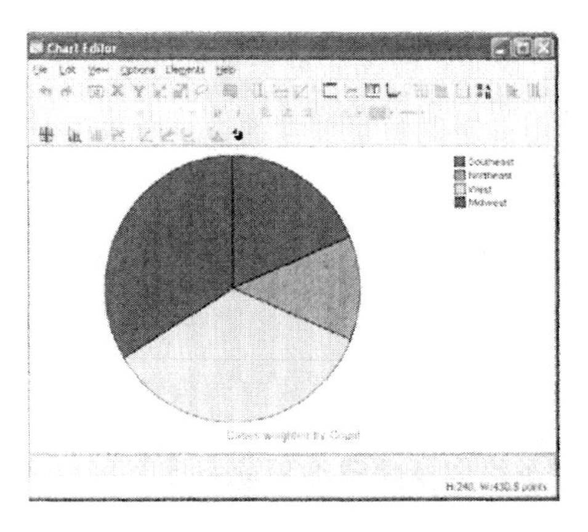

This Chart Editor is different than the editor for graphs made using the Interactive graphs options. For graphs made using the Legacy dialogs, the Chart Editor appears as a new window with a new version of the graph for you to change. To change the color of one of the slices, click once on that slice to highlight it and then double click on it to open a Properties dialog box. Choose the new color from the palette that appears and click on **Apply** and then **Close** to make the change.

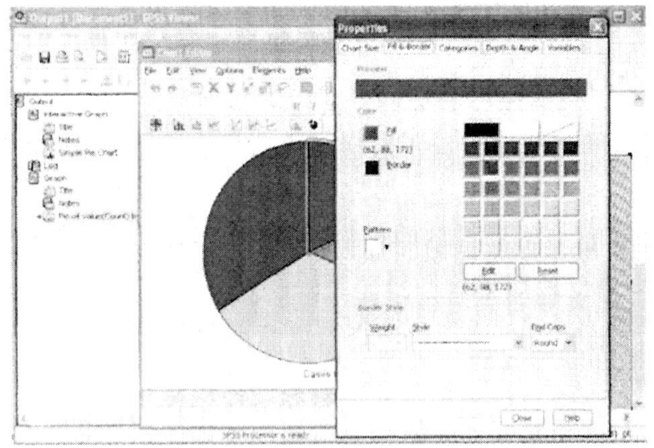

Close the Chart Editor when you are done making changes. The graph in the actual output will now show the changes you made in the Chart Editor.

Which option to use?

Both options are easy to use. Indicating that the slices represent the "count" in each category when you code the data numerically serves the same purpose as indicating that "count" is a weighting variable when a variable is entered as a string variable. If the category names are long, data entry might be easier using numerical coding. The graphs are nearly identical so the appearance of the output isn't really an issue. If you want the slices to appear in a non-alphabetical order that is very easy to do by using an appropriate coding scheme. For example, if you want the slices to appear largest to smallest clockwise around the circle, then you would code Midwest = 1, West = 2, Southeast = 3, and West = 4. Both data entry options include options for modifying the color, size and labeling of the resulting graph. However, the commands for some analyses, for example, for comparing the means of two groups, require that variables for defining the groups are coded numerically. You will learn more about this later.

Bar Charts for Tabulated Data

The command for making a bar chart operates exactly the same as the command for making a pie chart. If you enter a categorical variable using the string option choose the "Summaries of groups of cases" option and use the "Weight Cases" command to indicate which variable provides the recorded counts. If you code it numerically choose the "Values of individual cases" option. Let's make a bar chart for the data from Exercise 2.13 on the frequency of shark attacks worldwide. The data for this exercise is given in the table below:

Region	Frequency
Florida	365
Hawaii	60
California	40
Australia	94
Brazil	66
South Africa	76
Reunion Island	14
New Zealand	18
Japan	4
Hong Kong	6
Other	206

Enter the data set in SPSS using either option described above. To open another Data Editor, select **File → New → Data.** After you have entered the new data set (don't forget to select Frequency as a weighting variable using the Data menu), select **Graph → Interactive → Bar...** and then select the appropriate option for how you entered the data. For "Region" as a string variable, SPSS will produce the following output, with the categories listed alphabetically:

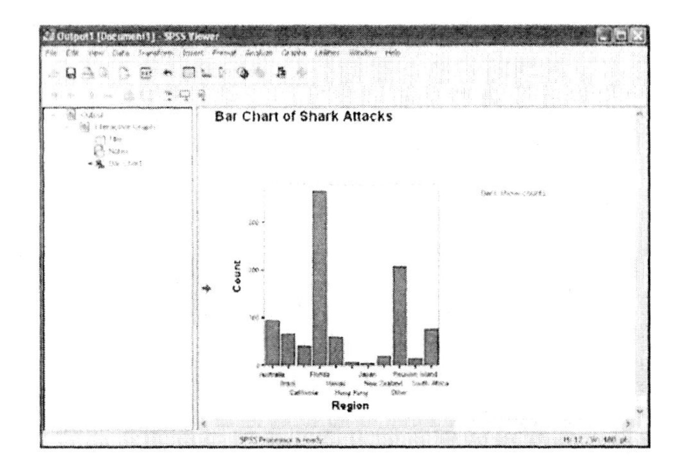

The size of the bar chart above was decreased to make the display fit the size of the window. Note that the vertical axis is labeled "Count." If you coded the data numerically, you will have to use the Legacy Dialog pie chart command. Select **Graph → Legacy Dialogs → Bar...** Click on the Values of individual cases option and then on **Define**.

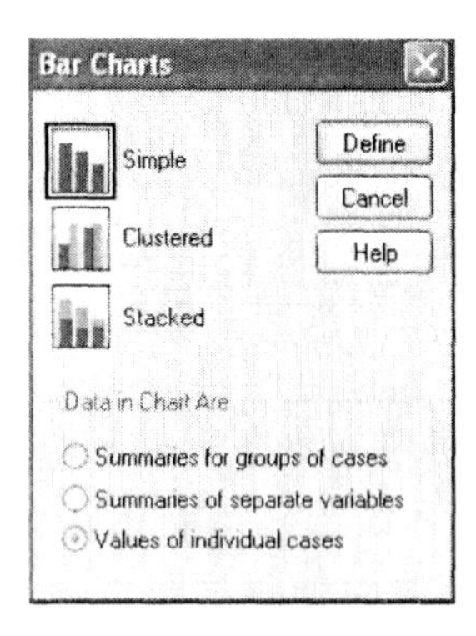

the vertical axis will be labeled "Value Frequency" and the order of the bars will depend on the coding scheme you used. Click on Variable underneath Category Labels and enter the region variable there. Enter the frequency variable in the Bars Represent box, as shown below.

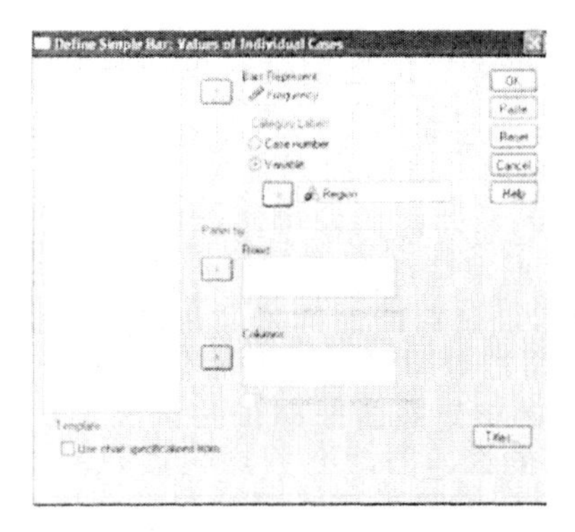

The output is shown below. Notice that the label on the vertical axis is Value Frequency and that the bars appear in the order of the numerical coding used.

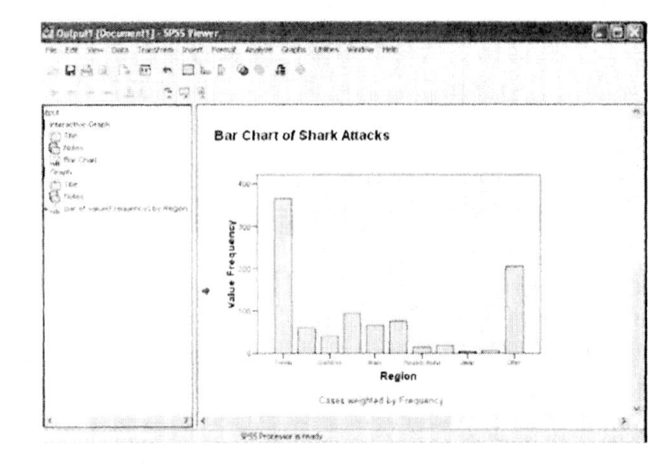

Frequency Tables and Graphs for Non-Tabulated Categorical Data

SPSS can also organize and display categorical data that hasn't already been tabulated. To demonstrate this, let's use the "gender" variable from the "fla_student_survey" data set from your text CD. Access this data file in SPSS by selecting **File → Open → Data** and locating your text CD in the "Look in" box. Find the file among those listed and double click on it. A portion of this data set is displayed below.

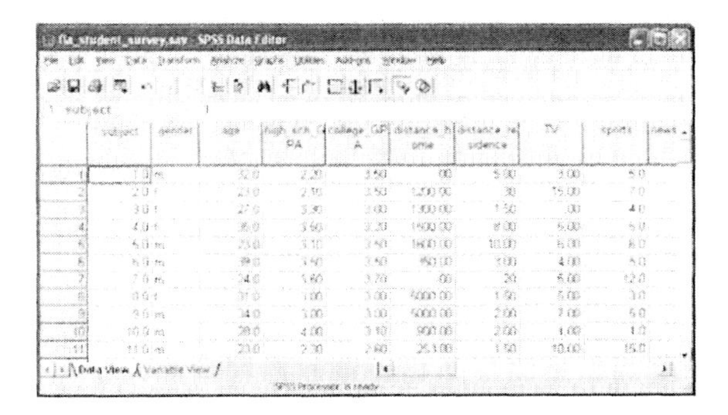

To find the frequency table for this variable, select **Analyze → Descriptive Statistics → Frequencies...**, then select "gender" and click on **OK**. The following output will appear:

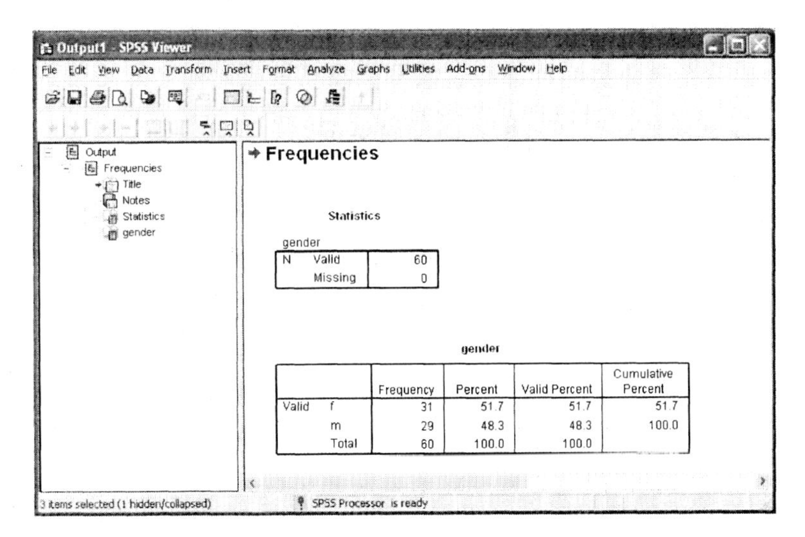

Statistics

gender

N	Valid	60
	Missing	0

gender

		Frequency	Percent	Valid Percent	Cumulative Percent
Valid	f	31	51.7	51.7	51.7
	m	29	48.3	48.3	100.0
	Total	60	100.0	100.0	

This output has three parts that are currently showing. The first is the title "Frequencies," which you should change to something like "Frequency Table for Gender." The second is a "Statistics" portion where SPSS reports how many "valid" or non-missing observations were analyzed. This part of the output can be omitted if there aren't any missing values in the data set. Or, you can temporarily "hide" it by double clicking on its icon in the Output Navigator. The last part of the output contains the actual frequency distribution. Here we see that 51.7% of the students surveyed were female. Since "gender" was entered as a String variable, the categories are listed alphabetically in this output.

You can make a pie or bar chart using the "Summaries for groups of cases" option of these graph commands, or you can use an option available in the "Frequencies" command. You might have noticed the "Charts" button in the Frequencies dialog box, shown below.

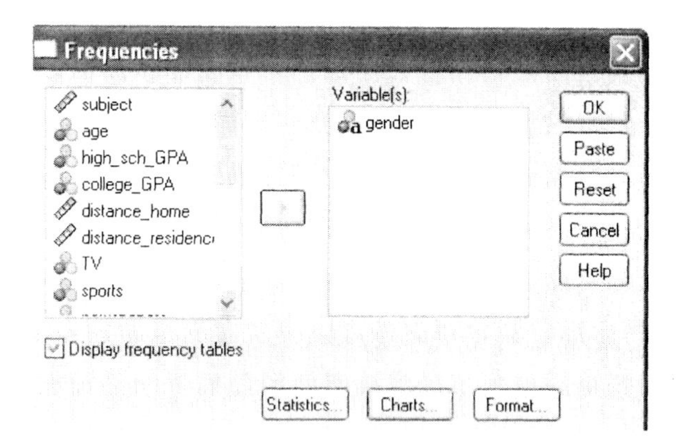

If you click on **Charts** a dialog box will appear that allows you to produce a bar chart or a pie chart along with the frequency table. Some of the commands under the "Analyze" menu also include options for producing graphs. The "Statistics" button in the "Frequencies" dialog box allows you to produce some statistics along with a frequency distribution of a data set.

Note that each of these commands will also work if the categorical variable is entered as a coded numeric variable.

Working with Quantitative Data

For quantitative data, you will now see how to make a dot plot, a stem-and-leaf plot, and histograms. If the data was collected over time you can also make a time plot.

Dot Plot

SPSS Version 15.0 includes a separate command for producing dot plots, but it is not the same kind of dot plot as in your text. Version 13.0 combined the dot plot with the scatter plot command. To demonstrate how to make a dot plot in SPSS we will use the "SODIUMmg" variable of the "cereal" data set from your text CD. Access this data and select **Graphs → Legacy Dialogs → Scatter/Dot...** In the dialog box that appears click on **Simple Dot** and then on **Define**.

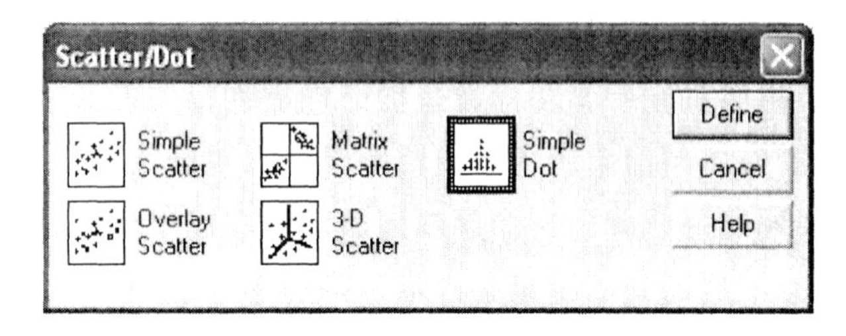

Enter "SODIUMmg" in the "X-Axis Variable" box.

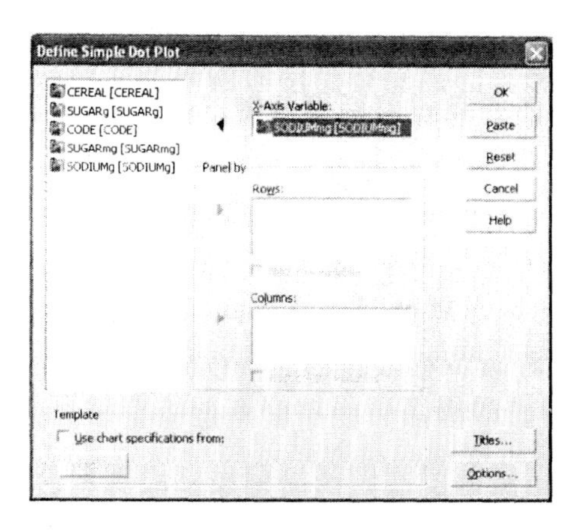

Click on **OK** to produce the output.

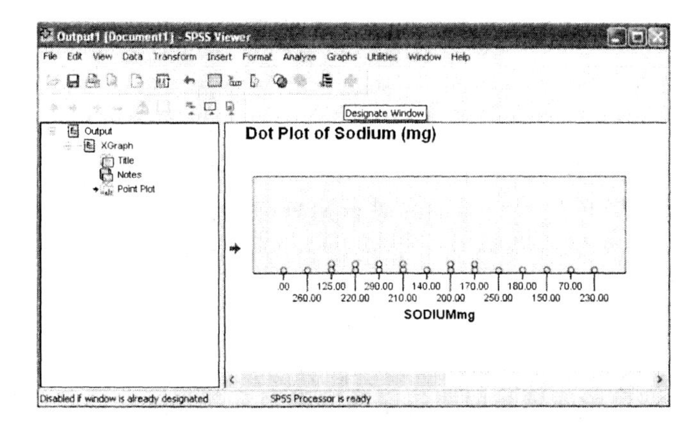

The size and fill of the circles can be modified using the Chart Editor. With the Chart Editor open, click once on one of the circles in the graph, then select **Edit → Properties**... Click on the "Marker" tab and enter a new size in the "Size" box. Click on the box next to "Fill" under "Color" and then on the black box to fill in the circles.

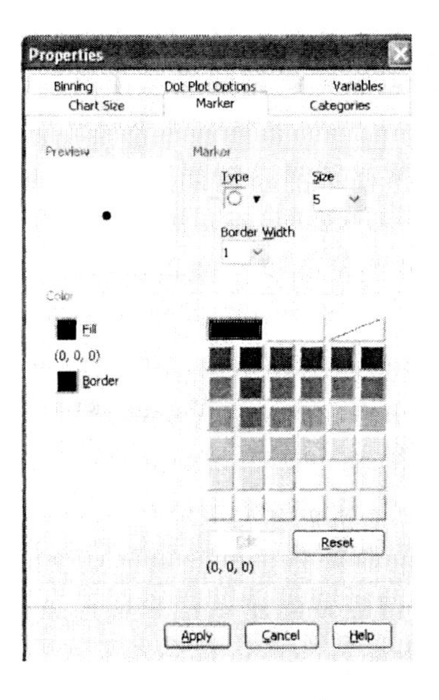

Click on **Apply** and then on **Close** to complete the change, then close the Chart Editor.

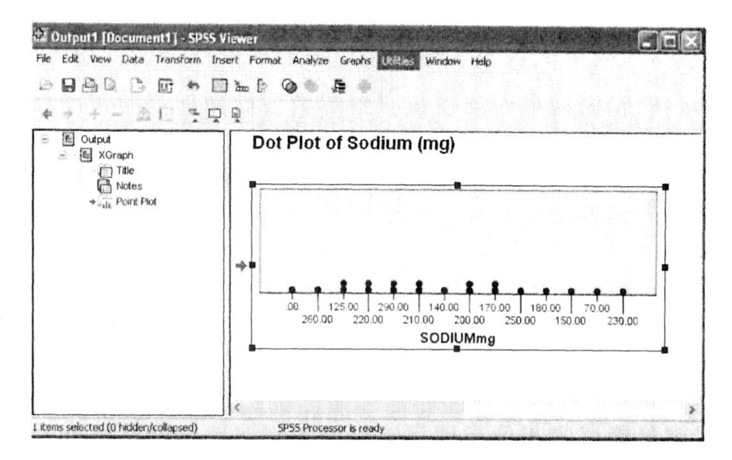

Stem-and-Leaf Plot

For this example we will use the data for Exercise 2.15 of final eBay prices (in dollars). The data is shown below and a portion of it is shown in the SPSS Data Editor below that.

235, 225, 225, 240, 250, 250, 210, 250, 249, 255, 200, 199,
240, 228, 255, 232, 246, 210, 178, 246, 240, 245, 225, 246, 225

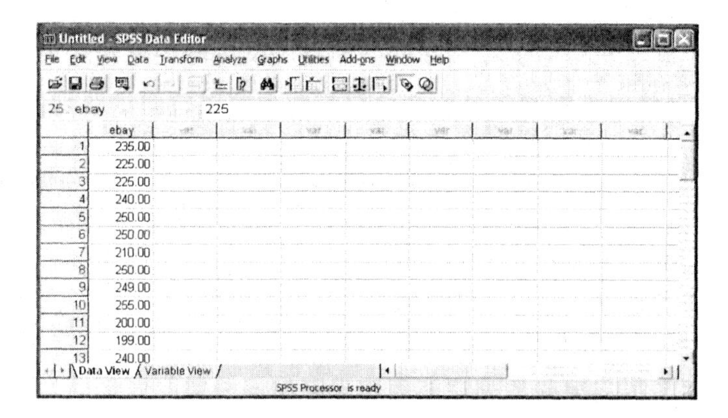

SPSS does not include stem-and-leaf plots with the other graphical methods under the "Graphs" menu. Instead it is an option under the "Explore" command under the "Analyze" menu. You will learn more about this very useful command later on.

To make a stem-and-leaf plot select **Analyze → Descriptive Statistics → Explore...** In the dialog box that appears, enter "eBay" in the "Dependent" list, and then click on **Plots** underneath "Display" in the lower left corner so that only graphical output will be produced.

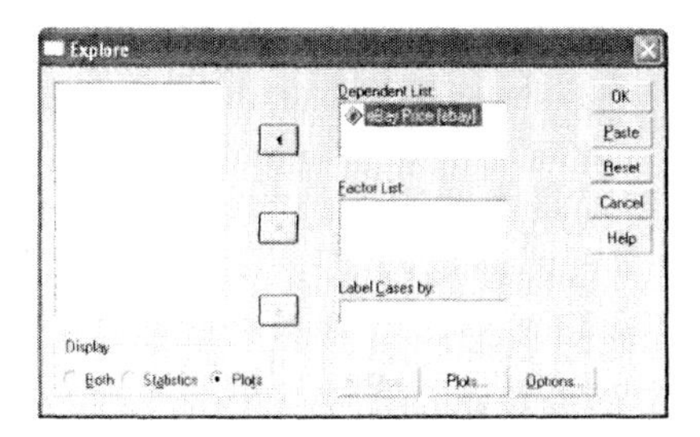

One of the default graphs produced by this command is a stem-and-leaf plot. The other default graph is a box plot. To produce only the stem-and-leaf plot, click on the "Plots" button on the lower right side of the dialog box and select "None" underneath "Boxplots."

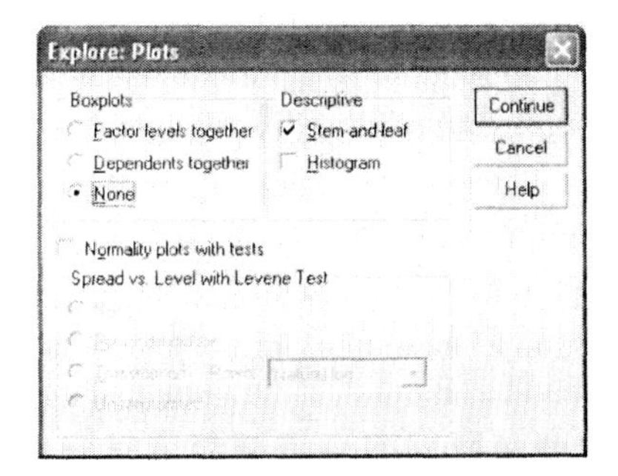

Click on **Continue** and then on **OK** to produce the stem-and-leaf plot. The following output will appear.

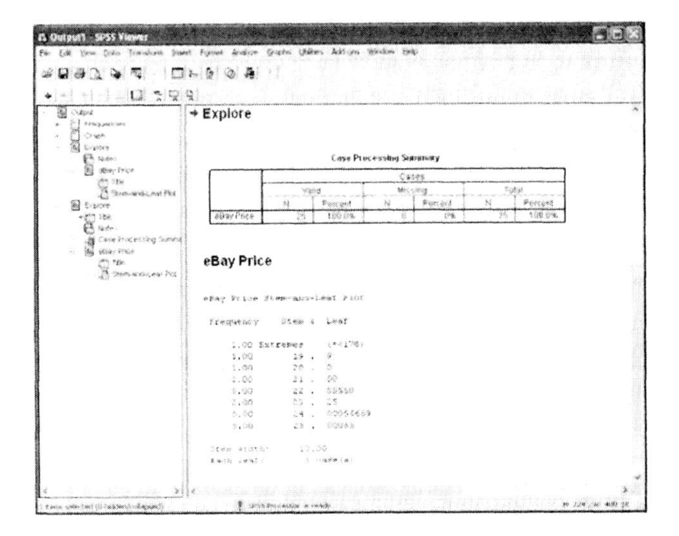

The "Explore" command includes a "Case Processing Summary" that reports the number of valid observations included in the analysis. Like the "Statistics" portion of the "Frequencies" command output, the "Case Processing Summary" may be omitted if there aren't any missing values in the data set. The stem-and-leaf plot is the last piece of output produced. The plot is actually text output that can be modified by double clicking on it to open a text editor. The first column on the left indicates how many "leaves" are associated with each "stem." Notice that the first stem is listed as "Extremes" with an observation "=< 178" and that the stem 18 is omitted since there aren't any observations between 180 and 189 in value.

Histograms

The command for making a histogram is one of the options available under the "Graphs" menu. You might have also noticed that it is an option under the "Frequencies" and "Explore" commands. Let's use the "Histogram" command under the "Graphs" menu to make a histogram of the New York annual average temperatures data for Exercise 2.27. The data for this exercise is the "central_park_yearly_temps" data set on the text CD, of which a portion is shown below.

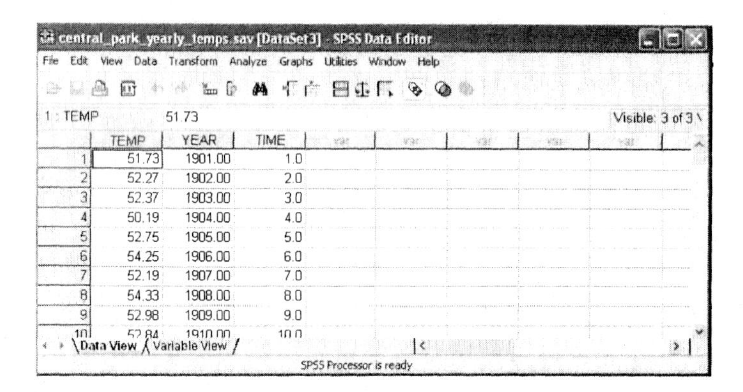

Select **Graphs → Interactive → Histogram...** and enter the "TEMP" variable in the box on the horizontal axis in the dialog box.

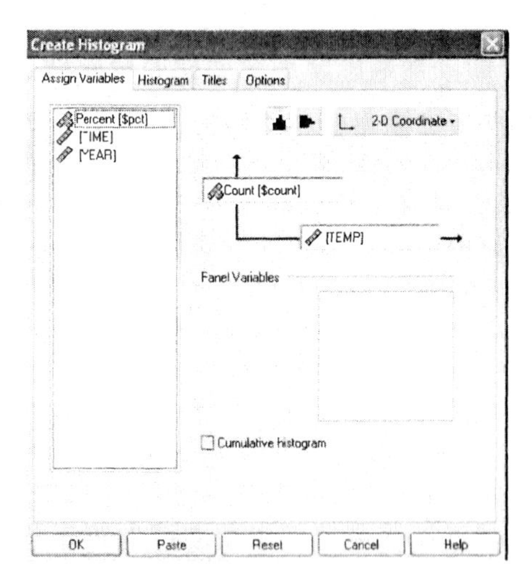

Click on **OK** to produce the graph. Notice that the graph is slightly different from the one in your text. More about this below.

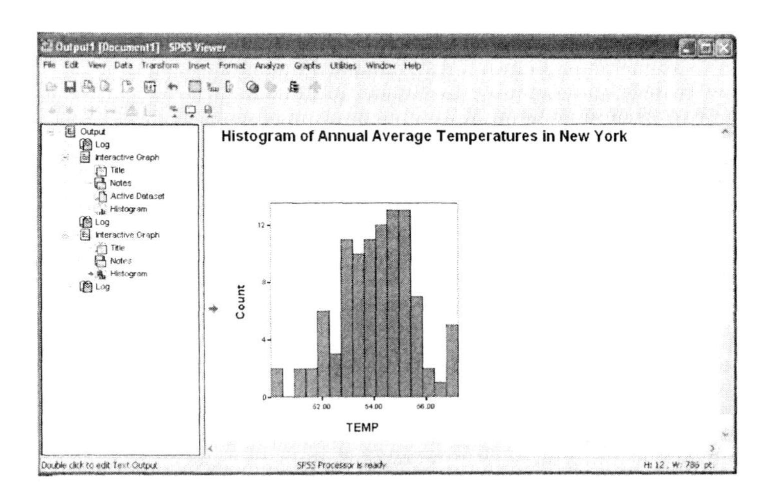

Changing the Number of Intervals and the Scale of a Histogram

The histogram above is an example of a "default" histogram made by SPSS. The "Histogram" command has a "rule" created by the SPSS programmers for determining how many intervals to produce, how wide they are, and where to start the first interval. It turns out that the "rule" used by SPSS didn't exactly match the "rule" used by the authors of your text book. The rule followed by SPSS may not produce the histogram you want or the "best" histogram for displaying your data. It is usually a good idea to make more than one histogram of the same data set to explore how changing the number of intervals and how wide they are affects the apparent shape of the distribution. The easiest way to make additional histograms is to make a copy of the default histogram. You can then modifying the copy using the Chart Editor. To make a copy of any piece of output click once on it and then select **Edit → Copy** and then **Edit → Paste After**.

You now have two copies of this histogram, the original one to leave as is and a second one to modify. Double click on the second graph to open the Chart Editor. To change the number of bars double click anywhere on the bars of the histogram. In the dialog box that appears, click on the "Interval Tool" button.

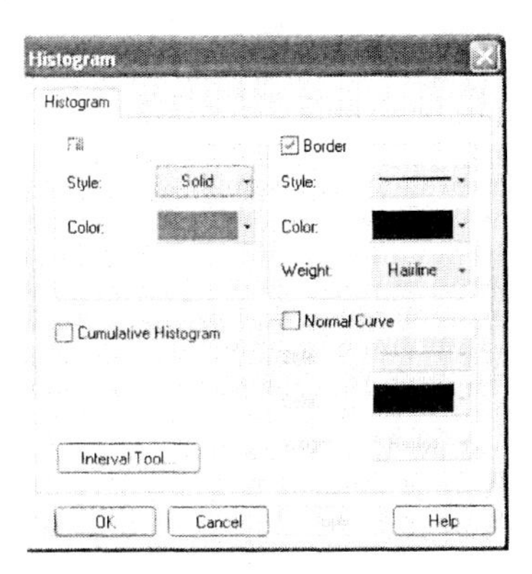

In the dialog box that appears you will see the default number of intervals that SPSS used, in this case 16. Grab the button on the axis to increase or decrease the number of intervals. The histogram will change automatically. You can also use the left and right arrow keys if you find it difficult to get the exact number of intervals you want by grabbing the button. If you click on the scroll bar beside the "Set" box you will see that you can also change the interval size and the starting point for the first interval.

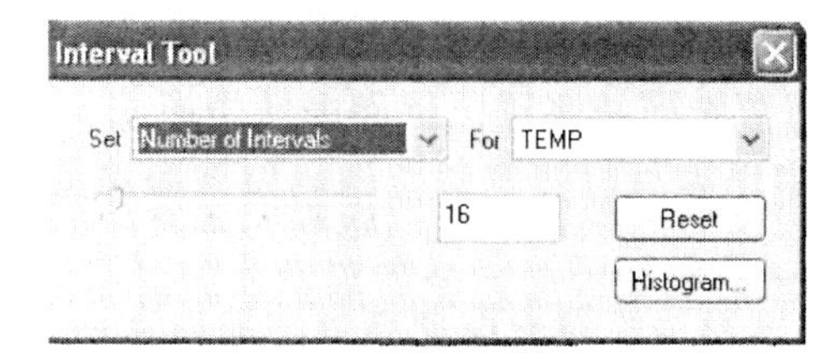

The same data set but with 8 intervals is shown below. By decreasing the number of intervals we are producing a histogram with a smoother shape, but be careful not to "over smooth" by using too few intervals!

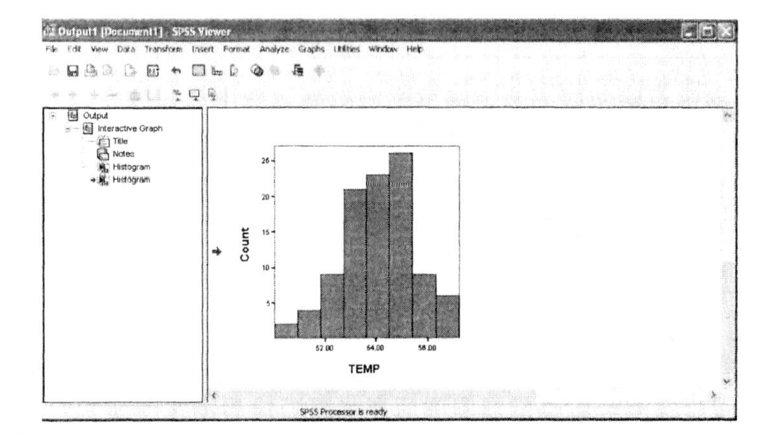

To change the scale on the horizontal axis, double click on one of the numbers below the x-axis while the Chart Editor is active. In the dialog box that appears, click on the check marks beside Minimum, Maximum, and Tick Interval and enter new values, as shown below.

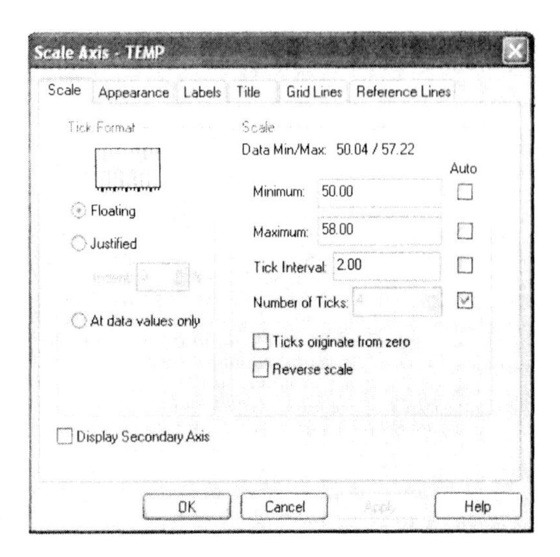

Try different values until you find ones that match up with the histogram bars. The modified histogram is shown below.

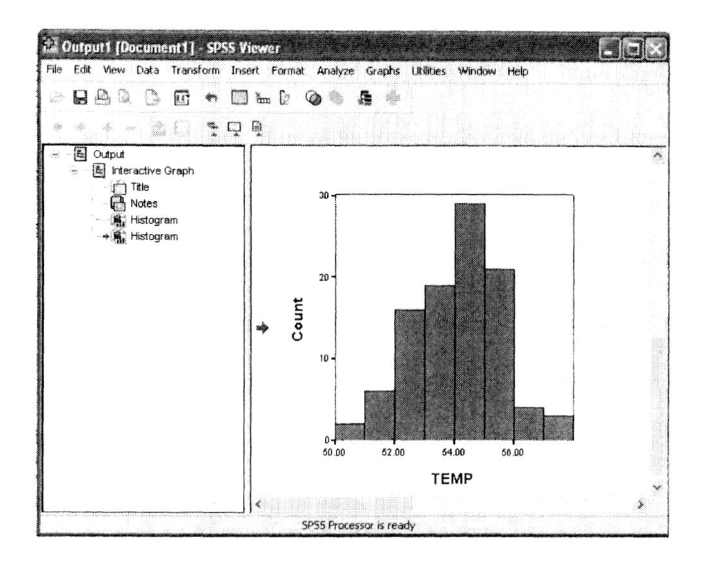

Time Plots

When data is collected over time (or some other dimension such as distance) a time or line plot can be useful for showing trends or other changes in the "process" that is producing the data. The "Line" command in the "Graphs" menu will produce a time plot. We will use the "newnan_temps" data file for Exercise 2.27 from your text CD for this demonstration. Access this data file and select **Graphs → Interactive → Line**... The following dialog box will appear.

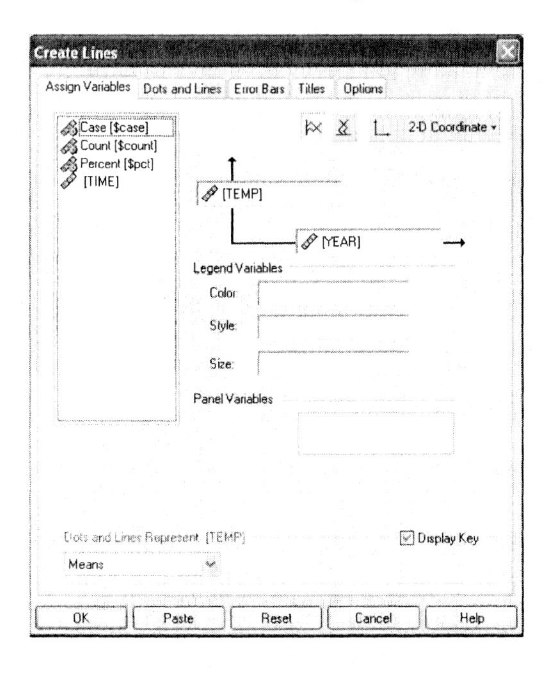

Replace the built-in variable "Count($count)" with the TEMP variable in the vertical axis box and place YEAR in the horizontal axis box, as shown above. Click on OK to produce the output.

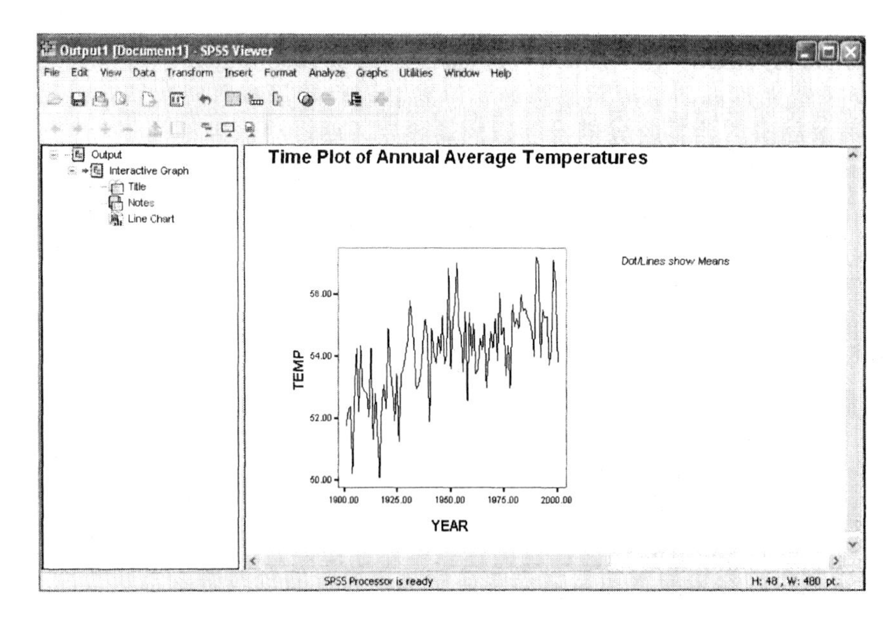

To change the scale on the horizontal axis so that every ten years is listed, rather than every twenty five years, double click on the graph to open the Chart Editor, then double click on one of the numbers below the x-axis. The same dialog box that appeared when you changed the number of bars in the histogram will appear. This time, increase the number of tick marks from 5 to 10 by clicking on the check mark next to the "Number of Ticks" box and entering the value 10. You can also use the scroll bar there to increase or decrease the number of tick marks. Change the minimum and maximum values as shown below.

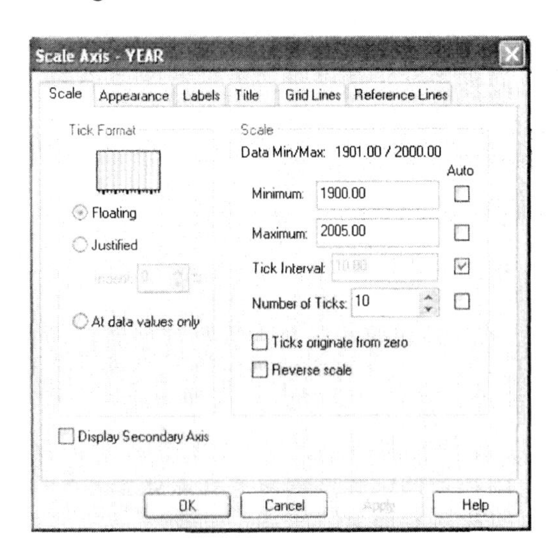

Click on **OK** to make the change, then close the Chart Editor.

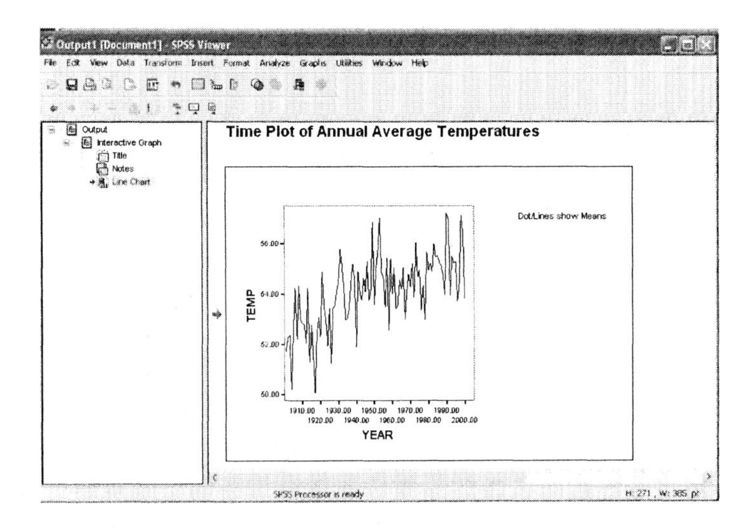

Descriptive Statistics

You have several options for finding descriptive statistics with SPSS using the "Descriptive Statistics" command under the "Analyze" menu: Frequencies, Descriptives, and Explore. "Descriptives" is the most basic command and was demonstrated in the first chapter; the default output for this command includes only the minimum, maximum, mean, and standard deviation. Under the "Options" button in the dialog box for this command you can produce some additional statistics.

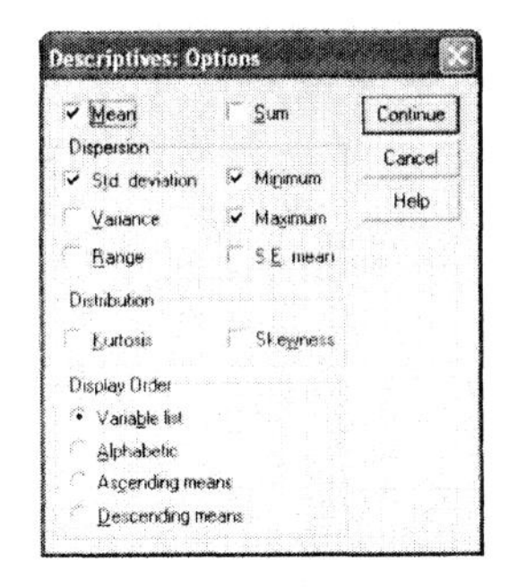

Click on any of the default statistics (the ones that are already checked) that you don't want to find to un-select them. The "Frequencies" and "Explore" commands offer more choices than "Descriptives" does.

Mean and Median

Let's demonstrate the "Frequencies" command using the data from Example 11 in Chapter 2. The data for this exercise consists of the per capita carbon dioxide emissions in 1999 by country. The values for the eight largest countries by population size are given below, in metric tons per person:

Country	China	India	U.S.	Indonesia	Brazil	Russia	Pakistan	Bangladesh
CO2	2.3	1.1	19.7	1.2	1.8	9.8	0.7	0.2

Enter this data in SPSS as shown below.

	CO2
1	2.3
2	1.1
3	19.7
4	1.2
5	1.8
6	9.8
7	.7
8	.2

Select **Analyze → Descriptive Statistics → Frequencies...** and enter CO2 in the "Variable(s)" box. Click on the "Statistics" button and select the statistics you want to produce. For this example, we will find just the mean and median for CO2.

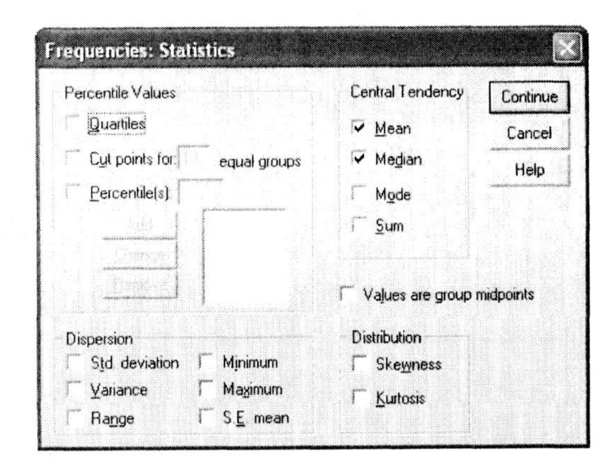

Click on **Continue** and then click on the "Display frequency tables" box so that the frequency distribution won't be produced. With only 7 observations, the frequency distribution will not be useful here.

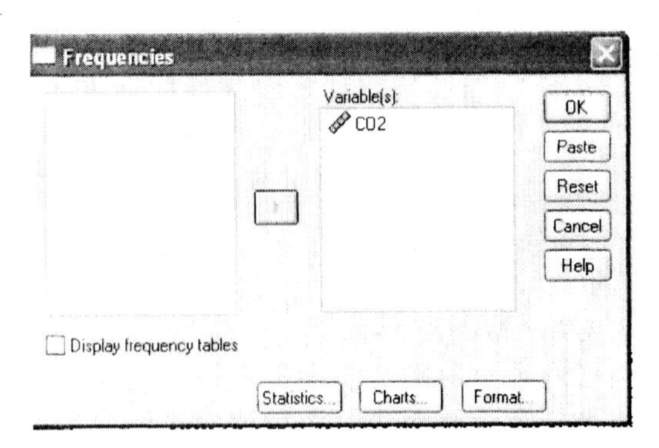

Note that you could click on the "Charts" button to also produce a histogram for this data set. Click on **OK** to produce the output. The following display shows the default output from the "Descriptives" command along with the mean and median output from the "Frequencies" command for comparison.

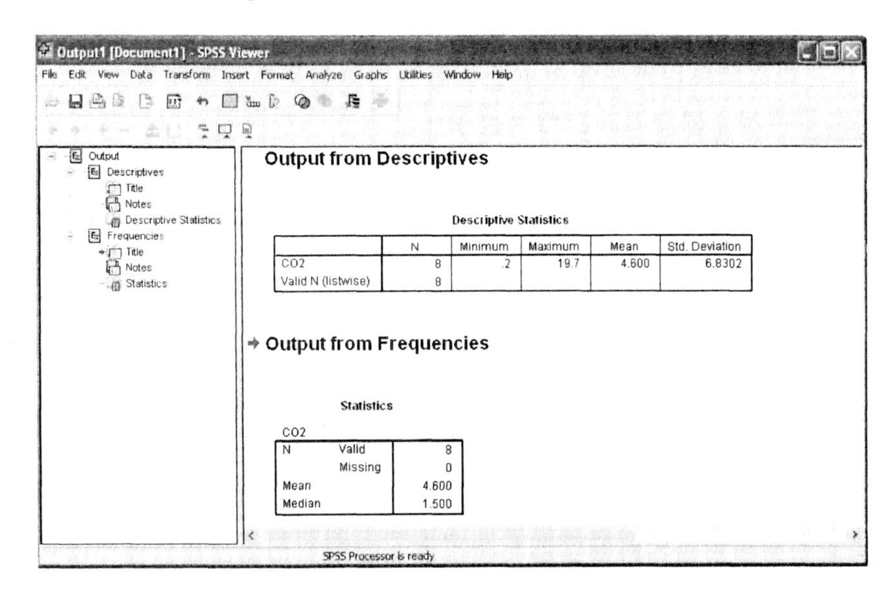

Standard Deviation and Range

The standard deviation and range can be found using any one of "Frequencies," "Descriptives," and "Explore." Let's find these statistics for the carbon dioxide data. Select **Analyze → Descriptive Statistics → Frequencies...** again. Click on the "Options" button and select "Standard Deviation" and "Range" from the list of statistics available under Dispersion.

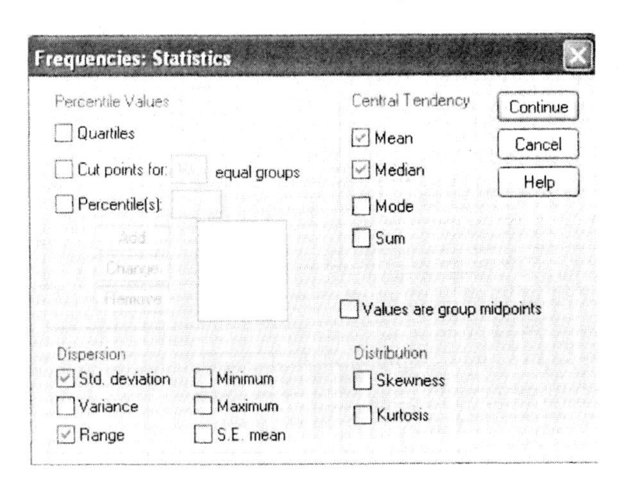

Click on **Continue** and then on **OK** to produce the output.

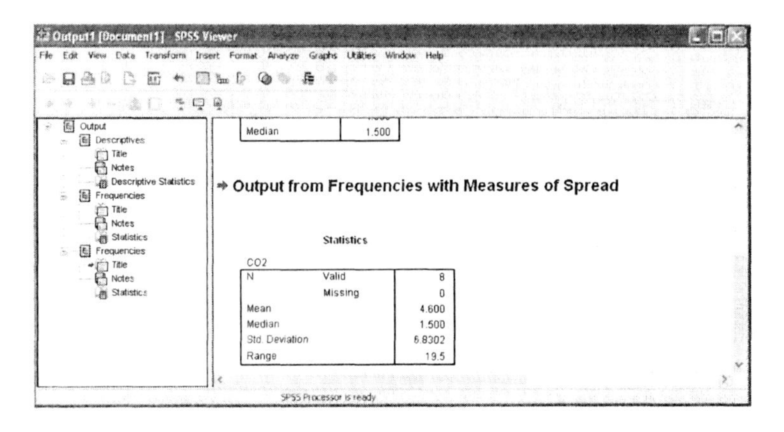

Quartiles and Other Percentiles

Both the "Frequencies" and "Explore" commands will produce quartiles and some percentiles. You actually have more options with the "Frequencies" command; the "Explore" command will only produce the 5th, 10th, 25th, 50th, 75th, 90th and 95th percentiles. Let's find the quartiles and the 30th and 70th percentiles for the data from Exercise 2.64 using "Frequencies." The data is the unemployment rate for the 15 nations of the European Union in 2003 and is shown in the following table:

Belgium	7.8	France	8.4	Italy	6.7
Denmark	3.2	Portugal	7.2	Finland	7.0
Germany	7.7	Netherlands	3.6	Austria	4.5
Greece	8.7	Luxembourg	5.0	Sweden	6.0
Spain	8.6	Ireland	4.4	U.K.	5.4

Enter the data in SPSS as shown below.

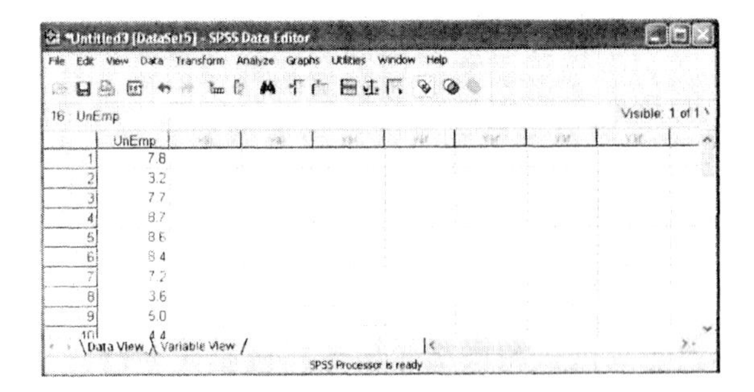

Select **Analyze → Descriptive Statistics → Frequencies...** and enter the unemployment rate variable in the "Variables" box, then click on the "Statistics" button. Click on "Quartiles" under "Percentile Values," then click on Percentiles. In the box next to Percentiles, enter "30" (without the quotes) and then click on the Add button. Enter 70 in that same box and click on Add again.

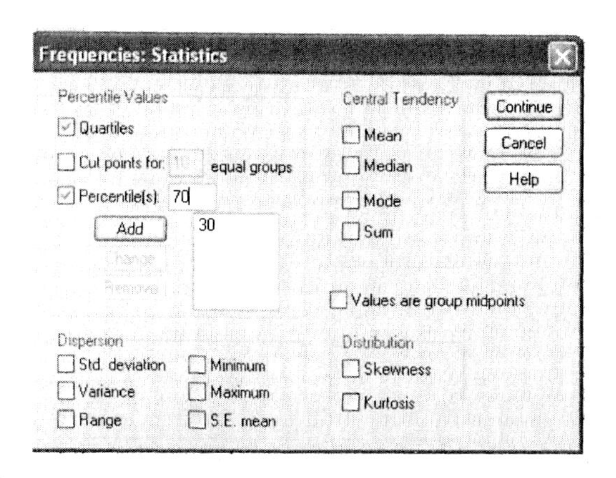

Click on **Continue** and then click on "Display frequency tables" in the Frequencies dialog box so that the frequency distribution will not be produced.

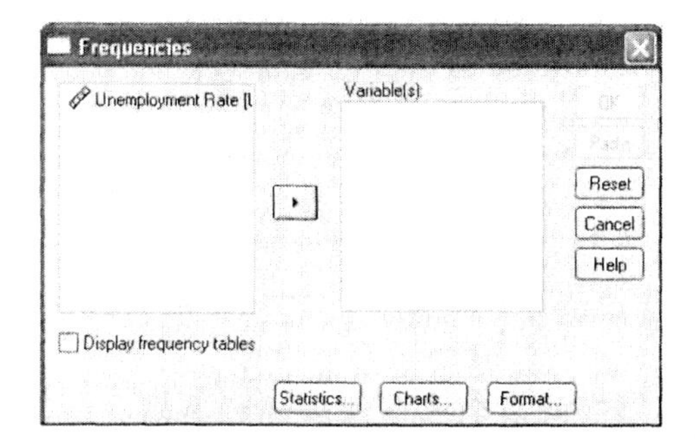

Click on **OK** to produce the output.

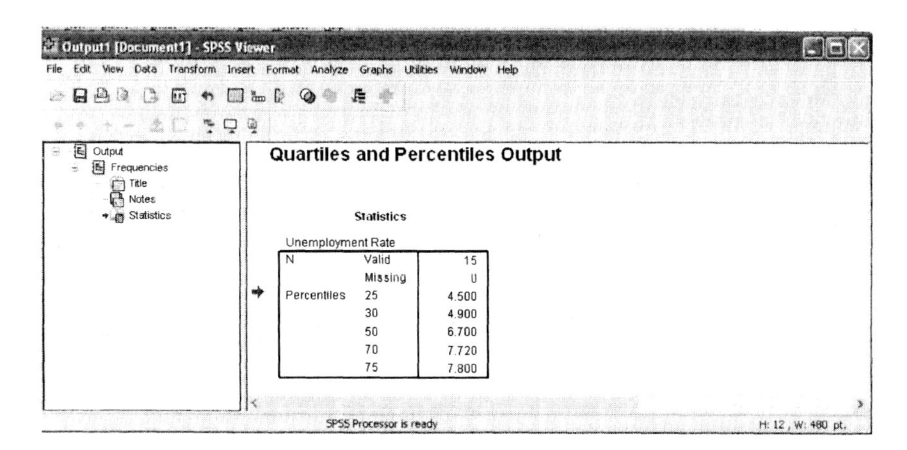

Using the Explore command to find percentiles will be demonstrated at the end of this chapter.

Box Plots

The "Graphs" menu includes a command for producing a box plot, but it is also an option under the "Explore" command. Let's produce a box plot for the unemployment rate data from the previous example. Select **Graphs → Interactive → Boxplot...** The following dialog box will appear:

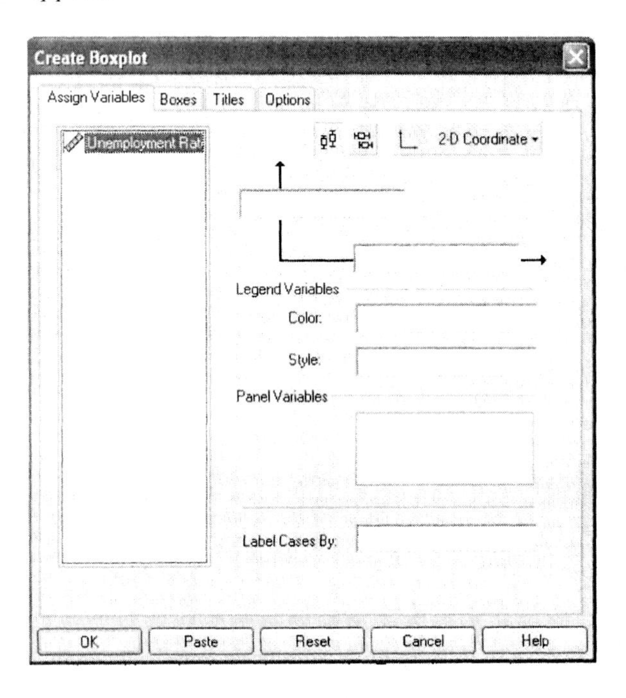

SPSS has options for producing box plots vertically or horizontally. In the dialog box above the vertical option is selected. If this is what you want, drag the unemployment variable to the box on the vertical axis. To change to horizontal box plots, click on the button that shows horizontal box plots. For a horizontal box plot, drag the unemployment variable to the box on the horizontal axis.

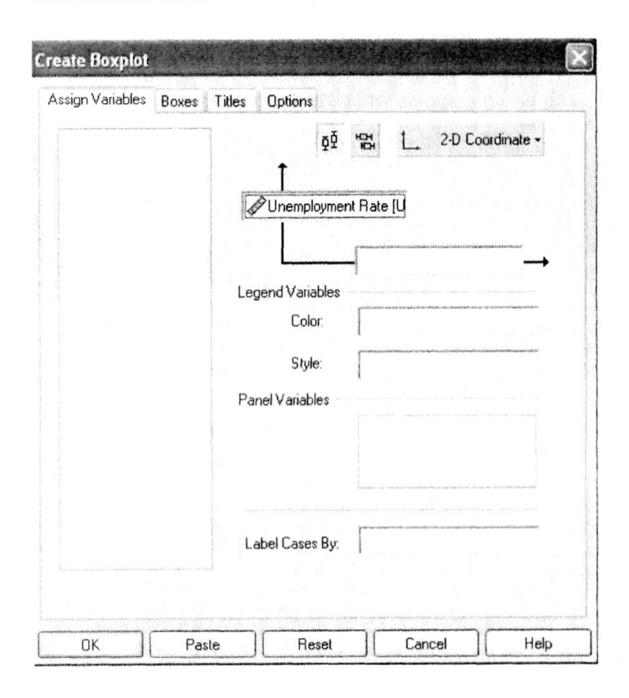

Click on **OK** to produce the graph.

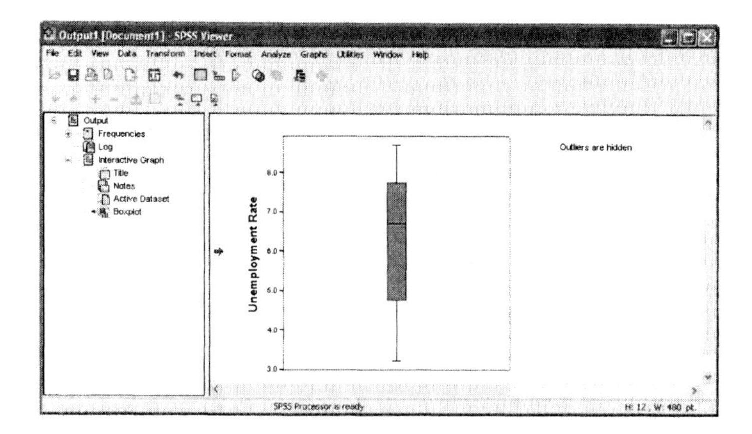

This is an example of a "simple" box plot, one that shows the five-number summary only (minimum, the three quartiles and the maximum). To produce a box plot that shows outliers, double click on the Boxes tab in the Boxplot dialog box. Click on the box next to Outliers, then on OK.

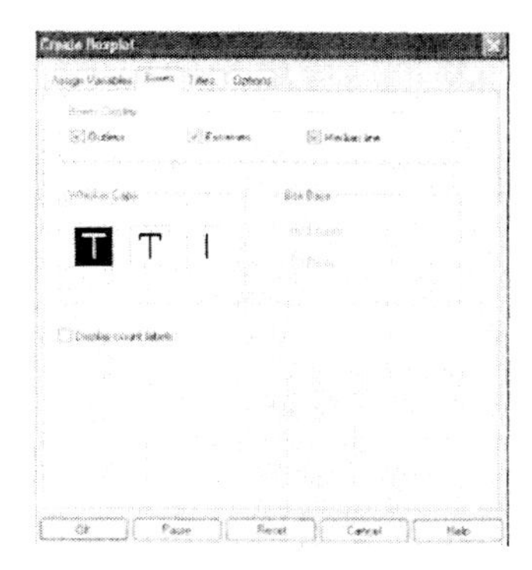

This data set doesn't have any outliers according to the rule SPSS uses when producing a box plot. To demonstrate this, though, change the first observation in the unemployment data set to 17.8 and make the box plot again. For the box plot below, the horizontal axis button was selected.

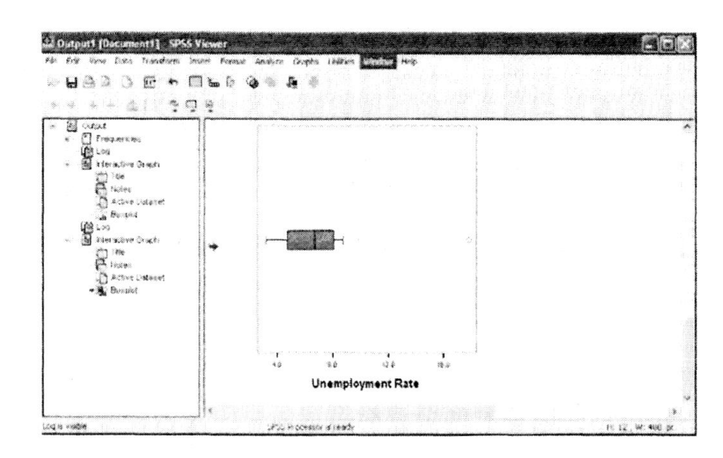

Notice that the text “Outliers are hidden” does not appear with this graph and that the right-hand “whisker” does not extend all the way out to the maximum value but is stopped at a “fence;” any observations beyond the fence are considered to be outliers. There is one outlier in this data set (the observation we changed), indicated by an open circle. Different programs have different rules for locating upper and lower fences and for identifying outliers. Two possible rules are to extend the upper whisker out to a distance of $1.5 \times$ IQR from the third quartile, or to the next larger data value below this distance. Any points beyond the fence are displayed as outliers. SPSS appears to be using the later rule. Similar rules are applied for the left-hand whisker if there are outliers on that side of the distribution. “Extreme” outliers (more than $3 \times$ IQR from the first or third quartile) are denoted by an asterisk (*).

Side-By-Side Box Plots

To see how to make side-by-side box plots of a quantitative variable for groups defined by a categorical variable, open the “fl_student_survey” data set from your text CD. Select **Graphs → Interactive → Boxplot...** and click on the horizontal box plots button. Then drag the age variable to the box on the horizontal axis and the gender variable to box on the vertical axis.

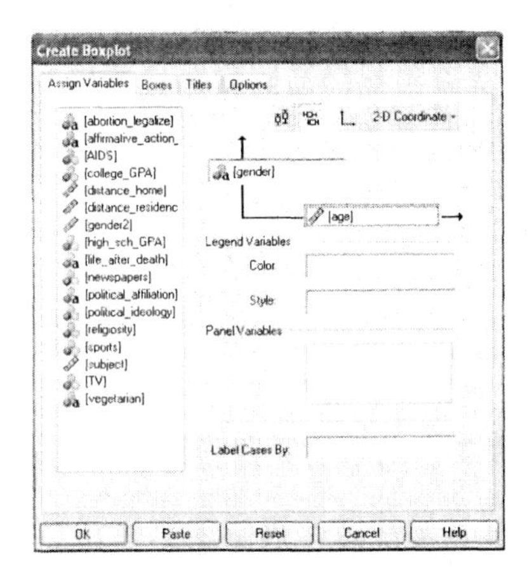

Click on **OK** to produce the output.

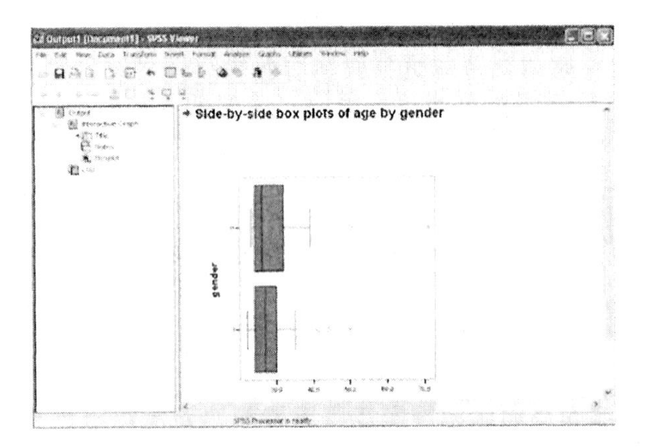

Side-by-side box plots are very useful for comparing the distributions of two or more groups.

The Explore Command

We used the "Explore" command above to produce a stem-and-leaf plot; however, we only produced a small portion of what this command can do all at once. With this one command you can not only find descriptive statistics and percentiles, but also produce a histogram, stem-and-leaf plot and a box plot! To produce all of this output, make sure that "Both" is selected under "Display" and that both "Descriptives" and "Percentiles" are selected under "Statistics." Click on the "Plots" button and select both "Histogram" and "Stem-and-leaf" under "Descriptive," along with one of the Boxplot options. Click on **Continue** and then on **OK** to see all the output this one command will produce!

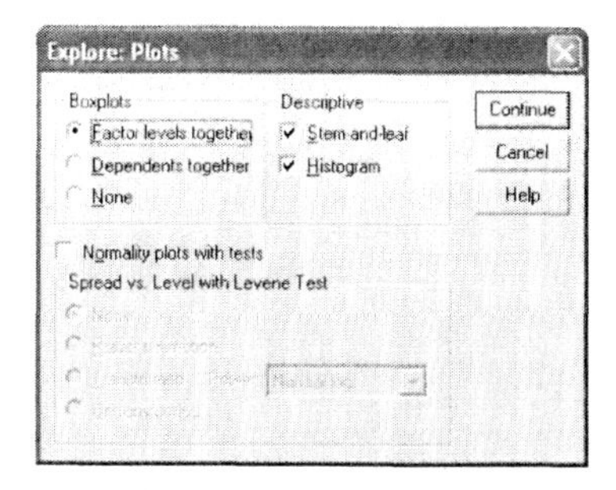

Comparison of Frequencies, Descriptives, and Explore

The following table will help you determine which of these three commands under "Descriptive Statistics" is the most useful for the analyses you want to do. Notice that "Frequencies" can be used on both categorical and quantitative data, while "Descriptives" and "Explore" only operate on quantitative data. Also, "Descriptives" is the only one of these commands that does not include the option to produce at least one type of graph.

Command	**Variable Type**	**Statistics**	**Graphs**
Frequencies	Categorical Quantitative	Frequency Table Mean, Median, Mode Standard Deviation, Variance, Range Minimum, Maximum, S.E. Mean Quartiles, Percentiles	Bar and pie charts Histogram
Descriptives	Quantitative	Mean Standard Deviation, Variance, Range Minimum, Maximum, S.E. Mean	None
Explore	Quantitative	Mean, Median Confidence Interval for the Mean Standard Deviation, Variance, Range, IQR Minimum, Maximum, S.E. Mean Some Percentiles	Stem-and-leaf Histogram Box plot

Chapter 3 Association: Contingency, Correlation and Regression

In the previous chapter you saw how to produce descriptive statistics and graphs for one categorical or quantitative variable. This chapter will introduce you to statistical methods for analyzing the association between two variables.

Two Categorical Variables

When working with two categorical variables, you will want to display the joint distribution in a contingency table. You might also want to produce some conditional (marginal) proportions or maybe a side-by-side bar chart to make comparisons.

Contingency Tables

To produce a contingency table from data that has already been tabulated, the process is similar to how we dealt with tabulated data for a single variable in the previous chapter. In that case, we entered two variables: one for the categories and one for the count or frequency of observations in each category. For a contingency table, we will need an additional column, one for the categories of the second categorical variable.

As an example, let's look at the data for Example 2. This data sct has two categorical variables, "Pesticide Status" and "Food Type," and is shown in the table below.

	Pesticide Status	
Food Type	Present	Not Present
Organic	29	98
Conventional	19485	7086

Open a new SPSS Data Editor and enter the data in three columns, as shown below. Notice that there are 4 "cases" in this table, one for each combination of the two categories for "Food Type" and the two categories for "Pesticide." You can enter "Food Type" and "Pesticide" as string-type variables or as numeric variables with value labels.

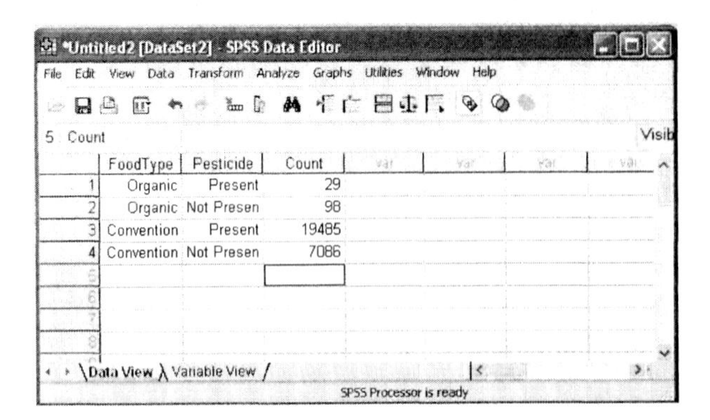

As with the examples from the previous chapter with one categorical variable, you will have to indicate to SPSS that the data in the variable "Count" represents the number of observations in each of the 4 different combinations of the categories of the "Food Type" and "Pesticide" variables. Select **Data → Weight Cases...** and enter "Count" as the "Frequency Variable" to "Weight cases by."

To produce a contingency table for this data using SPSS, select **Analyze → Descriptive Statistics → Crosstabs...** and enter "FoodType" as the "Row Variable" and "Pesticide" as the "Column Variable" and click on **OK**.

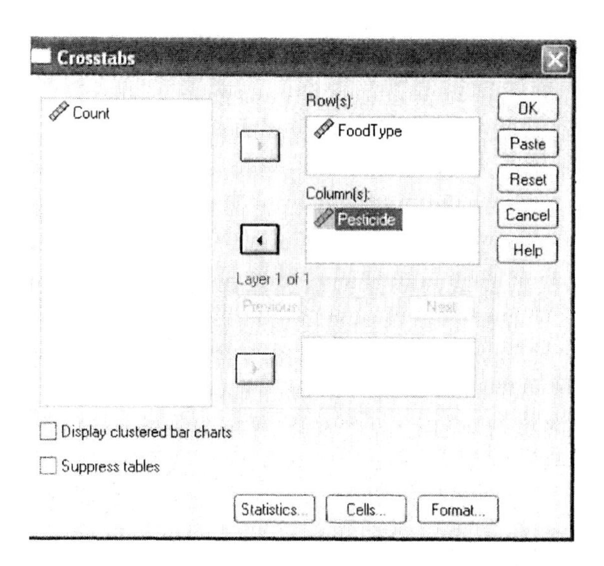

The output is shown below, without the "Case Processing Summary."

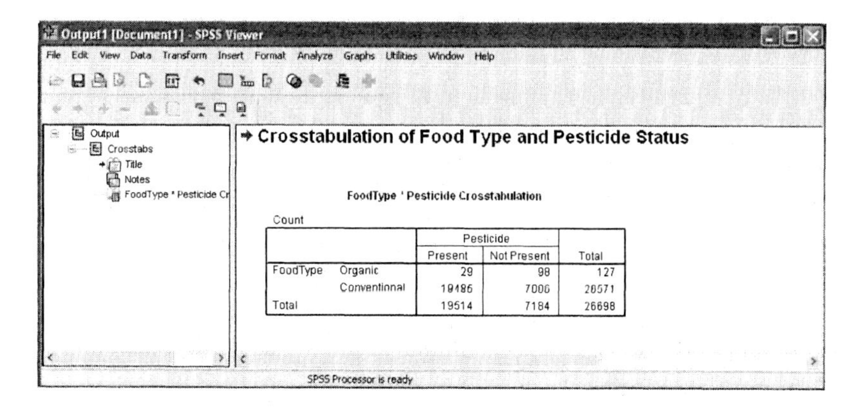

FoodType * Pesticide Crosstabulation

Count

		Pesticide		
		Present	Not Present	Total
FoodType	Organic	29	98	127
	Conventional	19486	7086	26571
Total		19514	7184	26698

Both of the categorical variables were entered as Numeric variables in this example. If you entered the variables as string variables, you will find that SPSS reorders the categories in the output produced. For this example, Conventional would appear above Organic and Not Present would be to the left of Present.

Conditional Proportions

If all you needed was the contingency table for this example, then there wouldn't be any need to have entered it in SPSS. But, with the data entered this way you can now find conditional proportions and produce a side-by-side bar chart allowing you to compare the "Pesticide" variable by "FoodType." To find the conditional proportions for "Pesticide" select **Analyze → Descriptive Statistics → Crosstabs...** again, but this time click on the "Cells" button. In the dialog box, click on "Row" underneath Percentages.

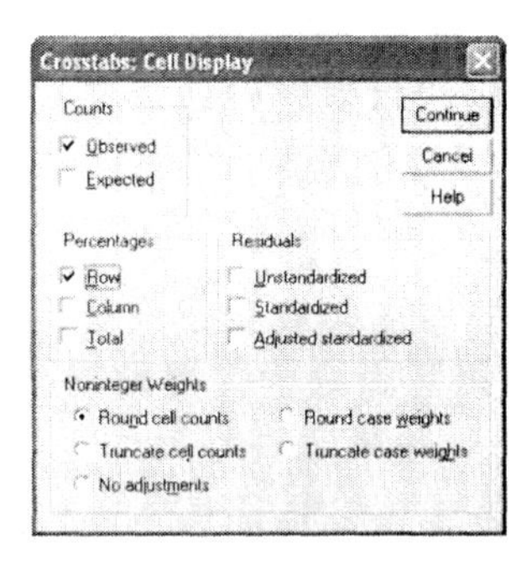

Click on **Continue** and then click on **OK** to produce the following output. If you also select "Column" and "Total" under "Percentages" then these results would be added to the output.

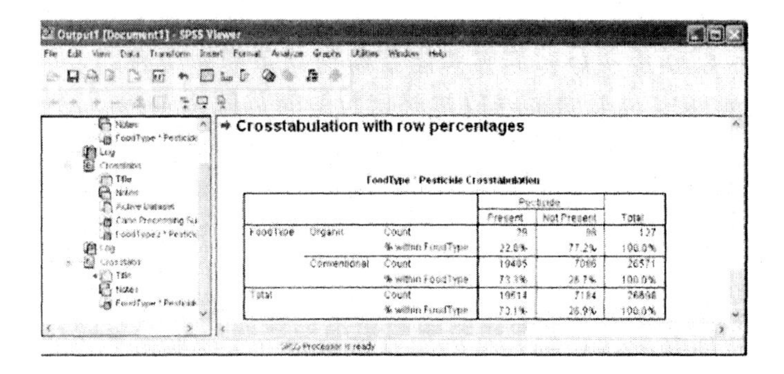
Crosstabulation with row percentages

FoodType * Pesticide Crosstabulation

			Pesticide		
			Present	Not Present	Total
FoodType	Organic	Count	29	98	127
		% within FoodType	22.8%	77.2%	100.0%
	Conventional	Count	19485	7086	26571
		% within FoodType	73.3%	26.7%	100.0%
Total		Count	19514	7184	26698
		% within FoodType	73.1%	26.9%	100.0%

Side-By-Side Bar Charts

To produce side-by-side bar charts for this data, you can use the Bar chart option from the Interactive Graphs menu. But, the Crosstabs command also has an option for producing bar charts. Select **Analyze → Descriptive Statistics → Crosstabs...** again and click on the "Display clustered bar charts" box.

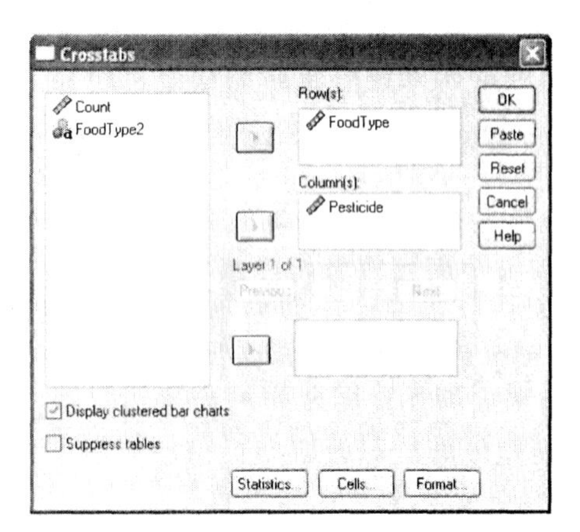

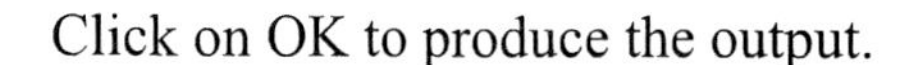
Click on OK to produce the output.

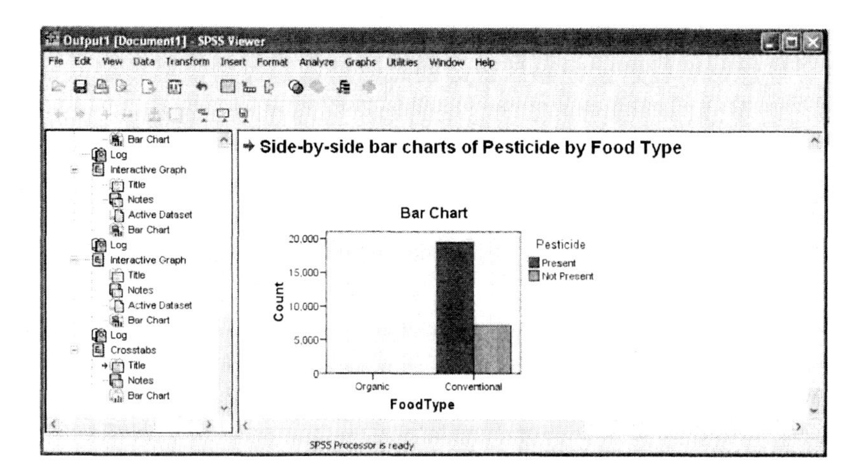

Because the number of Organic foods included in this data set is so much smaller than the number of Conventional foods, the bars for the Organic foods are barely visible. If you want to produce side-by-side bar charts of Food Type by Pesticide instead, make Food Type the Column variable and Pesticide the Row variable.

Just as in the case of one categorical variable, there is another option for producing bar charts for two categorical variables. For this option, do not weight the cases by the variable "Count." (If the cases are weighted, then select **Data → Weight Cases…** and click on "Do not weight cases.") Select **Graphs → Interactive → Bar…** and enter the Count variable in the box on the vertical axis (replacing the SPSS built-in Count($count) variable) and either FoodType or Pesticide in the box on the horizontal axis. Place whichever variable you did not put on the horizontal axis in the box below Panel Variables. When you place a numeric variable in the Panel Variable box SPSS will ask you if you want to convert the variable to a string variable. Indicate yes. The example below shows FoodType on the horizontal axis and Pesticide as a panel variable.

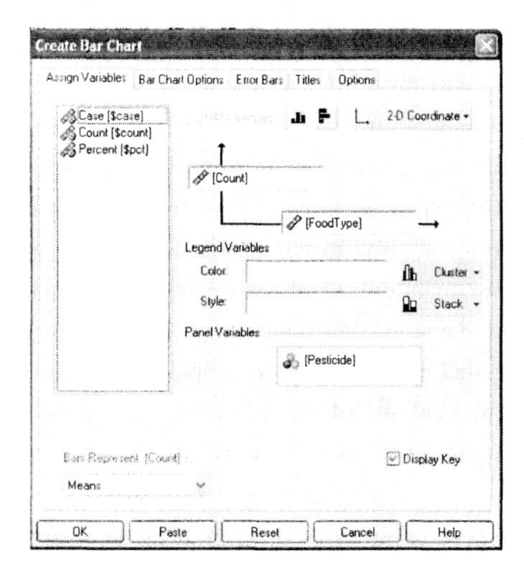

Click on **OK** to produce the output. The following output will be produced.

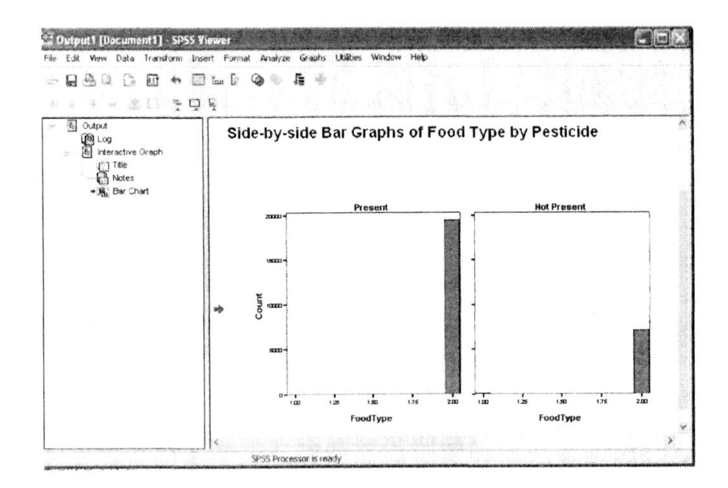

The graph is similar to the first side-by-side bar chart produced above except the bars are narrower and there is a separate graph for each level of Pesticide. If you switch the roles of Pesticide and Food Type the graph would have separate panels for Organic and Conventional foods and there would be bars for each level of Pesticide.

Contingency Tables for Non-Tabulated Data

What if the data hasn't been tabulated yet? This is actually an easier situation to deal with since you won't have to bother with a weighting variable. Let's demonstrate this using the variables "gender" and "vegetarian" from the "fla_student_survey" data set from your text CD. Once you have accessed this data file, select **Analyze → Descriptive Statistics → Crosstabs...** and enter "gender" as the row variable and "vegetarian" as the column variable.

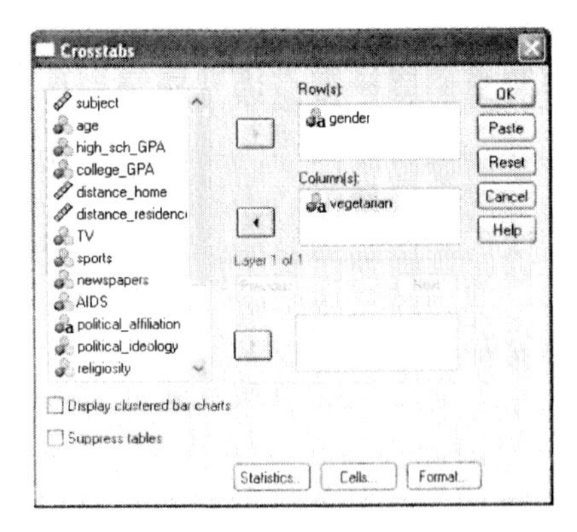

Click on **OK** to produce the contingency table.

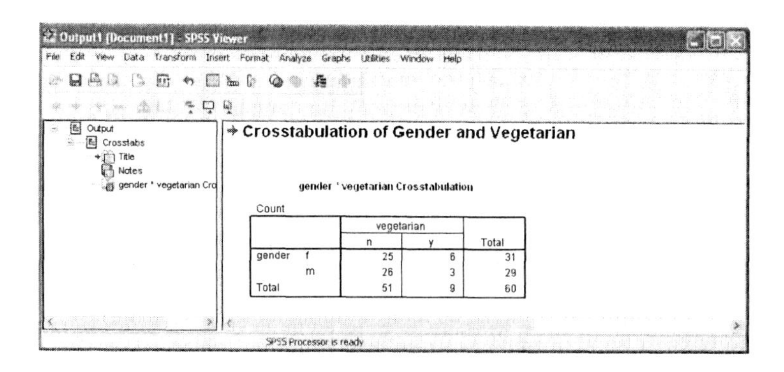

		vegetarian		Total
		n	y	
gender	f	25	6	31
	m	26	3	29
Total		51	9	60

Let's produce side-by-side bar charts using the same process as described above. For example, if you select **Graphs → Interactive → Bar...** and enter "gender" in the box on the horizontal axis and "vegetarian" under Panel Variables, the following graph will result.

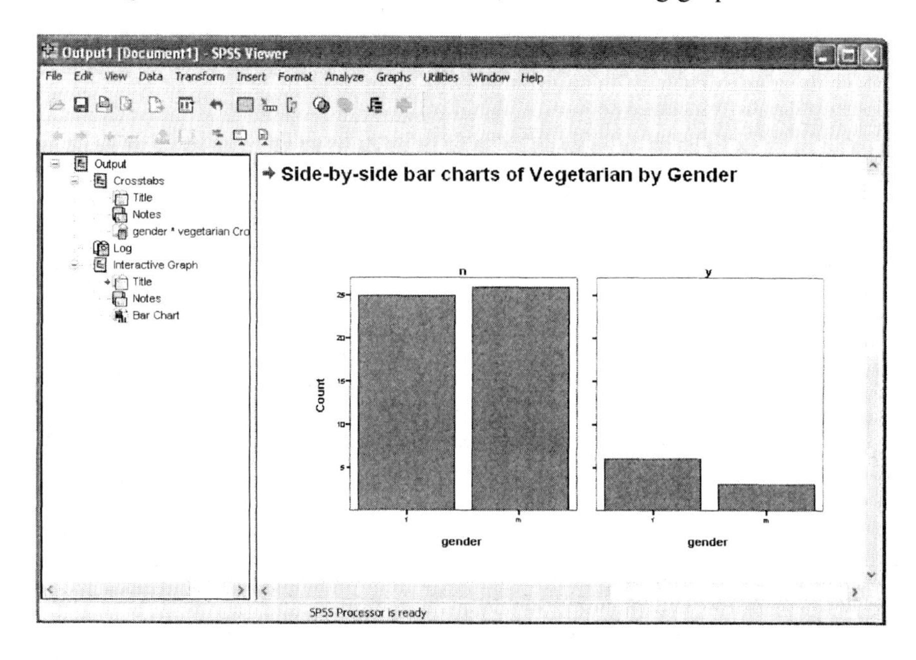

One Categorical Variable and One Quantitative Variable

For this situation you would want to produce descriptive statistics and graphs of the quantitative variable for each category of the categorical variable. The easiest way to do this is to use the "Explore" command. To demonstrate this, let's use the "fla_student_survey" data set.

Select **Analyze → Descriptive Statistics → Explore...** and enter "TV" in the "Dependent List" and "vegetarian" in the "Factor List." Click on **OK**. Among the output produced is a table with various descriptive statistics for average TV watching for the vegetarian and non-vegetarian students and side-by-side box plots, both shown below. Some of the descriptive statistics were omitted to allow the output to appear in the window without having to use the scroll bar. The box plots were also modified to be horizontal instead of vertical. To do this follow the directions given in the first chapter for modifying a bar chart produced using the Legacy Dialogs command and click on the "Transpose Chart" option from the Options menu in the Chart Editor.

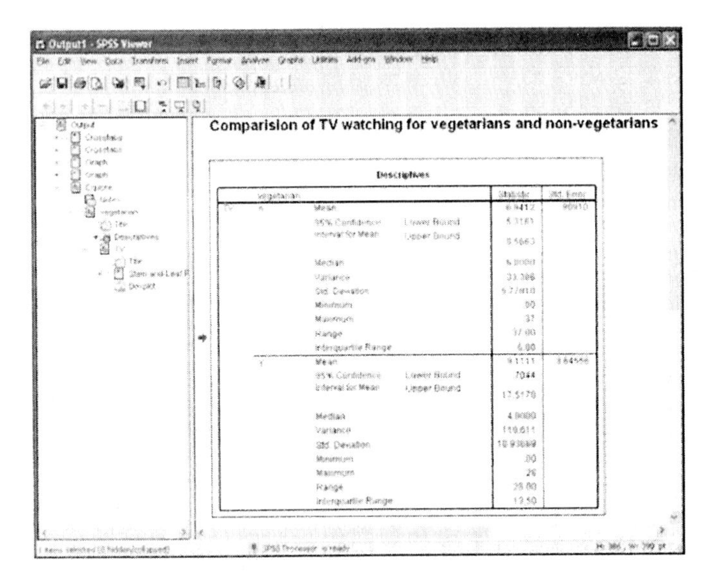

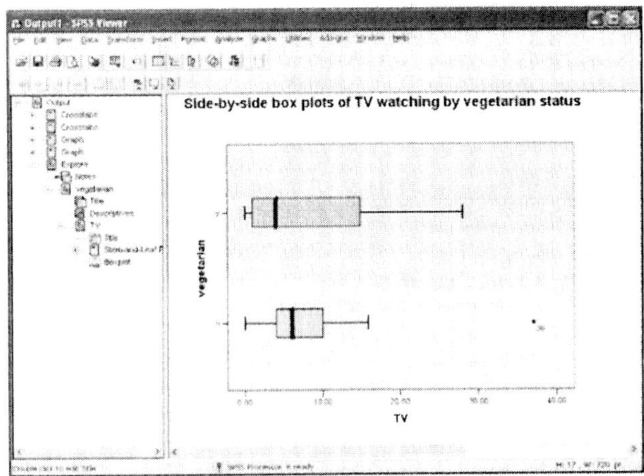

Another option for comparing a quantitative variable for groups defined by a qualitative variable is the "Compare Means" command. Let's demonstrate this command with the same data set as above. Select **Analyze → Compare Means → Means...** and enter the "TV" variable in the "Dependent List" and the "vegetarian" variable in the "Independent list."

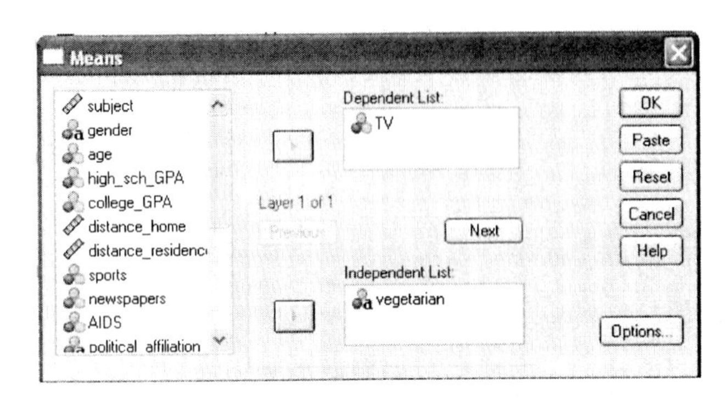

Click on **OK** to produce the output.

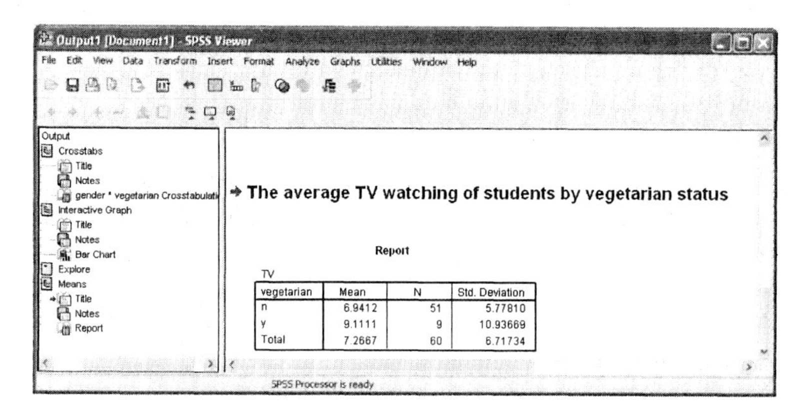

vegetarian	Mean	N	Std. Deviation
n	6.9412	51	5.77810
y	9.1111	9	10.93669
Total	7.2667	60	6.71734

The "Options" button in the "Means" dialog box has options for producing some additional statistics for each group.

A third alternative for producing statistics and graphs of a quantitative variable for groups defined by a categorical variable is to use the "split file" command from the "Data" menu. Continuing with the same data set, let's compare the TV watching habits for the groups defined by political affiliation. Select **Data → Split File...** and click on "Organize output by groups." Click on the "political affiliation" variable and then on the triangle button to enter that variable in the "Groups Base on:" box. Click on **OK**.

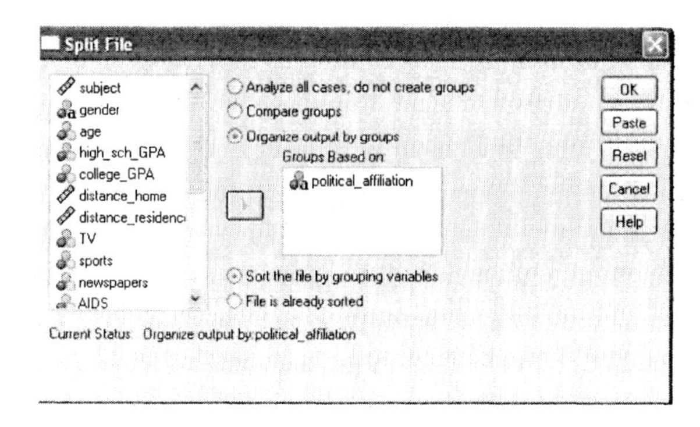

All of the commands you do after this will be done separately for each group defined by the grouping variable. The "Compare groups" option will do the same thing but it will organize the output to make comparisons easier. The output shown below is from the "Descriptives..." command for the TV watching variable after the "Split File..." command was executed. Only the first two levels of political affiliation appear in this screen.

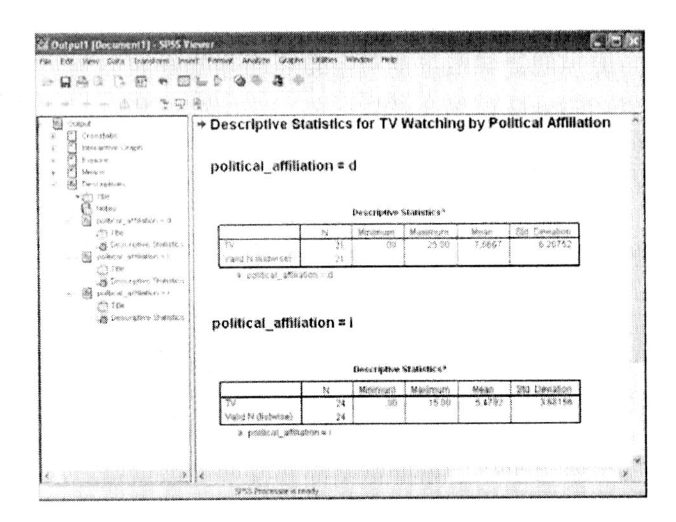

When you are done using the split file, select **Data → Split File...** again and select "Analyze all cases, do not create groups." Click on **OK** to complete the command. All subsequent commands done on this same data set will be done on the whole data set, rather than separately for each group defined by the political affiliation variable.

Two Quantitative Variables

For two quantitative variables you will want to display the association in a scatterplot. If the association appears linear, the correlation coefficient and regression line will also be useful.

Scatterplots

To produce a scatterplot to investigate the association between two quantitative variables, use the "Scatter/Dot" command under the "Graphs" menu. Let's demonstrate this using the "GDP" and "Internet Use" variables from the "human_development" data set on your text CD. Access this data file, then select **Graphs → Interactive → Scatterplot...** Enter "INTERNET" in the box on the vertical axis and "GDP" in the box on the horizontal axis, as shown below.

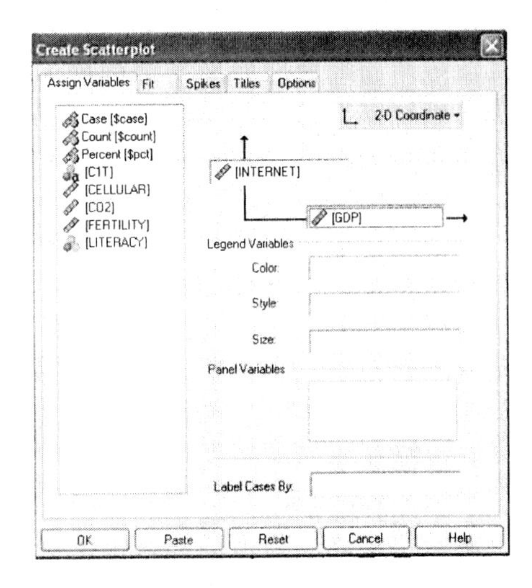

Click on **OK** to produce the output. As demonstrated in Chapter 2 with the dot plot, you can change the size, shape and color of the plotting symbol. This process will also be described again in the section below on how to add the regression line to a scatter plot.

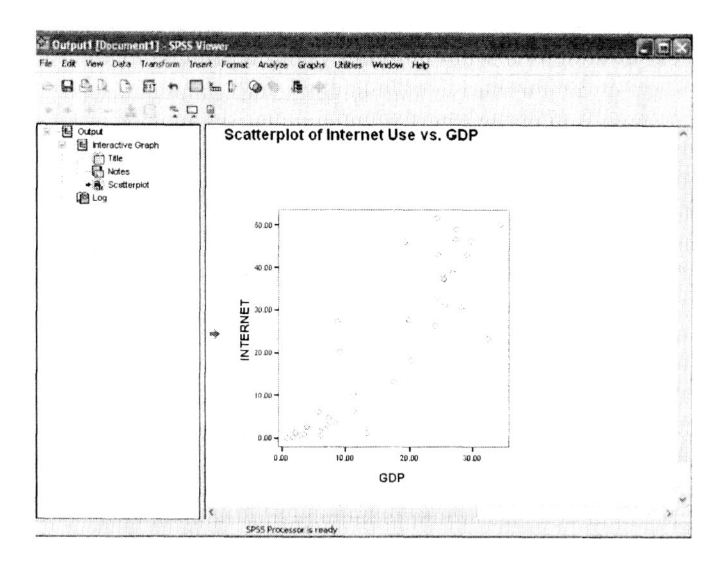

The Correlation Coefficient

To find the correlation coefficient for the association between these two variables select **Analyze → Correlate → Bivariate...** and enter both variables in the "Variables" box.

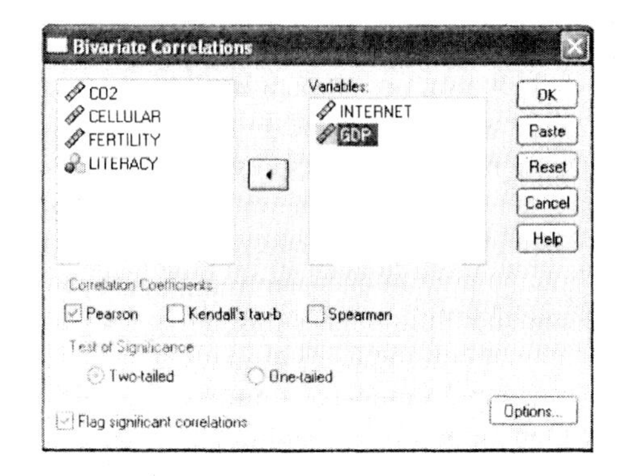

Click on **OK** to produce the output.

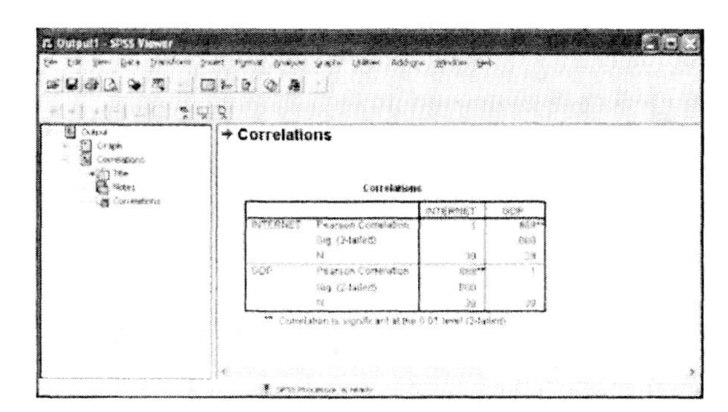

SPSS produces a "correlation matrix" that shows the correlation of each variable with itself (1) along with the correlation between "GDP" and "INTERNET" and the correlation between "INTERNET" and "GDP" (.888). SPSS also includes the p-value results ("Sig") of a hypothesis test to determine whether or not this correlation is "significantly different" from zero. Hypothesis tests and p-values will be discussed later on in this manual. Note that you can also obtain the correlation coefficient from the regression output that SPSS produces.

The Regression Line

To demonstrate this command, let's use the "Batting Average" and "Team Scoring" variables in Table 3.5 in your text. A portion of this data set and the scatter plot are shown below.

*Untitled1 [DataSet0] - SPSS Data Editor

File Edit View Data Transform Analyze Graphs Utilities Window Help

1 : TeamScore 5.1

	BatAve	TeamScore
1	.269	5.1
2	.284	5.0
3	.287	4.9
4	.271	4.7
5	.272	4.7
6	.285	5.7
7	.274	4.7
8	.277	4.7
9	.278	5.2
10	.255	4.3

Data View / Variable View

SPSS Processor is ready

Produce the scatterplot of Team Scoring vs. Batting Average, as shown below.

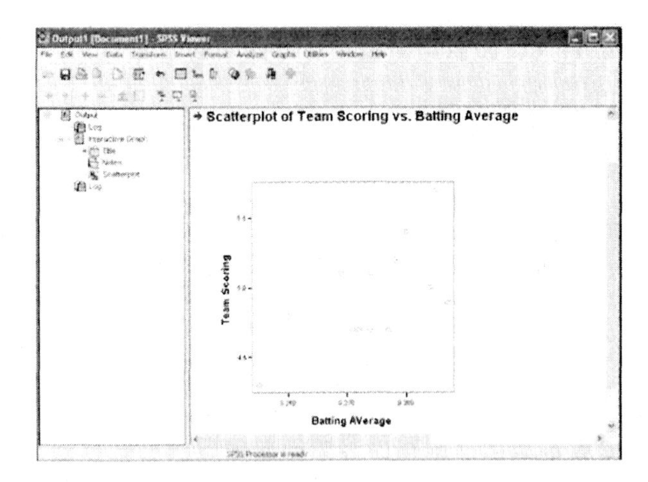

To find the best-fit regression line, select **Analyze → Regression → Linear...** Enter Batting Average as the Independent variable and Team Scoring as the Dependent variable, as shown in the dialog box below.

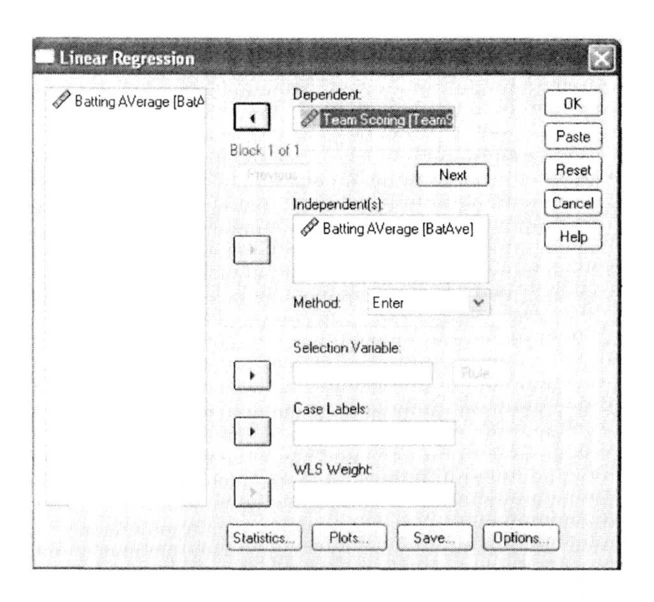

Click on **OK** to produce the output. The resulting output has five parts, some of which you will learn about later on. The parts that you will need right now are the title, the Model Summary, and the Coefficients. The other parts have been omitted from the output shown below.

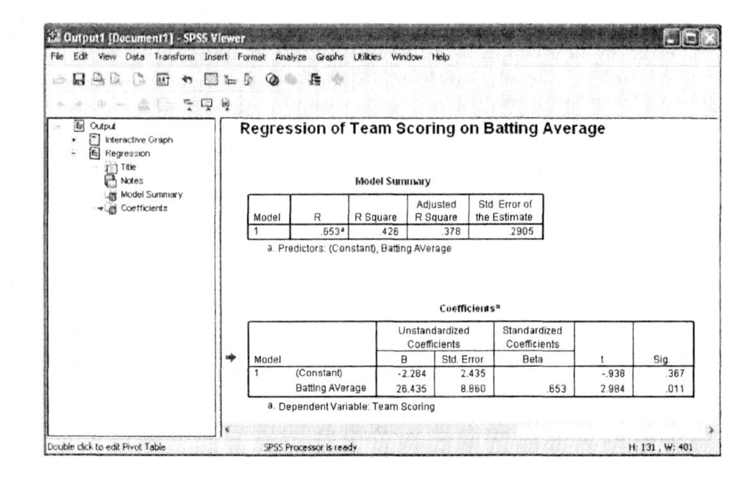
Regression of Team Scoring on Batting Average

Model Summary

Model	R	R Square	Adjusted R Square	Std. Error of the Estimate
1	.653a	.426	.378	.2905

a. Predictors: (Constant), Batting AVerage

Coefficientsa

Model		Unstandardized Coefficients B	Unstandardized Coefficients Std. Error	Standardized Coefficients Beta	t	Sig.
1	(Constant)	-2.284	2.435		-.938	.367
	Batting AVerage	26.435	8.860	.653	2.984	.011

a. Dependent Variable: Team Scoring

The "Model Summary" provides the absolute value of the correlation coefficient under "R." Since the association is positive we know that r is +.653. The intercept and slope can be found in the "B" column under "Unstandardized Coefficients" in the "Coefficients" part of the output. The intercept can be found next to "(Constant)" (value = -2.284) and the slope is next to the independent variable, "BAT_AVG" in this case (value = 26.453). So the regression line is $\hat{Y} = -2.284 + 26.453X$. Note: Most of the rest of the statistics included in the regression output are covered in a later chapter in your text and in this manual.

If you are interested in finding some descriptive statistics for the regression variables just click on the "Statistics" button before you click on **OK**. Click on "Descriptives," then on **Continue**, and finally on **OK**.

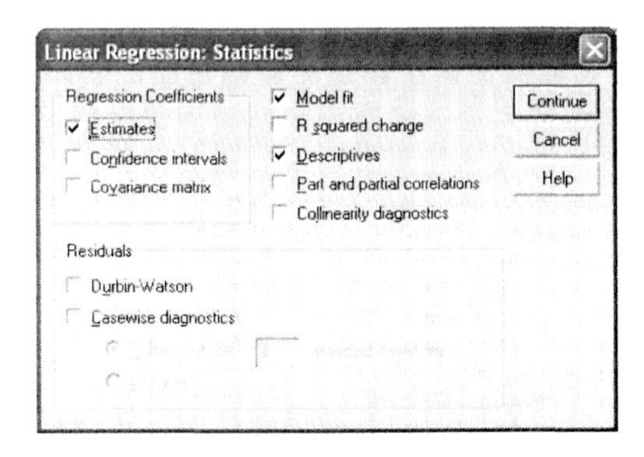

Another option for finding the best-fit line will be described in Chapter 12. This approach will be introduced for fitting an exponential model to a curved data set, but it could also be used to find a best-fit line.

Adding the Regression Line to a Scatterplot

The "Scatterplot..." command has an option for you to add the regression line to the plot. Select **Graphs → Interactive → Scatterplot...** again but click on the Fit tab. From the options available under Method, select "Regression," as shown below.

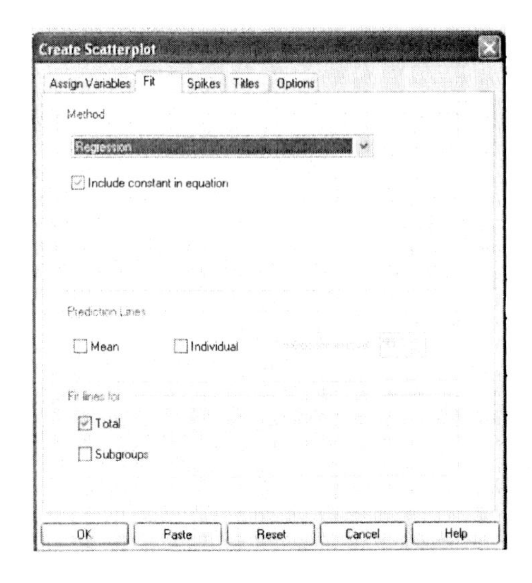

Click on OK to produce the output.

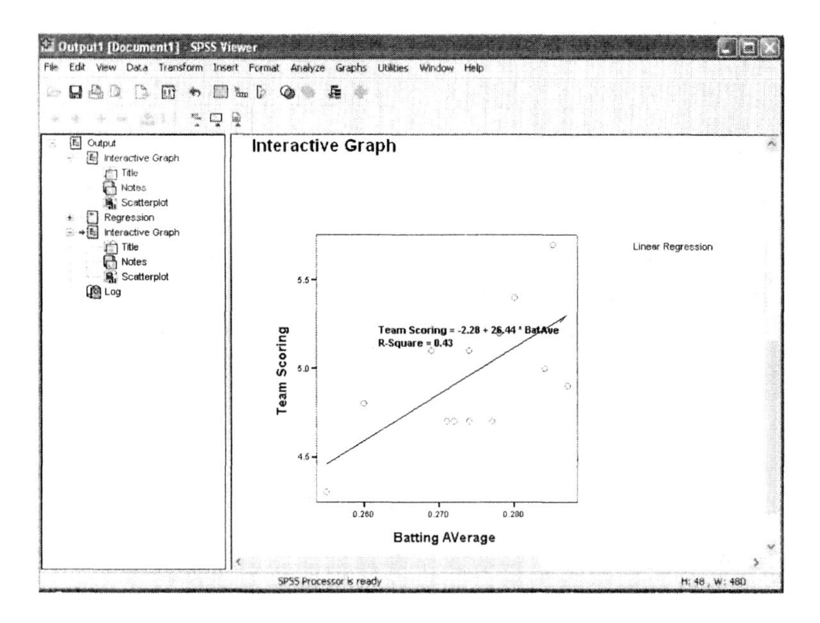

To change the plotting symbol, double click on the graph to open the Chart Editor then double click on one of the points. In the dialog box that appears, select a new symbol from those available under "Style." You can also change the color used, as shown below.

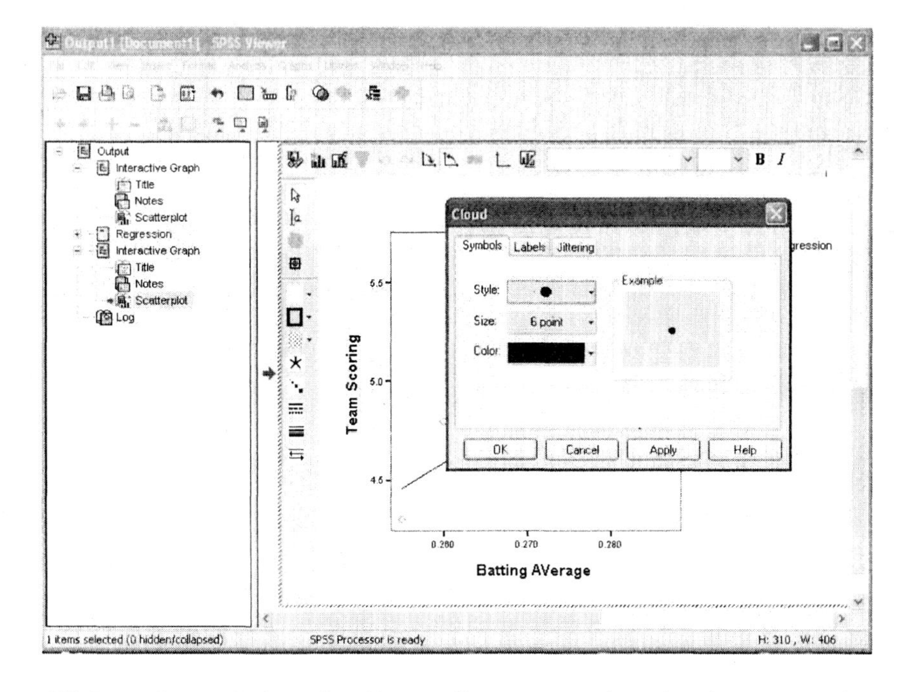

Click on OK and close the Chart Editor to complete the change. In the scatterplot below the regression line information was dragged to a position outside of the graph while the Chart Editor was active.

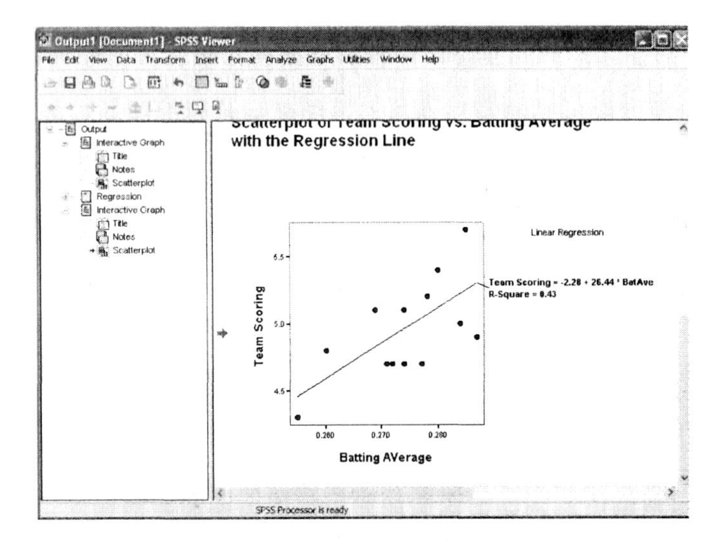

Regression lines using the Legacy Dialogs Scatter/Dot… command

The process for adding the regression line to a scatter plot made using the Legacy Dialogs version of a scatter plot is very different. Select **Graphs → Legacy Dialogs → Scatter/Dot…** Make sure that the **Simple Scatter** option is selected and click on **Define**.

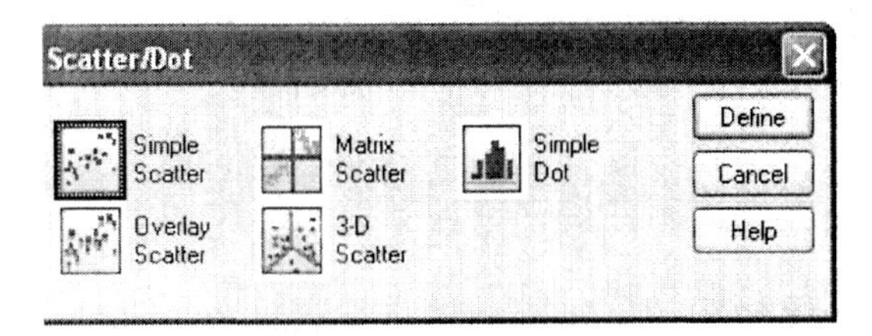

In the dialog box that appears, select Team Scoring as the Y-axis variable and Batting Average as the X-Axis variable, using the arrow buttons next to those boxes.

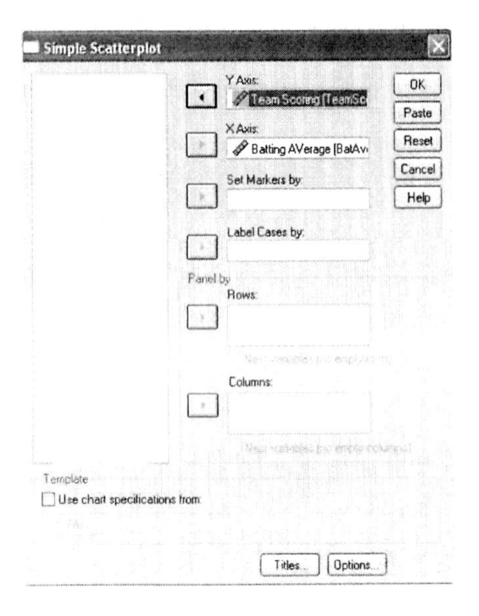

Click on **OK** to produce the output.

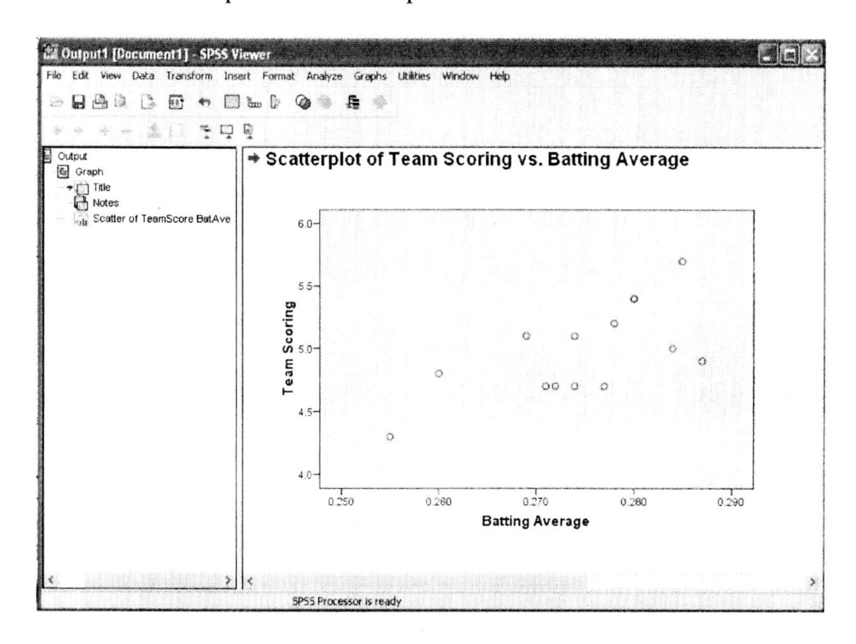

To add the regression line and change the plotting symbol, double click on the graph to open the Chart Editor.

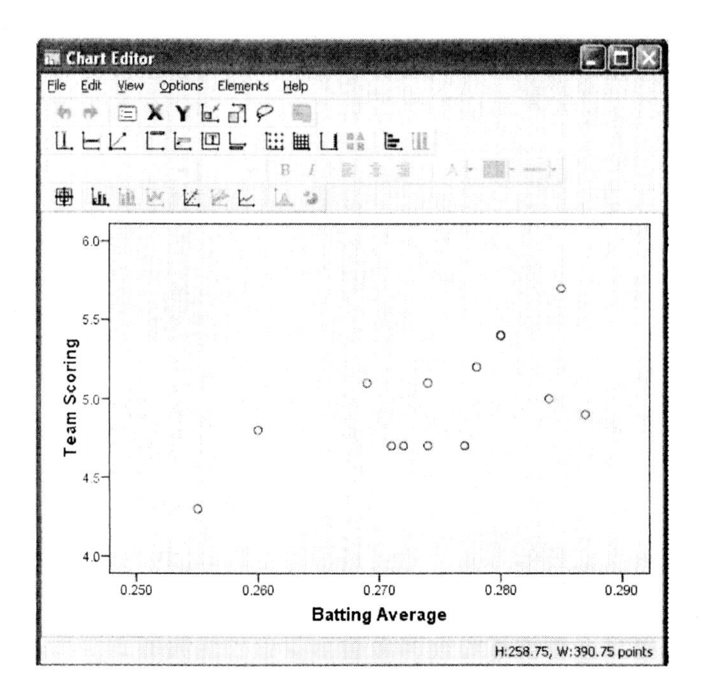

To add the regression line, select the "Fit line at total" option from the Elements menu.

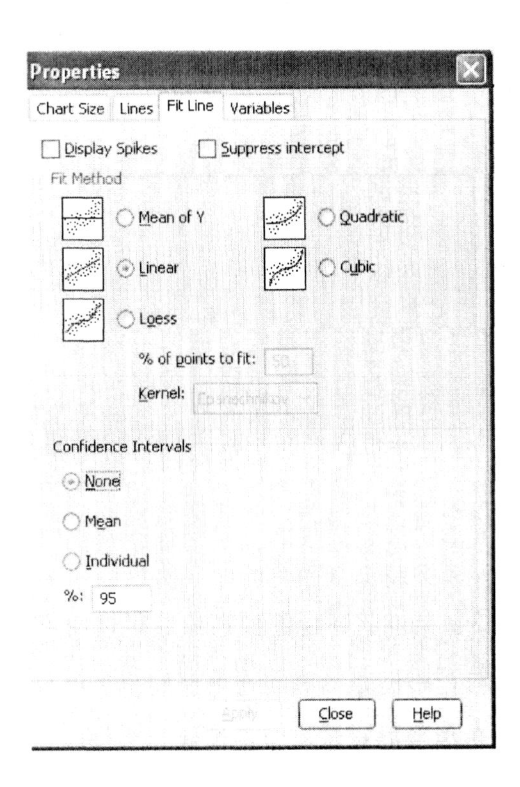

Click on **Close** and the line should be added to the graph.

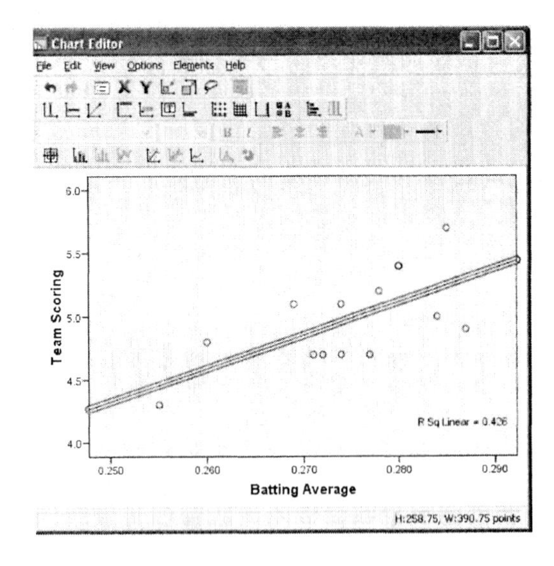

To change the plot symbol, double click on one of the circles in the graph. In the dialog box that appears, click on the box to the left of “Fill” below “Color” and then on the black box to the right of that.

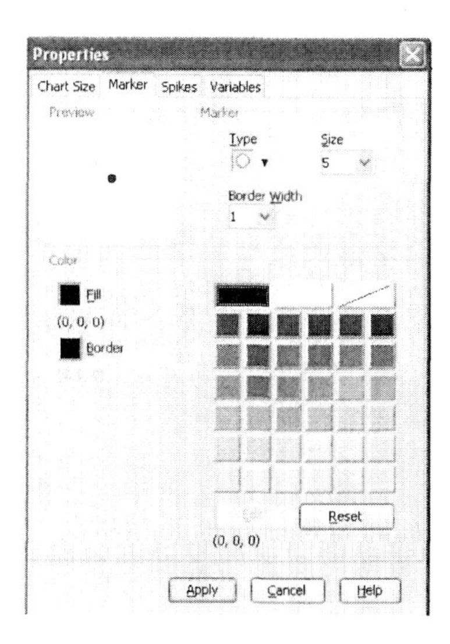

Click on **Apply** and then **Close** to make the change.

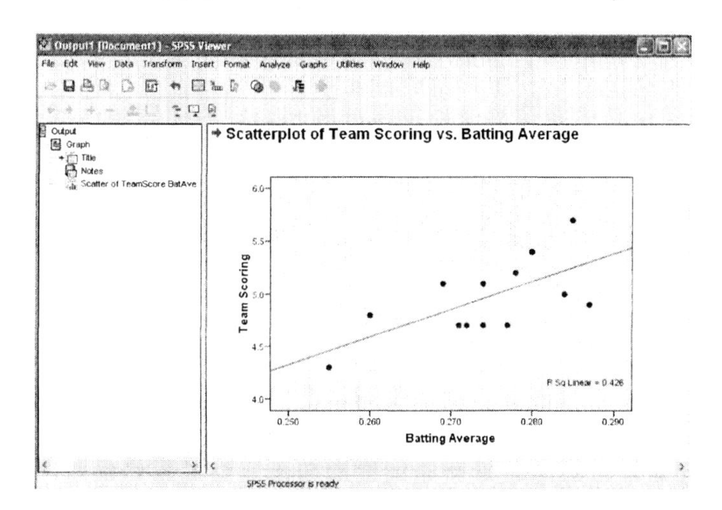

Saving and Graphing Regression Residuals

You have two options for producing a histogram of the residuals from a regression analysis. Residual plots can be useful for assessing the fit of the best-fit line and for assessing the presence of outliers. The first option is to select the Plots button at the bottom of the regression dialog box. To demonstrate this, let's use the "buchanan_and_the_butterfly_ballot" data set from Example 10 in Chapter 3, which can be found on the text CD. Access this data file and select **Analyze → Regression → Linear...** Enter "Perot" as the Independent Variable and "Buchanan" as the Dependent Variable. Click on the **Plots...** button at the bottom of the dialog box. In the new dialog box that appears, click on "Histogram" below "Standardized Residual Plots."

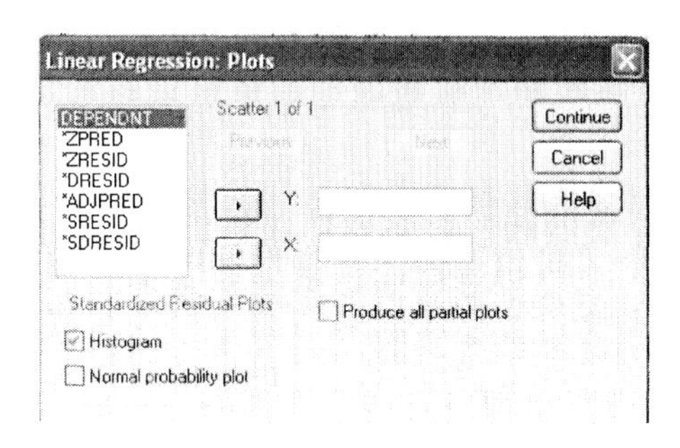

Click on **Continue** and then on **OK**. As part of the regression output, SPSS will produce the following graph.

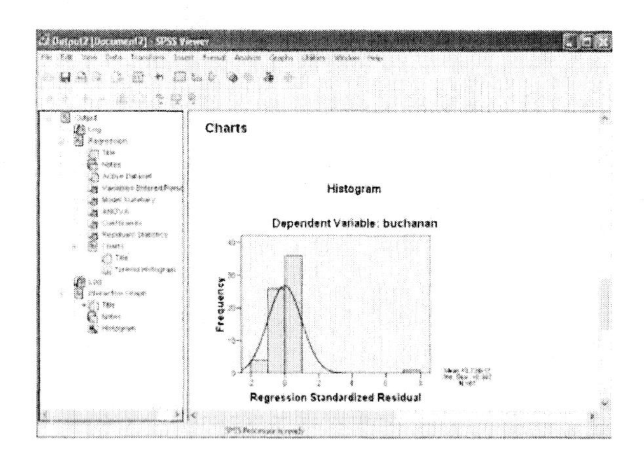

If you would prefer a histogram of the Unstandardized residuals, these have to be saved first and then graphed. Repeat the same regression command, but click on the **Save…** button at the bottom of the Regression dialog box. In the dialog box that appears, click on Unstandardized underneath "Residuals," as shown below.

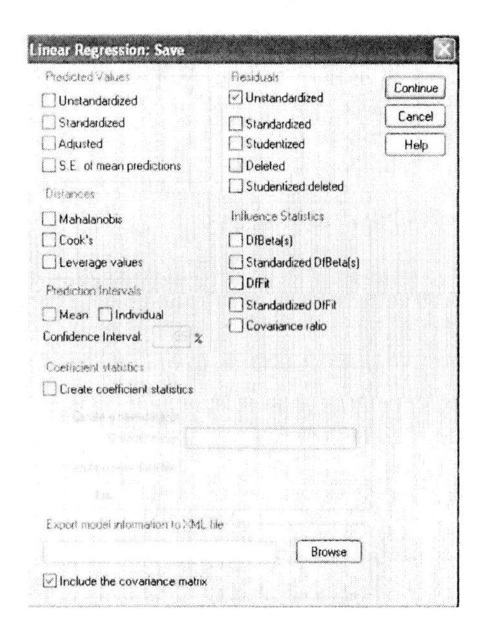

Click on **Continue** and then on the **Plots** button again to "unselect" Histogram so that this output won't be produced again. Click on **Continue** and then on **OK** to complete the regression analysis. The Unstandardized residuals will be saved to the data set with the variable name "RES_1."

*buchanan_and_the_butterfly_ballot.sav [DataSet1] - SPSS Data Editor

File Edit View Data Transform Analyze Graphs Utilities Window Help

1 : county Alachua Visible: 6 of

	county	perot	gore	bush	buchanan	RES_1
1	Alachua	8072.00	47300.00	34062.00	262.00	-27.53116
2	Baker	667.00	2392.00	5610.00	73.00	48.08359
3	Bay	5922.00	18850.00	38637.00	248.00	35.29824
4	Bradford	819.00	3075.00	5414.00	65.00	34.65193
5	Brevard	25249.00	97318.00	115185.0	570.00	-333.34446
6	Broward	38964.00	386518.0	177279.0	789.00	-604.44456
7	Calhoun	630.00	2155.00	2873.00	90.00	66.40577
8	Charlotte	7783.00	29645.00	35426.00	182.00	-97.20386
9	Citrus	7244.00	25525.00	29766.00	270.00	10.05709

Data View / Variable View

SPSS Processor is ready

Produce the histogram of the residuals as described in Chapter 2 of this manual.

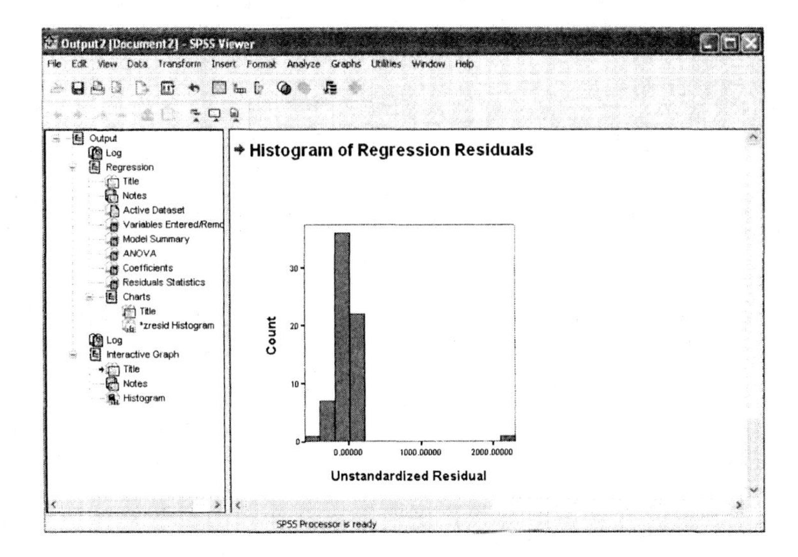

The histograms have the same basic shape – and the outlier appears in both graphs – but the scales are different and this second graph doesn't automatically include a normal plot overlay. These default histograms produced by SPSS are a little different from the one in your text but the outlier shows up just the same.

Influential Points

If you discover a potential influential point, it is very easy to temporarily eliminate this observation and repeat the analysis, to assess the impact of that one observation. To demonstrate how to do this, let's take a look at the "us_statewide_crime" data from Example 13 in Chapter 3, which is on your text CD. Access this data file and make a scatterplot of Y = murderrate vs. X = college.

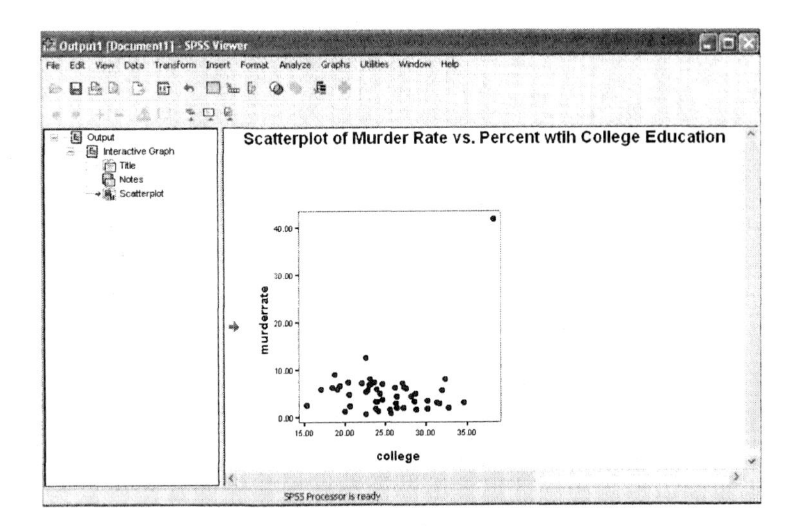

The plot above was resized to fit in the output window with the title, and the plot symbol and color were also changed. Notice the outlier in the upper right corner that is also a potential influential point. This is the observation for Washington DC. Make the plot again but this time add the regression line by clicking on the Fit tab and selecting Regression from the Method options.

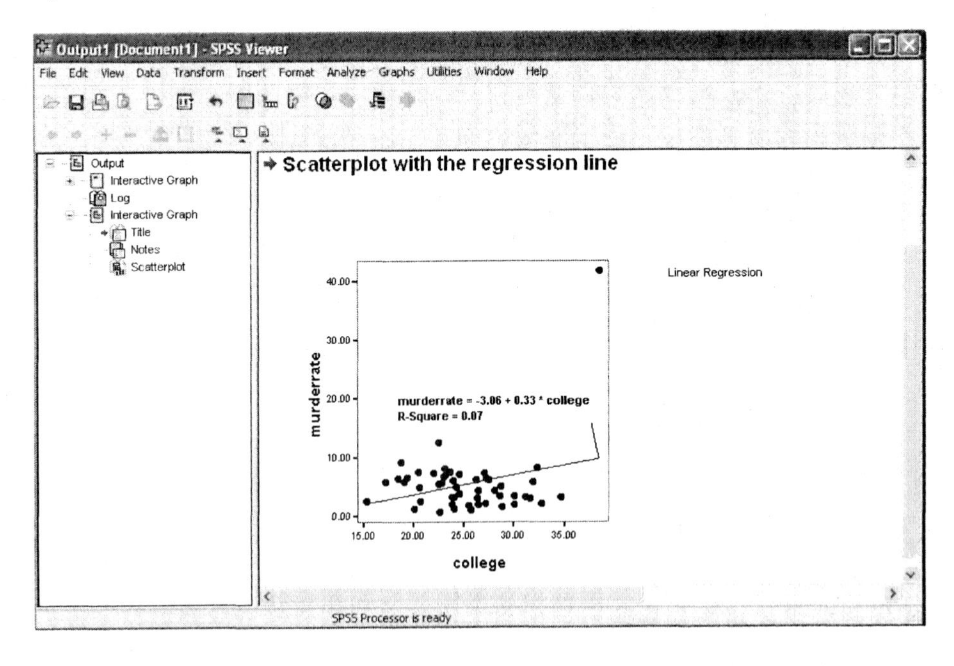

The line does appear to be "pulled up" toward the Washington DC data point. To produce the same graph without Washington DC, select **Data → Select Cases...** and click on "If condition is satisfied."

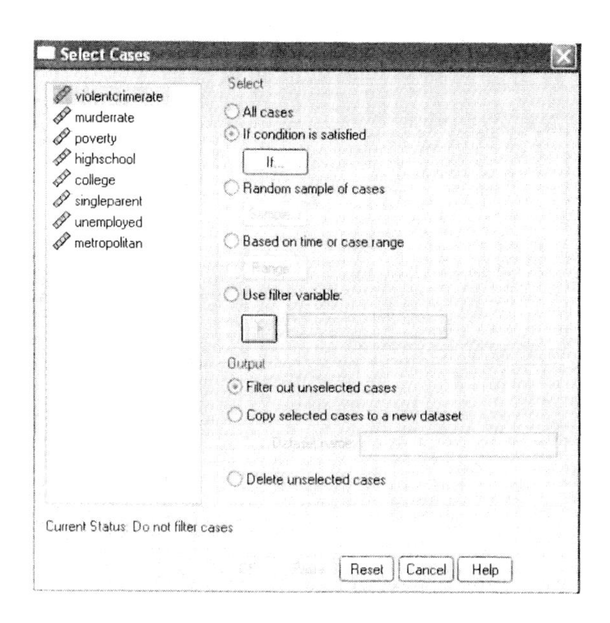

Click on the "If..." button and enter the expression "murderrate < 40" in the box at the top. Since the murder rate for DC is so much larger than it is for the other states, this is an easy way to eliminate the DC observation from the analysis without actually deleting the observation from the data set.

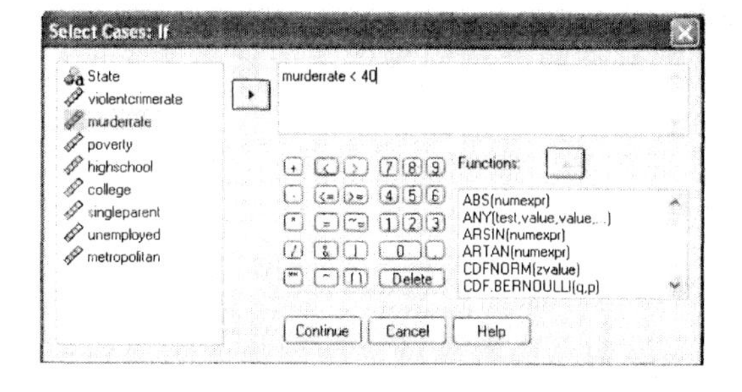

Click on **Continue** and then on **OK**. If you look at the Data Editor now you will see that the number to the left of "District of Columbia" is crossed out. SPSS also adds a "filter" variable to the data set with a value of 1 for cases to include in subsequent analyses and 0 for those to exclude. All of the cases except DC will have a value of 1 for the "filter," while DC has the value 0.

us_statewide_crime.sav - SPSS Data Editor

1 : State Alabama

	State	violentcrimerate	murderrate	poverty	highschool	college	singleparent	unemployed	metropolitan
1	Alabama	486.00	7.40	14.70	77.50	20.40	26.00	4.60	7
2	Alaska	567.00	4.30	8.40	90.40	28.10	23.20	6.60	4
3	Arizona	532.00	7.00	13.50	85.10	24.60	23.50	3.90	8
4	Arkansas	445.00	6.30	15.80	81.70	18.40	24.70	4.40	4
5	California	622.00	6.10	14.00	81.20	27.50	21.80	4.90	9
6	Colorado	334.00	3.10	8.50	89.70	34.60	20.80	2.70	8
7	Connecticut	325.00	2.90	7.70	88.20	31.60	22.90	2.30	9
8	Delaware	684.00	3.20	9.90	86.10	24.00	25.60	4.00	8
9	District of Columbia	1508.00	41.80	17.40	83.20	38.30	44.70	5.80	10
10	Florida	812.00	5.60	12.00	84.00	22.80	26.50	3.60	9
11	Georgia	505.00	8.00	12.50	82.60	23.10	25.50	3.70	6
12	Hawaii	244.00	2.90	10.60	87.40	26.30	19.10	4.30	7

Data View / Variable View

SPSS Processor is ready Filter On

Make the scatter plot of murder rate vs. college again with the regression line.

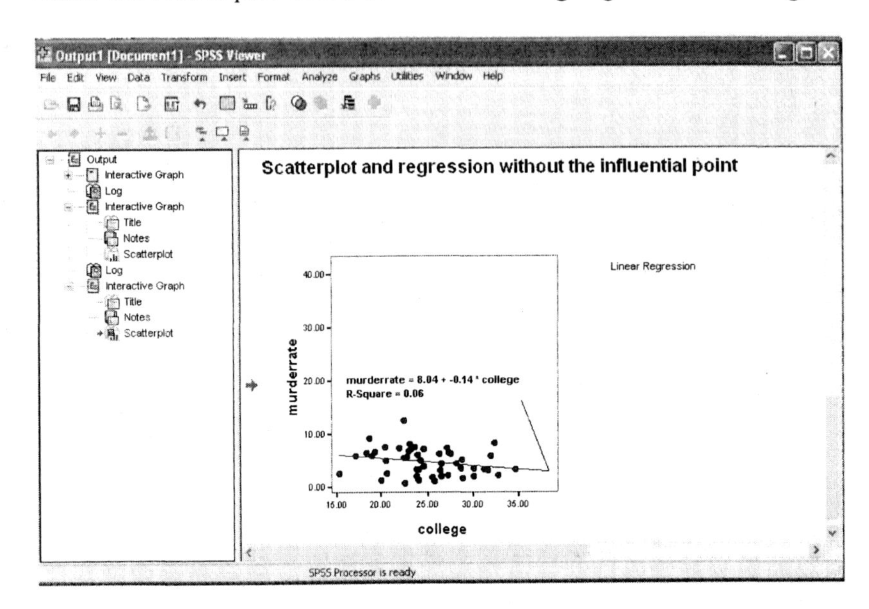

Instead of a positive slope, as in the plot with DC, this graph shows a regression line with a negative slope. The value of R-Square is also changed.

Chapter 4 Sampling

Most of the time that you will be using SPSS you will probably be doing data analyses similar to the examples given in the previous chapters of this manual. Occasionally, though, you might need to generate a random sample from a data set or from a list of numbers that represent the individuals in a population. These tasks are easy to do without SPSS, of course, but if you don't have access to a random number table, SPSS has a command for selecting samples that will do the job for you!

Random Samples

If you want to select a random sample from within a data set, you can do this very easily using SPSS. You can also use this feature to generate a sequence of random numbers. First we will demonstrate how to select a random sample from an existing data set. Access the "fla_student_survey" data set from your text CD. Suppose we want to select a random sample of 5 students from the 60 students in this data set. Select **Data → Select Cases...** and click on "Random sample of cases."

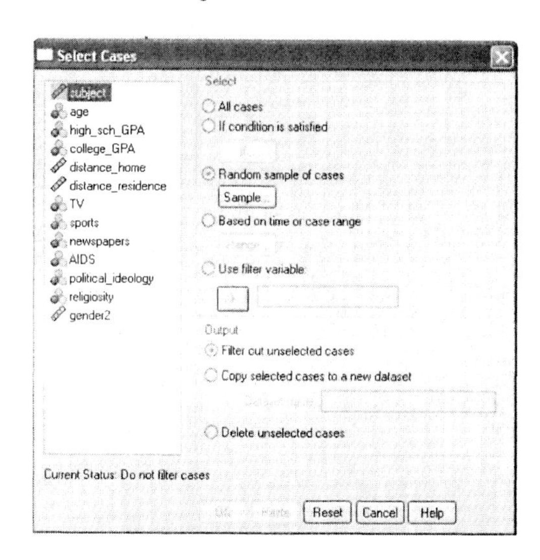

Click on the "Sample" button and in the dialog box that appears click on "Exactly" and enter "5" and "60" as in the example dialog box below (so that it reads "Exactly 5 cases from the first 60 cases").

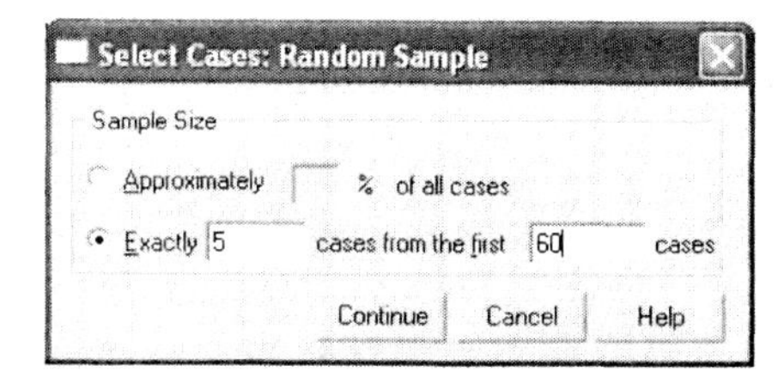

Click on **Continue** and then on **OK**. You will see that most of the case numbers now have a diagonal slash indicating that they are not included in the random sample.

*fla_student_survey.sav [DataSet2] - SPSS Data Editor

File Edit View Data Transform Analyze Graphs Utilities Window Help

1 : subject 2 Visible: 20 o

	subject	gender	age	high_sch_G	college_GP	distance_h	distance_re	TV
1	2.0	f	23.0	2.10	3.50	1200.00	.30	1
2	3.0	f	27.0	3.30	3.00	1300.00	1.50	
3	4.0	f	35.0	3.50	3.20	1500.00	8.00	
4	8.0	f	31.0	3.00	3.00	5000.00	1.50	
5	12.0	f	27.0	3.50	3.60	190.00	3.00	1
6	15.0	f	28.0	3.00	3.50	3500.00	1.00	
7	16.0	f	25.0	3.80	3.30	210.00	10.00	
8	17.0	f	41.0	4.00	3.00	1000.00	15.00	
9	20.0	f	28.0	3.00	3.80	120.00	1.00	2
10	21.0	f	26.0	3.70	3.70	8000.00	8.00	
11	22.0	f	27.0	4.00	3.70	2.00	2.50	
12	24.0	f	23.0	3.70	3.70	2.00	2.00	

Data View / Variable View

SPSS Processor is ready

In the example above we can see that the 9th observation corresponding to subject #20 was selected for the sample. To see which cases were selected you can scroll through the data set or print a list using the "Case Summaries" command. Select **Analyze → Reports → Case Summaries...** and enter the variables you want to review for your sample. In the example below, the variables "subject," "age," and "TV" have been selected.

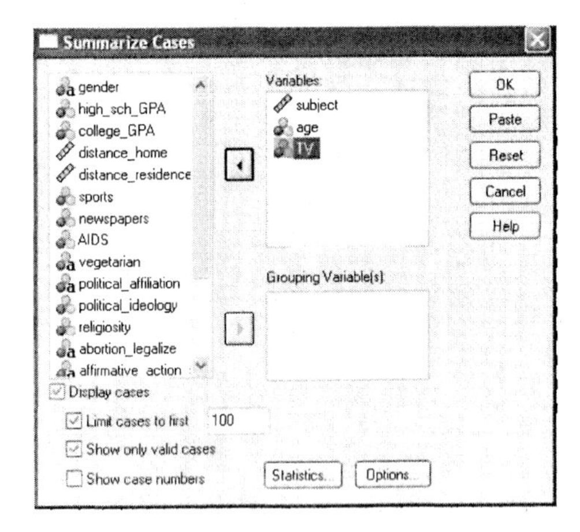

Click on **OK** to produce the output. In the output that is produced, we can see that observations 20, 46, 55, 29 and 58 were selected. If you use the "Select Cases" command again, you will get a different random sample each time.

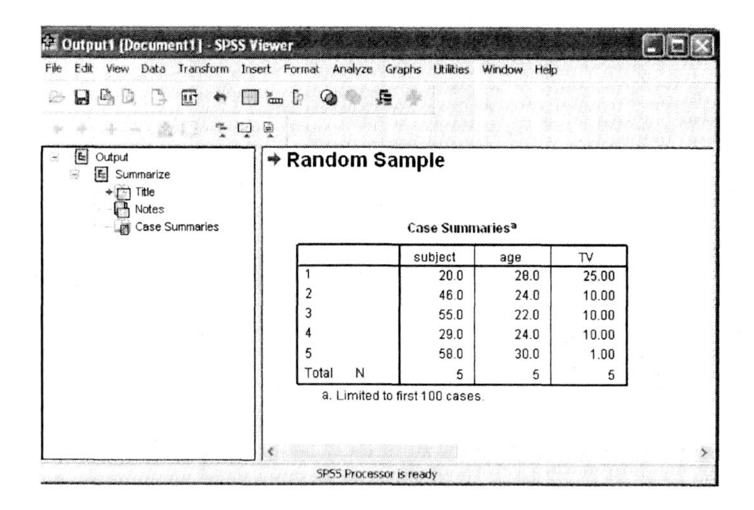

Generating a Sequence of Random Numbers

If you are not working with an existing SPSS data file and you would like to generate a sequence of random numbers to identify a random sample, as in Example 5 of Chapter 4 of your text, SPSS can generate the numbers for you. First, you will have to enter the numbers corresponding to the items you will be sampling from. In Example 5, the items are 60 account numbers. Enter the numbers 1 through 60 in the first column of the Data Editor. You can name the variable "account" if you want to, but it is not necessary.

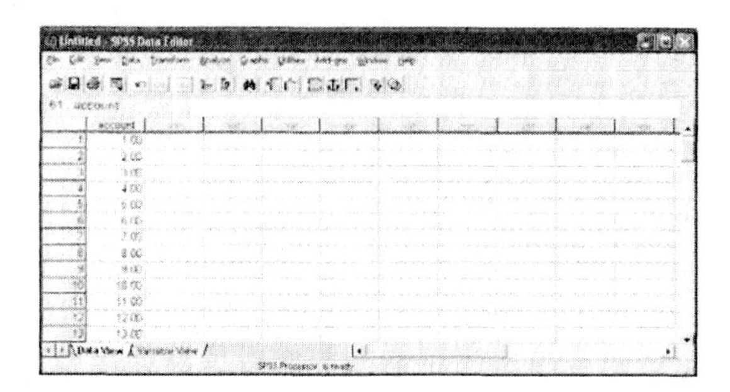

Select a random sample as above. Select **Data → Select Cases...** then click on "Random sample of cases." Click on "Sample" and enter 10 for the number of cases to select and 60 for the total number of cases. Click on "Continue" and then on "OK.'

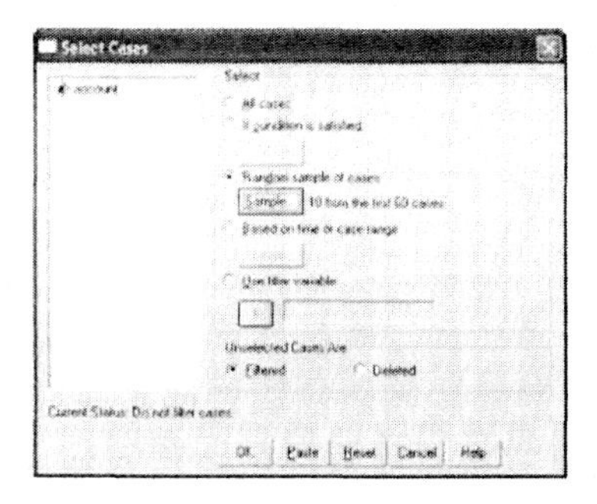

Scroll through the list to see which numbers were selected or use the "Case Summaries" command as above.

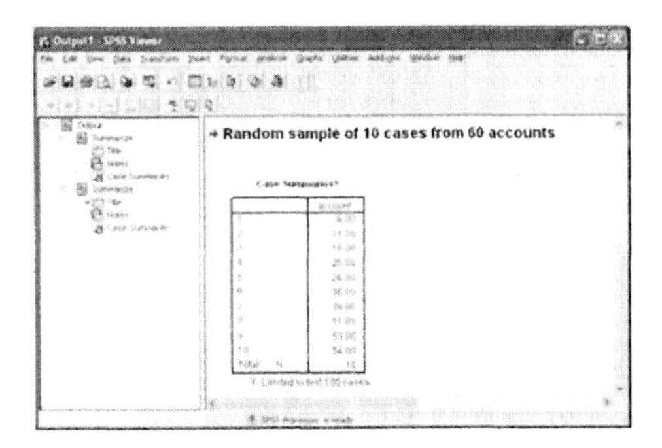

For this example, accounts 6, 11, 18, 25, 26, 36, 39, 51, 53 and 54 were selected for the random sample. Note: you can actually select the sample without having the numbers 1 through 60 entered into Data Editor. Open a new Data Editor and enter a 0 in the 40th row of the first column. SPSS will give this column the default name "VAR00001." All of the observations above this zero will appear as a ".", which is the "missing value" symbol in SPSS. Scroll down to the 60th row and enter another 0. Now, repeat the same process for selecting a random sample as above. SPSS will create the filter variable to select the sample. You will have to scroll through the data file to find out which "observations" were selected. In the example below you can see that the 12th observation was selected for the sample. The other four observations are 19, 37, 57 and 58.

*Untitled1 [DataSet0] - SPSS Data Editor

File Edit View Data Transform Analyze Graphs Utilities Window Help

1 : filter_$ 0 Visible: 2 of

	VAR00001	filter_$	var	var	var	var	var
1	.	0					
2	.	0					
3	.	0					
4	.	0					
5	.	0					
6	.	0					
7	.	0					
8	.	0					
9	.	0					
10	.	0					
11	.	0					
12	.	1					

Data View / Variable View

SPSS Processor is ready

Chapter 5 Probability Calculations

Your main goal for learning to use SPSS is to learn which commands produce the output you need, but you can also use SPSS to do probability calculations. It is fairly easy to find probabilities using tables, but it is nice to know that you can use SPSS to do this when you don't have access to the right table!

The binomial and normal distributions are among the discrete and continuous probability distributions available in SPSS. To find a probability, you will need a Data Editor with at least one variable and at least one observation already in it – the commands for calculating probabilities (or for producing any output) won't work in an empty Data Editor. Open a new Data Editor in SPSS and enter 0 (or any other value) in the first row of the first column, just so there is at least one observation entered. SPSS will name the variable "VAR00001" but there is no need to rename this variable because you are not going to use it for anything.

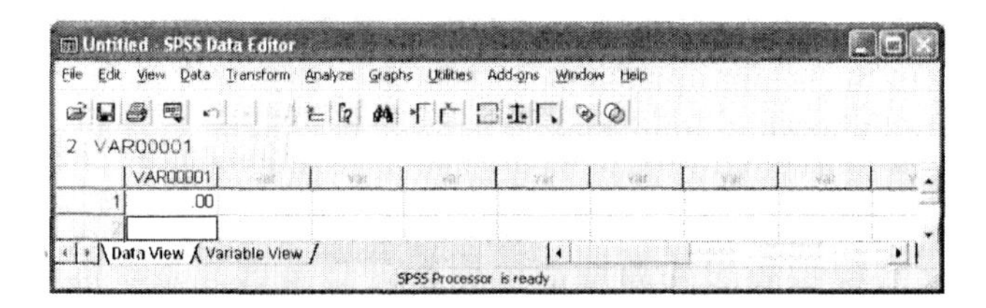

Binomial Probability Calculations

To find the binomial probability of no women in 10 employees chosen for managerial training, as in Example 12 of Chapter 6 in your text, select **Transform → Compute Variable...** and enter "prob1" (or some other variable name) in the "Target Variable" box in the dialog box that appears. Click on the box under "Numeric Expression" and scroll through the list under "Function Group" to find "PDF & Noncentral PDF." (PDF stands for "Probability Density Function.") Click on this to reveal the list of available probability distributions under "Functions and Special Variables." Scroll through this list to find "PDF.Binom" and double click on it.

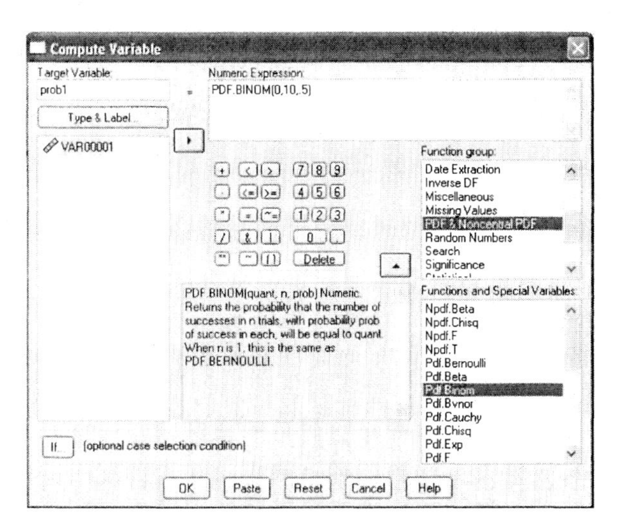

The expression "PDF. Binom(?,?,?) will appear in the "Numeric Expression" box. The first question mark is for the value that you want to calculate the probability for, in this case 0. The other two question marks are for the sample size (n) and the probability of a success (p), respectively. We want the probability of no (0) successes in a sample of n=10 when the probability of a success is p=0.50. This value of p represents the case when there are an equal proportion of men and women from the population from which the sample is taken. Edit the three question marks by clicking on the expression in the "Numeric Expression" box, then erasing the three questions marks and typing in "0,10,.5" between the parentheses. It should look like the example above.

Click on **OK** to complete the calculation. Notice that no output is produced from this command, other than a log statement. Instead, the probability will appear in the Data Editor in a new column with variable name "prob1." You will have to increase the number of decimal places and the width of the variable in the Variable View to see this probability value.

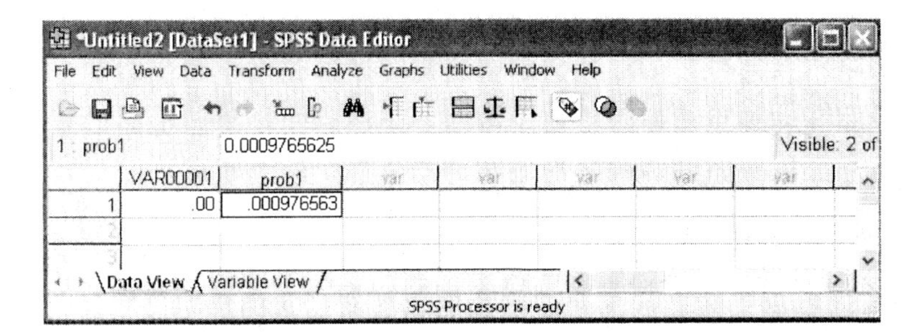

So, the probability of no women selected in a sample of 10 when the proportion of women is .5 is .000977, or approximately .001. To find the entire probability distribution for n=10 and p=0.50, just repeat this process and change the first value entered to the desired value. For example, to find the probability of selecting 3 women in 10 trials when the proportion of women is .50 the "Numeric Expression" would be "PDF.Binom(3,10,.5)." SPSS will replace the first probability calculation with this new result unless you enter a new variable name, such as prob2 in the "Target Variable" box. If you don't define a new variable for this calculation, SPSS will ask you if you want to replace the existing variable with the new calculation. Click on **OK** if you don't mind losing the first probability result; otherwise, click on **Cancel** and give the variable a new name in the Target Variable box.

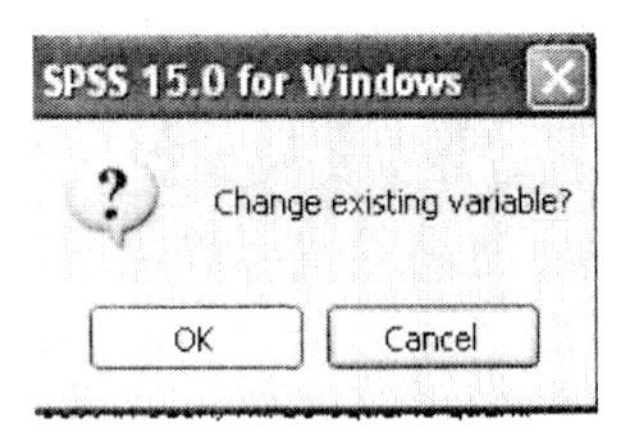

You could also find the enter distribution by creating a variable to contain each of the possible values of the binomial random variable and then replacing the first question mark with the name of this variable. The display below shows a Data Editor with a variable named X and the Compute Variable command dialog box with the same calculations as above, except the value for the first question mark is the variable X.

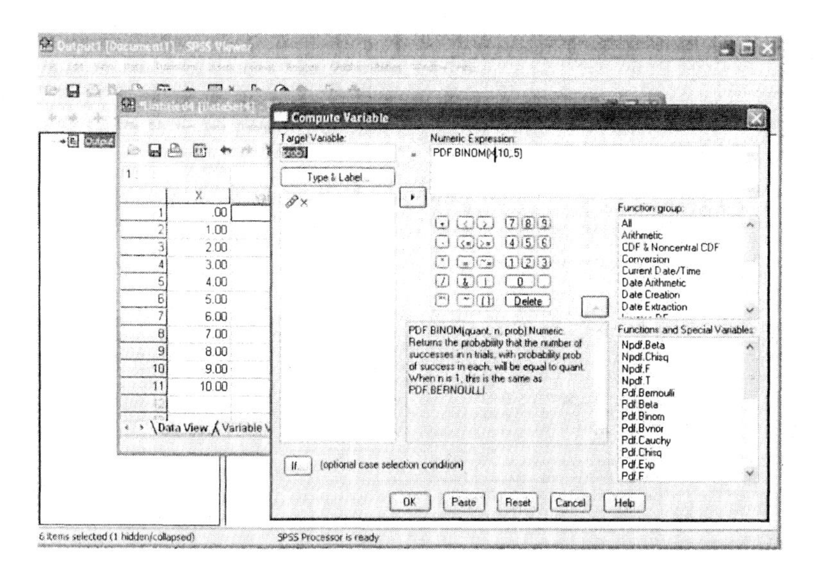

This command would produce the following results in the Data Editor. The number of decimal places was increased to 5.

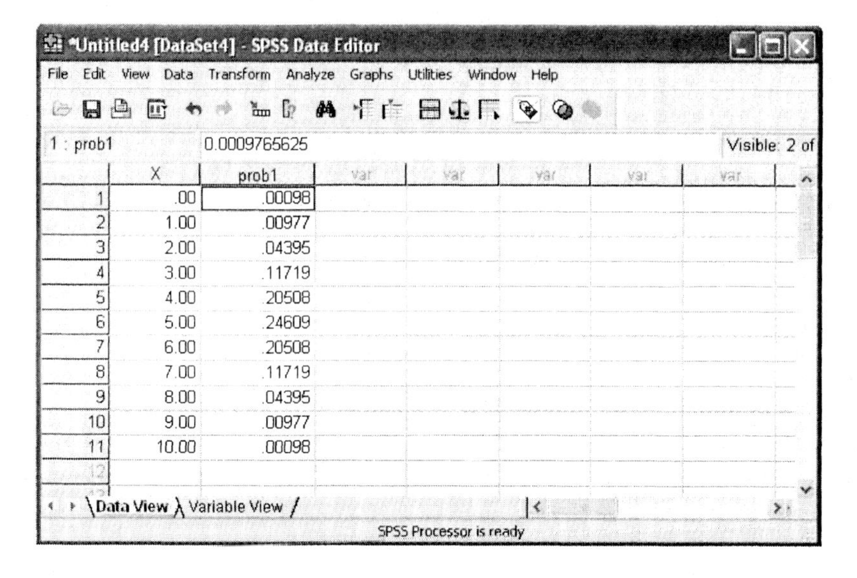

	X	prob1
1	.00	.00098
2	1.00	.00977
3	2.00	.04395
4	3.00	.11719
5	4.00	.20508
6	5.00	.24609
7	6.00	.20508
8	7.00	.11719
9	8.00	.04395
10	9.00	.00977
11	10.00	.00098

SPSS also has a function for finding the cumulative distribution function for the Binomial distribution. Select **Transform → Compute Variable...** again but click on "CDF & Noncentral CDF" under "Function group" instead of the PDF option. (CDF represents "Cumulative Distribution Function.") Double click on Cdf:Binom so that the expression CDF:BINOM(?,?,?) appears in the Numeric Expression box. Replace the three question marks with 3,10,5 and click on OK. This will calculate $P(X \leq 3)$ for a binomial random variable with $n=10$ and $p=.5$. The result is approximately 0.171875. The entire cumulative distribution can be found following the same process as above for the probability distribution. Select **Transform → Compute Variable...** and replace the first question mark with the variable X. Name the new variable something like "cum1."

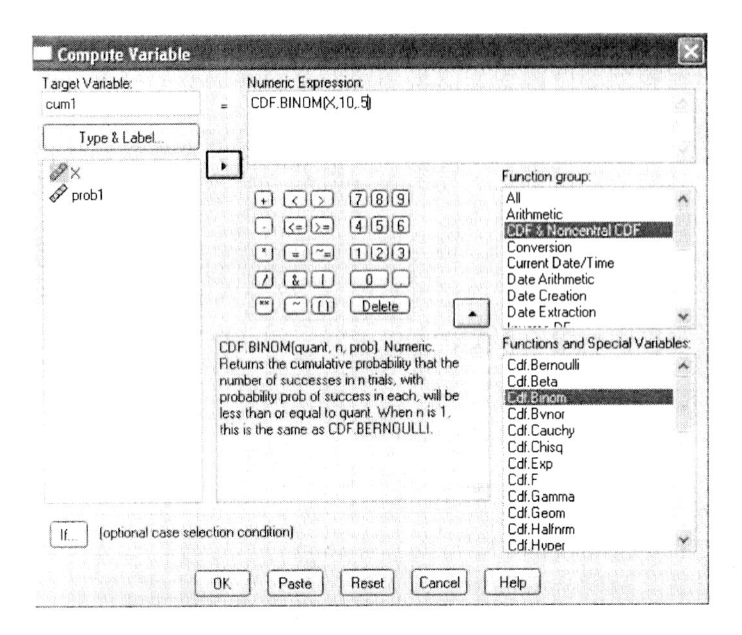

Click on **OK** to see the cumulative distribution.

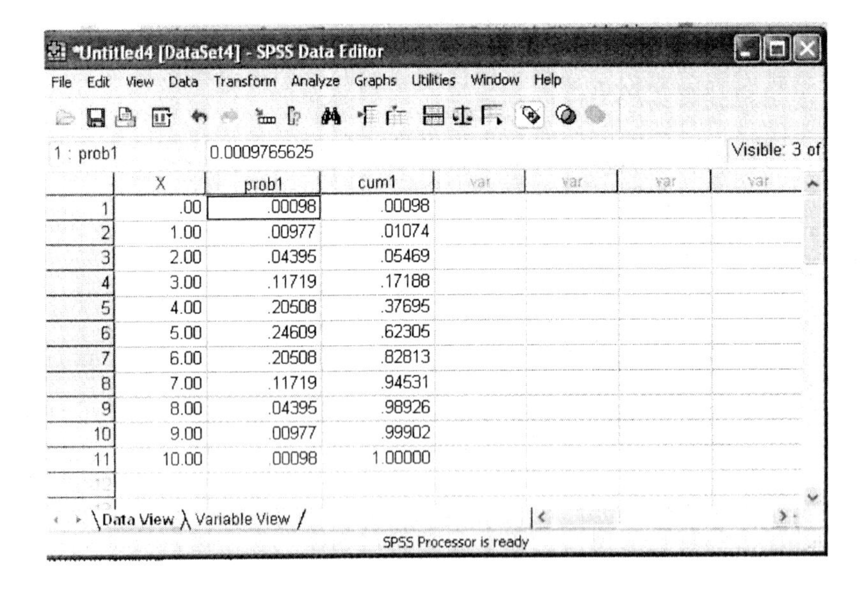

	X	prob1	cum1
1	.00	.00098	.00098
2	1.00	.00977	.01074
3	2.00	.04395	.05469
4	3.00	.11719	.17188
5	4.00	.20508	.37695
6	5.00	.24609	.62305
7	6.00	.20508	.82813
8	7.00	.11719	.94531
9	8.00	.04395	.98926
10	9.00	.00977	.99902
11	10.00	.00098	1.00000

Normal Probability Calculations

The process for calculating a probability for a normal distribution is the same as the process for finding a cumulative binomial probability. Select **Transform → Compute Variable...** and scroll through the list of available distributions under "Functions and Special Variables to find "Cdf.Normal" and double click on it.

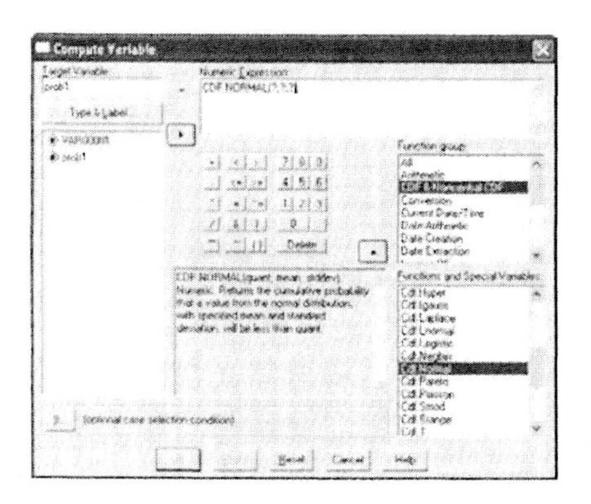

The expression "CDF.Normal(?,?,?)" will appear in the "Numeric Expression" box. Here the three question marks represent a value from a normal distribution that you want to find the cumulative probability for, and the mean and standard distribution of that distribution, respectively. SPSS will report the probability of observing a normal random variable at or below the value you enter. Let's use Example 8 of Chapter 6 in your text to demonstrate how to find normal probabilities. In this example, you are asked to find the probability of an SAT score higher than 650 when the mean SAT score is 500 and the standard deviation is 100. Replace the three question marks, as above, with "650,500,100" and click on **OK**.

The probability shown in the Data Editor is .933193, but this is the probability of an SAT score *below* 650, not above. You can find the probability of an SAT score larger than 650 by subtracting .933193 from 1, of course, or you could enter the expression "1 – CDF.Normal(650,500,100)" in the "Numeric Expression" box and let SPSS do the calculation for you.

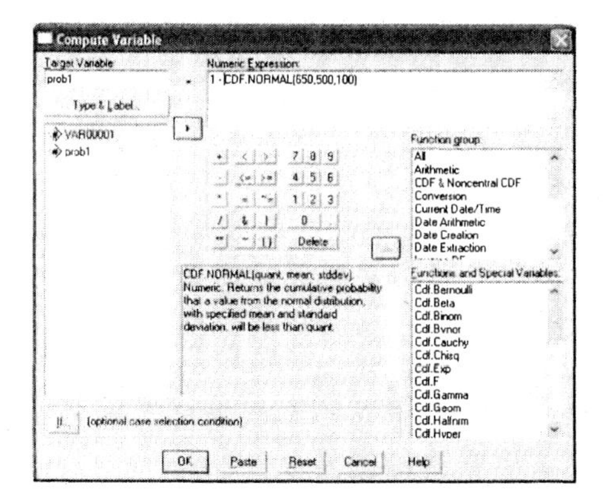

Finding a Standard Normal P-value

The function for finding normal probabilities may be useful for you later on when you need to find something called a "p-value" for a standard normal Z-score called a "test statistic." A p-value is a probability associated with an observed Z-score. For these calculations, the mean and standard deviation are 0 and 1, respectively, because the corresponding distribution is standard normal. For example, suppose we have an observed Z-score of 1.97 and the p-value we need corresponds to the probability of getting a value larger than 1.97. (Don't worry – the reason for needing this probability will be explained in your statistics course. Here you will just see how to use SPSS to find the probability.) Select **Transform → Compute Variable...**, enter "pvalue" as the "Target Variable" and enter the expression "1-CDF.Normal(1.97,0,1)" in the "Numeric Expression" box.

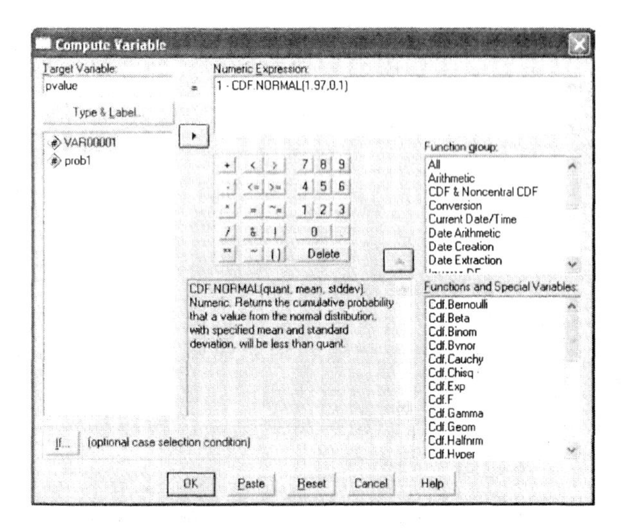

Click on **OK** to find the p-value.

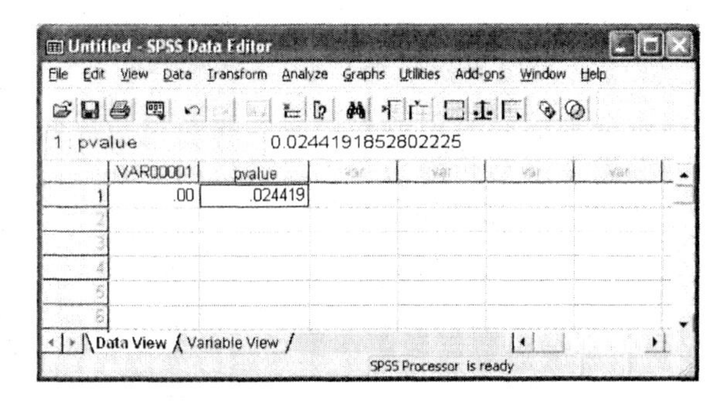

The p-value is .024419.

If you scroll through the list of Cdf functions available you will see some additional distribution functions that you may need to use to find p-values later on in your statistics course. These include the chi-square distribution (Cdf.chisq), the F distribution (Cdf.F), and the T distribution (Cdf.T).

Inverse Distribution Functions

In some situations you may need to find the z-score value for the standard normal distribution that has a specified proportion of the distribution below it in value. For example, to construct a 95% confidence interval for a proportion, as in Example 2 in Chapter 7, you need the z-score that has 97.5% of the distribution below it in value. The easiest way to find such a z-score is to use a standard normal table. But, in a situation where you don't have access to a table you can use SPSS to find the z-score you need. Select **Transform → Compute Variable...** and enter "zvalue" (or some other variable name) in the "Target Variable" box. Scroll through the list under "Function group" to find "Inverse DF" and click on it. (DF stands for "Distribution Function.") Scroll through the list under "Functions and Special Variables" to find "Idf.Normal" and double click on it. The expression "IDF.NORMAL(?,?,?)" will appear in the "Numeric Expression" box. This command works similarly to the normal distribution Cdf function above in that the last two question marks are for the mean and standard deviation of a normal distribution, respectively. The first question mark is for the appropriate probability value. If you want to find the zscore for a 95% confidence interval, replace the three question marks with ".975,0,1". You have to enter .975 instead of .95 because the "Inverse DF" is the inverse of the cumulative distribution function. So, to find the middle 95% you need the value that has 97.5% below it.

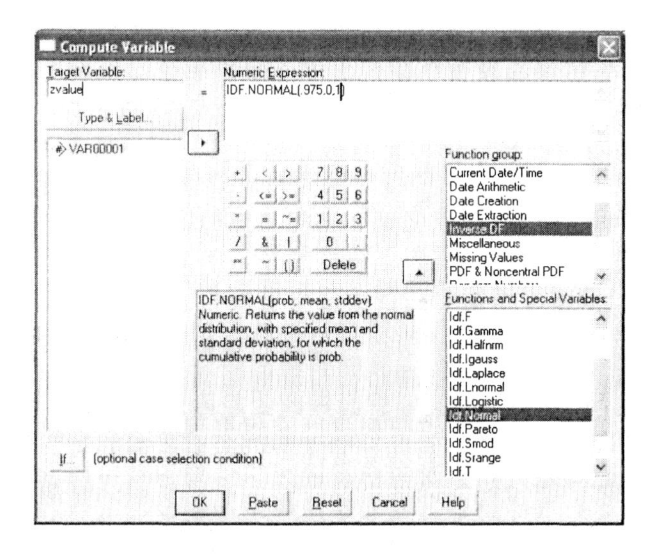

Click on **OK** to complete the calculation. Just like the Pdf and Cdf functions, no output will be produced. Instead the zscore value will appear as the value of a new variable created by the "Transform" command.

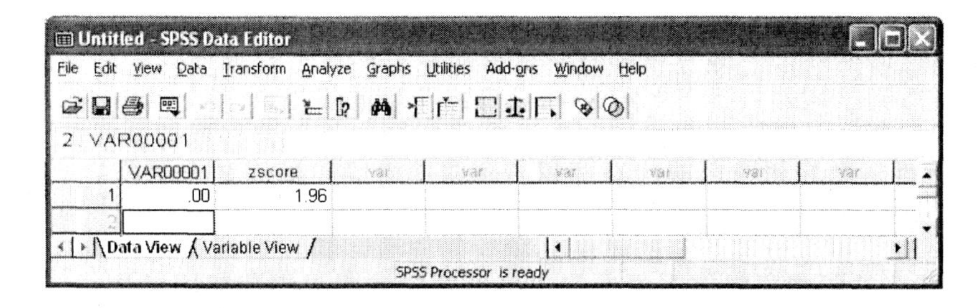

Chapter 6 Sampling Distributions

Simulating a Sampling Distribution

SPSS does not have a built-in command for simulating a sampling distribution but the process for doing this is fairly easy and takes only a few steps. Let's demonstrate this using the data for Exercise 7.63, the ages of all 50 heads of households in a small fishing village in Nova Scotia. This data is provided below.

<table>
<tr><td>50</td><td>45</td><td>23</td><td>28</td><td>67</td><td>62</td><td>41</td><td>68</td><td>37</td><td>60</td><td>41</td><td>70</td><td>47</td><td>66</td><td>51</td><td>57</td><td>40</td><td>36</td><td>38</td><td>81</td><td>27</td><td>37</td><td>56</td><td>71</td><td>39</td></tr>
<tr><td>46</td><td>49</td><td>30</td><td>28</td><td>31</td><td>45</td><td>43</td><td>43</td><td>54</td><td>62</td><td>67</td><td>48</td><td>32</td><td>42</td><td>33</td><td>36</td><td>25</td><td>29</td><td>57</td><td>39</td><td>50</td><td>64</td><td>76</td><td>63</td><td>29</td></tr>
</table>

Enter the ages in SPSS, then select **Data → Select Cases...** and click on "Random sample of Cases." Click on the "Sample" button and fill in the dialog box so that SPSS will select 9 of the 50 cases in the data set. Click on **Continue** and then on **OK**.

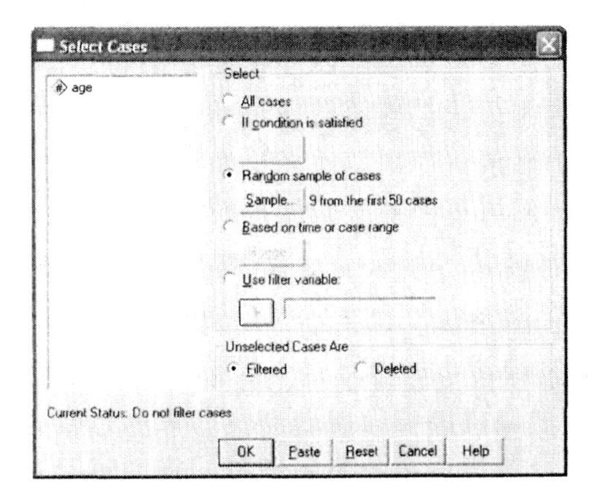

The next step is to create a new variable that has only the values of "age" for the 9 selected households for the first sample. To do this, select **Transform → Compute Variable...** and enter "sample1" in the "Target Variable" box. Enter the expression "age" in the "Numeric Expression" box, then click on the "If..." button in the lower left corner of the dialog box. Click on "Include if case satisfies condition" and enter "filter_$>0" in the box.

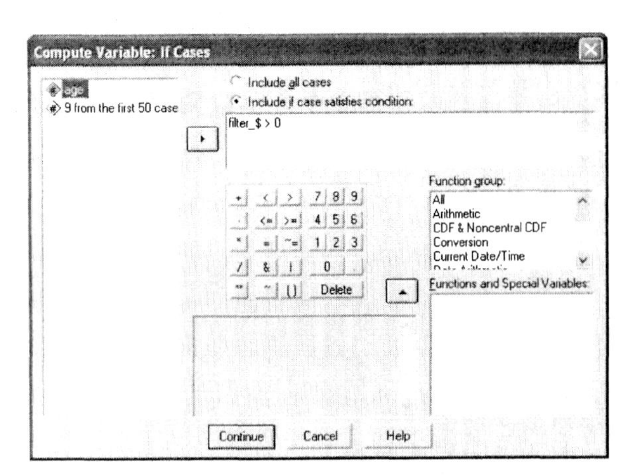

Click on **Continue**. Your dialog box should be filled in as the one below.

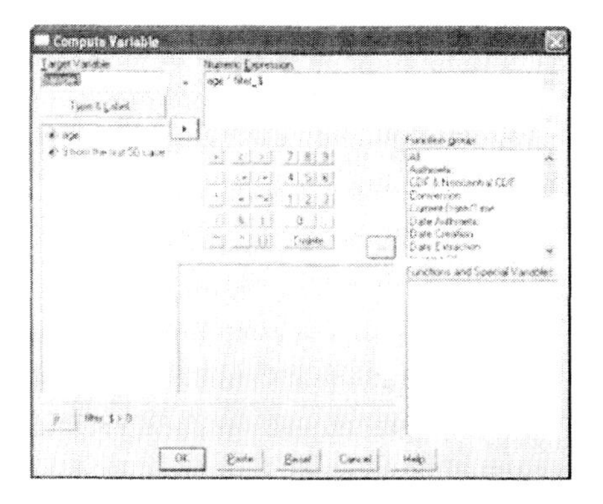

Click on **OK** to create the sample.

For this sample, the ages 23, 40, 36, 46, 45, 32, 39, 50, and 29 were selected. To create a second random sample, select **Data → Select Cases...** again and just click on **OK** since the dialog box is already set for selecting a new sample of 9 cases. Now select **Transform → Compute Variable...** again, but change the "Target Variable" to "sample2" and click on **OK**.

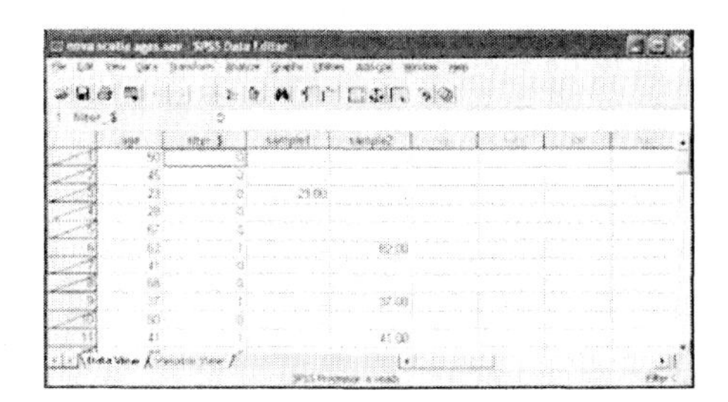

For this second sample, ages 62, 37, 41, 27, 56, 71, 39, 30, and 31 were selected. Repeat this process, changing the target variable name each time, for as many samples as you wish to create. Once you have the number of samples you need, you can find the mean of each sample by using one of the options under the "Descriptive Statistics" command. But, first you will need to "turn off" the "Select Cases" option under the Data menu. Select **Data → Select Cases...**, click on "All cases" under "Select," and then on **OK**. The following display shows the means for twenty samples of size nine chosen from this distribution.

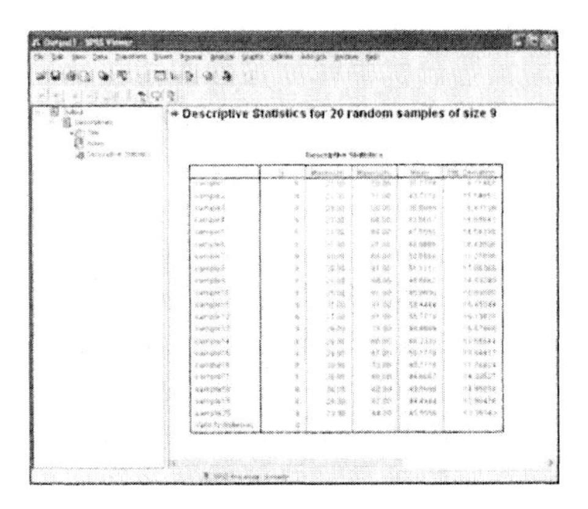

You can now enter these means in the same data file under a new variable named “mean.”

To investigate the sampling distribution, find the mean and standard deviation of the variable “mean” and produce a histogram, as shown below.

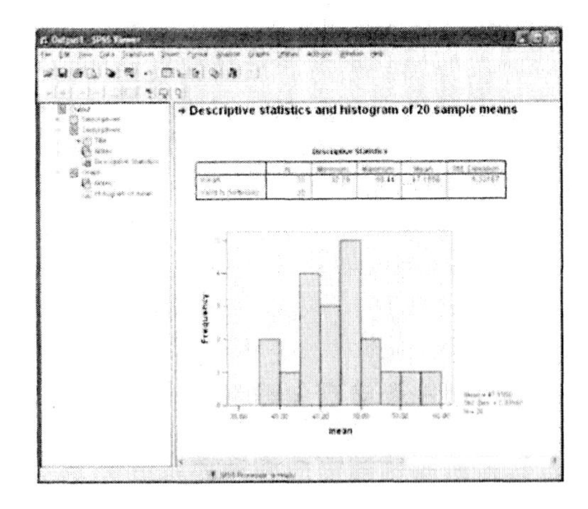

For these twenty samples, the mean of the sample means is 47.1556, which is very close to the population mean of 47.18. The standard deviation of the sample means is 5.33167, which is a little more than the expected value of 14.74/3, or 4.9133. The histogram also shows a relatively mounded and symmetric shape, as expected.

Chapter 7 Confidence Intervals

In this chapter we will demonstrate SPSS commands for producing a confidence interval for a mean. SPSS does not have a command for producing a confidence interval for a proportion. You can use SPSS to find the appropriate critical z-score for the level of confidence you are using if you don't have access to a normal distribution table. This process is described in Chapter 5 of this manual, in the section on Inverse Distribution Functions.

Confidence Interval for a Mean

Two of the commands under the "Analyze" menu will provide the confidence interval for a mean, "Explore" under the "Descriptive Statistics" command and "One-sample T Test" under the "Compare Means" command. We will demonstrate the "Explore" command in this section using the eBay auction data from Example 7 in Chapter 8, and then see how the "One-Sample T Test" command operates in the next chapter. The data is shown below and is the final prices (in dollars) of an item sold using two methods, "buy-it-now" and "bidding only."

Buy-it-now	235, 225, 225, 240, 250, 250, 210
Bidding only	250, 249, 255, 200, 199, 240, 228, 255, 232, 246, 210, 178, 246, 240, 245, 225, 246, 225

Data Entry Options

You have two options for entering the data in SPSS. One way would be to enter the data in two separate columns, one for each method used to sell the item. The data would appear as below.

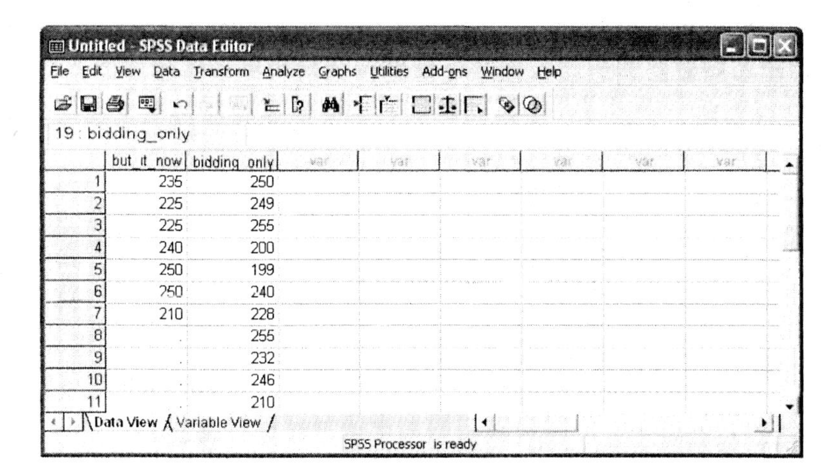

The other option would be to enter all of the selling prices in one variable and then to use a second variable to identify which selling method was used. The first variable would be a numeric variable, named something like "final_price," and the second one could be either numeric or string and named "selling_method," for example. Entered using this second method the data would appear as below.

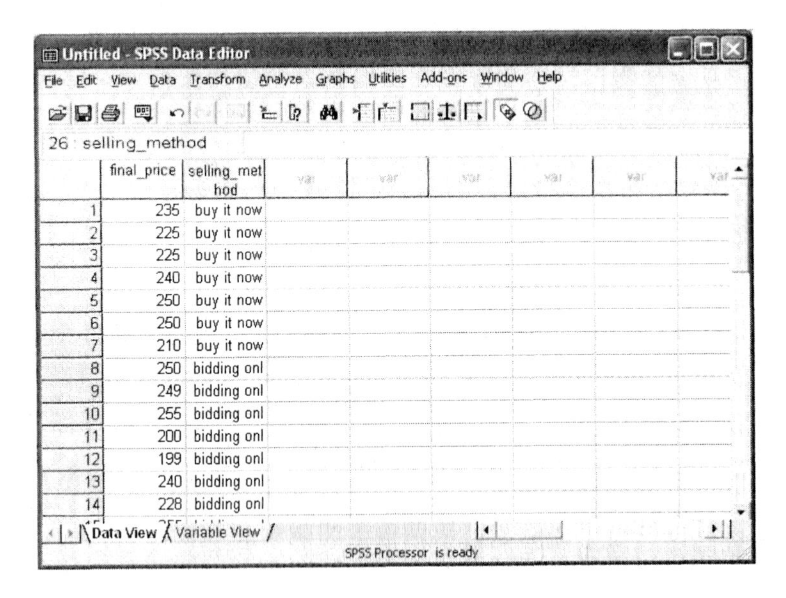

For finding the confidence interval for the mean of each sample you can use either data entry method. But, if you want to use a method for comparing the means of two or more independent samples, you will need to use the second data entry method. And, if you want to use a method for comparing the means of two or more dependent samples, you will need to use the first data entry method.

To find the confidence interval for a mean using the first data entry method, select **Analyze → Descriptive Statistics → Explore...** and enter both variable names in the box under "Dependent List." If you do not want any graphs, click on "Statistics" under "Display."

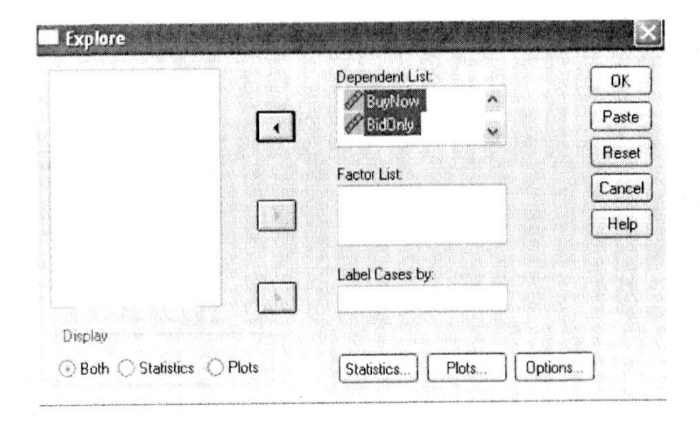

Click on **OK** to produce the output. Some of the statistics produced by SPSS have been deleted from the output display below.

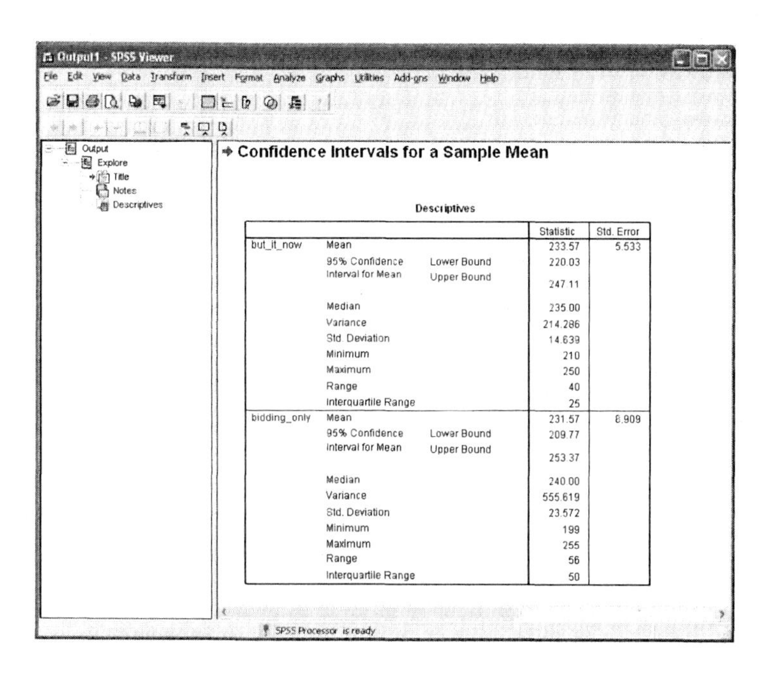

Confidence Intervals for a Sample Mean

Descriptives

			Statistic	Std. Error
but_it_now	Mean		233.57	5.533
	95% Confidence Interval for Mean	Lower Bound	220.03	
		Upper Bound	247.11	
	Median		235.00	
	Variance		214.286	
	Std. Deviation		14.639	
	Minimum		210	
	Maximum		250	
	Range		40	
	Interquartile Range		25	
bidding_only	Mean		231.57	8.909
	95% Confidence Interval for Mean	Lower Bound	209.77	
		Upper Bound	253.37	
	Median		240.00	
	Variance		555.619	
	Std. Deviation		23.572	
	Minimum		199	
	Maximum		255	
	Range		56	
	Interquartile Range		50	

The confidence interval for the "buy it now" method is (220.03, 247.11) and for the "bidding only" method it is (209.77, 253.37). If you wish to change the level of confidence, click on the **Statistics** button and in the dialog box that appears, enter a new level of confidence to replace 95 in the appropriate box. The example below shows the level of confidence changed to 90%.

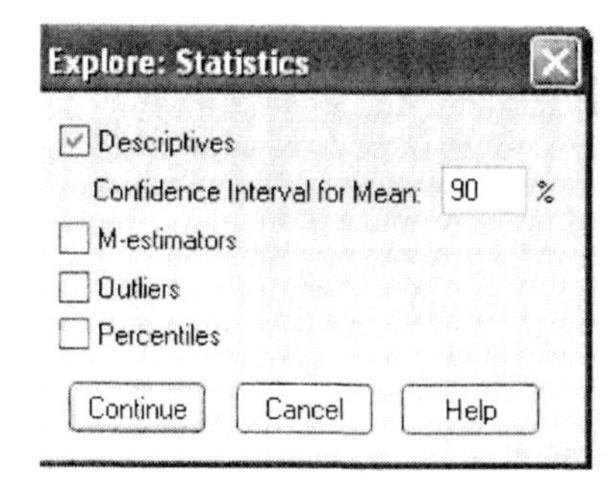

To find confidence intervals for a mean using the second data entry method select **Analyze → Descriptive Statistics → Explore...** and enter the variable with the prices in the "Dependent List" box and the variable with the selling method in the "Factor List" box. Click on **OK**. Except for some labels, the output will appear exactly the same as the output produced above.

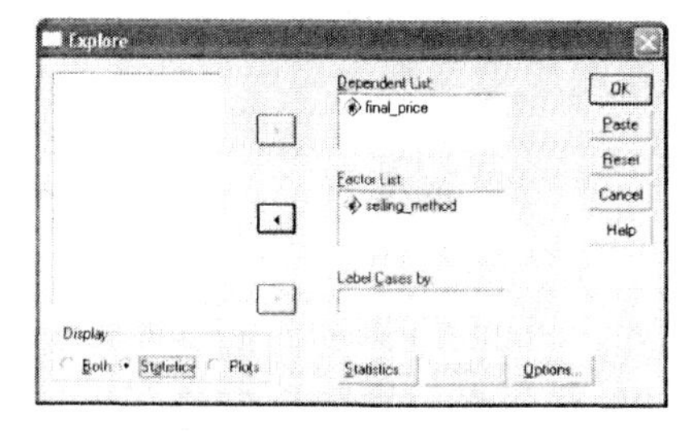

Chapter 8 Significance Tests about Hypotheses

In this section you will be introduced to a number of SPSS commands that you can use to conduct a hypothesis test.

Hypothesis Test for a Proportion

SPSS has two commands that you can use to do a test on the proportion of "successes" in a population. The first test is called the "Binomial Test" and it will give you approximately the same results as the Z-test in your text. The second test is a Chi-square "Goodness-of-fit" test; although the results for this test are equivalent to the Z-test for large samples, the process for doing this test is a little more complicated than it is for the Binomial Test. Both of these tests are available under the "Nonparametric Tests" command under the "Analyze" menu and both tests will be demonstrated below. You will learn about nonparametric tests in the last chapter in your text and in this manual. Remember, though, that the Z-test for a proportion is easy to calculate by hand and you can use the method described in Chapter 5 of this manual to find the corresponding standard normal p-value.

The Binomial Test

Let's demonstrate this test using the data from Example 4 in Chapter 9. In this example, dogs were trained to identify patients with bladder cancer, so p is the proportion of dogs that make the correct selection on a given trial. In this study, the dogs made the correct selection 22 times in 54 trials. Enter this data in SPSS using two variables, one to indicate the result of each trial ("correct" = 1 or "incorrect" = 0) and one to indicate the count for each result. The "Result" variable must be entered as a coded numeric variable and you must code the response that corresponds to the proportion of interest with the value 1. Don't forget to select **Data → Weight Cases...** and enter "Count" as the weighting variable.

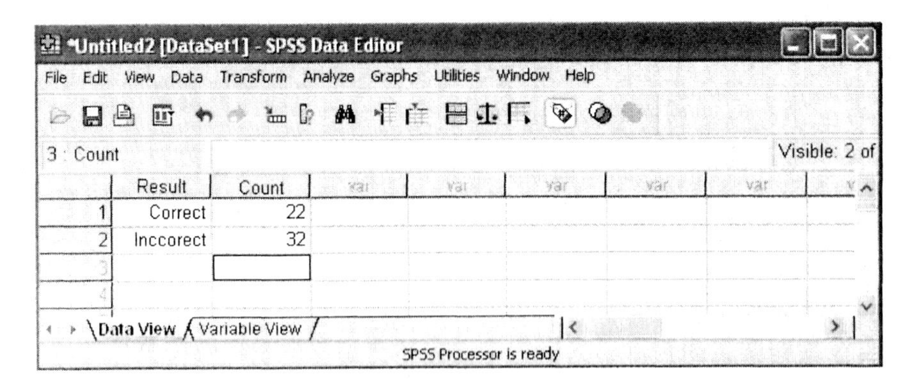

The null hypothesis is that the proportion of correct results is 1/7 and the alternate hypothesis is that it is different from 1/7. To conduct the test, select **Analyze → Nonparametric Tests → Binomial...** Enter "Result" in the "Test Variable List" and .143 (=1/7) in the "Test Proportion" box.

Click on **OK** to produce the output.

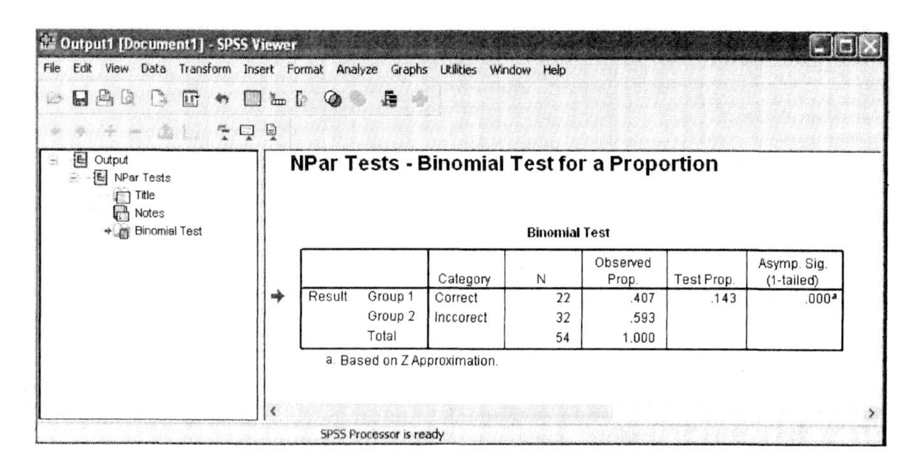

NPar Tests - Binomial Test for a Proportion

Binomial Test

		Category	N	Observed Prop.	Test Prop.	Asymp. Sig. (1-tailed)
Result	Group 1	Correct	22	.407	.143	.000[a]
	Group 2	Inccorect	32	.593		
	Total		54	1.000		

a. Based on Z Approximation.

SPSS reports the observed proportion of each result (correct and incorrect) as well as the test (null) proportion for the correct response. SPSS also reports the p-value for the two-sided alternative hypothesis, but not the actual value of the test statistic. The p-value isn't .000, however. SPSS only reports p-values to three decimal places, so we can tell that the p-value is less than .0005. With such a small p-value (less than .0005) you can't tell that the results for this Binomial Test differ slightly from the results for the Z-test. You will be able to see this, though, with the next example.

For Example 6 in Chapter 9, p is the proportion of correct predictions made by therapeutic touch practitioners. In this example the alternative hypothesis is that p is more than $p_0 = .50$. The sample results were that 70 out of 150 therapeutic touch practitioners correctly identified which of their hands was closer to the hand of a researcher. Enter this data in SPSS using the same format as above. (If you have just done the previous example, all you will have to do is replace the counts with the appropriate counts for this example.)

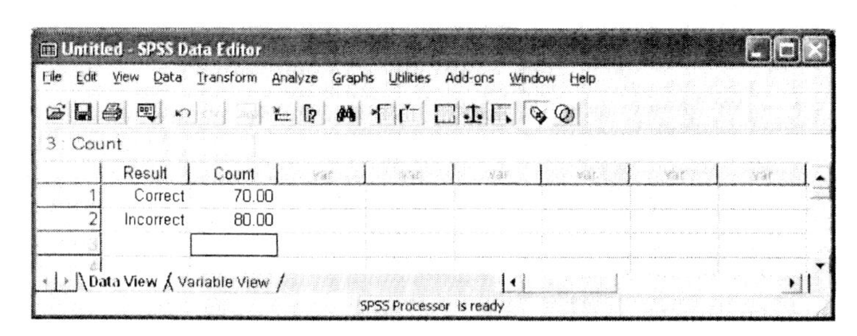

	Result	Count
1	Correct	70.00
2	Incorrect	80.00

To conduct the test, select **Analyze → Nonparametric Tests → Binomial…** Enter "Result" in the "Test Variable List" and 0.50 in the "Test Proportion" box.

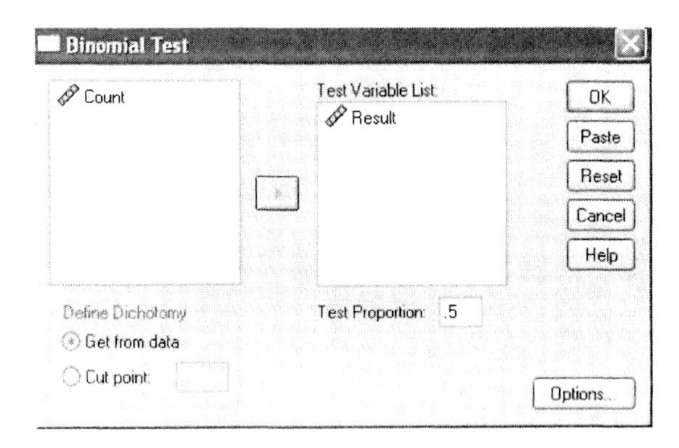

Click on **OK** to complete the test.

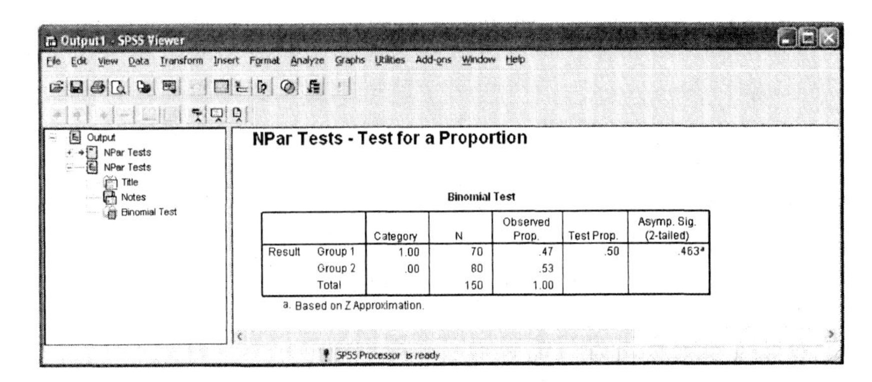
NPar Tests - Test for a Proportion

Binomial Test

		Category	N	Observed Prop.	Test Prop.	Asymp. Sig. (2-tailed)
Result	Group 1	1.00	70	.47	.50	.463a
	Group 2	.00	80	.53		
	Total		150	1.00		

a. Based on Z Approximation.

Since we are using the "greater than" alternate hypothesis and the observed proportion is less than the hypothesize proportion, the p-value for the Binomial Test is 1-0.463/2 = 0.7685, which is a little less than the p-value of 0.79 reported in your text. For most situations, the results for the Binomial Test and the Z-test will agree; that is, if you reject the null hypothesis using one test then you will also reject it using the other test.

Chi-Square Goodness-of-Fit Test

This test compares the observed frequencies of observations in two or more categories to the expected frequencies assuming a specified distribution. When there are just two categories, the chi-square test statistic is equal to the square of the Z-statistic for this situation. Although calculated using a chi-square distribution instead of the standard normal distribution, for large samples the p-value reported by SPSS for this test turns out to be the same as the p-value for the Z-statistic for the two-sided alternative hypothesis.

Let's try this approach using both of the examples we used above for the Binomial Test. Recall that for Example 4 there were 22 correct responses and 32 incorrect responses.

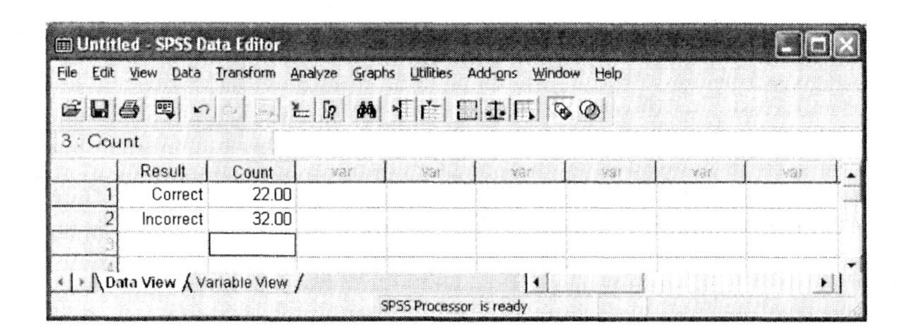

To conduct the test, select **Analyze → Nonparametric Tests → Chi-square...** Enter "Result" in the "Test Variable List" and then click on "Values" below "Expected Values." This test is based on comparing the actual counts to the expected counts, instead of comparing the sample proportion to the hypothesized proportion. To do this test, you have to tell SPSS what the expected frequencies would be for the two groups (correct and incorrect) under the null hypothesis. The hypothesized value for p is 1/7 and there were 54 trials, so we would expect 54/7 or 7.71 correct trials and 54 - 7.71 = 46.29 incorrect trials. Click on the box next to "Values" and enter "7.71" and click on "Add." Then click on that box again, enter "46.29" and click on "Add" again. We had to round these values because SPSS only allows a limited number of decimal places in the "Expected Values" box. At this point, the "OK" button should be highlighted.

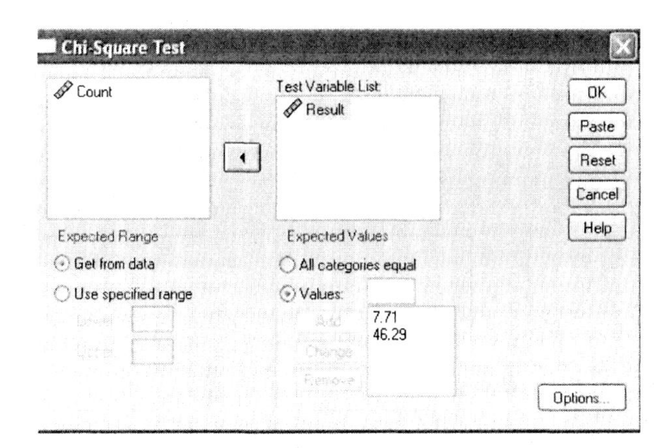

Click on **OK** to complete the command.

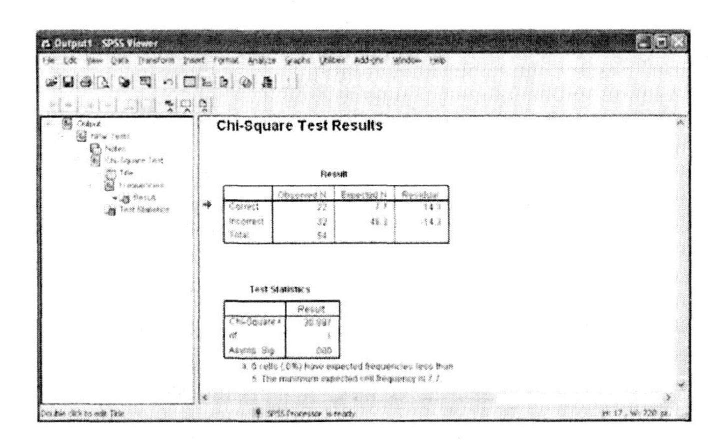

The value of the test statistic is 30.897. The Z test statistic in your text is calculated as $Z = \frac{.407 - .143}{\sqrt{\frac{.143(1-.143)}{54}}} = 5.555$, where $\hat{p} = \frac{22}{54} = .407$ and $p_0 = \frac{1}{7} = .143$. You can check that the square of 5.555 is roughly 30.897, the value of the Chi-Square test statistic reported by SPSS. The p-value for the test is the value in the output next to "Asymp. Sig." and is less than .0005.

Now let's try this with Example 6 where there were 70 correct responses and 80 incorrect, and the alternative hypothesis is one-sided. Enter the data in SPSS as shown below.

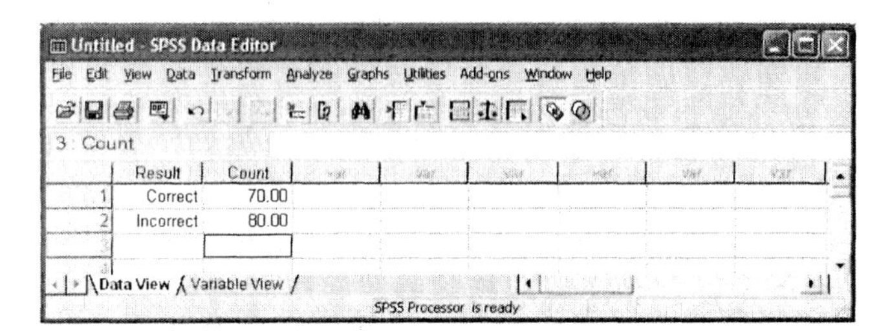

Select **Analyze → Nonparametric Tests → Chi-square...** Enter "Result" in the "Test Variable List" and then click on "All categories equal."

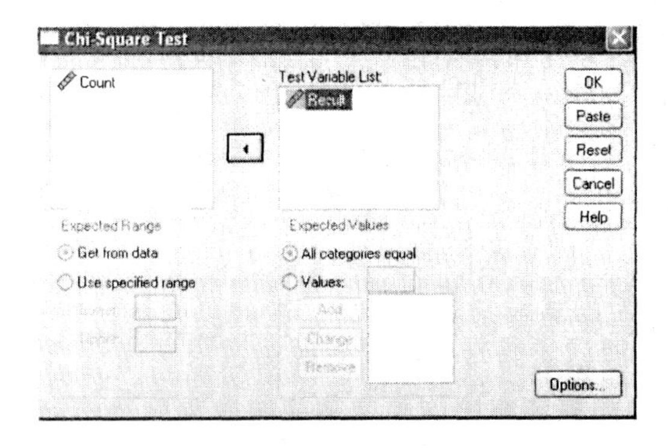

Click on **OK** to complete the command.

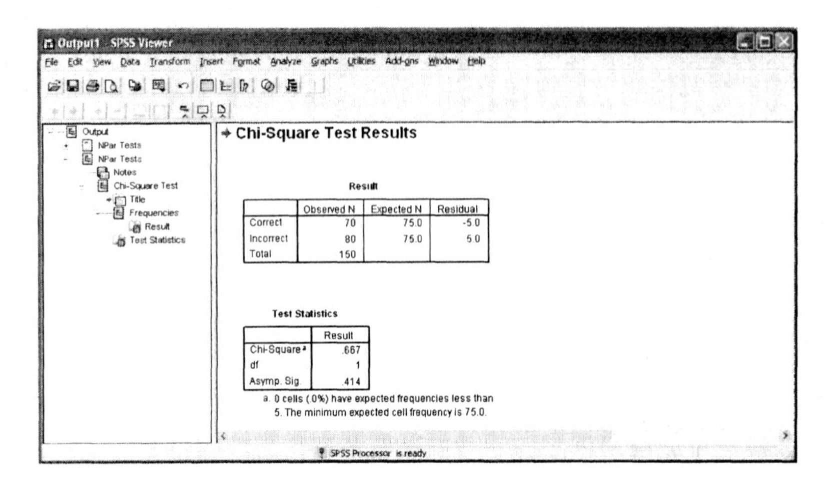

Since we know that Z = -0.82, we expect that the chi-square statistic will be $(-0.82)^2 = 0.67$. The chi-square statistic reported by SPSS is 0.667 and the p-value is .414. This p-value represents two times the "tail area" above 0.817, where 0.817 is the square root of 0.667. Since the sample proportion is less than the expected proportion, we know that the test statistic is -0.817 and the p-value should be 1 - .414/2, or .793. These results agree more closely with the results of the Z-test; however, you have to convert the sample and null proportions to sample and null (expected) counts and, although the p-values are the same, the value of the chi-square test statistic is the square of the Z-statistic, which might be a little confusing for the beginning statistician!

Hypothesis Test for a Mean

To demonstrate how to conduct a test for a mean, we will use the anorexia data set in Example 8 of Chapter 9. This data set consists of the weights of 29 anorexia patients both before and after a treatment program, along with their change in weights. This data is shown in the table below.

	Weight				Weight				Weight		
Girl	Before	After	Change	Girl	Before	After	Change	Girl	Before	After	Change
1	80.5	82.2	1.7	11	85.0	96.7	11.7	21	83.0	81.6	-1.4
2	84.9	85.6	0.7	12	89.2	95.3	6.1	22	76.5	75.7	-0.8
3	81.5	81.4	-0.1	13	81.3	82.4	1.1	23	80.2	82.6	2.4
4	82.6	81.9	-0.7	14	76.5	72.5	-4.0	24	87.8	100.4	12.6
5	79.9	76.4	-3.5	15	70.0	90.9	20.9	25	83.3	85.2	1.9
6	88.7	103.6	14.9	16	80.6	71.3	-9.3	26	79.7	83.6	3.9
7	94.9	98.4	3.5	17	83.3	85.4	2.1	27	84.5	84.6	0.1
8	76.3	93.4	17.1	18	87.7	89.1	1.4	28	80.8	96.2	15.4
9	81.0	73.4	-7.6	19	84.2	83.9	-0.3	29	87.4	86.7	-0.7
10	80.5	82.1	1.6	20	86.4	82.7	-3.7				

We would like to determine if the after weights are "significantly" different from the before weights, on average. Access the data set on your text CD, but notice that the data set also includes the same variables for a second group of anorexia patients. We want only the first 29 patients (those that were given the "cognitive" treatment"). To eliminate the other patients select **Data → Select Cases…** and select only those subjects in this group. Group is coded as "1" for the Cognitive Treatment patients and "2" for the control patients. Use the select cases command to select only those patients with a Group value of 1. Alternatively, you could just enter the data in the table above, using three columns, as in the display below. Note that for this demonstration we don't need the before and after weights, so you could just enter the data for the change variable.

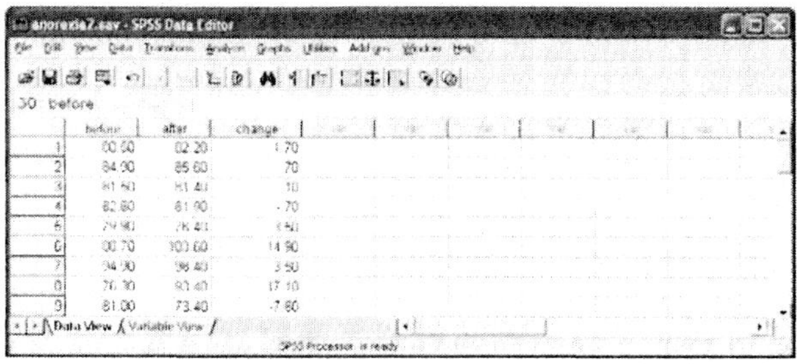

To do the test, select **Analyze → Compare Means → One-Sample T Test...** and enter the "change" variable in the "Test Variable(s)" box.

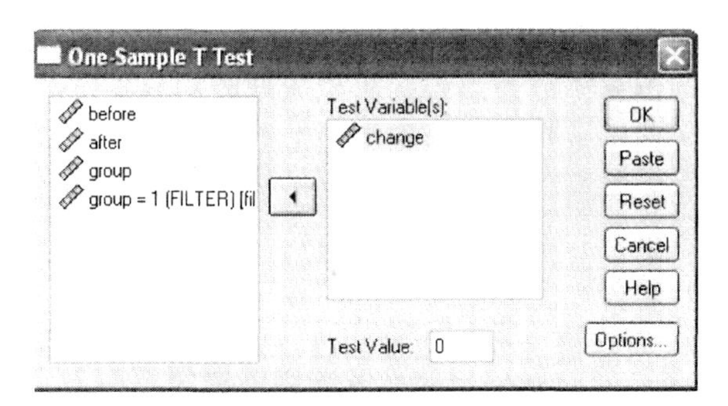

The dialog box includes a box for you to change the "null" or Test Value. For this example, we want to test if the change is different from zero, which is the default value. Click on **OK** to produce the output.

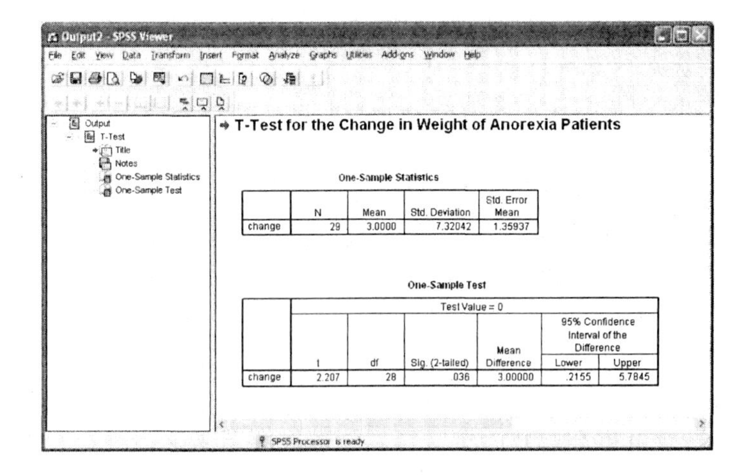

T-Test for the Change in Weight of Anorexia Patients

One-Sample Statistics

	N	Mean	Std. Deviation	Std. Error Mean
change	29	3.0000	7.32042	1.35937

One-Sample Test

	Test Value = 0					
					95% Confidence Interval of the Difference	
	t	df	Sig. (2-tailed)	Mean Difference	Lower	Upper
change	2.207	28	.036	3.00000	.2155	5.7845

The output consists of two parts. The first part provides some descriptive statistics for the "change" variable, including the mean and standard deviation. The second part provides the value of the test statistic, the degrees of freedom for the associated T-distribution, and the p-value for the two-sided (or two-tailed) alternate hypothesis. The p-values is listed as "Sig. (2-tailed)" in the output. SPSS does not include an option for finding one-sided p-values. If you are doing a one-sided test you can find the p-value as one-half of Sig or 1 minus one-half of Sig, depending on the direction of your alternate hypothesis and the value (positive or negative) of the T-statistic. The output also includes a confidence interval for the mean. For this particular example, we could have done a paired or dependent samples T-test instead, since the before and after weights are provided. This is demonstrated in the next chapter.

Chapter 9 Comparing Two Groups

In this chapter we will look at methods for comparing the means or proportions for two samples. While there are no commands for producing just confidence intervals for these situations, most of these commands either include a confidence interval as part of the default output or they include an option for producing a confidence interval.

Comparing the Proportions of Two Independent Samples

SPSS includes an option under the "Nonparametric Tests" command for doing a test to compare the proportions of "successes" from two independent samples. Unlike the Binomial Test, the results of this test will agree with the test shown in your test. SPSS also includes a command for conducting a chi-square "Contingency Table" test that is equivalent to the Z-test. The chi-square test will be discussed in the next chapter.

Let's demonstrate this test using the data from Example 5 in Chapter 10. In this example, 5 of 88 teenagers who watched TV for less than 1 hour per day reported an aggressive act, while 154 of 465 teenagers who watched TV at least 1 hour per day reported an aggressive act. Enter this data in SPSS using three variables, one for the response variable (aggressive act), one for the group (TV watching habits) and one for the counts in each group with each response. As we saw in the previous chapter, both the response variable and the group variable must be entered as coded numeric variables. Make sure to weight the cases using the "count" variable.

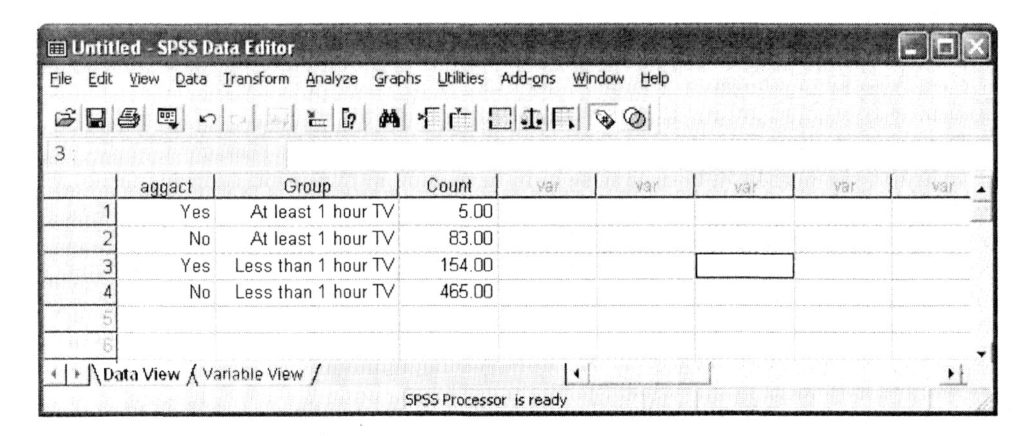

	aggact	Group	Count	var	var	var	var	var
1	Yes	At least 1 hour TV	5.00					
2	No	At least 1 hour TV	83.00					
3	Yes	Less than 1 hour TV	154.00					
4	No	Less than 1 hour TV	465.00					
5								
6								

To do the test, select **Analyze → Nonparametric Tests → 2 Independent Samples...** and enter the "aggressive act" variable in the "Test Variable List" box and the "group" variable in the "Grouping Variable:" box. Click on "Define Groups" and enter the two numeric values you used to define the groups. The data for this example was entered using "1" for the "less than 1 hour of TV" group and "2" for the "at least 1 hour of TV" group. Click on Mann-Whitney-U so that the OK button is highlighted, as shown below.

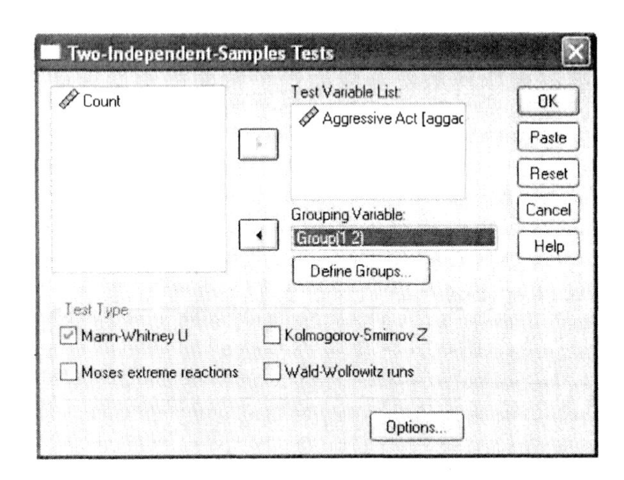

Click on **OK** to produce the output.

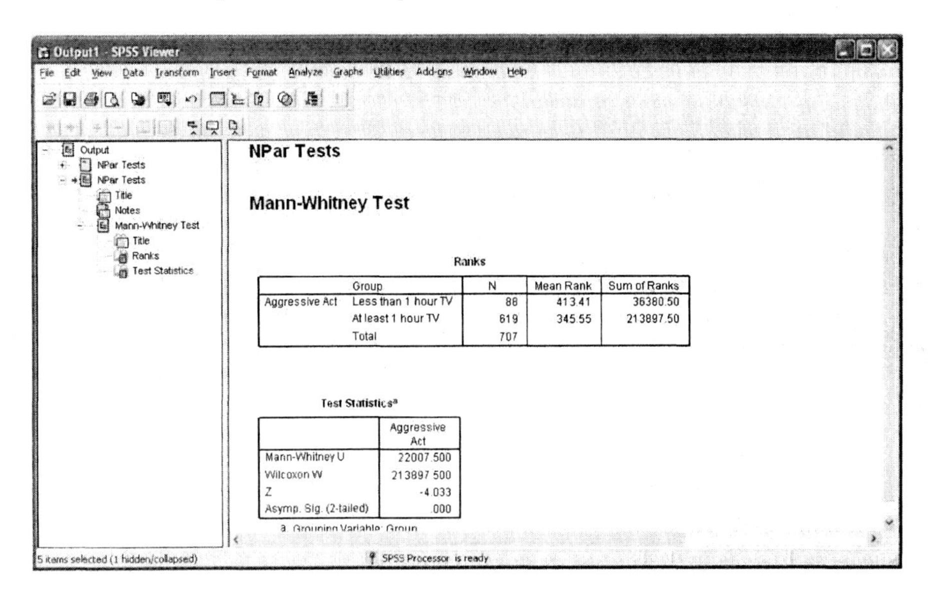

	Group	N	Mean Rank	Sum of Ranks
Aggressive Act	Less than 1 hour TV	88	413.41	36380.50
	At least 1 hour TV	619	345.55	213897.50
	Total	707		

	Aggressive Act
Mann-Whitney U	22007.500
Wilcoxon W	213897.500
Z	-4.033
Asymp. Sig. (2-tailed)	.000

The value listed next to "Z" in the output is the value of the two-sample Z-statistic shown in your text, and the value next to "Asymp. Sig" is the two-sided p-value for this test.

Comparing the Means of Two Independent Samples

The "Analyze" menu in SPSS includes both parametric and nonparametric methods for comparing means. When the samples to be compared are independent, the parametric methods are produced by "Independent-Samples T Test" under the "Compare Means" command. As mentioned in Chapter 7 of this manual, the data must be entered using the second data entry method; that is, you must have one variable for the variable of interest (the dependent variable) and a second variable to identify the sample or group (the independent variable). The "Independent-Samples T Test" command will work if the variable identifying the groups is a string variable or a numerically coded variable. The nonparametric commands will be discussed in the last chapter of this manual.

Let's demonstrate these commands using the "newspaper" variable from the "fla_student_survey" data set on your text CD. Access this data file, then select **Analyze → Compare Means → Independent-Samples T Test...** Enter "newspaper" in the "Test Variable(s)" box and "gender" in the "Grouping Variable" box. Because a variable selected as a grouping variable could have more than two groups, SPSS needs you to enter information about the groups to be compared. This is still necessary even with data set is like this one where the grouping variable has only two groups. Click on "Define Groups" and enter "m" in the "Group 1" box and "f" in the "Group 2" box. If you are working with a data set in which the variable that defines the groups is coded numerically, you would enter the appropriate numeric values instead of "m" and "f." Click on **Continue**.

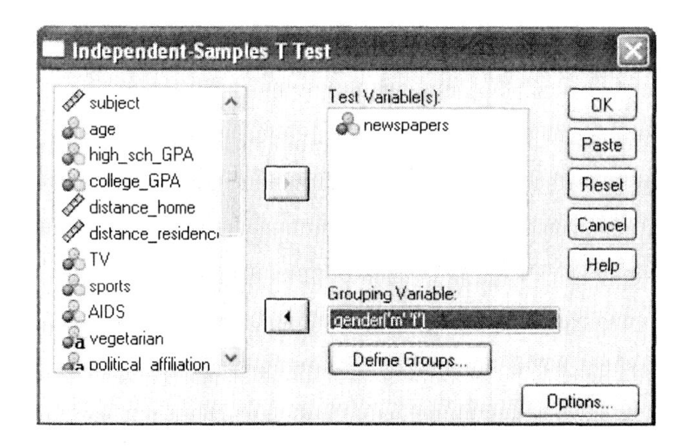

Click on **OK** to produce the output.

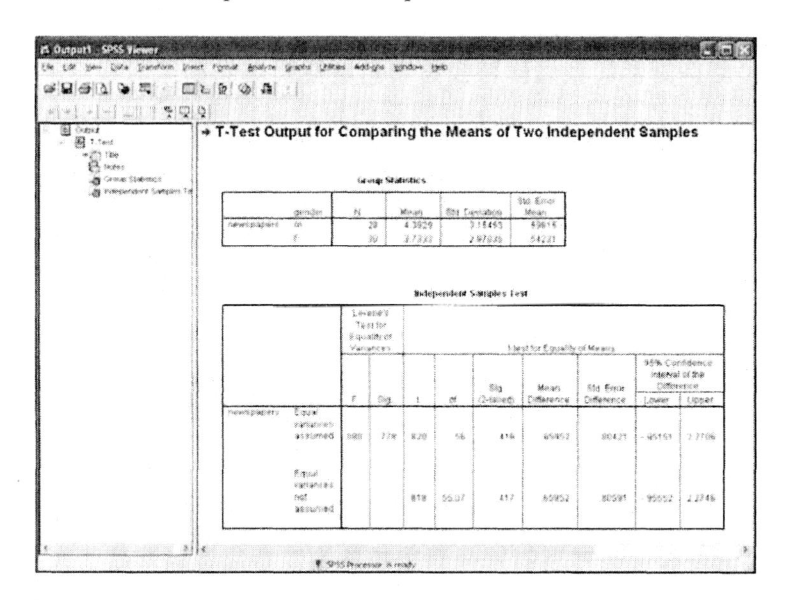

The first part of the output provides some descriptive statistics for the variable "newspaper" for each gender. The second part includes the results of a test to compare the variances of "newspaper" for the two genders (Levene's Test for Equality of Variances) and the results of tests to compare the means of "newspaper" for the two genders (t-test for Equality of Means).

This part of the output has been modified so that it will all fit on the screen at the same time. We are interested in the results of the test to compare two means. SPSS reports the results for both the equal variances and unequal variances versions of the test statistic for comparing means. Your text book uses the unequal variances test statistic, and presents the equal variances test statistic as an alternative that can be used when it is reasonable to assume, under the null hypothesis, that both the means and the variances are the same. For this data set, the t-statistic assuming equal variances is .82, with a two-sided p-value of .416, while the t-statistic assuming unequal variances is .818, with a two-sided p-value of .417. We would make the same conclusion regardless of which test statistic we used for this data set. The evidence suggests no difference in the average newspaper reading habits between this sample of male and female students.

Your text mentions an "F" test for comparing the variances of two groups that is not "robust" to the assumption of normal data. Levene's test, however, is a robust test and can be used to decide whether it is better to use the equal variances t-statistic or the unequal variances t-statistic. The hypothesis of equal variances would be rejected if the p-value is small. For this data set, the p-value ("Sig.") for Levene's test is .778, indicating that there is no evidence to support the conclusion that the variances are different. In this case, the equal variances test statistic could be used instead of the unequal variances test statistic. Since the unequal variances test statistic is always valid, many text book authors present only this statistic.

Comparing the Means of Two Dependent Samples

As indicated in Chapter 7 of this manual, to compare the means of two dependent samples, the data must be entered with each of the paired variables in a separate column. We will demonstrate the methods for comparing means of dependent samples using the cell phone data from Example 12 in Chapter 10. The reaction times are shown in the table below.

	Using cell phone?			Using cell phone?	
Student	No	Yes	Student	No	Yes
1	604	636	17	525	626
2	556	623	18	508	501
3	540	615	19	529	574
4	522	672	20	470	468
5	459	601	21	512	578
6	544	600	22	487	560
7	513	542	23	515	525
8	470	554	24	499	647
9	556	543	25	448	456
10	531	520	26	558	688
11	599	609	27	589	679
12	537	559	28	814	960
13	619	595	29	519	558
14	536	565	30	462	482
15	554	573	31	521	527
16	467	554	32	543	536

Enter the data in SPSS in two columns labeled "no" and "yes." We will use SPSS to calculate the differences for us, so you don't need to calculate them by hand or copy them from the example in the text. Select **Transform → Compute Variable...** and enter "difference" in the "Target Variable" box. Then, enter the expression "yes – no" in the "Numeric Expression" box.

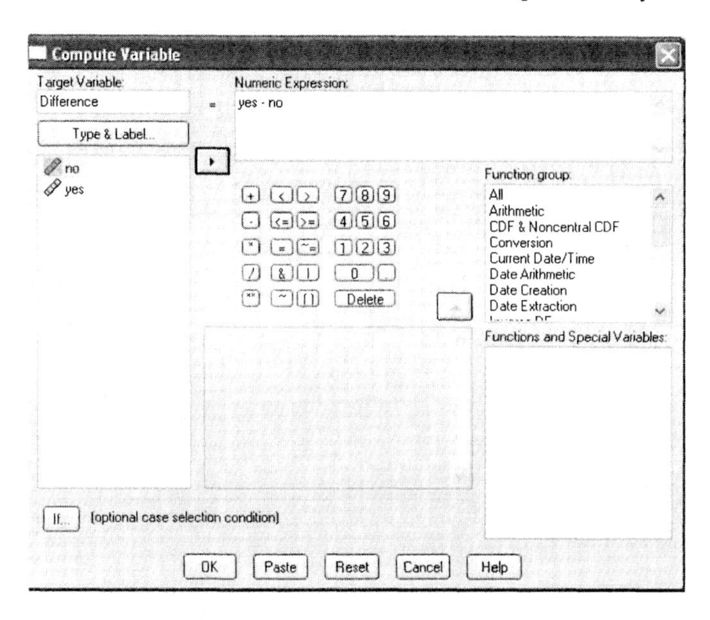

Click on **OK** and the new variable "Difference" will be added to the SPSS Data Editor. You can now find descriptive statistics about the difference variable, or produce a histogram or other graph. A histogram and box plot of the differences are shown below.

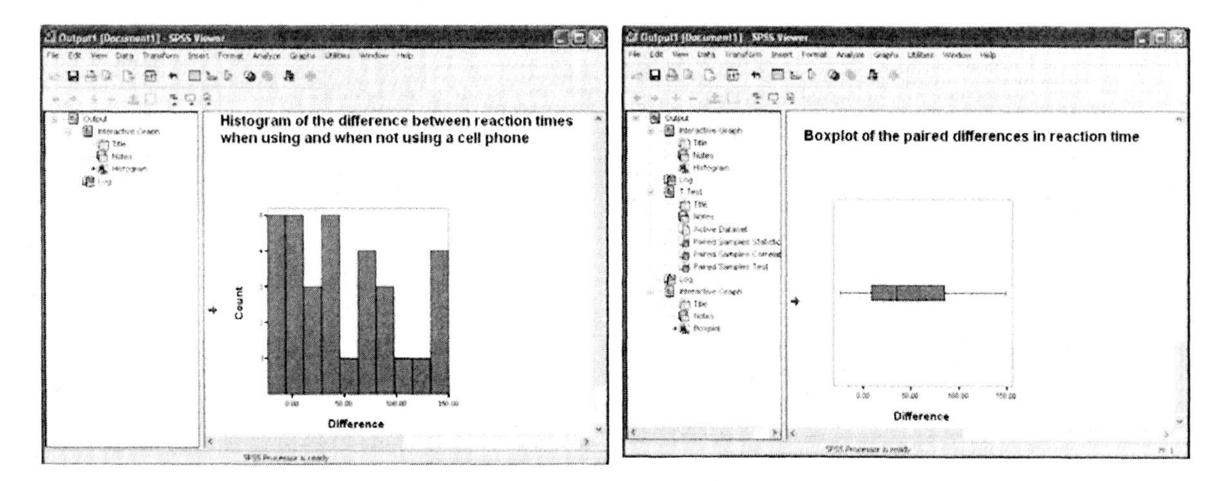

Paired-Samples T-Test

To conduct a test on the mean difference, select **Analyze → Compare Means → Paired-Samples T Test ...** and click once on "no" and once on "yes" in the variable list on the left. As you do this SPSS will enter these variables as "Variable 1" and "Variable 2" below "Current Selections."

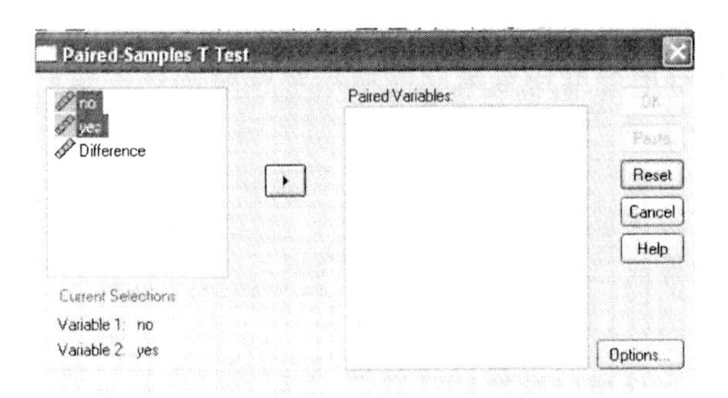

Click on the triangle button to enter the pair "no-yes" in the "Paired Variables" box.

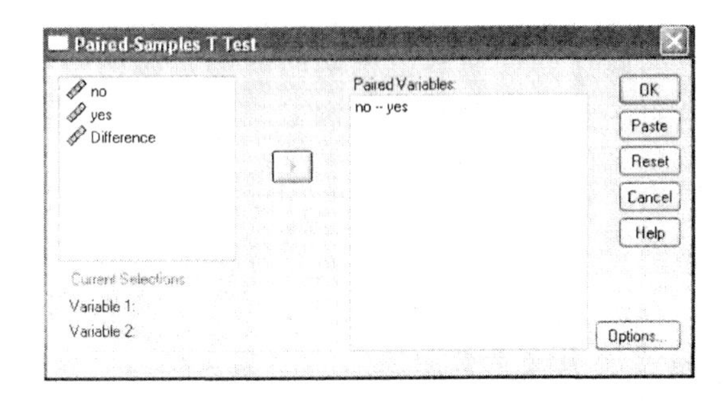

Click on **OK** to produce the output. SPSS will calculate the differences as "no – yes" instead of "yes – no." If you want the difference to be "yes – no" you will have to enter the "Yes" column to the left of the "No" column in the Data Editor. It doesn't make any difference, though, for the analysis because the results will just be the negative of the result if the differences were calculated as "yes – no."

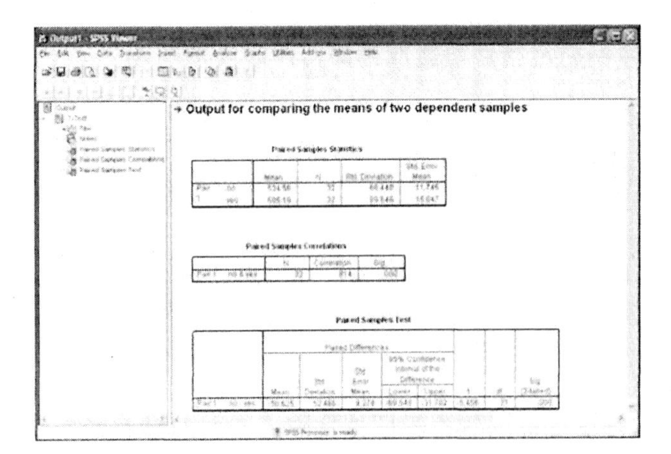

Similar to the output produced for the independent samples methods, SPSS produces some descriptive statistics of each sample and then the results for the paired t-test on the mean difference between the paired observations, including a confidence interval for the mean difference. SPSS also produces the correlation coefficient for the paired variables. Note that since we had SPSS calculate the differences you could have done a one-sample T-test on the differences instead of doing the paired T-test command, as demonstrated in the previous chapter.

If you use a test value of zero for the one-sample test, the results will be the same as for the paired t-test.

The paired T-test command in SPSS does not include any graphing options. To look at the distribution of the differences, you would have to enter the differences or have SPSS calculate them as we already did. Another useful graph for comparing the two variables is a side-by-side Boxplot. Select **Graphs → Legacy Dialogs → Boxplot...** and click on Summaries of separate variables and then on **Define**.

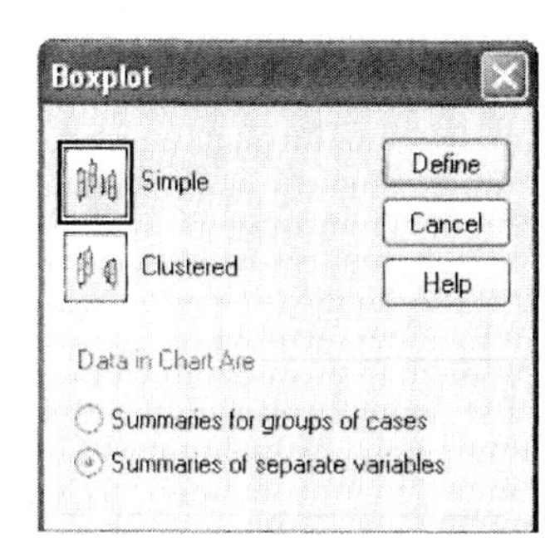

Enter both the before and after variables in the "Boxes Represent" box, as shown below.

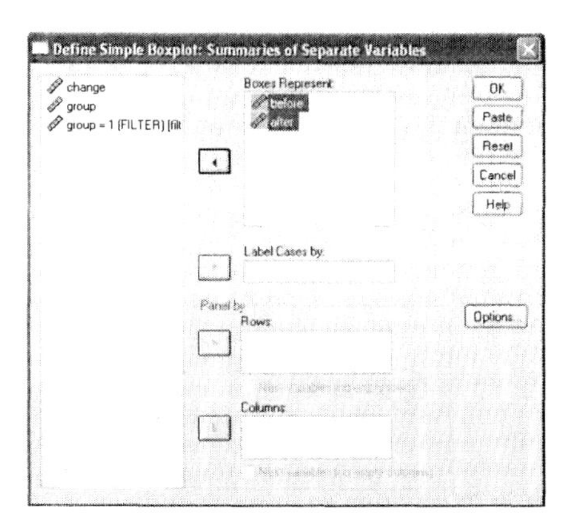

Click on **OK** to produce the output.

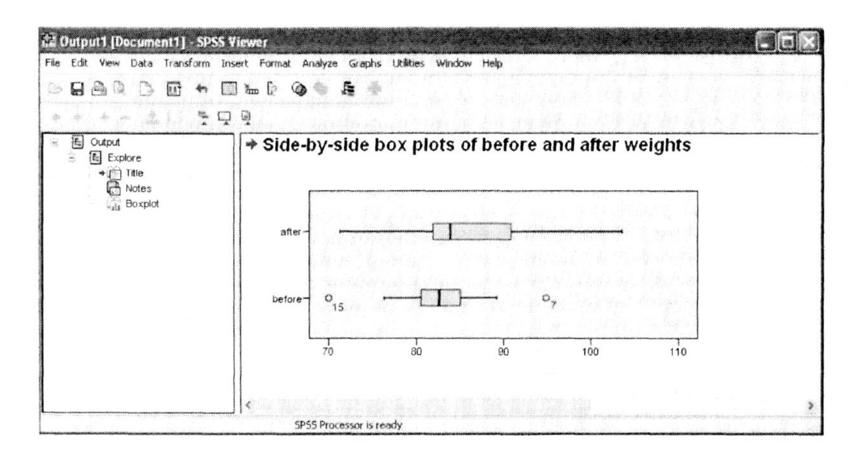

McNemar's Test

Although SPSS does not have a command for doing a large-sample Z-test to compare proportions between two independent samples, it does have a command for conducting McNemar's test on dependent samples when the data is categorical. To demonstrate this command, we will use the data from Example 14 in Chapter 10. This data is from the General Social Survey and is shown in the following table.

	Belief in Hell	
Belief in Heaven	Yes	No
Yes	833	125
No	2	160

Enter the data in SPSS using three variables, "heaven," "hell," and "count" and weight the cases by "count." Both "heaven" and "hell" must be entered as coded numeric variables in order to do this test.

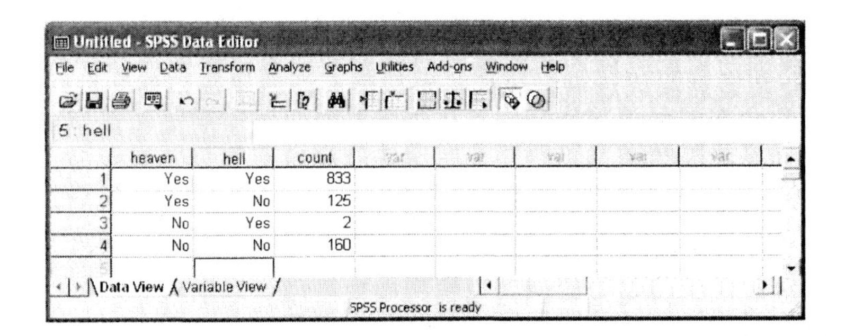

To complete the test, select **Analyze → Nonparametric Tests → 2 Related Samples...** Click on the paired variables as you did for the paired t-test above and then click on the arrow button. Click on "McNemar" to select that test and then on Wilcoxon to "unselect" that test.

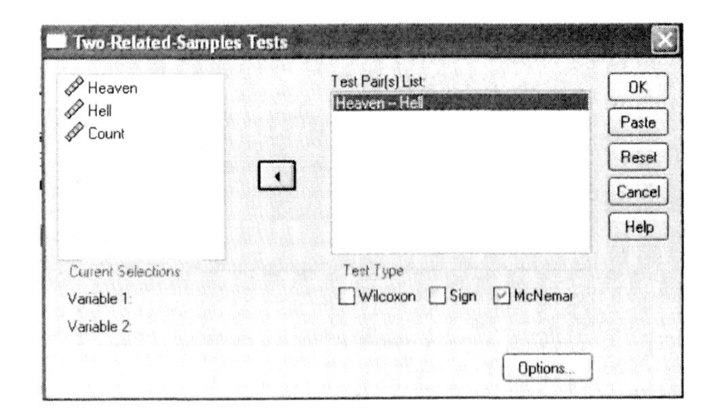

Click on **OK** to produce the output.

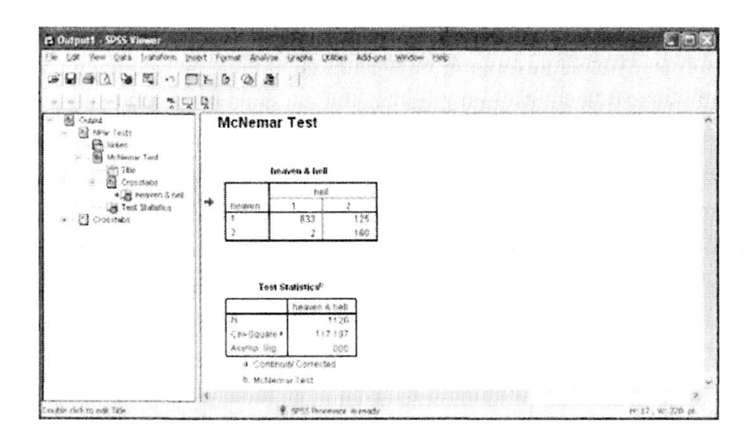

There is also an option for doing McNemar's test under the "Crosstabs" command. Select **Analyze → Descriptive Statistics → Crosstabs…** and click on the "Statistics" button. Click on "McNemar" and then on **Continue**.

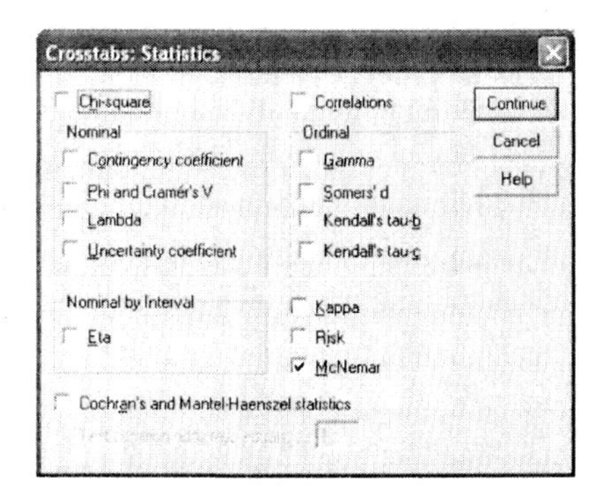

Enter "hell" as the "Column" variable and "heaven" as the "Row" variable. Click on **OK** to produce the output.

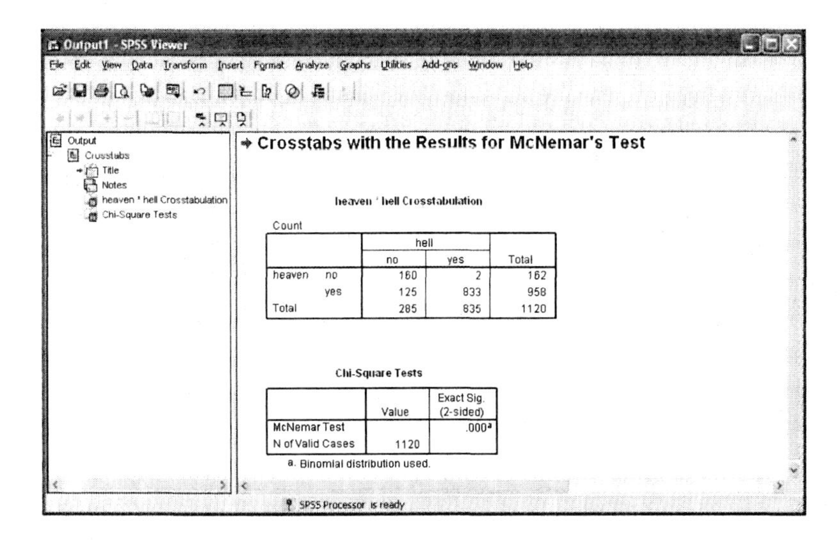

		hell no	hell yes	Total
heaven	no	160	2	162
	yes	125	833	958
Total		285	835	1120

	Value	Exact Sig. (2-sided)
McNemar Test		.000[a]
N of Valid Cases	1120	

SPSS reports only the p-value for this test.

Chapter 10 Analyzing the Association between Categorical Variables

In chapter 8 of this manual you saw how to use a chi-square test that is equivalent to a one-sample Z-test for a proportion. There is also a chi-square test that is equivalent to the Z-test for comparing a proportion between two independent samples. Depending on the situation, this test is sometimes called a test for independence (or association) or a test for homogeneity. This chi-square test can be thought of as an extension of the Z-test when there are more than two groups to be compared and/or the variable has more than two categories.

Contingency Table Tests

To demonstrate this chi-square test we will use the data provided in Example 3 in Chapter 11. This data is from the General Social Survey and shows the joint distribution of the variables "Happiness" and "Income." The data is shown in the following table.

		HAPPINESS	
INCOME	Not too Happy	Pretty Happy	Very Happy
Above average	21	159	110
Average	53	372	221
Below average	94	249	83

Enter the data in SPSS using three columns, one for each of the response variables, HAPPINESS and INCOME, and one for the count in each cell of the table. Remember that HAPPINESS and INCOME can be entered as string or numerically coded variables. Don't forget to select **Data → Weight Cases…** and enter "count" as a weighting variable.

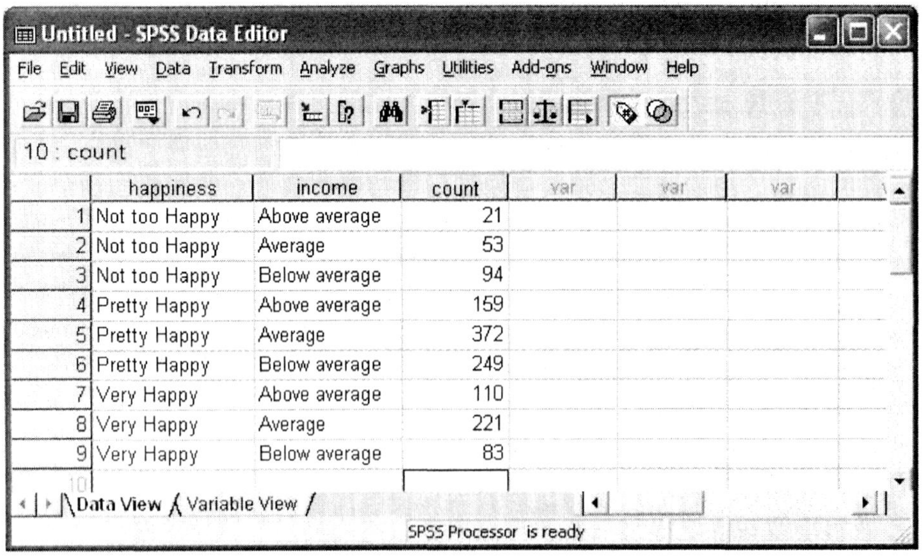

Untitled - SPSS Data Editor

File Edit View Data Transform Analyze Graphs Utilities Add-ons Window Help

10 : count

	happiness	income	count	var	var	var
1	Not too Happy	Above average	21			
2	Not too Happy	Average	53			
3	Not too Happy	Below average	94			
4	Pretty Happy	Above average	159			
5	Pretty Happy	Average	372			
6	Pretty Happy	Below average	249			
7	Very Happy	Above average	110			
8	Very Happy	Average	221			
9	Very Happy	Below average	83			
10						

Data View / Variable View

SPSS Processor is ready

To conduct the test of independence between the happiness and income variables, select **Analyze → Descriptive Statistics → Crosstabs**… and enter "income" as the "Row" variable and "happiness" as the "Column" variable.

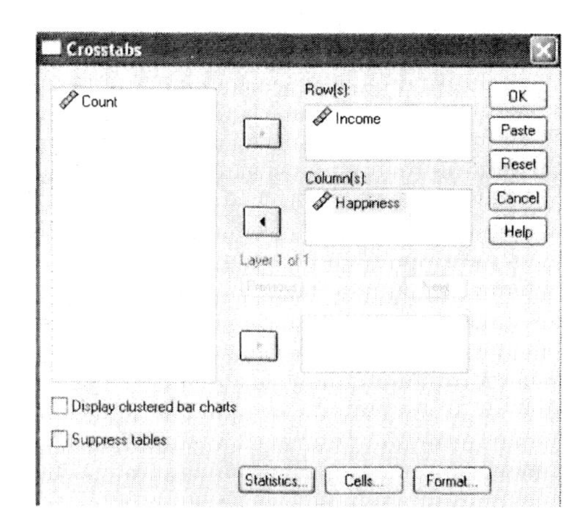

Click on the "Statistics" button and then on "Chi-square."

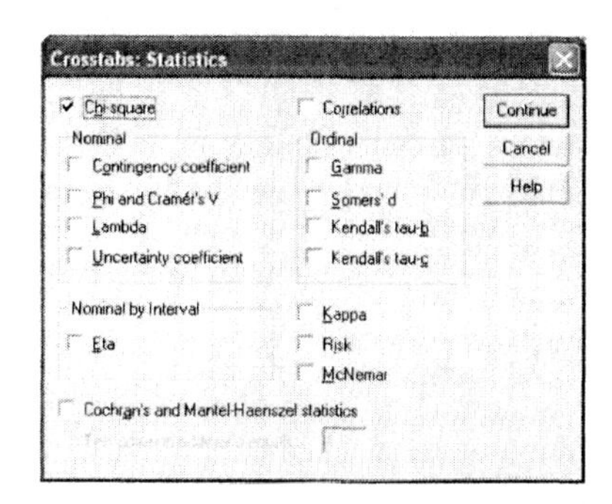

Click on **Continue** and then on **OK** to produce the output.

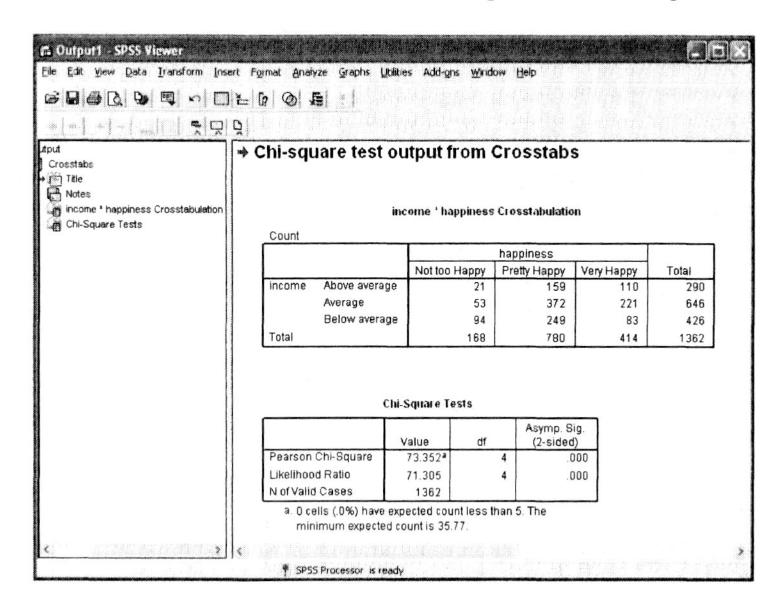

Chi-square test output from Crosstabs

income * happiness Crosstabulation

Count

		happiness			Total
		Not too Happy	Pretty Happy	Very Happy	
income	Above average	21	159	110	290
	Average	53	372	221	646
	Below average	94	249	83	426
Total		168	780	414	1362

Chi-Square Tests

	Value	df	Asymp. Sig. (2-sided)
Pearson Chi-Square	73.352[a]	4	.000
Likelihood Ratio	71.305	4	.000
N of Valid Cases	1362		

a. 0 cells (.0%) have expected count less than 5. The minimum expected count is 35.77.

In addition to the value of the chi-square statistic and its associated p-value, under "Asymp. Sig.", SPSS also reports the number of cells with expected cell count less than 5. Note that the results of the test would be the same if "income" was entered as the "Column" variable and "happiness" was the "Row" variable.

You may recall from chapter 2 of this manual that "Crosstabs" includes options for calculating the conditional (row or column) proportions for a contingency table. It also has an option for calculating the standardized residual for each cell in the table. Select **Analyze → Descriptive Statistics → Crosstabs**... again and click on the "Cells..." button. Click on "Expected" under "Counts" ("Observed" should already be selected) and also on "Adjusted standardized" under "Residuals." These are the residuals that your text refers to as "standardized."

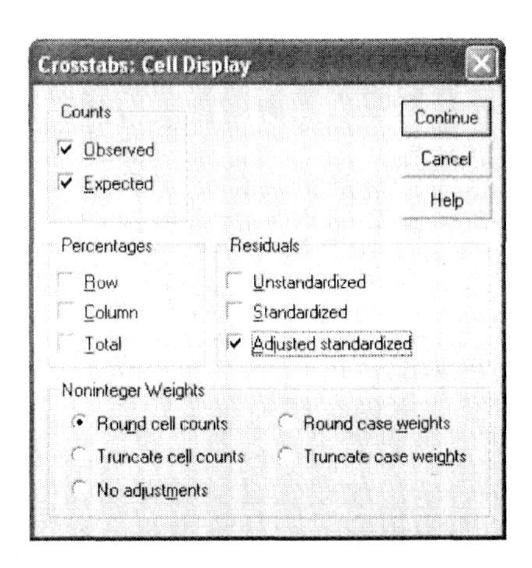

Click on **Continue** and then on **OK**. The residuals will be added to the contingency table in the output.

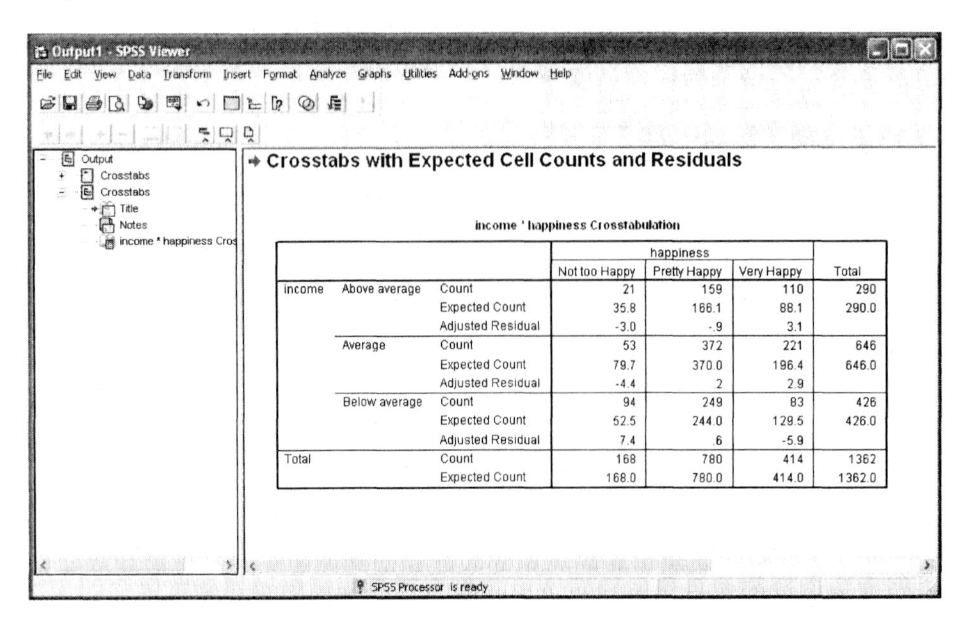

income * happiness Crosstabulation

			happiness			Total
			Not too Happy	Pretty Happy	Very Happy	
income	Above average	Count	21	159	110	290
		Expected Count	35.8	166.1	88.1	290.0
		Adjusted Residual	-3.0	-.9	3.1	
	Average	Count	53	372	221	646
		Expected Count	79.7	370.0	196.4	646.0
		Adjusted Residual	-4.4	.2	2.9	
	Below average	Count	94	249	83	426
		Expected Count	52.5	244.0	129.5	426.0
		Adjusted Residual	7.4	.6	-5.9	
Total		Count	168	780	414	1362
		Expected Count	168.0	780.0	414.0	1362.0

Comparing Two Proportions

Let's demonstrate this test using Example 5 from Chapter 10. In this example, p is the proportion that showed aggressive behavior and the two groups are determined by their amount of TV watching. The results are shown in the following table.

	Aggressive Act	
TV Watching	Yes	No
Less than 1 hour per day	5	83
At least 1 hour per day	154	465

Enter the data in SPSS using three variables, one for each of the categorical variables and one for the count in each cell of the table. Make sure to weight the cases by the "count" variable.

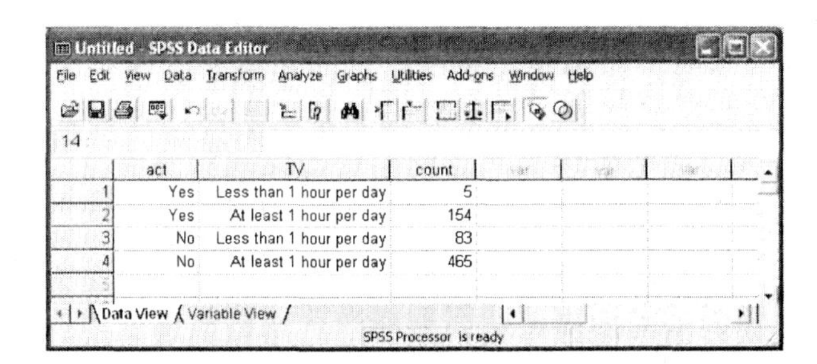

	act	TV	count
1	Yes	Less than 1 hour per day	5
2	Yes	At least 1 hour per day	154
3	No	Less than 1 hour per day	83
4	No	At least 1 hour per day	465

To do the test select **Analyze → Descriptive Statistics → Crosstabs**... and enter the "hours" variable in the "Row(s)" box and the "TV" variable in the "Column(s)" box. Click on the "Statistics" button and select the "Chi-Square" statistic.

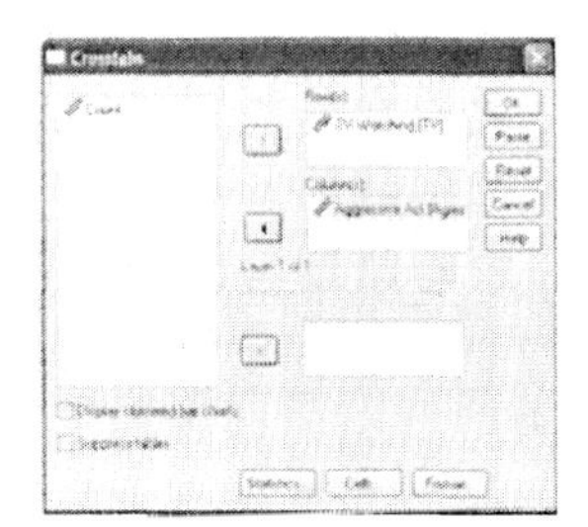

Click on **Continue** and then on **OK**. A portion of the output is shown below.

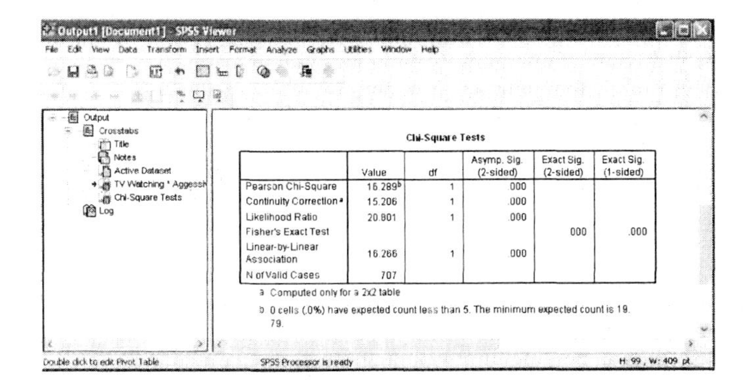

Chi-Square Tests

	Value	df	Asymp. Sig. (2-sided)	Exact Sig. (2-sided)	Exact Sig. (1-sided)
Pearson Chi-Square	16.289[b]	1	.000		
Continuity Correction[a]	15.206	1	.000		
Likelihood Ratio	20.801	1	.000		
Fisher's Exact Test				.000	.000
Linear-by-Linear Association	16.266	1	.000		
N of Valid Cases	707				

a. Computed only for a 2x2 table

b. 0 cells (.0%) have expected count less than 5. The minimum expected count is 19.79.

The value of the chi-square test statistic is 16.289, which is the same as the square of the value of the Z-statistic given in your text. (If you don't round your calculations at intermediate steps when calculating the Z-statistic, you will get a value of -4.036 and the square of -4.036 is 16.289.) The p-value for this test (and for the corresponding Z-statistic) is less than .0005.

Fisher's Exact Test

When the Chi-Square test is requested as part of the Crosstabs command, SPSS will automatically conduct Fisher's Exact Test when the table is two-by-two, which is useful when dealing with a small sample size. To demonstrate this test we will use the data from Example 10 in Chapter 11. This data is shown in the following table.

	PREDICTION	
ACTUAL	Milk	Tea
Milk	3	1
Tea	1	3

This data set is from an experiment where a taster made a prediction of whether tea or milk was poured in a cup first. The experiment consisted of 4 trials where milk was poured first and 4 where tea was poured first.

Enter the data in SPSS using three variables, "actual," "prediction" and "count," then select **Data → Weight Cases...** and enter "count" as the weighting variable. Select **Analyze → Descriptive Statistics → Crosstabs**..., then click on the **Statistics** button and select the Chi-square test. Click on **Continue** and enter "actual" as the "Row" variable and "predicted" as the "Column" variable. Click on **OK** to produce the output.

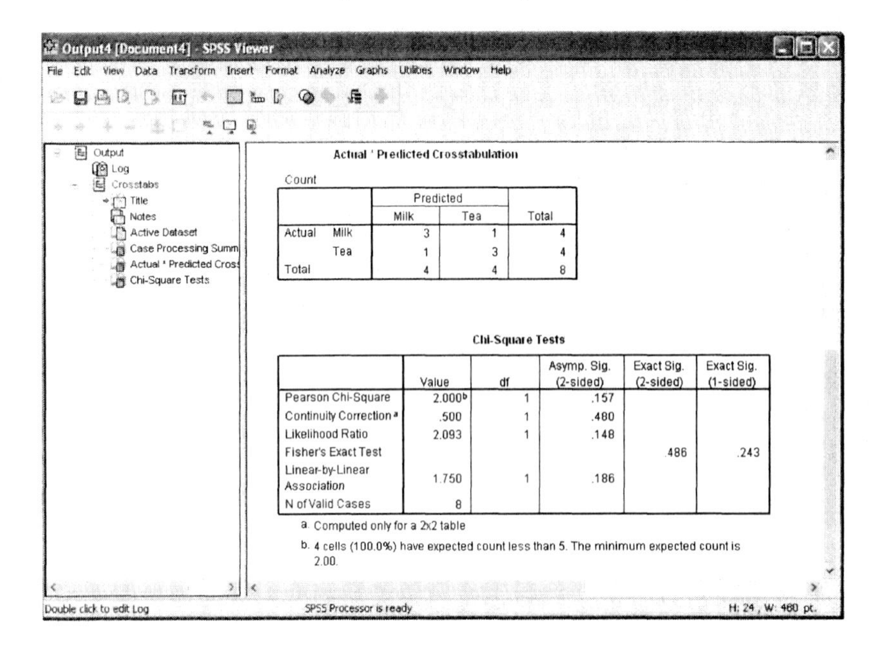

Actual * Predicted Crosstabulation

Count

		Predicted		
		Milk	Tea	Total
Actual	Milk	3	1	4
	Tea	1	3	4
Total		4	4	8

Chi-Square Tests

	Value	df	Asymp. Sig. (2-sided)	Exact Sig. (2-sided)	Exact Sig. (1-sided)
Pearson Chi-Square	2.000[b]	1	.157		
Continuity Correction[a]	.500	1	.480		
Likelihood Ratio	2.093	1	.148		
Fisher's Exact Test				.486	.243
Linear-by-Linear Association	1.750	1	.186		
N of Valid Cases	8				

a. Computed only for a 2x2 table

b. 4 cells (100.0%) have expected count less than 5. The minimum expected count is 2.00.

Fisher's Exact Test is included as part of the output.

Chi-Square Goodness-of-Fit Test

The chi-square test statistic can also be used to assess how well a data set follows a specified categorical distribution. This test can be done for distributions with two or more categories, but the example demonstrated here has just two categories. The data we'll use represents the results of Mendel's famous experiment conducted to evaluate his theory concerning the results of crossing a pure yellow strain of peas with a pure green strain. His theory predicted that the result would be 75% yellow and 25% green. The actual data is shown in the SPSS Data Editor below.

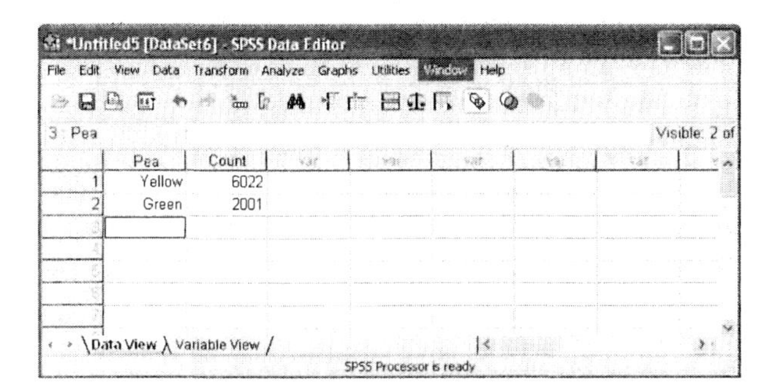

Enter the data in SPSS as shown above and make sure to indicate that Count is a weighting variable. To conduct the test, select **Analyze → Nonparametric Tests → Chi-Square...** Enter Pea in the Test Variable List box. Click on Values under "Expected Values," enter 6017 and click on the **Add** button. SPSS will not allow you to enter the actual expected value of 6017.25. It has a limit on the number of digits you can enter for the expected values. Enter 2006 in the same box and click on **Add** again.

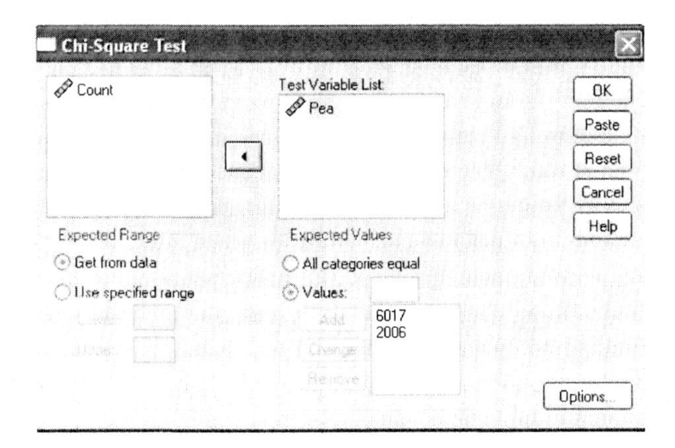

Click on **OK** to produce the output.

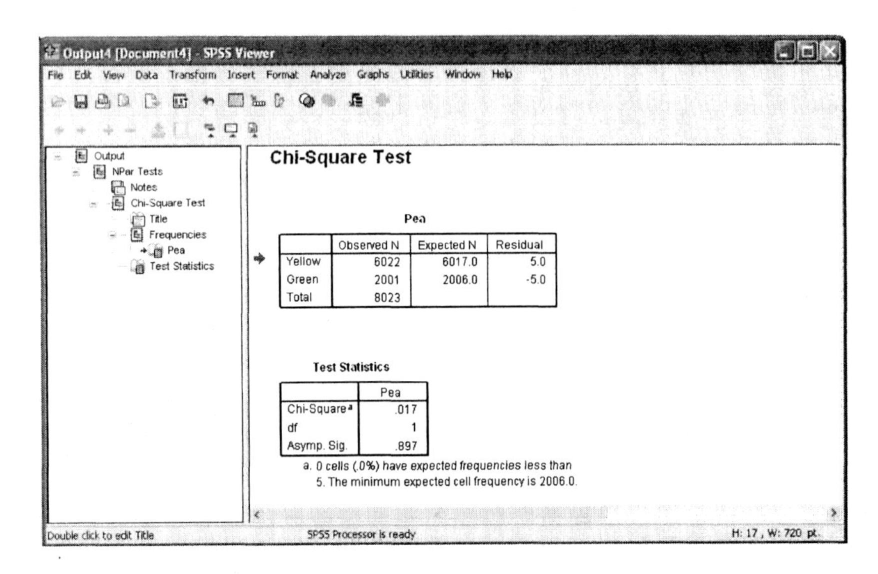

Pea

	Observed N	Expected N	Residual
Yellow	6022	6017.0	5.0
Green	2001	2006.0	-5.0
Total	8023		

Test Statistics

	Pea
Chi-Square[a]	.017
df	1
Asymp. Sig.	.897

a. 0 cells (.0%) have expected frequencies less than 5. The minimum expected cell frequency is 2006.0.

The value of the test statistic is .017, which is slightly different from the value calculated in your text because we couldn't enter the exact expected values of 6017.25 and 2005.75.

Chapter 11 Regression Analysis

You probably noticed in the regression examples in Chapter 3 of this manual that the regression command in SPSS does a lot more than just find the slope and intercept of the best fit line. Some of the output that SPSS produces for this command was omitted back in Chapter 3. We will explore that output more in this chapter

Testing for the significance of the slope

In Chapter 3 we used SPSS to find the correlation coefficient between two variables and to find the best fit line, but we didn't address whether or not the association was "significant" or "meaningful." In the regression output SPSS provides the results for a test to determine if the slope of the best fit line is significantly different from zero. We will explore this output using the "high school female athletes" data set from your text CD. Access this data set and select **Analyze → Regression → Linear...** Enter Maximum Bench Press (@1RMBENCHlbs) as the "Dependent" variable and Number of 60-Pound Bench Presses (BRTF_60) as the "Independent" variable.

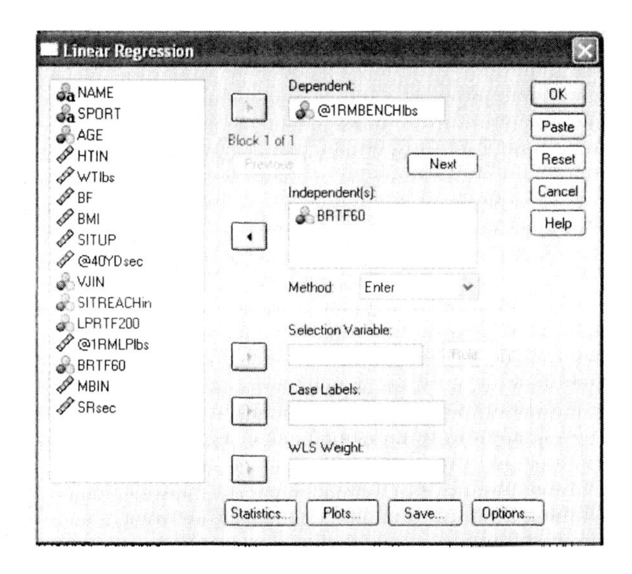

Click on **OK** to produce the output. The Model Summary and Coefficients portions of the regression output are shown below. The best fit line is $\hat{y} = 63.5 + 1.49x$, as shown in your text and verified by the "B" values under Unstandardized Coefficients in the output. Next to the estimates of the intercept and slope you will see their corresponding standard errors. On the far right of the "Coefficients" output you will see columns titled "t" and "Sig." These columns are for conducting two separate hypothesis tests, one to determine if the y-intercept is significantly different from zero and the other to determine if the slope is significantly different from zero. For this example we're only interested in the test concerning the slope.

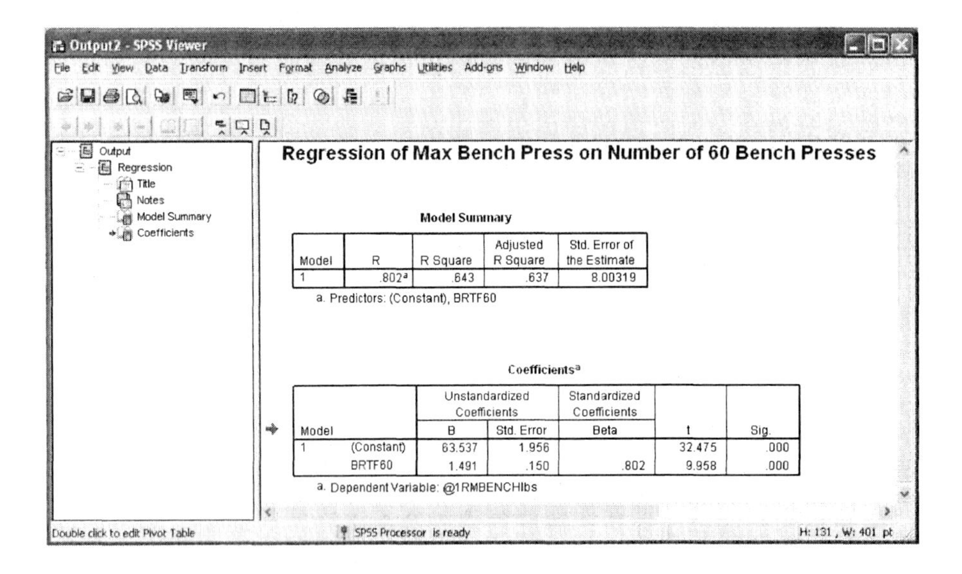

Regression of Max Bench Press on Number of 60 Bench Presses

Model Summary

Model	R	R Square	Adjusted R Square	Std. Error of the Estimate
1	.802[a]	.643	.637	8.00319

a. Predictors: (Constant), BRTF60

Coefficients[a]

Model		Unstandardized Coefficients		Standardized Coefficients	t	Sig.
		B	Std. Error	Beta		
1	(Constant)	63.537	1.956		32.475	.000
	BRTF60	1.491	.150	.802	9.958	.000

a. Dependent Variable: @1RMBENCHlbs

In your text book the test statistic for testing the significance of the slope is given as $t = \frac{b-0}{se}$, where b is the estimate of the slope and se is the standard error of b. For this example, $t = \frac{1.491-0}{.15} = 9.96$, which agrees with the t-statistic provided by SPSS. The p-value for the two-sided test is given by "Sig," which is listed as .0000. This result just means that the p-value is less than .00005.

The Squared Correlation and Residual Standard Deviation

Included in the Model Summary portion of the regression output is a value listed as "R-Square." This value tells us the proportion reduction in error in using $\hat{y}$ to predict y instead of $\bar{y}$. For a regression analysis with one Y-variable and one X-variable, the proportional reduction in error can easily be found by squaring the value of the correlation coefficient, r. We will explore the actual formula for doing this calculation in the next chapter, but for now it is easy to see for the example above that $(.802)^2 = .643$, so the error in predicting the maximum number of bench presses is 64.3% smaller using the regression equation than the error using the average number of bench presses.

The residual standard deviation is given in the Model Summary under "Std. Error of the Estimate." For this example, the residual standard deviation is 8.003. This value can be used to create a prediction interval for y as described in your text.

Confidence Interval for the Slope

The Regression command in SPSS includes an option for also calculating a confidence interval for the slope. Select **Analyze → Regression → Linear...** and click on the "Statistics" button. Click on "Confidence Intervals" under "Regression Coefficients" and then on **Continue**.

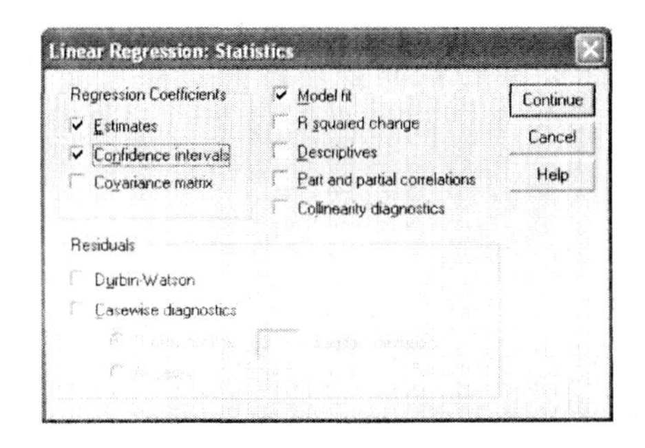

Click on **OK** to produce the output. Notice the additional columns for the lower and upper bounds in the "Coefficients" portion of the output. This portion of the output was modified so that all of it would fit on the screen.

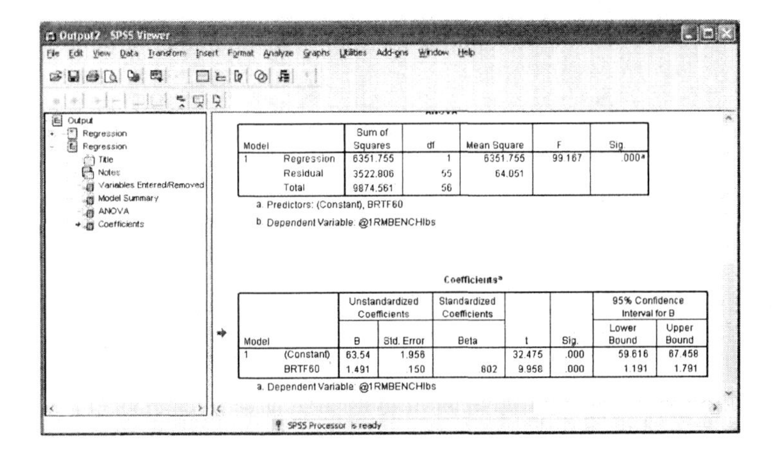

Model		Sum of Squares	df	Mean Square	F	Sig.
1	Regression	6351.755	1	6351.755	99.167	.000a
	Residual	3522.806	55	64.051		
	Total	9874.561	56			

a. Predictors: (Constant), BRTF60

b. Dependent Variable: @1RMBENCHlbs

Coefficientsa

Model		Unstandardized Coefficients		Standardized Coefficients	t	Sig.	95% Confidence Interval for B	
		B	Std. Error	Beta			Lower Bound	Upper Bound
1	(Constant)	63.54	1.956		32.475	.000	59.616	67.458
	BRTF60	1.491	.150	.802	9.956	.000	1.191	1.791

a. Dependent Variable: @1RMBENCHlbs

Detecting Unusual Observations

The Regression command in SPSS includes an option for printing a table of observations with large residuals. To print this table, select **Analyze → Regression → Linear...** and click on the "Statistics" button. Click on "Casewise Diagnostics" under "Residuals" and enter the value "2" in the "standard deviations" box. Click on **Continue** and then on **OK**.

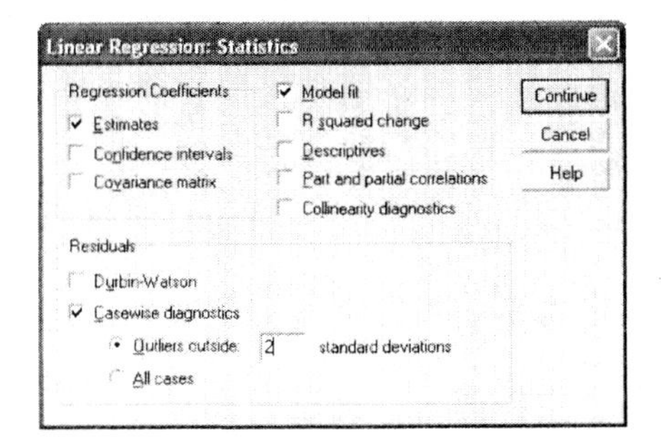

The output is shown below. This data set has three observations with standardized residuals larger than 2. Case #10, for example, represents a student whose residual is 2.386 standard errors above zero, indicating that the amount this student can actually bench press is far above the value predicted by the best-fit regression line.

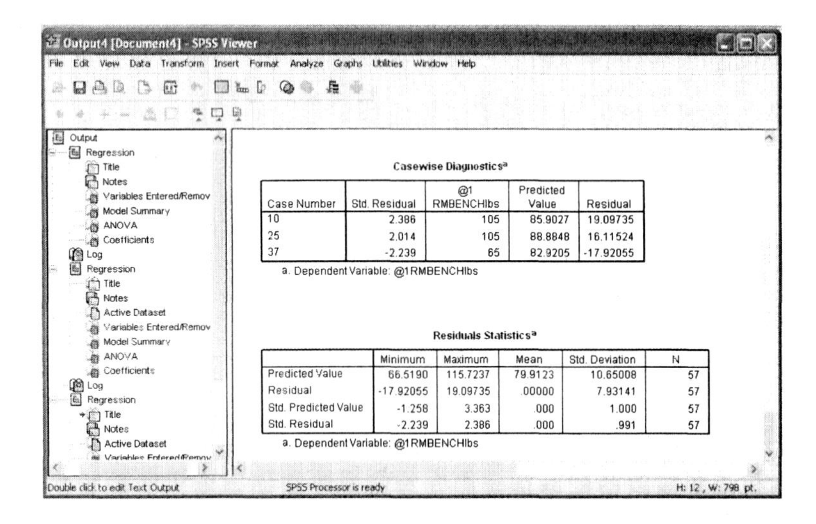

Casewise Diagnostics[a]

Case Number	Std. Residual	@1 RMBENCHlbs	Predicted Value	Residual
10	2.386	105	85.9027	19.09735
25	2.014	105	88.8848	16.11524
37	-2.239	65	82.9205	-17.92055

a. Dependent Variable: @1RMBENCHlbs

Residuals Statistics[a]

	Minimum	Maximum	Mean	Std. Deviation	N
Predicted Value	66.5190	115.7237	79.9123	10.65008	57
Residual	-17.92055	19.09735	.00000	7.93141	57
Std. Predicted Value	-1.258	3.363	.000	1.000	57
Std. Residual	-2.239	2.386	.000	.991	57

a. Dependent Variable: @1RMBENCHlbs

Exponential Regression

SPSS has an option for fitting an exponential model instead of a linear model. To demonstrate this we will use the data for Example 18 in Chapter 12 in your text. This data is shown in the table below:

Year	Years since 1995	Number of People
1995	0	16
1996	1	36
1997	2	76
1998	3	147
1999	4	195
2000	5	369
2001	6	513

The data has been entered into the SPSS Data Editor shown below.

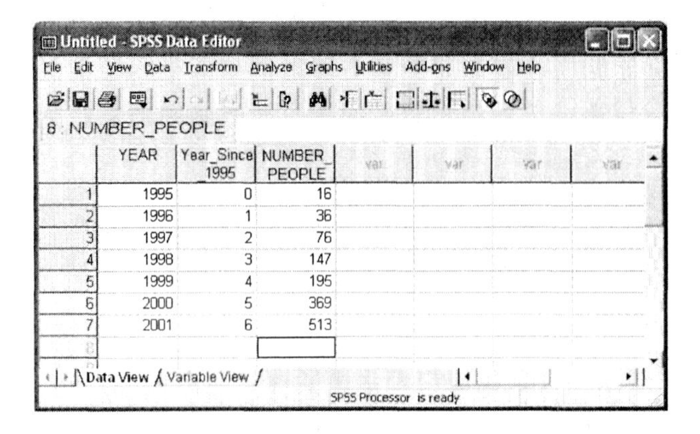

	YEAR	Year_Since 1995	NUMBER_ PEOPLE
1	1995	0	16
2	1996	1	36
3	1997	2	76
4	1998	3	147
5	1999	4	195
6	2000	5	369
7	2001	6	513

Select **Analyze → Regression → Curve Estimation…** and enter "NUMBER_PEOPLE" as the dependent variable and "Year_Since _995" as the independent variable. Click on "Exponential" under "Models." You will notice that "Linear" has already been selected. SPSS will produce both the linear and exponential fits to this data set. If you don't want to see the linear results, click on "Linear" to un-select it.

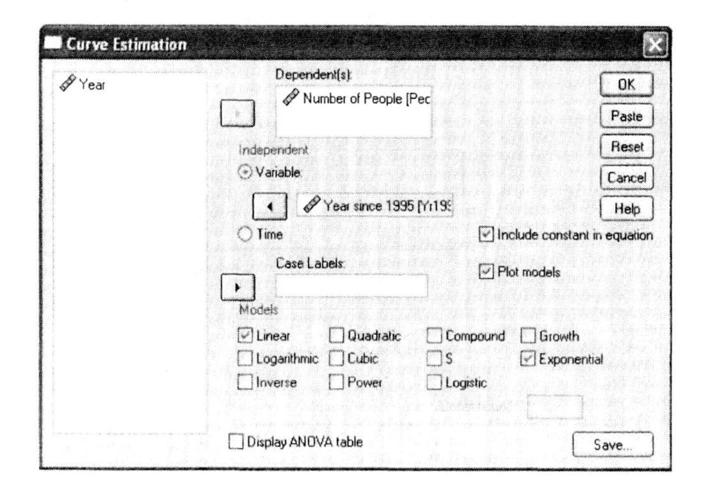

Click on **OK** to produce the output. Both the linear and exponential results are shown below. Once nice option of this command is that SPSS automatically produces a scatter plot showing the best fit curve (or line).

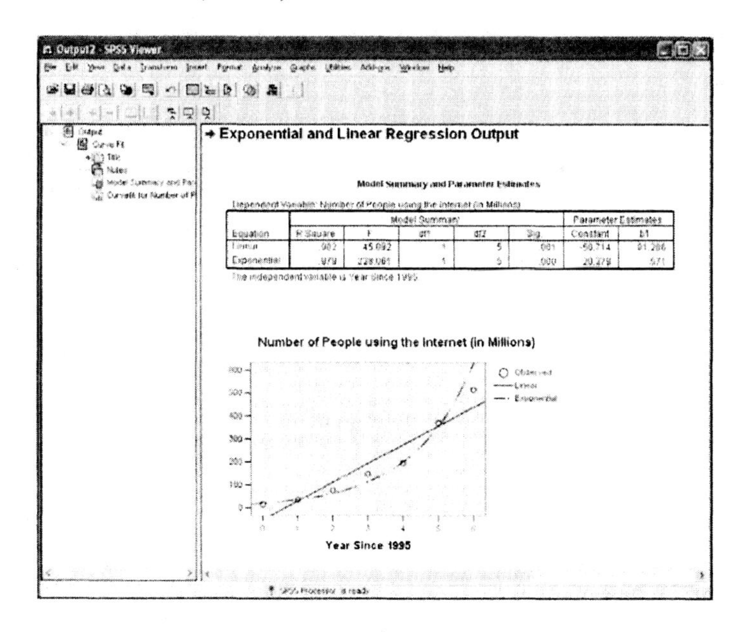

Some of the output produced by SPSS has been omitted in the display above. Unlike the example in your text, SPSS does the exponential regression using the logarithm with base e (instead of base 10), so the base for the exponent x in the exponential model appears different than it does in your text. To convert the coefficient b1 from SPSS to log base 10, raise e to the power given by the value of b1 from the output. That is, $e^{.571} = 1.77$. So the exponential regression model is $\hat{y} = 20.38 \times 1.77^{x}$.

Another way to find the exponential regression model is to use SPSS to calculate the logarithm of the number of people using the internet, as shown in your text. Select **Transform → Compute Variable...** and enter the variable name "LogPeople" in the Target Variable box. Then select "Arithmetic" under "Function Group" and scroll through the "Functions and Special Variables" list to find "Lg10." Double click on this and then replace the question mark in the expression "LG10(?)" with the people variable.

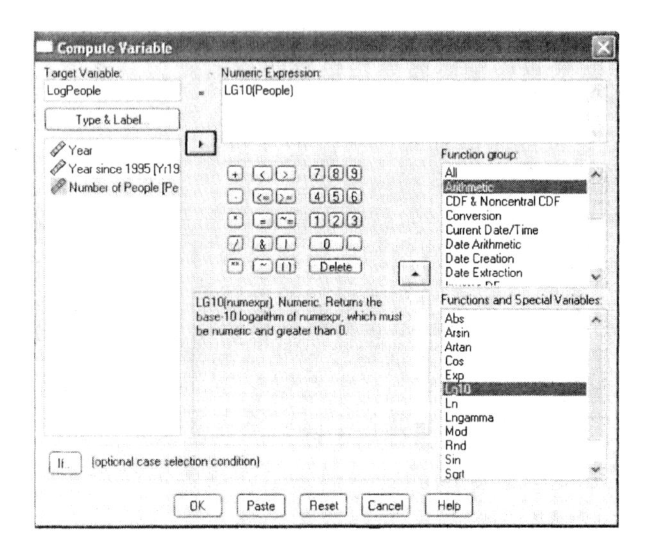

Click on **OK**. The data editor will now include the base 10 logarithm of the number of people using the Internet as a new variable.

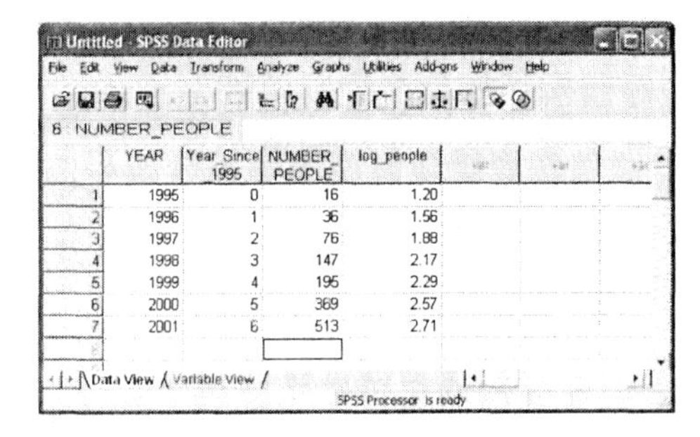

	YEAR	Year_Since 1995	NUMBER_PEOPLE	log_people
1	1995	0	16	1.20
2	1996	1	36	1.56
3	1997	2	76	1.88
4	1998	3	147	2.17
5	1999	4	195	2.29
6	2000	5	369	2.57
7	2001	6	513	2.71

To do the exponential regression using this approach, select **Analyze → Regression → Linear...** and enter "LogPeople" as the "Dependent" variable and "Year_Since_1995" as the "Independent" variable. Click on **OK**.

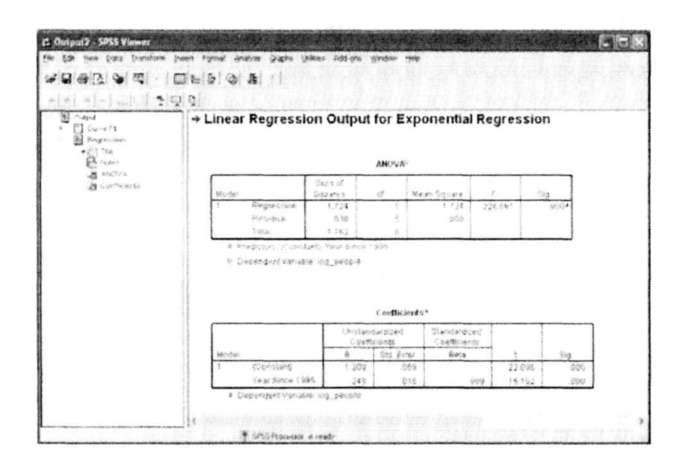

As your text describes, the logarithm of the mean is a linear function of x. From our output we have $\hat{\log} y = 1.309 + .248x$. To transform this to an exponential model, the alpha coefficient is 10 raised to the power of the intercept and the beta coefficient is 10 raised to the power of the slope. Or, $\hat{\alpha} = 10^{1.309} = 20.37$ and $\hat{\beta} = 10^{.248} = 1.77$. So, the exponential regression model is $\hat{y} = 20.37 \times 1.77^{x}$, which is the same result we obtained above using the first approach.

Chapter 12 Multiple Regression

When we explored regression analysis in the previous chapter and in Chapter 3 we used examples of "simple" linear regression; that is, the regression of one Y (dependent) variable on one X (independent) variable. The "Regression" command in SPSS allows you to also do multiple regression with more than one independent variable. To demonstrate how to do this and how to read the output that SPSS produces we will use the data for Example 2 in Chapter 13 on the selling prices of houses. This data set is available on your text CD in the file "house_selling_prices" and a portion is shown in the table below.

Home	Selling Price	House Size	Number of Bedrooms	Number of Bathrooms	Lot Size	Real Estate Tax	NW
1	145,000	1240	3	2	18,000	1360	Yes
2	69,000	1120	3	2	17,000	830	No
3	163,000	1710	3	2	14,000	2150	Yes

Access this data set on your text CD and select **Analyze → Regression → Linear...** and enter "price" as the "Dependent" variable and both "size" and "lot" as "Independent(s)" variables.

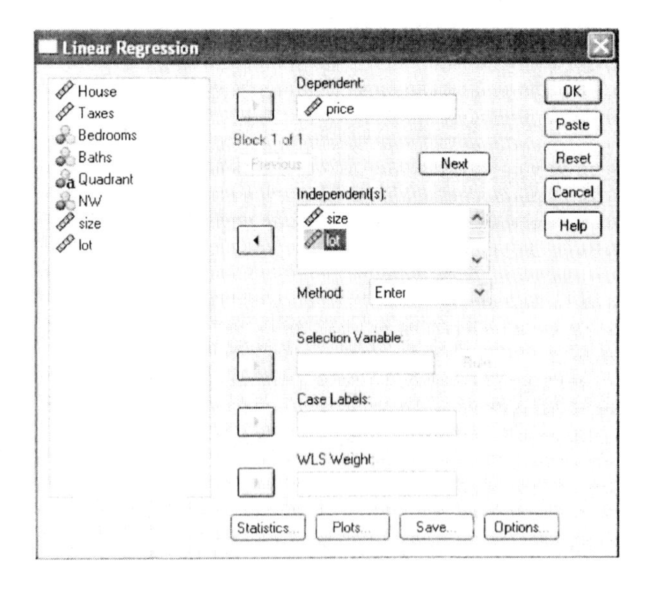

Click on **OK** to produce the output. The output is shown below. The "Coefficients" portion of the output provides us with the estimates of the parameters that describe the effects of house size and lot size on selling price. As shown in your text these values are 54.8 and 2.84, respectively.

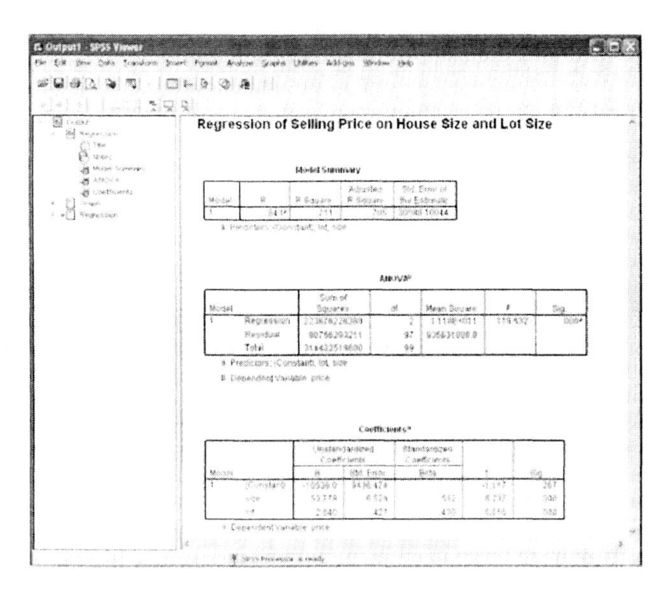

Regression of Selling Price on House Size and Lot Size

Model Summary

Model	R	R Square	Adjusted R Square	Std. Error of the Estimate
1	.843ª	.711	.705	30588.10044

a. Predictors: (Constant), lot, size

ANOVAᵇ

Model		Sum of Squares	df	Mean Square	F	Sig.
1	Regression	223676226380	2	1.118E+011	119.532	.000ª
	Residual	90756293211	97	935631008.8		
	Total	314432519600	99			

a. Predictors: (Constant), lot, size

b. Dependent Variable: price

Coefficientsª

Model		Unstandardized Coefficients B	Unstandardized Coefficients Std. Error	Standardized Coefficients Beta	t	Sig.
1	(Constant)	-10536.0	9436.474		-1.117	.267
	size	53.779	6.526	[illegible]	8.237	.000
	lot	2.840	.427	.430	6.656	.000

a. Dependent Variable: price

This portion of the output also includes T-tests for testing the significance of each of the independent variables, which will be discussed for another example below. The Model Summary portion of the output includes the values of the multiple correlation, R, and R-Squared. These values are .843 and .711, respectively. To see how to calculate R-Squared, the ANOVA portion of the output provides the value for the residual sum of squares and the total sum of squares. (This portion of the output is produced below with price recoded as the actual selling price divided by 1000, to match the output as given in the text. Transforming the dependent variable in this way does not affect the degree of association of selling price with the two independent variables.) The value of R^2 is $\dfrac{314{,}433 - 90{,}756}{314{,}433} = .711$.

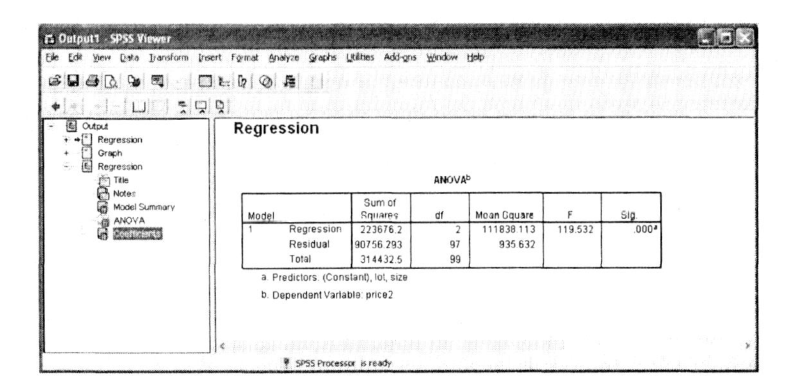

Regression

ANOVAᵇ

Model		Sum of Squares	df	Mean Square	F	Sig.
1	Regression	223676.2	2	111838.113	119.532	.000ª
	Residual	90756.293	97	935.632		
	Total	314432.5	99			

a. Predictors: (Constant), lot, size

b. Dependent Variable: price2

The ANOVA portion of the output also includes an F-test for the overall regression model; that is, it tests if at least one of the independent variables is significant. F-tests and Analysis of Variance (ANOVA) are discussed later in this chapter.

To see the associations between selling price with each of the independent variables select **Graphs → Legacy Dialogs → Scatter/Dot...** and select the "Matrix Scatter" option. Click on **Define**.

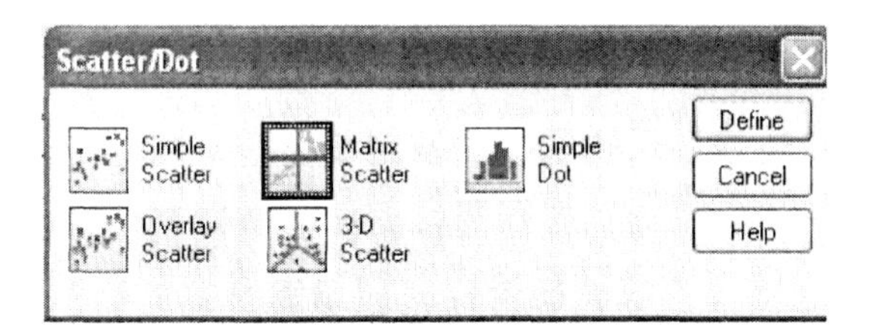

Enter the dependent variable and both independent variables in the "Matrix Variables" box.

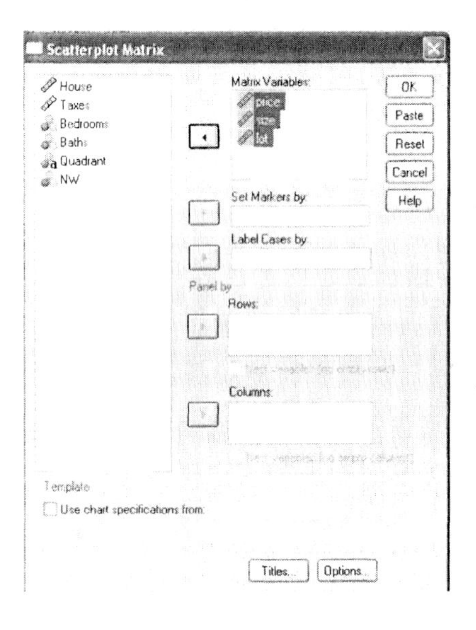

Click on **OK** to produce the output.

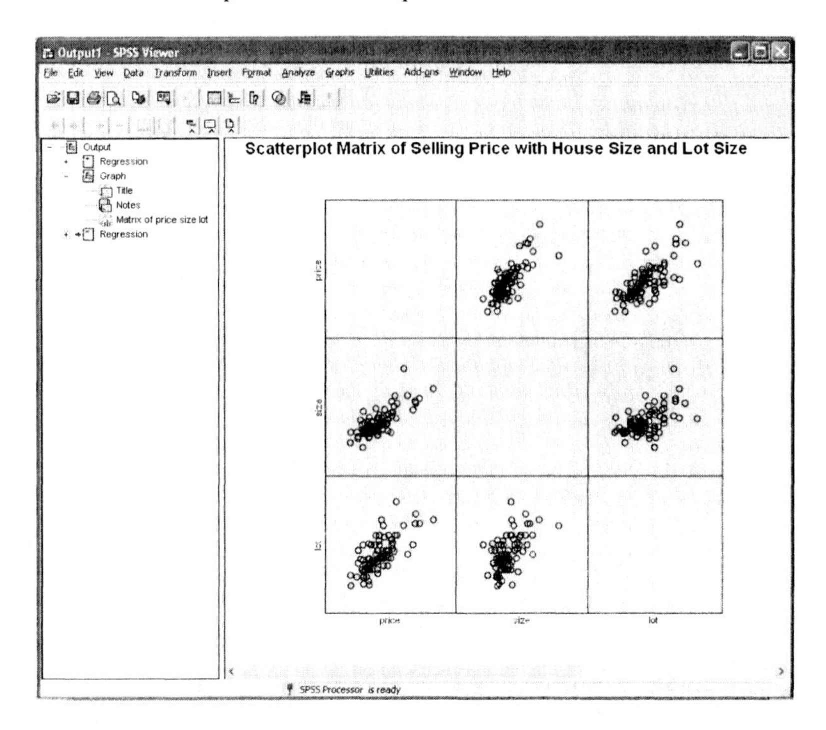

The top row of the matrix shows selling price on the Y-axis with house size on the X-axis in the second box and lot size on the X-axis in the third box. If you would like to see the corresponding correlation coefficients between each pair of variables in this matrix, select **Analyze → Correlate → Bivariate...** and enter all three variables in the Variables box.

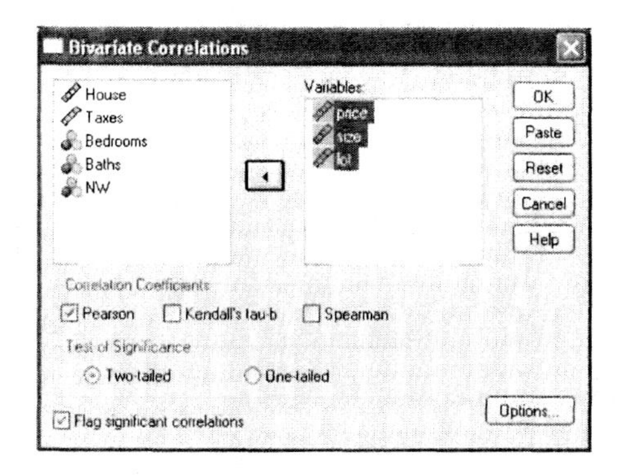

Click on **OK** to produce the output.

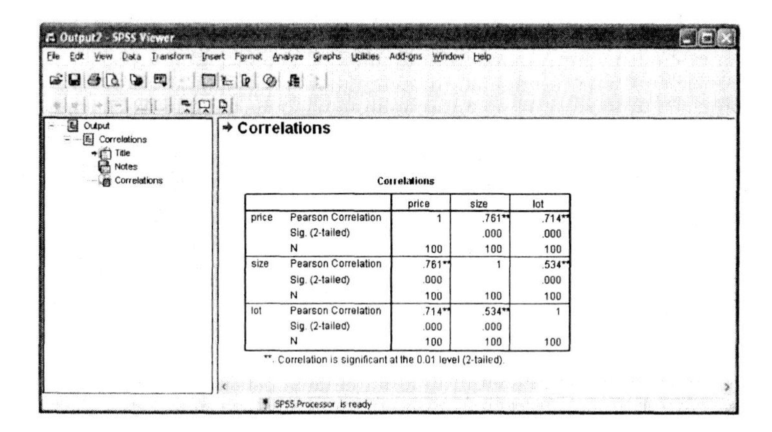

Correlations

		price	size	lot
price	Pearson Correlation	1	.761**	.714**
	Sig. (2-tailed)		.000	.000
	N	100	100	100
size	Pearson Correlation	.761**	1	.534**
	Sig. (2-tailed)	.000		.000
	N	100	100	100
lot	Pearson Correlation	.714**	.534**	1
	Sig. (2-tailed)	.000	.000	
	N	100	100	100

**. Correlation is significant at the 0.01 level (2-tailed).

Testing the Significance of the Regression Coefficients

To demonstrate how to use SPSS to test for the significance of the regression coefficients, we will use the "college_athletes" data set on your text CD. A portion of this data set is shown in the table below.

TBW	HGT	BF_%	AGE
96	62	13.0	23
130	65	16.3	17
107	64.5	17.3	21

Access this data set and select **Analyze → Regression → Linear...** Enter "TBW" as the "Dependent" variable and "HGT," "BF_Percent," and "AGE" as the "Independent(s)" variables.

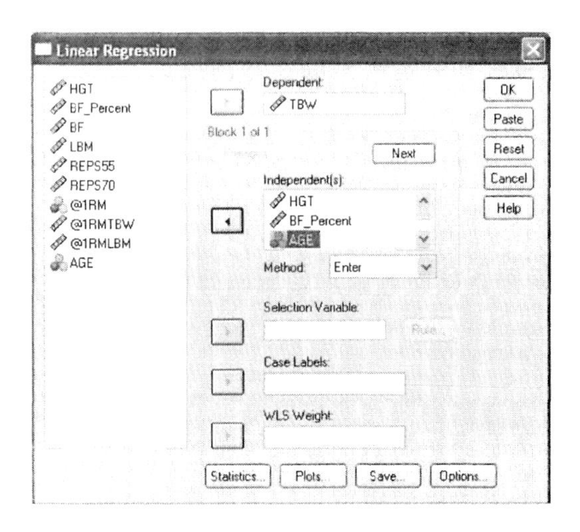

Click on **OK** to produce the output. The "Coefficients" portion of this output is shown below.

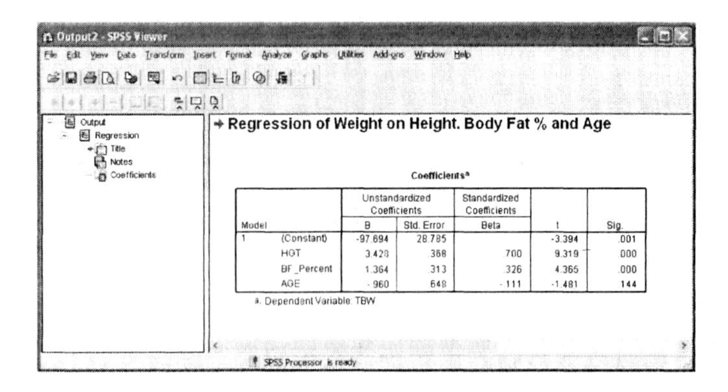

Regression of Weight on Height. Body Fat % and Age

Coefficients[a]

Model		Unstandardized Coefficients		Standardized Coefficients	t	Sig.
		B	Std. Error	Beta		
1	(Constant)	-97.694	28.785		-3.394	.001
	HGT	3.428	.368	.700	9.319	.000
	BF_Percent	1.364	.313	.326	4.365	.000
	AGE	-.960	.648	-.111	-1.481	.144

a. Dependent Variable: TBW

The t-statistics in the "Coefficients" portion of the output, and their corresponding p-values (Sig.) indicate that both height and body fat percent are significant but age is not.

The Regression ANOVA

To determine if a set of independent variables collectively have an effect on a dependent variable you will use the "ANOVA" portion of the regression output. We will demonstrate this by using the multiple regression output we just produced for the "college_athletes" data set. If you don't have this output then access this data file and select **Analyze → Regression → Linear...** Enter "TBW" as the "Dependent" variable and "HGT," "BF_Percent," and "AGE" as the "Independent(s)" variables. Click on **OK**. The "ANOVA" portion of the output is shown below.

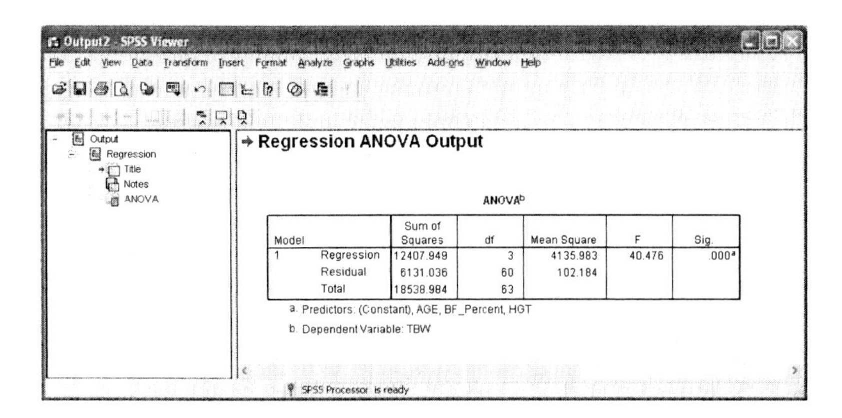

ANOVA[b]

Model		Sum of Squares	df	Mean Square	F	Sig.
1	Regression	12407.949	3	4135.983	40.476	.000[a]
	Residual	6131.036	60	102.184		
	Total	18538.984	63			

a. Predictors: (Constant), AGE, BF_Percent, HGT

b. Dependent Variable: TBW

The F-statistic is for testing that at least one of the three independent variables is significantly different from zero. The corresponding p-value is given under "Sig." and, as with previous examples, this result tells us that the p-value is less than .00005. You will learn more about ANOVA tables in the next chapter.

Another Residual Plot

In Chapter 3 of this manual we looked at a histogram of the residuals as a diagnostic tool for assessing the fit of the best fit line and for looking for unusual observations. Another graph that is useful is to plot the residuals vs. the independent variable. The Regression command in SPSS has options under the "Plots" button for graphing the standardized residuals (*ZRESID) vs. the dependent variable (DEPENDNT) and for plotting the dependent variable vs. each of the independent variables separately (click on "Produce all partial plots), but to graph the standardized residuals vs. one of the independent variables you will have to save the residuals and produce the plot using the "Scatter/Dot" command. Let's demonstrate this using the "house_ selling_prices" data set. Access this data set and select **Analyze → Regression → Linear...** Enter "price" as the "Dependent" variable and "house size" and "lot size" as the "Independent(s)" variables. Click on the "Save" button and select "Standardized" under "Residuals." Click on **Continue** and then on **OK** to produce the regression output. A new column with name "ZRE_1" will be added to the Data Editor with the standardized residual of each observation.

To make the residual plot select **Graphs → Interactive → Scatterplot...** and enter the residuals ("ZRE_1") as the variable for the Y-axis and house size as the variable for the X-axis. Click on **OK** to produce the plot.

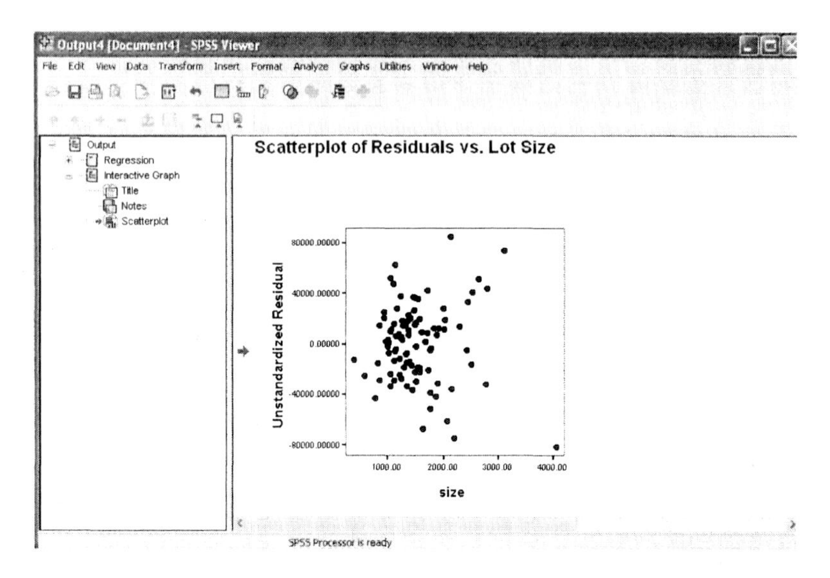

You can add a horizontal line to the graph by selecting the Regression method under the Fit tab. Make a similar graph with the other independent variable on the X-axis to check on its association with the residuals.

Categorical Predictor Variables

When a regression analysis is to be performed over groups identified by a categorical variable, the group to which a particular observation belongs can be identified in SPSS using an "indicator variable." An indicator variable takes just two values, 0 and 1, and so can be used when there are just two groups. If there are more than two groups, additional indicator variables can be used. The method for doing this is explained in your text.

Let's demonstrate the use of an indicator variable using the "high_jump2" data set on your text CD. This data set has three variables, "winning_height, "gender" and "year." The variable "gender" is an indicator variable, taking the value of "0" for females and "1" for males. A portion of this data set is shown below.

high jump 2.sav - SPSS Data Editor

	winning_height	gender	year
1	1.81	Male	1896.00
2	1.90	Male	1900.00
3	1.80	Male	1904.00
4	1.91	Male	1908.00
5	1.93	Male	1912.00
6	1.94	Male	1920.00
7	1.98	Male	1924.00
8	1.94	Male	1928.00

To do the regression, select **Analyze → Regression → Linear...** and enter the "winning height" variable in the "Dependent" box and both "gender" and the "year" variable in the "Independent(s)" box. You do not need to do anything special in SPSS to designate "gender" as an indicator variable.

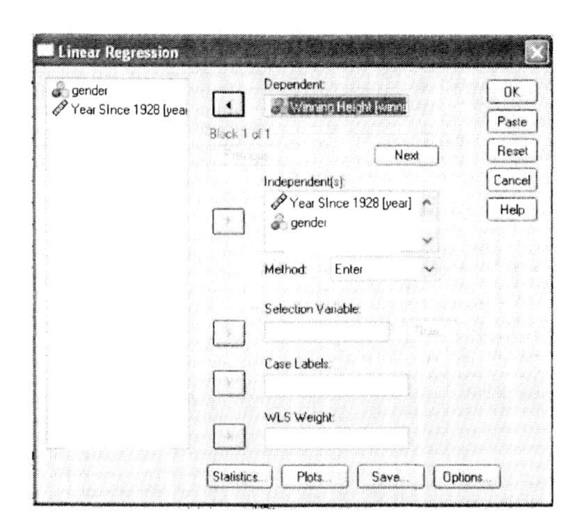

Click on **OK** to produce the output.

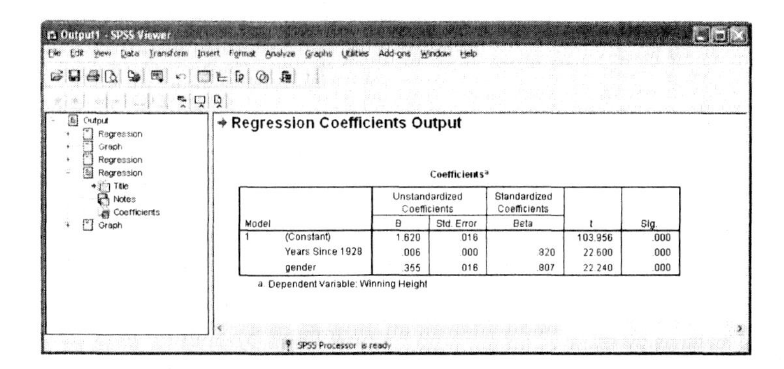

Coefficients[a]

Model		Unstandardized Coefficients		Standardized Coefficients	t	Sig.
		B	Std. Error	Beta		
1	(Constant)	1.620	.016		103.956	.000
	Years Since 1928	.006	.000	.820	22.600	.000
	gender	.355	.016	.807	22.240	.000

a. Dependent Variable: Winning Height

The best-fit regression line is $\hat{y} = 1.62 + .006x_1 + .355x_2$. For females $x_2 = 0$, so the best-fit line is $\hat{y} = 1.62 + .006x_1$. For males $x_2 = 1$, so the best-fit line is $\hat{y} = 1.62 + .006x_1 + .355(1)$, or $\hat{y} = 1.975 + .006x_1$. The result is two separate but parallel lines, one for the male data and the other for the female data.

To see if a model with two separate but parallel lines is a good "fit" to this data set, we can graph the data using a different symbol for the male and female data points. To do this, select **Graphs → Interactive → Scatterplot...,** and place the "height" variable in the box on the vertical axis and the "year" variable in the box on the horizontal axis. Then, enter the "gender" variable in the Style box under Legend Variables.

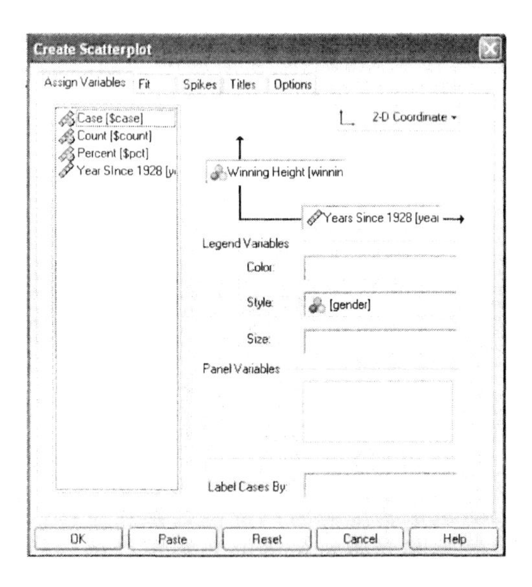

Click on **OK** to produce the plot.

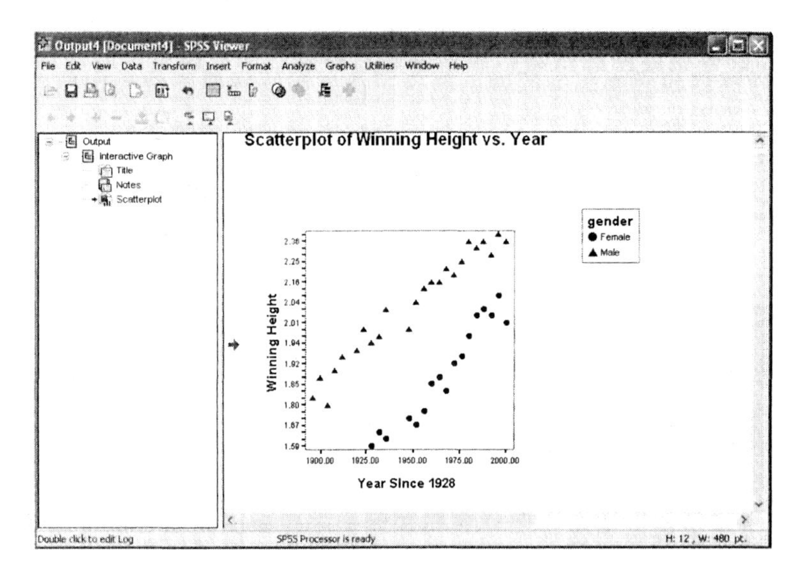

SPSS will use a different symbol for the two groups defined by the "gender" variable. You can use the Chart Editor to change the plot symbols and colors if you desire. The plot symbols in the graph above were changed from open to filled-in. We can see from this plot that two separate but parallel lines is a reasonable model for this data set.

Logistic Regression

SPSS can also do logistic regression on a binary response variable. To demonstrate this we will use the data from Example 12 in Chapter 13 on income (in 1000s of Euros) and whether or not the respondent has a travel credit card. The dependent variable Y is coded as 0 = "no" and 1 = "yes." Access the "credit_card_ and_income" data set from your text CD and select **Analyze →**

Regression → Binary Logistic... Enter the variable "y" as the "Dependent" variable and "income" as the "Covariate."

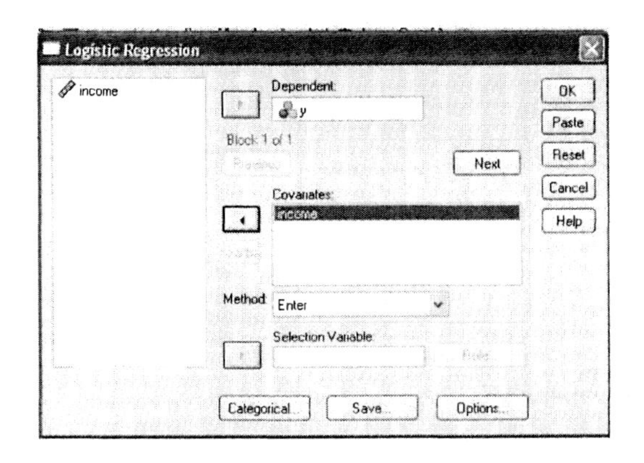

Click on **OK** to produce the output.

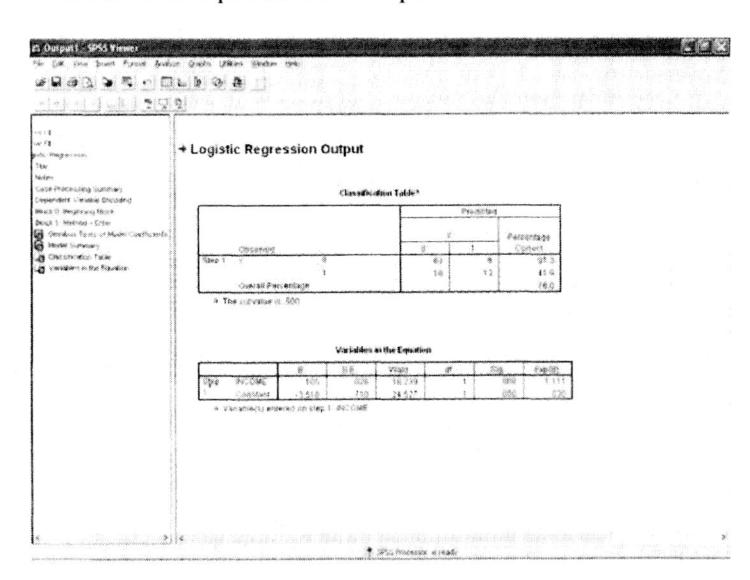

The coefficients for the logistic model are given under "B" in the bottom portion of the output.

The logistic equation would be $\hat{p} = \dfrac{e^{-3.52+0.105x}}{1+e^{-3.52+.105x}}$, which agrees with the result shown in your text. For example, for x=12 the estimate would be $\hat{p} = \dfrac{e^{-3.52+.105(12)}}{1+e^{-3.52+.105(12)}} = .0944$. So, the estimated proportion of those whose income is 12 thousand Euros that have a travel credit card is around 9%.

Chapter 13 Analysis of Variance

The extension of the independent samples T-test to compare more than two means is called Analysis of Variance. SPSS has two options for performing an ANOVA, "One-Way ANOVA" under the "Compare Means" command and "Univariate" under the "General Linear Model" command. We will demonstrate "One-Way ANOVA" in this section and "Univariate" for doing two-way and higher ANOVAs. For the One-way option under the Compare Means command, the grouping variable must be coded numerically, but for the Univariate option under the General Linear Model command, the grouping variable can be either numeric or a string variable. Unlike the two independent samples situation, for ANOVA we have to assume that the variances of the groups being compared are equal. We will show how to check this assumption using SPSS.

One-Way ANOVA

To demonstrate how to compare the means of more than two independent samples, we will use the telephone holding data from Example 2 in Chapter 14. This data is shown in the table below.

Recorded Message	Holding Time Observations
Advertisement	5, 1, 11, 2, 8
Muzak	0, 1, 4, 6, 3
Classical	13, 9, 8, 15, 7

Enter this data in SPSS using one variable for the holding times and another for the type of message. The variable for the type of music must be coded numerically. To do the ANOVA, select **Analyze → Compare Means → One-Way ANOVA...** and enter the holding time variable in the "Dependent List" and the message type variable in the "Factor" list.

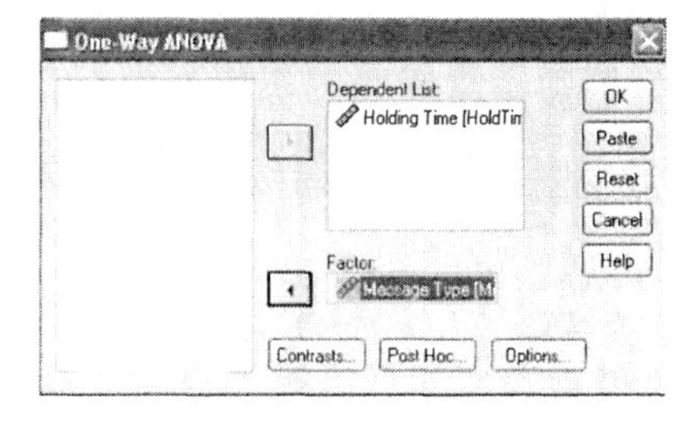

Click on **OK** to produce the output.

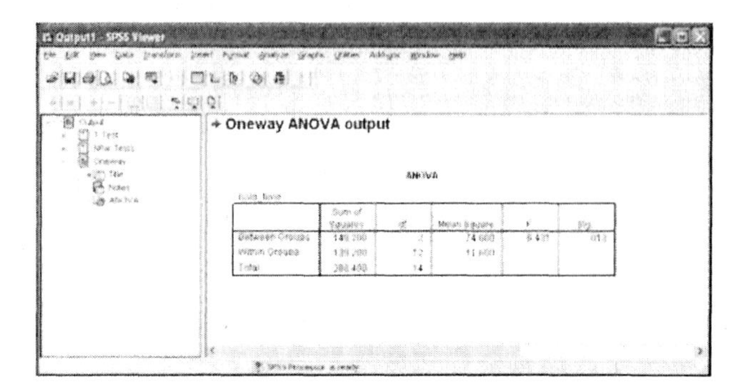

The p-value is small so this suggests that there is a difference in the average amount of time subjects were willing to hold based on the type of message they heard.

To find out which groups differ and if the equal variances assumption is met, we can use some of the options of this command. Select **Analyze → Compare Means → One-Way ANOVA...** again and click on the "Post Hoc..." button. Click on "Tukey" and then on **Continue.**

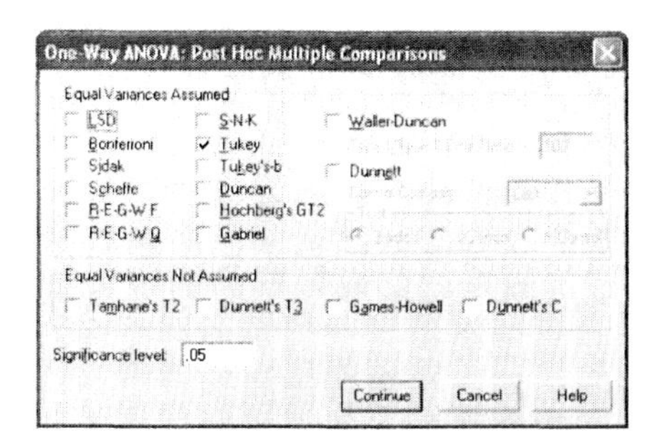

Click on the "Options" button and then on both "Descriptive" and "Homogeneity of variance test."

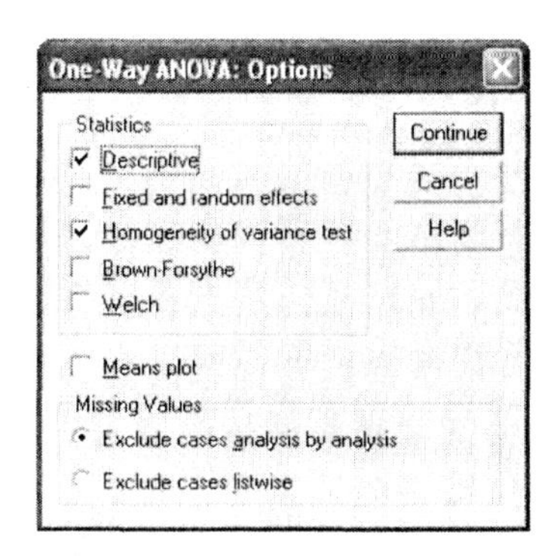

Click on **Continue** and then on **OK**. SPSS will produce a lot of additional output.

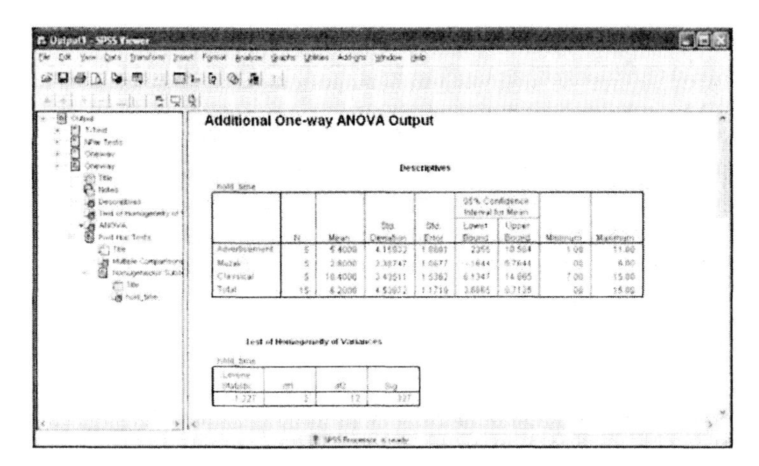

The output above shows the group means and standard deviations along with individual confidence intervals for the mean of each group. Levene's test for equal variances is also produced. The p-value for this test is .327, indicating that there is no evidence to suggest that the variances are unequal.

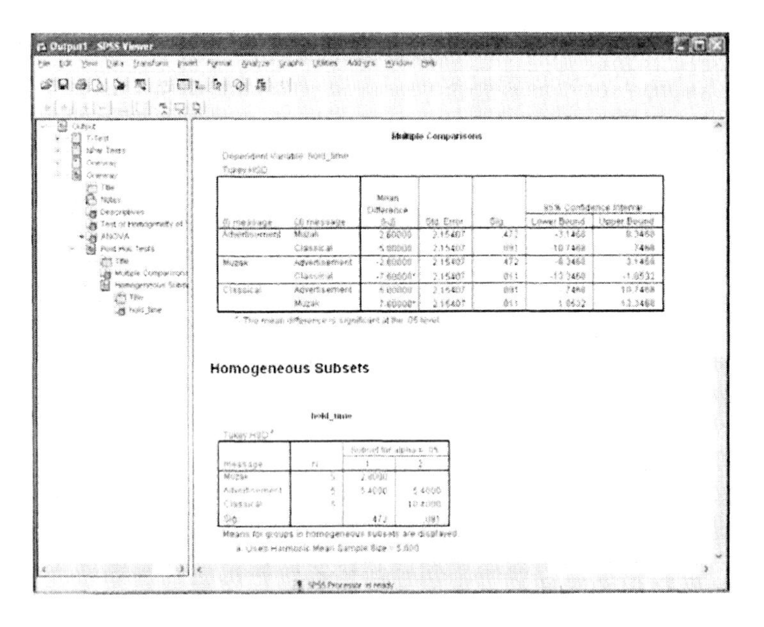

This output shows Tukey confidence intervals for simultaneous comparisons of each pair of groups. The last part of the output shows the conclusions of which groups differ based on the Tukey confidence intervals.

Two-Way ANOVA

As indicated at the beginning of this chapter, if you are doing a one-way ANOVA you can use either "One-Way ANOVA" under the "Compare Means" command or you can use "Univariate" under the "General Linear Model" command. To conduct an ANOVA with two or more factors you will need to use the "General Linear Model" command. To demonstrate this command we will use the agricultural data from Example 9 in Chapter 14. The data is provided below.

Fertilizer	Manure			Plot		
Level	Level	1	2	3	4	5
High	High	13.7	15.8	13.9	16.6	15.5
High	Low	16.4	12.5	14.1	14.4	12.2
Low	High	15.0	15.1	12.0	15.7	12.2
Low	Low	12.4	10.6	13.7	8.7	10.9

Enter the data in SPSS in three columns, one for the crop yield and one each for the two factors, fertilizer level and manure level. If coding the grouping values numerically, enter the value "1" for the "high" level of each factor and "0" for the "low" level. Use value labels under the Variable View to label the values appropriately. A portion of the data set is visible in the Data Editor shown below.

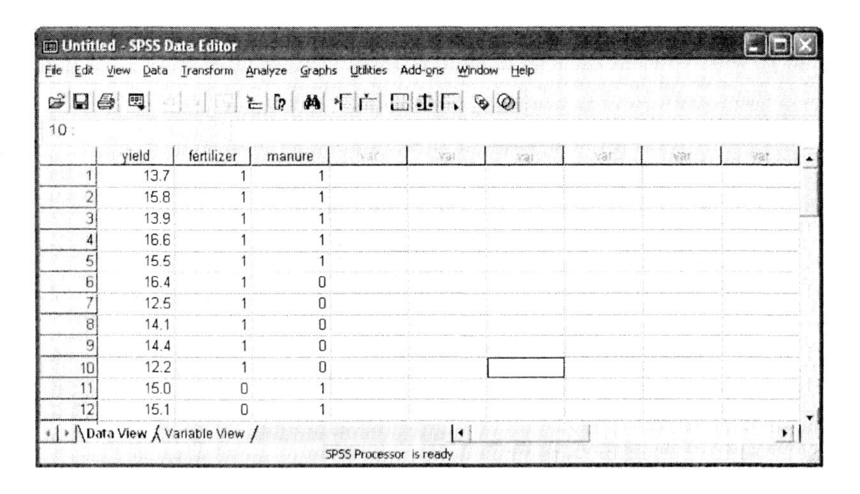

	yield	fertilizer	manure
1	13.7	1	1
2	15.8	1	1
3	13.9	1	1
4	16.6	1	1
5	15.5	1	1
6	16.4	1	0
7	12.5	1	0
8	14.1	1	0
9	14.4	1	0
10	12.2	1	0
11	15.0	0	1
12	15.1	0	1

To do the ANOVA select **Analyze → General Linear Model → Univariate...** and enter "yield" as the "Dependent Variable" and both "fertilizer" and "manure" as "Fixed Factor(s)."

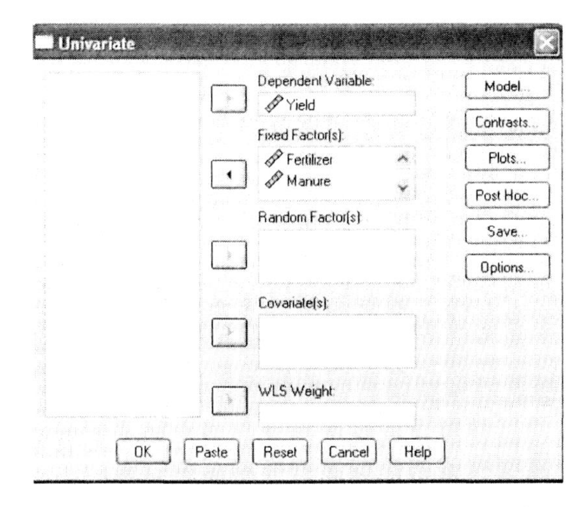

The default model in SPSS includes an interaction term. To run the ANOVA without this term, click on the "Model" button and then on "Custom." Click once on "fertilizer" under "Factors & Covariates" and then once on the arrow button underneath "Build Term(s)." Repeat this process with "manure." Your dialog box should look like the one shown below.

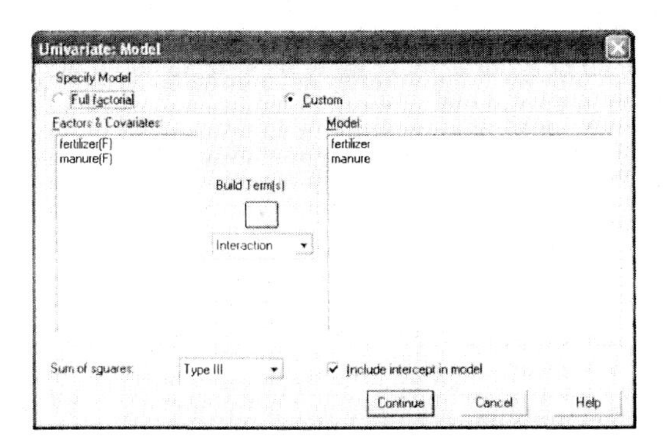

Click on **Continue** and then on **OK**.

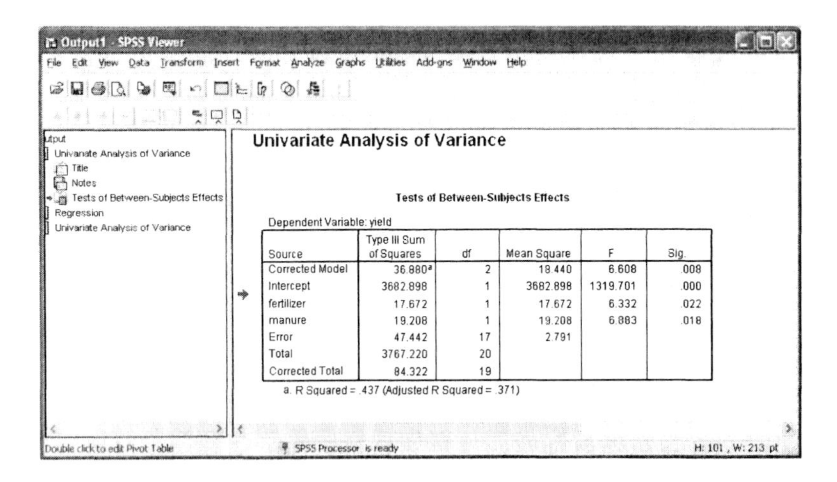

Univariate Analysis of Variance

Tests of Between-Subjects Effects

Dependent Variable: yield

Source	Type III Sum of Squares	df	Mean Square	F	Sig.
Corrected Model	36.880ᵃ	2	18.440	6.608	.008
Intercept	3682.898	1	3682.898	1319.701	.000
fertilizer	17.672	1	17.672	6.332	.022
manure	19.208	1	19.208	6.883	.018
Error	47.442	17	2.791		
Total	3767.220	20			
Corrected Total	84.322	19			

a. R Squared = .437 (Adjusted R Squared = .371)

This ANOVA output includes some additional terms under "Source" that are not in the ANOVA table in your text, but the results for "fertilizer" and "manure" are the same.

The "Options" button on the "Univariate" dialog box includes a number of useful options. For example, you can request the estimated marginal means, additional descriptive statistics and a test for equal variances (homogeneity).

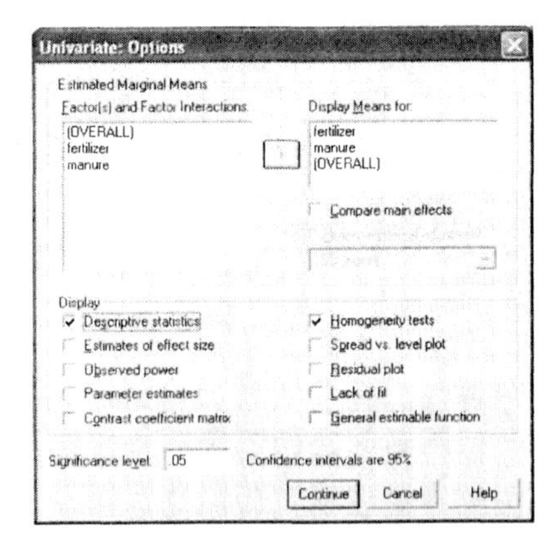

The Regression Approach to ANOVA

For factors that have just two categories, such as "fertilizer" and "manure" in the previous example, you can also do ANOVA using the "Regression" command. For factors with more than two categories it is easier to just use the "General Linear Model" command (or the "Compare Means" command for a one-way ANOVA). Using the same agricultural data set select **Analyze → Regression → Linear...** and enter "yield" as the "Dependent" variable and both "fertilizer" and "manure" as "Independent(s)" variables.

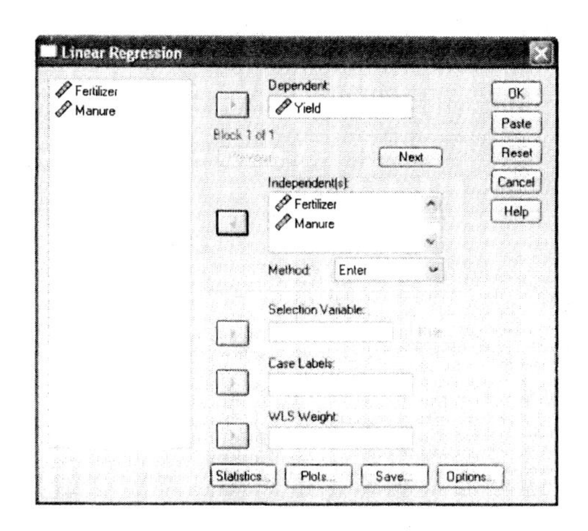

Click on **OK** to produce the output.

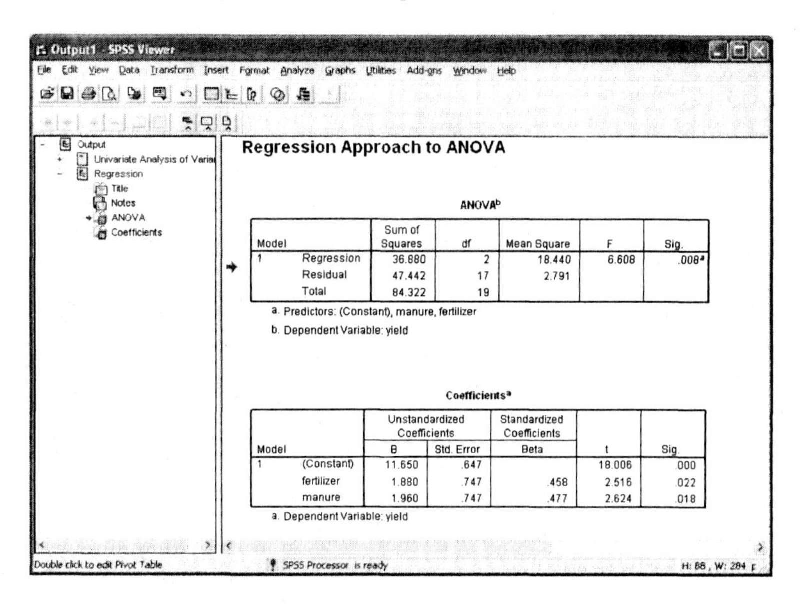

Regression Approach to ANOVA

ANOVA[b]

Model		Sum of Squares	df	Mean Square	F	Sig.
1	Regression	36.880	2	18.440	6.608	.008[a]
	Residual	47.442	17	2.791		
	Total	84.322	19			

a. Predictors: (Constant), manure, fertilizer

b. Dependent Variable: yield

Coefficients[a]

Model		Unstandardized Coefficients		Standardized Coefficients	t	Sig.
		B	Std. Error	Beta		
1	(Constant)	11.650	.647		18.006	.000
	fertilizer	1.880	.747	.458	2.516	.022
	manure	1.960	.747	.477	2.624	.018

a. Dependent Variable: yield

The ANOVA table from the "Regression" command does not separate out the effects of each factor separately, as in the first ANOVA table we produced for this data. But, the t-tests provided in the "Coefficients" portion of the output will allow you to test for the significance of each factor separately. The "slopes" associated with each factor under the "Coefficients" portion of the output provide the values you can use to predict the size of the yield for each combination of the levels of the two factors. For example, for a plot in which the high level of each factor was used we would predict an average yield of 11.65 + 1.88 + 1.96 = 15.49.

Two-Way ANOVA with Interaction

The default model for two-way and higher ANOVA in SPSS includes all interaction terms. For two-way ANOVA there is just one interaction term. To do the ANOVA with the interaction term for the agricultural data set, select **Analyze → General Linear Model → Univariate...** again and click on the "Model" button. This time, make sure that "Full factorial" is selected.

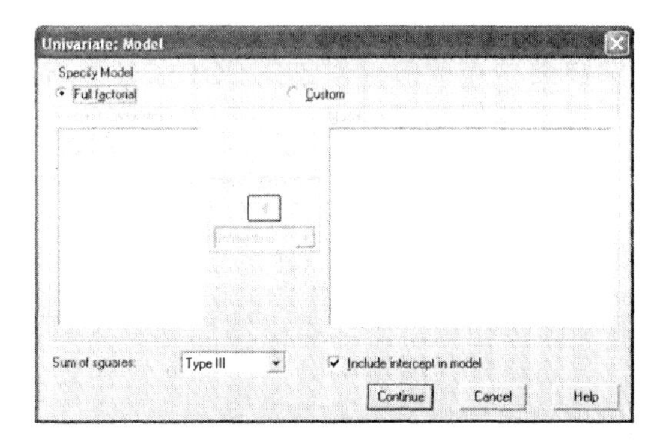

Click on **Continue** and then on **OK**.

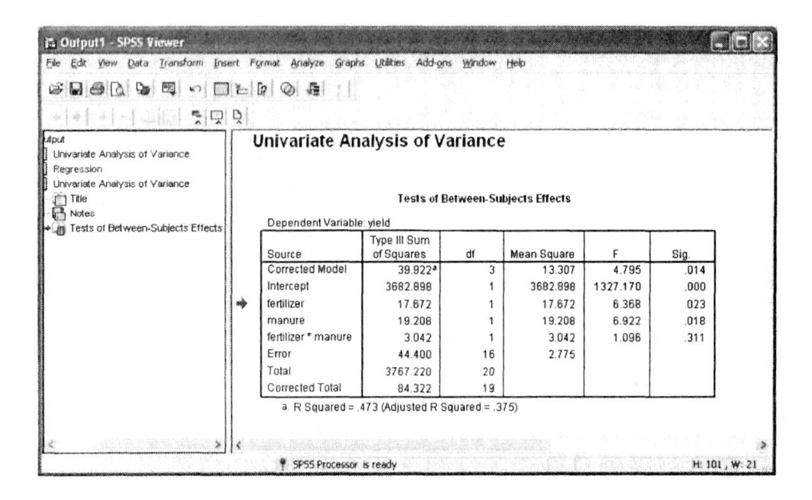

Univariate Analysis of Variance

Tests of Between-Subjects Effects

Dependent Variable: yield

Source	Type III Sum of Squares	df	Mean Square	F	Sig.
Corrected Model	39.922[a]	3	13.307	4.795	.014
Intercept	3682.898	1	3682.898	1327.170	.000
fertilizer	17.672	1	17.672	6.368	.023
manure	19.208	1	19.208	6.922	.018
fertilizer * manure	3.042	1	3.042	1.096	.311
Error	44.400	16	2.775		
Total	3767.220	20			
Corrected Total	84.322	19			

a. R Squared = .473 (Adjusted R Squared = .375)

Here the interaction term is not significant.

To include the interaction term using the regression approach, the grouping variables must be coded numerically and you will have to create an interaction variable in the data set. Select **Transform → Compute Variable...** and enter the variable name "interaction" under "Target Variable" and the expression "fertilizer*manure" in the "Numeric Expression" box. Click on **OK**. This new variable will be added to your data set.

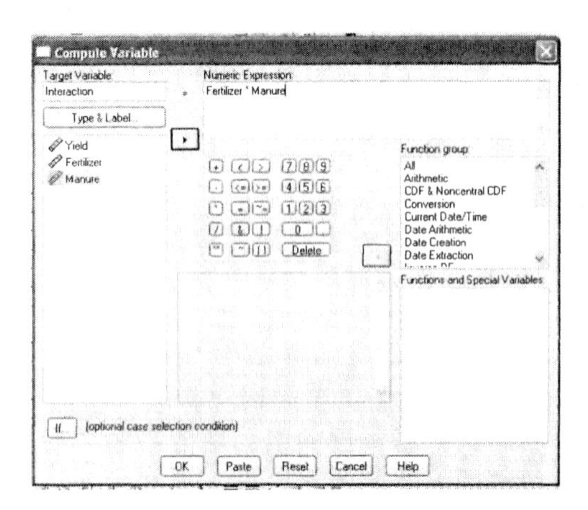

To do the ANOVA, select **Analyze → Regression → Linear...** and add the interaction variable to the "Independent(s)" variables list. Click on **OK**.

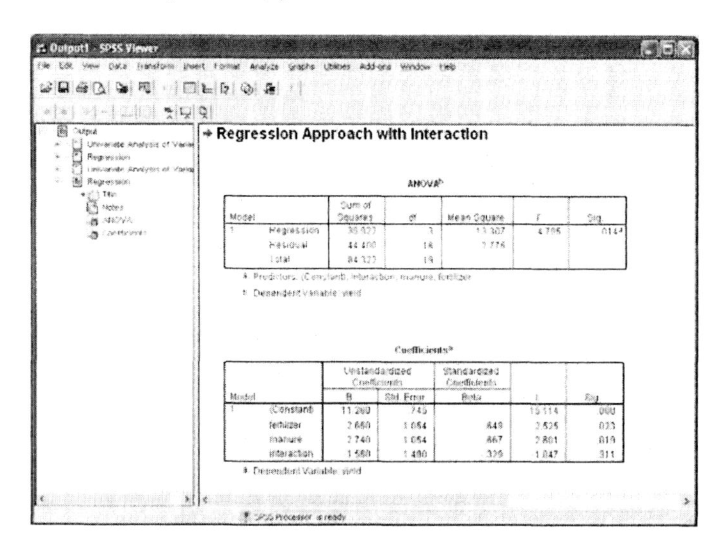

As before, SPSS does not separate out the effect of the interaction term in the "ANOVA" portion of the output, but you can test for its significance using the T-test under "Coefficients."

Chapter 14 Nonparametric Statistics

All of the methods we've looked at so far for comparing means have an underlying normality assumption. It turns out that these methods are fairly "robust" when it comes to working with non-normal data; that is, the methods perform well in detecting a difference in means even when the data is not exactly normally distributed. For data from highly skewed distributions or from populations that differ substantially in the amount of variation present, these normal-based methods do not perform well and in these situations we need some alternate methods that do not involve a normality assumption.

The Wilcoxon Test

The nonparametric test for this situation is called the Wilcoxon test (or the Wilcoxon rank sum test or the Mann-Whitney test). As your text explains, it is based on replacing the actual data values with their associated ranks among all observations. To demonstrate this test, we will use the respiratory ventilation data from Exercise 15.4. This data set consists of respiratory ventilation measurements made on two groups, a control group and a treatment group that would undergo hypnosis. The data is shown in the following table.

Controls	3.99 4.19 4.21 4.54 4.64 4.69 4.84 5.48
Treated	4.36 4.67 4.78 5.08 5.16 5.20 5.52 5.74

Enter the data in SPSS using two variables, one for the ventilation measurements and one for the group. For this analysis, the grouping variable must be entered as a numeric variable. Recall that for the "Compare Means" and "Univariate" commands you could enter the grouping variable as either a numeric of string variable.

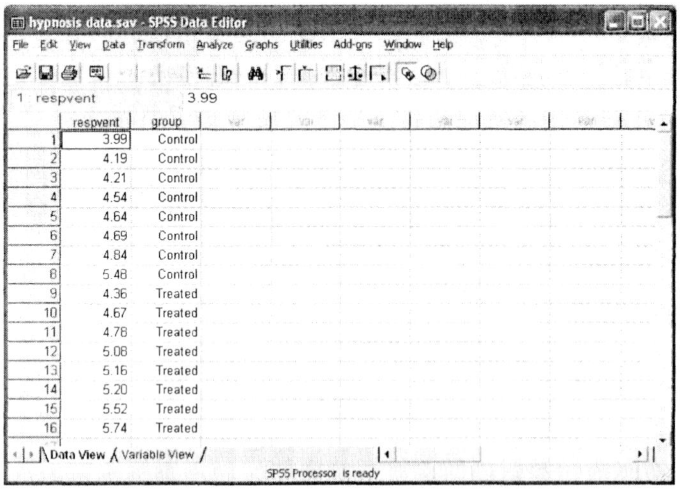

hypnosis data.sav - SPSS Data Editor

File Edit View Data Transform Analyze Graphs Utilities Add-ons Window Help

1 : respvent 3.99

	respvent	group
1	3.99	Control
2	4.19	Control
3	4.21	Control
4	4.54	Control
5	4.64	Control
6	4.69	Control
7	4.84	Control
8	5.48	Control
9	4.36	Treated
10	4.67	Treated
11	4.78	Treated
12	5.08	Treated
13	5.16	Treated
14	5.20	Treated
15	5.52	Treated
16	5.74	Treated

Data View / Variable View

SPSS Processor is ready

To do the nonparametric test, select **Analyze → Nonparametric Tests → 2 Independent Samples…** and enter the "respiratory ventilation" variable under "Test Variable List" and the "group" variable in the box under "Grouping Variable." Click on **Define Groups** and enter the values you used to define the two groups. In the example below you can see that the groups were defined using the values 1 and 2.

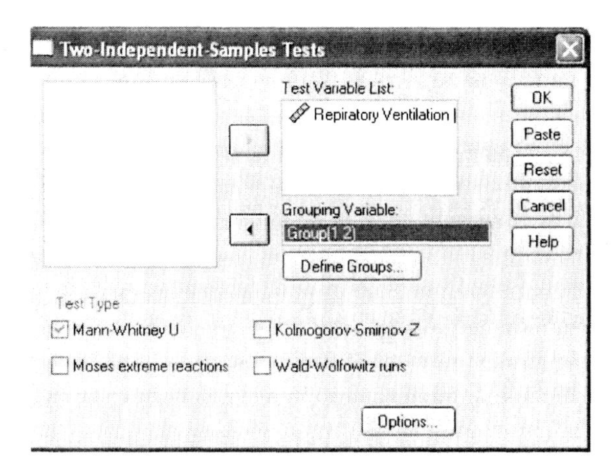

Click on **Continue** and then on **OK**.

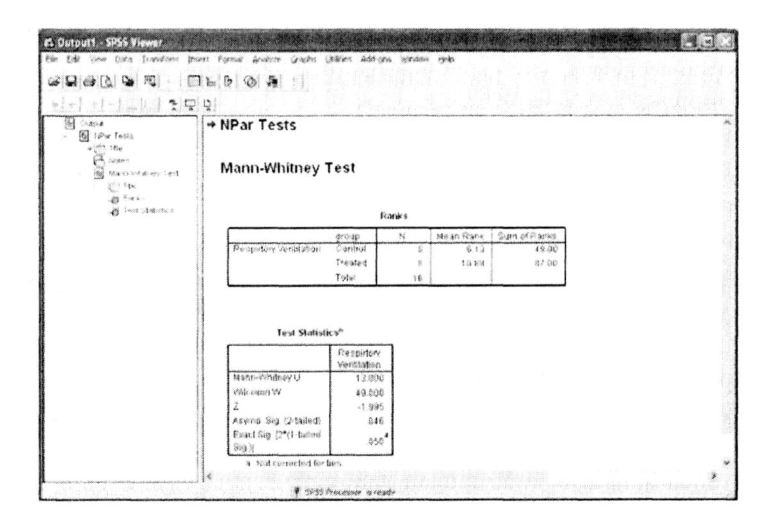

SPSS will assign a rank to each observation, but only the mean rank and sum of ranks are reported for each group. The results for this data set show an exact two-sided p-value of .050, but for this situation we need the one-sided p-value. Since we want to know if the treated group (those who would be hypnotized) ventilated more, and we can see that the mean ranks for the treated group is larger for the treated group, the p-value will be half of .050 or .025.

The Kruskal-Wallis Test

The extension of the Wilcoxon test for comparing the means of more than two samples is called the Kruskal-Wallis Test. In this test, as in the Wilcoxon test, the observations are replaced by their ranks among all observations. To demonstrate this test we'll use the data from Example 5 in Chapter 15. This data, shown in the table below, is from a study to investigate the impact of dating on GPA.

Dating Group	GPA observations						
Rare	1.75	3.15	3.50	3.68			
Occasional	2.00	3.20	3.44	3.50	3.60	3.71	3.80
Regular	2.40	2.95	3.40	3.67	3.70	4.00	

Enter this data in SPSS with one variable for the GPA and another for the dating group. The grouping variable must be coded numerically.

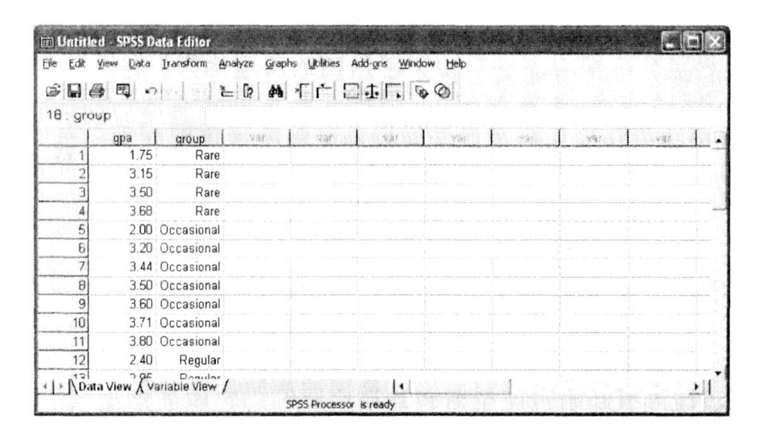

To conduct the Kruskal Wallis nonparametric test on this data set select **Analyze → Nonparametric Tests → K Independent Samples...** Enter "GPA" in the "Test Variable List" box and the "group" variable in the "Grouping Variable" box. The dialog box is similar to the ANOVA dialog box except that you need to click on the "Define Range..." button and enter the minimum and maximum values of the "Grouping Variable" that identify the groups to be compared. For this example, group was coded with Rare = 1, Occasional = 2, and Regular = 3, so the values 1 and 3 appear in the range for Group in the dialog box below.

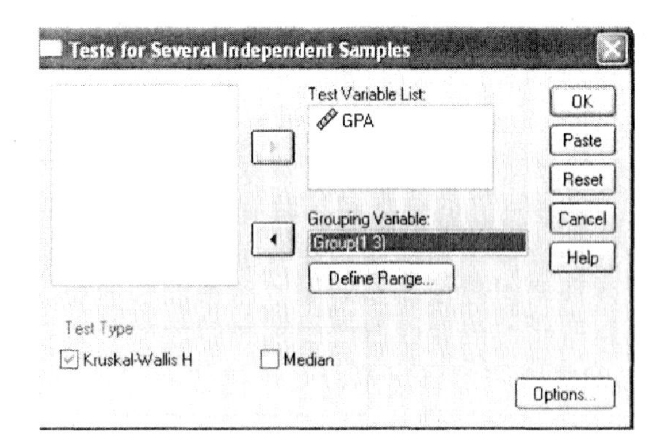

The Kruskal-Wallis test is the default option for this command. Click on **OK** to produce the output.

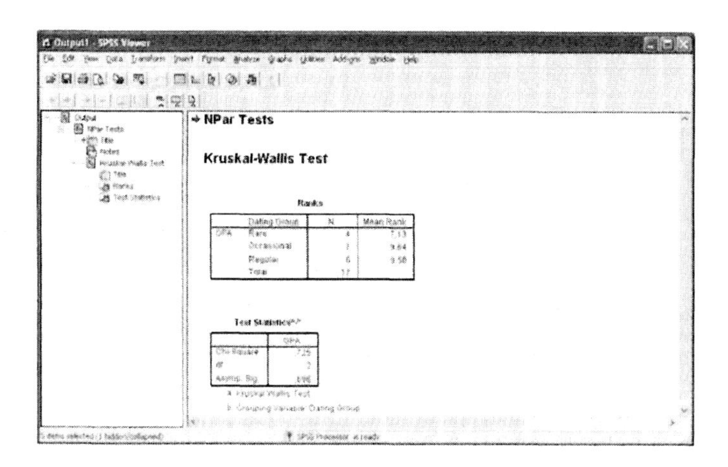

As in the previous test, SPSS will assign a rank to each observation but the output reports just the mean rank for each group. The test statistic and its associated p-value are reported, as well. The large p-value for this data set suggests that GPA is independent of dating group.

The Sign Test

For analyzing data from paired or dependent samples, your text describes two separate tests, the Sign Test and the Wilcoxon Signed-Ranks Test. The difference between these two tests is that, in the sign test only the number of positive differences is recorded while in the Wilcoxon Signed-Rank the differences are ranked in absolute value and the sum of the positive ranks is recorded. To demonstrate the Sign Test we will use data from the "georgia_student_suvey" file from your text CD. This example is described in Example 6 from Chapter 15 of your text. The paired variables are the time (in minutes per day) spent watching TV and browsing the Internet. Access this data file and select **Analyze → Nonparametric Tests → 2 Related Samples…** and enter the paired variables in the same way as for the Paired T-test described in Chapter 9 of this manual. Click once on the "internet" variable and once on the "TV" variable and then click on the triangle button to enter the pair in the "Test Pair(s) List" box. Click on the "Sign" option to select it. The Wilcoxon Test (demonstrated in the next example) is the default option. You will have to click once on it to un-select it.

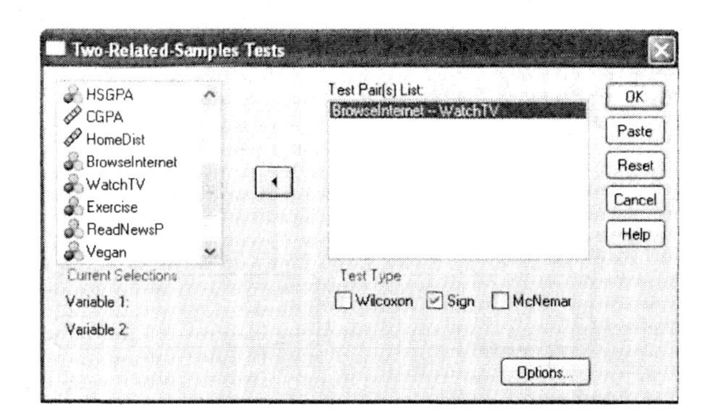

Click on **OK** to produce the output.

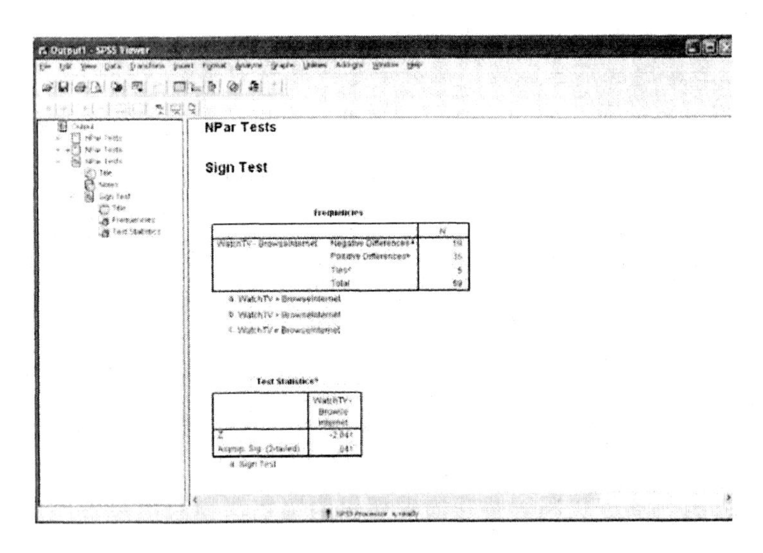

The output shows that there were 19 students who spent more time browsing the Internet (so the difference was negative) and 35 students who spent more time watching TV 9. All students who spent the same amount of time on these activities are not included in the analysis. SPSS reports both the exact value of the test statistic and is associated p-value.

The Wilcoxon Signed-Ranks Test

Let's use the same data set to demonstrate this test. Select **Analyze → Nonparametric Tests → 2 Related Samples...** again except this time select the Wilcoxon test (and un-select the Sign test).

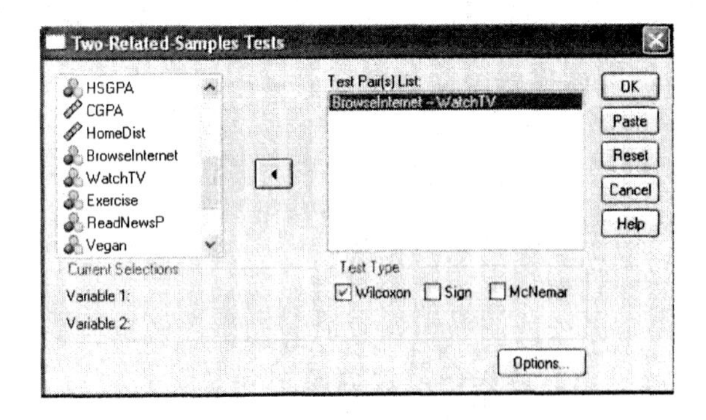

Click on **OK** to produce the output.

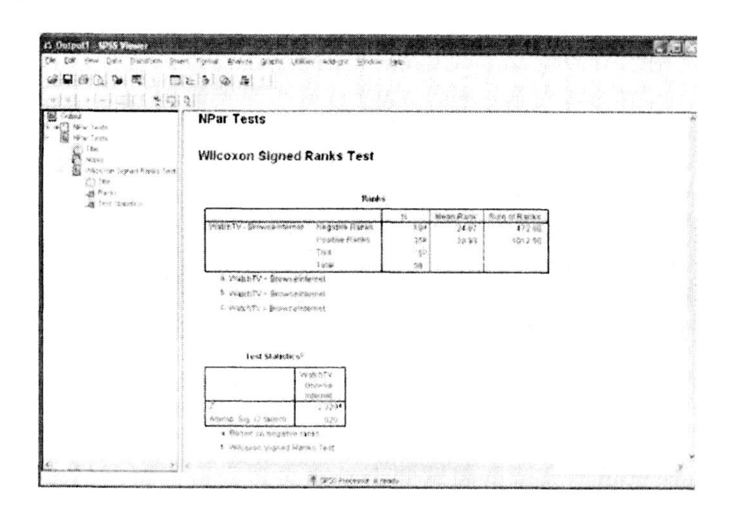

SPSS reports both the mean ranks and sum of ranks for the negative and positive differences. The test statistic is the sum of the positive ranks, for which SPSS reports a two-sided p-value.

Appendix A: Summary of SPSS Commands

The following is a list of frequently used SPSS procedures using the convention:

Select **Menu name→ Command name → Subcommand** or **option**

In the dialog box that appears, select one or two response (dependent) variables and explanatory (independent) variables where appropriate. When you have entered enough information for SPSS to be able to complete the command, the "OK" button will become highlighted.

Operating SPSS

Creating a new data set

Start SPSS and select "Type in new data" or select **File → New → Data**. (Page 20)

Opening an existing data file

Select **File → Open → Data...** and locate the file from options under "Look in:" (Page 27)

Changing the window display

Select **Window→ SPSS Data Editor** or **SPSS Viewer** (Page 8)

Saving data

Select **File → Save as...** when the Data Editor is the active window. (Page 7)

Saving output

Select **File → Save as...** when the Output Viewer is the active window. (Page 16)

Printing output

Select the output to be printed, then select **File → Print.** (Page 15)

Exiting SPSS

Select **File→ Exit.** (Page 17)

Data Management

Naming a variable

Click on the **Variable View** tab and enter a new name under "Name." (Page 4)

Labeling a variable

Click on the **Variable View** tab and enter a label under "Label." (Page 5)

Selecting a variable type

Click on the Variable View tab, click on the box under "Type" and select "Numeric" or "String." (Page 4)

Labeling the values of a numerically coded qualitative variable

Click on the **Variable View** tab then on the box under "Values," enter a value in the "Value" box and a label in the "Label" box, then click on "Add." (Page 5)

Calculating a new variable from existing variables

Select **Transform → Compute Variable...** enter a variable name in the "Target Variable" box and an expression in the "Numeric Expression" box. (Pages 73, 80, 97, 114, and 132)

Selecting a random sample

Select **Data → Select Cases...** then select "**Random sample of cases...**" (Pages 69)

Omitting specific cases

Select **Data → Select Cases...** then select "**If condition is Satisfied...**" and enter an expression that will eliminate specific cases; to analyze all cases for subsequent commands, Select **Data → Select Cases...** again and select "**All cases**." (Page 66)

Weighting cases for tabulated data

Select **Data → Weight Cases...** and enter a count variable in the "Frequency Variable" box. (Page 19)

Repeating an analysis of a quantitative variable for each value of a qualitative variable

Select **Data → Split File...** and select an option for comparing groups, enter the qualitative variable in the "Groups Base on" box; to analyze all cases for subsequent commands select **Data → Split File...** again, select "**Analyze all cases**..." (Pages 53 and 54)

Descriptive Statistics and Graphs

Listing data

Select **Analyze → Reports → Case summaries...** (Page 7)

Frequency distributions

Univariate: Select **Analyze → Descriptive Statistics → Frequencies...** (Pages 27 and 45)

Bivariate: **Analyze → Descriptive Statistics → Crosstabs...** (Pages 47 and 50)

Descriptive Statistics

Select **Analyze → Descriptive Statistics → Descriptives...** (Pages 10, 30, 42 and 45)

Select **Analyze → Descriptive Statistics → Frequencies...** and request optional statistics. (Pages 38, 39, 40 and 45)

Select **Analyze → Compare means → Means...** (Page 52)

Select **Analyze → Descriptive Statistics → Explore...** and request optional statistics. (Pages 45, 51, and 84)

Univariate Graphs

Categorical Data

Pie Chart: Select **Graphs → Interactive → Pie...** (Page 20) or select **Graphs → Legacy Dialogs → Pie...** (Page 23)

Bar Chart: Select **Graphs → Interactive → Bar...** (Page 25) or select **Graphs → Legacy Dialogs → Bar** (Page 26)

Quantitative Data

Dot Plot: Select **Graphs → Legacy Dialogs → Dot...** (Page 28)

Histogram: Select **Graphs → Interactive → Histogram...** (Pages 11 and 32) or select **Graphs → Legacy Dialogs → Histogram...**

Boxplot: Select **Graphs → Interactive → Boxplot…** (Pages 46 and 47) or select **Graphs → Legacy Dialogs → Boxplot…** (Page 100)

Bar Chart, Pie Chart or Histogram: Select **Analyze→ Descriptive Statistics → Frequencies…** and request optional Charts. (Page 33 and 45)

Boxplot, Histogram or Stem-and-Leaf Plot: Select **Analyze→ Descriptive Statistics → Explore…** and request optional Plots. (Page 31 and 45)

Line Plot: Select **Graphs → Interactive → Line…** (Page 35)

Bivariate Graphs

Two Categorical Variables

Select **Graphs → Interactive → Bar…** (Page 51) or select **Descriptive Statistics → Crosstabs…** and click on the Clustered Bar Charts option (Page 50)

One Categorical Variable and One Quantitative Variable

Select **Graphs → Interactive → Boxplots…** (Page 44) or select **Graphs → Legacy Dialogs → Boxplots…** and select the Summaries for groups of cases option

Select **Analyze → Descriptive Statistics → Explore…** (Page 51)

See also **Data → Split File…** above under Data Management .

Two Quantitative Variables

Scatterplot: Select **Graphs → Interactive → Scatterplot…** (Pages 54, 58 and 123) for a graph of Y vs. X or select **Graphs → Legacy Dialogs → Scatter/Dot…→ Matrix Scatter** (Page 117) for a graph of Y vs. several X variables.

Investigating Associations

Two Categorical Variables

Select **Analyze → Descriptive Statistics → Crosstabs…** (Pages 48, 101, 103 – 106)

Two Quantitative Variables

Select **Analyze → Correlate → Bivariate…** (Pages 55 and 119)

Select **Analyze → Regression → Linear…** (Pages 63, 109 – 111 and 114)

Select **Analyze → Regression → Curve Estimation…** (Page 112)

Select **Analyze → Regression → Binary Logistic…** (Page 125)

One Quantitative Variable with Several Quantitative Variables

Select **Analyze → Regression → Linear…** (Pages 116, 119 – 122)

Comparing Means and Proportions

Comparing Means

One Sample

Select **Analyze → Compare Means → One-Sample T-Test…** (Page 92)

Two Independent Samples

Select **Analyze → Compare Means → Means…** (Page 52)

Select **Analyze → Compare Means → Independent Samples T-test…** (Page 95)

Select **Analyze → Nonparametric Tests → 2 Independent Samples…** (Page 134)

Two Dependent Samples

Select **Analyze → Compare Means → Paired-Samples T-Test…** (Page 97)

Select **Analyze → Nonparametric Tests → 2 Related Samples…** (Pages 137 and 138)

Two or More Independent Samples

Select **Analyze → Compare Means → One-Way ANOVA…** (Page 126 and 127)

Select **Analyze → General Linear Model → Univariate…** (Pages 129 and 132)

Select **Analyze → Nonparametric Tests → K Independent Samples…** (Page 136)

Select **Analyze → Regression → Linear…** (Pages 130 and 133)

Comparing Proportions

One Sample

Select **Analyze → Nonparametric Tests → Binomial…** (Page 86)

Select **Analyze → Nonparametric Tests → Chi-Square...** (Pages 89 and 107)

Two Independent Samples

Select **Analyze →Descriptive Statistics → Crosstabs...** and select the Chi-square statistics under "Statistics." (Pages 103 and 104)

Two Dependent Samples

Select **Analyze → Nonparametric Tests → 2 Related Samples...** (Pages 100, 137 – 138)

Managing Output

Hiding Output

In the Output Navigator, double click on the icon of the output; double click on the icon again to "un-hide" (Page 10)

Making Copies of Output

Select **Edit → Copy** and then **Edit → Paste After** (Page 33)

Modifying Output

Titles (Page 14)

Text (Page 11)

Graphs

Resizing (Pages 11 and 15)

Pie Charts – changing the items displayed and the color of the plot (Pages 21 and 24)

Dot Plots – changing plot symbol (Page 29)

Histograms – changing the number of intervals and their width (Page 34)

Time Plots – changing the scale on the time axis (Page 36)

Boxplots – rotating the plot (Pages 42 and 51)

Scatter Plots –changing the plot symbol (Pages 59 and 63); adding the regression line (Pages 59 and 61)

StatCrunch Manual

Michael Kowalski

University of Alberta

SECOND EDITION

STATISTICS

THE ART AND SCIENCE OF LEARNING FROM DATA

Agresti • Franklin

Upper Saddle River, NJ 07458

Table of Contents

Chapter 1 - Introduction to StatCrunch

StatCrunch is web-based statistical software package. All you will need to use StatCrunch is a Java-capable web browser, an Internet connection and a password. The purpose of this chapter is to introduce you to the basic features of StatCrunch and the StatCrunch windows. Specifically, we will outline how to open and close StatCrunch, load data, enter data, manipulate data and save data files.

Starting StatCrunch

If you are accessing StatCrunch through the web server at your institution go to the designated website provided by your institution. Your starting page may look like this:

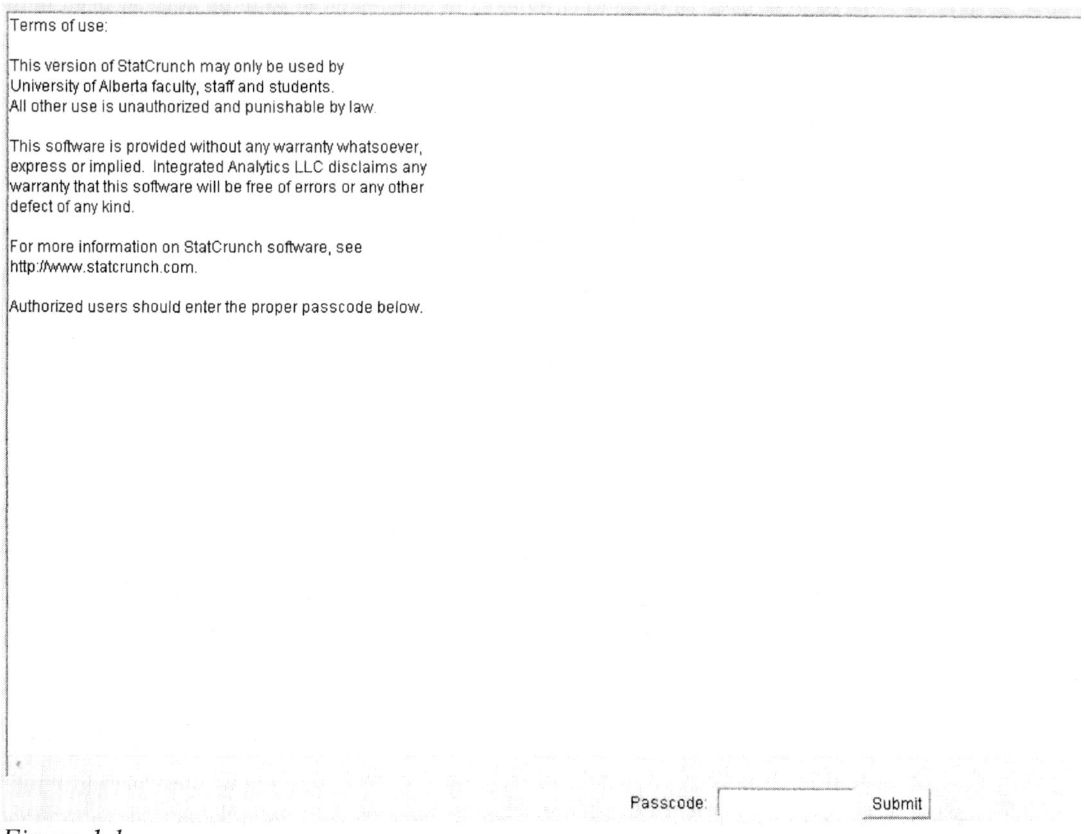

Terms of use:

This version of StatCrunch may only be used by
University of Alberta faculty, staff and students.
All other use is unauthorized and punishable by law.

This software is provided without any warranty whatsoever, express or implied. Integrated Analytics LLC disclaims any warranty that this software will be free of errors or any other defect of any kind.

For more information on StatCrunch software, see http://www.statcrunch.com.

Authorized users should enter the proper passcode below.

Passcode: Submit

Figure 1.1

Enter your password and then click **Submit**. This will begin your session. If you have an account at www.statcrunch.com, sign in and click **Open StatCrunch** on the main page.

Ending StatCrunch

In order to end your session simply shut down your web browser. Make sure that when you decide to quit that you have saved everything you need, as you will not be prompted to do so.

StatCrunch Data Window

After you have entered your password the main StatCrunch window will open. The main StatCrunch window is a single spreadsheet where the data is displayed. Columns and rows arrange the spreadsheet. The columns generally represent variables and the rows generally represent your observations.

On the top you will see your menu bar. The menus are:

StatCrunch – commands to modify the display or navigate between data and output
Data – commands for opening, organizing, modifying and saving data
Stat – commands for producing output for basic descriptive and inferential statistics
Graphics – commands to produce graphical displays
Help – a help guide to using StatCrunch

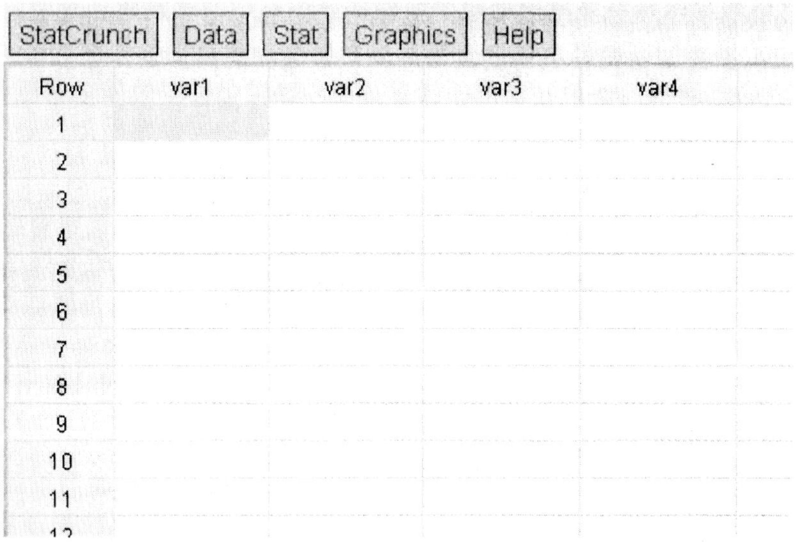

Figure 1.2

The title of your data file is displayed directly to the right of the help menu.

By default, output that you generate will be displayed in separate pop-up windows. You have the option to change this and create split a split screen where the output is displayed. To change the way you would like the output to be displayed click on **StatCrunch➔Display in➔** and then choose your option.

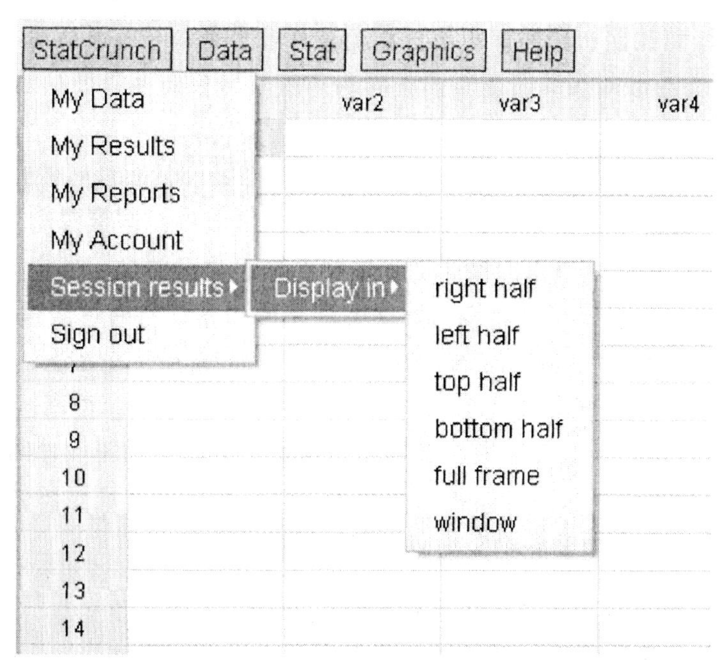

Figure 1.3

Depending on your preference you can display the output in the right half, left half, top half, and bottom half or as a separate full frame.

For example, the **right-half** display will look like this:

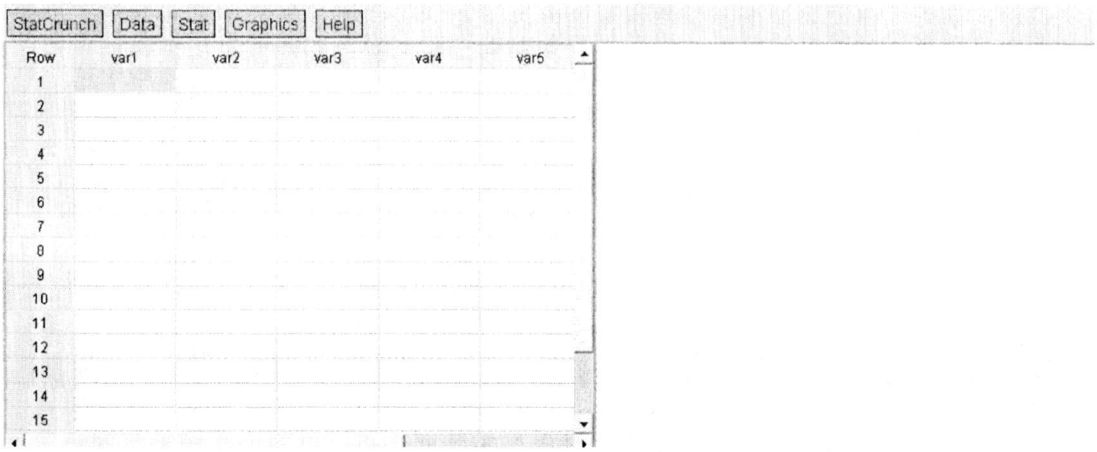

Figure 1.4

The main data page is on the left and output would be displayed on the right.

The **window** option is the default, producing separate pop-up windows with output. Choosing the **full frame** option will create two tabs at the bottom. You can navigate between the data and output by clicking on the tabs. The **full frame** display is displayed below:

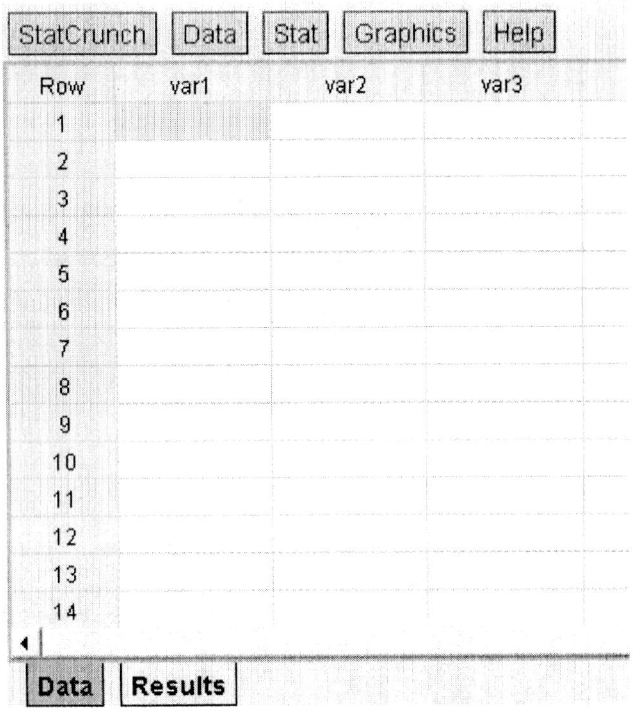

Figure 1.5

Before we begin to generate output we need some data. Although StatCrunch has several options, the two most common ways to obtain data is to enter it manually or to load it from a file.

Entering Data

Refer to the data file in Figure 1.2 in your text. This is a portion of the data from the University of Florida student survey, which is available on your CD, "fla_student_survey". First, let's enter the variable names. To change the first variable (the first column) "*var1*" to "*Student*", click on the column header "*var1*", click delete and then type in "*Student*". Note that, unlike other programs such as Excel, you cannot simply type in the new content of a cell. In order to change the contents of a cell (variable header or observation) you must delete the content first and then type in the new content. You can navigate between cells using the arrow butting on your keyboard. Hit the right button to highlight "*var2*", and then click delete and type in "*Gender*". Continue this until all the variables names are entered. You cannot use the tab button to move to the right.

Once the variable names are entered we can begin entering the data. Click on the cell where you wish to enter the data and type in the appropriate value. Hitting enter on your keyboard will move you to the cell immediately below. Otherwise you can navigate around using the arrow keys. At any time if you wish to change a value that is entered, you can highlight the cell, click delete, and then enter the new value. Again, remember to hit delete first to modify the contents of a cell. Enter the data displayed in Figure 1.2 in your text so that your data file looks like this one.

Untitled

StatCrunch | Data | Stat | Graphics | Help

Row	Student	Gender	Race	Age	GPA	TV	Veg	PolParty	Married?
1	1	m	w	32	3.5	3	no	rep	1
2	2	f	w	23	3.5	15	yes	dem	0
3	3	f	w	27	3	0	yes	dem	0
4	4	f	h	35	3.2	5	no	ind	1
5	5	m	w	23	3.5	6	no	ind	0
6	6	m	w	39	3.5	4	yes	dem	1
7	7	m	b	24	3.7	4	no	ind	0
8	8	f	h	31	3	4	no	ind	1
9									

Figure 1.6

Before we look at loading data from a file, let's save this first.

Saving Your Data File

If you are accessing StatCrunch through a server then follow these steps to save data to your hard drive. In order to save your data file, select **Data➔Save data.** A pop-up window will appear. Click on **Browse** to define the file name and the location where you want the data to be saved. To save your data as a text file, make sure the "text with delimiter" button is chosen. You can choose the type of delimiter you want. You will need to remember this when loading the data file. For now you can just leave it as "space". When you are ready to save the file, click **Export** to save the file. After you have clicked the export button the file name directly to the right of the help menu should change to the file name that you have specified. Make a note as to where you have saved the file, as we will use this data file it in further illustrations.

You can also choose the save you file as an html file. If you do so, the saved file will open as a table in your web browser.

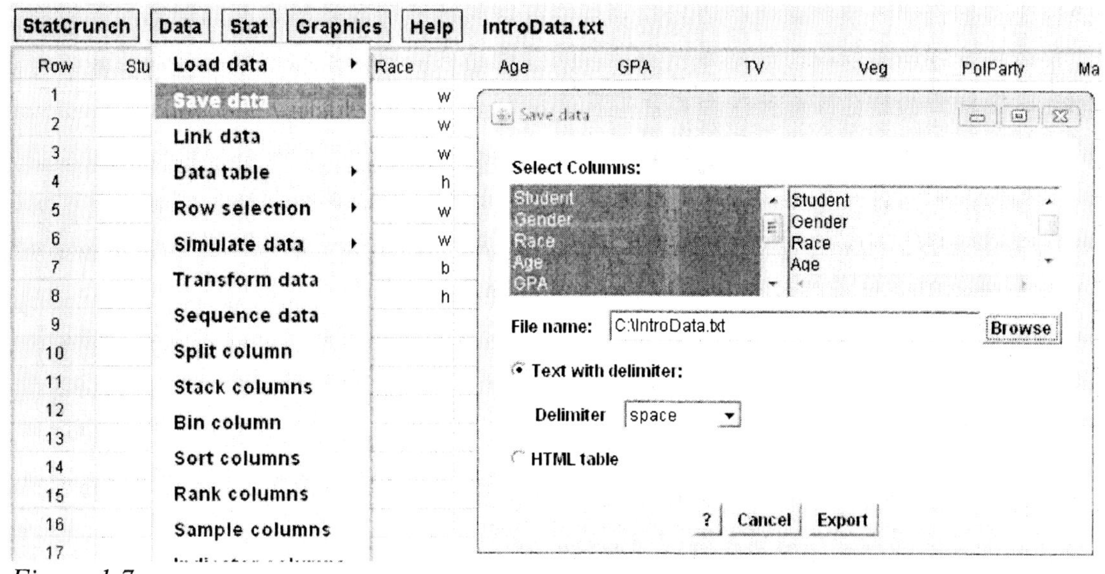

Figure 1.7

Note: If you are accessing StatCrunch though the StatCrunch website, please see their instructions on how to save data to your hard drive.

Loading Data from a File

If you are accessing StatCrunch through a server then follow these steps to load data from your hard drive. To illustrate how to load a data file, re-start StatCrunch so that your main data screen is blank. Go to **Data➔Load data➔from file.** In the pop-up window enter the location of the file or click **Browse** to find the file and then click **Okay**. The data file should open in the data window and the file name should change from "Untitled" to whatever you saved it as.

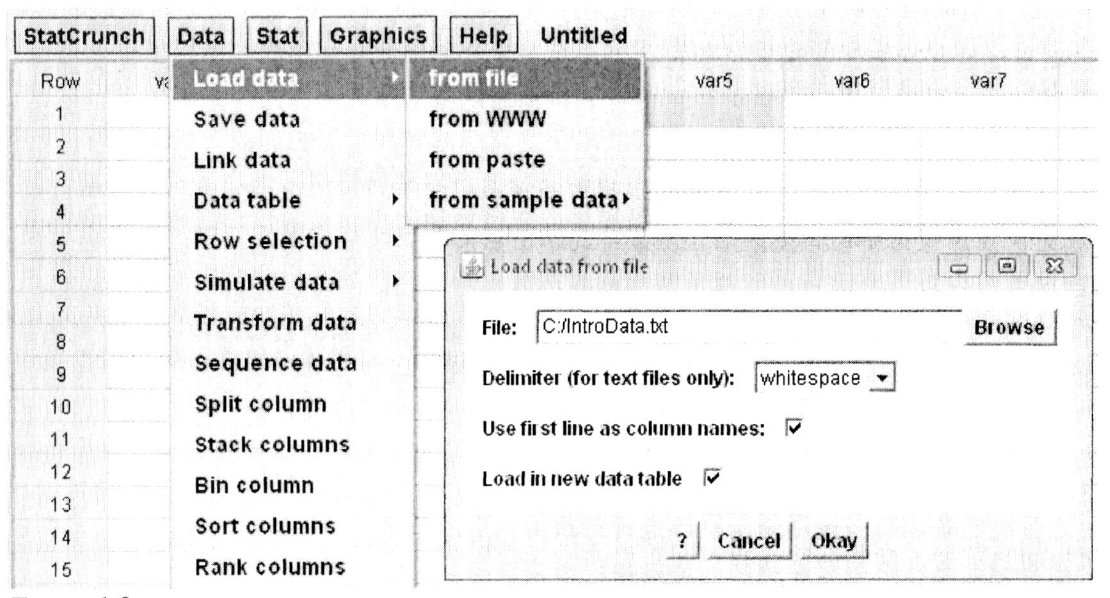

Figure 1.8

Note: If you are accessing StatCrunch though the StatCrunch website, please see their instructions on how to load data.

Editing Your Data File

StatCrunch offers a variety of editing and organizing options. I will go over just a few of the more common features.

Deleting columns

To delete entire columns go to **Data➔Data table➔Delete columns.** A pop-up window will appear where you can select the columns you wish to delete. Click on the columns you wish to delete then click **Delete**. The columns selected should disappear. *CAUTION: StatCrunch does not have an "undo" option. For example, if you accidentally delete a column you will have to re-start your session to get it back.*

IntroData.txt

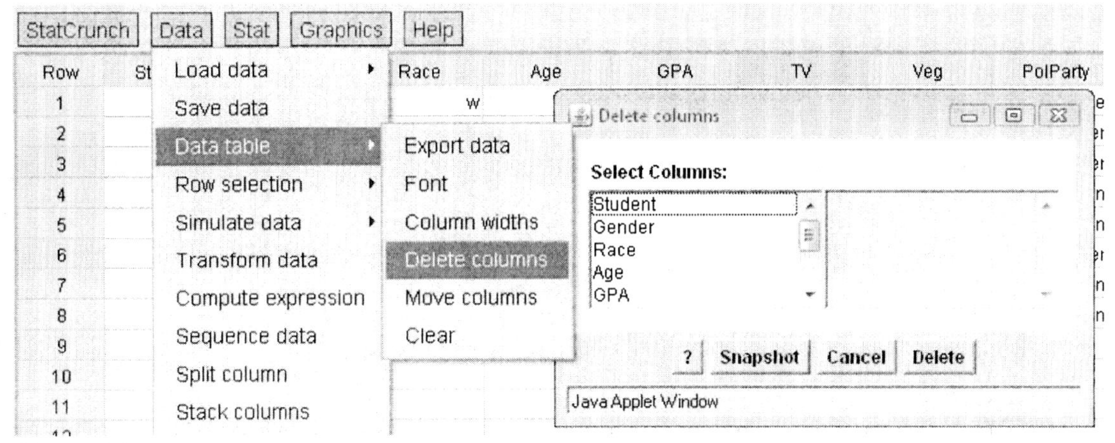

Figure 1.9

Other features in the **Data table** menu include:

Font – change the font type and size
Column widths – change the column widths (You can also change the column widths by clicking the right edge of the column and dragging it to the desired width.)
Move columns – move around columns for organizational benefit.
Clear – this option clears the entire data sheet (remember there is no "undo" function)

Transforming Data

Let's suppose you are interested in age in months. To transform the variable Age, which is in years, to months you need to multiply each value by 12. You could do this manually although it would be quite tedious. Instead, StatCrunch can do this for you. Click **Data ➔Transform data.**

IntroData.txt

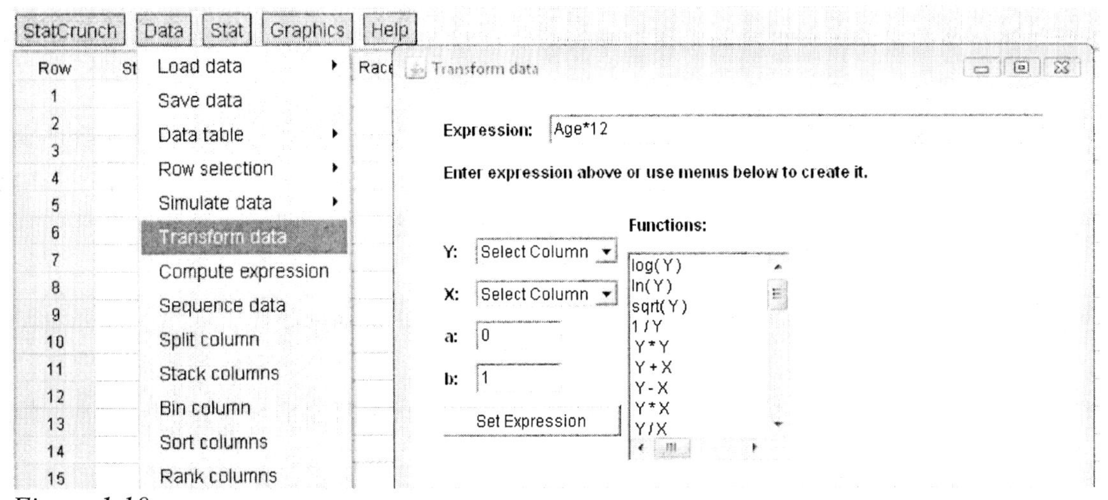

Figure 1.10

A pop-up will appear. To transform Age to age in months, you can define the transformation to be Age*12. Fill in "Age*12" in the "Expressions" line and click **Compute.** This will create an additional variable titled "Age*12" which will be the age in months. Verify the values are in fact the Ages multiplied by 12. StatCrunch also has many built in functions for more complicated transformations.

Other Editing Features

Sorting Columns (Data➔Sort columns): This option allows you to sort your data file by a specified column. You can sort all or just some of your columns.

Rank columns (Data➔Rank columns): This option will create a variable that will rank the values of a specified column. The ranks of tied observations will be averaged.

The other function I will leave you to explore, as they are not that common.

Now that you are familiar with the StatCrunch environment as well as the more common functions in the data menu we are ready to start producing output. In the next chapter we will look at how to generate descriptive statistics and how to generate graphical displays.

General Comment on StatCrunch Output

As a general comment, when producing output in StatCrunch a pop-up window will appear after you choose the appropriate function. On this pop-up you will see a "**next**" and a "**back**" button. Clicking these buttons will let you navigate and edit your options for the chosen function. I will refer to the first screen as "**Screen 1**". Clicking next will open "**Screen 2**" and so on. Not all screens are needed and many are repeated across different functions so we won't always go over each and every screen within each function.

Chapter 2 - Exploring Data with Graphs and Numerical Summaries

Descriptive statistics and graphical displays are very important in understanding the properties of a dataset. They help us better understand the properties of the data we are dealing with and give us insight in choosing an appropriate inferential method. This chapter will go over many of the common graphical techniques available in StatCrunch. As well we will show how to obtain descriptive statistics for quantitative variables.

Graphs for Categorical Variables (Tabulated Data)

This section will explore graphical methods for categorical variables with tabulated data (summarized in a frequency distribution). We will go over two graphical techniques: Pie Charts and Bar Charts.

Pie Charts for Tabulated Data

Example 2: "Shark Attacks around the World"

This example wishes to compare the number of shark attacks, "*frequency*" for several geographical regions, "*region*". In Table 2.1 of your text the number of shark attacks is summarized in a frequency distribution (tabulated) for each of 11 regions. Enter the data as it appears in Table 2.1.

To produce a Pie Chart for the number of shark attacks by region select **Graphics→Pie Chart→with summary.**

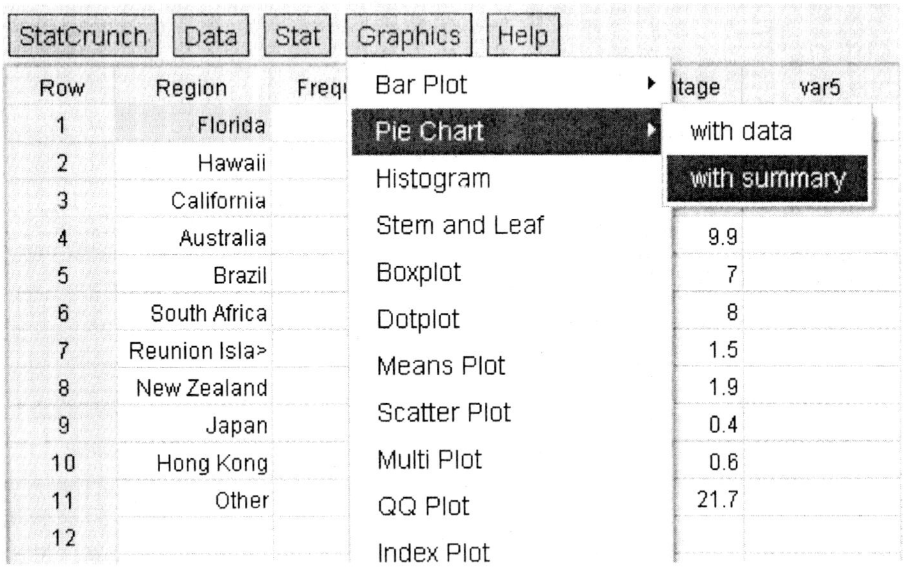

Figure 2.1a

*Note: This is just a reminder about the StatCrunch functions. For each function a pop-up will appear. Clicking "**next**" and "**back**" will let you navigate through the "**Screens**" of the chosen function.*

In **Screen 1** select "*Region*" as the "Categories in" and "*Frequency*" as the "Counts in". At this point you could click **Create Graph** to produce your Pie Chart with the default display settings. Otherwise you can click **Next** to go through the rest of the screens to modify the default display setting.

*Note: The **"Where:"** option will be available in almost all StatCrunch functions. It allows you to specify a portion of the dataset (specific rows) to be used in the analysis.*

Screen 1

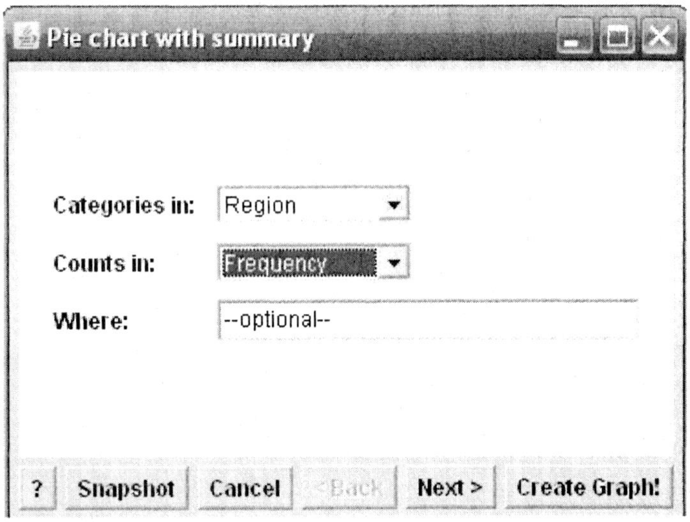

Pie chart with summary

Categories in: Region

Counts in: Frequency

Where: --optional--

? Snapshot Cancel < Back Next > Create Graph!

Figure 2.1b

Screen 2

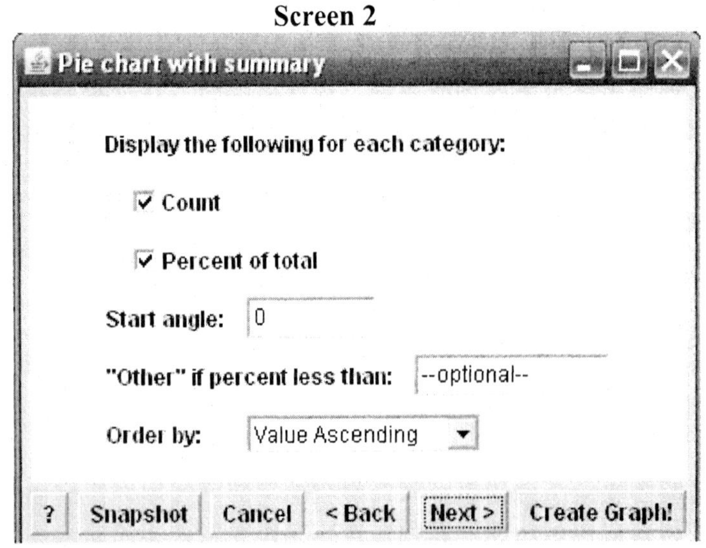

Figure 2.1c

In **Screen 2** you have the option to display the counts and/or the percentages of the total. As well you can modify the start angle as well as the order of the counts (these options are purely cosmetic). The "Other if percent less than" option allows you to create a category called "Other" which will group together all categories with a relative frequency of observations less than your specified amount. This is beneficial if there are many categories with relatively small counts. Considering our variable already has an "Other" category this is not applicable. For now I will just leave **Screen 2** as is.

Clicking **Next** to **Screen 3** will give you the option to add a title for your graph. I will title my graph appropriately "*Pie Chart of Shark Attacks*".

Screen 3

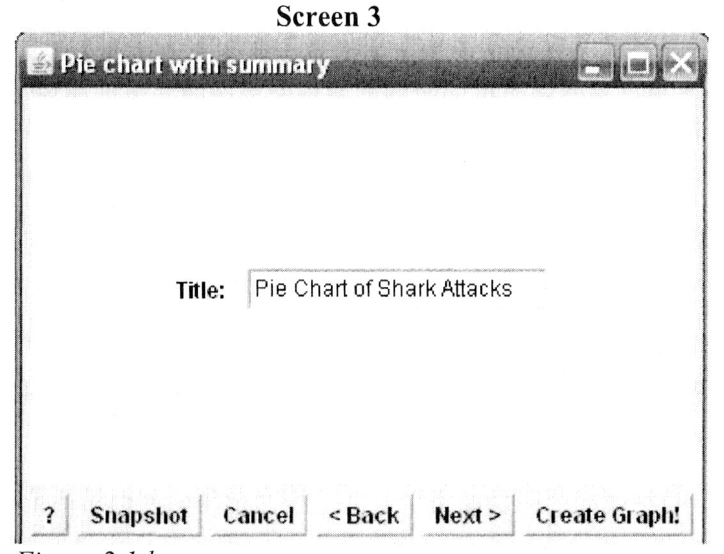

Figure 2.1d

Finally if you click **Next** to **Screen 4** you can modify the colour scheme. I will leave this screen as is.

Note: At your convenience you are encouraged to experiment with different setting to see how they affect your graph.

When you are ready to create your Pie Chart click **Create Graph**! Your Pie Chart should look like this:

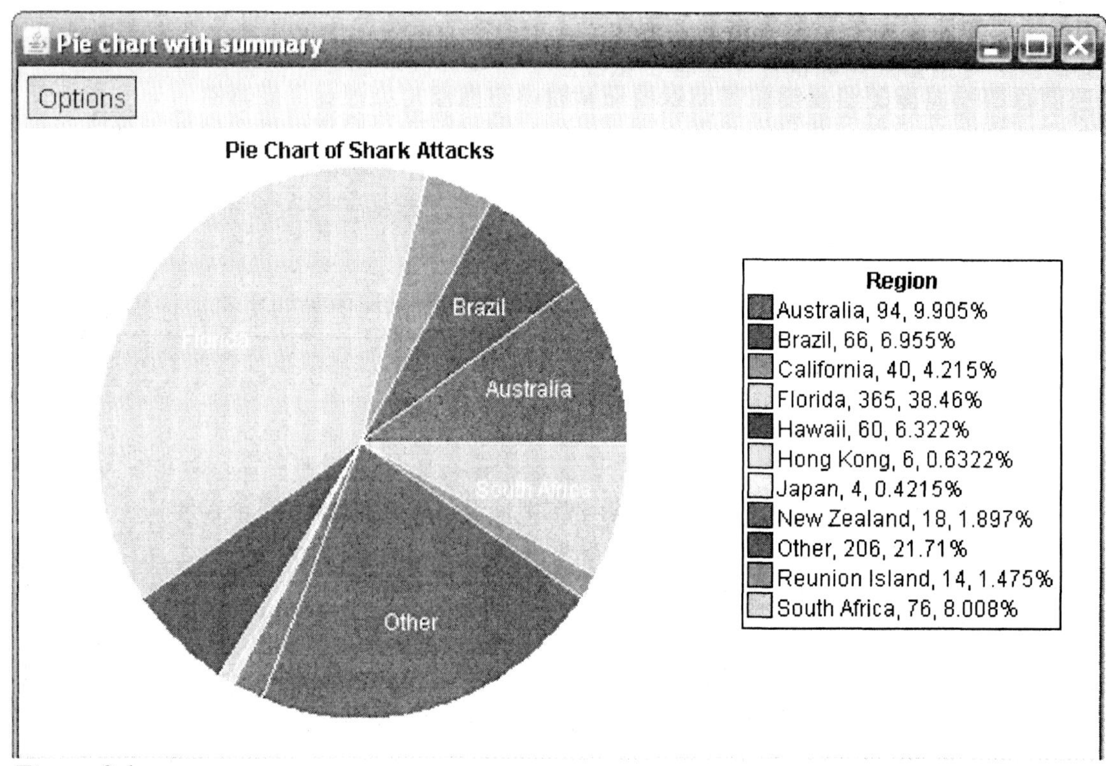

Figure 2.1e

Bar Charts for Tabulated Data

Example 2, Exercise 2.13: "Shark Attacks around the World"

To produce a Bar Chart for the number of shark attacks by region select **Graphics➔Bar Chart➔with summary.**

In **Screen 1** select "*Region*" as the "Categories in" and "*Frequency*" as the "Counts in". At this point you could click **Create Graph** to produce your Bar Chart with the default display settings. Otherwise you can click **Next** to modify the default display settings.

Screen 1

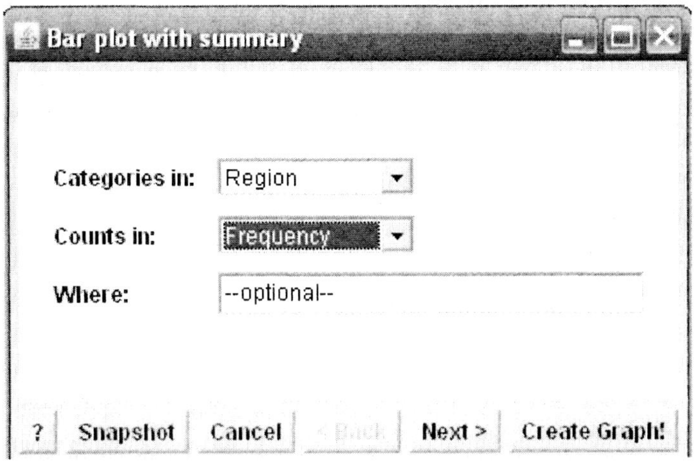

Bar plot with summary

Categories in: Region

Counts in: Frequency

Where: --optional--

? Snapshot Cancel < Back Next > Create Graph!

Figure 2.2a

Screen 2

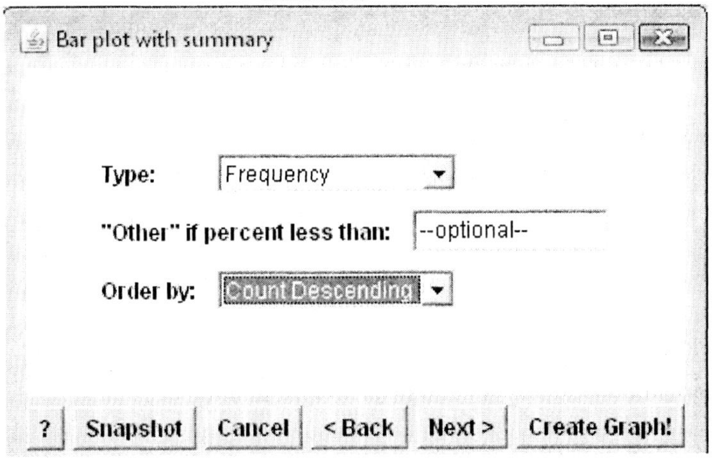

Bar plot with summary

Type: Frequency

"Other" if percent less than: --optional--

Order by: Count Descending

? Snapshot Cancel < Back Next > Create Graph!

Figure 2.2b

In **Screen 2** you have the option to have the bar heights scaled using the frequencies or the relative frequencies. Again StatCrunch gives you the option to create an "Other" category for those groups with relative frequencies lower than some specified value. In **Screen 2** you can also choose how the bars will be ordered. For this illustration I will choose to order them from highest frequency to lowest frequency (Count Descending).

Clicking **Next** to **Screen 3** will give you the option to create labels for your axis as well as a title for your graph. In addition, you can add horizontal and/or vertical grid lines.

Screen 3

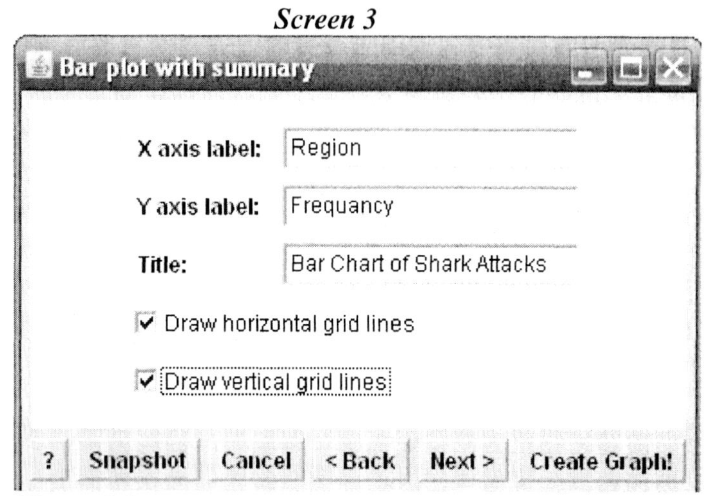

Figure 2.2c

Finally if you click **Next** to **Screen 4** you can modify the colour scheme. I will leave this screen as is.

When you are ready to create your Bar Chart click **Create Graph**! Your Bar Chart should look like this:

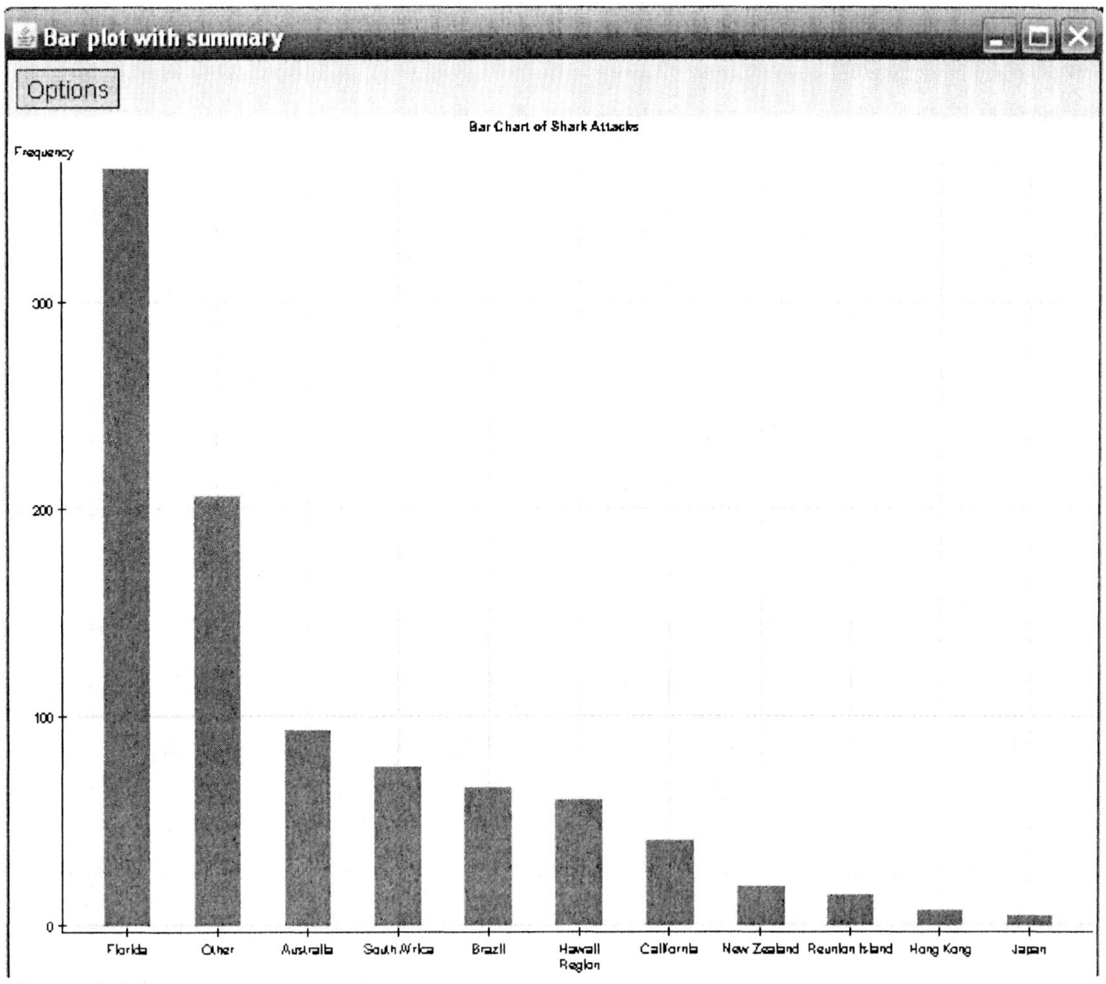

Figure 2.2d

As you can see above, the category with the largest frequency is displayed first followed by the next largest and so on (Order by: Count Descending, in Screen 2). As well the vertical axis is scaled to the actual number of shark attacks (Type: Frequency, in Screen 2).

Again if you are unhappy with your display you can always go back and modify your settings. You can click **Options** at any time on the top left to go back and edit your display settings. Suppose you wanted to display your bars in alphabetical order corresponding to their region. In **Screen 2,** change the "Order by" option from "Count Descending" to "Value Ascending". Also, suppose you want you bar heights to be the relative frequencies rather than the frequencies. In **Screen 2,** change the "Type" to "Relative Frequency". Now click **Create Graph!** The following should be displayed:

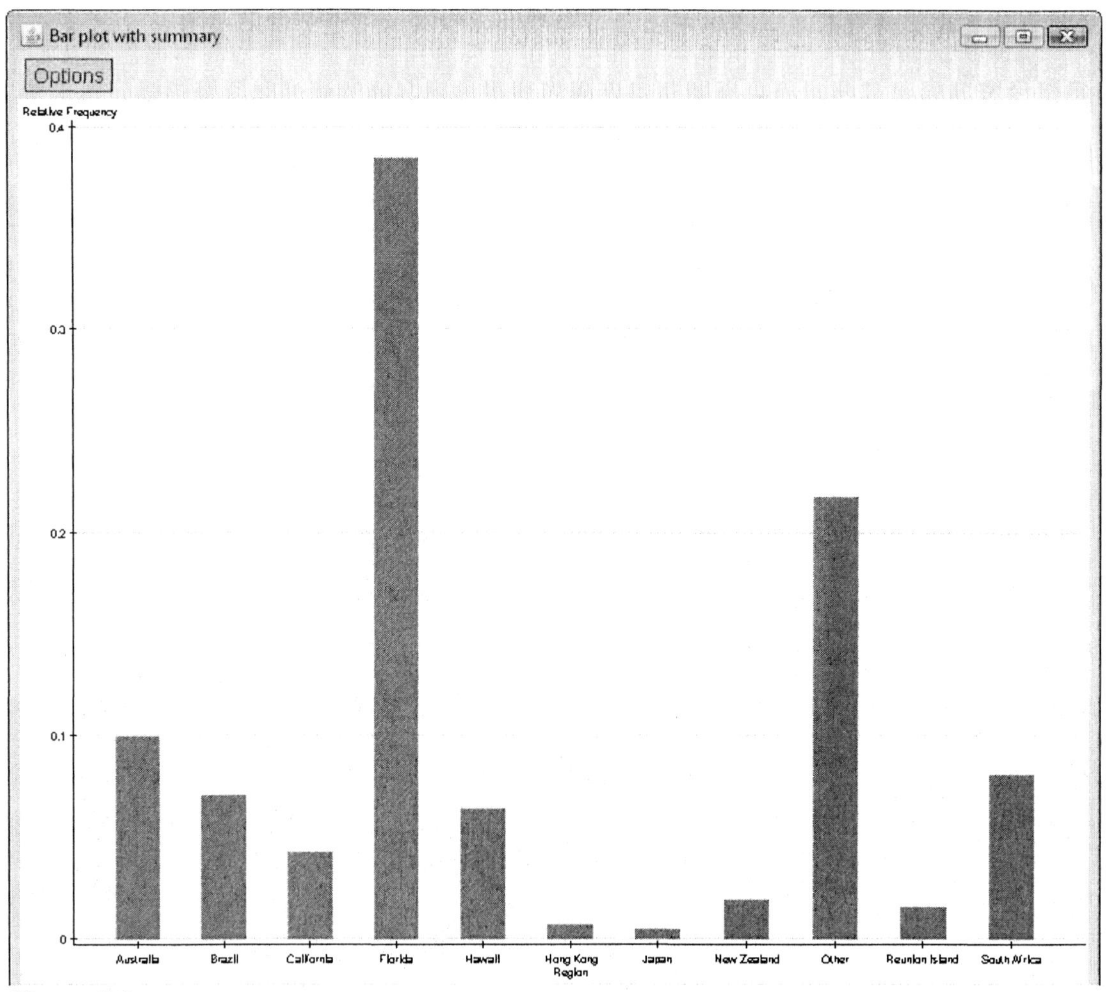

Figure 2.2e

Note that the regions are now displayed in alphabetical order. As well the vertical axis is now scaled to the relative frequency (percentage) of shark attacks. For example, Florida had about 38% of the Shark attacks.

Graphs for Categorical Variables (Non-Tabulated Data)

It is also possible to produce Pie Charts and Bar Charts when the data isn't summarized in a frequency distribution.

Example: For example, refer to the partial data set in Figure 1.2 in your text.

StatCrunch | Data | Stat | Graphics | Help | Untitled

Row	Student	Gender	Race	Age	GPA	TV	Veg	PolParty	Married?	var10
1	1	m	w	32	3.5	3	no	rep	1	
2	2	f	w	23	3.5	15	yes	dem	0	
3	3	f	w	27	3	0	yes	dem	0	
4	4	f	h	35	3.2	5	no	ind	1	
5	5	m	w	23	3.5	6	no	ind	0	
6	6	m	w	39	3.5	4	yes	dem	1	
7	7	m	b	24	3.7	4	no	ind	0	
8	8	f	h	31	3	4	no	ind	1	
9										
10										
11										

Figure 2.3a

One of the categorical variables is "*Political Party*" with 3 possible values: *Republican (rep), Democrat (dem)* or *Independent (ind)*. In the dataset above the results of this variable are given in non-tabulated or "raw" form. That is, we see the value of the variable for each individual. In any case, we can produce Pie Charts and Bar Charts for data in this form by using the **Graphics➔Pie Chart➔with data** and **Graphics➔Bar Chart➔with data** functions, respectively. As seen in the last section, the Pie Chart and Bar Chart functions are very similar so I will discuss them simultaneously.

In either case, in **Screen 1**, simply select "*PolParty*" as the variable in "Select Columns" and click **Create Graph!**

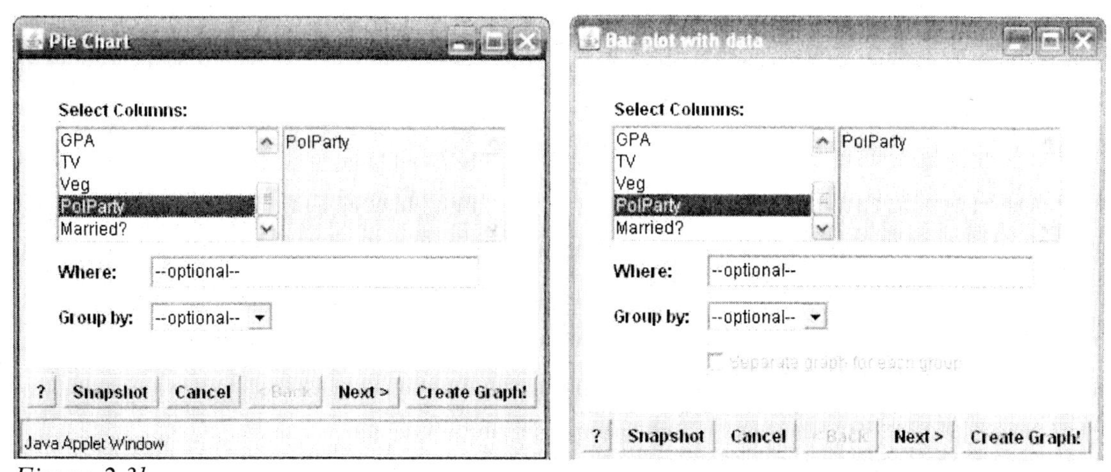

Figure 2.3b

Once again you can click "next" to scroll through the rest of the screens to modify the display settings. The rest of the screens are very similar to those using the "with summary" option.

Here is the Pie Chart and Bar Chart for political party from the partial dataset in Chapter 1.

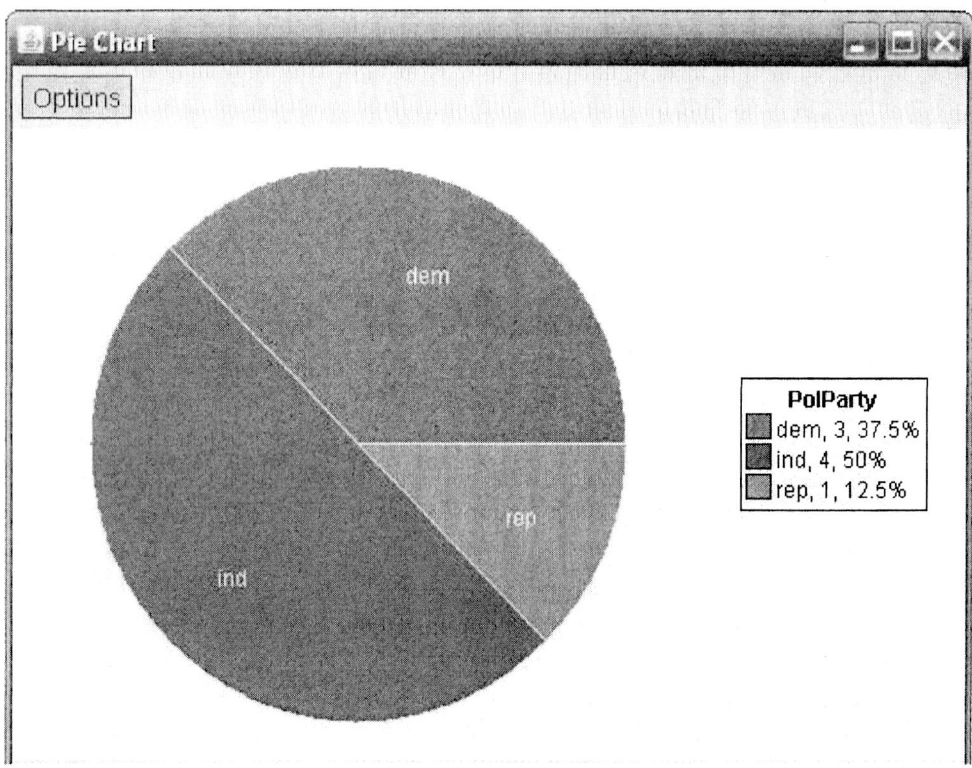

Figure 2.3c

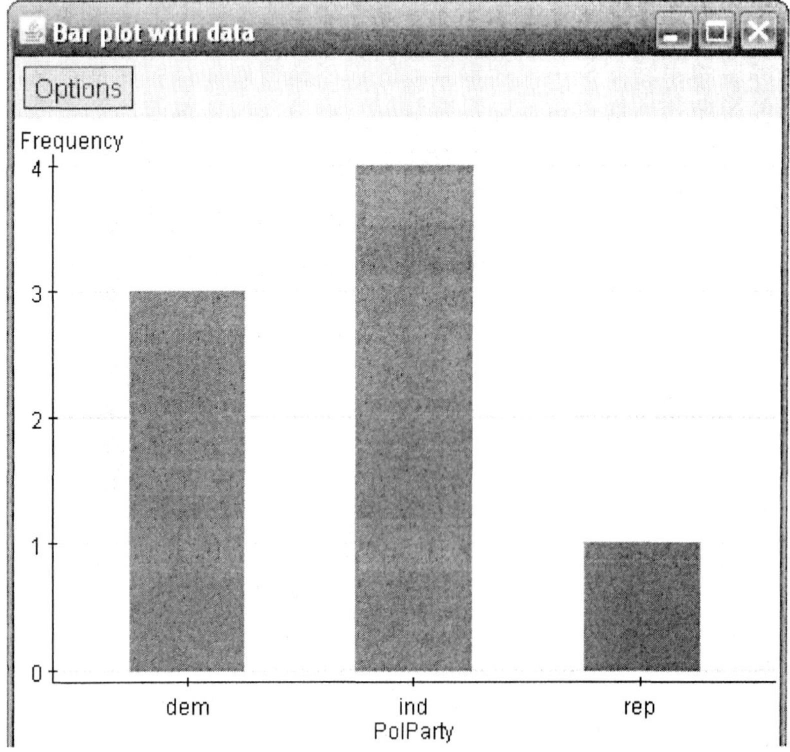

Figure 2.3d

Note: Unfortunately StatCrunch doesn't allow you to move around the labels in your Pie Chart or change the font of the labels and titles.

Example 3: "How Much Electricity Comes from Renewable Energy Sources"

Refer to the data in Table2.2 in your text. The categorical variable is the source of electricity. Below are the charts for percentage use of each source for the U.S.:

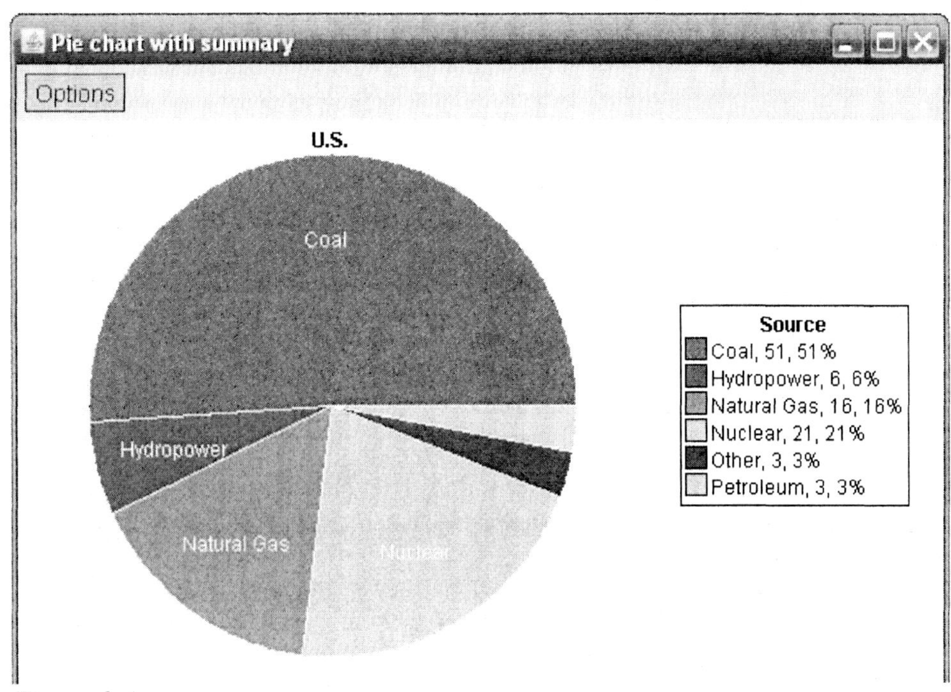

Figure 2.4a

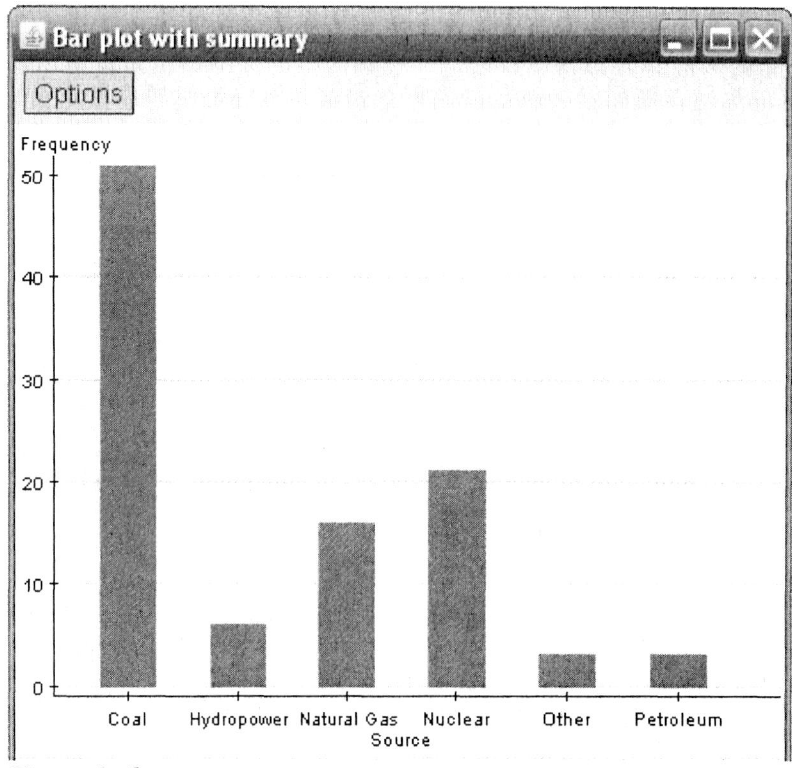

Figure 2.4b

Graphs for Quantitative Variables

This section will explore graphical methods for data from quantitative variables. We will discuss five graphical techniques: Dot Plots, Stem and Leaf Plots, Histograms, Box Plots, and Time Plots.

Dot Plots

Example 4: "Exploring Health Value of Cereals"

Refer to Table 2.3 on page 34 of your text for the data. The table lists the Sodium content (in mg), Sugar content (in g) for 20 popular breakfast cereals. As well, each cereal is classified as an adult's cereal (Type A) or a children's cereal (Type C). Enter the data as it appears in Table 2.3 on page 34 of your text or load the "cereal" dataset from your text CD. For now we will focus on the variable sodium content, "*SODIUM(mg)*".

To produce a Dot Plot, select **Graphics→Dotplot.**

In **Screen 1** choose the variable(s) for which you wish to produce the plot by clicking on them on the left. Click on "*SODIUM(mg)*" to create a Dot Plot for sodium content. The variable(s) you click should appear on the right. To remove a variable, click on that variable on the left again and it should disappear. The "Group by:" option will be discussed later. Click **Next** to go to **Screen 2**.

Screen 1

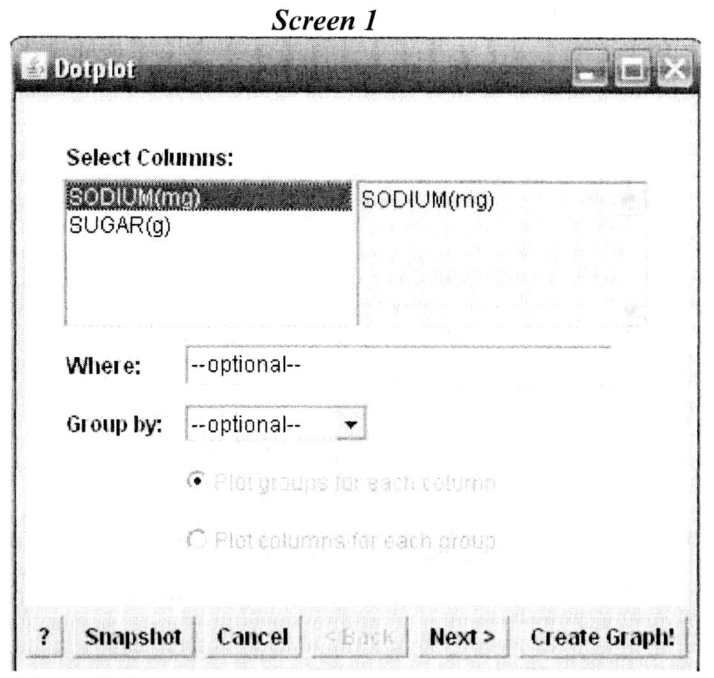

Figure 2.5a

In **Screen 2** you can edit the labels and the title. The X-axis label will be the variable name if left blank. You can also add grid lines if you prefer.

Screen 2

Dotplot

X axis label: Sodium Content

Y axis label:

Title: Dotplot of Sodium Content

Draw horizontal grid lines

Draw vertical grid lines

For multiple graphs:

Use same Y axis

Use same X axis

? | Snapshot | Cancel | < Back | Next > | Create Graph!

Figure 2.5b

When you are finished, click **Create Graph!**

Your Dot Plot for sodium content should look like this:

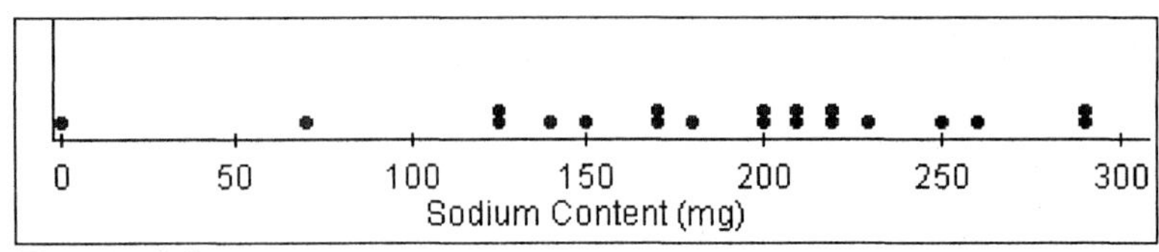

Figure 2.5c

More on Dot Plots

StatCrunch also has the ability to produce several graphs simultaneously. As well, it can produce several plots for each level of a categorical variable. For example, suppose we wish to produce a Dot Plot for each of the variables, "*SODIUM(mg)*" and "*SUGAR(g)*". In addition, suppose we want separate Dot Plots for the two different types of cereals, Type A for adults and Type C for children. Open the Dot Plot function again or click **Options** on the top left of the graph you just obtained.

In **Screen 1** select both variables by clicking on them on the left. They should both appear on the right. In the "Group by" drop down menu, select "*CODE*". This is the categorical variable categorizing each cereal as either Adult or Children.

Click **Next** if you with to edit the labels or title, otherwise click **Create Graph!**

Screen 1

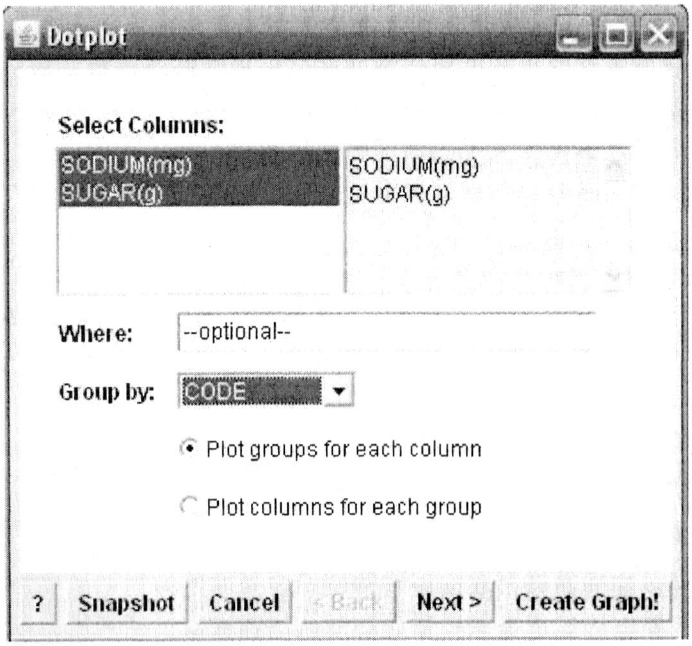

Dotplot

Select Columns:

SODIUM(mg)
SUGAR(g)

SODIUM(mg)
SUGAR(g)

Where: --optional--

Group by: CODE

Plot groups for each column

Plot columns for each group

? | Snapshot | Cancel | < Back | Next > | Create Graph!

Figure 2.5d

The first screen to open has the Dot Plots of sodium content for each of the cereal types. If you click **Next**, you will see the Dot Plots of sugar content for each of the cereal types. The graphs are displayed below.

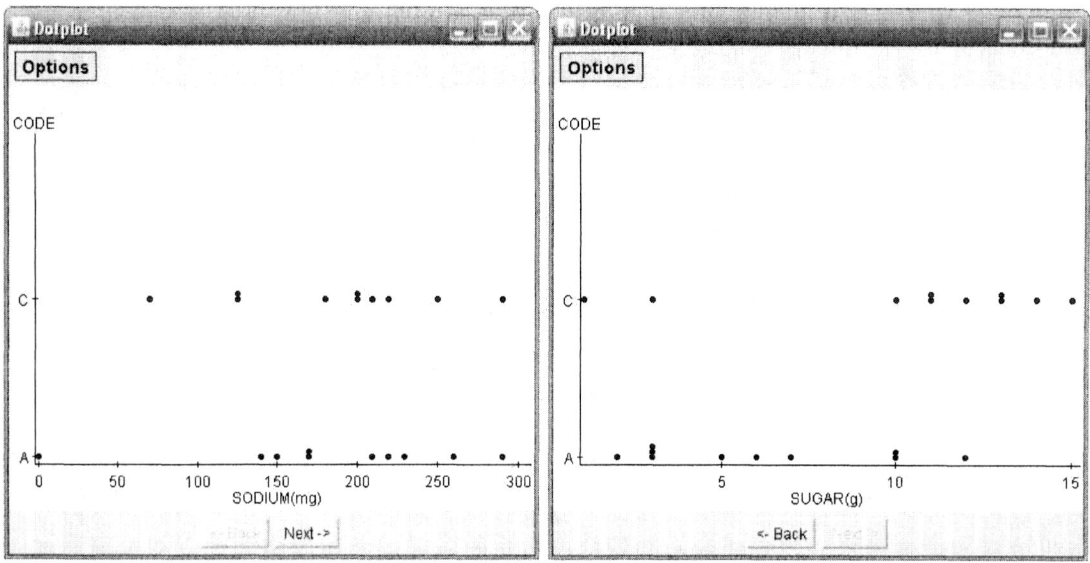

Figure 2.5e

Stem-and-Leaf Plot

Example 5: "Exploring the Health Value of Cereals"

To produce a Stem-and-Leaf Plot, select **Graphics➔Stem and Leaf.**

In **Screen 1** choose the variable(s) for which you wish to produce the plot by clicking on it on the left. Click on "*SODIUM(mg)*" to create a Stem-and-Leaf plot for sodium content.

Screen 1

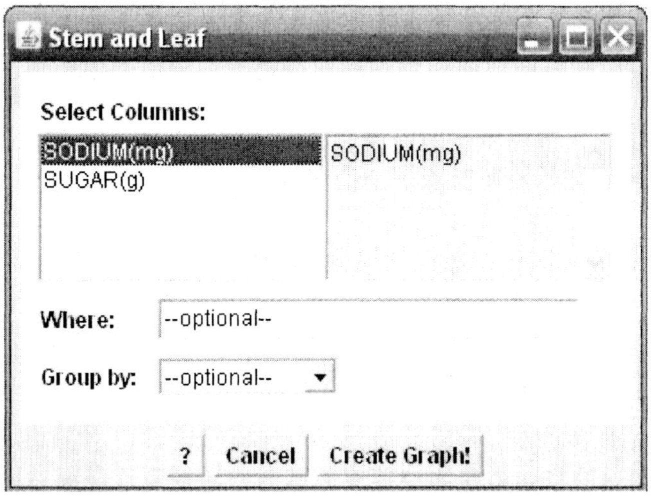

Stem and Leaf

Select Columns:

SODIUM(mg) | SODIUM(mg)
SUGAR(g)

Where: --optional--

Group by: --optional--

? Cancel Create Graph!

Figure 2.6a

When you are finished, click **Create Graph!** Your Stem-and-Leaf Plot should look like this:

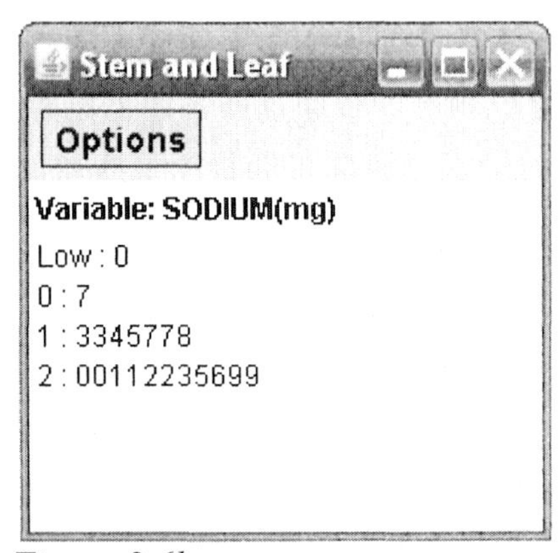

Figure 2.6b

NOTE: Unfortunately StatCrunch does not allow you to modify the "stems".

If you had chosen both variables in **Screen 1** you would end up with the following:

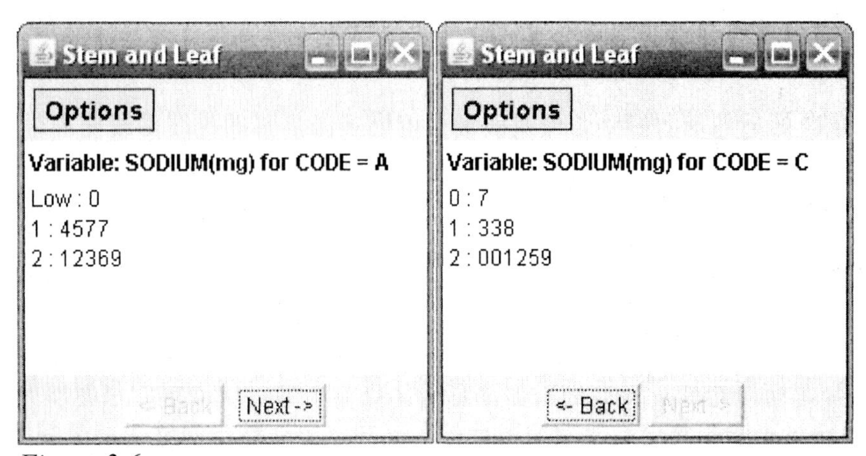

Figure 2.6c

As well you could display separate plots for the two types of cereals by selecting "*CODE*" in the "Group by:" drop down in **Screen 1**.

Histogram

Example 7: "Exploring the Health Value of Cereals"

To produce a Histogram, select **Graphics➔Histogram.**

In **Screen 1** choose the variable(s) for which you wish to produce the plot by clicking on it on the left. Click on "*SODIUM(mg)*" to create a Histogram for sodium content. Again you have the option to display separate histograms for each type of cereal by selecting "*CODE*" in the "Group by:" drop down menu. Click **Next** to go to **Screen 2.**

Screen 1

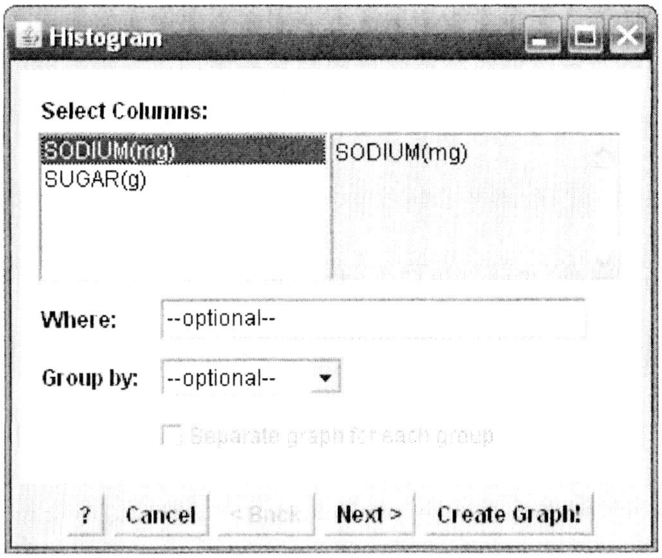

Figure 2.7a

In **Screen 2** you can choose to display the bar heights as either the frequencies or the relative frequencies. You can also modify the starting point of the lowest bar ("Start bins at") as well as the bar widths ("Binwidth"). For now let's choose bar widths of 40 to mimic the histogram in your text. Click **Next** twice to go to **Screen 4 (**I will come back to Screen 3 at the end).

Screen 2

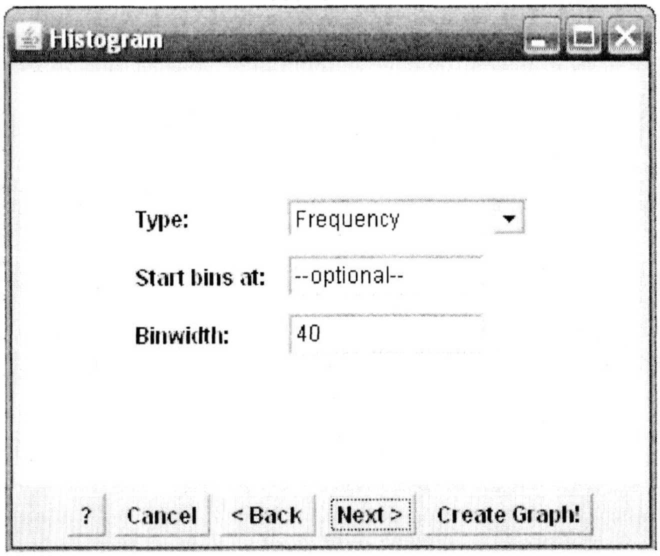

Figure 2.7b

In **Screen** 4 you can edit the axis titles as well as the graph title. The default X-axis title is the variable of interest. The default Y-axis title will be either "Frequency" or "Relative Frequency", depending on what you chose in **Screen 2**. You can also choose to include gridlines. Let's include horizontal gridlines.

Screen 4

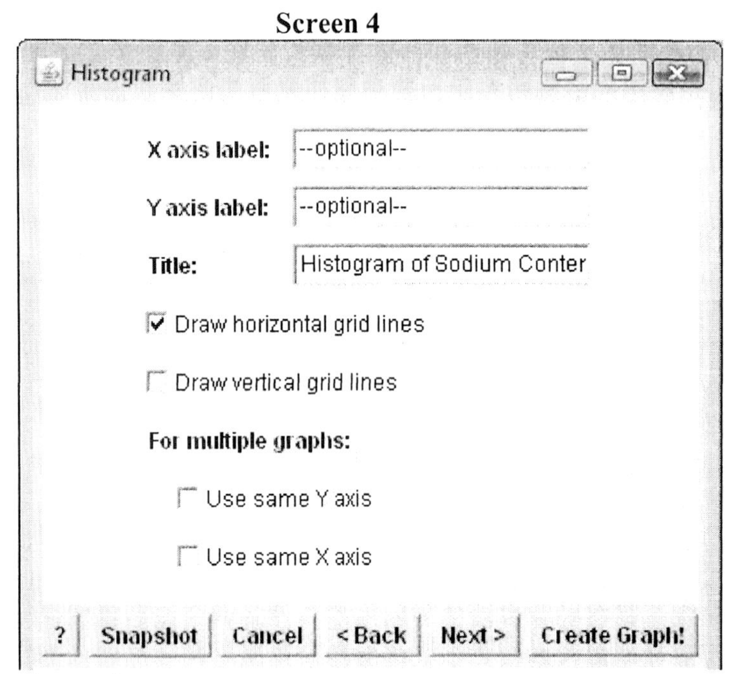

Figure 2.7c

Click **Create Graph**! Your Histogram should look like this:

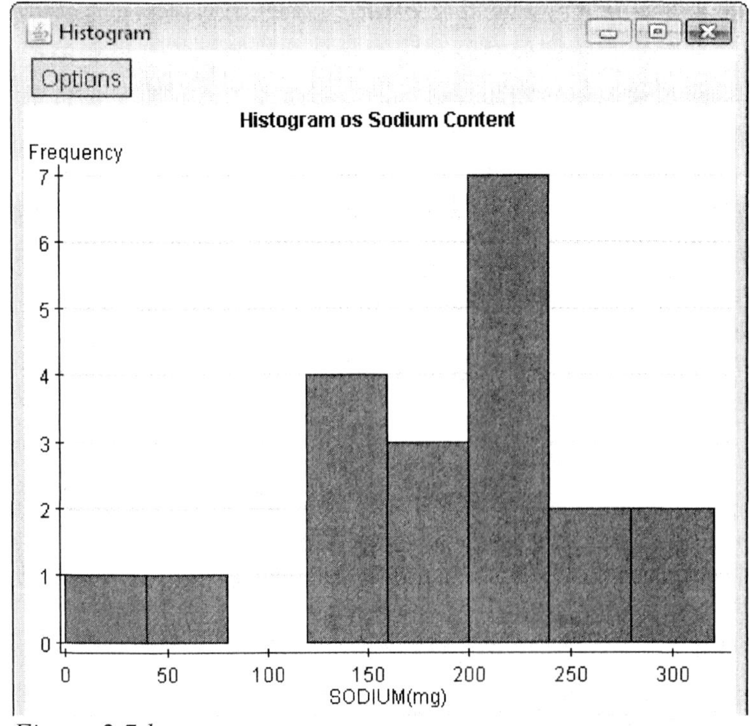

Figure 2.7d

With bar widths of 40 the first bar goes from 0 to 39, the second from 40 to 79, etc…

Suppose you want bar widths of 20? If you wish to modify any of the settings simply click **Options** to bring up your Histogram function. Navigate to **Screen 2** and change the "Binwidth" to 20. Click **Create Graph!** The following graph should be displayed:

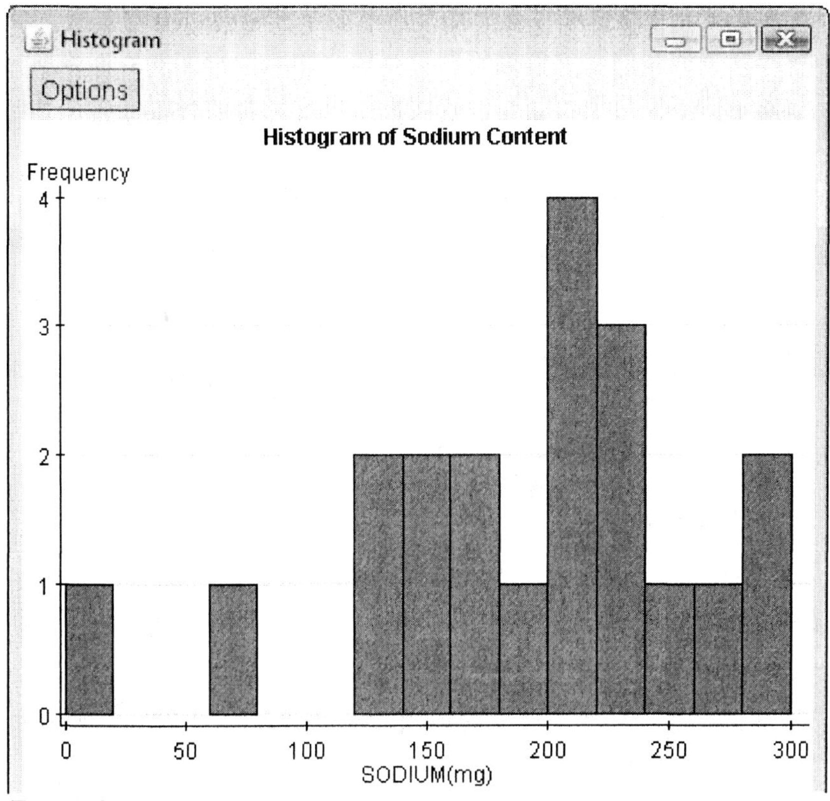

Figure 2.7e

Here the first bar goes from 0 to 19, the second from 20 to 39 etc…

Screen 3 (Optional): You may elect to come back to this after you have read Chapters 5 and 6. Suppose you wish display a common Probability Distribution[1] over your histogram. The purpose of doing this is to allow you to better visualize if your histograms follows approximately a certain common probability distribution. For example, does sodium content follow approximately a normal distribution? We can display an overlay of the normal distribution be selecting "Normal" in the "Overlay Density" drop down in Screen 3. You then have the option to choose the mean and standard deviation. If you leave them blank StatCrunch will just use the sample mean and sample standard deviation of your variable. If you click "Create Graph!" you should end up with the following:

[1] Probability distributions aren't formally discussed until chapter 5. However the "Normal" distribution or "bell" distribution is introduced briefly at the end of Chapter 2.

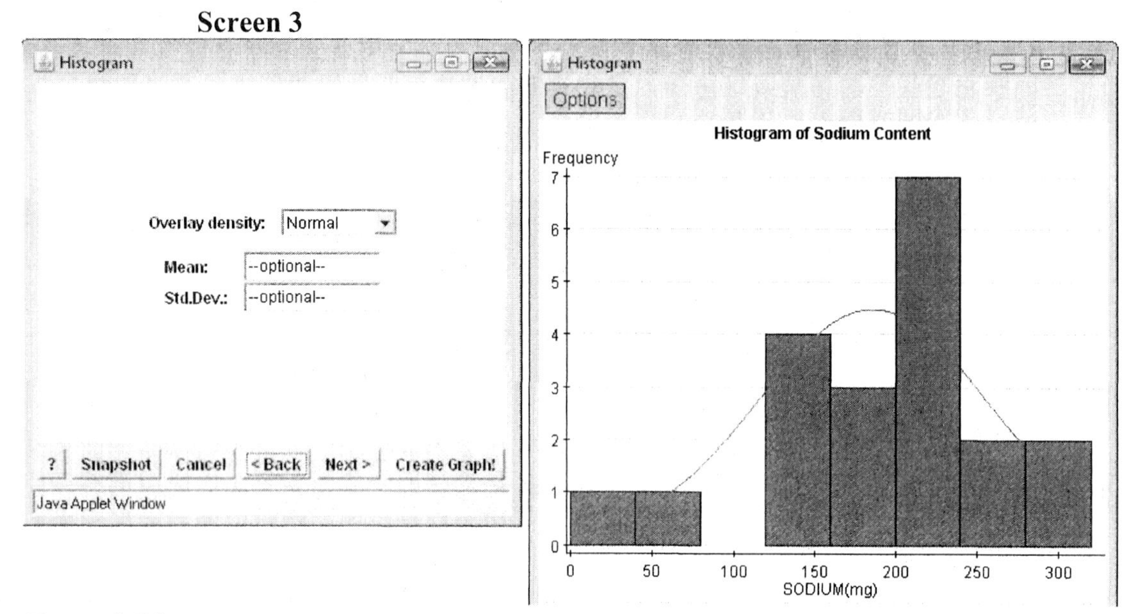

Figure 2.7f

Here you can better visualize if the distribution of sodium content is roughly Normal.

Box Plots

Example 17: "Box Plot for the Breakfast Cereal Sodium Content"

To produce a Box Plot, select **Graphics➔Boxplots.**

In **Screen 1** choose the variable(s) for which you wish to produce the plot by clicking on it on the left. Click on "*SODIUM(mg)*" to create a Box Plot for sodium content. Again you have the option to display separate Box Plots for each type of cereal by selecting "*CODE*" in the "Group by:" drop down menu. Click **Next** to go to **Screen 2**.

Screen 1

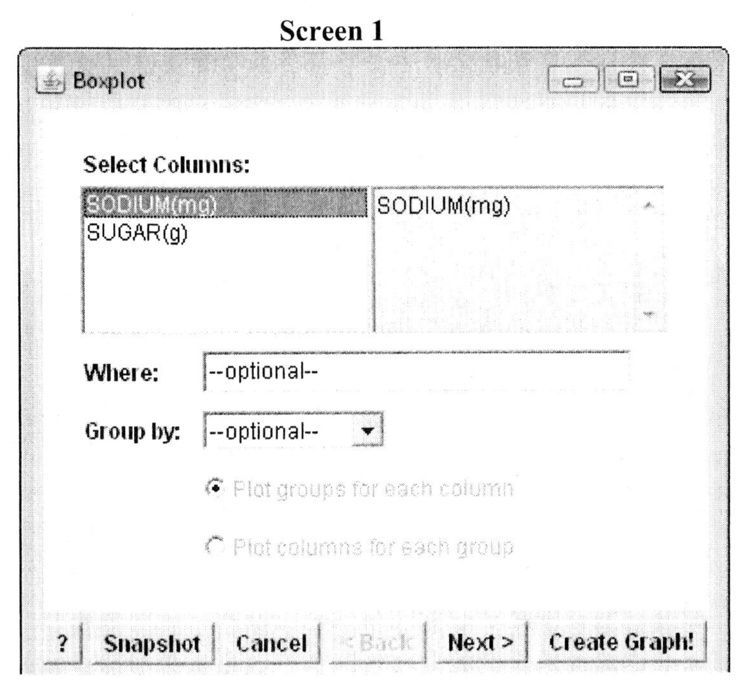

Figure 2.8a

In **Screen 2** you can choose to identify outliers on you plot. If you wish to display outliers check "Use fences to identify outliers". The box plot in your text has outliers displayed so let's check the box. You can also choose to display the plot horizontally. Click **Next** to go to **Screen 3**.

Screen 2

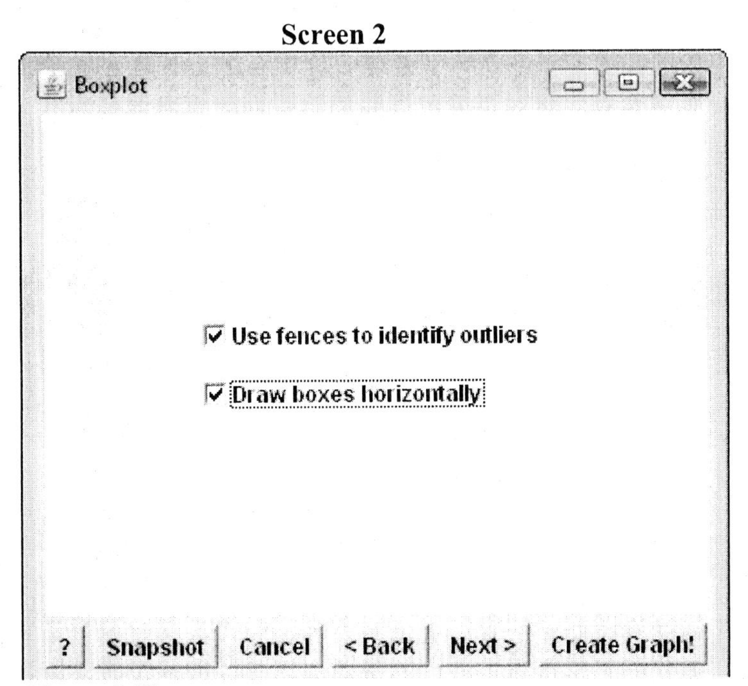

Figure 2.8b

Screen 3 allows you to edit axis labels, titles, and include gridlines. When you are done click **Create Graph**! Your Box Plot should look like this:

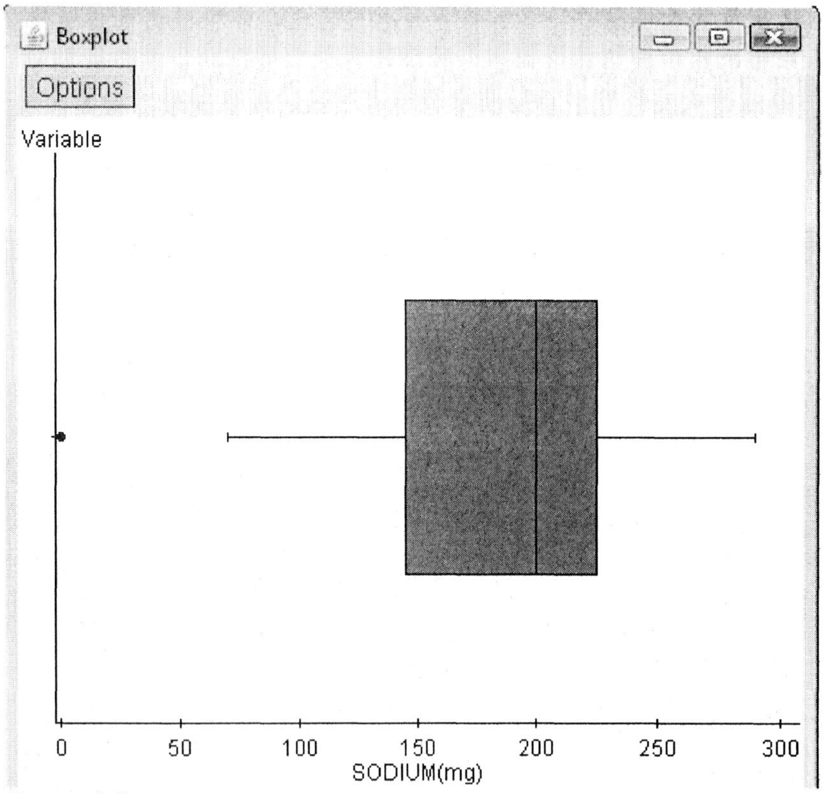

Figure 2.8c

Note that the cereal with 0 sodium content (Frosted Mini Wheat's) is identified as an outlier. It is displayed as a separate point outside the stems of the box plot. Had you chose not to display outliers (not checked the "Use fences to identify outliers" box) your box plot would look like:

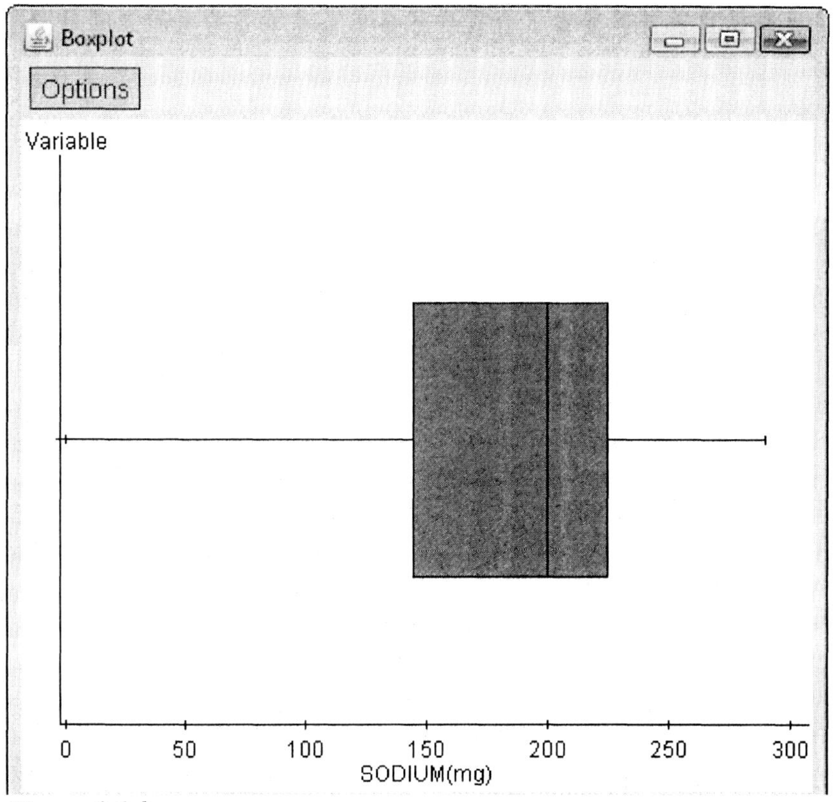

Figure 2.8d

Side-by-side Box Plots

A common graphical display used to compare the distributions of two groups is to look at Side-by-Side Box Plots. To produce Side-by-Side Box Plots to compare the distributions of sodium content for the two groups of cereal types simply choose "*CODE*" in the "Group by" drop down in **Screen 1**. Your Side-by-Side Box Plots should look like:

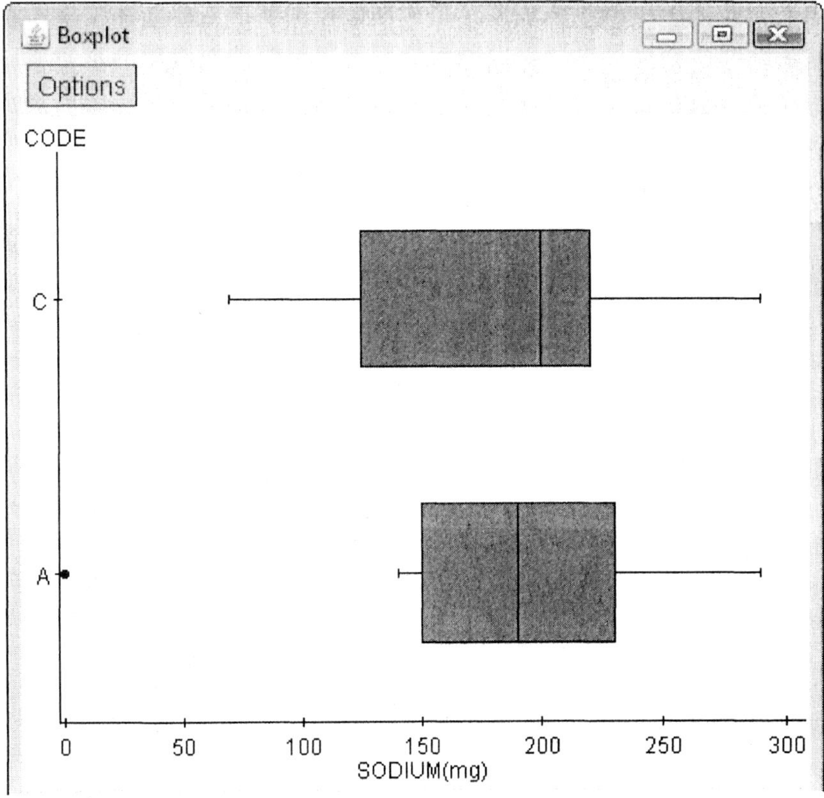

Figure 2.8e

Here you can compare the centres and spreads of sodium content for the two types of cereals.

Time Plots

StatCrunch does not have an actual Time Plot function. However, Time Plots are essentially Scatter Plots[2] with the X-variable being "*time*" and the dots connected by lines. The purpose is to try and visualize a trend in the data over time.

Example 9: "Is There a Trend Towards Warming in New York City?"

The data for Example 9 can be obtained from your text CD as "Central Park Yearly Temps". The data file contains three variables, "*TEMP*" (average annual temperature), "*YEAR*" and "*Time*".

To produce a Time Plot, select **Graphics➔Scatter Plot.**

In **Screen 1** select "*YEAR*" as the X-variable. You could also choose "*TIME*". Select "*TEMP*" as the Y-variable. There is no categorical variable that groups the observations, so click **Next** to go to **Screen 2**.

---

[2] Scatterplot are normally discussed in Chapter 3, but we will introduce them here for the purpose of illustrating Time Plots.

Screen 1

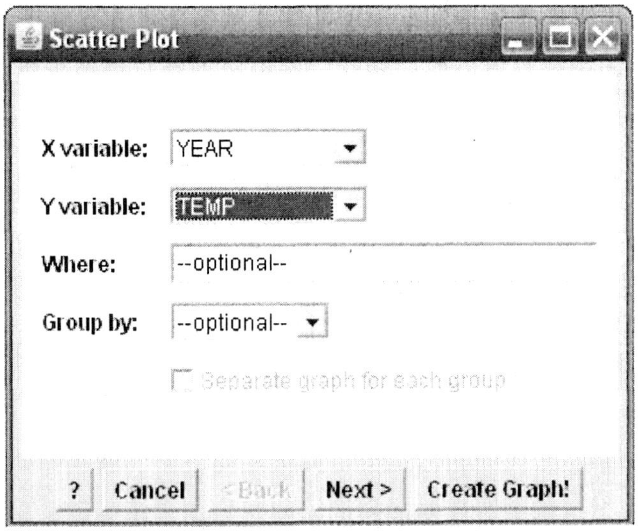

Scatter Plot

X variable: YEAR

Y variable: TEMP

Where: --optional--

Group by: --optional--

Separate graph for each group

? Cancel < Back Next > Create Graph!

Figure 2.9a

In **Screen 2,** check both boxes. For a conventional Scatter Plot we would not connect the points. In Time Plots it is easier to see a trend if the points are connected. Click **Next** to go to **Screen 3**.

Screen 2

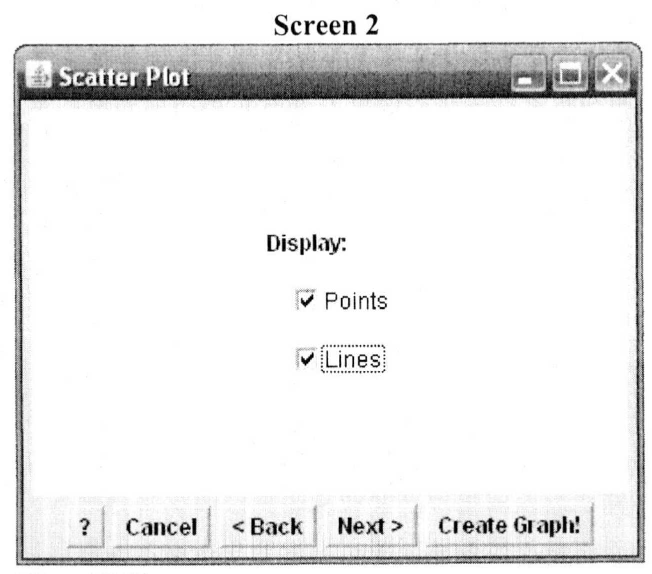

Figure 2.9b

Screen 3 allows you to edit axis labels and titles. When you are done click **Create Graph!** Your Time Plot for average annual temperature should look like this:

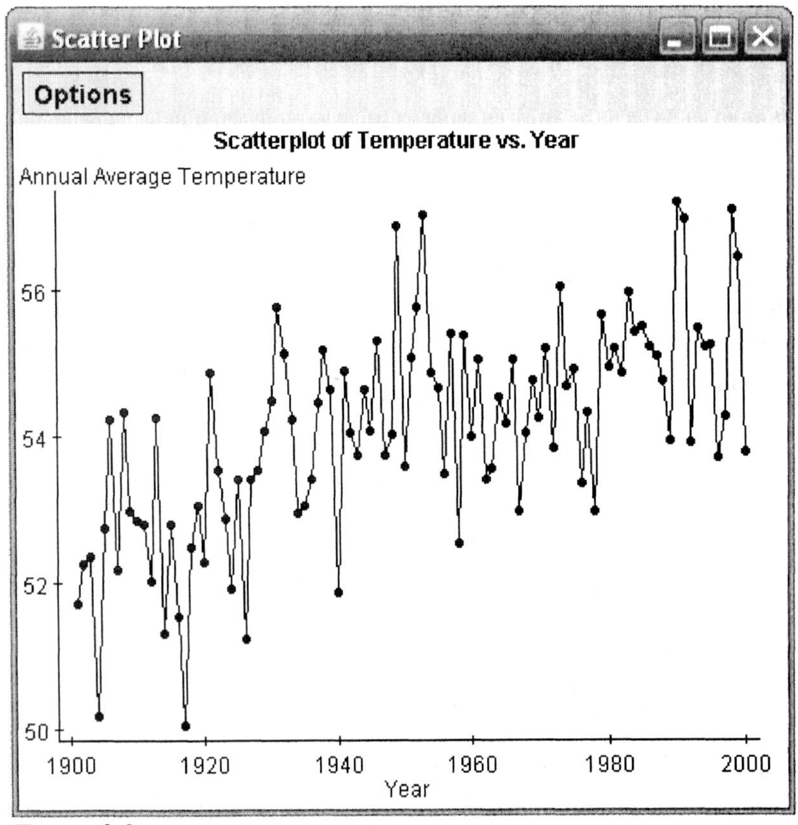

Figure 2.9c

Descriptive Statistics for Quantitative Variables

The most common descriptive statistics are the **Center** and the **Spread.** This section will discuss how to obtain these types of descriptive statistics and others for quantitative variables.

Examples 10 and 16: "Center, Spread and Quartiles for the Cereal Sodium Data"

To obtain the descriptive statistics for the sodium content of the cereals select **Stat➔Summary Stats➔Columns.** We choose "Columns" as the Sodium contents are listed in a column.

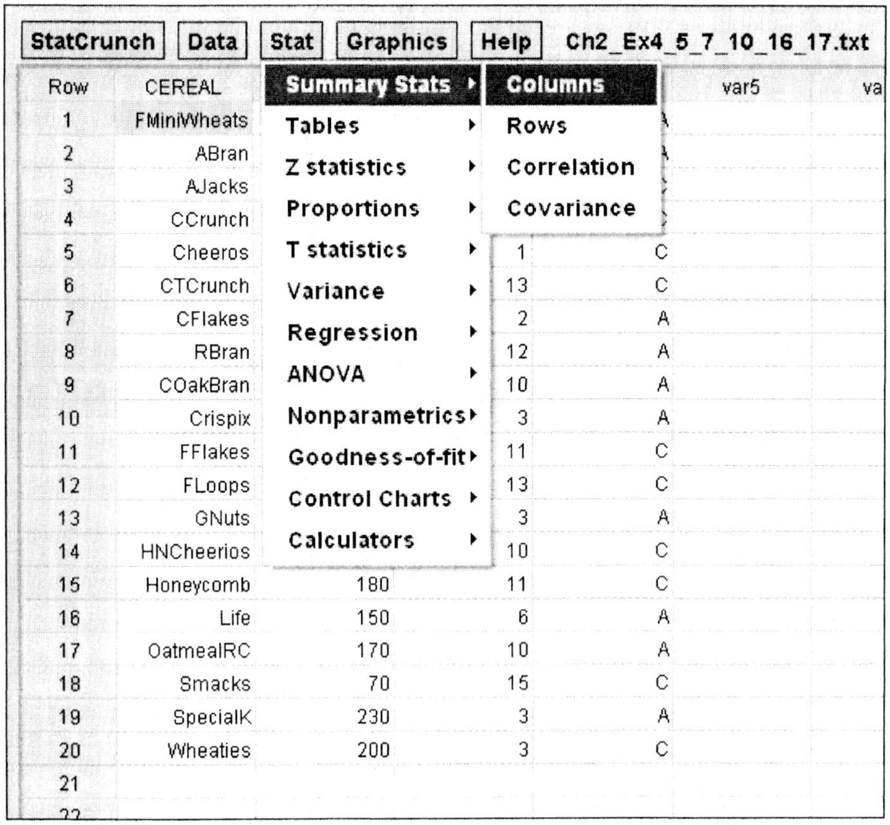

StatCrunch | Data | Stat | Graphics | Help | Ch2_Ex4_5_7_10_16_17.txt

Summary Stats ▸	Columns
Tables ▸	Rows
Z statistics ▸	Correlation
Proportions ▸	Covariance
T statistics ▸
Variance ▸
Regression ▸
ANOVA ▸
Nonparametrics ▸
Goodness-of-fit ▸
Control Charts ▸
Calculators ▸

Row	CEREAL				var5	va
1	FMiniWheats			A		
2	ABran			A		
3	AJacks			C		
4	CCrunch			C		
5	Cheeros		1	C		
6	CTCrunch		13	C		
7	CFlakes		2	A		
8	RBran		12	A		
9	COakBran		10	A		
10	Crispix		3	A		
11	FFlakes		11	C		
12	FLoops		13	C		
13	GNuts		3	A		
14	HNCheerios		10	C		
15	Honeycomb	180	11	C		
16	Life	150	6	A		
17	OatmealRC	170	10	A		
18	Smacks	70	15	C		
19	SpecialK	230	3	A		
20	Wheaties	200	3	C		
21						

Figure 2.10a

In **Screen 1** choose the variable(s) for which you wish to obtain descriptive statistics by clicking on them on the list on the left. To obtain the descriptive for sodium content, click on "*SODIUM(mg)*". Again you have the option to obtain results separately for each type of cereal by selecting "*CODE*" in the "Group by:" drop down menu. Click **Next** to go to **Screen 2**.

Screen 1

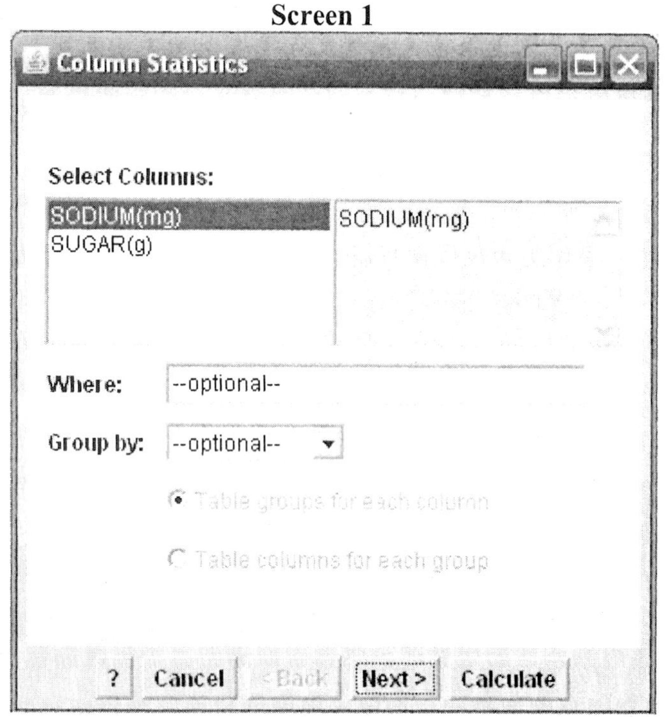

Figure 2.10b

The available descriptive statistics are listed on the left. Click on the ones you want. If chosen they will be displayed on the right. The ones highlighted below are the defaults.

Screen 2

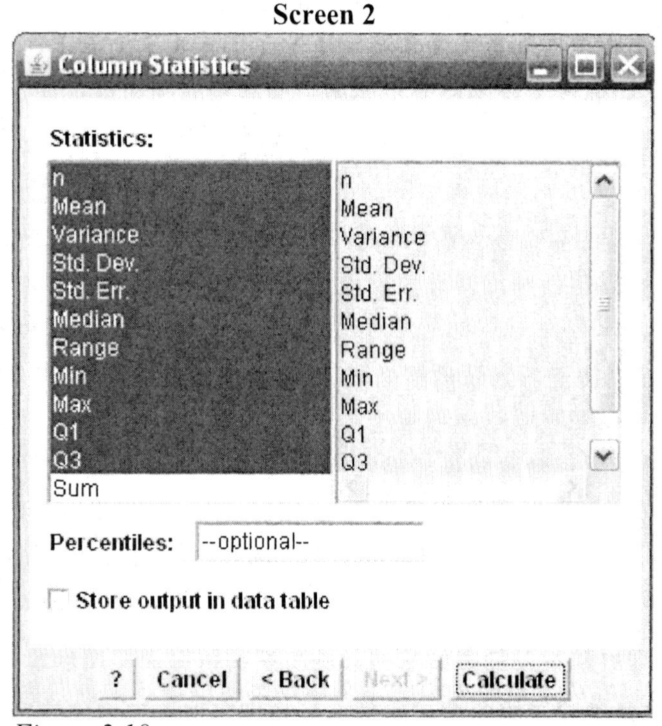

Figure 2.10c

Click **Calculate.** You should obtain the following table.

Column Statistics

Options

Summary statistics:

Column	n	Mean	Variance	Std. Dev.	Std. Err.	Median	Range	Min	Max	Q1	Q3
SODIUM(mg)	20	185.5	5076.0527	71.24642	15.931184	200	290	0	290	145	225

Figure 2.10d

Center: There are two common measures for center, the **Mean** and the **Median**. The Mean (or arithmetic average) sodium content for the 20 (n) cereals is 185.5mg. The Median (middle value or 50th percentile) sodium content is 200mg.

Spread: Spread can be described using the **Range**, the **Variance**, the **Standard Deviation** and/or **Interquartile Range**, depending on the situation. The Variance of sodium content is 5076.05, and the Standard Deviation is 71.246 (the positive square root of the variance). The Range (largest – smallest observation) is 290-0=290. The Interquartile range must be calculated by hand as the difference between the third quartile (Q3 or 75th percentile) and the first quartile (Q1 or the 25th percentile). Thus, the Interquartile range for sodium content is 225-145=80. The standard error is also a measure for spread, which will be discussed later.

Recall Box Plots: Box Plots are typically displayed using a Five-Number Summary composed of the Minimum, the First Quartile (Q1), the Median, the Third Quartile (Q3), and the Maximum. The "box" is drawn extending from Q1 to Q3. A line is drawn in the box at the median. Then stems are extended from the box to the minimum and maximum observations. Here was our Box Plot for the Sodium content of the cereals without using fences to identify outliers. As you can see the lower stem extends to the left to 0 (minimum). The left edge of the box is 145 (Q1). The right edge of the box is 225 (Q3). The line in the box is at 200 (the median). The upper stem extends to the right to 290 (maximum).

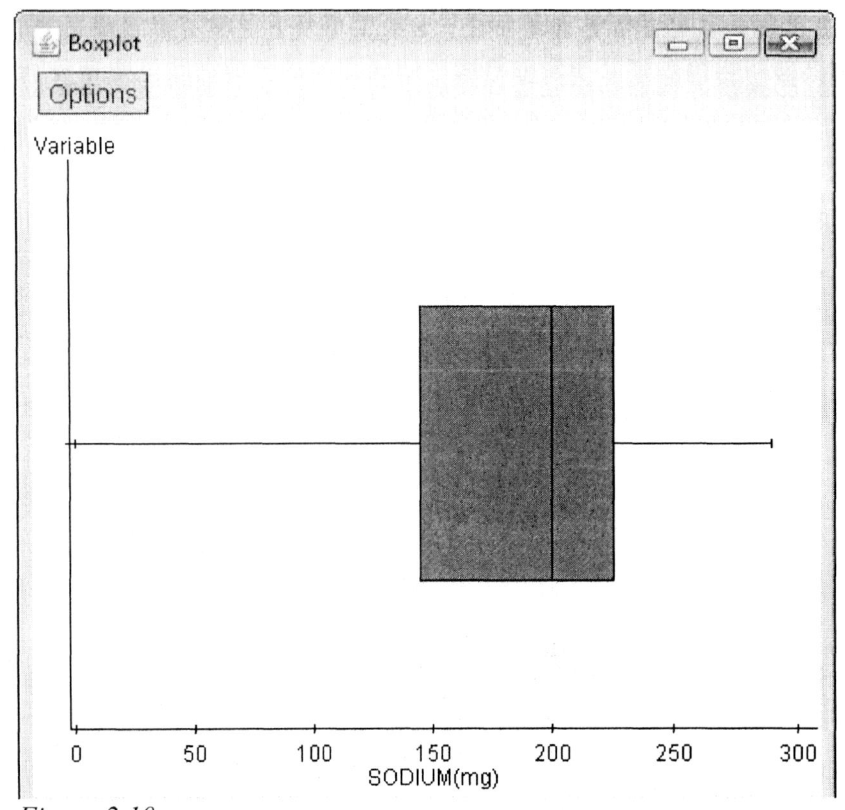

Figure 2.10e

Note: We saw another Box Plot that identified outliers as points separate from the Box Plot. In these cases the stems extend to the minimum or maximum observations that are not identified as outliers.

Earlier we mentioned that the most common descriptive statistics are the **Center** and the **Spread.** Another descriptive measure that might be important is the position of a particular number in the distribution. Percentiles are used to describe positions in the distribution. We have already seen the 25th percentile (Q1), the 50th percentile (median) and the 75th percentile (Q3). Suppose we want to know where the 80th or 90th percentiles lie? Click **Options** in your Column Statistics output to re-open the function. In **Screen 2** you will see a "Percentiles" blank. To obtain the values that correspond to the 80th and 90th percentiles, enter "80" and "90" in the blank. Also, select only n, Mean, Variance and Std. Dev. Click on the other values to deselect them.

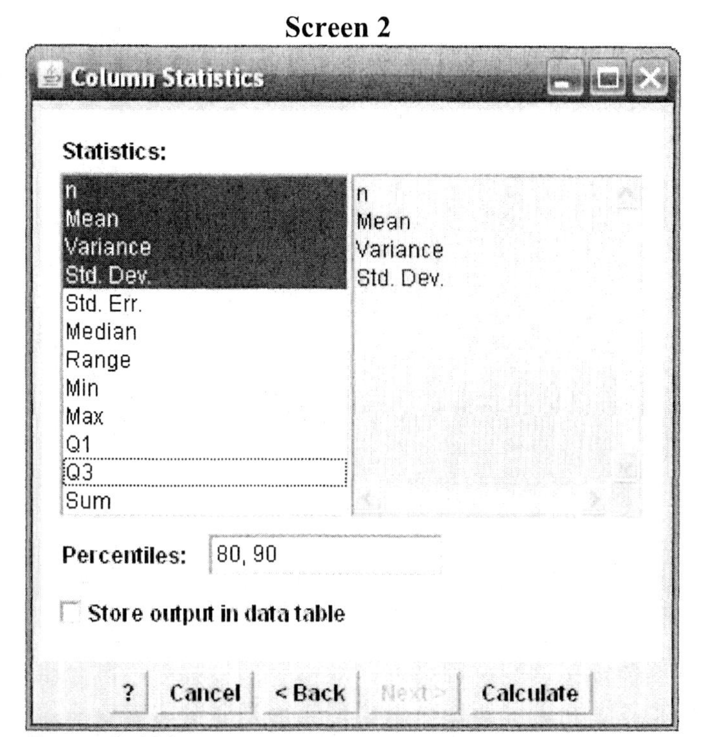

Figure 2.10f

Click **Calculate**. You should obtain the following:

Column Statistics

Options

Summary statistics:

Column	n	Mean	Variance	Std. Dev.	80th Per.	90th Per.
SODIUM(mg)	20	185.5	5076.0527	71.24642	240	275

Figure 2.10g

The 80th percentile of Sodium content is 240 and the 90th percentile is 275.

Like the other functions we can display descriptive statistics separately for different groups. Click **Options** in the table above. In the **Screen 1**, select "*CODE*" in the "Group by" drop down. Click **Calculate.** You should obtain the table below:

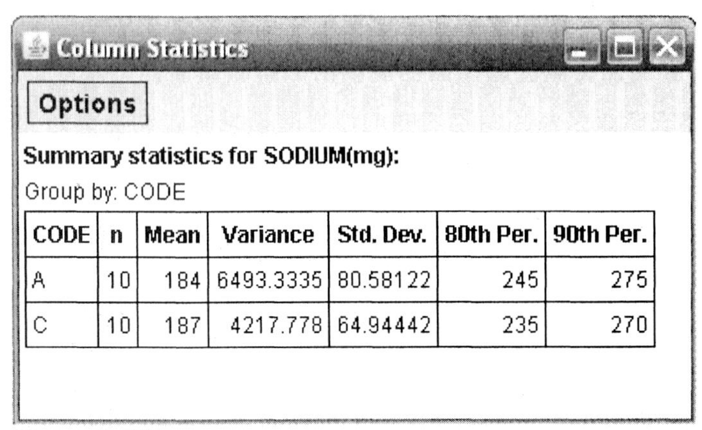
Column Statistics

Options

Summary statistics for SODIUM(mg):

Group by: CODE

CODE	n	Mean	Variance	Std. Dev.	80th Per.	90th Per.
A	10	184	6493.3335	80.58122	245	275
C	10	187	4217.778	64.94442	235	270

Figure 2.10h

Additional Examples

Example 11: "CO_2 Pollution of the Eight Largest Nations"

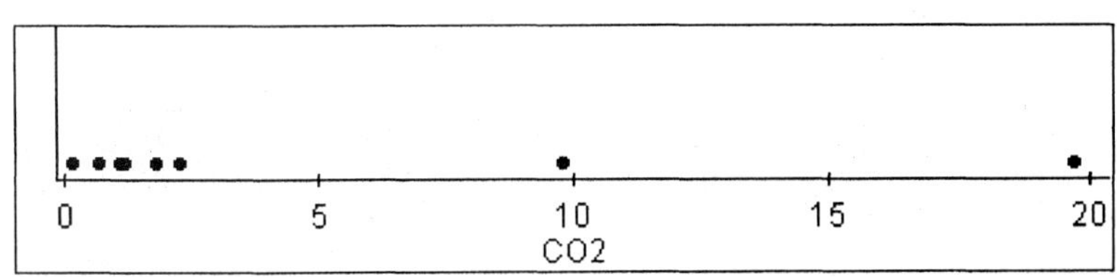

Figure 2.11a

Column Statistics

Options

Summary statistics:

Column	n	Mean	Variance	Std. Dev.	Std. Err.	Median	Range	Min	Max	Q1	Q3
CO2	8	4.6	46.65143	6.830185	2.4148352	1.5	19.5	0.2	19.7	0.9	6.05

Figure 2.11b

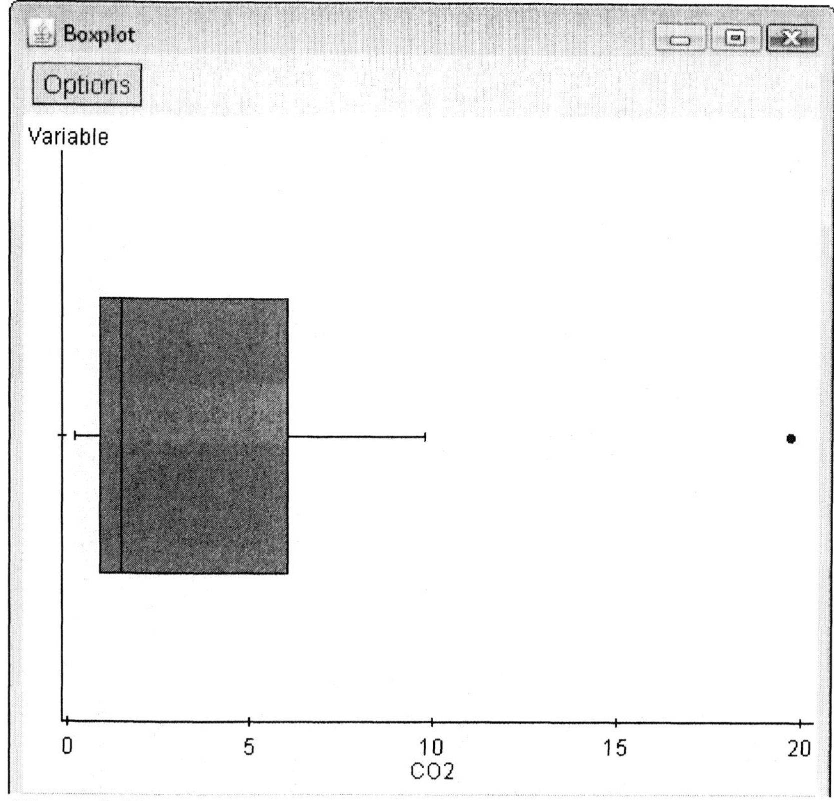

Figure 2.11c

Example 13: "Comparing Women's and Men's Ideal Number of Children"

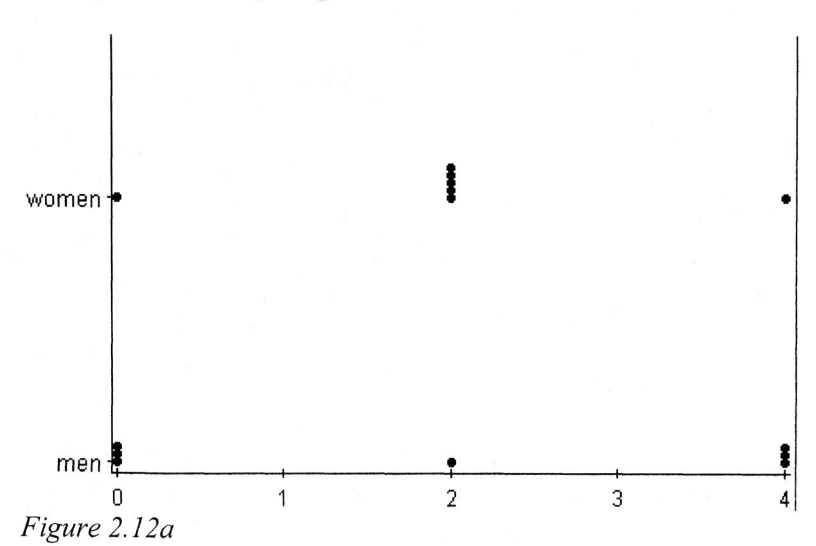

Figure 2.12a

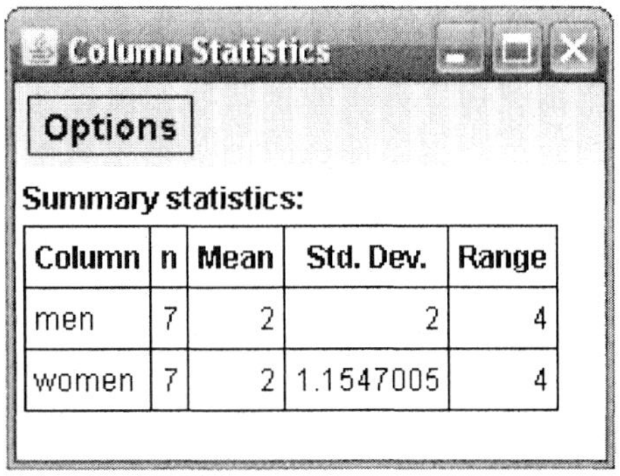

Column Statistics

Options

Summary statistics:

Column	n	Mean	Std. Dev.	Range
men	7	2	2	4
women	7	2	1.1547005	4

Figure 2.12b

Example 15: "Describing Female College Student Heights"

If you select "*GENDER*" in the "Group by" drop down you will obtain a histogram with the different genders color-coded. To produce separate histograms, one for males (*GENDER*=0) and one for females (*GENDER*=1), check the box "Separate graph for each group".

Histogram

Select Columns:

HEIGHT
GENDER

HEIGHT

Where: --optional--

Group by: GENDER

Separate graph for each group

? Cancel < Back Next > Create Graph!

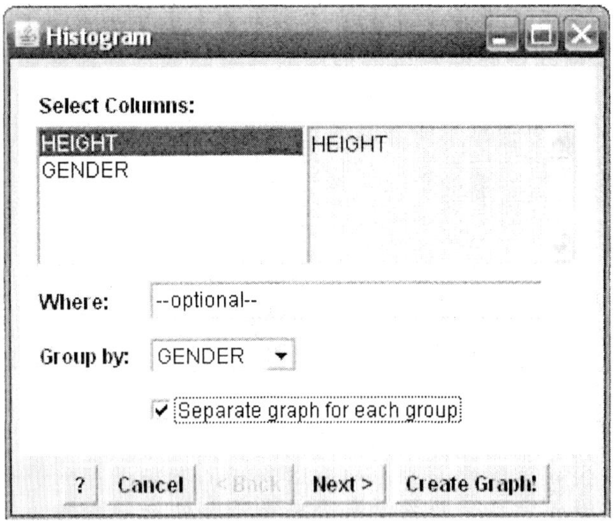

Figure 2.13a

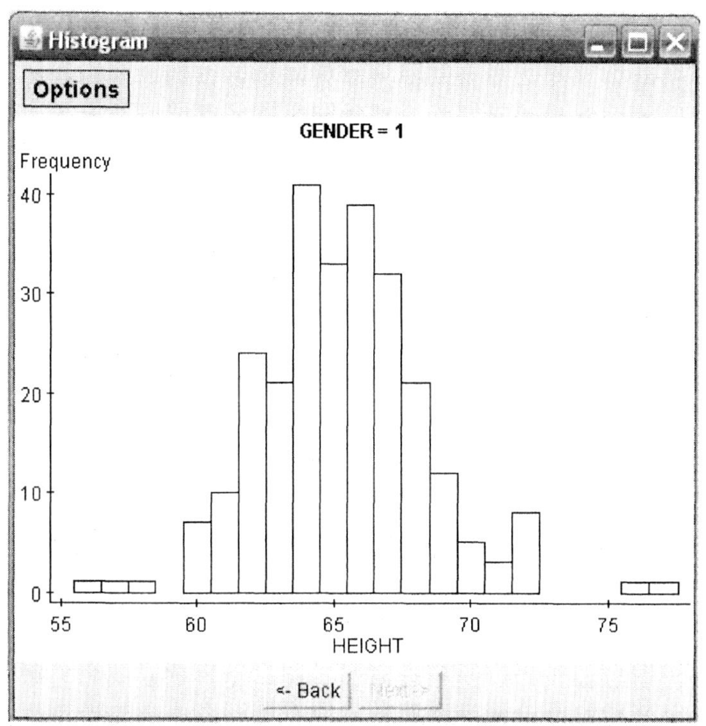

Figure 2.13b

If you chose to overlay the normal distribution you would obtain:

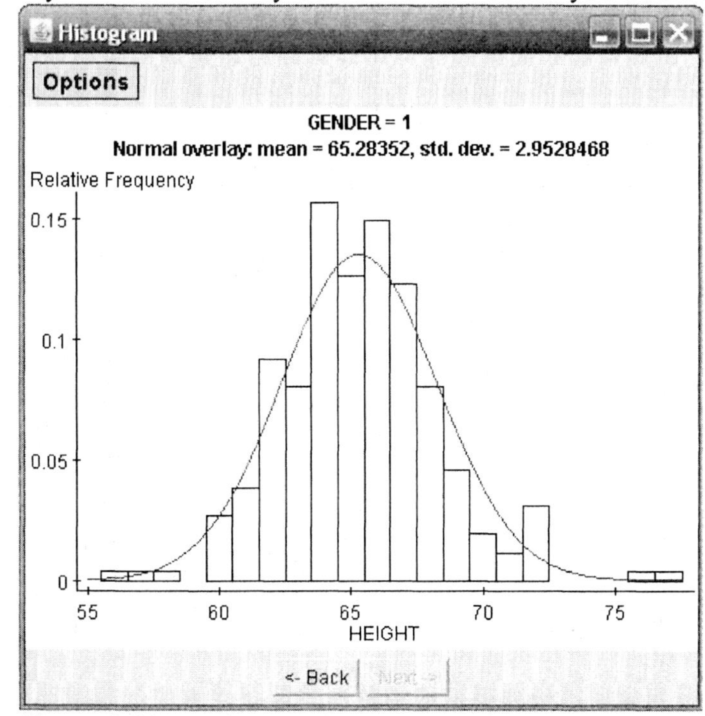

Figure 2.13c

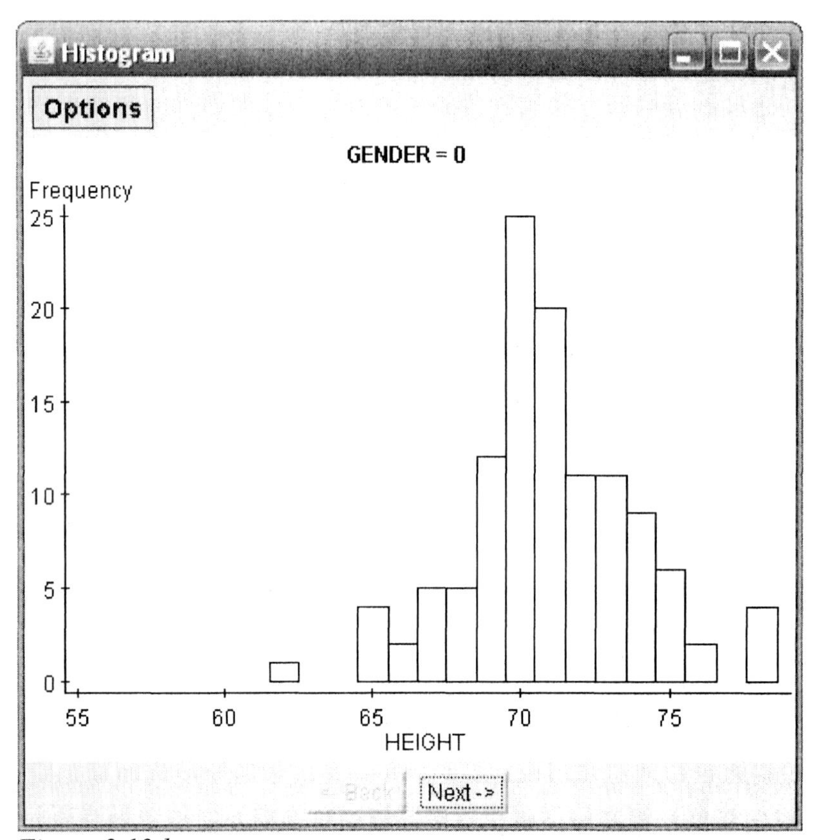

Figure 2.13d

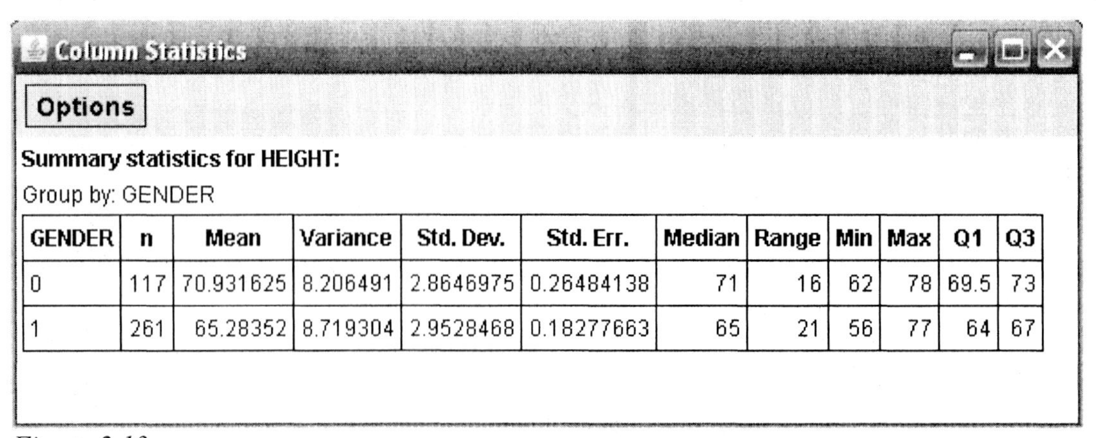

Column Statistics

Options

Summary statistics for HEIGHT:

Group by: GENDER

GENDER	n	Mean	Variance	Std. Dev.	Std. Err.	Median	Range	Min	Max	Q1	Q3
0	117	70.931625	8.206491	2.8646975	0.26484138	71	16	62	78	69.5	73
1	261	65.28352	8.719304	2.9528468	0.18277663	65	21	56	77	64	67

Figure 2.13e

Chapter 3 - Association: Contingency, Correlation and Regression

In Chapter 2 we discussed graphical techniques and numerical summaries for a single categorical or quantitative variable. In Chapter 3 we will discuss methods for exploring relationships between two variables.

Two Categorical Variables

Although we will formally discuss how to explore relationships between two categorical variables in Chapter 11, here we will discuss exploratory techniques here. When dealing with data grouped by two categorical variables we are often interested in the joint distribution of the two variables. The joint distribution is defined by the proportion of observations that fall into each combination of categories. As well we can explore the conditional distribution of each variable by looking at conditional proportions. All these proportions can be summarized in a contingency table. A useful visual display that we will look at is a Side-by-side Bar Chart.

Contingency Tables and Side-by-side Bar Charts for Tabulated Data

Recall that tabulated data is data that is summarized in a frequency distribution.

Examples 2 and 3: "Are Pesticides Present Less Often in Organic Foods?"

Table 3.1 in your text displays the frequency of foods sampled that fall into all possible combinations of two categorical variables, "*Food Type*" with categories "Organic" and "Conventional", and "*Pesticide Status*" with categories "Present" and "Not Present". The table displayed is called a Contingency Table. In order to replicate the table we must enter the data into StatCrunch as displayed below.

StatCrunch | Data | Stat | Graphics | Help

Row	Food Type	Present	Not Present
1	Organic	29	98
2	Conventional	19485	7086

Figure 3.1a

Contingency Tables

To produce a Contingency Table select **Stat→Tables→Contingency→with summary:**

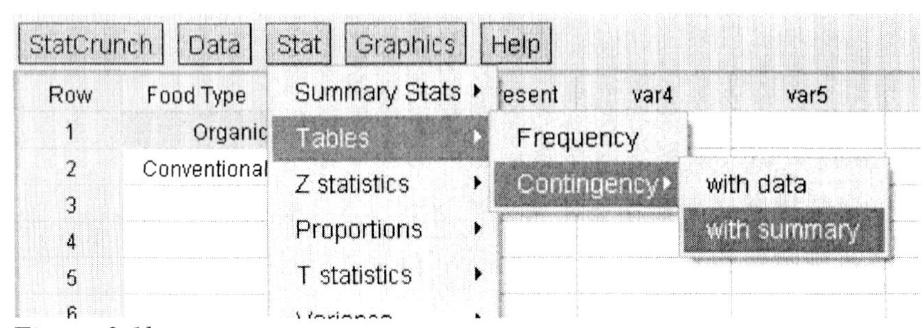

Figure 3.1b

In **Screen 1** select the columns for which the frequencies are entered, "Present" and "Not Present". Under Row Labels select "*Food Type*".

Screen 1

Contingency Table with summary

Select columns for table:

Present
Not Present

Present
Not Present

Row labels in: Food Type

Column variable: --optional--

? | Snapshot | Cancel | < Back | Next > | Calculate

Figure 3.1c

Click **Calculate.** You should obtain the following:

Contingency Table with su...

Options

Contingency table results:

Rows: Food Type

Columns: None

	Present	Not Present	Total
Organic	29	98	127
Conventional	19485	7086	26571
Total	19514	7184	26698

Statistic	**DF**	**Value**	**P-value**
Chi-square	1	163.87474	<0.0001

Figure 3.1d

Notice that StatCrunch automatically calculates the row and column totals. Please compare this to Table 3.1 in your text. For now we will ignore the bottom table.

As mentioned earlier we are often interested in the proportions (relative frequencies) rather than the frequencies. If you click **Options➔Edit** then click **Next** to go to **Screen 2** you will have the option to display the proportions and conditional proportions for each variable. Click on "Row percent" to display the conditional proportions for the row variable, "*Food Type*". Click on "Column percent" to display the conditional proportions for the column variable, "*Pesticide Status*". Click on "Percent of Total" to display the joint distribution (overall percentage) of each combination of the two variables.

Screen 2

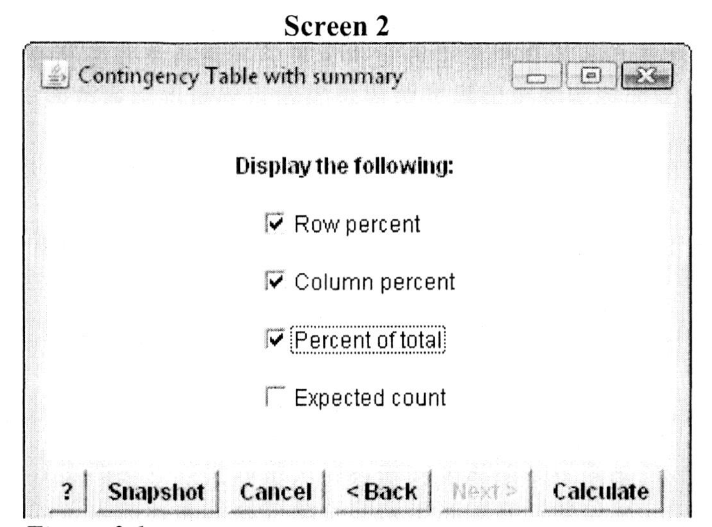

Figure 3.1e

Click **Calculate.**

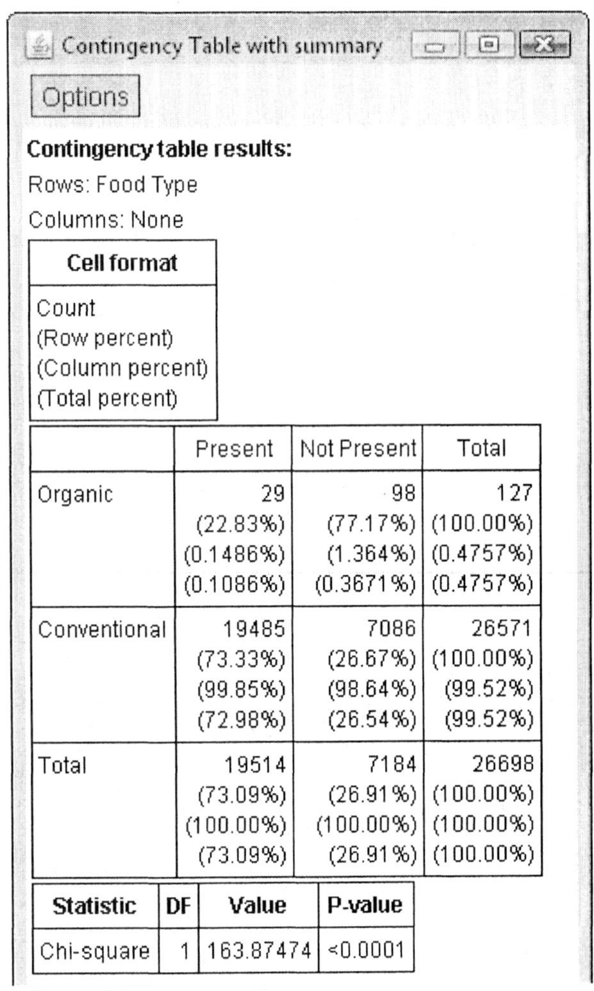

Contingency Table with summary

Options

Contingency table results:

Rows: Food Type

Columns: None

Cell format
Count (Row percent) (Column percent) (Total percent)

	Present	Not Present	Total
Organic	29 (22.83%) (0.1486%) (0.1086%)	98 (77.17%) (1.364%) (0.3671%)	127 (100.00%) (0.4757%) (0.4757%)
Conventional	19485 (73.33%) (99.85%) (72.98%)	7086 (26.67%) (98.64%) (26.54%)	26571 (100.00%) (99.52%) (99.52%)
Total	19514 (73.09%) (100.00%) (73.09%)	7184 (26.91%) (100.00%) (26.91%)	26698 (100.00%) (100.00%) (100.00%)

Statistic	DF	Value	P-value
Chi-square	1	163.87474	<0.0001

Figure 3.1f

The Cell Format table displays what the numbers represent. The first row in each cell is the frequency.

The second row represents the conditional distribution of "*Pesticide Status*" given "*Food Type*" (compare these values to Table 3.2 in your text). For example, the value 22.83% in the "Organic/Present" cell is the proportion of organic foods that contain pesticides = 29/127 (the proportion of foods that contain pesticides given that they are organic). Notice that the row totals are equal to 100%.

The third row represents the conditional distribution of "*Food Type*" given "*Pesticide Status*". For example, the value 0.1486% in the "Organic/Present" cell is the proportion of foods with pesticides present that are organic = 29/19514 (the proportion of foods that are organic given that they contain pesticides). Notice that the column totals are equal to 100%.

The fourth row represents the joint distribution of the two variables. These values are the proportion of foods falling in each combination of the two variables. For example, the value 0.1086% in the "Organic/Present" cell is the proportion of foods that are organic and contain pesticides (29/26698).

Side-by-side Bar Charts

This option is not available in StatCrunch when dealing with summary data. You can however make separate bar charts for each level of one of the variables. You will however have to change the way the data is entered. See the section for Bar Charts in Chapter 2.

Contingency Tables and Side-by-side Bar Charts for Non-Tabulated Data

If data is given in "raw" form it is called non-tabulated data.

Example: Refer to the data "fla_student_survey" on your text CD.

This data set is non-tabulated as we have the observations for each individual. Let's consider the relationship between "*gender*" (male (m) or female (f)) and "*political affiliation*" (republican (r), democrat (d) and independent (i)).

Contingency Tables

Select **Stat→Tables→Contingency→with data.**

In **Screen 1** select the "Row" and "Column" variables. It doesn't really matter which is which. Let's choose "*gender*" as the "Row variable" and "*political affiliation*" as the "Column variable".

Screen 1

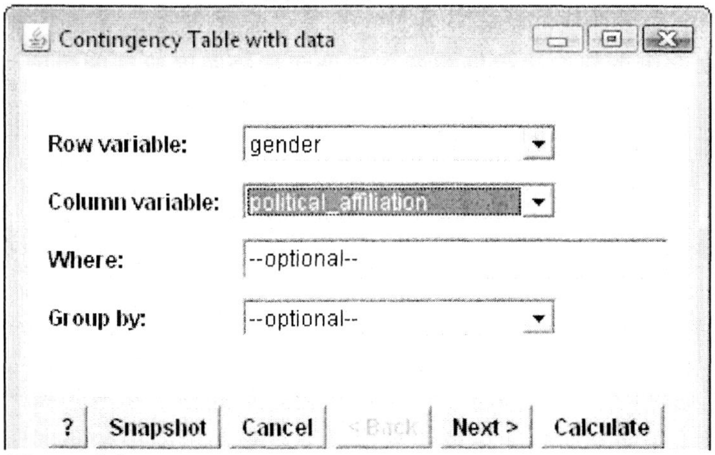

Contingency Table with data

Row variable: gender

Column variable: political_affiliation

Where: --optional--

Group by: --optional--

? | Snapshot | Cancel | < Back | Next > | Calculate

Figure 3.2a

Just like the results for tabulated data you display the joint and conditional proportions. If you go to **Screen 2** you can add these values to your table. For now click **Calculate** to produce the Contingency Table.

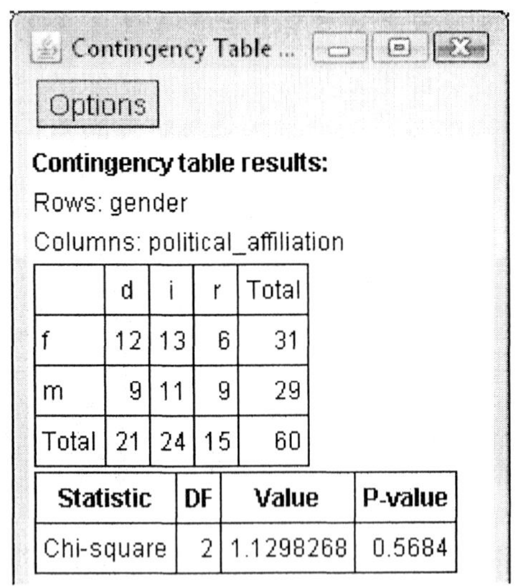

	d	i	r	Total
f	12	13	6	31
m	9	11	9	29
Total	21	24	15	60

Statistic	DF	Value	P-value
Chi-square	2	1.1298268	0.5684

Figure 3.2b

Suppose you want separate contingency tables for different levels of a third categorical variable. You can do this by choosing that variable in the "Group by" menu. Suppose we want separate contingency tables for "vegetarians" and "non-vegetarians". In the "Group by" menu select "*vegetarian*". Click **Calculate.** The tables are displayed below:

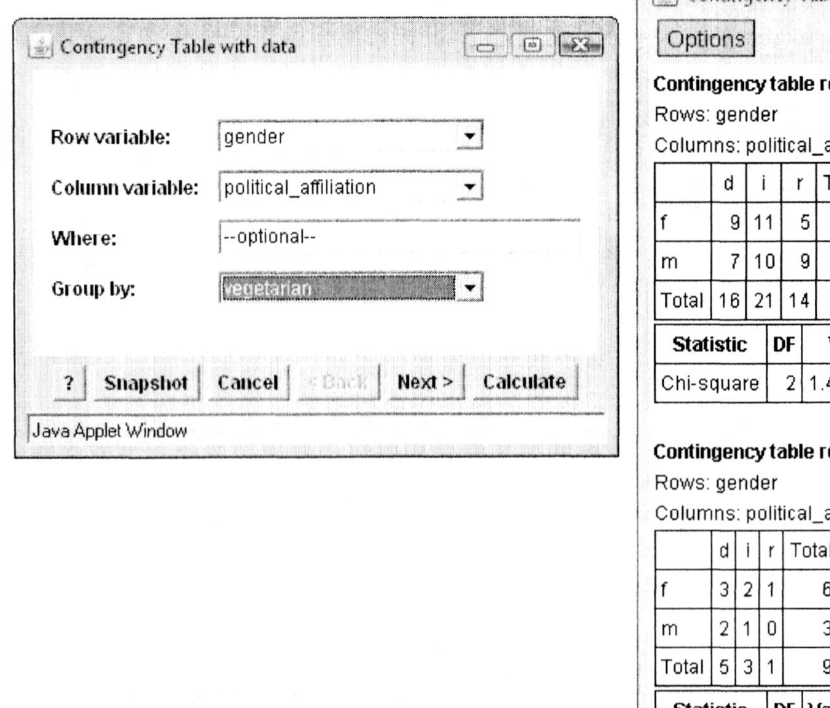

	d	i	r	Total
f	9	11	5	25
m	7	10	9	26
Total	16	21	14	51

Statistic	DF	Value	P-value
Chi-square	2	1.4214149	0.4913

	d	i	r	Total
f	3	2	1	6
m	2	1	0	3
Total	5	3	1	9

Statistic	DF	Value	P-value
Chi-square	2	0.6	0.7408

Figure 3.2c

Side-by-side Bar Charts

When we have non-tabulated data we can produce Side-by-Side Bar Charts. Select **Graphics➔Bar Plot➔with data.**

Screen 1

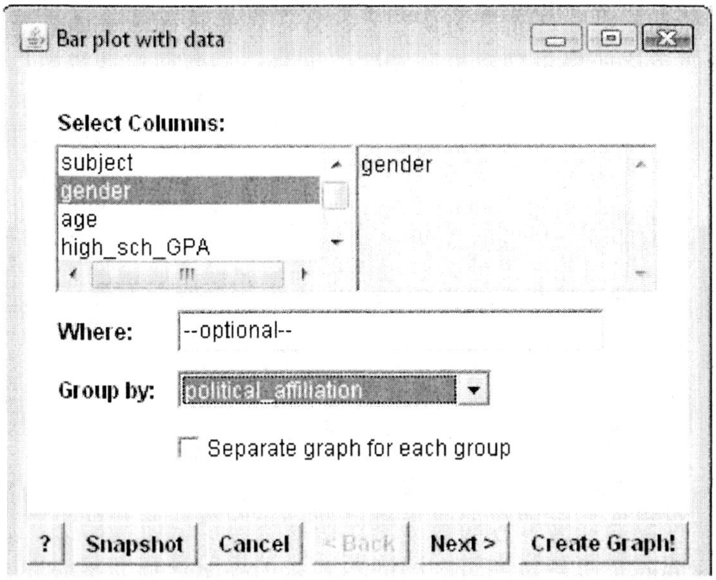

Bar plot with data

Select Columns:

subject
gender
age
high_sch_GPA

gender

Where: --optional--

Group by: political_affiliation

Separate graph for each group

? | Snapshot | Cancel | < Back | Next > | Create Graph!

Figure 3.2d

If we want to display Bar Charts for gender grouped by political affiliation, select the column "*gender*" and choose "*political affiliation*" in the "Group by" drop down menu. See Bar Charts in Chapter 2 for other options. Click **Create Graph.**

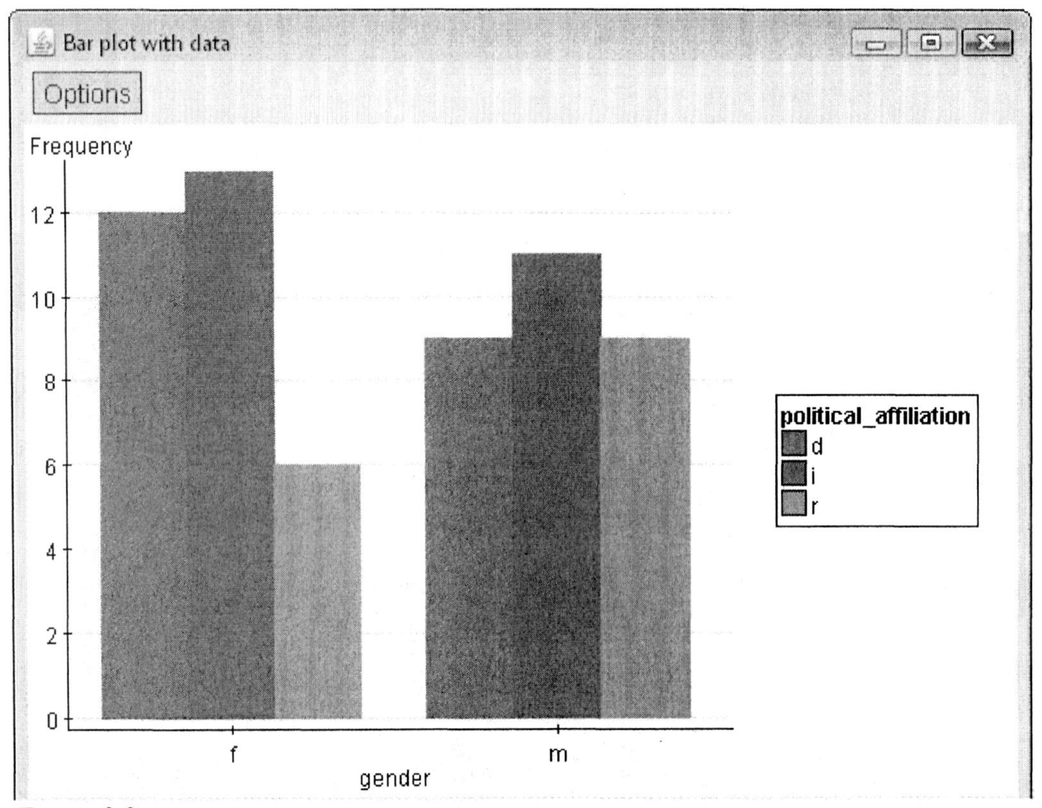

Figure 3.2e

Alternatively you can display three separate Bar Charts for each level of political affiliation. In **Screen 1** click "Separate graph for each group." Here you would have to scroll back and forth to see each graph.

If you want to display three Box Plots side-by-side, scroll to **Screen 4.** In "Columns per page" select 3 (as there are 3 levels of political affiliation). Click **Create Graph!** Here we have the Bar Chart for gender displayed separately for each political affiliation.

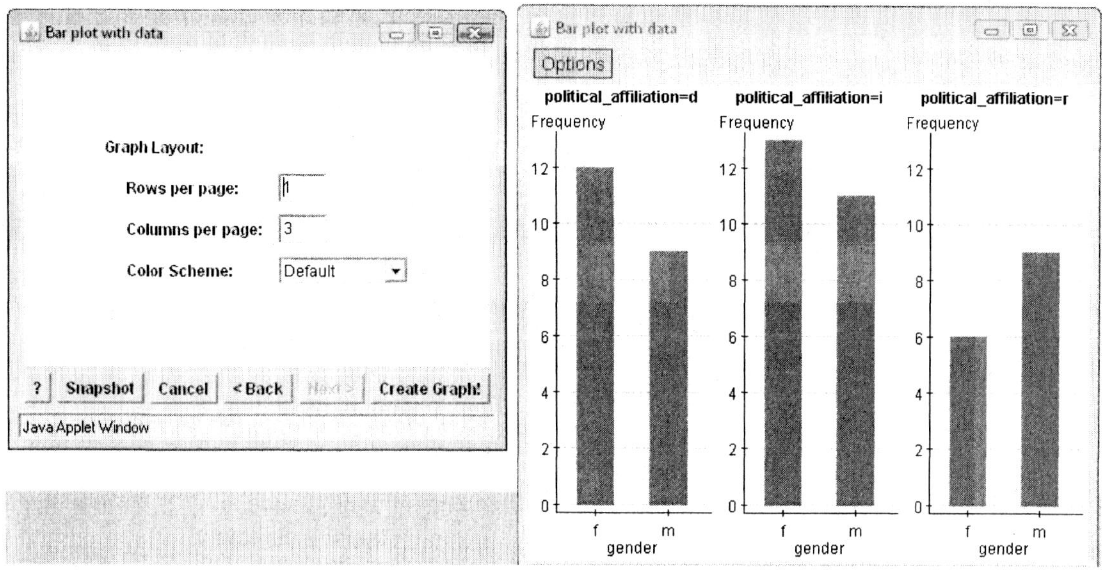

Figure 3.2f

One Categorical and One Quantitative Variable

To explore the relationship between one quantitative and one categorical variable you can look at the numerical summaries (means and standard deviations) or graphs (such as Box Plots) for the quantitative variable for each level of the categorical variable. We already explored an example of this in Chapter 2 when we looked at sodium content (quantitative variable) and cereal type (categorical: adult and children).

Two Quantitative Variables

Now let's consider exploring the relationship between two quantitative variables. Here we will look at a Scatterplot to visualize the relationship or trend between the two variables. When talking about the relationship between two quantitative variables we often are interested in how one variable (the response or dependant variable) depends on another (the explanatory or independent variable). It is common to classify the response variable as the Y-variable and the explanatory variable as the X-variable. If the trend or association seems to be linear (a straight line) then numerical summaries such as correlation and the regression line apply.

Scatterplots

Recall, we already saw an example of Scatterplot when we introduced Time Plots.

Example 5: "Constructing a Scatterplot for Internet Use and GDP"

The data for this example can be found in the "Human Development" file on your CD.

This example wishes to explore the relationship between percentage of adults that use the Internet and GDP for 39 industrial nations. Specifically we are interested in how Internet use depends on GDP. Therefore, we classify Internet use as the response or Y-variable and GDP as the explanatory or X-variable.

To create a Scatterplot to visualize the relationship, select **Graphics➔Scatter Plot.** Select "*INTERNET*" in the "Y-variable" menu and "*GDP*" in the "X-variable" menu. You can also choose to display separate graphs for each level of a categorical variable using the "Group by" menu. Let's leave this as is. Click **Next** to go to **Screen 2**.

Screen 1

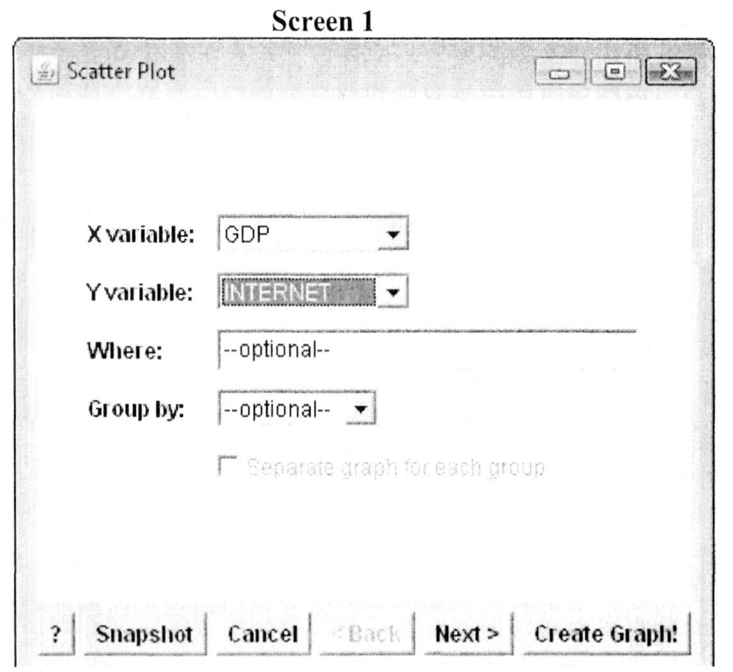

Figure3.3a

In **Screen 2** you have the option to display the actual points and connect them by lines. A conventional Scatterplot displays only the points not connected by line (so make sure the "Points" box is checked and the "Lines" box is unchecked). **Screen 3** allows you to add a title for your graph as well as change the axis labels (by default they will be labelled using the variable names). You can also include grid lines if you wish. When you are ready, click **Create Graph!**

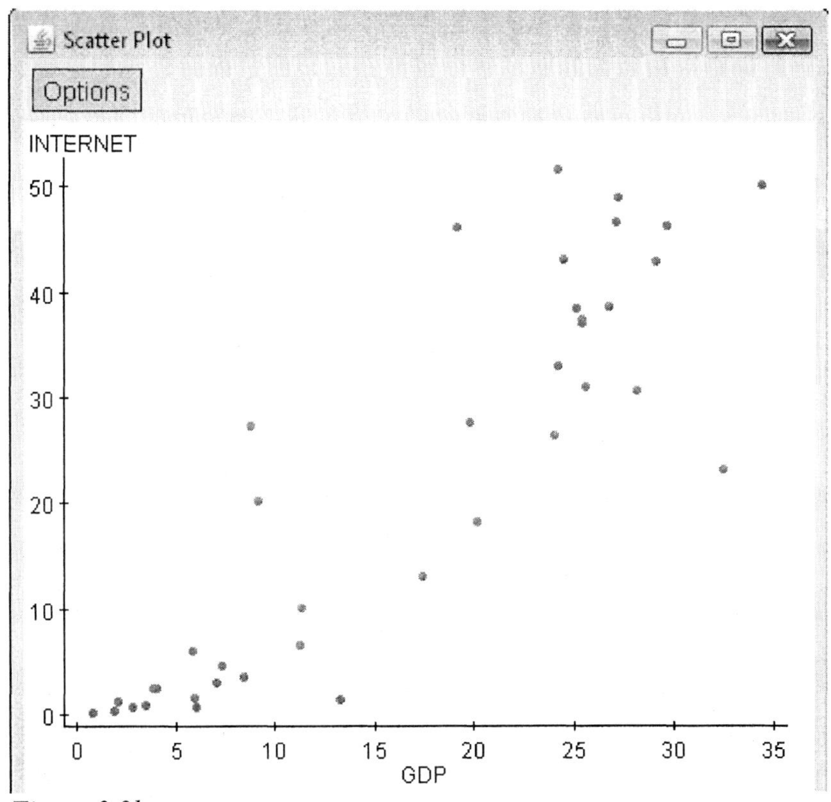

Figure 3.3b

There appears to be an increasing linear trend.

Correlation

If there appears to be a **linear** association between two quantitative variables we can calculate the correlation coefficient to describe the strength of the linear association.

Example 7: "What's the Correlation between Internet Use and GDP?"

To calculate the correlation coefficient between Internet use and GDP select **Stat➔Summary Stats➔Correlation.** Select the variables for which you want their correlation, so "*INTERNET*" and "*GDP*". Note: The correlation coefficient is independent of which is the explanatory variable and which is the response variable. Click **Calculate.**

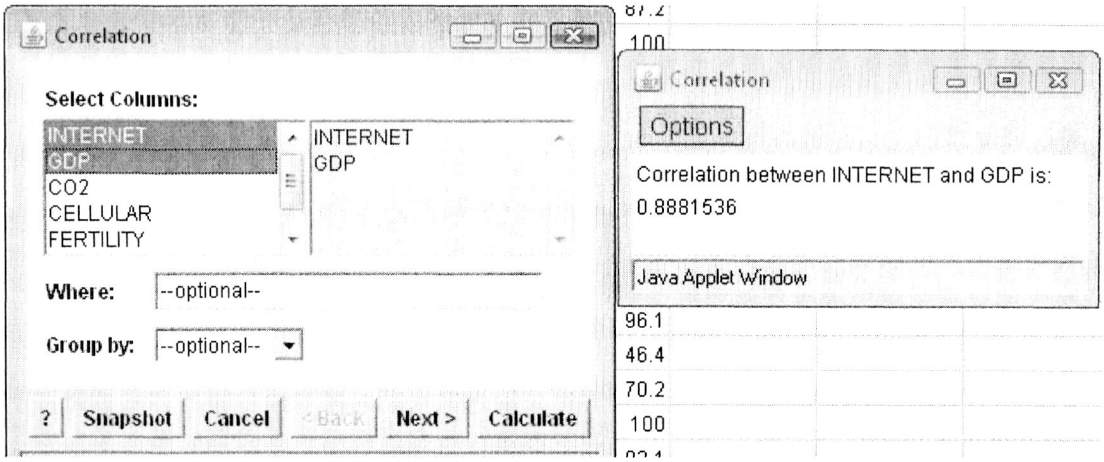

Correlation

Select Columns:

INTERNET
GDP
CO2
CELLULAR
FERTILITY

Where: --optional--

Group by: --optional--

? Snapshot Cancel < Back Next > Calculate

Correlation

Options

Correlation between INTERNET and GDP is:
0.8881536

Java Applet Window

Figure 3.4a

The output above shows that the correlation between Internet use and GDP is 0.8881536. The correlation of positive 0.8881536 indicates a fairly strong positive (or increasing) linear trend. This is consistent with the Scatterplot obtained earlier.

If you want several correlation coefficients you can choose several variables then click **Calculate**. For example select all the quantitative variables. The table below on the right displays all bivariate correlations. **NOTE:** The correlation coefficient only describes the strength of the linear association between two quantitative variables. Thus, it is always wise to look at a Scatterplot before interpreting the correlation coefficient.

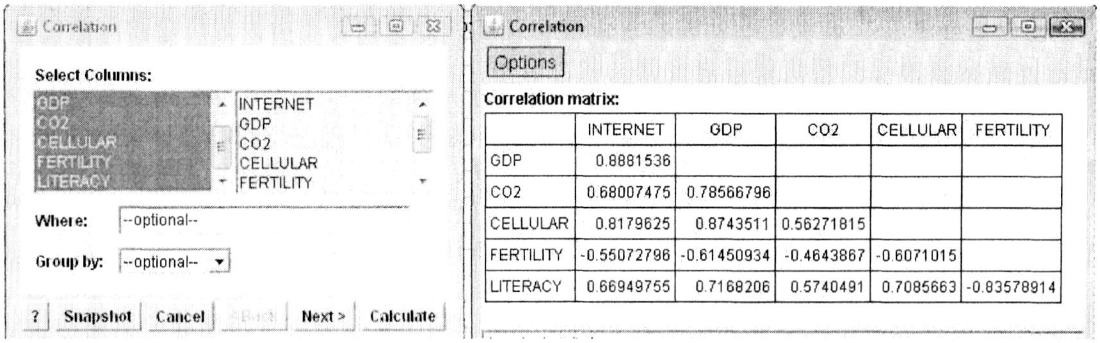

Correlation

Select Columns:

GDP
CO2
CELLULAR
FERTILITY
LITERACY

Where: --optional--

Group by: --optional--

? Snapshot Cancel < Back Next > Calculate

Correlation

Options

Correlation matrix:

	INTERNET	GDP	CO2	CELLULAR	FERTILITY
GDP	0.8881536				
CO2	0.68007475	0.78566796			
CELLULAR	0.8179625	0.8743511	0.56271815		
FERTILITY	-0.55072796	-0.61450934	-0.4643867	-0.6071015	
LITERACY	0.66949755	0.7168206	0.5740491	0.7085663	-0.83578914

Figure 3.4b

The Regression Line and Regression Function

If there appears to be a linear association between two quantitative variables (see the Scatterplot) we can find the linear regression equation. This is the equation for the straight line that best describes the linear pattern of the data. With the regression equation we can estimate or predict values of the response variable from values of the explanatory variable.

Exercise 3.29: "Regression Line for Internet Use and GDP"

In order to calculate the regression equation for Internet use on GDP, select **Stat➔Regression➔Simple Linear** ("simple" means that there is exactly one explanatory variable). In **Screen 1** define the X and Y variables. Select "*INTERNET*" in the "Y-variable" menu and "*GDP*" in the "X-variable" menu. Click **Next** to go to **Screen 2.**

Screen 1

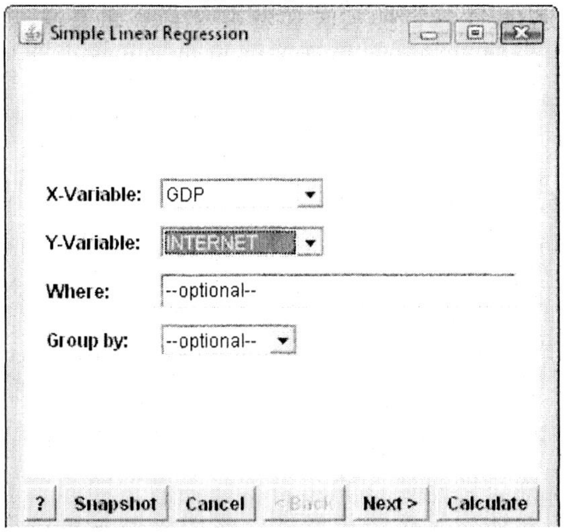

Simple Linear Regression

X-Variable: GDP

Y-Variable: INTERNET

Where: --optional--

Group by: --optional--

? Snapshot Cancel < Back Next > Calculate

Figure 3.5a

In **Screen 2** we can ask StatCrunch to calculate a predicted value of Internet use given some specified value of GDP. Although we have the actual observation for the United States ("*GDP*"=34.3 and "*INTERNET*"=50.2) let's ask StatCrunch to calculate the predicted Internet use for GDP = 34.3. We can also save the residuals for each observation (observed–predicted value) as well as the fitted values (predicted values).

Screen 2

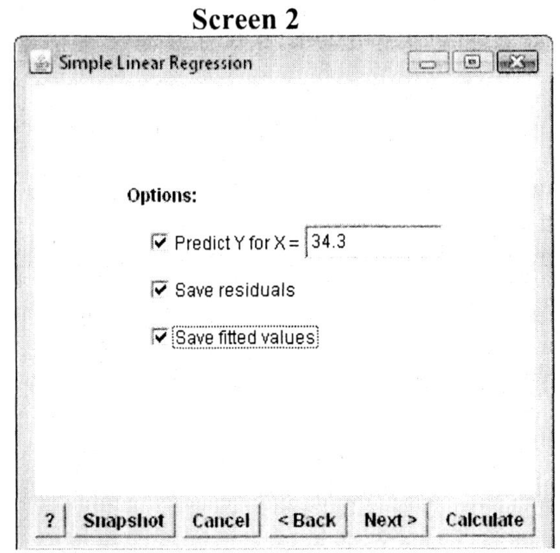

Figure 3.5b

For now click **Calculate.** The Regression output is displayed below:

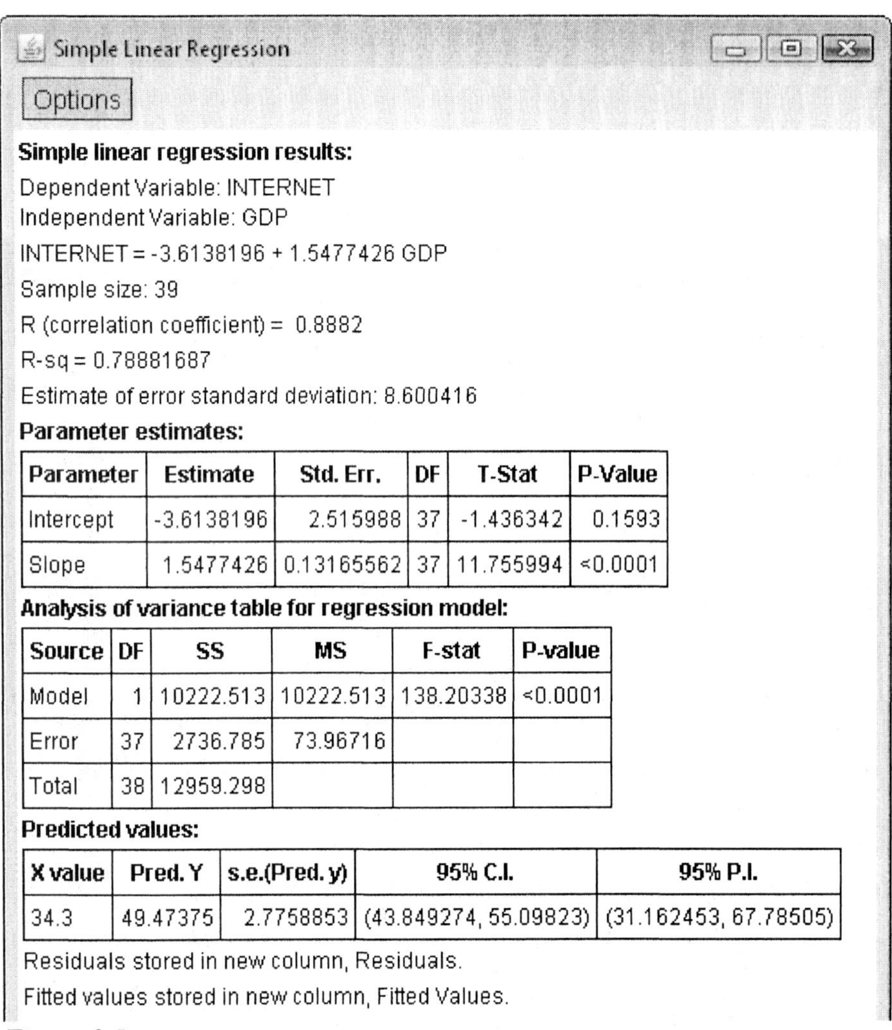

Simple Linear Regression

Options

Simple linear regression results:

Dependent Variable: INTERNET
Independent Variable: GDP
INTERNET = -3.6138196 + 1.5477426 GDP
Sample size: 39
R (correlation coefficient) = 0.8882
R-sq = 0.78881687
Estimate of error standard deviation: 8.600416

Parameter estimates:

Parameter	Estimate	Std. Err.	DF	T-Stat	P-Value
Intercept	-3.6138196	2.515988	37	-1.436342	0.1593
Slope	1.5477426	0.13165562	37	11.755994	<0.0001

Analysis of variance table for regression model:

Source	DF	SS	MS	F-stat	P-value
Model	1	10222.513	10222.513	138.20338	<0.0001
Error	37	2736.785	73.96716		
Total	38	12959.298			

Predicted values:

X value	Pred. Y	s.e.(Pred. y)	95% C.I.	95% P.I.
34.3	49.47375	2.7758853	(43.849274, 55.09823)	(31.162453, 67.78505)

Residuals stored in new column, Residuals.
Fitted values stored in new column, Fitted Values.

Figure 3.5c

In the output above the linear regression equation is given as

INTERNET = -3.6138196+1.5477426GDP.

Specifically the slope is calculated to be 1.5477426 (which indicates an increasing trend) and the y-intercept is calculated to be -3.6138196.

We can also find these values in the **Parameter estimates** table under the **Estimate** column. This output also gives us the correlation coefficient (which we already calculated separately) of 0.8882.

In addition, in the **Predicted values** table, we see that the predicted Internet use, "Pred. Y", for GDP = 34.3 (X=34.3) is 49.47375. This could easily be calculated by hand as -3.6138196+1.5477426(34.3)=49.47375.

Plotting the Regression Line on the Scatterplot

Sometimes it is nice to see the Regression Line on the Scatterplot. In the Regression function, go to **Screen 3** and click on "Plot the fitted line". Click **Calculate.** The first screen to appear should be the regression output. Click **Next** and the following should be displayed:

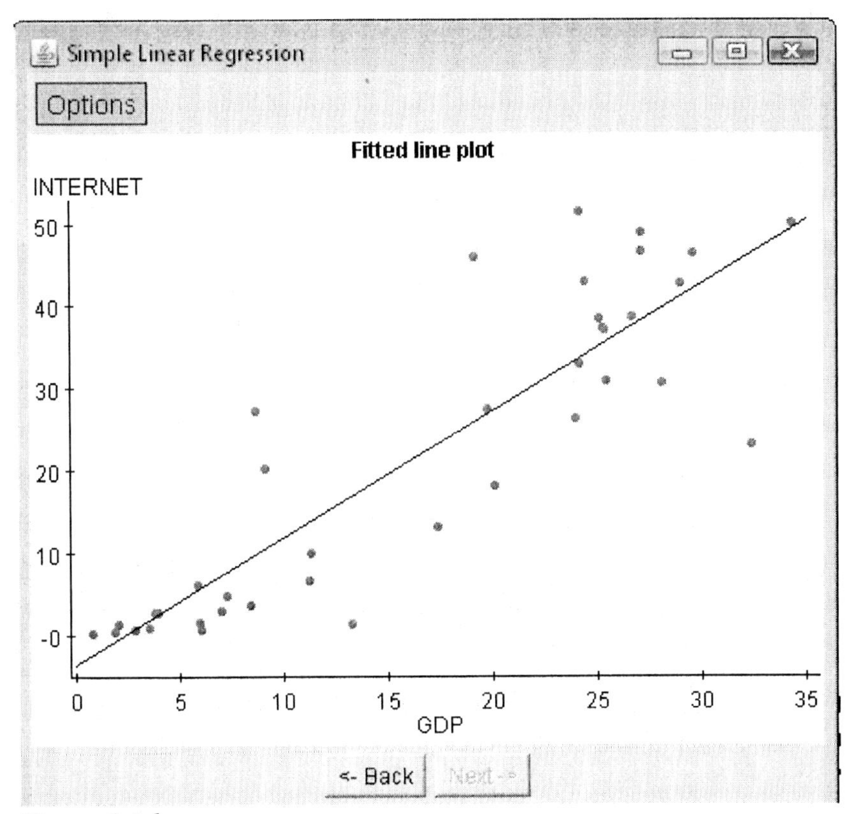

Figure 3.5d

Other Regression Results

This regression output displays several other statistics most of which will be discussed on Chapter 12. For now here are a few others we can discuss.

1) R-sq=0.78881678: This can be interpreted as the proportion of variation in Y that is explained by the linear regression on X. Or, we can explain about 78.88% of the variation in Internet use using a linear regression on GDP.

2) Residuals and Fitted Values: Recall in **Screen 2** we clicked on "Save residuals" and "Save fitted values". These numbers are automatically added as two new columns in your data file.

StatCrunch | Data | Stat | Graphics | Help

Row	C1-T	INTERNET	GDP	CO2	CELLULAR	FERTILITY	LITERACY	Residuals	Fitted Values
1	Algeria	0.65	6.09	3	0.3	2.8	58.3	-5.161933	5.811933
2	Argentina	10.08	11.32	3.8	19.3	2.4	96.9	-3.8266265	13.906627
3	Australia	37.14	25.37	18.2	57.4	1.7	100	1.4875901	35.65241
4	Austria	38.7	26.73	7.6	81.7	1.3	100	0.94266015	37.75734

Figure 3.5e

For example the fitted Internet use for **Algeria** is 5.811933. This is the same as the predicted value calculated using the regression equation:

$$\mathbf{-3.6138196+1.5477426(6.09) = 5.81933.}$$

The residual for each observation is the difference between the actual observed response and the fitted or predicted response. For example the residual for Algeria is 0.65 - 5.811933 = -5.161933.

3) Unusual Points or Outliers: As residuals describe the prediction we can identify unusual observations as those with high absolute residuals. We can look at a histogram of the residuals to try and identify whether or not we have any unusual observations. You can either use the histogram function if you have already saved the residuals as we did above or you can click on "Histogram of Residuals" in **Screen 3** of the regression function. Using the regression function you will obtain the following which automatically overlays the normal distribution on the histogram. Note: Depending on what else you have checked you might have to click next a few times.

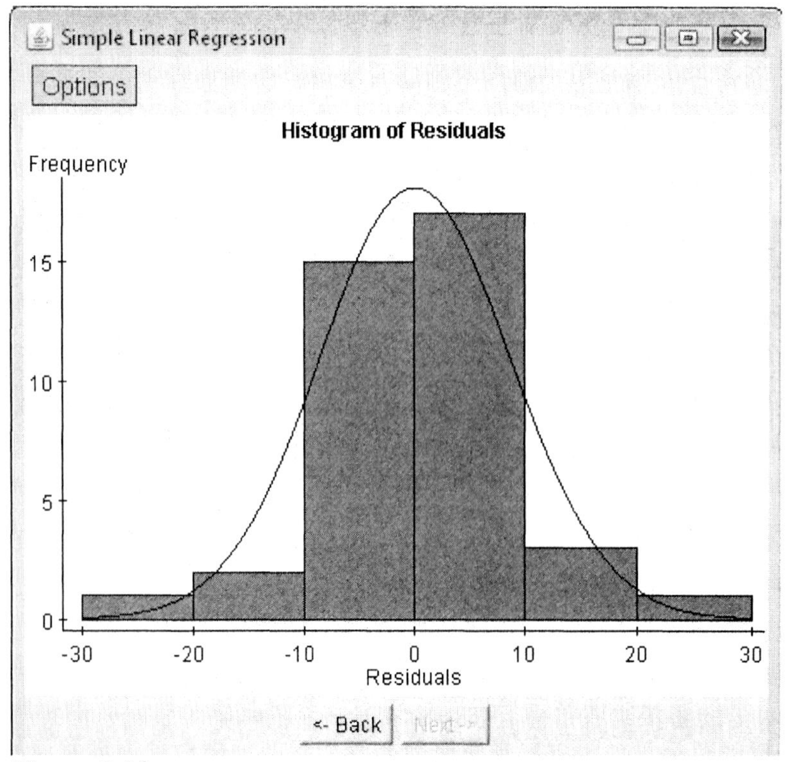

Figure 3.5f

The histogram of residuals does not indicate that there are any unusual observations.

Additional Examples

Example 6 and 10: "Did the Butterfly Ballot Cost Al Gore the 2000 Presidential Election?"

The data can be found in "buchanan_and_the_butterfly_ballot" on you text CD. We are interested in the relationship between the number of votes cast in the 67 Florida counties for Pat Buchanan in the 2000 presidential election vs. the number of votes cast in the 67 Florida counties for Ross Perot in the 1996 presidential election. Specifically we are trying to determine if there are any unusual differences.

The Scatterplot for the number of votes for Buchanan vs. the number of votes for Perot is:

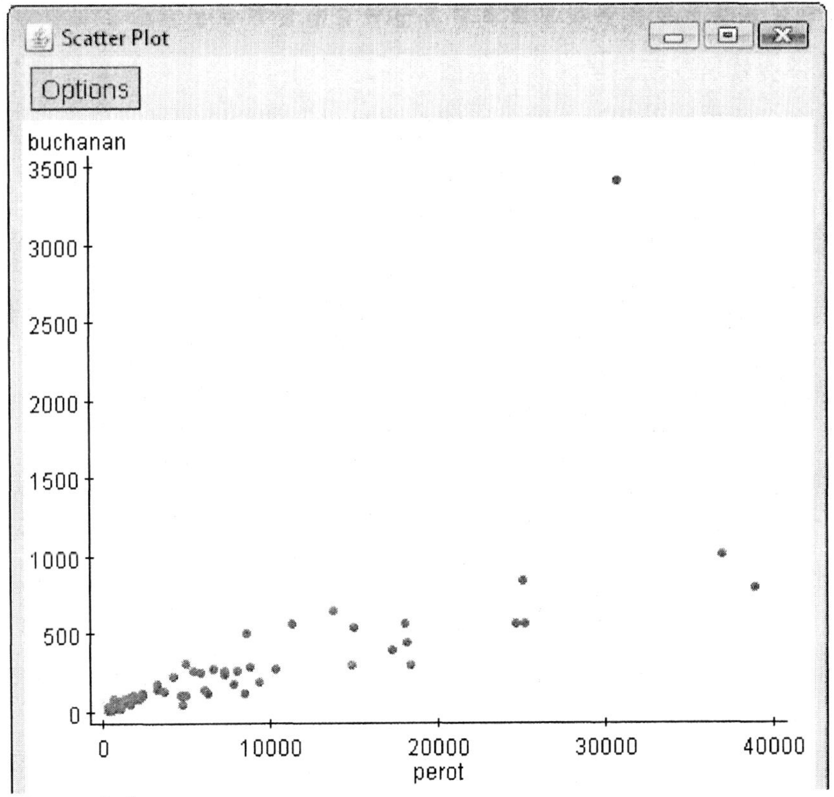

Figure 3.6a

The histogram shows one unusual observation. Upon inspection of the data set this observation is from Palm Beach County. Besides Palm Beach County, there does appear to be a strong linear relationship between the number of votes between the two candidates.

The histogram of residuals using the regression function is:

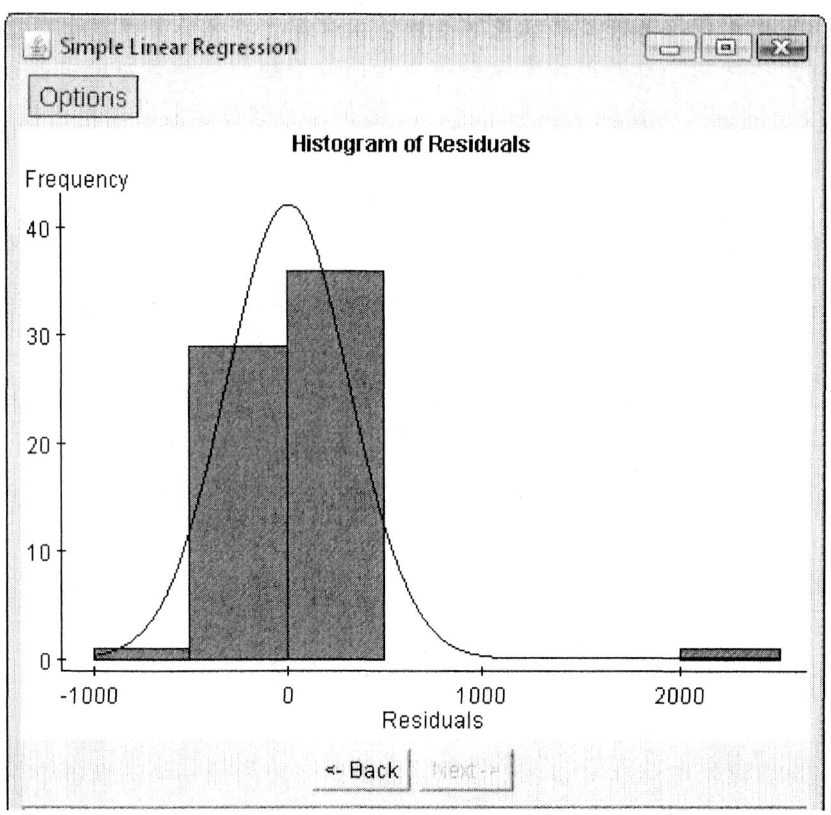

Figure 3.6b

You'll note that this does not appear to be the same as the histogram displayed in Figure 3.15 in your text. This is mainly because the bar widths are not the same size. Unfortunately, we cannot modify the bar width in the Regression function. Nonetheless, we can still see that the observation for Palm Beach County has an extremely high residual.

In order to produce the same histogram as in Figure 3.15 in your text you will have to first **save the residuals** (**Screen 2** in your regression function). Then you will have to use the **Histogram** function. In **Screen 2 of the Histogram** function enter -675 in the "Start bins at" blank and 250 as the "Binwidth". You should get the following:

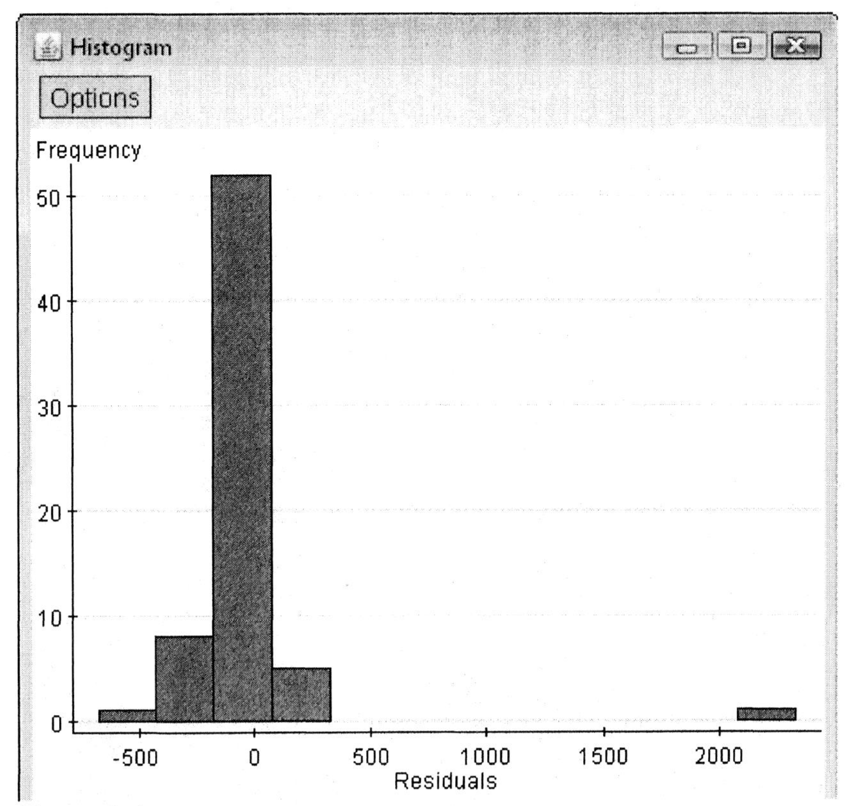

Figure 3.6c

Example 9: "How Can We Predict Baseball Scoring Using Batting Average?"

The data can be found in Table 3.5 in your text or "al team statistics" on you text CD. We are interested in the relationship between team batting average and team scoring for American League teams in 2006.

The Scatterplot of team scoring vs. batting average is:

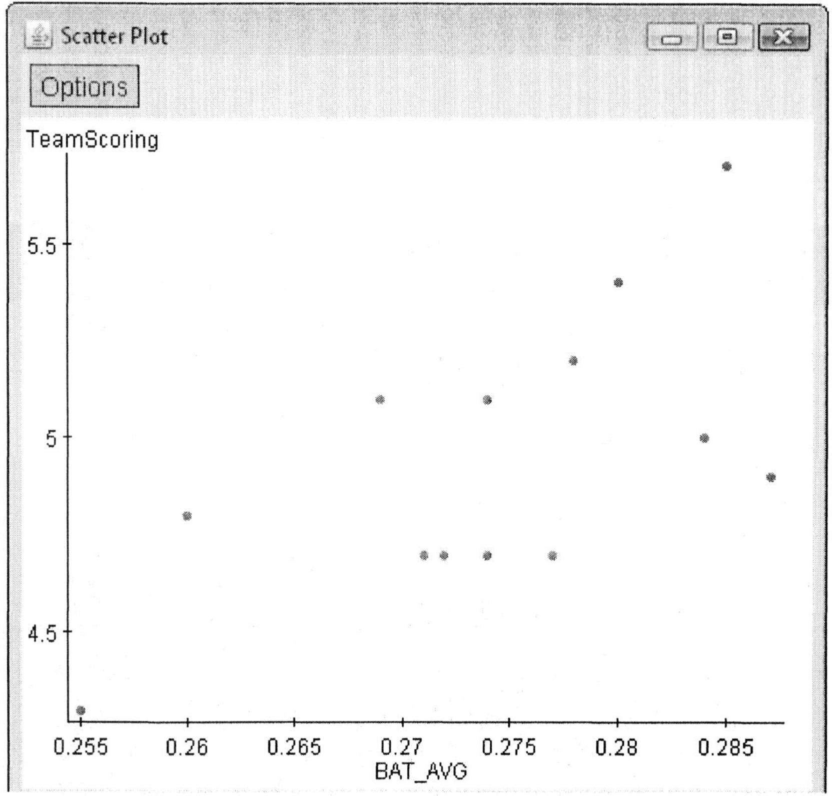

Figure 3.7a

There appears to be a moderate increasing linear pattern.

The correlation between batting average and team scoring is:

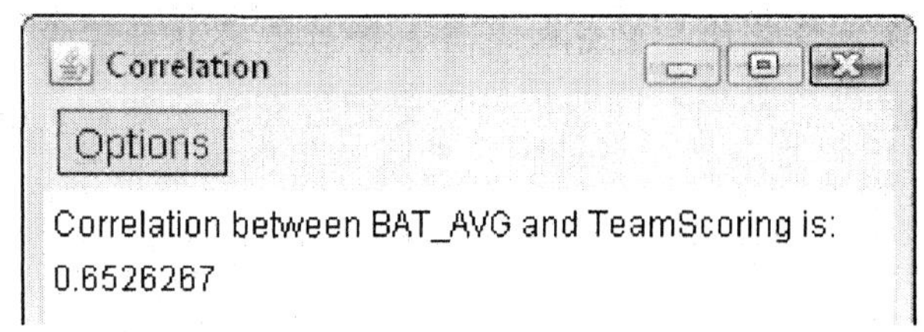

Figure 3.7b

Example 11: "What's the Gender Difference in Winning Olympic High Jumps?"

The data can be found in "high jump" on you text CD. We are interested in the relationship between the heights of winning jumps and time. As well we wish to compare these heights between males and females. Here we have three variables of interest: the year (quantitative), the height (quantitative), and gender (categorical). In order to display the data on one graph in StatCrunch we need to change how the data is entered. We need three columns, one for each variable. Unfortunately StatCrunch doesn't have a copy and paste function so we need to manually enter the data as follows (ordered by year):

StatCrunch | Data | Stat | Graphics | Help

Row	Year	Height	Gender
1	1896	1.81	male
2	1900	1.9	male
3	1904	1.8	male
4	1908	1.905	male
5	1912	1.93	male
6	1920	1.935	male
7	1924	1.98	male
8	1928	1.94	male
9	1928	1.59	female
10	1932	1.97	male
11	1932	1.657	female
12	1936	2.03	male
13	1936	1.6	female
14	1948	1.98	male

Figure 3.8a

In the Scatterplot function, select "*Year*" as the X-variable and "*Height*" as the Y-variable. In order to display the points differently for males and females, select "*gender*" in the "Group by" drop down.

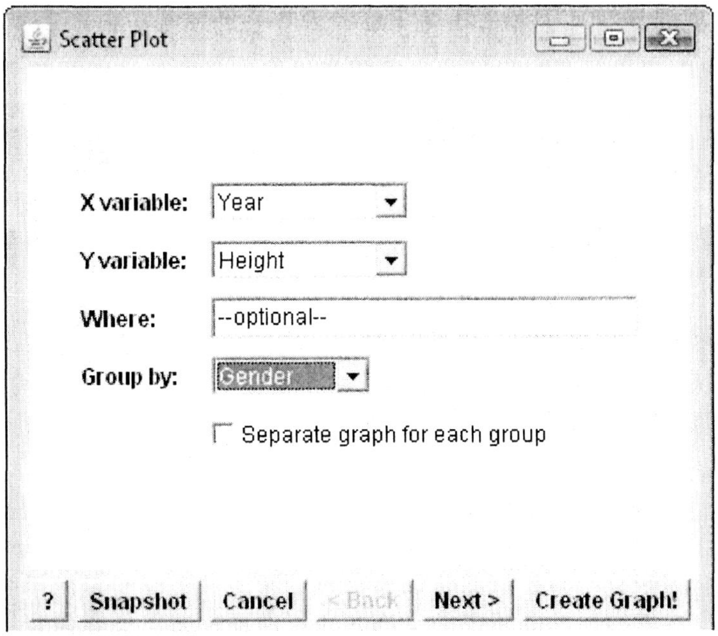

Scatter Plot

X variable: Year

Y variable: Height

Where: --optional--

Group by: Gender

☐ Separate graph for each group

? | Snapshot | Cancel | < Back | Next > | Create Graph!

Figure 3.8b

Click **Create Graph!**

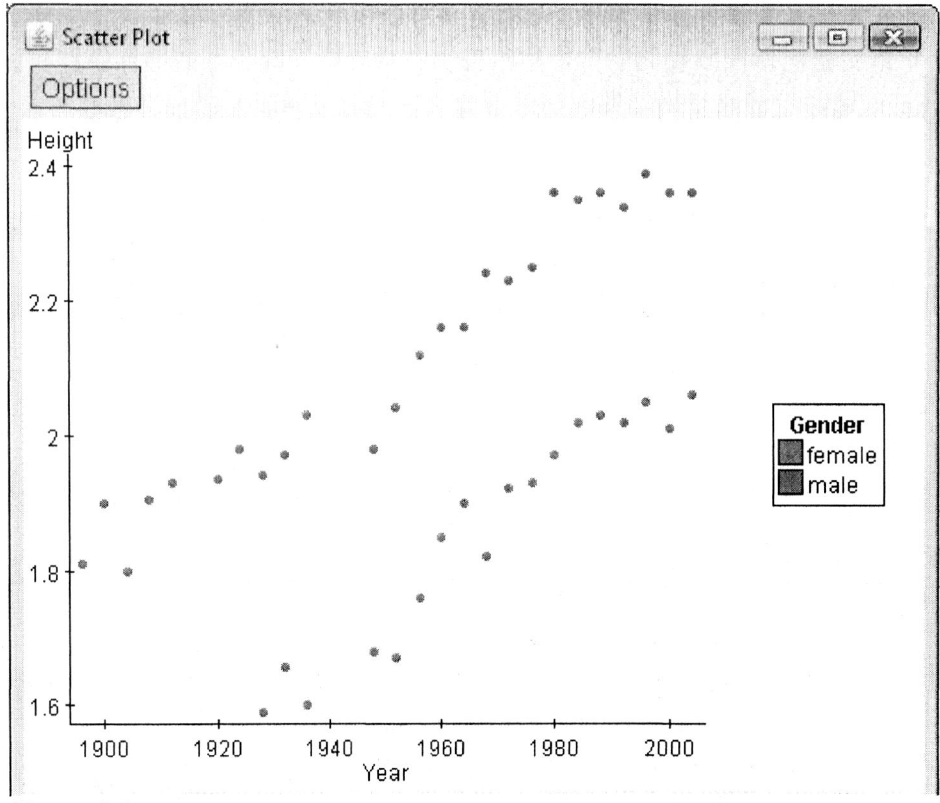

Figure 3.8c

The Scatterplot above shows a definite increasing linear trend in winning height over time. In addition it shows that men have consistently jumped higher than women for any given year. You can find the correlation between height and year separately as follows:

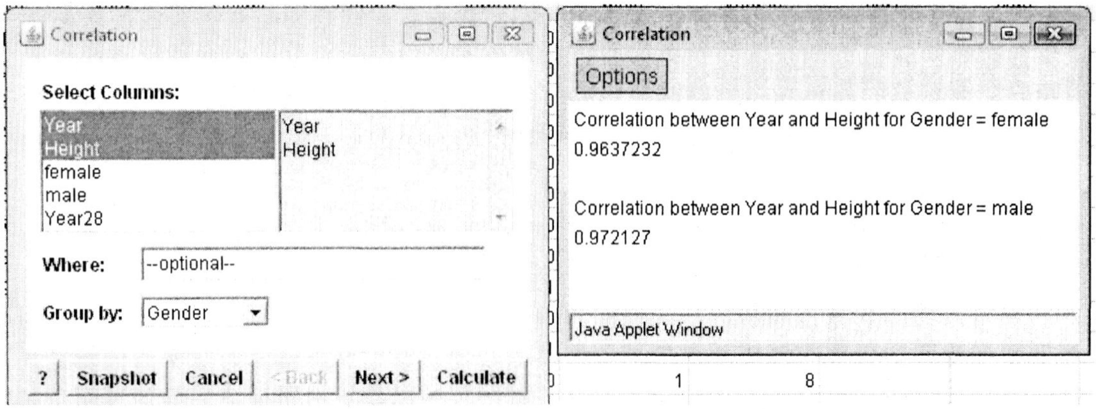

Correlation
Select Columns:
Year
Height
female
male
Year28
Year
Height
Where: --optional--
Group by: Gender
? Snapshot Cancel < Back Next > Calculate

Correlation
Options
Correlation between Year and Height for Gender = female
0.9637232
Correlation between Year and Height for Gender = male
0.972127
Java Applet Window

Figure 3.8d

Example 12: "How can we Forecast Future Global Warming?"

Open the file "central_park_yearly_temp.txt" from your CD. This example explores the trend in average annual temperature for Central Park in New York City from 1901 to 2000.

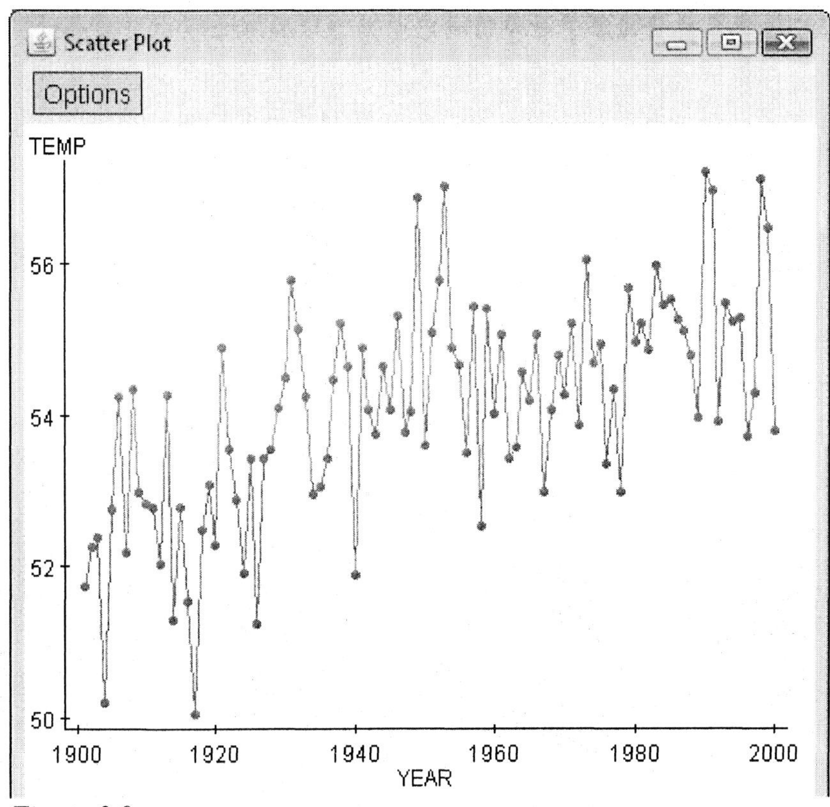

Figure 3.9a

The linear regression line is *TEMP* = 52.52 + 0.031*TIME*. The slope indicates that average annual temperature increases on average by 0.031 degrees per year. Although 0.031 degrees per year may seem insignificant, this equates to 3.1 degrees over the course of the century, which may seem significant.

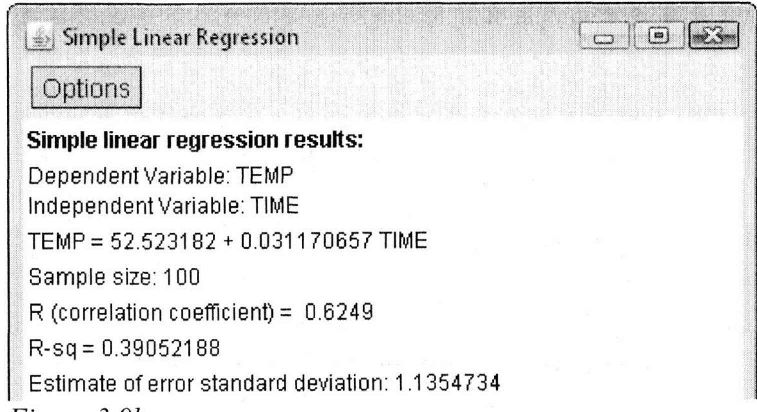

Figure 3.9b

Here is the Scatterplot with the regression line added:

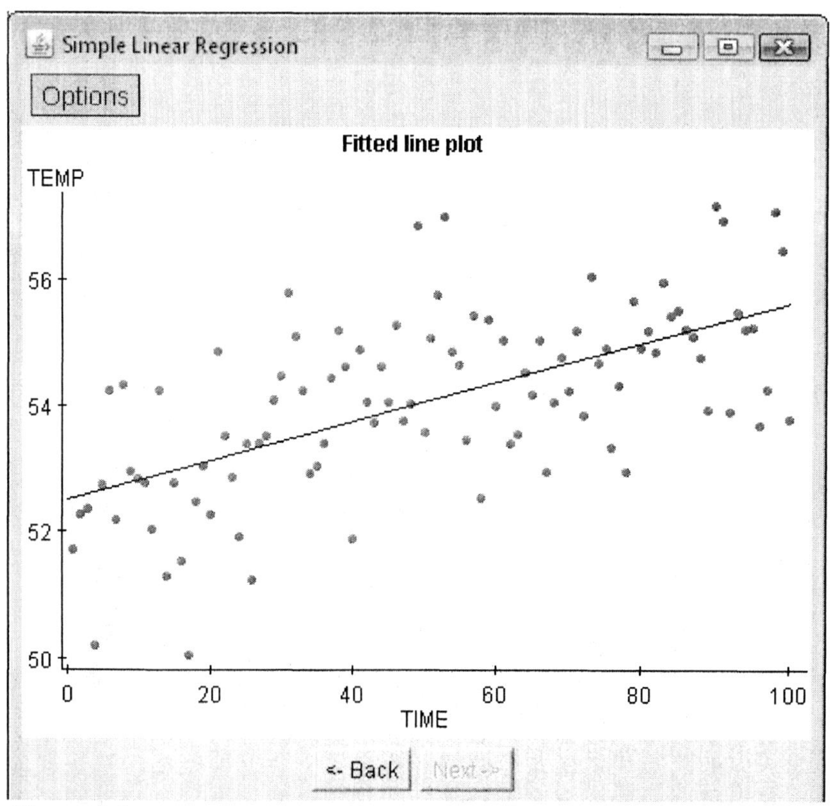

Figure 3.9c

Example 13: "Is Higher Education Associated with Higher Murder Rates?"

Open the file "us_statewide_crime" from your CD. This example wishes to explore the relationship between murder rates (annual number of murders per 100,000 people) and percentage of adult residents with a college education for the 50 U.S. states plus the District of Columbia.

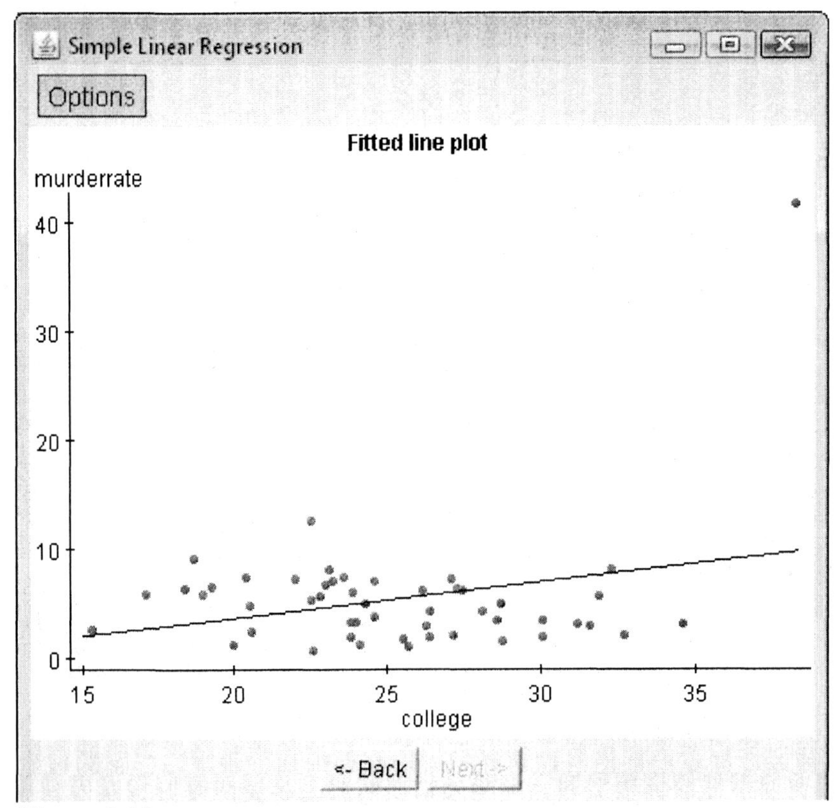

Figure 3.10a

The Scatterplot above shows one clear observation that falls outside the general trend. This is the observation for the District of Columbia. If we were to only look at the regression line

MurderRate = -3.058 + 0.33307College

We might conclude that there is a positive association between college education and murder rates. However, as we can see from the Scatterplot with the District of Columbia included a regression line is not appropriate and the resulting equation is misleading.

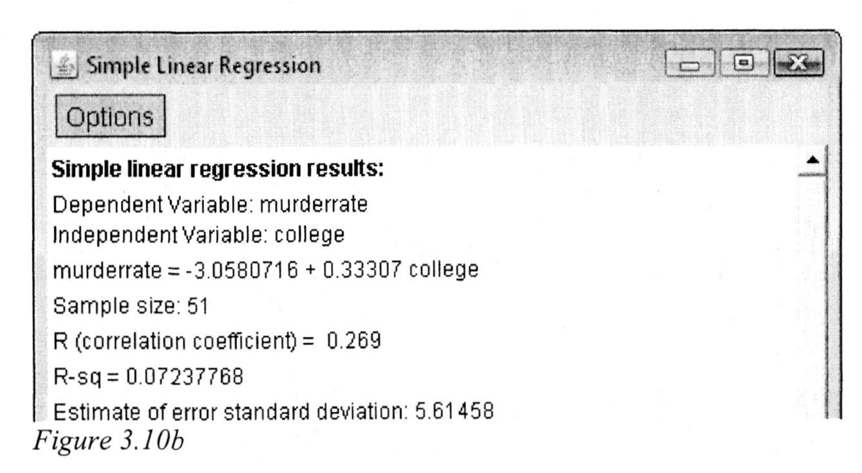

Figure 3.10b

If we want to see the regression line without the District of Columbia re-open the Regression function and run the analysis without that observation. To ignore this observation enter "State<>"District of Columbia"" in the "Where" blank. This tells StatCrunch to exclude this observation.

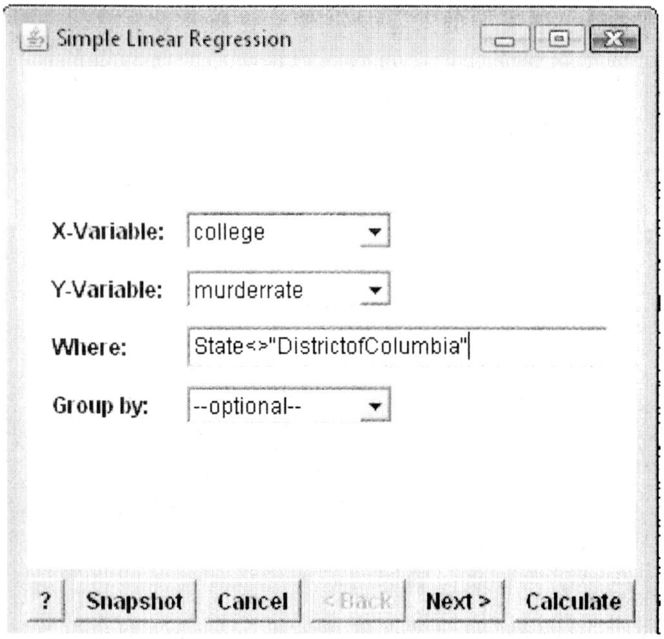

Simple Linear Regression

X-Variable: college

Y-Variable: murderrate

Where: State<>"DistrictofColumbia"

Group by: --optional--

? Snapshot Cancel < Back Next > Calculate

Figure 3.10c

The regression line for only the 50 states is

MurderRate = 8.042 – 0.138College.

This equation now shows a decreasing relationship.

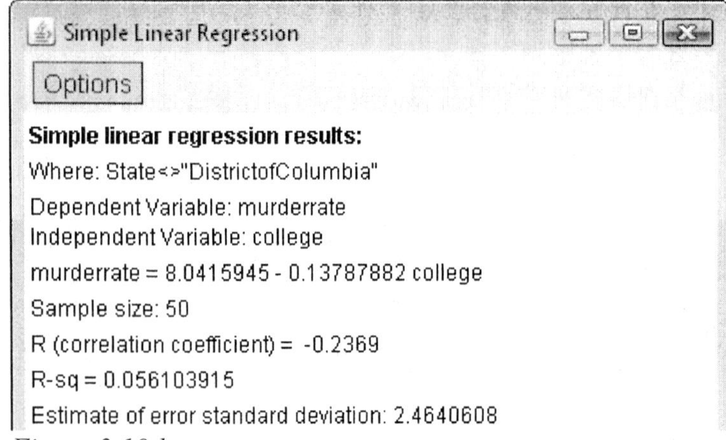

Figure 3.10d

Based on the Scatterplot below this is definitely a more appropriate overall conclusion.

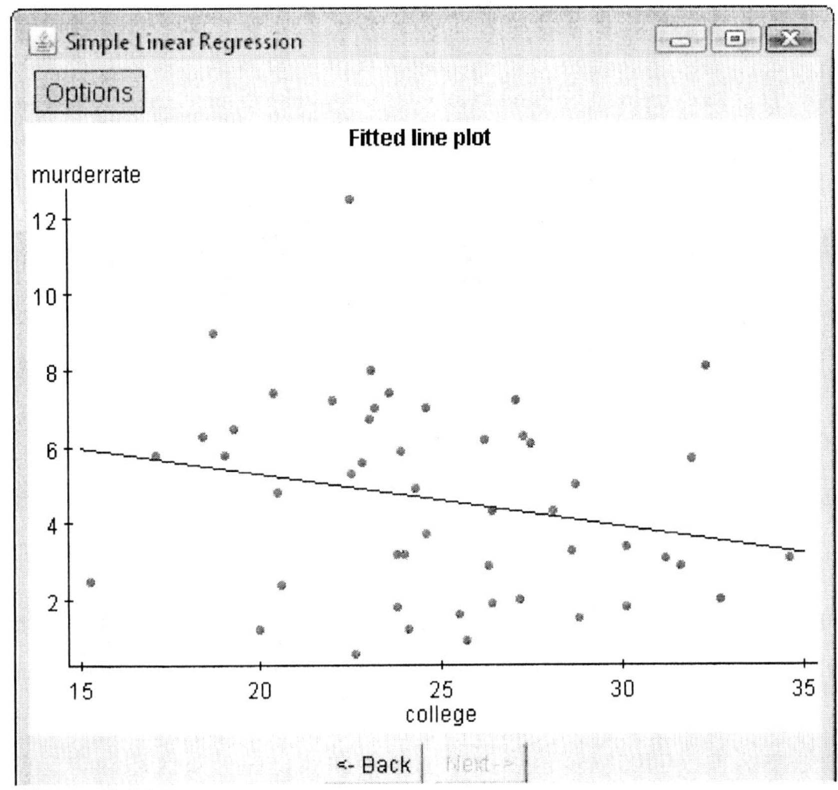

Simple Linear Regression
Options
Fitted line plot
murderrate
12
10
8
6
4
2
15
20
25
30
35
college
<- Back
Next->

Figure 3.10e

Chapter 4 - Gathering Data

This chapter is concerned with the fundamentals of experiment design and sampling data. In experiments we have to assign individuals to different treatment groups. For example, a scientist may be conducting a clinical trial to test the effectiveness of a new treatment. In this case, the scientist will need to choose some of the participants to take the new treatment and some to take a placebo. When conducting a survey we need to select a sample of individuals from a population. For example, a dataset for potential political candidate may include names of thousands or even millions of potential voters and they want to conduct a telephone poll to measure their level of support. As it is extremely time consuming (or simply impossible) to call all of them, the candidate will call only a portion of them.

How does the political candidate select who should be interviewed to obtain a representative estimate? In order to obtain a representative estimate the candidate must obtain a **random sample**.

How does the scientist choose who will be administered the new treatment and who the placebo to properly measure if the new treatment is more effective? To properly measure the effectiveness of the new drug and to draw a cause-and-effect conclusion the scientist must **randomly assign** individuals to the different treatment groups.

Although these tasks can be accomplished with the use of a random number table (usually found in the appendix of most statistics textbooks), this chapter will illustrate how StatCrunch can do the work for you!

Please refer to the dataset "fla_student_survey" on your text CD for illustrations.

Random Sampling

Sample Columns

Although there are different ways to choose a random sample from a list of individuals or observations, StatCrunch has a function that will do the work for you. Select **Data→Sample Columns.**

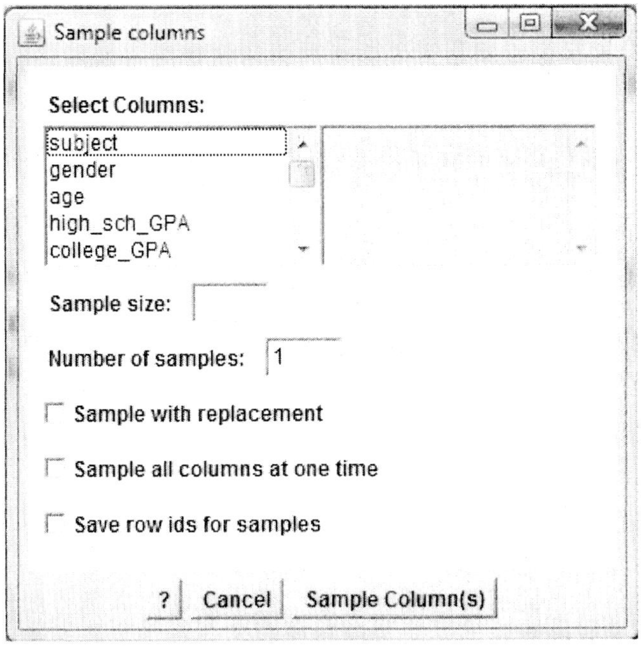

Sample columns

Select Columns:

subject
gender
age
high_sch_GPA
college_GPA

Sample size:

Number of samples: 1

Sample with replacement

Sample all columns at one time

Save row ids for samples

? Cancel Sample Column(s)

Figure 4.1a

This function allows you to select random observations from one or more columns. Suppose we want to select a sample of 10 subjects from the list. Since the first column is the subject number we want to sample this column. Click on "*subject*". Enter 10 as the sample size. We want 10 distinct individuals so leave the "Sample with replacement" unchecked. Since we are only sampling from one column and the "*subject*" column corresponds to the row id we can leave the last two boxes unchecked.

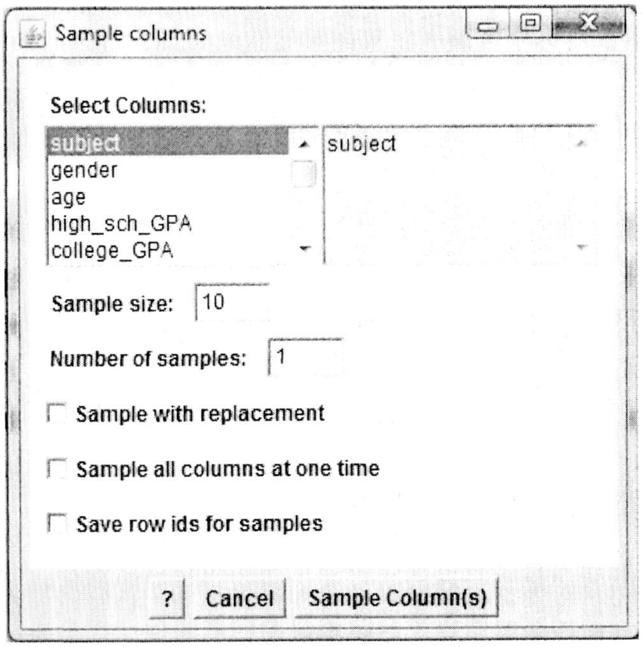

Sample columns

Select Columns:

subject
gender
age
high_sch_GPA
college_GPA

subject

Sample size: 10

Number of samples: 1

Sample with replacement

Sample all columns at one time

Save row ids for samples

? Cancel Sample Column(s)

Figure 4.1b

Click **Sample Column(s).** The list of subjects that were chosen will appear in a new column titled "*Sample(subject)*".

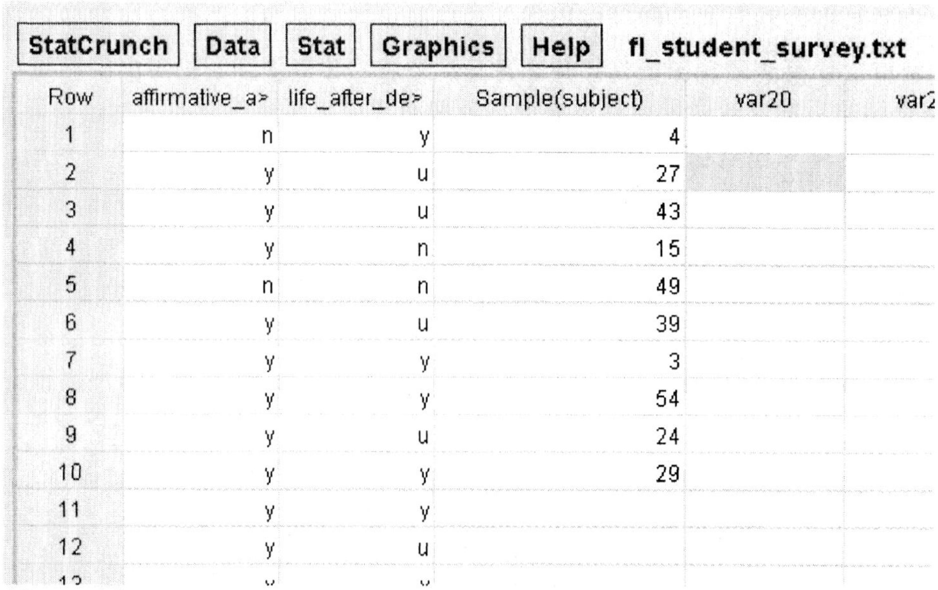

StatCrunch | Data | Stat | Graphics | Help | **fl_student_survey.txt**

Row	affirmative_a>	life_after_de>	Sample(subject)	var20	var2
1	n	y	4		
2	y	u	27		
3	y	u	43		
4	y	n	15		
5	n	n	49		
6	y	u	39		
7	y	y	3		
8	y	y	54		
9	y	u	24		
10	y	y	29		
11	y	y			
12	y	u			

Figure 4.1c

This time StatCrunch chose subjects (or observations) 4, 27, 43, 15, 49, 39, 3, 54, 24 and 29. The purpose of this illustration was to simply select the individuals. If you wanted to analyze data from your sample you could select all the columns to be sample and click "Sample all columns at one time". This will display the data from all columns for the subjects selected. Then, use the tools from the rest of this manual to analyze the data.

Random Number Generation

Another technique that can be used to choose a random sample is to assign a random number to each individual. Then according to some (pre-determined) rule about these random numbers you can select your sample. For example, suppose we want to select 10 observations from the dataset. You can ask StatCrunch to randomly generate numbers for each individual. Then an easy way to select your sample is to choose the subjects corresponding to lowest (or equivalently, highest) 10 numbers.

Select **Data→Simulate data.** This menu lists many different ways to generate numbers according to different probability distributions. The most common distribution used in simulation is the Uniform distribution with lower bound 0 and upper bound 1. This is commonly referred to as the Uniform(0,1) distribution. This distribution function will randomly generate numbers between 0 and 1. Click on **Uniform.**

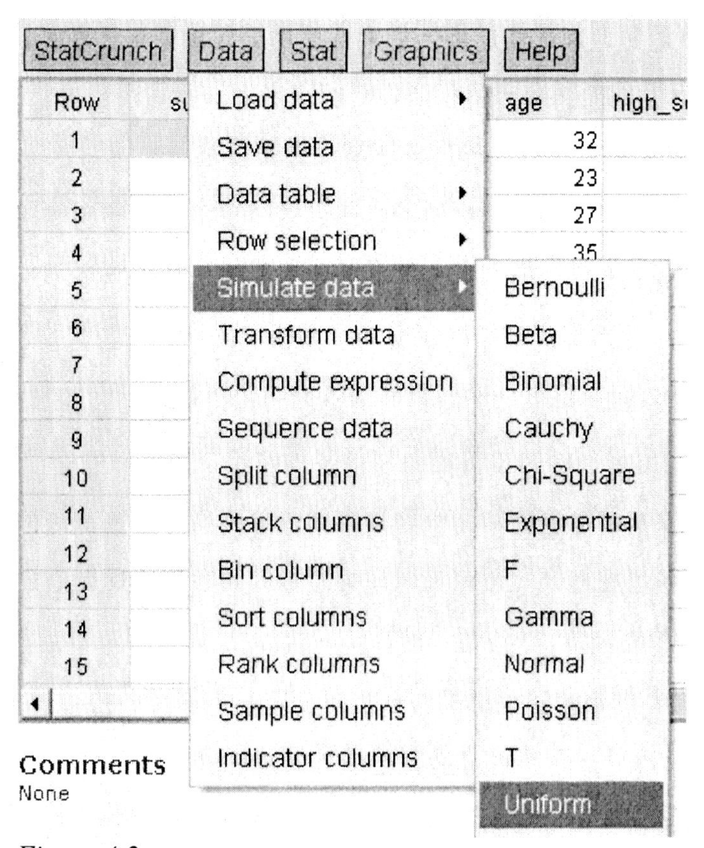

Figure 4.2a

The blank "Rows" corresponds to the number or random numbers to be generated. We want a number generated for each of the 60 individuals, so enter 60 in "Rows". We only want one list of numbers so enter 1 in "Columns". The "a" refers to the lower bound of the distribution, which we want to be 0. The "b" refers to the upper bound, which we want to be 1. Then you can choose your generator type. The default is a preset generator that is determined by two seeds when you open StatCrunch. A dynamic generator chooses the seed using the clock on your computer. Otherwise you can fix the seed yourself. You can experiment with these later, for now just leave this as the default. Click **Simulate.**

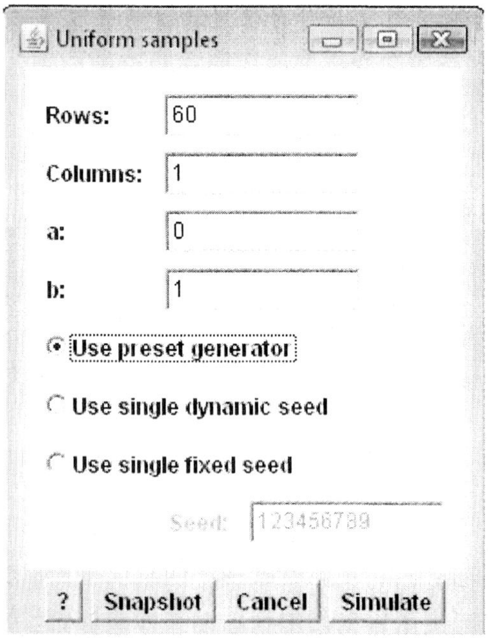

Uniform samples

Rows: 60

Columns: 1

a: 0

b: 1

Use preset generator

Use single dynamic seed

Use single fixed seed

Seed: 123456789

? Snapshot Cancel Simulate

Figure 4.2b

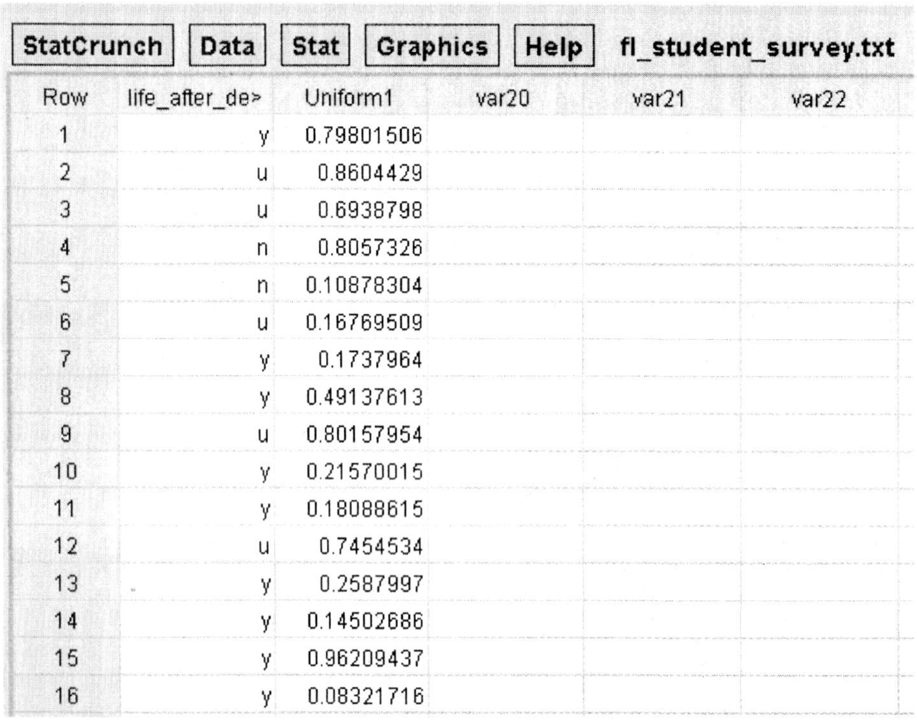

StatCrunch Data Stat Graphics Help fl_student_survey.txt

Row	life_after_de>	Uniform1	var20	var21	var22
1	y	0.79801506			
2	u	0.8604429			
3	u	0.6938798			
4	n	0.8057326			
5	n	0.10878304			
6	u	0.16769509			
7	y	0.1737964			
8	y	0.49137613			
9	u	0.80157954			
10	y	0.21570015			
11	y	0.18088615			
12	u	0.7454534			
13	y	0.2587997			
14	y	0.14502686			
15	y	0.96209437			
16	y	0.08321716			

Figure 4.2c

The function will create another column titled "*Uniform1*". If you scroll down you'll see that all numbers are between 0 and 1. In order to select you sample of 10 find the subjects that correspond to the smallest 10 numbers in "*Uniform1*". To speed up the process you can use the "Sort columns" or "Rank columns" function.

Although using random number generation is a bit more tedious than the Sample Columns function, it has other benefits. One benefit is that it provides a way to randomly rank the individuals. This is beneficial if you are choosing a sample of individuals to interview, say in a political poll. Suppose you want to obtain the opinions of exactly 10 individuals. In telephone polls it is common for people to not respond because they either do not answer or simply hang up. In these cases you'll need to obtain another observation to make your quota. Since each individual can be ranked according to his or her random number you can simply select the next lowest numbered individual to be sample.

Random Assignment

Like sampling there are many different ways to randomly assign individuals to different treatment groups. One of the easiest ways to assign individuals to two treatment groups is to simply flip a coin for each individual. This method will allocate approximately 50% of the individuals to each group. If you must have an equal number in each group then you can simply make up slips of paper where half say "1" and the other half say "2". Then for each individual you can draw a number out of a hat and assign them to that treatment group. Although trivial, these types of techniques (and many other creative techniques) are random and legitimate. However, if you have a large sample of individuals they can be tedious. The good news is that we've already discussed how StatCrunch can help. Regardless of how many treatment groups we have or whether we want exact numbers or approximate proportions allocated to each group, we can do this using our Uniform random numbers.

Assigning Exact Numbers to Each Group

Suppose we want to randomly assign the 60 individuals to three treatment groups with exactly 20 in each group. One method to accomplish this is to generate a Uniform(0,1) random number for each individual. Then, rank the individuals on order of their number. Select the lowest 20 for the first group, the next lowest 20 for the second group and the highest 20 for the third group. You're done!

Assigning Approximate Proportions to Each Group

Suppose we want to randomly assign to 60 individuals to two groups where each individual has a 50/50 chance of being in either group. Again, generate Uniform(0,1) random numbers for each individual. Then (by the properties of the Uniform(0,1) distribution) assign an individual to group 1 if their number is between 0 and just below 0.5 and to group 2 if their number is between 0.5 and 1. This will ensure approximately 50% in each group and is the equivalent to flipping a coin for each individual.

If you had 4 groups then assign those with Uniform(0,1) random numbers between 0 and just below 0.25 to group 1, between 0.25 and just below 0.5 to group 2, between 0.5 and just below 0.75 to group 3 and between 0.75 and 1 to group 4.

Chapter 6 - Probability Distributions

This chapter will discuss how to calculate probabilities from a few common probability distributions, notably the continuous Normal distribution and the discrete Binomial distribution. Although tables are available for some of these calculations it is also possible to calculate these probabilities using StatCrunch. The use of these calculators will be illustrated using several examples in your text.

Normal Calculations

The Normal Probability calculator can be found by selecting **Stat→Calculators→Normal.**

Example 7: "What IQ Do You Need to Get into MENSA?"

In order to qualify as a member of MENSA your IQ must be in the 98-th percentile. It is given that IQ scores follow approximately a Normal distribution with mean of 100 and standard deviation of 16. It is convenient to denote this distribution as $N(100, 16)$.

How many standard deviations above the mean is the 98th percentile?

In order to find the number of standard deviation above the mean that correspond to the 98th percentile we can refer to the Standard Normal Distribution (the Normal distribution with mean 0 and standard deviation 1). Open the **Normal calculator**. Enter 0 as the Mean and 1 as the Standard Deviation, and 0.98 as displayed. Select "<=" to indicate that the left tail probability should be 0.98. Click **Compute.**

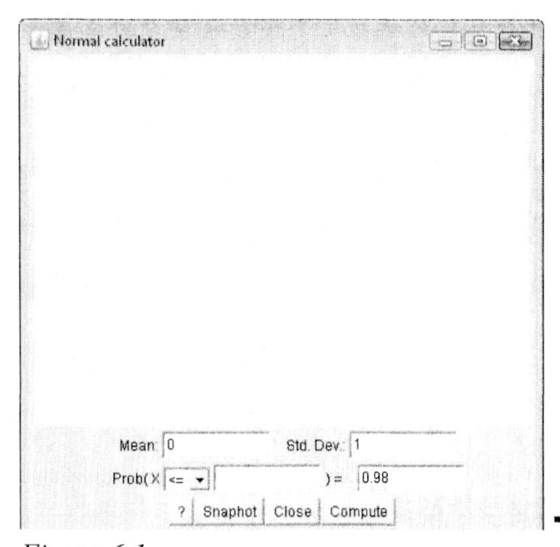

Figure 6.1a

→

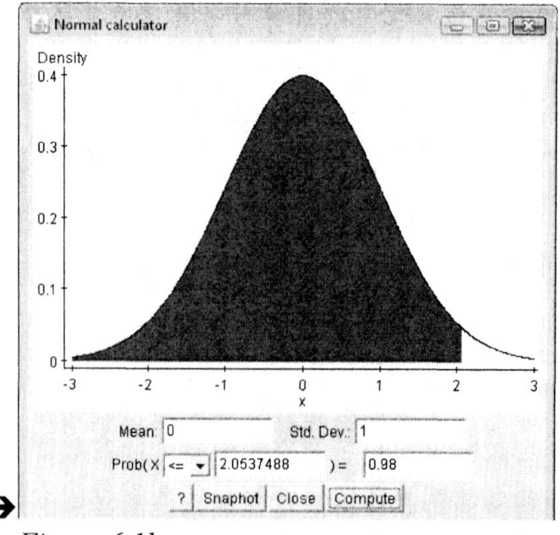

Figure 6.1b

The diagram should then appear as above indicating that the 98th percentile on the standard normal distribution is 2.054. Or, equivalently the 98th percentile falls about 2.054 standard deviations above the mean on any normal distribution.

What is the IQ score for the 98th percentile?

From part (a) we can find this value directly as 100+2.054(16)=132.864. Otherwise we can use the normal probability calculator as follows:

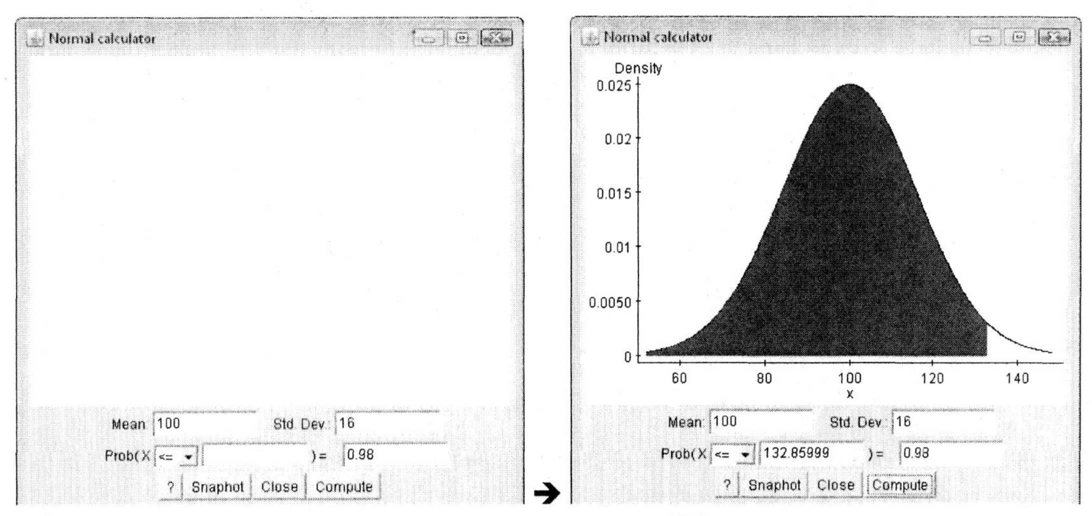

Figure 6.2a *Figure 6.2b*

The calculator tells us that you would need an IQ of 132.86 to fall in the 98th percentile. The difference between the two calculations is simply due to slight rounding error in the first.

Example 8: "Finding Your Relative Standing on the SAT"

The scores on the three components of the SAT (Critical Reading, Mathematics and Writing) each deviate approximately according to the Normal distribution with mean of 500 and standard deviation of 100. Scores from any component range from 200 to 800.

If one of your SAT scores was 650, how many standard deviations from the mean was it?

This can be found directly by standardizing the value of 650, $\frac{(650-500)}{100}=1.5$. Or, the score of 650 is 1.5 standard deviations above the mean.

What percentage of SAT scores was higher than yours?

Open the **Normal calculator**. You need to find the probability that a $N(500,100)$ random variable takes on a value greater (or equal) than 650. Fill in the following on the left and click **Compute.**

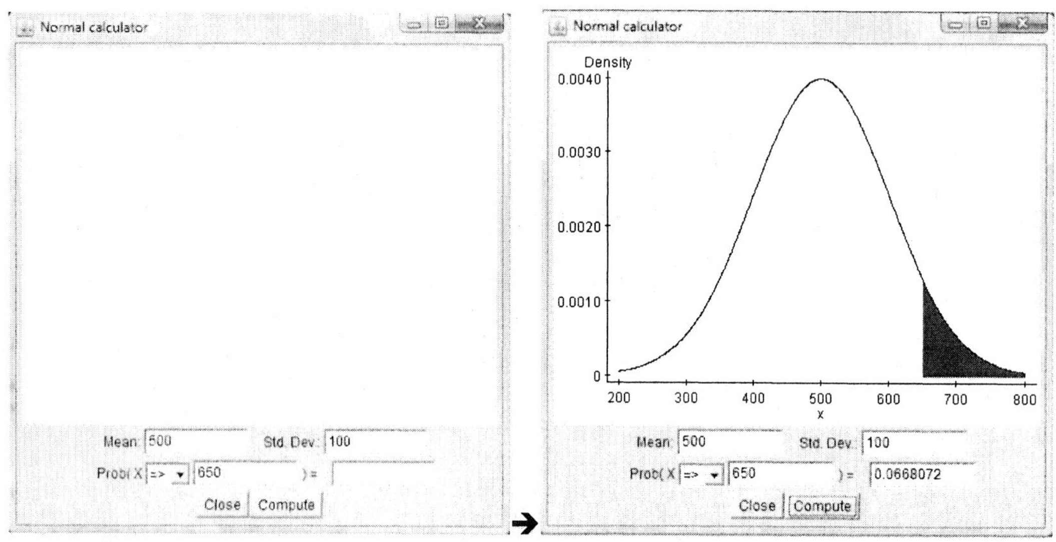

Figure 6.3a *Figure 6.3b*

The right panel indicates that about 6.68% of the SAT scores are above 650. Or equivalently, the score of 650 is in the 93.32th percentile (1 – 0.0668 = 0.9332).

Example 9: "What Proportion of Students Get a Grade of B?"

An instructor in introductory statistics always gives a grade of B to those who score between 80 and 90. If the scores deviate approximately according to a $N(83,5)$ probability distribution, what proportion will receive a B?

If we let X denote a midterm score then the probability of scoring between 80 and 90 can be expressed as:

$$P(80 \leq X \leq 90) = P(X \leq 90) - P(X \leq 80)$$

We can use the **Normal calculator** to find $P(X \leq 90)$ and $P(X \leq 80)$ as follows:

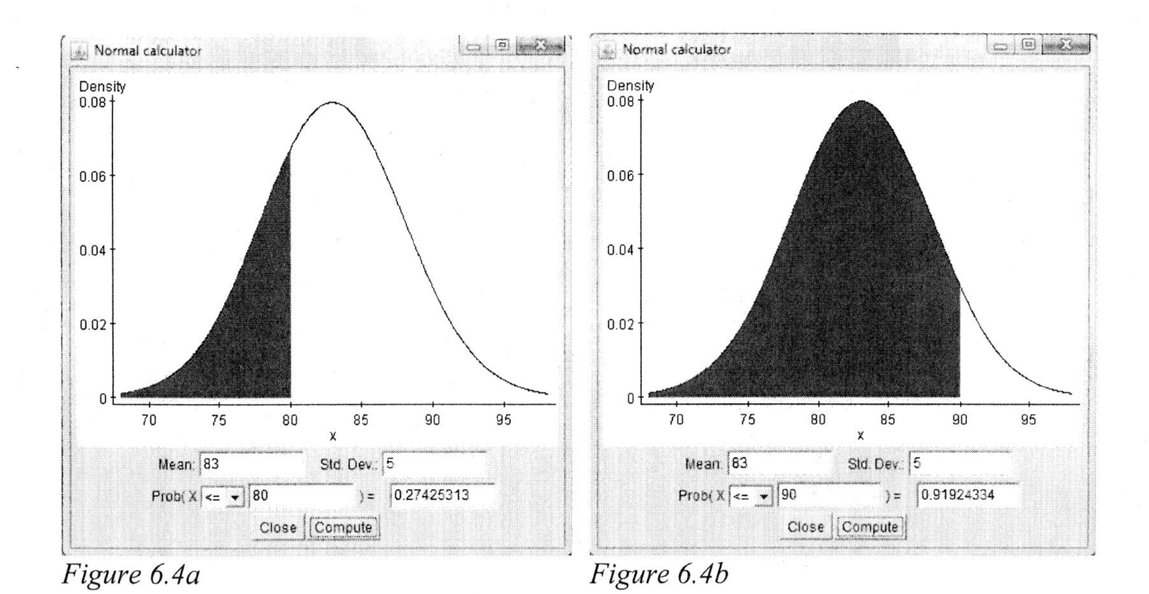

Figure 6.4a *Figure 6.4b*

From Figure 6.4a we find that $P(X \leq 80) = 0.2743$. From Figure 6.4b we find that $P(X \leq 90) = 0.9192$. Thus, the probability of scoring between 80 and 90 is 0.9192 – 0.2743 = 0.6449. About 64.5% of the students will get a B.

Binomial Calculation

The Binomial Probability calculator can be found by selecting **Stat➔Calculators➔Binomial.**

Example 12: "Are Women Passed Over for Managerial Training?'

Out of a large employee pool of more than a thousand individuals (about half male and half female with similar qualifications) eligible for management training 10 were selected. Out of the 10 selected, none were female.

What is the probability of selecting 0 females if there was in fact no gender bias?

If we let X denote the number of females selected, we are interested in $P(X = 0)$. In the Binomial calculator select n=10 (the sample size) and p=0.5 (the probability of randomly selecting a female). Then accordingly find the "=" sign in the drop down and enter 0 in the next blank. Click **Compute.**

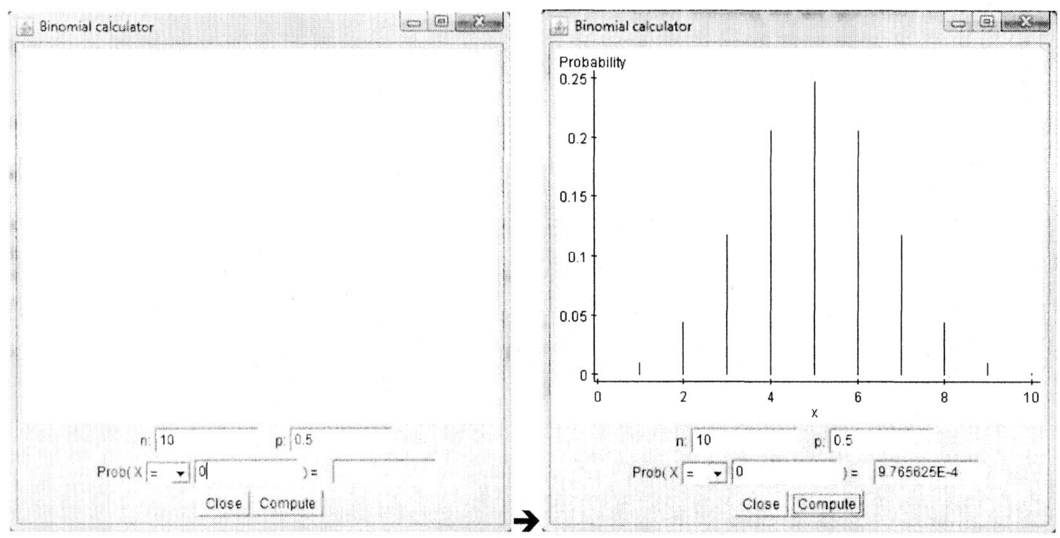

Figure 6.5a *Figure 6.5b*

The result in Figure 6.5 b shows that $P(X=0)$=0.000977. Or equivalently if they selected 10 individuals from a large sample with roughly the same number of males and females there is about a 1 in 1000 chance (actually 1 in 1024) that no females would be selected.

The next part is not in you text.

What is the probability that at least 2 females are selected?

Here we find that $P(X \geq 2)$=0.9893.

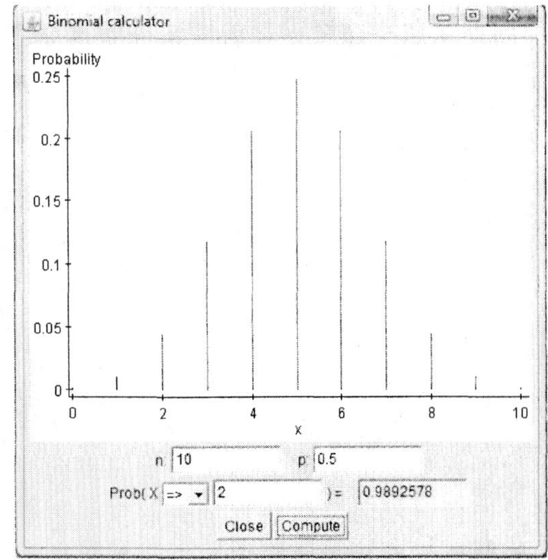

Figure 6.5c

Equivalently $P(X \geq 2) = 1 - P(X < 2)$ where below we find that $P(X < 2)$=0.0107. Thus, $P(X \geq 2) = 1 - P(X < 2)$ = 1 – 0.0107 = 0.9893.

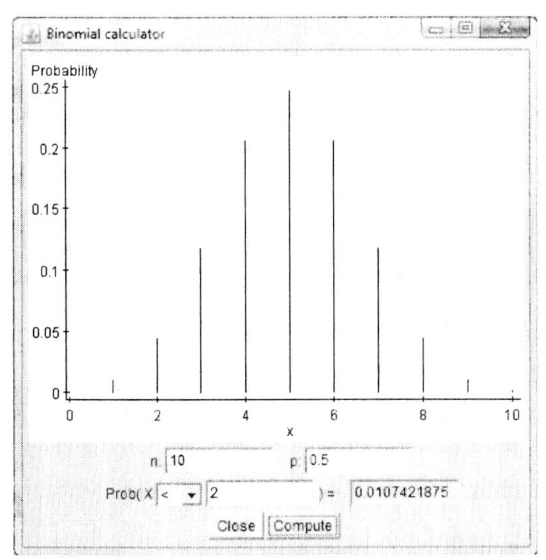

Figure 6.5d

Example 14: "How Can We Check for Racial Profiling?"

A study conducted by the American Civil Liberties Union analyzed whether Africa-American drivers were more likely than others to be targeted by police for traffic stops. The study showed that out of 262 stops during one week in 1997, 207 of the drivers were African-American. During that time, 42.2% of the population in that area were African-American.

Does the number of African-Americans stopped suggest possible bias? To answer this question we can calculate the probability that out of the 262 stops, at least 207 were African-American if in fact there was a probability of 0.422 that any stopped driver was African-America.

If we let X denote the number of African-American stopped, we are interested in $P(X \geq 207)$ where X is a Binomial random variable with n = 262 and p = 0.422.

From the Binomial calculator below we find that $P(X \geq 207)$ is roughly 0.

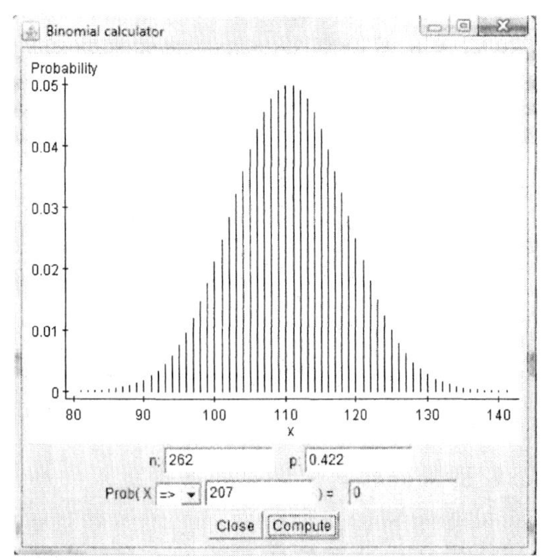

Figure 6.6a

Normal Approximation

We can approximate the probability above as $P(Y \geq 207)$ where Y is approximately a normal distribution with mean = np = 111 and standard deviation = $\sqrt{np(1-p)}$ = 8. Below we see again that this probability is virtually 0.

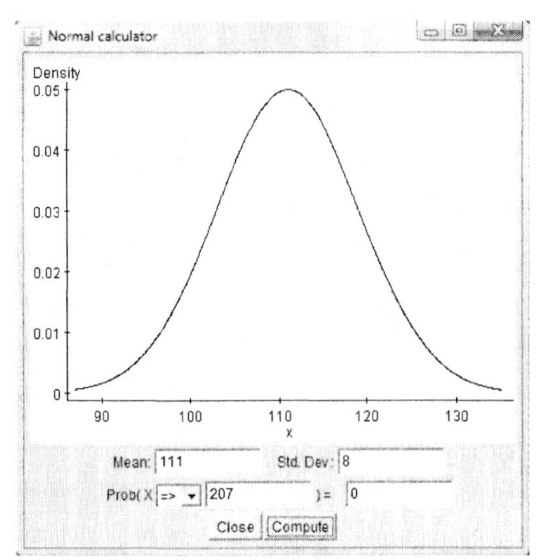

Figure 6.6b

Chapter 8 - Statistical Inference: Confidence Intervals

In this chapter we will illustrate how to obtain confidence intervals for a single proportion and a single mean using StatCrunch.

Confidence Intervals for a Population Proportion

With Summary

StatCrunch can calculate confidence intervals for a single proportion with only two pieces of information, the number of "successes" and the total number of observation. Select **Stat➔Proportion➔One sample➔with summary.**

Screen 1 **Screen 2**

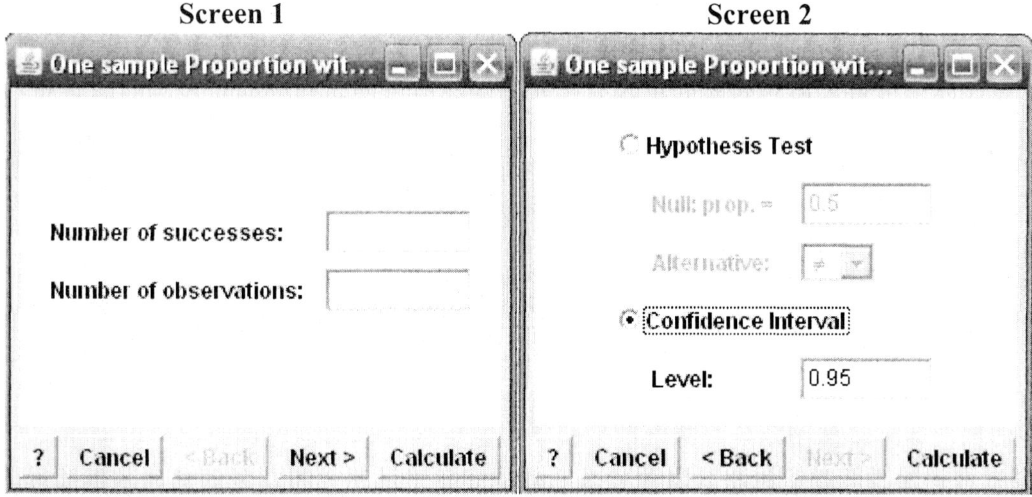

Figure 8.1a *Figure 8.1b*

In **Screen 1** enter the number of successes and the total number of observations in the blanks. In **Screen 2**, select "Confidence Interval" and type in the desired confidence level (the default is 0.95 for a 95% confidence interval). If you want a 99% confidence interval then change the level to 0.99. Once everything is specified, click **Calculate** and StatCrunch will calculate your confidence interval.

Example 2: "Should a Wife Sacrifice Her Career for Her Husbands?"

One question on the General Social Survey asked whether or not respondents agreed with the following statement: "It is more important for a wife to help her husband's career than to have one herself." The example states that 19% of the 1823 respondents agreed. In order for StatCrunch to calculate a confidence interval for the population proportion with that opinion we need to specify the number of respondents that agreed. The exact number isn't given in the question. However, 19% of 1823 (0.19*1823) is about 346. In **Screen 1** enter 346 as the number of successes and 1823 as the number of observations. Click **Next** to go to **Screen 2** and select "Confidence Interval" button. Make sure 0.95 is entered as the "level". Click **Calculate.** You should obtain the following:

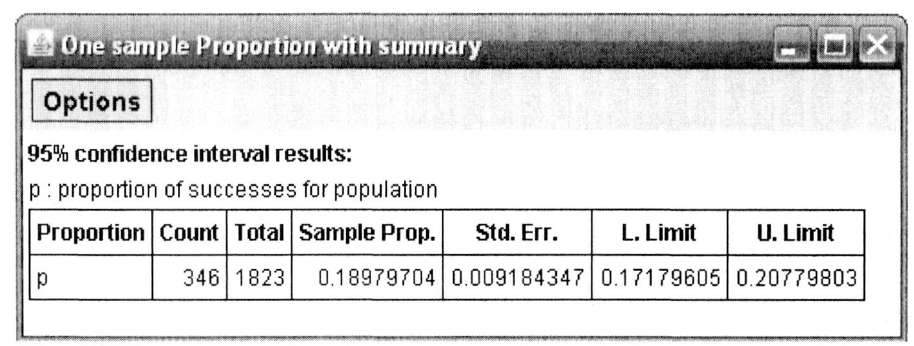
One sample Proportion with summary

Options

95% confidence interval results:

p : proportion of successes for population

Proportion	Count	Total	Sample Prop.	Std. Err.	L. Limit	U. Limit
p	346	1823	0.18979704	0.009184347	0.17179605	0.20779803

Figure 8.2

The output above states that the lower limit of the confidence interval is 0.1718 and that the upper limit is 0.2078. Thus, the 95% C.I. for the proportion of respondents that agreed with the statement is about (0.17, 0.21). It also states that the sample proportion is 0.1898 and that the standard error is 0.0092.

Example 3: "Would You Pay Higher Prices to Protect the Environment?"

In 2000, the GSS found that 518 out of 1154 said they would pay higher prices to protect the environment. From the output below the 95% C.I. for the proportion that said they would is (0.42, 0.48). The estimate is 0.45 with a standard error of 0.015.

One sample Proportion with summary

Options

95% confidence interval results:

p : proportion of successes for population

Proportion	Count	Total	Sample Prop.	Std. Err.	L. Limit	U. Limit
p	518	1154	0.4488735	0.014641471	0.4201767	0.47757024

Figure 8.3

Example 5: "Should a Husband Be Forced to Have Children?"

Out of 598 respondents, 366 said that it is all right for a husband to refuse children even if the wife wants them. From the output below the 99% C.I. for the proportion that said it is all right is (0.56, 0.66). The estimate is 0.61 with a standard error of 0.02.

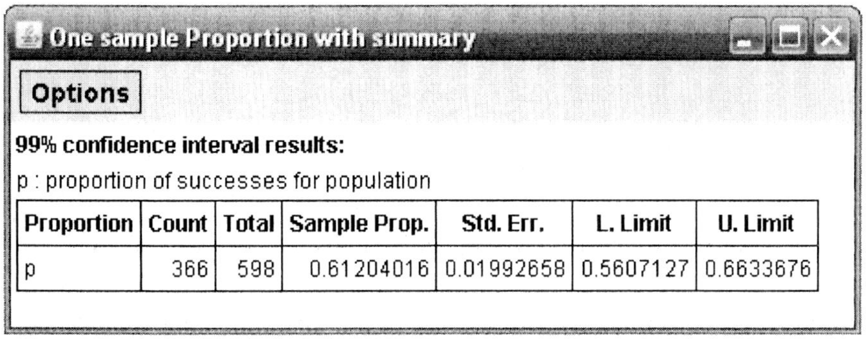

One sample Proportion with summary

Options

99% confidence interval results:

p : proportion of successes for population

Proportion	Count	Total	Sample Prop.	Std. Err.	L. Limit	U. Limit
p	366	598	0.61204016	0.01992658	0.5607127	0.6633676

Figure 8.4

With Data

StatCrunch can also find confidence intervals for a proportion from raw data.

Example: Recall the data from the Florida Student Survey, "fla_student_survey".

Let's calculate a 90% confidence interval for the proportion of Florida Students that say they are Republican. Open the dataset and select **Stat➔Proportion➔One sample➔with data.** In **Screen 1** choose the variable "*political_affiliation*". Then identify what constitutes a success. Here it is "r" for republican. Click **Next** to go to **Screen 2.**

Screen 1

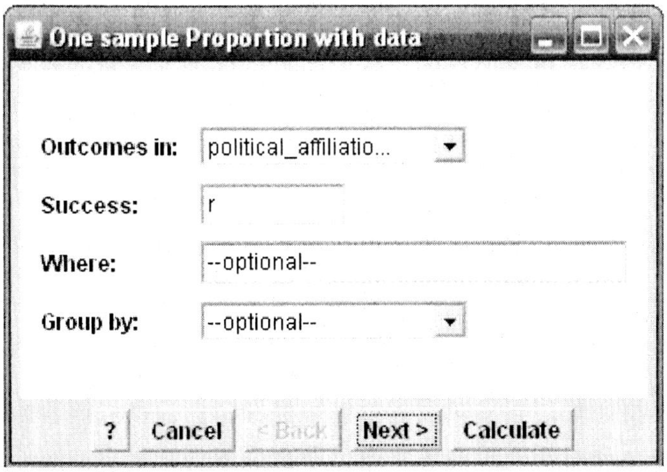

One sample Proportion with data

Outcomes in: political_affiliatio...

Success: r

Where: --optional--

Group by: --optional--

? Cancel < Back Next > Calculate

Figure 8.5a

In **Screen 2,** select "Confidence Interval" and change the confidence level to 0.9. Click **Calculate.** The output below states that 15 out the 60 students identified themselves as Republican (you can check this by going and counting them). The sample proportion is 0.25 and the standard error is 0.0559. A 90% C.I. for the proportion of Florida Students that are Republican is then (0.158, 0.342).

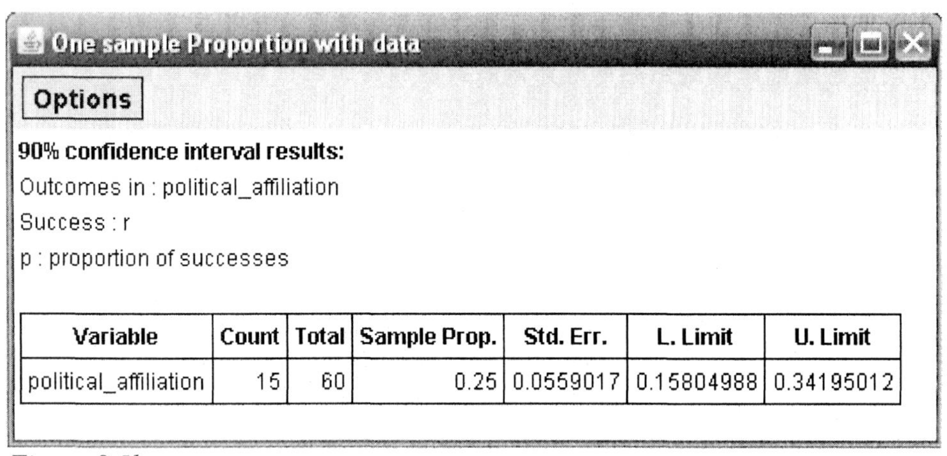

One sample Proportion with data

Options

90% confidence interval results:

Outcomes in : political_affiliation

Success : r

p : proportion of successes

Variable	Count	Total	Sample Prop.	Std. Err.	L. Limit	U. Limit
political_affiliation	15	60	0.25	0.0559017	0.15804988	0.34195012

Figure 8.5b

Confidence Interval for a Population Mean

There are two functions in StatCrunch that calculate confidence intervals for single population means. One uses the Z-distribution (**Stat➔Z statistics➔One sample)** and the other uses the T-distribution (**Stat➔T statistics➔One sample)**. Both functions are similar except the Z statistics requires that you know the population standard deviation. We will go over the T statistics confidence interval.

With Summary

StatCrunch can calculate confidence intervals for a single mean with only three pieces of information, the sample mean, the sample standard deviation and the sample size. Select **Stat➔T statistics➔One sample➔with summary.**

Screen 1

One sample T statistics with s...

Sample mean:

Sample std. dev.:

Sample size:

? | Snapshot | Cancel | < Back | Next >

Figure 8.6a

In **Screen 1** enter the sample mean, the sample standard deviation and the sample size. In **Screen 2**, select "Confidence Interval" and type in the desired confidence level (the default is

0.95 for a 95% confidence interval). If you want a 99% confidence interval then change the level to 0.99. Once everything is specified, click **Calculate** and StatCrunch will calculate your confidence interval.

Example 6: "On the Average, How Much Television Do You Watch?"

You are given that out of a sample of 899 individuals, the average time in a day they watch television is 2.865 hours. The sample standard deviation is also given to 2.617. To find a 95% Confidence Interval for the average time, open **Stat→T statistics→One sample→with summary** and fill in **Screen 1** as follows:

One sample T statistics with summary

Sample mean: 2.865

Sample std. dev.: 2.617

Sample size: 899

? Snapshot Cancel < Back Next > Calculate

Figure 8.6b

Click **Next** to go to **Screen 2.** In **Screen 2** select "Confidence Interval" with level 0.95 and click **Calculate**.

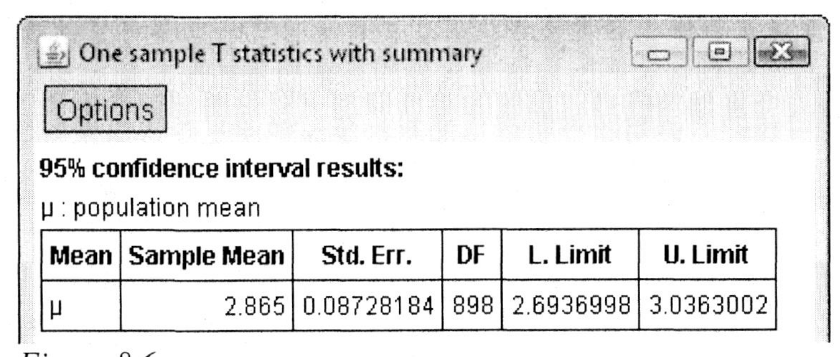
One sample T statistics with summary

Options

95% confidence interval results:

μ : population mean

Mean	Sample Mean	Std. Err.	DF	L. Limit	U. Limit
μ	2.865	0.08728184	898	2.6936998	3.0363002

Figure 8.6c

The output above shows that the 95% Confidence Interval is (2.69, 3.04).

With Data

Example 7: "eBay Auctions of Palm Handled Computers"

This example looks at the selling prices for a certain type of PDA using differing selling options, "buy-it-now" options and "bidding" option. First, either load the data from the "ebay_auctions" file on your CD or enter the data as it appears in your text. Enter the data into StatCrunch in two separate columns, one for "*Buy-it-now*" results and one for "*Bidding only*" results.

StatCrunch | Data | Stat | Graphic

Row	Buy-it-now	Bidding
1	235	250
2	225	249
3	225	255
4	240	200
5	250	199
6	250	240
7	210	228
8		255
9		232
10		246
11		210
12		178

Figure 8.7a

Open the one sample T statistics C.I. by selecting **Stat➔T statistics➔One sample➔with Data.** To obtain a C.I. for the mean of each sample select both variables by clicking them on the left. Click **Next.**

Screen 1

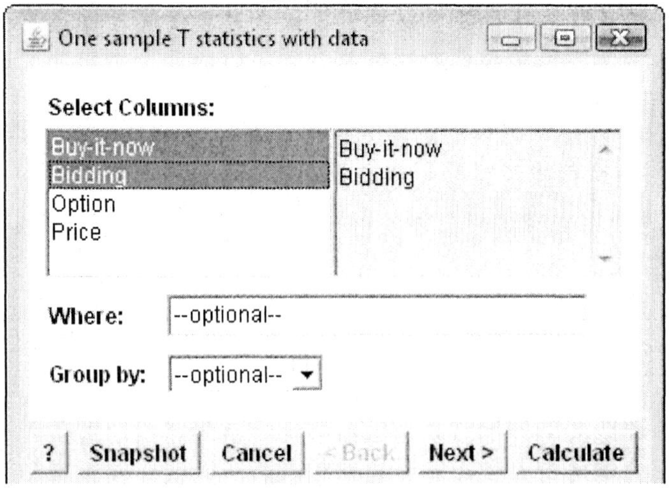

One sample T statistics with data

Select Columns:

Buy-it-now
Bidding
Option
Price

Buy-it-now
Bidding

Where: --optional--

Group by: --optional--

? | Snapshot | Cancel | < Back | Next > | Calculate

Figure 8.7b

In **Screen 2** select "Confidence Interval" and then type in the confidence level. The example asks for 95% confidence intervals so type in 0.95 for the level. Click **Calculate.**

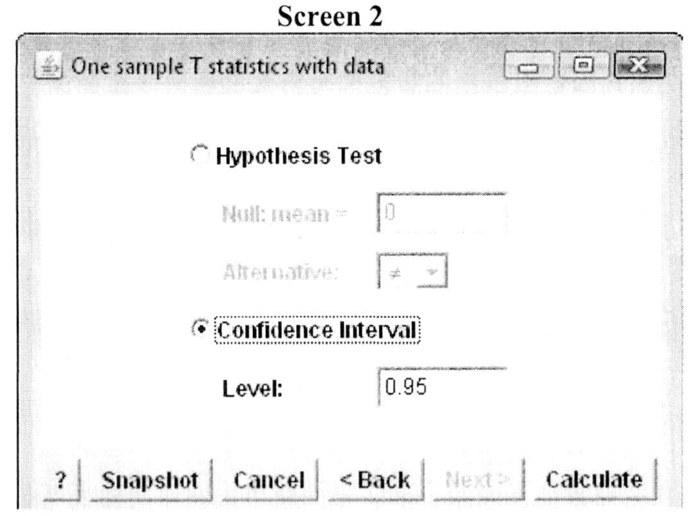

Figure 8.7c

One sample T statistics with data

Options

95% confidence interval results:

μ : mean of Variable

Variable	Sample Mean	Std. Err.	DF	L. Limit	U. Limit
Buy-it-now	233.57143	5.5328336	6	220.03308	247.10979
Bidding	231.61111	5.17039	17	220.70255	242.51968

Figure 8.7d

The table above shows the results for both samples. The 95% confidence interval for the average selling price using the "buy-it-now" option is ($220.03, $247.11). The 95% confidence interval for the average selling price using the "bidding" option is ($220.70, $242.52).

Alternate Data Entry

Another common way to enter data of this type is to enter all the prices in one column and enter the selling options in another. In the table below option 1 represents the selling prices for the "buy-it-now" option and option 2 represents the selling prices for the "bidding" option.

Option	Price
1	235
1	225
1	225
1	240
1	250
1	250
1	210
2	250
2	249
2	255
2	200
2	199
2	240

Figure 8.8a

Again select **Stat➔T statistics➔One sample➔with data.** Now choose the variable "*Price*" and group by "*Option*". In **Screen 2** select "Confidence Interval" and then type in 0.95 for the confidence level. Click **Calculate.**

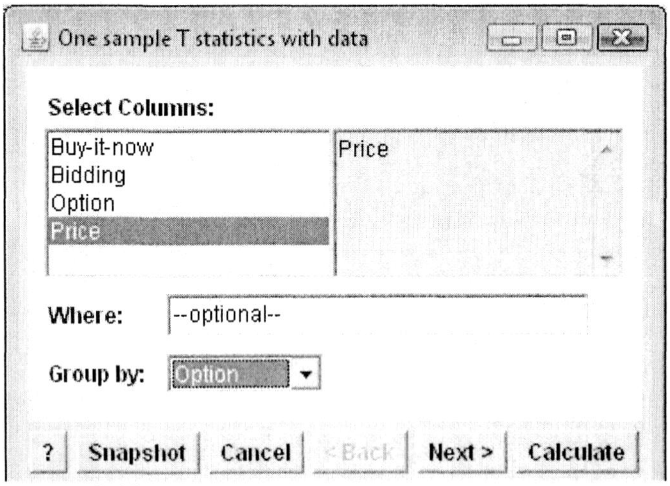

One sample T statistics with data

Select Columns:

Buy-it-now
Bidding
Option
Price

Price

Where: --optional--

Group by: Option

? Snapshot Cancel < Back Next > Calculate

Figure 8.8b

The following table lists the results for both groups and as you can see they are the same as before.

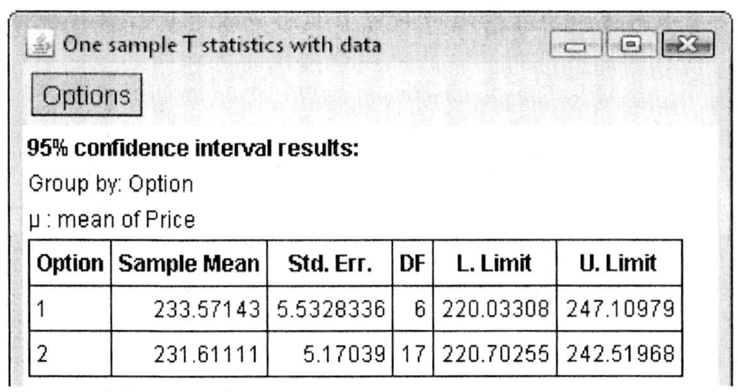
One sample T statistics with data

Options

95% confidence interval results:

Group by: Option

μ : mean of Price

Option	Sample Mean	Std. Err.	DF	L. Limit	U. Limit
1	233.57143	5.5328336	6	220.03308	247.10979
2	231.61111	5.17039	17	220.70255	242.51968

Figure 8.8c

Chapter 9 - Statistical Inference: Significance Tests about Hypothesis

In this chapter we will illustrate how to carry out hypothesis tests for a single proportion and a single mean using StatCrunch.

Hypothesis Test for a Population Proportion

With Summary

Just like confidence intervals for a single proportion StatCrunch can carry out hypothesis tests for a single proportion with only two pieces of information, the number of "successes" and the total number of observation. Select **Stat➔Proportion➔One sample➔with summary.**

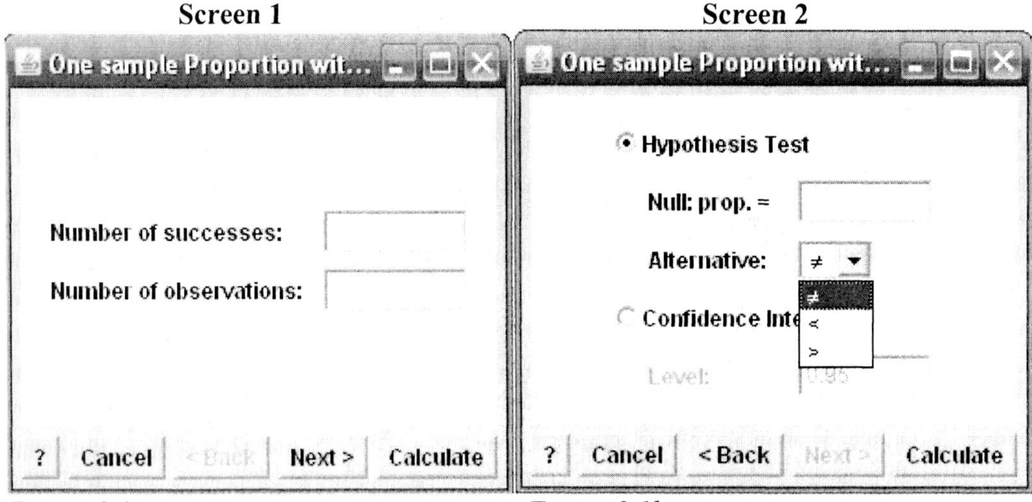

Figure 9.1a *Figure 9.1b*

In **Screen 1** enter the number of successes and the total number of observations in the blanks. In **Screen 2** select "Hypothesis Test", type in the desired value under the Null Hypothesis, and select the type of alternative. You can choose from a two-tailed alternative (not equal or "≠"), a left-tailed alternative (less than or "<"), or a right-tailed alternative (greater than or ">"). Once everything is specified, click **Calculate** and StatCrunch will calculate the results (estimate, standard error, test-statistic and p-value).

Examples 1 and 3: "Are Astrologers' Predictions Better than Guessing?"

Examples 1 and 3 describe a study that attempt to test whether an astrologers predictions are any better than pure guessing. Each prediction involved choosing the individual (out of three) that matched the horoscope they were given. For more information read Examples 1 and 3 in your text. Random guessing would lead to about 1 out of 3 correct predictions. Out of the 116 replications, the astrologers were correct in 40 of their predictions. Does this imply that astrologers' predictions are better than guessing?

In **Screen 1** enter 40 as the number of successes and 116 as the number of observations. Click **next** to go to **Screen 2** and select "Hypothesis Test". Under the null hypothesis we want the value that would be true if they we're simply guessing, so 1/3 or 0.33333 (you have to enter this in decimal form). If the astrologers' predictions were better than pure guessing then the proportion of correct predictions would be greater than 1/3. In which case choose the right-tailed alternative (greater than or ">"). Click **Calculate.** You should obtain the following:

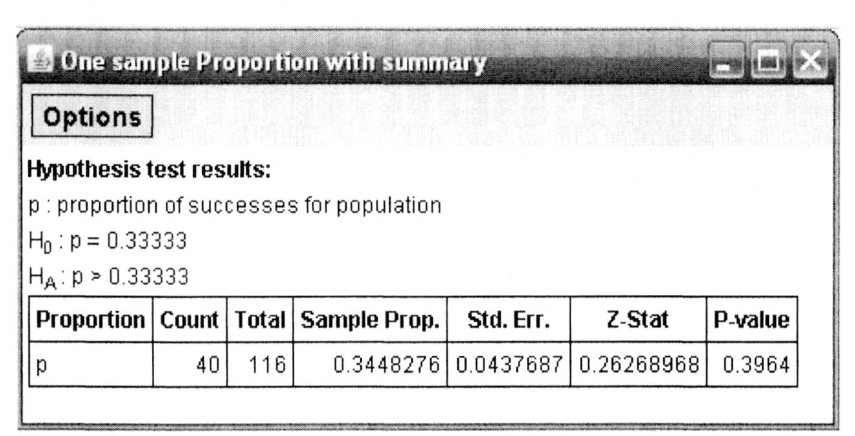
One sample Proportion with summary

Options

Hypothesis test results:

p : proportion of successes for population

$H_0 : p = 0.33333$

$H_A : p > 0.33333$

Proportion	Count	Total	Sample Prop.	Std. Err.	Z-Stat	P-value
p	40	116	0.3448276	0.0437687	0.26268968	0.3964

Figure 9.2

The output above states that the sample proportion of correct predictions is 0.3448 and the standard error is 0.0438. The test-statistic is 0.262 and the p-value is 0.3964 (rounded to 0.40 in your text).

Example 4: "Dr. Dog: Can Dogs Detect Cancer by Smell?"

In a total of 54 trials (using 6 dogs of different breeds), the dogs chose correctly 22 times. Each time the dog was to choose 1 urine sample from 7. Based on pure guessing the dogs would be correct 1 out of 7 times.

In **Screen 1** enter 22 as the number of successes and 54 as the number of observations. Click **next** to go to **Screen 2** and select "Hypothesis Test". Under the null hypothesis we want the value that would be true if they we're simply guessing, so 1/7 or 0.14286. If the dogs' predictions were better or worse than pure guessing then the proportion of correct predictions would not equal to 1/7. In which case choose the two-tailed alternative (not equal to or "≠"). Click **Calculate.** You should obtain the following:

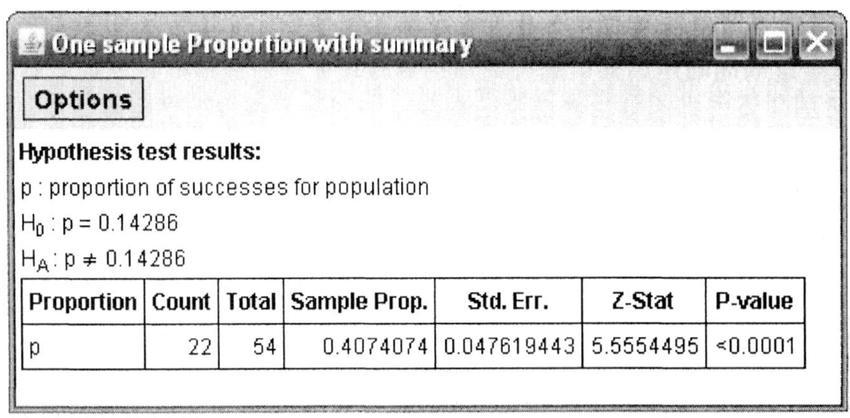
One sample Proportion with summary

Options

Hypothesis test results:

p : proportion of successes for population

$H_0 : p = 0.14286$

$H_A : p \neq 0.14286$

Proportion	Count	Total	Sample Prop.	Std. Err.	Z-Stat	P-value
p	22	54	0.4074074	0.047619443	5.5554495	<0.0001

Figure 9.3

The output above states that the sample proportion of correct predictions is 0.4074 and the standard error is 0.0476. The test-statistic is 5.555 and the p-value is less than 0.0001.

With Data

StatCrunch can also carry out hypothesis tests for a single proportion from raw data.

Example: Recall the data from the Florida Student Survey, "fla_student_survey".

Suppose there is a claim that the population proportion of Republicans is less than 30%. Do the results in the survey indicate that less than 30% of the students are Republican?

Open the dataset and select **Stat→Proportion→One sample→with data.** In **Screen 1** below choose the variable "*political_affiliation*". Then identify what constitutes a success. Here it is "r" for Republican. In **Screen 2,** select "Hypothesis Test" and type in 0.3 (30%) for the "Null:prop." (This is the null hypothesized value). We are interested in assessing the evidence that less than 30% of the students are Republicans so select the left-tailed (less than or "<") alternative. Click **Calculate.**

Screen 2

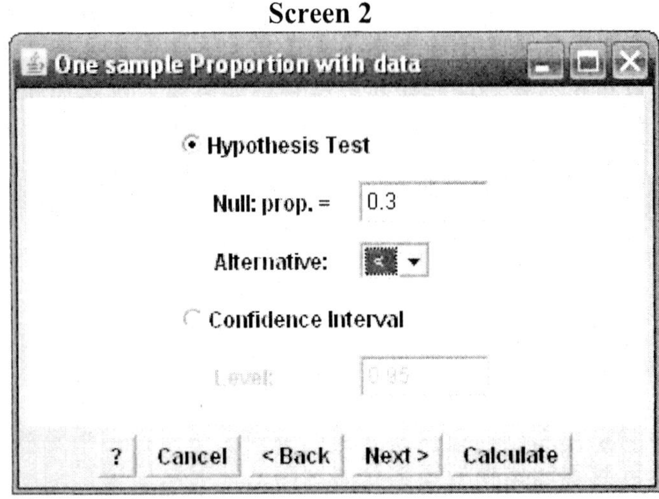

Figure 9.4a

The output below states that 15 out the 60 students identified themselves as Republican (you can check this by going and counting them). The sample proportion is 0.25 and the standard error is 0.0592. The test-statistic is -0.845 and the p-value is 0.199.

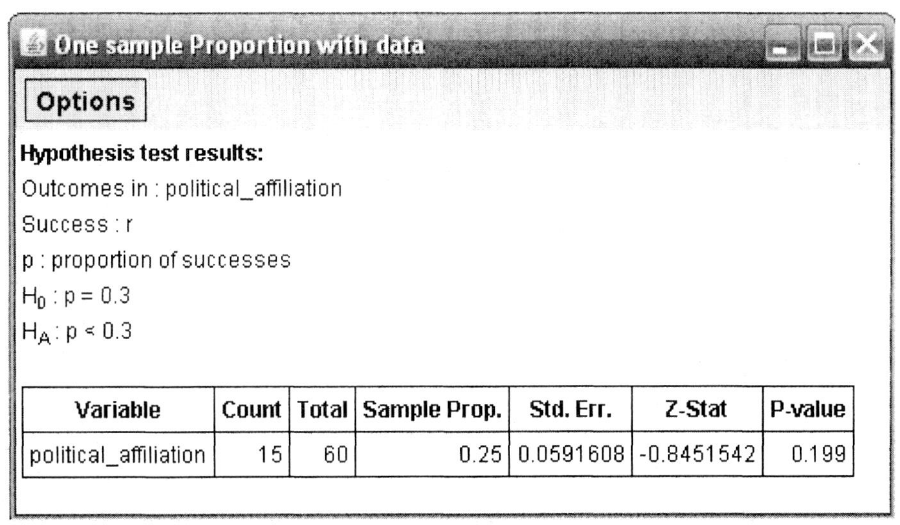
One sample Proportion with data

Options

Hypothesis test results:

Outcomes in : political_affiliation

Success : r

p : proportion of successes

$H_0 : p = 0.3$

$H_A : p < 0.3$

Variable	Count	Total	Sample Prop.	Std. Err.	Z-Stat	P-value
political_affiliation	15	60	0.25	0.0591608	-0.8451542	0.199

Figure 9.4b

Hypothesis Test for a Population Mean

Similar to Confidence Interval for population means, there are two functions in StatCrunch that carry out Hypothesis Tests for single population means. One uses the Z-distribution (**Stat➔Z statistics➔One sample**) and the other uses the T-distribution (**Stat➔T statistics➔One sample**). Again we will go over the T statistics hypothesis test only.

With Summary

StatCrunch can calculate confidence intervals for a single mean with only three pieces of information, the sample mean, the sample standard deviation and the sample size. Select **Stat➔T statistics➔One sample➔with summary.**

Screen 1

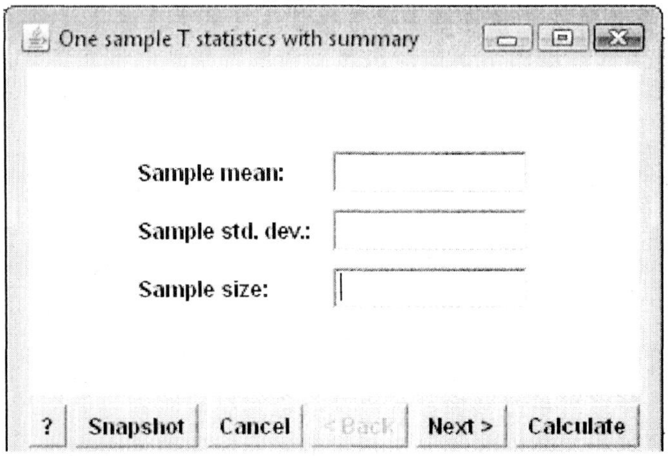

One sample T statistics with summary

Sample mean:

Sample std. dev.:

Sample size:

? Snapshot Cancel < Back Next > Calculate

Figure 9.5a

In **Screen 1** enter the sample mean, the sample standard deviation and the sample size. In **Screen 2** select "Hypothesis Test", type in the desired value under the Null Hypothesis, and select the type of alternative. Once everything is specified, click **Calculate** and StatCrunch will carry out your test.

Example 7, Exercise 9.31: "Do Americans Work a 40 Hour Week?"

From a sample of 895 men, their average workweek reported is 45.3 hours with a standard deviation of 14.8. Is this evidence that the workweek for men isn't 40 hours? To carry out this test, open **Stat→T statistics→One sample→with summary** and fill in **Screen 1** as follows:

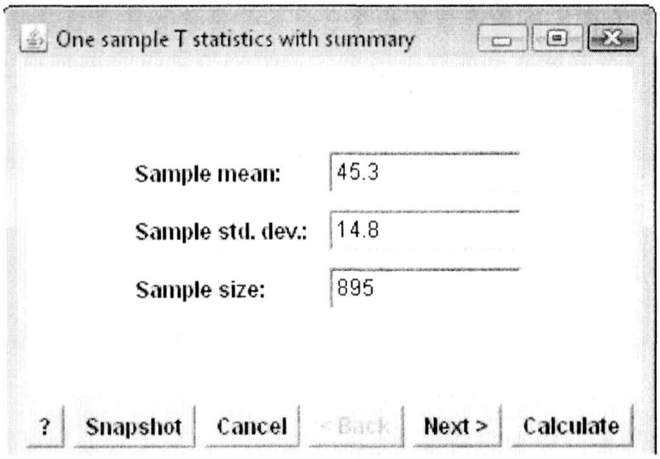

One sample T statistics with summary

Sample mean: 45.3

Sample std. dev.: 14.8

Sample size: 895

? Snapshot Cancel < Back Next > Calculate

Figure 9.5b

Click **Next** to go to **Screen 2.** In **Screen 2** select "Hypothesis Test" with "Null: mean" equal to 40. Select the "not equal" alternative and click **Calculate**.

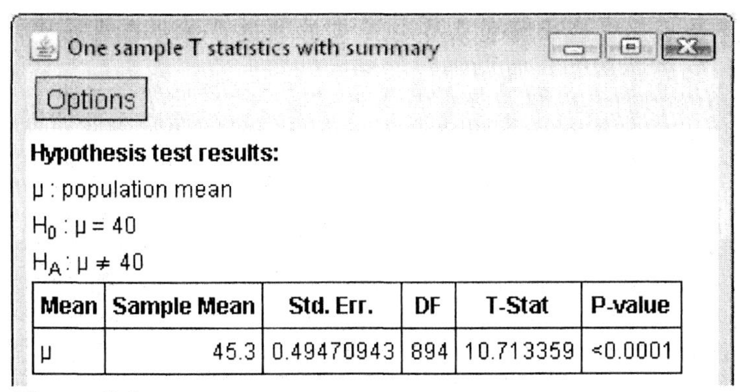
One sample T statistics with summary

Options

Hypothesis test results:

μ : population mean

$H_0 : \mu = 40$

$H_A : \mu \neq 40$

Mean	Sample Mean	Std. Err.	DF	T-Stat	P-value
μ	45.3	0.49470943	894	10.713359	<0.0001

Figure 9.5c

The output above shows a test statistic of 10.7 with a p-value less than 0.0001.

With Data

Example 9: "Testing Whether Software Operates Properly"

Earlier we showed how StatCrunch could randomly generate numbers. Let's test whether StatCrunch does this properly. Let's **Simulate** (see Chapter 4) 100 Uniform random numbers between 0 and 1.

Uniform samples

Rows: 100

Columns: 1

a: 0

b: 1

Use preset generator

Use single dynamic seed

Use single fixed seed

Seed: 123456789

? Cancel Simulate

Figure 9.6a

If StatCrunch were doing this properly we'd expect our numbers to vary uniformly from 0 to 1. The average should be 0.5. Let's test whether the average is any different from 0.5.

Open the one sample T-statistics C.I. by selecting **Stat→T statistics→One sample→with data.** In **Screen 1** select the variable "*Uniform1*".

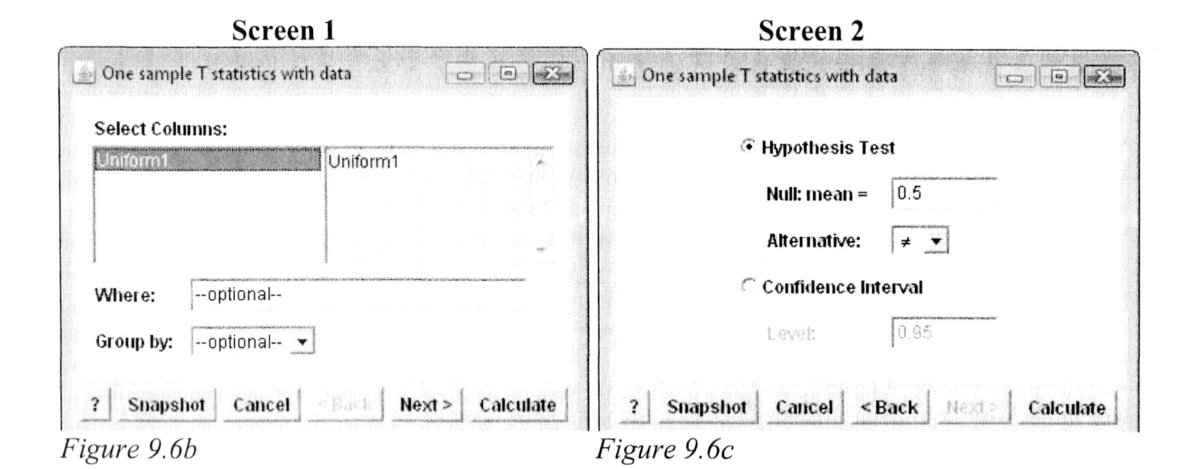

Figure 9.6b *Figure 9.6c*

In **Screen 2** select "Hypothesis Test" and type in 0.5 as the null hypothesized value. Select the two-sided alternative and click **Calculate.**

The output below shows that the average value generated was 0.469 with a standard error of 0.0297. The test-statistic is -1.06 with a p-value of 0.2917. The degrees of freedom are 99 (n-1). Based on this p-value there is very little evidence to indicate that StatCrunch is not generating uniform number properly.

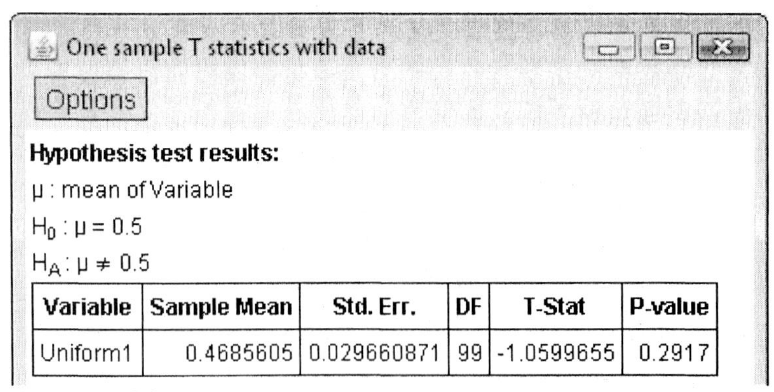
One sample T statistics with data

Options

Hypothesis test results:

μ : mean of Variable

$H_0 : \mu = 0.5$

$H_A : \mu \neq 0.5$

Variable	Sample Mean	Std. Err.	DF	T-Stat	P-value
Uniform1	0.4685605	0.029660871	99	-1.0599655	0.2917

Figure 9.6d

Example 8: "Mean Weight Change in Anorexic Girls"

This example wishes to explore the weight change in girls suffering from anorexia after undergoing a certain type of therapy designed to help weight gain. The file "anorexia" on your CD lists each girls weight before therapy and then at the end of the study. Did the therapy help with weight gain? There are two ways to carry out the test. The first method is to calculate the differences first and use the one-sample procedures. The second is to use the paired procedures (also discussed in chapter 10). Both procedures are equivalent.

Calculating the Differences First

Select **Data→Transform data.** You want to create a column that is the difference in weight after – before ("*cog2* – *cog1*"). Select "*cog2*" as "Y" and "*cog1*" as "X" then choose "Y-X" as the "Function" and click **Set Expression**. "cog2"-"cog1" should appear in the "Expression" line at the top. Click **Compute**.

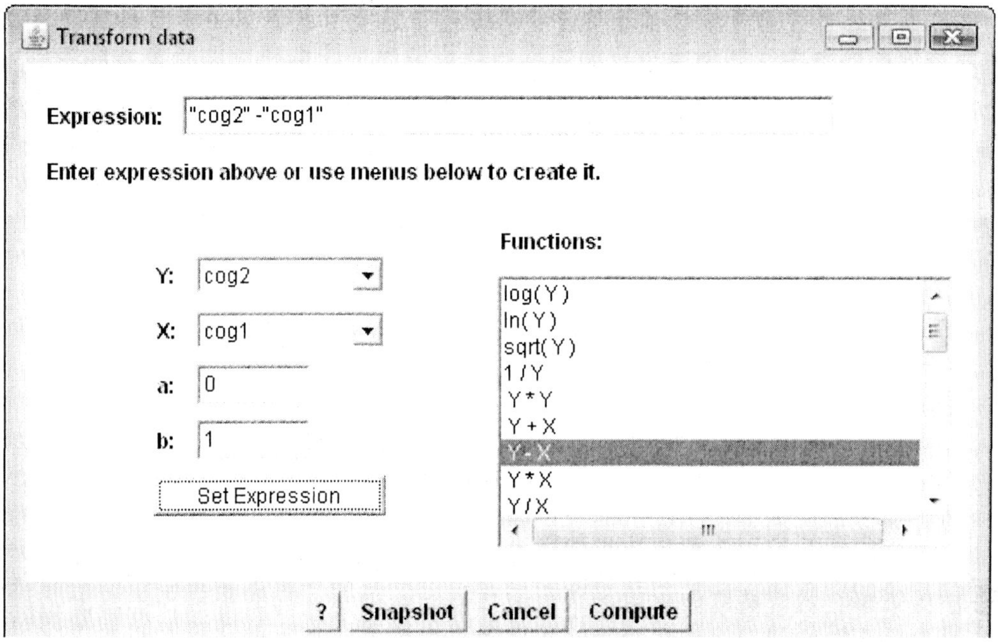

Figure 9.7a

In the data table you should see a new variable called "*cog2-cog1*" with the observations being the differences. In order to carry out a test to determine if the therapy helped in weight gain you want to carry out a one-sample T-test for this new variable. Select **Stat→T statistics→One sample→with data.** Click on the new variable "*cog2-cog1*". Click **Next.**

Screen 1

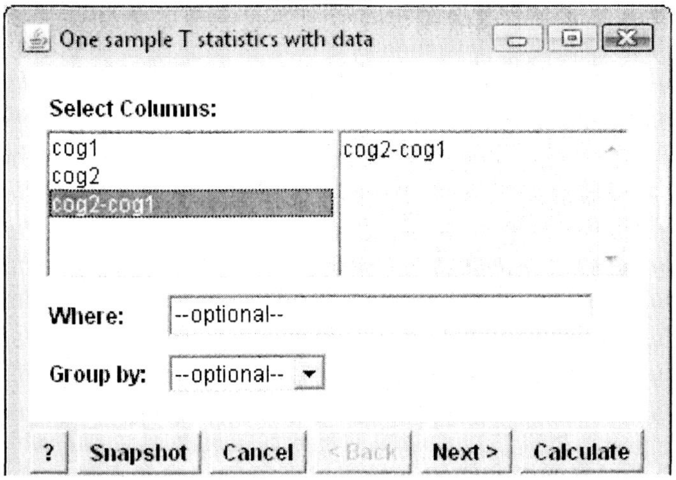

Figure 9.7b

If the therapy helped then the weight after should be higher or "cog2-cog1" should be positive (or simply greater than 0). Select "Hypothesis Test" with the Null mean being 0 and choose the right-sided or ">" option. Click **Calculate.**

Screen 2

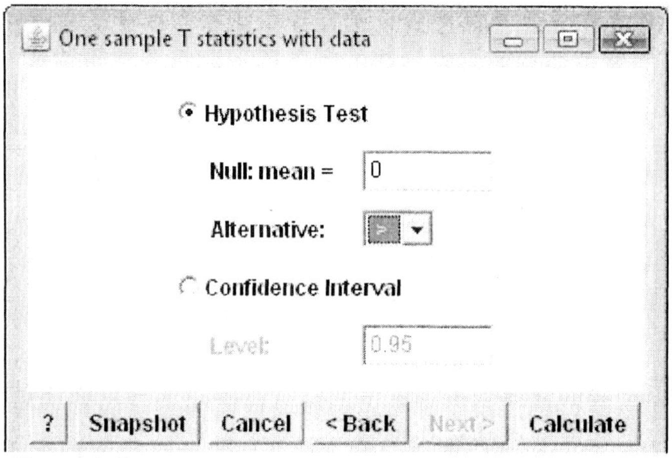

One sample T statistics with data

Hypothesis Test

Null: mean = 0

Alternative: >

Confidence Interval

Level: 0.95

? Snapshot Cancel < Back Next > Calculate

Figure 9.7c

The results below show that average weight gain was 3 with a standard error of 1.359. The test-statistic is 2.207 with a p-value of 0.0178. The degrees of freedom are 28 (n (the number of differences) – 1).

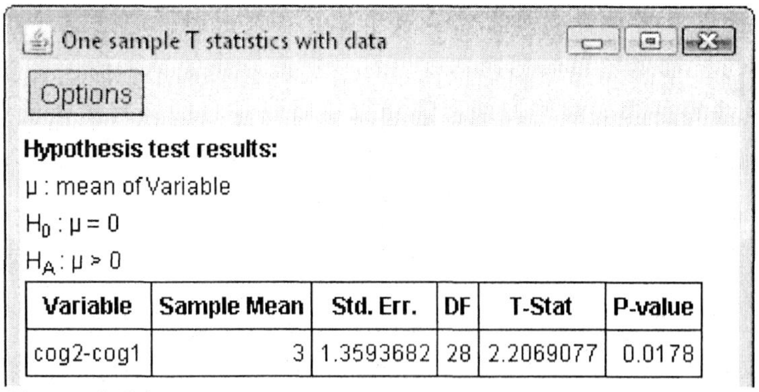

One sample T statistics with data

Options

Hypothesis test results:

μ : mean of Variable

$H_0 : \mu = 0$

$H_A : \mu > 0$

Variable	Sample Mean	Std. Err.	DF	T-Stat	P-value
cog2-cog1	3	1.3593682	28	2.2069077	0.0178

Figure 9.7d

Paired Procedure

Example 8 is actually a **Paired Two-sample** problem. We have two-samples that are dependent. Each sample contains an observation from the same individual. In the end, we focus on only the differences (in the pairs). Two-sample procedures are actually discussed in Chapter 10. However, the paired two-sample procedure is actually the same as the one-sample procedure on the differences, in which case the paired procedure will be illustrated here in order to make the comparison. Just like the one sample procedure we can carry out a test and calculate confidence intervals. Let's carry out the test to determine if the therapy aided in weight gain. Select **Stat➔T statistics➔Paired.** We are interested in the difference after – before (or equivalently "cog2" – "cog1"). Thus, select sample 1 to be "*cog2*" and sample 2 to be "*cog1*". You can ask StatCrunch to save the differences if you haven't done so already (in the transformation). Click **Next.**

Screen 1

Paired T statistics

Sample 1 - Sample 2:

Sample 1 in: cog2

Sample 2 in: cog1

Where: --optional--

Group by: --optional--

Save differences

? Cancel < Back Next > Calculate

Figure 9.8a

Fill in **Screen 2** just as you did before and click **Calculate.** The output below shows the same results as the output we saw earlier, as expected!

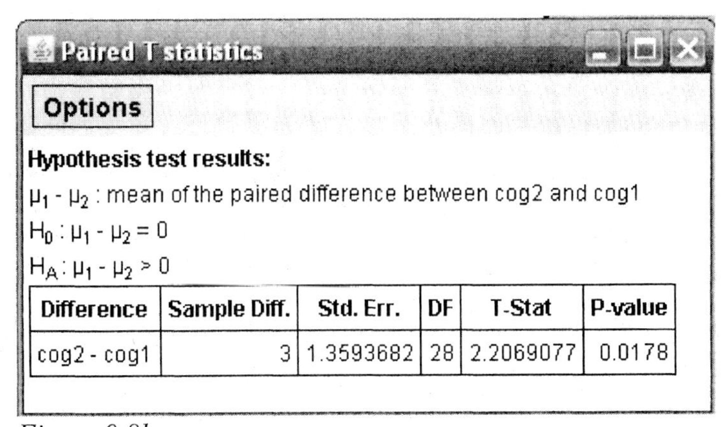

Paired T statistics

Options

Hypothesis test results:

$\mu_1 - \mu_2$: mean of the paired difference between cog2 and cog1

$H_0 : \mu_1 - \mu_2 = 0$

$H_A : \mu_1 - \mu_2 > 0$

Difference	Sample Diff.	Std. Err.	DF	T-Stat	P-value
cog2 - cog1	3	1.3593682	28	2.2069077	0.0178

Figure 9.8b

Chapter 10 - Comparing Two Groups

In this chapter we will illustrate how to use StatCrunch to carry out Hypothesis Tests and calculate Confidence Intervals for the difference in two population proportions and the difference in two population means.

Comparing Two Population Proportions (Independent Samples)

Although it is possible to carry out tests and calculate confidence intervals for the difference in two proportions with raw data it is much easier to use summary data (same as the one-sample case). Even if you have raw data, you can easily find the summaries first and then carry out the analysis. The following will illustrate how to carry out tests and calculate confidence intervals for the difference between two population proportions with summary data.

With Summary

Confidence Intervals

Example 4: "C.I. Comparing Heart Attack Rates for Aspirin and Placebo"

In this example we want to compare the heart attack rates between individuals taking aspirin vs. those taking a placebo. In particular, we want to calculate a 95% confidence interval for the difference in the proportion of heart attacks between the two groups.

Select **Stat→Proportions→Two Samples→with summary.** In **Screen 1** enter the number of successes and the total number of observations in the blanks for each of the two groups. In this example, 189 out of 11034 individuals taking the placebo (sample 1) experienced a heart attack and 104 out of 11037 individuals taking Aspirin (sample 2) experienced a heart attack. Click **Next** to go to **Screen 2**.

Screen 1

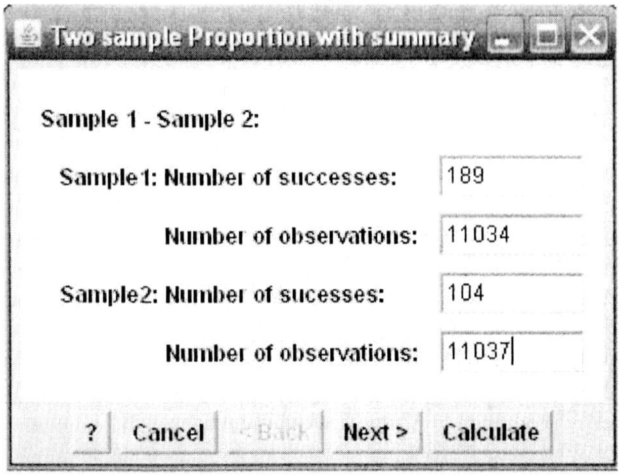

Two sample Proportion with summary

Sample 1 - Sample 2:

Sample1: Number of successes: 189

Number of observations: 11034

Sample2: Number of sucesses: 104

Number of observations: 11037

? Cancel < Back Next > Calculate

Figure 10.1a

In **Screen 2**, select "Confidence Interval" and type in the desired confidence level. In this example we want a 95% confidence interval so you would enter 0.95 (the default value). Click **Calculate.**

Screen 2

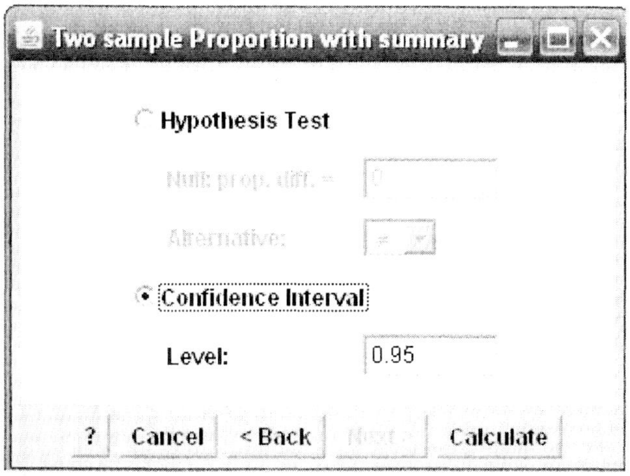

Two sample Proportion with summary

Hypothesis Test

Null: prop. diff. = 0

Alternative: ≠

Confidence Interval

Level: 0.95

? Cancel < Back Next > Calculate

Figure 10.1b

The observed (estimated) difference in the proportion of heart attacks between the two groups is 0.00771. The standard error of the difference is 0.00154. Finally the 95% confidence interval for the difference in the proportion of heart attacks between those taking a placebo vs. those taking Aspirin is (0.0047, 0.0107).

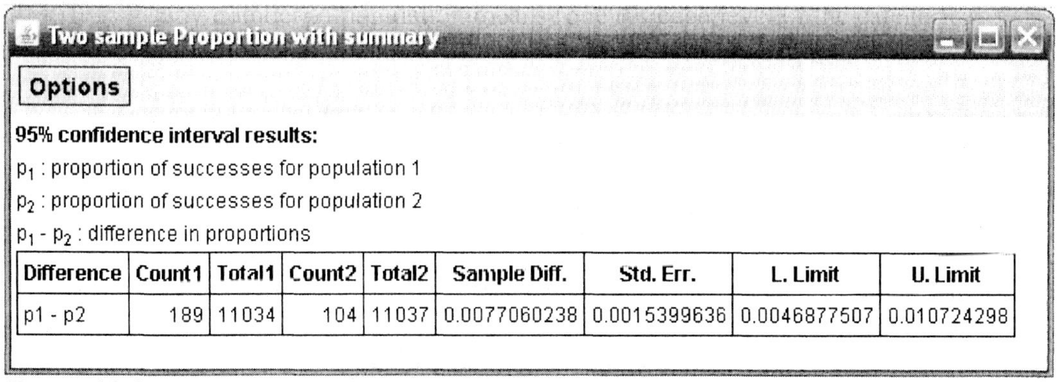

Two sample Proportion with summary

Options

95% confidence interval results:

p_1 : proportion of successes for population 1

p_2 : proportion of successes for population 2

$p_1 - p_2$: difference in proportions

Difference	Count1	Total1	Count2	Total2	Sample Diff.	Std. Err.	L. Limit	U. Limit
p1 - p2	189	11034	104	11037	0.0077060238	0.0015399636	0.0046877507	0.010724298

Figure 10.1c

If you wanted a 99% confidence interval you would enter 0.99 as your confidence level in **Screen 2**. The output below shows that a 99% confidence interval for the difference in the proportion of heart attacks between those taking a placebo vs. those taking Aspirin is (0.0037, 0.0117).

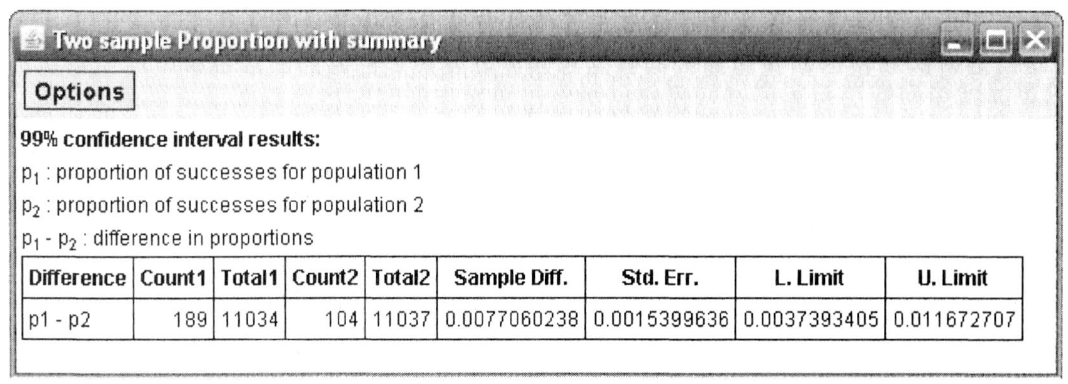

Difference	Count1	Total1	Count2	Total2	Sample Diff.	Std. Err.	L. Limit	U. Limit
p1 - p2	189	11034	104	11037	0.0077060238	0.0015399636	0.0037393405	0.011672707

Figure 10.1d

Hypothesis Tests

Example 5: "Is TV Watching Associated with Aggressive Behavior?"

In this example we want to explore the proportion of teenagers that commit aggressive acts. In particular this example wishes to compare the proportion of aggressive acts between those who watch less than 1 hour a day of television to those that watch at least 1 hour a day of television. It is hypothesized that the likelihood of aggressive behaviour may be somehow different depending on the level of television watching by the individual. In which case, we will carry out a two-sided hypothesis test for any difference in the two population proportions.

Select **Stat➔Proportions➔Two Samples➔with summary.** In **Screen 1** enter the number of successes and the total number of observations in the blanks for each of the two groups. In this example, 5 out of 88 teenagers who watched less than 1 hour a day of television reported committing an aggressive act, whereas 154 out of 619 teenagers who watched less than 1 hour a day of television reported committing an aggressive act. Click **Next** to go **Screen 2.**

Screen 1

Two sample Proportion with summary

Sample 1 - Sample 2:

Sample1: Number of successes: 5

Number of observations: 88

Sample2: Number of sucesses: 154

Number of observations: 619

? | Cancel | < Back | Next > | Calculate

Figure 10.2a

In **Screen 2,** select "Hypothesis Test" with the null proportion difference being 0. The test is looking for any difference, thus select the "not equal" as the alternative. Click **Calculate.**

Screen 2

Two sample Proportion with summary

Hypothesis Test

Null: prop. diff. = 0

Alternative: ≠

Confidence Interval

Level: 0.95

? Cancel < Back Next > Calculate

Figure 10.2b

The observed (estimated) difference in the proportion of aggressive acts between the teenagers who watched less than 1 hour of television a day vs. those who watched at least 1 hour of television a day is -0.192. The standard error of the difference is 0.0.0476. Finally, the test-statistic is -4.04 and the two-sided p-value is less than 0.0001. This tells us that there is extremely strong evidence against the null hypothesis in favor of the alternative that there exists a difference.

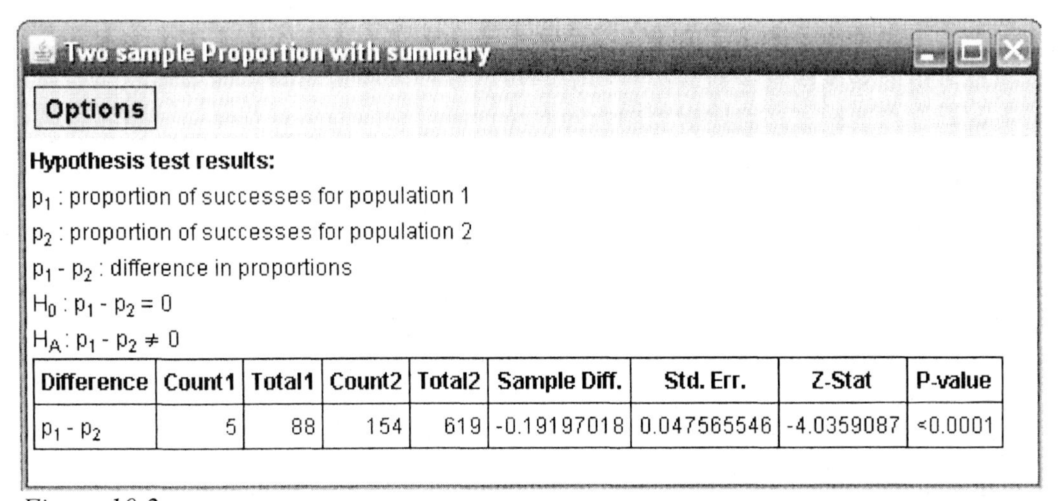

Two sample Proportion with summary

Options

Hypothesis test results:

p_1 : proportion of successes for population 1

p_2 : proportion of successes for population 2

$p_1 - p_2$: difference in proportions

$H_0 : p_1 - p_2 = 0$

$H_A : p_1 - p_2 \neq 0$

Difference	Count1	Total1	Count2	Total2	Sample Diff.	Std. Err.	Z-Stat	P-value
$p_1 - p_2$	5	88	154	619	-0.19197018	0.047565546	-4.0359087	<0.0001

Figure 10.2c

To estimate the magnitude of the difference you could click **Options** and go back and produce a 95% confidence interval. Note: The Standard Error is different for the confidence interval. The 95% confidence interval is (-0.25, -0.13).

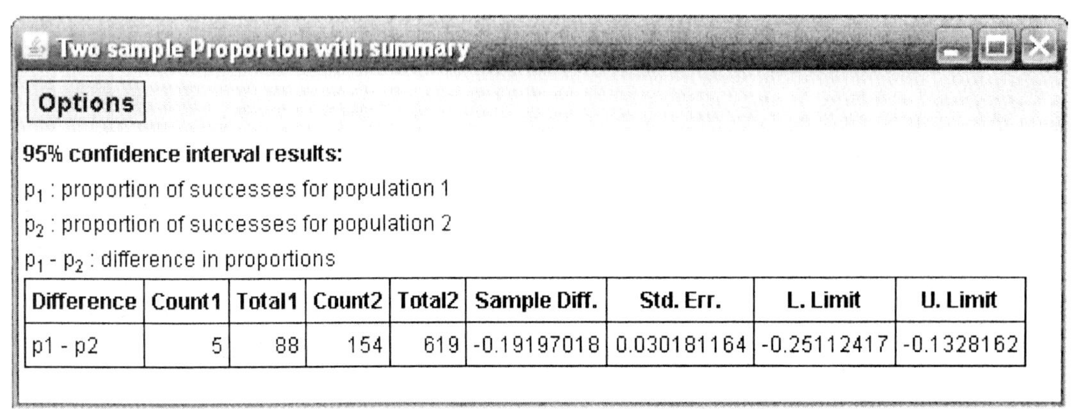

Difference	Count1	Total1	Count2	Total2	Sample Diff.	Std. Err.	L. Limit	U. Limit
p1 - p2	5	88	154	619	-0.19197018	0.030181164	-0.25112417	-0.1328162

Figure 10.2d

Comparing Two Population Means (Independent Samples)

Similar to the One Sample T procedures StatCrunch can carry out the Two Sample T procedures using **summary data (Stat➔T statistics➔Two sample➔with summary)** or using **raw data (Stat➔T statistics➔Two sample➔with data).**

With Summary

Here you'll need to simply enter the sample means, sample standard deviation and sample sizes of the two groups in **Screen 1** and the desired procedure in **Screen 2** and click **Calculate** to obtain the results.

Confidence Intervals

Example 8: "Nicotine - How Much More Addicted Are Smokers then Ex-Smokers?"

It is given that out of 75 smokers, their average *HONC* score was 5.9 with standard deviation 3.3. Out of 257 ex-smokers their average *HONC* score was 1.0 with a standard deviation of 2.3. To find a 95% Confidence Interval for the difference in average *HONC* score between smokers and ex-smokers, open **Stat➔T statistics➔Two sample➔with summary** and fill in **Screen 1** as follows:

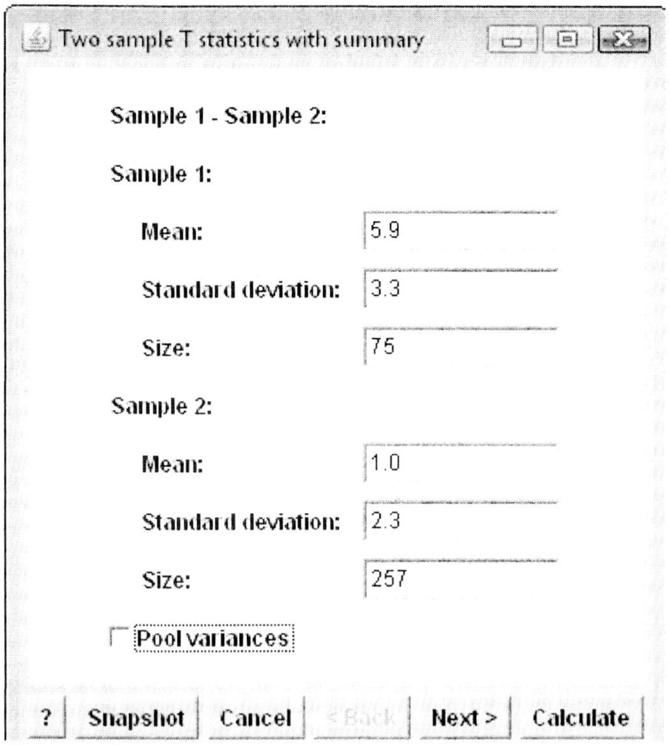

Two sample T statistics with summary

Sample 1 - Sample 2:

Sample 1:

Mean: 5.9

Standard deviation: 3.3

Size: 75

Sample 2:

Mean: 1.0

Standard deviation: 2.3

Size: 257

Pool variances

? Snapshot Cancel < Back Next > Calculate

Figure 10.3a

Make sure "Pool variances" is unchecked. Click **Next** to go to **Screen 2.** Select "Confidence Interval" with level 0.95 and click **Calculate.**

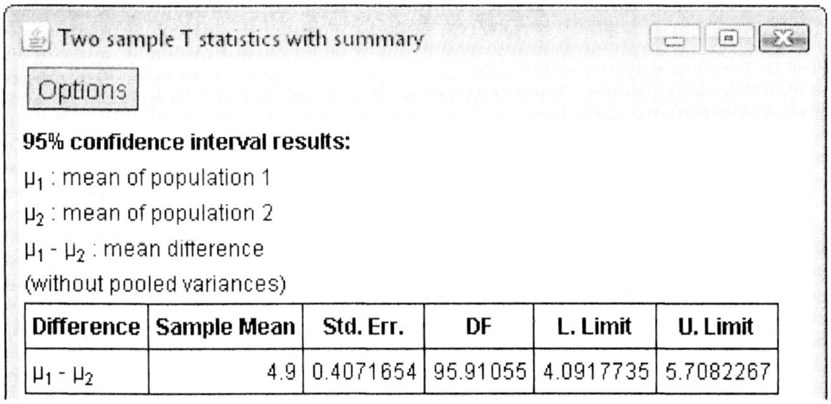

Two sample T statistics with summary

Options

95% confidence interval results:

μ_1 : mean of population 1

μ_2 : mean of population 2

$\mu_1 - \mu_2$: mean difference

(without pooled variances)

Difference	Sample Mean	Std. Err.	DF	L. Limit	U. Limit
$\mu_1 - \mu_2$	4.9	0.4071654	95.91055	4.0917735	5.7082267

Figure 10.3b

The 95% Confidence Interval is (4.1, 5.7).

Hypothesis Test

Example: "Is There a Real Difference in Average HONC scores?"

To carry out a test for the difference in average *HONC* score between smokers and ex-smokers, open **Stat➔T statistics➔Two sample➔with summary** and fill in **Screen 1** we did earlier. Make sure "Pool variances" is unchecked. Click **Next** to go to **Screen 2.** Select "Hypothesis Test" with "Null: mean diff" of 0 and the "not equal" alternative. Click **Calculate.**

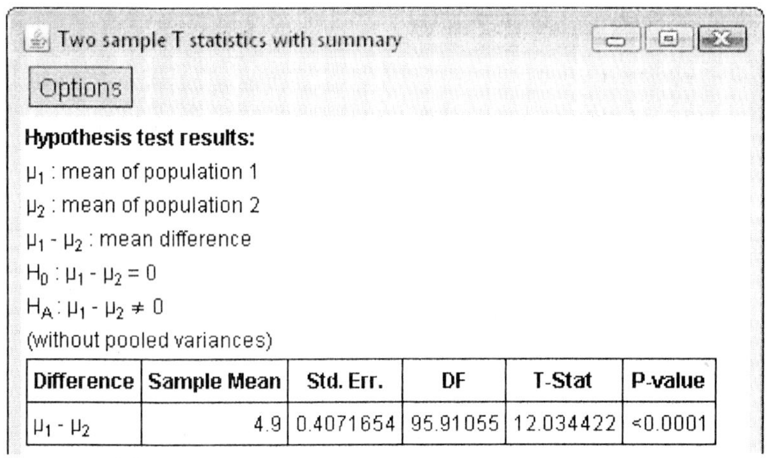

Two sample T statistics with summary

Options

Hypothesis test results:

μ_1 : mean of population 1

μ_2 : mean of population 2

$\mu_1 - \mu_2$: mean difference

$H_0 : \mu_1 - \mu_2 = 0$

$H_A : \mu_1 - \mu_2 \neq 0$

(without pooled variances)

Difference	Sample Mean	Std. Err.	DF	T-Stat	P-value
$\mu_1 - \mu_2$	4.9	0.4071654	95.91055	12.034422	<0.0001

Figure 10.3c

The output gives a test-statistic of 12.03 and a p-value less than 0.0001.

With Data

In order to carry out analysis for the difference in two population means **with data** the data must be entered in two separate columns, one for each group.

Confidence Intervals

Exercise 10.20: Ebay Auctions

In Chapter 8 we looked at separate One-Sample Confidence Intervals for the average selling price of PDA's using the "*Buy-it-now*" option and the "*Bidding*" methods in Ebay. Now let's focus on the difference in selling price between the two methods. Specifically let's find a 95% confidence interval for the difference in average selling price between the two methods. First, either load the data from the "ebay_auctions" file on your CD or enter the data as it appears in your text. The data should be in two separate columns, one for "*Buy-it-now*" results and one for "*Bidding*". The Summary Statistics are:

Column Statistics

Options

Summary statistics:

Column	n	Mean	Variance	Std. Dev.
Buy-it-now	7	233.57143	214.28572	14.638501
Bidding	18	231.61111	481.1928	21.936108

Figure 10.4a

Select **Stat➔T Statistics➔Two Sample➔with data.** In **Screen 1**, select "*Buy-it-now*" as Sample 1 and "*Bidding*" as Sample 2. For now, make sure the "Pool variances" is unchecked. Click **Next** to go to **Screen 2.**

Screen 1

Two sample T statistics with data

Sample 1 - Sample 2:

Sample 1 in: Buy-it-now

Where : --optional--

Sample 2 in: Bidding

Where: --optional--

Pool variances

? | Snapshot | Cancel | < Back | Next > | Calculate

Figure 10.4b

In **Screen 2**, select "Confidence Interval" and 0.95 as the confidence level. Click **Calculate.**

The estimated difference in average selling price between the "*Buy-it-now*" and "*Bidding*" options is $1.96. The standard error of the difference is 7.57. Finally, the 95% confidence interval is ($-14.05, $17.97). This tells us that we are 95% confidence that the average selling price of the PDA's in question using the "*Buy-it-now*" option is anywhere between $17.97 more to $14.05 less than the average selling price using the "*Bidding*" option.

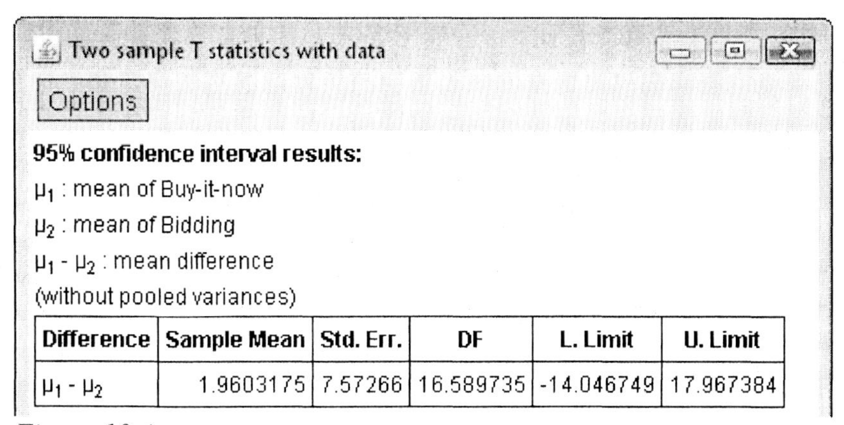

Two sample T statistics with data

Options

95% confidence interval results:

μ_1 : mean of Buy-it-now

μ_2 : mean of Bidding

$\mu_1 - \mu_2$: mean difference

(without pooled variances)

Difference	Sample Mean	Std. Err.	DF	L. Limit	U. Limit
$\mu_1 - \mu_2$	1.9603175	7.57266	16.589735	-14.046749	17.967384

Figure 10.4c

Note: The degrees of freedom are calculated to 16.59. You can check this using the formula given in your text.

Hypothesis Tests

Exercise 10.21: Ebay Data

Suppose we wanted to carry out a test to determine if there is any difference in average selling price between the two selling methods. In **Screen 2** in **Stat➔T Statistics➔Two Sample➔with data,** select "Hypothesis Test" with the null mean difference being 0. The test is looking for any difference, thus select "not equal" as the alternative. Click **Calculate.**

The test-statistic is 0.259 and the p-value is 0.7989. This indicates virtually no evidence of a difference.

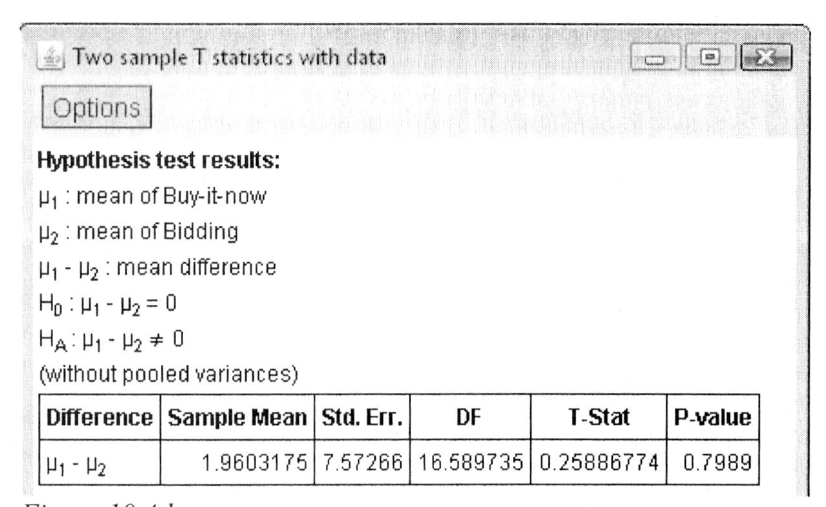

Difference	Sample Mean	Std. Err.	DF	T-Stat	P-value
$\mu_1 - \mu_2$	1.9603175	7.57266	16.589735	0.25886774	0.7989

Figure 10.4d

C.I. and Tests Assuming Equal Standard Deviations

Sometimes it is reasonable to assume that the population standard deviations (or variances) are the same for the groups in question. If this is a reasonable assumption then we "pool" the two sample standard deviations to calculate an estimate for the common population standard deviation.

Exercise 10.37: "Comparing Clinical Therapies"

This exercise wishes to compare the average improvement between two therapies for treating depression. The improvement is measured as the change in depression score (the higher the better the improvement). Enter the data as it appears in Exercise 10.37. In the first column enter the observations from "*Therapy 1*" and in the second column enter the observations from "*Therapy 2*". The descriptive statistics are:

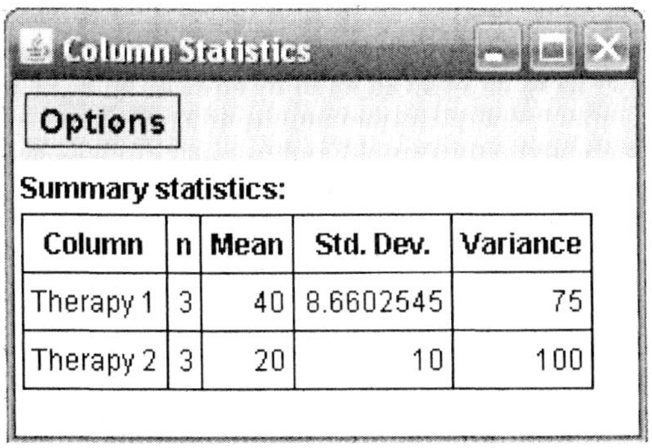

Column	n	Mean	Std. Dev.	Variance
Therapy 1	3	40	8.6602545	75
Therapy 2	3	20	10	100

Figure 10.5a

For inference about the difference in average depression scores select **Stat➔T Statistics➔Two Sample➔with data.** Select "*Therapy 1*" as Sample 1 and "*Therapy 2*" as Sample 2. Since we are assuming equal standard deviations (or equal variances) check the "Pool variances" box. Click **Next** to go to **Screen 2.**

Screen 1

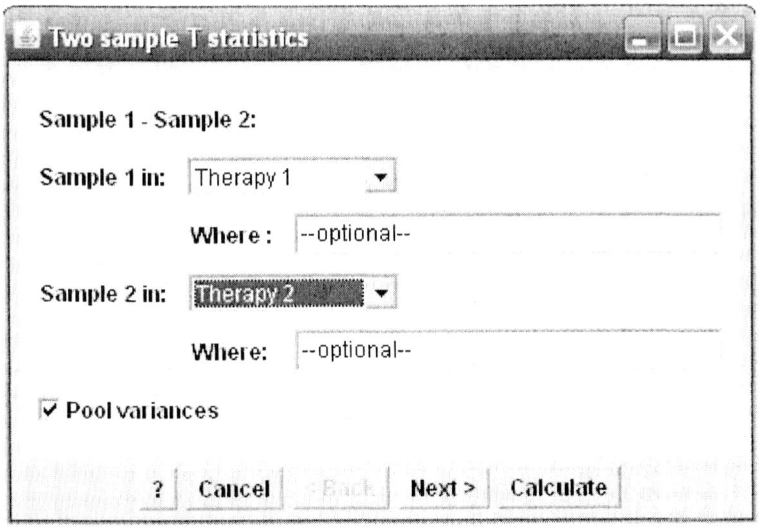

Figure 10.5b

First, let's look at a test for any difference in average depression score between the two therapies. In **Screen 2,** select "Hypothesis Test" with the null proportion difference being 0. The test is looking for any difference, thus select the "not equal" as the alternative. Click **Calculate.**

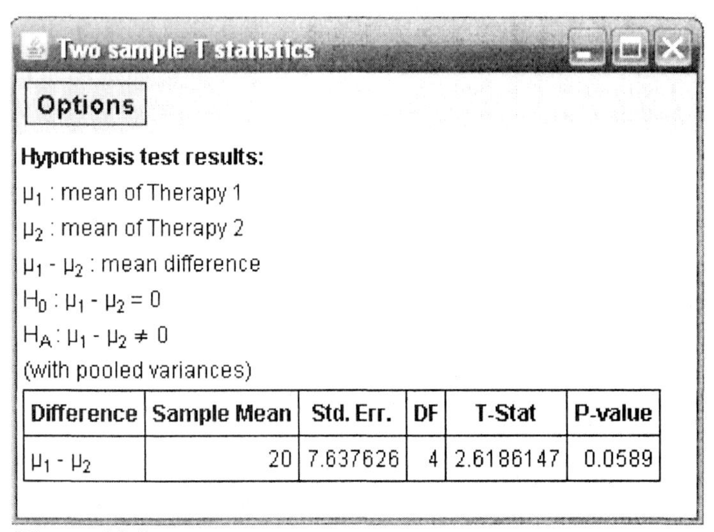

Figure 10.5c

The estimated difference is 20 with standard error being 7.64. The test-statistic is 2.62 with a p-value of 0.0589.

Note 1: When assuming equal standard deviation the degrees of freedom are calculated as $n_1 + n_2 - 2 = 3 + 3 - 2 = 4$.

Note 2: The standard error is calculates after finding the pooled standard deviation (which is not reported in the output).

Click **Options** and go back to **Screen 2.** Select "Confidence Interval" with level 0.95. Click **Calculate.**

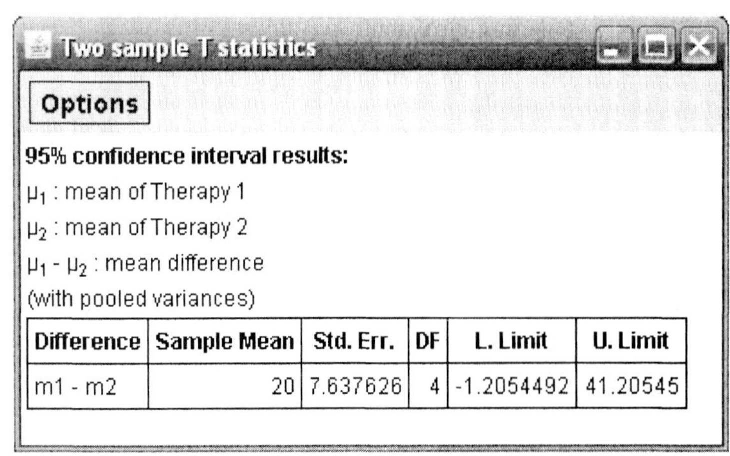

Figure 10.5d

The 95% confidence interval for the difference in average score between the Therapy 1 vs. Therapy 2 is (-1.21, 41.21).

Comparing Two Population Means (Dependent or Paired Samples)

We already discussed these methods in Chapter 9. When analyzing the difference in means between observations from a Paired or Dependent sample we basically analyze the average difference for each pair of observations. In which case, the paired procedures are equivalent to the One-Sample procedures for the average difference. Let's look at another example.

Example 12 and 13: "Cell Phones and Driving"

This example is comparing average reaction times while driving when not using a cell phone to those when using a cell phone. Each subject was tested without a cell phone and with a cell phone. Thus, we have dependent samples. Enter the data as it appears in Table 10.9 in your text. You only need to enter the "*No*" and "*Yes*" columns. Select **Stat➔T Statistics➔Paired.** Choose "*Yes*" as Sample 1 and "*No*" as Sample 2. Check the "Save differences" box if you wish to save the differences. Click **Next** to go to **Screen 2.**

Screen 1

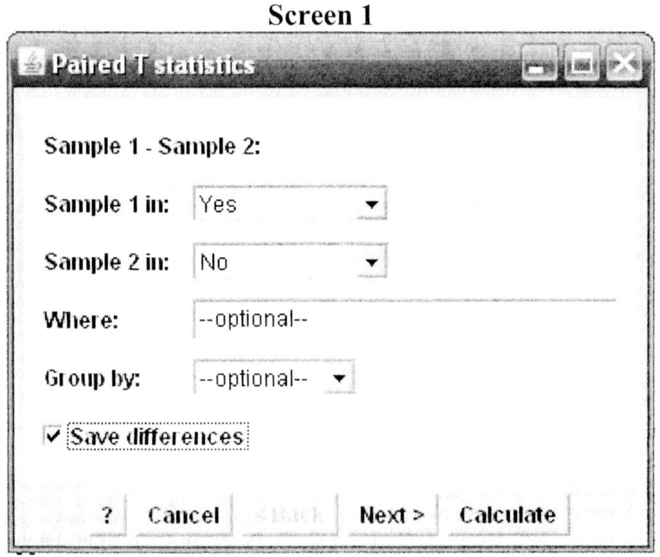

Figure 10.6a

Are the reaction times slower when using a cell phone?

In **Screen 2,** select "Hypothesis Test" with the null proportion difference being 0. If reaction times were slower when using a cell phone, we would expect a positive difference (for "*Yes*" – "*No*"). Thus, thus select ">" as the alternative. Click **Calculate.** .

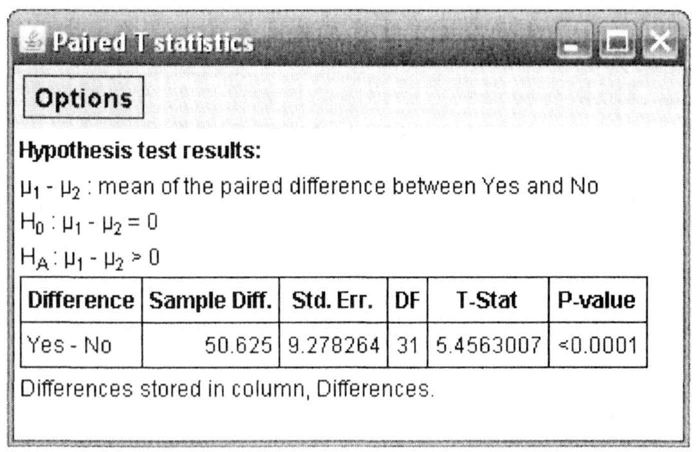

Paired T statistics

Options

Hypothesis test results:

$\mu_1 - \mu_2$: mean of the paired difference between Yes and No

$H_0 : \mu_1 - \mu_2 = 0$

$H_A : \mu_1 - \mu_2 > 0$

Difference	Sample Diff.	Std. Err.	DF	T-Stat	P-value
Yes - No	50.625	9.278264	31	5.4563007	<0.0001

Differences stored in column, Differences.

Figure 10.6b

The estimate average difference is 50.625 with a standard error of 9.278. The test-statistic is 5.46 and the p-value is less than 0.0001

If you want a 95% confidence interval then click **Options** and go back to **Screen 2.** Select "Confidence Interval" with level 0.95. Click **Calculate.**

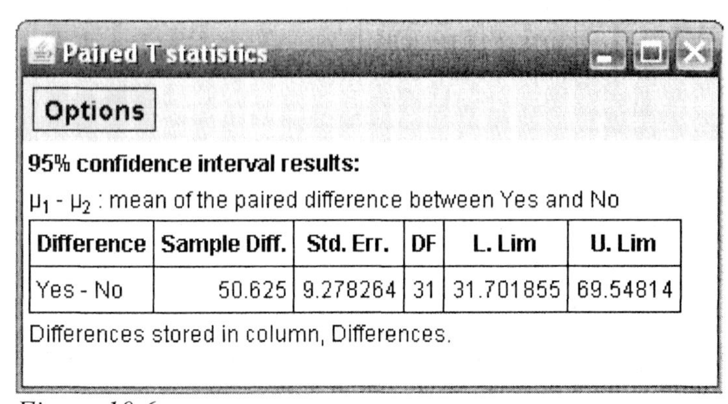

Paired T statistics

Options

95% confidence interval results:

$\mu_1 - \mu_2$: mean of the paired difference between Yes and No

Difference	Sample Diff.	Std. Err.	DF	L. Lim	U. Lim
Yes - No	50.625	9.278264	31	31.701855	69.54814

Differences stored in column, Differences.

Figure 10.6c

The 95% confidence interval is reported to be (31.70, 69.55).

Chapter 11 - Analyzing the Association between Categorical Variables

In this chapter we will illustrate how to analyze associations between two categorical variables. In particular we will look at two Chi-Square tests, one to test for dependence (or association) between two categorical variables, and the other to test a particular distribution of a categorical variable.

Testing for Dependence (or Association)

Example 1, 3 and 4: "Happiness and Family Income"

This example explores the relationship between happiness and family income. Both variables are categorical with three levels each. Happiness measured as not happy, pretty happy or very happy. Family income is measured as above average, average or below average. The observed frequencies are summarized in your text as displayed below. Enter the data in StatCrunch as it appears below.

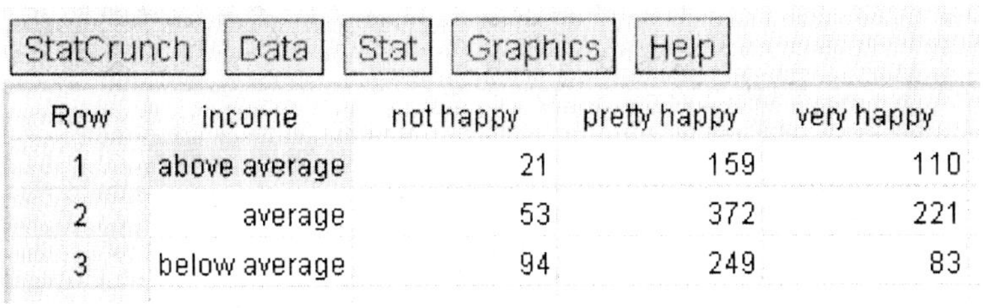

StatCrunch | Data | Stat | Graphics | Help

Row	income	not happy	pretty happy	very happy
1	above average	21	159	110
2	average	53	372	221
3	below average	94	249	83

Figure 11.1a

To carry out the test for some dependence select **Stat➔Tables➔Contingency➔with summary.** Select the three columns for the levels of happiness. In "Row labels in:" select "*income*" from the drop down menu. This is all you need to obtain the test results. Otherwise you can click **Next** to obtain other summaries. Click **Next** to go to **Screen 2.**

Screen 1

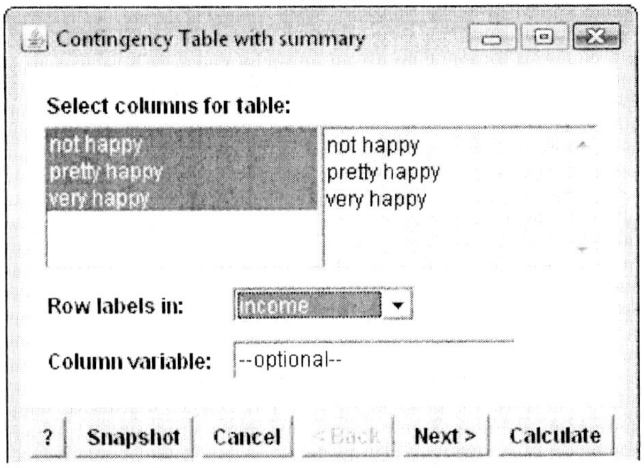

Contingency Table with summary

Select columns for table:

not happy
pretty happy
very happy

Row labels in: income

Column variable: --optional--

? | Snapshot | Cancel | < Back | Next > | Calculate

Figure 11.1b

Here you have the option to obtain cell summaries. The Row percents will give the observed conditional distribution for happiness given income level. The Column percents give the observed conditional distribution for income level given happiness. The Percentage of total is the observed joint distribution for the two variables. Lastly, the expected counts are expected number out of the total falling into each of the nine categories if the two variables are independent (or not associated in any way). Select them all and click **Calculate.**

Screen 2

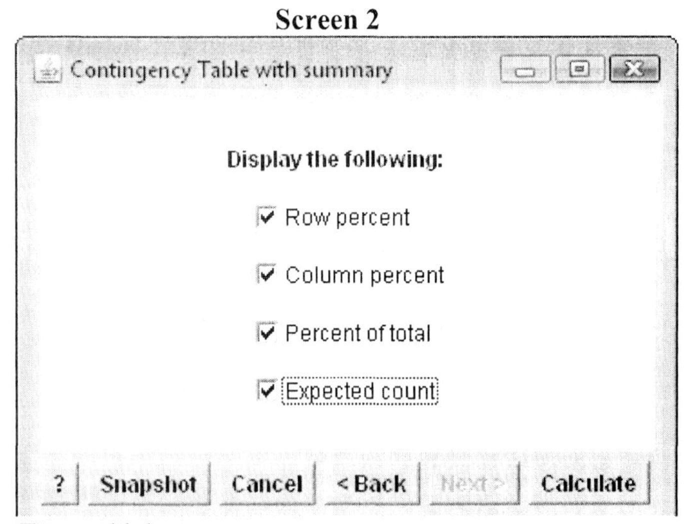

Figure 11.1c

The output below summarizes each value chosen above, as well as the observed counts for each cell. In addition it reports the Chi-Square test-statistic, the degrees of freedom and the p-value for the test of dependence between the two variables.

e.g. Pretty Happy with Average Income

In the cell for individuals that are pretty happy and have an average income they observed 372 individuals. The observed conditional proportion of being pretty happy given that you have an average income is 372/646 = 0.5759 or 57.59%. Compare all of the values under "row percent" to those in Table 11.2 in you text. The observed condition proportion of having an average income given that you are pretty happy is 372/780 = 0.4769 or 47.69%. The observed proportion of pretty happy individuals with average income is 372/1362 = 0.2731 or 27.31%. Lastly, if the two variables were independent we would expect to see 370 out 1362 to be pretty happy with an average income. The value "370" is calculated using the formula on the bottom of page 491 in your text as:

$$\frac{(\text{Row total}) \times (\text{Column total})}{\text{Total sample size}} = \frac{646 \times 780}{1362} = 370.0 \text{ (rounded to 1 decimal place)}$$

Compare all the values under "Expected count" to those in Table 11.5 in your text.

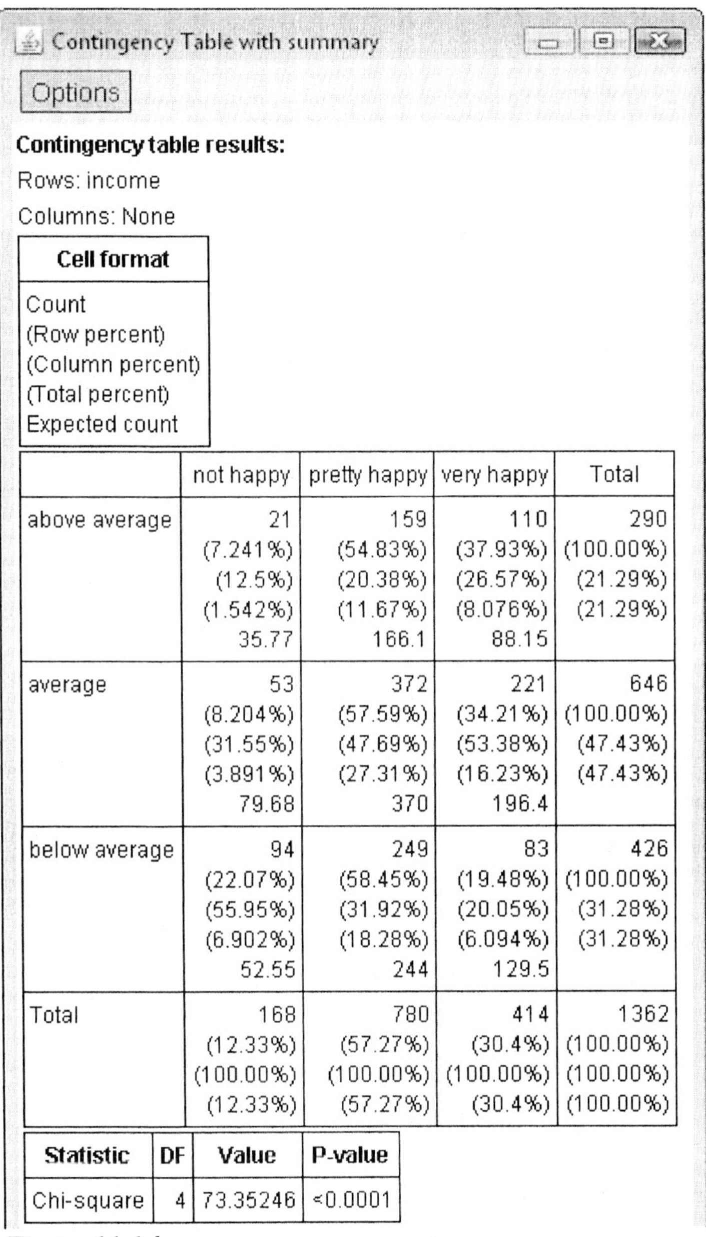
Contingency Table with summary

Options

Contingency table results:

Rows: income

Columns: None

Cell format
Count (Row percent) (Column percent) (Total percent) Expected count

	not happy	pretty happy	very happy	Total
above average	21 (7.241%) (12.5%) (1.542%) 35.77	159 (54.83%) (20.38%) (11.67%) 166.1	110 (37.93%) (26.57%) (8.076%) 88.15	290 (100.00%) (21.29%) (21.29%)
average	53 (8.204%) (31.55%) (3.891%) 79.68	372 (57.59%) (47.69%) (27.31%) 370	221 (34.21%) (53.38%) (16.23%) 196.4	646 (100.00%) (47.43%) (47.43%)
below average	94 (22.07%) (55.95%) (6.902%) 52.55	249 (58.45%) (31.92%) (18.28%) 244	83 (19.48%) (20.05%) (6.094%) 129.5	426 (100.00%) (31.28%) (31.28%)
Total	168 (12.33%) (100.00%) (12.33%)	780 (57.27%) (100.00%) (57.27%)	414 (30.4%) (100.00%) (30.4%)	1362 (100.00%) (100.00%) (100.00%)

Statistic	DF	Value	P-value
Chi-square	4	73.35246	<0.0001

Figure 11.1d

For the test of any dependence or association the Chi-Square test-statistic is 73.35 and the p-value is less than 0.0001. There are 4 degrees of freedom. This is extremely strong evidence of some dependence or some association between the two variables.

Goodness-of-Fit

Another Chi-Square test involves testing an assumed distribution for one categorical variable. This test is called the Goodness-of-fit test.

Suppose you are interested in determining if a single six-sided die is unfair (the probability of the values occurring on a single roll are some how different). In order to do this you could collect a sample (roll the die over and over again). For this experiment let's roll the die 60 times. I did this and obtained the following results:

VALUE	1	2	3	4	5	6
Frequency	*6*	*14*	*10*	*12*	*8*	*10*

Is there evidence above that the die is unfair? Under the null hypothesis we would assume that the die is fair (the proportion of each value is the same = 1/6). The alternative would state that the proportions are not all equal to 1/6 (at least one is different). In StatCrunch you will need to enter the "Observed Count" as well as the "Expected Count" for each value under the null hypothesis. If the null hypothesis were true, we would expect that out of 60 rolls we would observe 10 of each value. Enter the following into StatCrunch:

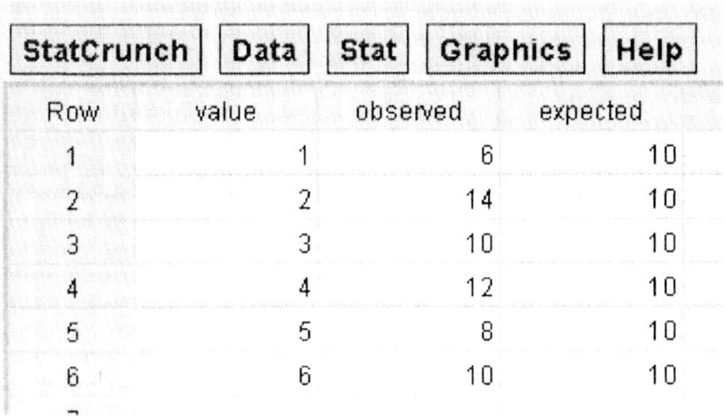

StatCrunch | Data | Stat | Graphics | Help

Row	value	observed	expected
1	1	6	10
2	2	14	10
3	3	10	10
4	4	12	10
5	5	8	10
6	6	10	10

Figure 11.2a

Select **Stat→Goodness-of-fit→Chi-Square test.** Select the appropriate variables for the observed and expected counts and click **Calculate.**

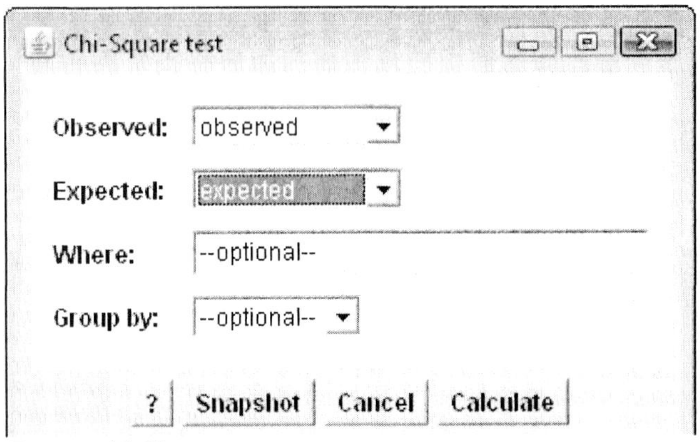

Chi-Square test

Observed: observed

Expected: expected

Where: --optional--

Group by: --optional--

? | Snapshot | Cancel | Calculate

Figure 11.2b

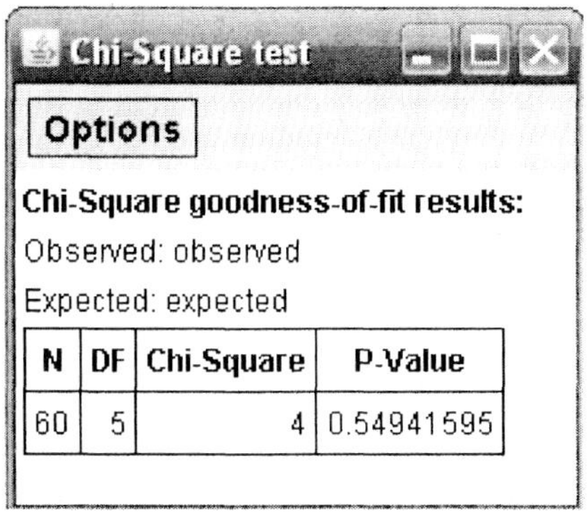

Chi-Square test

Options

Chi-Square goodness-of-fit results:

Observed: observed

Expected: expected

N	DF	Chi-Square	P-Value
60	5	4	0.54941595

Figure 11.2c

The output above gives a test-statistic of 4 and a p-value of 0.549 indicating virtually no evidence against the null hypothesis. In other words, there is virtually no evidence that based on the observed 60 rolls that the die is unfair.

Chapter 12 - Analyzing the Association between Quantitative Variables: Regression Analysis

We first introduced associations between quantitative variables in Chapter 3. There we discussed how to obtain Scatterplots to visualize the association between two quantitative variables, Correlation and R-squared as descriptive measures for the strength of linear associations, and finally, the Simple Linear Regression line. There is a lot more to SLR than this! In this chapter we look at inference for SLR. In particular, we will show how to use StatCrunch to test for the significance of the linear association, as well as estimate or predict the response with confidence.

Review of Chapter 3

Let's start with a review of some of the descriptive procedures from Chapter 3.

Example 1, 2, 3, 6 and 9: "Bench Press"

Open the file "high school female athletes". We are interested in the association between maximum bench press ("*maxBP*") and the number of 60-punds bench presses performed before fatigue ("*BP(60)*"). To obtain a Scatterplot of "*maxBP*" vs. "*BP(60)*" select **Graphics➔Scatter Plot** and then specify the Y and X variables. The Scatterplot should look like this:

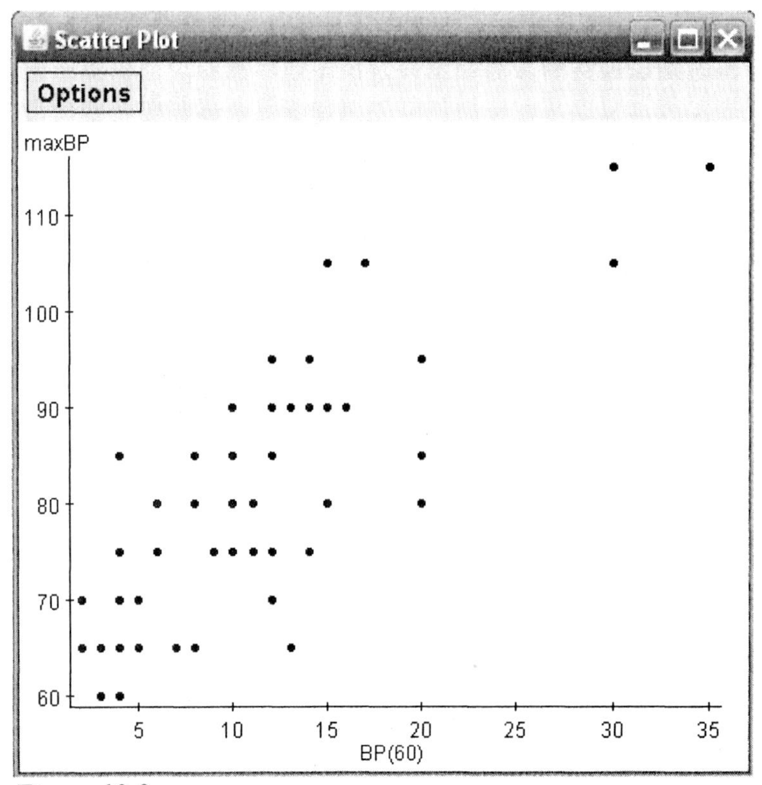

Figure 12.0a

To obtain the regression output, select **Stat➔Regression➔Simple Linear** and then specify the Y and Y variables. The output is:

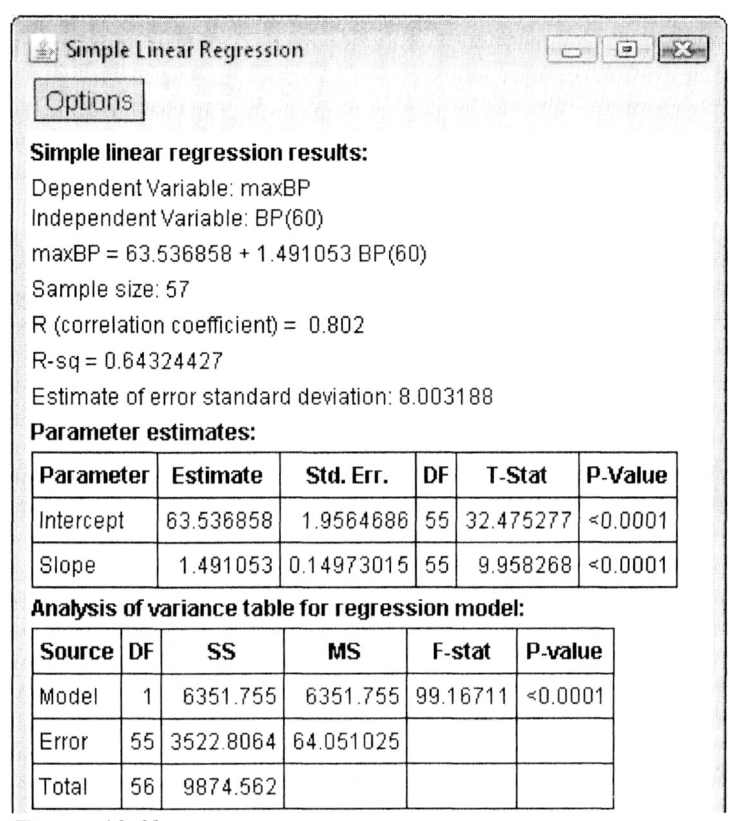

Simple Linear Regression

Options

Simple linear regression results:

Dependent Variable: maxBP
Independent Variable: BP(60)
maxBP = 63.536858 + 1.491053 BP(60)
Sample size: 57
R (correlation coefficient) = 0.802
R-sq = 0.64324427
Estimate of error standard deviation: 8.003188

Parameter estimates:

Parameter	Estimate	Std. Err.	DF	T-Stat	P-Value
Intercept	63.536858	1.9564686	55	32.475277	<0.0001
Slope	1.491053	0.14973015	55	9.958268	<0.0001

Analysis of variance table for regression model:

Source	DF	SS	MS	F-stat	P-value
Model	1	6351.755	6351.755	99.16711	<0.0001
Error	55	3522.8064	64.051025		
Total	56	9874.562			

Figure 12.0b

In the output above we find the Simple Linear Regression line to be

$$maxBP = 63.54 + 1.49 * BP(60)$$

For example, an athlete that can do 35, 60-pound bench presses has a predicted maximum bench press of 63.54+(1.49*35) = 115.69 pounds.

The Linear Correlation between the two variables is given to be R = 0.802.

Another descriptive measure used in SLR is r-squared given to be R-sq = 0.643 (which is the Correlation squared). This value can be interpreted as the proportional reduction in error in the response after fitting the regression. Thus, we could say that the regression on the number of 60-pound bench presses explains 64.3% of the error in predicting the maximum bench press.

Now that we've reviewed some of the topics from Chapter 3, let's move on to inference for SLR.

Inference for Regression Coefficients

In the "Parameter Estimates" table we are given the parameter estimate, the standard error of the estimate, the test statistic testing the null hypothesis that the parameter is 0 (this is the default hypothesized value), the degrees of freedom for the test (n – 2 for SLR) and finally the p-value for the alternative hypothesis that the parameter is not 0 (the two-sided p-value).

Test for Linear Significance

First, we may be interested in significance test for the linear association. Is there a significant linear association? In SLR, the test for linear significance is a test for the slope being 0 (the null hypothesis) vs. the slope being not 0 (the alternative hypothesis). Thus, the results of the test will be in the row corresponding to the slope.

Example 11: "Is Strength Associated with 60-pound Bench Presses?"

Parameter estimates:

Parameter	Estimate	Std. Err.	DF	T-Stat	P-Value
Intercept	63.536858	1.9564686	55	32.475277	<0.0001
Slope	1.491053	0.14973015	55	9.958268	<0.0001

Figure 12.0c

In the "Parameter estimates" table above you'll find the estimated slope to 1.49 with a standard error of 0.15. The test-statistic is also given (calculated as the ratio of the slope divided by the standard error) to be 9.96 with a p-value less than 0.0001.

One-Sided Tests: If you were only interested in testing whether or not there is a positive linear association (a right-sided alternative) then you divide the p-value in the output by 2.

Confidence Interval for the Slope

Unfortunately StatCrunch does not give confidence intervals for the regression coefficients. In order to calculate a confidence interval for the slope you will have to obtain the critical value off your t-table and manually calculate the interval.

Example 12: "Estimating the Slope for Predicting Maximum Bench Press"

From StatCrunch we know the estimated slope is 1.49 with a standard error of 0.15. If we wanted a 95% confidence interval we find our critical value off the t-table with n-2 = 55 degrees of freedom with right tail probability of 0.025. The critical value is 2.000. Thus, the 95% confidence interval is

$$1.49 \pm 2.000(0.15)$$
$$\Rightarrow (1.19, 1.79).$$

Residuals and Residual Standard Deviation

How do the observations vary about the line? Are there any unusual observations? To answer these questions we look at the residuals. A residual is the difference between an observed response and what it would be predicted to be using some model. For simple linear regression we calculate a residual as the difference between an observed response and what it would be predicted to be on the line.

Detecting Outliers using Residuals

We can use the residuals to help identify unusual observations. Abnormally large or small residuals may indicate an unusual observation. Recall, a Residual is the difference between an observed response and a predicted response.

Example 13: "Detecting an Underachieving College Student"

Open the data file "Georgia student survey". Suppose we are interested in the association between high school GPA ("*HSGPA*") and college GPA ("*CPGA*"). The regression output is displayed below:

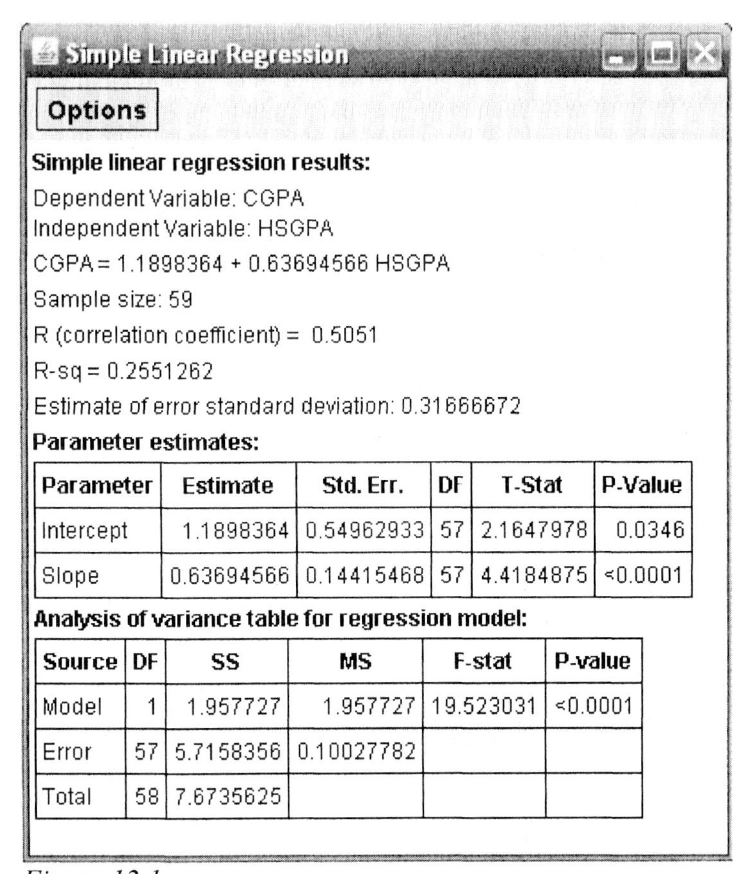
Simple Linear Regression

Options

Simple linear regression results:

Dependent Variable: CGPA
Independent Variable: HSGPA
CGPA = 1.1898364 + 0.63694566 HSGPA
Sample size: 59
R (correlation coefficient) = 0.5051
R-sq = 0.2551262
Estimate of error standard deviation: 0.31666672

Parameter estimates:

Parameter	Estimate	Std. Err.	DF	T-Stat	P-Value
Intercept	1.1898364	0.54962933	57	2.1647978	0.0346
Slope	0.63694566	0.14415468	57	4.4184875	<0.0001

Analysis of variance table for regression model:

Source	DF	SS	MS	F-stat	P-value
Model	1	1.957727	1.957727	19.523031	<0.0001
Error	57	5.7158356	0.10027782		
Total	58	7.6735625			

Figure 12.1a

The estimated regression equation is CPGA = 1.190 + 0.637HSGPA.

What is the residual for individual 59?

If you look at the data file, individual 59 had a high school GPA of 3.6 and a college GPA of 2.5. Based on the regression equation, the estimated (or predicted or fitted) college GPA of an individual with a high school GPA of 3.6 is:

CPGA = 1.190 + (0.637*3.6) = 3.48.

The **residual** for this individual is then 2.5-3.48 = -0.98.

To obtain the residuals for each response go to **Screen 2** and click on "Save residuals".

Screen 2

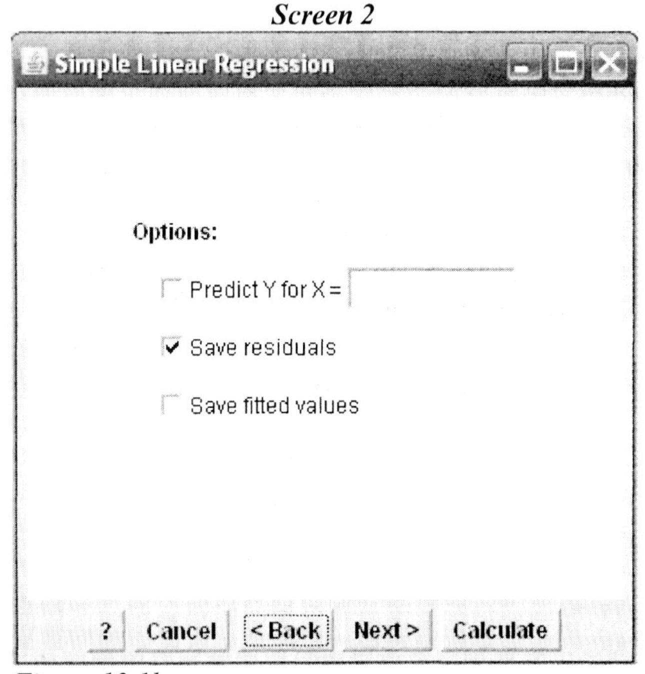

Figure 12.1b

The residuals will be saved in a new column on your data sheet called "*Residuals*".

In **Screen 3** you can ask StatCrunch to obtain a Histogram of Residuals.

Screen 3

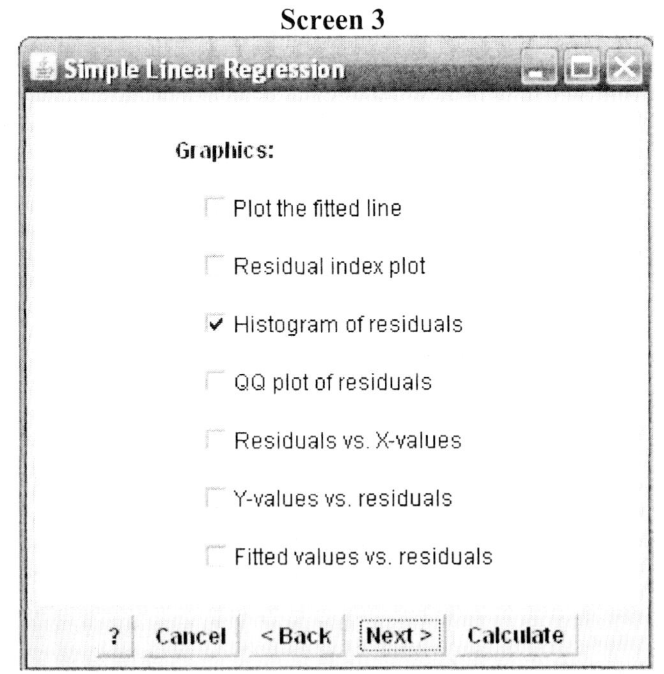

Figure 12.1c

The histogram of residuals will have the normal density superimposed and should look like:

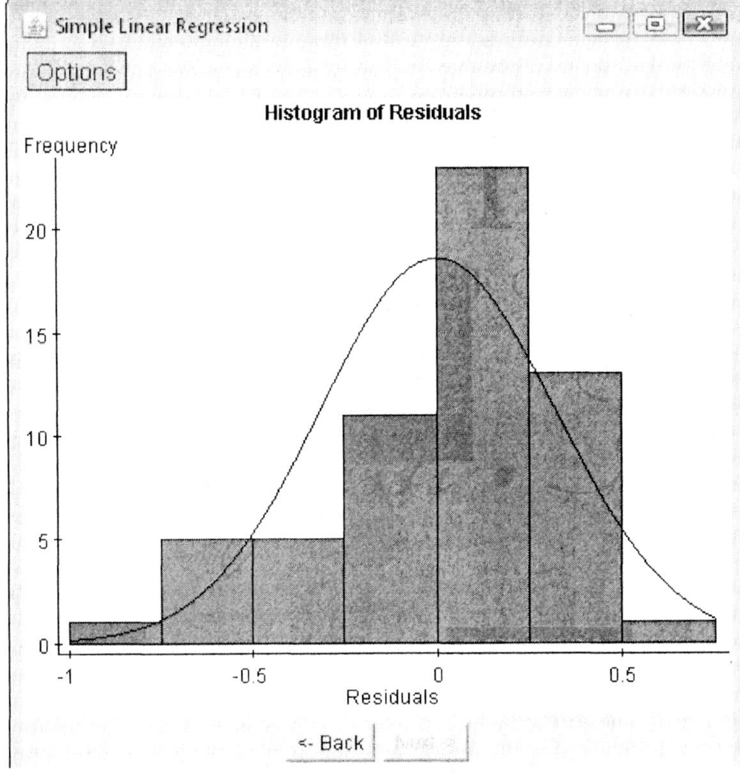

Figure 12.1d

Standardized Residuals

Sometimes it is better to look at Standardized Residuals. A Standardized Residual is calculated by dividing the residual by the standard deviation of the residuals. The Standardized Residuals are similar to z-scores. In which case, we can identify possible outliers or unusual observations as perhaps those with absolute residuals greater than 3. Although StatCrunch doesn't have a direct function to calculate the standardized residuals, we can calculate them by using the "Transform data" function.

Select **Data ➔ Transform Data**. You want to divide the column "*Residual*" by the standard error of the model which is the "Estimate of error standard deviation" = 0.31666672. Or equivalently, you can multiply "*Residual*" by 1/0.31666672 = 3.157894. Fill in the "Transform data" function as displayed below and click **Compute.**

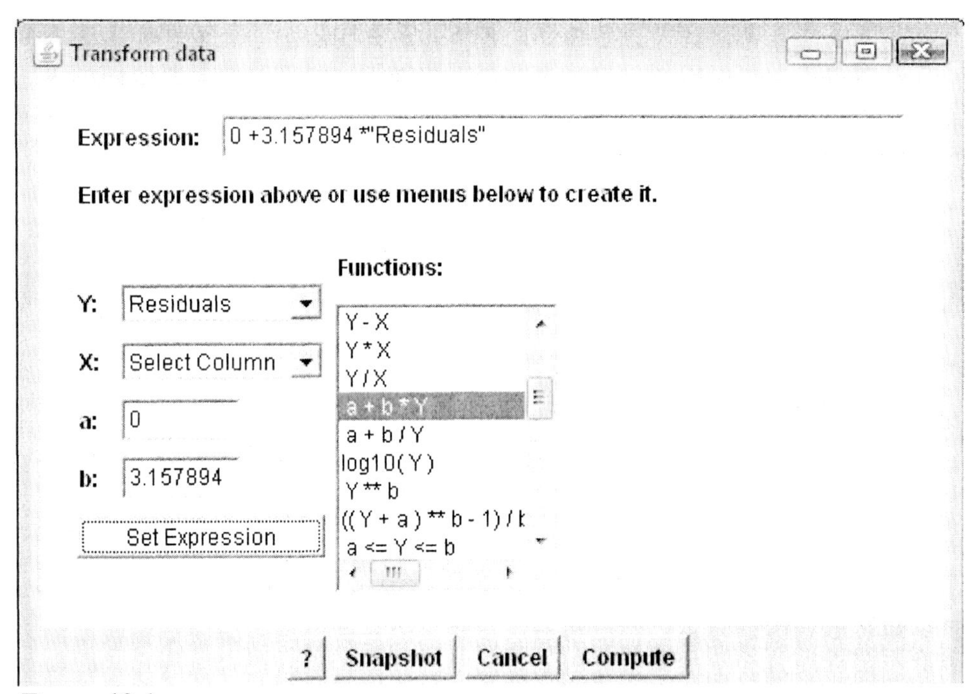

Figure 12.1e

There should be a new column labelled "0+3.157894*Residuals". You can change the title to "Std. Residuals". Note that the standardized residual of observation 59 is -3.10.

A histogram of the standardized residuals is displayed below (with bins starting at -3.25 and "binwidth" being 0.5):

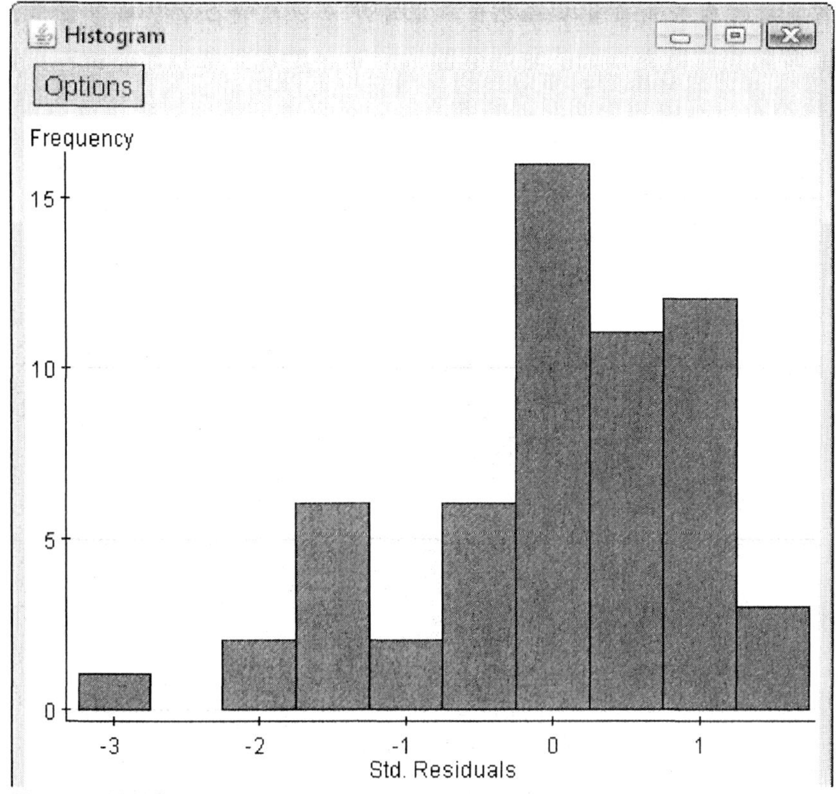

Figure 12.1f

Residual Standard Deviation

The residual standard deviation is called the "Estimate of error standard deviation" in StatCrunch.

Example 15: "Bench Press"

The residual standard deviation is given to be 8.003188. This tells us that typical deviation from the line is about 8.00. This value can also be calculated as the square root of the mean square error (MSE) found in the ANOVA table:

$$s = \sqrt{MSE} = \sqrt{64.051} = 8.0032$$

Confidence Intervals for the Mean and Predicted Response

We can use our regression model to predict responses of individuals as well as average responses for a group of individuals with 95% confidence. Unfortunately StatCrunch does not allow us to modify the confidence levels for these intervals.

Example 16: "Predicting Maximum Bench Press and Estimating Its Mean"

Open the file "high school female athletes ".

What is a 95% **Prediction Interval** for the maximum bench press of an individual that can do x = 11, 60-pound bench presses? What is a 95% **Confidence Interval** for the average maximum bench press of all individuals that can do x = 11 60-pound bench presses? Open the simple linear regression function (**Stat→Regression→Simple Linear), select** the appropriate x and y variables and click **Next** to go to **Screen 2.** Fill in Screen 2 as follows:

Screen 2

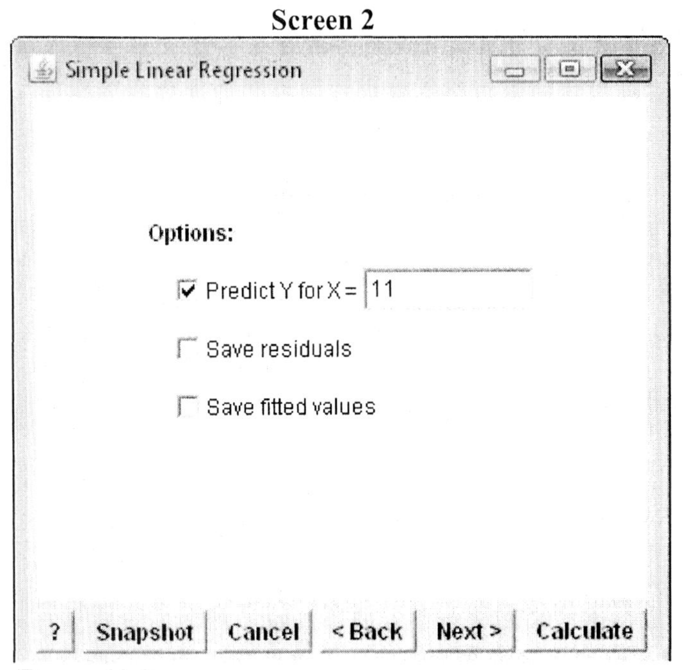

Figure 12.2a

Click **Calculate.** Here is a portion of the regression output (the bottom):

Predicted values:

X value	Pred. Y	s.e.(Pred. y)	95% C.I.	95% P.I.
11	79.93844	1.0600514	(77.81405, 82.06283)	(63.759613, 96.11726)

Figure 12.2b

The table above tells us that the predicted maximum bench press or estimated average maximum bench press at x = 11 is 79.94. The value "s.e.(Pred.y)" is the standard error for the estimated mean response. The 95% Confidence Interval for the estimated average maximum bench press for all individuals that can do 11, 60-pound bench presses is (77.8, 82.06). This is calculated as $79.94 \pm 2.00(1.06) \Rightarrow (77.8, 82.06)$.

Although the standard error for the predicted response isn't given, the 95% Prediction Interval for the maximum bench press of a single individuals that can do 11 60-pound bench presses is given to be (63.8, 96.1).

Analysis of Variance (ANOVA) for SLR

In general, the ANOVA for Regression allows us to test for the significance of ANY of the explanatory variables. Since SLR only has the one explanatory variable this test is equivalent (in result) to the t-test for the slope being 0 vs. not being 0. It can be shown that the F-statistic is the square of the t-statistic and that the p-values are equal.

Exercise 12.49: "Bench Press"

In the regression output the ANOVA table is:

Analysis of variance table for regression model:

Source	DF	SS	MS	F-stat	P-value
Model	1	6351.755	6351.755	99.16711	<0.0001
Error	55	3522.8064	64.051025		
Total	56	9874.562			

Figure 12.3

The F-statistic is given to be 99.16711 with a p-value less than 0.0001. If you recall, the t-statistic was given to be 9.958 which squared is equal to F-statistic.

The Total Sum-of-Squares (SS in the Total row) is the sum-of-squared residuals in the single data set of responses. This is given to be 9874.562. Essentially, the total sum-of-squares is broken up into two parts:

1) The amount explained by the regression: This is the reduction in sum-of-squares by fitting the regression, SS for the Model = 6351.755. This row is sometimes labelled "Regression".
2) The amount unexplained by the regression: This is the left over sum-of-squares in the regression model, SS Error = 3522.8064.

The "mean-squares" (MS) are calculated as the sum-of-squares divided by their degrees of freedom. Then, the F-statistic is the MS(Model) / MSE.

Exponential Regression: A Model for Non-linearity

In a linear model the rate of change in Y is constant (linear). Another common association between two variables is an exponential association. This type of association occurs when the rate of change in Y either increases or decreases as X increases. The model to describe this association is: $\mu_y = \alpha\beta^x$

We can find the estimates for the model parameters by fitting a simple linear regression on the *logarithm* of the response.

Example 18: "Explosion in Number of People Using the Internet"

Enter the data for the first three columns as they appear in Table 12.9 in your text into StatCrunch. In this example we are looking at the association between time and number of people using the Internet (in millions). If a linear association were appropriate we'd expect the change in the number of users to be constantly changing over time. Upon inspection of the Scatterplot it appears the change from year to year seems to be increasing over time. Thus, perhaps an Exponential model is appropriate. In order to pursue this type of association we can first transform the response to their logarithms. In the end it doesn't matter what base you use so let's use base 10. In the "Transform data" function calculate the logs base 10 as follows:

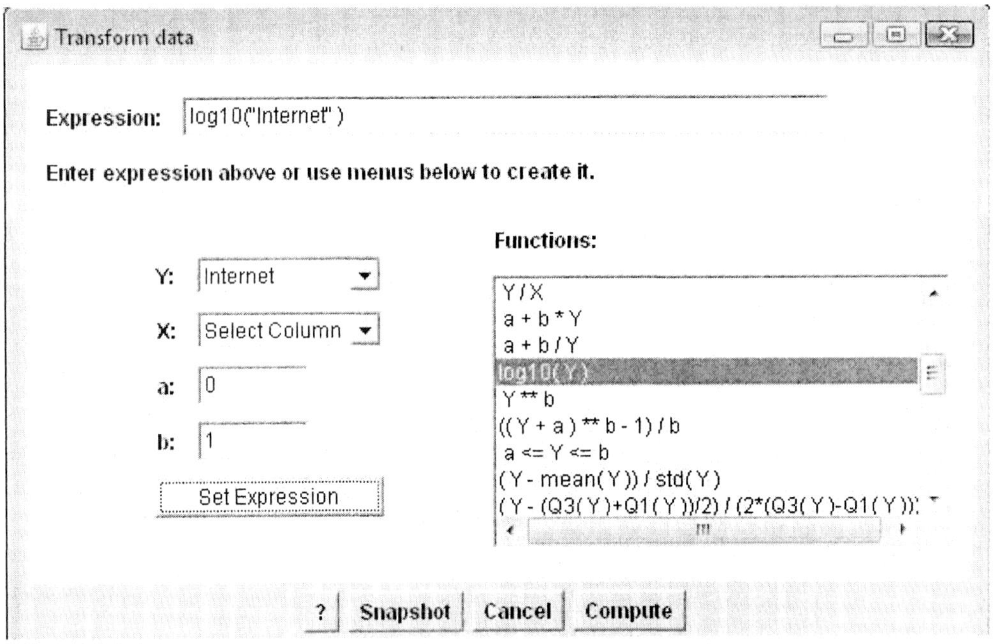

Transform data

Expression: log10("Internet")

Enter expression above or use menus below to create it.

Y: Internet

X: Select Column

a: 0

b: 1

Set Expression

Functions:

Y / X
a + b * Y
a + b / Y
log10(Y)
Y ** b
((Y + a) ** b - 1) / b
a <= Y <= b
(Y - mean(Y)) / std(Y)
(Y - (Q3(Y)+Q1(Y))/2) / (2*(Q3(Y)-Q1(Y));

? Snapshot Cancel Compute

Figure 12.4a

Click **Compute**. You should have a new column in your data file.

Row	YEAR	X	Internet	log10(Internet)
1	1995	0	16	1.20412
2	1996	1	36	1.5563025
3	1997	2	76	1.8808136
4	1998	3	147	2.1673174
5	1999	4	195	2.2900345
6	2000	5	369	2.5670264
7	2001	6	513	2.7101173

Figure 12.4b

The Scatterplot of the "*log10(Internet)*" vs. "*X*" with the Regression Line is:

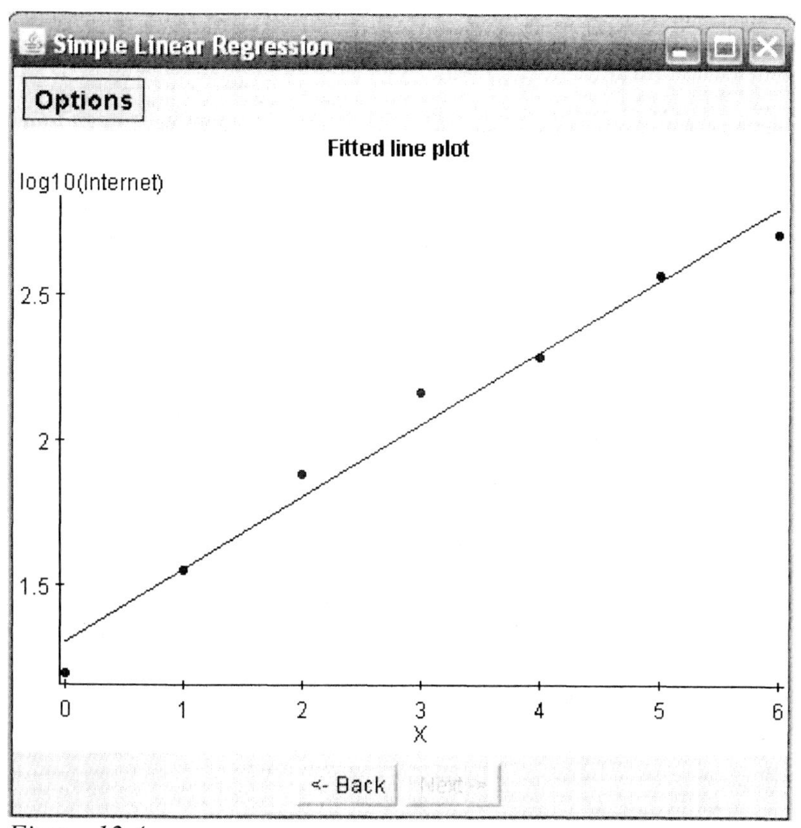

Figure 12.4c

Now, run the SLR function for Y = "*log10(Internet)*" vs. "*X*". The output is displayed below:

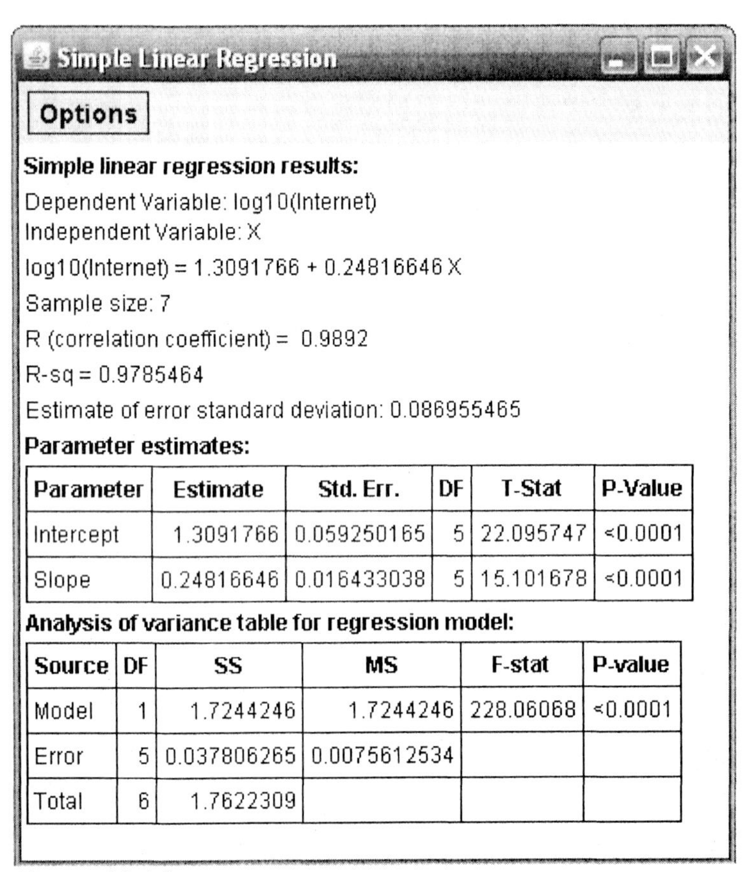

Simple Linear Regression

Options

Simple linear regression results:

Dependent Variable: log10(Internet)

Independent Variable: X

log10(Internet) = 1.3091766 + 0.24816646 X

Sample size: 7

R (correlation coefficient) = 0.9892

R-sq = 0.9785464

Estimate of error standard deviation: 0.086955465

Parameter estimates:

Parameter	Estimate	Std. Err.	DF	T-Stat	P-Value
Intercept	1.3091766	0.059250165	5	22.095747	<0.0001
Slope	0.24816646	0.016433038	5	15.101678	<0.0001

Analysis of variance table for regression model:

Source	DF	SS	MS	F-stat	P-value
Model	1	1.7244246	1.7244246	228.06068	<0.0001
Error	5	0.037806265	0.0075612534		
Total	6	1.7622309			

Figure 12.4d

From the output above the estimated simple linear regression equation is:

$$Log_{10}(Internet) = 1.3092 + 0.2482X$$

The estimates for the parameters in the exponential model are then found by taking the antilog of their respective estimates in the equation above. This is done as follows:

$$a = 10^{1.3092} = 20.38 \quad \text{and} \quad b = 10^{0.2482} = 1.7708$$

Thus, the estimated exponential regression equation is:

$$Internet = 20.38 \times 1.7708^{x}$$

Chapter 13 - Multiple Regression

In Chapters 3 and 12 we looked at examples of Simple Linear Regression. "Simple" means that there is exactly one quantitative explanatory variable. Quite often however, we have more than one quantitative explanatory variable. In these cases it is beneficial to consider a single model that includes all of them. This chapter will explore these types of models, called Multiple Linear Regression models (models with more than one explanatory variable).

First we will discuss how to obtain the estimated regression equation as well as other numerical summaries, such as multiple correlation and R-squared, using StatCrunch. Then, we will look at how StatCrunch can help us with some inferential procedures used in MLR. Finally, we will look at regression models that include categorical variables as explanatory variables, as well as a regression model for categorical response variable.

The Multiple Regression Equation

Example 2: "Predicting Selling Price Using House and Lot Size"

Open the file "house_selling_prices". We are interested in the association between selling price (in dollars) of a house and the house and lot sizes (in square feet). To obtain the regression output select **Stat➔Regression➔Multiple Linear.** In general, select the explanatory variables you wish to include in your model by clicking them in the "X Variables" list and select the response variable from the "Y Variable" drop down menu. In this example we want a regression model with house size ("*size*") and lot size ("*lot*") as the X (explanatory) variables and "*price*" as the Y (response) variable. Fill in **Screen 1** as follows and click **Calculate.**

Screen 1

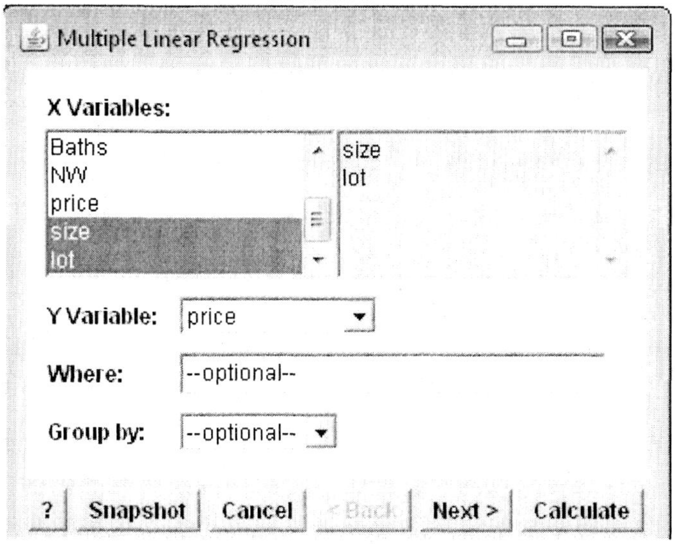

Figure 13.1a

The Multiple Linear Regression output is displayed below.

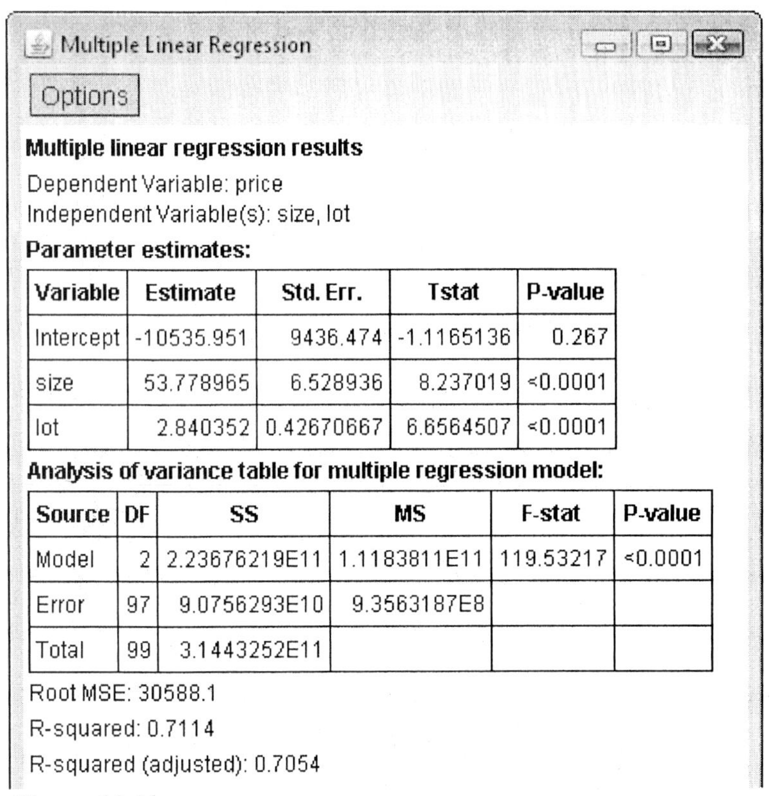
Multiple Linear Regression

Options

Multiple linear regression results

Dependent Variable: price
Independent Variable(s): size, lot

Parameter estimates:

Variable	Estimate	Std. Err.	Tstat	P-value
Intercept	-10535.951	9436.474	-1.1165136	0.267
size	53.778965	6.528936	8.237019	<0.0001
lot	2.840352	0.42670667	6.6564507	<0.0001

Analysis of variance table for multiple regression model:

Source	DF	SS	MS	F-stat	P-value
Model	2	2.23676219E11	1.1183811E11	119.53217	<0.0001
Error	97	9.0756293E10	9.3563187E8		
Total	99	3.1443252E11			

Root MSE: 30588.1
R-squared: 0.7114
R-squared (adjusted): 0.7054

Figure 13.1b

In the "Parameter Estimates" table you'll find a row for each parameter in the model (the intercept and a coefficient for each explanatory variable). The second column gives the least-squares estimates for the model coefficients. From the table above, the estimated intercept is -10535.951, the estimate coefficient on the variable "*size*" is 53.779, and the estimated coefficient on the variable "*lot*" is 2.840.

Thus, the estimate regression equation is

$$\hat{y} = -10535.951 + 53.779 size + 2.840 lot.$$

Just like a SLR model, you can estimate an average response or predict a single response by substituting in the chosen values of the X variables. For example, the predicted selling price of a 1240 square foot house ("*size*"=1240) on an 18000 square foot lot ("*lot*"=18000) is

$$\hat{y} = -10535.951 + 53.779(1240) + 2.840(18000) = 107270,$$

or $107,270.

Confidence Intervals for the Mean and Predicted Response

Confidence intervals for the mean response of Prediction intervals for a single response in MLR are not available in StatCrunch.

Correlation and R-Squared

In Chapter 3 we discussed the Linear Correlation coefficient, which is used to describe the strength of a linear association between the two variables. We also discussed R-squared which measures the proportional reduction in error by using the SLR model vs. using the sample mean to predict y. These concepts can be expanded to MLR.

In MLR we have the Multiple Correlation coefficient that describes the strength of the association between Y and a set of explanatory variables, and R-squared that measures the proportional reduction in error by using the MLR model vs. using the sample mean to predict y.

Example 3: "How Well Can We Predict House Selling Prices?"

Analysis of variance table for multiple regression model:

Source	DF	SS	MS	F-stat	P-value
Model	2	2.23676219E11	1.1183811E11	119.53217	<0.0001
Error	97	9.0756293E10	9.3563187E8		
Total	99	3.1443252E11			

Root MSE: 30588.1

R-squared: 0.7114

R-squared (adjusted): 0.7054

Figure 13.1c

In the MLR output above, StatCrunch gives us R-squared = 0.7114. Thus, we can reduce error by 71.14% by predicting selling price using the MLR model on house and lot size.

This value can also be calculated using the ANOVA table above as follows:

$$R^2 = \frac{223676}{314433}$$

StatCrunch doesn't give us Multiple Correlation coefficient; however it can be calculated as the positive square root of R-squared. Thus, the Multiple Correlation is equal to $\sqrt{0.7114} = 0.84$. Unlike the Linear Correlation (which ranges from -1 to +1); the Multiple Correlation is always positive between 0 and 1. The closer it is to 1 the stronger the association.

Sometimes we may wish to see individual linear associations between the response and each of the explanatory variables. You can visualize these associations by producing several

Scatterplots, one for the response variable vs. each of the explanatory variables. Also, you might want to know how strong these linear associations are by looking at their individual Linear Correlation coefficients. Select **Stats➔Summary Stats➔Correlation.** Click on the variables on the left that you are interested in. For our example, choose "*price*", "*size*", and "*lot*". Then click **Calculate**.

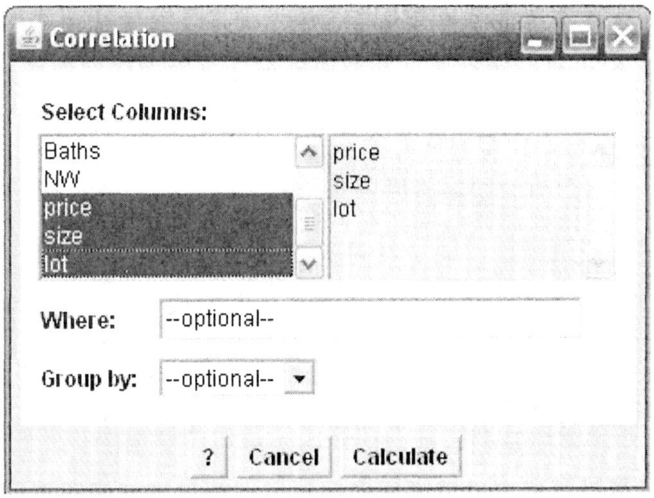

Correlation

Select Columns:

Baths
NW
price
size
lot

price
size
lot

Where: --optional--

Group by: --optional--

? Cancel Calculate

Figure 13.2a

The output below is the Correlation Matrix for the variables selected. It gives the Linear Correlation for each pair of variables (including each pair of explanatory variables).

Correlation

Options

Correlation matrix:

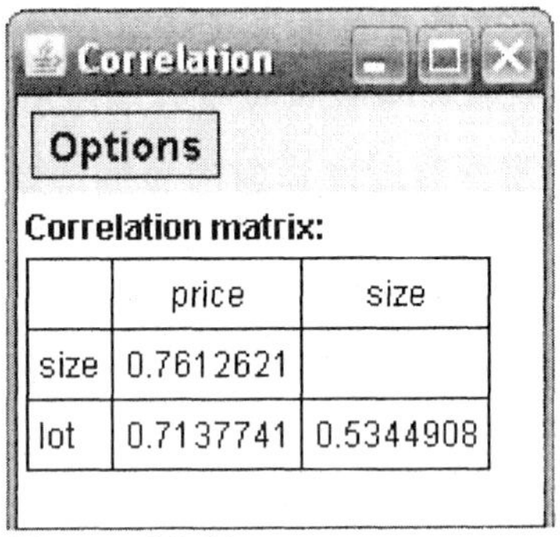

	price	size
size	0.7612621	
lot	0.7137741	0.5344908

Figure 13.2b

For example, the linear correlation between "*price*" and "*size*" is 0.761.

Inference for Regression Coefficients

Very similar to the SLR output, in the Parameter Estimates table, we are given the parameter estimates, the standard error of these estimates, the test statistics for the null hypothesis that the parameter is 0, and the p-value for the 2-sided alternative that the parameter is not 0. The only thing different is that the degrees of freedom are missing. For all t-tests for an individual coefficient the degree of freedom can be found in the Error row in the ANOVA table (calculated as df = n – the number of coefficients in the model).

Below is the **Parameter estimates** table for the housing price example.

Parameter estimates:

Variable	Estimate	Std. Err.	Tstat	P-value
Intercept	-10535.951	9436.474	-1.1165136	0.267
size	53.778965	6.528936	8.237019	<0.0001
lot	2.840352	0.42670667	6.6564507	<0.0001

Figure 13.3

The tests of significance for the explanatory variables have very small p-values, both less than 0.001. This indicates that both house size and lot size appear to be significant in describing average selling price.

Example 4: "What Helps Predict a Female Athlete's Weight?"

This example is exploring how certain physical characteristics (such as height ('*HGT*"), body fat percentage ("*%BF*"), and "*AGE*") can predict the body weight of a female athlete ("*TBW*"). Open the file "college athletes". Obtain the MLR output with "*HGT*", "*%BF*" and "*AGE*" as the X variables and "*TBW*" as the Y variable. The output is displayed below:

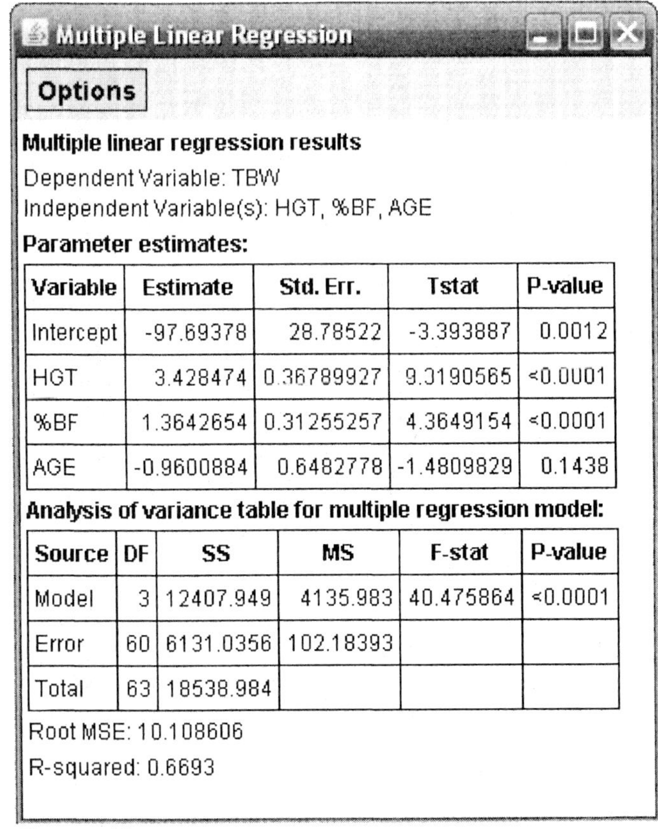
Multiple Linear Regression

Options

Multiple linear regression results

Dependent Variable: TBW

Independent Variable(s): HGT, %BF, AGE

Parameter estimates:

Variable	Estimate	Std. Err.	Tstat	P-value
Intercept	-97.69378	28.78522	-3.393887	0.0012
HGT	3.428474	0.36789927	9.3190565	<0.0001
%BF	1.3642654	0.31255257	4.3649154	<0.0001
AGE	-0.9600884	0.6482778	-1.4809829	0.1438

Analysis of variance table for multiple regression model:

Source	DF	SS	MS	F-stat	P-value
Model	3	12407.949	4135.983	40.475864	<0.0001
Error	60	6131.0356	102.18393		
Total	63	18538.984			

Root MSE: 10.108606

R-squared: 0.6693

Figure 13.4

From the output above we find the regression equation is

$$\hat{y} = -97.693 + 3.429HGT + 1.364\%BF - 0.960AGE$$

In this example we want to test the significance of "*AGE*" as an explanatory variable for average "*TBW*", in the model with "*HGT*" and "*%BF*". The estimated coefficient on "*AGE*" is -0.960 with a standard error of 0.648.

The estimated coefficient (or the estimated effect of "*AGE*") is the estimated change in average "*TBW*" by aging 1 year, for fixed "*HGT*" and "*%BF*". We'd expect "*TBW*" to go down by 0.96, on average, when aging 1 year while maintaining the same "*HGT*" and "*%BF*".

The test statistic for the test that the coefficient for "*AGE*" is 0 (null hypothesis) vs. not 0 (alternative hypothesis) is -1.48 (-0.96/0.648). This results in a p-value of 0.14 (calculated off the t-distribution with 60 degrees of freedom). The large p-value shows little evidence that "*AGE*" helps to predict average "*TBW*", after accounting for "*HGT*" and "*%BF*".

Example 5: "What's Plausible for the Effect of Age on Weight?"

Here we want to estimate the effect of "*AGE*". This is simply the coefficient, -0.96. If you want to produce a confidence interval for the effect you will have to calculate it by hand as StatCrunch doesn't produce confidence intervals for regression parameters. StatCrunch does however give us the estimate (-0.96) and standard error (0.648). For a 95% confidence interval (based on 60 degrees of freedom) the critical value is 2.00 (from your t-table). Thus, the 95% confidence interval for the effect of "*AGE*" on average "*TBW*" (in the model accounting for "*HGT*" and "*%BF*") is:

$$-0.96 \pm 2.00(0.648)$$
$$\Rightarrow (-2.3, 0.3).$$

Variability around the Regression Equation

The estimated variability around the regression equation (otherwise known as the Residual Variability) is found as MSE. The estimated standard deviation around the regression equation (otherwise known as the Residual Standard Deviation) is then found as the positive square root of MSE, given in the bottom of MLR output as "Root MSE".

Example 6: "Estimating Variability of Female Athletes' Weight"

The estimated variability of female athletes' weight (about the regression equation) is 102.18 ("MSE"). The estimated standard deviation of female athletes' weight (about the regression equation) is 10.11 ("Root MSE").

Analysis of Variance (ANOVA) for MLR

The ANOVA for Regression allows us to test for the significance of ANY of the explanatory variables.

Example 7: "The F-Test for Predictors of Athletes' Weight"

Here is the ANOVA table from the regression output we saw earlier:

Analysis of variance table for multiple regression model:

Source	DF	SS	MS	F-stat	P-value
Model	3	12407.949	4135.983	40.475864	<0.0001
Error	60	6131.0356	102.18393		
Total	63	18538.984			

Figure 13.5

The F-statistic is 40.476 with a p-value less than 0.0001. The p-value is calculated off the F-distribution with df1 = 3 and df2 = 60. The small p-value indicates very strong evidence that at least one of the explanatory variables is useful in predicting average "*TBW*".

Residual Plots

Residual plots can help assess the validity of the model assumptions such as normality and equal variability.

Example 8: "Residuals for House Selling Price"

Open the MLR function and select the appropriate variables in **Screen 1**. Click **Next** to go to **Screen 2** and click "Save residuals". Click **Calculate.** The residuals will be saved in a new column in your data file called "*Residuals*". If you want to standardize the residuals you'll need to transform the column by dividing it by the standard deviation of residuals ("Root MSE"). Follow the steps outlined in Chapter 12 to standardize the residuals. The histogram of standardized residuals is:

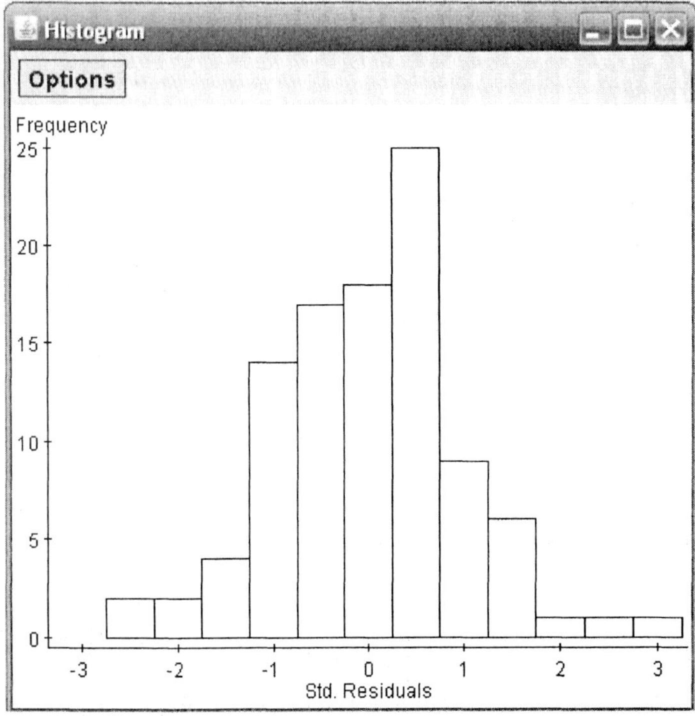

Figure 13.6a

Example 9: "Another Residual Plot for House Selling Prices"

Here is a Scatterplot of the standardized residuals vs. house size:

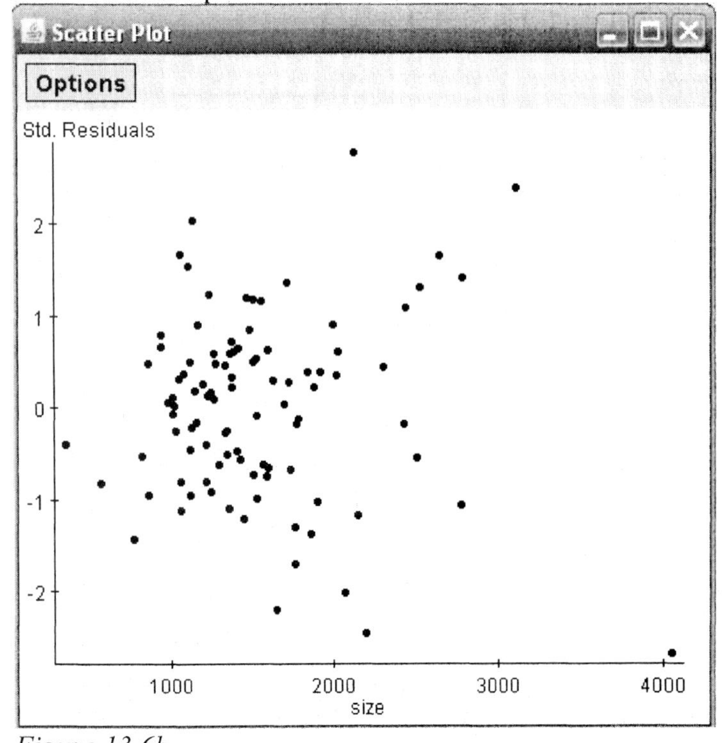

Figure 13.6b

Categorical Explanatory Variables

In order to include categorical variables in a regression model you include a set of Indicator variables. For each level of a categorical variable an Indicator variable can be created. Each Indicator variable equals 1 if the observation belongs to the level of the observation; otherwise the Indicator variable equals 0. For each observation, exactly one indicator equals 1, all others are zero. In general, if the categorical variable has k levels then you would include k-1 of the indicator variables in the regression model.

Example 10: "Including Region in Regression for House Selling Price"

Another variable in this example is "*Quadrant*" with 4 levels: NW, NE, SW, and SE. Suppose we are only interested in the difference in average house price between the NW and all others, while accounting for house size. First, create another variable, say "*region*", which has two levels, either "*NW*" (north west) or "*Other*". Since the variable "*region*" only has two levels we need one indicator variable. In your data set you have the following indicator for "*region*":

$$NW = \begin{cases} 1, & \text{if region} = \text{NW} \\ 0, & \text{else} \end{cases}$$

To fit a regression model with house size and region as your explanatory variables, select the MLR function and choose "*size*" and "*NW*" as your X variables ("*NW*" is the variable used to describe the effects of region) and "*price*" as your Y variable. Click **Calculate.** You should obtain the following output:

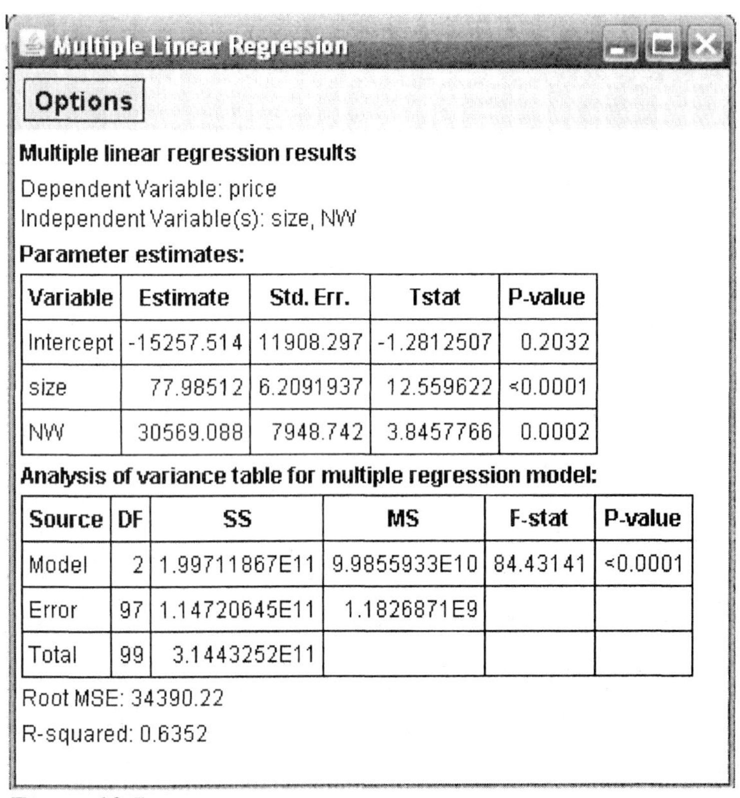
Multiple Linear Regression

Options

Multiple linear regression results

Dependent Variable: price

Independent Variable(s): size, NW

Parameter estimates:

Variable	Estimate	Std. Err.	Tstat	P-value
Intercept	-15257.514	11908.297	-1.2812507	0.2032
size	77.98512	6.2091937	12.559622	<0.0001
NW	30569.088	7948.742	3.8457766	0.0002

Analysis of variance table for multiple regression model:

Source	DF	SS	MS	F-stat	P-value
Model	2	1.99711867E11	9.9855933E10	84.43141	<0.0001
Error	97	1.14720645E11	1.1826871E9		
Total	99	3.1443252E11			

Root MSE: 34390.22

R-squared: 0.6352

Figure 13.7

The regression equation is

$$\hat{y} = -15257.51 + 77.99size + 30569.09NW$$

When including categorical variables in a regression model (by including a set of indicators) you are able to compare conventional regression models (with quantitative explanatory variables only) across the levels of the categorical. If we fit the model above for both levels of region we have two SLR models for average price given house size

i) The estimated SLR equation for houses in the NW is:
$$\hat{y} = -15257.51 + 77.99size + 30569.09(1) = 15311.58 + 77.99size$$

ii) The estimated SLR equation for houses in the other areas is:
$$\hat{y} = -15257.51 + 77.99size + 30569.09(0) = -15257.51 + 77.99size$$

Indicator Columns

Suppose you wanted to include "*Quadrant*" as the explanatory variable. First, you would have to create additional indicator variables. The four indicator variables for "*Quadrant*" are:

$$NW = \begin{cases} 1, & \text{if region} = \text{NW} \\ 0, & \text{else} \end{cases}, \quad SW = \begin{cases} 1, & \text{if region} = \text{SW} \\ 0, & \text{else} \end{cases},$$

$$NE = \begin{cases} 1, & \text{if region} = \text{NE} \\ 0, & \text{else} \end{cases}, \quad SE = \begin{cases} 1, & \text{if region} = \text{SE} \\ 0, & \text{else} \end{cases}$$

In StatCrunch, select **Data→Indicator Columns.** Then, click on "*Quadrant*" and hit **Create Indicators.**

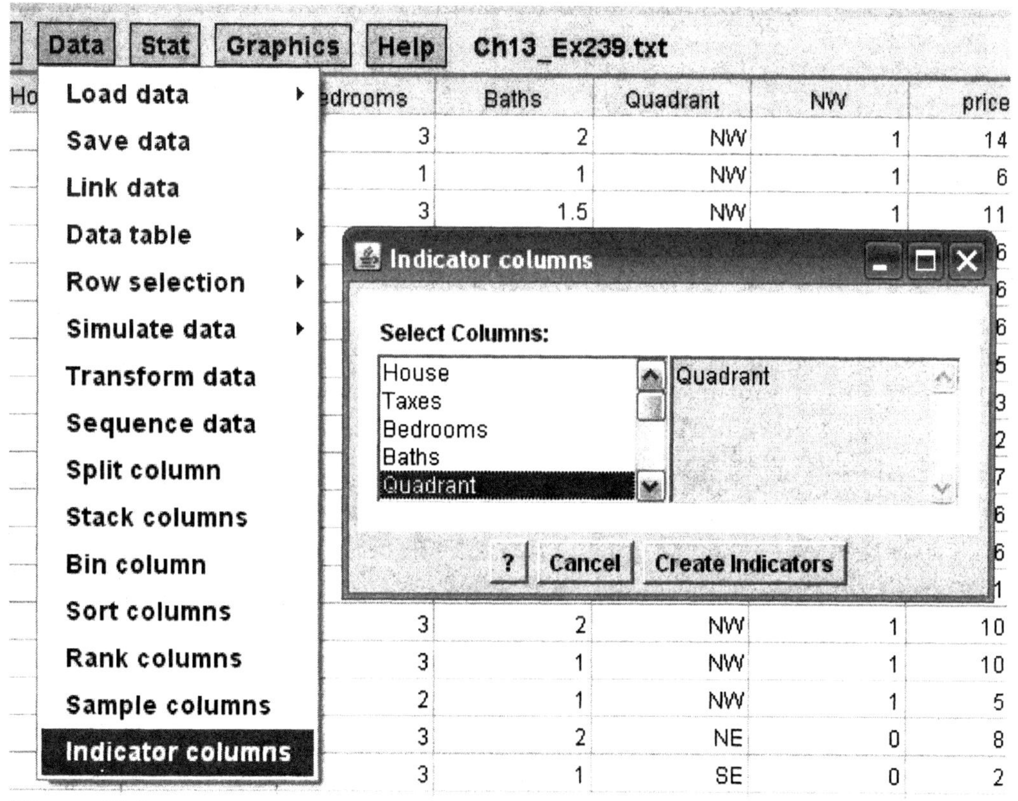

Figure 13.8a

An indicator variable will be created for each unique level of "*Quadrant*". Here is a portion of the data (I changed the titles for the indicators):

	Quadrant	NW	price	size	lot	NE	NW	SE	SW
2	NW	1	145000	1240	18000	0	1	0	0
1	NW	1	68000	370	25000	0	1	0	0
.5	NW	1	115000	1130	25000	0	1	0	0
2	SW	0	69000	1120	17000	0	0	0	1
2	NW	1	163000	1710	14000	0	1	0	0
2	NW	1	69900	1010	8000	0	1	0	0
2	NW	1	50000	860	15300	0	1	0	0
2	NW	1	137000	1420	18000	0	1	0	0
2	NW	1	121300	1270	16000	0	1	0	0
2	NW	1	70000	1160	8000	0	1	0	0
2	NE	0	64500	1220	12000	1	0	0	0
2	NW	1	167000	1690	30000	0	1	0	0
2	NW	1	114600	1380	15500	0	1	0	0
2	NW	1	103000	1590	16800	0	1	0	0
1	NW	1	101000	1050	16000	0	1	0	0
1	NW	1	50000	770	22100	0	1	0	0
2	NE	0	85000	1410	12000	1	0	0	0
1	SE	0	22500	1060	3500	0	0	1	0
2	NW	1	90000	1300	17500	0	1	0	0

Figure 13.8b

In order to fit "*Quadrant*" as an explanatory you will need to include three of the indicators (any 3) in your regression model. For illustration, I will use "*NE*", "*NW*" and "*SE*". Open the MLR function and choose "*size*", "*NE*", "*NW*" and "*SE*" as your X variables and "*price*" as your Y variable. Click **Calculate.** You should obtain the following output:

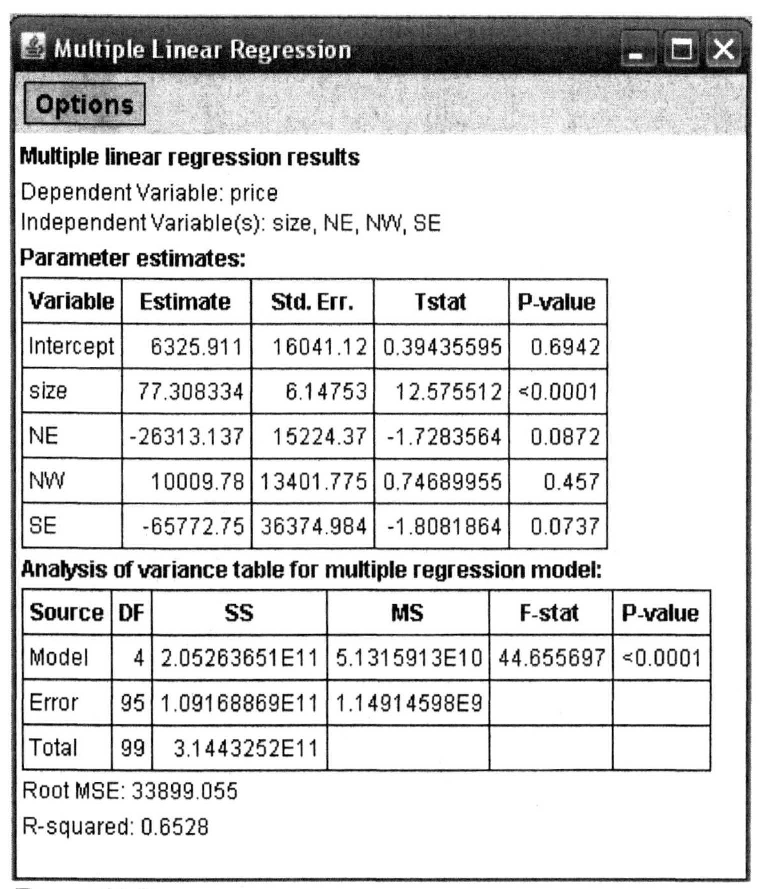
Multiple Linear Regression

Options

Multiple linear regression results

Dependent Variable: price

Independent Variable(s): size, NE, NW, SE

Parameter estimates:

Variable	Estimate	Std. Err.	Tstat	P-value
Intercept	6325.911	16041.12	0.39435595	0.6942
size	77.308334	6.14753	12.575512	<0.0001
NE	-26313.137	15224.37	-1.7283564	0.0872
NW	10009.78	13401.775	0.74689955	0.457
SE	-65772.75	36374.984	-1.8081864	0.0737

Analysis of variance table for multiple regression model:

Source	DF	SS	MS	F-stat	P-value
Model	4	2.05263651E11	5.1315913E10	44.655697	<0.0001
Error	95	1.09168869E11	1.14914598E9		
Total	99	3.1443252E11			

Root MSE: 33899.055

R-squared: 0.6528

Figure 13.8c

The multiple regression equation is

$$\hat{y} = 6325.91 + 77.31 size - 26313.14 NE + 10009.78 NW - 65772.75 SE$$

Or

i) $\hat{y} = -19987.23 + 77.31 size$ for houses in thc NE
ii) $\hat{y} = 16335.69 + 77.31 size$ for houses in the NW
iii) $\hat{y} = -59446.84 + 77.31 size$ for houses in the SE
iv) $\hat{y} = 6325.91 + 77.31 size$ for houses in the SW

Example 11: "Comparing Winning High Jumps for Men and Women"

This example wants to compare high jump heights over time (a quantitative explanatory variable), accounting for gender (a categorical explanatory variable). Open the file "high jump". You will have to modify the data so that there are three variables, "*height*", "*year*" and "*gender*". For example, here is a portion of the data how it should look (sorted by "*year*"):

Row	Year	Height	Gender	female	male	Year28
1	1896	1.81	male	0	1	-32
2	1900	1.9	male	0	1	-28
3	1904	1.8	male	0	1	-24
4	1908	1.905	male	0	1	-20
5	1912	1.93	male	0	1	-16
6	1920	1.935	male	0	1	-8
7	1924	1.98	male	0	1	-4
8	1928	1.94	male	0	1	0
9	1928	1.59	female	1	0	0
10	1932	1.97	male	0	1	4
11	1932	1.657	female	1	0	4
12	1936	2.03	male	0	1	8
13	1936	1.6	female	1	0	8
14	1948	1.98	male	0	1	20
15	1948	1.68	female	1	0	20

Figure 13.9a

Also included in the table are "*Year28*" (= Year-1928) and the two possible indicators ("*male*" and "*female*") for "*gender*'. I will use the indicator "*male*".

Open the MLR function and choose "*Year28*", and "*male*" as your X variables and "*height*" as your Y variable. Click **Calculate.** You should obtain the following output:

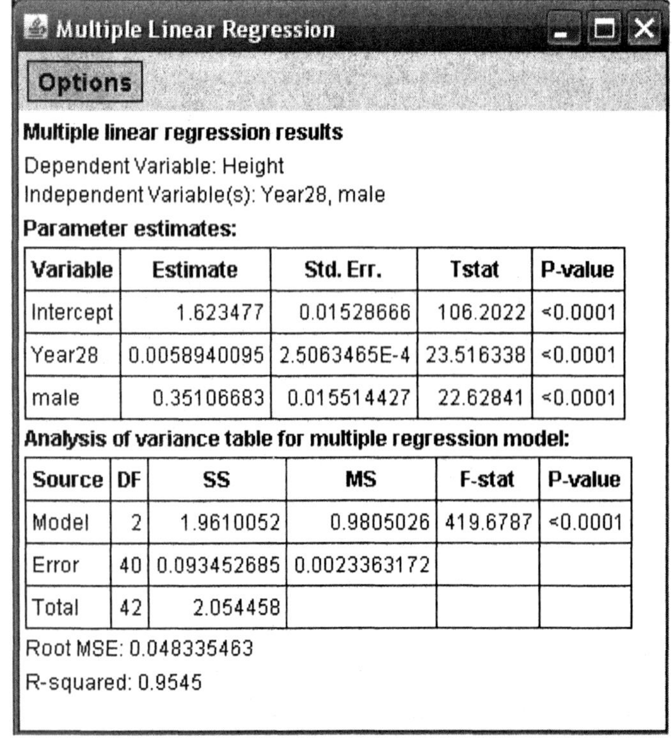
Multiple Linear Regression

Options

Multiple linear regression results

Dependent Variable: Height

Independent Variable(s): Year28, male

Parameter estimates:

Variable	Estimate	Std. Err.	Tstat	P-value
Intercept	1.623477	0.01528666	106.2022	<0.0001
Year28	0.0058940095	2.5063465E-4	23.516338	<0.0001
male	0.35106683	0.015514427	22.62841	<0.0001

Analysis of variance table for multiple regression model:

Source	DF	SS	MS	F-stat	P-value
Model	2	1.9610052	0.9805026	419.6787	<0.0001
Error	40	0.093452685	0.0023363172		
Total	42	2.054458			

Root MSE: 0.048335463

R-squared: 0.9545

Figure 13.9b

The multiple regression equation is

$$\hat{y} = 1.6234 + 0.0059(year - 1928) + 0.3511 male$$

Or

i) $\hat{y} = 1.9745 + 0.0059(year - 1928)$ for males
ii) $\hat{y} = 1.6234 + 0.0059(year - 1928)$ for females

Below is the Scatterplot of "*height*" vs. "*year*" grouped by gender:

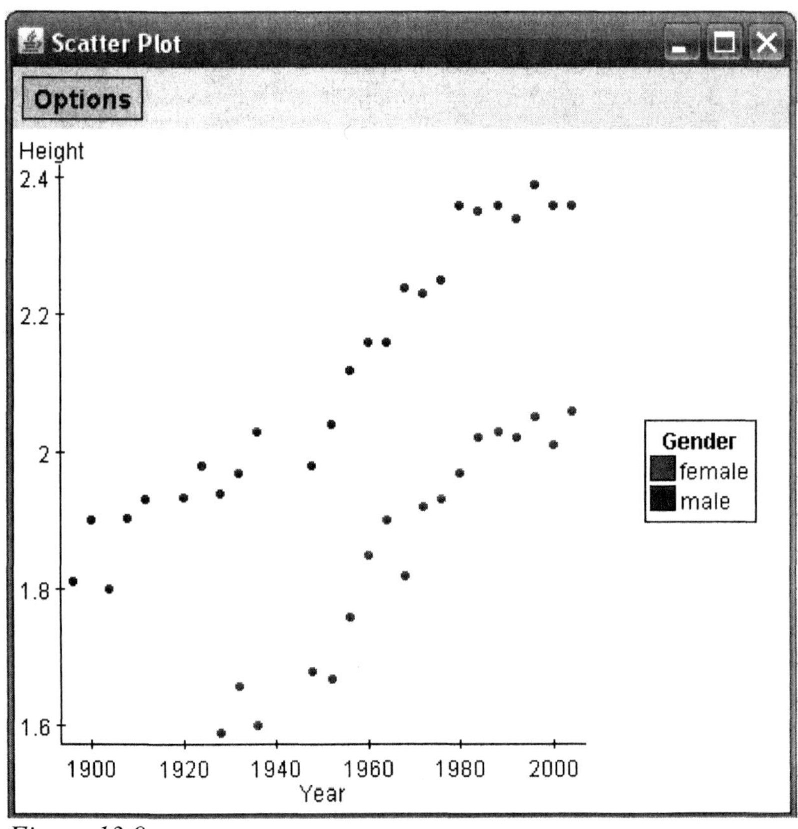

Figure 13.9c

Logistic Regression

So far all of our regression models have been for quantitative response variables. If your response variable is categorical we use a logistic regression model. The logistic regression model then can be used to estimate the probability that the categorical response takes on a particular level. If your response variable has two levels we call this **binary logistic regression**.

Example 12: "Annual Income and Having a Travel Credit Card"

Open the "Credit card and income" data file from your CD.

Is there an association with annual income and having a credit card? In this example, the response variable y, whether or not a person has a credit card, is categorical with two levels, either they do have a credit card (y=1) or they don't (y=0). The explanatory variable is annual income in thousands or Euros.

To obtain the logistic regression equation, select **Stat➔Regression➔with Data.** We use the "with Data" option in this example as we have the values of each variable listed for each of the 100 sampled individuals.

Select "*income*" as your X variable, "*y*" as your Y variable, and indicate that "*y*"=1 is a success. Click **Calculate.**

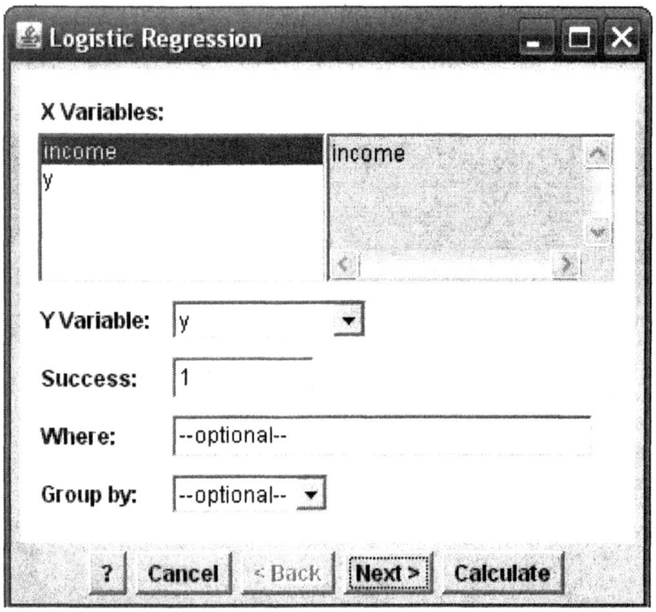

Logistic Regression

X Variables:
income
y
income

Y Variable: y
Success: 1
Where: --optional--
Group by: --optional--

? Cancel < Back Next > Calculate

Figure 13.10a

The following is a portion of the logistic regression output.

Logistic regression results

Dependent Variable: y (Success = 1)
Independent Variable(s): income

Parameter estimates

Variable	Estimate	Std. Err.	Zstat	P-value	Odds Ratio	95% Low. Lim.	95% Up. Lim.
Intercept	-3.517947	0.7103348	-4.9525194	<0.0001			
income	0.105408914	0.026157392	4.029794	<0.0001	1.1111649	1.0556328	1.1696182

Figure 13.10b

In the table above we find the estimated coefficients to be $\hat{\alpha} = -3.518$ and $\hat{\beta} = 0.105$. Thus, the estimated logistic regression equation is

$$\hat{p} = \frac{e^{-3.518+0.105 income}}{1+e^{-3.518+0.105 income}}$$

Example 14: "Estimating Proportion of Students who've Used Marijuana"

Is there an association between alcohol and tobacco use and marijuana use? A survey was conducted and each individual was asked whether or not they have ever used each of alcohol, cigarettes, or marijuana. A multiple logistic regression model can be used to estimate the probability of marijuana use given whether or not an individual used alcohol and/or cigarettes.

Here both explanatory variables are categorical. Define the following indicators:

$$alcohol = \begin{cases} 1, & \text{if they said yes to alcohol use} \\ 0, & \text{if they said no to alcohol use} \end{cases}$$

$$cigarettes = \begin{cases} 1, & \text{if they said yes to cigarette use} \\ 0, & \text{if they said no to cigarette use} \end{cases}$$

The counts are given for all combinations of levels of the three variables. You'll have to edit the dataset as follows:

Row	alcohol	cigarette	marijuana	total
1	1	1	911	1449
2	1	0	44	500
3	0	1	3	46
4	0	0	2	281
5				

Figure 13.11a

For example, out of 1449 individuals that used alcohol and cigarettes, 911 said they used marijuana.

In order to obtain the logistic regression equation, select **Stat➔Regression➔with Summary.** First, select "*alcohol*" and "*cigarettes*" as your X-variables. Select "*marijuana*" as the frequency of successes (here a success means that an individual used marijuana). You also have to indicate the total from each of the category combinations for the explanatory variables. Click **Calculate.**

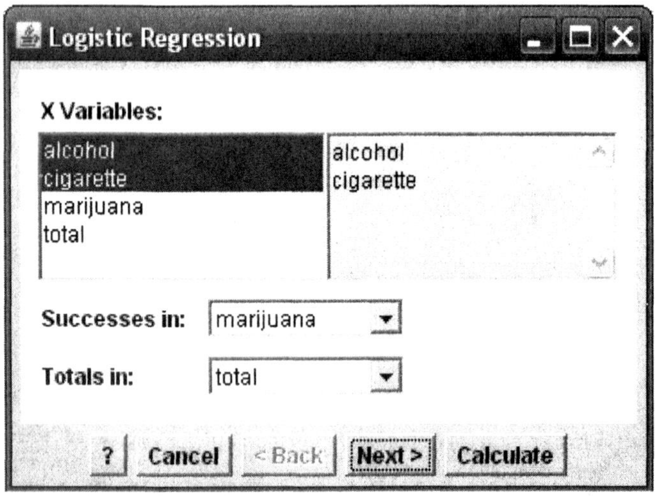

Logistic Regression

X Variables:

alcohol
cigarette
marijuana
total

alcohol
cigarette

Successes in: marijuana

Totals in: total

? Cancel < Back Next > Calculate

Figure 13.11b

The following is a portion of the logistic regression output.

Parameter estimates

Variable	Estimate	Std. Err.	Zstat	P-value	Odds Ratio	95% Low. Lim.	95% Up. Lim.
Intercept	-5.3090425	0.47519702	-11.1722975	<0.0001			
alcohol	2.9860144	0.464678	6.4259863	<0.0001	19.806585	7.9665027	49.24379
cigarette	2.8478892	0.1638394	17.3822	<0.0001	17.25133	12.512948	23.78403

Figure 13.11c

In the table above we find the estimated coefficients to be $\hat{\alpha} = -5.309$, $\hat{\beta}_1 = 2.986$ and $\hat{\beta}_2 = 2.848$. Thus, the estimated logistic regression equation is

$$\hat{p} = \frac{e^{-5.309+2.986alcohol+2.848cigarettes}}{1+e^{-5.309+2.986alcohol+2.848cigarettes}}$$

Chapter 14 - Comparing Groups: Analysis of Variance Methods

In Chapter 10 we looked at methods to determine how a quantitative response variable depends on a categorical explanatory (that is, we compared two independent population means). There we carried out a t-test to determine if the two means differ and found confidence intervals for the mean difference. Now, suppose we have a categorical explanatory variable with more than two levels (or, we have more than two means to compare). Or, what if we have two categorical explanatory variables?

In this chapter we will look at methods to determine how a quantitative response variable depends on one or two categorical explanatory variables with any number of levels. First we will look at an Analysis of Variance (ANOVA) F-test to determine if the quantitative response depends on a single categorical explanatory variable (One-Way ANOVA F-test for any mean differences). Here, we will also consider follow-up analysis to explore specific differences between the means. Last, we will look at Two-Way ANOVA F-tests to determine if the response variable depends on each of the explanatory variables (tests for main factor effects). In addition we will consider a test to determine if the main factors interact.

One-Way Classification

In the One-Way classification we have a single quantitative response variable with a single categorical explanatory variable with any number of levels. Another way to look at this is that we have any number of population means (one for each level of the categorical explanatory variable).

One-Way ANOVA F-Test

The first step in this type of analysis involves a general test to determine if the response depends on the explanatory variable. In other words, are there any differences among any of the population means (one for each level of the categorical explanatory variable)? This test is called the One-Way ANOVA F-Test.

Examples 3: "ANOVA for Customers' Telephone Holding Times"

This example wishes to explore the average time until a caller hangs up while waiting on hold after calling an airline to make a reservation. In particular they are interested in the effects of different recordings (ads, Muzak, or classical music) on average holding time. Is there a difference in average holding times between the different recordings? If so, how do they differ? The quantitative response variable is holding time (in minutes). The categorical explanatory variable is the type of recording with three levels: ads, Muzak or classical music. The data for this example can be found on page 627 in your text.

There are two ways you can enter the data. The first is to create a column of times for each type of recording. The second is to have a column for type of recording and another for the time. Both are illustrated below:

Row	ads	muzak	classical	recording	time
1	5	0	13	ads	5
2	1	1	9	ads	1
3	11	4	8	ads	11
4	2	6	15	ads	2
5	8	3	7	ads	8
6				muzak	0
7				muzak	1
8				muzak	4
9				muzak	6
10				muzak	3
11				classical	13
12				classical	9
13				classical	8
14				classical	15
15				classical	7

Figure 14.1a

The One-Way ANOVA F-test tests the following:

$H_0 : \mu_{muzak} = \mu_{ads} = \mu_{classical}$ (all means are equal)

H_a : At least two means are different (not all means are equal)

To carry out the One-Way ANOVA F-test, select **Stat➔ANOVA➔One Way.** Then, choose the columns you wish to compare (if you have entered the data in separate columns) and click **Calculate.**

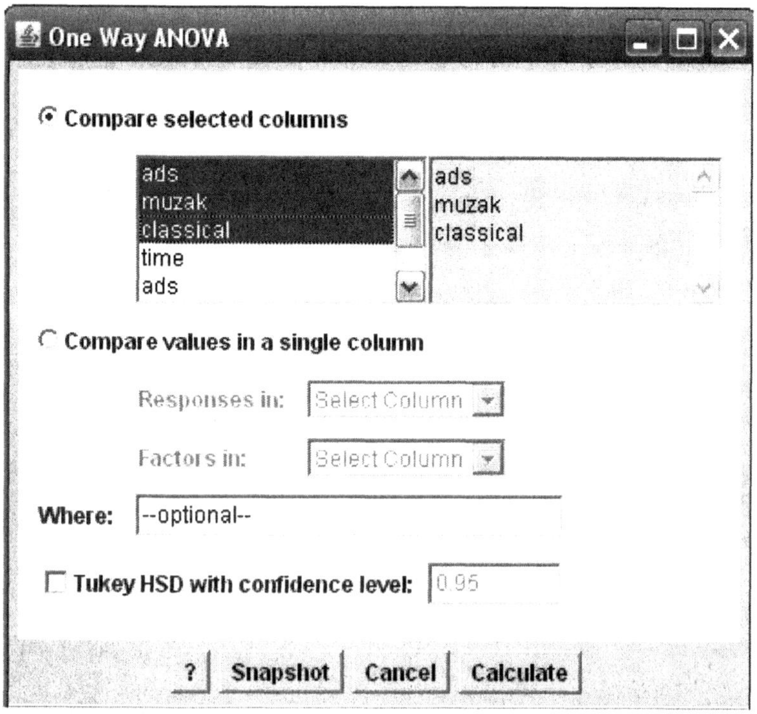

One Way ANOVA

Compare selected columns

ads
muzak
classical
time
ads

ads
muzak
classical

Compare values in a single column

Responses in: Select Column

Factors in: Select Column

Where: --optional--

Tukey HSD with confidence level: 0.95

? Snapshot Cancel Calculate

Figure 14.1b

Or, choose "*time*" as the response variable and "*recording*" as the factor (explanatory variable) then click **Calculate.**

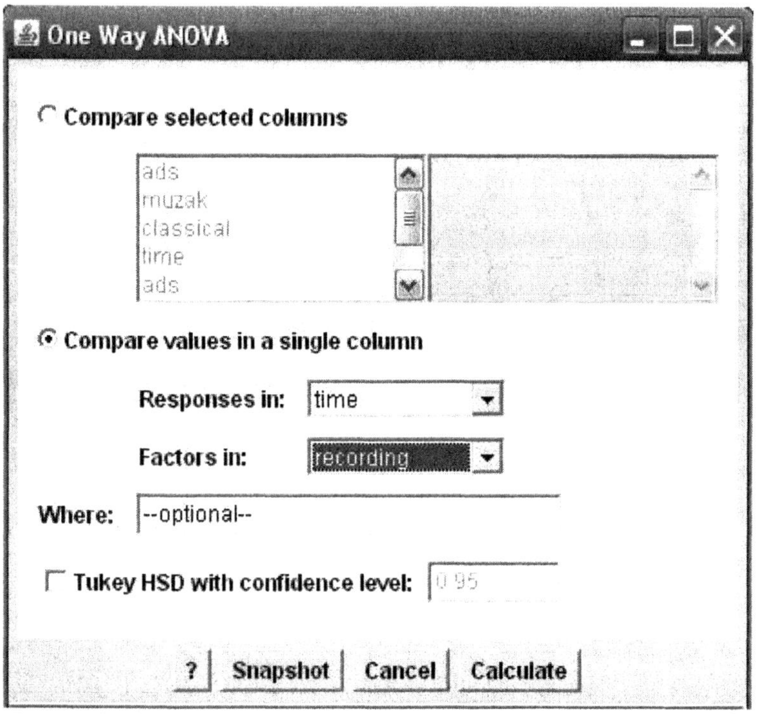

One Way ANOVA

Compare selected columns

ads
muzak
classical
time
ads

Compare values in a single column

Responses in: time

Factors in: recording

Where: --optional--

Tukey HSD with confidence level: 0.95

? Snapshot Cancel Calculate

Figure 14.1c

You will end up with one of the following outputs (which are essentially the same):

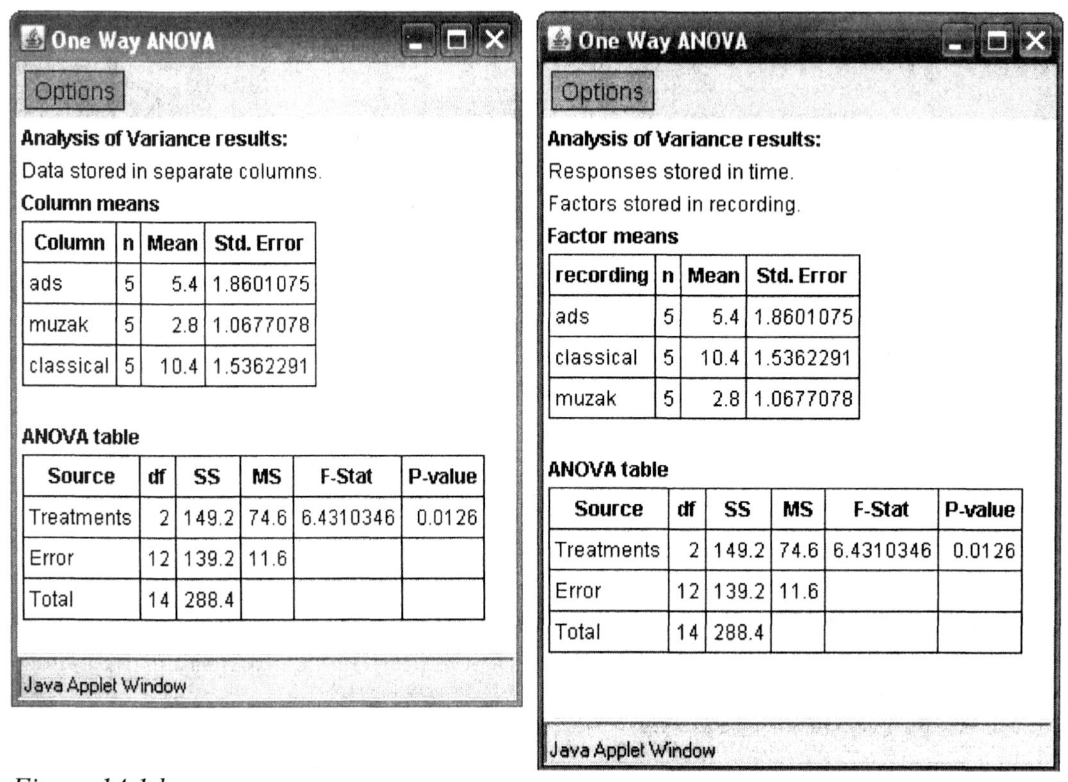

One Way ANOVA

Options

Analysis of Variance results:

Data stored in separate columns.

Column means

Column	n	Mean	Std. Error
ads	5	5.4	1.8601075
muzak	5	2.8	1.0677078
classical	5	10.4	1.5362291

ANOVA table

Source	df	SS	MS	F-Stat	P-value
Treatments	2	149.2	74.6	6.4310346	0.0126
Error	12	139.2	11.6		
Total	14	288.4			

Java Applet Window

One Way ANOVA

Options

Analysis of Variance results:

Responses stored in time.

Factors stored in recording.

Factor means

recording	n	Mean	Std. Error
ads	5	5.4	1.8601075
classical	5	10.4	1.5362291
muzak	5	2.8	1.0677078

ANOVA table

Source	df	SS	MS	F-Stat	P-value
Treatments	2	149.2	74.6	6.4310346	0.0126
Error	12	139.2	11.6		
Total	14	288.4			

Java Applet Window

Figure 14.1d

The first table in the outputs (Column means or Factor Means) shows the descriptive statistic; sample size (n), sample mean, and standard error for each group. Recall, the standard error is the sample standard deviation divided by the square root of the sample size.

The second table is the One-Way ANOVA table. You can compare this table to Table 14.2 in your text (note that the "Treatments" row in the StatCrunch table is equivalent to the "Group" row in your text). The test-statistic is 6.43 and the p-value is 0.013 (which is found off the F-distribution with dfl = 2 and df2 = 12). This tells us that there is fairly strong evidence that at least two of the means are different.

Follow-Up Analysis (Multiple Comparisons)

Part of the follow-up analysis after the One-Way ANOVA F-test is to compare all means two at a time and to summarize significant differences. This procedure is called multiple comparisons and fixes the overall confidence level of all comparisons simultaneously (also called simultaneous inference). There are many methods for multiple comparisons, one of which is the "Tukey" method. In the One-Way ANOVA procedure there is the option to produce the Tukey confidence intervals. You also have the option to choose the overall confidence level. You can summarize significant differences (at the overall confidence level that you designate) as those intervals that do not contain 0.

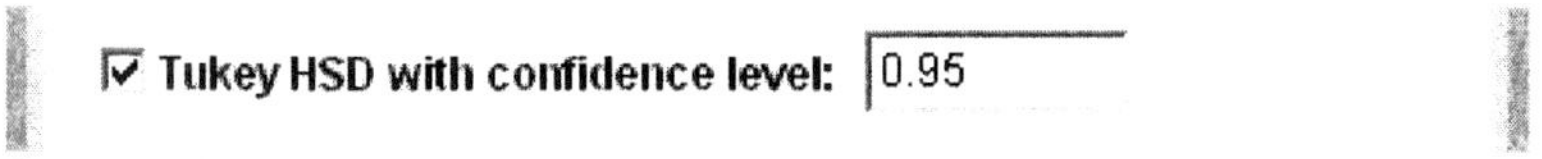

☑ Tukey HSD with confidence level: 0.95

Figure 14.2a

Exercise 14.15: "Tukey holding time comparison."

Below are the Tukey confidence intervals with overall confidence level of 95%.

Tukey 95% Simultaneous Confidence Intervals

ads subtracted from

	Lower	Upper
muzak	-8.346768	3.146768
classical	-0.7467682	10.746768

muzak subtracted from

	Lower	Upper
classical	1.8532318	13.346768

Figure 14.2b

To summarize, the Tukey confidence intervals are:

$$\mu_{muzak} - \mu_{ads} : (-8.35, 3.15)$$

$$\mu_{classical} - \mu_{ads} : (-0.75, 10.75)$$

$$\mu_{classical} - \mu_{muzak} : (1.85, 13.35)$$

According to the Tukey method the only significant difference is between the classical and Muzak groups.

ANOVA and Regression

As we saw in the last chapter, it is possible to include categorical explanatory variables in a regression model by including a set of categorical variables (one less then the number of levels f that variable).

Example 7: "Regression Analysis of Telephone Holding Times"

In order to find the regression model for average holding time given the type of recording we must include two indicator variables for recording type (**Data➔Indicator columns**). As well, the response variable must be entered in one column. The date should be as displayed below:

time	recording	ads	classical	muzak
5	ads	1	0	0
1	ads	1	0	0
11	ads	1	0	0
2	ads	1	0	0
8	ads	1	0	0
0	muzak	0	0	1
1	muzak	0	0	1
4	muzak	0	0	1
6	muzak	0	0	1
3	muzak	0	0	1
13	classical	0	1	0
9	classical	0	1	0
8	classical	0	1	0
15	classical	0	1	0
7	classical	0	1	0

Figure 14.3a

To produce the regression output, select **Stat➔Regression➔Multiple Linear** and choose any two of the indicators as your "X variables". Let's use "*ads*" and "*Muzak*" as the indicators for recording type. Select "*time*" as your "Y variable" and click **Calculate.** The regression output is:

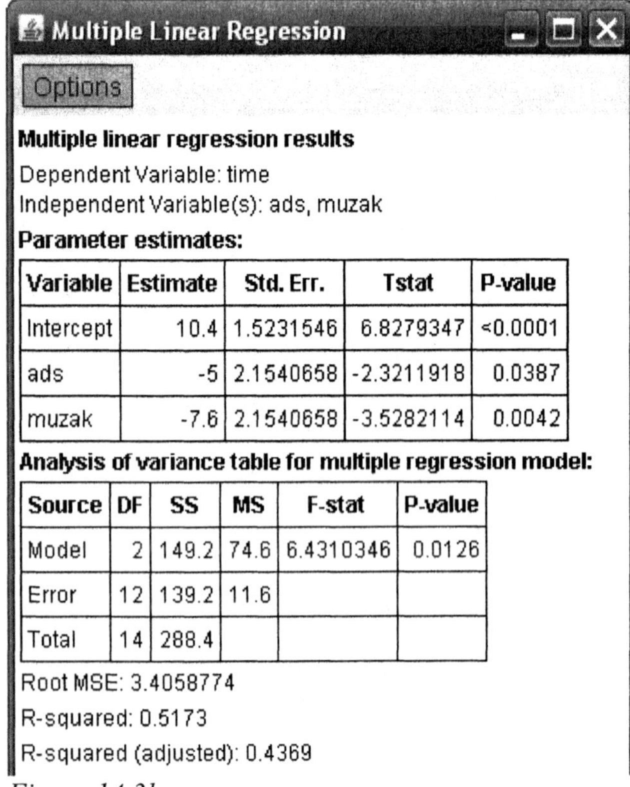
Multiple Linear Regression

Options

Multiple linear regression results

Dependent Variable: time

Independent Variable(s): ads, muzak

Parameter estimates:

Variable	Estimate	Std. Err.	Tstat	P-value
Intercept	10.4	1.5231546	6.8279347	<0.0001
ads	-5	2.1540658	-2.3211918	0.0387
muzak	-7.6	2.1540658	-3.5282114	0.0042

Analysis of variance table for multiple regression model:

Source	DF	SS	MS	F-stat	P-value
Model	2	149.2	74.6	6.4310346	0.0126
Error	12	139.2	11.6		
Total	14	288.4			

Root MSE: 3.4058774

R-squared: 0.5173

R-squared (adjusted): 0.4369

Figure 14.3b

Compare this output to Table 14.6 in your text. The estimated regression equation is:

$$\hat{y} = 10.4 - 5ads - 7.6muzak$$

If you fit this model for all levels of recording type you get the sample means.

$$\hat{y}_{ads} = 10.4 - 5(1) - 7.6(0) = 5.4$$
$$\hat{y}_{muzak} = 10.4 - 5(0) - 7.6(1) = 2.8$$
$$\hat{y}_{classical} = 10.4 - 5(0) - 7.6(0) = 10.4$$

You should note that these fitted means are the sample means.

The ANOVA table gives the F-test for any linear significance, which is equivalent the One-Way ANOVA F-test for any mean differences. This regression F-test tests the following:

$$H_0 : \beta_1 = \beta_2 = 0$$
$$H_a : \text{At least one is different from 0}$$

This is equivalent to the One-Way ANOVA F-test for any mean differences. Upon inspection you'll see that the ANOVA tables (for the One-Way approach and for regression) are equivalent.

Two-Way Classification

Suppose there are two categorical explanatory variables. Here we will look at tests for the main factor effects as well as interaction between the two explanatory variables.

Example 9: "Testing the Main Effects for Corn Yield"

This example wishes to explore how different levels (low or high) of a nitrogen-based fertilizer and manure affect corn yield. The response variable is corn yield. The two explanatory variables are fertilizer level and manure level, both with levels low or high. The data are displayed in Table 14.9 in your text. You'll have to enter the data in three columns, one for the level of each explanatory variable (h - high or l – low), and the other for the value of corn yield. A portion of the data is displayed below:

Row	yield	fertilizer	manure
1	13.7	h	h
2	15.8	h	h
3	13.9	h	h
4	16.6	h	h
5	15.5	h	h
6	16.4	h	l
7	12.5	h	l
8	14.1	h	l
9	14.4	h	l
10	12.2	h	l
11	15	l	h
12	15.1	l	h
13	12	l	h

Figure 14.4a

To obtain the Two-Way ANOVA table select **Stat➔ANOVA➔Two Way.** Select "*yield*" as the response variable and then "*fertilizer*" and "*manure*" as the row and column factors (it doesn't matter which is which). Click **Next** to go to **Screen 2.**

Screen 1

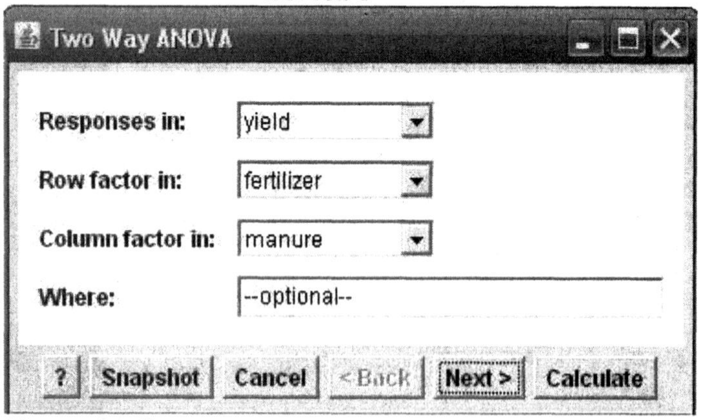

Two Way ANOVA

Responses in:	yield
Row factor in:	fertilizer
Column factor in:	manure
Where:	--optional--

? | Snapshot | Cancel | < Back | Next > | Calculate

Figure 14.4b

By default, StatCrunch will fit the model that includes interaction (called the Non-Additive Fit). For now we won't consider interaction. In **Screen 2** select the "Fit Additive Model" option to fit the model without interaction (called the Additive Fit). Let's choose to display the means table as well. Click **Calculate.**

Screen 2

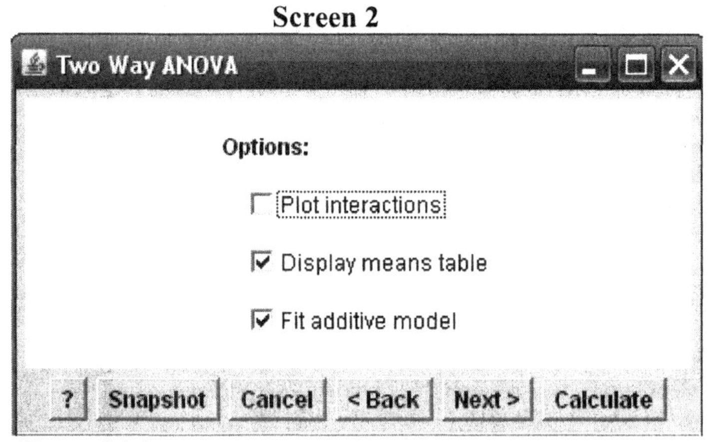

Figure 14.4c

The Two-Way ANOVA output is displayed below. Compare this table to Table 14.10 in your text. Also, compare the Means table in the output to Table 14.9 in your text.

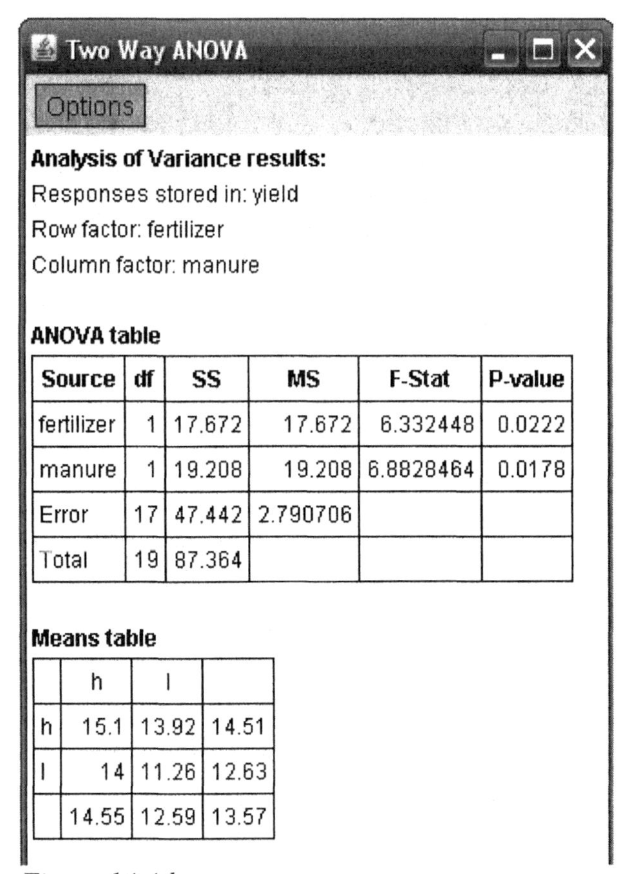
Two Way ANOVA

Options

Analysis of Variance results:

Responses stored in: yield

Row factor: fertilizer

Column factor: manure

ANOVA table

Source	df	SS	MS	F-Stat	P-value
fertilizer	1	17.672	17.672	6.332448	0.0222
manure	1	19.208	19.208	6.8828464	0.0178
Error	17	47.442	2.790706		
Total	19	87.364			

Means table

	h	l	
h	15.1	13.92	14.51
l	14	11.26	12.63
	14.55	12.59	13.57

Figure 14.4d

The first row, labelled "fertilizer", gives a test statistic of 6.33 and a p-value 0.0222. These are the results for the test if fertilizer level has an effect on average yield, after accounting for manure level.

The second row, labelled "manure", gives s test statistic of 6.88 and a p-value of 0.0178. These results are for the test if manure level has an effect on average yield, after accounting for fertilizer level.

Example 10: "Regression Modeling to Estimate and Compare Mean Corn Yields"

We can model a Two-Way situation in a similar fashion to that of the One-Way situation. We will simply have to create indicator variables for both factors. We can define:

$$fert\_h = \begin{cases} 1, \text{ if fertilizer level is high} \\ 0, \text{ else} \end{cases} \quad , \quad manu\_h = \begin{cases} 1, \text{ if manure level is high} \\ 0, \text{ else} \end{cases}$$

$$fert\_l = \begin{cases} 1, \text{ if fertilizer level is low} \\ 0, \text{ else} \end{cases} \quad , \quad manu\_l = \begin{cases} 1, \text{ if manure level is low} \\ 0, \text{ else} \end{cases}$$

The data should look like this (we'll discuss the last column later):

Row	fertilizer	manure	fert_h	fert_l	manu_h	manu_l	fert_h*manu_h
1	h	h	1	0	1	0	1
2	h	h	1	0	1	0	1
3	h	h	1	0	1	0	1
4	h	h	1	0	1	0	1
5	h	h	1	0	1	0	1
6	h	l	1	0	0	1	0
7	h	l	1	0	0	1	0
8	h	l	1	0	0	1	0
9	h	l	1	0	0	1	0
10	h	l	1	0	0	1	0
11	l	h	0	1	1	0	0
12	l	h	0	1	1	0	0
13	l	h	0	1	1	0	0
14	l	h	0	1	1	0	0

Figure 14.5a

For each factor we will need to include one indicator (let's use the level high). To fit the regression, select **Stat➔Regression➔Multiple Linear.** In **Screen 1,** select "*yield*" as the "Y variable" and "*fert_h*" and "*manu_h*" as the "X variables". Click **Calculate.** Below is the regression output.

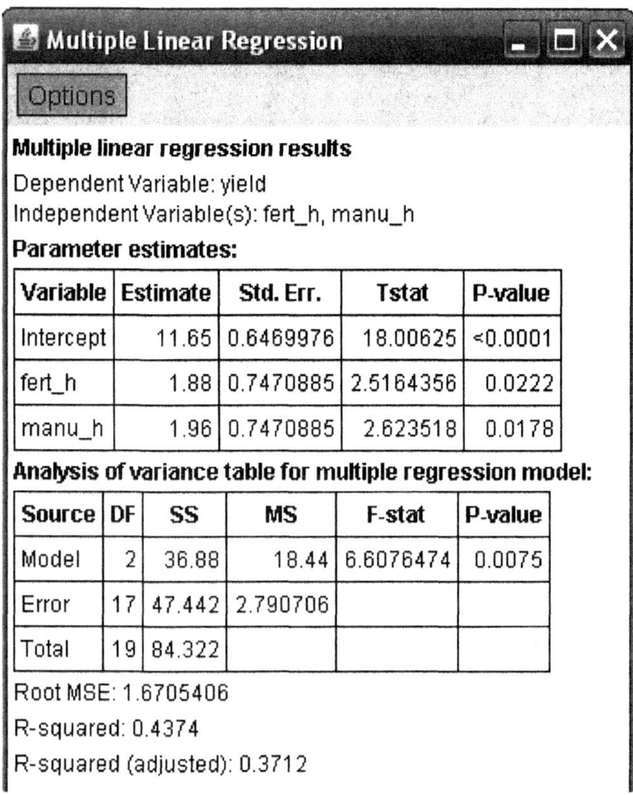
Multiple Linear Regression

Options

Multiple linear regression results

Dependent Variable: yield

Independent Variable(s): fert_h, manu_h

Parameter estimates:

Variable	Estimate	Std. Err.	Tstat	P-value
Intercept	11.65	0.6469976	18.00625	<0.0001
fert_h	1.88	0.7470885	2.5164356	0.0222
manu_h	1.96	0.7470885	2.623518	0.0178

Analysis of variance table for multiple regression model:

Source	DF	SS	MS	F-stat	P-value
Model	2	36.88	18.44	6.6076474	0.0075
Error	17	47.442	2.790706		
Total	19	84.322			

Root MSE: 1.6705406

R-squared: 0.4374

R-squared (adjusted): 0.3712

Figure 14.5b

The estimated regression equation is:

$$\hat{y} = 11.65 + 1.88 fert\_h + 1.96 manu\_h$$

If you fit this model for all levels of the two factors you get the following fitted means.

$$\hat{y}_{fert\_h,manu\_h} = 11.65 + 1.88 + 1.96 = 15.49$$
$$\hat{y}_{fert\_h,manu\_l} = 11.65 + 1.88 = 13.53$$
$$\hat{y}_{fert\_l,manu\_h} = 11.65 + 1.96 = 13.61$$
$$\hat{y}_{fert\_l,manu\_l} = 11.65$$

You should note that these fitted means are NOT the same as the sample means. The reason for this is that these fitted means are *smoothed* over so that the effect for either factor is the same at all levels of the other factor (there is no interaction in the fitted model).

The next example will show how to test for interaction using StatCrunch.

Example 11: "Testing for Interaction with Corn Yield Data"

In order to test for interaction we must include it in the model. Open the Two-Way function. Make sure **Screen 1** is filled in correctly. In **Screen 2** make sure the "Fit Additive Model" is NOT selected. Click **Calculate.** You should then get the following table. You'll note that this table includes a row for interaction.

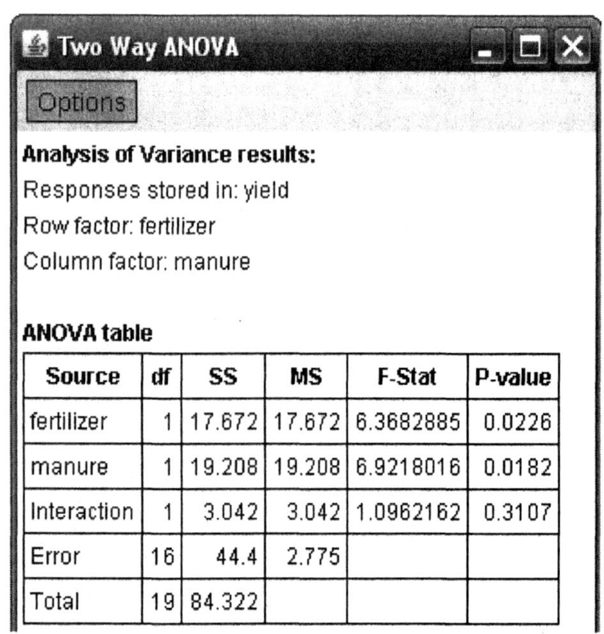

Analysis of Variance results:

Responses stored in: yield

Row factor: fertilizer

Column factor: manure

ANOVA table

Source	df	SS	MS	F-Stat	P-value
fertilizer	1	17.672	17.672	6.3682885	0.0226
manure	1	19.208	19.208	6.9218016	0.0182
Interaction	1	3.042	3.042	1.0962162	0.3107
Error	16	44.4	2.775		
Total	19	84.322			

Figure 14.6

Compare this table to Table 14.14 in your text. The test-statistic for interaction is 1.096 with a p-value of 0.31. This is an indication that there is very little evidence that the two explanatory variables interact. Thus, we are justified to remove the interaction term and carry out further analysis of main factor effects using the additive fit (basically the test results in Example 9 are justified).

Interaction Plots

In **Screen 2** of your Two-Way function there is an option to "Plot Interaction". If you click this and click **Calculate** you'll end up with the following plot:

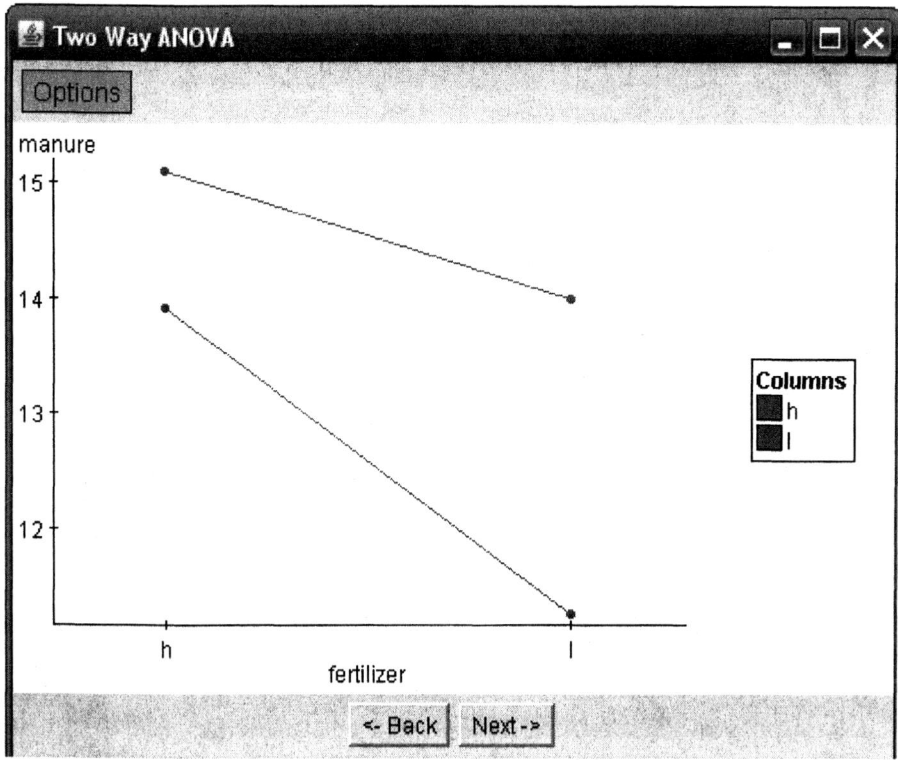

Figure 14.7

The plot displays the 4 sample means (for each combination of the two factors). These plots can help to assess additivity. If the lines are parallel (or roughly parallel) this is an indication that the two factors do not interact and that an additive model may be appropriate.

Regression with Interaction

If you want to fit the regression model that includes interaction you must produce an interaction term between the two indicators "*fert_h*" and "*manu_h*". To create the interaction variable you can multiply the two columns "*fert_h*" and "*manu_h*" using the transform function. The interaction column is displayed in *Figure 14.5a*. Include this interaction variable as another "X variable" in your regression. The regression output is displayed below.

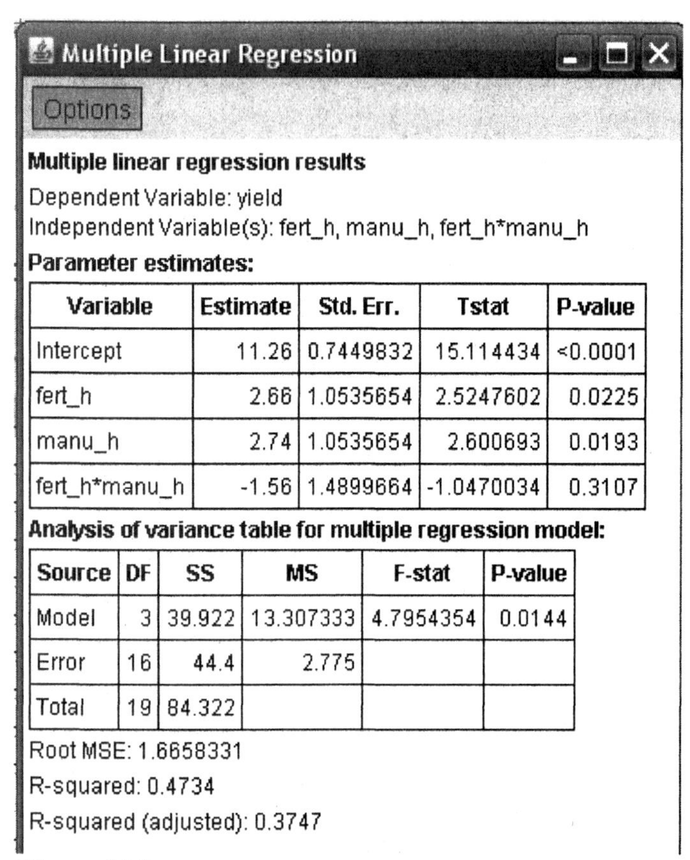
Multiple Linear Regression

Options

Multiple linear regression results

Dependent Variable: yield

Independent Variable(s): fert_h, manu_h, fert_h*manu_h

Parameter estimates:

Variable	Estimate	Std. Err.	Tstat	P-value
Intercept	11.26	0.7449832	15.114434	<0.0001
fert_h	2.66	1.0535654	2.5247602	0.0225
manu_h	2.74	1.0535654	2.600693	0.0193
fert_h*manu_h	-1.56	1.4899664	-1.0470034	0.3107

Analysis of variance table for multiple regression model:

Source	DF	SS	MS	F-stat	P-value
Model	3	39.922	13.307333	4.7954354	0.0144
Error	16	44.4	2.775		
Total	19	84.322			

Root MSE: 1.6658331

R-squared: 0.4734

R-squared (adjusted): 0.3747

Figure 14.8

The estimated regression equation is:

$$\hat{y} = 11.26 + 2.66 fert\_h + 2.74 manu\_h - 1.56 fert\_h \times manu\_h$$

If you fit this model for all levels of the two factors you get the following fitted means:

$$\hat{y}_{fert\_h,manu\_h} = 11.26 + 2.66 + 2.74 - 1.56 = 15.10$$
$$\hat{y}_{fert\_h,manu\_l} = 11.26 + 2.66 = 13.92$$
$$\hat{y}_{fert\_l,manu\_h} = 11.26 + 2.74 = 14.00$$
$$\hat{y}_{fert\_l,manu\_l} = 11.26$$

You should note that (other than minor rounding error) these fitted means are the same as the sample means.

Chapter 15 - Nonparametric Statistics

So far all of the methods we have discussed require the assumption of normality. That is, we assume the data are samples from normal population. Thankfully, these methods perform quite well in situations where our data is not normal (from the Central Limit Theorem). However, there are situations when non-normal data is problematic (such as highly skewed data and data with extreme outliers). Thus, alternative methods must be considered. This chapter will discuss a few alternative methods that can be used when the normality assumption in violated. These methods are also particularly useful when dealing with small sample sizes.

Wilcoxon or Mann-Whitney Test

This section discusses an alternative to the Two-Independent Sample T-tools to compare two mean. The Nonparametric test used to compare two independent groups is called the Wilcoxon Rank-Sum test or the Mann-Whitney Test. StatCrunch calls it the Mann-Whitney Test.

Example 3: "Do Drivers Using Cell Phones Have Slower Reaction Times?"

This example wishes to compare the reaction times between drivers that are using cell phones and those that aren't. Open the file "cell phone" on your CD. To carry out the Mann-Whitney Test select **Stat➔Nonparametrics➔Mann-Whitney.** Select the variables for groups 1 and 2 as displayed below and click **Next** to go to **Screen 2.**

Screen 1

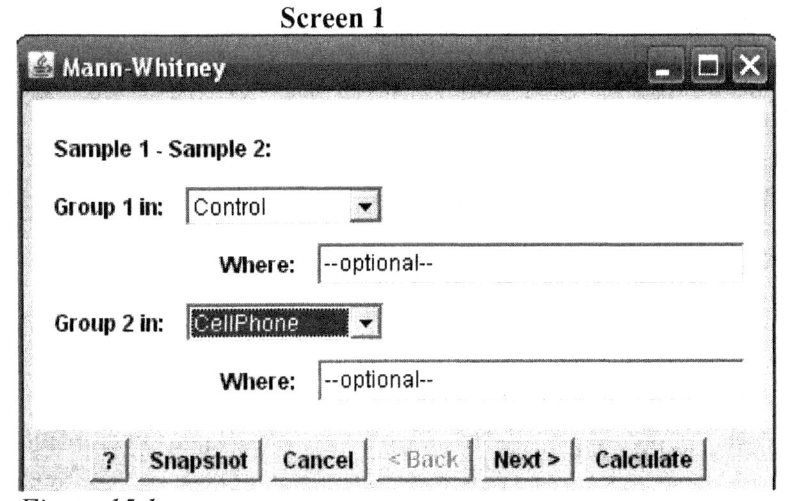

Figure 15.1a

Screen 2 gives you the same options as in the T-tools procedures we went over earlier. You can choose to carry out a test (and specify the alternative) or you can produce a confidence interval with your specified level. Let's carry out a test for any difference. Specifically we are testing for any median difference in responses.

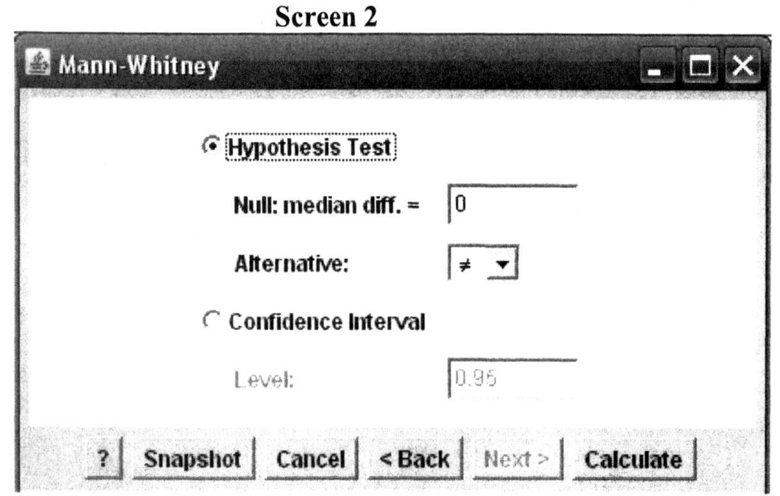

Figure 15.1b

Click **Calculate.** The output is displayed below.

Mann-Whitney

Options

Hypothesis test results:

m1 = median of Control

m2 = median of CellPhone

Parameter : m1 - m2

H_0 : Parameter = 0

H_A : Parameter ≠ 0

Difference	n1	n2	Diff. Est.	Test Stat.	P-value	Method
m1 - m2	32	32	-44.5	864	0.0184	Norm. Approx.

Figure 15.1c

The test statistic given in the output is the "Wilcoxon W" test statistic, as described in your text. See Table 15.5 in your text. The two-sided p-value is reported to be 0.0184, found using the large-sample (normal) approximation.

Confidence Intervals

Example 4: "Estimating the Difference between Median Reaction Times"

In **Screen 2** you could have chosen to produce a 95% confidence interval. Specifically these are confidence intervals for the difference in the median responses. The output for the 95% confidence interval for the difference in the median response is displayed below:

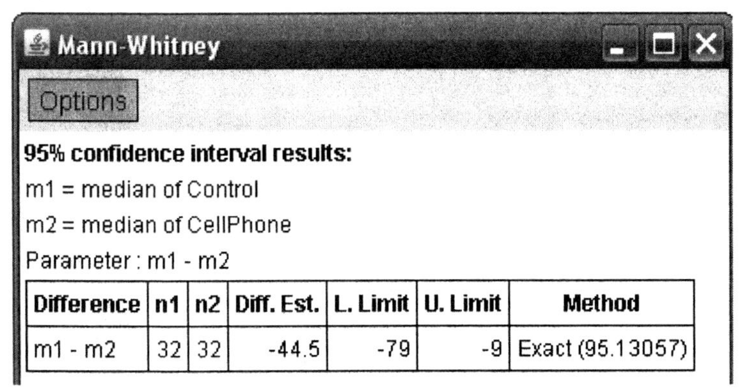

Difference	n1	n2	Diff. Est.	L. Limit	U. Limit	Method
m1 - m2	32	32	-44.5	-79	-9	Exact (95.13057)

Figure 15.1d

Compare this to Table 15.6 in your text. This states that the actual confidence level used is 95.1%. Thus, the 95.1% confidence interval for the difference in median response time between the control group vs. the cell phone group is (-79, -9).

Or, as stated in your text, the 95.1% confidence interval for the difference in median response time between the cell phone group vs. the control group is (9, 79).

Kruskal-Wallis Test

This section discusses an alternative to the One-Way ANOVA F-test to compare several means. The Nonparametric test used to compare any number of independent groups is called the Kruskal-Wallis Test.

Example 5: "Does Heavy Dating Affect College GPA?"

This example wishes to compare the GPA's between students with different dating habits (rare, occasional, or regular). The data are given in Table 15.7 in your text. Similar to the One-Way ANOVA procedure, you can either enter the data in three separate columns one for each Dating Group, or you can enter the data in two columns one for GPA and the other for Dating Group. I have entered the data in two columns as displayed below:

Row	Dating	GPA
1	rare	1.75
2	rare	3.15
3	rare	3.5
4	rare	3.68
5	occasional	2
6	occasional	3.2
7	occasional	3.44
8	occasional	3.5
9	occasional	3.6
10	occasional	3.71
11	occasional	3.8
12	regular	2.4
13	regular	2.95
14	regular	3.4
15	regular	3.67
16	regular	3.7
17	regular	4

Figure 15.2a

To run the test, select **Stat➔Nonparametrics➔Kruskal Wallis.** Select "*GPA*" as the response variable and "*Dating*" as the factor (or explanatory variable) and click **Calculate.**

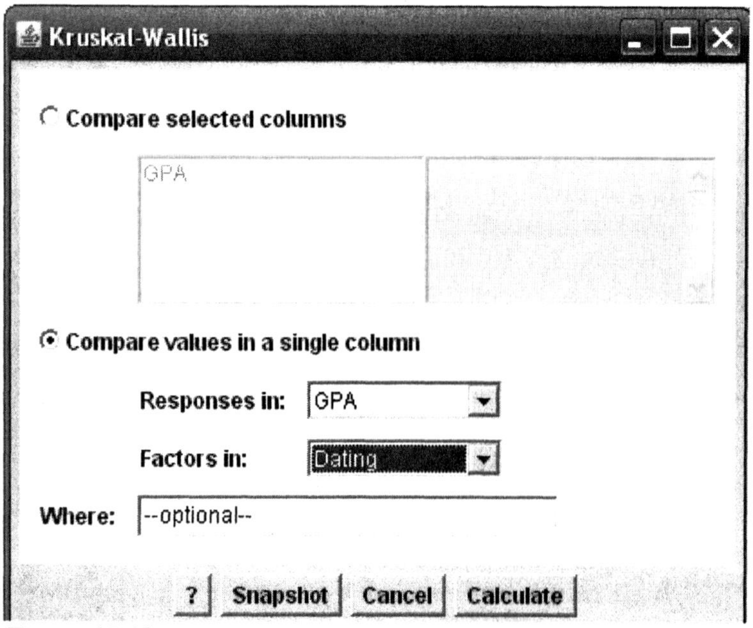

Kruskal-Wallis

Compare selected columns

GPA

Compare values in a single column

Responses in: GPA

Factors in: Dating

Where: --optional--

? Snapshot Cancel Calculate

Figure 15.2b

The output is displayed below:

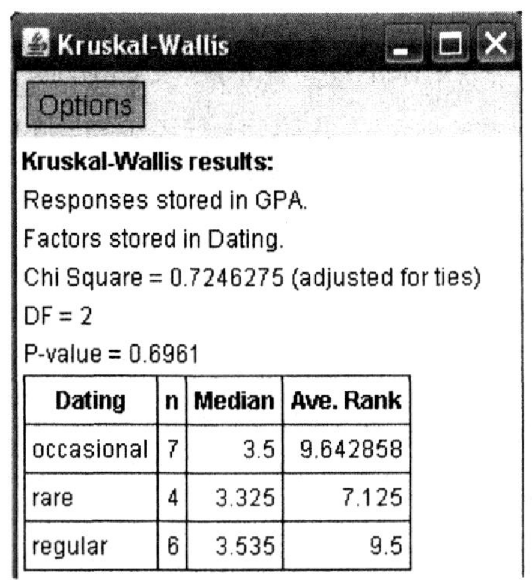
Kruskal-Wallis

Options

Kruskal-Wallis results:

Responses stored in GPA.

Factors stored in Dating.

Chi Square = 0.7246275 (adjusted for ties)

DF = 2

P-value = 0.6961

Dating	n	Median	Ave. Rank
occasional	7	3.5	9.642858
rare	4	3.325	7.125
regular	6	3.535	9.5

Figure 15.2c

Compare this output to Table 15.8 in your text. The test-statistic for the test of any median differences is $H = 0.72$ with a p-value of 0.696. The p-value is based on a Chi-Squared distribution with 2 (# of groups – 1) degrees of freedom. The output also displays the sample medians and average ranks.

Sign Test and Wilcoxon Signed-Rank Test

Your text discusses two Nonparametric methods for Paired Samples: the Sign Test and the Wilcoxon Signed-Rank Test. These are two alternatives to the Paired T-tools.

Sign Test

The Sign Test for any difference is basically a test to determine if the proportion of positive differences in the paired observations is different from 0.5.

Example 6: "Spend More Time Browsing the Internet or Watching TV?"

This example wishes to compare the time individuals watch television vs. the time they browse the Internet. Each individual records his or her two times. Thus, this is a paired sample. The data can be found in the "Georgia Student survey". The column for Internet time is "*BrowseInternet*" and the column for TV time is "*WatchTV*". In order to carry out the Sign Test we must first calculate the column of differences. Use the transform function to calculate the difference between "*WatchTV*" and "*BrowseInterent*". The first three observations should be 60, 100 and 30. I relabelled the difference column "*TV-Internet*".

Let p = the proportion of students that spend more time watching television than they do browsing the Internet. The Sign Test for any difference is testing the following:

$$H_0: p = 0.5$$
$$H_a: p \neq 0.5$$

To carry out the Sign Test, select **Stat➔Nonparametrics➔Sign Test.** Select the column "*TV-Internet*" and click **Next** to go to **Screen 2.**

Screen 1

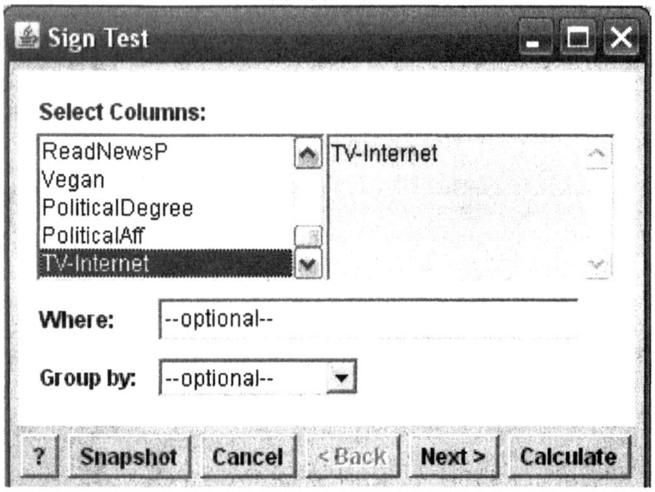

Sign Test

Select Columns:

ReadNewsP
Vegan
PoliticalDegree
PoliticalAff
TV-Internet

TV-Internet

Where: --optional--

Group by: --optional--

? | Snapshot | Cancel | < Back | Next > | Calculate

Figure 15.3a

Screen 2 gives you the same options as in the T-tools procedures we went over earlier. You can choose to carry out a test (and specify the alternative) or you can produce a confidence interval with your specified level. Let's carry out a test for any difference. The output is displayed below:

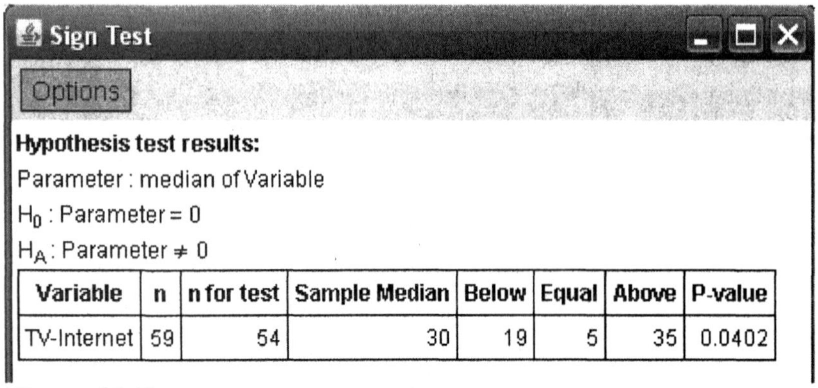

Sign Test

Options

Hypothesis test results:

Parameter : median of Variable

H_0 : Parameter = 0

H_A : Parameter ≠ 0

Variable	n	n for test	Sample Median	Below	Equal	Above	P-value
TV-Internet	59	54	30	19	5	35	0.0402

Figure 15.3b

The table tells us that out of 59 students, 19 (below) spent more time on the Internet, 35 (above) spent more time watching TV and 5 (equal) spent the same amount of time on each. The sample size used in the calculation of the test-statistic (which isn't given) is 54 (ignoring the ties). Thus, the proportion of positive differences is $\hat{p}$ =35/54=0.648. The p-value given is not the same as the one in your text. From the results above you can manually calculate the standard error and test statistic given in Example 6 in your text. Otherwise you can run **Stat→Proportion→One Sample→with summary**. In **Screen 1** specify the "Number of Successes" to be 35 and the "Number of observations" to be 54. In **Screen 2,** select "Hypothesis Test" with the alternative being not equal to 0.5. Click **Calculate.**

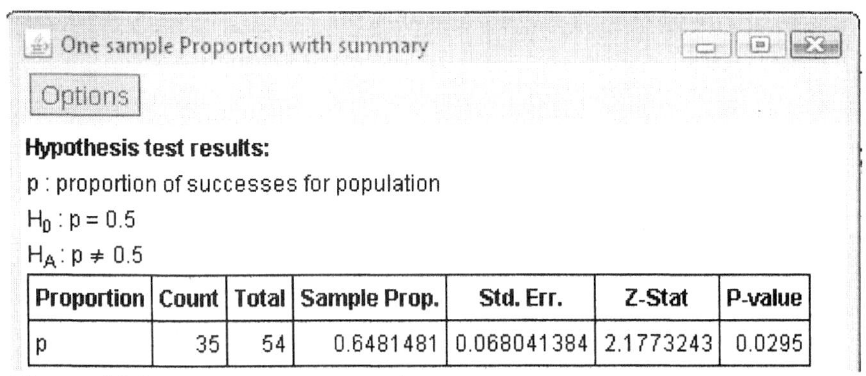
One sample Proportion with summary

Options

Hypothesis test results:

p : proportion of successes for population

$H_0 : p = 0.5$

$H_A : p \neq 0.5$

Proportion	Count	Total	Sample Prop.	Std. Err.	Z-Stat	P-value
p	35	54	0.6481481	0.068041384	2.1773243	0.0295

Figure 15.3c

Wilcoxon Signed-Rank Test

The Wilcoxon Signed-Rank Test for any difference tests to see if the median difference is different from 0. To illustrate we will use the same example.

To carry out the Wilcoxon Signed-Rank Test **Stat→Nonparametrics→Wilcoxon Signed Ranks.** Select the column "*TV-Internet*" and click **Next** to go to **Screen 2.**

Screen 2 gives you the same options as in the T-tools procedures we went over earlier. You can choose to carry out a test (and specify the alternative) or you can produce a confidence interval with your specified level. Let's carry out a test for any difference. The output is displayed below:

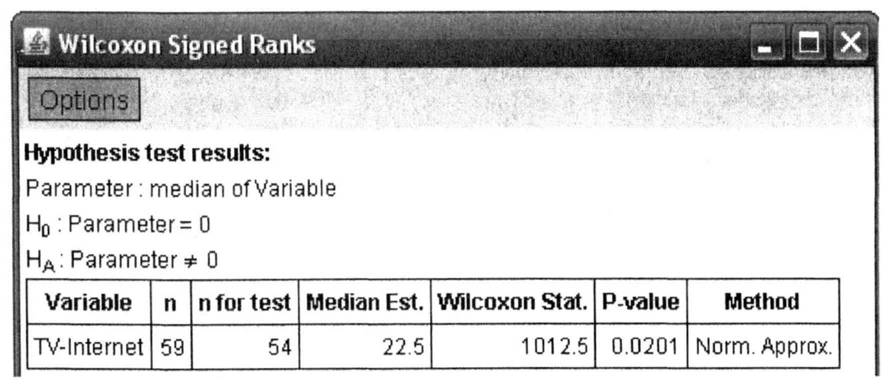
Wilcoxon Signed Ranks

Options

Hypothesis test results:

Parameter : median of Variable

H_0 : Parameter = 0

H_A : Parameter $\neq$ 0

Variable	n	n for test	Median Est.	Wilcoxon Stat.	P-value	Method
TV-Internet	59	54	22.5	1012.5	0.0201	Norm. Approx.

Figure 15.4a

Compare this output to Table 15.10 in your text. The Wilcoxon Test-Statistic is 1012.5 with a p-value of 0.0201. The estimated median difference is 22.5. This p-value indicates strong evidence that the population median difference is different from 0.

JMP MANUAL

Jennifer Lewis Priestley

Kennesaw State University

SECOND EDITION

STATISTICS

THE ART AND SCIENCE OF LEARNING FROM DATA

Agresti • Franklin

Upper Saddle River, NJ 07458

Table of Contents for Chapter Examples and Exercises

INTRODUCTION TO JMP

This lab session is designed to introduce you to JMP (pronounced "Jump"), a product of the SAS Institute. JMP has been designed to be a smaller, more accessible tool for analysis than the much larger SAS System.

This session was developed using JMP version 7.0.

During this session you will learn how to enter and exit JMP, how to enter data, open files, and how to execute commands to conduct basic statistical analysis. As with any new skill, using this software will require practice, patience and more practice. JMP is based on a spreadsheet format, similar to other familiar packages like Microsoft Excel, and can be easily imported from and exported to other programs. It offers a wide variety of statistical functions and graphics, which are easy to use, while offering slightly more sophisticated options relative to more generic applications like Excel.

Chapter 2. Example 3 – How Much Electricity Comes from Renewable Energy Sources?

After launching JMP, the initial JMP Starter screen will appear:

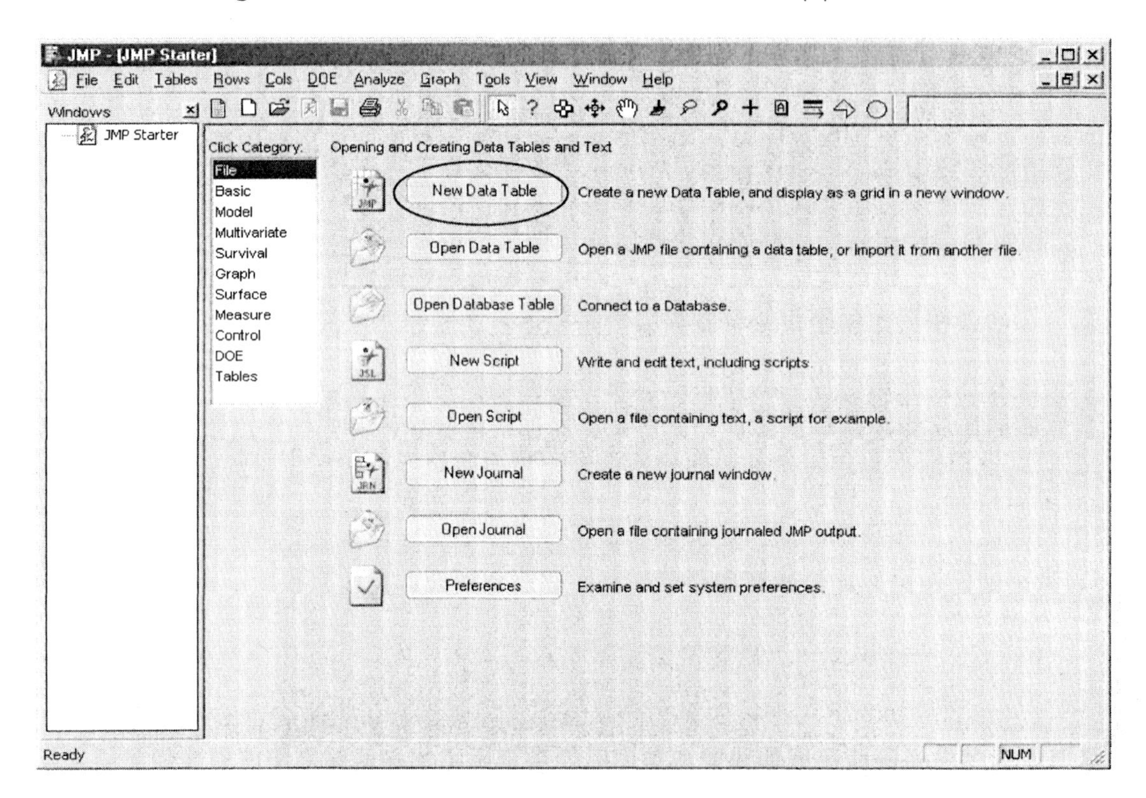

For this problem, we will enter data into a new table. Choose **New Data Table** as indicated above.

The New Data Table screen looks like this:

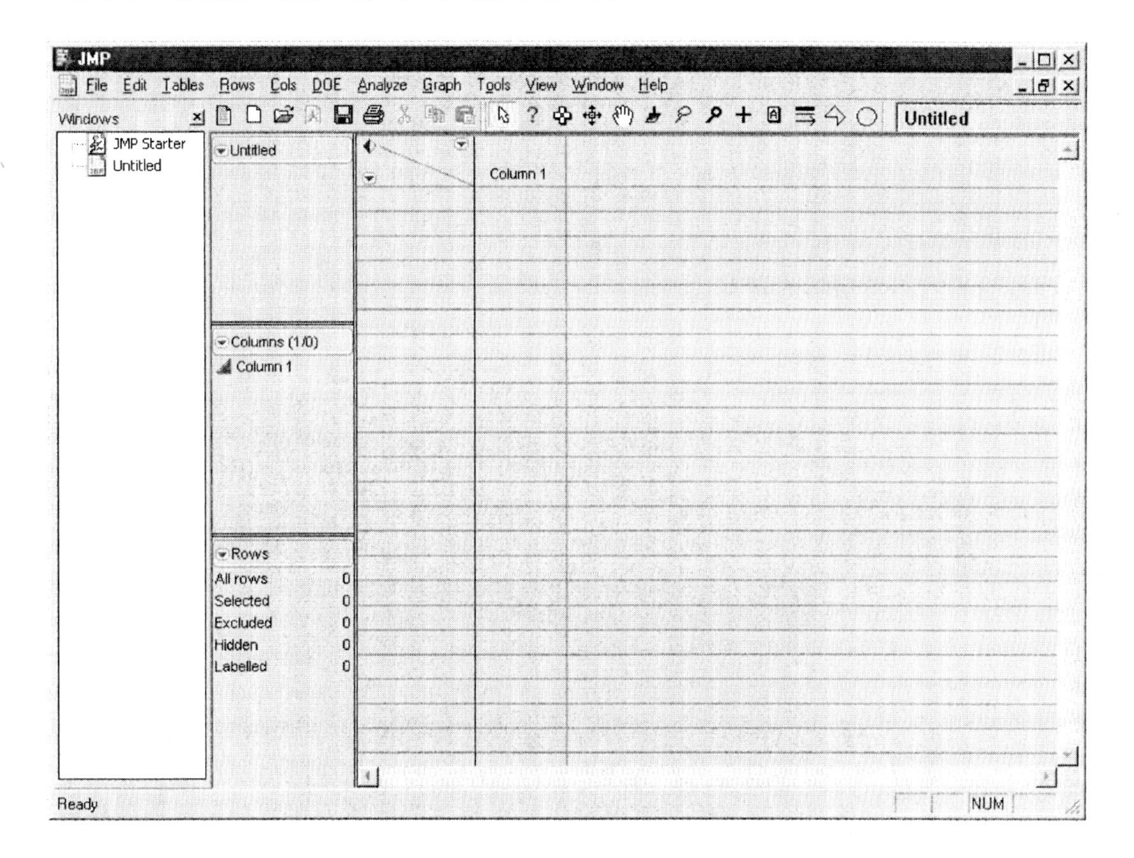

Place your cursor in the cell currently titled "**Column 1**" and type "***Source***". Then, place your cursor in the large space just to the right and double click. This will create "**Column 2**". Title this column "***U.S. Percentage***". Finally, create "**Column 3**", and title this column "***Canada Percentage***".

Your screen should now look like this:

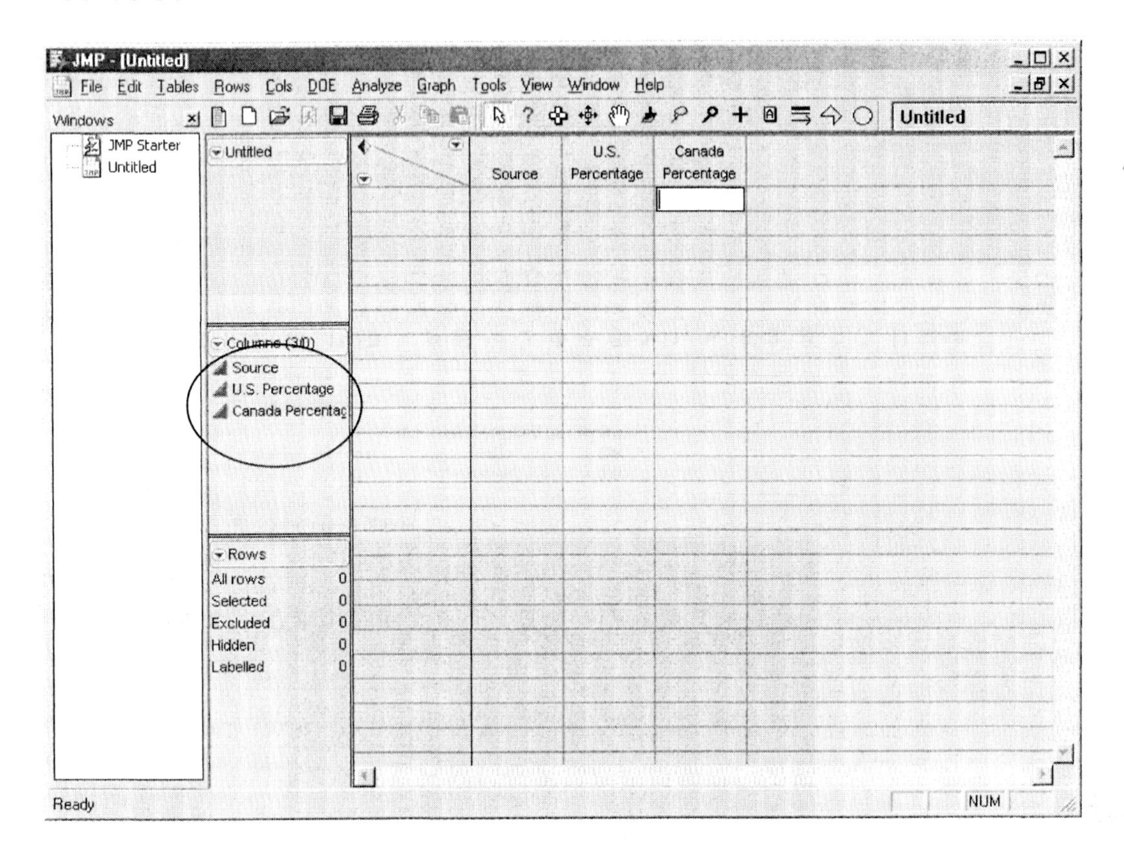

Now, type in the rest of the data as shown in the table. You will notice that as you type the **Sources** of electricity, JMP will generate a pop up box asking "***Non numeric data entered. Change column to character column?***" As you will learn, some data is numeric (typically numbers), and some data is character (typically letters). It is important for the program (and you) to know what kind of data you are working with. When you get this warning from JMP, select **Change**. You will notice that the symbols on the left side of the screen (circled above) will change from blue triangles to red vertical bars. These symbols are important, because they remind you how JMP "sees" your data. The blue triangles indicate numeric data, while the red bars indicate character data.

If you ever need to change how your data is "seen" by JMP, just click inside the first row of the column (variable) you need to change. Once the column is highlighted, right click and choose **Column Info**. Then change the **Data Type** and **Modeling Type** as needed. Note that if you change the **Data Type** from **Character** to **Numeric**, you will also need to change the **Modeling Type** from **Nominal** to **Continuous** and vice versa.

At this point your screen should look like this:

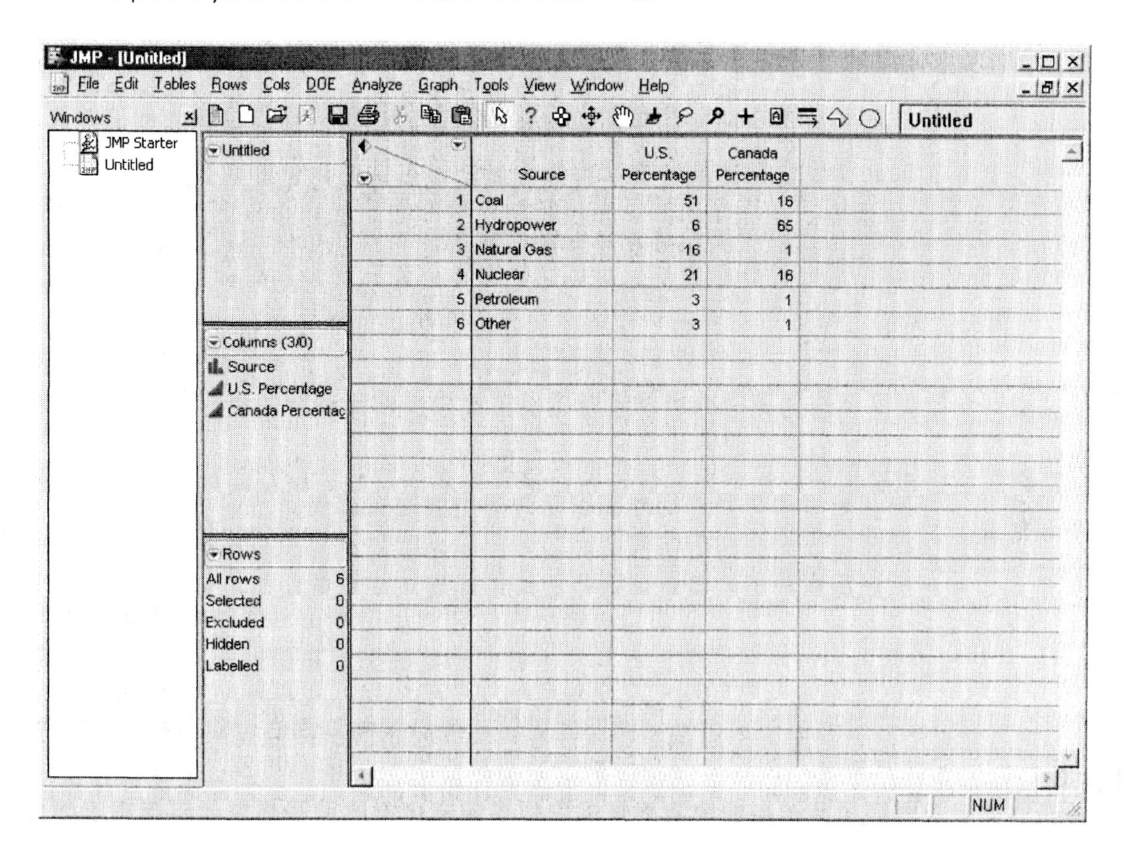

To create a pie chart of the **U.S. Percentage**, select **Graph>Chart**. In the **Options Box**, change the **Bar Chart** option to **Pie Chart**. Then, in the **Select Columns Box**, highlight ***Source***. Then select the **Categories,X,Levels** button. This will determine the "slices" of the pie. Finally, in the **Select Columns Box**, highlight ***U.S. Percentage***. Then select the **Statistics** button and select **% of Total**. This will determine the values for each slice.

At this point, your screen should look like this:

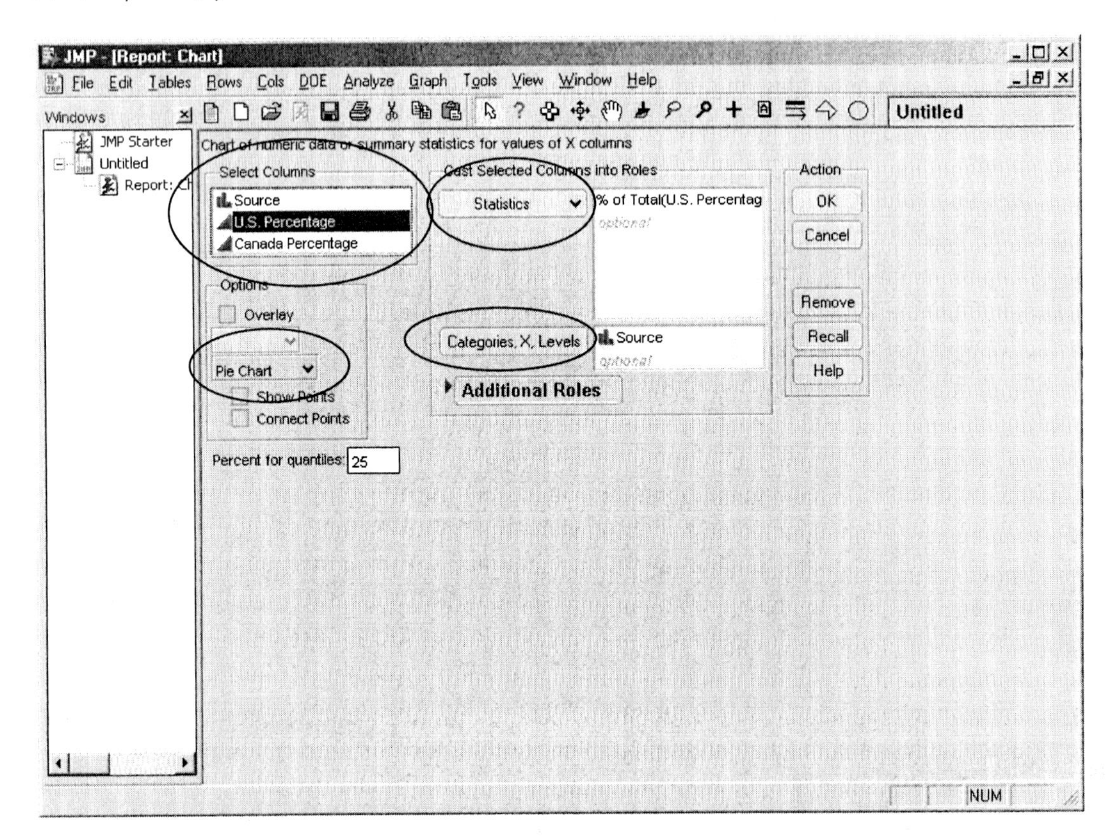

Now select **OK**.

Your screen should look like this:

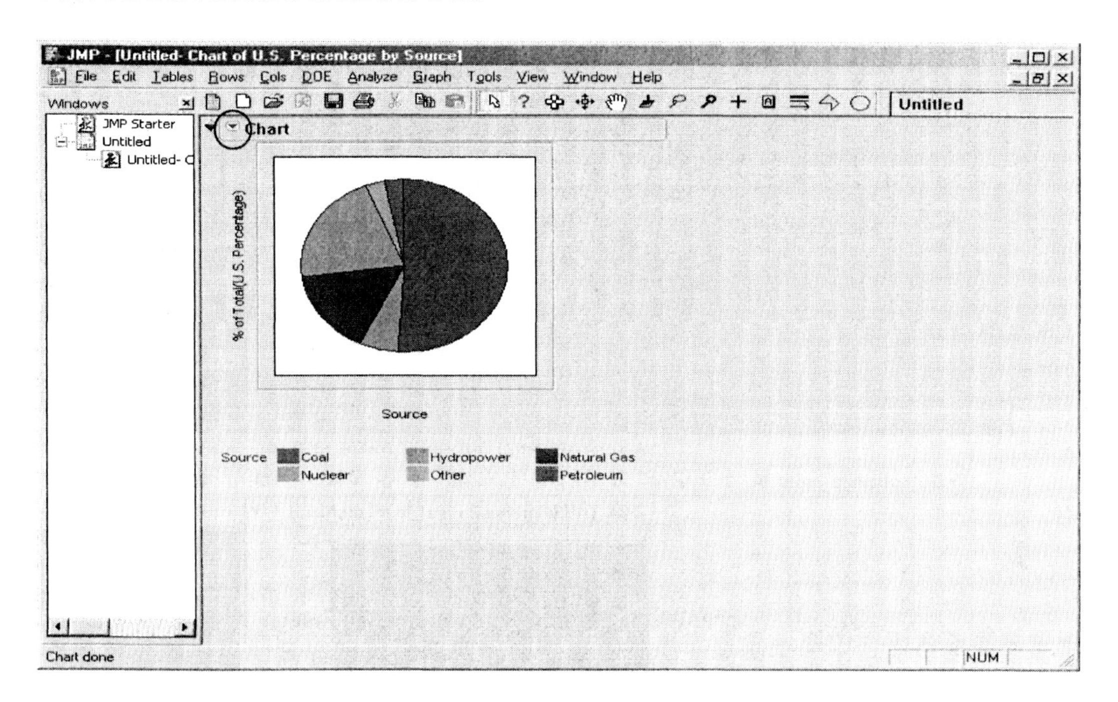

There are a few changes we can make to improve the readability of our pie chart. Lets change a few colors and add some labels. Change the "***Coal***" color from red (which may be difficult to distinguish from orange) to yellow. To do this, place your cursor inside the red box next to "***Coal***" and right click. Select **Yellow**. Change other colors in your graphic as needed. To add labels, click on the little red triangle in the upper left corner of the Pie Chart (see circle above). Select **Label Options>Label by Percent of Total Values**. Now, place your cursor on the bottom right corner and drag it until the pie chart is the desired size.

Your screen should look like this:

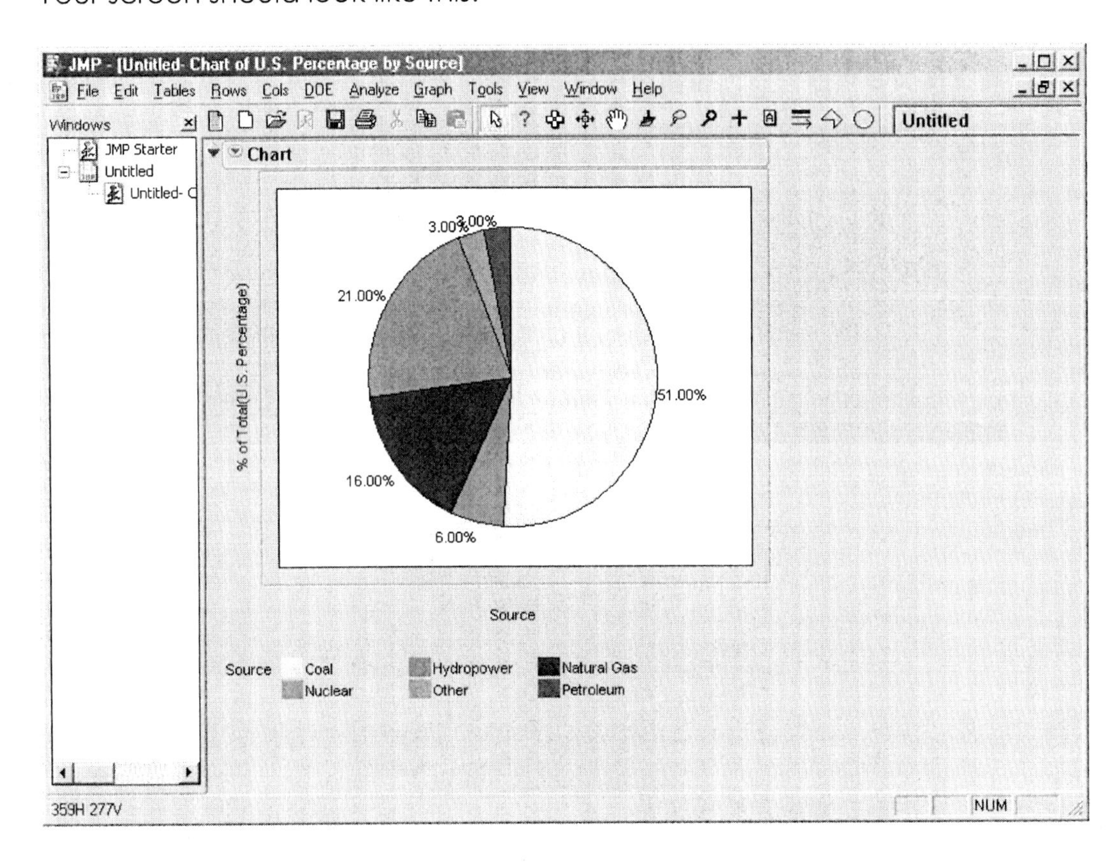

To create the bar chart, simply click on the same red triangle as before, and select **Vertical Chart**:

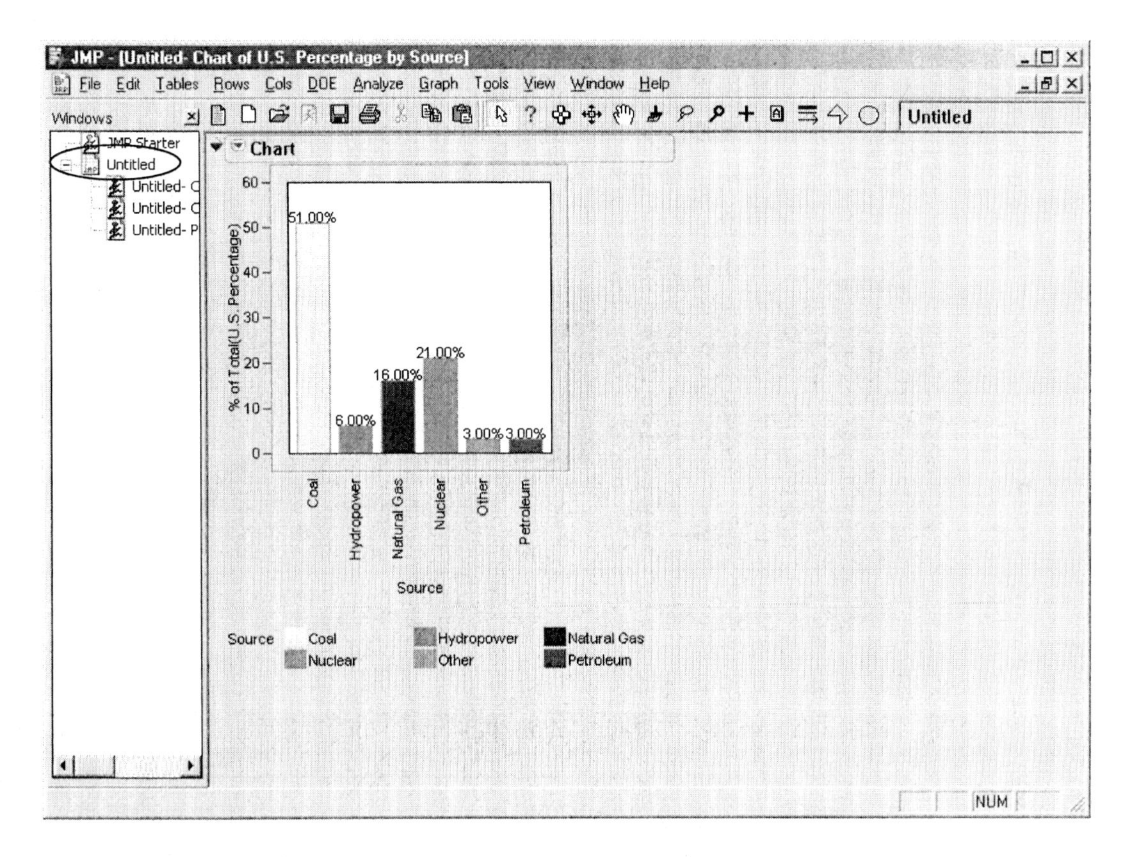

To reorder the bars from most frequently occurring to least frequently occurring, we will use a special version of a bar chart – the Pareto chart.

In the Results Window (circled above), click on the worksheet "Untitled", which will return us to the original data.

Select **Graph** > **Pareto Plot**. You should see this screen:

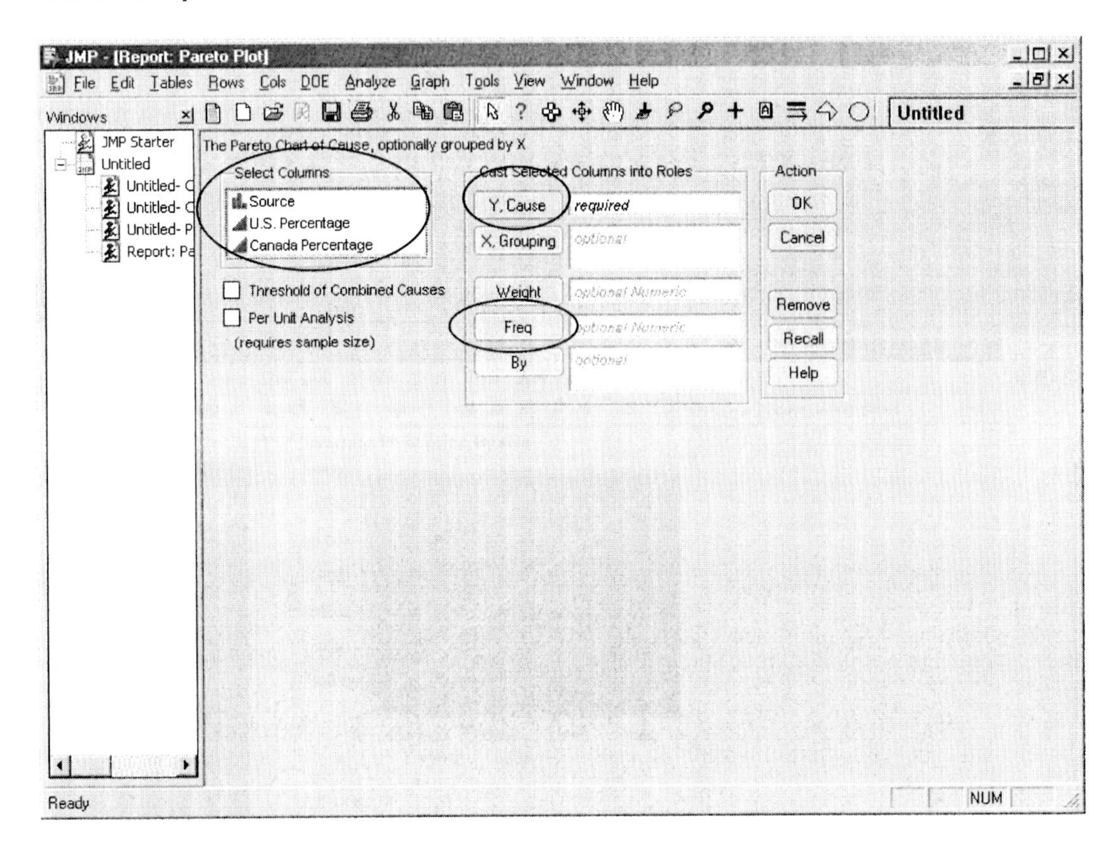

In the **Select Columns** box, select ***Source***. Then click on the **Y, Cause** button (this process will be referred to as "assigning a role"). Then, in the **Select Columns** box, select ***U.S. Percentage***. Then click on the **Freq** button. Select **OK**.

You should see the following screen:

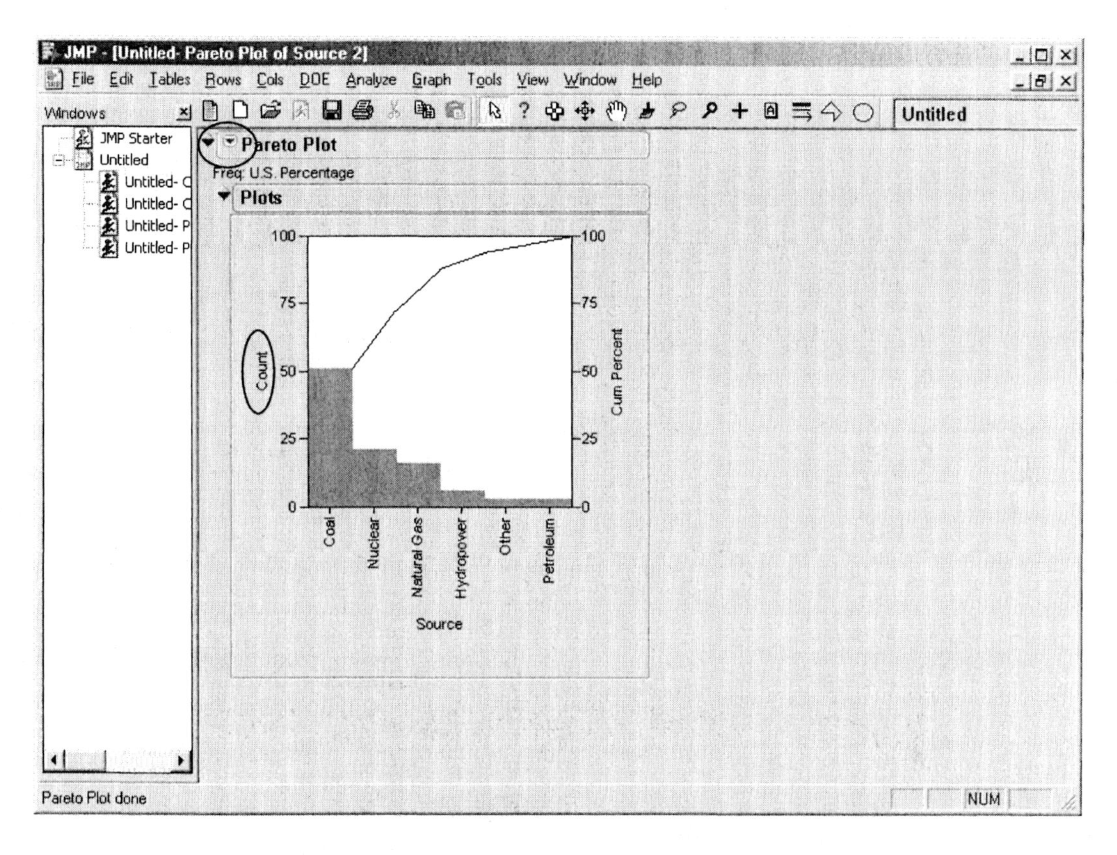

To change the axis label from "**Count**" to "**Percent**", click on the little red triangle and select the **Percent Scale** option.

If you need to show only the bars, without the cumulative line, click on the little red triangle and deselect **Show Cum Percent Curve**. To rescale the y-axis, maneuver the cursor over the y-axis until the little hand appears (in JMP this is referred to as a "grabber"). Using the grabber, pull the axis up or down as needed.

Chapter 2. Exercise 2.11. Weather Stations

Open the JMP Starter Screen, and select **New Data Table**. Enter the data as shown. Ensure that the ***Region*** column is a character variable and the ***Frequency*** column is a numeric variable:

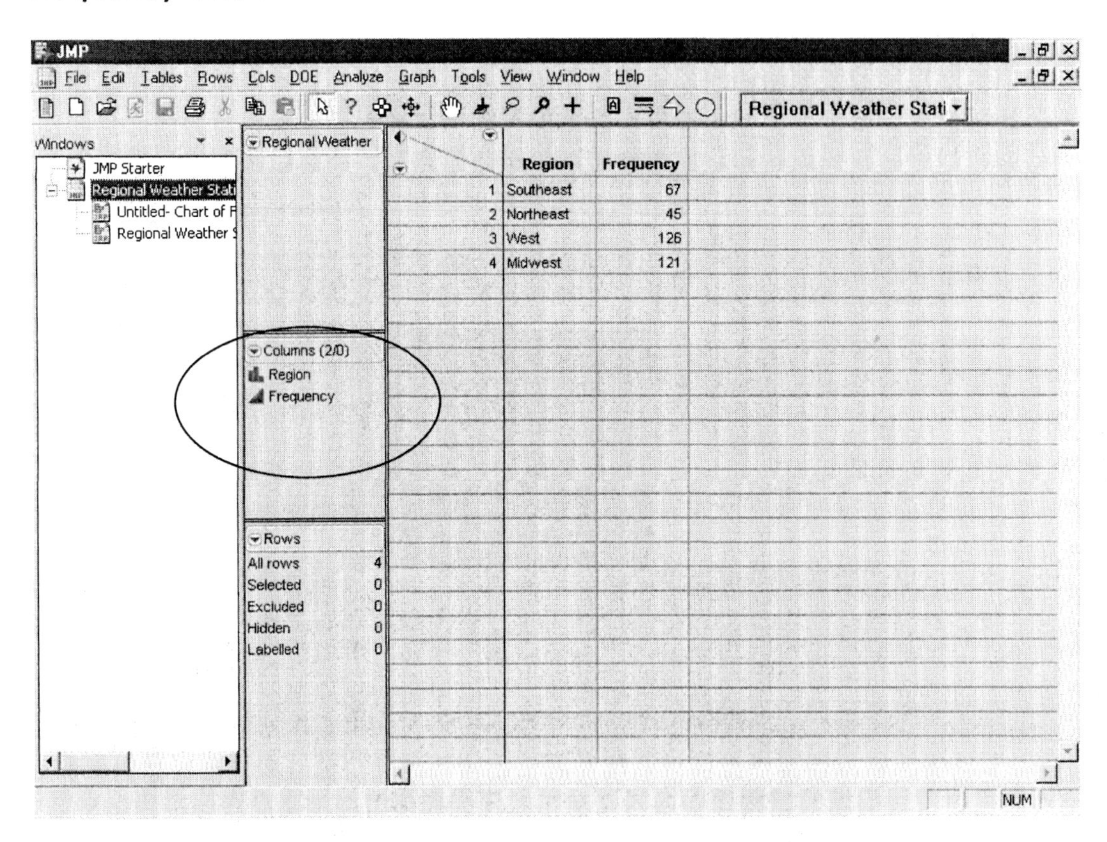

To create a pie chart of the **Frequencies**, select **Graph>Chart**. In the **Options Box**, change the **Bar Chart** option to **Pie Chart**. Then, in the **Select Columns Box**, highlight ***Region***. Then select the **Categories,X,Levels** button. This will determine the "slices" of the pie. Finally, in the **Select Columns Box**, highlight ***Frequency***. Then select the **Statistics** button and select **% of Total**. This will determine the values for each slice. The Pie Chart will be generated. To add labels, click on the little red triangle in the upper left corner of the Pie Chart. Select **Label Options>Label by Percent of Total Values**. Now, place your cursor on the bottom right corner and drag it until the pie chart is the desired size.

You should see the following screen:

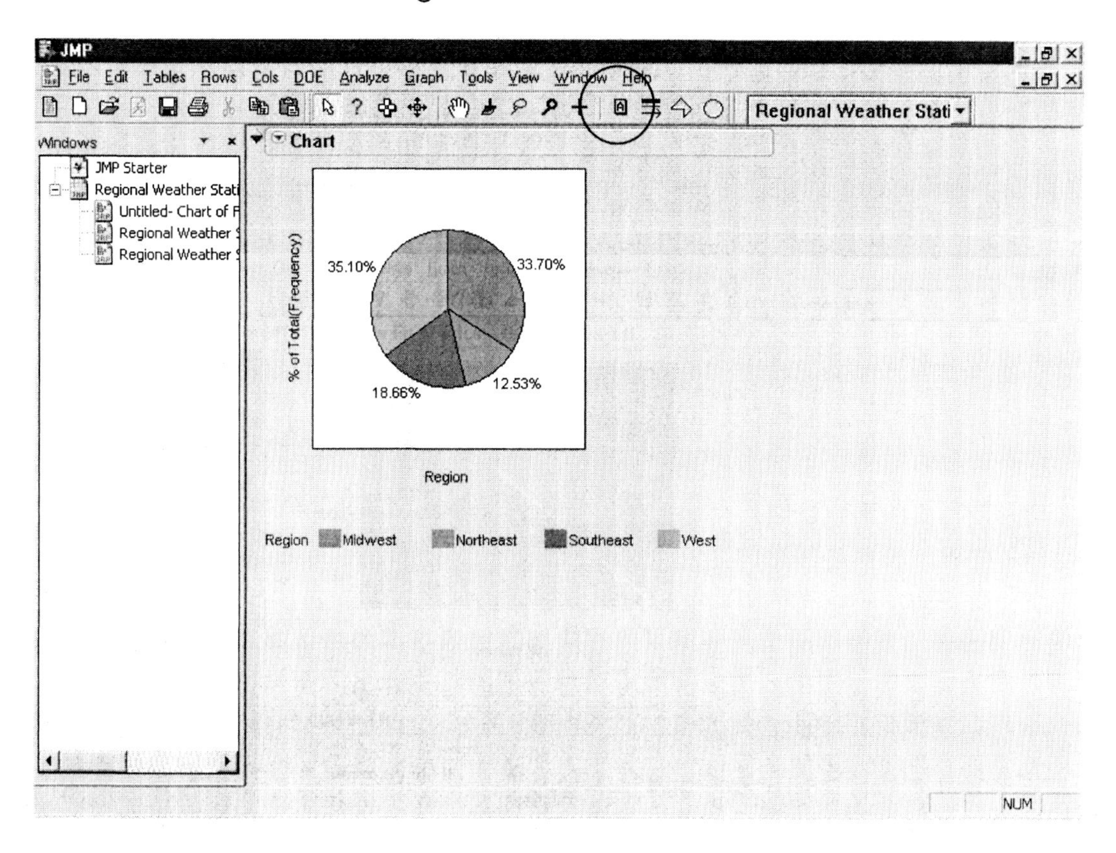

To add a title to the Pie Chart, click on the "annotate" button as indicated above. Place the annotate box inside the pie chart. Type "***Regional Distribution of Weather Stations***", then click outside the box.

You should see the following screen:

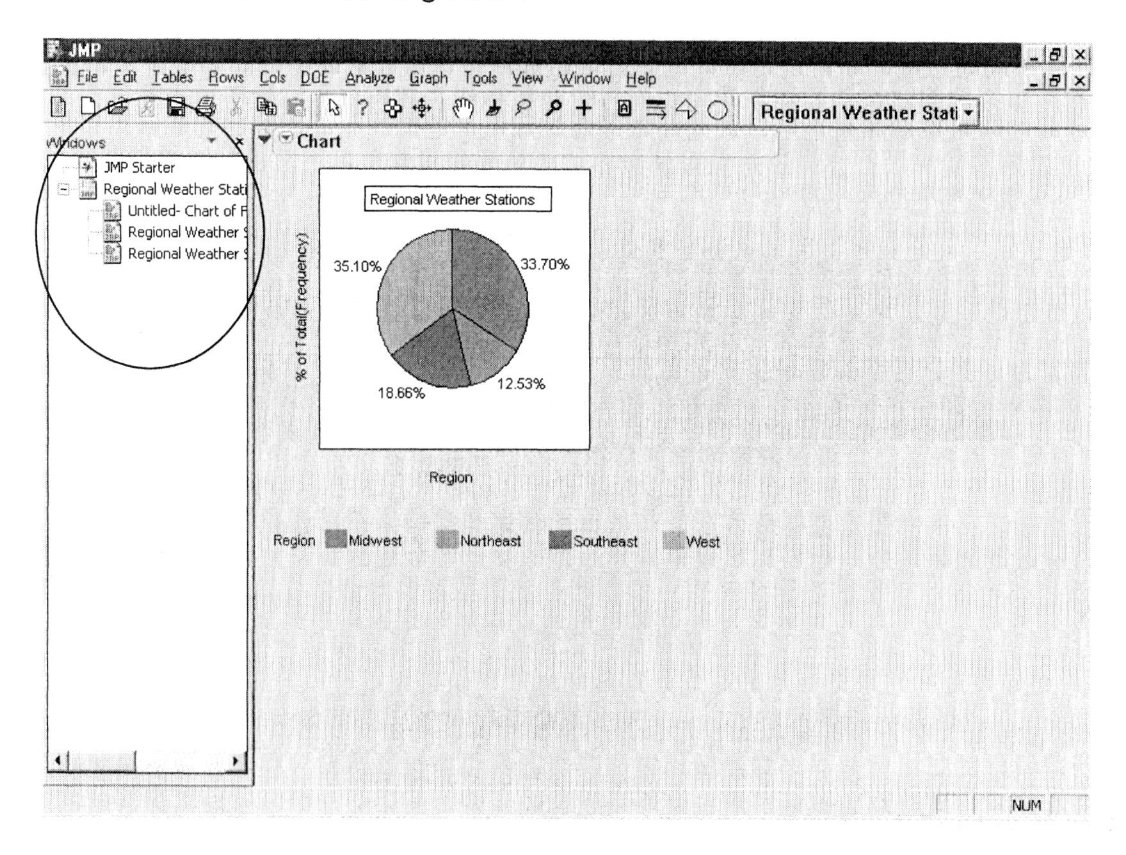

You can change the background color of the annotation box by right clicking inside the box and then select **Background Color**.

To generate the corresponding bar chart, return to the datasheet (you can jump from output screens to the datasheet by clicking inside the results window as circled above).

Select **Graph>Chart**. In the **Options** box, select **Vertical** and **Bar Chart**. Assign ***Frequency*** to the **Statistics** role and select "**Data**". Assign ***Region*** to the **Categories,X,Levels** role. Select **OK**. Annotate a title as before.

You should see the following screen:

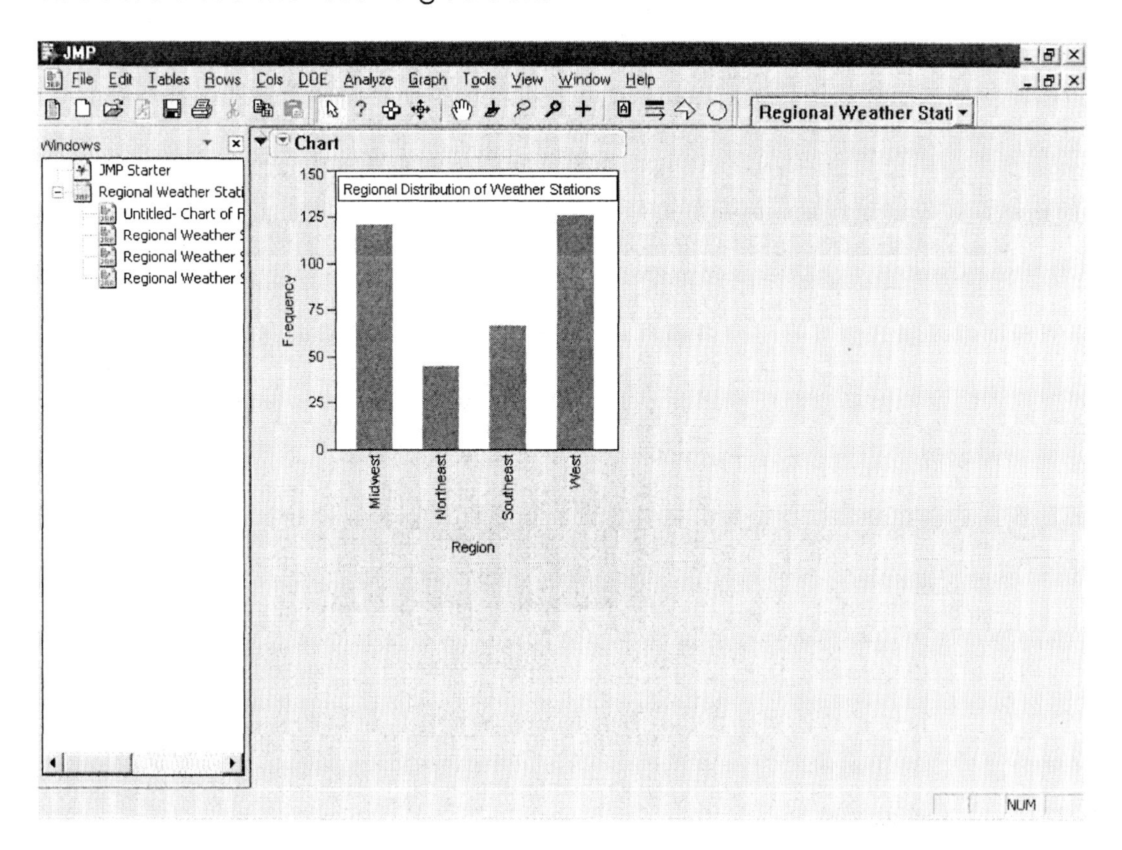

Chapter 2. Example 5. Exploring the Health Value of Cereals: Stem and Leaf Plot.

Stem and Leaf plots (and most other forms of single variable exploration) are easy to generate by selecting **Analyze>Distribution**.

First, return to the JMP Starter screen:

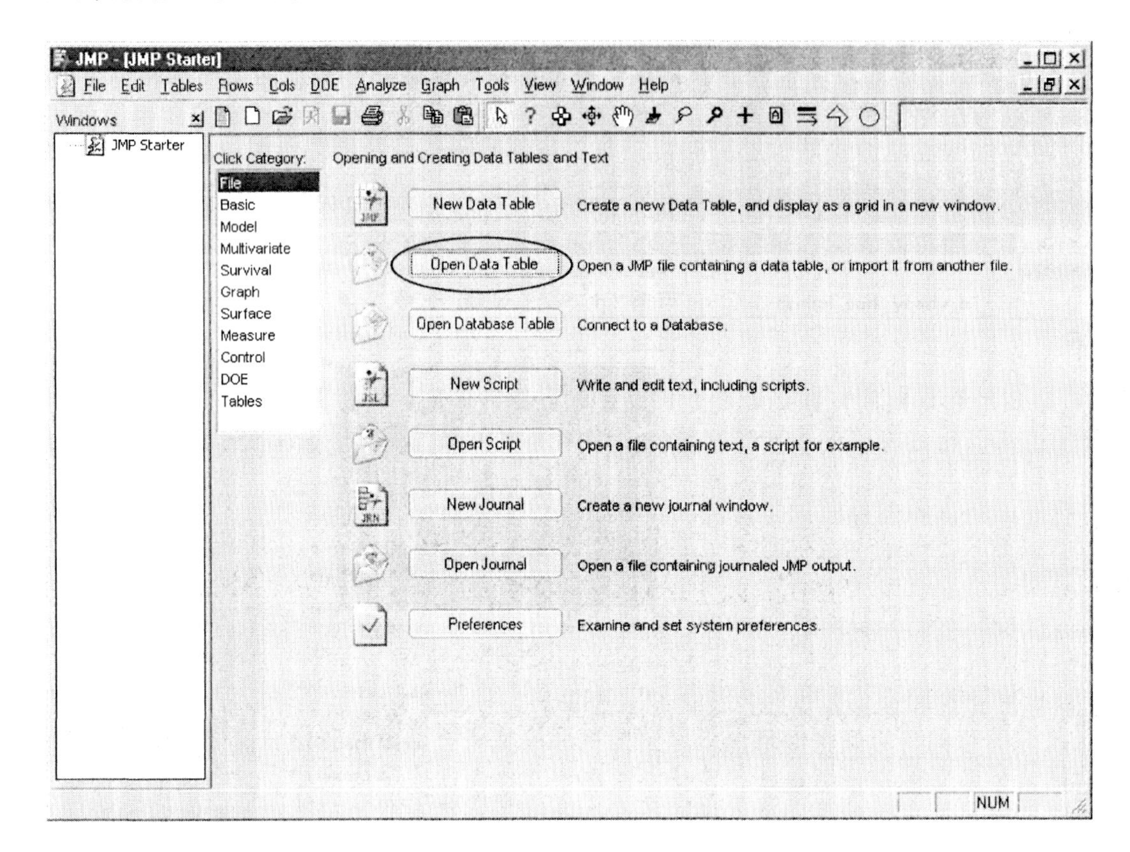

Open the cereal data from the JMP folder of the data disk.

Select **Analyze** > **Distribution**.

Assign ***Sodium(mg)*** to the **Y, Columns** role. Select **OK**.

The default output for this option produces a variety of univariate plots and descriptive statistics:

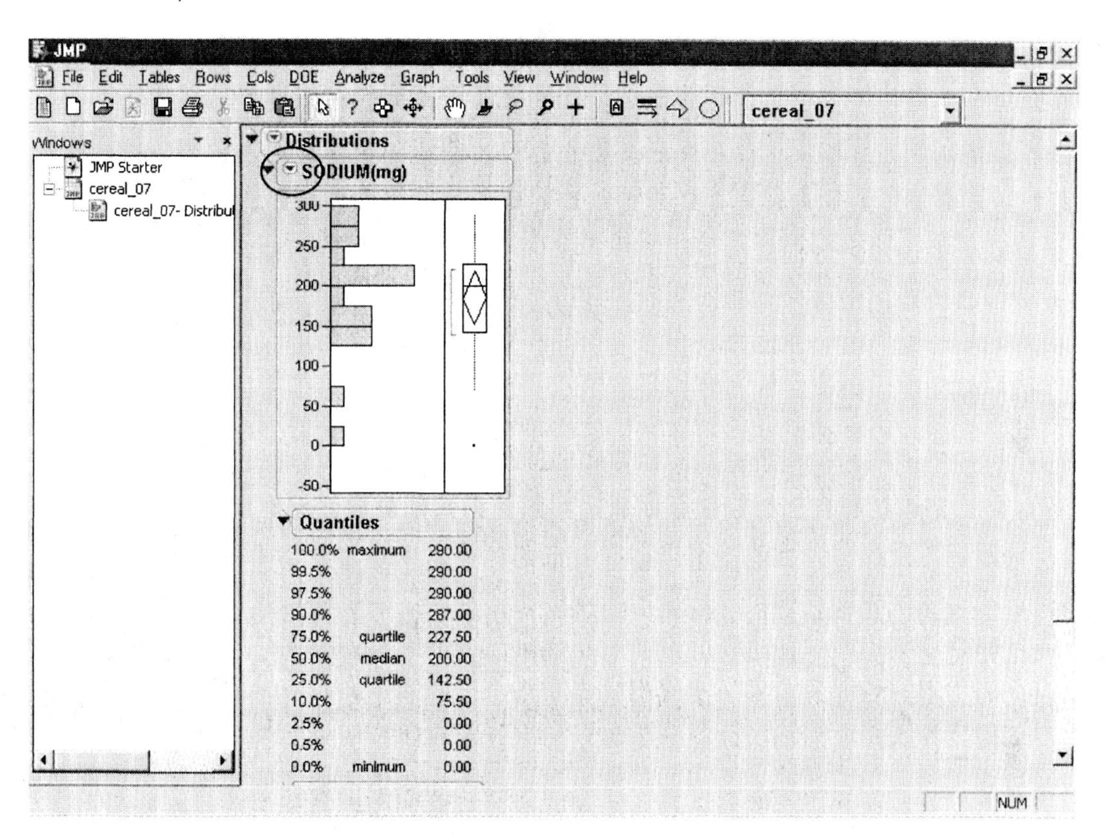

The Stem and Leaf plot does not appear automatically. To generate it, click on the red triangle as indicated above. Select **Stem and Leaf**.

You should see the following screen:

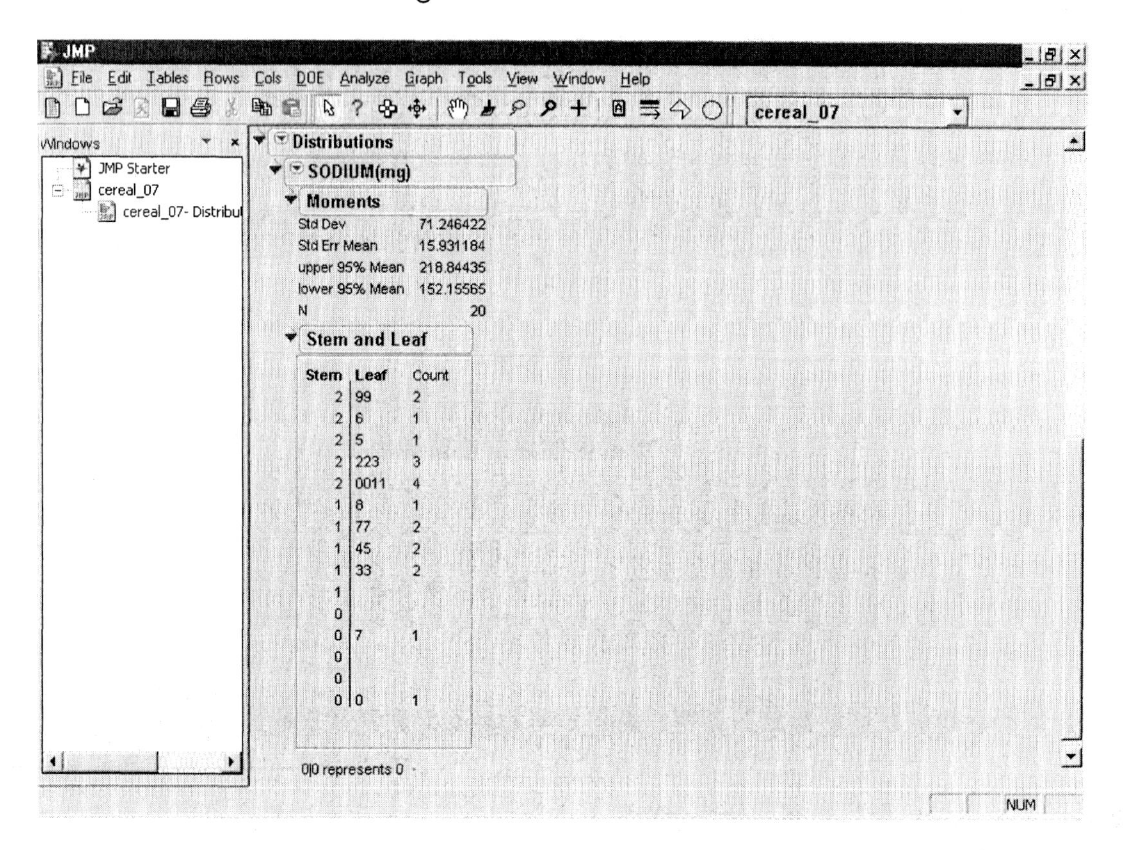

Chapter 2. Exercise 2.15. eBay Prices.

Open the eBay data from the JMP folder of the data disk.

To generate the Stem and Leaf plot of the price, select **Analyze>Distribution**. Assign ***price*** to the **Y, Columns** role. Select **OK**. On the output screen, click on the red triangle next to "***price***" and select **Stem and Leaf**.

You should see the following screen:

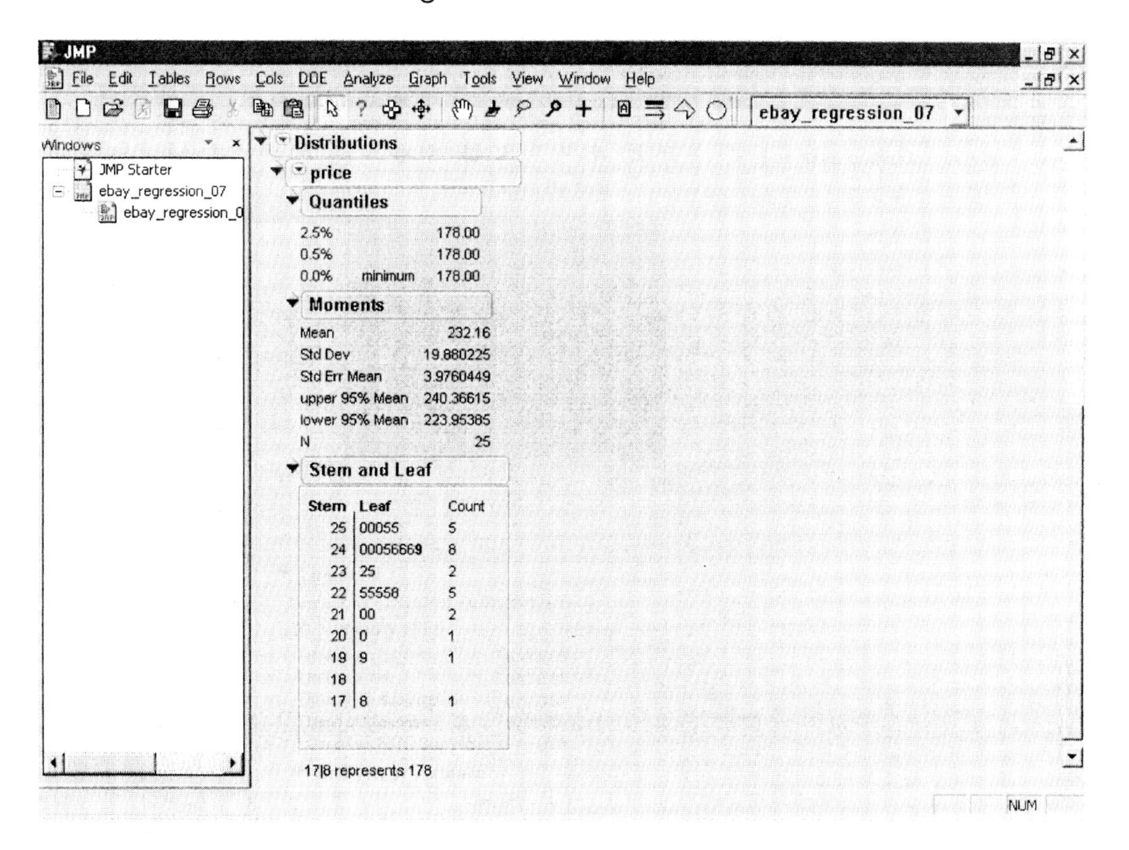

Chapter 2. Example 7. Exploring the Health Value of Cereals.

Open the cereal data from the JMP folder of the data disk.

Select **Analyze>Distribution**. Assign ***sodium(mg)*** to the **Y, Columns** role. Select **OK**.

You should see the following screen:

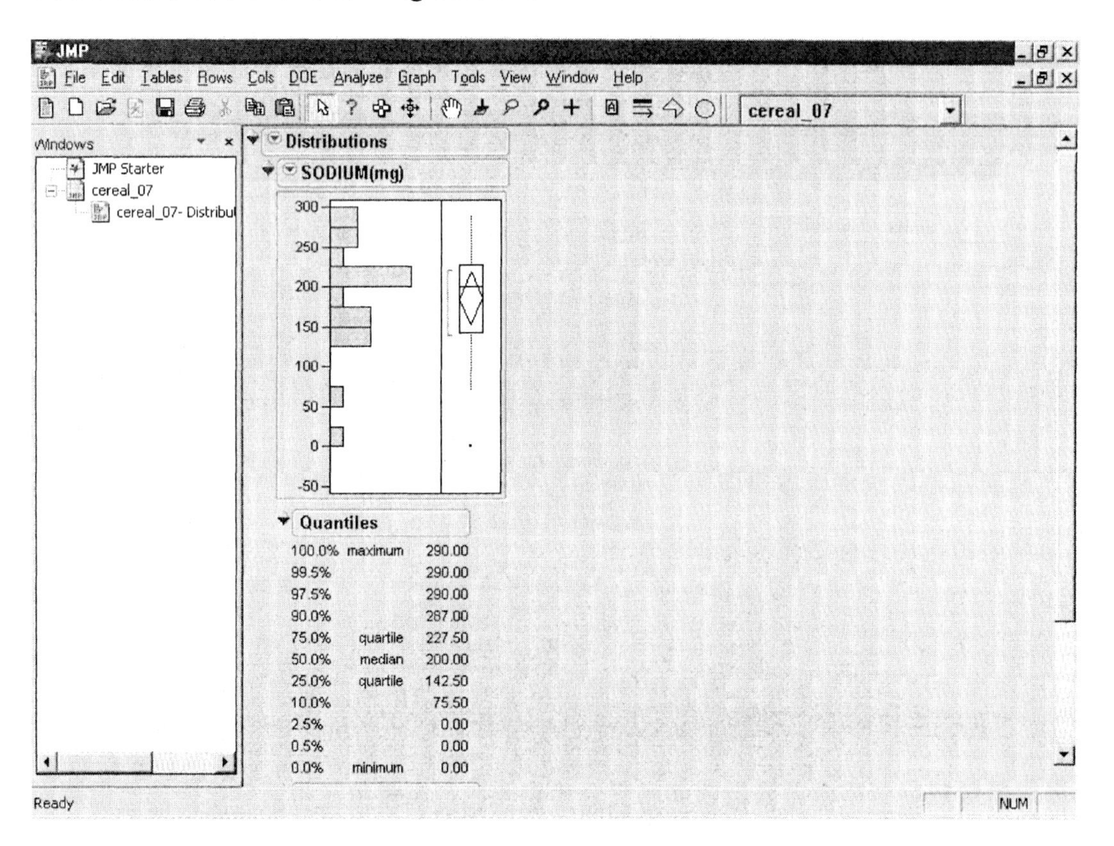

The default output produces a histogram on its side. To pivot the histogram so the graphic is horizontal, click on the red triangle circled above, and select **Histogram Options** and then deselect **Vertical**.

You should see the following screen:

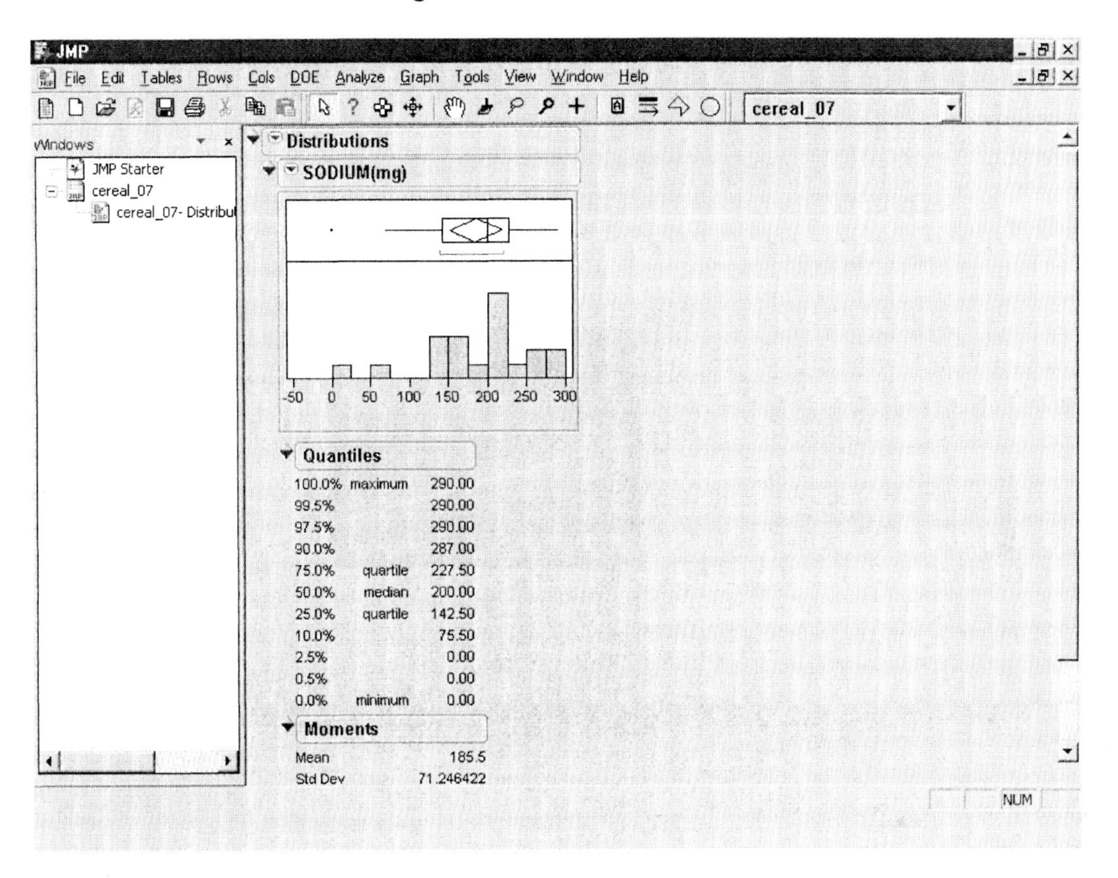

Chapter 2, Exercise 2.28. Is whooping cough close to being eradicated?

Open the whooping_cough data from the JMP folder of the data disk.

Select **Analyze** > **Modeling** > **Time Series.** Assign the ***Year*** variable as the **X, Time ID** and assign ***Incid Rate*** to the **Y, Time Series** role.

You should see the following popup appear:

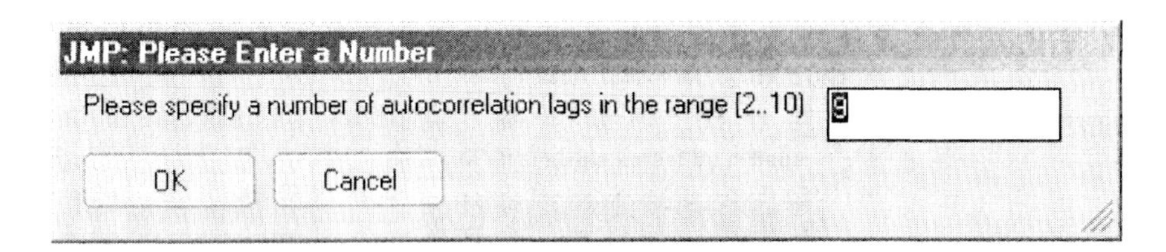

Whatever is indicated in the box, just select **OK**.
You should see the following screen:

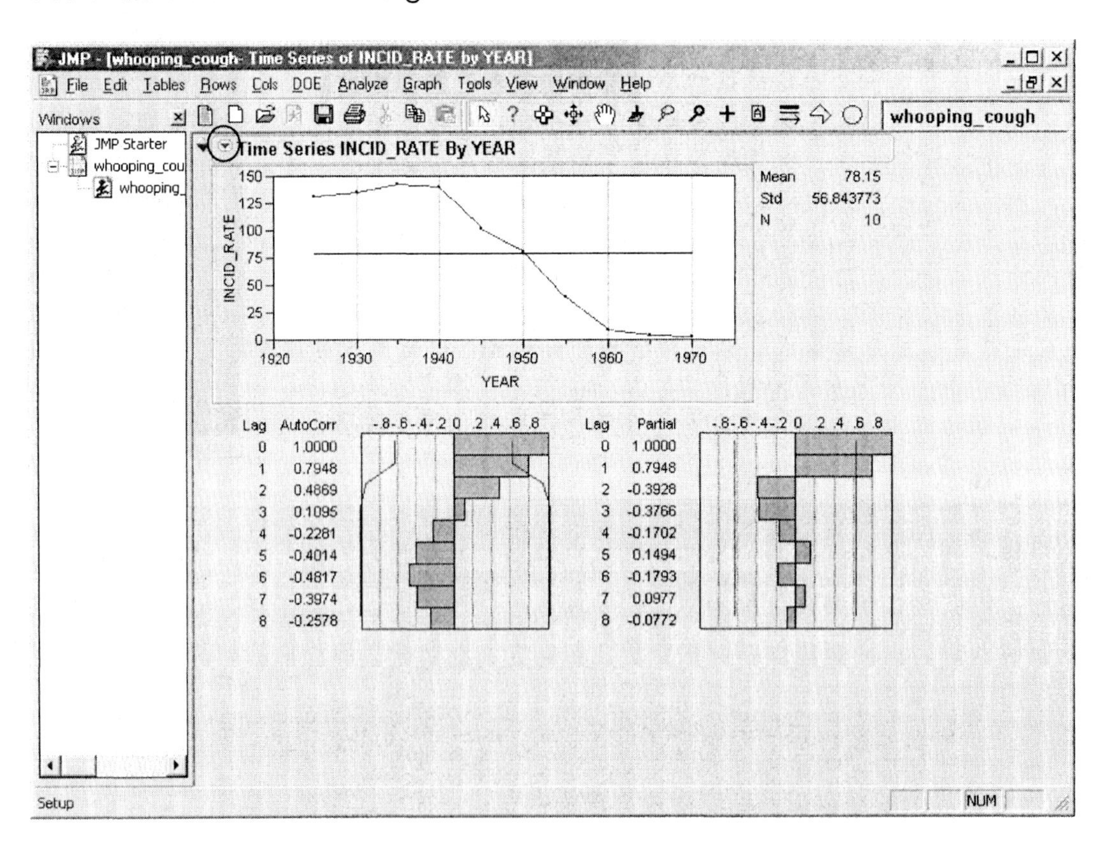

To hide the two boxes at the bottom of the screen, click on the red triangle noted above and deselect **Autocorrelation** and **Partial Autocorrelation**.

Chapter 2. Exercise 2.29. Warming in Newnan, Georgia?

Open the newnan_ga_temps data from the JMP folder of the data disk.

Select **Analyze** > **Modeling** > **Time Series.** Assign the ***Year*** variable to the **X, Time ID** role and ***Temp*** to the **Y, Time Series role**.

You should see the following screen:

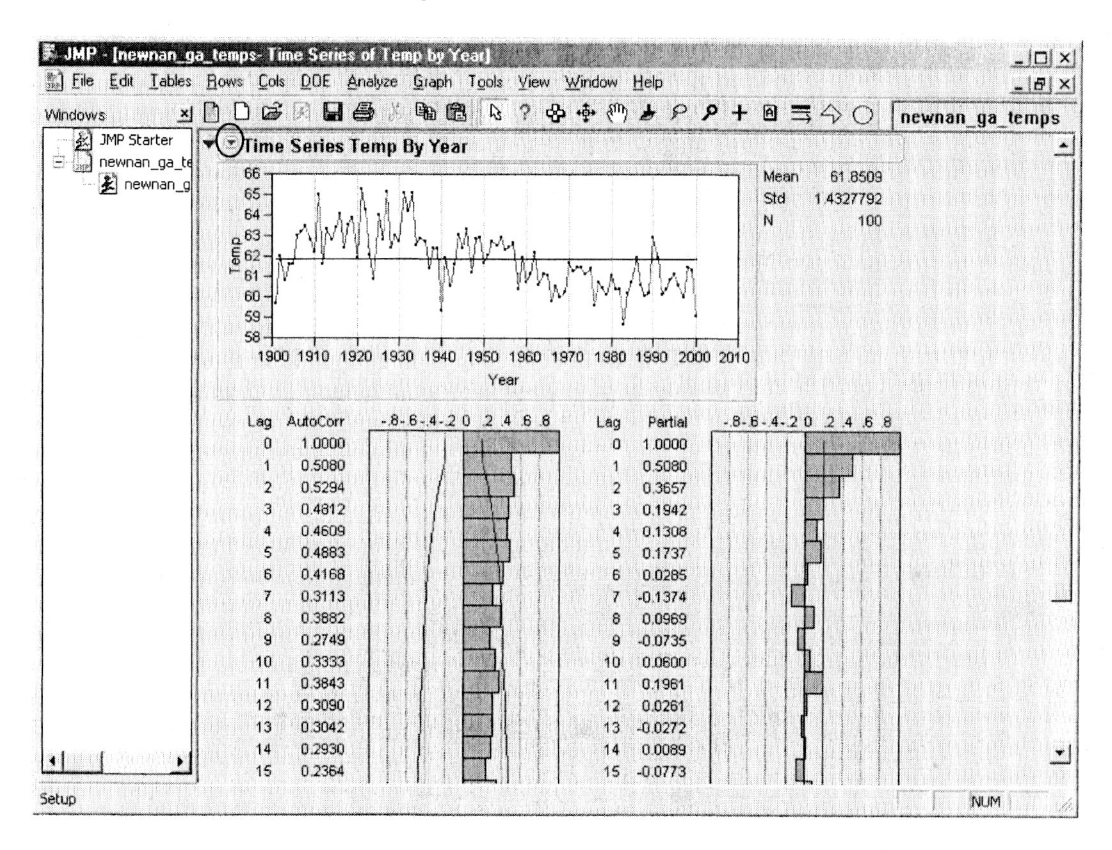

As before, to hide the two boxes at the bottom of the screen, click on the red triangle noted above and deselect **Autocorrelation** and **Partial Autocorrelation**.

Chapter 2. Example 10. What's the Center of the Cereal Sodium Data?

Open the cereal data from the JMP folder of the data disk.

There are two ways to generate information regarding the "center" of the sodium data.

One way is to select **Tables>Summary**. Select ***Sodium*** from the **Select Columns** list. Click on **Statistics** > **Mean**. Click on **Statistics** again, > **Median** (you could select any number of statistics from this list).

You should see the following screen:

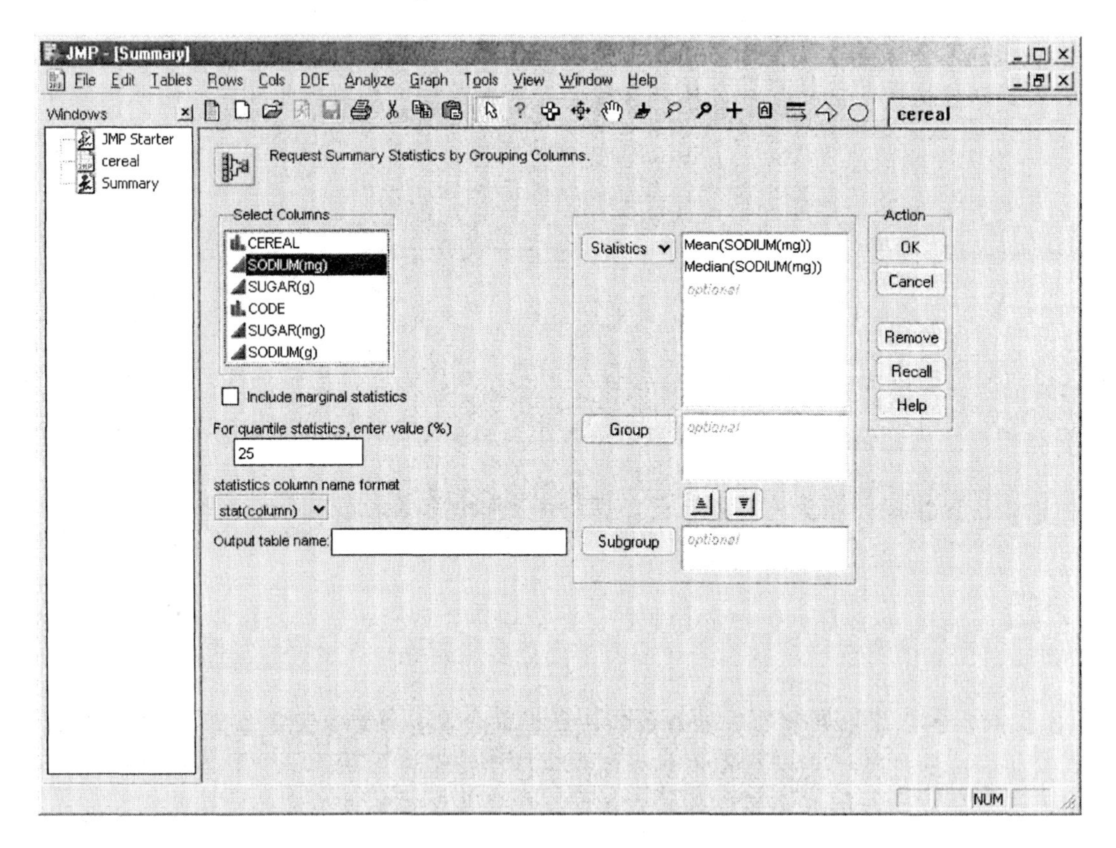

Select **OK**.

You should see the following screen:

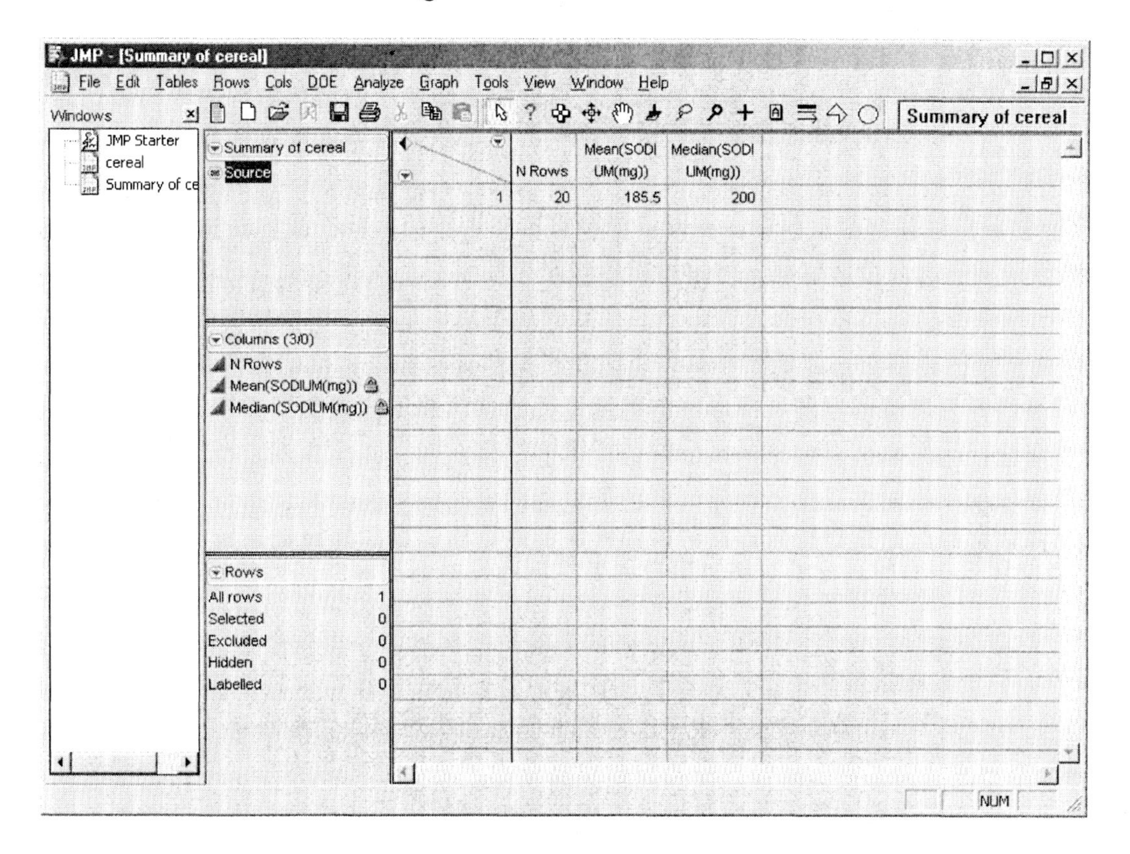

Note that you can size the width of the columns by placing your cursor in between the columns and then dragging the lines wider or narrower as needed.

A second method of generating information regarding the center of the dataset is to select **Analyze>Distribution**. Assign ***sodium(mg)*** to the **Y, Columns** role. Select **OK**. When the output screen is generated, toggle to the bottom.

You should see the following screen:

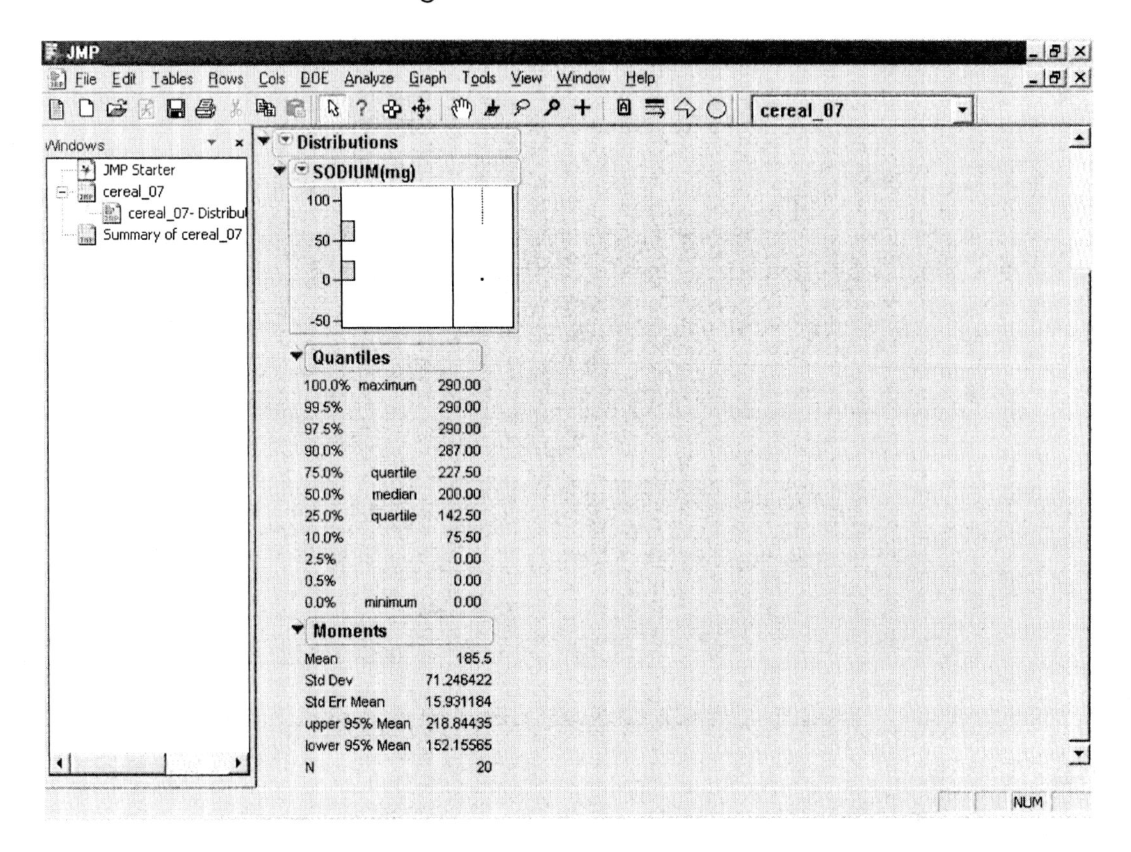

The Median value can be found in the **Quantiles** section. The Mean is included in the **Moments** section.

Chapter 2. Example 15. Describing Female College Student Heights.

Open the heights data from the JMP folder of the data disk.

Select **Analyze>Distribution**. Assign ***Height*** to the **Y, Columns** role and ***Gender*** to the **By** role. Select **OK**. When the output screen is generated, click on the red triangle next to Height in each section of the output. Select **Display Options>Horizontal**.

You should see the following screen:

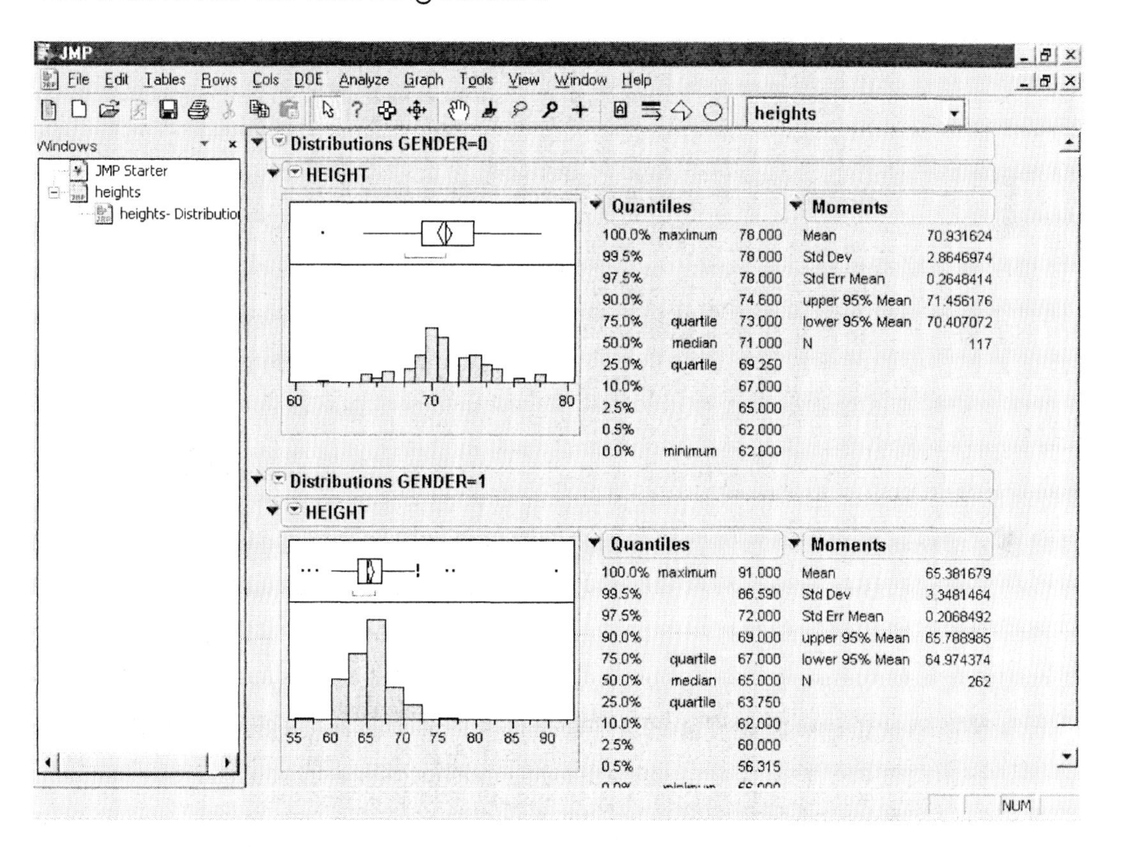

We are only interested in the women, who are designated as Gender = 1.

To use the empirical rule to better understand the heights, return to the dataset. Select **Tables>Sort**. Assign ***Gender*** first and then ***Height*** to the **By** role. To ensure that the women are at the top, click on the triangle option. This will ensure that we execute a descending sort of gender and an ascending sort of height. Select **OK**.

You should see the following screen:

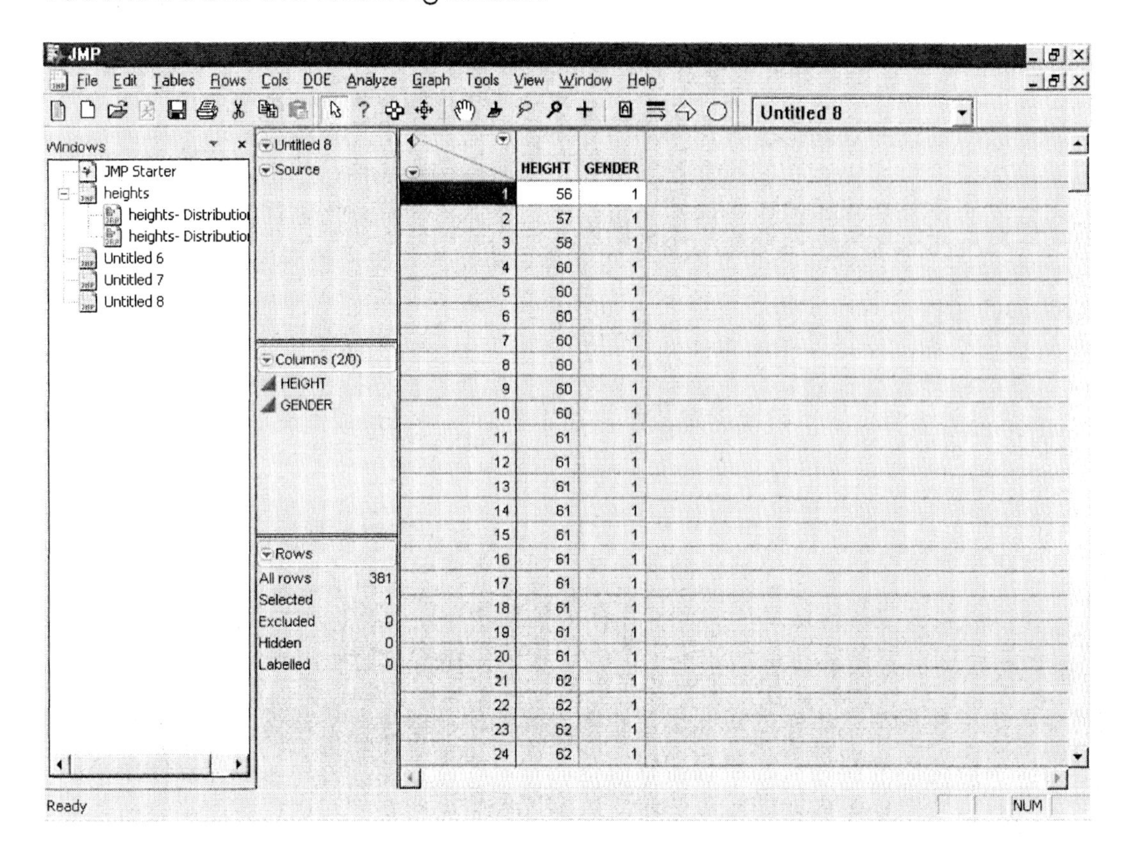

Since we have a total of 262 women, the center 68% would be the center 178 values. If you take 262 minus 178, and then divide by 2, you have 42. So, we take observation number 42 – which is 62 – and observation number 220 – which is 68 – we can say, using the empirical rule, that approximately 68% of all observations lie between 62 and 68. Using the same approach for 95% and 99%, we would come up with 59 and 71, and 56 and 74, respectively.

Chapter 2. Exercise 2.61. EU Data File.

Open the european_union_unemployment data from the JMP folder of the data disk.

To construct a graph to describe these values, select **Analyze>Distribution**. Assign ***unemployment*** to the **Y, Columns** role. On the output screen, to rotate the histogram, click on the red triangle in the upper left corner, select **Histogram Options** and deselect **Vertical.**

You should see the following screen:

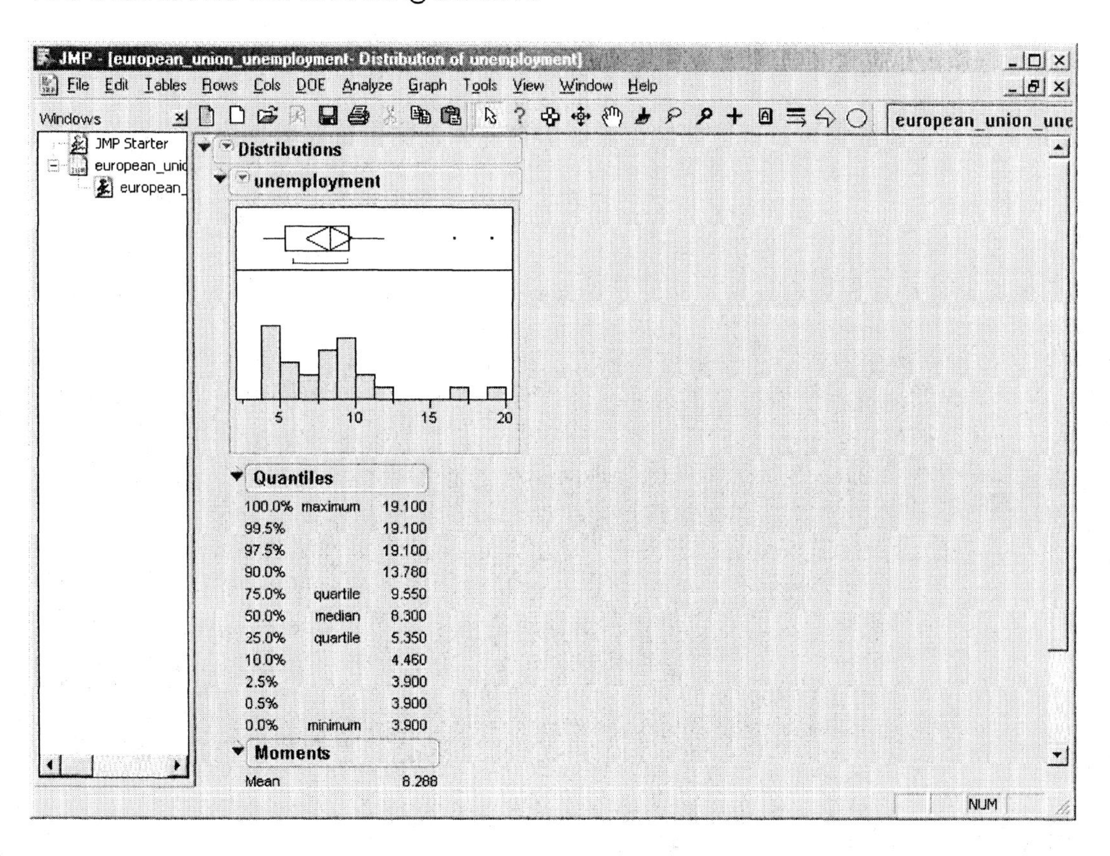

Scrolling to the bottom of the output, you will see the standard deviation as well.

An alternative chart to display this data, is a simple vertical bar chart, showing each of the country names. A bar chart is generated by returning to the dataset, and selecting **Graph>Chart**. Assign ***countries*** as the **Categories, X, Levels**. Then select ***unemployment*** and **Statistics>Data**. Click on **OK.**

When the chart appears, it will be "bunched" together. Place your cursor in the bottom right hand corner and drag the frame until the chart is sized more appropriately:

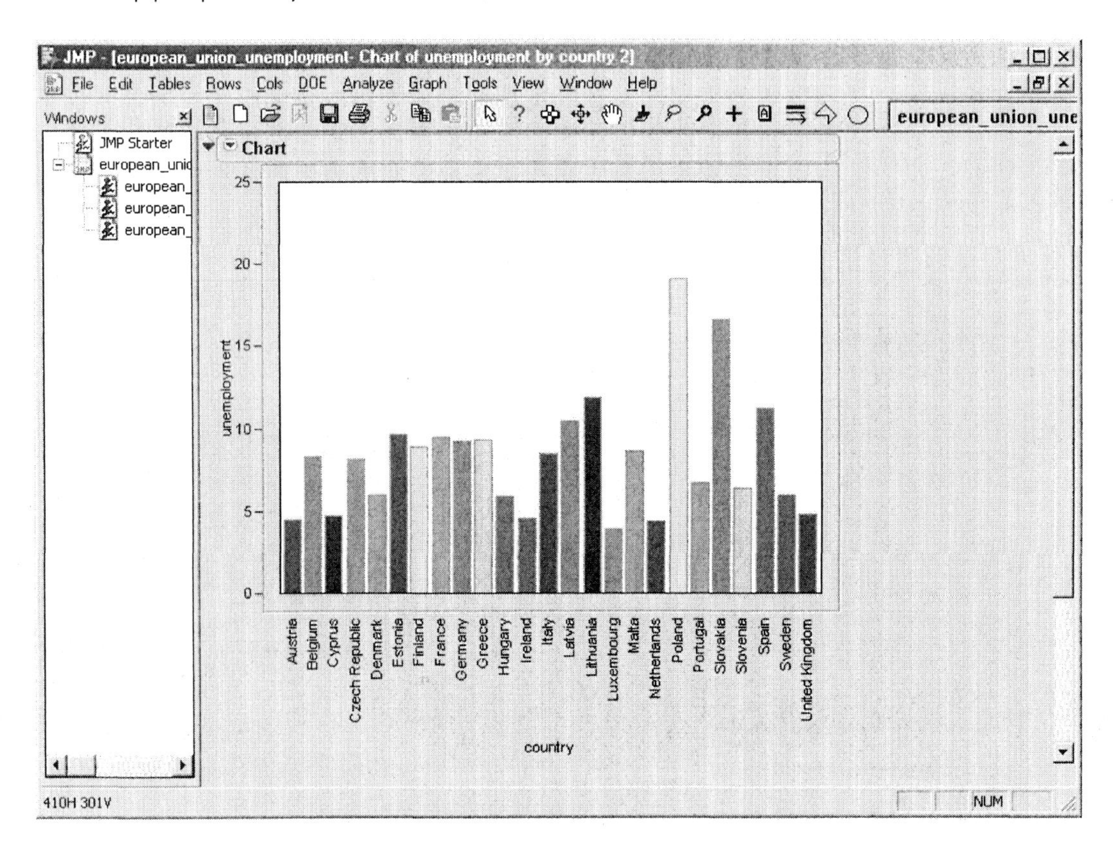

Chapter 2. Example 16. What Are the Quartiles for the Cereal Sodium Data?

Open the cereal data from the JMP folder of the data disk.

Select **Analyze>Distribution.** Assign ***Sodium*** to the **Y, Columns** role. Select **OK**.

You should see the following screen:

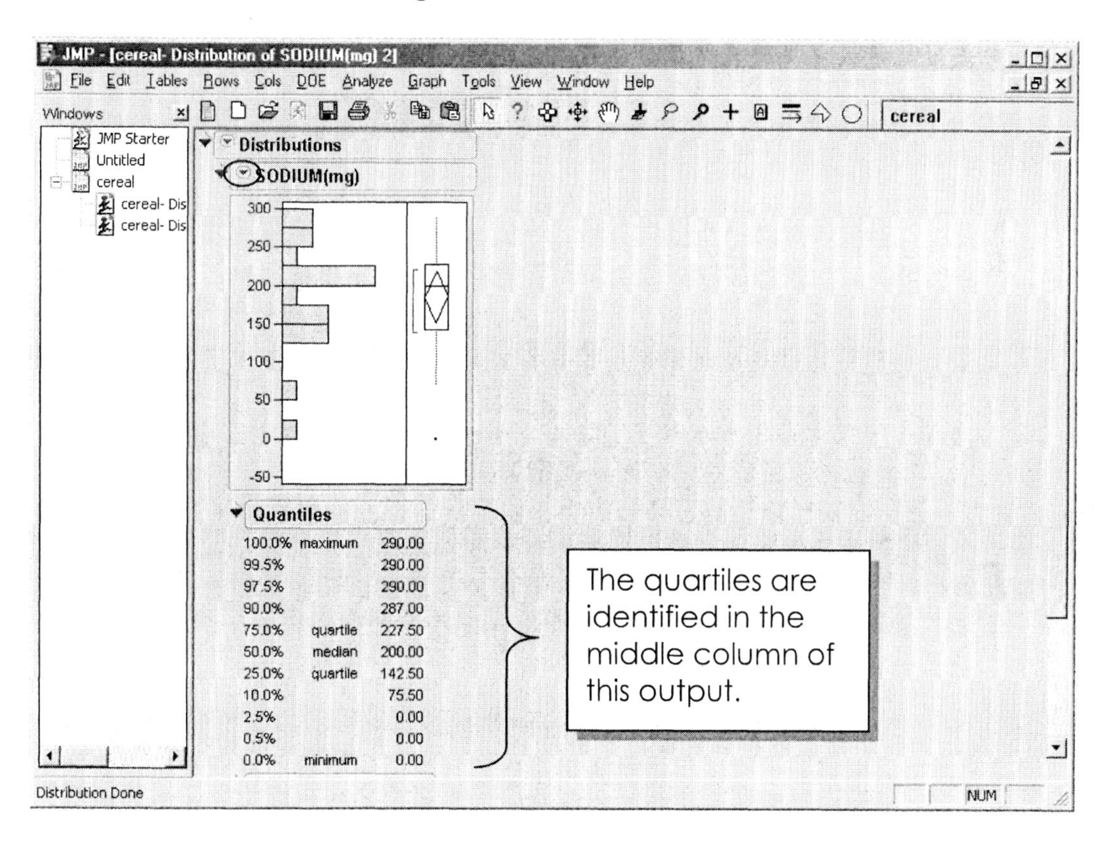

Note that the boxplot is provided as part of the histogram graphic, when this graphic is displayed as above (this is the default). In JMP, the "box" represents Q1 to Q3, with the median identified as the line in the middle. Observations which are more than 1.5*(Q3-Q1 or IQR) are identified with an asterisk ("*") and are considered to be outliers.

You can easily change this display to read horizontally rather than vertically by clicking on the red triangle identified above and selecting **Display Options>Horizontal Layout**:

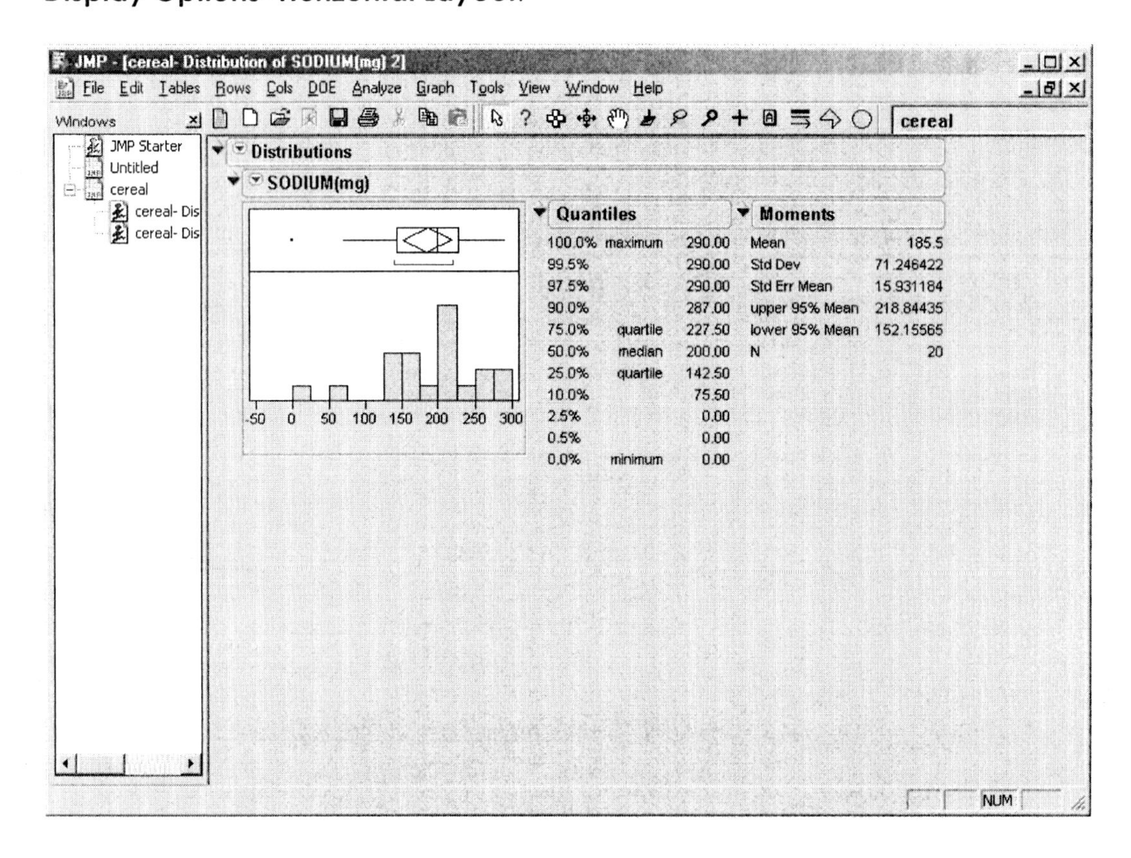

Chapter 2. Exercise 2.76. Energy Statistics.

Open the energy_eu data from the JMP folder of the data disk.

Select **Analyze>Distribution**. Assign ***energy*** to the **Y, Columns** role. Select **OK**.

You should see the following screen:

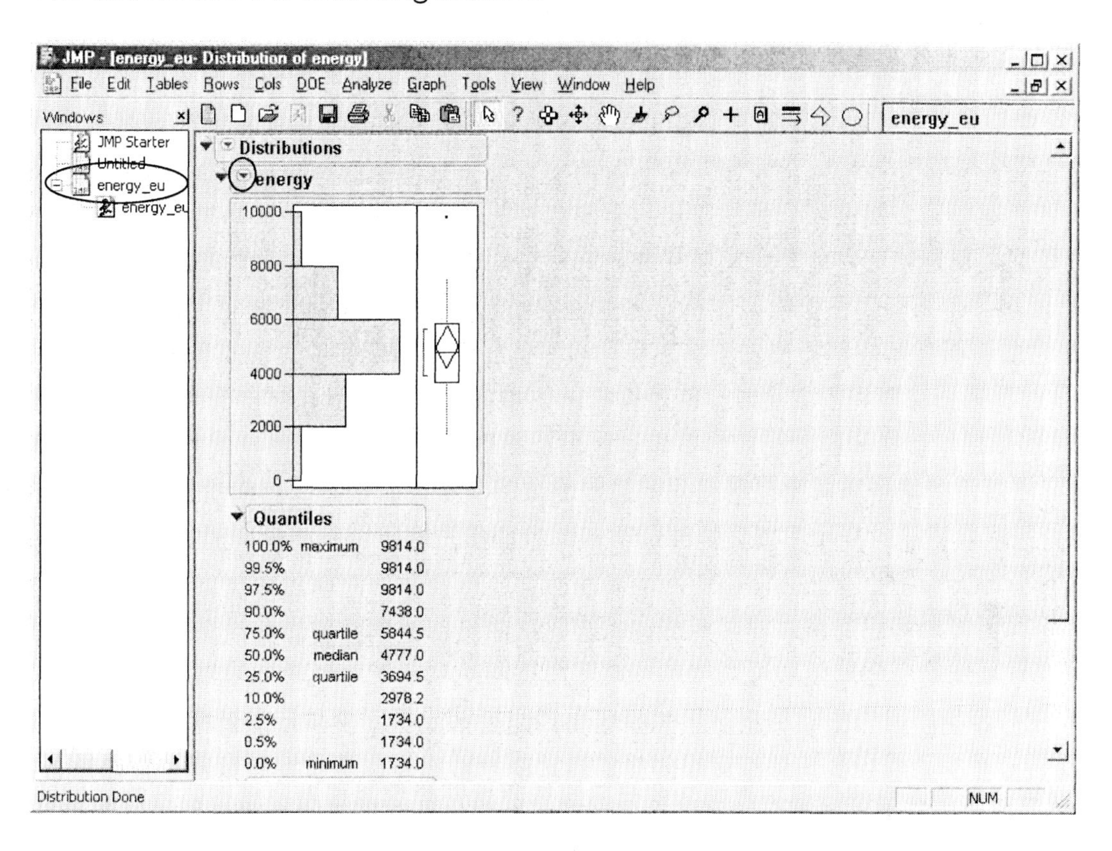

Note that the moments (specifically, the mean and the standard deviation) can be found by toggling to the bottom of this screen.

The five number summary (Minimum, Q1, Median, Q3 and Maximum) can be found in the **Quantiles** section above.

To produce the standard deviations for each country in the dataset, click on the red triangle as circled above, select **Save>Standardized**. Now, return to the original dataset by selecting the energy_eu report icon in the report window as circled above.

You should now see this screen:

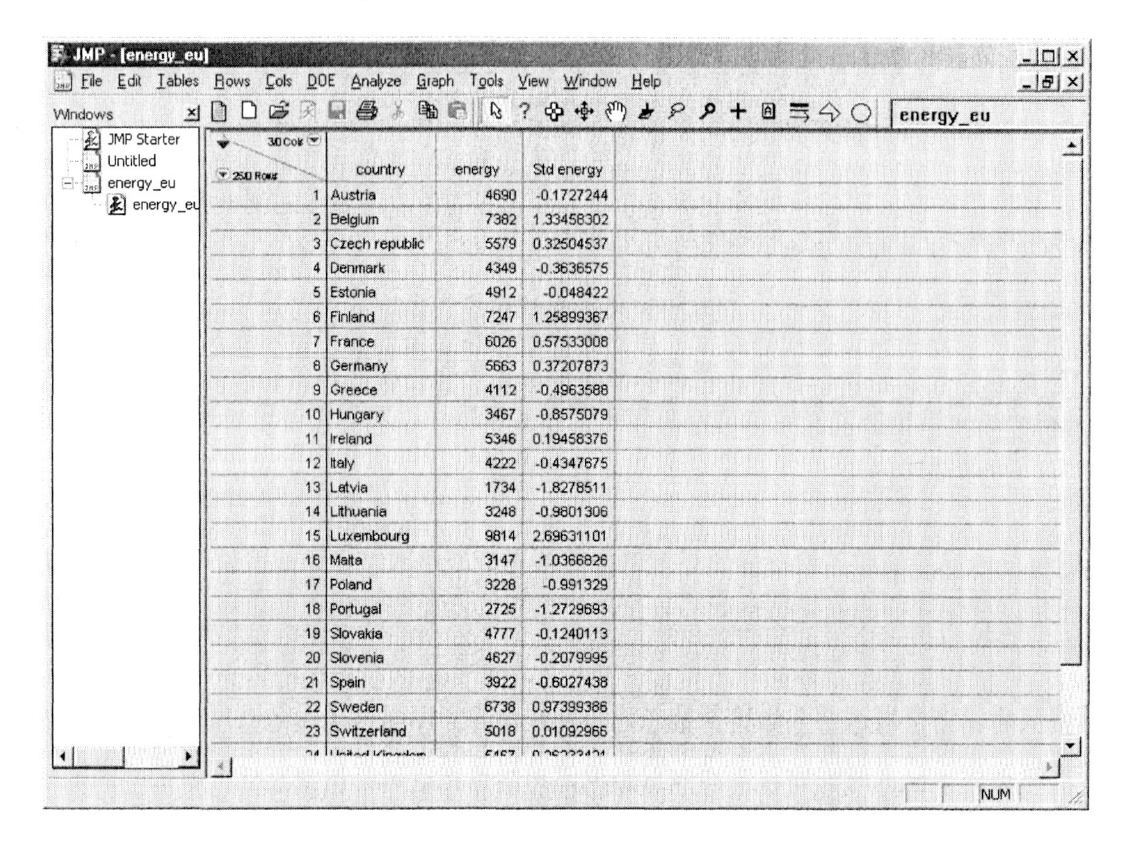

	country	energy	Std energy
1	Austria	4690	-0.1727244
2	Belgium	7382	1.33458302
3	Czech republic	5579	0.32504537
4	Denmark	4349	-0.3636575
5	Estonia	4912	-0.048422
6	Finland	7247	1.25899367
7	France	6026	0.57533008
8	Germany	5663	0.37207873
9	Greece	4112	-0.4963588
10	Hungary	3467	-0.8575079
11	Ireland	5346	0.19458376
12	Italy	4222	-0.4347675
13	Latvia	1734	-1.8278511
14	Lithuania	3248	-0.9801306
15	Luxembourg	9814	2.69631101
16	Malta	3147	-1.0366826
17	Poland	3228	-0.991329
18	Portugal	2725	-1.2729693
19	Slovakia	4777	-0.1240113
20	Slovenia	4627	-0.2079995
21	Spain	3922	-0.6027438
22	Sweden	6738	0.97399386
23	Switzerland	5018	0.01092966

The "***Std energy***" column has converted all of the "energy" values to z-scores (number of standard deviations from the mean).

From this table, Italy's energy value of 4222 was .4347675 standard deviations BELOW (because it has a negative sign) the mean of 4998.

Chapter 2. Exercise 2.81. Florida Students Again.

Open the fl_student_survey data from the JMP folder of the data disk.

Select **Analyze>Distribution**. Assign ***TV*** to the **Y, Columns** role and select **OK**.

You should see the following screen:

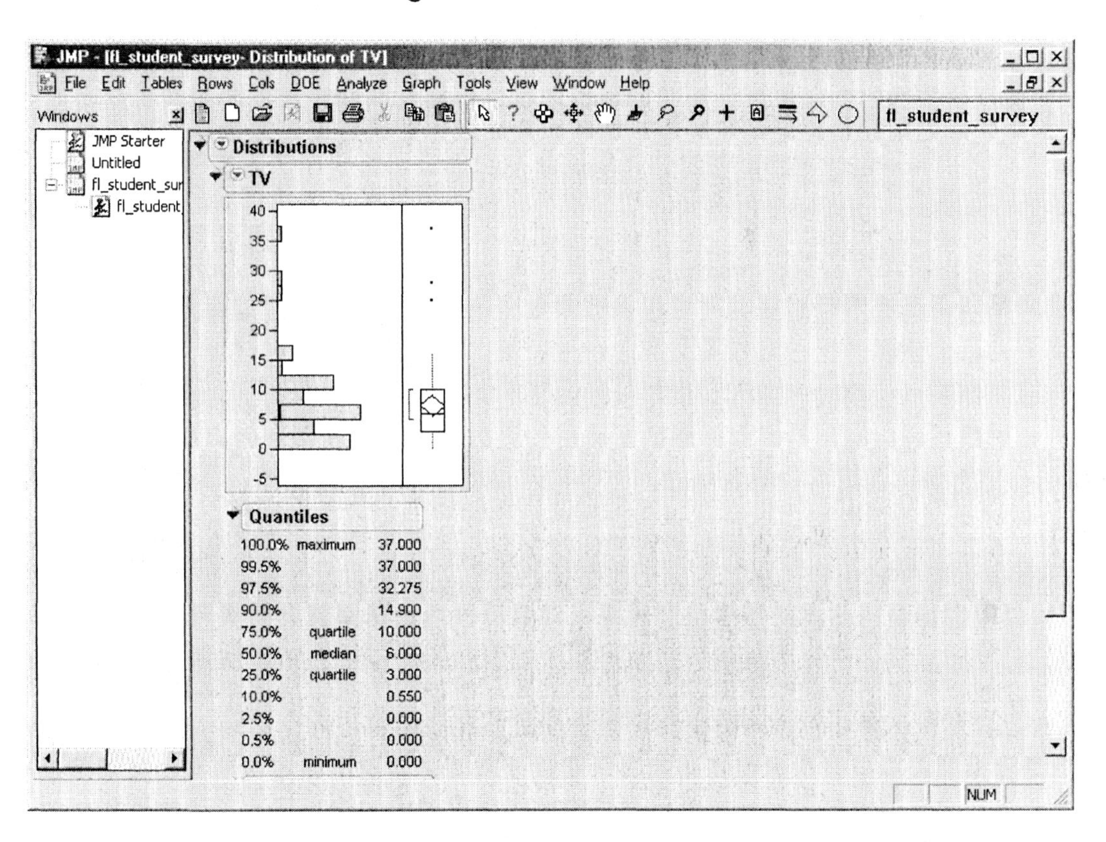

The histogram and the boxplot shown above indicate that three outliers appear to exist, as is evident from the asterisks. To check the z-scores, click on the red triangle as circled above, then select **Save>Standardized**.

When you return to the dataset, an additional column would have been created titled "**Std TV**". These are the z-scores for the TV variable. To easily see where the outliers are (those observations with the highest and lowest z-scores), you can sort the dataset by the z-scores.

When viewing the dataset, select **Tables>Sort**. Select the new ***Std TV*** variable as the **By** variable:

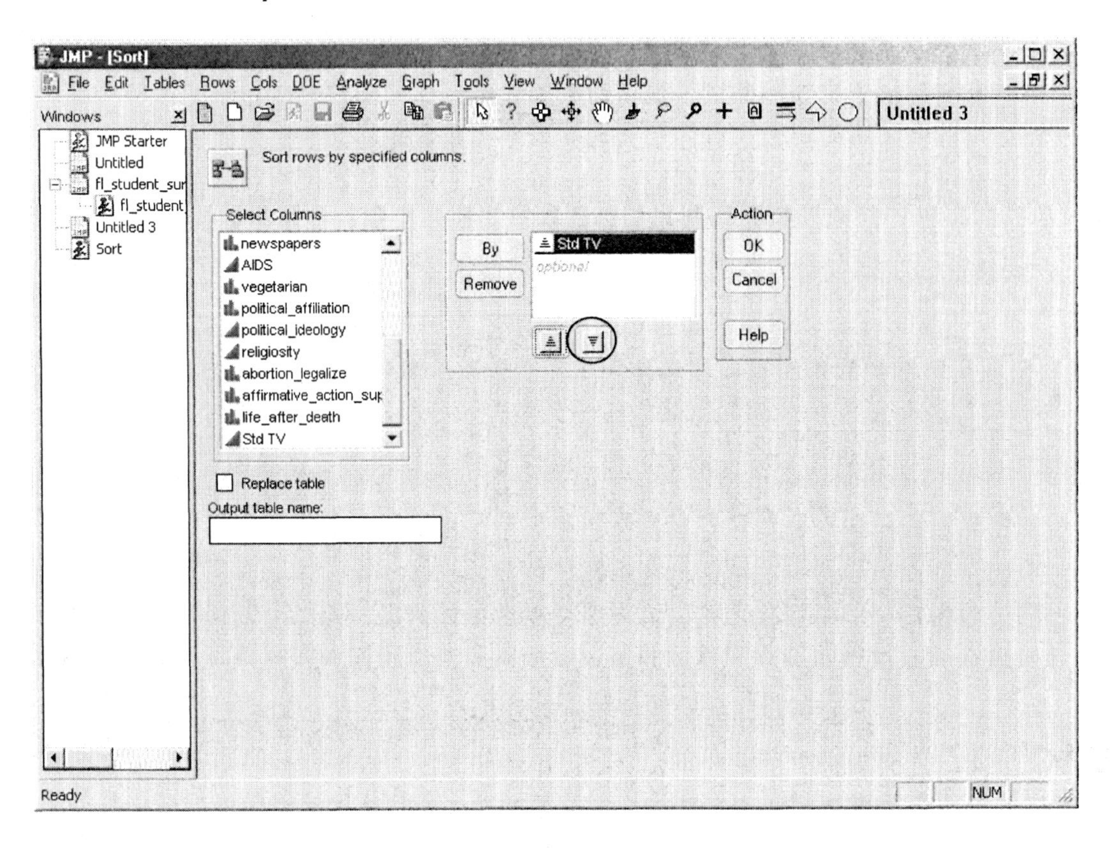

You can determine if you want an ascending sort (the default) or a descending sort. Since we are interested in the largest values, select the descending sort as identified above. Select **OK**.

After you toggle to the right to see the ***Std TV*** variable, you should see the following screen:

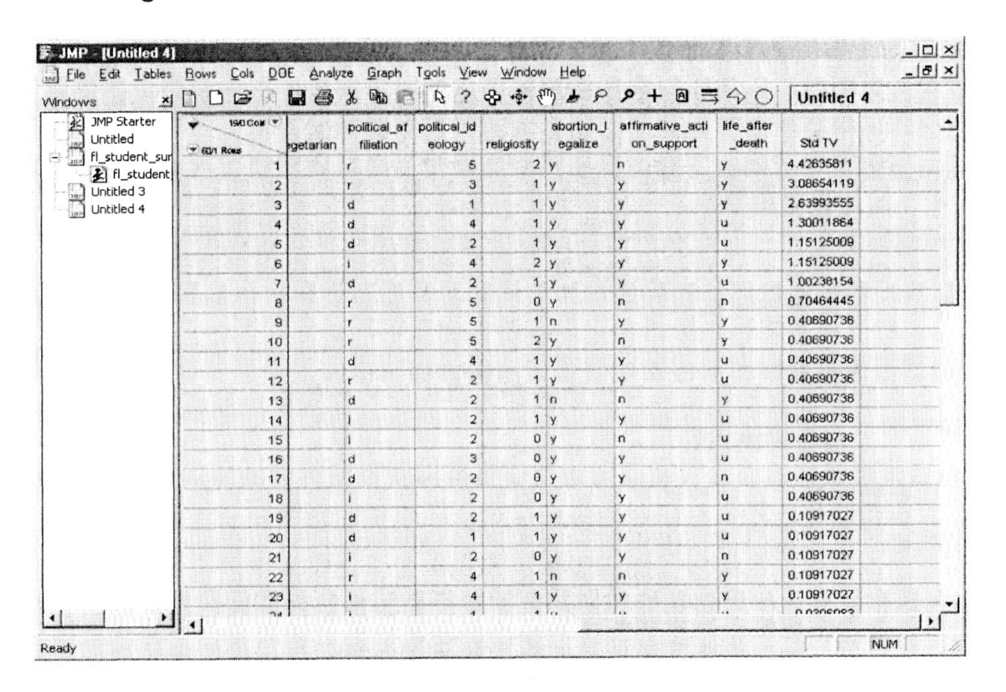

	getarian	political_affiliation	political_ideology	religiosity	abortion_legalize	affirmative_action_support	life_after_death	Std TV
1		r	5	2	y	n	y	4.42635811
2		r	3	1	y	y	y	3.08654119
3		d	1	1	y	y	y	2.63993555
4		d	4	1	y	y	u	1.30011864
5		d	2	1	y	y	u	1.15125009
6		i	4	2	y	y	y	1.15125009
7		d	2	1	y	y	u	1.00238154
8		r	5	0	y	n	n	0.70464445
9		r	5	1	n	y	y	0.40690736
10		r	5	2	y	n	y	0.40690736
11		d	4	1	y	y	u	0.40690736
12		r	2	1	y	y	u	0.40690736
13		d	2	1	n	n	y	0.40690736
14		i	2	1	y	y	u	0.40690736
15		i	2	0	y	n	u	0.40690736
16		d	3	0	y	y	u	0.40690736
17		d	2	0	y	y	n	0.40690736
18		i	2	0	y	y	u	0.40690736
19		d	2	1	y	y	u	0.10917027
20		d	1	1	y	y	u	0.10917027
21		i	2	0	y	y	n	0.10917027
22		r	4	1	n	n	y	0.10917027
23		i	4	1	y	y	y	0.10917027

From this sort, subjects 39, 47 and 20, respectively watch the most TV, and appear to be outliers.

Chapter 2. Exercise 2.82. Females or males watch more TV?

Open the fl_student_survey data from the JMP folder of the data disk.

Select **Analyze>Distribution**. Assign ***TV*** to the **Y, Columns** role and ***gender*** to the **By** role.

Your screen should look like this:

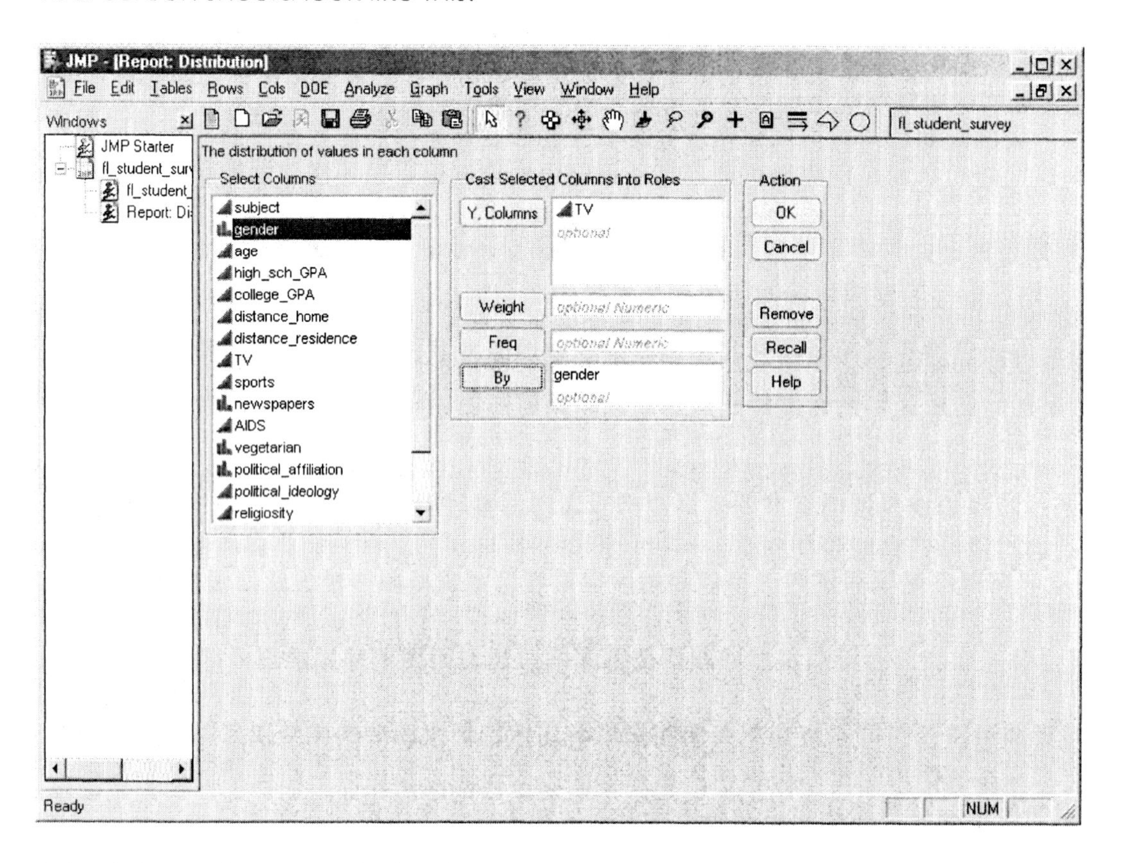

Select **OK**. When the Distribution Platform is generated, you will notice that the analysis of the females is indicated first, followed by the males.

To see this information simultaneously, change the layouts for both sets of analysis to a "**horizontal display**" (click on the red triangle next to "***TV***", select **Display Options>Horizontal Layout**).

You should see the following screen:

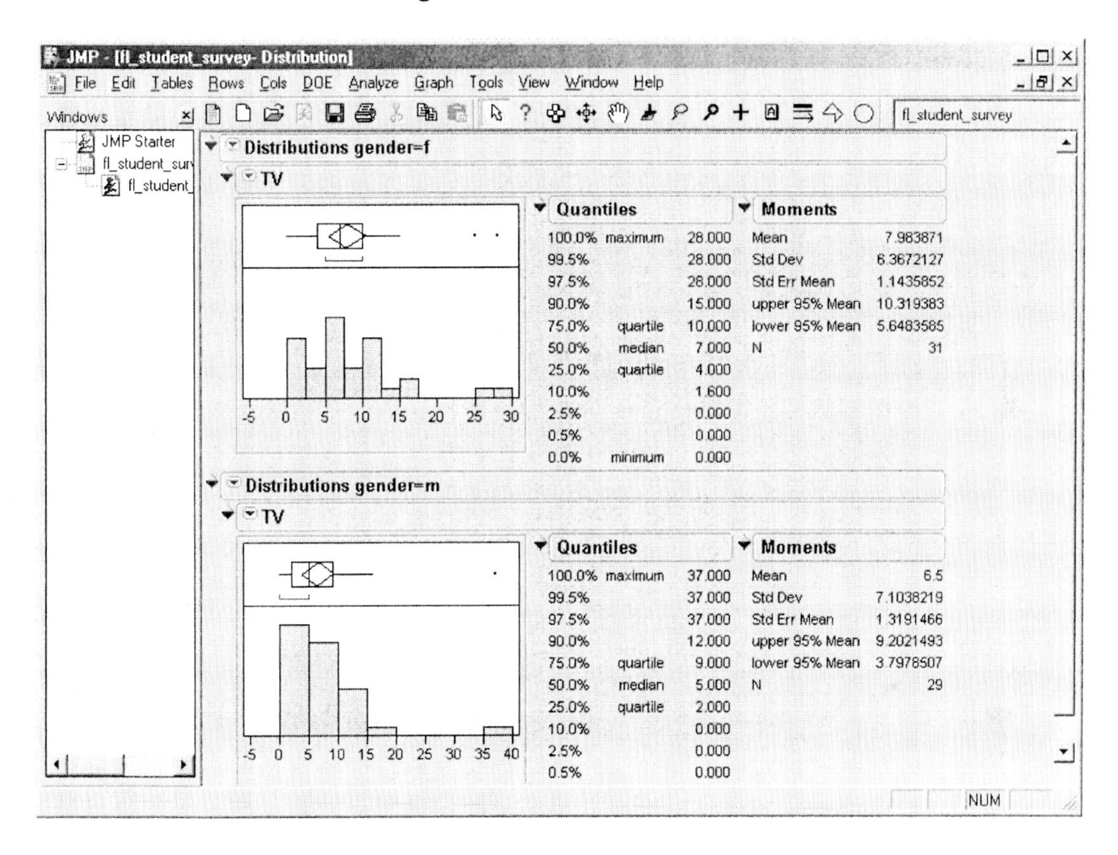

Chapter 2. Exercise 2.86. Enrollment Trends.

Input the data as provided. To create a Line Graph of the African American enrollment, select **Graph>Chart**. Select ***African American*>Statistics>Data**. Select ***Year*>Categories,X,Levels**. Finally, on the left side, change **Bar Chart** (the default) to **Line Chart**. Click **OK**.

You should see the following screen:

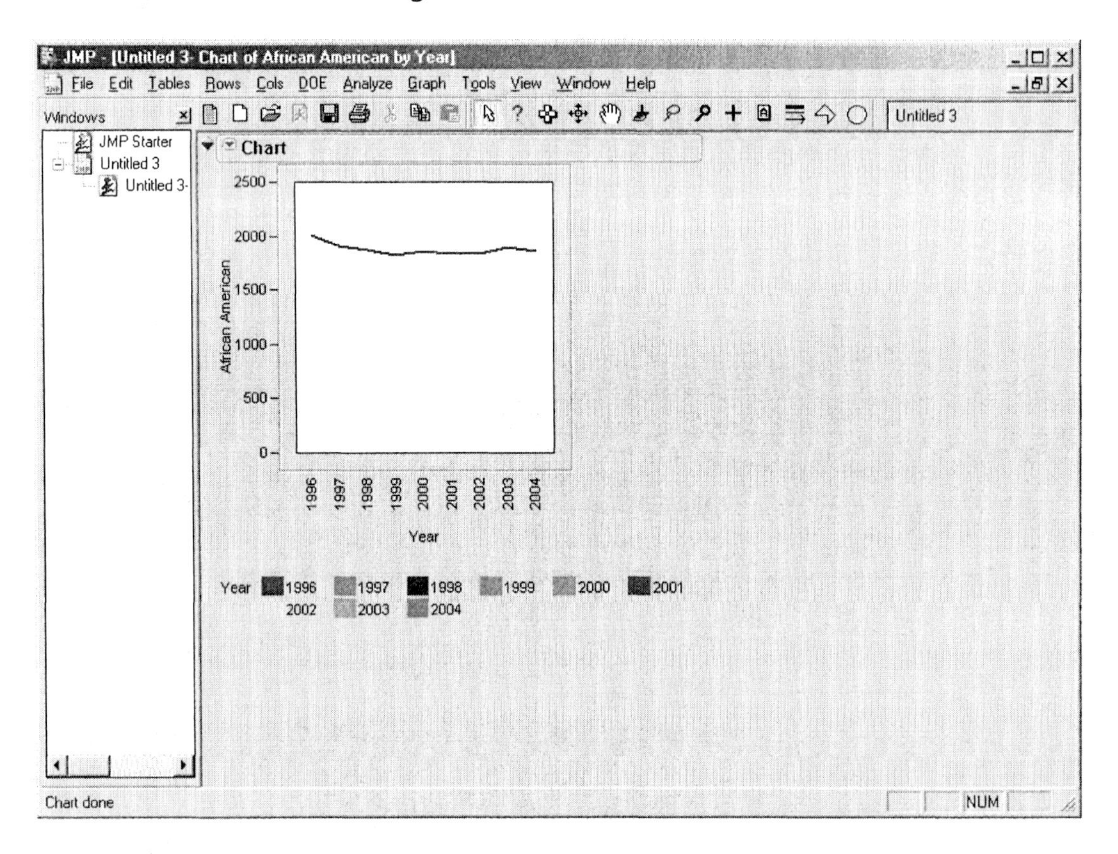

Although technically correct, this graphic can be improved. Specifically, the y-axis contains too much unused space. To scale the y-axis, double click on the y-axis.

You will see the following screen:

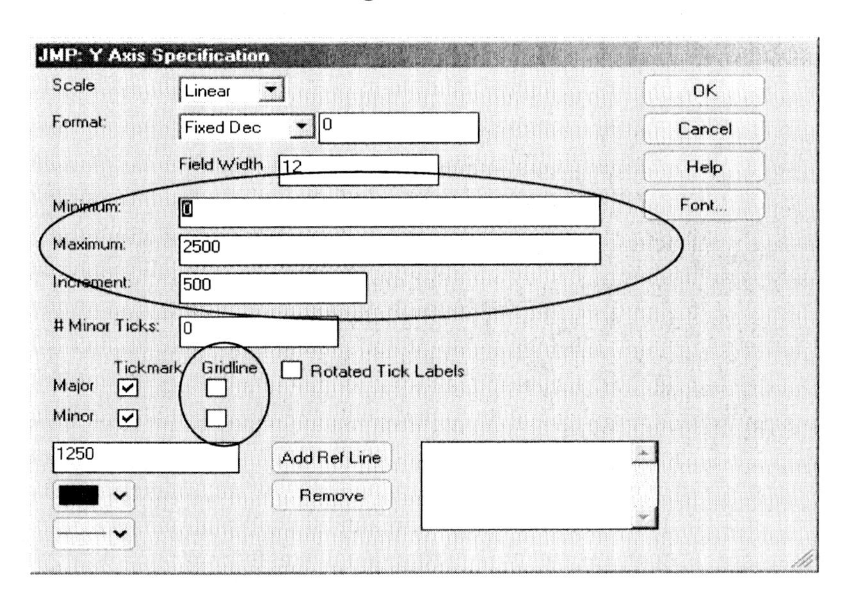

In this screen, change the **Minimum** value to 1500 and change the **Maximum** value to 2100. Change the Increment to 100. Finally, check the **Gridline** boxes as circled above. The chart should now look like this:

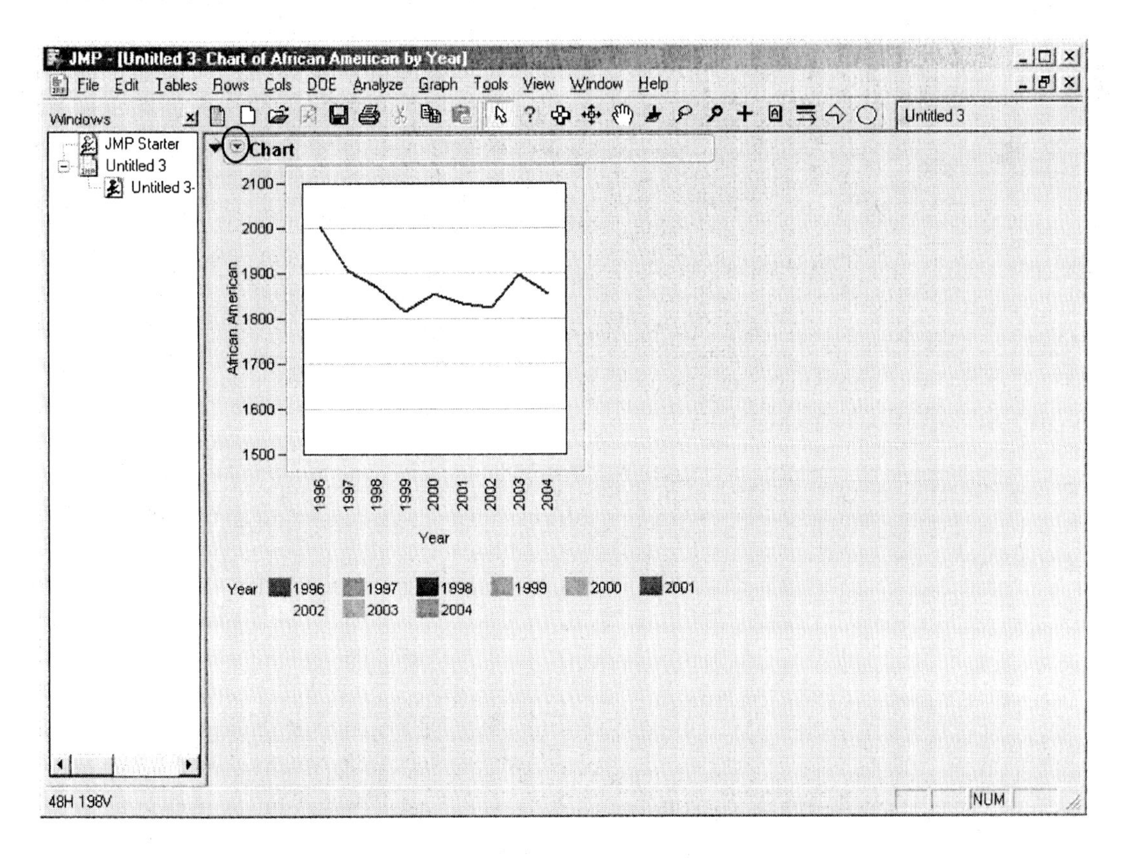

The legend for the years is unnecessary in this graphic. To omit the legend, click on the red triangle, and deselect **Show Level Legend**.

Now, the graphic should appear like this:

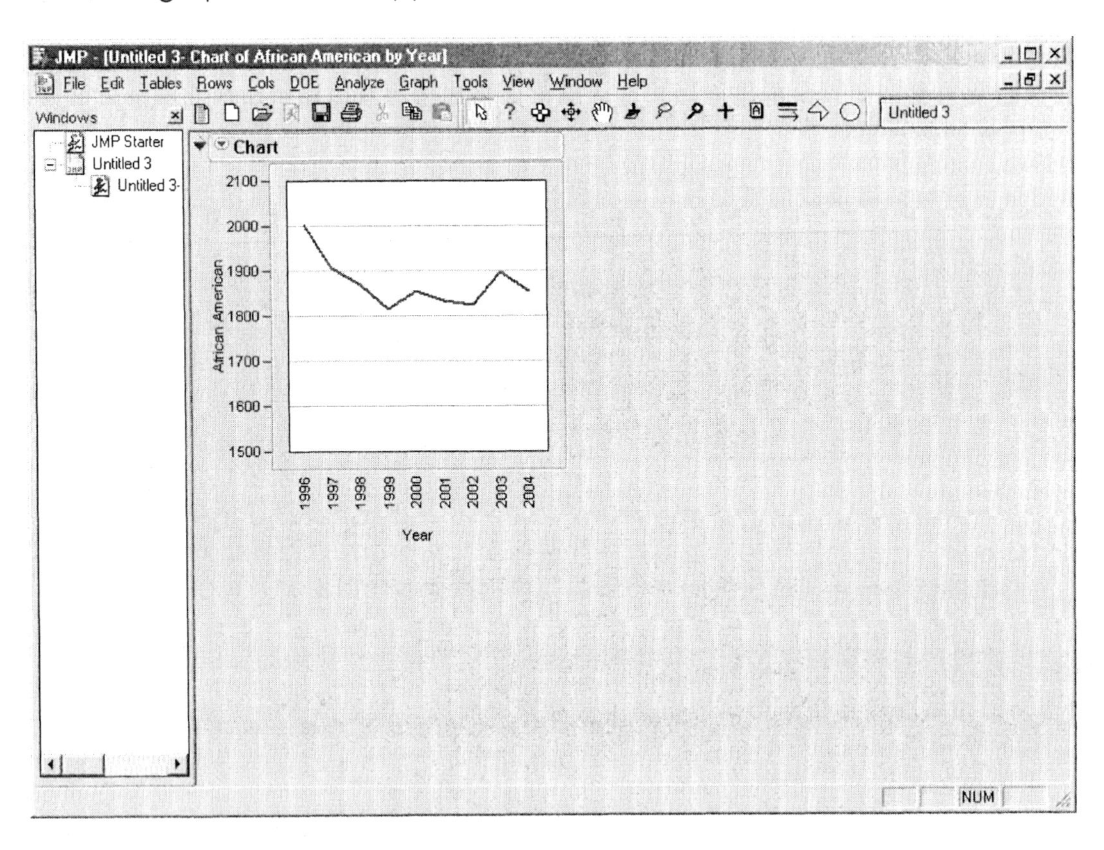

At this point, you can interpret the trend.

To evaluate and graph the percentage of total students who are African American, we need to begin by creating an additional column or variable in our dataset.

Return to the original data. Double click on the first available column – column 4. Title this column "***Ratio of African American Students to Total***". After the column has been titled, double click on the title box.

You should see the following screen:

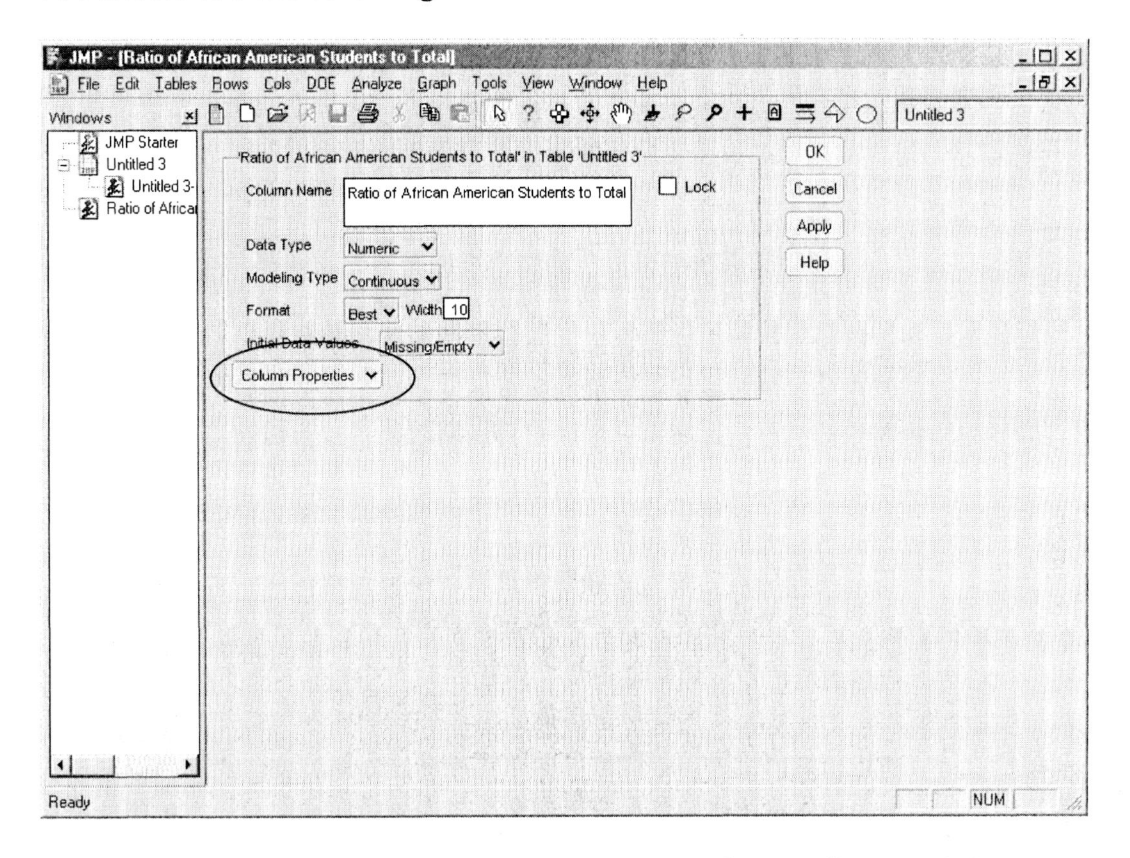

Select **Column Properties>Formula**. Click on the box "**Edit Formula**".

You should see the following screen:

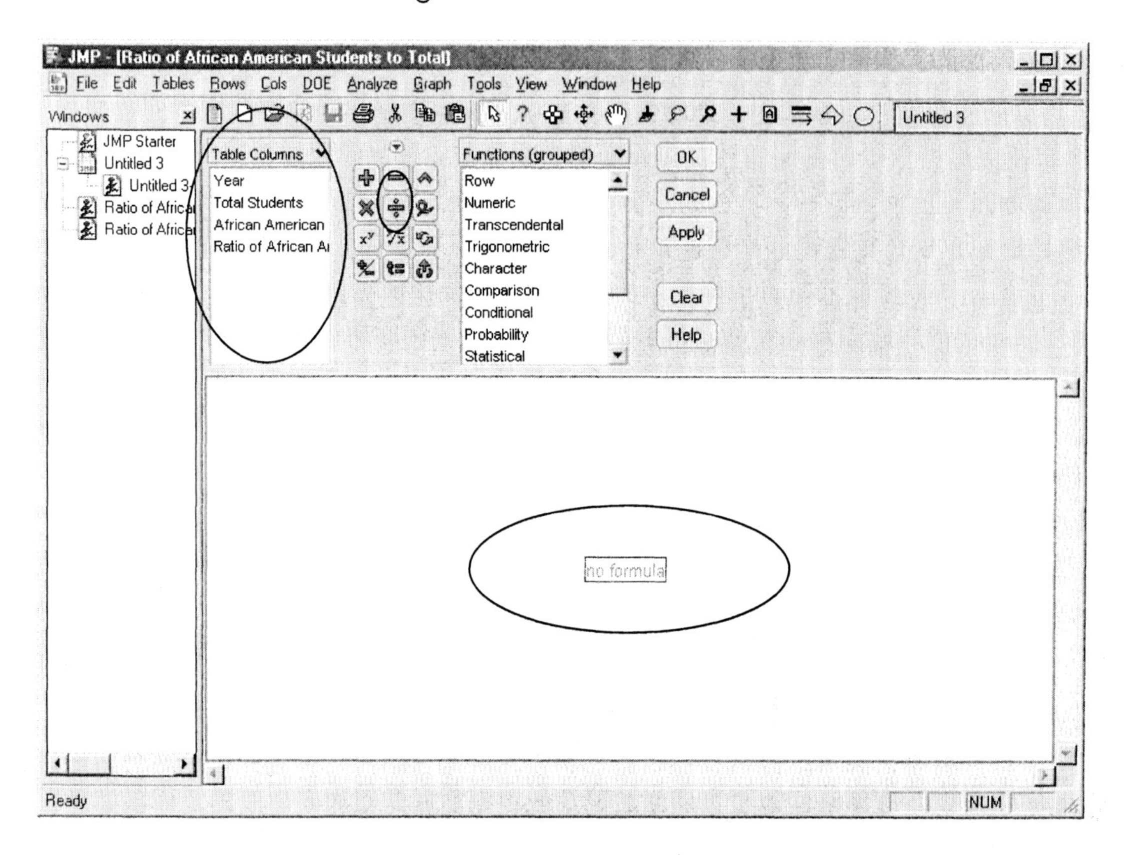

We want to write a formula that will divide ***African American Students*** by the ***Total Number of Students***. From the **Table Columns** list above, select "***African American Students***". Then click on the division button circled above. Then select "***Total Students***" from the **Table Columns** list. Within the red box circled above, there should now be a visual representation of ***African American Students*** divided by ***Total Students***. Select **OK>OK**.

You should now see the following screen:

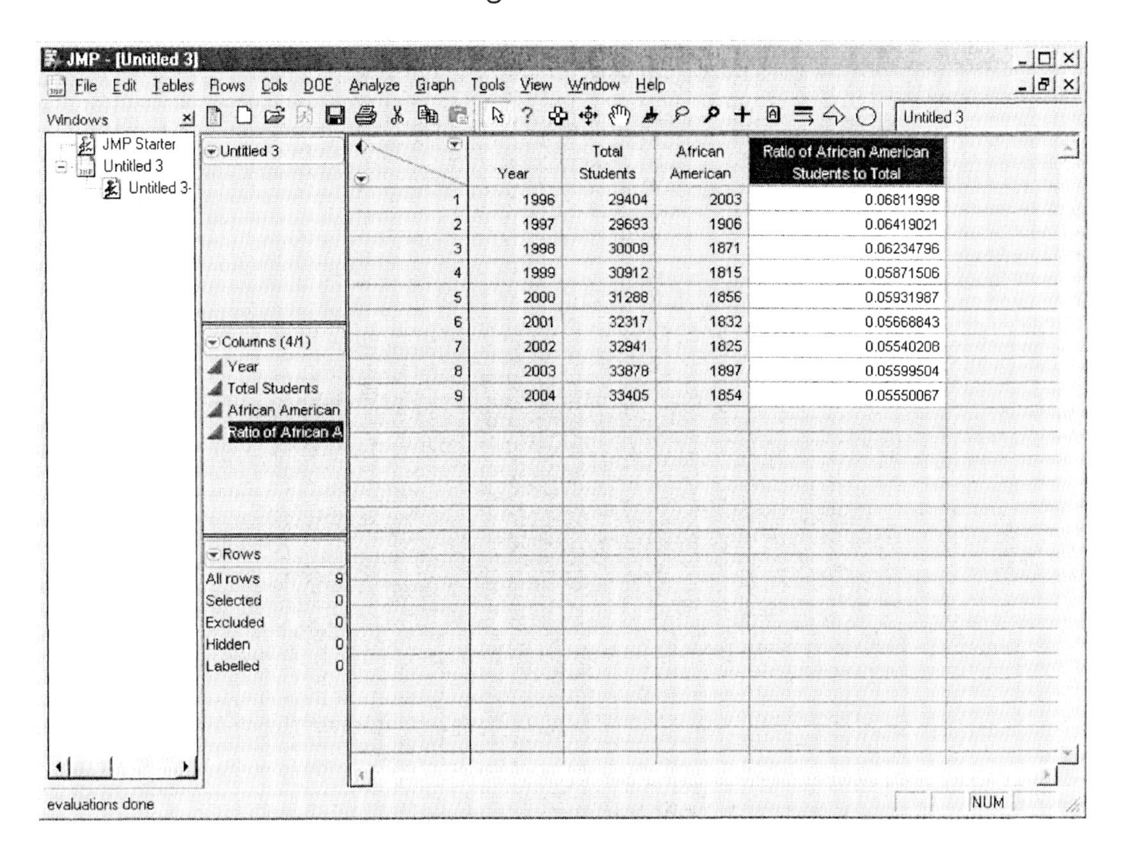

	Year	Total Students	African American	Ratio of African American Students to Total
1	1996	29404	2003	0.06811998
2	1997	29693	1906	0.06419021
3	1998	30009	1871	0.06234796
4	1999	30912	1815	0.05871506
5	2000	31288	1856	0.05931987
6	2001	32317	1832	0.05668843
7	2002	32941	1825	0.05540208
8	2003	33878	1897	0.05599504
9	2004	33405	1854	0.05550067

The new ratio variable has been created and can now be graphed just like the African American variable was graphed above (indicating a very clear trend):

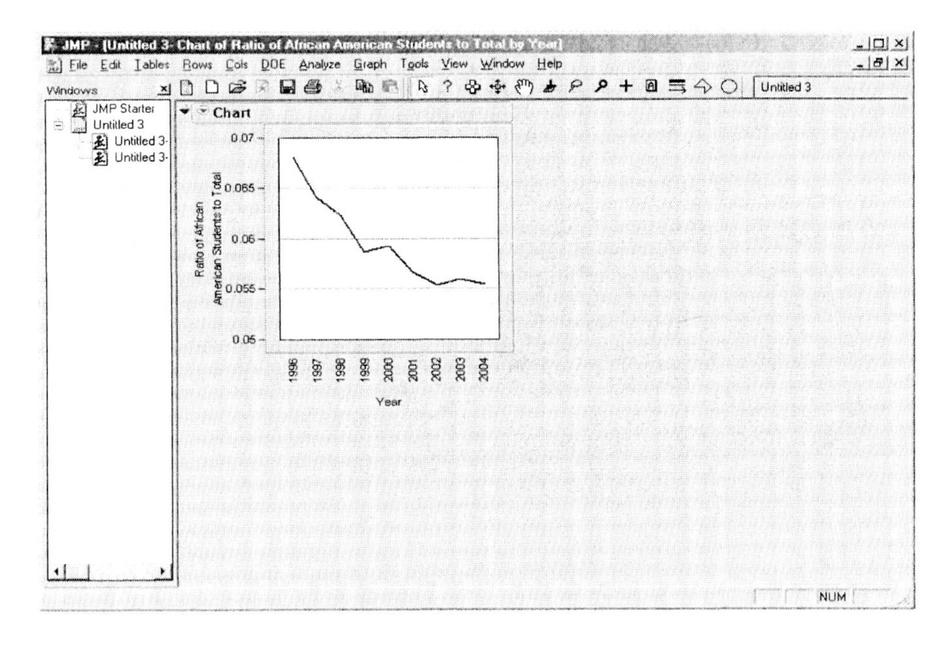

Chapter 2. Exercise 2.119 Temperatures in Central Park.

Open the central_park_yearly_temps data from the JMP folder of the data disk.

Select **Analyze>Distribution**. Assign ***Temp*** to the **Y,Columns** role and select **OK**.

You should see the following screen:

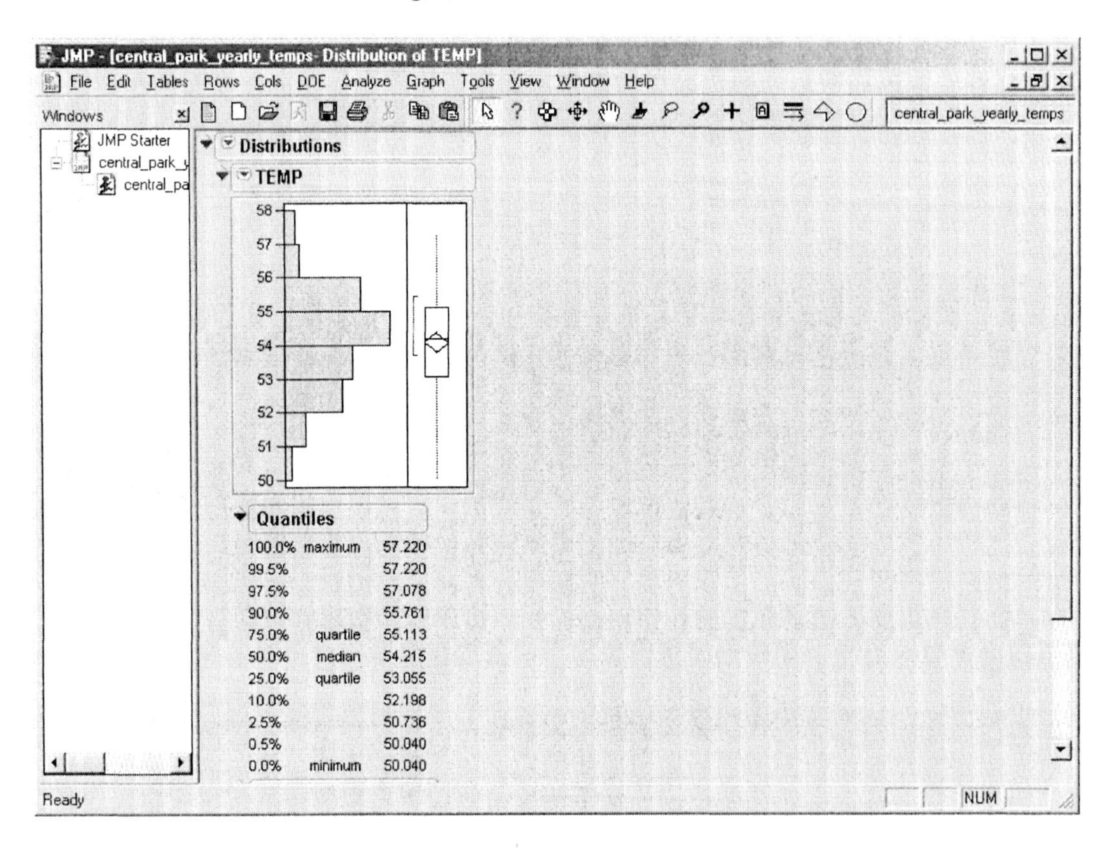

All of the information required for questions a) through d) can be found on this screen.

Chapter 2. Exercise 2.128. Baseball's great home run hitters.

Open the baseballs_hr_hitters data from the JMP folder of the data disk.

To compare performances, one option is to look at the distributions and the boxplots for each hitter on the same page. To do this, select **Analyze>Distribution**. Then, select ***BRHR, RMHR, HAHR, MMHR, SSHR*** and ***BBHR*** (for Babe Ruth, Roger Maris, Hank Aaron, Mark McGwire, Sammy Sosa, and Barry Bonds, respectively) and assign each of these variables to the **Y,Columns** role. You should see the following screen:

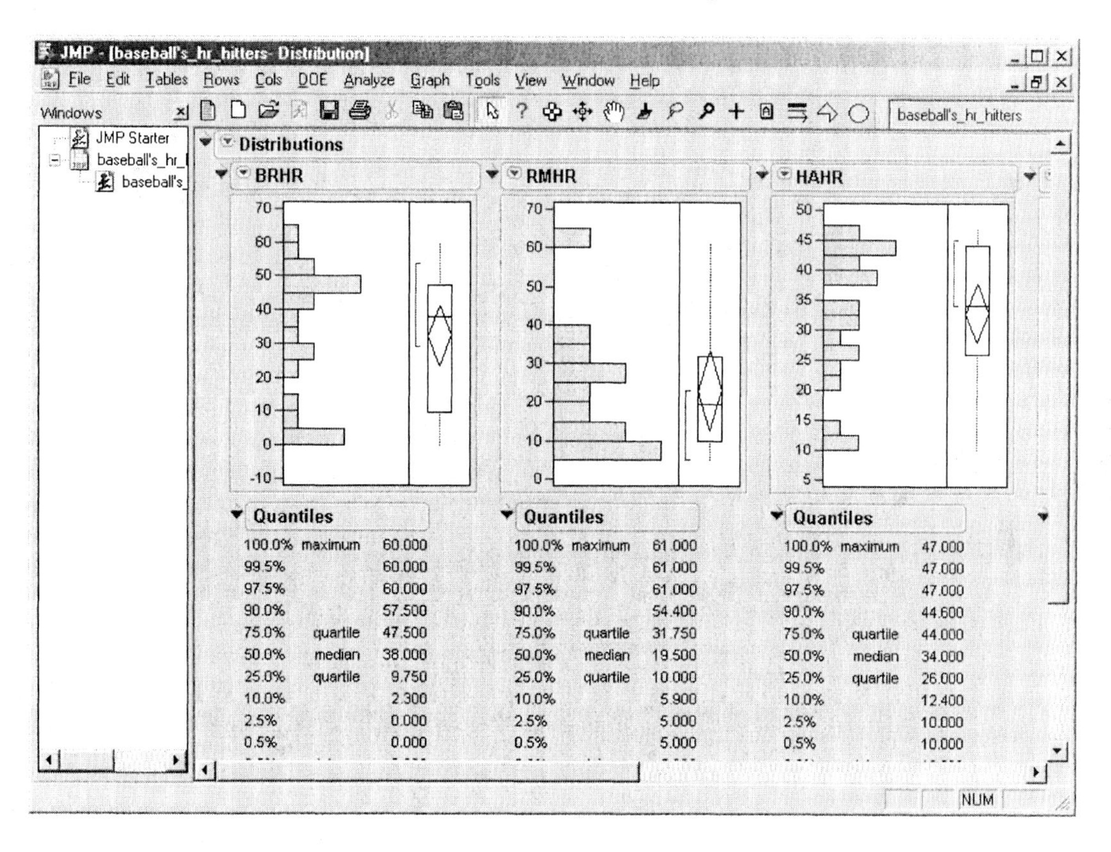

Chapter 2. Exercise 2.129. How much spent on haircuts?

Open the georgia_student_survey data from the JMP folder of the data disk.

One option to compare the price paid for a haircut by gender is to look at the histograms and the boxplots. To do this, select **Analyze>Distribution**. Then assign ***haircut*** to the **Y,Columns** role and ***gender*** to the **By** role. Recall that the default layout will be vertical (one output will follow the next). To rotate to have a horizontal layout, click on the red triangle next to "Haircut" and select **Display Options>Horizontal Layout**.

You should see the following screen:

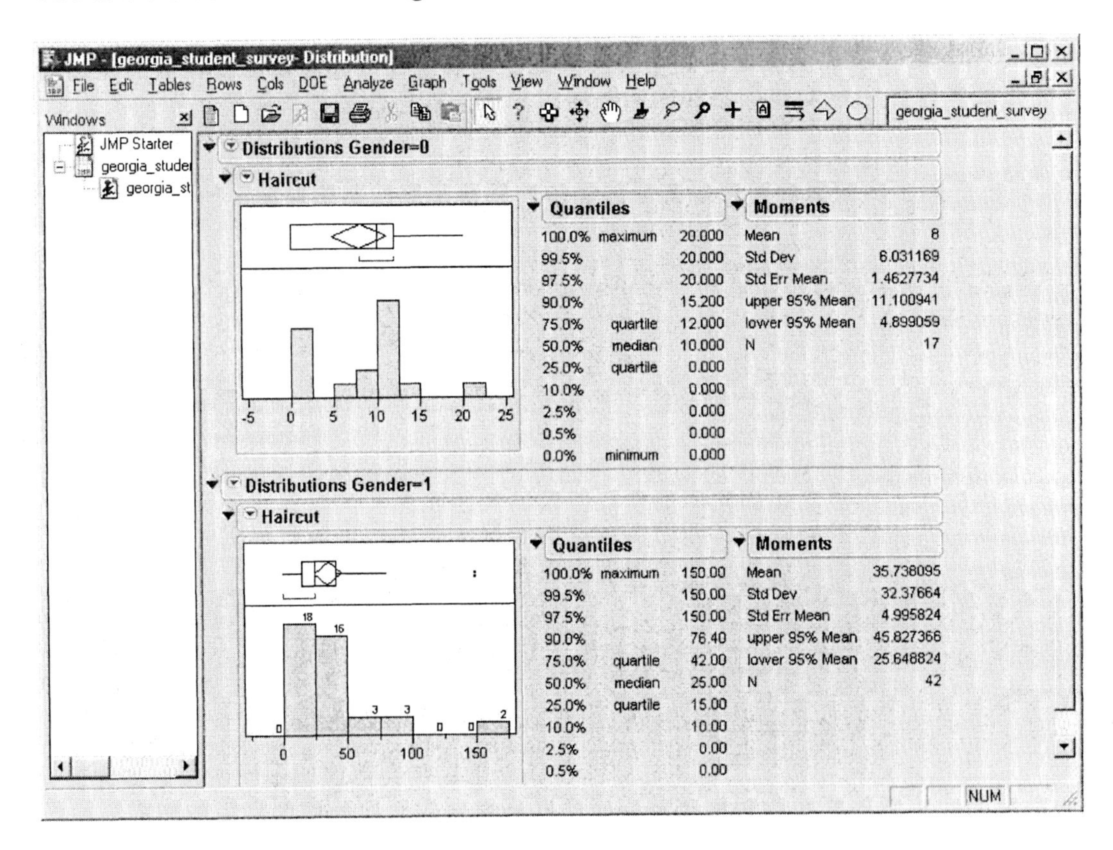

Chapter 3. Example 5. Constructing a Scatterplot for Internet Use and GDP.

Open the human_development data from the JMP folder of the data disk.

Select **Analyze>Fit Y by X**. Assign ***Internet*** for the **Y, Response** role and ***GDP*** for the **X,Factor** role. Select **OK.**

You should see the following screen:

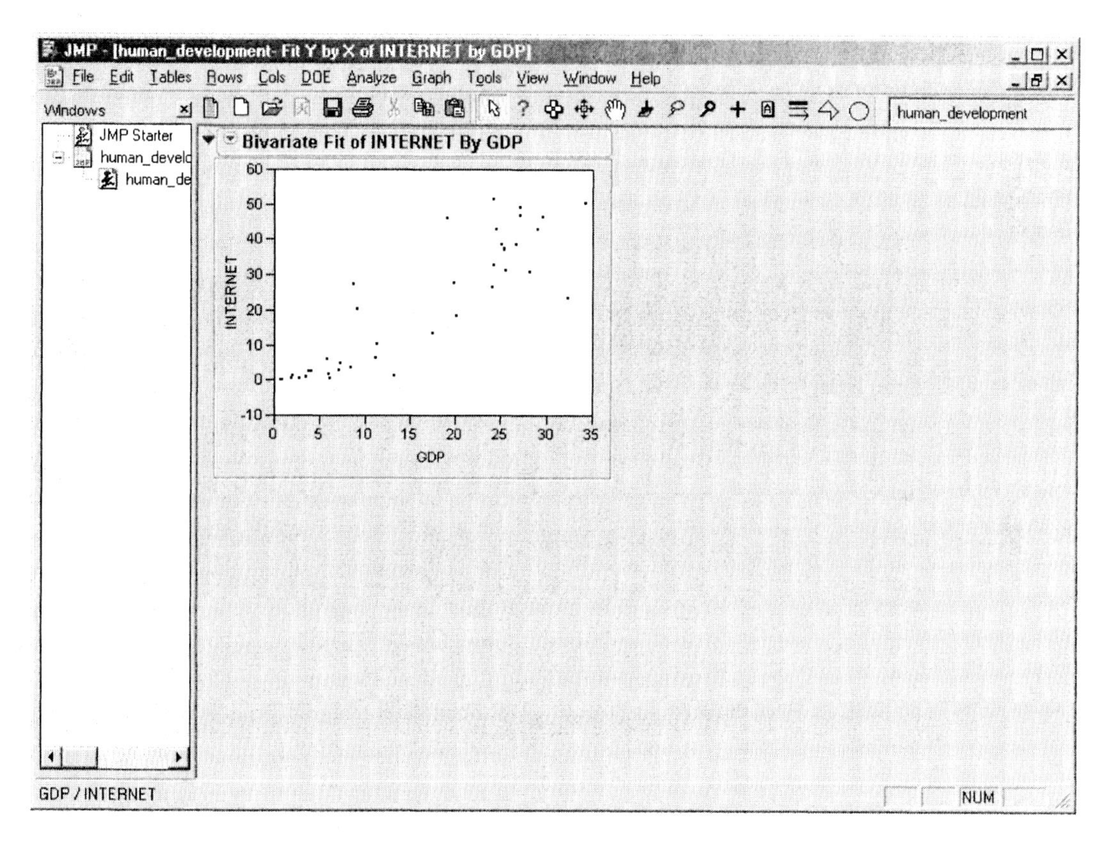

To rescale the y-axis to have a minimum value of zero, double click on the y-axis and enter a value of 0 for the Minimum.

Chapter 3. Example 7. What's the Correlation Between Internet Use and GDP?

Open the human_development data from the JMP folder of the data disk.

Select **Analyze>Multivariate Methods>Multivariate**. Assign BOTH ***Internet*** and ***GDP*** for the **Y, Columns** role. Select **OK**.

You should see the following screen:

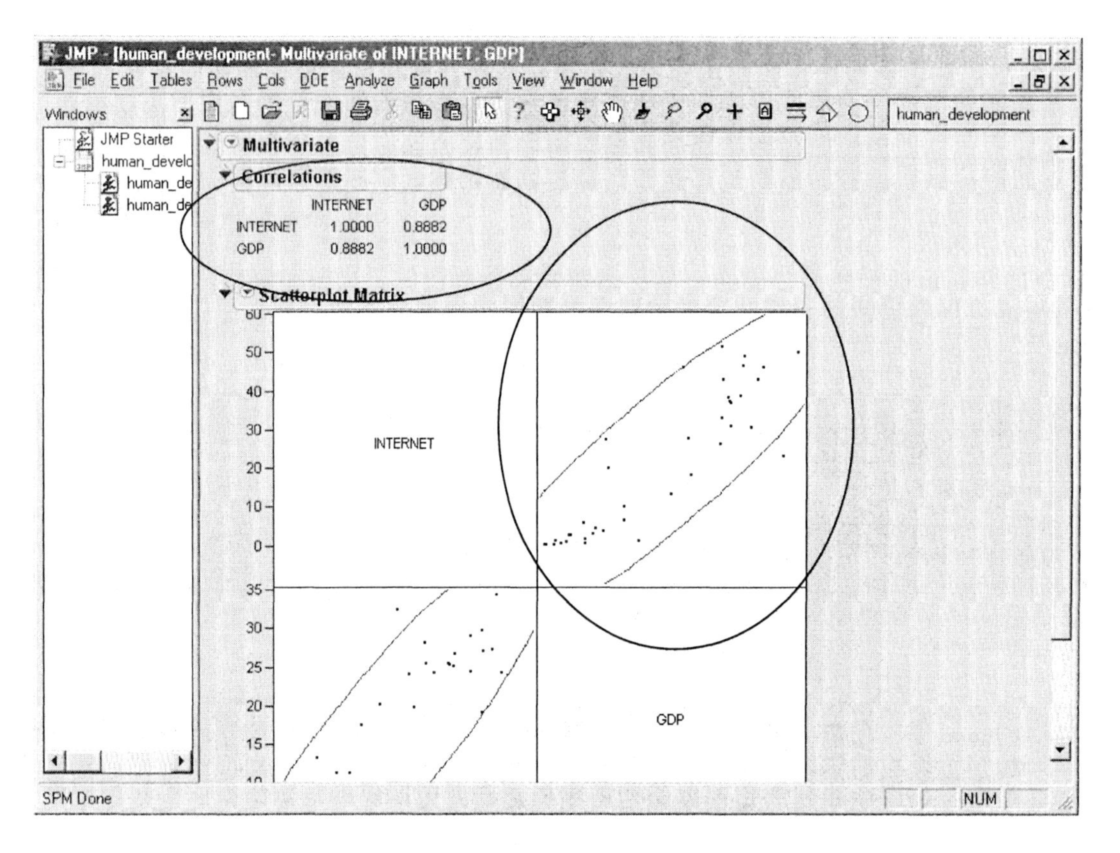

The correlation matrix is identified above. The scatterplot identified above, has ***Internet*** on the y-axis and ***GDP*** on the x-axis. The scatterplot on the opposite diagonal has ***Internet*** on the x-axis and ***GDP*** on the y-axis.

Chapter 3. Exercise 3.21 Which mountain bike to buy?

Open the mountain_bike data from the JMP folder of the data disk.

Select **Analyze>Multivariate Methods>Multivariate**. Assign the variables ***price*** and ***weight*** to the **Y, Columns** role. Select **OK.**

You should see the following screen:

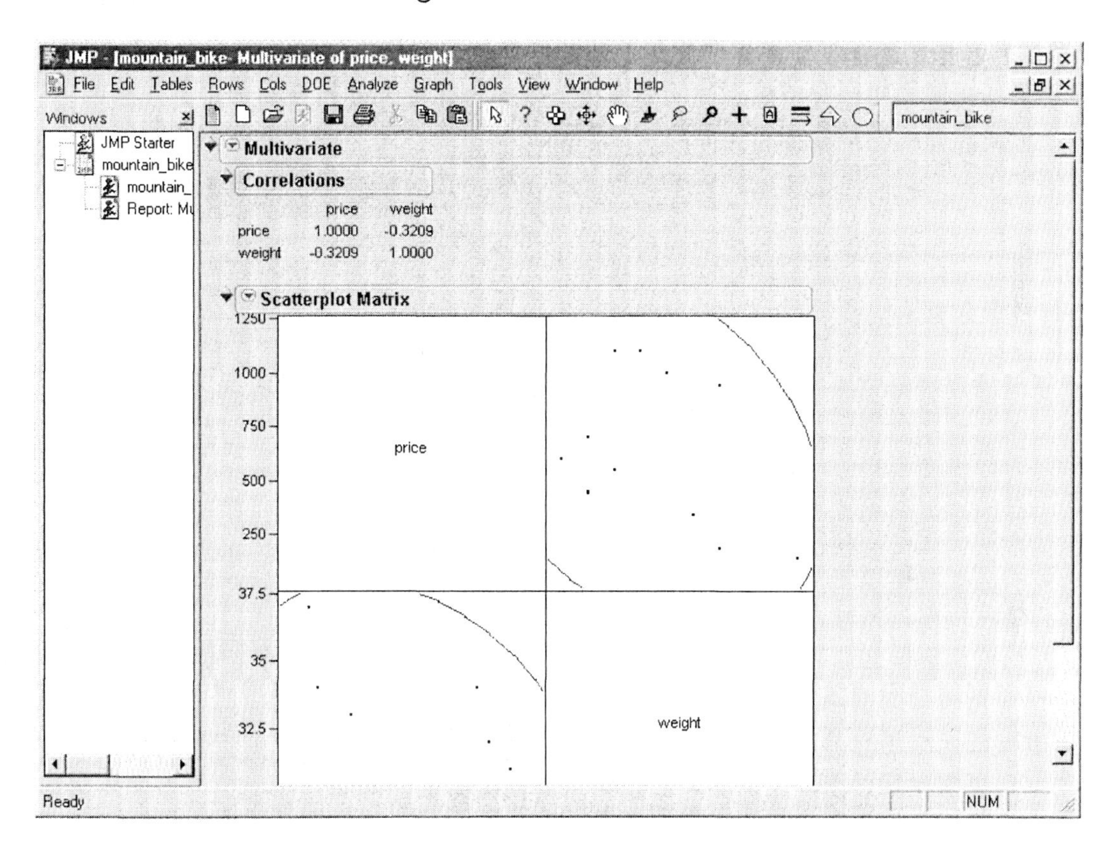

The scatterplot in the NE corner has ***price*** on the y-axis and ***weight*** on the x-axis. The scatterplot in the SW corner has the two variables in the opposite positions.

Chapter 3. Exercise 3.22. Enchiladas and sodium-revisited.

Open the enchiladas data from the JMP folder of the data disk.

Select **Analyze>Multivariate Methods>Multivariate**. Assign the variables ***sodium content*** and ***cost*** to the **Y, Columns** role. Select **OK**.

You should see the following screen:

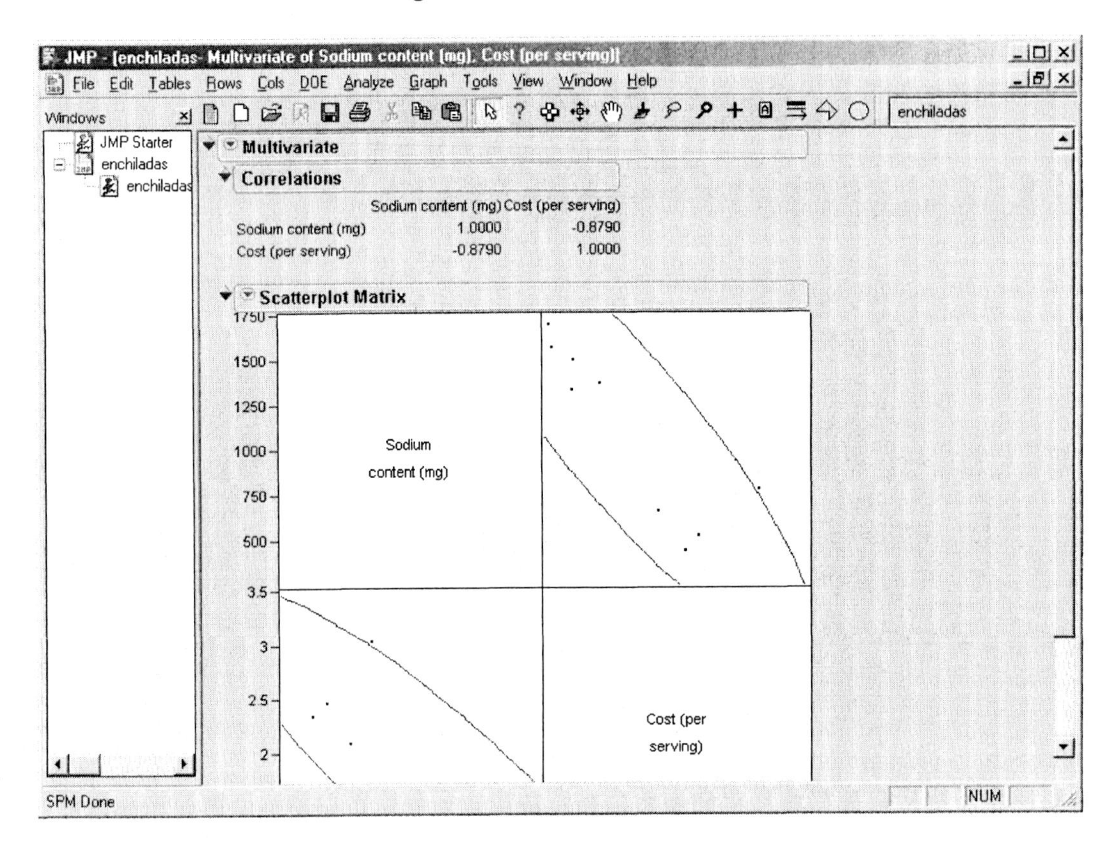

Chapter 3. Exercise 3.23. Buchanan vote.

Open the Buchanan_and_the_butt#218AA data from the JMP folder of the data disk.

To construct the boxplots, select **Analyze>Distribution**. Then assign ***perot, gore, bush*** and ***buchanan*** for the **Y, Columns** role. Select **OK**.

You should see the following screen:

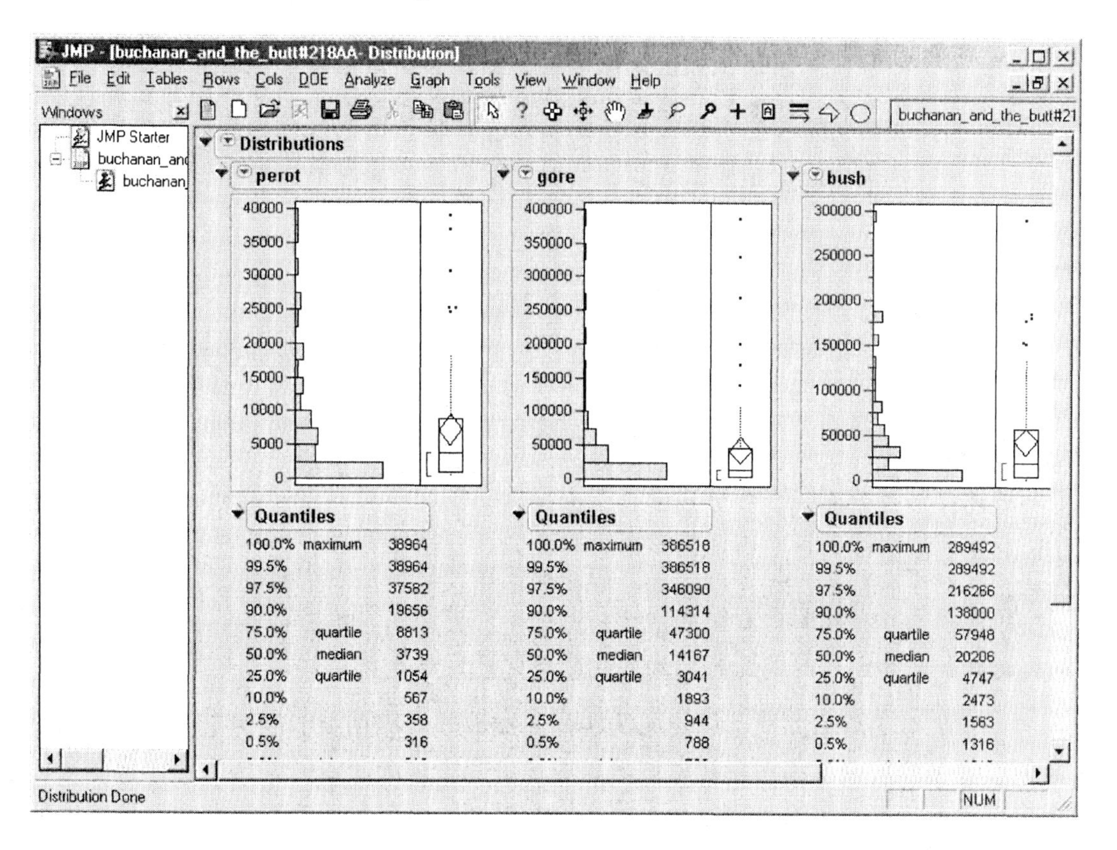

To construct a scatterplot evaluating the Buchanan vote versus the Gore vote, select **Analyze>Multivariate Methods>Multivariate**. Assign ***buchanan*** and ***gore*** for the **Y, Columns** role. Select **OK**.

You should see the following screen:

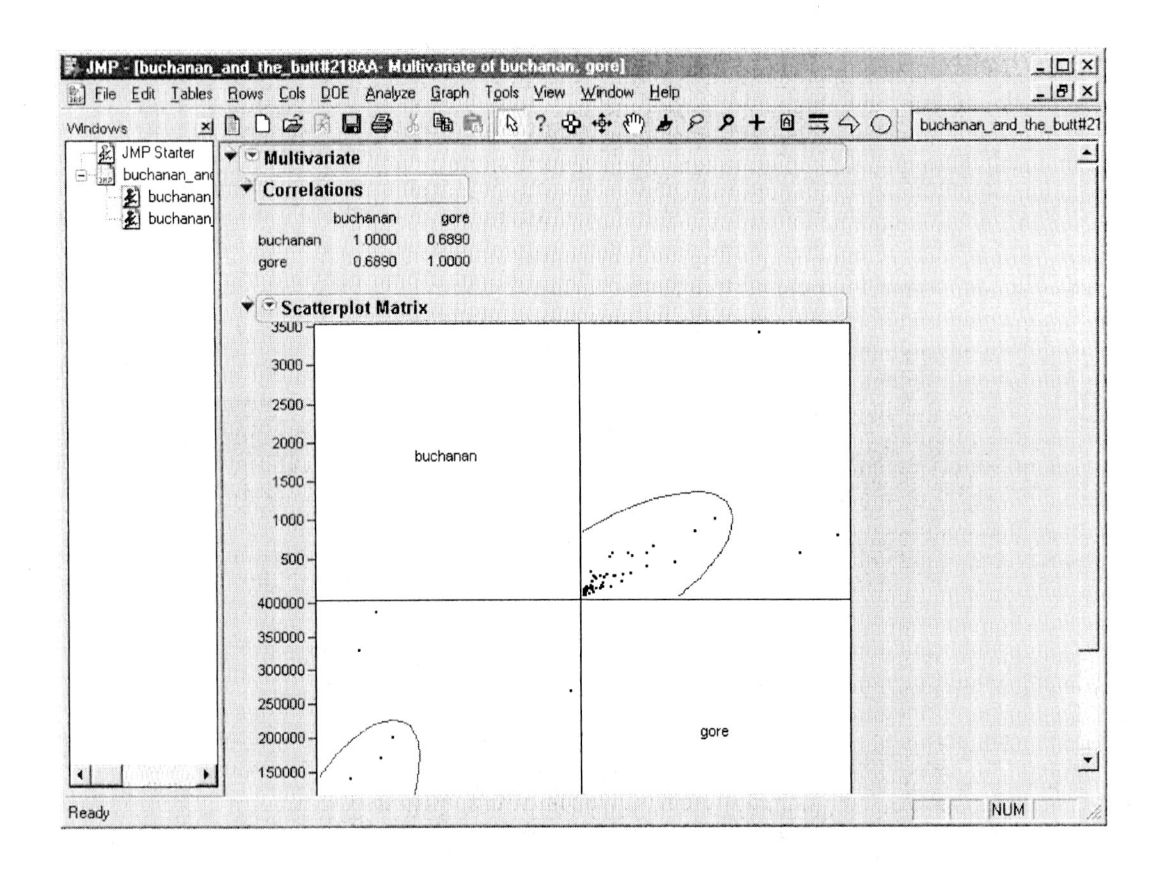
JMP - [buchanan_and_the_butt#218AA- Multivariate of buchanan, gore]
File Edit Tables Rows Cols DOE Analyze Graph Tools View Window Help
Windows
buchanan_and_the_butt#21
JMP Starter
buchanan_an
buchanan
buchanan
Multivariate
Correlations
buchanan gore
buchanan 1.0000 0.6890
gore 0.6890 1.0000
Scatterplot Matrix
3500
3000
2500
2000
1500
1000
500
buchanan
400000
350000
300000
250000
200000
150000
gore
Ready
NUM

Chapter 3. Example 9. How Can We Predict Baseball Scoring Using Batting Average?

Open the al_team_statistics data from the JMP folder of the data disk.

To create the scatterplot and generate the regression equation, select **Analyze>Fit Y By X**. Assign ***Team Scoring*** to the **Y, Columns** role and ***BAT_AVG*** to the **X, Factor** role. Select **OK**. Then, click on the red triangle next to "**Bivariate Fit...**". Select **Fit Line**.

You should see the following screen:

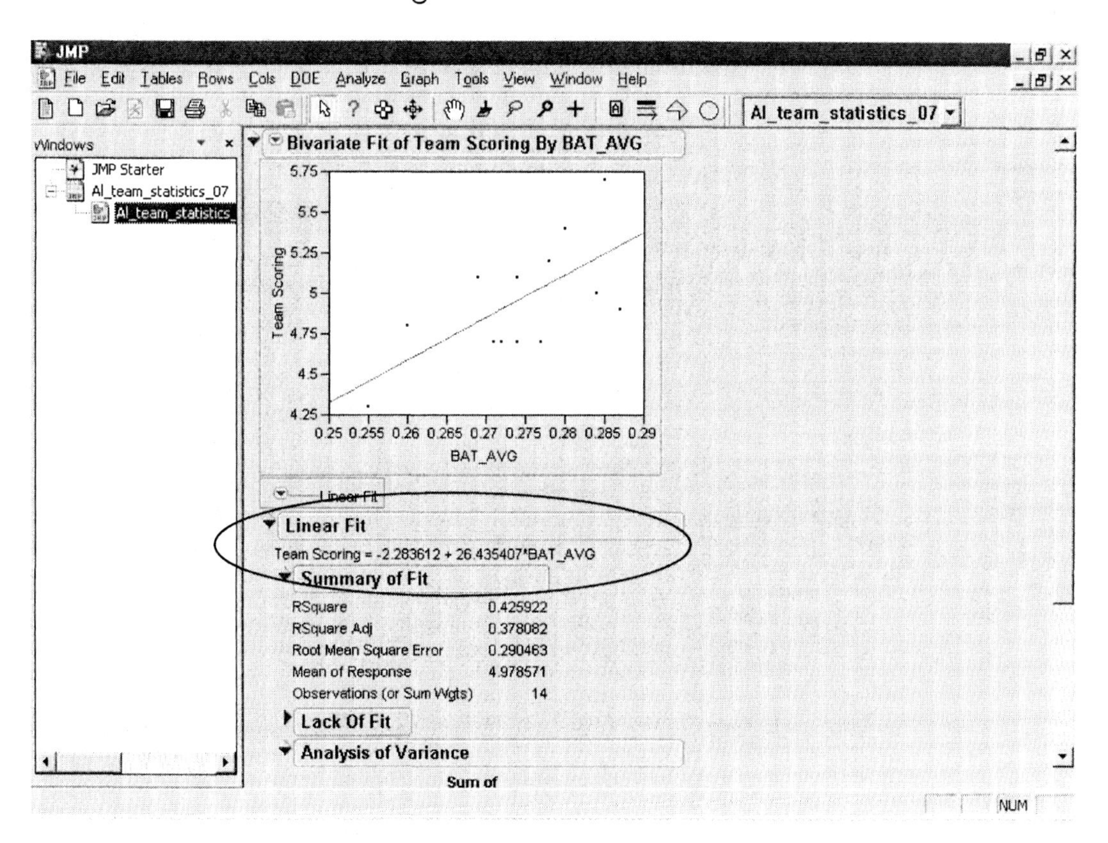

The regression equation information is circled above.

Chapter 3. Example 10. How Can We Detect an Unusual Vote Total?

Open the Buchanan_and_the_butt#218AA data from the JMP folder of the data disk.

To evaluate the Buchanan vote versus the Perot vote, select **Analyze>Fit Y By X**. Assign ***buchanan*** to the **Y, Columns** role and ***perot*** to the **X, Factor** role. Then, click on the red triangle next to "Bivariate Fit...". Select **Fit Line**.

You should see the following screen:

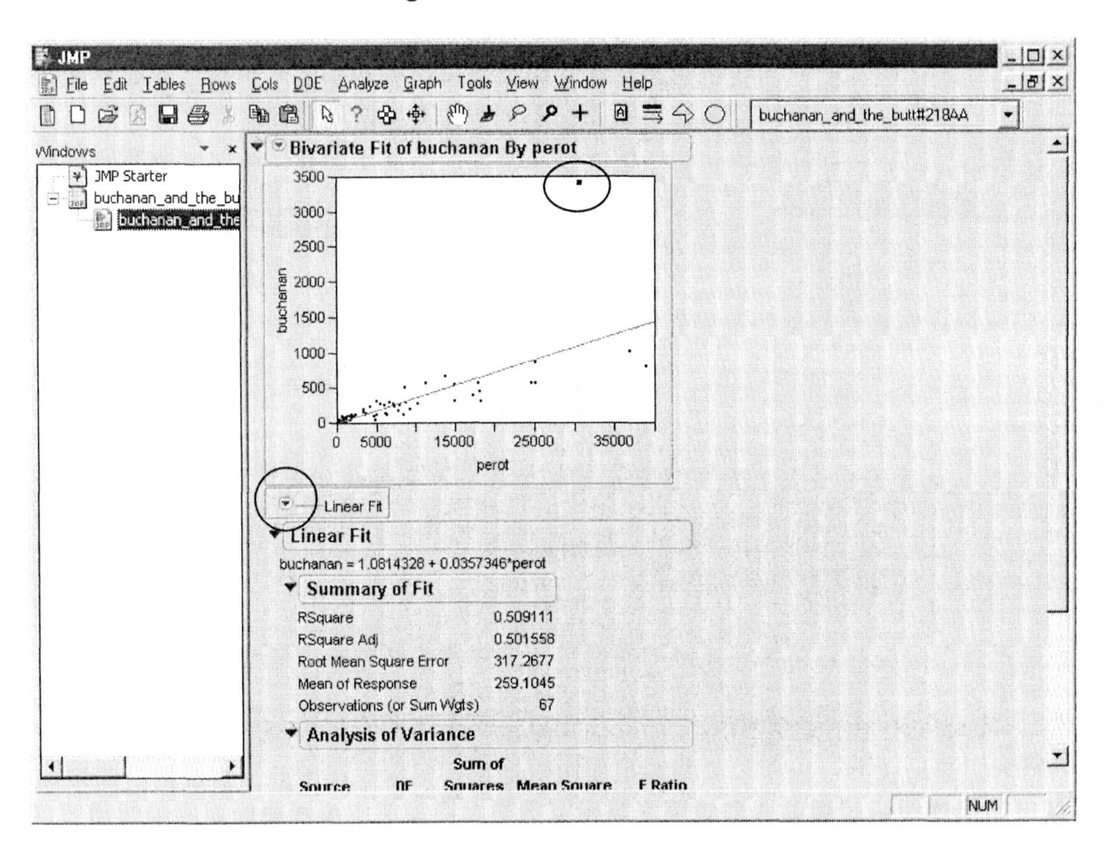

Click on the outlying observation circled above. You will see the number "50" appear. This indicates that the outlier is observation number 50 in the dataset. If you return to the datasheet, you will see that observation number 50 is Palm Beach County.

To evaluate the residuals, and to create a histogram of the residuals, click on the red triangle circled above. Select **Save Residuals**. This selection will create a new variable in the datasheet called "***Residuals Buchanan***". Returning to the dataset, create a histogram using this new variable.

Select **Analyze>Distribution**. Assign ***Residuals Buchanan*** in the **Y, Columns** role. Select **OK.** Click on the red triangle next to "**Residuals Buchanan**". Change the **Display Option** to **Horizontal Layout**.

You should see the following screen:

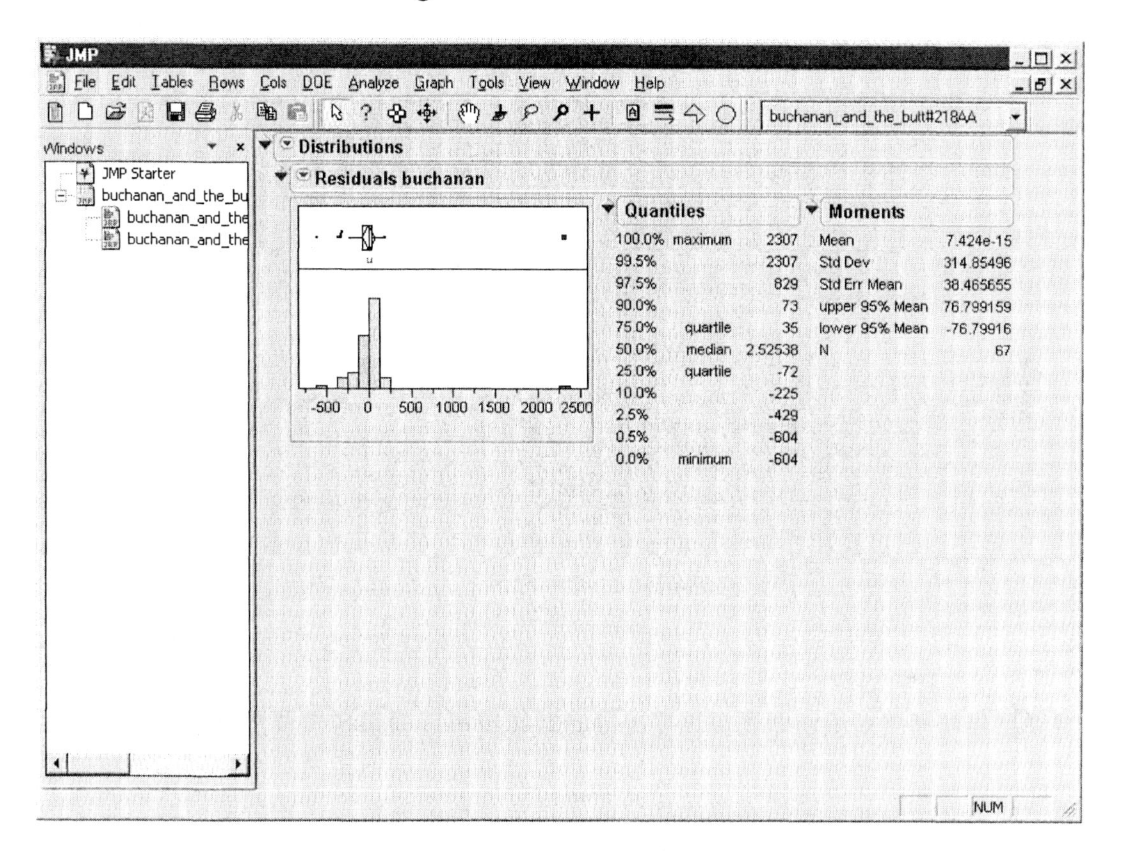

Chapter 3. Exercise 3.26. Home selling prices.

Open the house_selling_prices data from the JMP folder of the data disk.

To create the scatterplot and the linear equation, select **Analyze>Fit Y By X**. Assign ***price*** to the **Y, Columns** role and ***size*** to the **X, Factor** role. Select **OK**. Click on the red triangle next to "Bivariate Fit..." and select **Fit Line**.

You should see the following screen:

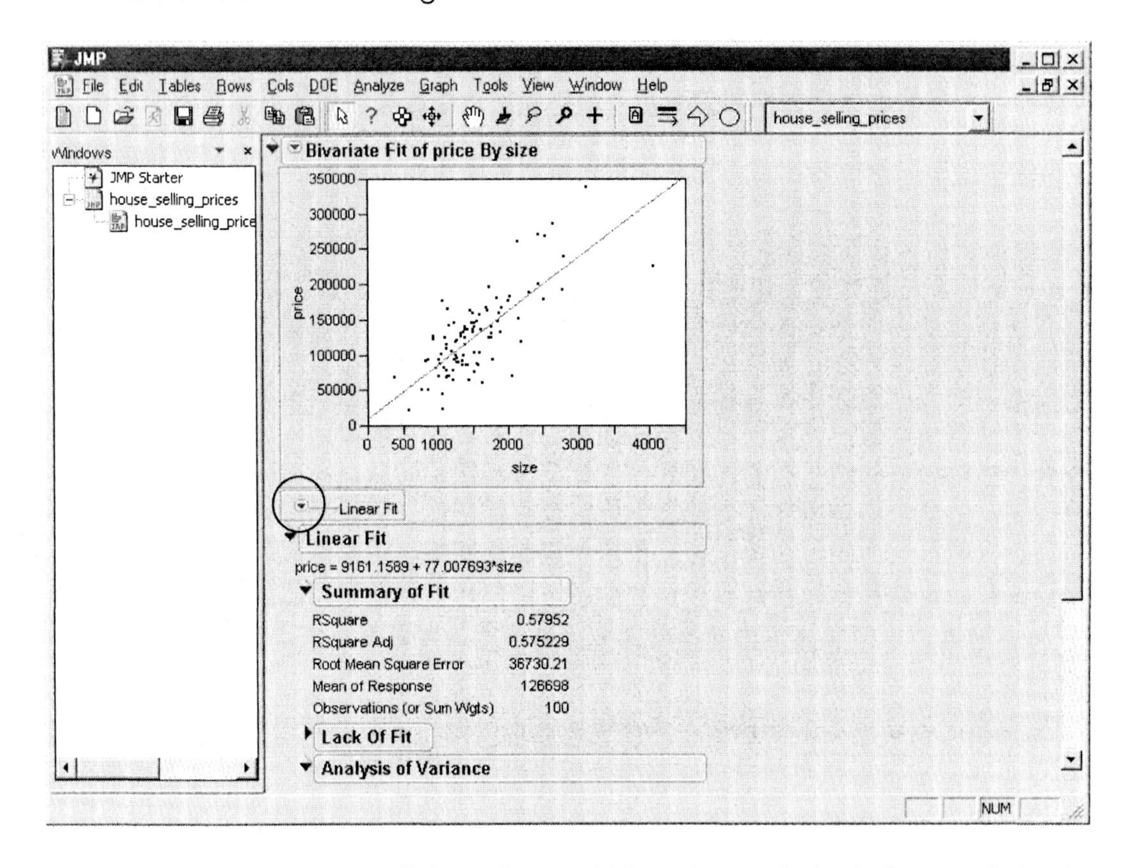

To identify the residuals, click on the red triangle as circled above. Select **Save Residuals**. This will create a new variable in the datasheet titled "***Residuals Price***".

Chapter 3. Exercise 3.41. Mountain Bikes Revisited.

Open the mountain_bike data from the JMP folder of the data disk.

Select **Analyze> Fit Y By X.** Assign ***price*** to the **Y, Columns** role and ***weight*** to the **X, Factor** role. Select **OK.** Click on the red triangle next to "**Bivariate Fit**...". Select **Fit Line.**

You should see the following screen:

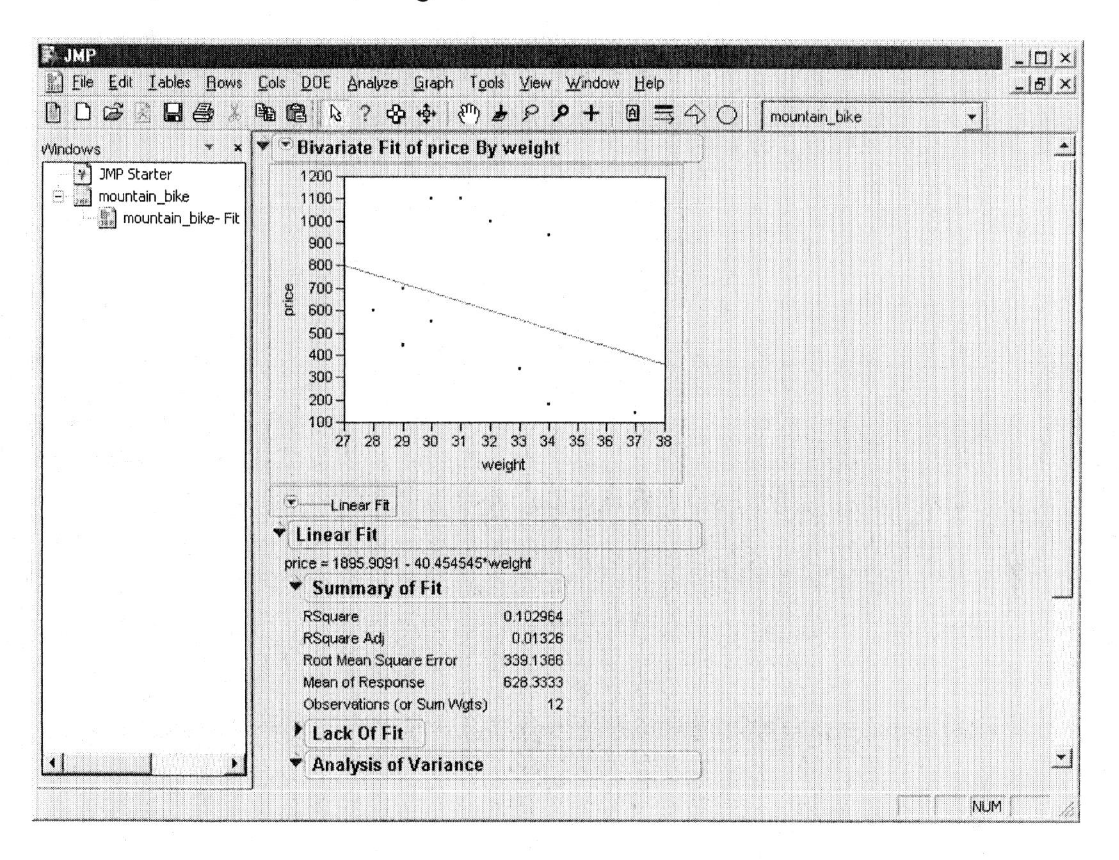

Chapter 3. Exercise 3.42. Mountain bike and suspension type.

To build separate regression lines for the two suspension types, return to the datasheet. Select **Analyze>Fit Y By X**. Assign ***price_FE*** to the **Y, Columns** role and ***weight_FE*** to the **X, Factor** role. Select **OK**. Click on the red triangle next to "**Bivariate Fit**..." and select **Fit Line**.

You should see the following screen:

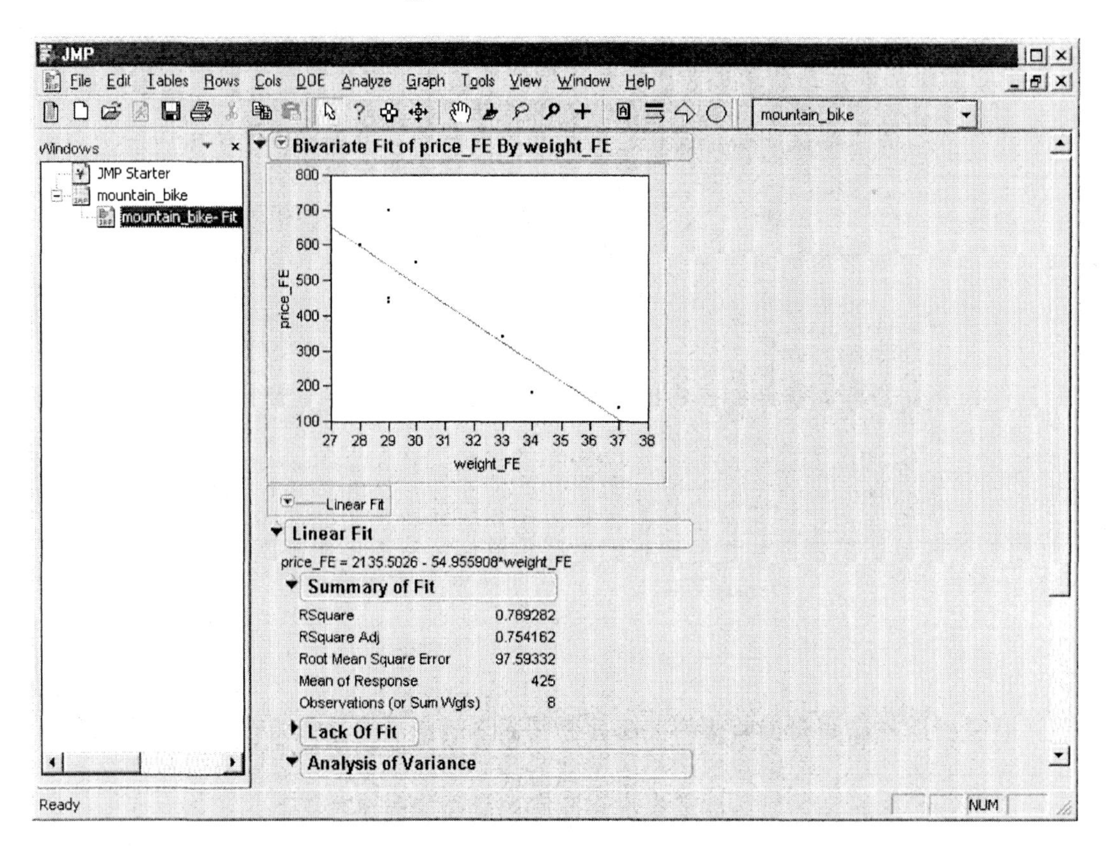

To create a similar graphic and equation for the second suspension type, execute the same commands, but assign ***price_FU*** to the **Y, Columns** role and ***weight_FU*** to the **X, Factor** role.

You should see the following screen:

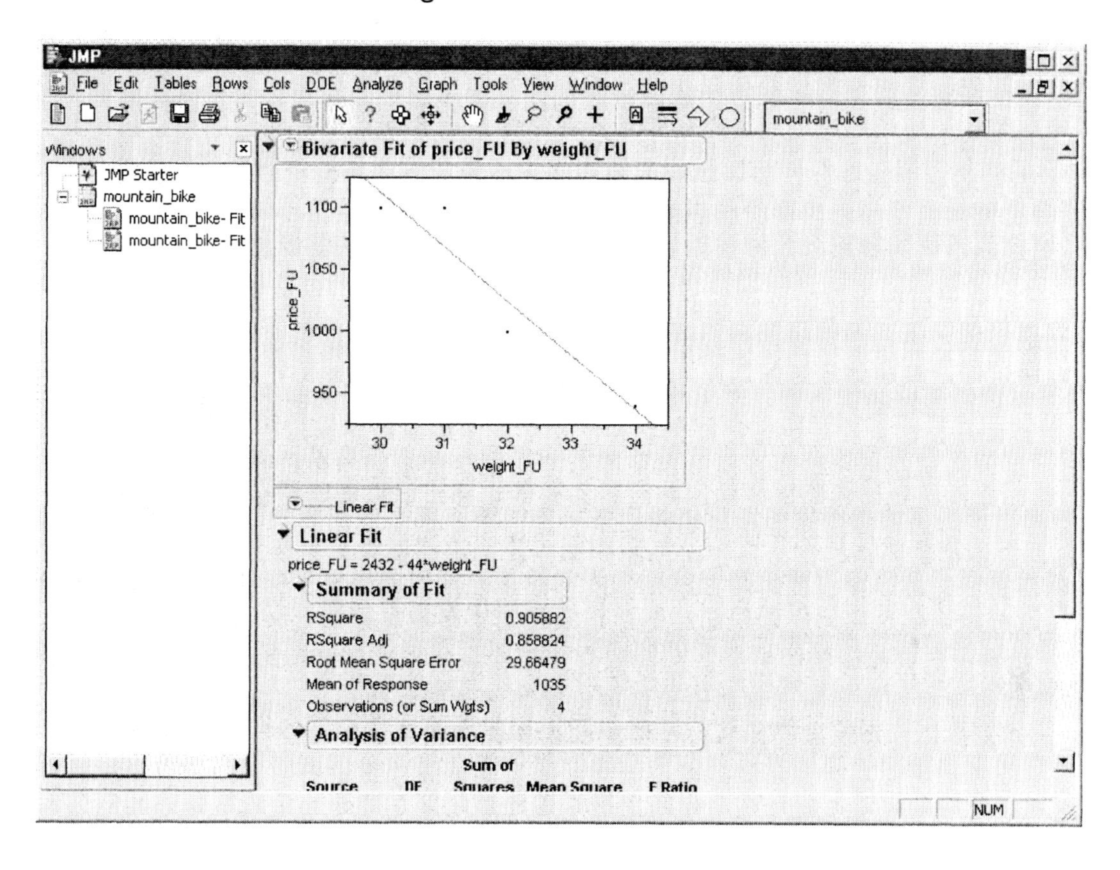

Chapter 3. Exercise 3.43. Olympic High Jump.

Open the high_jump data from the JMP folder of the data disk.

To generate the scatterplot and regression equation for the men, select **Analyze>Fit Y By X**. Assign ***Men_Meters*** to the **Y, Columns** role and ***Year_Men*** to the **X, Factor** role. Select **OK**. Click on the red triangle next to "Bivariate Fit...". Select **Fit Line**.

You should see the following screen:

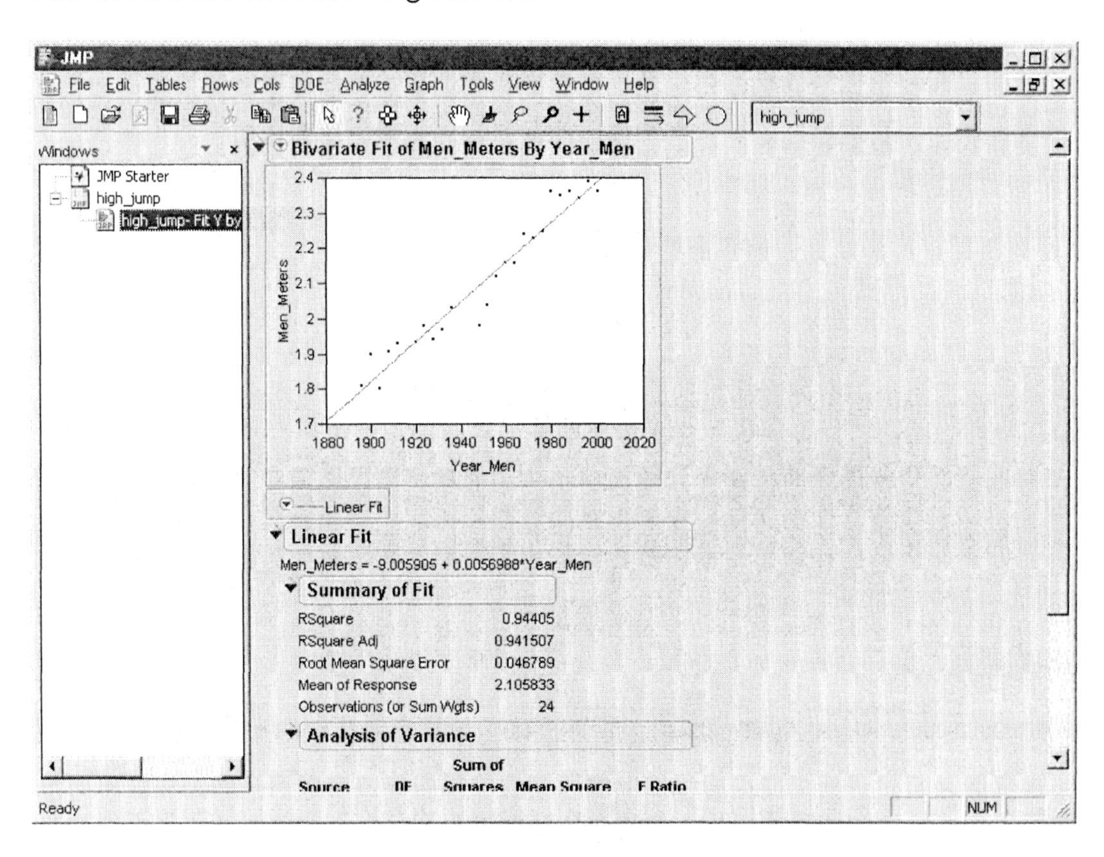

To generate the same information for the women, execute the same commands, but assign ***Women_Meters*** to the **Y, Columns** role and ***Year_Women*** to the **X, Factor** role.

You should see the following screen:

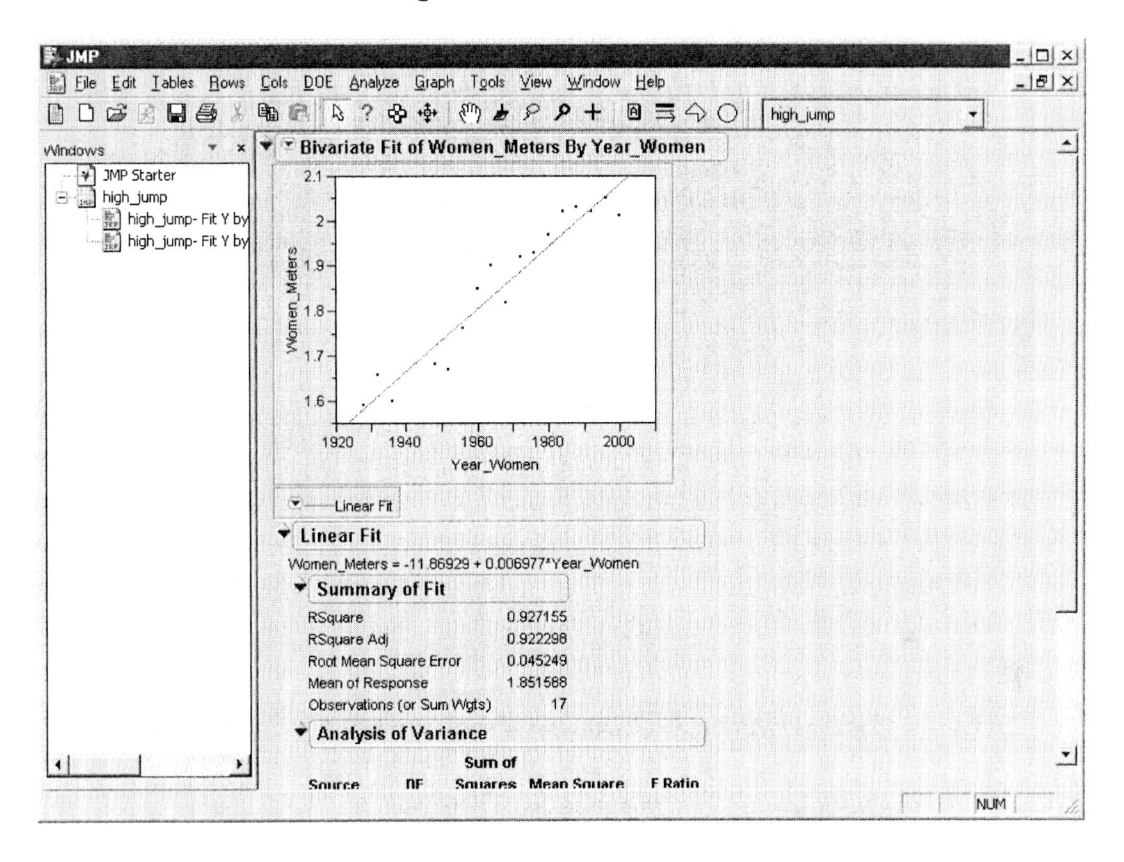

Chapter 3. Example 12. How Can We Forecast Future Global Warming?

Open the central_park_yearly_temps data from the JMP folder of the data disk.

To generate the scatterplot and regression line for the time series data, select **Analyze>Fit Y By X.** Assign ***TEMP*** to the **Y, Columns** role and ***TIME*** to the **X, Factor** role. Click on the red triangle next to "**Bivariate Fit...**" Select **Fit Line**. Then select **Fit Each Value.**

You should see the following screen:

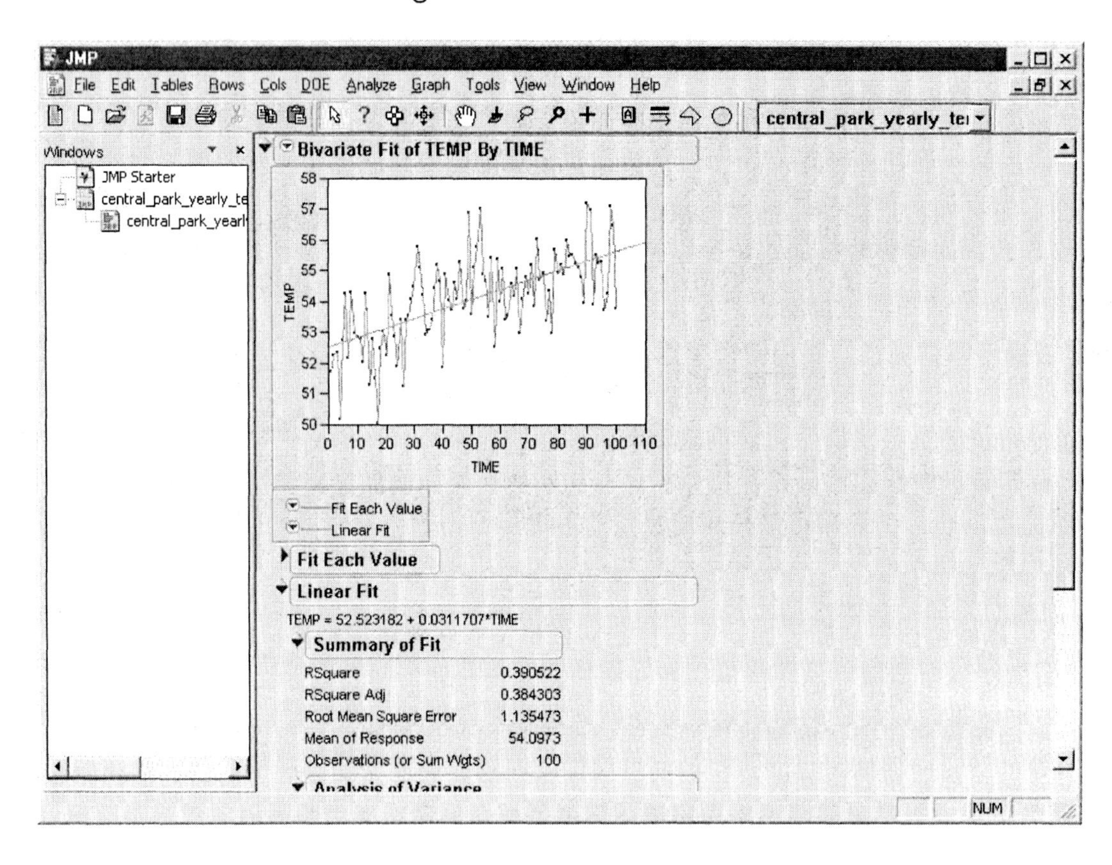

Chapter 3. Example 13. Is Higher Education Associated with Higher Murder Rates?

Open the us_statewide_crime data from the JMP folder of the data disk.

To construct the scatterplot and the regression line, select **Analyze> Fit Y By X**. Assign ***murder rate*** to the **Y, Column** role and ***college*** to the **X, Factor** role. Select **OK**. Click on the red triangle next to "**Bivariate Fit**...". Select **Fit Line**.

You should see the following screen:

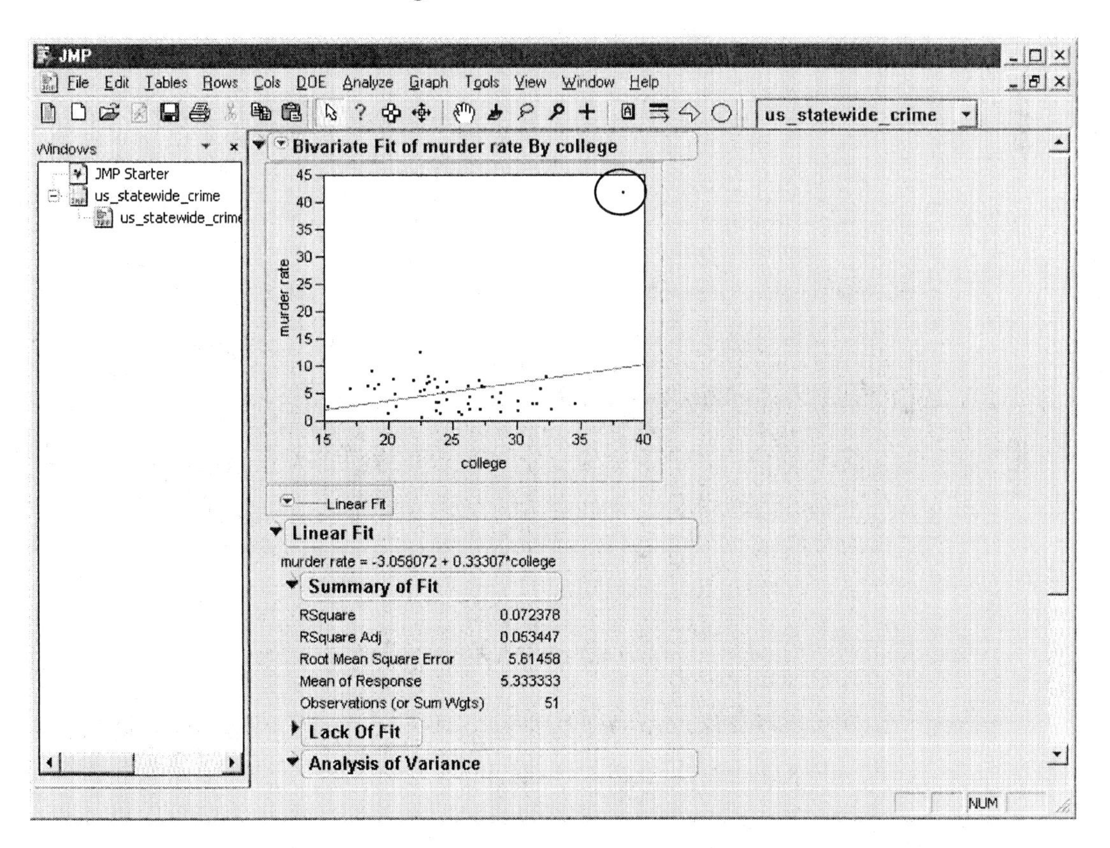

The observation indicated above is an outlier. Place the cursor on the dot and click. The number "9" should appear, indicating that this is observation number 9. To evaluate the same data, without the outlier identified above (D.C.), go back to the datasheet. Observation number 9 (row 9) should be highlighted (if not, click on this row). Select **Rows>Exclude/Unexclude**. You will see a ⊘ symbol next to row 9, indicating that it will now be excluded. Execute the exact same operations as before, with this observation excluded.

You should see the following screen:

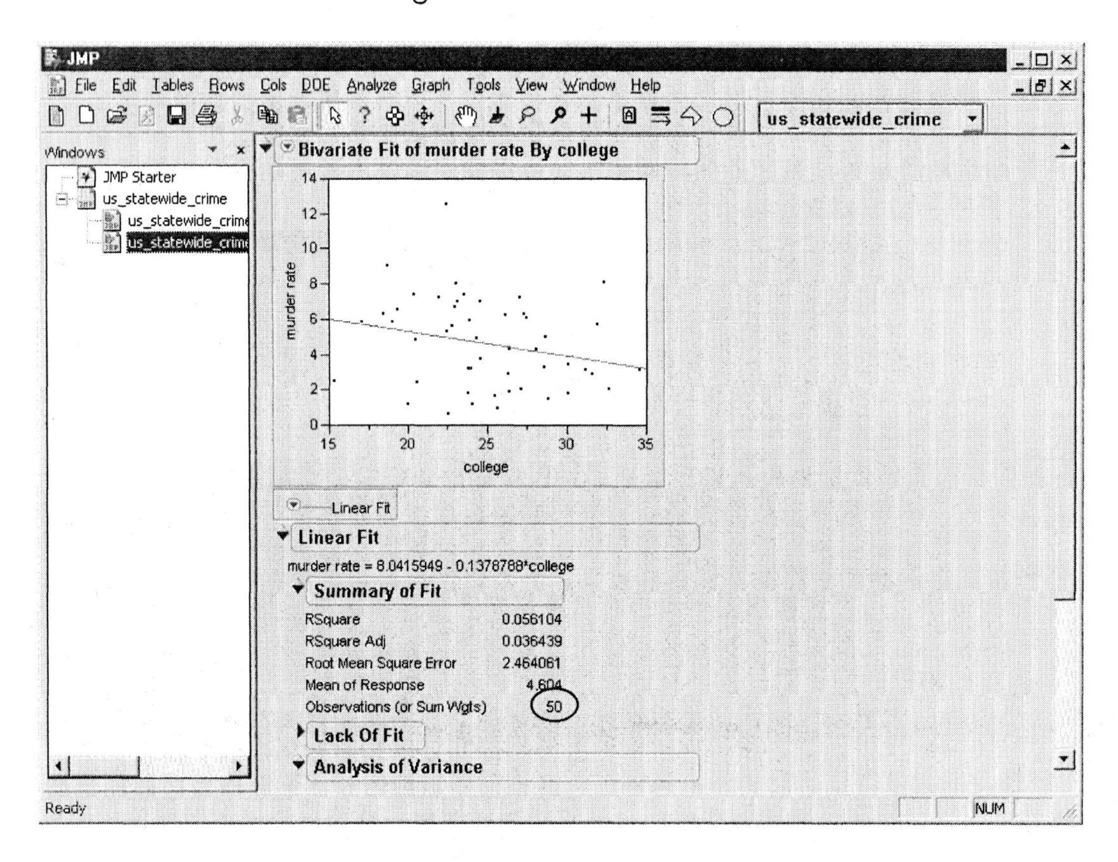

Note that the number of observations decreased from 51 in the previous screen to 50, indicating that the observation was, in fact, excluded.

Chapter 3. Exercise 3.46. U.S. average annual temperatures.

Open the us_temperatures data from the JMP folder of the data disk.

To create the scatterplot and the trend line (and regression equation), select **Analyze>Fit Y By X**. Assign ***Temperature*** to the **Y, Column** role and ***Year*** to the **X, Factor** role. Select **OK**. Click on the red triangle next to "**Bivariate Fit**..." Select **Fit Line**.

You should see the following screen:

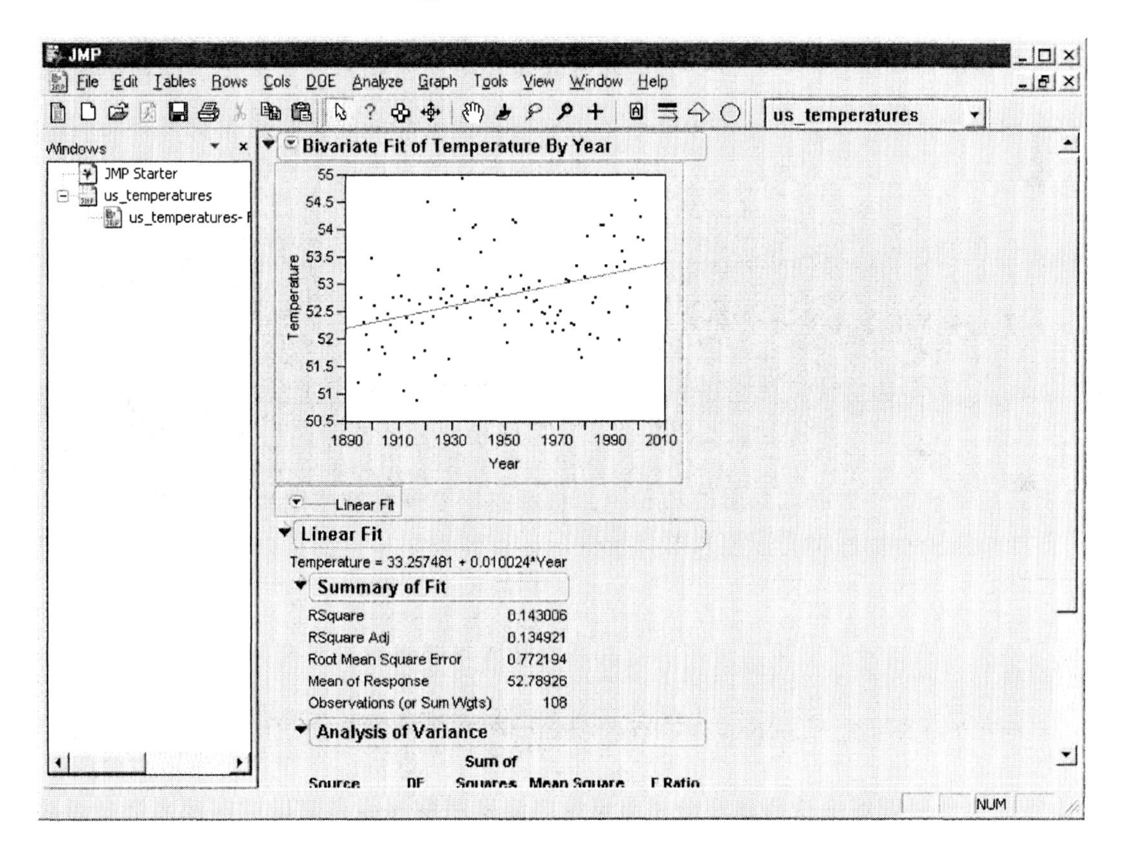

Chapter 3. Exercise 3.48. Murder and Poverty.

Open the us_statewide_crime data from the JMP folder of the data disk.

To construct the scatterplot and the trend line, select **Analyze> Fit Y By X.** Assign ***murder rate*** to the **Y, Column** role and ***poverty*** to the **X, Factor** role. Select **OK.** Click on the red triangle next to "**Bivariate Fit...**". Select **Fit Line.**

You should see the following screen:

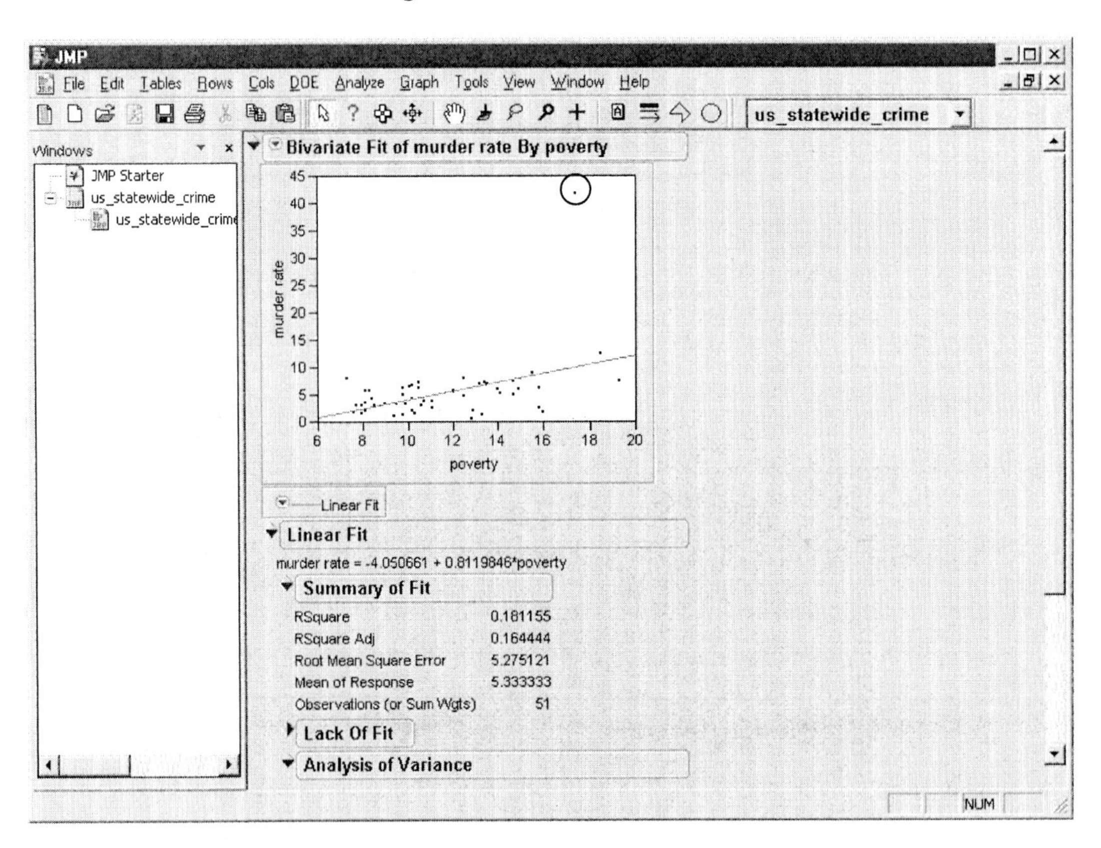

Click on the outlier observation identified above. Then return to the original datasheet. The row for the selected observation (D.C.) should be highlighted.

To rerun the analysis without this observation, select **Row>Exclude/Unexclude**. Then, execute the same operations as before, with D.C. excluded.

You should see the following screen:

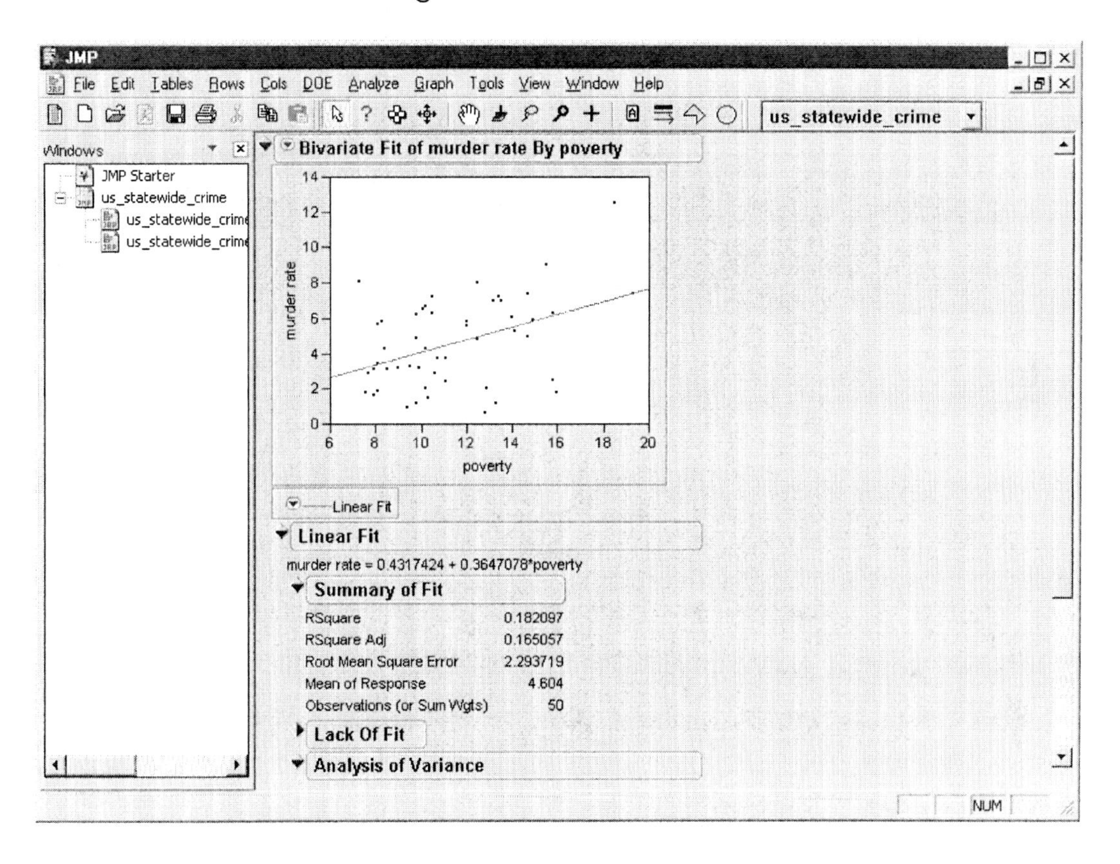

Chapter 3. Exercise 3.50. Looking for outliers.

Open the us_statewide_crime data from the JMP folder of the data disk.

To construct the scatterplot, select **Analyze>Fit Y By X**. Assign ***single parent*** to the **Y, Column** role and ***college*** to the **X, Factor** role. Select **OK**.

You should see the following screen:

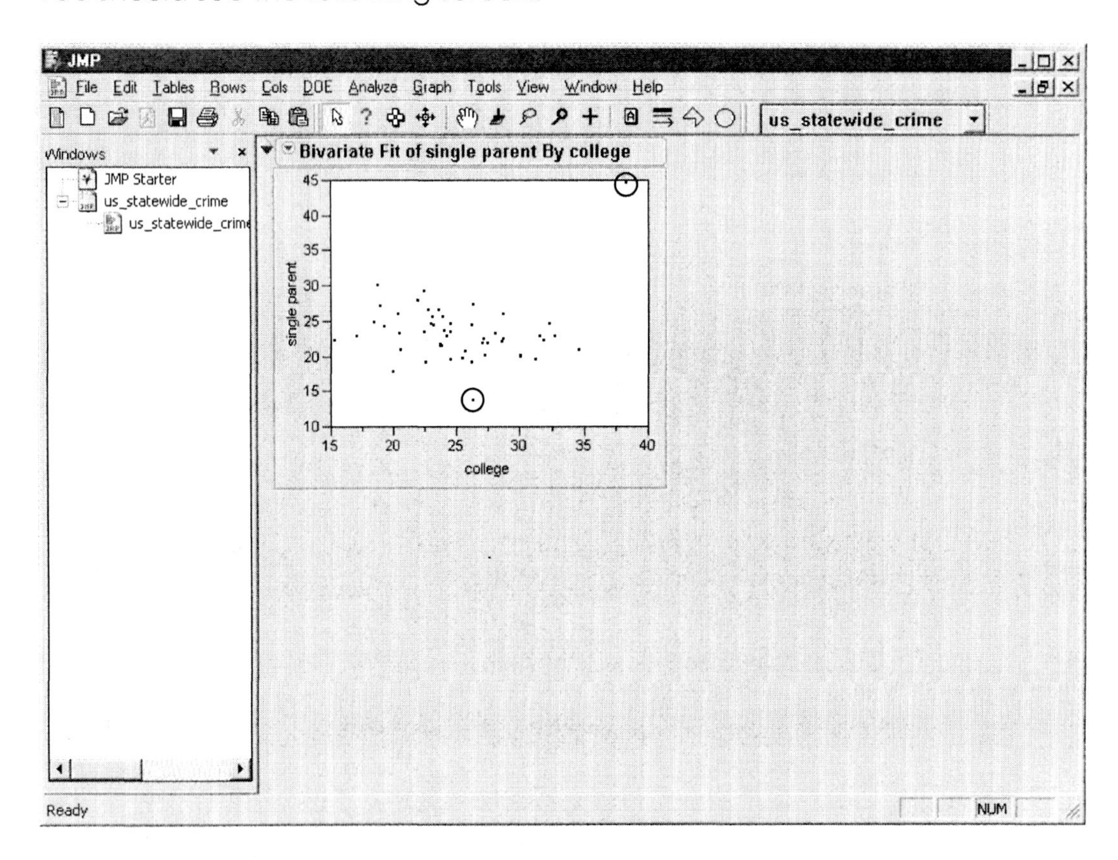

To find the regression equation for the entire dataset, click on the red triangle next to "Bivariate Fit...". Select **Fit Line**. The information for the regression equation for the entire dataset will be provided.

Click on the outlier at the top of the screen. The number 9 should appear, indicating that this is the 9th observation (row) in the dataset. Return to the datasheet. Select **Rows>Exclude/Unexclude**. Then, execute the same commands as above, to generate the regression equation for the dataset without this observation (D.C.).

You should see the following screen:

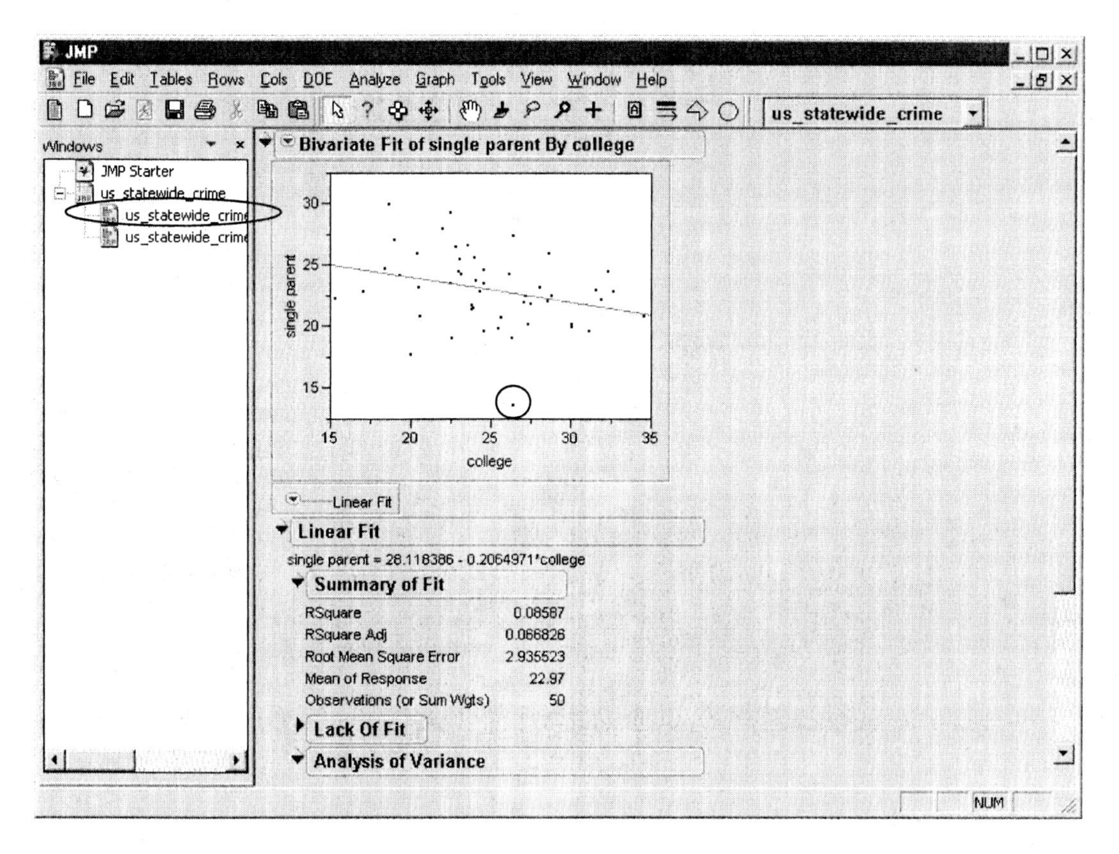

To reinstate D.C., return to the datasheet. D.C. should still be highlighted with the symbol ⊘ on the row. Select **Rows>Exclude/Unexclude**. The symbol should disappear, reinstating D.C.

To delete the other outlier, identified at the bottom of the plot, return to the original graphic (this will most likely be the first report in the results tree as indicated above). Click on the datapoint at the bottom of the plot (the other outlier). The number 45 should appear, indicating, the 45th observation, which is Utah. Return to the datasheet. This observation (row) should be highlighted. The ⊘ symbol should appear on this row. Execute the same operations as before, with Utah excluded.

You should see the following screen:

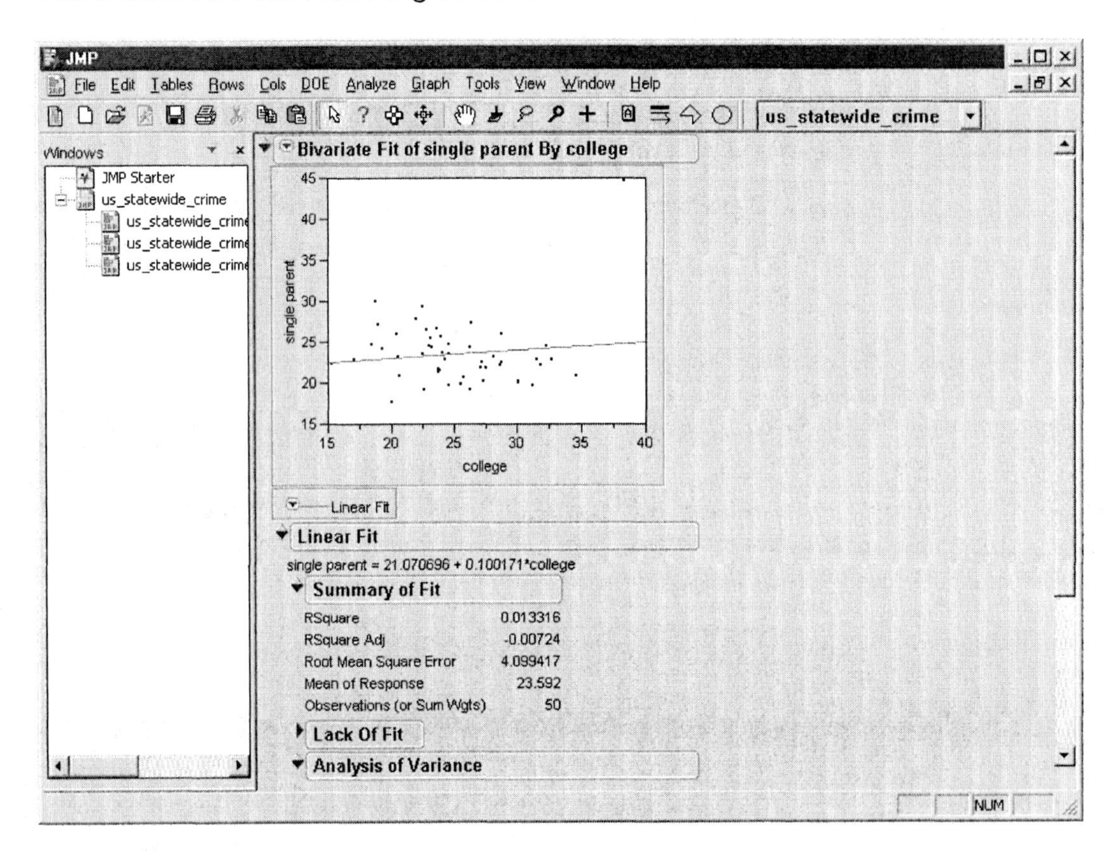

Chapter 3. Exercise 3.51. Regression between cereal sodium and sugar.

Open the cereal data from the JMP folder of the data disk.

To create a scatterplot and regression line, select **Analyze>Fit Y By X**. Assign ***sugar*** to the **Y, Columns** role and ***sodium*** to the **X, Factor** role. Select **OK**. On the scatterplot screen, click on the red triangle next to "Bivariate Fit..." and select **Fit Line**.

You should see the following screen:

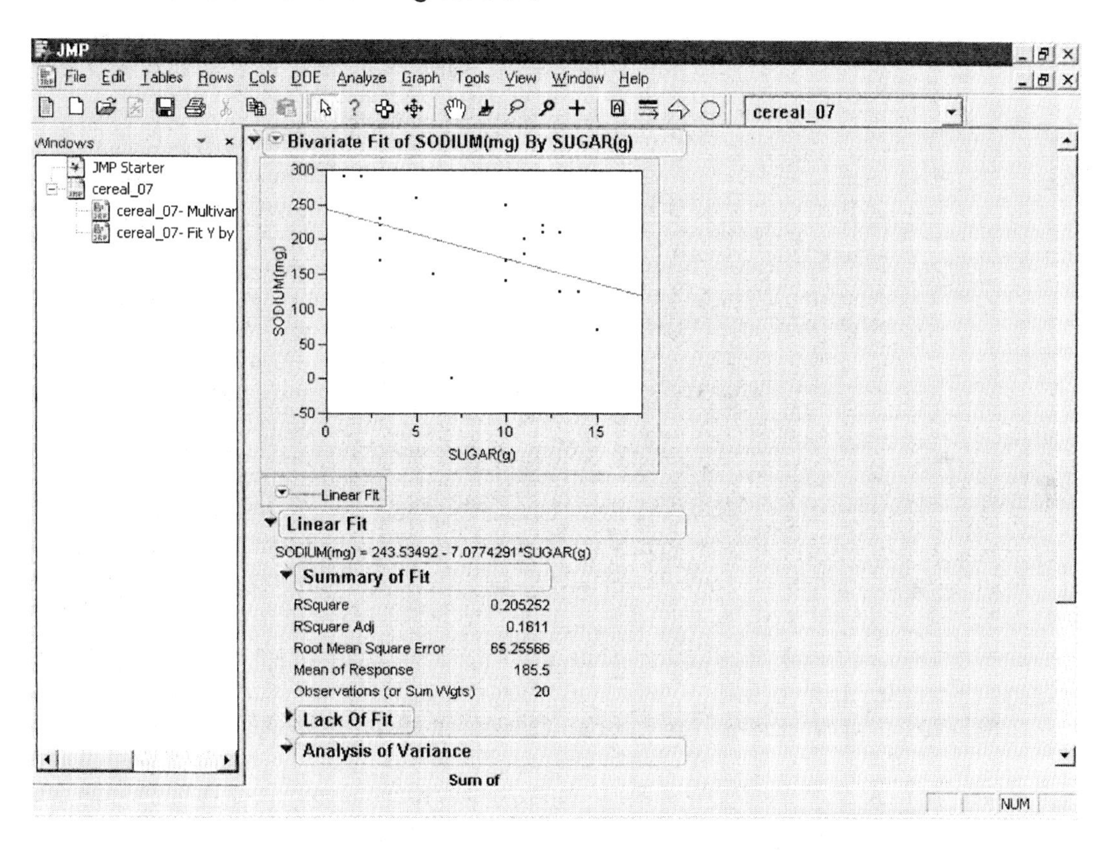

To identify the observation number of potential outliers, click on the observation of interest on the scatterplot.

To generate the correlation value, either just take the square root of the Rsquare value above (and make the value negative, since the regression line goes down), or, return to the datasheet, select **Analyze>Multivariate Methods>Multivariate**. Assign BOTH ***sodium*** and ***sugar*** to the **Y, Columns** role. Select **OK.**

The correlation matrix will be displayed (toggle to the top of the output screen):

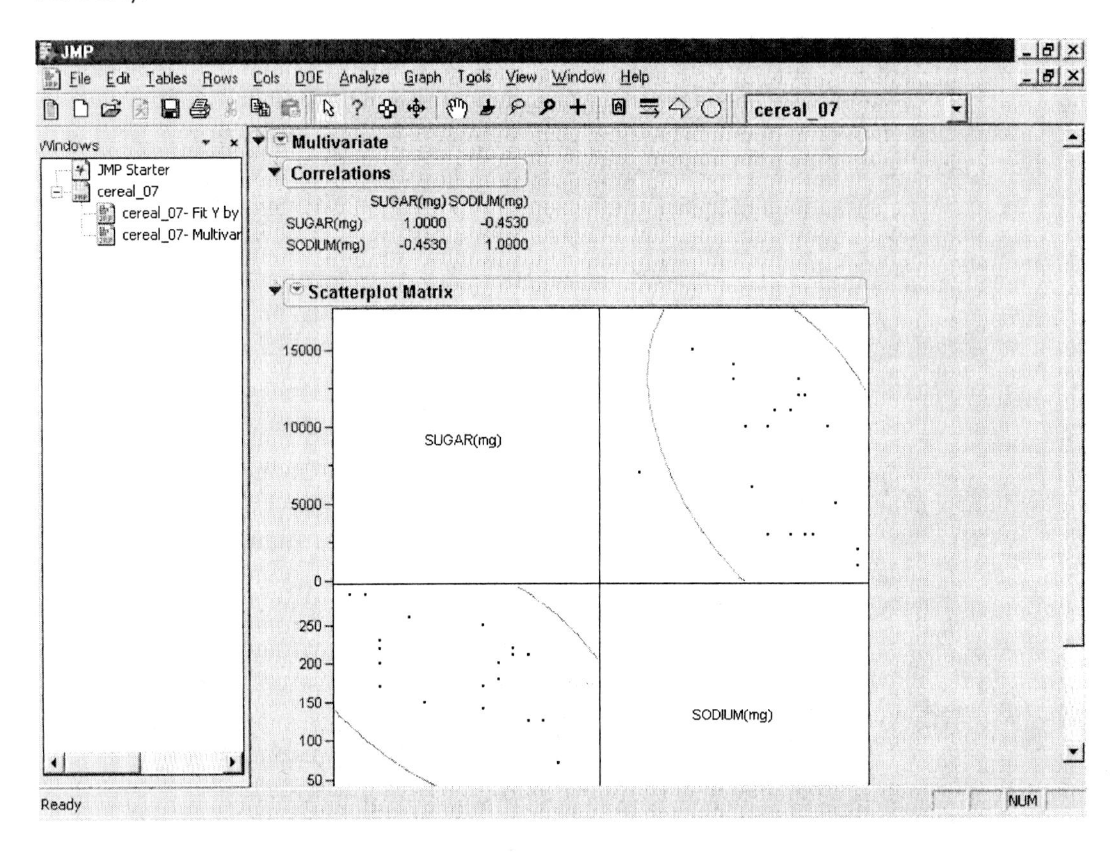

To execute the analysis above with particular observations excluded, identify the observation(s) to be excluded from the scatterplot. Then, return to the datasheet, select the row to be excluded and select **Rows>Exclude/Unexclude**. The⊘ symbol will appear next to that observation. Rerun the same executions as above.

Chapter 3. Exercise 3.52. TV in Europe.

Open the tv_europe data from the JMP folder of the data disk.

To create a scatterplot and regression line, select **Analyze>Fit Y By X**. Assign ***%Imports*** to the **Y, Columns** role and ***%Private*** to the **X, Factor** role. Select **OK**. On the scatterplot screen, click on the red triangle next to "**Bivariate Fit**..." and select **Fit Line**.

You should see the following screen:

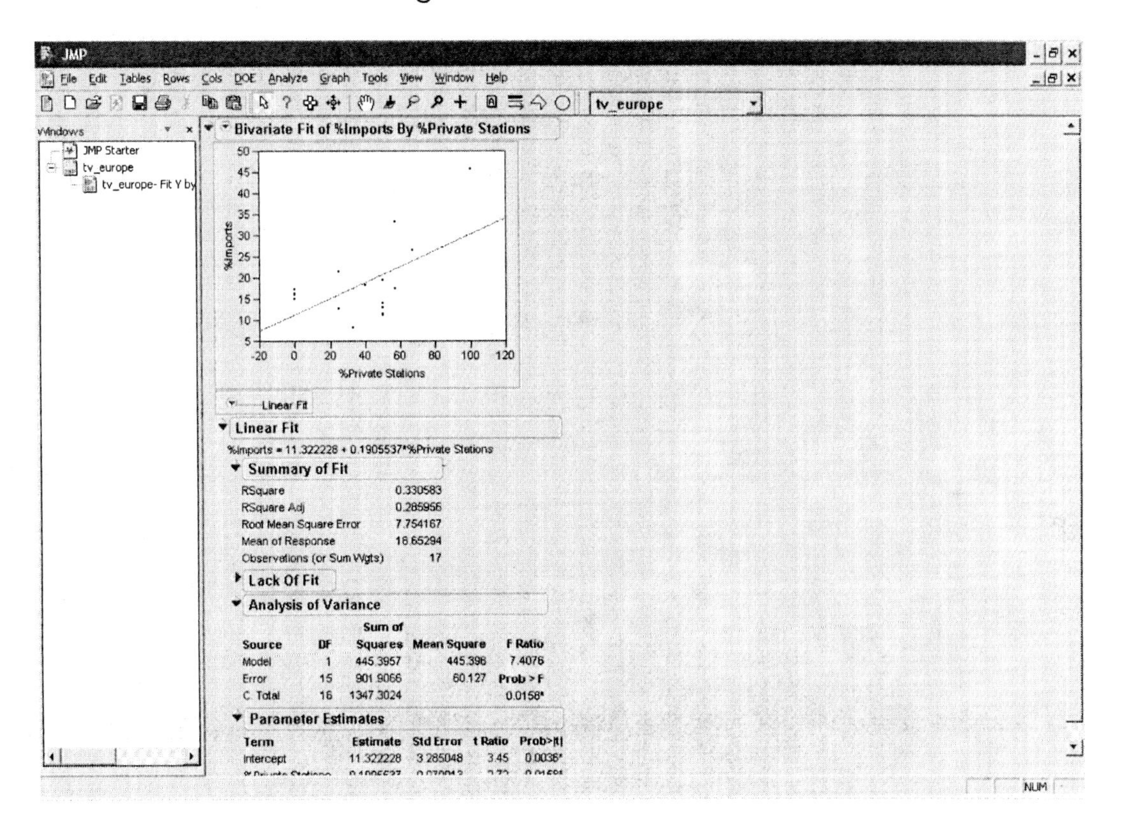

To generate the correlation value, either just take the square root of the Rsquare value above, or, return to the datasheet, select **Analyze>Multivariate Methods>Multivariate**. Assign BOTH ***%Import*** and ***%Private*** to the **Y, Columns** role. Select **OK**.

The correlation matrix will be displayed:

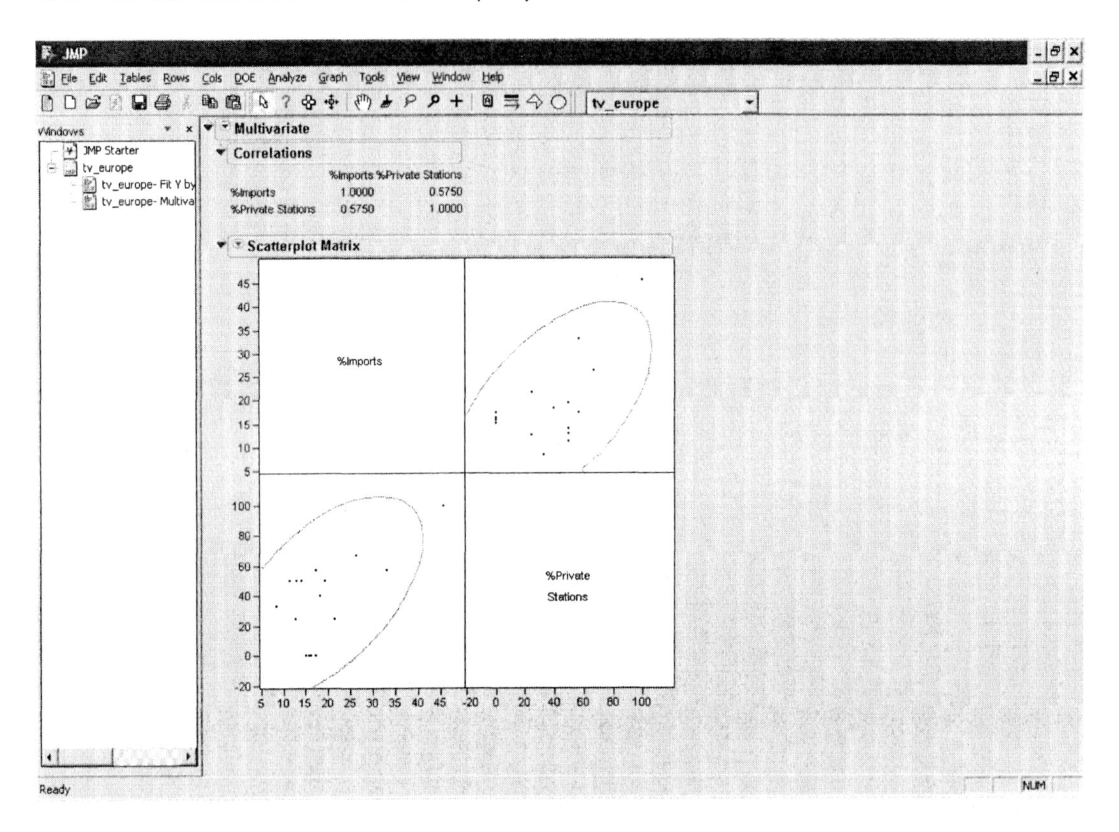

To execute the analysis above with particular observations excluded, identify the observation(s) to be excluded from the scatterplot. Then, return to the datasheet, select the row to be excluded and select **Rows>Exclude/Unexclude**. The⊘ symbol will appear next to that observation. Rerun the same executions as above.

Chapter 3. Exercise 3.80. Fertility and GDP.

Open the human_development data from the JMP folder of the data disk.

To create a scatterplot and regression line, select **Analyze>Fit Y By X**. Assign ***fertility*** to the **Y, Columns** role and ***GDP*** to the **X, Factor** role. Select **OK**. On the scatterplot screen, click on the red triangle next to "Bivariate Fit..." and select **Fit Line**.

You should see the following screen:

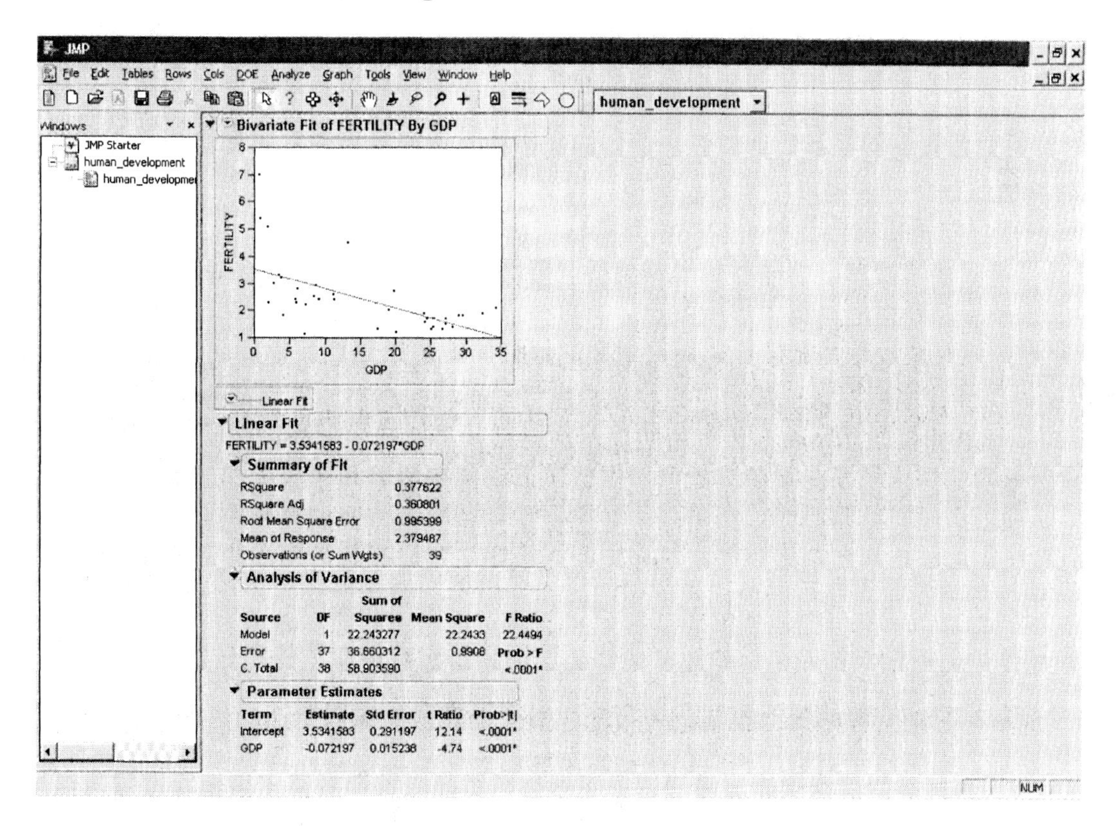

To generate the correlation value, either just take the square root of the Rsquare value above, or, return to the datasheet, select **Analyze>Multivariate Methods>Multivariate**. Assign BOTH ***GDP*** and ***Fertility*** to the **Y, Columns** role. Select **OK**.

The correlation matrix will be displayed:

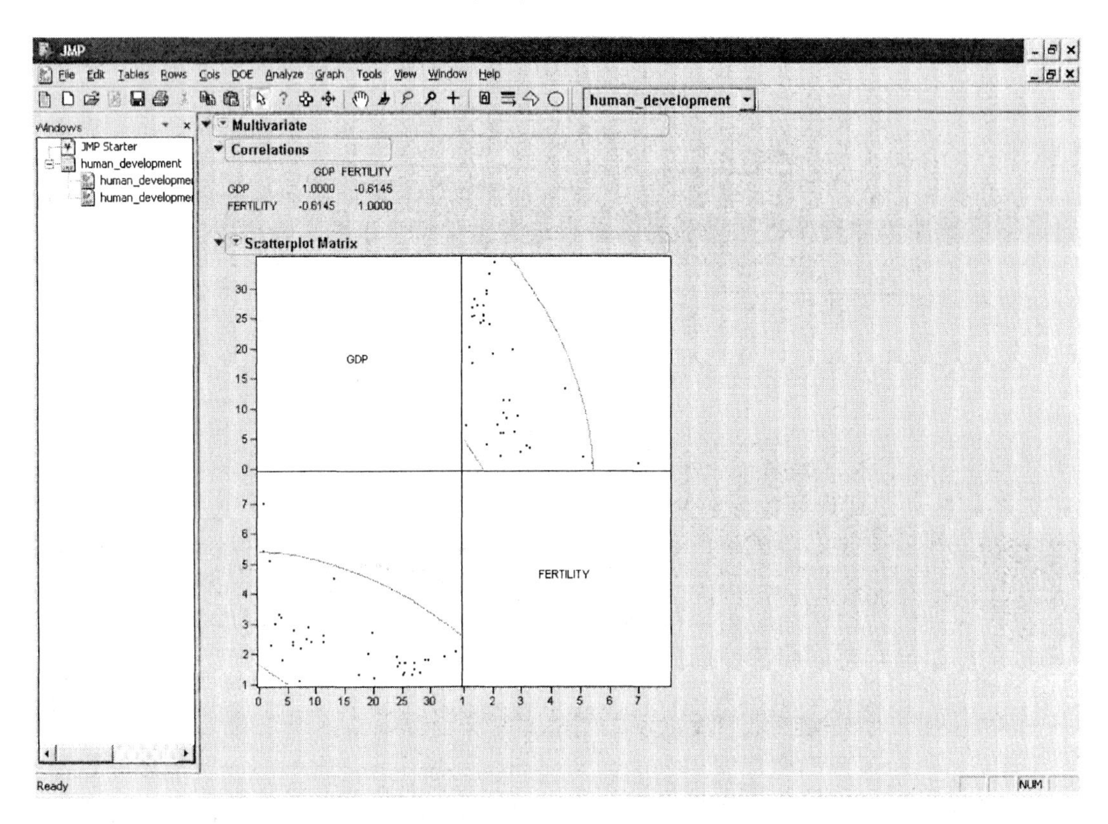

To evaluate the association of contraception with fertility, execute the same operations, substituting contraception for GDP.

Chapter 3. Exercise 3.84. Changing units for cereal data.

Open the cereal data from the JMP folder of the data disk.

To create a scatterplot and regression line in grams (the original units), select **Analyze>Fit Y By X**. Assign ***sodium* (mg)** to the **Y, Columns** role and ***sugar* (mg)** to the **X, Factor** role. Select **OK**. On the scatterplot screen, click on the red triangle next to "Bivariate Fit..." and select **Fit Line**.

You should see the following screen:

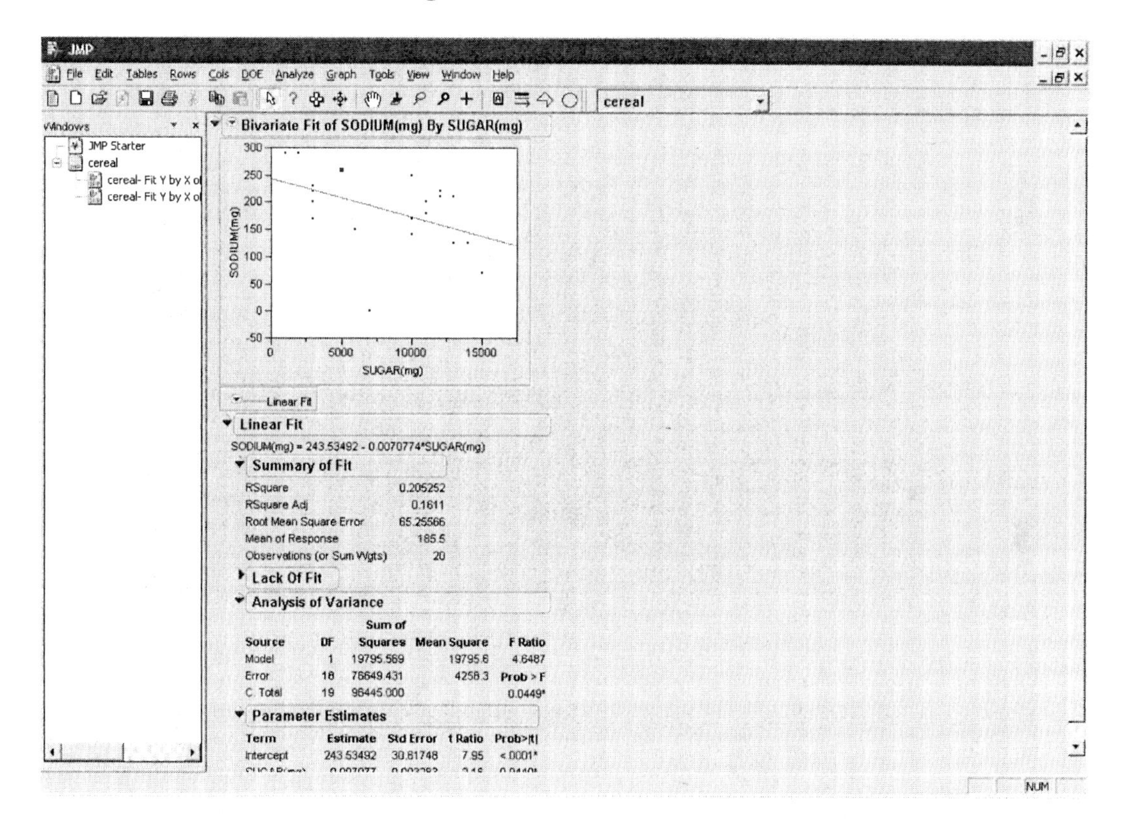

To convert the analysis from milligrams to grams, return to the datasheet. Execute the same commands as before, but use the ***sodium (g)*** and ***sugar (g)*** variables.

You should see the following screen:

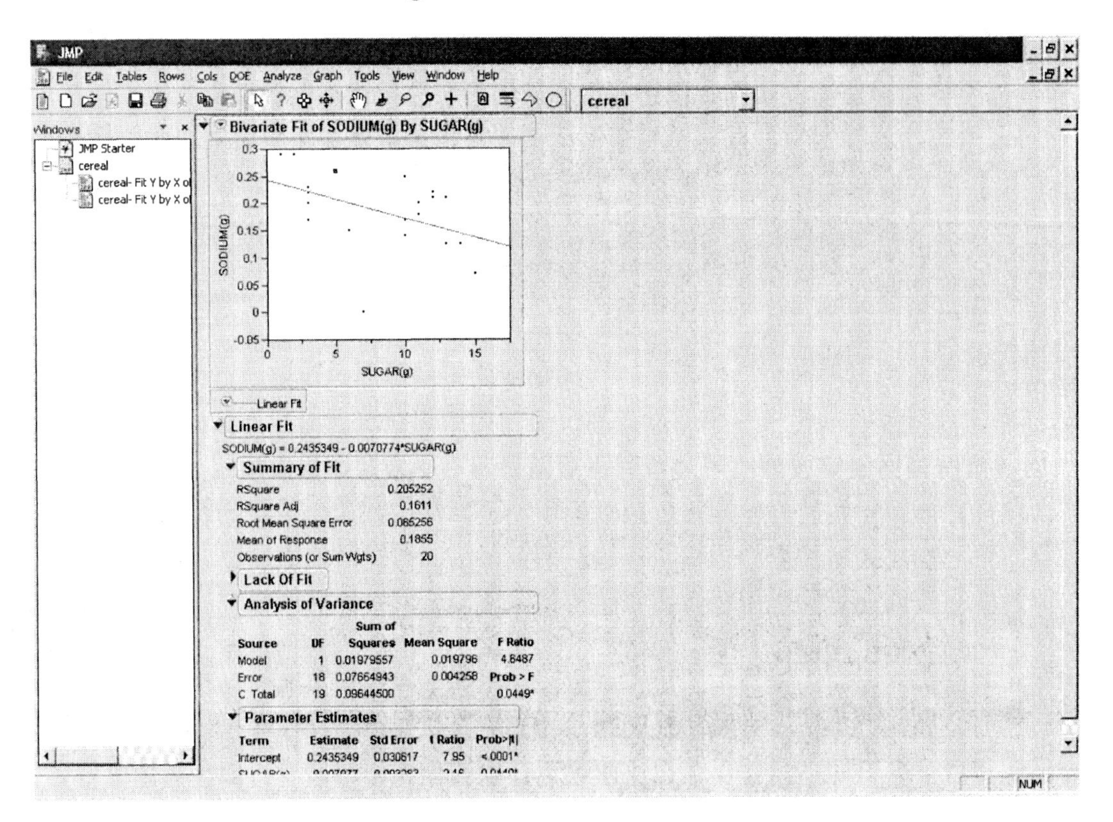

Notice that the only difference in the regression equations is the location of the decimal points.

Chapter 3. Exercise 3.86. Violent crime and college education.

Open the us_statewide_crime data from the JMP folder of the data disk.

To create a scatterplot and regression line, select **Analyze>Fit Y By X**. Assign ***violent crime rate*** to the **Y, Columns** role and ***college*** to the **X, Factor** role. Select **OK**. On the scatterplot screen, click on the red triangle next to "**Bivariate Fit**..." and select **Fit Line**.

You should see the following screen:

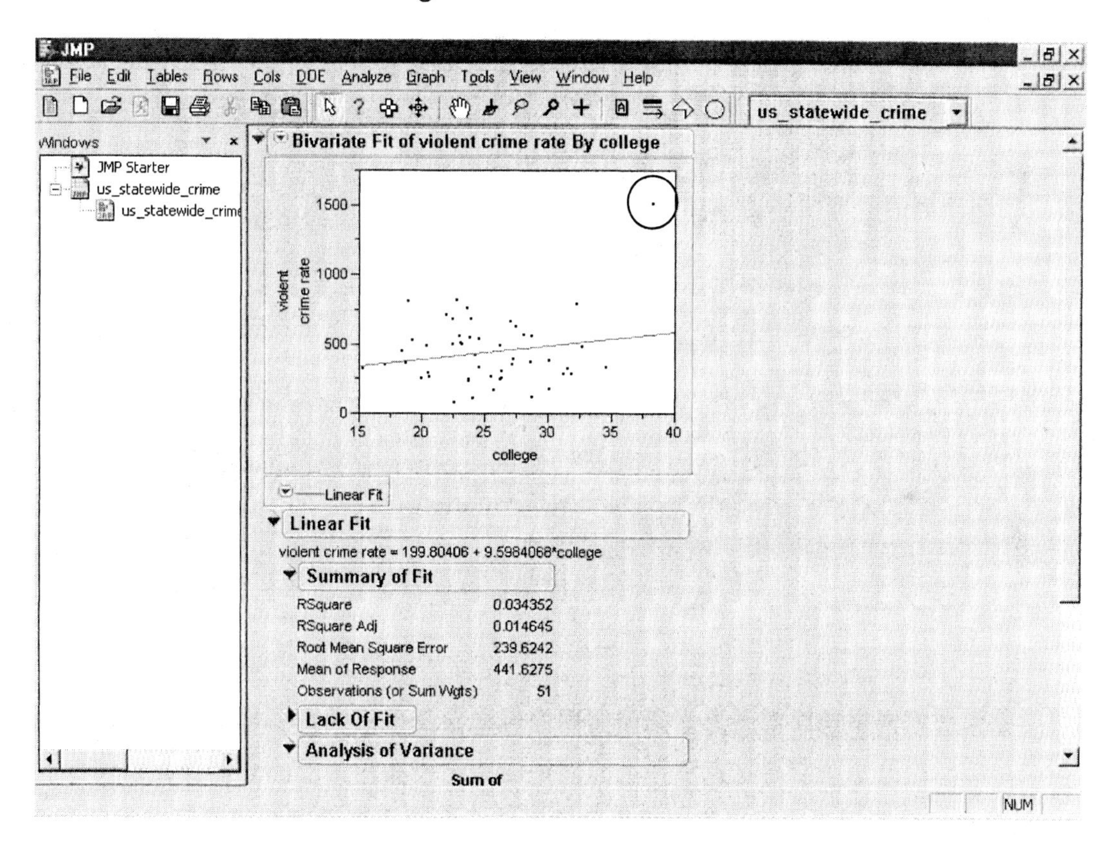

To identify any observation numbers for potential outliers – such as the one identified above – click on the observation of interest in the scatterplot. The observation above is number 9.

To evaluate the scatterplot and the regression line without this observation, return to the datasheet. Highlight row number 9 (if it is not already highlighted). Select **Rows>Exclude/Unexclude**. This action will exclude observation number 9 (D.C.) from any analysis. Rerun the same executions as before, with this observation excluded.

You should see the following screen:

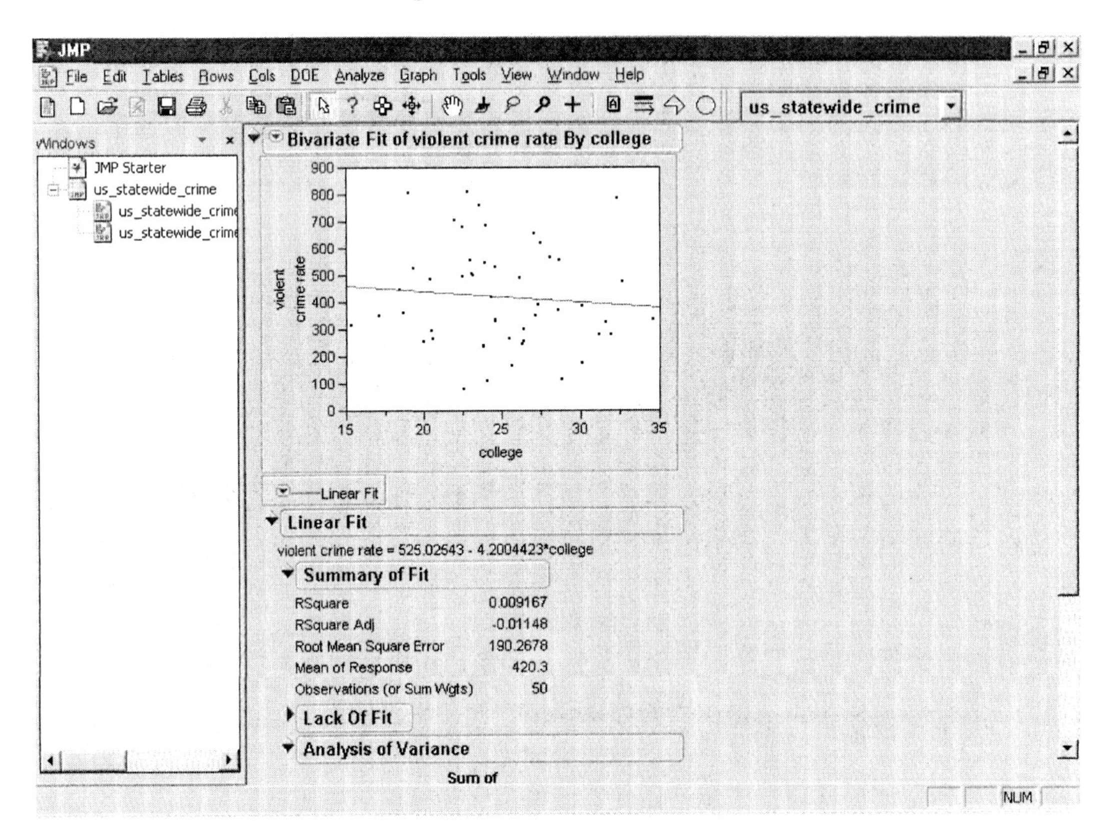

Chapter 3. Exercise 3.88. Crime and urbanization.

Open the us_statewide_crime data from the JMP folder of the data disk.

To generate the five number summary and the boxplot for the violent crime variable, select **Analyze>Distribution**. Assign ***violent crime*** to the **Y, Columns** role. Select **OK**.

You should see the following screen:

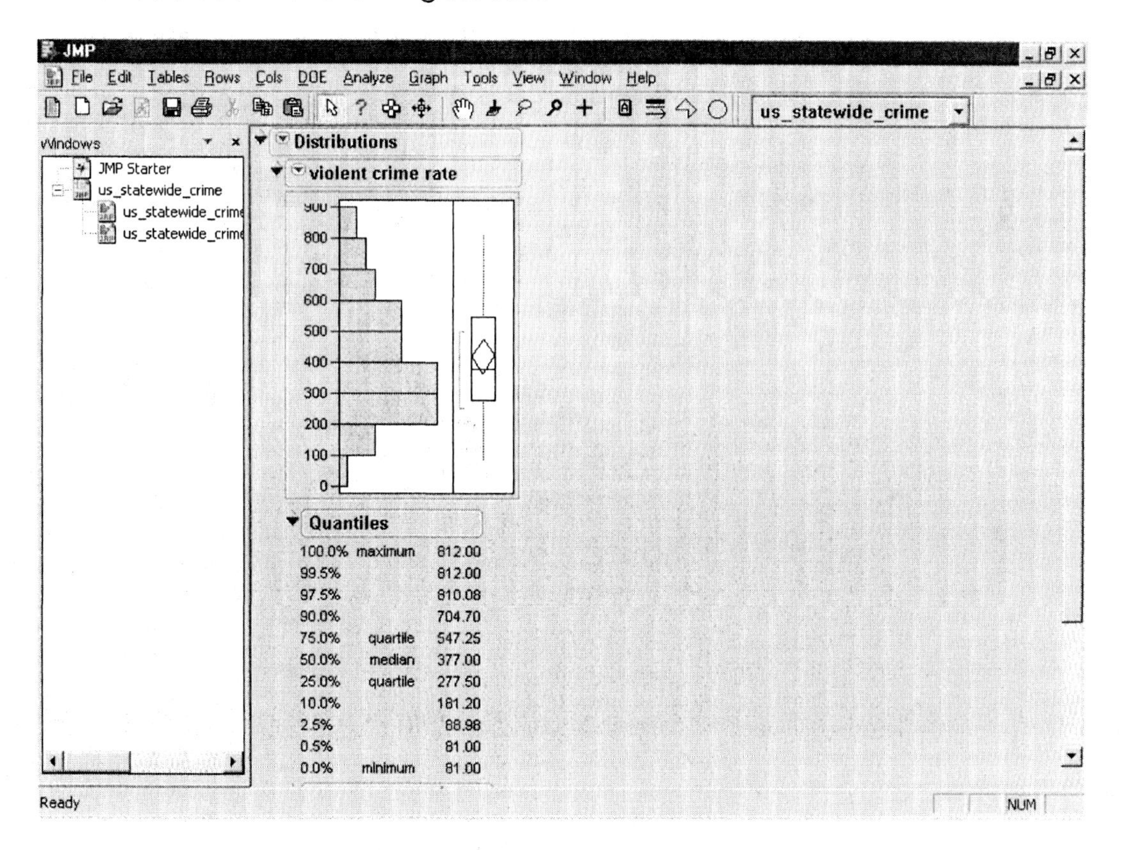

Note that the **Moments** section at the bottom of the screen contains the summary figures.

To generate the scatterplot and the regression line, select **Analyze>Fit Y By X**. Assign ***violent crime rate*** to the **Y, Columns** role and ***metropolitan*** to the **X, Factor** role. Select **OK**. On the scatterplot screen, click on the red triangle next to "**Bivariate Fit...**" and select **Fit Line**.

You should see the following screen:

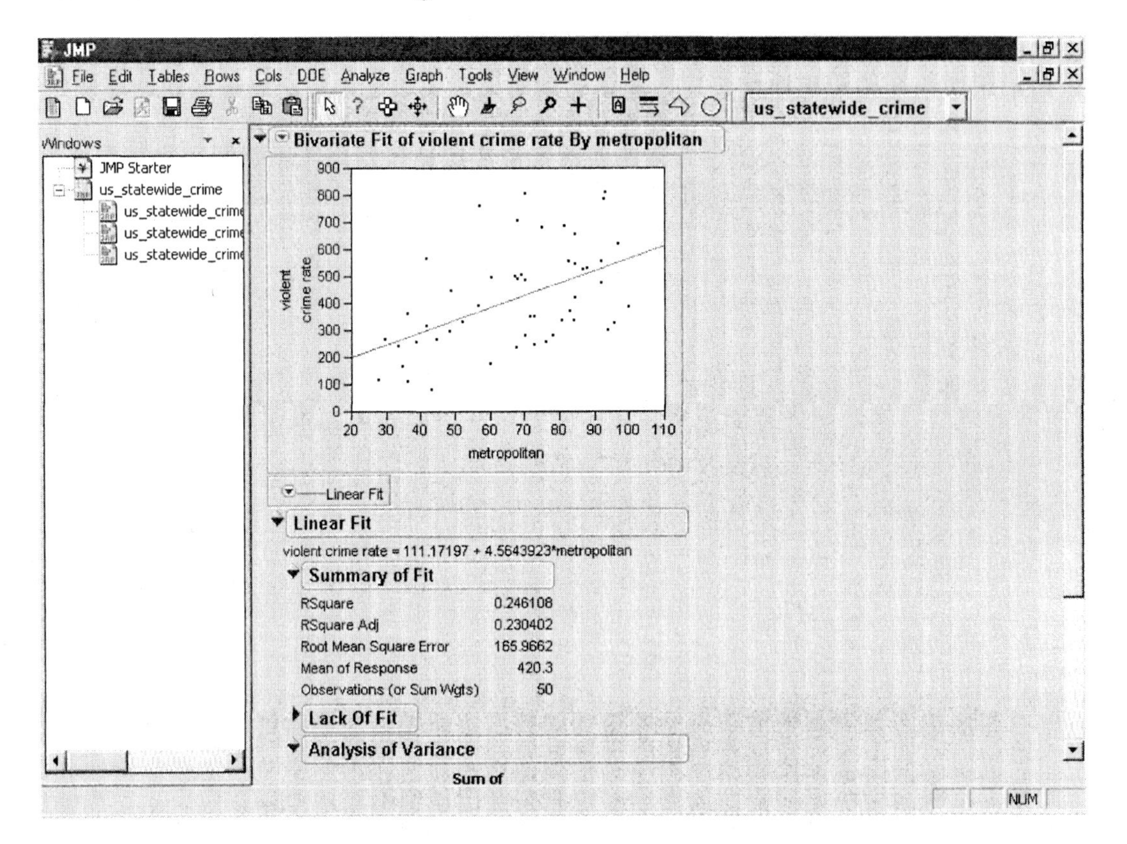

Chapter 3. Exercise 3.89. High School graduation rates and health insurance.

Open the hs_graduation_rates data from the JMP folder of the data disk.

To generate the scatterplot and regression line, select **Analyze>Fit Y By X**. Assign ***health ins*** to the **Y, Columns** role and ***HS Grad Rate*** to the **X, Factor** role. Select **OK**. On the scatterplot screen, click on the red triangle next to "**Bivariate Fit**..." and select **Fit Line**.

You should see the following screen:

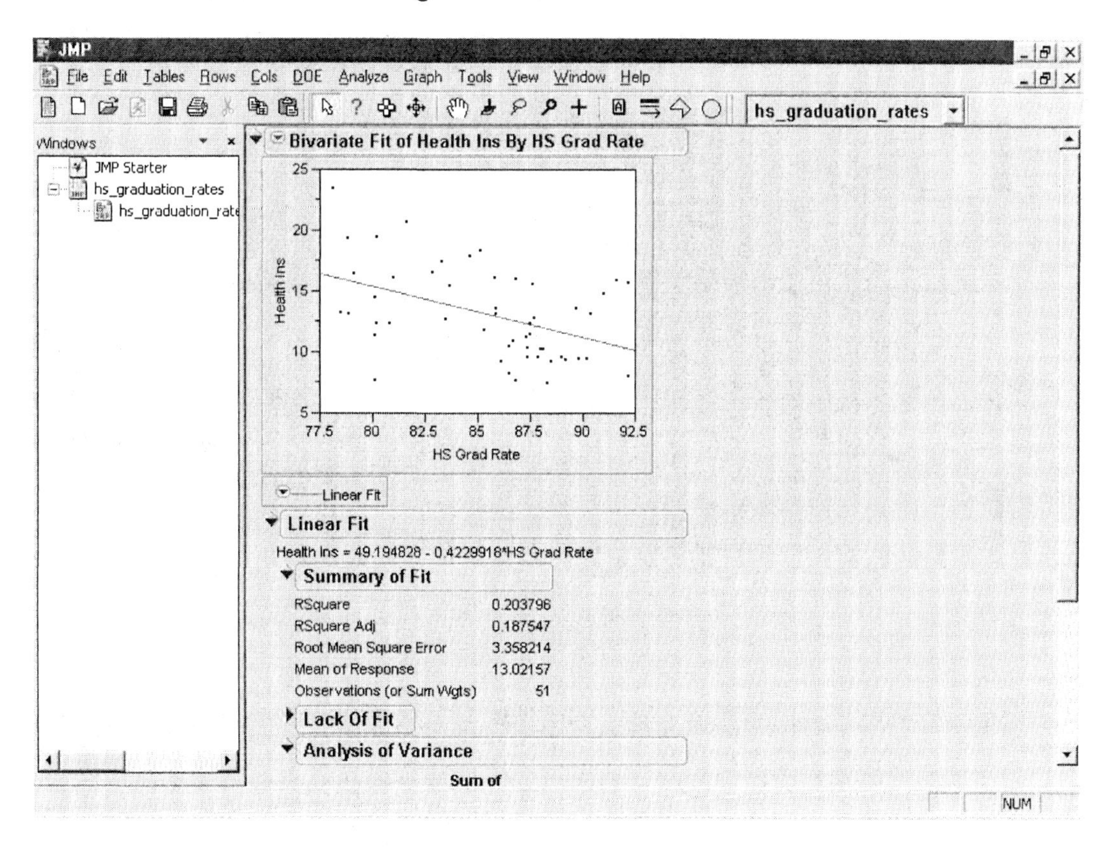

To generate the correlation value, either just take the square root of the Rsquare value above, or, return to the datasheet, select **Analyze>Multivariate Methods>Multivariate**. Assign BOTH ***Health Ins*** and ***HS Grad Rate*** to the **Y, Columns** role. Select **OK**.

The correlation matrix will be displayed:

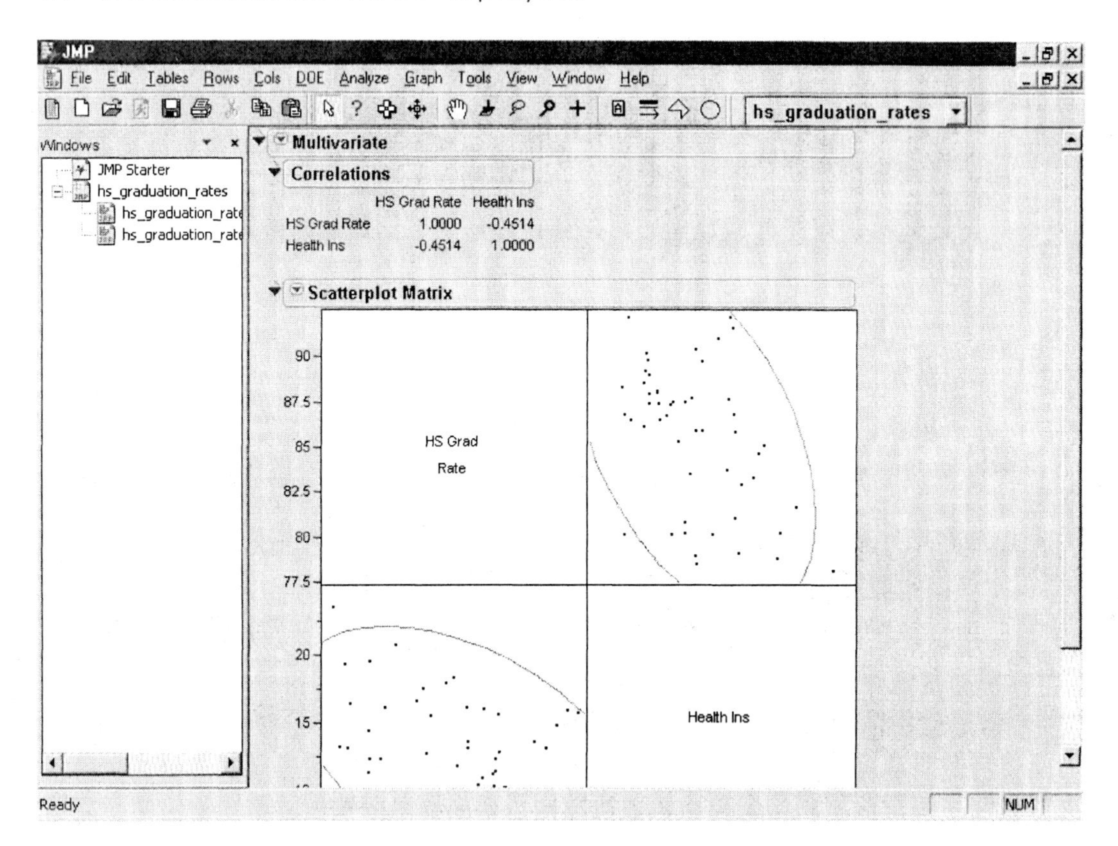

Chapter 3. Exercise 3.97. Warming in Newnan, GA.

Open the newnan_ga_temps data from the JMP folder of the data disk.

To generate the scatterplot and regression line, select **Analyze>Fit Y By X**. Assign ***Temp*** to the **Y, Columns** role and ***Year*** to the **X, Factor** role. Select **OK**. On the scatterplot screen, click on the red triangle next to "**Bivariate Fit**..." and select **Fit Line.**

You should see the following screen:

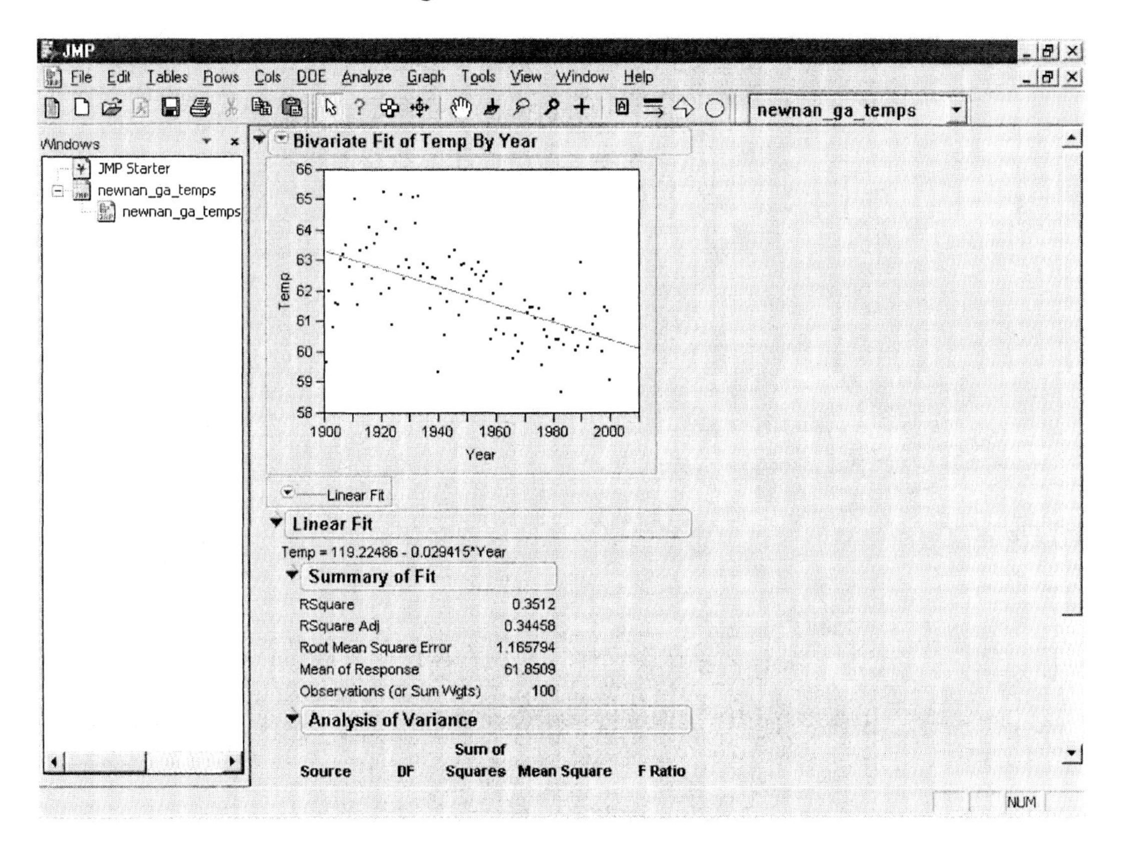

To assign random numbers to a dataset, select Cols>New Column. You will see the following screen:

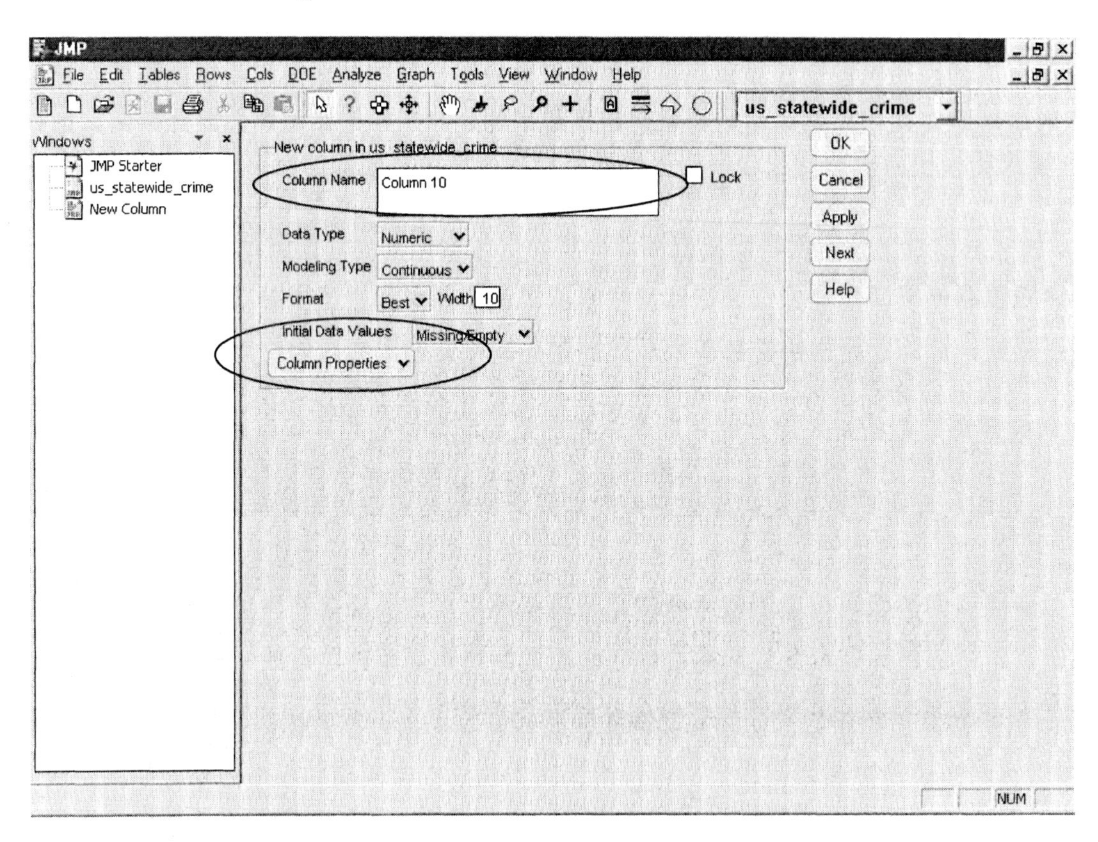

Give the column a name (like "Random"). Select **Column Properties>Formula**.

You will see the following screen:

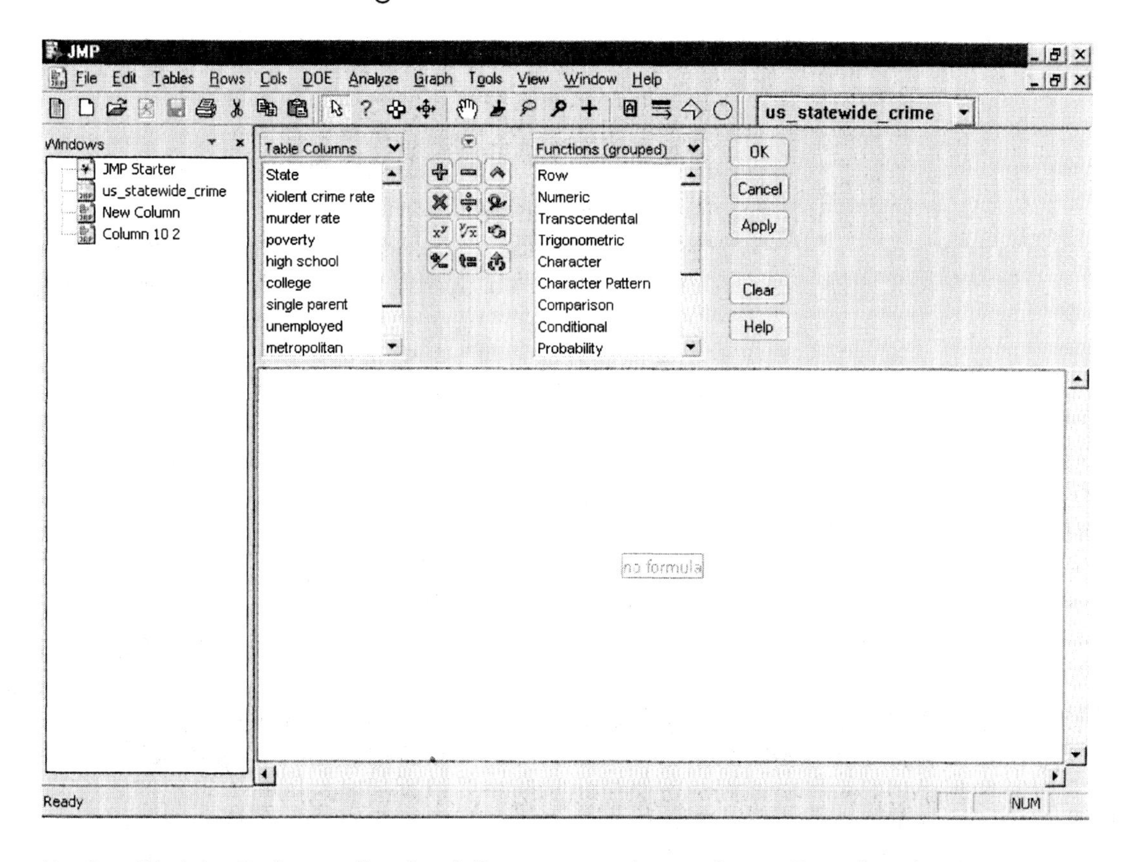

Under "Table Columns" select the new column (e.g., Random). Then, under "Functions (grouped)" toggle down until you see "Random". A long list of distributions will appear. Unless you have a strong understanding of probability distributions, you should select "Random Uniform". A random uniform distribution will generate random numbers between 0 and 1, with all numbers having the same probability of occurrence. Once Random Uniform is selected, the red box above will read "Random Uniform ()". Select **OK>OK**.

To have JMP randomly select a sample of observations from a dataset, select **Rows>Row Selection>Select Randomly**. JMP will generate a dialogue box, asking for the percentage of the dataset that needs to be excluded from the analysis. Enter a number between 0 and 1. For example, if you require a sample of 40% of the dataset, enter .60. Select **OK**. The datasheet will now have 60% of the observations randomly selected. Then, select **Rows>Select/Unselect**. Now, those observations representing 60% of the dataset have been excluded, leaving the remaining 40% available as a sample for analysis.

Chapter 6. Example 7. What IQ Do You Need to Get into MENSA?

There are several ways to approach this question using JMP. One way, is to generate a long string of numbers (observations) which following the distribution in question – IQ s are normally distributed with a mean of 100 and a standard deviation of 16. Then, using this set of simulated observations, you can examine the distribution, including the percentiles.

To do this, open a new data table in JMP.

Double click on Column 1. In the Column Info box, name the column "***IQ Scores***". Then click on **Column Properties>Formula**. Then click on **Edit Formula**. Scroll to the bottom of the **Functions**, to **Random**. Select **Random>Random Normal.**

You should see the following screen:

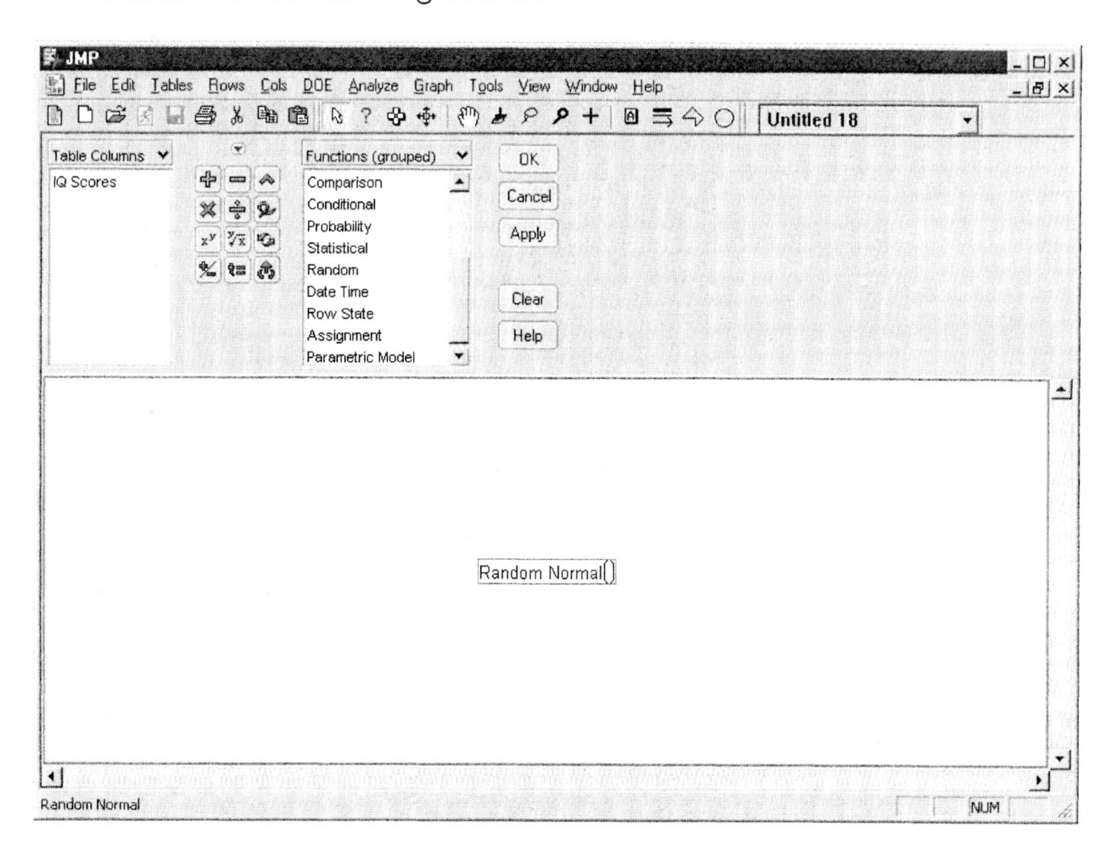

Inside the parentheses after Normal, you need to enter the mean and the standard deviation of the distribution in question. In this case, the values are 100 and 16. After you have entered these values inside the parentheses (they need to be entered with a comma separating them), click **OK>OK**. When you return to the datasheet, there will be no observations in the ***IQ Scores*** column. Select **Rows>Add Rows**. Enter the number "1000" and select **OK.**

Now that you have 1000 observations, which follow a normal distribution with a mean of 100 and a standard deviation of 16, select **Analyze>Distribution**. Assign the single variable to the **Y, Columns** role and select **OK.**

On the output screen, click on the red triangle next to "***IQ Scores***" and select **Fit Distribution>Normal**. Also, it might be helpful to change the layout to horizontal (**red triangle>display options>horizontal**).

You should see the following screen:

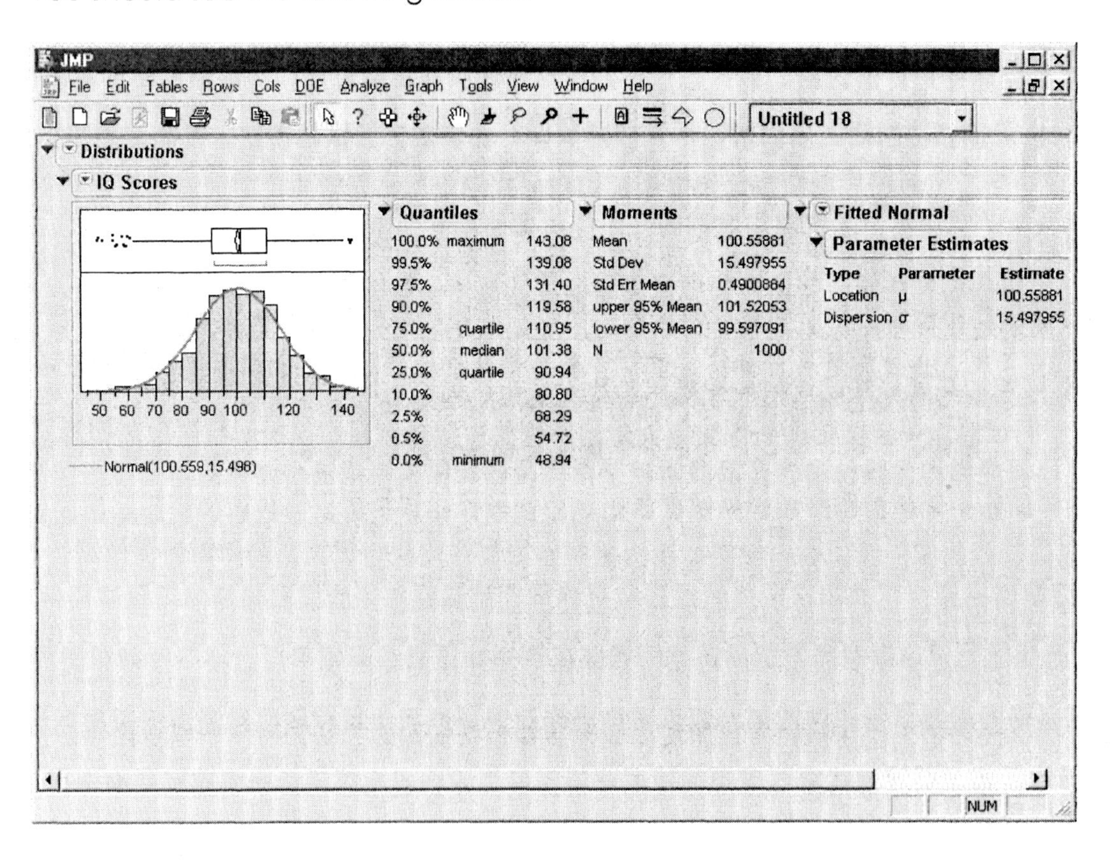

Because we were using a random number generator, your numbers may be slightly different, but they should be close.

You can see from the percentiles information that the 97.5 percentile value is 131.40. This would indicate that 97.5% of all people have an IQ of below 131.40. If MENSA was interested in the score that separated the bottom 98% from the top 2%, we would expect that it would be slightly higher than 131.40, but lower than the 99.5 percentile value of 139.08.

Chapter 6. Exercise 6.39 NBA Shooting.

Similar to the previous exercise, we will be generating a column of random numbers to answer the questions from this exercise. To begin, return to the JMP Starter menu and select **New Data Table**.

Double click on Column 1. In the Column Info box, name the column "***Free Throws***". Then click on **Column Properties>Formula**. Then click on **Edit Formula**. Scroll to the bottom of the **Functions**, to **Random**. Select **Random>Random Binomial**.

Inside the parentheses after Binomial, you need to enter the Number of Trials (in this case 10) and the probability of success of any one trial (in this case .90). After you have entered these values inside the parentheses (they need to be entered with a comma separating them), click **OK>OK**. When you return to the datasheet, there won't be any observations. To create the observations, select **Rows>Add Rows**. Enter "100", for 100 rows. Select **OK**.

After you have created a single variable with 100 observations, which following a binomial distribution, select **Analyze>Distribution**. Assign the single variable to the **Y, Columns** role. Select **OK**.

You should see the following screen:

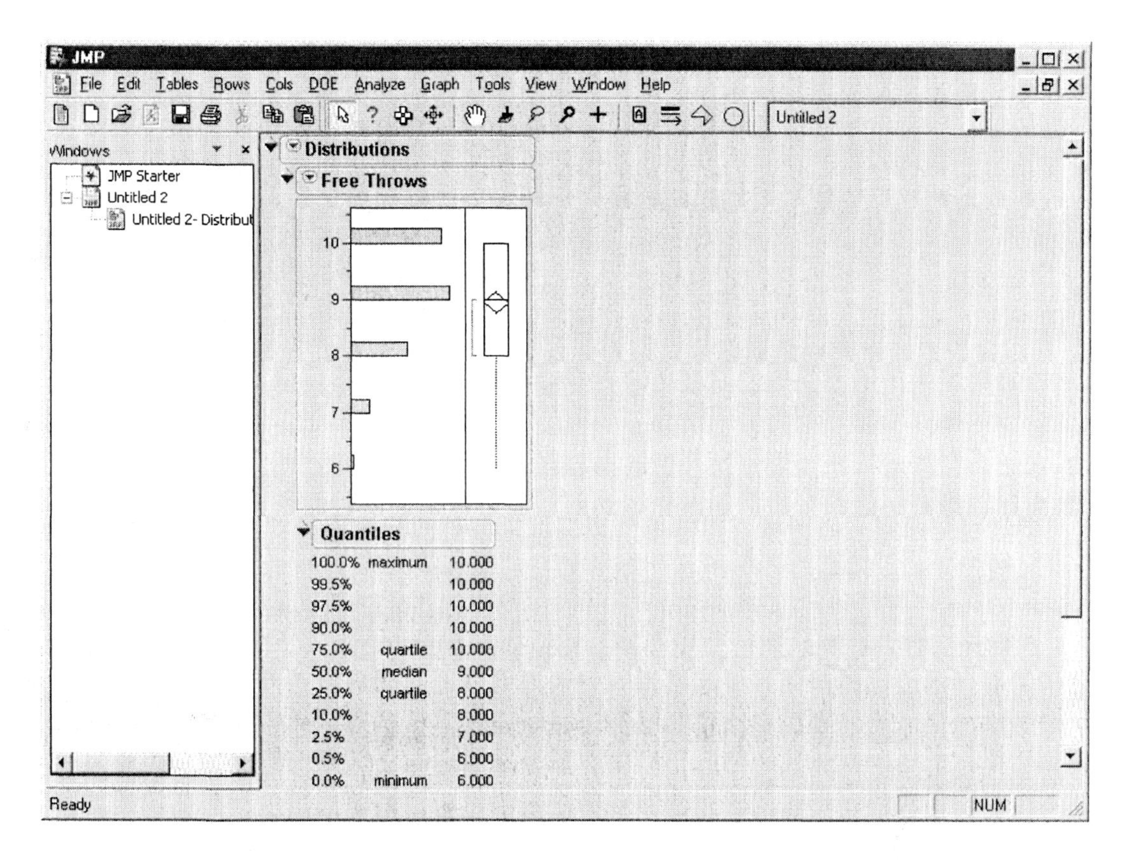

Click on the red triangle next to "**Free Throws**". Select **Save>Prob Scores**. This action will save the probability of each outcome of free throws to be saved to the original dataset.

You should see the following screen:

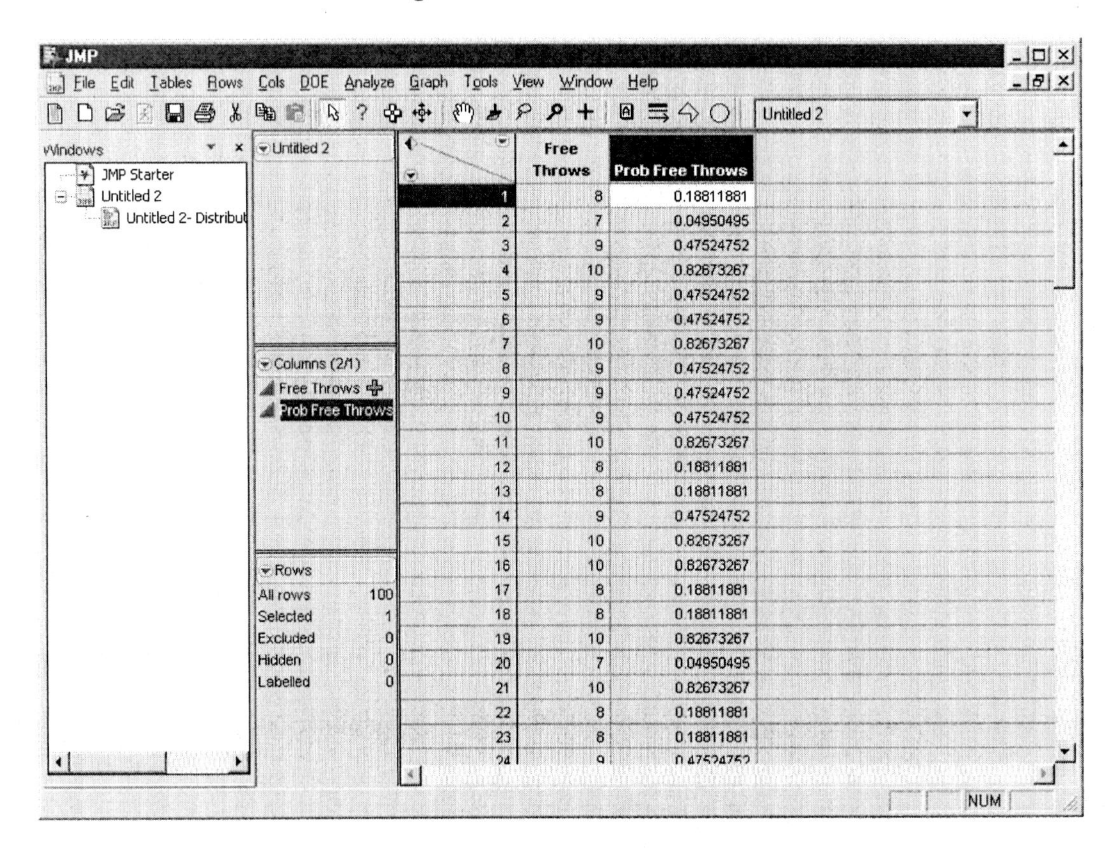

From this output, the probability of making 10 free throws is .8267 or 82.67%. The probability of making 9 free throws is .4752 or 47.52%.

Chapter 8. Exercise 8.5. Watching TV.

To create the dataset for analysis, open the JMP Starter page, and select **New Data Table**.

In the blank data set, right click on the top of column 1. Select **Column Info**. For **Column Name**, enter "***Hours***". Ensure that the **Data Type** is identified as **Numeric** and the **Modeling Type** is **Continuous**. Select **OK**.

After you enter the data in the table, select **Analyze>Distribution**. Assign the ***Hours*** variable to the **Y, Columns** role. Select **OK**.

From the output screen, click on the red triangle next to the variable name ***Hours***. Select **Confidence Interval**. Select **OK**.

You should see the following screen:

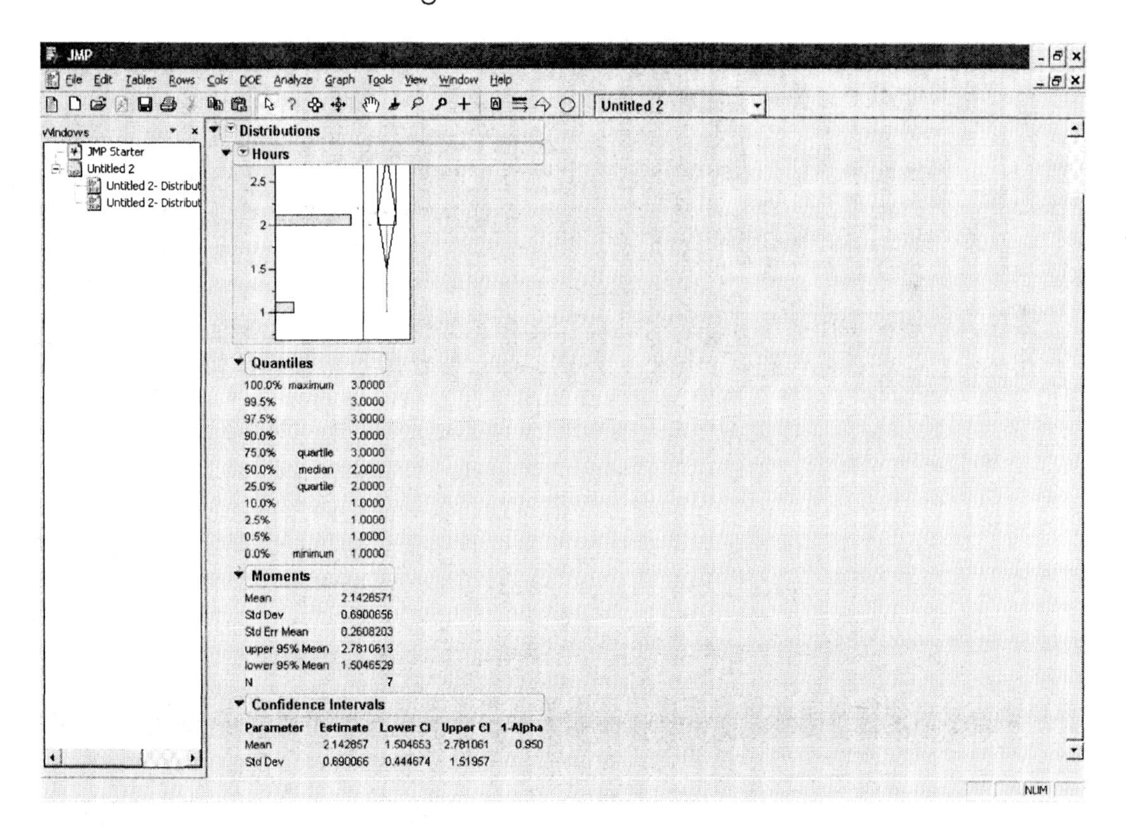

The point estimate is the mean as identified above. The Margin of Error is the difference between the mean estimate and the Lower (or Upper) CL: 2.14-1.50 =.64.

Chapter 8. Exercise 8.32. New Graduate's Income.

To create the dataset for analysis, open the JMP Starter page, and select **New Data Table.**

In the blank data set, right click on the top of column 1. Select **Column Info**. For **Column Name**, enter "***Salaries***". Ensure that the **Data Type** is identified as **Numeric** and the **Modeling Type** is **Continuous**. Select **OK.**

After you enter the data in the table, select **Analyze>Distribution**. Assign the ***Salaries*** variable to the **Y, Columns** role. Select **OK.**

From the output screen, click on the red triangle next to the variable name ***Salaries***. Select **Confidence Interval.** Select **OK.**

You should see the following screen:

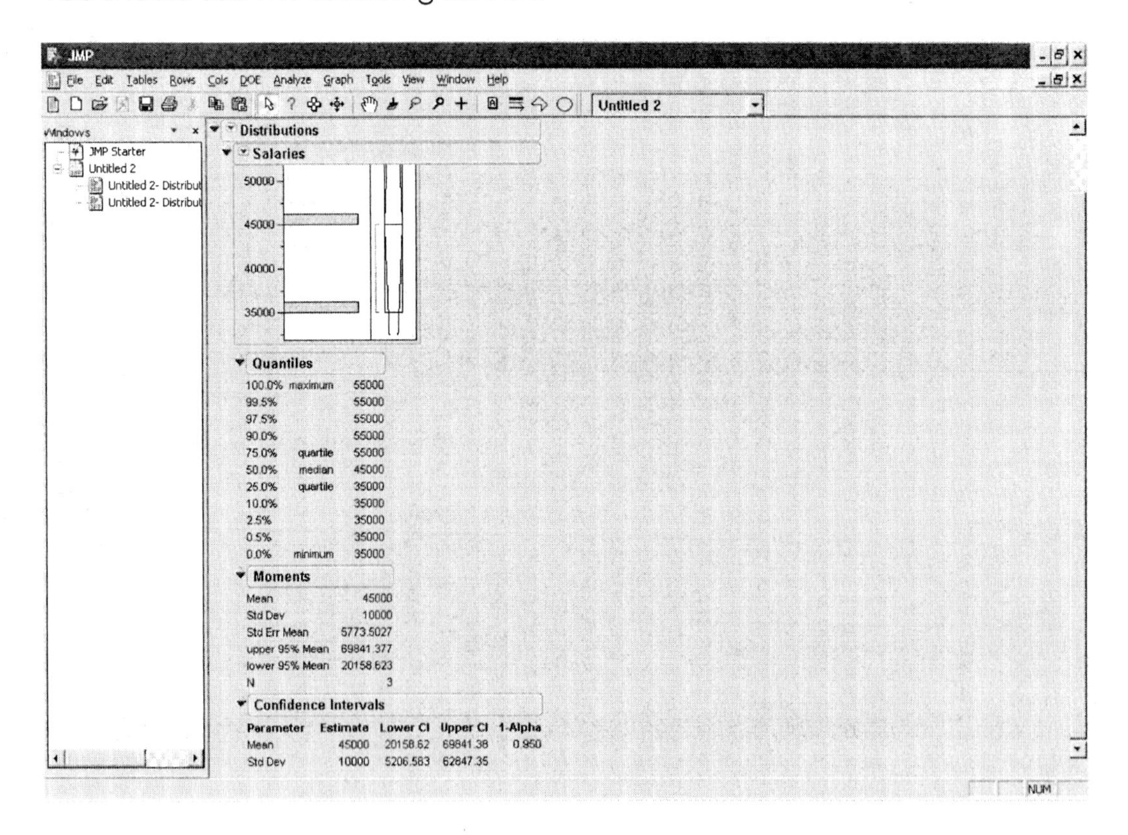

To generate the 99% confidence interval, click on the red triangle next to the variable name "***Salaries***", select **Confidence Interval** again. This time, make the confidence level 99%.

Chapter 8. Exercise 8.33 TV Watching for Muslims.

To create the dataset for analysis, open the JMP Starter page, and select **New Data Table**.

In the blank data set, right click on the top of column 1. Select **Column Info**. For **Column Name**, enter "***Hours***". Ensure that the **Data Type** is identified as **Numeric** and the **Modeling Type** is **Continuous**. Select **OK**.

After you enter the data in the table, select **Analyze>Distribution**. Assign the ***Hours*** variable to the **Y, Columns** role. Select **OK**.

From the output screen, click on the red triangle next to the variable name ***Hours***. Select **Confidence Interval**. Select **OK**.

You should see the following screen:

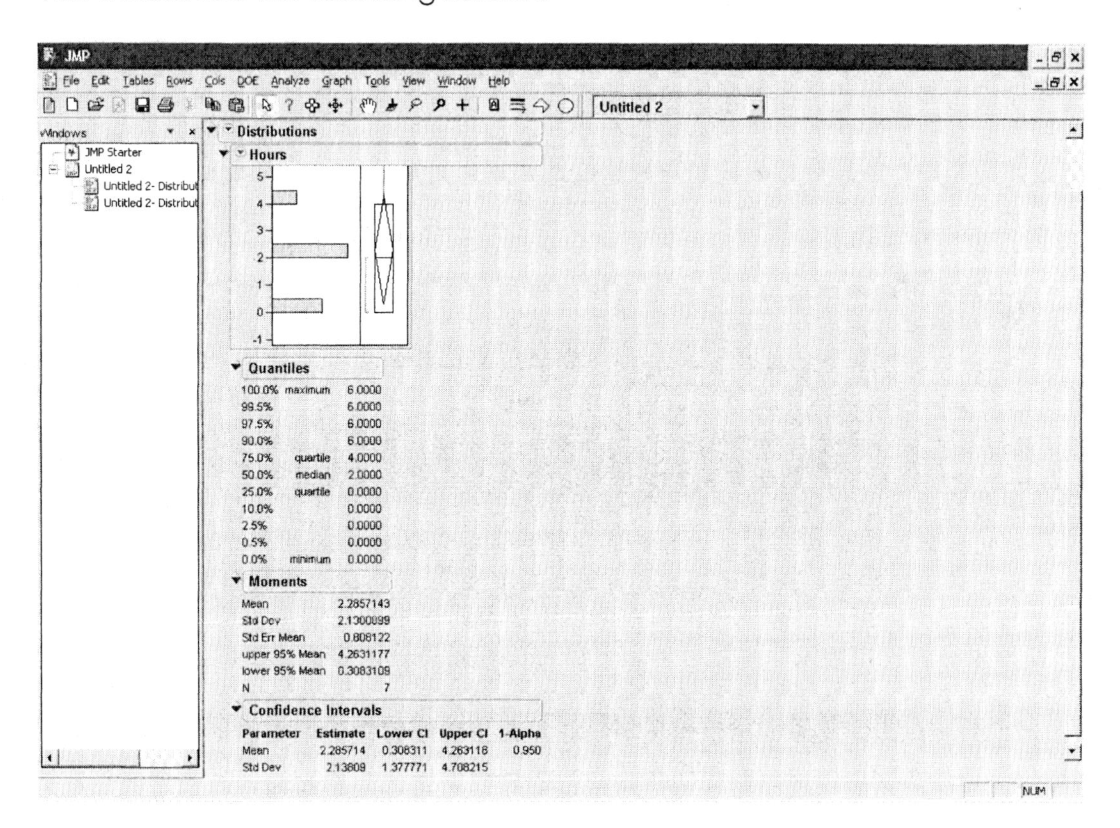

Chapter 8. Example 7. eBay Auctions of Palm Handheld Computers.

Open the ebay_auctions JMP file from the data disk.

To obtain the descriptive statistics for both purchase options, select **Analyze>Distribution**. Assign both options, ***Bidding*** and ***Buy-It-New*** to the **Y, Columns** role. Select **OK**.

You should see the following screen:

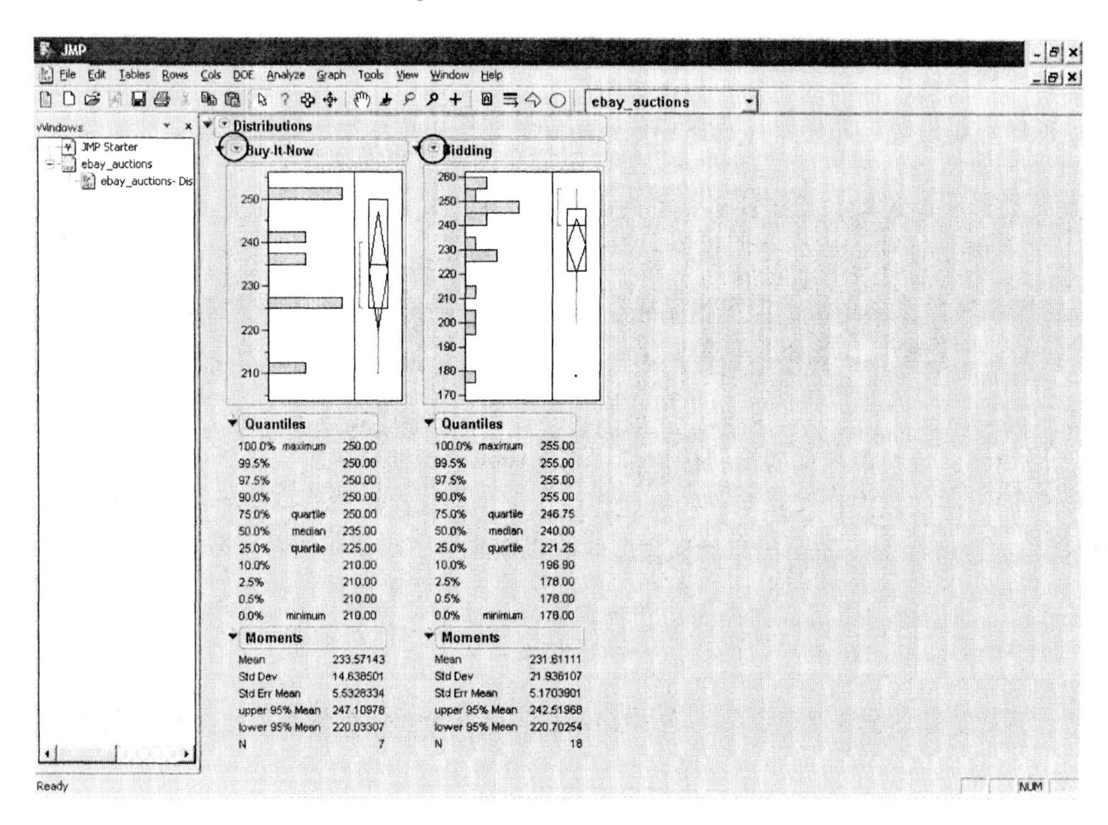

To generate the 95% confidence intervals, click on the red triangle next to each variable name as identified above. Then select **Confidence Interval**.

You should see the following screen:

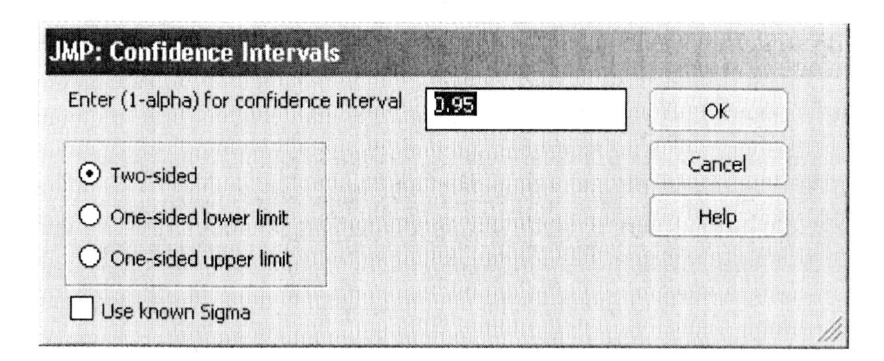

The interval requested is 95% - which is the default (if a 99% interval is needed, enter 0.99 in the box above). And, the interval is two sided – which is also the default. Select **OK**.

You should see the following screen:

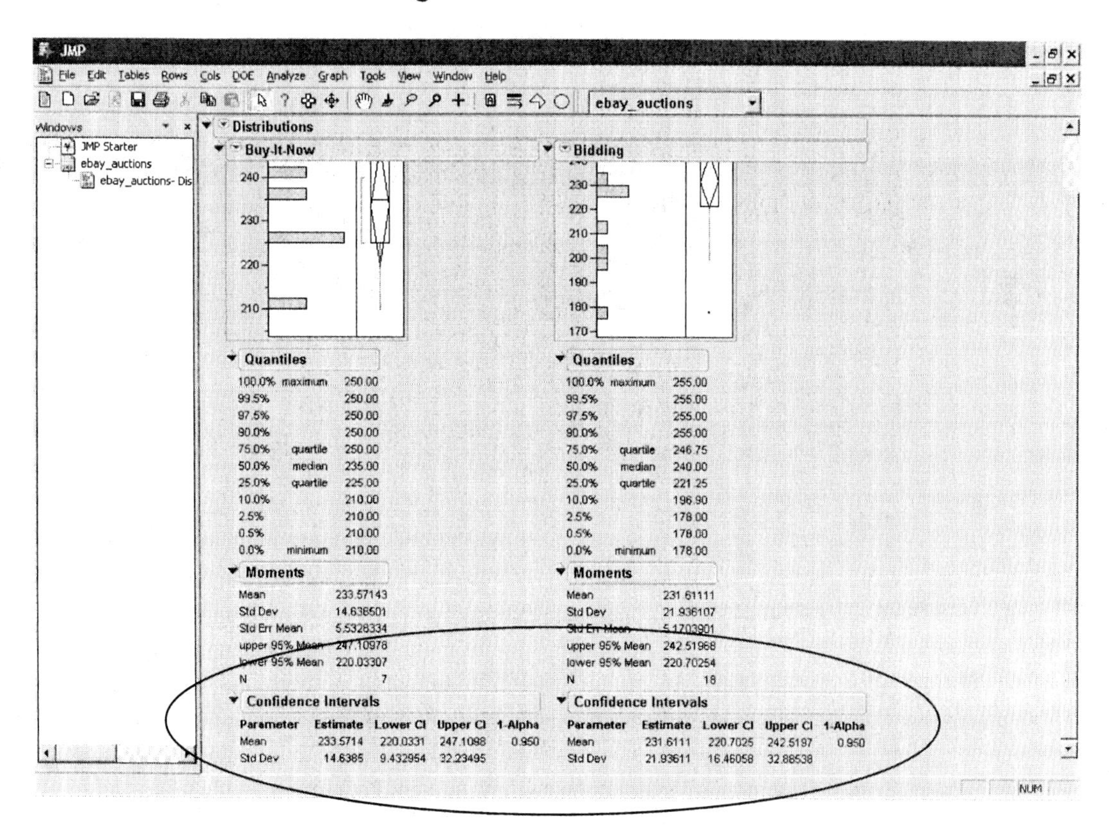

The confidence intervals for both options have been included in the output above.

Chapter 8. Exercise 8.38. How often read a newspaper?

Before beginning the analysis of the newspaper variable, notice that the icon next to the variable is a series of red bars. Recall that this indicates a qualitative variable. Since the variable is actually quantitative (the number of newspapers students read), it needs to be converted into a quantitative variable. To do this, right click on the variable name newspapers. Select **Column Info**. Change the **Data Type** to **Numeric** and the **Modeling Type** to **Continuous**. Select **Apply>OK**. Now, the variable icon for newspapers should be a blue triangle, indicating that the variable is quantitative.

To generate the descriptive statistics and the 95% confidence interval, select **Analyze>Distribution**. Assign the ***newspapers*** variable to the **Y, Columns** role. Select **OK.**

From the output screen, click on the red triangle next to the variable name ***newspapers***. Select **Confidence Interval**. Select **OK.**

You should see the following screen:

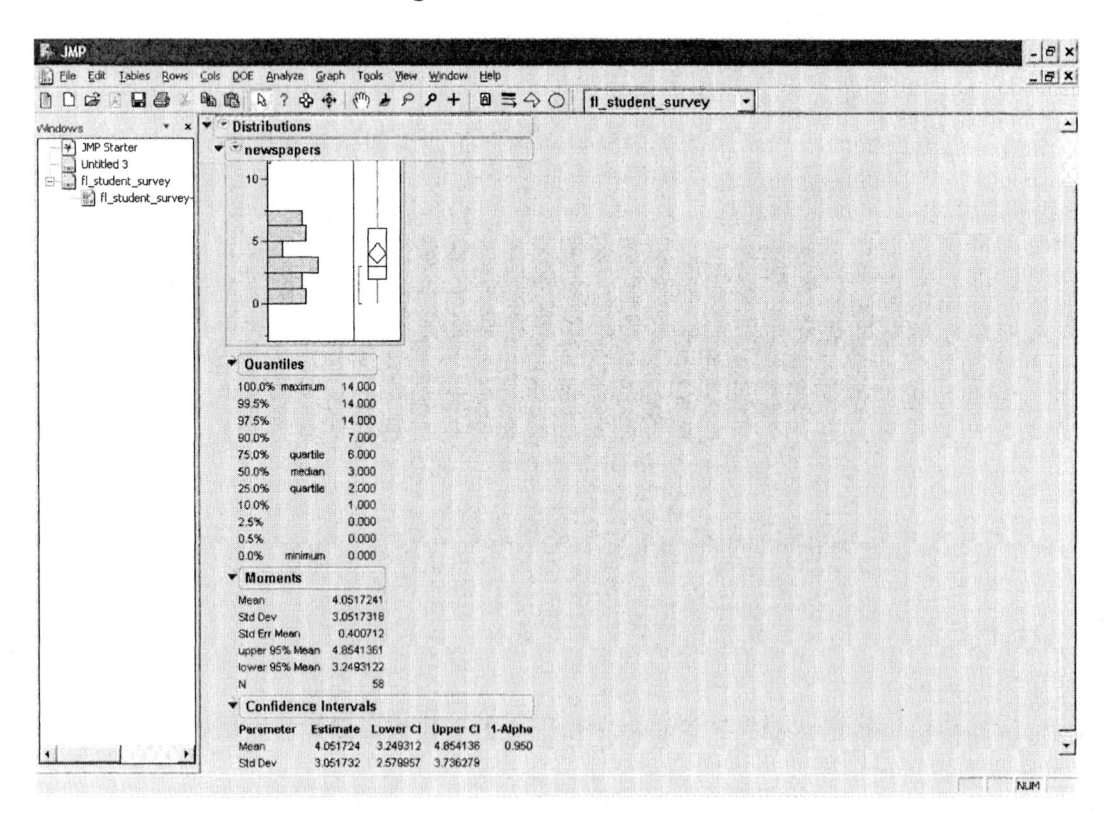

Chapter 8. Exercise 8.85. Revisiting Mountain Bikes.

Open the mountain_bike file from the JMP data disk.

Select **Analyze>Distribution**. Assign ***Price*** to the **Y, Columns** role. Select **OK**. Click on the red triangle next to the variable ***price***, and select **Confidence Interval**. Select **OK**.

You should see the following screen:

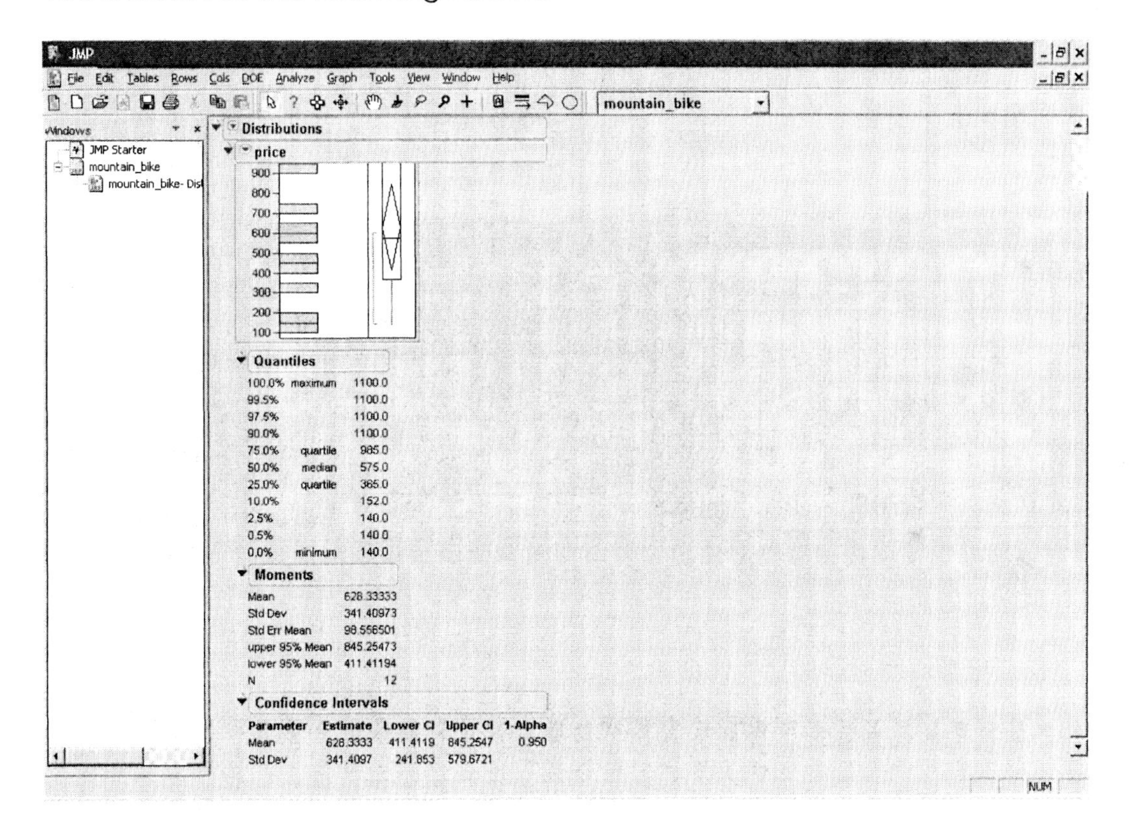

Examine the histogram and the boxplot regarding the distribution of the data before interpreting the confidence interval.

Chapter 8. Exercise 8.87. Income for families in public housing.

From the JMP Starter screen, select "**New Data Table**". Enter the **Income** data as provided.

You screen should look like this:

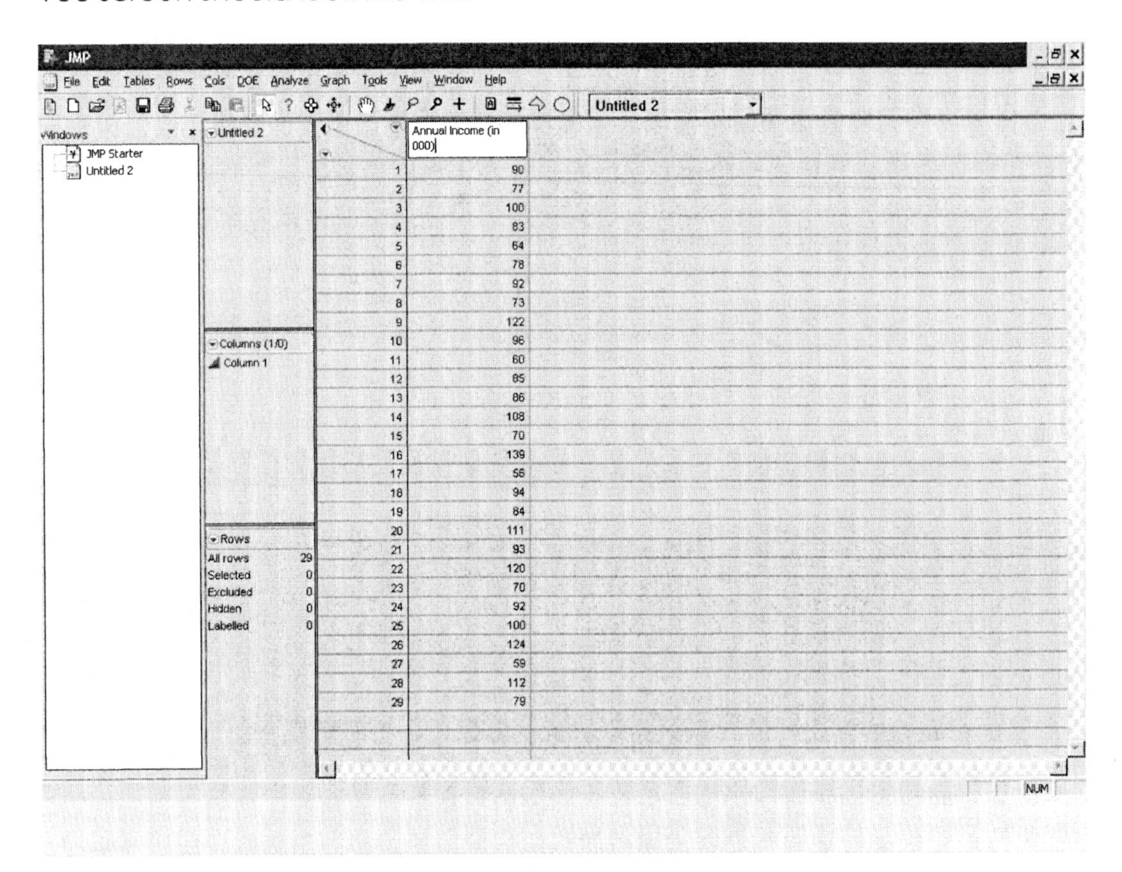

To generate the descriptive statistics, select **Analyze>Distribution**. Assign ***Annual Income*** to the **Y, Columns** role. Select **OK**.

To generate the 95% confidence interval, click on the triangle next to "Annual Income...". Select **Confidence Interval**. Set the **Confidence Interval** to **0.95**, and ensure that **Two-sided** is checked. Select **OK**.

You should see the following screen:

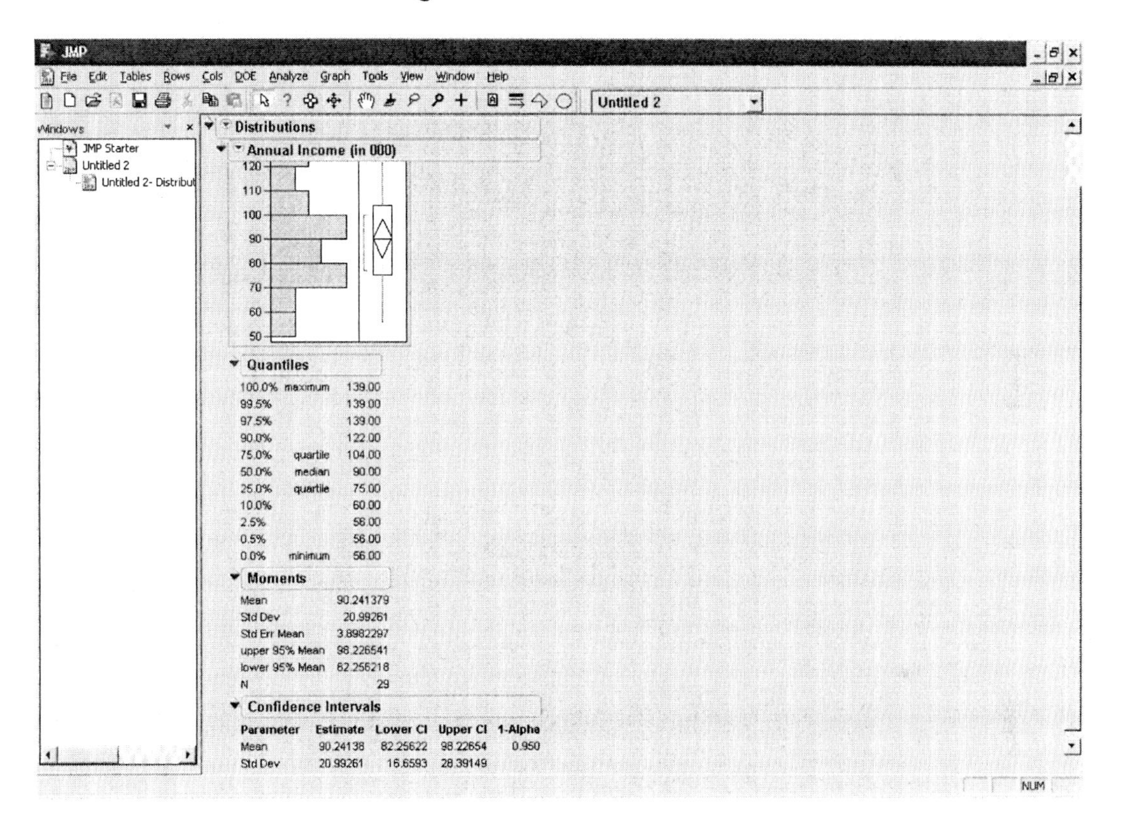

Chapter 9. Example 8. Mean Weight Change in Anorexic Girls.

Open the anorexia file from the JMP data disk.

To determine if the change in weight for girls was different from 0, select **Analyze>Distribution**. Assign the ***Change*** variable to the **Y, Columns** role. Select **OK**. Click on the red triangle next to the variable name "***Change***". Select **Test Mean.** In this instance, the test is to determine if the mean is statistically different from zero, which is the default value. Select **OK**.

You should see the following screen:

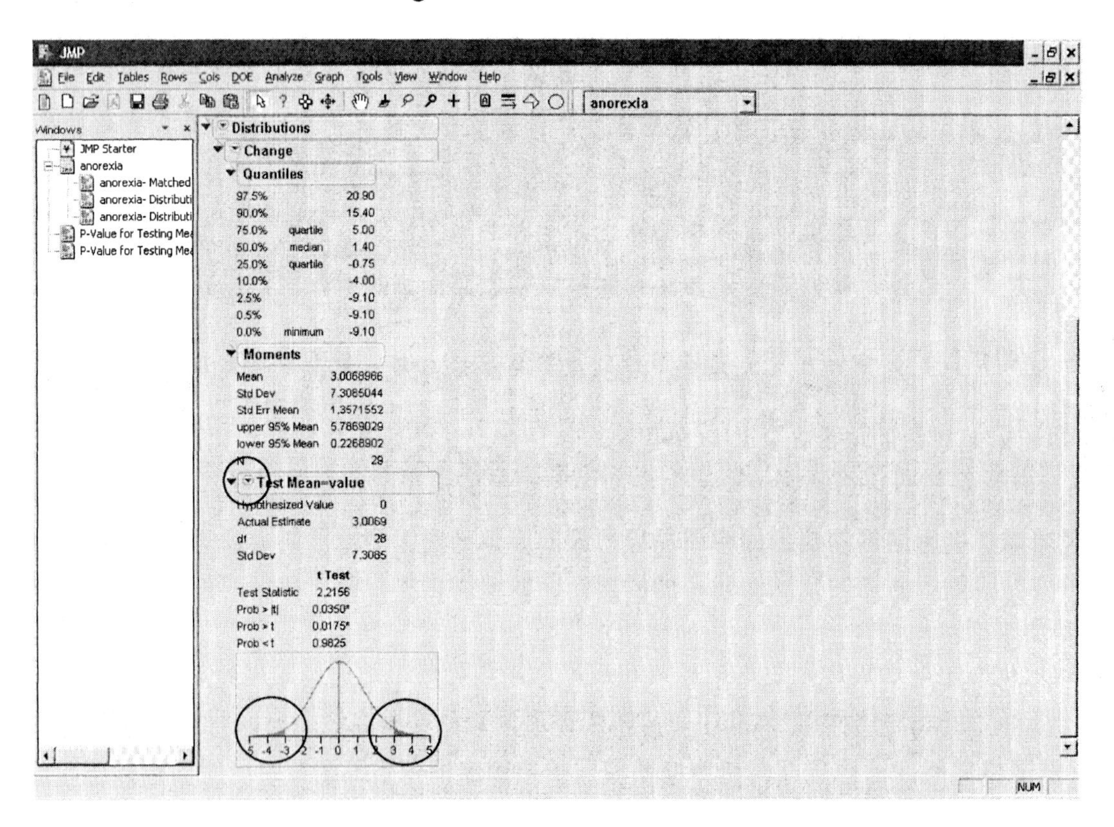

As can be seen in the visual above, a two-tailed test was conducted – the test was for Ha: $\mu \neq 0$. From the p-value above, we would conclude that the Null hypothesis should be rejected, and there is evidence to conclude that the mean weight change is different from 0. To execute a one-tailed test, such as Ha: $\mu > 0$, click on the red triangle next to "Test Mean = value" as identified above, select **PValue Animation**.

You should see the following screen:

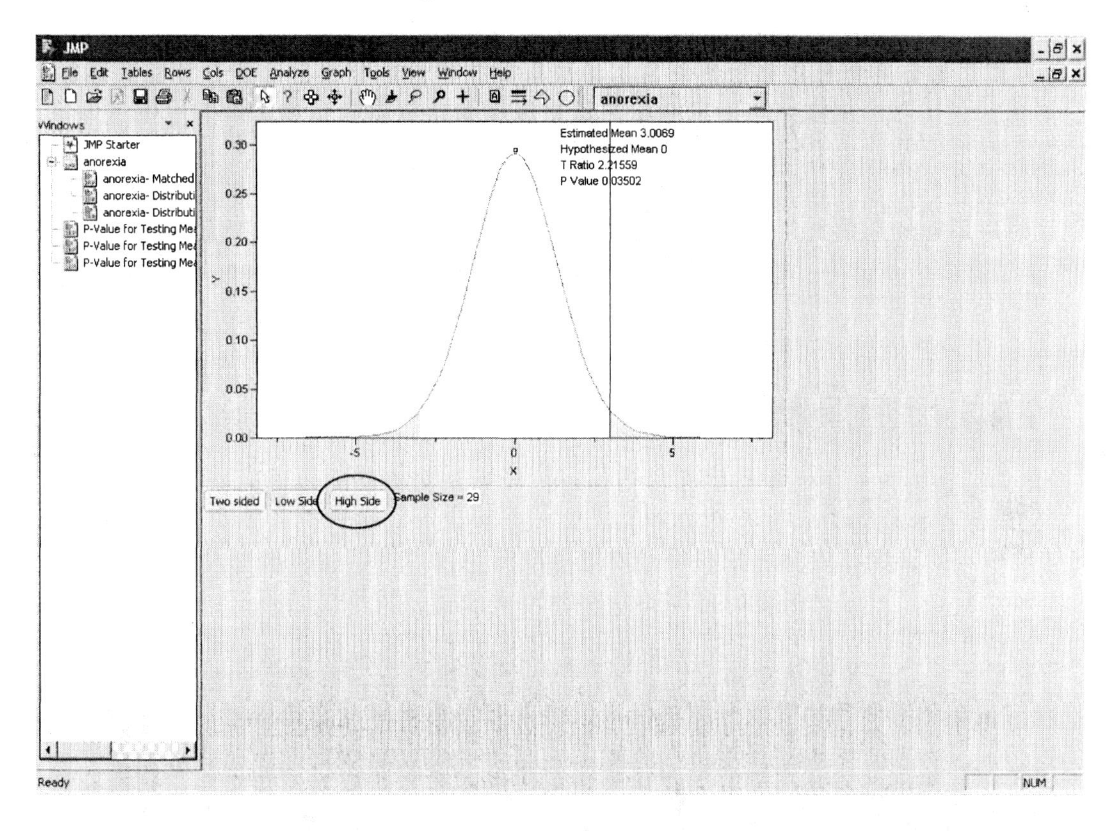

To execute the one tailed test - Ha: μ > 0, click on the **High Side** button as identified above. The results for this one tailed test will be generated.

You should see the following screen:

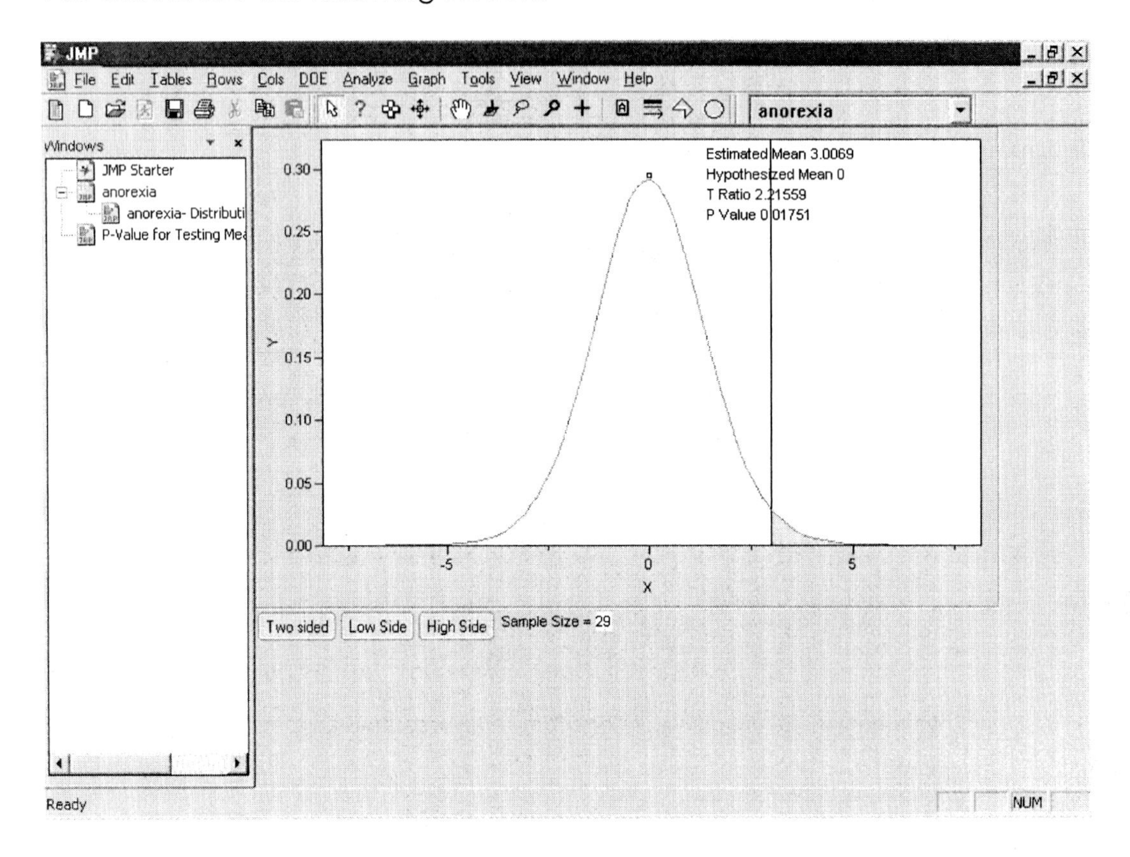

The new p-value of .01751 would indicate that we would still reject the Null Hypothesis that there is no difference (unless the alpha value was set to .01, at which point we would fail to reject the Null).

Chapter 9. Exercise 9.33. Lake pollution.

Open a New Data Table from the JMP Starter page. Title the first column "***Gallons Wastewater***". Enter the data as provided.

Select **Analyze>Distribution**. Assign the single variable to the **Y, Columns** role. Select **OK**. The mean, standard deviation and standard error will be provided.

To test Ha: μ>1000, click on the red triangle next to the variable name "***Gallons of Wastewater***". Select **Test Mean**. In the box **Specify Hypothesized Mean**, enter the value 1000. Select **OK**.

You should see the following screen:

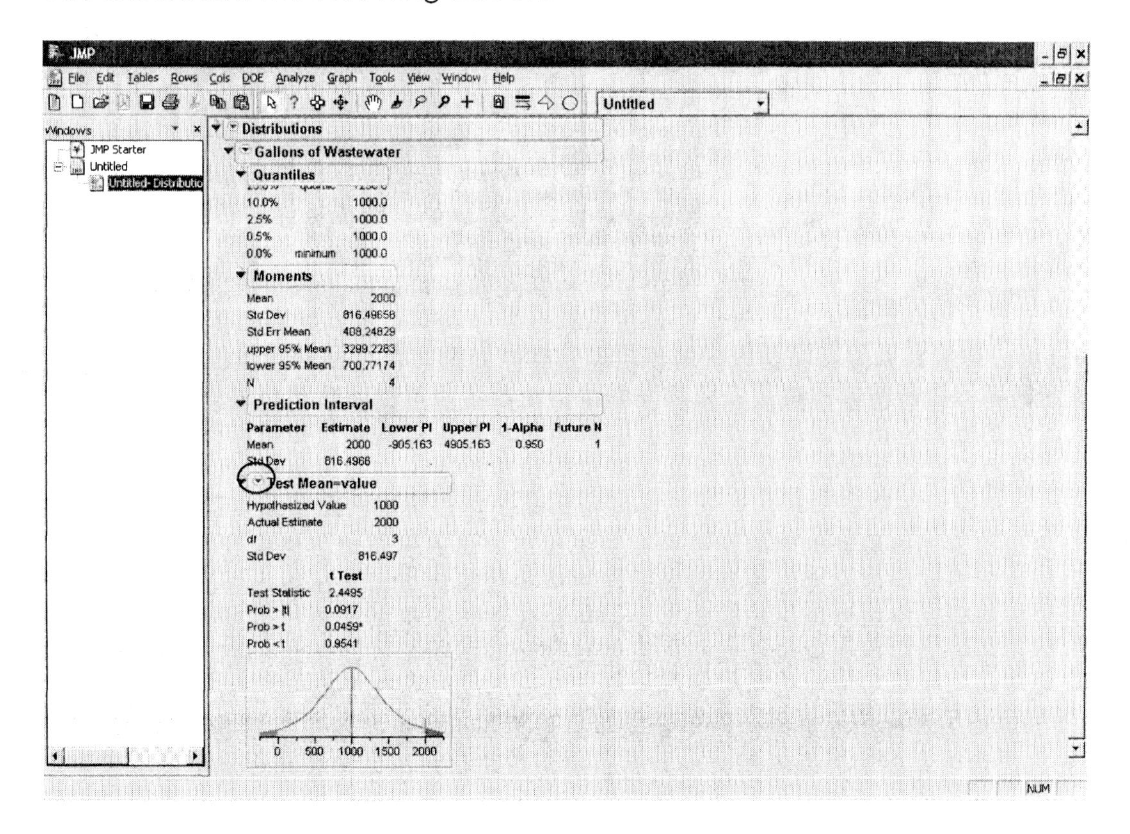

The information provided in the section Test Mean = value, is for a two tailed test. To obtain the values for a one tailed test, click on the red triangle as indicated above. Select **Pvalue Animation> High Side**.

You should see the following screen:

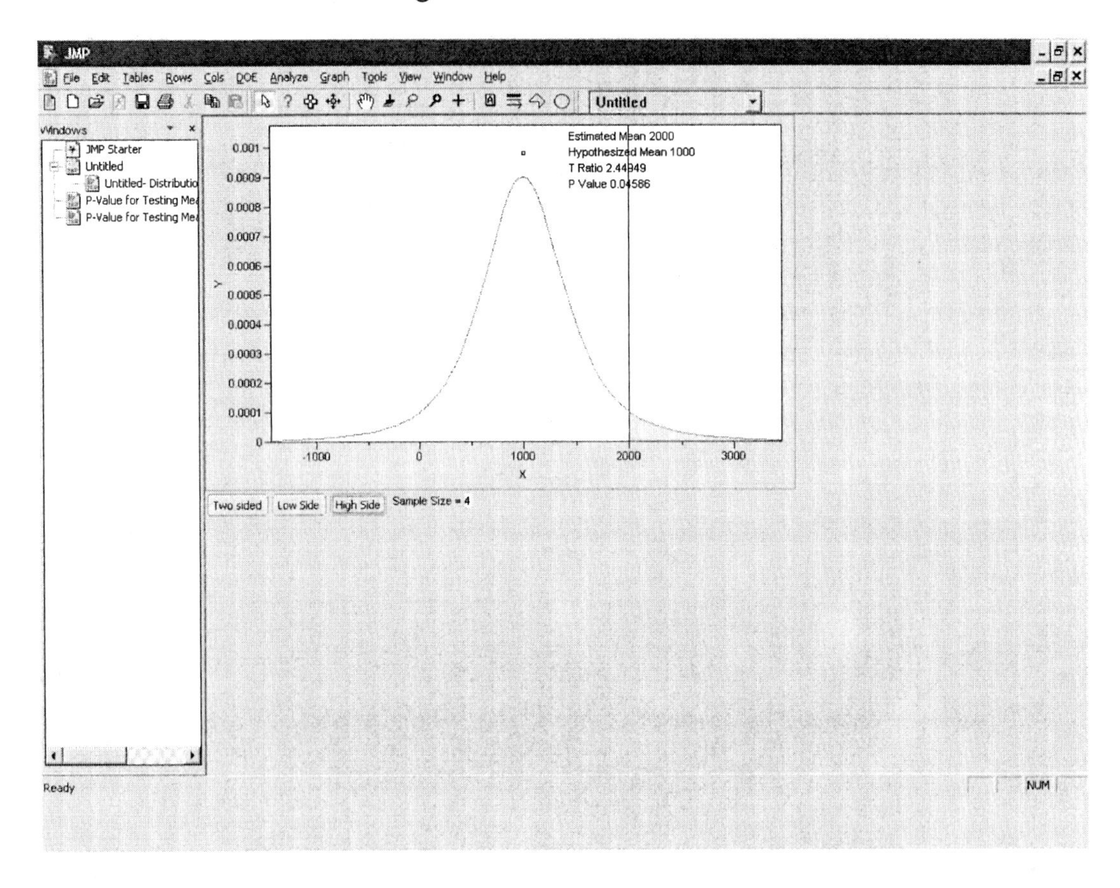

From the p-value of .04586, we would reject the Null Hypothesis (at alpha = .05) and conclude that the population mean is greater than 1000.

Chapter 9. Exercise 9.36. Too little or too much wine?

To create the dataset for analysis, open the JMP Starter page, and select **New Data Table.**

In the blank data set, right click on the top of column 1. Select **Column Info**. For **Column Name**, enter "***ml***" for milliliters. Ensure that the **Data Type** is identified as **Numeric** and the **Modeling Type** is **Continuous**. Select **OK.**

After you enter the data in the table, select **Analyze>Distribution**. Assign the ***ml*** variable to the **Y, Columns** role. Select **OK**.

You should see the following screen:

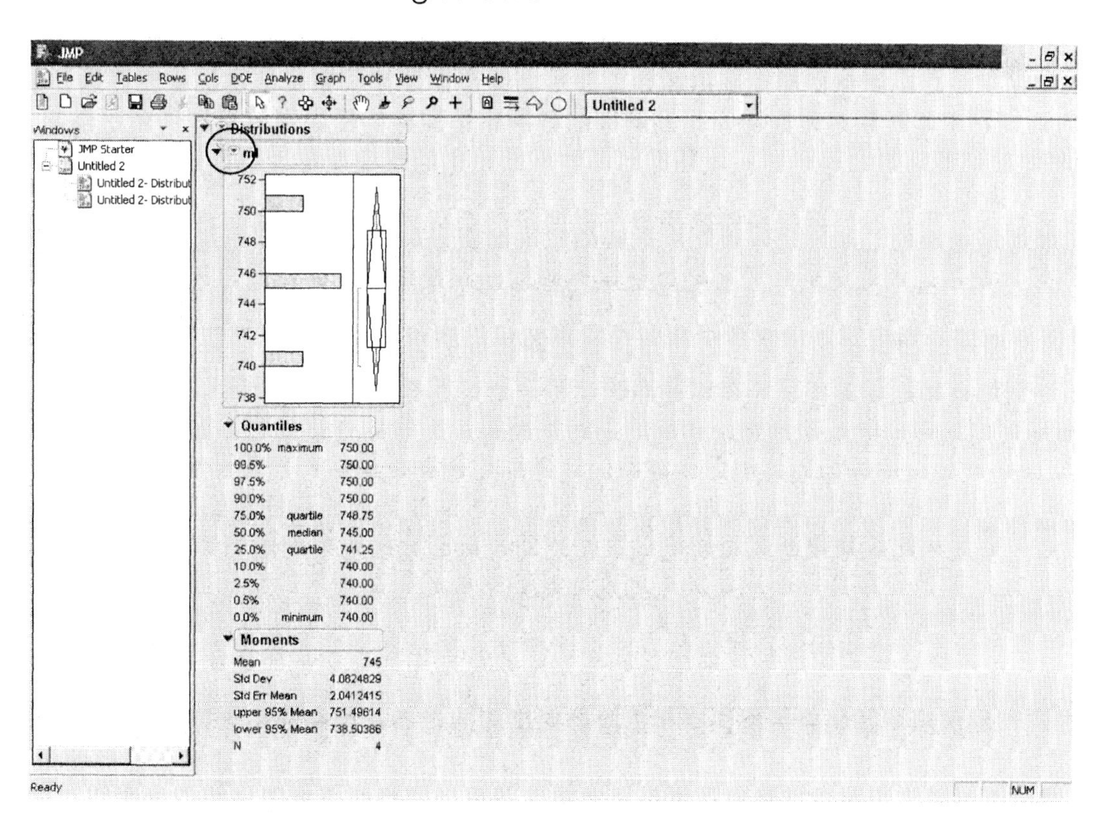

To test the hypothesis that the true mean is different from 752 (a two-tailed test), click on the red triangle as identified above. Select **Test Mean**. For the specified **Hypothesis Mean**, enter *752*. Select **OK**. When the output is generated, it reflects a two-tailed test.

Chapter 9. Exercise 9.40. Sensitivity Study.

Open the anorexia file from the JMP data disk.

To evaluate the effect of the outlier of 20.9 lbs of lost weight, first, generate the two-sided confidence interval with the observation in tact. To do this, select **Analyze>Distribution**. Assign ***cogchange*** to the **Y, Columns** role. Select **OK**. Click on the red triangle next to the variable name "***cogchange***". Select **Confidence Interval**. Select **OK**.

You should see the following screen:

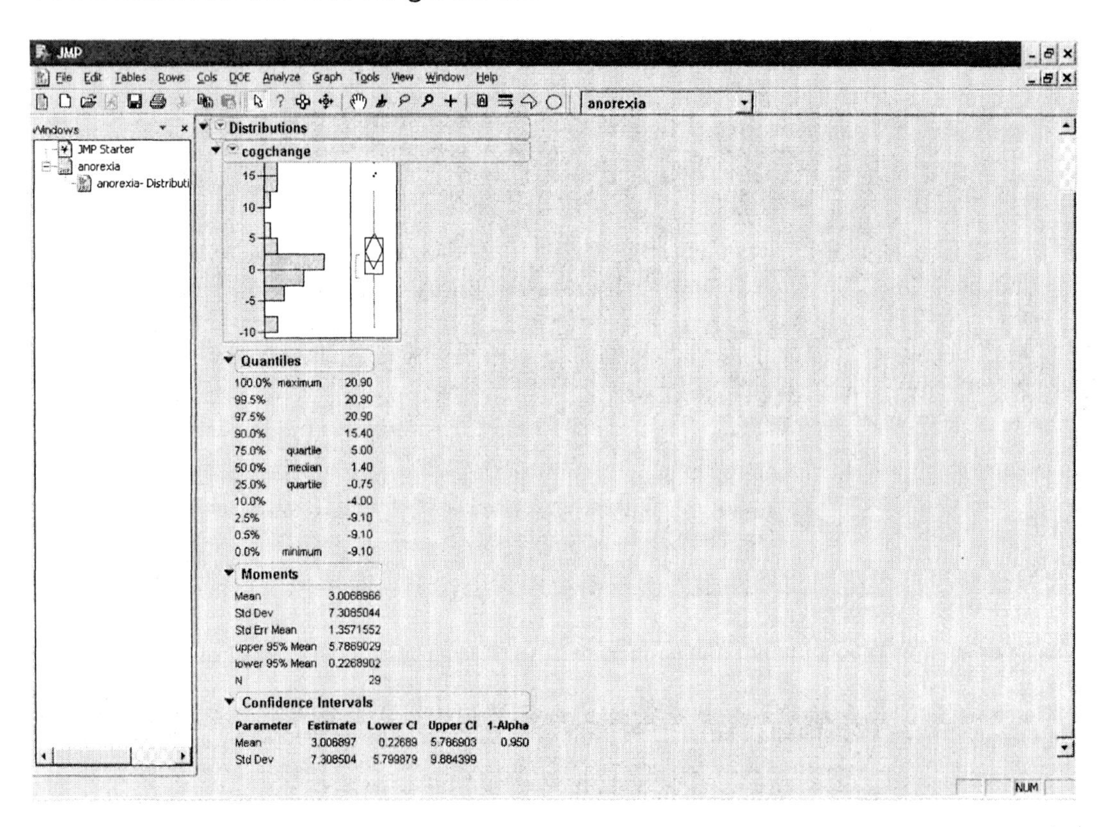

Note the size of the Confidence Interval – it ranges from a low of .22689 to a high of 5.786903.

To evaluate the sensitivity of these results to the outlier of 20.9, return to the datasheet. The outlier is observation number 15. Edit this value to be 2.9 instead of 20.9. (You may get a warning message from JMP. Just select OK). Repeat the same executions as above.

The new output screen should look like this:

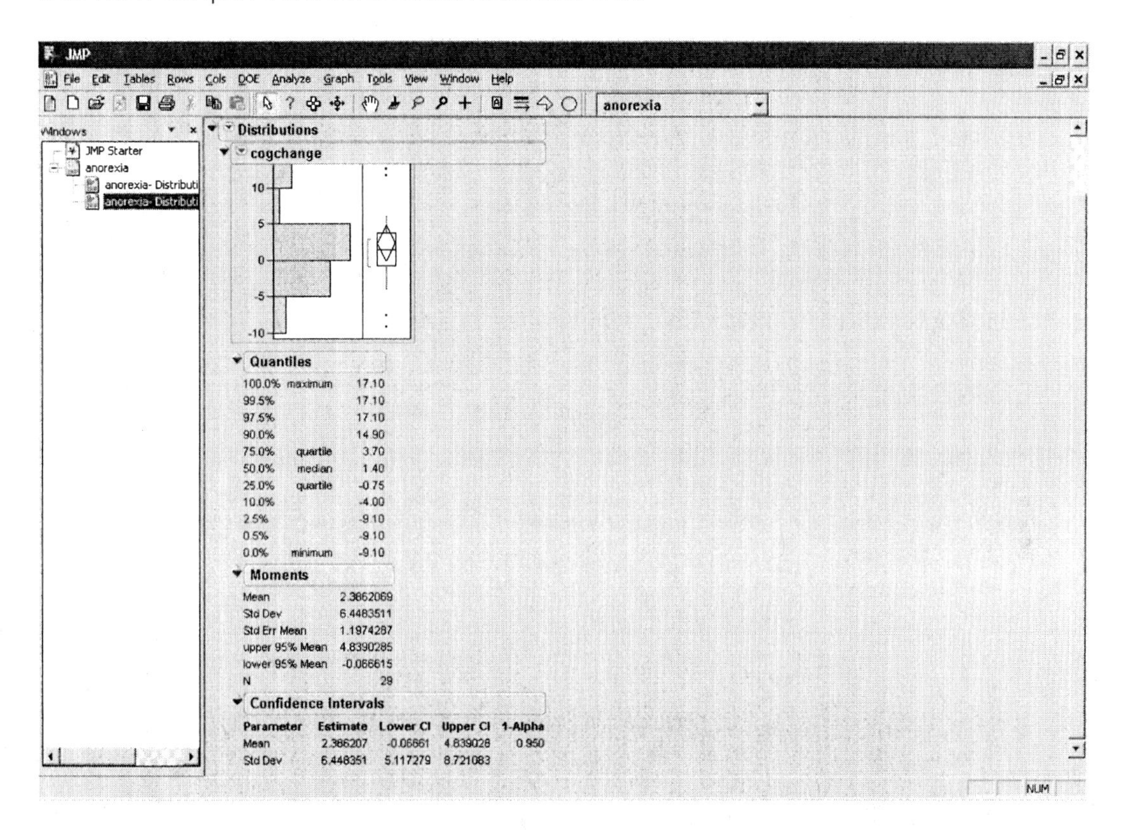

Note how the Interval changed. It is now -.06661 to 4.839028.

Chapter 9. Exercise 9.34. Weight change for controls.

Open the dataset anorexia from the JMP data disk.

To determine if the change in weight for girls in the control group was different from 0, select **Analyze>Distribution**. Assign the ***Conchange*** variable to the **Y, Columns** role. Select **OK**. Click on the red triangle next to the variable name "***Conchange***". Select **Test Mean**. In this instance, the test is to determine if the mean is statistically different from zero, which is the default value. Select **OK**.

You should see the following screen:

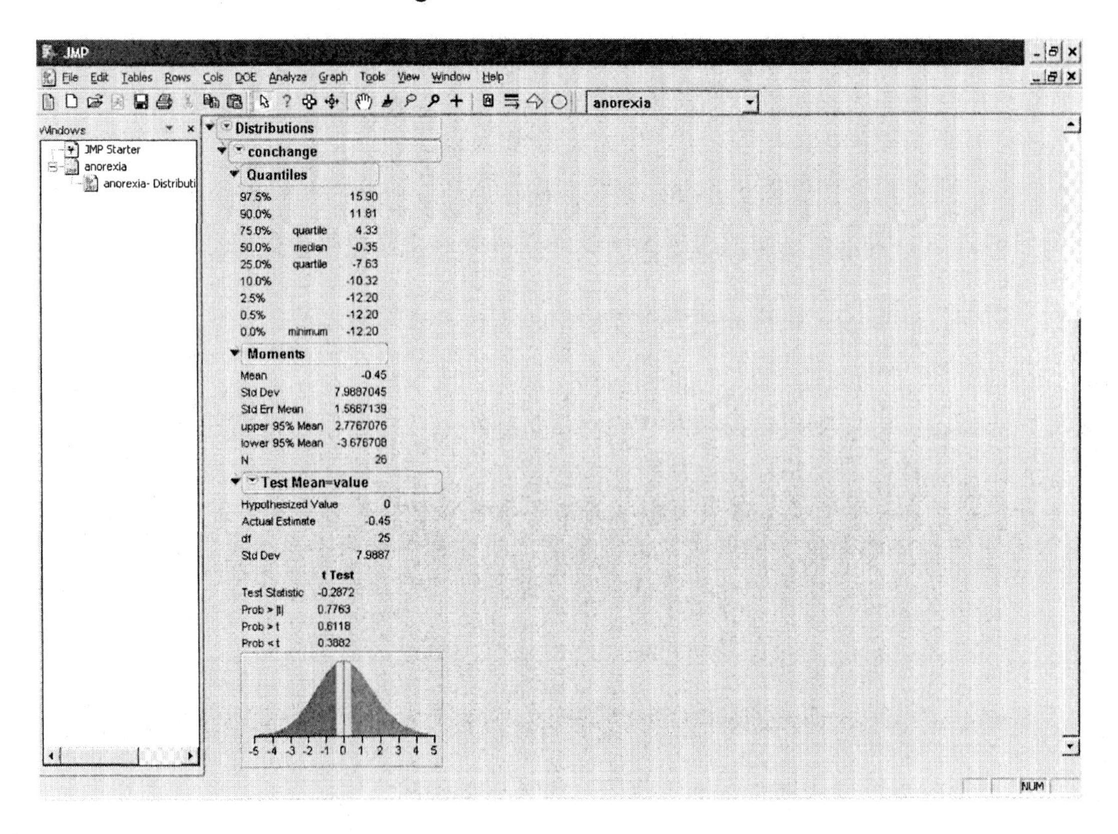

Chapter 9. Exercise 9.39. Anorexia in teenage girls.

Open the dataset anorexia from the JMP data disk.

To determine if the change in weight for girls in family therapy was different from 0, either enter the data as provided in a new column, or generate the same data by selecting **Cols>New Column**. Enter the name of the new column (FamChange). Then **Column Properties>Formula**. For the formula, select **Fam2-Fam1**. Select **OK**.

You should see the following screen:

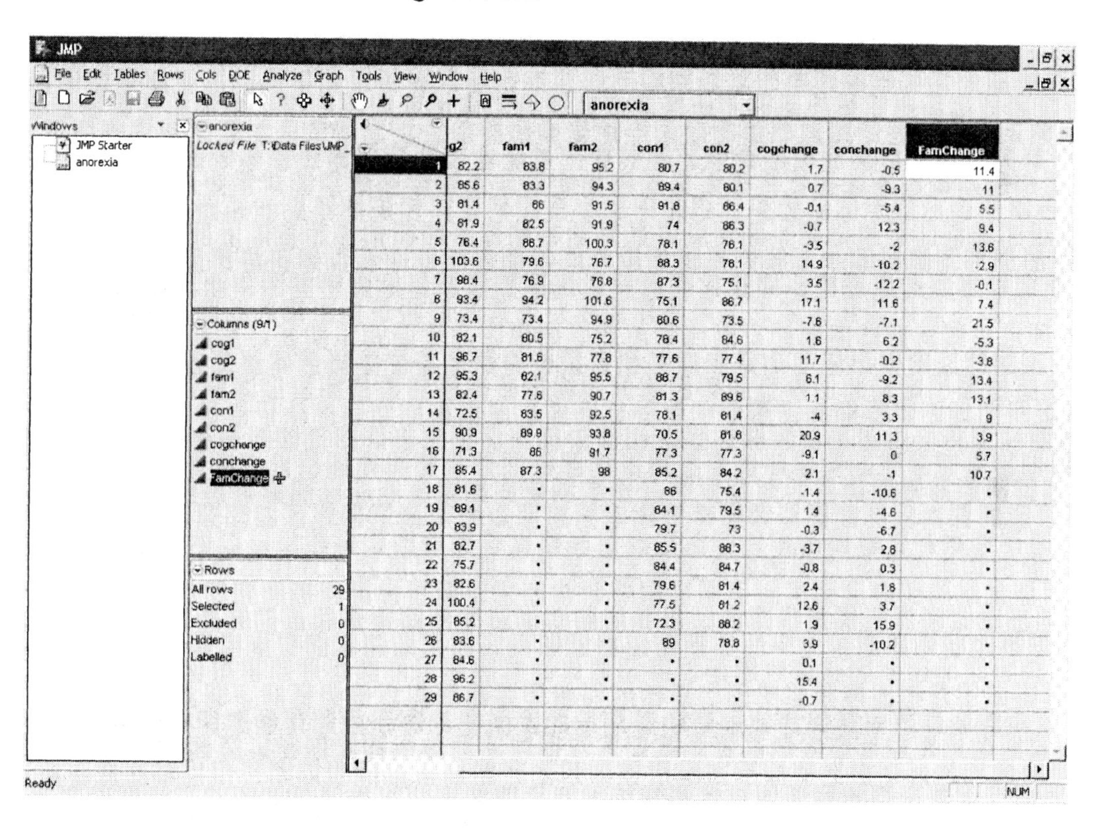

	g2	fam1	fam2	con1	con2	cogchange	conchange	FamChange
1	82.2	83.8	95.2	80.7	80.2	1.7	-0.5	11.4
2	85.6	83.3	94.3	89.4	80.1	0.7	-9.3	11
3	81.4	86	91.5	91.8	86.4	-0.1	-5.4	5.5
4	81.9	82.5	91.9	74	86.3	-0.7	12.3	9.4
5	76.4	86.7	100.3	78.1	76.1	-3.5	-2	13.6
6	103.6	79.6	76.7	88.3	76.1	14.9	-10.2	-2.9
7	98.4	76.9	76.8	87.3	75.1	3.5	-12.2	-0.1
8	93.4	94.2	101.6	75.1	86.7	17.1	11.6	7.4
9	73.4	73.4	94.9	80.6	73.5	-7.6	-7.1	21.5
10	82.1	80.5	75.2	78.4	84.6	1.6	6.2	-5.3
11	96.7	81.6	77.8	77.6	77.4	11.7	-0.2	-3.8
12	95.3	82.1	95.5	88.7	79.5	6.1	-9.2	13.4
13	82.4	77.6	90.7	81.3	89.6	1.1	8.3	13.1
14	72.5	83.5	92.5	78.1	81.4	-4	3.3	9
15	90.9	89.9	93.8	70.5	81.8	20.9	11.3	3.9
16	71.3	86	91.7	77.3	77.3	-9.1	0	5.7
17	85.4	87.3	98	85.2	84.2	2.1	-1	10.7
18	81.6	•	•	86	75.4	-1.4	-10.6	•
19	89.1	•	•	84.1	79.5	1.4	-4.6	•
20	83.9	•	•	79.7	73	-0.3	-6.7	•
21	82.7	•	•	85.5	88.3	-3.7	2.8	•
22	75.7	•	•	84.4	84.7	-0.8	0.3	•
23	82.6	•	•	79.6	81.4	2.4	1.8	•
24	100.4	•	•	77.5	81.2	12.6	3.7	•
25	85.2	•	•	72.3	88.2	1.9	15.9	•
26	83.6	•	•	89	78.8	3.9	-10.2	•
27	84.6	•	•	•	•	0.1	•	•
28	96.2	•	•	•	•	15.4	•	•
29	86.7	•	•	•	•	-0.7	•	•

Select **Analyze>Distribution**. Assign the ***Famchange*** variable to the **Y, Columns** role. Select **OK**. Click on the red triangle next to the variable name "**Famchange**". Select **Test Mean**. In this instance, the test is to determine if the mean is statistically different from zero, which is the default value. Select **OK**.

You should see the following screen:

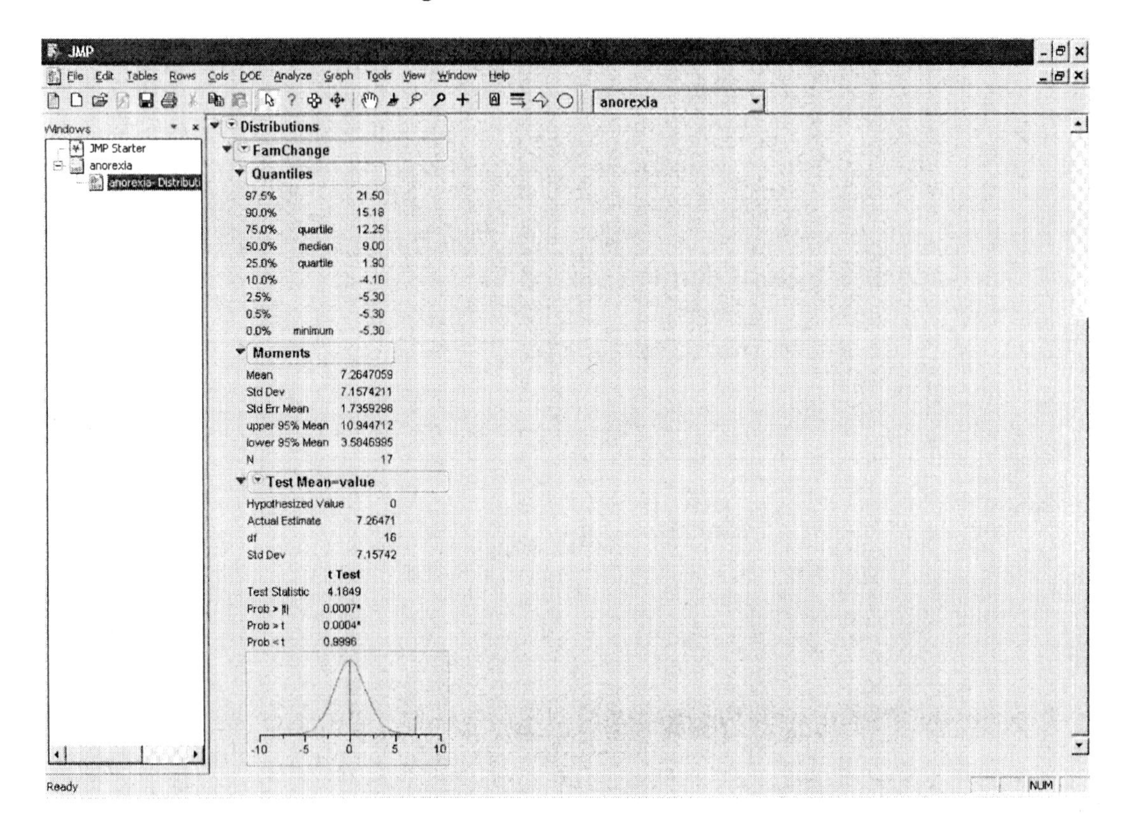

From the p-value above of .0004, we would reject the Null Hypothesis, and conclude that the mean difference is significantly different from 0.

Chapter 9. Exercise 9.85. Tennis balls in control?

Open a new data table from the JMP Starter menu.

Enter the data as provided for the eight tennis balls.

Select **Analyze>Distribution**. Assign the single variable (***Ball Weight***) to the **Y, Columns** role. Select **OK**. Click on the red triangle next to the variable name "**Ball Weight**". Select **Test Mean.** In this instance, the test is to determine if the mean is statistically different from 57.6. Select **OK.**

You should see the following screen:

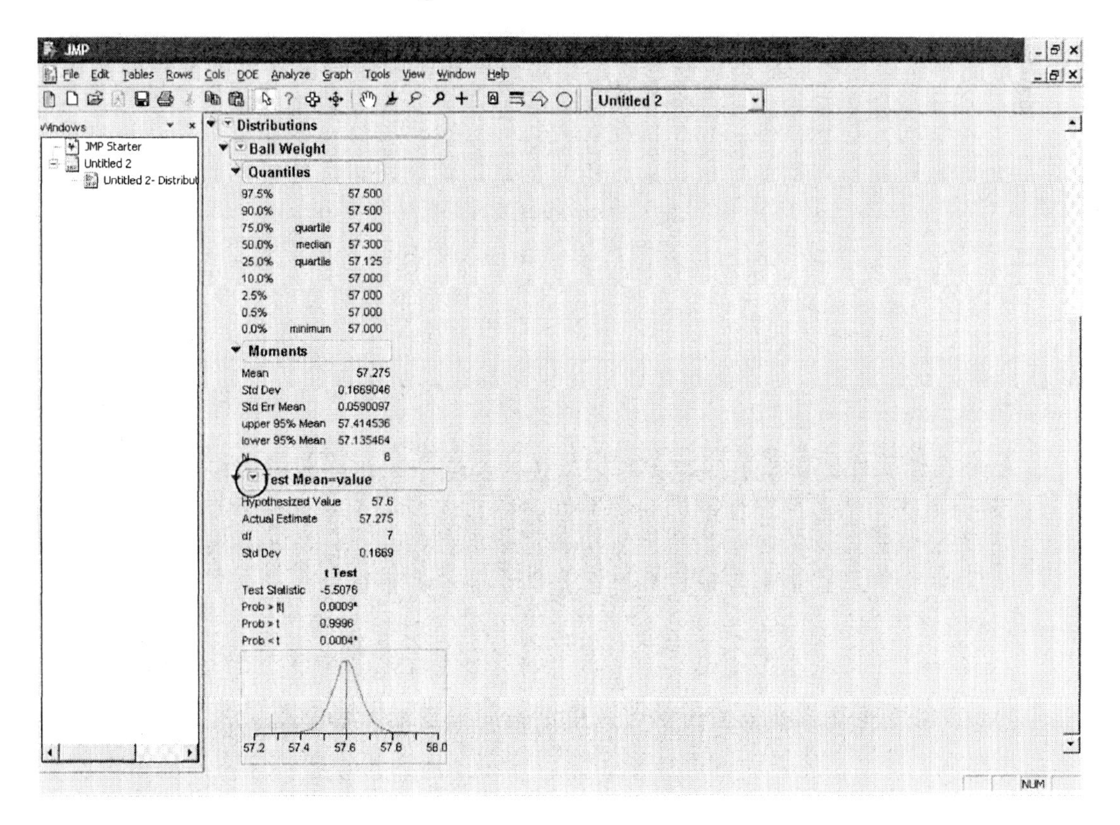

To determine the power of this test, click on the red triangle next to "Test Mean = value" as identified above. Select **Power Animation.**

You should see the following screen:

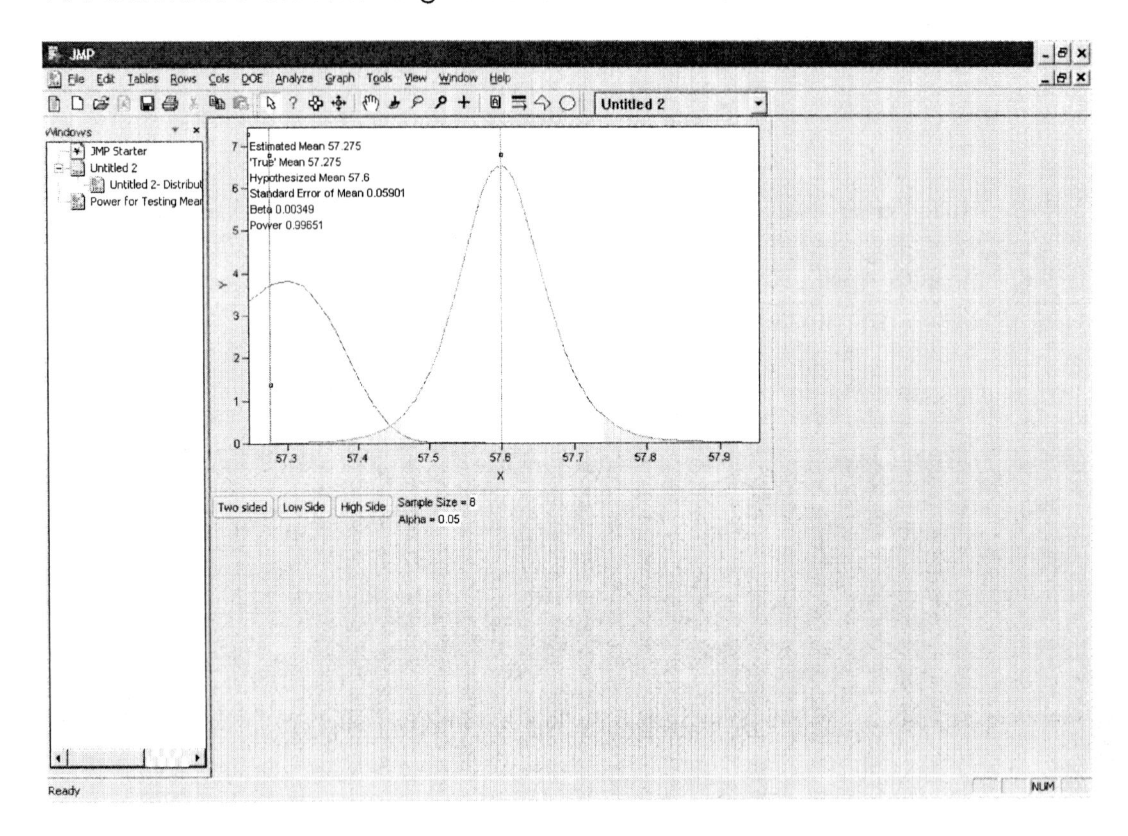

The Power of this test is .99651 – indicating that if a difference existed, there is a 99.651% chance that we would have detected it.

Chapter 10. Exercise 10.20. eBay Auctions.

Open the ebay_auctions JMP file from the data disk.

To construct the 95% confidence intervals for the two options, select **Analyze>Distribution**. Assign both variables to the **Y, Columns** role. Select **OK**.

Click on the red triangle next to each variable name and select **Confidence Interval**. Ensure that the interval is set to 95% and two tails. Select **OK**.

You should see the following screen:

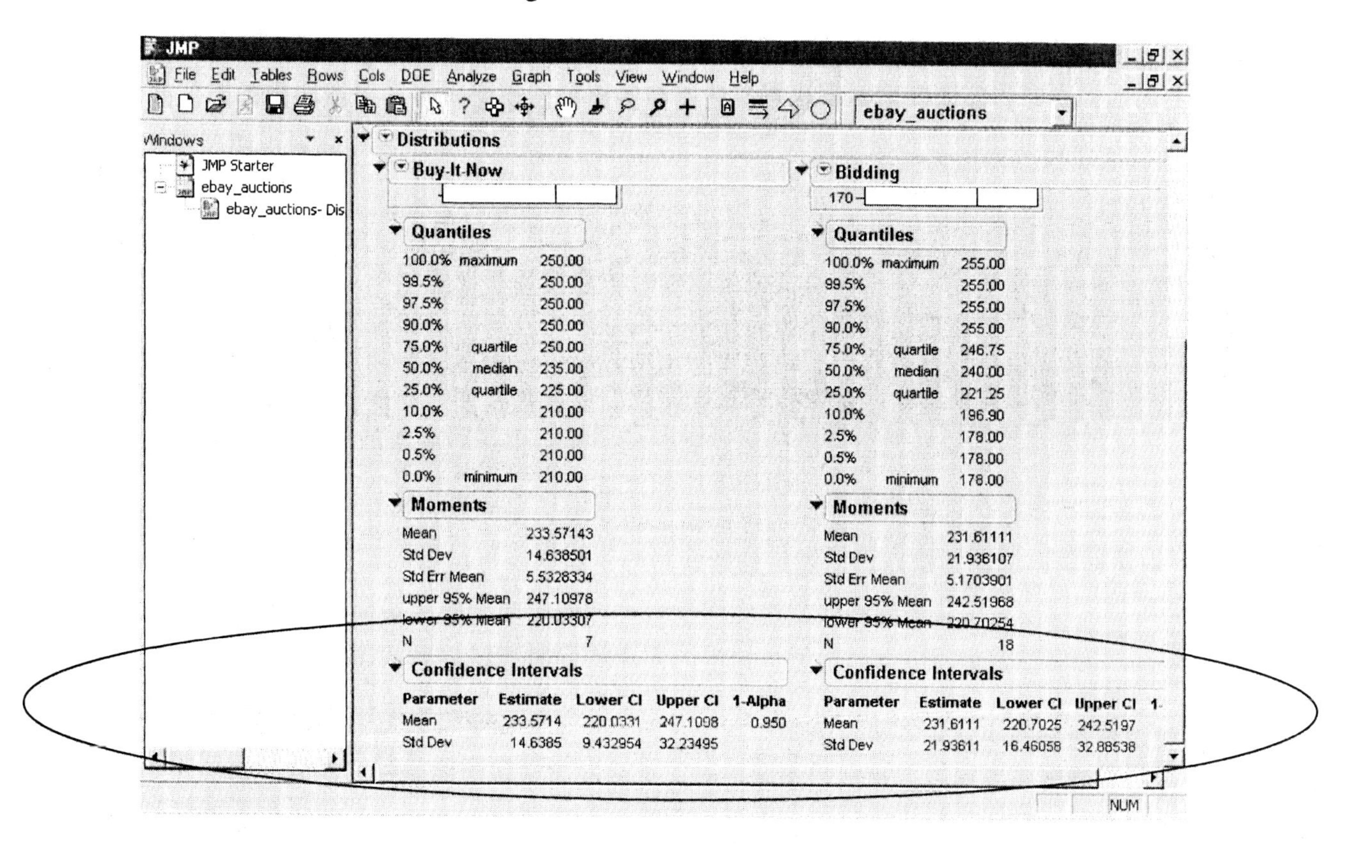

The confidence intervals for the two options is circled above.

Chapter 10. Exercise 10.21. Test comparing auction prices.

Continuing from Exercise 10.20 above, begin by "stacking" the data into a single column. To do this, select **Tables>Stack**. Assign both the ***Buy-it-Now*** and the ***Bidding*** columns to the **Stack Columns** role. Select **OK**.

You should see the following screen:

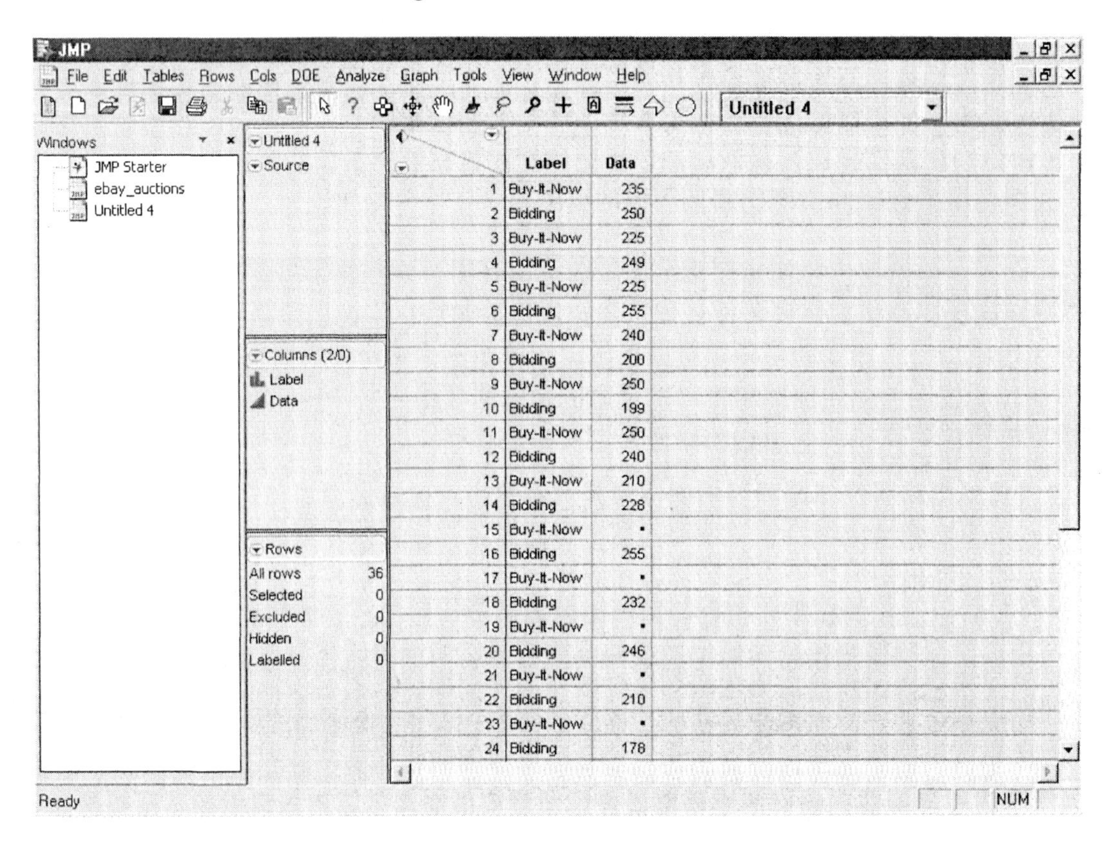

Change the title of the column "***Label***" to "***Option***" and change the title of the column "***Data***" to "***Price***".

To generate the plots (the histogram and box plot), select **Analyze>Distribution**. Assign ***Price*** to the **Y, Columns** role and ***Option*** to the **By** role. Select **OK**. The plots and descriptive statistics will be displayed. Recall that it might be helpful to change the display layout from vertical to horizontal to see the plots together.

To test Ha: $\mu_1 \neq \mu_2$, return to the JMP Starter menu. Under the "**Click Category**" list, select **Basic>Two Sample T test**. Assign ***Option*** to the **X, Grouping** role and ***Price*** to the **Y, Response** role. Select **OK**.

You should see the following screen:

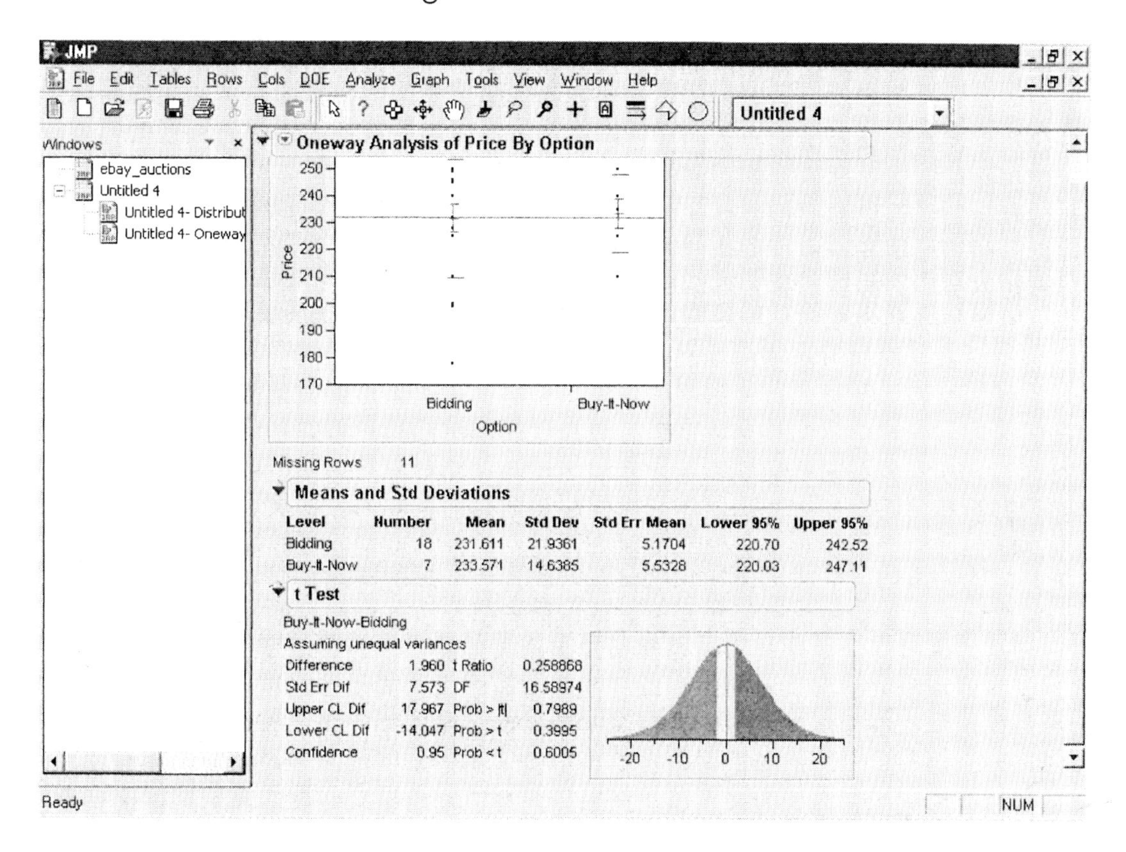

The p-value of .6005 above, would lead us to fail to reject the Null Hypothesis, and conclude that there is no difference in price between the two options.

Chapter 10. Exercise 10.28. Student Survey.

Open the fl_student_survey datafile from the JMP folder of the data disk.

To analyze newspaper reading habits by gender, we will first need to redefine the newspaper variable in the dataset from character to numeric. To do this, double click on the column heading ***newspapers***. Identify the "Data Type" as **Numeric** and the "Modeling Type" as **Continuous**. Select **OK**. When you return to the dataset, the variable newspapers should now be identified with a blue triangle – indicating numeric data – rather than with red bars.

To construct plots for the males and females for newspaper reading, select **Analyze>Distribution**. Assign ***newspapers*** to the **Y, Columns** role and ***gender*** to the **By** role. Select **OK**.

Remember that by selecting the horizontal display option, both graphics can be more easily seen on a single screen.

To generate the 95% confidence intervals, simply click on the red triangle next to ***newspapers*** and select **Confidence Interval**. The defaults should indicate a 95% two-sided interval. Select **OK**.

You should see the following screen:

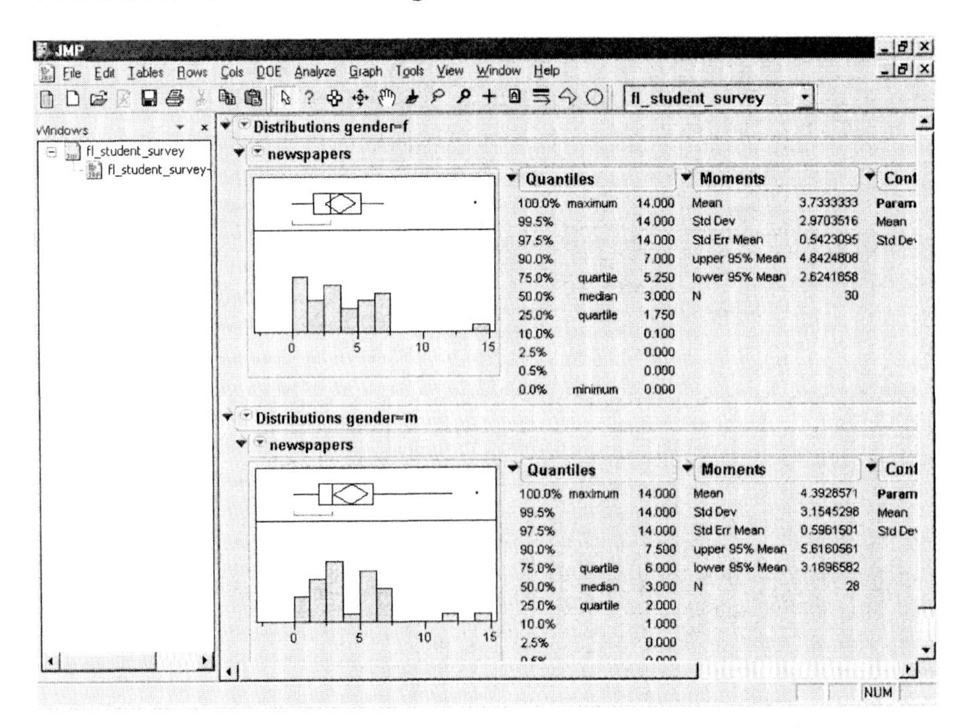

Chapter 10. Exercise 10.29. Study Time.

From the JMP Starter Screen, select New Data Table.

Enter the data into the new table as follows:

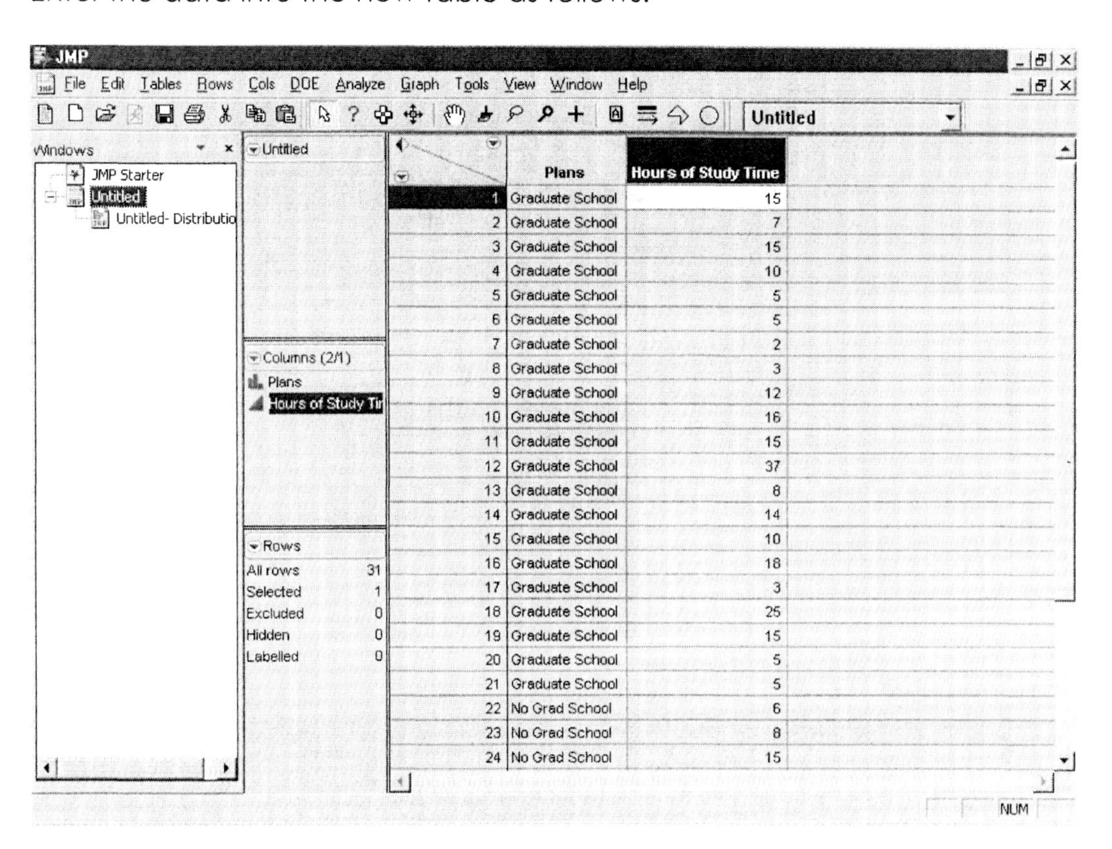

Note: you can type "graduate school" one time and then select **Edit>Copy>Edit>Paste**, just as you would in EXCEL.

To find the mean and sample standard deviation for each group, select **Analyze>Distribution**. Assign ***Hours of Study Time*** to the **Y, Columns** role and ***Plans*** to the **By** role. Select **OK**.

When the output screen appears, click on the red triangle next to "Hours of Study Time" and select **Display Options>Horizontal Display**. Do this for both sections.

You should see the following screen:

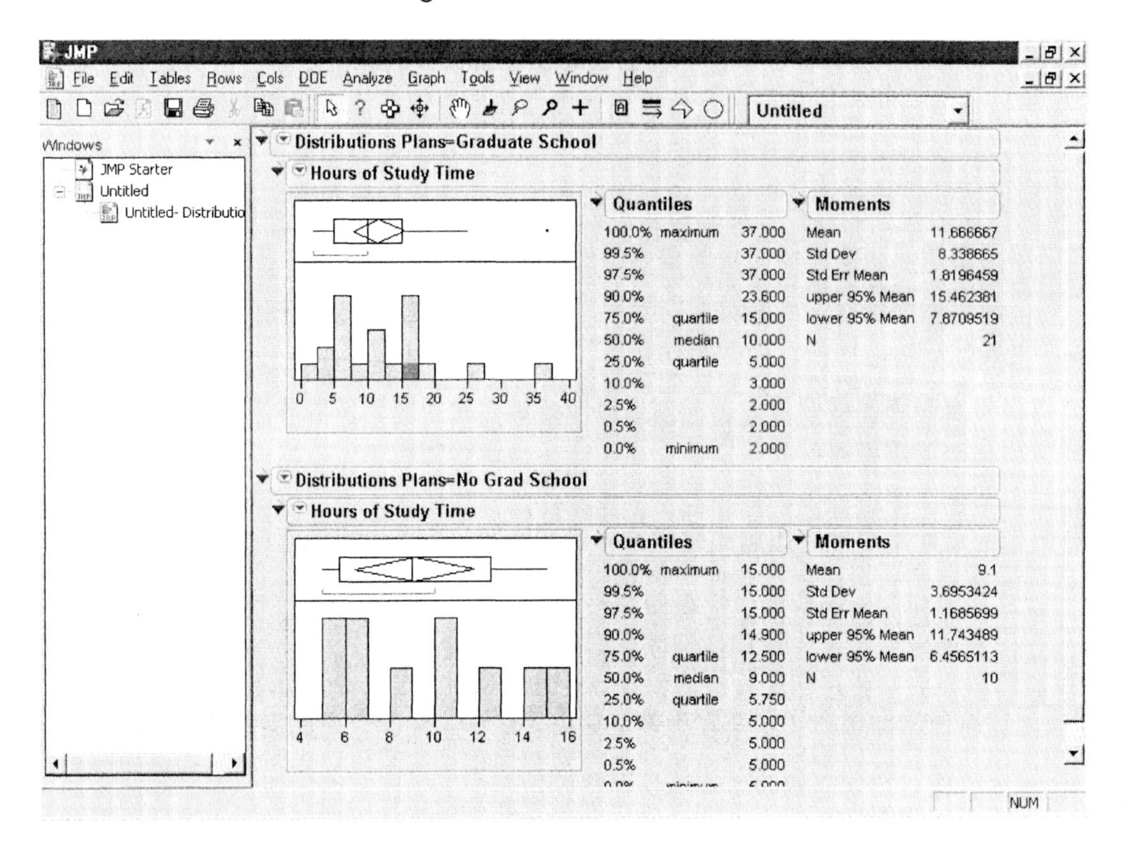

The mean and standard deviation for those planning to go to graduate school are 11.66 and 8.33, respectively. The mean and standard deviation for those not planning to go to graduate school are 9.1 and 3.69, respectively. This indicates that those with graduate school plans study more frequently than those with no graduate school plans.

The standard errors are also reported in the Moments sections. The standard error for those planning to go to graduate school is higher than the standard error for those with no graduate school plans – 1.82 versus 1.68.

To generate the confidence intervals for both samples, click on the red triangle as before, and select **Confidence Interval>OK**. Do this for both sections.

You should see the following screen:

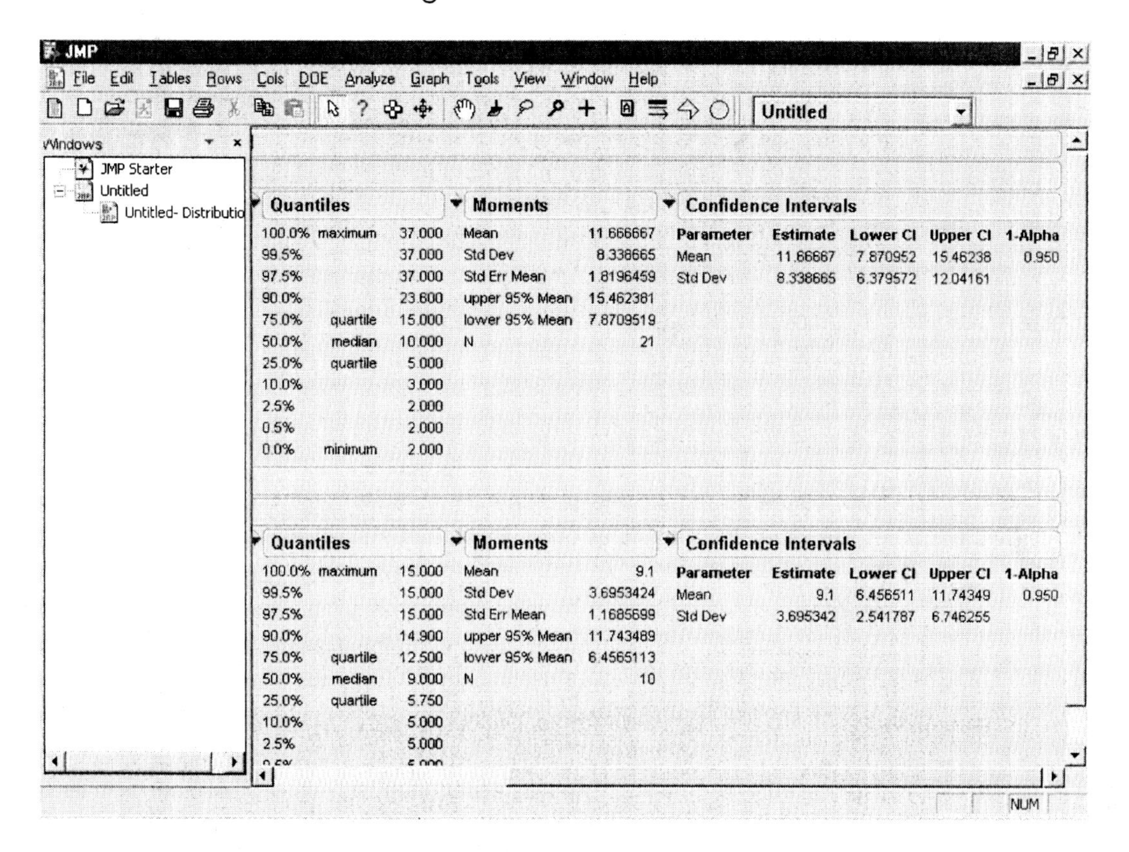

Quantiles		
100.0%	maximum	37.000
99.5%		37.000
97.5%		37.000
90.0%		23.600
75.0%	quartile	15.000
50.0%	median	10.000
25.0%	quartile	5.000
10.0%		3.000
2.5%		2.000
0.5%		2.000
0.0%	minimum	2.000

Moments	
Mean	11.666667
Std Dev	8.338665
Std Err Mean	1.8196459
upper 95% Mean	15.462381
lower 95% Mean	7.8709519
N	21

Parameter	Estimate	Lower CI	Upper CI	1-Alpha
Mean	11.66667	7.870952	15.46238	0.950
Std Dev	8.338665	6.379572	12.04161	

Quantiles		
100.0%	maximum	15.000
99.5%		15.000
97.5%		15.000
90.0%		14.900
75.0%	quartile	12.500
50.0%	median	9.000
25.0%	quartile	5.750
10.0%		5.000
2.5%		5.000

Moments	
Mean	9.1
Std Dev	3.6953424
Std Err Mean	1.1685699
upper 95% Mean	11.743489
lower 95% Mean	6.4565113
N	10

Parameter	Estimate	Lower CI	Upper CI	1-Alpha
Mean	9.1	6.456511	11.74349	0.950
Std Dev	3.695342	2.541787	6.746255	

The 95% interval for those with plans for graduate school is much larger than the interval for those without plans for graduate school – (7.87-15.46) versus (6.46 – 11.74).

Chapter 10. Exercise 10.39. Vegetarians more Liberal?

Open the fl_student_survey datafile from the JMP folder of the data disk.

To generate the ttest assuming unequal variances, from the **JMP Starter Screen**, select **Basic** from the **Category List** and then **Two Sample t Test.** Assign ***political_ideology*** to the **Y, Response** role and ***vegetarian*** to the **X, Grouping** role. Select **OK.**

You should see the following screen:

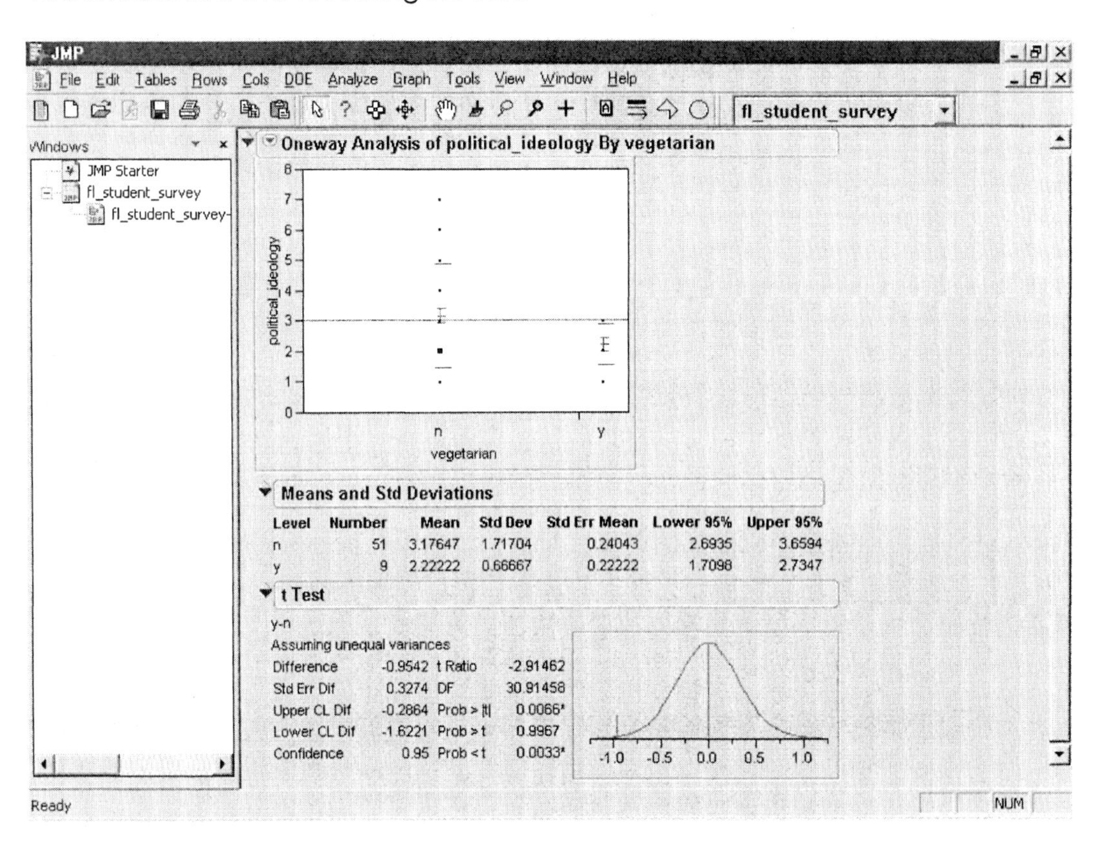

The information regarding the t Ratio is –2.91 (this value should be interpreted as an absolute value), with a corresponding p-value is .0066, indicating a highly significant difference between the two groups.

To generate the same information assuming equal variances, click on the red triangle next to "Oneway"... and select **Means/ANOVA/Pooled t**".

You should see the following screen:

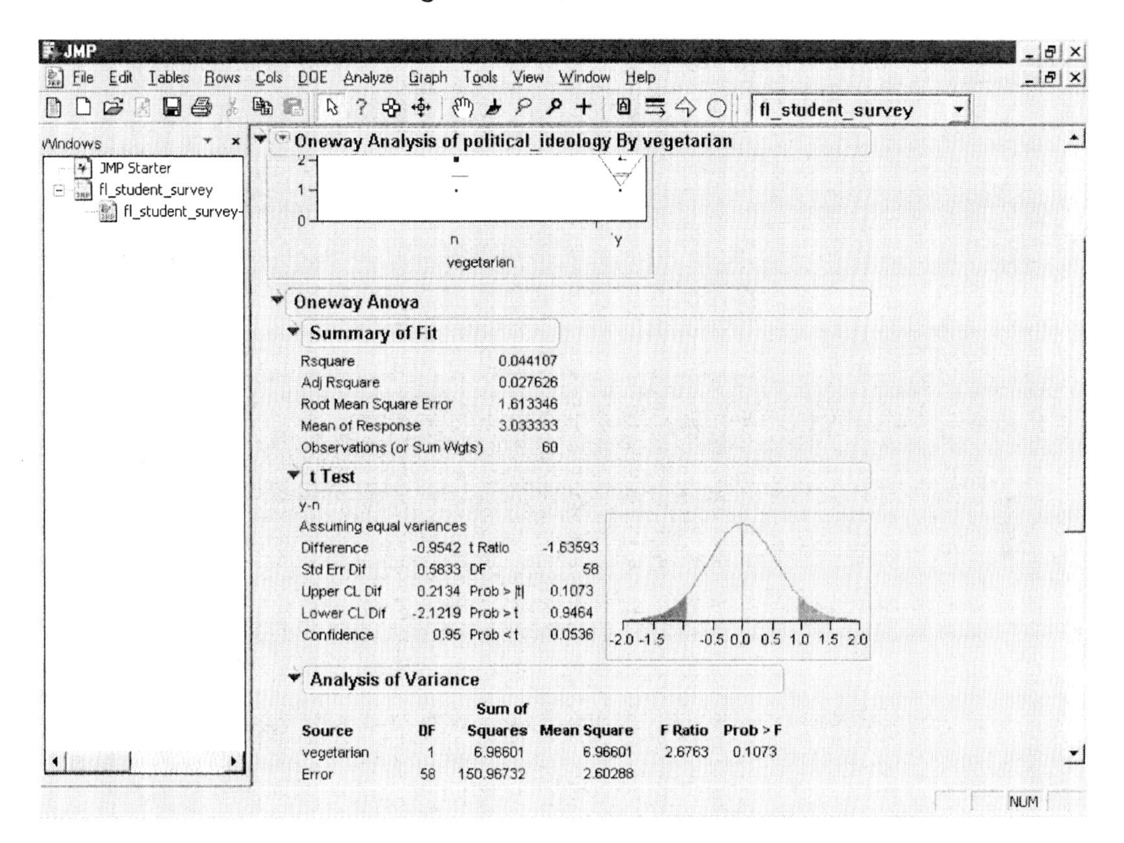

The t Ratio value is less compelling at –1.64, with a corresponding p-value of .1073. At this level of significance, we would fail to reject the Null Hypothesis and conclude that there is no difference in the populations.

Because the sample sizes are so different (although probably representative of the proportions of vegetarian/not vegetarian), and the standard deviations are so different, it would be inappropriate to assume equal variance. The first set of results is more likely.

Chapter 10. Exercise 10.47. Does Exercise help blood pressure?

From the JMP Starter Screen, select New Data Table.

Enter the data into the new table as follows:

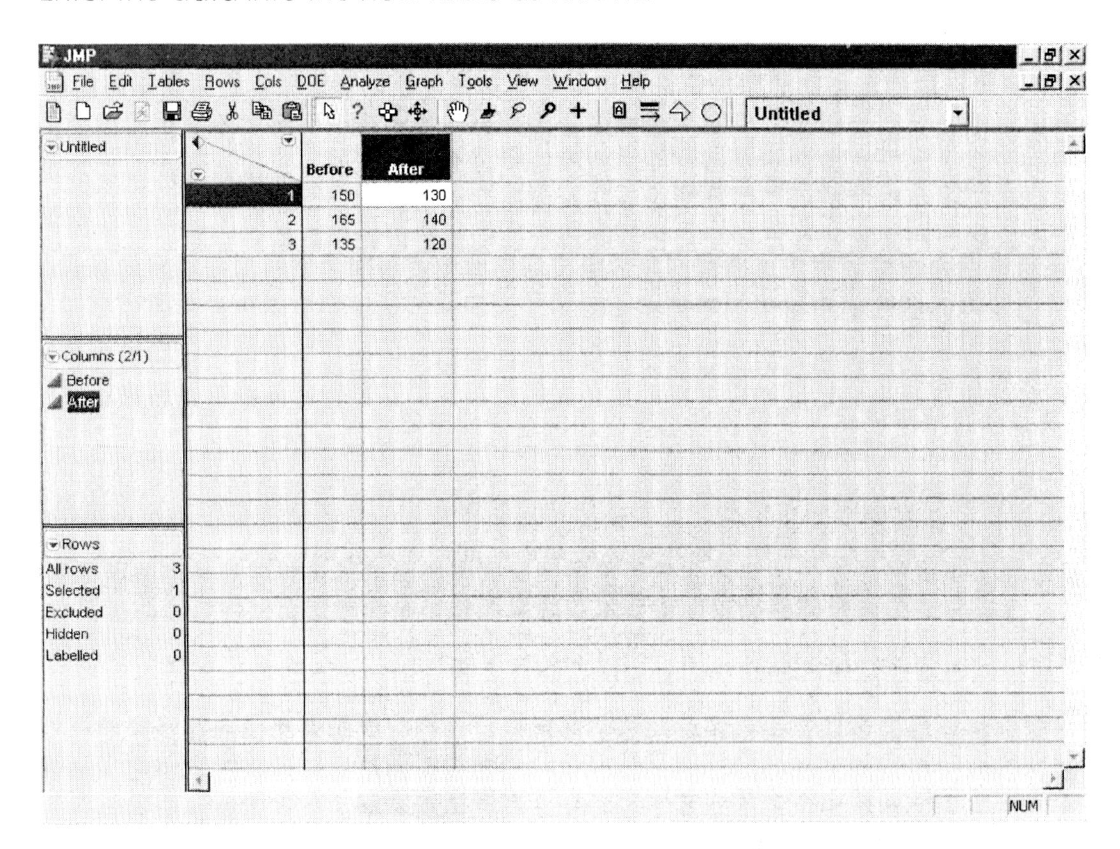

Return to the JMP Starter Screen, select **Basic** from the **Category List** and then **Matched Pairs**. Assign both variables to the **Y, Paired Response** role. Select **OK**.

You should see the following screen:

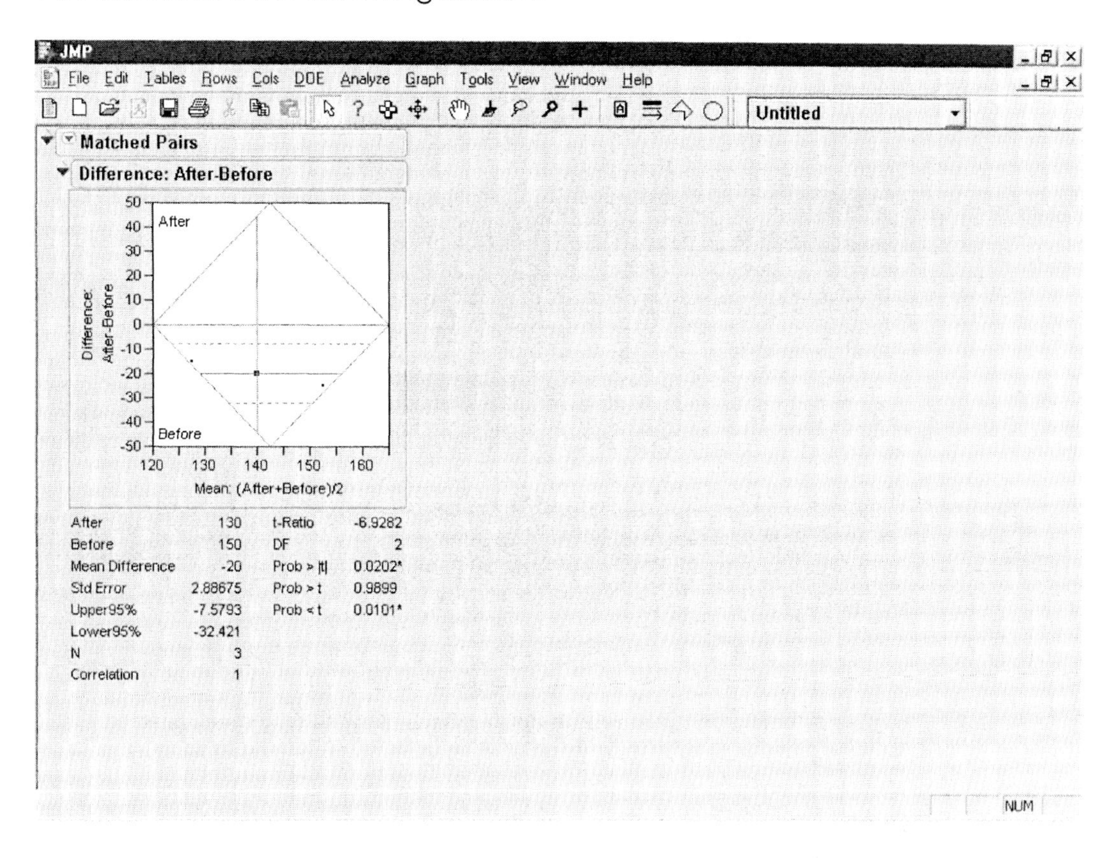

The Before and After values represent dependent samples because they come from the same population of people.

From this screen, the mean value of the After BP is 130, while the mean value of the Before is 150, and the calculated mean difference is –20 (you can treat this as an absolute value since this is a two-tailed test). The t-Ratio statistic is –6.9282, with a corresponding p-value of .0202. These results would indicate that there is evidence to reject the Null Hypothesis and conclude that a difference does exist.

The 95% confidence interval is included in the output above, and indicates an interval of the difference as –7.57 to –32.42. The difference could be as great as 32.42 BP points, or as low as 7.57 BP points. It is important to note that the interval does not include 0.

Chapter 10. Exercise 10.49. Social activities for students.

From the JMP Starter Screen, select New Data Table.

Enter the data into the new table as follows:

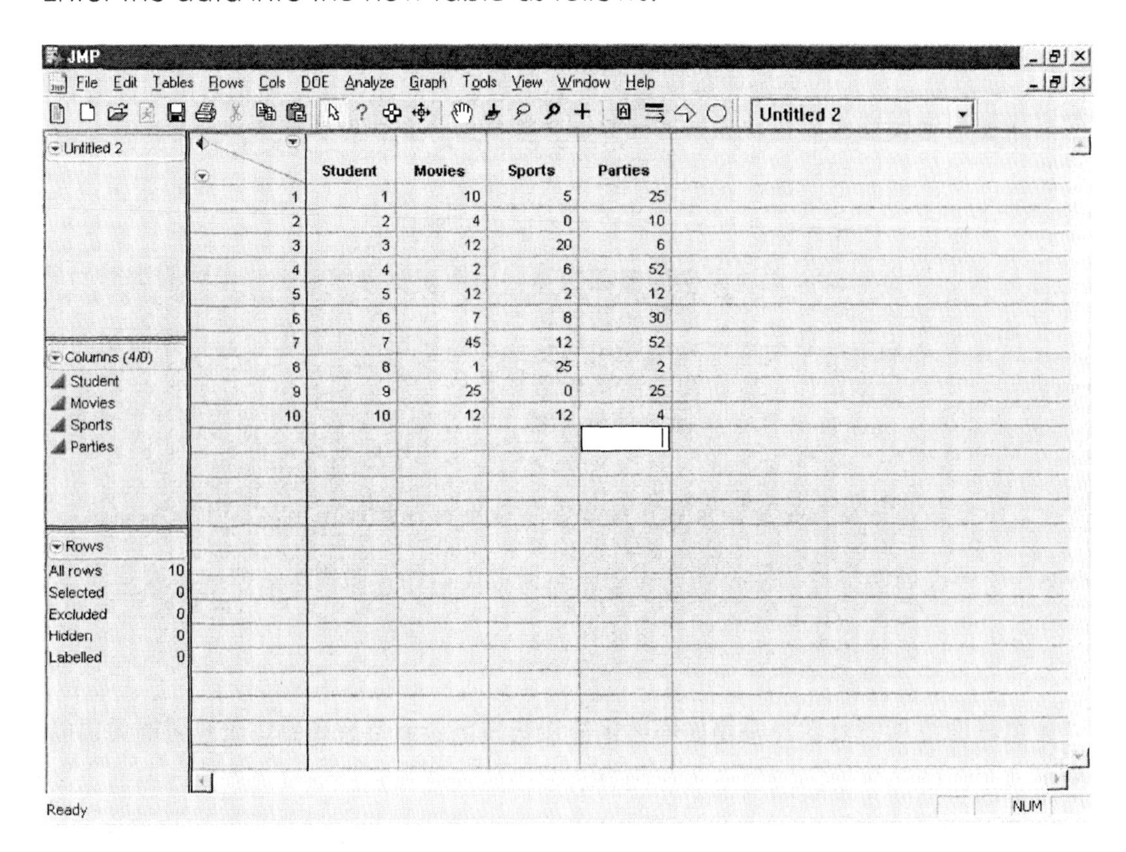

	Student	Movies	Sports	Parties
1	1	10	5	25
2	2	4	0	10
3	3	12	20	6
4	4	2	6	52
5	5	12	2	12
6	6	7	8	30
7	7	45	12	52
8	8	1	25	2
9	9	25	0	25
10	10	12	12	4

To compare the means of movie attendance and sports attendance, treat the samples as dependent.

To generate the results of a paired ttest, return to the JMP Starter Screen, select **Basic** from the **Category List** and then **Matched Pairs**. Assign both the ***movies*** and ***sports*** variables to the **Y, Paired Response** role. Select **OK**.

You should see the following screen:

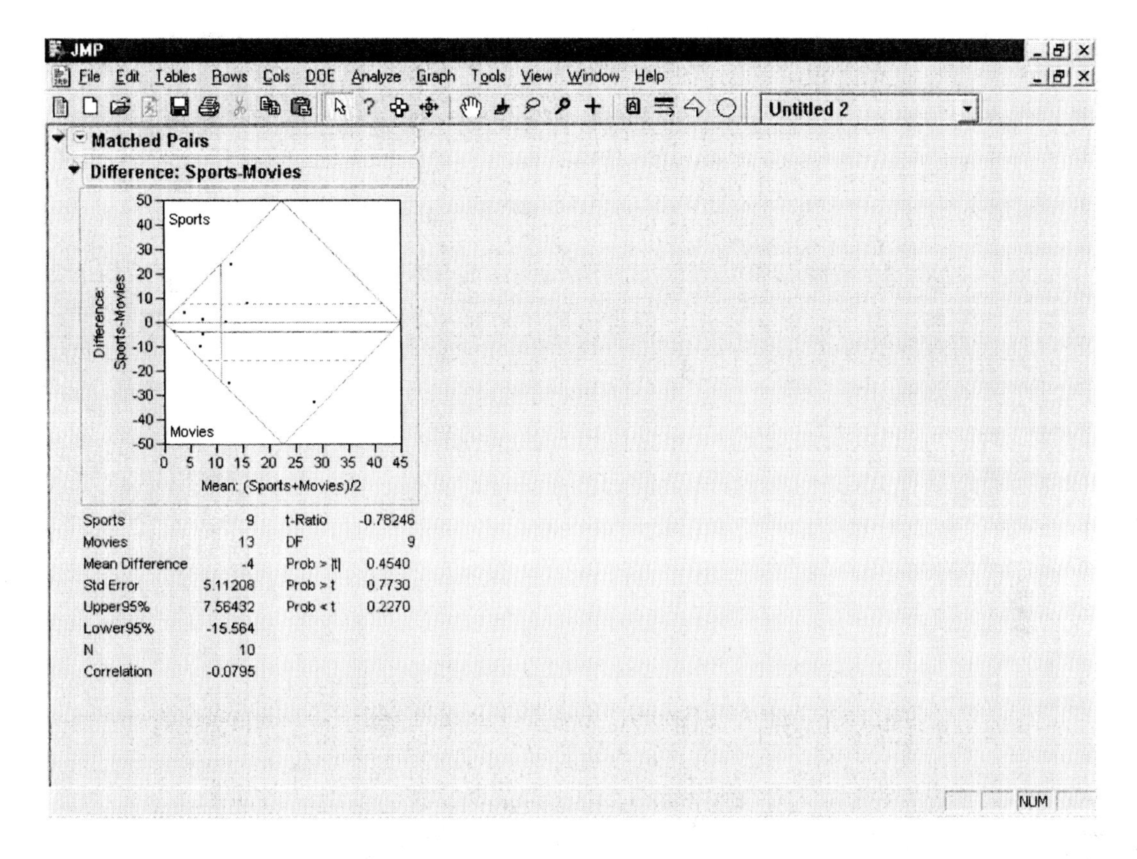

The information above includes the 95% confidence interval for the mean difference. This interval ranges from a high of 7.58 to a low of –15.56. Note that the interval includes 0.

The t-Ratio is -.78246, with a corresponding p-value of .4540. This p-value would lead us to fail to reject the Null Hypothesis, and conclude that there is no difference in the number of times students go to the movies versus attend sports activities.

Chapter 11. Exercise 11.1. Gender gap in politics?

From the JMP Starter Screen, select **New Data Table**.

Enter the data into the new table as follows:

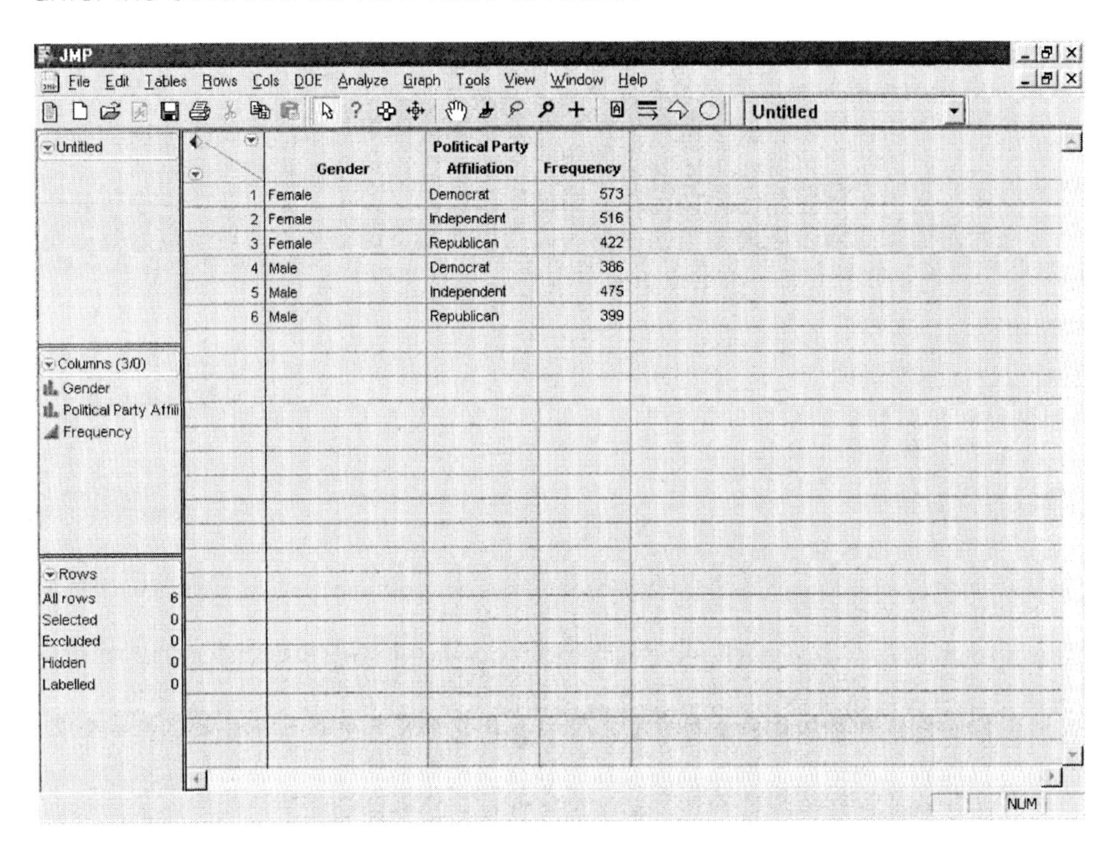

To construct a contingency table, showing the conditional distributions, select **Analyze>Fit Y By X**. Assign ***Political Party Affiliation*** to the **Y, Response** role, ***Gender*** to the **X, Factor** role and ***Frequency*** to the **Weight** role. Select **OK**.

You should see the following screen:

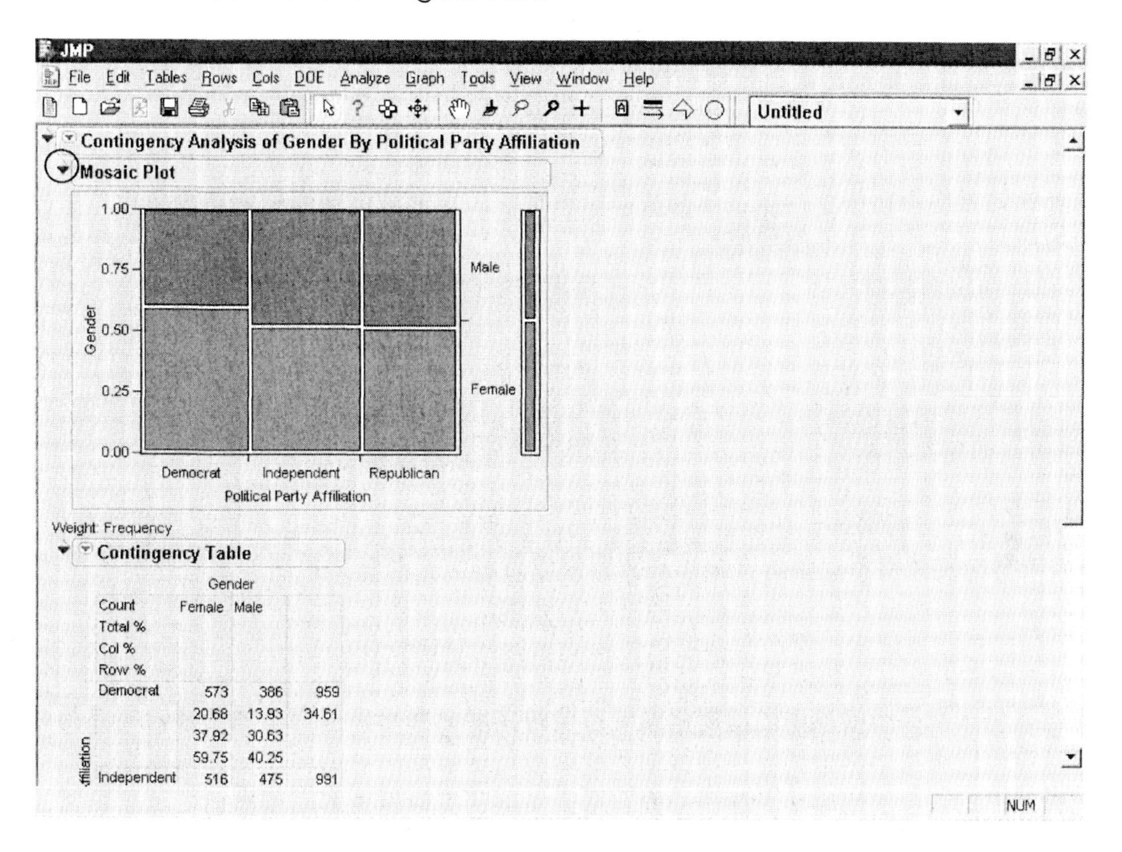

To hide the mosaic plot (if you find it distracting), click on the blue triangle as indicated above.

You should see the following screen:

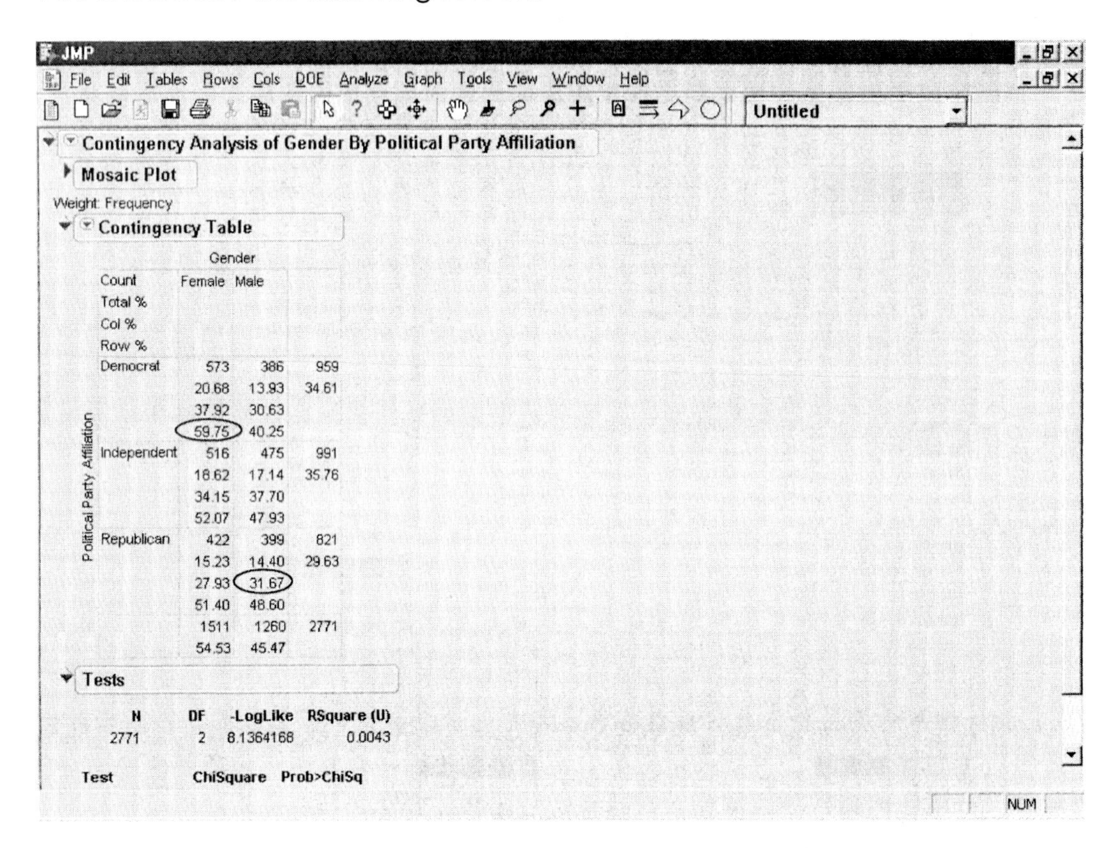

The contingency table provides the information regarding the conditional distributions. For example, of those who identified themselves as Democrats, 59.75% were Female. Of the Males, 31.67% identified themselves as Republican.

To construct Bar Graphs of this data, select **Graph>Chart**. Assign ***Frequency*** to the **Statistics** role (set **Statistics** to **Data**), and assign both ***Gender*** and ***Political Party Affiliation*** to the **Categories,X,Levels** role. Select **OK**.

You should see the following screen:

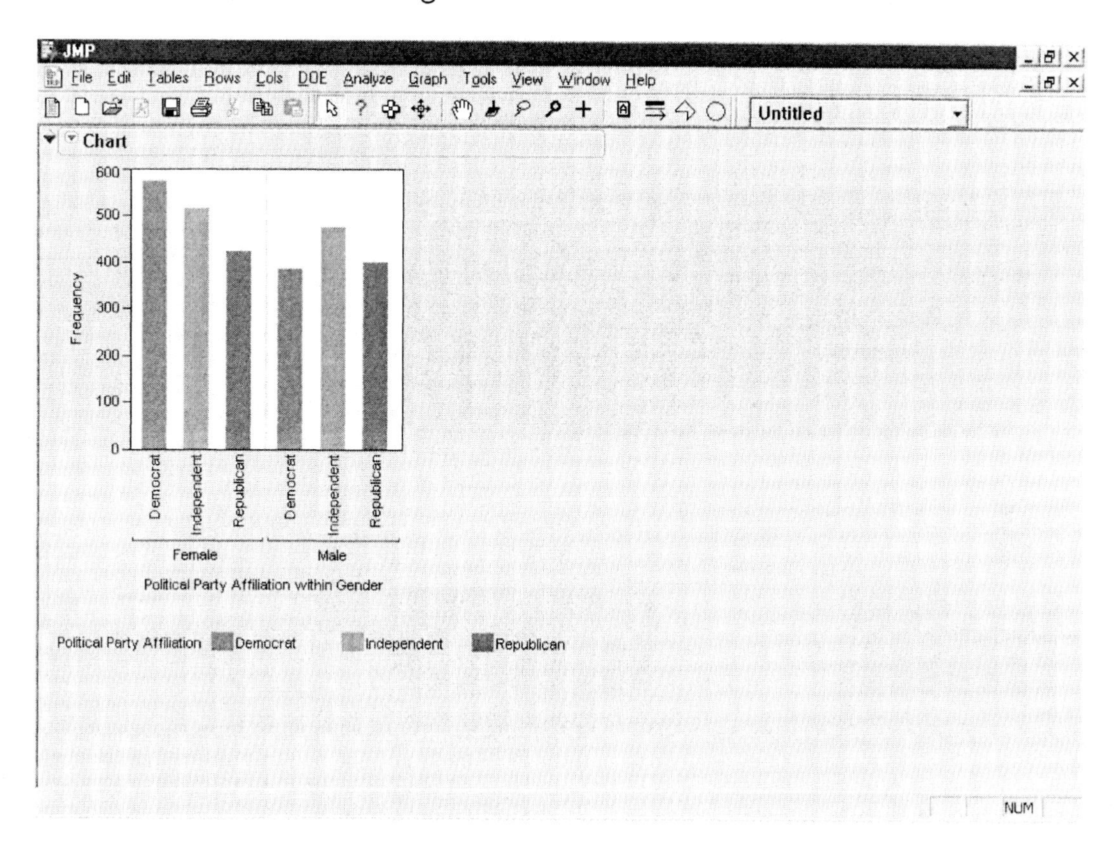

Chapter 11. Example 3. Chi-Squared for Happiness and Family Income.

From the JMP Starter Screen, select **New Data Table**.

Enter the data into the new table as follows:

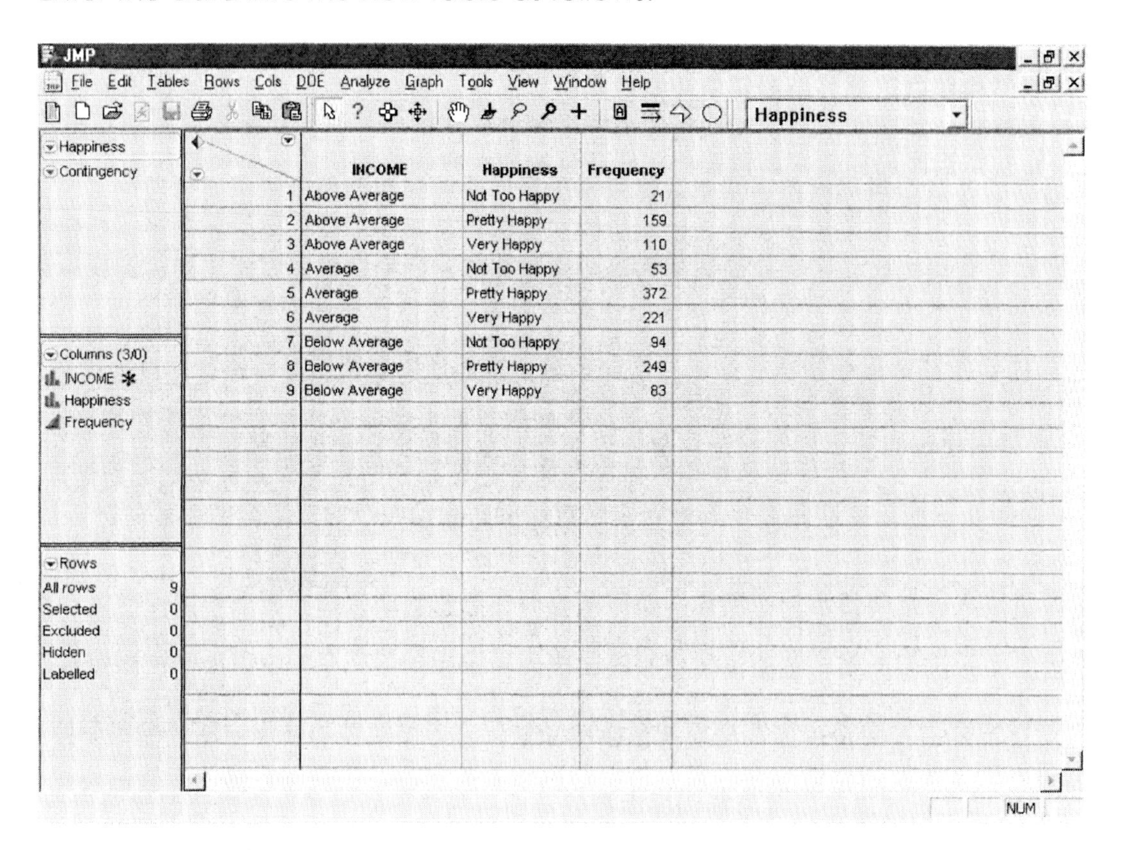

	INCOME	Happiness	Frequency
1	Above Average	Not Too Happy	21
2	Above Average	Pretty Happy	159
3	Above Average	Very Happy	110
4	Average	Not Too Happy	53
5	Average	Pretty Happy	372
6	Average	Very Happy	221
7	Below Average	Not Too Happy	94
8	Below Average	Pretty Happy	249
9	Below Average	Very Happy	83

To perform a Chi-Square test of independence, select **Analyze>Fit Y By X**. Assign ***Happiness*** to the **Y, Response** role, ***Income*** to the **X, Factor** role, and ***Frequency*** to the **Weight** role. Select **OK**. Toggle to the bottom of the output screen.

You should see the following screen:

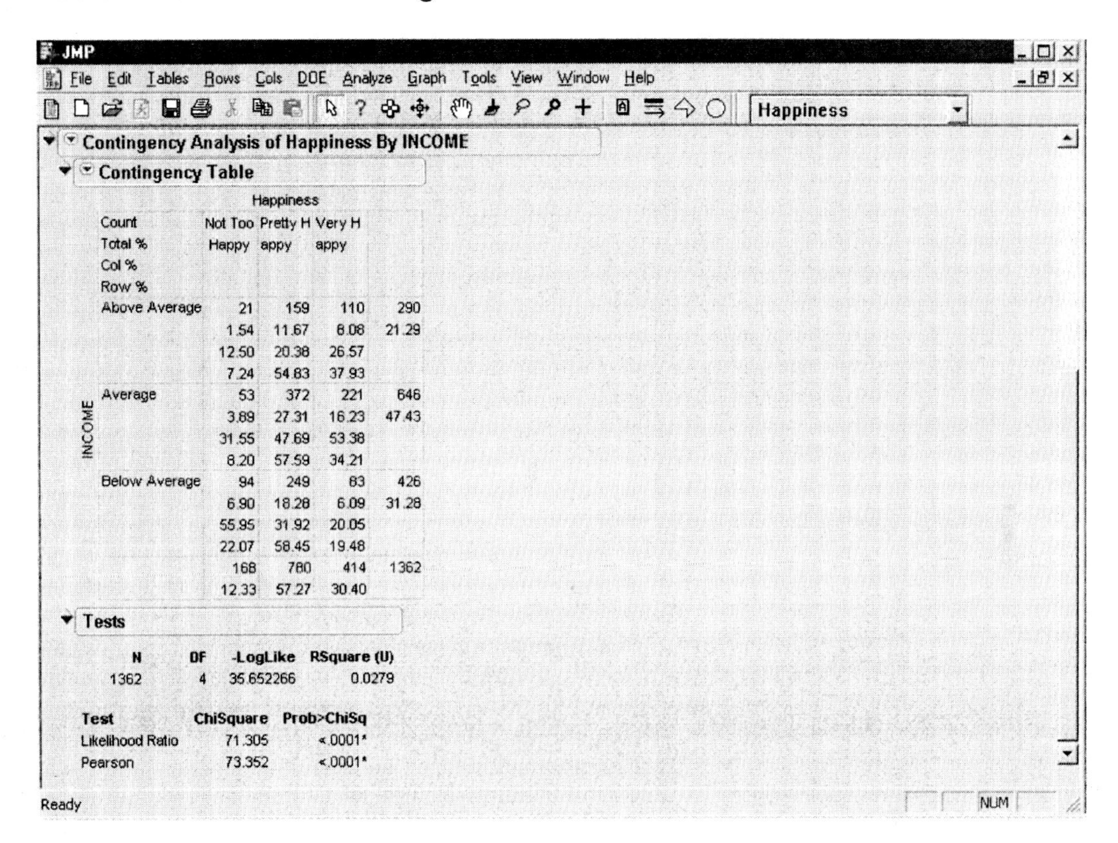

INCOME	Count / Total % / Col % / Row %	Happiness: Not Too Happy	Pretty Happy	Very Happy	
	Above Average	21	159	110	290
		1.54	11.67	8.08	21.29
		12.50	20.38	26.57	
		7.24	54.83	37.93	
	Average	53	372	221	646
		3.89	27.31	16.23	47.43
		31.55	47.69	53.38	
		8.20	57.59	34.21	
	Below Average	94	249	83	426
		6.90	18.28	6.09	31.28
		55.95	31.92	20.05	
		22.07	58.45	19.48	
		168	780	414	1362
		12.33	57.27	30.40	

N	DF	-LogLike	RSquare (U)
1362	4	35.652266	0.0279

Test	ChiSquare	Prob>ChiSq
Likelihood Ratio	71.305	<.0001*
Pearson	73.352	<.0001*

The ChiSquare value of 73.352, would indicate that the Null Hypothesis should be rejected, and we should conclude that Income and Happiness are related.

Chapter 11. Example 4. Are Happiness and Income Independent?

The output screen from Example 3 was:

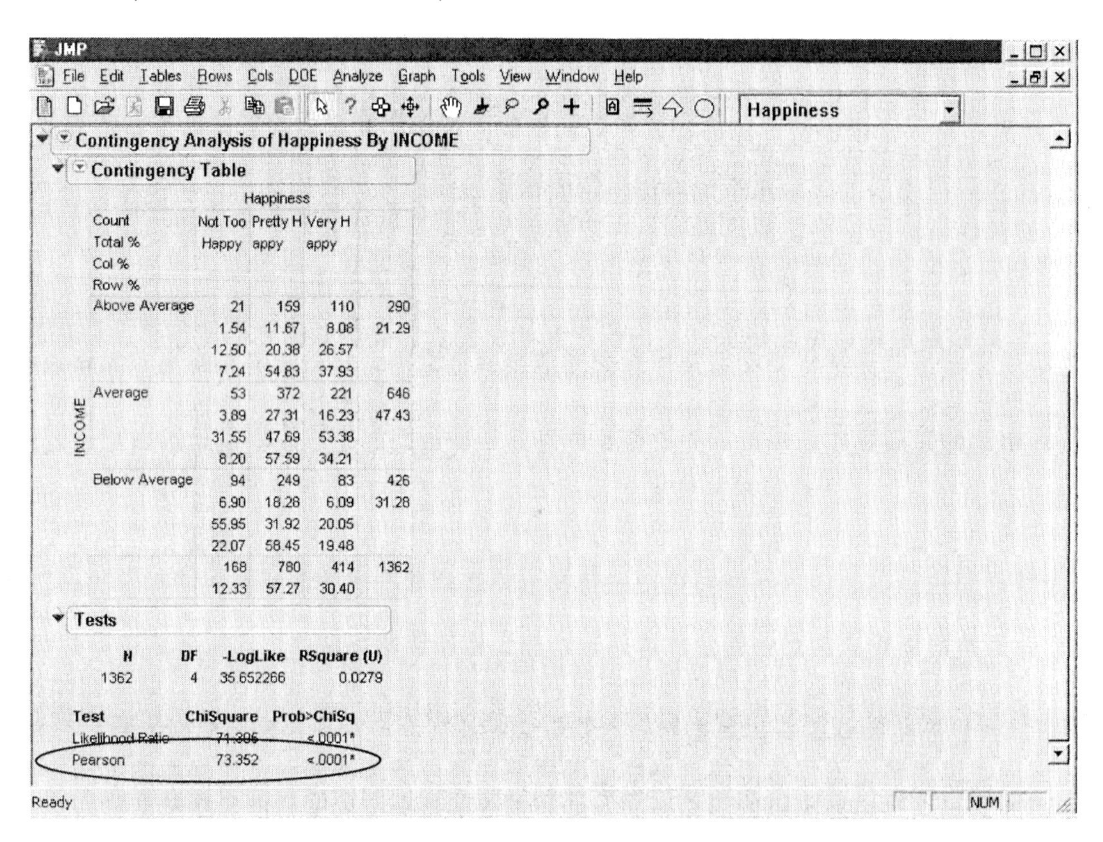

From this output, the p-value associated with the Chi-Square statistic of 73.352, is <.0001. This p-value is effectively 0. Alpha was selected to be .05 – this is the highest probability we are willing to accept of rejecting the Null Hypothesis when the Null is actually True (a Type 1 Error). The p-value represents the calculated probability of committing this same error, based upon the actual data. Since the calculated probability (0) is lower than what we said we were willing to accept (.05), we feel comfortable rejecting the Null and concluding that the two variables are related.

Chapter 11. Exercise 11.11. Life after death and gender.

From the JMP Starter Screen, select **New Data Table**.

Enter the data into the new table as follows:

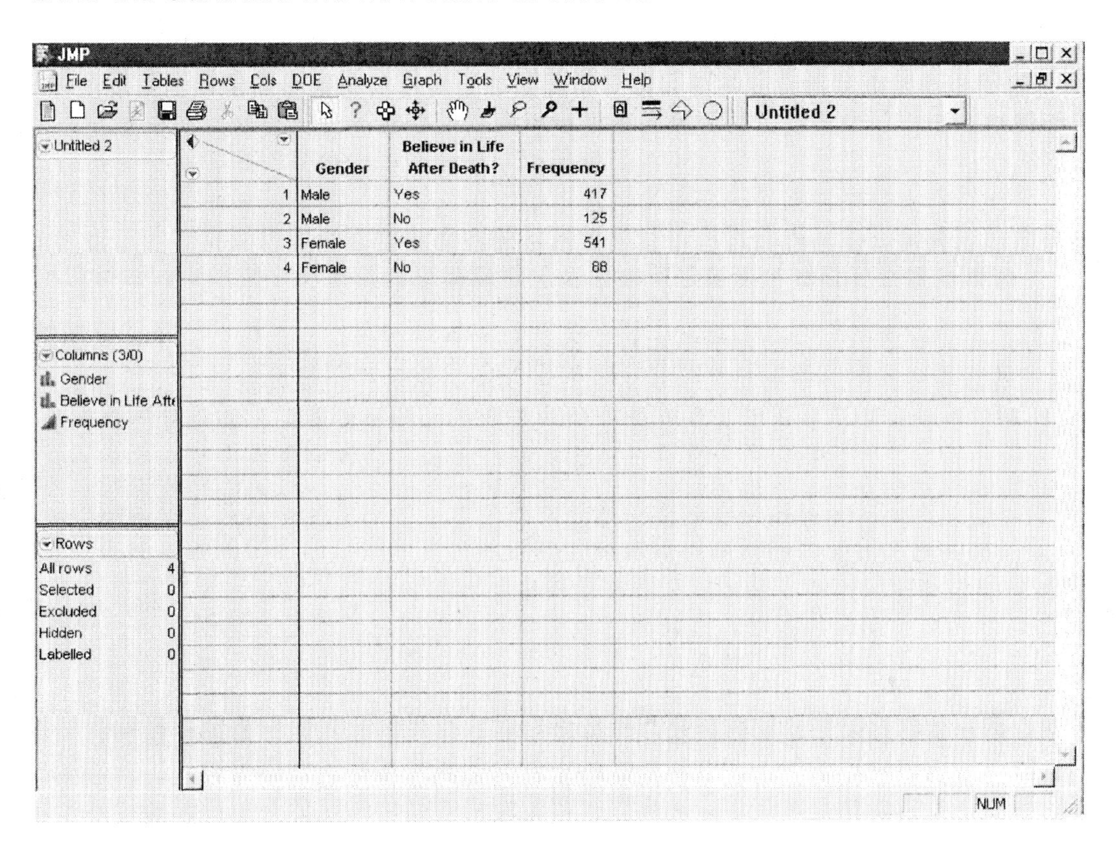

To generate the contingency table, including the expected values and the Chi-Square statistics, select **Analyze>Fit Y By X**. Assign Believe in Life After Death? to the **Y, Response** role, Gender to the **X, Factor** role and Frequency to the **Weight** role. Select **OK**. Toggle to the bottom of the output screen. Click on the red triangle next to "Contingency Table" and select **Expected**.

You should see the following screen:

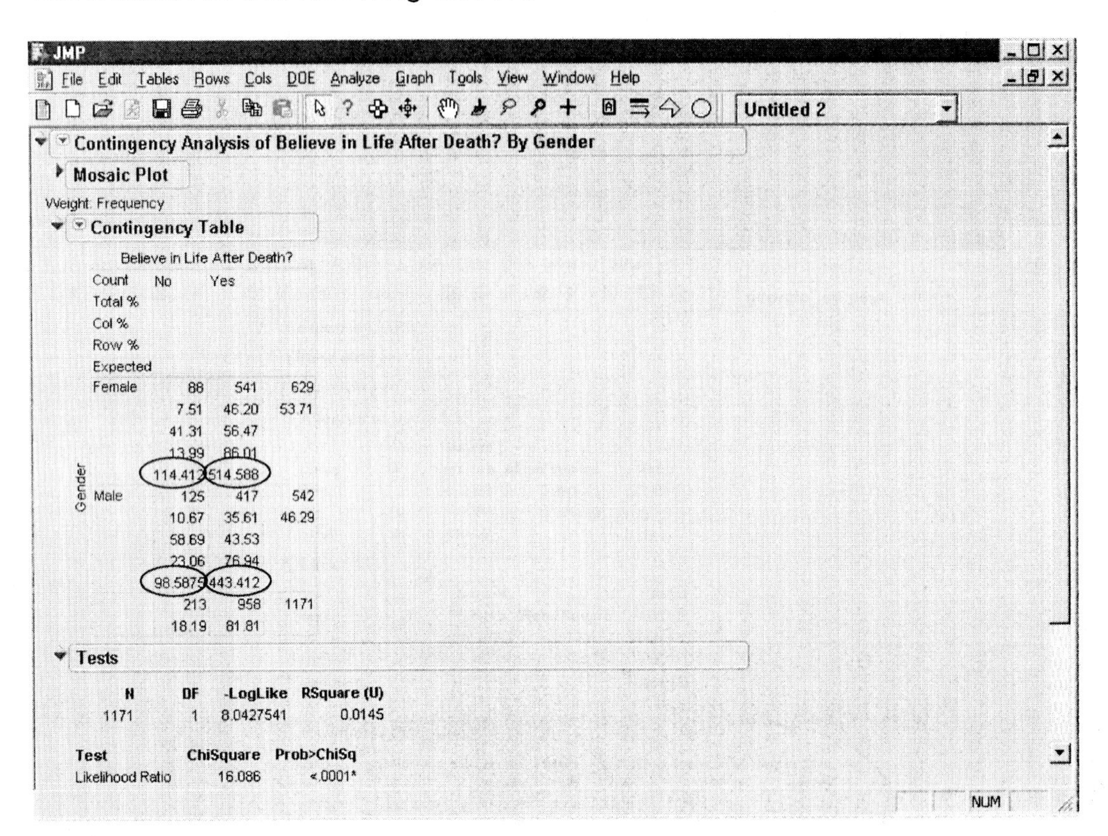

Notice in the contingency table, that the expected values have been included, indicating what the expected frequencies would have been if the two variables had no association. The actual counts are higher than expected for the Females indicating Yes and the Males indicating No.

The Chi-Square test results at the bottom of the screen indicate that the two variables are, in fact, related:

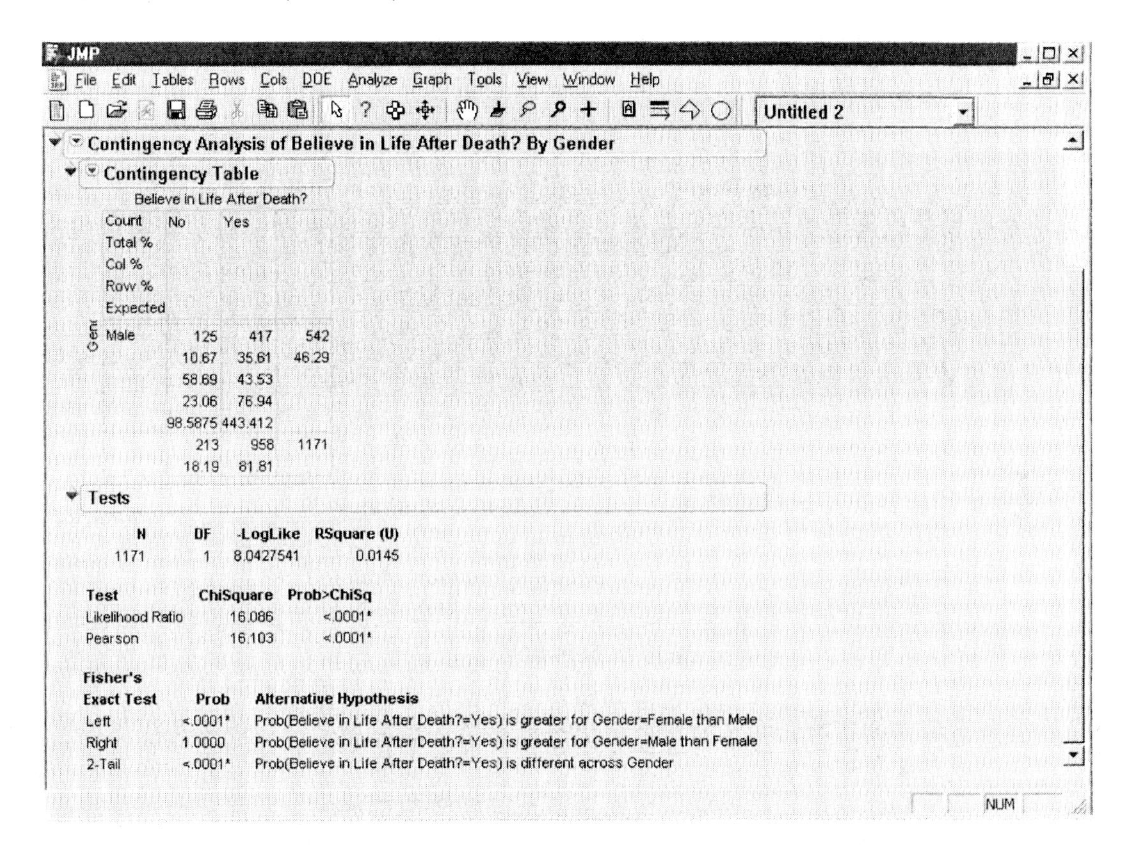

Chapter 11. Exercise 11.15. Help the Environment.

From the JMP Starter Screen, select **New Data Table**.

Enter the data into the new table as follows:

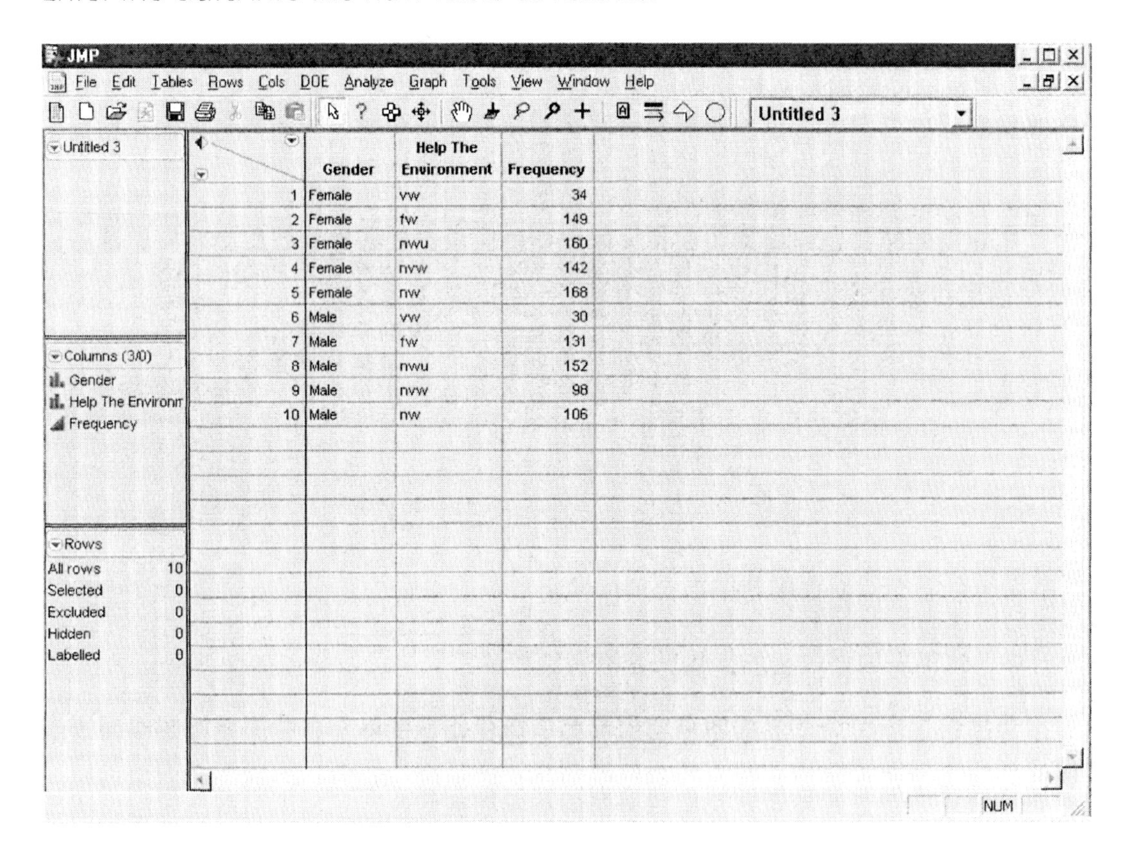

	Gender	Help The Environment	Frequency
1	Female	vw	34
2	Female	fw	149
3	Female	nwu	160
4	Female	nvw	142
5	Female	nw	168
6	Male	vw	30
7	Male	fw	131
8	Male	nwu	152
9	Male	nvw	98
10	Male	nw	106

The Hypothesis Statements for this test would be:

H_o: Gender and Willingness to Help the Environment are NOT Related
H_a: Gender and Willingness to Help the Environment ARE Related

To test the claim (H_a), select **Analyze>Fit Y By X**. Assign Help the Environment to the **Y, Response** role, Gender to the **X, Factor** role and Frequency to the **Weight** role. Select **OK**. Toggle to the bottom of the screen.

You should see the following screen:

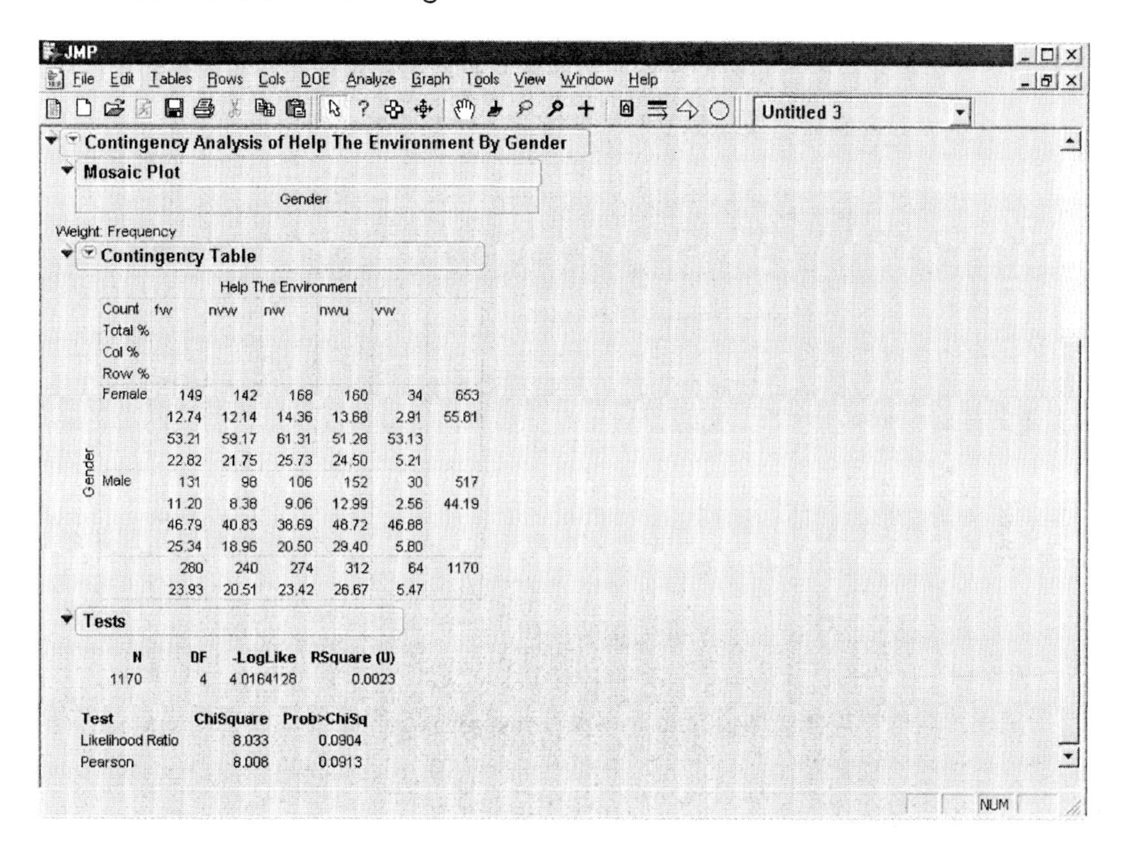

From the output, there are two rows, five columns, and four degrees of freedom ((2-1)*(5-1)) = 4.

The p-value is less than alpha = .1, but not lower than alpha = .05. Therefore, at .05, we would fail to reject the Null Hypothesis, but at .1, we would reject the Null Hypothesis.

Chapter 11. Exercise 11.41. Claritin and nervousness.

From the JMP Starter Screen, select **New Data Table**.

Enter the data into the new table as follows:

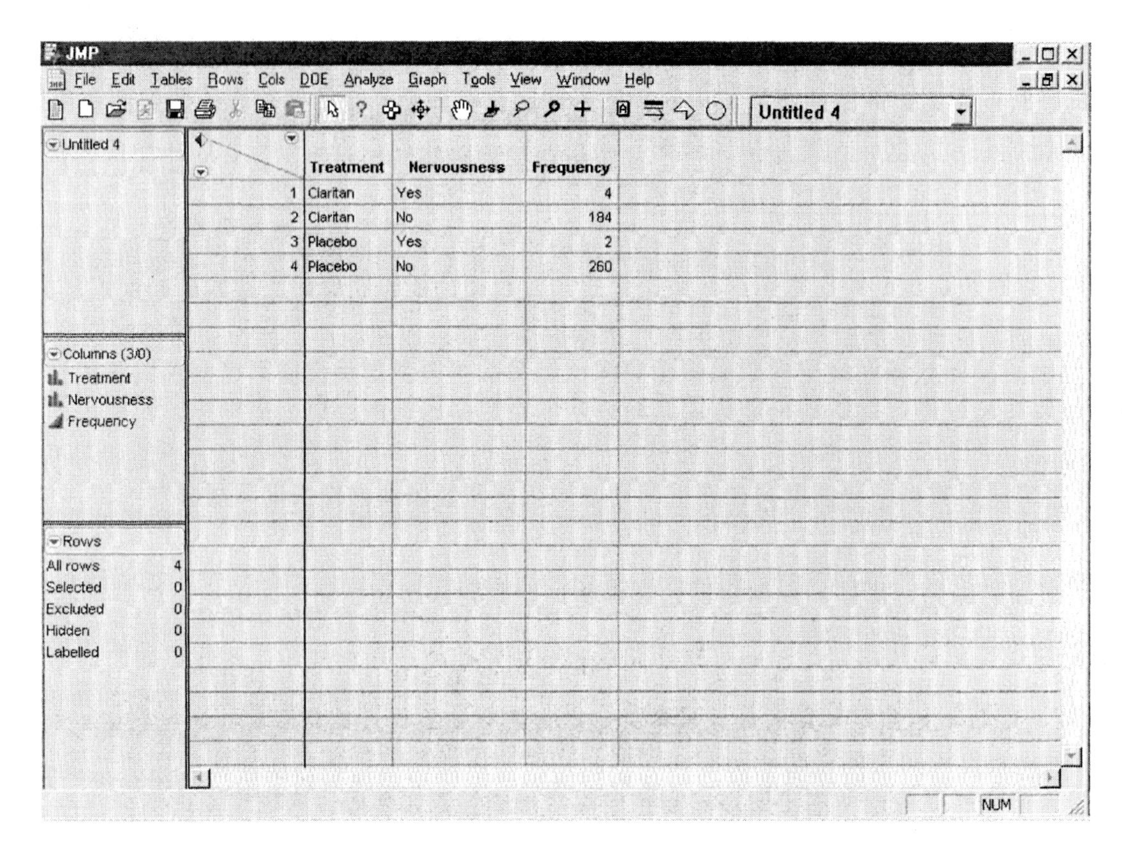

Select **Analyze>Fit Y By X**. Assign Treatment to the **X, Factor** role, Nervousness to the **Y, Response** role and Frequency to the **Freq** role. Select **OK**. Toggle to the bottom of the screen.

You should see the following screen:

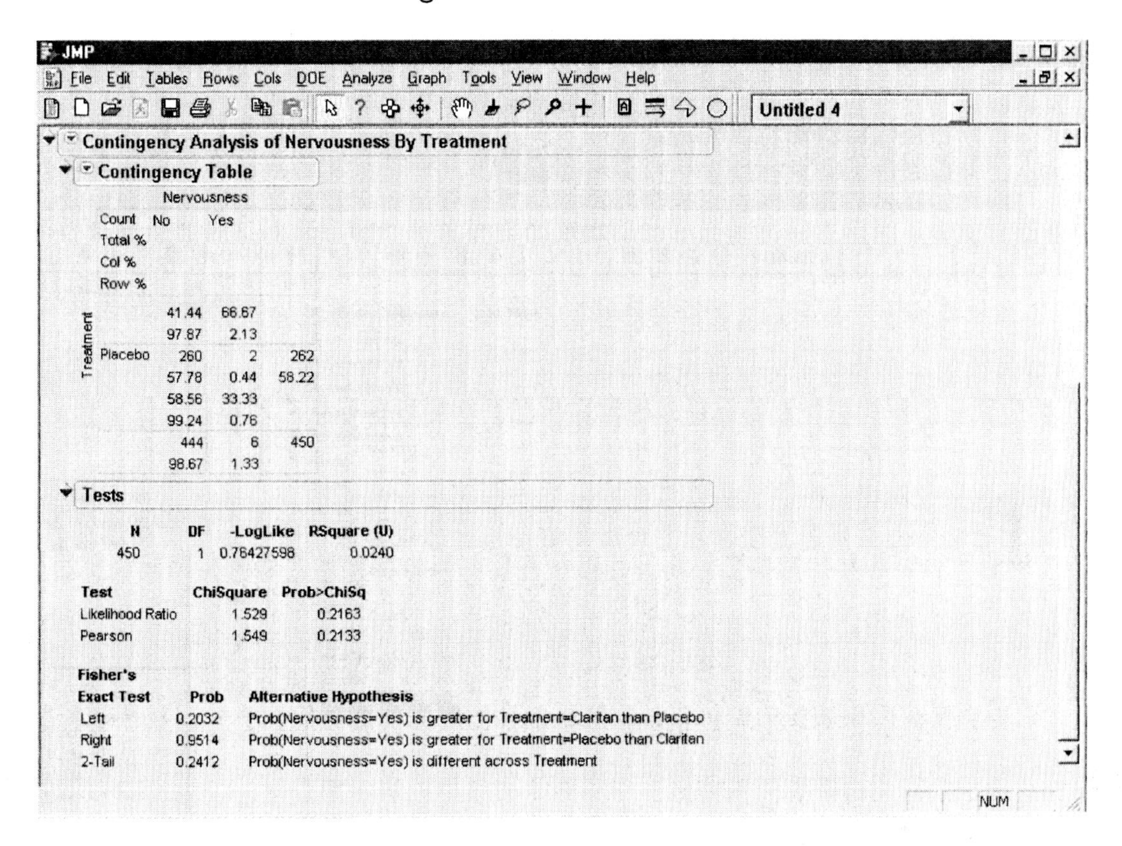

The p-value using Fisher's Exact Test is .2412, indicating that we would fail to reject the Null Hypothesis.

Chapter 11. Exercise 11.59. Student Data.

Open the fl_student_survey dataset from the JMP file.

We will be conducting a Chi-Square test to determine if religiosity is related to beliefs regarding life after death.

When you open the file, you should see the following screen:

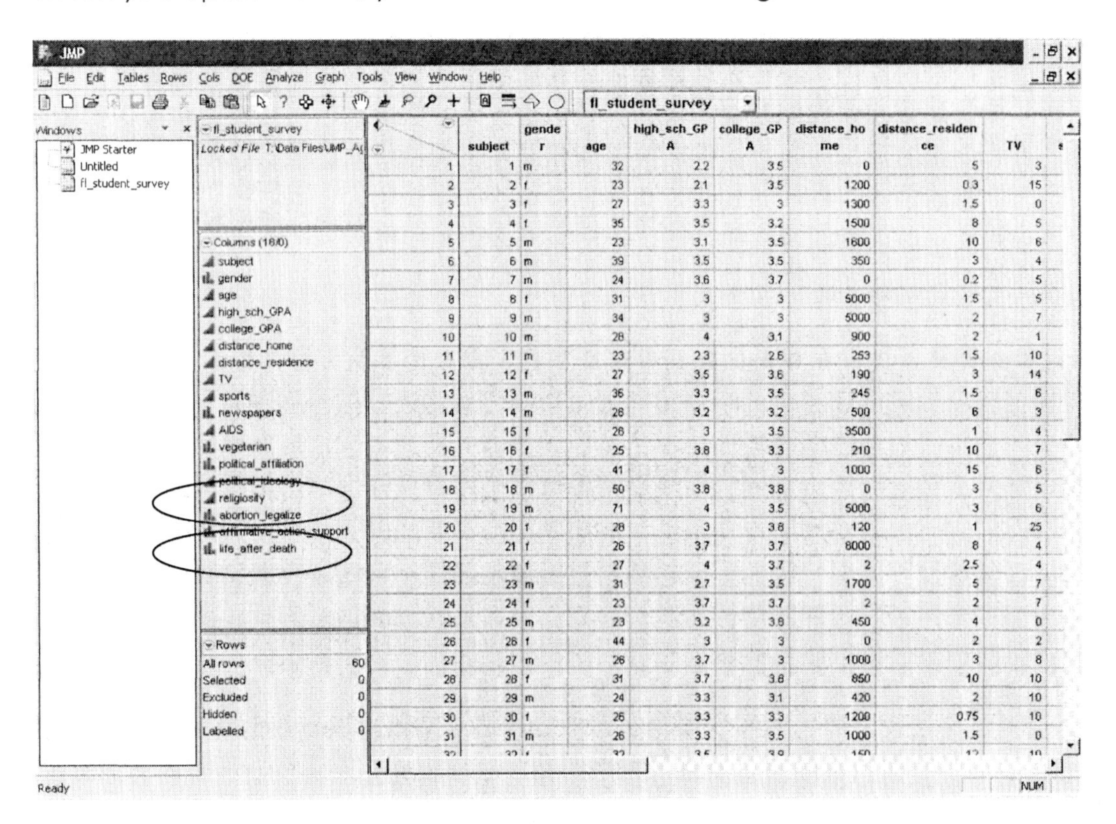

As you have learned, a Chi-Square test evaluates the relationship between two categorical variables. Recall in JMP that categorical variables are indicated with a series of small red bars, while quantitative data is indicated by a blue triangle. Notice on the screen that although the life after death variable is categorical, the religiosity variable has been read as quantitative. This is simply because the information was entered as numbers. We need to change the variable classification from quantitative to categorical to execute a Chi-Square test.

To change religiosity from quantitative to categorical, place your cursor on the column name "religiosity" in the list as circled above, and RIGHT click. Select **Column Info**. Set **Data Type** to **Character**, and **Modeling**

Type to **Nominal**. Select **OK**. Now, the religiosity variable should be identified as categorical:

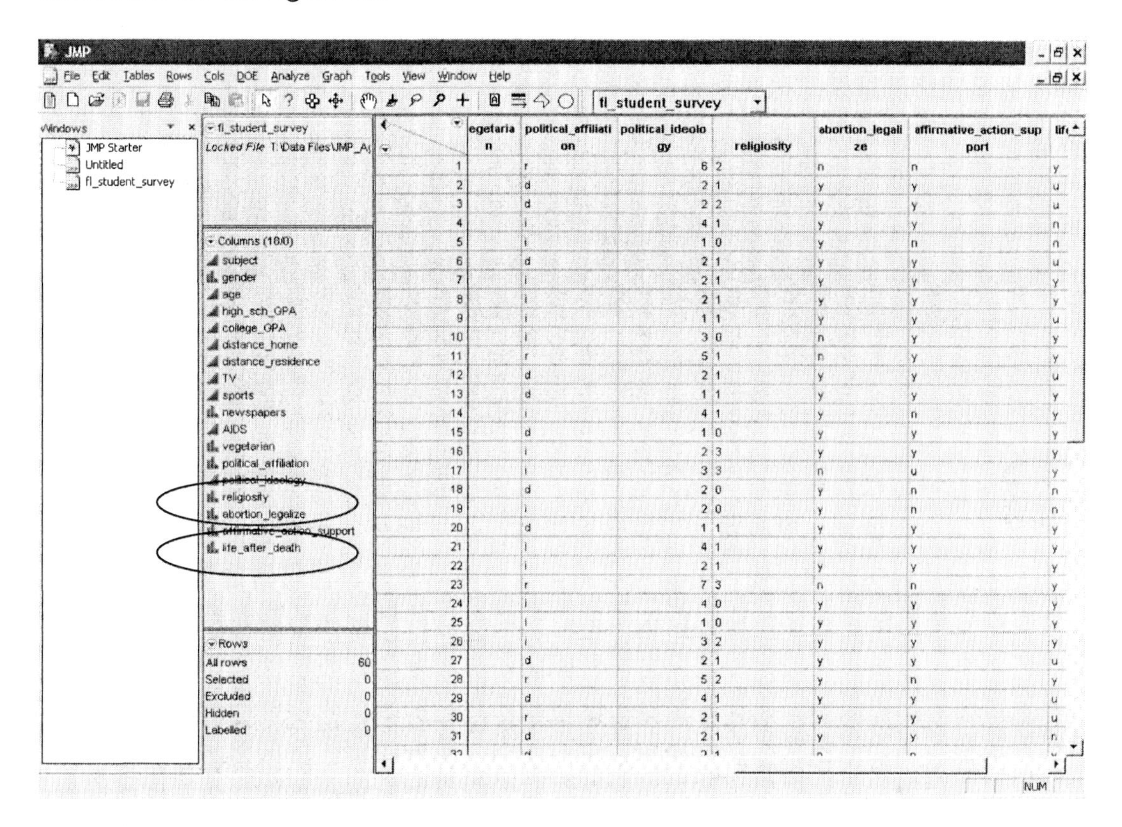

The appropriate hypothesis statements for this test are:

H_o: Religiosity and views on Life After Death are NOT Related
H_a: Religiosity and views on Life After Death ARE Related

To execute a Chi-Square test to determine if these two variables are related, select **Analyze>Fit Y by X**. Assign ***religiosity*** to the **Y, Response** role and ***life_after_death*** as the **X, Factor** role (for a Chi-Square test, these assignments could be reversed without any impact to the results). Select **OK**.

The information that we need is in the **Contingency Table**. If the colorful Mosaic Plot is confusing or distracting, hide it by clicking on the blue triangle next to the Plot heading.

The relevant information can be seen in the Contingency Table section:

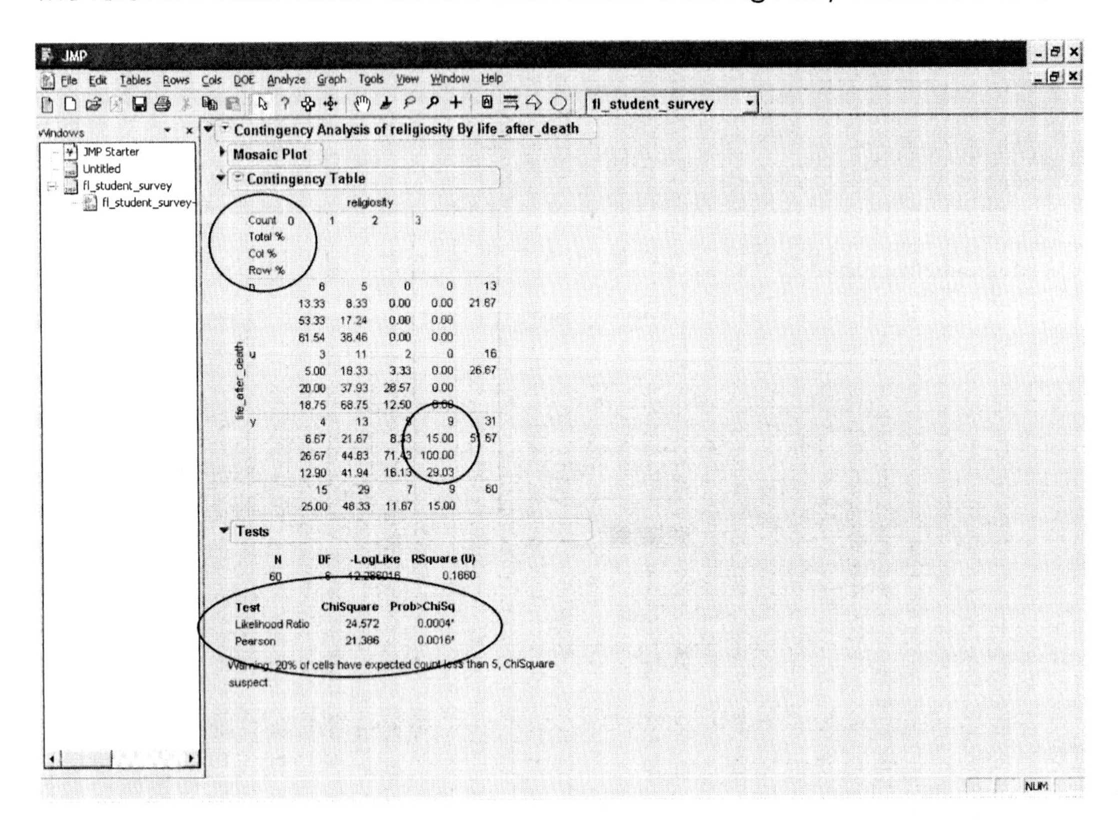

The Chi-Square statistic for this test is 21.386, corresponding to a p-value of .0016, indicating that the null hypothesis should be rejected and that a relationship between the variables appears to exist.

The cell in the upper left corner of the Contingency Table, provides a legend for the data. For example, the bottom right cell (circled above), indicates that 15% of the total population of students consider themselves to be highly religious (religiosity = 3) and they believe in life after death. Of those who indicated religiosity = 3, 100% believe in life after death. Of those who believe in life after death, 29.03% consider themselves to be highly religious.

Chapter 12. Example 2. What Do We Learn from a Scatterplot in the Strength Study?

Open the high_school_female_athlete data from the JMP folder of the data disk.

To generate the descriptive statistics for the number of 60 pound bench presses and the maximum bench press pounds, select **Analyze>Distribution**. Assign ***BP*** and ***BP60*** to the **Y, Columns** role. Select **OK**. Toggle to the bottom of the output screen.

You should see the following screen:

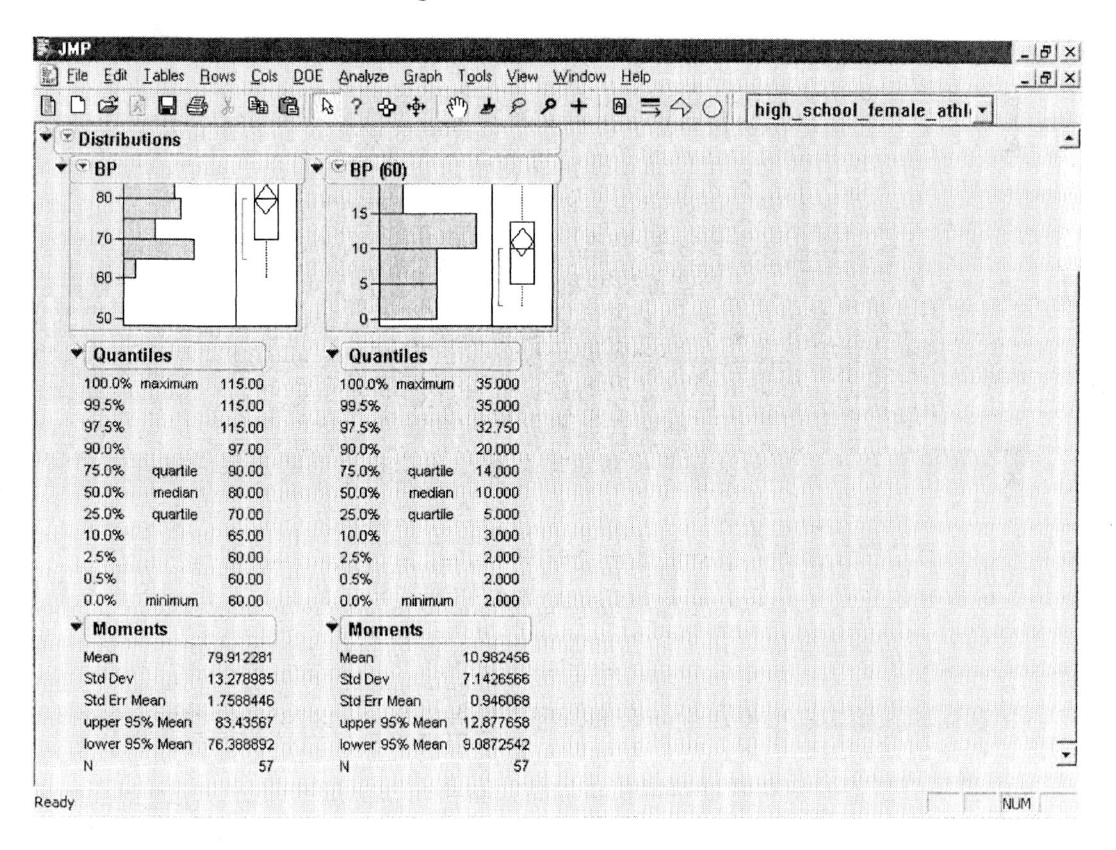

To generate the scatterplot of the two variables, select **Analyze> Fit Y by X**. Assign ***BP*** to the **Y, Response** role and assign ***BP60*** to the **X, Factor** role. Select **OK**.

You should see the following screen:

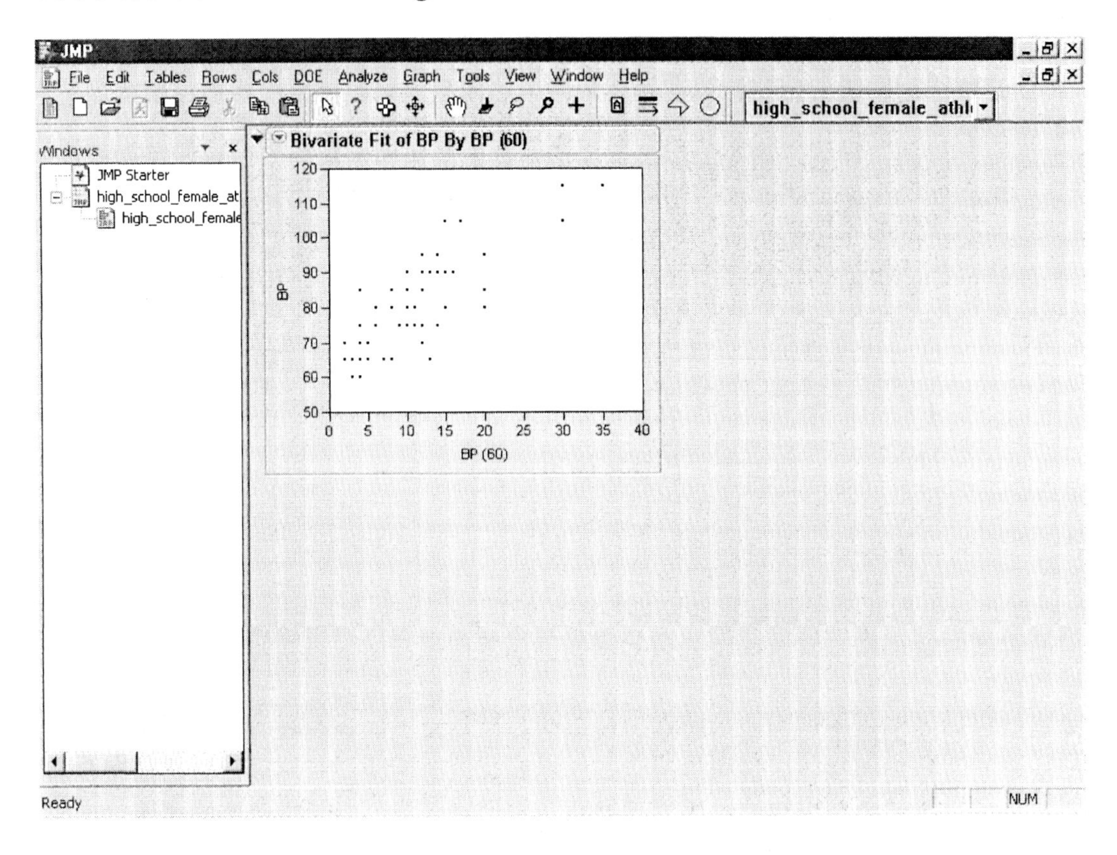

Chapter 12. Example 3. Which Regression Line Predicts Maximum Bench Press?

Open the high_school_female_athlete data from the JMP folder of the data disk.

To generate the regression line for y= maximum bench press and x= number of 60-pound bench presses, select **Analyze>Fit Y By X**. Assign ***BP*** to the **Y, Response** role and ***BP60*** to the **X, Factor** role. Select **OK**.

When the output screen is generated with the scatterplot, click on the red triangle next to "Bivariate Fit...". Select Fit Line.

You should see the following screen:

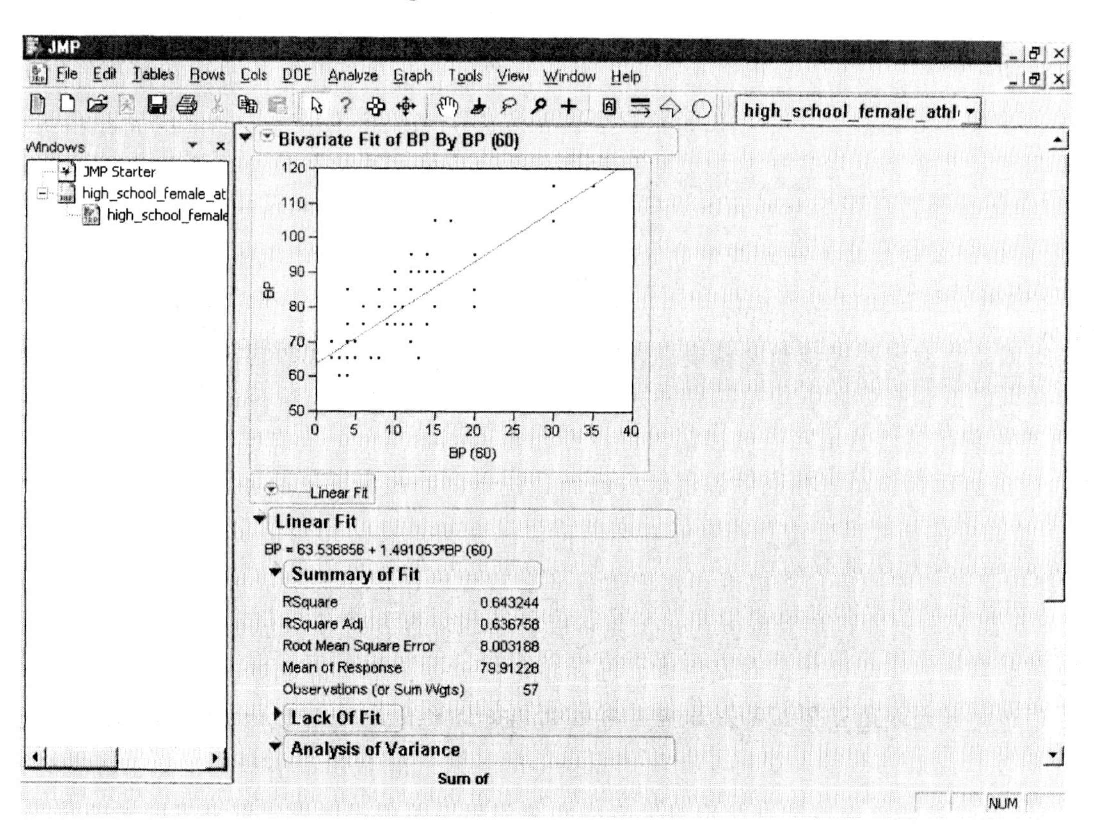

Chapter 12. Example 6. What's the Correlation for Predicting Strength?

Open the high_school_female_athlete data from the JMP folder of the data disk.

To determine the correlation for y= maximum bench press and x= number of 60-pound bench presses, select **Multivariate Methods>Multivariate**. Assign both ***BP*** and ***BP60*** to the **Y, Columns** role. Select **OK**. Toggle to the top of the output screen.

You should see the following screen:

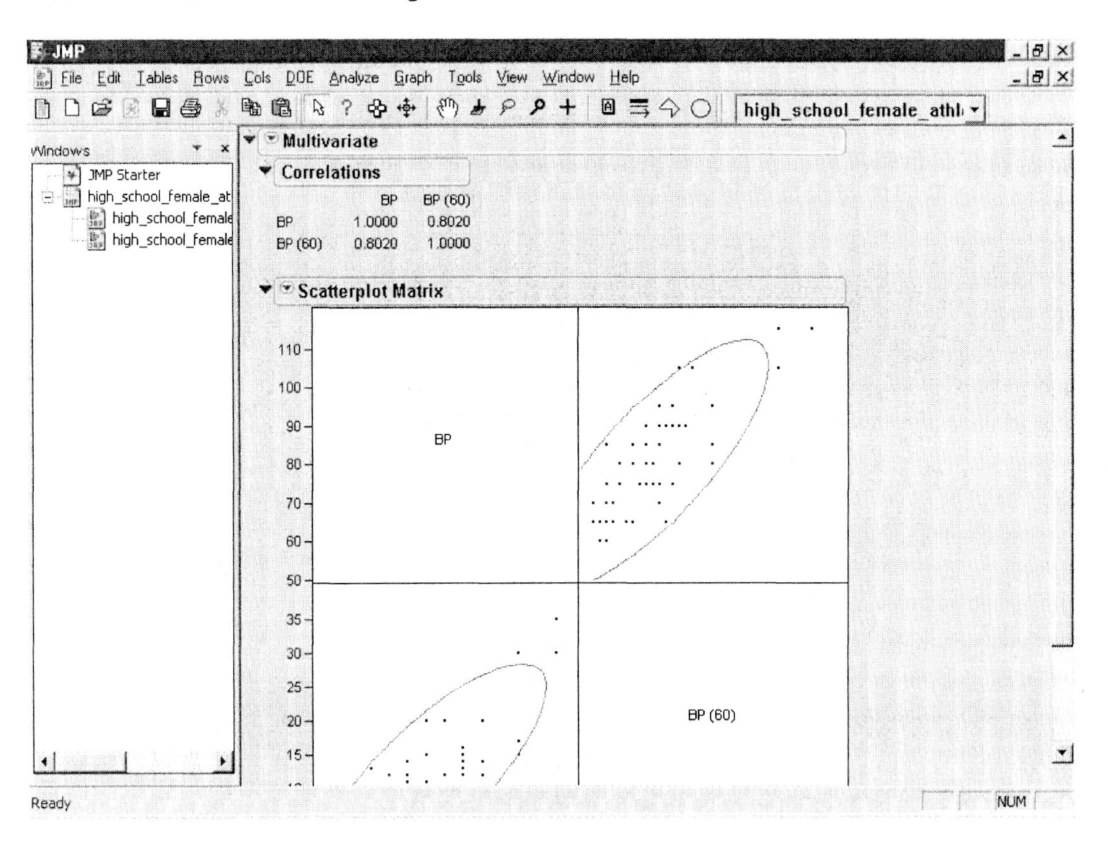

Chapter 12. Example 9. What Does R^2 Tell Us in the Strength Study?

Open the high_school_female_athlete data from the JMP folder of the data disk.

To generate the regression line for y= maximum bench press and x= number of 60-pound bench presses, select **Analyze>Fit Y By X**. Assign ***BP*** to the **Y, Response** role and ***BP60*** to the **X, Factor** role. Select **OK**.

When the output screen is generated with the scatterplot, click on the red triangle next to "Bivariate Fit...". Select Fit Line.

You should see the following screen:

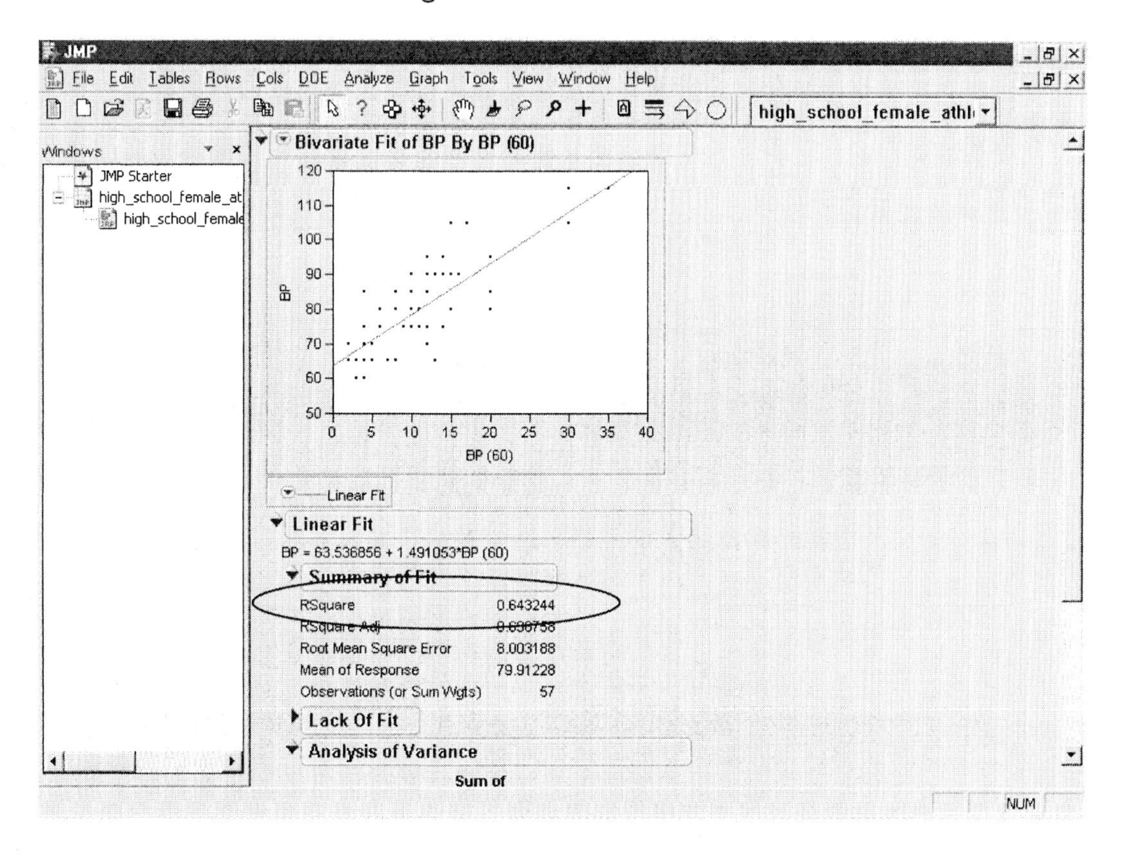

Chapter 12. Exercise 12.1. Car Mileage and Weight.

Open the car_weight_and_mileage data from the JMP folder of the data disk.

To generate the regression of weight on mileage, select **Analyze>Fit Y by X**. Assign ***mileage*** to the **Y, Response** role and ***weight*** as the **X, Factor** role. Select **OK**. After the scatterplot is generated, click on the red triangle next to "Bivariate Fit..." and select **Fit Line**. Toggle to the bottom of the screen.

You should see the following screen:

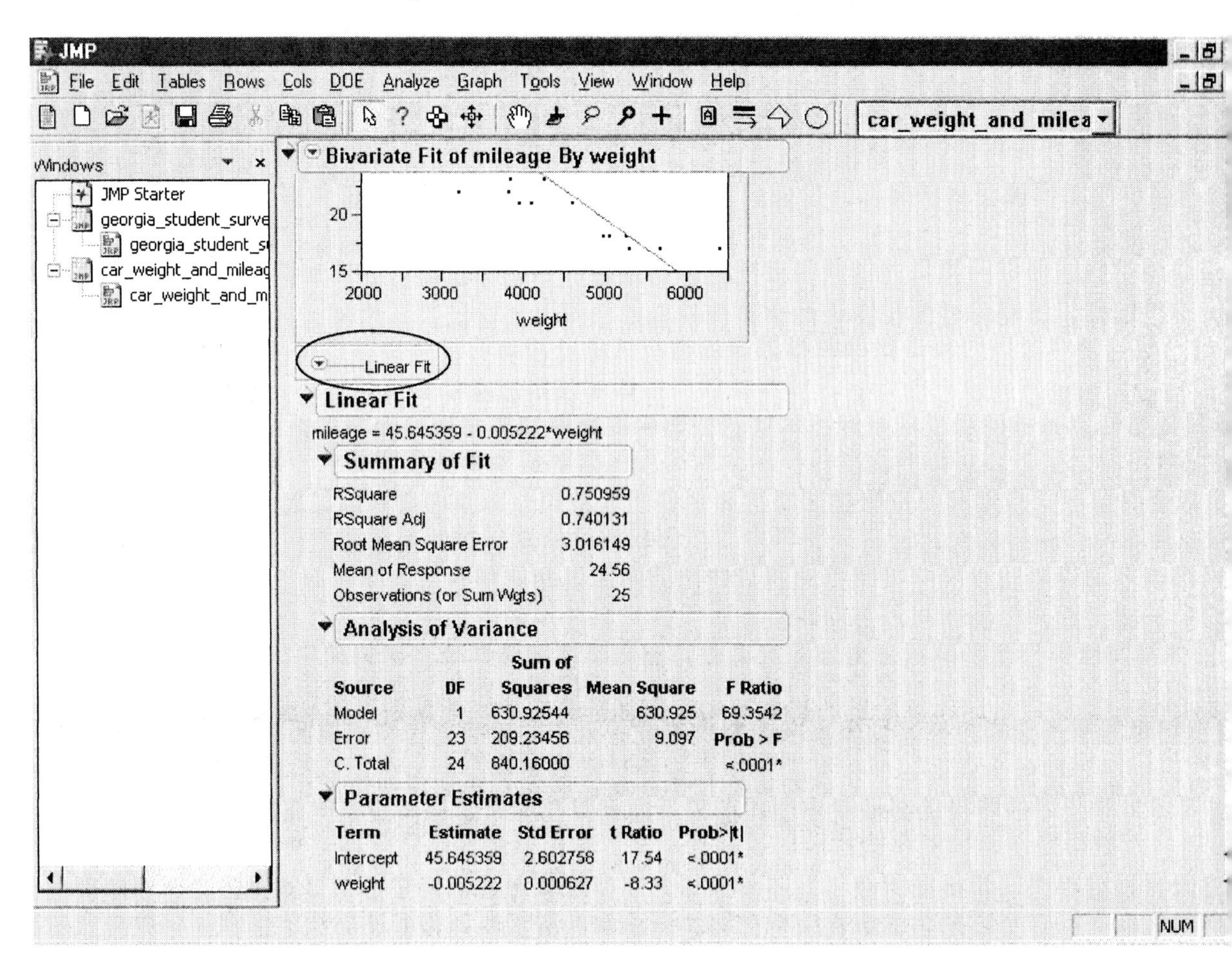

If you need the residuals (the differences between the predictions and the actuals), simply click on the red triangle next to "Linear Fit" as indicated above. Select **Save Residuals**. The residual values will now be included as a variable in the dataset. Once the residual values are part of the dataset, they can be graphed or analyzed like any other variable.

Chapter 12. Exercise 12.9. Predicting college GPA.

Open the georgia_student_survey data from the JMP folder of the data disk.

To generate the regression of High School GPA on College GPA, select **Analyze>Fit Y by X**. Assign ***CGPA*** to the **Y, Response** role and ***HSGPA*** as the **X, Factor** role. Select **OK**. After the scatterplot is generated, click on the red triangle next to "Bivariate Fit..." and select **Fit Line**. Toggle to the bottom of the screen.

You should see the following screen:

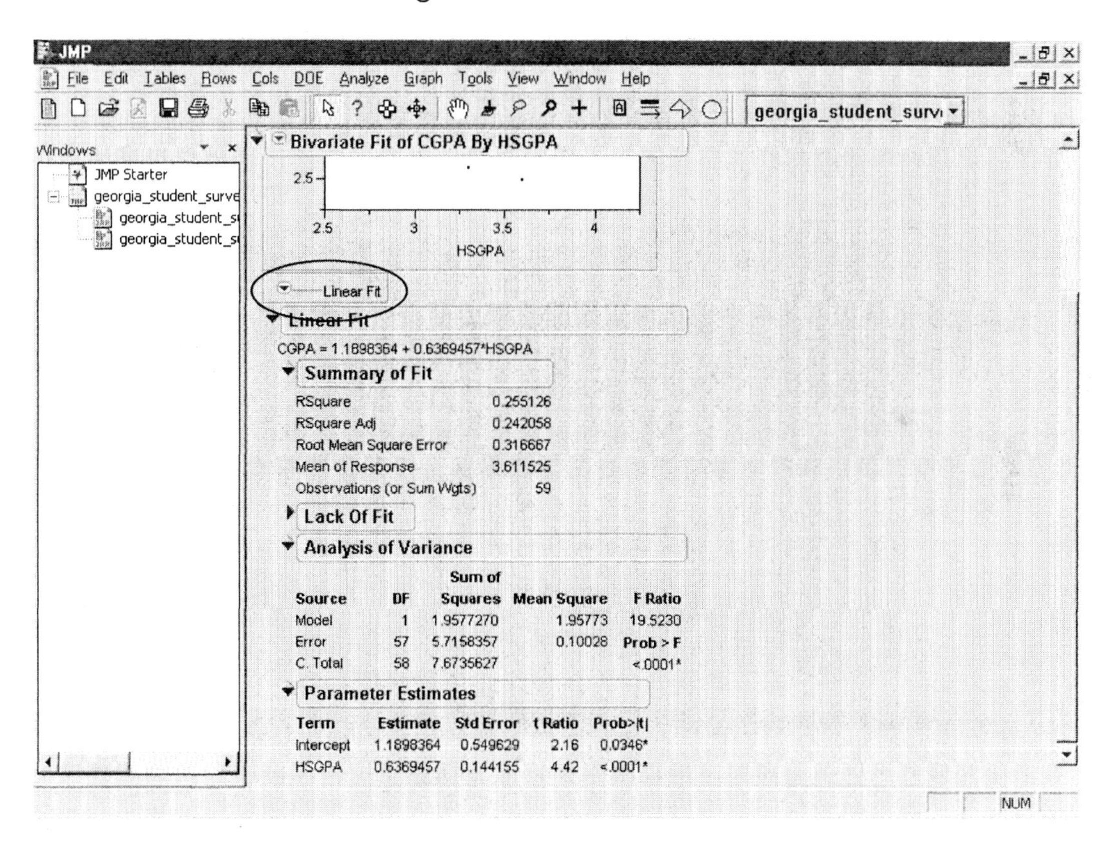

If you need the residuals (the differences between the predictions and the actuals), simply click on the red triangle next to "Linear Fit" as indicated above. Select **Save Residuals**. The residual values will now be included as a variable in the dataset. Once the residual values are part of the dataset, they can be graphed or analyzed like any other variable.

Chapter 12. Exercise 12.10. Exercise and Watching TV.

Open the georgia_student_survey data from the JMP folder of the data disk.

To generate the regression of watching TV on exercise, select **Analyze>Fit Y by X**. Assign ***exercise*** to the **Y, Response** role and ***watchtv*** as the **X, Factor** role. Select **OK**. After the scatterplot is generated, click on the red triangle next to "Bivariate Fit…" and select **Fit Line**.

You should see the following screen:

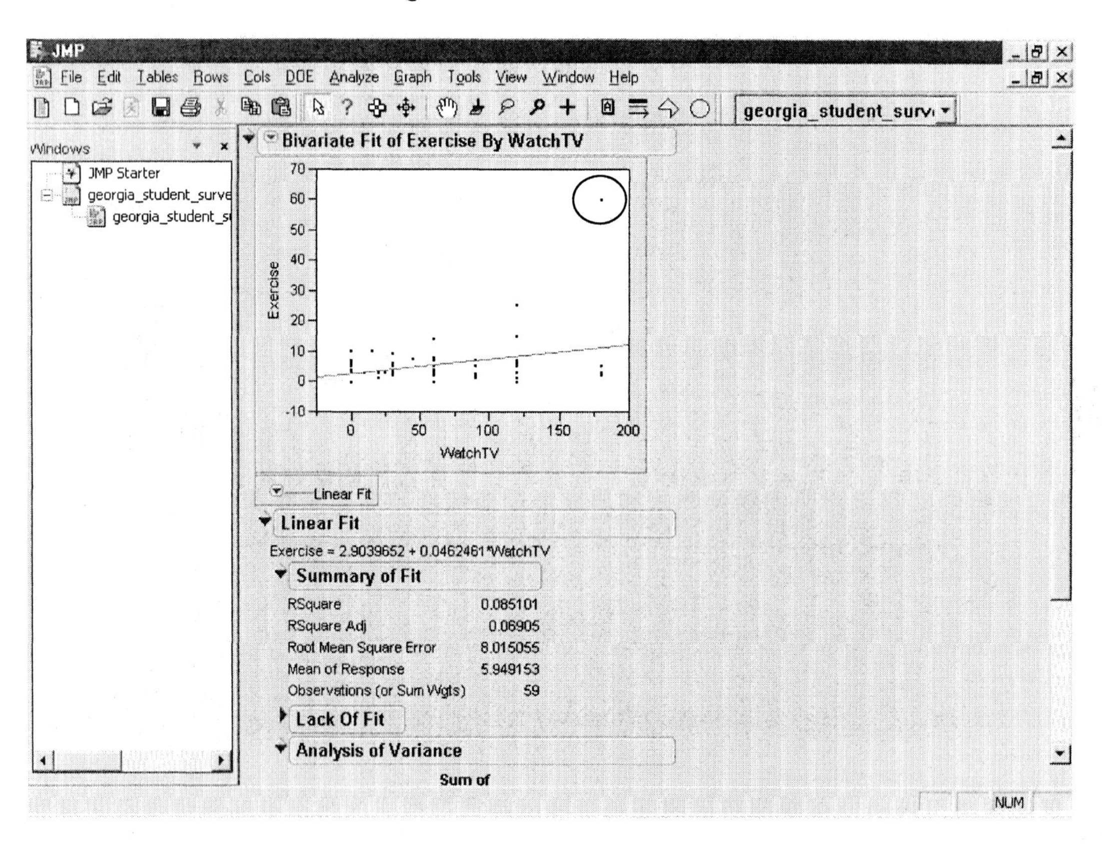

After examining the parameter estimates and the regression equation for the entire dataset, go back to the plot. The identified observation is an outlier. Click on it. You should see the number "14" appear, indicating that this is the 14th observation in the dataset. To rerun the analysis with this observation removed, return to the dataset. The 14th observation should be highlighted. If not, highlight it and select **Rows>Exclude/Unexclude**. The 14th observation will now be excluded from further analysis.

Rerun the regression analysis as before.

You should see the following screen:

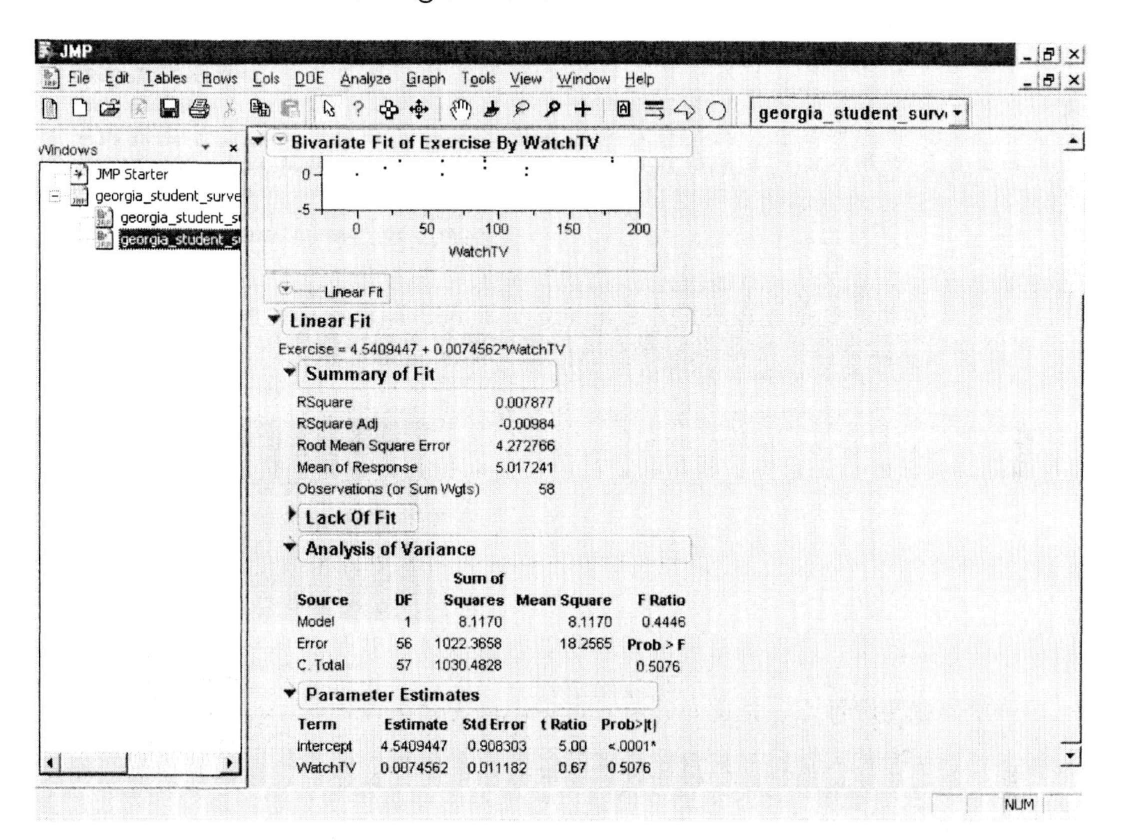

Chapter 12. Example 5. Correlations with Worldwide Internet Use.

Open the human_development data from the JMP folder of the data disk.

To generate the correlation matrix of internet use with GDP, Cellular, Fertility, and CO_2, select **Analyze>Multivariate Methods>Multivariate**. Assign all of the variables listed to the **Y, Columns** role. Select **OK**.

When the output screen is generated, toggle to the top.

You should see the following screen:

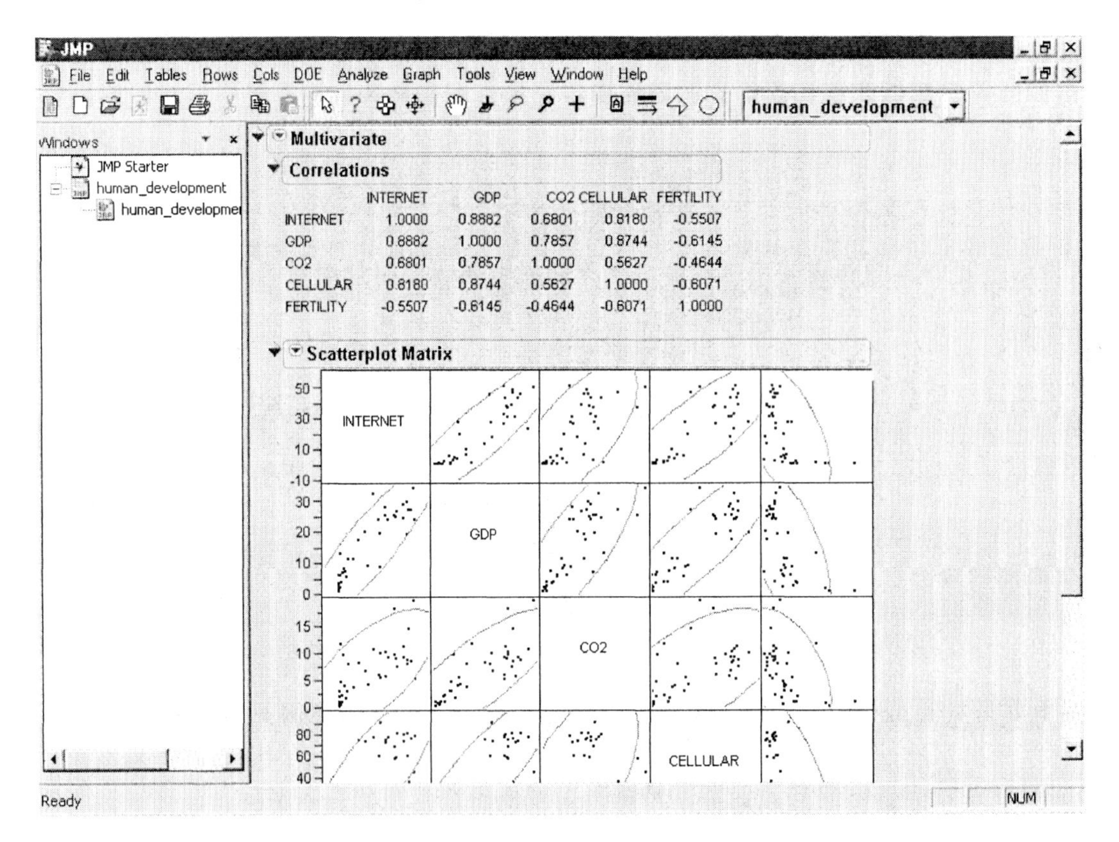

Note that the correlation matrix above is a "complete" matrix, such that the intersections in the upper diagonal are the same as the intersections in the bottom diagonal (e.g., GDP and Internet have the same correlation value of .8882 as Internet and GDP).

Chapter 12. Exercise 12.13. When can you compare slopes?

Open the human_development data from the JMP folder of the data disk.

To generate the regression of GDP on Cell Phone Usage, select **Analyze>Fit Y by X**. Assign ***Cellular*** to the **Y, Response** role and ***GDP*** as the **X, Factor** role. Select **OK**. After the scatterplot is generated, click on the red triangle next to "Bivariate Fit..." and select **Fit Line**.

You should see the following screen:

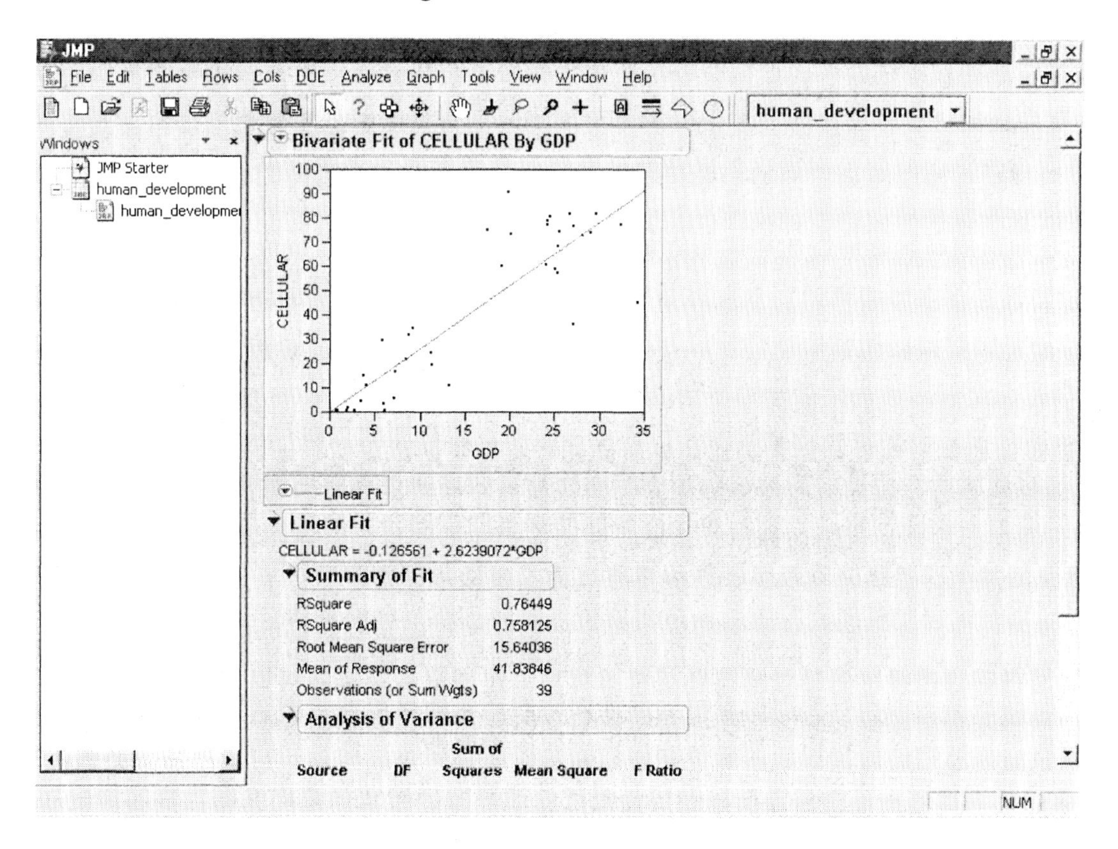

To generate the regression of GDP on Internet usage, select **Analyze>Fit Y by X**. Assign ***Internet*** to the **Y, Response** role and ***GDP*** as the **X, Factor** role. Select **OK**. After the scatterplot is generated, click on the red triangle next to "Bivariate Fit..." and select **Fit Line.**

You should see the following screen:

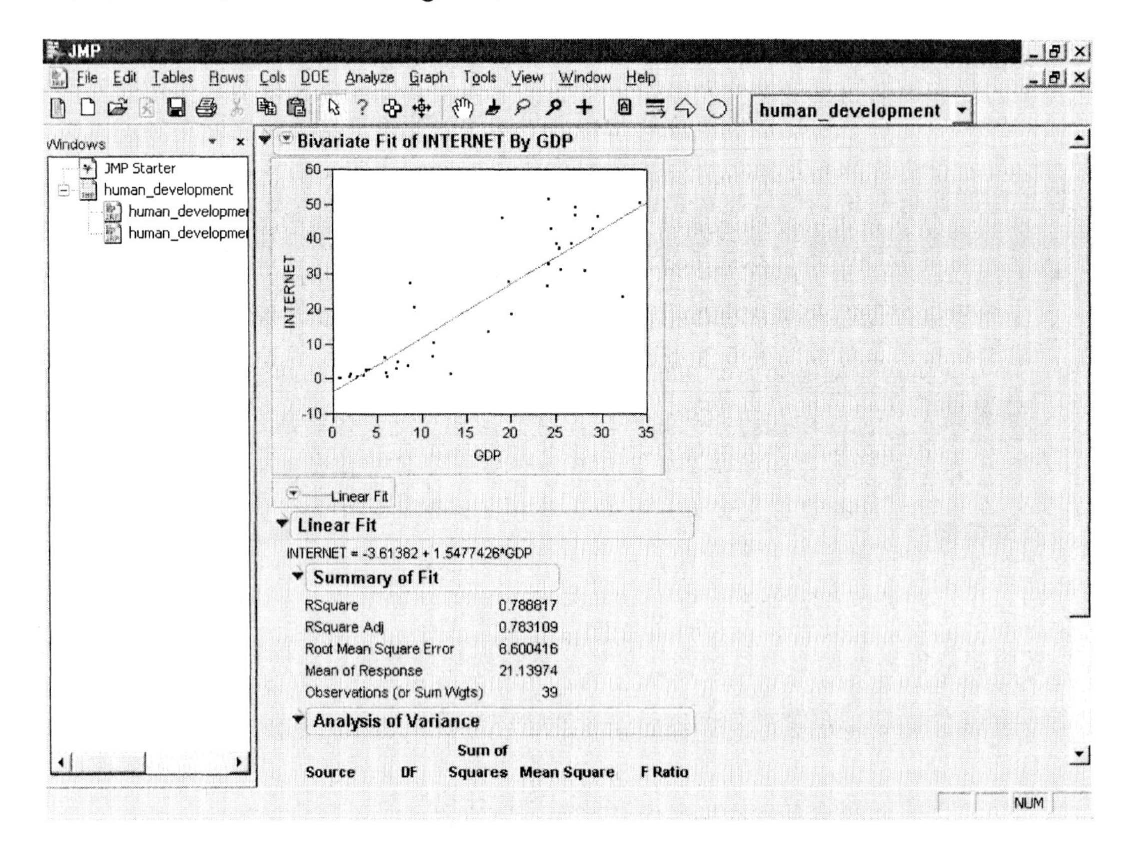

Chapter 12. Exercise 12.17. Student ideology.

Open the fl_student_survey data from the JMP folder of the data disk.

Correlation analysis requires two quantitative ratio scale variables. Looking at the dataset, you will notice that the newspapers variable has a series of red vertical bars as its icon, indicating that it is currently coded as a qualitative nominal scale variable. Since the data in this variable is truly quantitative (number of newspapers read), it needs to be recoded as a quantitative ratio scale variable. To do this, right click on the variable name "newspapers" (you can also right click on the newspapers column). Select **Column Info**. Set the **Data Type** to **Numeric** and the **Modeling Type** to **Continuous**. Select **Apply** and then **OK**. At this point, the variable newspapers should now have a blue triangle as its icon, indicating that it is now recognized as a quantitative ratio scale variable.

To generate the correlation between newspapers and political ideology, select **Analyze>Multivariate Methods>Multivariate**. Assign both variables to the **Y, Columns** role. Select **OK**. Toggle to the top.

You should see the following screen:

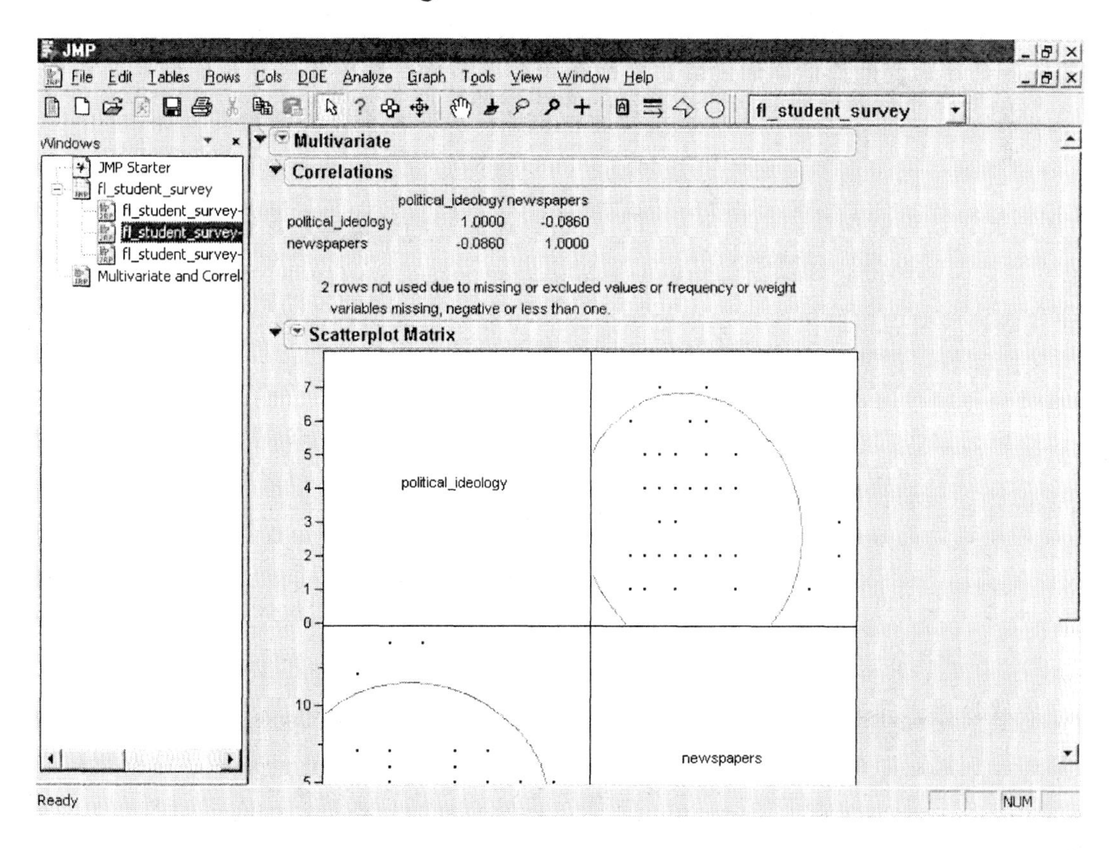

Chapter 12. Exercise 12.20. GPAs and TV Watching.

Open the georgia_student_survey data from the JMP folder of the data disk.

To generate the correlation matrix, select **Analyze>Multivariate Methods>Multivariate**. Assign ***HSGPA, CGPA*** and ***WatchTV*** to the **Y, Columns** role. Select **OK**. Toggle to the top of the output.

You should see the following screen:

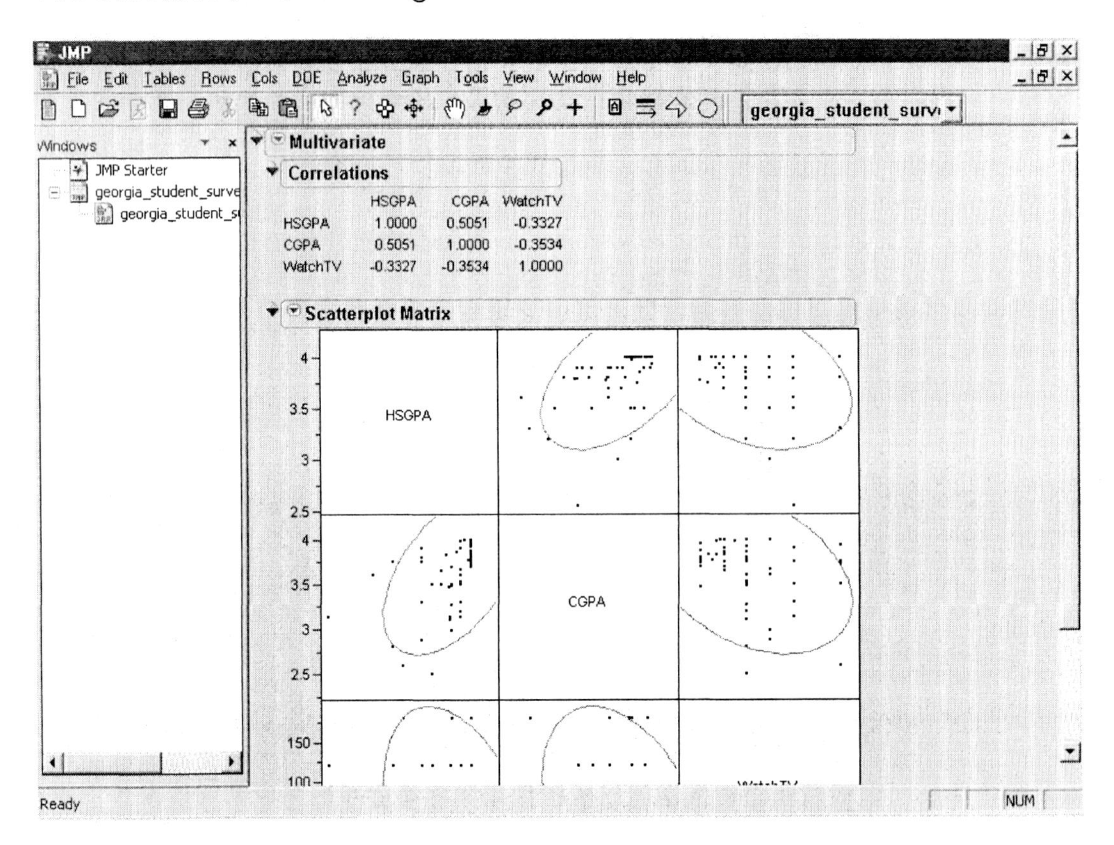

Recall that r^2 values can be generated by squaring the relevant terms in the matrix above.

Chapter 12. Exercise 12.30. Violent crime and single-parent families.

Open the us_statewide_crime data from the JMP folder of the data disk.

To analyze the effect of single parent families on the violent crime rate, select **Analyze>Fit Y by X**. Assign ***violent crime rate*** to the **Y, Columns** role and ***single parent*** to the **X, Factor** role. Select **OK**. To generate the regression line, click on the red triangle next to "Bivariate Fit...". Select **Fit Line**.

You should see the following screen:

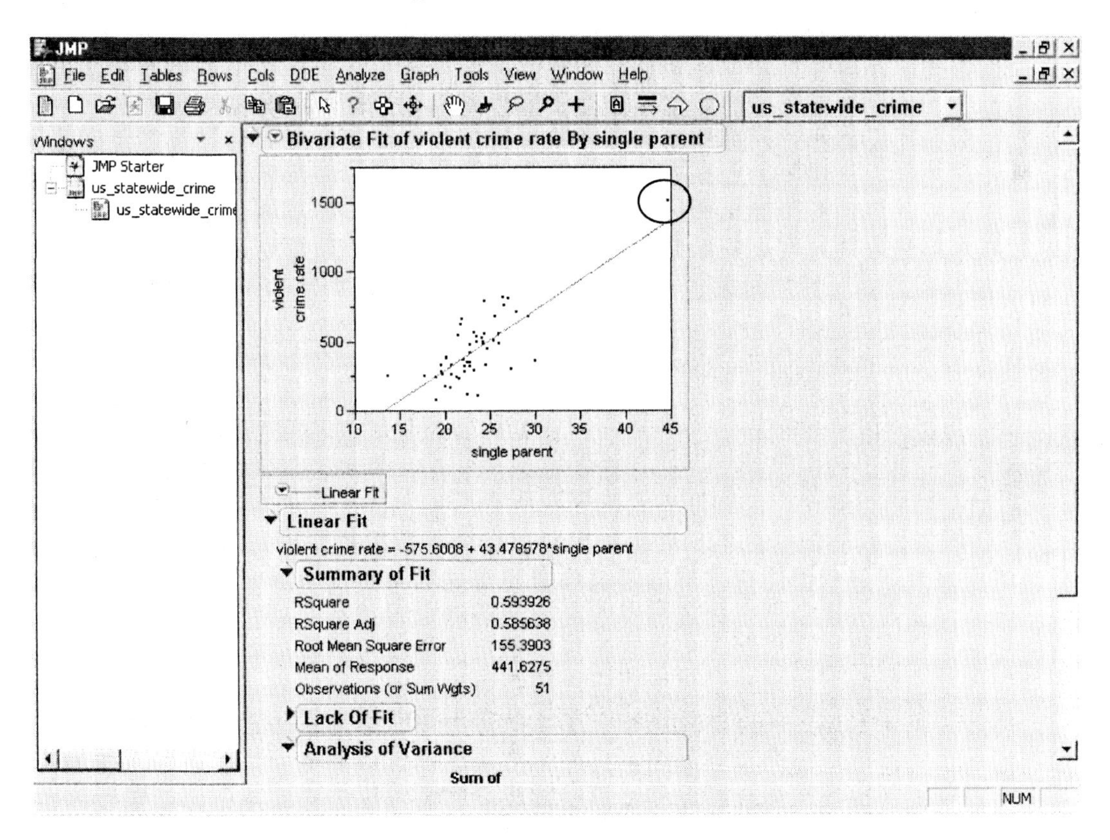

To rerun the analysis without the outlier identified above, click on the outlier. You should see the number '9' appear, indicating that this is the 9th observation in the dataset. Return to the dataset. The 9th observation (D.C.) should be highlighted. Select **Rows>Exclude/Unexclude**. Rerun the same regression analysis as before.

You should see the following screen:

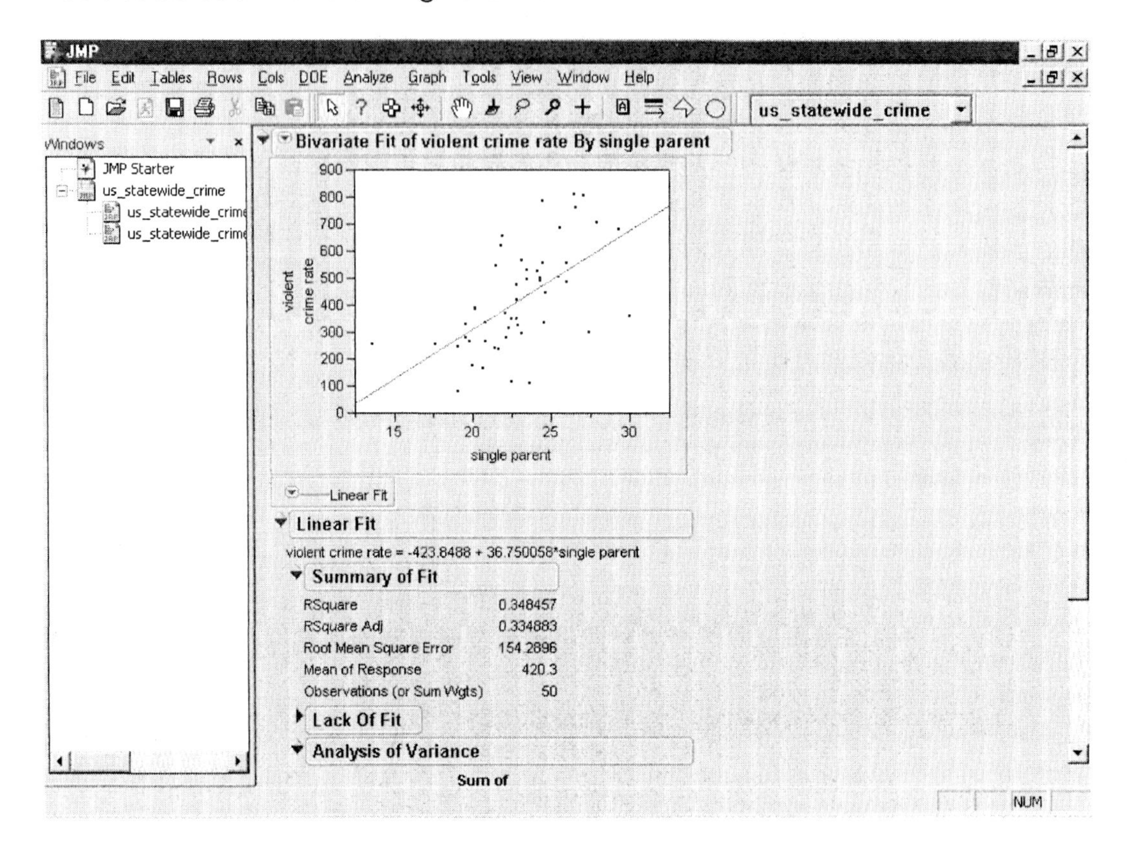

Chapter 12. Example 11. Is Strength Associated with 60-Pound Bench Press?

Open the high_school_female_athletes data from the JMP folder of the data disk.

To analyze whether the number of times an athlete can lift a 60 pound bench press helps to predict the maximum number of pounds the athlete can bench press, select **Analyze>Fit Y by X**. Assign ***BP*** to the **Y, Columns** role and ***BP60*** to the **X, Factor** role. Select **OK**. To generate the regression line and parameter estimates, click on the red triangle next to "Bivariate Fit..." and select **Fit Line**.

You should see the following screen:

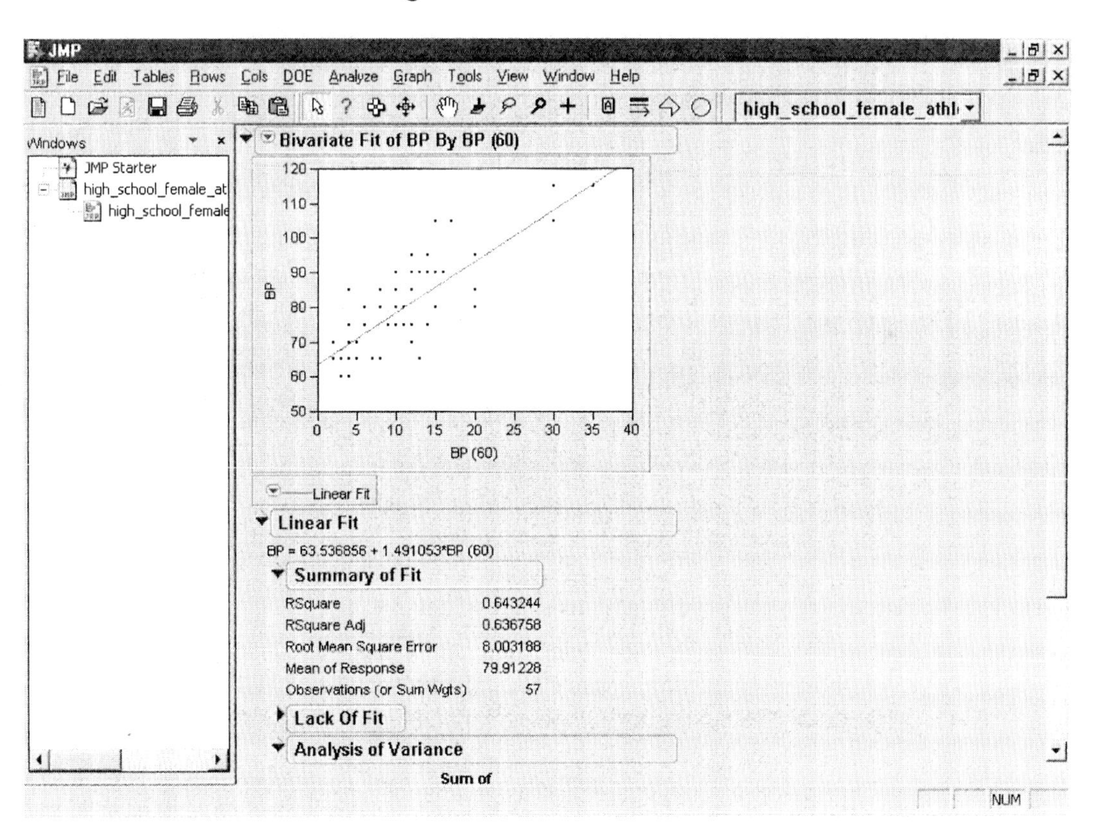

Toggle to the bottom of the screen to see the t-statistic associated with the test of H_0: β=0:

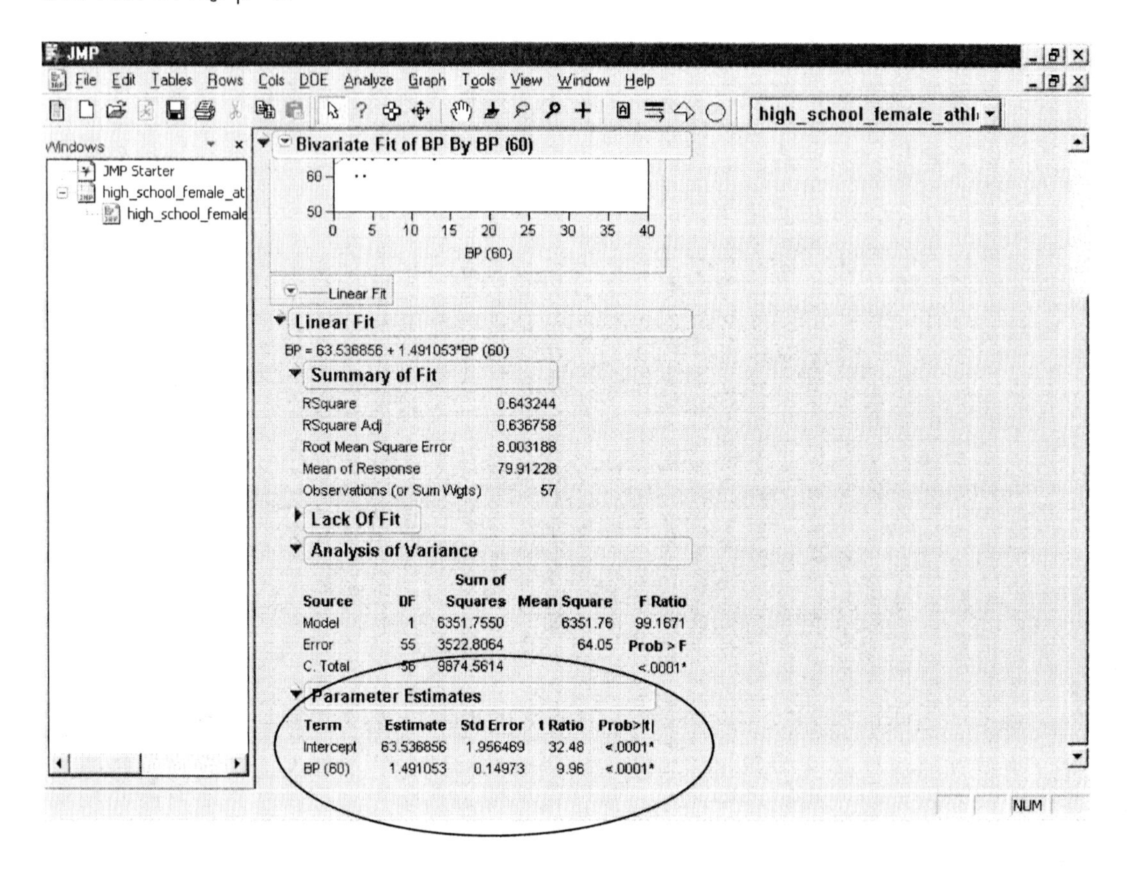

Chapter 12. Exercise 12.33. Predicting house prices.

Open the housing_prices data from the JMP folder of the data disk.

To evaluate the effect of size on housing price, select **Analyze>Fit Y By X**. Assign ***price*** to the **Y, Response** role and ***size*** to the **X, Factor** role. Select **OK**. When the output is generated, click on the red triangle next to "Bivariate Fit…". Select **Fit Line.**

You should see the following screen:

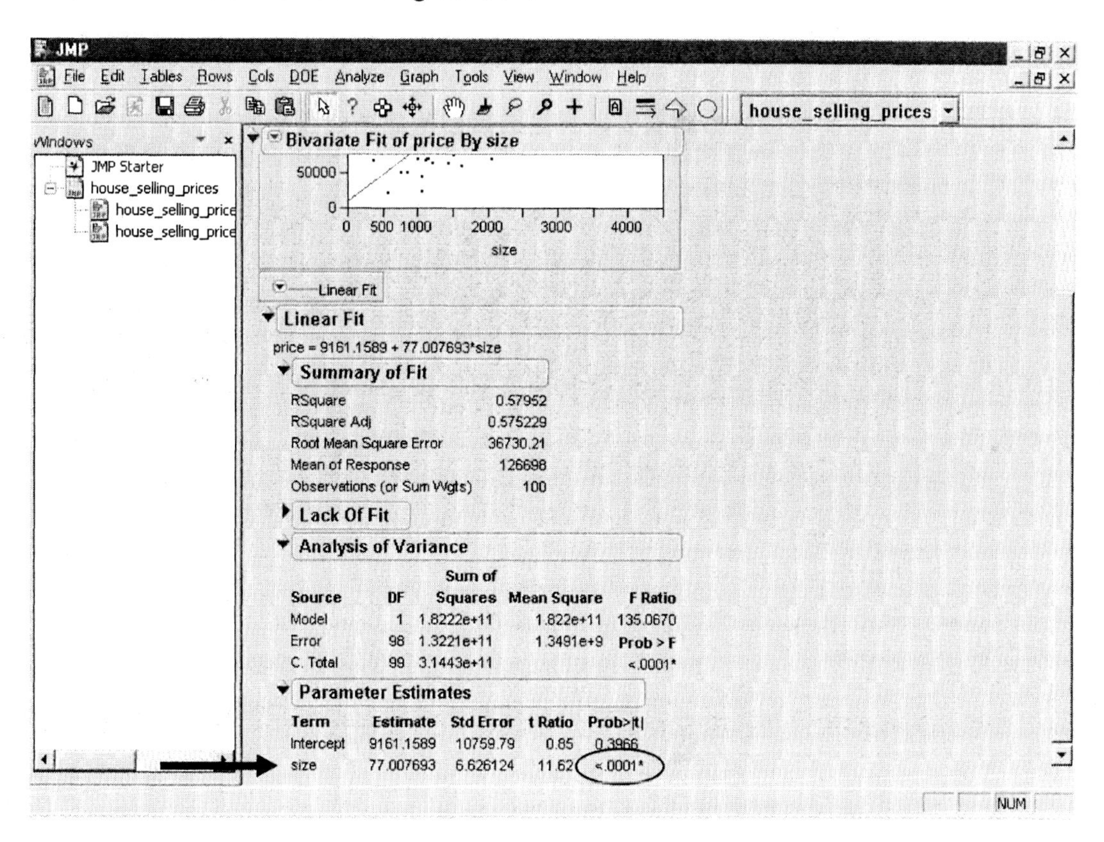

The information needed to assess the sample association (part a) can be found circled above. The confidence interval for the slope (i.e., parameter estimate for the predictor size) can be found by placing the cursor on the word "size" as listed in the terms of the Parameter Estimates section. Right click. Select **Columns>Upper 95%** and then **Columns>Lower 95%.**

You should see the following screen:

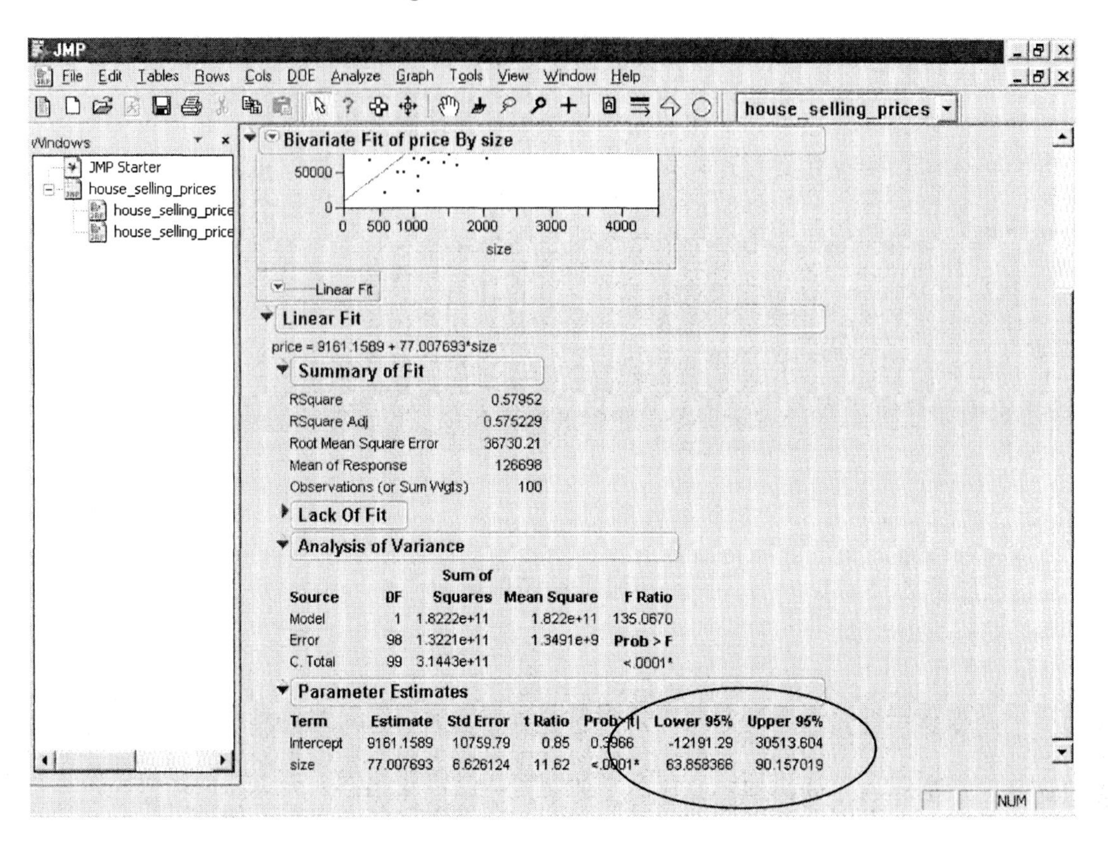

Chapter 12. Exercise 12.39. Advertising and Sales.

From the JMP Starter menu, select **New Data Table**.

In the new table, enter the ***Advertising*** data into column 1 and the ***Sales*** data into column 2.

You should see the following screen:

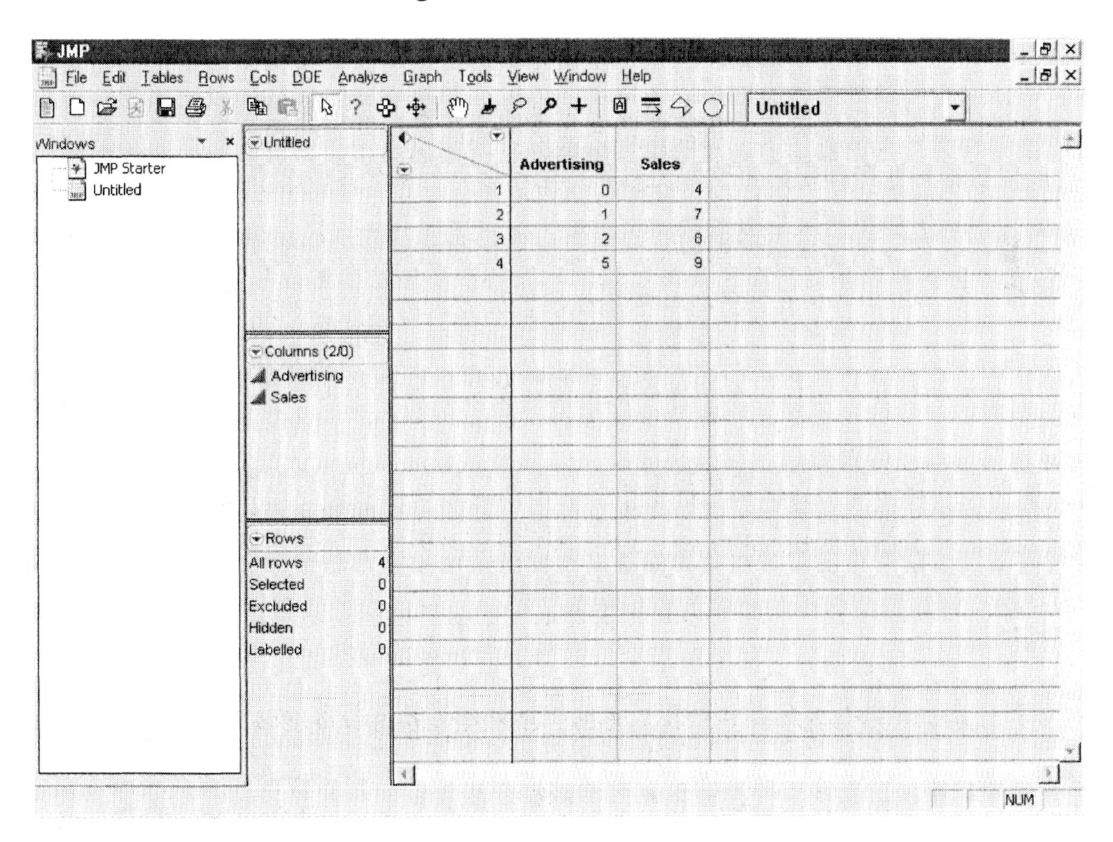

To determine the mean and standard deviation for the two variables, select **Analyze>Distribution**. Assign both variables to the **Y, Columns** role. Select **OK**. The descriptive statistics can be found at the bottom of the page.

To generate a regression equation and evaluate the significance of the slope, return to the datasheet. Select **Analyze>Fit Y By X**. Assign ***Sales*** to the **Y, Response** role and ***Advertising*** to the **X, Factor** role. Select **OK**. On the output page, click on the red triangle. Select **Fit Line**.

You should see the following screen:

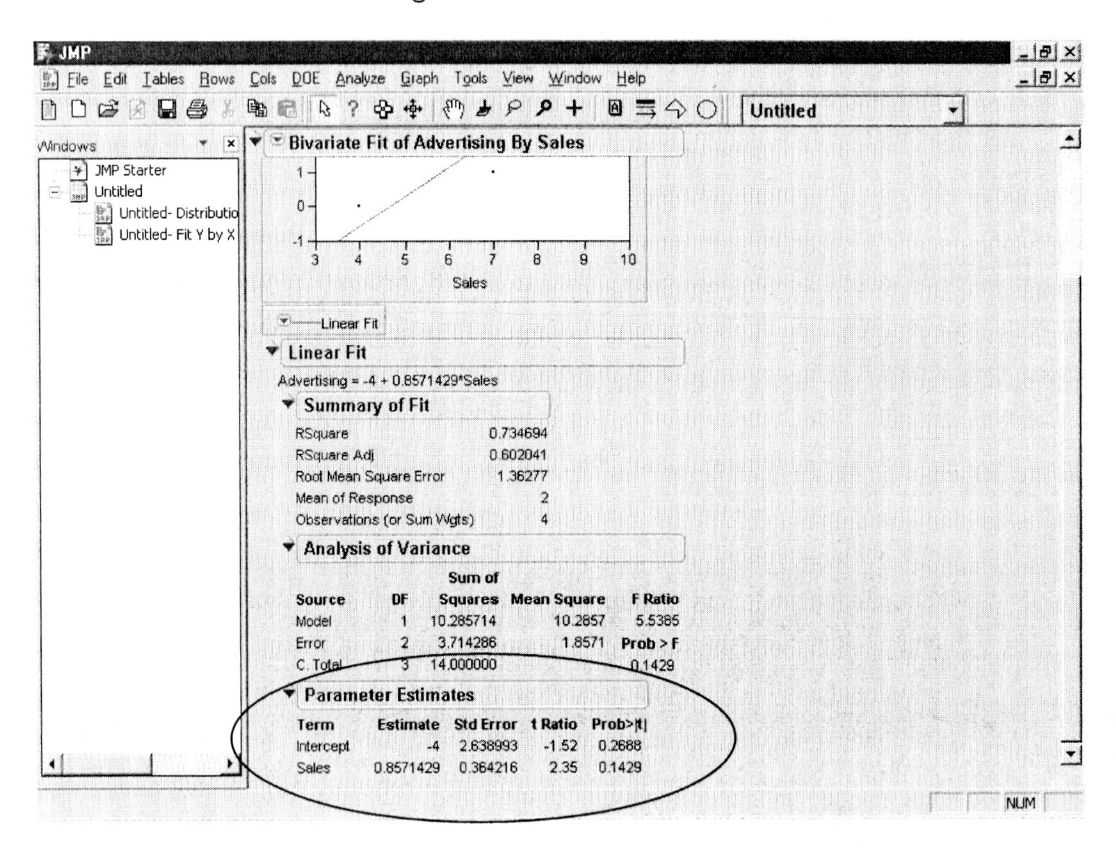

Chapter 12. Example 13. Detecting an Underachieving College Student.

Open the georgia_student_survey data from the JMP folder of the data disk.

To generate the regression equation for the effect of HS GPA on College GPA, select **Analyze>Fit Y By X**. Assign ***CGAP*** to the **Y, Response** role and **HSGPA** to the **X, Factor** role. Select **OK**. On the output page, click on the red triangle next to "Bivariate Fit...". Select **Fit Line**.

You should see the following screen:

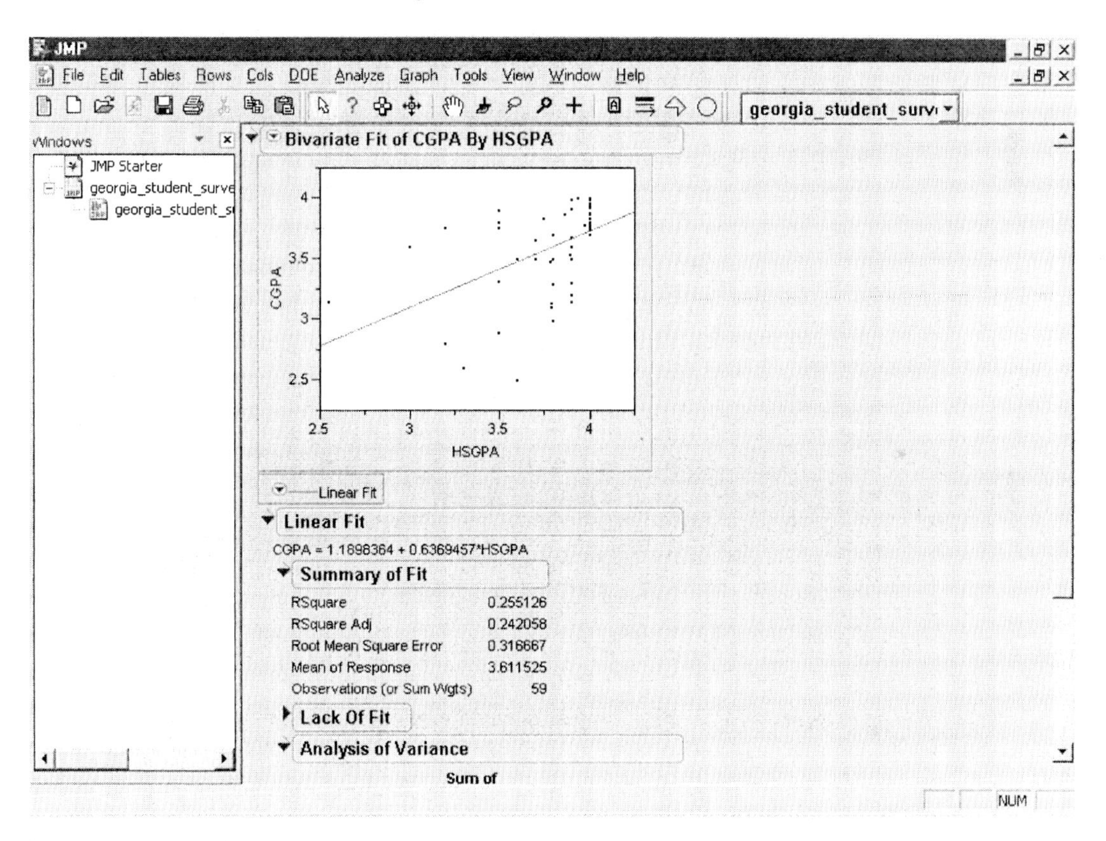

To evaluate the residuals of the model, click on the red triangle next to "Linear Fit". Select **Save Residuals**. Return to the datasheet. You should see a new column titled "***Residuals CGPA***". The values in this column represent the differences between the actual College GPA values and the predicted College GPA values. To identify any unusual observations, these values need to be standardized.

From the datasheet, select **Analyze>Distribution**. Assign ***Residuals CGPA*** to the **Y, Columns** role. Select **OK**. On the output page, click on the red triangle next to "Residuals CGPA". Select **Save>Standardized**. Return to the datasheet. You should see an additional column titled "***Std Residuals CGPA***". The values in this column represent the standardized values of the residuals (i.e. the z-scores). To sort the data so the largest values are at the top, select **Tables>Sort**. Assign ***Std Residuals CGPA*** as the **By** variable. Select **OK**.

You should see the following screen:

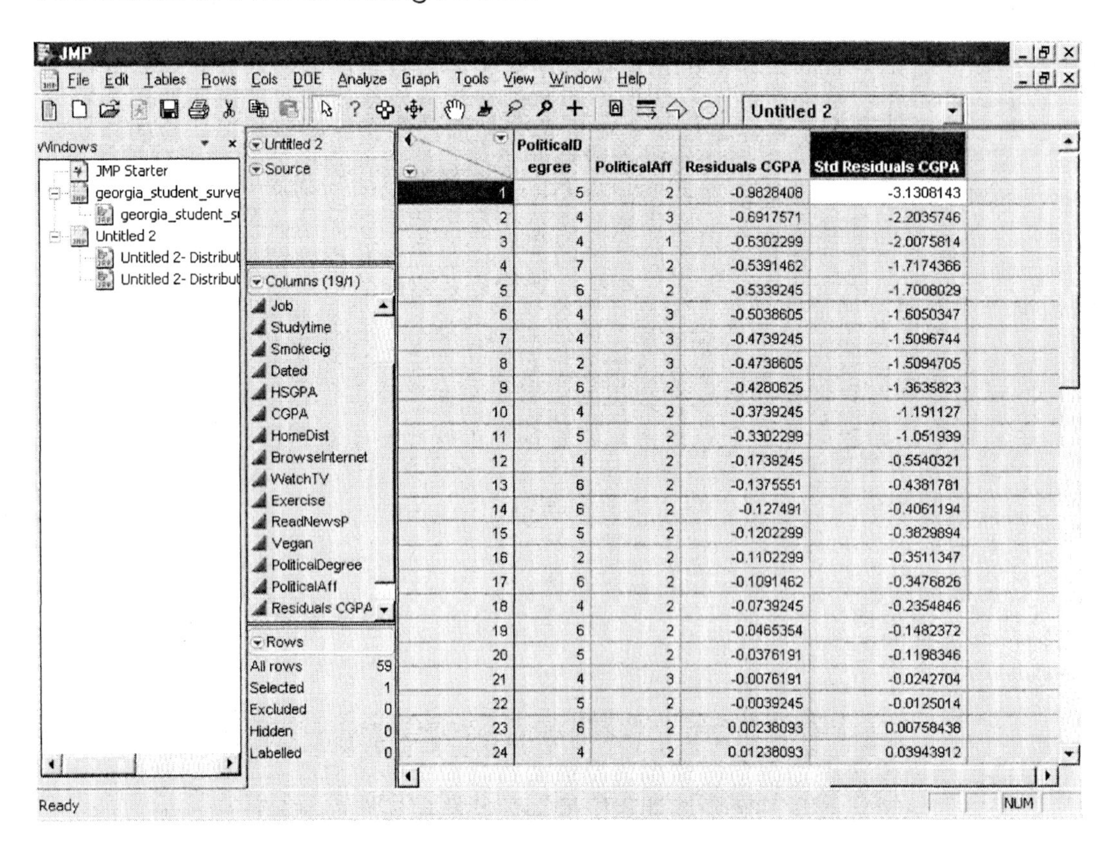

Chapter 12. Example 16. Predicting Maximum Bench Press and Estimating Its Mean.

Open the high_school_female_athletes data from the JMP folder of the data disk.

In this example, we require some additional information on the regression model that we cannot obtain by using the **Analyze>Fit Y By X** option. To obtain the confidence intervals for the mean and for the individual predictions, return to the JMP Starter menu. Select **Model>Fit Model.** Assign ***BP*** to the **Y** role and add ***BP60*** to the **Construct Model Effects** section. Select **Run Model.**

You should see the following screen:

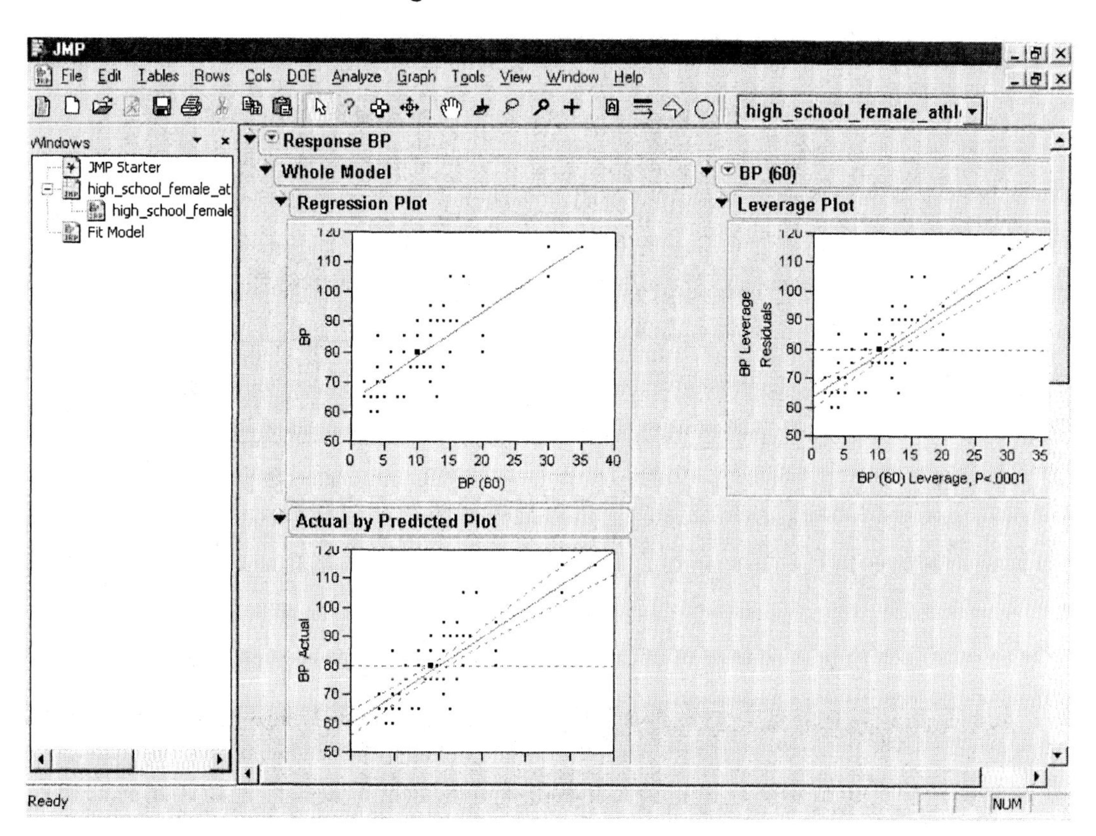

At the bottom of the screen are the same regression parameter estimates that we would have generated had we used **Analyze>Fit Y By X**. To generate the 95% confidence interval for the mean and for the individual predictions, click on the red triangle next to "Response BP". Select **Save Columns>Mean Confidence Interval** and **Save Columns>Ind Confidence Interval.**

Return to the datasheet. At the end of the variables (on the far right) you should see four additional columns:

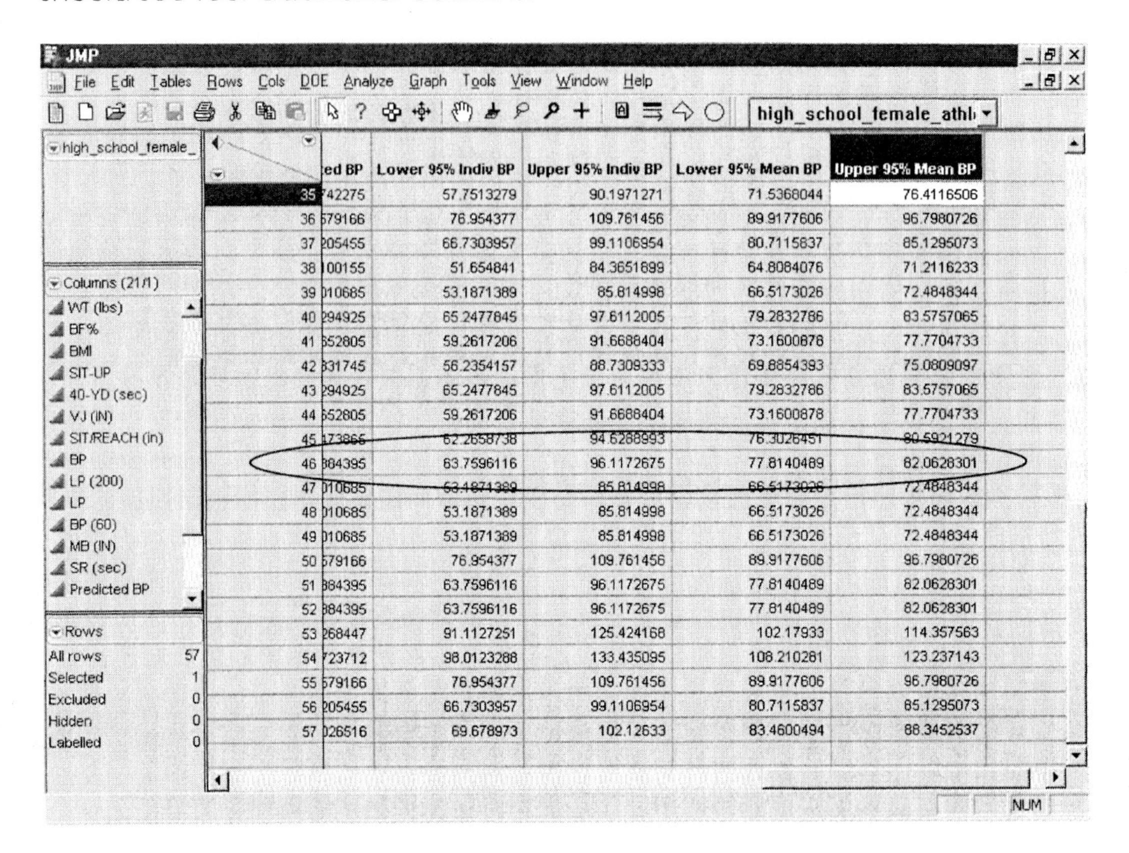

	ed BP	Lower 95% Indiv BP	Upper 95% Indiv BP	Lower 95% Mean BP	Upper 95% Mean BP
35	742275	57.7513279	90.1971271	71.5368044	76.4116506
36	579166	76.954377	109.761456	89.9177606	96.7980726
37	205455	66.7303957	99.1106954	80.7115837	85.1295073
38	100155	51.654841	84.3651699	64.8084076	71.2116233
39	010685	53.1871389	85.814998	66.5173026	72.4848344
40	294925	65.2477845	97.6112005	79.2832786	83.5757065
41	652805	59.2617206	91.6688404	73.1600878	77.7704733
42	831745	56.2354157	88.7309333	69.8854393	75.0809097
43	294925	65.2477845	97.6112005	79.2832786	83.5757065
44	652805	59.2617206	91.6688404	73.1600878	77.7704733
45	173866	62.2658738	94.6288993	76.3026451	80.5921279
46	884395	63.7596116	96.1172675	77.8140489	82.0628301
47	010685	53.1871389	85.814998	66.5173026	72.4848344
48	010685	53.1871389	85.814998	66.5173026	72.4848344
49	010685	53.1871389	85.814998	66.5173026	72.4848344
50	579166	76.954377	109.761456	89.9177606	96.7980726
51	884395	63.7596116	96.1172675	77.8140489	82.0628301
52	884395	63.7596116	96.1172675	77.8140489	82.0628301
53	268447	91.1127251	125.424168	102.17933	114.357563
54	723712	98.0123288	133.435095	108.210281	123.237143
55	579166	76.954377	109.761456	89.9177606	96.7980726
56	205455	66.7303957	99.1106954	80.7115837	85.1295073
57	026516	69.678973	102.12633	83.4600494	88.3452537

You can scroll down in the dataset to observation number 46, where BP60=11, and see that the 95% CI for the Mean = (77.81,82.06) and the 95% CI for the prediction Interval = (63.76,96.12).

Chapter 12. Exercise 12.43. Poor predicted strength.

Open the high_school_female_athletes data from the JMP folder of the data disk.

To generate the regression equation for the effect of ***BP60*** on **BP**, select **Analyze>Fit Y By X**. Assign ***BP*** to the **Y, Response** role and ***BP60*** to the **X, Factor** role. Select **OK**.

On the output page, click on the red triangle next to "Bivariate Fit...". Select **Fit line**. Click on the red triangle next to "Linear Fit". Select **Save Predicteds** and then select **Save Residuals**. When you return to the datasheet, you should see two additional columns:

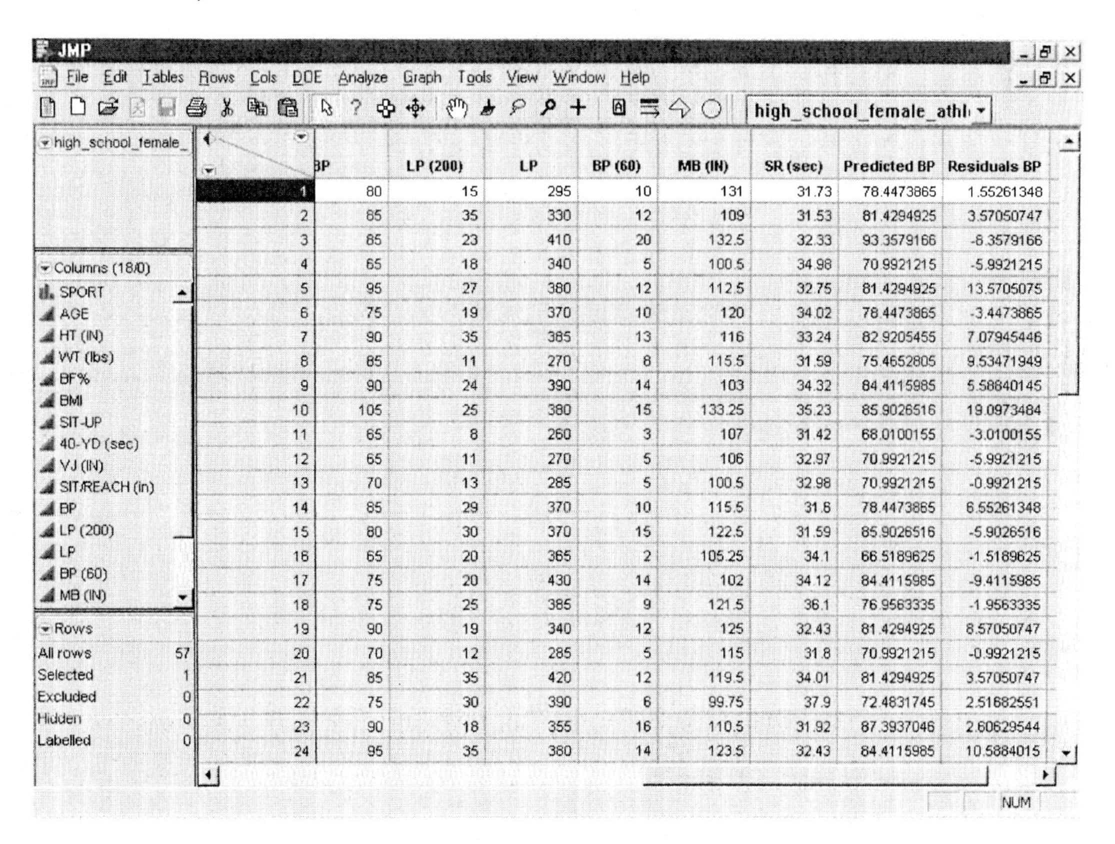

	BP	LP (200)	LP	BP (60)	MB (IN)	SR (sec)	Predicted BP	Residuals BP
1	80	15	295	10	131	31.73	78.4473865	1.55261348
2	85	35	330	12	109	31.53	81.4294925	3.57050747
3	85	23	410	20	132.5	32.33	93.3579166	-8.3579166
4	65	18	340	5	100.5	34.98	70.9921215	-5.9921215
5	95	27	380	12	112.5	32.75	81.4294925	13.5705075
6	75	19	370	10	120	34.02	78.4473865	-3.4473865
7	90	35	385	13	116	33.24	82.9205455	7.07945446
8	85	11	270	8	115.5	31.59	75.4652805	9.53471949
9	90	24	390	14	103	34.32	84.4115985	5.58840145
10	105	25	380	15	133.25	35.23	85.9026516	19.0973484
11	65	8	260	3	107	31.42	68.0100155	-3.0100155
12	65	11	270	5	106	32.97	70.9921215	-5.9921215
13	70	13	285	5	100.5	32.98	70.9921215	-0.9921215
14	85	29	370	10	115.5	31.6	78.4473865	6.55261348
15	80	30	370	15	122.5	31.59	85.9026516	-5.9026516
16	65	20	365	2	105.25	34.1	66.5189625	-1.5189625
17	75	20	430	14	102	34.12	84.4115985	-9.4115985
18	75	25	385	9	121.5	36.1	76.9563335	-1.9563335
19	90	19	340	12	125	32.43	81.4294925	8.57050747
20	70	12	285	5	115	31.8	70.9921215	-0.9921215
21	85	35	420	12	119.5	34.01	81.4294925	3.57050747
22	75	30	390	6	99.75	37.9	72.4831745	2.51682551
23	90	18	355	16	110.5	31.92	87.3937046	2.60629544
24	95	35	380	14	123.5	32.43	84.4115985	10.5884015

To generate an additional column of the standardized residuals, select **Analyze>Distribution**. Assign ***Residuals BP*** to the **Y, Columns** role. Select **OK**. From the output page, click on the red triangle next to "***Residuals BP***". Select **Save>Standardized**. When you return to the datasheet, there will be an additional column titled "***Std Residuals BP***".

Chapter 12. Exercise 12.49. ANOVA table for leg press.

Open the high_school_female_athletes data from the JMP folder of the data disk.

To evaluate the effect of the number of 200 pound leg presses on the maximum leg press weight, select **Analyze>Fit Y By X**. Assign ***LP*** to the **Y, Response** role and ***LP (200)*** to the **X, Factor** role. Select **OK**. On the output page, click on the red triangle next to "**Bivariate Fit...**". Select **Fit Line**. To generate the residuals, click on the red triangle next to "**Linear Fit**". Select **Save Residuals**. Return to the datasheet. There will be a new column titled "***Residuals LP***". To evaluate the standard deviation of these residuals, select **Analyze>Distribution**. Assign ***Residuals LP*** to the **Y, Columns** role. Select **OK**. The descriptive statistics will appear at the bottom of the page.

To generate the confidence interval around the x (LP200) values, select **Analyze>Dsitribution**. Assign ***LP (200)*** to the **Y, Columns** role. Select **OK**.

You should see the following screen:

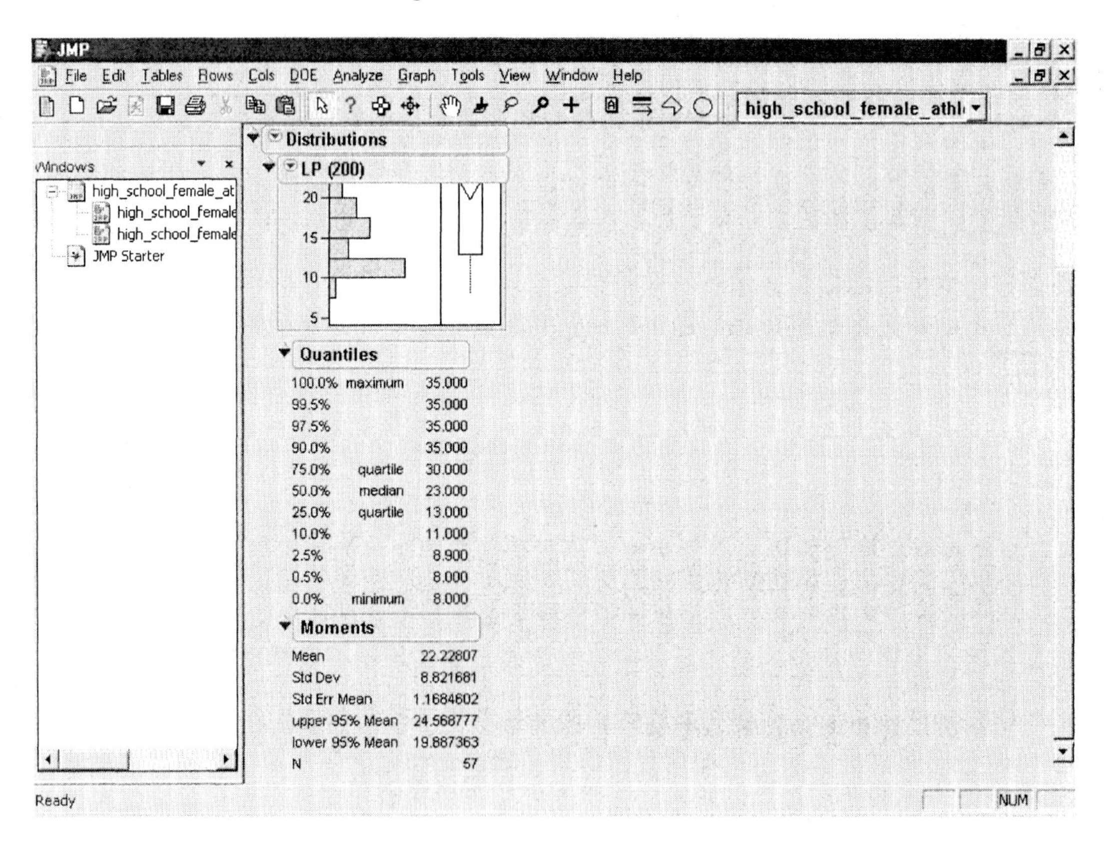

To generate the 95% Confidence Interval around the mean of 22.22, click on the red triangle next to LP (200) and select Confidence Interval. To generate the 95% Confidence Interval for the y (LP) values, execute the same steps as for the LP (200) values.

You should see the following screen:

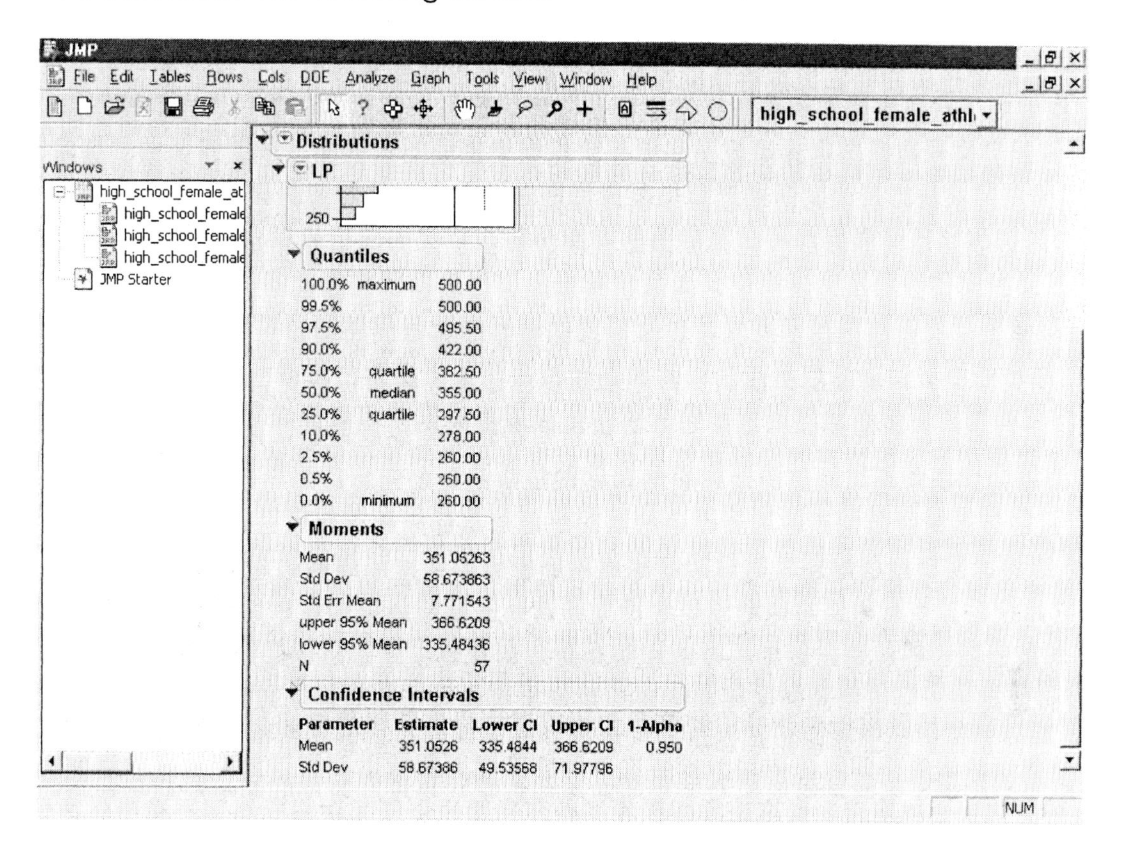

Chapter 13. Example 2. Predicted Selling Price Using House and Lot Size.

Open the house_selling_prices data from the JMP folder of the data disk.

To develop a multiple regression model with price as the response variable and house size and lot size as the explanatory variables, select **Analyze>Fit Model**.

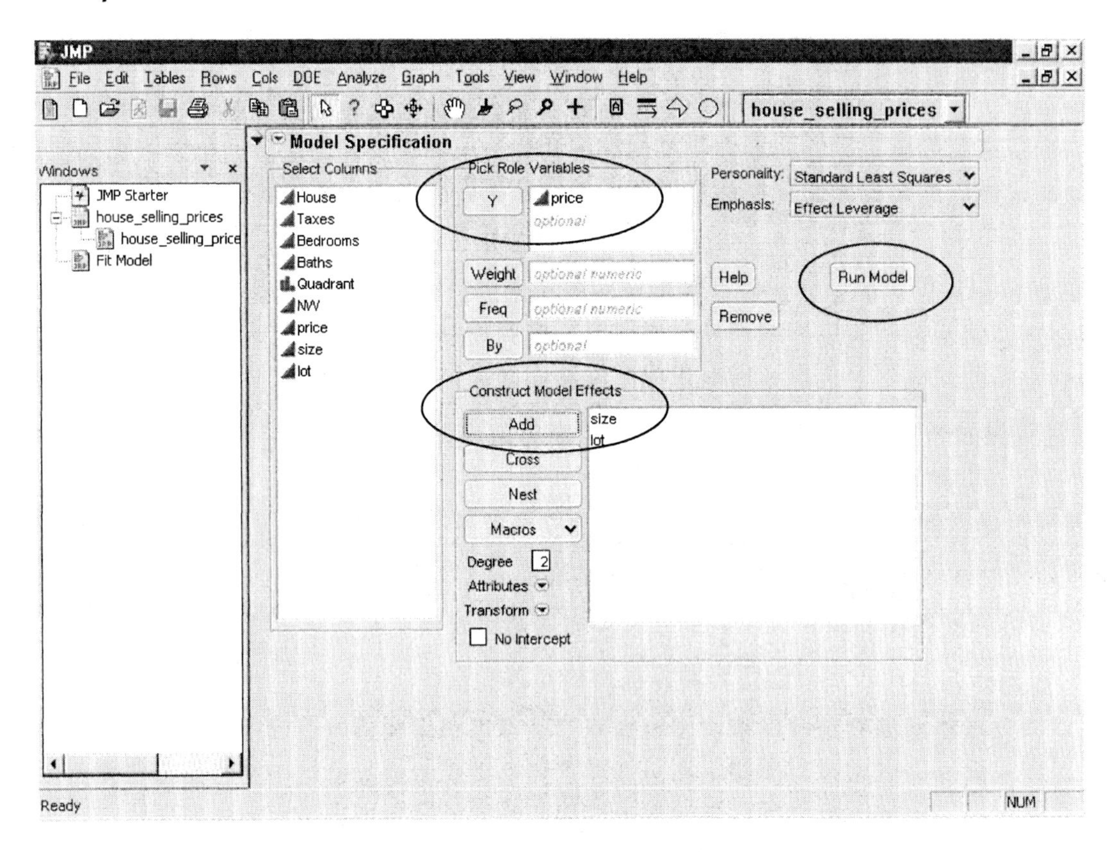

Assign price to the Y role and the Add size and lot as the Construct Model Effects. Select Run Model.

You should see the following screen:

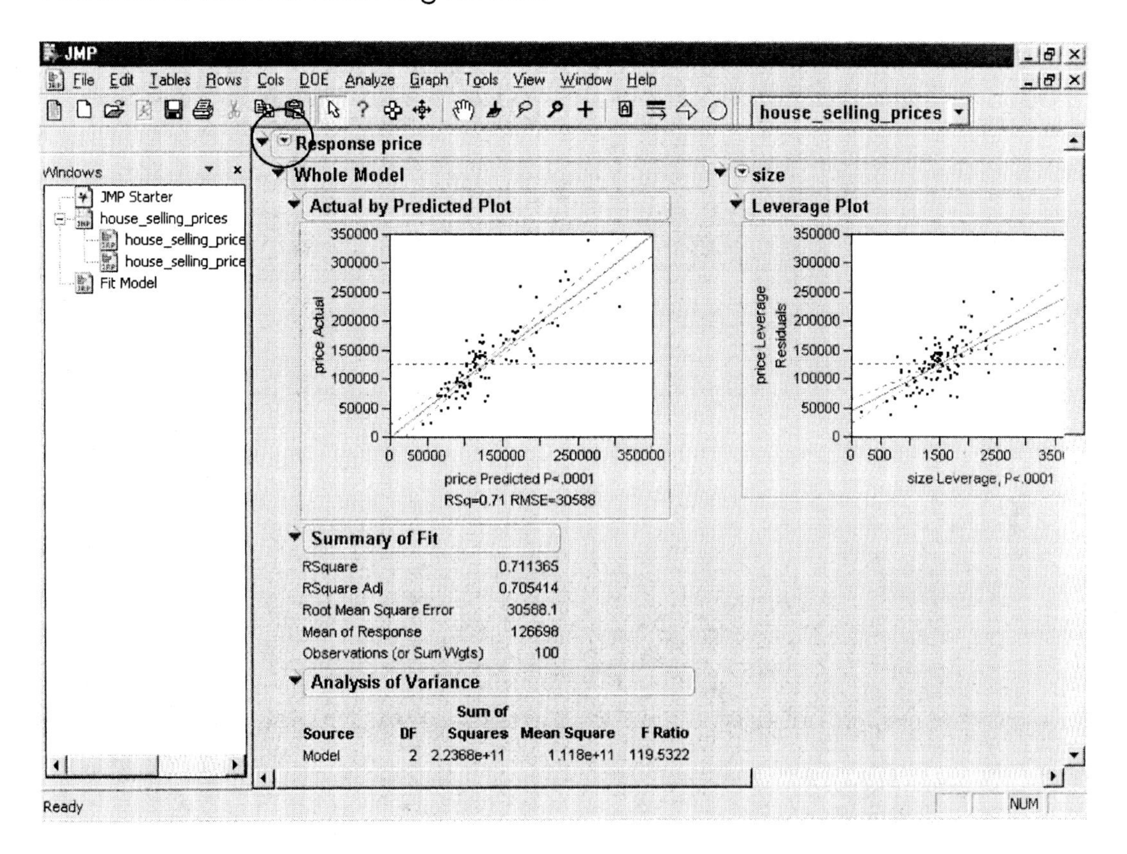

The parameter estimates for the model can be found at the bottom of the page.

To save the predictions and the residuals, click on the red triangle as identified above. Select **Save Columns>Predicted Values** and then **Save Columns>Residuals**. When you return to the datasheet, there will be two additional columns on the right titled "Predicted Price" and "Residual Price".

Chapter 13. Exercise 13.5. Does more education cause more crime?

Open the florida_crime data from the JMP folder of the data disk.

To assess the correlations among the variables crime, education and urbanization, select **Analyze>Multivariate Methods>Multivariate**. Assign ***crime, education and urbanization*** to the **Y, Columns** role. Select **OK**.

You should see the following screen:

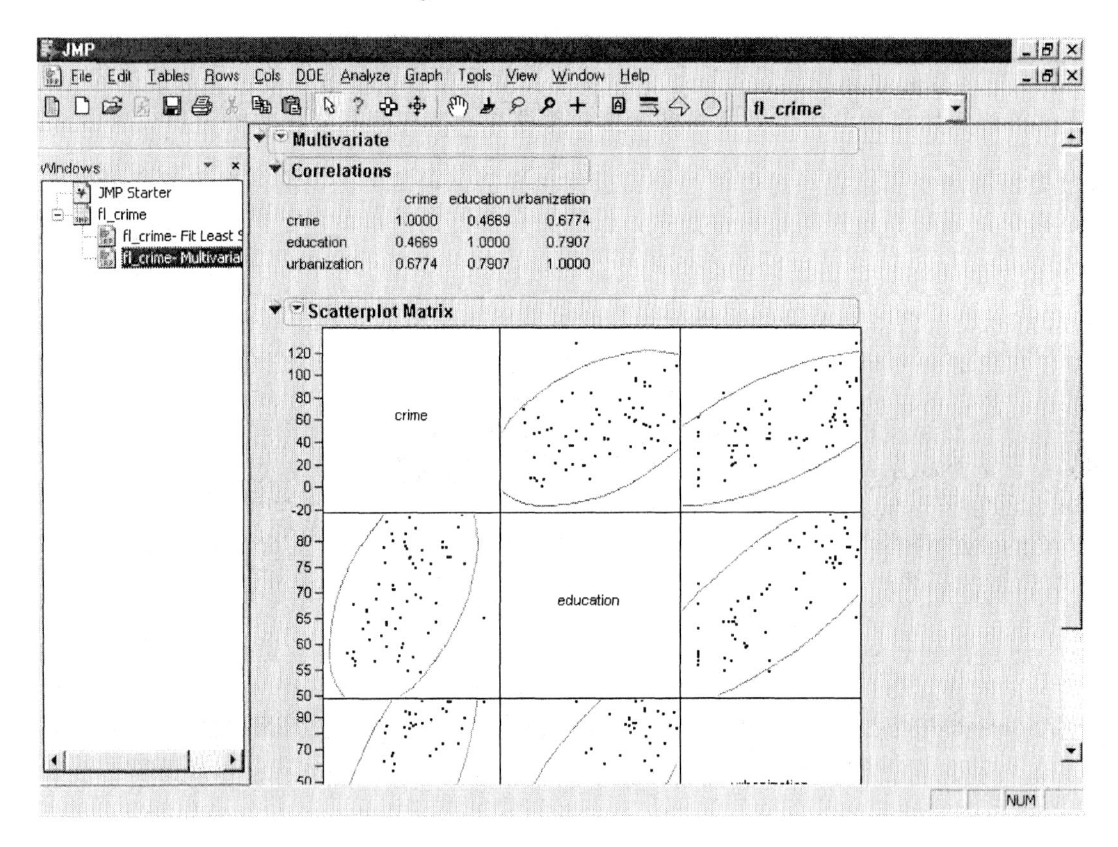

To develop a predictive model, select **Analyze>Fit Model**. Assign ***crime*** to the **Y role** and Add ***education*** and ***urbanization*** to the **Construct Model Effects**. Select **Run Model**.

You should see the following screen:

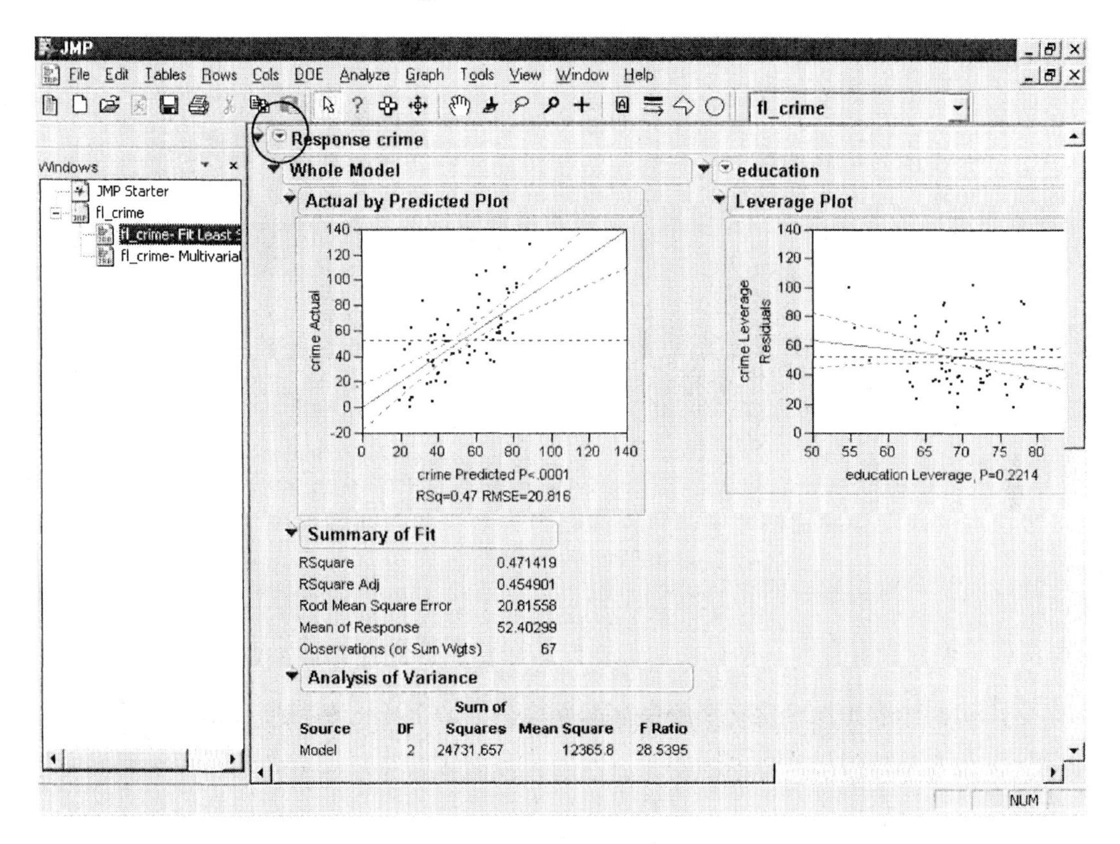

The parameter estimates for the model are at the bottom of the page.

To generate the prediction for 0% urbanization and 70% high school education, click on the red triangle as indicated above. Select **Save Columns>Prediction Formula**. This will create a new column in the dataset that has the prediction equation entered for each observation in the dataset (you can check this by clicking on the new column "***Pred formula crime***" and then selecting **Formula**. You should see the regression equation as the column formula). To generate the prediction as stated above, create a new observation, with 0 in the ***urbanization column*** and 70 in the ***education column***.

You should see the following screen:

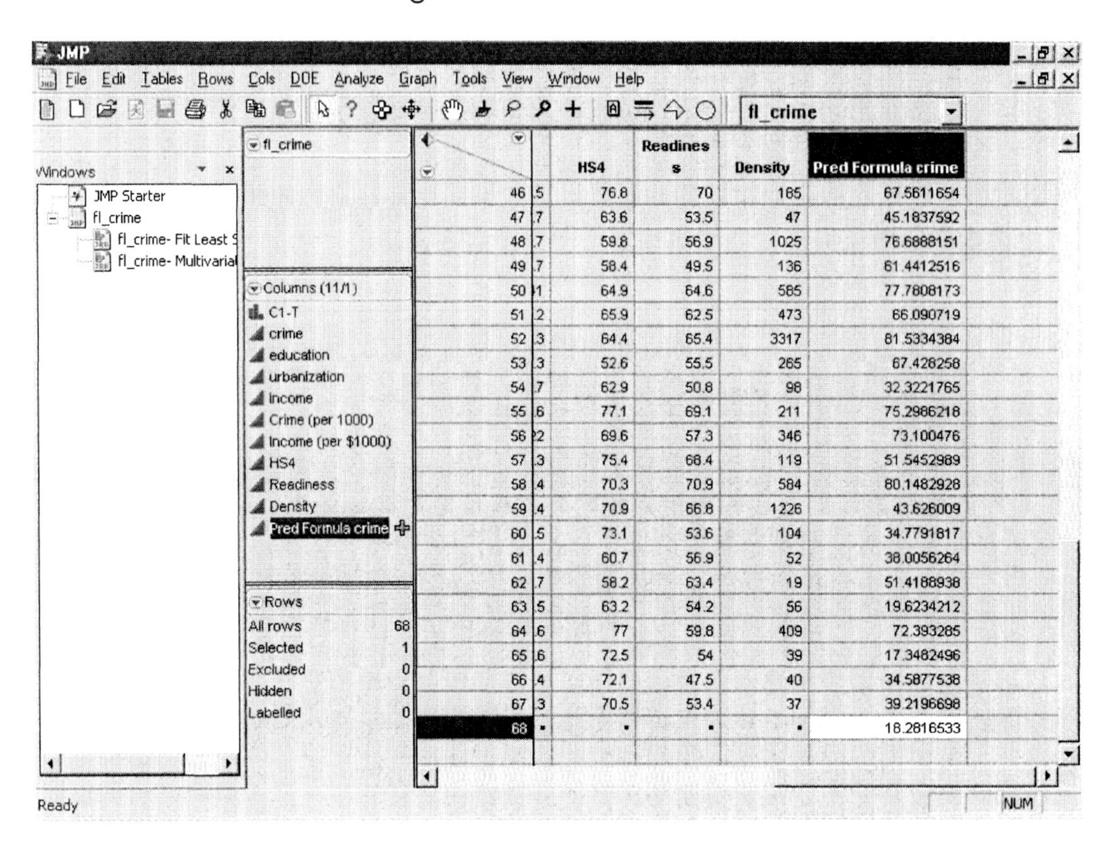

To generate additional predictions, create additional rows (observations).

Chapter 13. Example 3. How Well Can We Predict House Selling Prices?

Open the house_selling_prices data from the JMP folder of the data disk.

To generate the ANOVA table and the Multiple Regression Model where y=selling price and x1 is house size and x2 is lot size, select **Analyze>Fit Model**. Assign ***price*** to the **Y** role and Add ***size*** and ***lot*** to the **Construct Model Effects**. Select **Run Model**.

You should see the following screen:

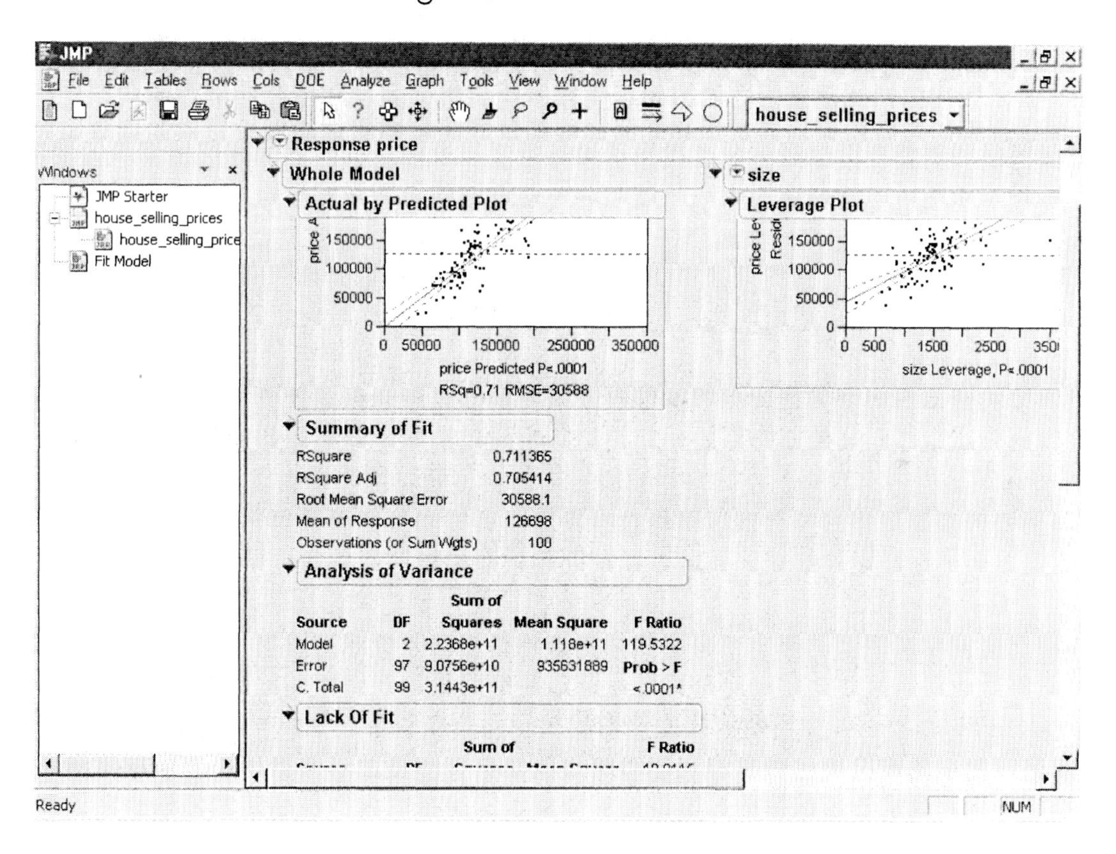

The information for the ANOVA and the Fit and Correlation statistics can be seen here.

Chapter 13. Example 4. What Helps Predict a Female Athlete's Weight?

Open the college_athletes data from the JMP folder of the data disk.

To generate the descriptive statistics for total body weight (TBW), height in inches (HGT), percent of body fat (%BF) and age, select **Analyze>Distribution**. Assign all four of these variables to the **Y, Columns** role. Select **OK**. All of the descriptive statistics will be available on the output screen.

To generate the regression equation, using weight as the response variable, select **Analyze>Fit Model**. Assign ***TBW*** as the **Y** variable and Add ***HGT, %BF*** and ***age*** as the **Construct Model Effects**. Select **Run Model.**

You should see the following screen:

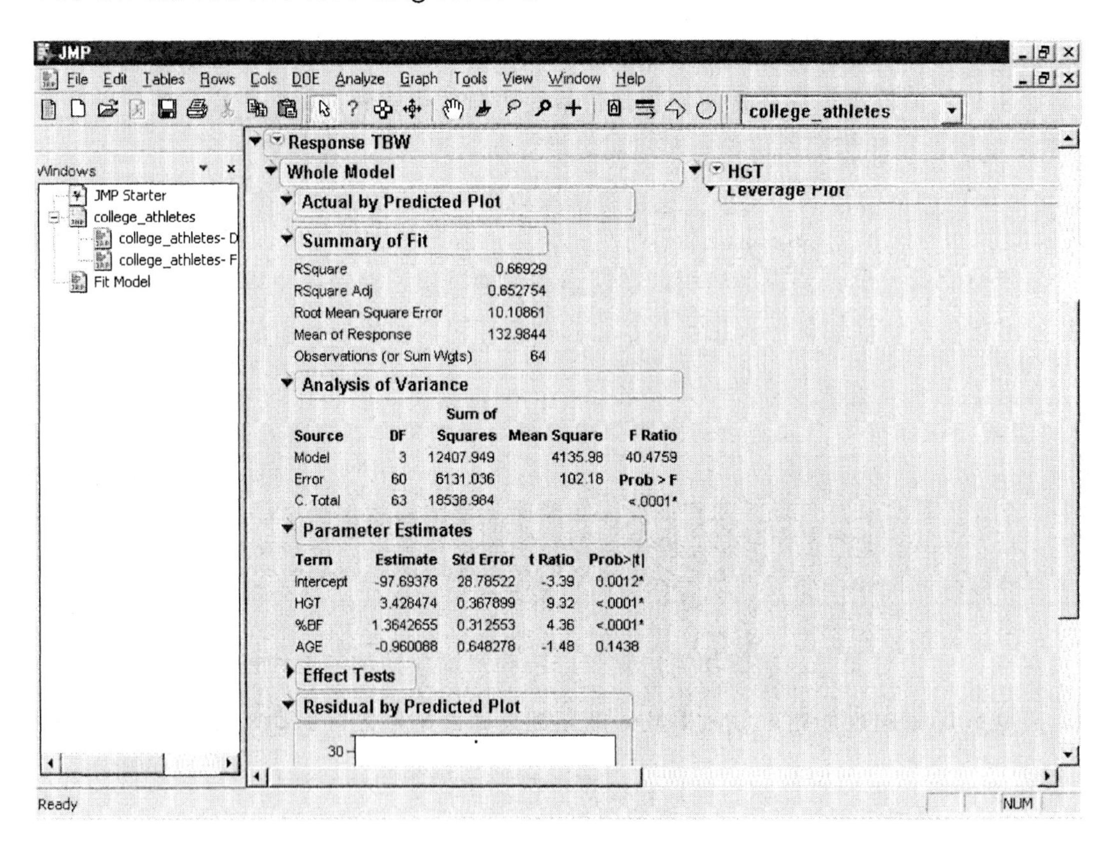

Chapter 13. Example 5. What's Plausible for the Effect of Age on Weight?

Open the college_athletes data from the JMP folder of the data disk.

To generate the regression equation, using weight as the response variable, select **Analyze>Fit Model**. Assign ***TBW*** as the **Y variable** and Add ***HGT, %BF*** and ***age*** as the Construct Model Effects. Select **Run Model**.

To generate the 95% Confidence Interval for the beta coefficient for age, place your cursor in the Parameter Estimates section. Right click. Select **Columns>Lower 95%** and then **Columns>Upper 95%.**

You should see the following screen:

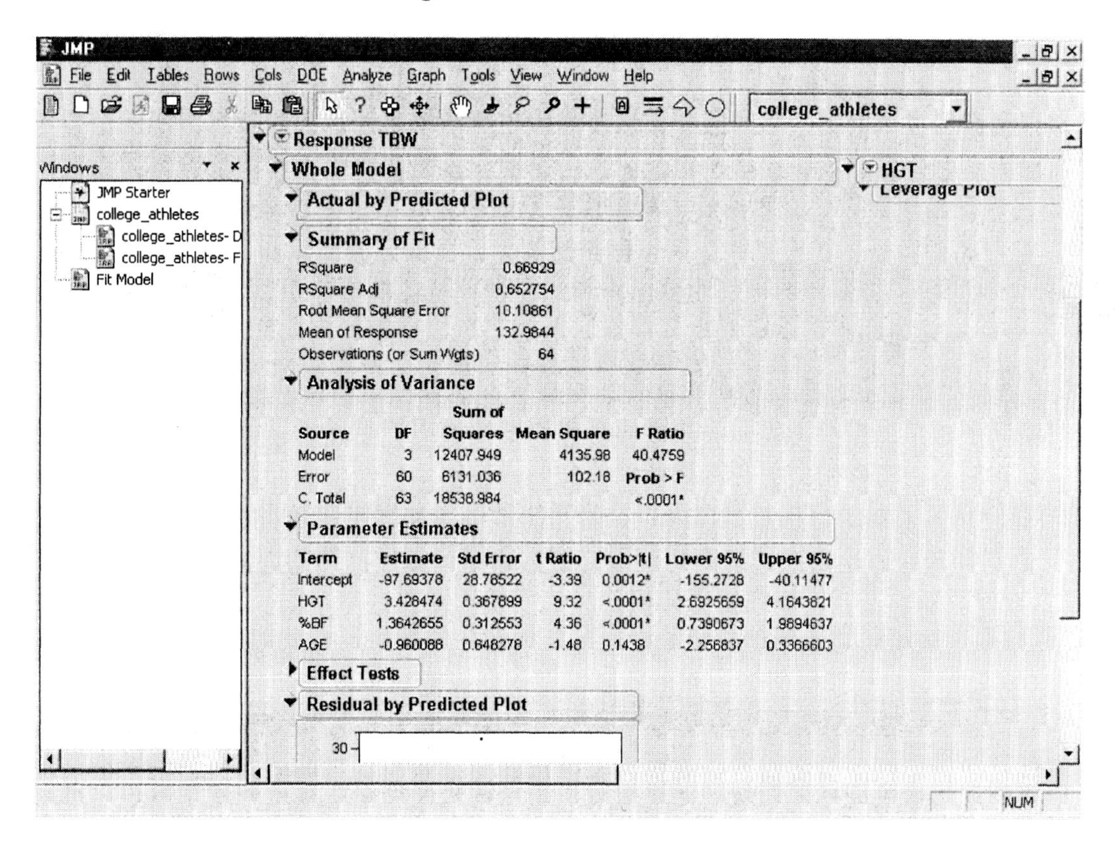

Chapter 13. Example 7. The F Test for Predictors of Athletes' Weight.

Open the college_athletes data from the JMP folder of the data disk.

To generate the ANOVA Table, select **Analyze>Fit Model**. Assign ***TBW*** as the **Y variable** and Add ***HGT, %BF*** and ***age*** as the **Construct Model Effects**. Select **Run Model**.

You should see the following screen:

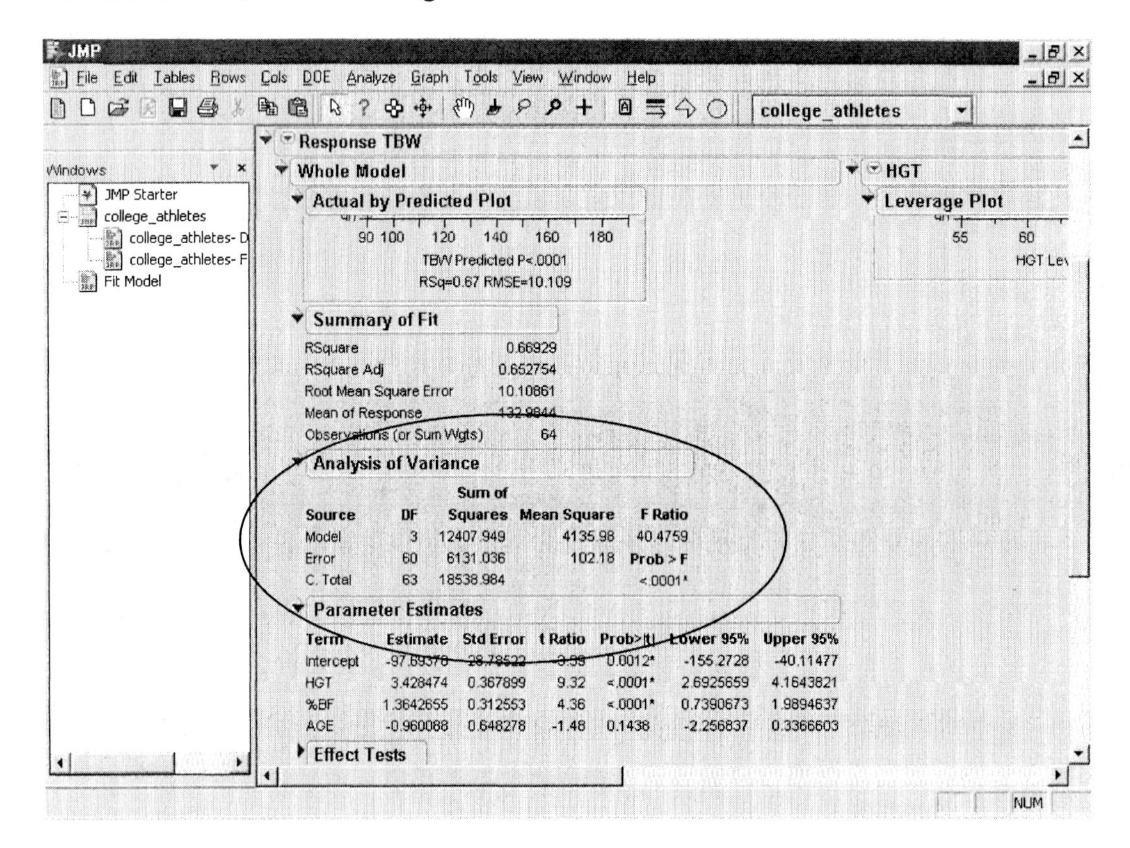

Chapter 13. Exercise 13.20. Study time help GPA?

Open the georgia_student_survey data from the JMP folder of the data disk.

To generate the regression equation where y=CGPA, x1 = HSGPA and x2 = Study time, select **Analyze>Fit Model**. Assign ***CGPA*** to the **Y,** role and Add ***HSGPA*** and ***Studytime*** to the **Construct Model Effects**. Select **Run Model**.

You should see the following screen:

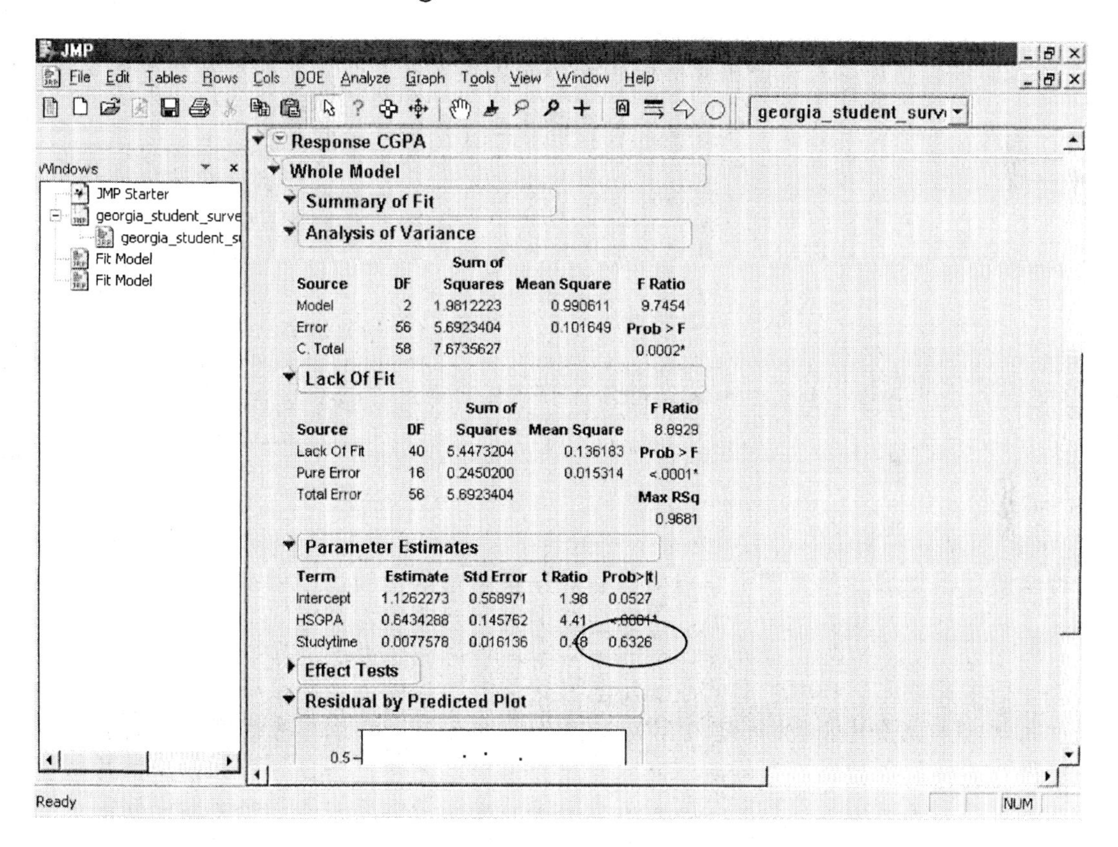

The p-value for the test of the slope coefficient for studytime being 0 is .6326 as circled above. The high p-value indicates that we would fail to reject the null hypothesis.

To find the 95% Confidence Interval around the slope estimate for study time, place your cursor inside the Parameter Estimates section and right click. Select **Columns>Lower 95%** and then **Columns>Upper 95%.**

You should see the following screen:

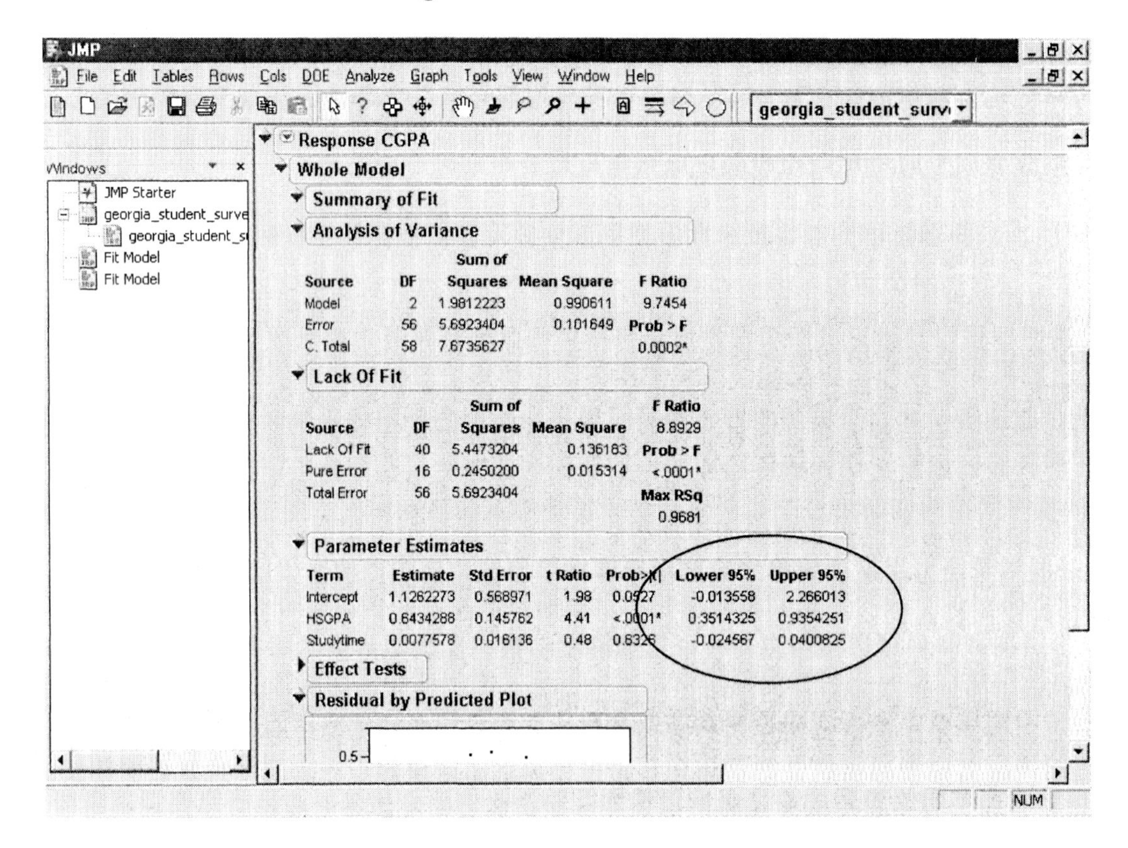

The 95% interval for the slope estimate is -.02 to .04, where the value of 0 is included in the interval. This is consistent with the finding from above.

At a fixed value of high school GPA, we infer that the population mean college GPA changes very little (if at all), again providing evidence for the conclusion that the estimate for the coefficient of studytime may not be different from 0.

Chapter 13. Exercise 13.29. More predictors for house price.

Open the house_selling_prices data from the JMP folder of the data disk.

To generate the parameter estimates using y=price, x_1 = size, x_2 = lot size and x_3 = real estate tax, select **Analyze>Fit Model**. Assign ***price*** to the **Y**, role and Add ***size, lot*** and ***Taxes*** to the **Construct Model Effects**. Select **Run Model**.

You should see the following screen:

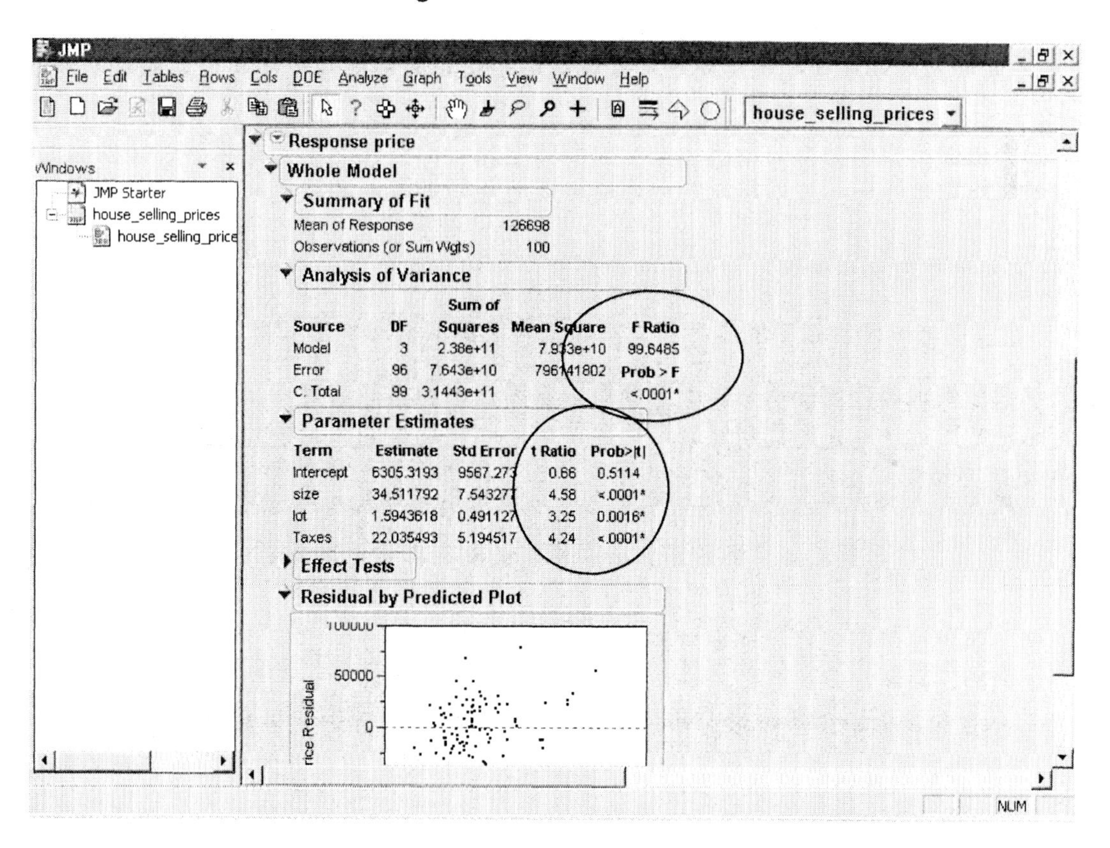

From the output, the null hypothesis would be that all of the beta values equals 0. Given the F-statistic of 99 and the associated p-value of <.0001, we can conclude that at least one of the beta values is not 0, indicating that at least one of the variables has an effect on housing price.

The t-statistics (and associated p-values) indicate that all of the variables in the model have an effect on housing price.

Chapter 13. Example 9. Another Residual Plot for House Selling Prices.

Open the house_selling_prices data from the JMP folder of the data disk.

To generate the residual plot from the regression model where y=price, x_1 = size and x_2 = lot size, select **Analyze>Fit Model**. Assign ***price*** to the **Y**, role and **Add *size*** and ***lot*** to the **Construct Model Effects**. Select **Run Model.**

You should see the following screen (toggle to the bottom):

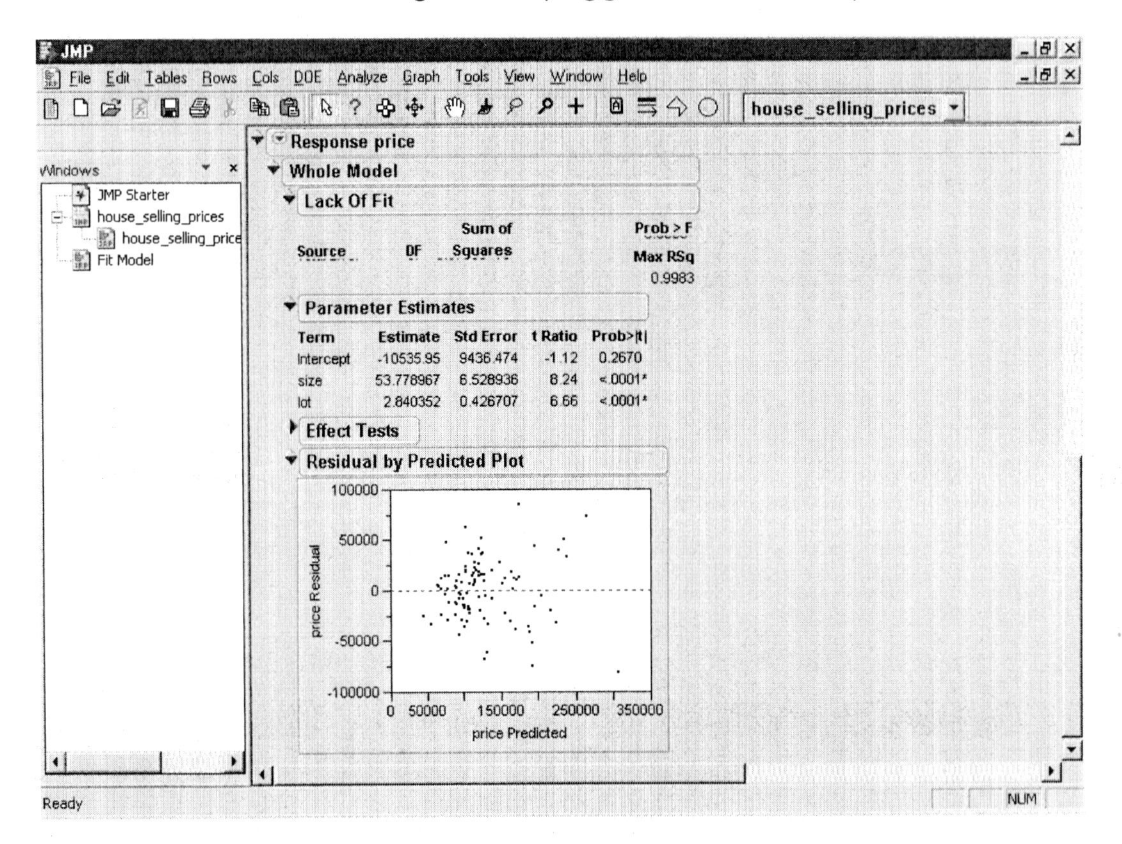

Chapter 13. Exercise 13.45. Houses, tax and NW.

Open the house_selling_prices data from the JMP folder of the data disk.

To generate the residual plot from the regression model where y=price, x_1 = real estate tax and x_2 = NW, select **Analyze>Fit Model**. Assign ***price*** to the **Y**, role and **Add *Taxes*** and ***NW*** to the **Construct Model Effects**. Select **Run Model**.

You should see the following screen:

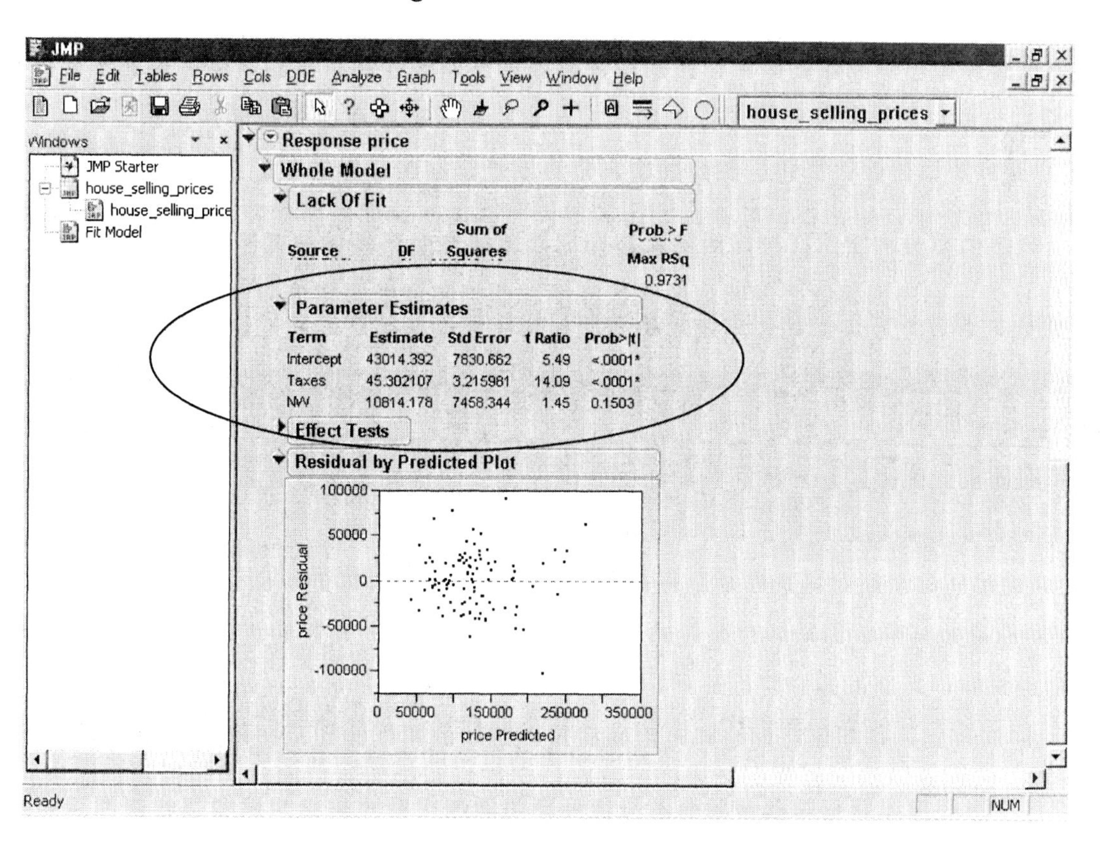

The regression equation can be derived from the Parameter Estimates section above: price = 43014+45.30(taxes)+10814(NW). Since the parameter estimate for NW is positive, this would indicate that houses in the NW section are more valuable than those outside of this section. Specifically, a house in the NW, is $10,814 more valuable than a house which is not in this section.

Chapter 13. Example 12. Annual Income and Having a Travel Credit Card.

Open the credit_card_and_income data from the JMP folder of the data disk.

Because Logistic Regression requires a nominal (1/0) dependent variable, we need to redefine the y variable in this dataset from numeric/continuous to numeric/nominal. To do this, right click on the y column. Select **Column Info**. Change the **Modeling Typ**e from **Continuous** to **Nominal**. After you do this, the icon designating the y variable will change from a blue triangle to a series of red bars – indicating that it is qualitative variable.

To generate a Logistic Regression model, where y=possession of a credit card and x=income, select **Analyze>Fit Y By X**. Assign ***Y*** to the **Y, Response** role and ***income*** to the **X, Factor** role. Select **OK**.

You should see the following screen:

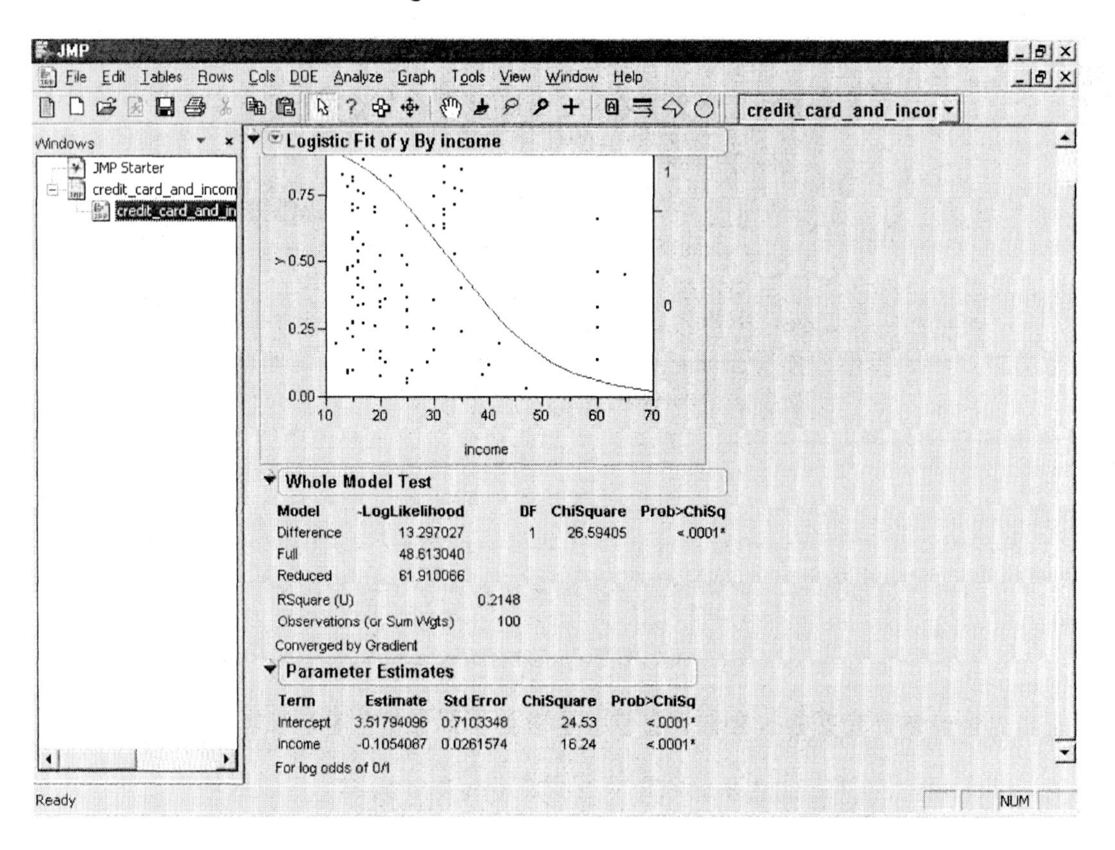

Chapter 14. Example 2. How Long Will You Tolerate Being Put On Hold?

From the JMP Starter Menu, select **New Data Table.**

Enter the data on hold times as follows:

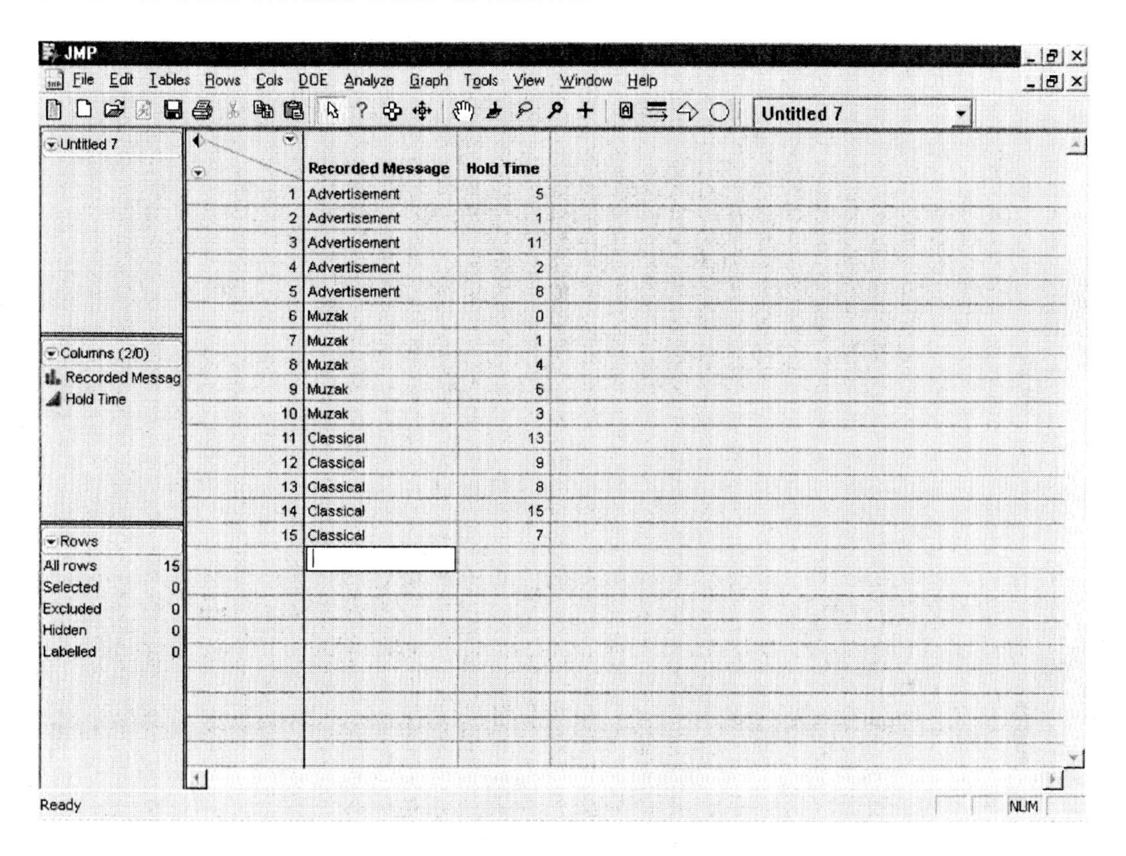

	Recorded Message	Hold Time
1	Advertisement	5
2	Advertisement	1
3	Advertisement	11
4	Advertisement	2
5	Advertisement	8
6	Muzak	0
7	Muzak	1
8	Muzak	4
9	Muzak	6
10	Muzak	3
11	Classical	13
12	Classical	9
13	Classical	8
14	Classical	15
15	Classical	7

To determine the means by recorded message, select **Analyze>Distribution**. Assign ***Hold Time*** to the **Y, Columns** role and ***Recorded Message*** to the **By** role. Select **OK**.

You should see the following screen:

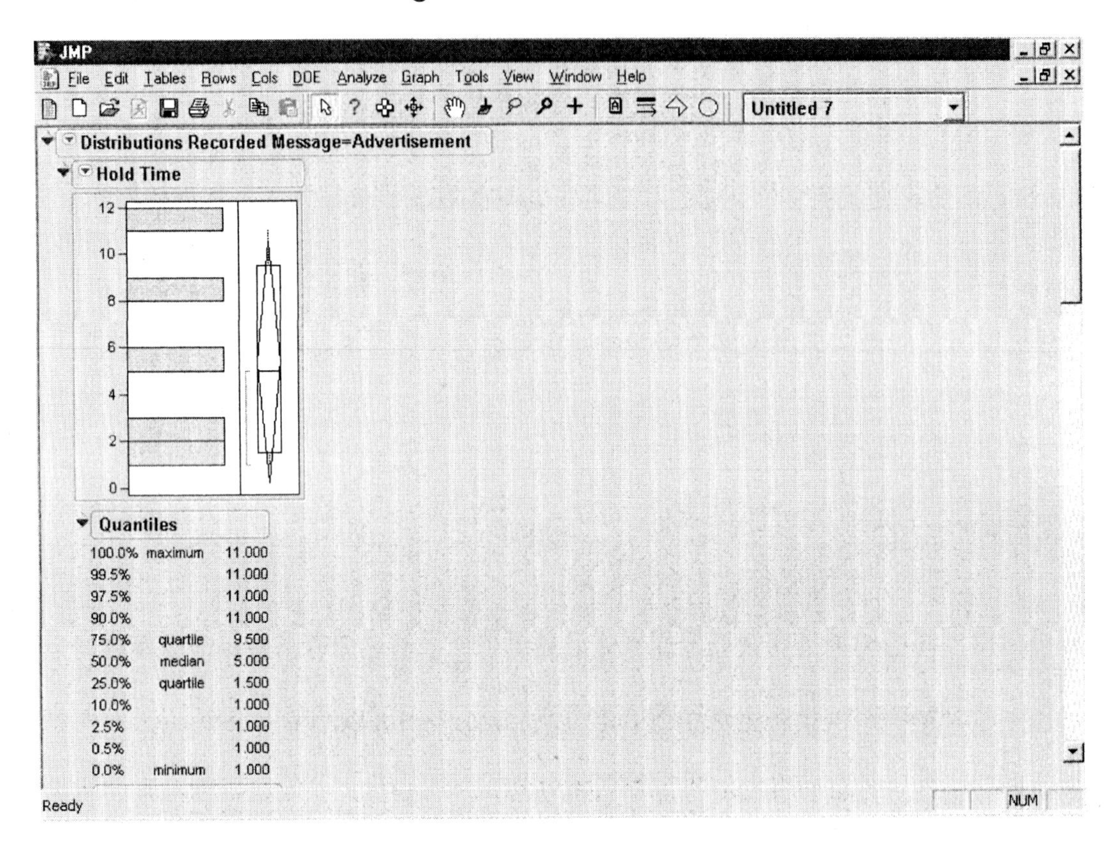

All of the descriptive statistics for each type of recorded message can be viewed on this output screen.

Chapter 14. Example 3. ANOVA for Customers' Telephone Holding Times.

To generate the ANOVA table from Example 2, return to the datasheet. Select **Analyze>Fit Y By X**. Assign ***Hold Time*** to the **Y, Response** role and ***Recorded Message*** to the **X, Factor** role. Select **OK**. In the output screen, click on the red triangle next to "Oneway Analysis...". Select **Means/ANOVA**.

You should see the following screen:

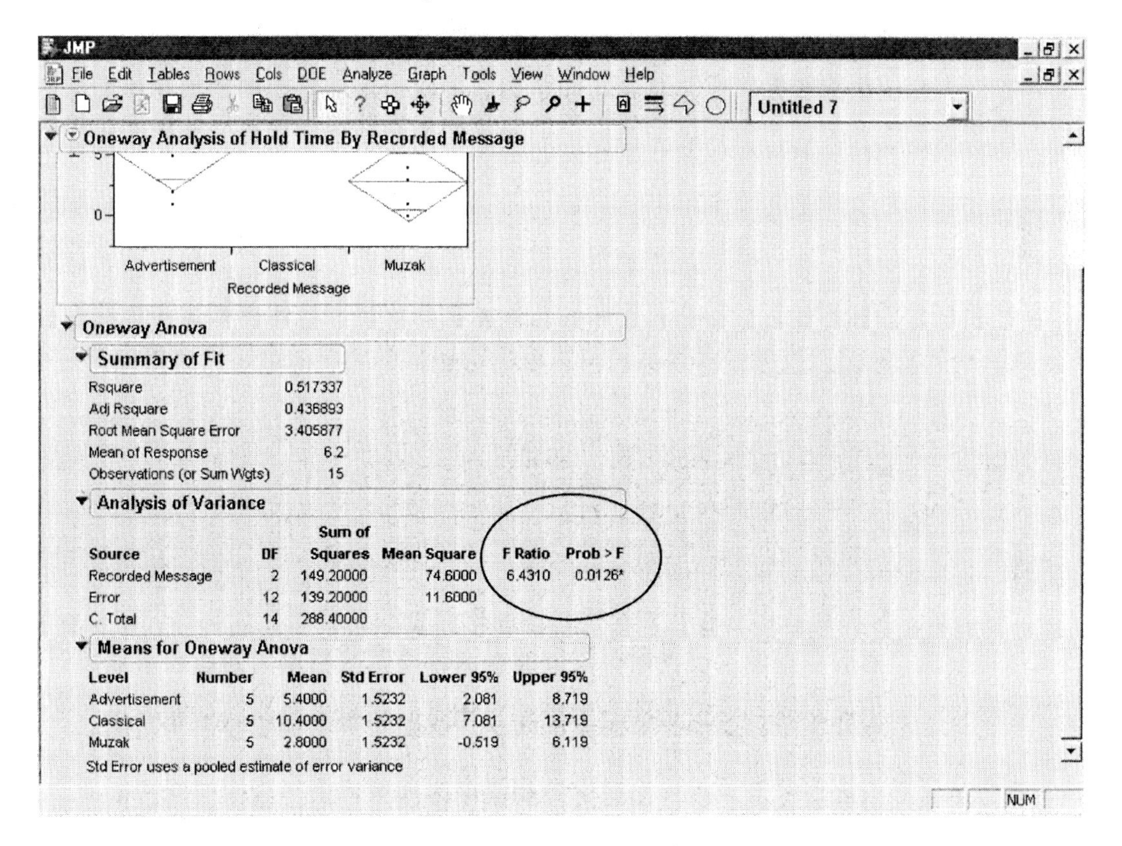

Chapter 14. Exercise 14.3. What's the best way to learn French?

From the JMP Starter Menu, select **New Data Table**.

Enter the data as below:

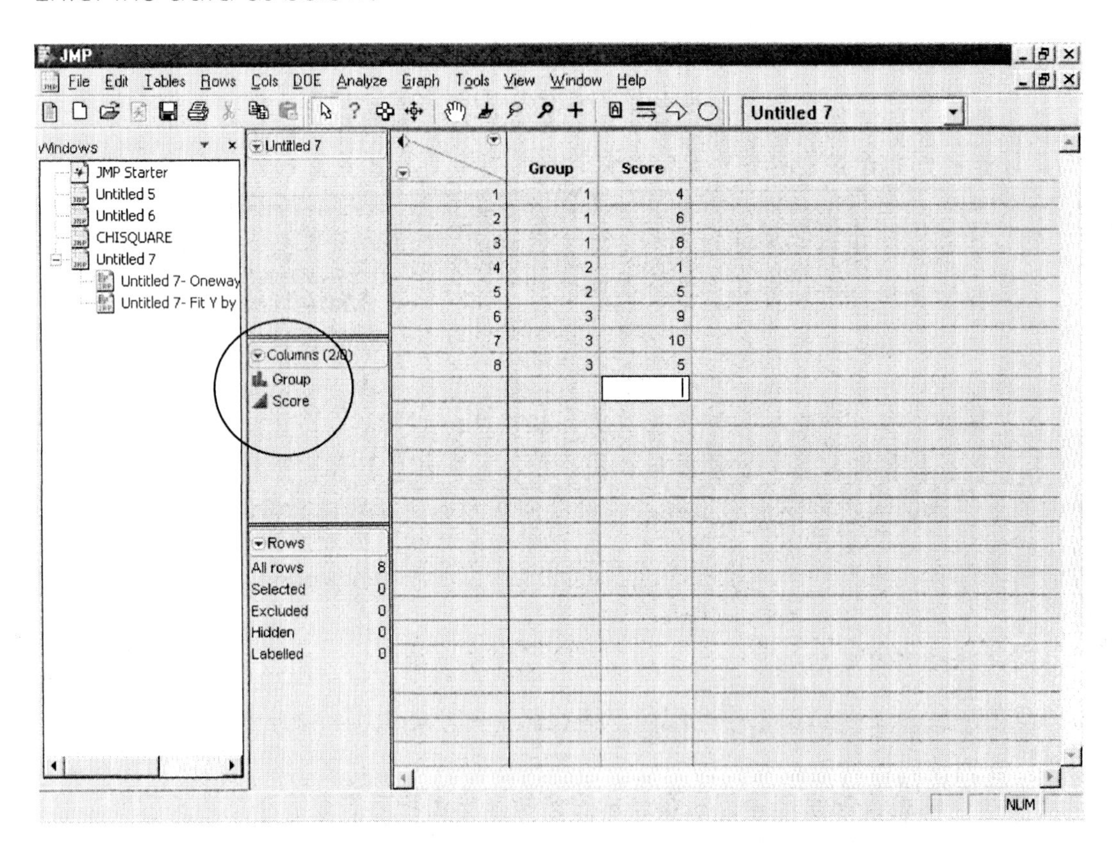

Note: ensure that the Group variable is coded as a Nominal variable.

To generate an ANOVA table, select **Analyze>Fit Y By X**. Assign ***Score*** to the **Y, Response** role and ***Group*** to the **X, Factor** role. Select **OK**.

Click on the red triangle next to "**Oneway Analysis…**". Select **Means/ANOVA**.

You should see the following screen (toggle down):

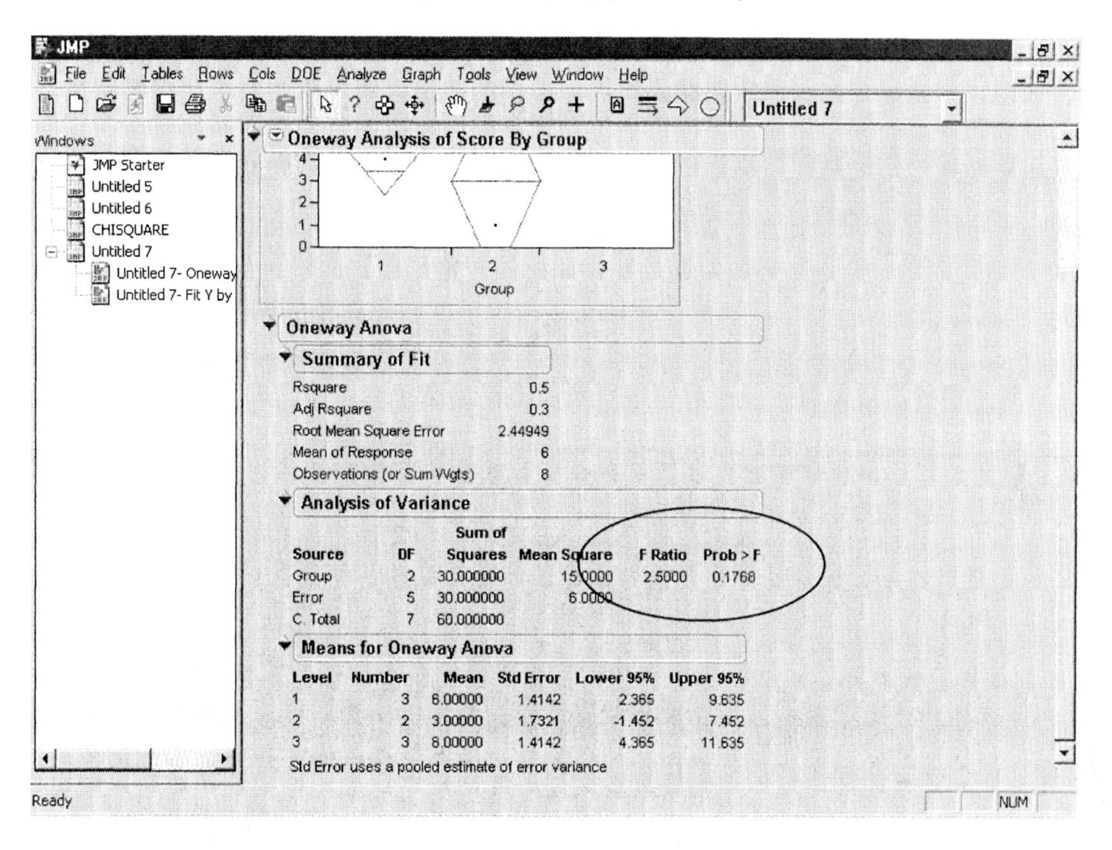

The five steps for the ANOVA test are:

Step 1: Check assumptions for independent samples, normal populations and equal standard deviations (assumed).

Step 2: H_0: $\mu_1 = \mu_2 = \mu_3$

H_a: at least 2 of the means are not equal

Step 3: F = 2.5 from output above.

Step 4: p=.1769 from output above.

Step 5: Fail to Reject the Null (assuming alpha <.1).

Since the F-statistic is calculated from the DF (which are based on sample size), the small sample size contributed to the low F-statistic.

Chapter 14. Exercise 14.14. Comparing telephone holding times.

You will need the data that was entered for Example 2 of this chapter.

The datasheet should look like this:

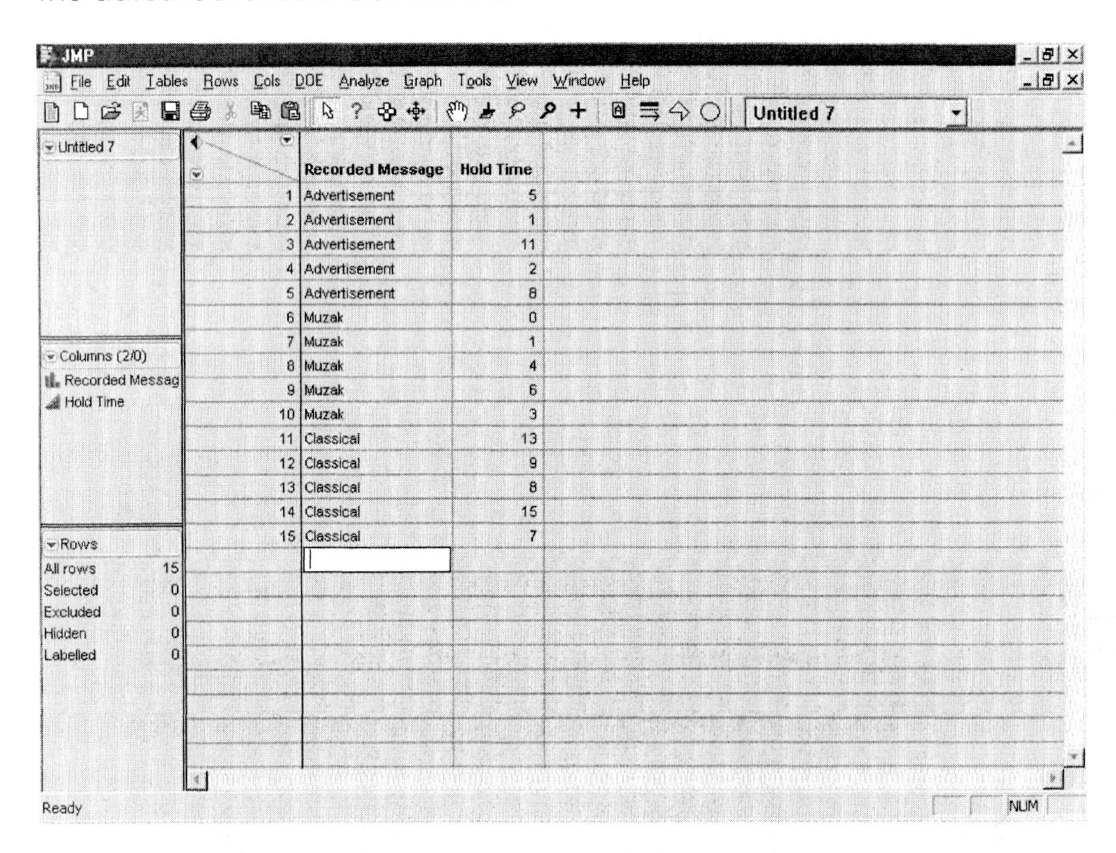

	Recorded Message	Hold Time
1	Advertisement	5
2	Advertisement	1
3	Advertisement	11
4	Advertisement	2
5	Advertisement	8
6	Muzak	0
7	Muzak	1
8	Muzak	4
9	Muzak	6
10	Muzak	3
11	Classical	13
12	Classical	9
13	Classical	8
14	Classical	15
15	Classical	7

To execute the group means, the ANOVA and the means comparisons, select **Analyze>Fit Y By X**. Assign ***Recorded Message*** to the **X, Factor** role and ***Hold Time*** to the **Y, Response** role. Select **OK**.

On the output screen, click on the red triangle next to "Oneway Analysis...". Select **Means/ANOVA**. Click on the red triangle again, and select **Compare Means>Each Pair, Students t**. Toggle to the bottom of the screen.

You should see the following screen:

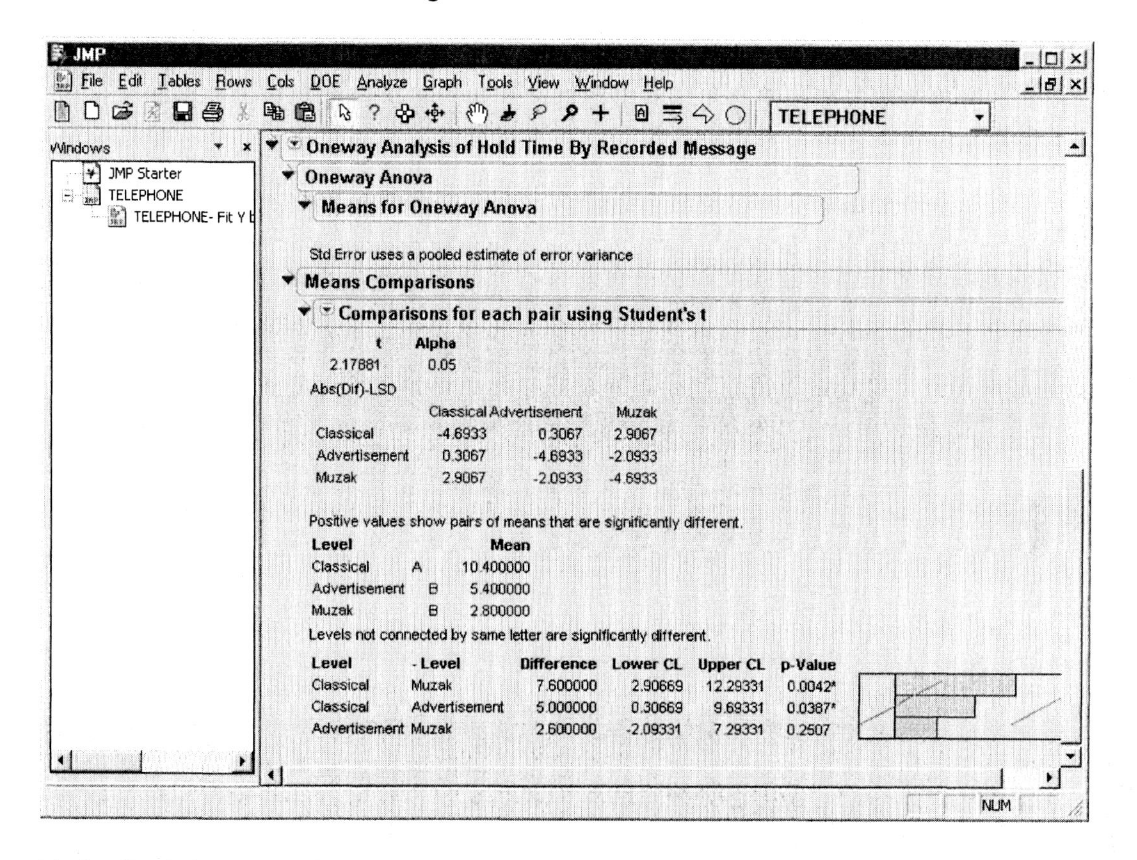

Note that the margin of error can be found by subtracting the lower CL from the upper CL and dividing by 2: 12.29331 – 2.90669 = 9.38662/2 = 4.69331.

Chapter 14. Exercise 14.15. Tukey holding times comparisons.

Continuing from the previous exercise, click on the red triangle next to "Oneway Analysis...". Select **Compare Means>All Pairs, Tukey HSD**. Toggle to the bottom of the screen.

You should see the following screen:

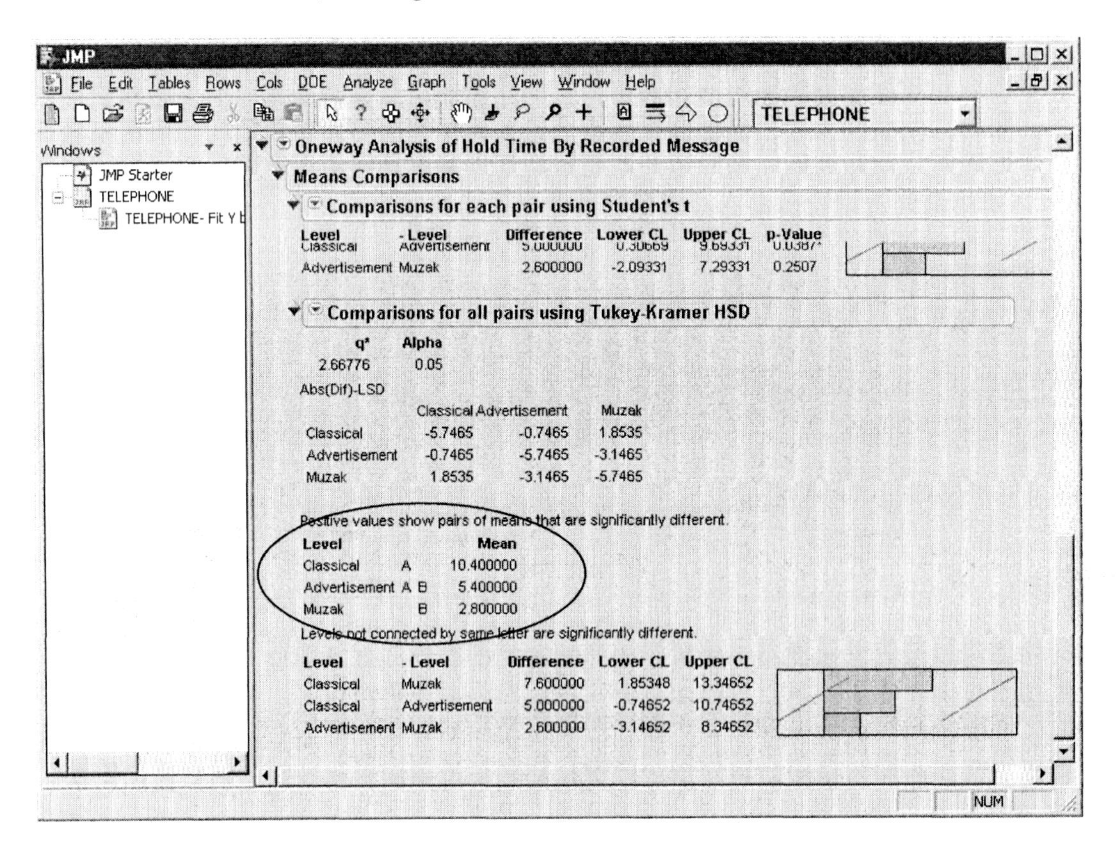

From the output above, the groups which are significantly different are designated with different letters. For example, Classical has a letter "A" next to it, Advertisement has an "A" and a "B", while Muzak has a "B". This would indicate that Classical and Muzak are significantly different, because they do not share a letter.

The margin of error for the Confidence Intervals is found by subtracting the Lower CL from the Upper CL and dividing by 2.

Chapter 14. Exercise 14.21. French ANOVA.

Continuing from Exercise 14.3, you should have the following dataset:

To generate the separate 95% confidence intervals for the comparisons of the means, select **Analyze>Fit Y By X**. Assign ***Group*** to the **X, Factor** role, and ***Score*** to the **Y, Response** role. Select **OK.**

On the output screen, click on the red triangle next to "Oneway Analysis...". Select **Means/ANOVA.** Click on the red triangle again and select **Compare Means>Each Pair, Student's t.**

You should see the following screen:

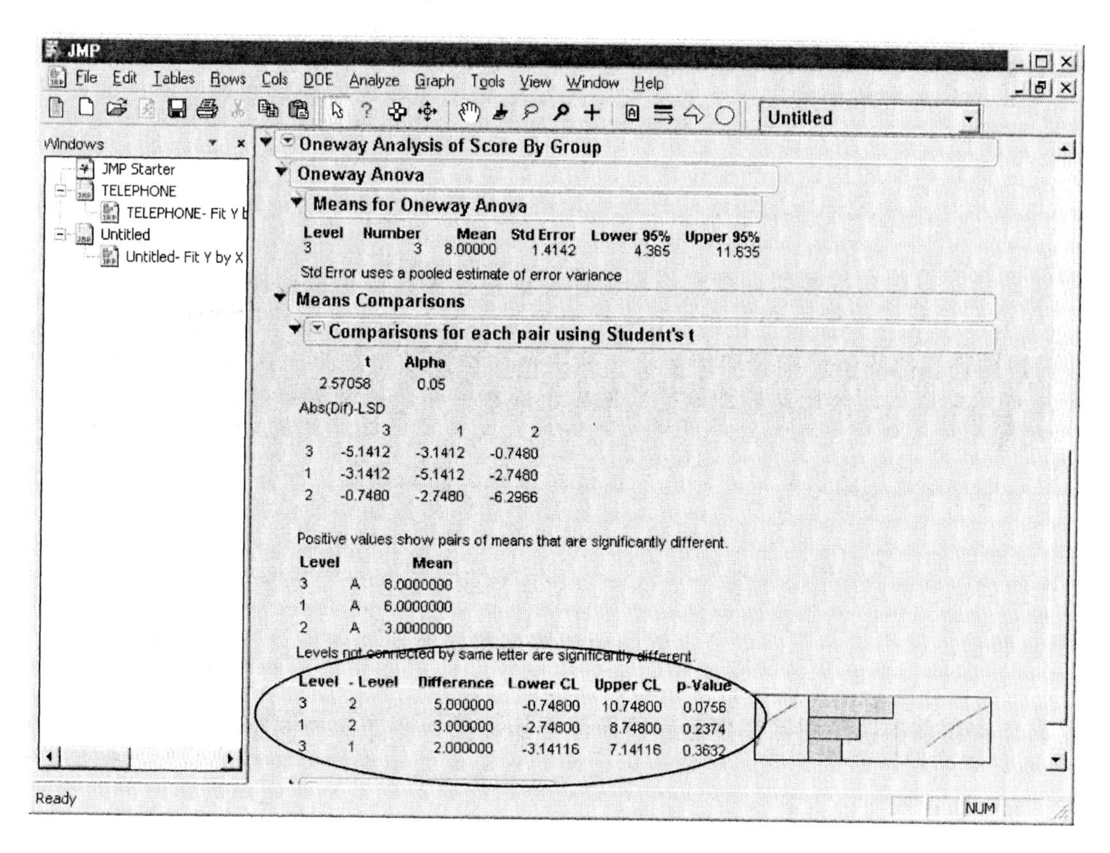

The results for the different comparisons are circled above.

To execute the Tukey comparisons, click on the red triangle again, and select **Compare Means>All Pairs, Tukey HSD.**

You should see the following screen:

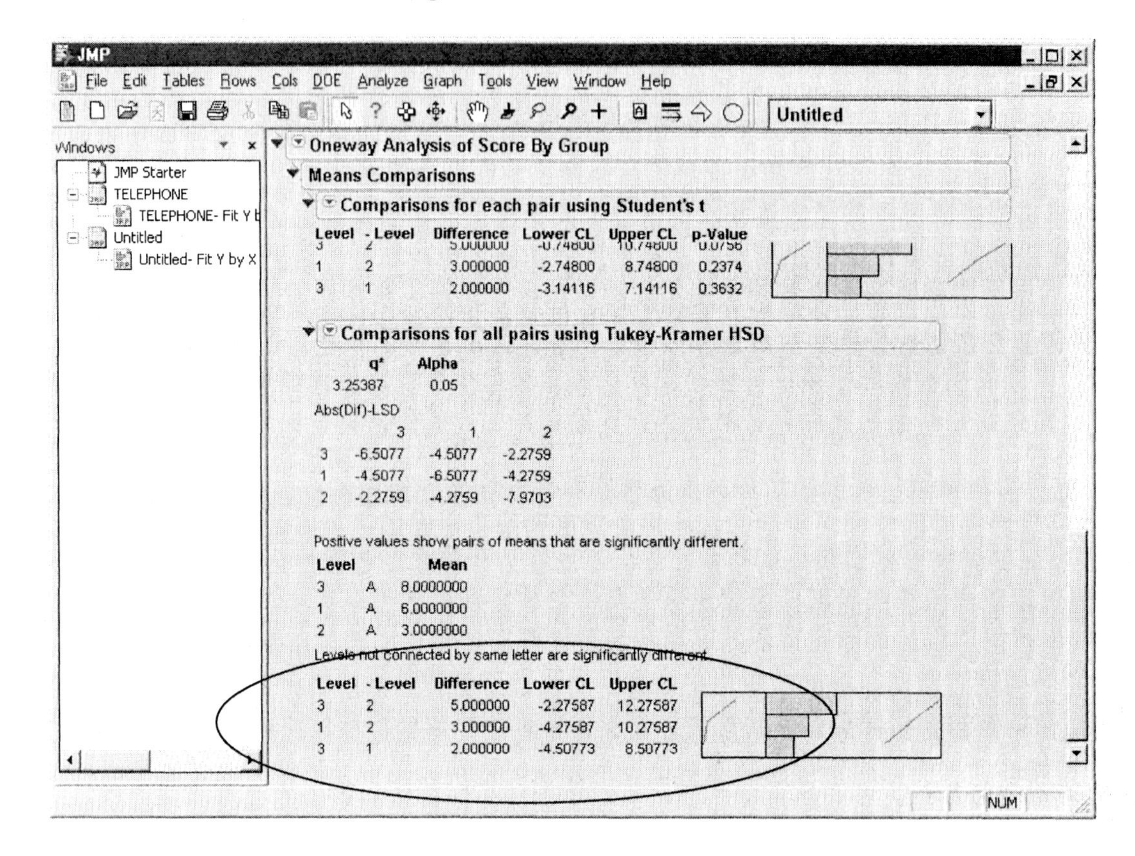

The results for the different comparisons are circled above.

Chapter 14. Example 9. Testing the Main Effects for Corn Yield.

From the JMP Starter Menu, select **New Data Table**.

Enter the data as below:

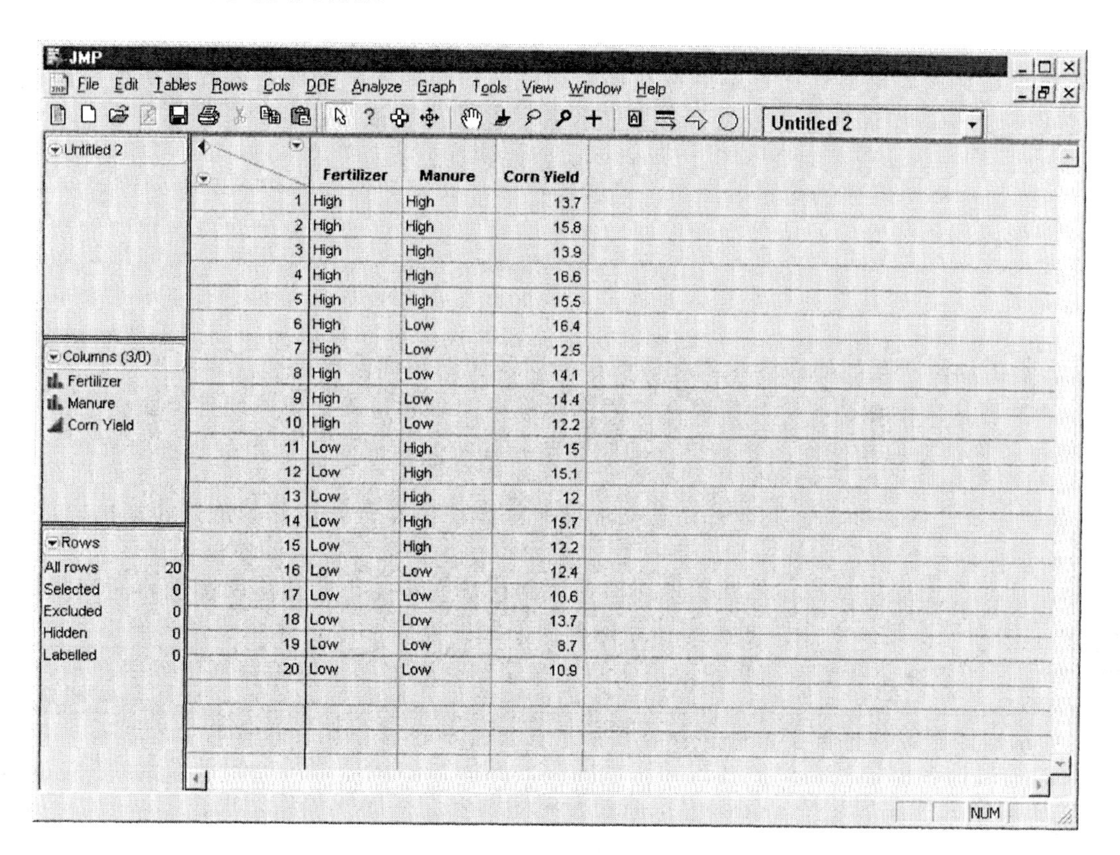

	Fertilizer	Manure	Corn Yield
1	High	High	13.7
2	High	High	15.8
3	High	High	13.9
4	High	High	16.6
5	High	High	15.5
6	High	Low	16.4
7	High	Low	12.5
8	High	Low	14.1
9	High	Low	14.4
10	High	Low	12.2
11	Low	High	15
12	Low	High	15.1
13	Low	High	12
14	Low	High	15.7
15	Low	High	12.2
16	Low	Low	12.4
17	Low	Low	10.6
18	Low	Low	13.7
19	Low	Low	8.7
20	Low	Low	10.9

To generate the descriptive statistics for Corn Yield by Fertilizer Level and by Manure Level, select **Analyze>Distribution**. Assign ***Corn Yield*** to the **Y, Columns** role and ***Fertilizer*** and ***Manure*** to the **By** role. Select **OK**.

The descriptive statistics (i.e., mean, standard deviation) can be seen in the Moments section of each part of the output screen:

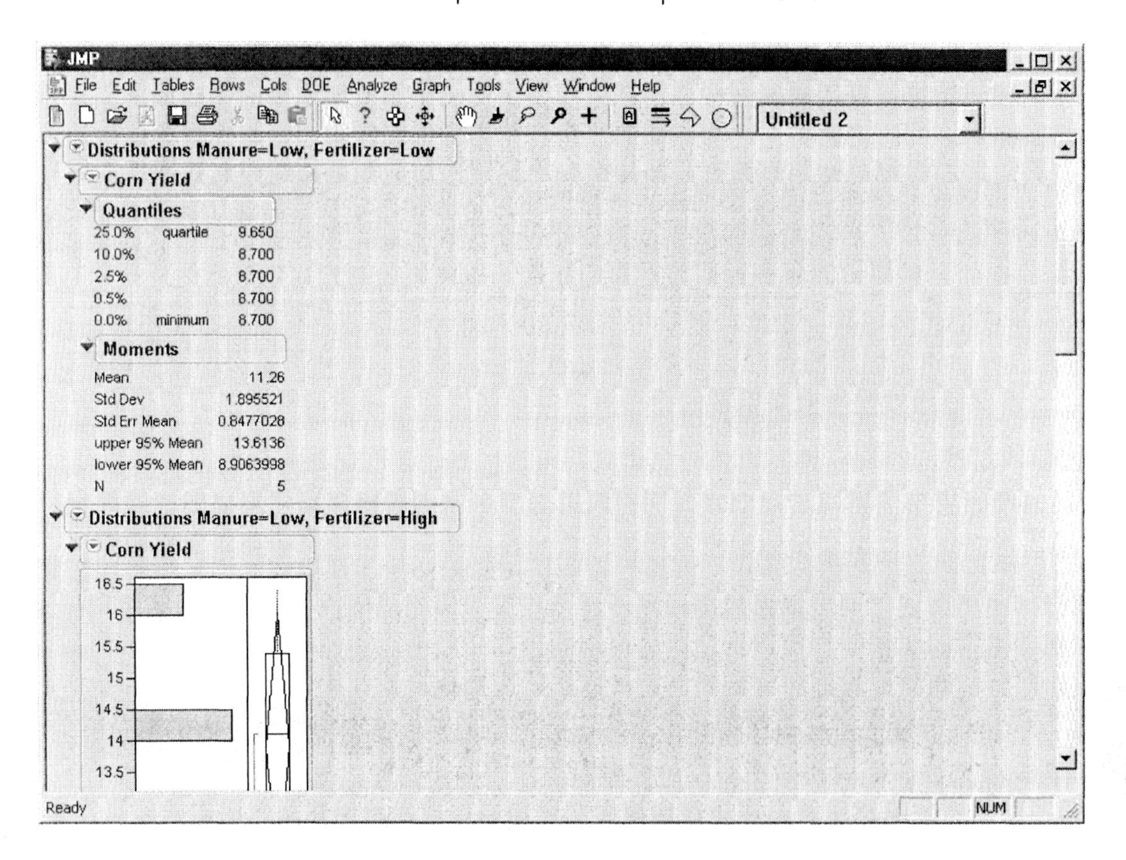

To test the hypotheses of the effects of fertilizer and manure on Corn Yield, select **Analyze>Fit Model**. Assign ***Corn Yield*** to the **Y** role and add ***Fertilizer*** and ***Manure*** for the **Construct Model Effects**. Select **Run Model**. Toggle to the bottom of the screen.

You should see the following screen:

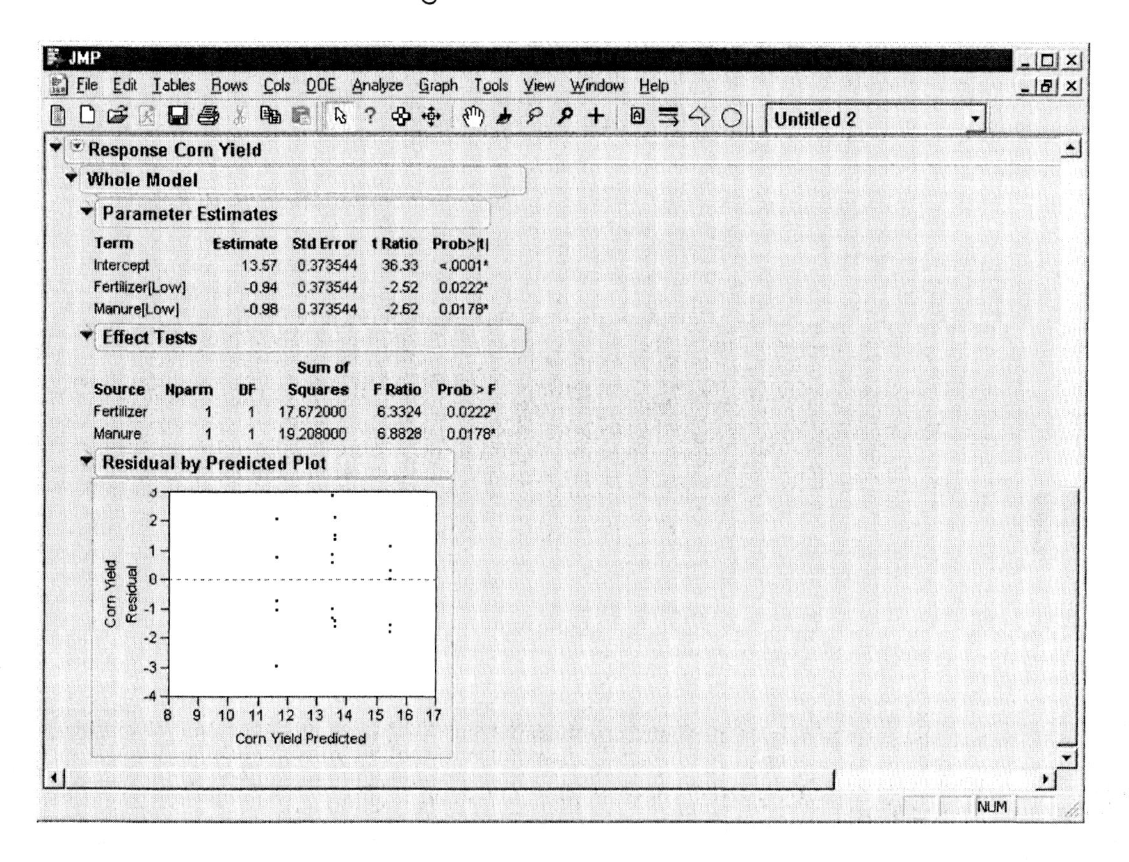

The F statistics for the two variables Fertilizer and Manure are indicated above. Both F statistics, and the associated p-values, would indicate that the Null Hypotheses should be rejected and that there is evidence to suggest that Fertilizer and Manure do affect Corn Yield.

Chapter 15. Example 5. Does Heavy Dating Affect College GPA?

From the JMP Starter Menu, select **New Data Table**.

Enter the data as below:

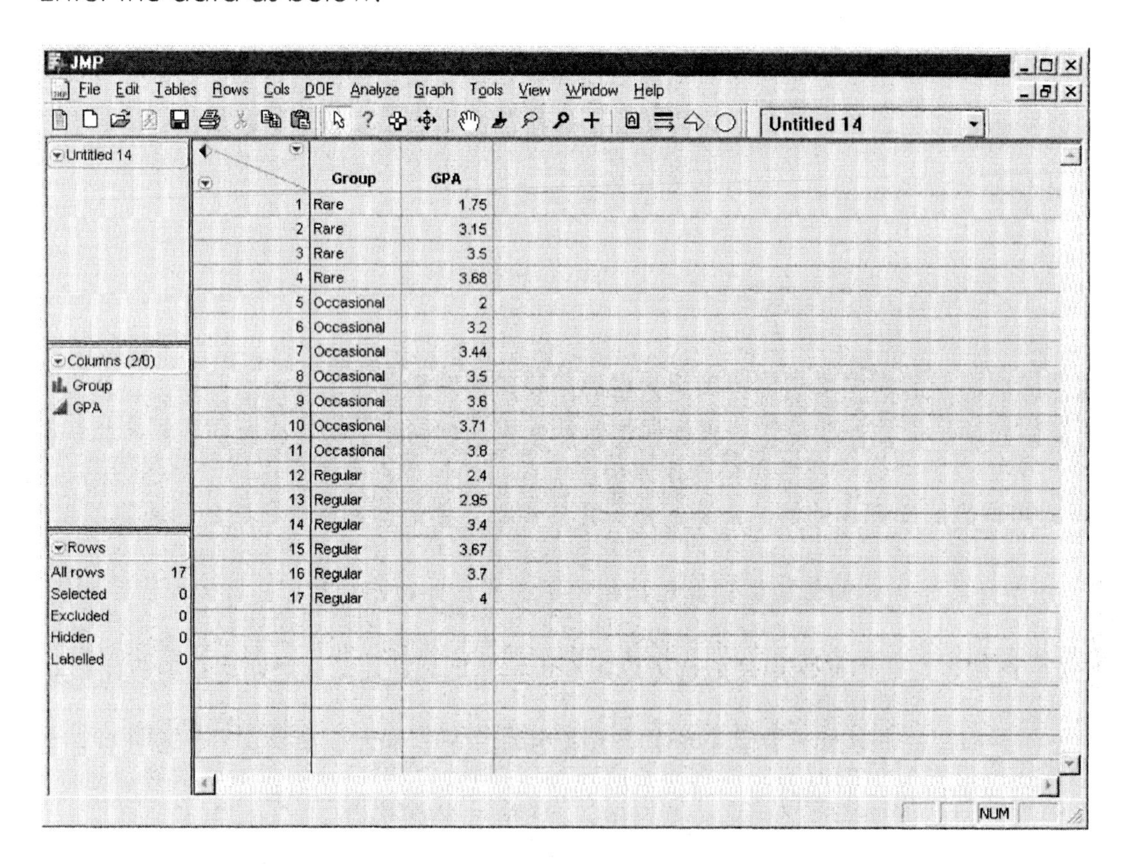

	Group	GPA
1	Rare	1.75
2	Rare	3.15
3	Rare	3.5
4	Rare	3.68
5	Occasional	2
6	Occasional	3.2
7	Occasional	3.44
8	Occasional	3.5
9	Occasional	3.6
10	Occasional	3.71
11	Occasional	3.8
12	Regular	2.4
13	Regular	2.95
14	Regular	3.4
15	Regular	3.67
16	Regular	3.7
17	Regular	4

To execute the Kruskal-Wallis Test, select **Analyze>Fit Y By X**. Assign ***GPA*** to the **Y, Response** role and **Group** to the ***X, Factor*** role. Select **OK**.

On the output screen, click on the red triangle next to "Oneway Analysis...". Select **Nonparametric>Wilcoxon Test**.

You should see the following screen:

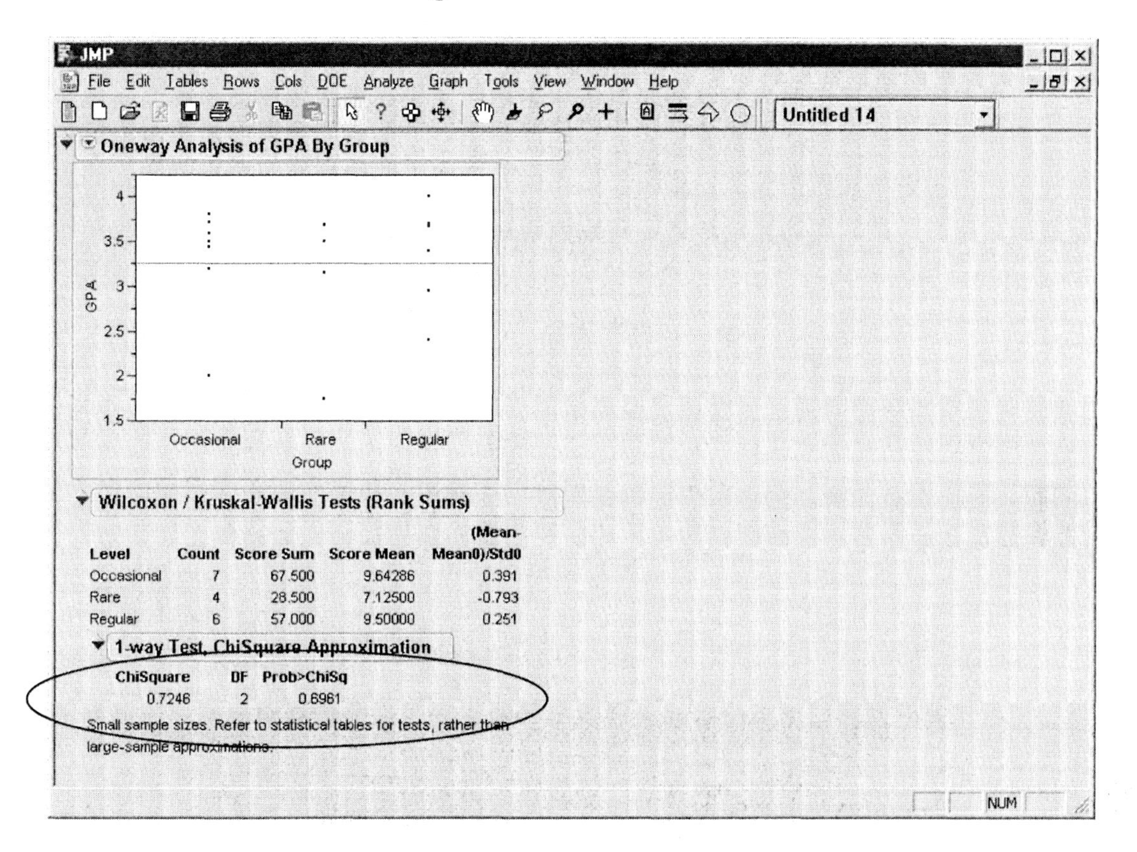

The test statistic is reported as .7246, and the associated p-value is reported as .6961. This information would lead us to fail to reject the Null Hypothesis and conclude that Dating does not affect College GPA.

Chapter 15. Exercise 15.22. Comparing Therapies for Anorexia.

Open the anorexia file from the JMP Folder of the data disk.

You will need to create the third change column – for the Family Therapy.

To do this, double click on first empty column. In the **Column Info** dialogue box, name the column "***Famchange***". Select **Column Properties>Formula**. Click on "**Edit Formula**". In the Formula box, click on Fam1. Then click on the "minus" symbol and then click on Fam2. Select **OK>OK**.

You should see the following screen:

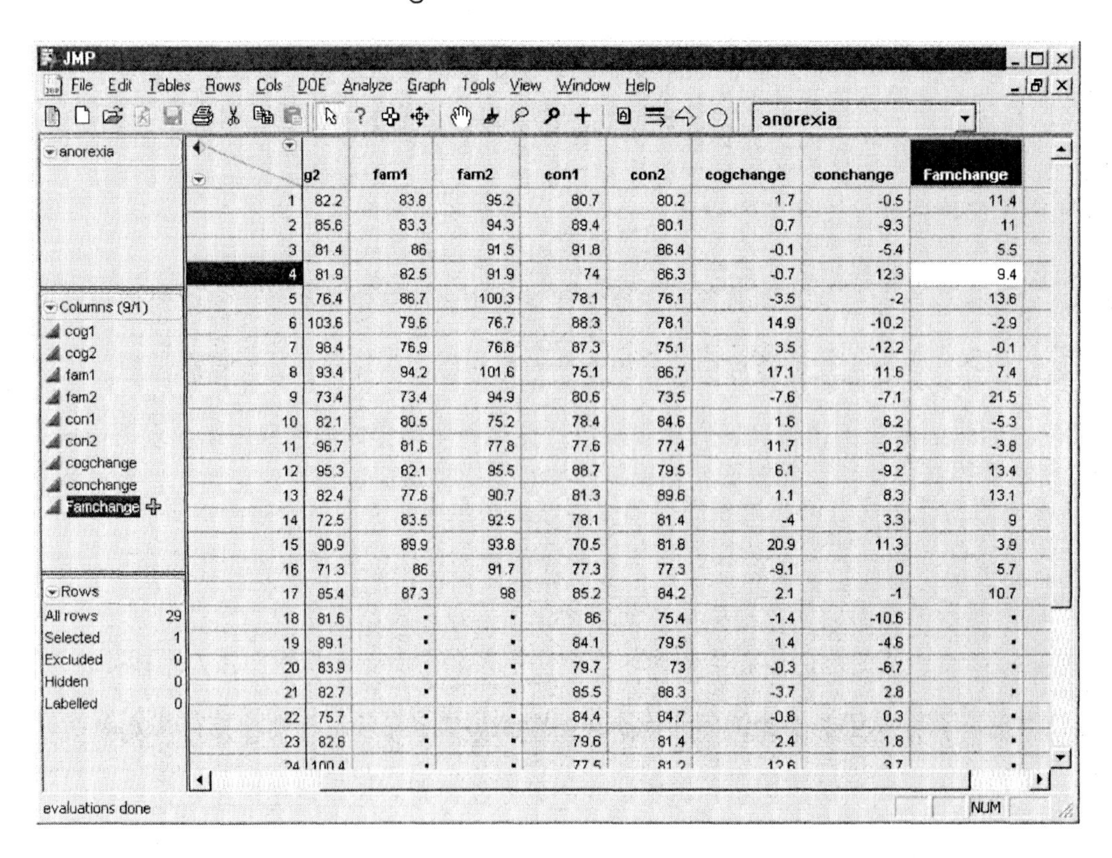

	g2	fam1	fam2	con1	con2	cogchange	conchange	Famchange
1	82.2	83.8	95.2	80.7	80.2	1.7	-0.5	11.4
2	85.6	83.3	94.3	89.4	80.1	0.7	-9.3	11
3	81.4	86	91.5	91.8	86.4	-0.1	-5.4	5.5
4	81.9	82.5	91.9	74	86.3	-0.7	12.3	9.4
5	76.4	86.7	100.3	78.1	76.1	-3.5	-2	13.6
6	103.6	79.6	76.7	88.3	78.1	14.9	-10.2	-2.9
7	98.4	76.9	76.8	87.3	75.1	3.5	-12.2	-0.1
8	93.4	94.2	101.6	75.1	86.7	17.1	11.6	7.4
9	73.4	73.4	94.9	80.6	73.5	-7.6	-7.1	21.5
10	82.1	80.5	75.2	78.4	84.6	1.6	6.2	-5.3
11	96.7	81.6	77.8	77.6	77.4	11.7	-0.2	-3.8
12	95.3	82.1	95.5	88.7	79.5	6.1	-9.2	13.4
13	82.4	77.6	90.7	81.3	89.6	1.1	8.3	13.1
14	72.5	83.5	92.5	78.1	81.4	-4	3.3	9
15	90.9	89.9	93.8	70.5	81.8	20.9	11.3	3.9
16	71.3	86	91.7	77.3	77.3	-9.1	0	5.7
17	85.4	87.3	98	85.2	84.2	2.1	-1	10.7
18	81.6	•	•	86	75.4	-1.4	-10.6	•
19	89.1	•	•	84.1	79.5	1.4	-4.6	•
20	83.9	•	•	79.7	73	-0.3	-6.7	•
21	82.7	•	•	85.5	88.3	-3.7	2.8	•
22	75.7	•	•	84.4	84.7	-0.8	0.3	•
23	82.6	•	•	79.6	81.4	2.4	1.8	•
24	100.4	•	•	77.5	81.2	12.6	3.7	•

Now, we need to "stack" the values for the three change categories into one column. To do this, select **Tables>Stack**. Assign ***cogchange, conchange*** and ***Famchange*** into the **Stack Columns** box. Select **OK**.

You should see the following screen:

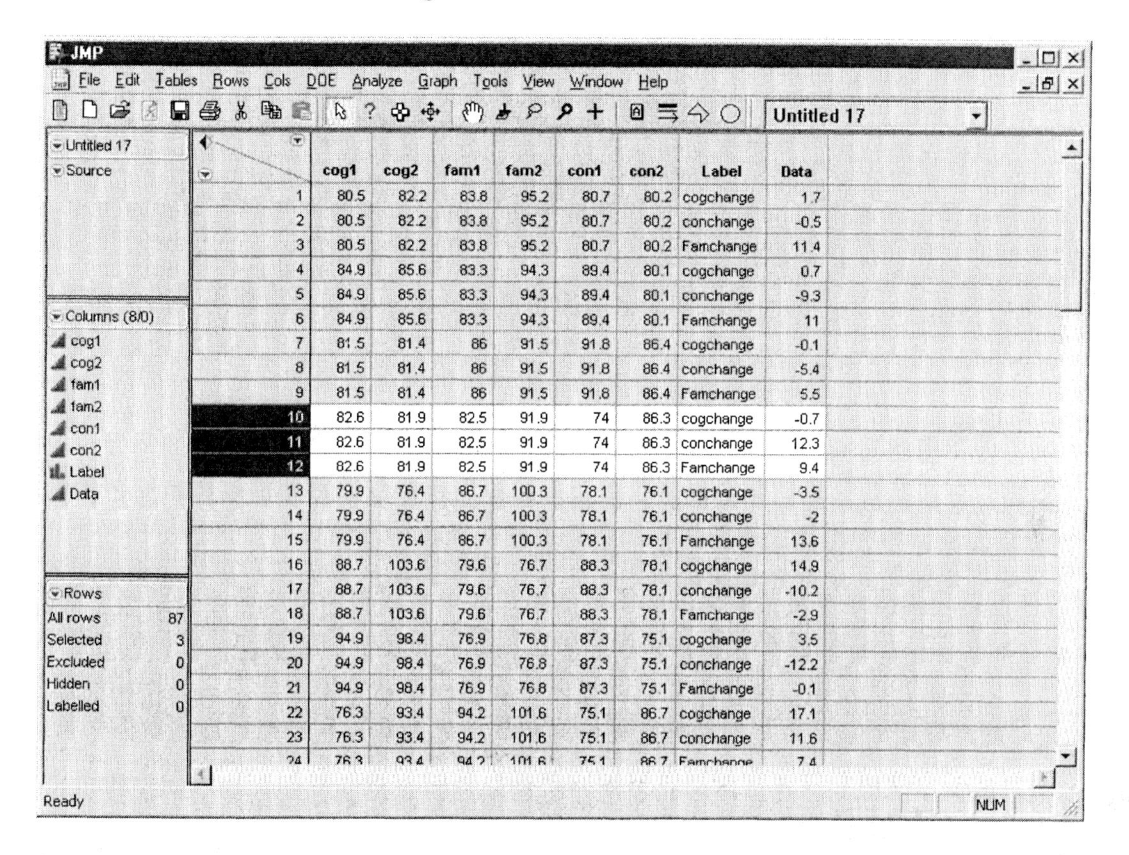

	cog1	cog2	fam1	fam2	con1	con2	Label	Data
1	80.5	82.2	83.8	95.2	80.7	80.2	cogchange	1.7
2	80.5	82.2	83.8	95.2	80.7	80.2	conchange	-0.5
3	80.5	82.2	83.8	95.2	80.7	80.2	Famchange	11.4
4	84.9	85.6	83.3	94.3	89.4	80.1	cogchange	0.7
5	84.9	85.6	83.3	94.3	89.4	80.1	conchange	-9.3
6	84.9	85.6	83.3	94.3	89.4	80.1	Famchange	11
7	81.5	81.4	86	91.5	91.8	86.4	cogchange	-0.1
8	81.5	81.4	86	91.5	91.8	86.4	conchange	-5.4
9	81.5	81.4	86	91.5	91.8	86.4	Famchange	5.5
10	82.6	81.9	82.5	91.9	74	86.3	cogchange	-0.7
11	82.6	81.9	82.5	91.9	74	86.3	conchange	12.3
12	82.6	81.9	82.5	91.9	74	86.3	Famchange	9.4
13	79.9	76.4	86.7	100.3	78.1	76.1	cogchange	-3.5
14	79.9	76.4	86.7	100.3	78.1	76.1	conchange	-2
15	79.9	76.4	86.7	100.3	78.1	76.1	Famchange	13.6
16	88.7	103.6	79.6	76.7	88.3	78.1	cogchange	14.9
17	88.7	103.6	79.6	76.7	88.3	78.1	conchange	-10.2
18	88.7	103.6	79.6	76.7	88.3	78.1	Famchange	-2.9
19	94.9	98.4	76.9	76.8	87.3	75.1	cogchange	3.5
20	94.9	98.4	76.9	76.8	87.3	75.1	conchange	-12.2
21	94.9	98.4	76.9	76.8	87.3	75.1	Famchange	-0.1
22	76.3	93.4	94.2	101.6	75.1	86.7	cogchange	17.1
23	76.3	93.4	94.2	101.6	75.1	86.7	conchange	11.6
24	76.3	93.4	94.2	101.6	75.1	86.7	Famchange	7.4

Rename the "***Label***" column "***Group***" and rename the "***Data***" column, "***Weight Change***".

To generate the nonparametric ANOVA, select **Analyze>Fit Y By X**. Assign ***Weight Change*** to the **Y, Response** role and ***Group*** to the **X, Factor** role. Select **OK**. On the output screen, click on the red triangle next to "Oneway Analysis..." and select **Nonparametric>Wilcoxon**.

You should see the following screen:

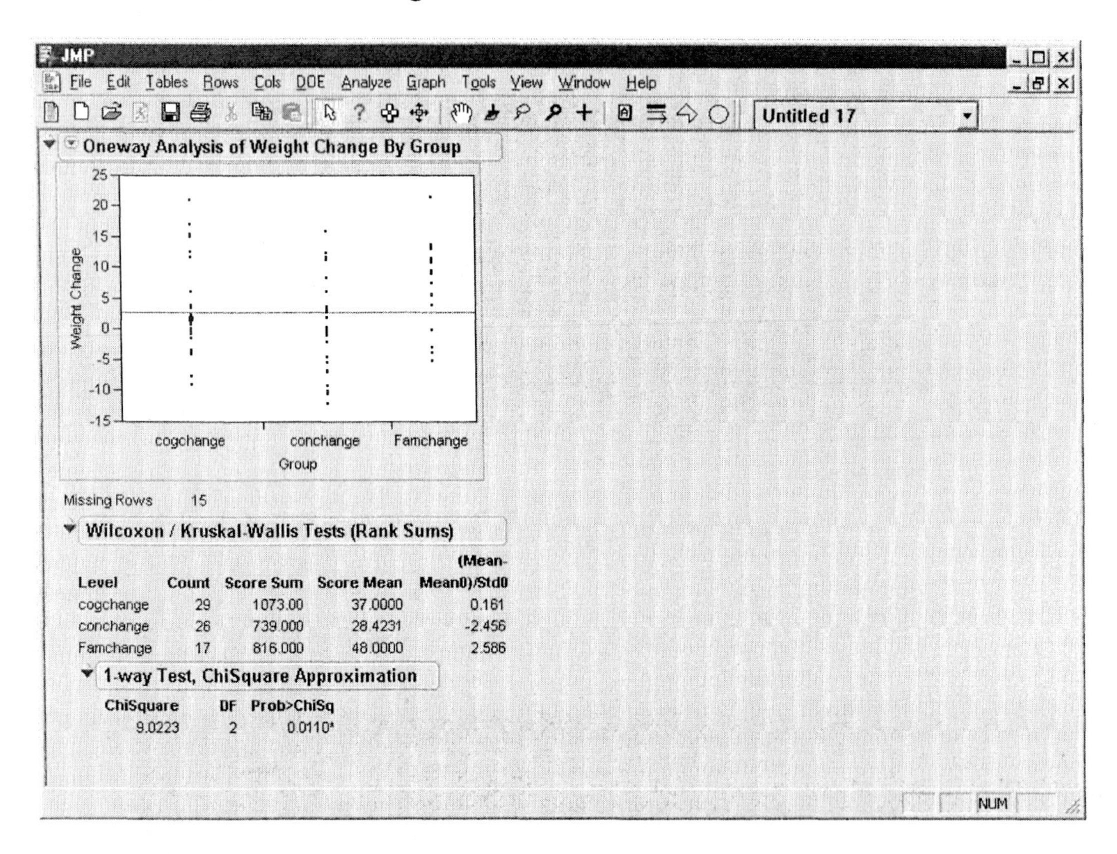

The Hypothesis Statements for this test would be:

H_o: the Weight Change means of cogchange, conchange and Famchange are equal (therapy type makes no difference)
H_a: At least two of the means are not equal (therapy types does make a difference)

From the output above, the test statistic of 9.0223, and the associated p-value of .0110 would lead us to reject the Null Hypothesis and conclude that the type of therapy does make a difference in weight change for anorexia patients.

PHStat 2.7 with Data Files for use with the Technology Manual to accompany Statistics The Art and Science of Learning from Data, 2e
Alan Agresti/Christine Franklin
ISBN 0136014593
CD License Agreement

Pearson Prentice Hall
Pearson Education, Inc.
Upper Saddle River, NJ 07458

READ THIS LICENSE CAREFULLY BEFORE OPENING THIS PACKAGE. BY OPENING THIS PACKAGE, YOU ARE AGREEING TO THE TERMS AND CONDITIONS OF THIS LICENSE. IF YOU DO NOT AGREE, DO NOT OPEN THE PACKAGE. PROMPTLY RETURN THE UNOPENED PACKAGE AND ALL ACCOMPANYING ITEMS TO THE PLACE YOU OBTAINED THEM. THESE TERMS APPLY TO ALL LICENSED SOFTWARE ON THE DISK EXCEPT THAT THE TERMS FOR USE OF ANY SHAREWARE OR FREEWARE ON THE DISKETTES ARE AS SET FORTH IN THE ELECTRONIC LICENSE LOCATED ON THE DISK:

1. **GRANT OF LICENSE and OWNERSHIP:** The enclosed CD-ROM ("Software") is licensed, not sold, to you by Pearson Education, Inc. publishing as Pearson Prentice Hall ("We" or the "Company") in consideration of your adoption of the accompanying Company textbooks and/or other materials, and your agreement to these terms. You own only the disk(s) but we and/or our licensors own the Software itself. This license allows instructors and students enrolled in the course using the Company textbook that accompanies this Software (the "Course") to use and display the enclosed copy of the Software on up to one computer of an educational institution, for academic use only, so long as you comply with the terms of this Agreement. You may make one copy for back up only. We reserve any rights not granted to you.
2. **USE RESTRICTIONS:** You may not sell or license copies of the Software or the Documentation to others. You may not transfer, distribute or make available the Software or the Documentation, except to instructors and students in your school who are users of the adopted Company textbook that accompanies this Software in connection with the course for which the textbook was adopted. You may not reverse engineer, disassemble, decompile, modify, adapt, translate or create derivative works based on the Software or the Documentation. You may be held legally responsible for any copying or copyright infringement that is caused by your failure to abide by the terms of these restrictions.
3. **TERMINATION:** This license is effective until terminated. This license will terminate automatically without notice from the Company if you fail to comply with any provisions or limitations of this license. Upon termination, you shall destroy the Documentation and all copies of the Software. All provisions of this Agreement as to limitation and disclaimer of warranties, limitation of liability, remedies or damages, and our ownership rights shall survive termination.
4. **DISCLAIMER OF WARRANTY: THE COMPANY AND ITS LICENSORS MAKE NO WARRANTIES ABOUT THE SOFTWARE, WHICH IS PROVIDED "AS-IS." IF THE DISK IS DEFECTIVE IN MATERIALS OR WORKMANSHIP, YOUR ONLY REMEDY IS TO RETURN IT TO THE COMPANY WITHIN 30 DAYS FOR REPLACEMENT UNLESS THE COMPANY DETERMINES IN GOOD FAITH THAT THE DISK HAS BEEN MISUSED OR IMPROPERLY INSTALLED, REPAIRED, ALTERED OR DAMAGED. THE COMPANY DISCLAIMS ALL WARRANTIES, EXPRESS OR IMPLIED, INCLUDING WITHOUT LIMITATION, THE IMPLIED WARRANTIES OF MERCHANTABILITY AND FITNESS FOR A PARTICULAR PURPOSE. THE COMPANY DOES NOT WARRANT, GUARANTEE OR MAKE ANY REPRESENTATION REGARDING THE ACCURACY, RELIABILITY, CURRENTNESS, USE, OR RESULTS OF USE, OF THE SOFTWARE.**
5. **LIMITATION OF REMEDIES AND DAMAGES: IN NO EVENT, SHALL THE COMPANY OR ITS EMPLOYEES, AGENTS, LICENSORS OR CONTRACTORS BE LIABLE FOR ANY INCIDENTAL, INDIRECT, SPECIAL OR CONSEQUENTIAL DAMAGES ARISING OUT OF OR IN CONNECTION WITH THIS LICENSE OR THE SOFTWARE, INCLUDING, WITHOUT LIMITATION, LOSS OF USE, LOSS OF DATA, LOSS OF INCOME OR PROFIT, OR OTHER LOSSES SUSTAINED AS A RESULT OF INJURY TO ANY PERSON, OR LOSS OF OR DAMAGE TO PROPERTY, OR CLAIMS OF THIRD PARTIES, EVEN IF THE COMPANY OR AN AUTHORIZED REPRESENTATIVE OF THE COMPANY HAS BEEN ADVISED OF THE POSSIBILITY OF SUCH DAMAGES.** SOME JURISDICTIONS DO NOT ALLOW THE LIMITATION OF DAMAGES IN CERTAIN CIRCUMSTANCES, SO THE ABOVE LIMITATIONS MAY NOT ALWAYS APPLY.
6. **GENERAL:** THIS AGREEMENT SHALL BE CONSTRUED IN ACCORDANCE WITH THE LAWS OF THE UNITED STATES OF AMERICA AND THE STATE OF NEW YORK, APPLICABLE TO CONTRACTS MADE IN NEW YORK, EXCLUDING THE STATE'S LAWS AND POLICIES ON CONFLICTS OF LAW, AND SHALL BENEFIT THE COMPANY, ITS AFFILIATES AND ASSIGNEES. This Agreement is the complete and exclusive statement of the agreement between you and the Company and supersedes all proposals, prior agreements, oral or written, and any other communications between you and the company or any of its representatives relating to the subject matter. If you are a U.S. Government user, this Software is licensed with "restricted rights" as set forth in subparagraphs (a)-(d) of the Commercial Computer-Restricted Rights clause at FAR 52.227-19 or in subparagraphs (c)(1)(ii) of the Rights in Technical Data and Computer Software clause at DFARS 252.227-7013, and similar clauses, as applicable. Should you have any questions concerning this agreement or if you wish to contact the Company for any reason, please contact in writing: Pearson Education, Inc., One Lake Street, Upper Saddle River, New Jersey 07458 "AS IS" LICENSE

****
SYSTEM REQUIREMENTS
****
* Windows 98, Windows NT, Windows 2000, Windows ME, or Windows XP
*In addition to the minimum processor requirements for the operating system your computer is running, this CD requires a Pentium II, 200 MHz or higher processor
*In addition to the RAM required by the operating system your computer is running, this CD requires 64 MB RAM
*Microsoft PowerPoint 97, 2000 or 2002
*Microsoft Word 97, 2000 or 2002
*Microsoft Excel 97, 2000, 2002, 2003, 2007 (Excel 97 users must apply the SR-2 or a later free update from Microsoft in order to user PHStat2. Excel 2000 and 2002 must have the macro security level set to Medium)
*Microsoft Excel Data Analysis ToolPak and Analysis ToolPak VBA installed (supplied on the Microsoft Office/Excel program CD)
*CD-ROM or DVD-ROM drive
*Mouse and keyboard
*Color monitor (256 or more colors and screen resolution settings set to 800 by 600 pixels or 1024 by 748 pixels)
*Internet browser (Netscape 4.x, 6.x or 7.x, or Internet Explorer 5.x or 6.x) and Internet connection suggested but not required

*This CD-ROM is intended for stand-alone use only. It is not meant for use on a network.

****
CD-ROM CONTENTS
****
-- PHStat 2 version 2.7 (PHSTAT2 folder)
PHStat2 is Windows software that assists you in learning the concepts of statistics while using Microsoft Excel. PHStat2 allows you to perform many common types of statistical analyses working with and using the familiar Microsoft Excel interface. For additional information about PHStat2, including technical, navigational and functional information, read the PHStat2_readme.rtf file in the Install\PHStat2 folder on this CD.
-- Data Files (Agresti Data_Sets folder)
-- Readme.txt

****
TECHNICAL SUPPORT
****
Prentice Hall does not support 3rd party products such as Microsoft Excel. For technical support options, visit http://support.microsoft.com on the Internet.
If you are having problems with PHStat software, first re-consult the detailed information contained in this HELP file and the PHStat2 page. PHStat2 technical support information can be found at http://www.prenhall.com/phstat.
If you continue to experience difficulties, visit Pearson Education's Technical Support Web site at http://247pearsoned.custhelp.com.
Our technical staff will need to know certain things about your system in order to help us solve your problems more quickly and efficiently. If possible, please be at your computer when you call for support. You should have the following information ready:
-Product title and ISBN
-Computer, make and model
-Operating system
-CD-ROM drive, make and model
-RAM available
-Hard disk space available
-Graphics card type
-Printer, make and model
-Detailed description of the problem, including the exact wording of any error messages.

For assistance with third-party software, please visit:
JMP Support (for JMP): http://www.jmp.com/support/techsup/index.shtml
MINITAB Support: http://www.minitab.com/support/index.htm
SPSS Support: http://www.spss.com/tech/spssdefault.htm
TI-8x Support: http://education.ti.com/us/support/main.html
Excel Support: http://support.microsoft.com/
PHStat Support: http://www.prenhall.com/phstat/phstat2/phstat2(main).htm
----